Fundamentals of Microelectronics

with Robotics and Bioengineering Applications

Third Edition

Behzad Razavi
University of California, Los Angeles

WILEY

SENIOR VP	Smita Bakshi
SENIOR DIRECTOR	Don Fowley
SENIOR EDITOR	Jennifer Brady
SENIOR MANAGING EDITOR	Judy Howarth
PRODUCTION EDITOR	Vinolia Benedict Fernando
COVER PHOTO CREDIT	© Just_Super/E+/Getty Images

This book was set in 10/12 TimesTenLTStd by SPi Global.

Founded in 1807, John Wiley & Sons, Inc. has been a valued source of knowledge and understanding for more than 200 years, helping people around the world meet their needs and fulfill their aspirations. Our company is built on a foundation of principles that include responsibility to the communities we serve and where we live and work. In 2008, we launched a Corporate Citizenship Initiative, a global effort to address the environmental, social, economic, and ethical challenges we face in our business. Among the issues we are addressing are carbon impact, paper specifications and procurement, ethical conduct within our business and among our vendors, and community and charitable support. For more information, please visit our website: www.wiley.com/go/citizenship.

ISBN: 978-1-119-69514-1 (PBK)
ISBN: 978-1-119-69487-8 (EVALC)

Library of Congress Cataloging-in-Publication Data

Names: Razavi, Behzad, author.
Title: Fundamentals of microelectronics : with robotics and bioengineering
applications / Behzad Razavi, University of California, Los Angeles.
Description: Third edition. | Hoboken : Wiley, [2021] | Includes index.
Identifiers: LCCN 2021000237 (print) | LCCN 2021000238 (ebook) | ISBN
9781119695141 (paperback) | ISBN 9781119694854 (adobe pdf) | ISBN
9781119694397 (epub)
Subjects: LCSH: Microelectronics—Textbooks.
Classification: LCC TK7874 .R395 2021 (print) | LCC TK7874 (ebook) | DDC
621.381—dc23
LC record available at https://lccn.loc.gov/2021000237
LC ebook record available at https://lccn.loc.gov/2021000238

The inside back cover will contain printing identification and country of origin if omitted from this page. In addition, if the ISBN on the back cover differs from the ISBN on this page, the one on the back cover is correct.

Printed in Singapore
M003814_030622

To Angelina and Jahan,
for their love and patience

About the Author

Behzad Razavi received the BSEE degree from Sharif University of Technology in 1985 and the MSEE and PhDEE degrees from Stanford University in 1988 and 1992, respectively. He was with AT&T Bell Laboratories and Hewlett-Packard Laboratories until 1996. Since 1996, he has been Associate Professor and subsequently Professor of electrical engineering at University of California, Los Angeles. His current research includes wireless transceivers, frequency synthesizers, phase-locking and clock recovery for high-speed data communications, and data converters.

Professor Razavi was an Adjunct Professor at Princeton University from 1992 to 1994, and at Stanford University in 1995. He served on the Technical Program Committees of the International Solid-State Circuits Conference (ISSCC) from 1993 to 2002 and VLSI Circuits Symposium from 1998 to 2002. He has also served as Guest Editor and Associate Editor of the IEEE Journal of Solid-State Circuits, IEEE Transactions on Circuits and Systems, and International Journal of High Speed Electronics.

Professor Razavi received the Beatrice Winner Award for Editorial Excellence at the 1994 ISSCC, the best paper award at the 1994 European Solid-State Circuits Conference, the best panel award at the 1995 and 1997 ISSCC, the TRW Innovative Teaching Award in 1997, the best paper award at the IEEE Custom Integrated Circuits Conference in 1998, and the McGraw-Hill First Edition of the Year Award in 2001. He was the co-recipient of both the Jack Kilby Outstanding Student Paper Award and the Beatrice Winner Award for Editorial Excellence at the 2001 ISSCC. He received the Lockheed Martin Excellence in Teaching Award in 2006, the UCLA Faculty Senate Teaching Award in 2007, and the CICC Best Invited Paper Award in 2009 and 2012. He was the co-recipient of the 2012 VLSI Circuits Symposium Best Student Paper Award. He was also recognized as one of the top 10 authors in the 50-year history of ISSCC. Professor Razavi received the IEEE Donald Pederson Award in Solid-State Circuits in 2011.

Professor Razavi is a Fellow of IEEE, has served as an IEEE Distinguished Lecturer, and is the author of *Principles of Data Conversion System Design, RF Microelectronics* (translated to Chinese, Japanese, and Korean), *Design of Analog CMOS Integrated Circuits* (translated to Chinese, Japanese, and Korean), *Design of Integrated Circuits for Optical Communications, and Fundamentals of Microelectronics* (translated to Korean and Portuguese). He is also the editor of *Monolithic Phase-Locked Loops and Clock Recovery Circuits* and *Phase-Locking in High-Performance Systems.*

Preface to Third Edition

The second edition of this book was published seven years ago. Since then, I have been developing new materials that can enhance the readers' learning experience. In particular, the third edition is introduced with two significant additions:

Companion Videos Students generally prefer to watch a lecture before reading the related concepts in a book. With the proliferation of online videos, this preference has become so dominant that students rarely consider reading as their first option. I therefore began to ponder how this book could benefit from video lectures.

In 2014, I produced a video lecture series, called Electronics 1, that paralleled roughly the first nine chapters of the book. The series spanned 45 one-hour lectures recorded in front of a smart board, including segments called "Frontiers in Electronics," which highlighted real-life applications. The videos were posted on YouTube and quickly garnered a wide audience. This motivated me to develop another 45 lectures, called Electronics 2, to cover Chapters 10–12. The 90 lectures have been collectively viewed 3 million times and for a total of 500,000 hours. With thousands of positive messages that the viewers have sent me, I am convinced that the videos can serve as an effective means of learning for the readers of this book as well. The video titles and links are shown at the beginning of each chapter.

Robotics and Bioengineering Applications In addition to communications and computing, a number of other fields rely heavily on microelectronics. Chief among them are robotics and bioengineering. The principal challenge in teaching such applications in a fundamental electronics course is how to distill the concepts in a language that the students can appreciate. I have collected about two dozen robotics and bioengineering examples that portray the use of various electronic devices and circuits in real-life systems.

Another change in the third edition is that the end-of-chapter problems have been rearranged, and various typos have been corrected.

Behzad Razavi
September 2020

Preface to Second Edition

The first edition of this book was published in 2008 and has been adopted by numerous universities around the globe for undergraduate microelectronics education. In response to the feedback received from students and instructors, this second edition entails a number of revisions that enhance the pedagogical aspects of the book:

1. Numerous sidebars have been added throughout the text on the history and applications of electronic devices and circuits, helping the reader remain engaged and motivated and allowing the instructor to draw upon real-life examples during the lecture. The sidebars are intended to demonstrate the impact of electronics, elevate the reader's understanding of the concepts, or provide a snapshot of the latest developments in the field.
2. A chapter on oscillators has been added. A natural descendent of feedback circuits, discrete and integrated oscillators have become indispensible in most devices and hence merit a detailed study.
3. The end-of-chapter problems have been rearranged to better agree with the progression of the chapter. Also, to allow the reader to quickly find the problems for each section, the corresponding section titles have been added. Moreover, the challenging problems have been ranked in terms of their difficulty level by one or two stars.
4. Since students often ask for the answers to problems so as to check the validity of their approach, the answers to even-numbered problems have been posted on the book's website.
5. Various typographical errors have been corrected.

I wish to thank all of the students and instructors who have provided valuable feedback in the past five years and helped me decide on the revisions for this edition.

Behzad Razavi
January 2013

Preface to First Edition

With the advances in the semiconductor and communication industries, it has become increasingly important for electrical engineers to develop a good understanding of microelectronics. This book addresses the need for a text that teaches microelectronics from a modern and intuitive perspective. Guided by my industrial, research, and academic experience, I have chosen the topics, the order, and the depth and breadth so as to efficiently impart analysis and design principles that the students will find useful as they enter the industry or graduate school.

One salient feature of this book is its synthesis- or design-oriented approach. Rather than pulling a circuit out of a bag and trying to analyze it, I set the stage by stating a problem that we face in real life (e.g., how to design a cellphone charger). I then attempt to arrive at a solution using basic principles, thus presenting both failures and successes in the process. When we do arrive at the final solution, the student has seen the exact role of each device as well as the logical thought sequence behind synthesizing the circuit.

Another essential component of this book is "analysis by inspection." This "mentality" is created in two steps. First, the behavior of elementary building blocks is formulated using a "verbal" description of each analytical result (e.g., "looking into the emitter, we see $1/g_m$"). Second, larger circuits are decomposed and "mapped" to the elementary blocks to avoid the need for writing KVLs and KCLs. This approach both imparts a great deal of intuition and simplifies the analysis of large circuits.

The two articles following this preface provide helpful suggestions for students and instructors. I hope these suggestions make the task of learning or teaching microelectronics more enjoyable.

A set of Powerpoint slides, a solutions manual, and many other teaching aids are available for instructors.

Behzad Razavi
November 2007

Acknowledgments

This book has taken four years to write and benefited from contributions of many individuals. I wish to thank the following for their input at various stages of this book's development: David Allstot (University of Washington), Joel Berlinghieri, Sr. (The Citadel), Bernhard Boser (University of California, Berkeley), Charles Bray (University of Memphis), Marc Cahay (University of Cincinnati), Norman Cox (University of Missouri, Rolla), James Daley (University of Rhode Island), Tranjan Farid (University of North Carolina at Charlotte), Paul Furth (New Mexico State University), Roman Genov (University of Toronto), Maysam Ghovanloo (North Carolina State University), Gennady Gildenblat (Pennsylvania State University), Ashok Goel (Michigan Technological University), Michael Gouzman (SUNY, Stony Brook), Michael Green (University of California, Irvine), Sotoudeh Hamedi-Hagh (San Jose State University), Reid Harrison (University of Utah), Payam Heydari (University of California, Irvine), Feng Hua (Clarkson University), Marian Kazmierchuk (Wright State University), Roger King (University of Toledo), Edward Kolesar (Texas Christian University), Ying-Cheng Lai (Arizona State University), Daniel Lau (University of Kentucky, Lexington), Stanislaw Legowski (University of Wyoming), Philip Lopresti (University of Pennsylvania), Mani Mina (Iowa State University), James Morris (Portland State University), Khalil Najafi (University of Michigan), Homer Nazeran (University of Texas, El Paso), Tamara Papalias (San Jose State University), Matthew Radmanesh (California State University, Northridge), Angela Rasmussen (University of Utah), Sal R. Riggio, Jr. (Pennsylvania State University), Ali Sheikholeslami (University of Toronto), Kalpathy B. Sundaram (University of Central Florida), Yannis Tsividis (Columbia University), Thomas Wu (University of Central Florida), Darrin Young (Case Western Reserve University).

I am grateful to Naresh Shanbhag (University of Illinois, Urbana-Champaign) for test driving a draft of the book in a course and providing valuable feedback. The following UCLA students diligently prepared the solutions manual: Lawrence Au, Hamid Hatamkhani, Alireza Mehrnia, Alireza Razzaghi, William Wai-Kwok Tang, and Ning Wang. Ning Wang also produced the Powerpoint slides for the entire book. Eudean Sun (University of California, Berkeley) and John Tyler (Texas A&M University) served as accuracy checkers. I would like to thank them for their hard work.

I thank my publisher, Catherine Shultz, for her dedication and exuberance. Lucille Buonocore, Carmen Hernandez, Dana Kellogg, Madelyn Lesure, Christopher Ruel, Kenneth Santor, Lauren Sapira, Daniel Sayre, Gladys Soto, and Carolyn Weisman of Wiley and Bill Zobrist (formerly with Wiley) also deserve my gratitude. In addition, I wish to thank Jessica Knecht and Joyce Poh for their hard work on the second edition.

My wife, Angelina, typed the entire book and kept her humor as this project dragged on. My deepest thanks go to her.

Behzad Razavi

Suggestions for Students

You are about to embark upon a journey through the fascinating world of microelectronics. Fortunately, microelectronics appears in so many facets of our lives that we can readily gather enough motivation to study it. The reading, however, is not as easy as that of a novel; we must deal with *analysis* and *design*, applying mathematical rigor as well as engineering intuition every step of the way. This article provides some suggestions that students may find helpful in studying microelectronics.

Rigor and Intuition Before reading this book, you have taken one or two courses on basic circuit theory, mastering Kirchoff's Laws and the analysis of RLC circuits. While quite abstract and bearing no apparent connection with real life, the concepts studied in these courses form the foundation for microelectronics—just as calculus does for engineering.

Our treatment of microelectronics also requires rigor but entails two additional components. First, we identify many applications for the concepts that we study. Second, we must develop *intuition*, i.e., a "feel" for the operation of microelectronic devices and circuits. Without an intuitive understanding, the analysis of circuits becomes increasingly more difficult as we add more devices to perform more complex functions.

Analysis by Inspection We will expend a considerable effort toward establishing the mentality and the skills necessary for "analysis by inspection." That is, looking at a complex circuit, we wish to decompose or "map" it to simpler topologies, thus formulating the behavior with a few lines of algebra. As a simple example, suppose we have encountered the resistive divider shown in Fig. (a) and derived its Thevenin equivalent. Now, if given the circuit in Fig. (b), we can readily replace V_{in}, R_1, and R_2 with a Thevenin equivalent, thereby simplifying the calculations.

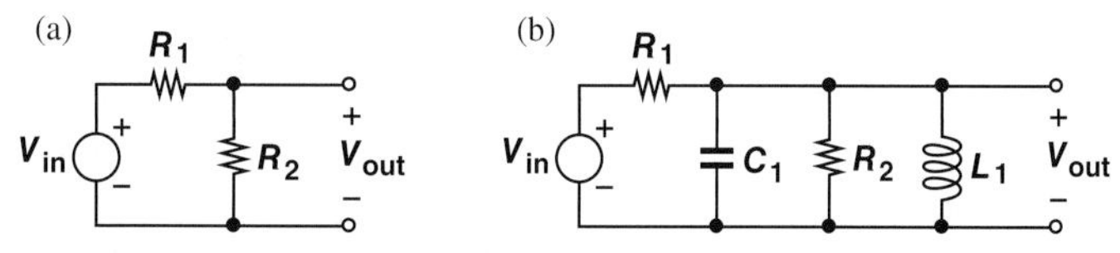

Example of analysis by inspections.

40 Pages per Week While taking courses on microelectronics, you will need to read about 40 pages of this book every week, with each page containing many new concepts, derivations, and examples. The lectures given by the instructor create a "skeleton" of each chapter, but it rests upon you to "connect the dots" by reading the book carefully and understanding each paragraph before proceeding to the next.

Reading and understanding 40 pages of the book each week requires concentration and discipline. You will face new material and detailed derivations on each page and should set aside two- or three-hour distraction-free blocks of time (no phone calls, TV, email, etc.) so that you can follow the *evolution* of the concepts while honing your analytical skills. I also suggest that you attempt each example before reading its solution.

40 Problems per Week After reading each section and going through its examples, you are encouraged to evaluate and improve your understanding by trying the corresponding end-of-chapter problems. The problems begin at a relatively easy level and gradually become more challenging. Some problems may require that you return to the section and study the subtle points more carefully.

The educational value provided by each problem depends on your *persistence*. The initial glance at the problem may be discouraging. But, as you think about it from different angles and, more importantly, re-examine the concepts in the chapter, you begin to form a path in your mind that may lead to the solution. In fact, if you have thought about a problem extensively and still have not solved it, you need but a brief hint from the instructor or the teaching assistant. Also, the more you struggle with a problem, the more appealing and memorable the answer will be.

Attending the lecture and reading the book are examples of "passive learning:" you simply receive (and, hopefully, absorb) a stream of information provided by the instructor and the text. While necessary, passive learning does not *exercise* your understanding, thus lacking depth. You may highlight many lines of the text as important. You may even summarize the important concepts on a separate sheet of paper (and you are encouraged to do so). But, to *master* the material, you need practice ("active learning"). The problem sets at the end of each chapter serve this purpose.

Homeworks and Exams Solving the problems at the end of each chapter also prepares you for homeworks and exams. Homeworks, too, demand distraction-free periods during which you put your knowledge to work and polish your understanding. An important piece of advice that I can offer here is that doing homeworks with your fellow students is a *bad* idea! Unlike other subject matters that benefit from discussions, arguments, and rebuttals, learning microelectronics requires quiet concentration. (After all, you will be on your own during the exam!) To gain more confidence in your answers, you can discuss the results with your fellow students, the instructor, or the teaching assistants *after* you have completed the homework by yourself.

Time Management Reading the text, going through the problem sets, and doing the homeworks require a time commitment of at least 10 hours per week. Due to the fast pace of the course, the material accumulates rapidly, making it difficult to keep up with the lectures if you do not spend the required time from the very first week. In fact, the more you fall behind, the less interesting and useful the lectures become, thus forcing you to simply write down everything that the instructor says while not understanding much. With your other courses demanding similar time commitments, you can soon become overwhelmed if you do not manage your time carefully.

Time management consists of two steps: (1) partitioning your waking hours into solid blocks, and (2) using each block *efficiently*. To improve the efficiency, you can take the following measures: (a) work in a quiet environment to minimize distractions; (b) spread the work on a given subject over the week, e.g., 3 hours every other day, to avoid saturation and to allow your subconscious to process the concepts in the meantime.

Prerequisites Many of the concepts that you have learned in the circuit theory courses prove essential to the study of microelectronics. Chapter 1 gives a brief overview to refresh your memory. With the limited lecture time, the instructor may not cover this material in the class, leaving it for you to read at home. You can first glance through the chapter and see which concepts "bother" you before sitting down to concentrate.

Suggestions for Instructors

Teaching undergraduate courses proves quite challenging—especially if the emphasis is on thinking and deduction rather than on memorization. With today's young minds used to playing fast-paced video games and "clicking" on the Internet toward their destination, it has become increasingly more difficult to encourage them to concentrate for long periods of time and deal with abstract concepts. Drawing upon more than one decade of teaching, this article provides suggestions that instructors of microelectronics may find helpful.

Therapy The students taking the first microelectronics course have typically completed one or two courses on basic circuit theory. To many, that experience has not been particularly memorable. After all, the circuit theory textbook is most likely written by a person *not* in the field of circuits. Similarly, the courses are most likely taught by an instructor having little involvement in circuit design. For example, the students are rarely told that node analysis is much more frequently used in hand calculations than mesh analysis is. Or, they are given little intuition with respect to Thevenin's and Norton's theorems.

With the foregoing issues in mind, I begin the first course with a five-minute "therapy session." I ask how many liked the circuit theory courses and came out with a "practical" understanding. Very few raise their hands. I then ask, "But how about your calculus courses? How many of you came out of these courses with a "practical" understanding?" Subsequently, I explain that circuit theory builds the foundation for microelectronics just as calculus does for engineering. I further mention that some abstractness should also be expected in microelectronics as we complete the foundation for more advanced topics in circuit analysis and design. I then point out that (1) microelectronics is very heavily based on intuitive understanding, requiring that we go *beyond* simply writing KVLs and KCLs and interpret the mathematical expressions intuitively, and (2) this course offers many applications of microelectronic devices and circuits in our daily lives. In other words, microelectronics is not as dry as arbitrary RLC circuits consisting of 1-Ω resistors, 1-H inductors, and 1-F capacitors.

First Quiz Since different students enter each course with different levels of preparation, I have found it useful to give a 10-minute quiz in the very first lecture. Pointing out that the quiz does not count towards their grade but serves as a gauge of their understanding, I emphasize that the objective is to test their knowledge rather than their intelligence. After collecting the quizzes, I ask one of the teaching assistants to assign a binary grade to each: those who would receive less than 50% are marked with a red star. At the end of the lecture, I return the quizzes and mention that those with a red star need to work harder and interact with the teaching assistants and myself more extensively.

The Big Picture A powerful motivational tool in teaching is the "big picture," i.e., the "practical" application of the concept under study. The two examples of microelectronic systems described in Chapter 1 serve as the first step toward creating the context for the material covered in the book. But, the big picture cannot stop here. Each new concept may merit an application—however brief the mention of the application may be—and most of this burden falls on the lecture rather than on the book.

The choice of the application must be carefully considered. If the description is too long or the result too abstract, the students miss the connection between the concept and the application. My general approach is as follows. Suppose we are to begin Chapter 2 (Basic Semiconductor Physics). I ask either "What would our world look like without semiconductors?" or "Is there a semiconductor device in your watch? In your cellphone? In your laptop? In your digital camera?" In the ensuing discussion, I quickly go over examples of semiconductor devices and where they are used.

Following the big picture, I provide additional motivation by asking, "Well, but isn't this stuff *old*? Why do *we* need to learn these things?" I then briefly talk about the challenges in today's designs and the competition among manufacturers to lower both the power consumption and the cost of portable devices.

Analysis versus Synthesis Let us consider the background of the students entering a microelectronics course. They can write KVLs and KCLs efficiently. They have also seen numerous "random" RLC circuits; i.e., to these students, all RLC circuits look the same, and it is unclear how they came about. On the other hand, an essential objective in teaching microelectronics is to develop specific circuit topologies that provide certain characteristics. We must therefore change the students' mentality from "Here's a circuit that you may never see again in your life. Analyze it!" to "We face the following problem and we must create (synthesize) a circuit that solves the problem." We can then begin with the simplest topology, identify its shortcomings, and continue to modify it until we arrive at an acceptable solution. This step-by-step synthesis approach (a) illustrates the role of each device in the circuit, (b) establishes a "design-oriented" mentality, and (c) engages the students' intellect and interest.

Analysis by Inspection In their journey through microelectronics, students face increasingly more complex circuits, eventually reaching a point where blindly writing KVLs and KCLs becomes extremely inefficient and even prohibitive. In one of my first few lectures, I show the internal circuit of a complex op amp and ask, "Can we analyze the behavior of this circuit by simply writing node or mesh equations?" It is therefore important to instill in them the concept of "analysis by inspection." My approach consists of two steps. (1) For each simple circuit, formulate the properties in an intuitive language; e.g., "the voltage gain of a common-source stage is given by the load resistance divided by $1/g_m$ plus the resistance tied from the source to ground." (2) Map complex circuits to one or more topologies studied in step (1).

In addition to efficiency, analysis by inspection also provides great intuition. As we cover various examples, I emphasize to the students that the results thus obtained reveal the circuit's dependencies much more clearly than if we simply write KVLs and KCLs without mapping.

"What If?" Adventures An interesting method of reinforcing a circuit's properties is to ask a question like, "What if we tie this device between nodes C and D rather than between nodes A and B?" In fact, students themselves often raise similar questions. My answer to them is "Don't be afraid! The circuit doesn't bite if you change it like this. So go ahead and analyze it in its new form."

For simple circuits, the students can be encouraged to consider several possible modifications and determine the resulting behavior. Consequently, the students feel much more comfortable with the original topology and understand why it is the only acceptable solution (if that is the case).

Numeric versus Symbolic Calculations In the design of examples, homeworks, and exams, the instructor must decide between numeric and symbolic calculations. The students may, of course, prefer the former type as it simply requires finding the corresponding equation and plugging in the numbers.

What is the value in numeric calculations? In my opinion, they may serve one of two purposes: (1) make the students comfortable with the results recently obtained, or (2) give the students a feel for the typical values encountered in practice. As such, numeric calculations play a limited role in teaching and reinforcing concepts.

Symbolic calculations, on the other hand, can offer insight into the behavior of the circuit by revealing dependencies, trends, and limits. Also, the results thus obtained can be utilized in more complex examples.

Blackboard versus Powerpoint This book comes with a complete set of Powerpoint slides. However, I suggest that the instructors carefully consider the pros and cons of blackboard and Powerpoint presentations.

I can offer the following observations. (1) Many students fall asleep (at least mentally) in the classroom if they are not writing. (2) Many others feel they are missing something if they are not writing. (3) For most people, the act of writing something on paper helps "carve" it in their mind. (4) The use of slides leads to a fast pace ("if we are not writing, we should move on!"), leaving little time for the students to digest the concepts. For these reasons, even if the students have a hardcopy of the slides, this type of presentation proves quite ineffective.

To improve the situation, one can leave blank spaces in each slide and fill them with critical and interesting results in real time. I have tried this method using transparencies and, more recently, tablet laptops. The approach works well for graduate courses but leaves undergraduate students bored or bewildered.

My conclusion is that the good old blackboard is still the best medium for teaching undergraduate microelectronics. The instructor may nonetheless utilize a hardcopy of the Powerpoint slides as his/her own guide for the flow of the lecture.

Discrete versus Integrated How much emphasis should a microelectronics course place on discrete circuits and integrated circuits? To most of us, the term "microelectronics" remains synonymous with "integrated circuits," and, in fact, some university curricula have gradually reduced the discrete design flavor of the course to nearly zero. However, only a small fraction of the students taking such courses eventually become active in IC products, while many go into board-level design.

My approach in this book is to begin with general concepts that apply to both paradigms and gradually concentrate on integrated circuits. I also believe that even board-level designers must have a basic understanding of the integrated circuits that they use.

Bipolar Transistor versus MOSFET At present, some controversy surrounds the inclusion of bipolar transistors and circuits in undergraduate microelectronics. With the MOSFET dominating the semiconductor market, it appears that bipolar devices are of little value. While this view may apply to graduate courses to some extent, it should be borne in mind that (1) as mentioned above, many undergraduate students go into board-level and discrete design and are likely to encounter bipolar devices, and (2) the contrasts and similarities between bipolar and MOS devices prove extremely useful in understanding the properties of each.

The order in which the two species are presented is also debatable. (Extensive surveys conducted by Wiley indicate a 50–50 split between instructors on this matter.)

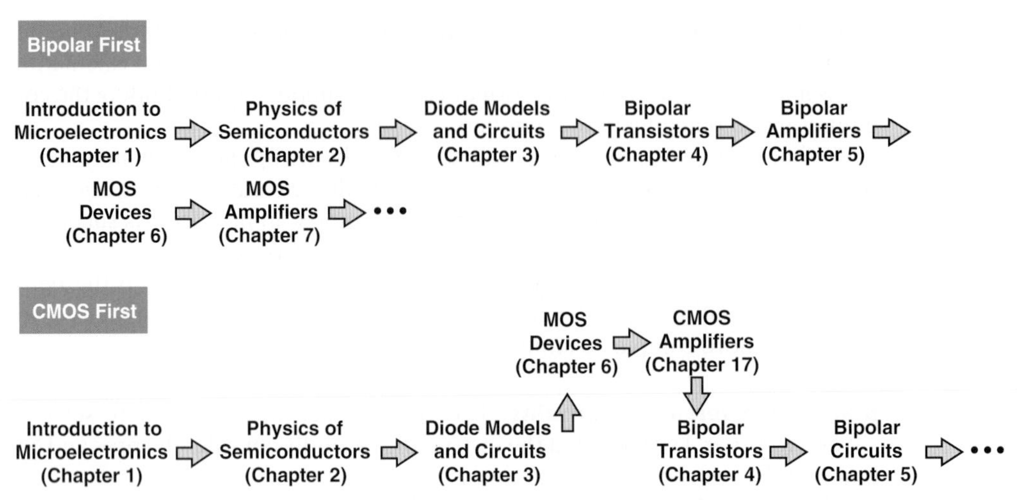

Course sequences for covering bipolar technology first or CMOS technology first.

Some instructors begin with MOS devices to ensure enough time is spent on their coverage. On the other hand, the natural flow of the course calls for bipolar devices as an extension of *pn* junctions. In fact, if diodes are immediately followed by MOS devices, the students see little relevance between the two. (The *pn* junctions in MOSFETs do not come into the picture until the device capacitances are introduced.)

My approach in this book is to first cover bipolar devices and circuits while building the foundation such that the MOS counterparts are subsequently taught with greater ease. As explained below, the material can comfortably be taught even in one quarter with no sacrifice of details of either device type.

Nonetheless, the book is organized so as to allow covering CMOS circuits first if the instructor so wishes. The sequence of chapters for each case is shown below. Chapter 16 is written with the assumption that the students have not seen any amplifier design principles so that the instructor can seamlessly go from MOS device phyics to MOS amplifier design without having covered bipolar amplifiers.

Course Syllabi This book can be used in a two-quarter or two-semester sequence. Depending on the instructor's preference, the courses can follow various combinations of the chapters. Figure illustrates some possibilities.

I have followed Syllabus I for the quarter system at UCLA for a number of years.[1] Syllabus II sacrifices op amp circuits for an introductory treatment of digital CMOS circuits.

In a semester system, Syllabus I extends the first course to current mirrors and cascode stages and the second course to output stages and analog filters. Syllabus II, on the other hand, includes digital circuits in the first course, moving current mirrors and cascodes to the second course and sacrificing the chapter on output stages.

Figure shows the approximate length of time spent on the chapters as practiced at UCLA. In a semester system, the allotted times are more flexible.

[1]We offer a separate undergraduate course on digital circuit design, which the students can take only *after* our first microelectronics course.

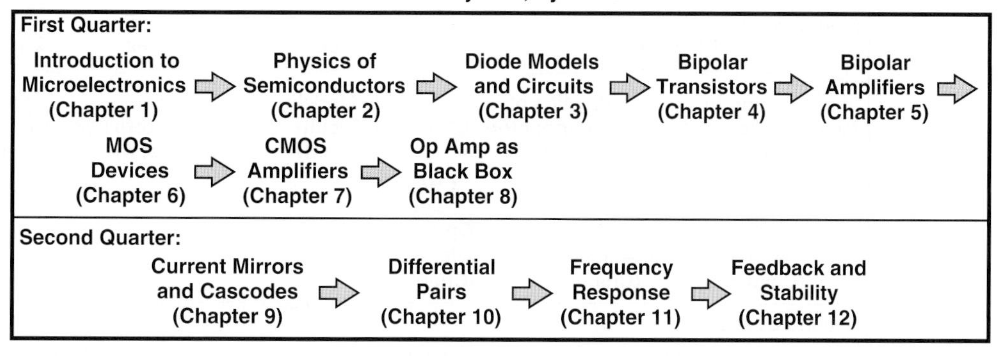

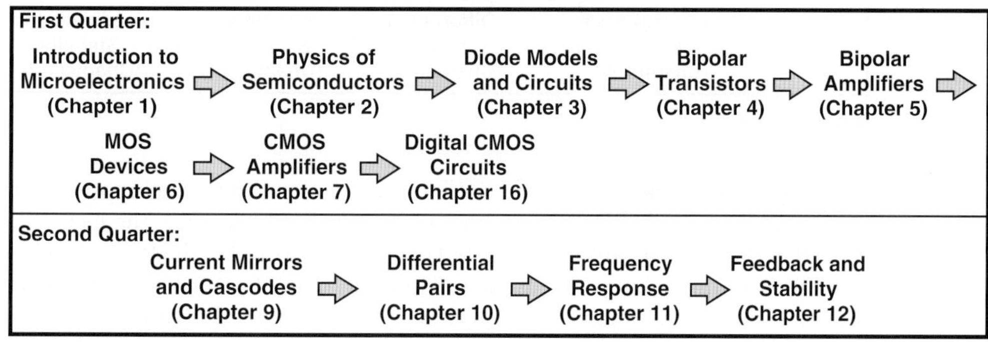

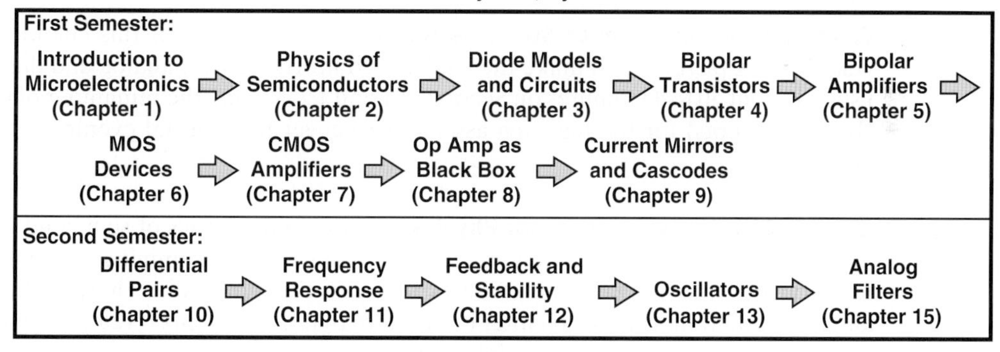

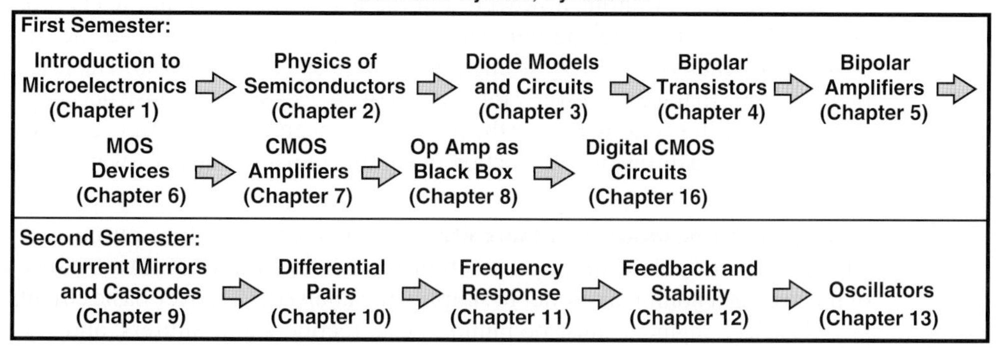

Different course structures for quarter and semester systems.

Quarter System, Syllabus I

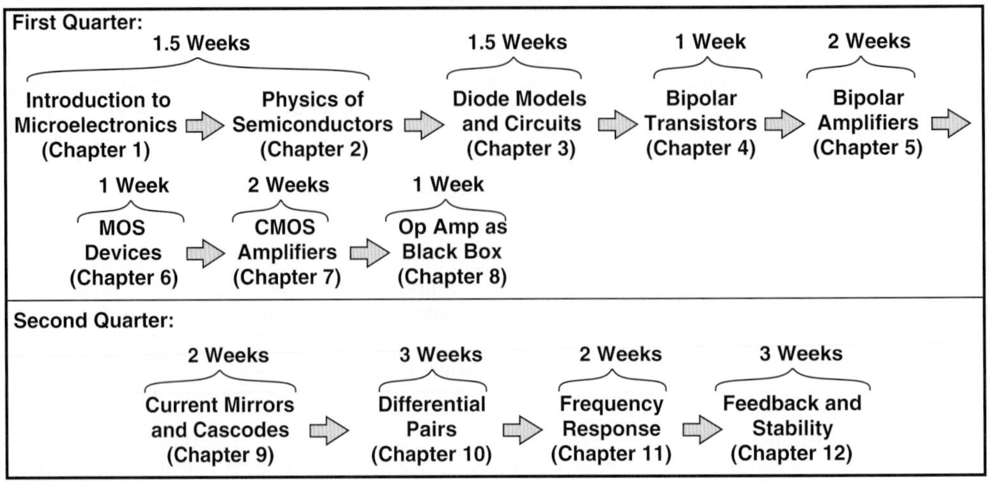

First Quarter:

1.5 Weeks		1.5 Weeks	1 Week	2 Weeks
Introduction to Microelectronics (Chapter 1) ⇨	Physics of Semiconductors (Chapter 2) ⇨	Diode Models and Circuits (Chapter 3) ⇨	Bipolar Transistors (Chapter 4) ⇨	Bipolar Amplifiers (Chapter 5) ⇨

1 Week	2 Weeks	1 Week
MOS Devices (Chapter 6) ⇨	CMOS Amplifiers (Chapter 7) ⇨	Op Amp as Black Box (Chapter 8)

Second Quarter:

2 Weeks	3 Weeks	2 Weeks	3 Weeks
Current Mirrors and Cascodes (Chapter 9) ⇨	Differential Pairs (Chapter 10) ⇨	Frequency Response (Chapter 11) ⇨	Feedback and Stability (Chapter 12)

Timetable for the two courses.

Coverage of Chapters The material in each chapter can be decomposed into three categories: (1) essential concepts that the instructor should cover in the lecture, (2) essential skills that the students must develop but cannot be covered in the lecture due to the limited time, and (3) topics that prove useful but may be skipped according to the instructor's preference.[2] Summarized below are overviews of the chapters showing which topics should be covered in the classroom.

Chapter 1: Introduction to Microelectronics The objective of this chapter is to provide the "big picture" and make the students comfortable with analog and digital signals. I spend about 30 to 45 minutes on Sections 1.1 and 1.2, leaving the remainder of the chapter (Basic Concepts) for the teaching assistants to cover in a special evening session in the first week.

Chapter 2: Basic Semiconductor Physics Providing the basics of semiconductor device physics, this chapter deliberately proceeds at a slow pace, examining concepts from different angles and allowing the students to digest the material as they read on. A terse language would shorten the chapter but require that the students reread the material multiple times in their attempt to decipher the prose.

It is important to note, however, that the instructor's pace in the classroom need not be as slow as that of the chapter. The students are expected to read the details and the examples on their own so as to strengthen their grasp of the material. The principal point in this chapter is that we must study the physics of devices so as to construct circuit models for them. In a quarter system, I cover the following concepts in the lecture: electrons and holes; doping; drift and diffusion; *pn* junction in equilibrium and under forward and reverse bias.

Chapter 3: Diode Models and Circuits This chapter serves four purposes: (1) make the students comfortable with the *pn* junction as a nonlinear device; (2) introduce the concept of linearizing a nonlinear model to simplify the analysis; (3) cover basic circuits with which any electrical engineer must be familiar, e.g., rectifiers and limiters; and (4) develop the

[2]Such topics are identified in the book by a footnote.

skills necessary to analyze heavily-nonlinear circuits, e.g., where it is difficult to predict which diode turns on at what input voltage. Of these, the first three are essential and should be covered in the lecture, whereas the last depends on the instructor's preference. (I cover it in my lectures.) In the interest of time, I skip a number of sections in a quarter system, e.g., voltage doublers and level shifters.

Chapter 4: Physics of Bipolar Transistors Beginning with the use of a voltage-controlled current source in an amplifier, this chapter introduces the bipolar transistor as an extension of *pn* junctions and derives its small-signal model. As with Chapter 2, the pace is relatively slow, but the lectures need not be. I cover structure and operation of the bipolar transistor, a very simplified derivation of the exponential characteristic, and transistor models, mentioning only briefly that saturation is undesirable. Since the T-model of limited use in analysis and carries little intuition (especially for MOS devices), I have excluded it in this book.

Chapter 5: Bipolar Amplifiers This is the longest chapter in the book, building the foundation necessary for all subsequent work in electronics. Following a bottom-up approach, this chapter establishes critical concepts such as input and output impedances, biasing, and small-signal analysis.

While writing the book, I contemplated decomposing Chapter 5 into two chapters, one on the above concepts and another on bipolar amplifier topologies, so that the latter could be skipped by instructors who prefer to continue with MOS circuits instead. However, teaching the general concepts does require the use of transistors, making such a decomposition difficult.

Chapter 5 proceeds slowly, reinforcing, step-by-step, the concept of synthesis and exploring circuit topologies with the aid of "What if?" examples. As with Chapters 2 and 4, the instructor can move at a faster pace and leave much of the text for the students to read on their own. In a quarter system, I cover all of the chapter, frequently emphasizing the concepts illustrated in Figure 5.7 (the impedance seen looking into the base, emitter, or collector). With about two (perhaps two and half) weeks allotted to this chapter, the lectures must be precisely designed to ensure the main concepts are imparted in the classroom.

Chapter 6: Physics of MOS Devices This chapter parallels Chapter 4, introducing the MOSFET as a voltage-controlled current source and deriving its characteristics. Given the limited time that we generally face in covering topics, I have included only a brief discussion of the body effect and velocity saturation and neglected these phenomena for the remainder of the book. I cover all of this chapter in our first course.

Chapter 7: CMOS Amplifiers Drawing extensively upon the foundation established in Chapter 5, this chapter deals with MOS amplifiers but at a faster pace. I cover all of this chapter in our first course.

Chapter 8: Operational Amplifier as a Black Box Dealing with op-amp-based circuits, this chapter is written such that it can be taught in almost any order with respect to other chapters. My own preference is to cover this chapter *after* amplifier topologies have been studied, so that the students have some bare understanding of the internal circuitry of op amps and its gain limitations. Teaching this chapter near the end of the first course also places op amps closer to differential amplifiers (Chapter 10), thus allowing the students to appreciate the relevance of each. I cover all of this chapter in our first course.

Chapter 9: Cascodes and Current Mirrors This chapter serves as an important step toward integrated circuit design. The study of cascodes and current mirrors here also provides the necessary background for constructing differential pairs with active loads or cascodes in Chapter 10. From this chapter on, bipolar and MOS circuits are covered together and various similarities and contrasts between them are pointed out. In our second microelectronics course, I cover all of the topics in this chapter in approximately two weeks.

Chapter 10: Differential Amplifiers This chapter deals with large-signal and small-signal behavior of differential amplifiers. The students may wonder why we did not study the large-signal behavior of various amplifiers in Chapters 5 and 7; so I explain that the differential pair is a versatile circuit and is utilized in both regimes. I cover all of this chapter in our second course.

Chapter 11: Frequency Response Beginning with a review of basic concepts such as Bode's rules, this chapter introduces the high-frequency model of transistors and analyzes the frequency response of basic amplifiers. I cover all of this chapter in our second course.

Chapter 12: Feedback and Stability Most instructors agree the students find feedback to be the most difficult topic in undergraduate microelectronics. For this reason, I have made great effort to create a step-by-step procedure for analyzing feedback circuits, especially where input and output loading effects must be taken into account. As with Chapters 2 and 5, this chapter proceeds at a deliberately slow pace, allowing the students to become comfortable with each concept and appreciate the points taught by each example. I cover all of this chapter in our second course.

Chapter 13: Oscillators This new chapter deals with both discrete and integrated oscillators. These circuits are both important in real-life applications and helpful in enhancing the feedback concepts taught previously. This chapter can be comfortably covered in a semester system.

Chapter 14: Output Stages and Power Amplifiers This chapter studies circuits that deliver higher power levels than those considered in previous chapters. Topologies such as push-pull stages and their limitations are analyzed. This chapter can be covered in a semester system.

Chapter 15: Analog Filters This chapter provides a basic understanding of passive and active filters, preparing the student for more advanced texts on the subject. This chapter can also be comfortably covered in a semester system.

Chapter 16: Digital CMOS Circuits This chapter is written for microelectronics courses that include an introduction to digital circuits as a preparation for subsequent courses on the subject. Given the time constraints in quarter and semester systems, I have excluded TTL and ECL circuits here.

Chapter 17: CMOS Amplifiers This chapter is written for courses that cover CMOS circuits before bipolar circuits. As explained earlier, this chapter follows MOS device physics and, in essence, is similar to Chapter 5 but deals with MOS counterparts.

Problem Sets In addition to numerous examples, each chapter offers a relatively large problem set at the end. For each concept covered in the chapter, I begin with simple, confidence-building problems and gradually raise the level of difficulty. Except for the device physics chapters, all chapters also provide a set of design problems that encourage students to work "in reverse" and select the bias and/or component values to satisfy certain requirements.

SPICE Some basic circuit theory courses may provide exposure to SPICE, but it is in the first microelectronics course that the students can appreciate the importance of simulation tools. Appendix A of this book introduces SPICE and teaches circuit simulation with the aid of numerous examples. The objective is to master only a *subset* of SPICE commands that allow simulation of most circuits at this level. Due to the limited lecture time, I ask the teaching assistants to cover SPICE in a special evening session around the middle of the quarter—just before I begin to assign SPICE problems.

Most chapters contain SPICE problems, but I prefer to introduce SPICE only in the second half of the first course (toward the end of Chapter 5). This is for two reasons: (1) the students must first develop their basic understanding and analytical skills, i.e., the homeworks must exercise the fundamental concepts; and (2) the students appreciate the utility of SPICE much better if the circuit contains a relatively large number of devices (e.g., 5–10).

Homeworks and Exams In a quarter system, I assign four homeworks before the midterm and four after. Mostly based on the problem sets in the book, the homeworks contain moderate to difficult problems, thereby requiring that the students first go over the easier problems in the book on their own.

The exam questions are typically "twisted" versions of the problems in the book. To encourage the students to solve *all* of the problems at the end of each chapter, I tell them that one of the problems in the book is given in the exam verbatim. The exams are open-book, but I suggest to the students to summarize the important equations on one sheet of paper.

Happy Teaching!

Contents

1

Introduction to Microelectronics

Over the past five decades, microelectronics has revolutionized our lives. While beyond the realm of possibility a few decades ago, cellphones, digital cameras, laptop computers, and many other electronic products have now become an integral part of our daily affairs.

Learning microelectronics *can* be fun. As we learn how each device operates, how devices comprise circuits that perform interesting and useful functions, and how circuits form sophisticated systems, we begin to see the beauty of microelectronics and appreciate the reasons for its explosive growth.

This chapter gives an overview of microelectronics so as to provide a context for the material presented in this book. We introduce examples of microelectronic systems and identify important circuit "functions" that they employ. We also provide a review of basic circuit theory to refresh the reader's memory.

1.1 ELECTRONICS VERSUS MICROELECTRONICS

The general area of electronics began about a century ago and proved instrumental in the radio and radar communications used during the two world wars. Early systems incorporated "vacuum tubes," amplifying devices that operated with the flow of electrons between plates in a vacuum chamber. However, the finite lifetime and the large size of vacuum tubes motivated researchers to seek an electronic device with better properties.

The first transistor was invented in the 1940s and rapidly displaced vacuum tubes. It exhibited a very long (in principle, infinite) lifetime and occupied a much smaller volume (e.g., less than 1 cm^3 in packaged form) than vacuum tubes did.

But it was not until 1960s that the field of microelectronics, i.e., the science of integrating many transistors on one chip, began. Early "integrated circuits" (ICs) contained only a handful of devices, but advances in the technology soon made it possible to dramatically increase the complexity of "microchips."

Example 1-1 Today's microprocessors contain about 100 million transistors in a chip area of approximately 3 cm × 3 cm. (The chip is a few hundred microns thick.) Suppose integrated circuits were not invented and we attempted to build a processor using 100 million "discrete" transistors. If each device occupies a volume of 3 mm × 3 mm × 3 mm, determine the minimum volume for the processor. What other issues would arise in such an implementation?

Solution The minimum volume is given by $27\ \text{mm}^3 \times 10^8$, i.e., a cube 1.4 m on each side! Of course, the wires connecting the transistors would increase the volume substantially.

In addition to occupying a large volume, this discrete processor would be extremely *slow*; the signals would need to travel on wires as long as 1.4 m! Furthermore, if each discrete transistor costs 1 cent and weighs 1 g, each processor unit would be priced at one million dollars and weigh 100 tons!

Exercise How much power would such a system consume if each transistor dissipates 10 μW?

This book deals mostly with microelectronics while providing sufficient foundation for general (perhaps discrete) electronic systems as well.

1.2 EXAMPLES OF ELECTRONIC SYSTEMS

At this point, we introduce two examples of microelectronic systems and identify some of the important building blocks that we should study in basic electronics.

1.2.1 Cellular Telephone

Cellular telephones were developed in the 1980s and rapidly became popular in the 1990s. Today's cellphones contain a great deal of sophisticated analog and digital electronics that lie well beyond the scope of this book. But our objective here is to see how the concepts described in this book prove relevant to the operation of a cellphone.

Suppose you are speaking with a friend on your cellphone. Your voice is converted to an electric signal by a microphone and, after some processing, transmitted by the antenna. The signal produced by your antenna is picked up by your friend's receiver and, after some processing, applied to the speaker [Fig. 1.1(a)]. What goes on in these black boxes? Why are they needed?

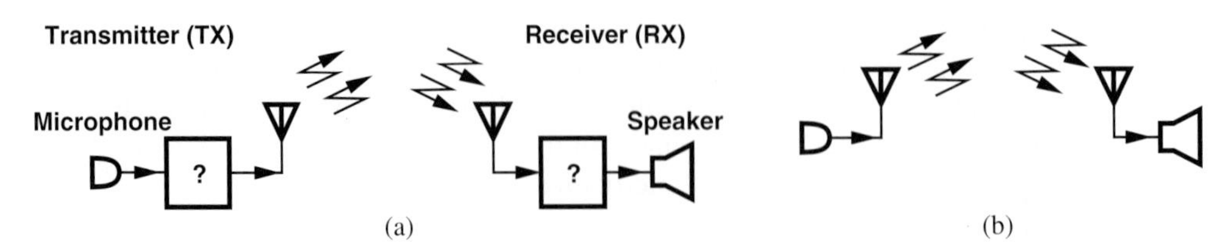

Figure 1.1 (a) Simplified view of a cellphone, (b) further simplification of transmit and receive paths.

Let us attempt to omit the black boxes and construct the simple system shown in Fig. 1.1(b). How well does this system work? We make two observations. First, our voice contains frequencies from 20 Hz to 20 kHz (called the "voice band"). Second, for an antenna to operate efficiently, i.e., to convert most of the electrical signal to electromagnetic radiation, its dimension must be a significant fraction (e.g., 25%) of the wavelength. Unfortunately, a frequency range of 20 Hz to 20 kHz translates to a wavelength[1] of 1.5×10^7 m to 1.5×10^4 m, requiring gigantic antennas for each cellphone. Conversely, to obtain a reasonable antenna length, e.g., 5 cm, the wavelength must be around 20 cm and the frequency around 1.5 GHz.

How do we "convert" the voice band to a gigahertz center frequency? One possible approach is to multiply the voice signal, $x(t)$, by a sinusoid, $A\cos(2\pi f_c t)$ [Fig. 1.2(a)]. Since multiplication in the time domain corresponds to convolution in the frequency domain, and since the spectrum of the sinusoid consists of two impulses at $\pm f_c$, the voice spectrum is simply shifted (translated) to $\pm f_c$ [Fig. 1.2(b)]. Thus, if $f_c = 1$ GHz, the output occupies a bandwidth of 40 kHz centered at 1 GHz. This operation is an example of "amplitude modulation."[2]

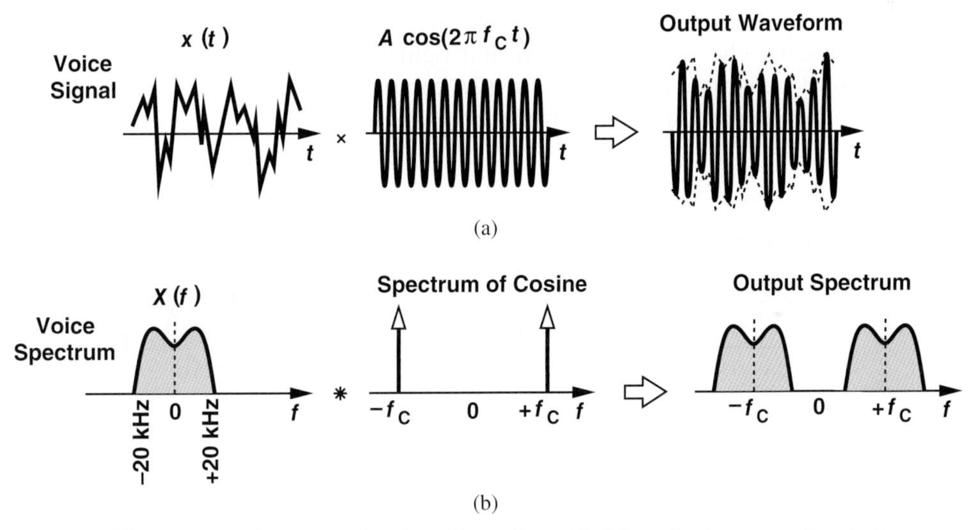

Figure 1.2 (a) Multiplication of a voice signal by a sinusoid, (b) equivalent operation in the frequency domain.

We therefore postulate that the black box in the transmitter of Fig. 1.1(a) contains a multiplier,[3] as depicted in Fig. 1.3(a). But two other issues arise. First, the cellphone must deliver a relatively large voltage swing (e.g., 20 V_{pp}) to the antenna so that the radiated power can reach across distances of several kilometers, thereby requiring a "power amplifier" between the multiplier and the antenna. Second, the sinusoid, $A\cos 2\pi f_c t$, must be produced by an "oscillator." We thus arrive at the transmitter architecture shown in Fig. 1.3(b).

[1]Recall that the wavelength is equal to the (light) velocity divided by the frequency.
[2]Cellphones in fact use other types of modulation to translate the voice band to higher frequencies.
[3]Also called a "mixer" in high-frequency electronics.

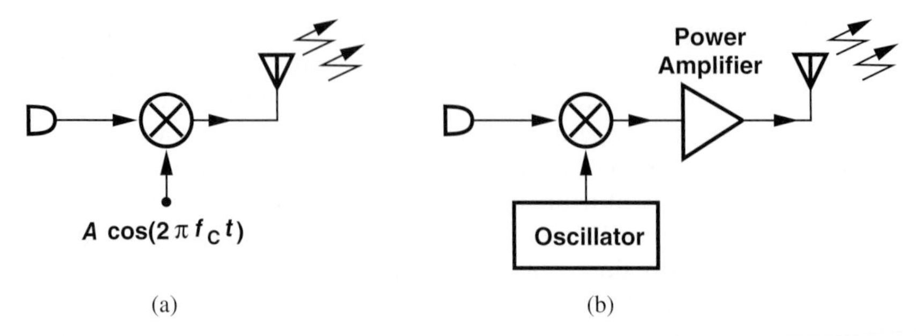

Figure 1.3 (a) Simple transmitter, (b) more complete transmitter.

Let us now turn our attention to the receive path of the cellphone, beginning with the simple realization illustrated in Fig. 1.1(b). Unfortunately, this topology fails to operate with the principle of modulation: if the signal received by the antenna resides around a gigahertz center frequency, the audio speaker cannot produce meaningful information. In other words, a means of translating the spectrum back to zero center frequency is necessary. For example, as depicted in Fig. 1.4(a), multiplication by a sinusoid, $A\cos(2\pi f_c t)$, translates the spectrum to left and right by f_c, restoring the original voice band. The newly-generated components at $\pm 2f_c$ can be removed by a low-pass filter. We thus arrive at the receiver topology shown in Fig. 1.4(b).

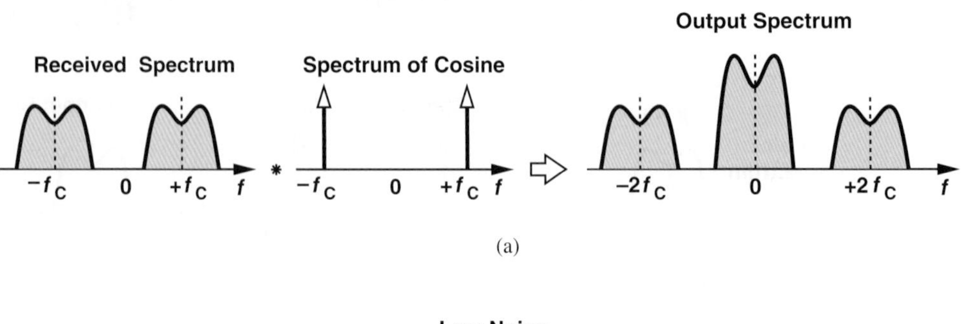

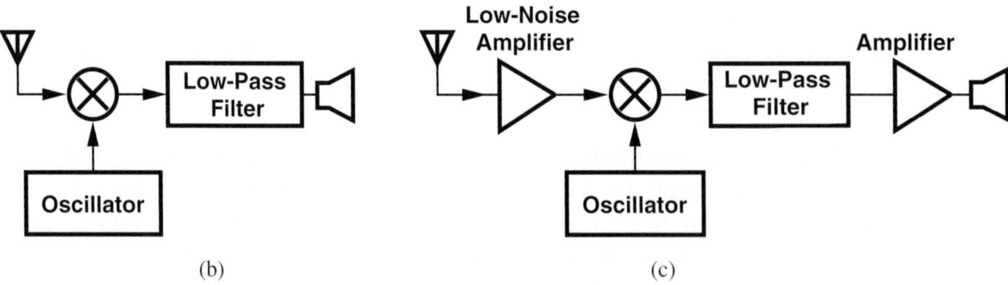

Figure 1.4 (a) Translation of modulated signal to zero center frequency, (b) simple receiver, (c) more complete receiver.

Our receiver design is still incomplete. The signal received by the antenna can be as low as a few tens of microvolts whereas the speaker may require swings of several tens or hundreds of millivolts. That is, the receiver must provide a great deal of amplification ("gain") between the antenna and the speaker. Furthermore, since multipliers typically suffer from a high "noise" and hence corrupt the received signal, a "low-noise amplifier" must precede the multiplier. The overall architecture is depicted in Fig. 1.4(c).

Today's cellphones are much more sophisticated than the topologies developed above. For example, the voice signal in the transmitter and the receiver is applied to a digital signal processor (DSP) to improve the quality and efficiency of the communication. Nonetheless, our study reveals some of the *fundamental* building blocks of cellphones, e.g., amplifiers, oscillators, and filters, with the last two also utilizing amplification. We therefore devote a great deal of effort to the analysis and design of amplifiers.

Having seen the necessity of amplifiers, oscillators, and multipliers in both transmit and receive paths of a cellphone, the reader may wonder if "this is old stuff" and rather trivial compared to the state of the art. Interestingly, these building blocks still remain among the most challenging circuits in communication systems. This is because the design entails critical *trade-offs* between speed (gigahertz center frequencies), noise, power dissipation (i.e., battery lifetime), weight, cost (i.e., price of a cellphone), and many other parameters. In the competitive world of cellphone manufacturing, a given design is never "good enough" and the engineers are forced to further push the above trade-offs in each new generation of the product.

1.2.2 Digital Camera

Another consumer product that, by virtue of "going electronic," has dramatically changed our habits and routines is the digital camera. With traditional cameras, we received no immediate feedback on the quality of the picture that was taken, we were very careful in selecting and shooting scenes to avoid wasting frames, we needed to carry bulky rolls of film, and we would obtain the final result only in printed form. With digital cameras, on the other hand, we have resolved these issues and enjoy many other features that only electronic processing can provide, e.g., transmission of pictures through cellphones or ability to retouch or alter pictures by computers. In this section, we study the operation of the digital camera.

The "front end" of the camera must convert light to electricity, a task performed by an array (matrix) of "pixels."[4] Each pixel consists of an electronic device (a "photodiode") that produces a current proportional to the intensity of the light that it receives. As illustrated in Fig. 1.5(a), this current flows through a capacitance, C_L, for a certain period of time, thereby developing a proportional voltage across it. Each pixel thus provides a voltage proportional to the "local" light density.

Now consider a camera with, say, 6.25 million pixels arranged in a 2500×2500 array [Fig. 1.5(b)]. How is the output voltage of each pixel sensed and processed? If each pixel contains its own electronic circuitry, the overall array occupies a very large area, raising the cost and the power dissipation considerably. We must therefore "time-share" the signal processing circuits among pixels. To this end, we follow the circuit of Fig. 1.5(a) with a simple, compact amplifier and a switch (within the pixel) [Fig. 1.5(c)]. Now, we connect a wire to the outputs of all 2500 pixels in a "column," turn on only one switch at a time, and apply the corresponding voltage to the "signal processing" block outside the column. The overall array consists of 2500 of such columns, with each column employing a dedicated signal processing block.

[4]The term "pixel" is an abbreviation of "picture cell."

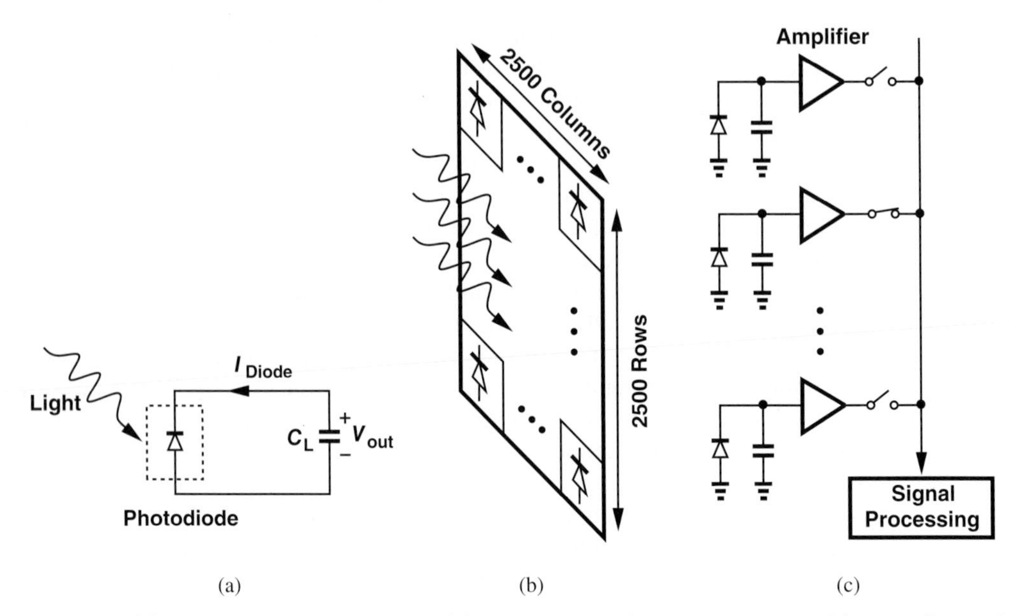

Figure 1.5 (a) Operation of a photodiode, (b) array of pixels in a digital camera, (c) one column of the array.

Example 1-2 A digital camera is focused on a chess board. Sketch the voltage produced by one column as a function of time.

Solution The pixels in each column receive light only from the white squares [Fig. 1.6(a)]. Thus, the column voltage alternates between a maximum for such pixels and zero for those receiving no light. The resulting waveform is shown in Fig. 1.6(b).

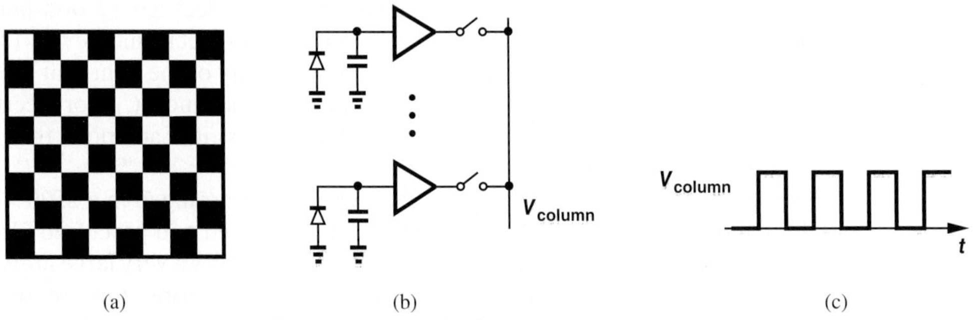

Figure 1.6 (a) Chess board captured by a digital camera, (b) sense circuits, and (c) voltage waveform on one column.

Exercise Plot the voltage if the first and second squares in each row have the same color.

What does each signal processing block do? Since the voltage produced by each pixel is an analog signal and can assume all values within a range, we must first "digitize" it by means of an "analog-to-digital converter" (ADC). A 6.25 megapixel array must thus incorporate 2500 ADCs. Since ADCs are relatively complex circuits, we may time-share one ADC between every two columns (Fig. 1.7), but requiring that the ADC operate twice as fast (why?). In the extreme case, we may employ a single, very fast ADC for all 2500 columns. In practice, the optimum choice lies between these two extremes.

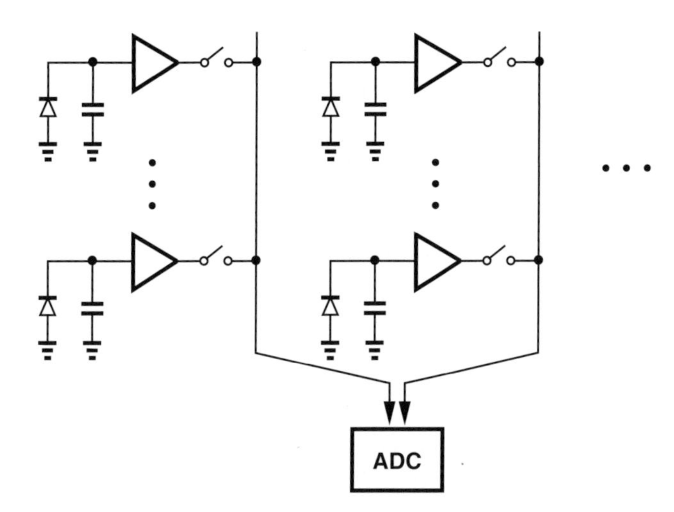

Figure 1.7 Sharing one ADC between two columns of a pixel array.

Once in the digital domain, the "video" signal collected by the camera can be manipulated extensively. For example, to "zoom in," the digital signal processor (DSP) simply considers only a section of the array, discarding the information from the remaining pixels. Also, to reduce the required memory size, the processor "compresses" the video signal.

The digital camera exemplifies the extensive use of both analog and digital microelectronics. The analog functions include amplification, switching operations, and analog-to-digital conversion, and the digital functions consist of subsequent signal processing and storage.

1.2.3 Analog Versus Digital

Amplifiers and ADCs are examples of analog functions, circuits that must process each point on a waveform (e.g., a voice signal) with great care to avoid effects such as noise and "distortion." By contrast, digital circuits deal with binary levels (ONEs and ZEROs) and, evidently, contain no analog functions. The reader may then say, "I have no intention of working for a cellphone or camera manufacturer and, therefore, need not learn about analog circuits." In fact, with digital communications, digital signal processors, and every other function becoming digital, is there any future for analog design?

Well, some of the assumptions in the above statements are incorrect. First, not every function can be realized digitally. The architectures of Figs. 1.3 and 1.4 must employ low-noise and low-power amplifiers, oscillators, and multipliers regardless of whether the actual communication is in analog or digital form. For example, a 20-μV signal (analog or digital) received by the antenna cannot be directly applied to a digital gate. Similarly, the video signal collectively captured by the pixels in a digital camera must be processed with low noise and distortion before it appears in the digital domain.

Second, digital circuits require analog expertise as the speed increases. Figure 1.8 exemplifies this point by illustrating two binary data waveforms, one at 100 Mb/s and another at 1 Gb/s. The finite risetime and falltime of the latter raises many issues in the operation of gates, flipflops, and other digital circuits, necessitating great attention to each point on the waveform.

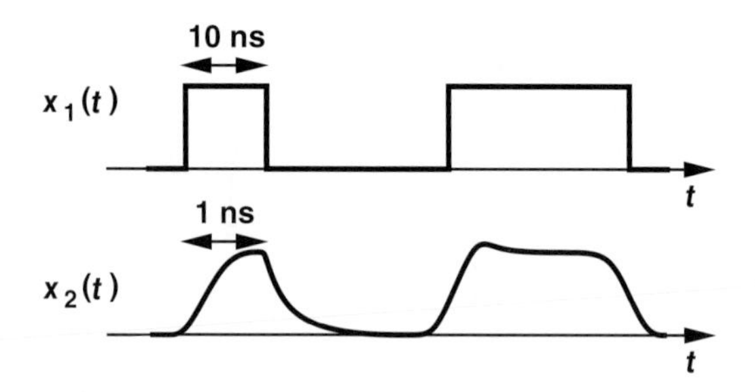

Figure 1.8 Data waveforms at 100 Mb/s and 1 Gb/s.

1.3 BASIC CONCEPTS*

Analysis of microelectronic circuits draws upon many concepts that are taught in basic courses on signals and systems and circuit theory. This section provides a brief review of these concepts so as to refresh the reader's memory and establish the terminology used throughout this book. The reader may first glance through this section to determine which topics need a review or simply return to this material as it becomes necessary later.

1.3.1 Analog and Digital Signals

An electric signal is a waveform that carries information. Signals that occur in nature can assume all values in a given range. Called "analog," such signals include voice, video, seismic, and music waveforms. As shown in Fig. 1.9(a), an analog voltage waveform swings through a "continuum" of values and provides information at each instant of time.

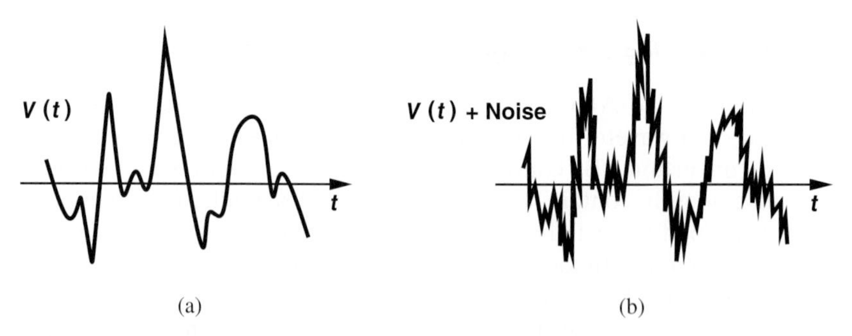

Figure 1.9 (a) Analog signal, (b) effect of noise on analog signal.

While occurring all around us, analog signals are difficult to "process" due to sensitivities to such circuit imperfections as "noise" and "distortion."[5] As an example, Fig. 1.9(b)

*This section serves as a review and can be skipped in classroom teaching.
[5]Distortion arises if the output is not a linear function of input.

illustrates the effect of noise. Furthermore, analog signals are difficult to "store" because they require "analog memories" (e.g., capacitors).

By contrast, a digital signal assumes only a finite number of values at only certain points in time. Depicted in Fig. 1.10(a) is a "binary" waveform, which remains at only one of two levels for each period, T. So long as the two voltages corresponding to ONEs and ZEROs differ sufficiently, logical circuits sensing such a signal process it correctly—even if noise or distortion create some corruption [Fig. 1.10(b)]. We therefore consider digital signals more "robust" than their analog counterparts. The storage of binary signals (in a digital memory) is also much simpler.

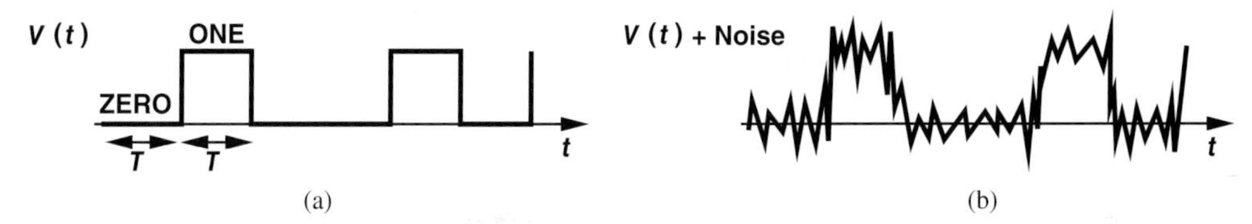

Figure 1.10 (a) Digital signal, (b) effect of noise on digital signal.

The foregoing observations favor processing of signals in the digital domain, suggesting that inherently analog information must be converted to digital form as early as possible. Indeed, complex microelectronic systems such as digital cameras, camcorders, and compact disk (CD) recorders perform some analog processing, "analog-to-digital conversion," and digital processing (Fig. 1.11), with the first two functions playing a critical role in the quality of the signal.

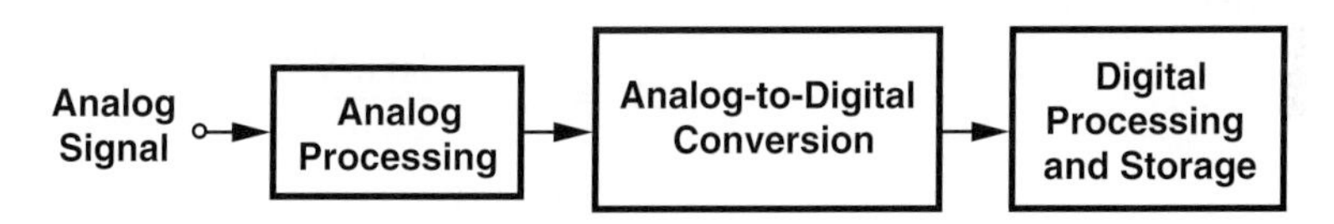

Figure 1.11 Signal processing in a typical system.

It is worth noting that many digital binary signals must be viewed and processed as analog waveforms. Consider, for example, the information stored on a hard disk in a computer. Upon retrieval, the "digital" data appears as a distorted waveform with only a few millivolts of amplitude (Fig. 1.12). Such a small separation between ONEs and ZEROs proves inadequate if this signal is to drive a logical gate, demanding a great deal of amplification and other analog processing before the data reaches a robust digital form.

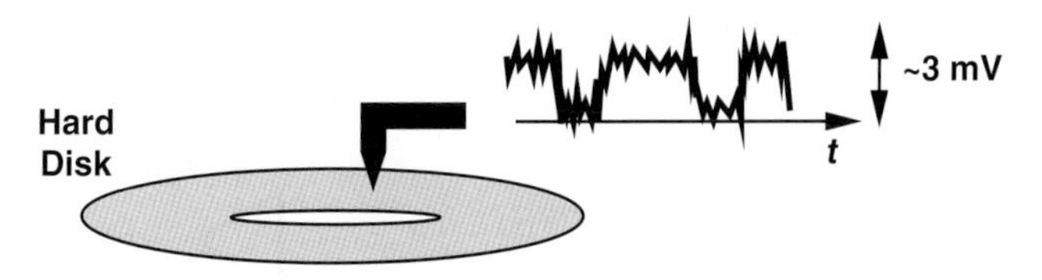

Figure 1.12 Signal picked up from a hard disk in a computer.

1.3.2 Analog Circuits

Today's microelectronic systems incorporate many analog functions. As exemplified by the cellphone and the digital camera studied above, analog circuits often limit the performance of the overall system.

The most commonly-used analog function is amplification. The signal received by a cellphone or picked up by a microphone proves too small to be processed further. An amplifier is therefore necessary to raise the signal swing to acceptable levels.

The performance of an amplifier is characterized by a number of parameters, e.g., gain, speed, and power dissipation. We study these aspects of amplification in great detail later in this book, but it is instructive to briefly review some of these concepts here.

A voltage amplifier produces an output swing greater than the input swing. The voltage gain, A_v, is defined as

$$A_v = \frac{v_{out}}{v_{in}}. \tag{1.1}$$

In some cases, we prefer to express the gain in decibels (dB):

$$A_v|_{dB} = 20 \log \frac{v_{out}}{v_{in}}. \tag{1.2}$$

For example, a voltage gain of 10 translates to 20 dB. The gain of typical amplifiers falls in the range of 10^1 to 10^5.

Example 1-3 A cellphone receives a signal level of 20 μV, but it must deliver a swing of 50 mV to the speaker that reproduces the voice. Calculate the required voltage gain in decibels.

Solution We have

$$A_v = 20 \log \frac{50\ \text{mV}}{20\ \mu\text{V}} \tag{1.3}$$

$$\approx 68\ \text{dB}. \tag{1.4}$$

Exercise What is the output swing if the gain is 50 dB?

In order to operate properly and provide gain, an amplifier must draw power from a voltage source, e.g., a battery or a charger. Called the "power supply," this source is typically denoted by V_{CC} or V_{DD} [Fig. 1.13(a)]. In complex circuits, we may simplify the notation to that shown in Fig. 1.13(b), where the "ground" terminal signifies a reference point with zero potential. If the amplifier is simply denoted by a triangle, we may even omit the supply terminals [Fig. 1.13(c)], with the understanding that they are present. Typical amplifiers operate with supply voltages in the range of 1 to 10 V.

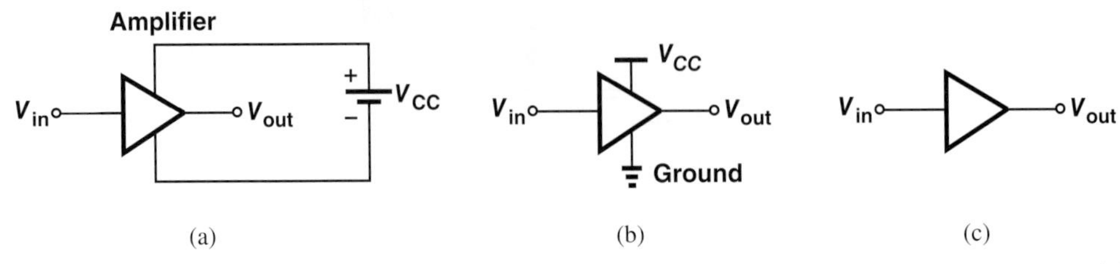

Figure 1.13 (a) General amplifier symbol along with its power supply, (b) simplified diagram of (a), (c) amplifier with supply rails omitted.

What limits the *speed* of amplifiers? We expect that various capacitances in the circuit begin to manifest themselves at high frequencies, thereby lowering the gain. In other words, as depicted in Fig. 1.14, the gain rolls off at sufficiently high frequencies, limiting the (usable) "bandwidth" of the circuit. Amplifiers (and other analog circuits) suffer from trade-offs between gain, speed, and power dissipation. Today's microelectronic amplifiers achieve bandwidths as large as tens of gigahertz.

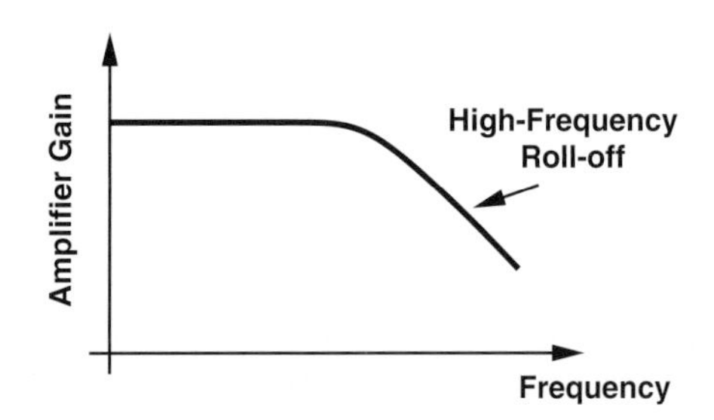

Figure 1.14 Roll-off of an amplifier's gain at high frequencies.

What other analog functions are frequently used? A critical operation is "filtering." For example, an electrocardiograph measuring a patient's heart activities also picks up the 60-Hz (or 50-Hz) electrical line voltage because the patient's body acts as an antenna. Thus, a filter must suppress this "interferer" to allow meaningful measurement of the heart.

1.3.3 Digital Circuits

More than 80% of the microelectronics industry deals with digital circuits. Examples include microprocessors, static and dynamic memories, and digital signal processors. Recall from basic logic design that gates form "combinational" circuits, and latches and flipflops constitute "sequential" machines. The complexity, speed, and power dissipation of these building blocks play a central role in the overall system performance.

In digital microelectronics, we study the design of the internal circuits of gates, flipflops, and other components. For example, we construct a circuit using devices such as transistors to realize the NOT and NOR functions shown in Fig. 1.15. Based on these implementations, we then determine various properties of each circuit. For example, what limits the speed of a gate? How much power does a gate consume while running at a certain speed? How *robustly* does a gate operate in the presence of nonidealities such as noise (Fig. 1.16)?

Figure 1.15 NOT and NOR gates.

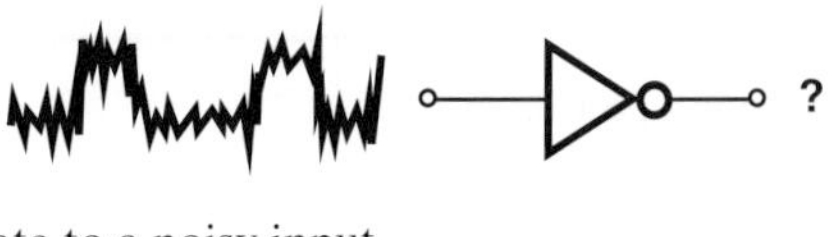

Figure 1.16 Response of a gate to a noisy input.

Example 1-4 Consider the circuit shown in Fig. 1.17, where switch S_1 is controlled by the digital input. That is, if A is high, S_1 is on and vice versa. Prove that the circuit provides the NOT function.

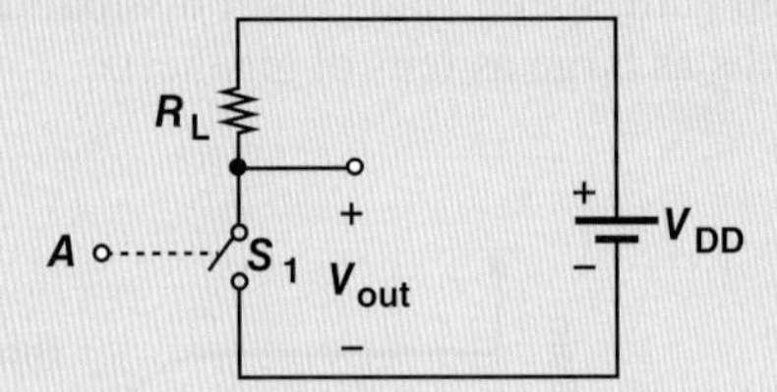

Figure 1.17

Solution If A is high, S_1 is on, forcing V_{out} to zero. On the other hand, if A is low, S_1 remains off, drawing no current from R_L. As a result, the voltage drop across R_L is zero and hence $V_{out} = V_{DD}$; i.e., the output is high. We thus observe that, for both logical states at the input, the output assumes the opposite state.

Exercise Determine the logical function if S_1 and R_L are swapped and V_{out} is sensed across R_L.

The above example indicates that *switches* can perform logical operations. In fact, early digital circuits did employ mechanical switches (relays), but suffered from a very limited speed (a few kilohertz). It was only after "transistors" were invented and their ability to act as switches was recognized that digital circuits consisting of millions of gates and operating at high speeds (several gigahertz) became possible.

1.3.4 Basic Circuit Theorems

Of the numerous analysis techniques taught in circuit theory courses, some prove particularly important to our study of microelectronics. This section provides a review of such concepts.

Kirchoff's Laws The Kirchoff Current Law (KCL) states that the sum of all currents flowing *into* a node is zero (Fig. 1.18):

$$\sum_j I_j = 0. \tag{1.5}$$

KCL in fact results from conservation of charge: a nonzero sum would mean that either some of the charge flowing into node X *vanishes* or this node *produces* charge.

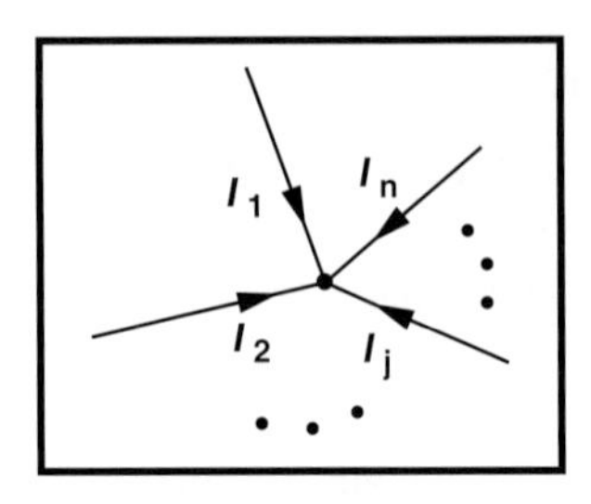

Figure 1.18 Illustration of KCL.

The Kirchoff Voltage Law (KVL) states that the sum of voltage drops around any closed loop in a circuit is zero [Fig. 1.19(a)]:

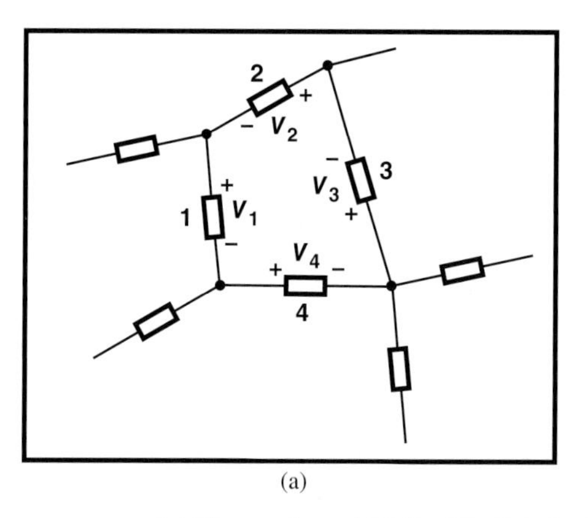

(a)

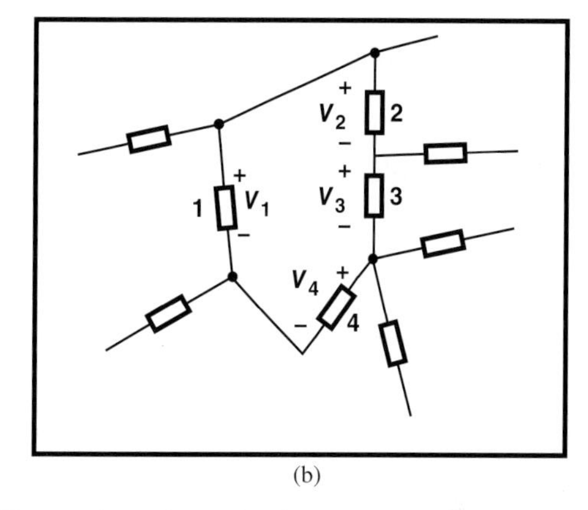

(b)

Figure 1.19 (a) Illustration of KVL, (b) slightly different view of the circuit.

$$\sum_j V_j = 0, \tag{1.6}$$

where V_j denotes the voltage drop across element number j. KVL arises from the conservation of the "electromotive force." In the example illustrated in Fig. 1.19(a), we may sum the voltages in the loop to zero: $V_1 + V_2 + V_3 + V_4 = 0$. Alternatively, adopting the modified view shown in Fig. 1.19(b), we can say V_1 is *equal* to the sum of the voltages across elements 2, 3, and 4: $V_1 = V_2 + V_3 + V_4$. Note that the polarities assigned to V_2, V_3, and V_4 in Fig. 1.19(b) are different from those in Fig. 1.19(a).

In solving circuits, we may not know a priori the correct polarities of the currents and voltages. Nonetheless, we can simply assign arbitrary polarities, write KCLs and KVLs, and solve the equations to obtain the actual polarities and values.

Example 1-5

The topology depicted in Fig. 1.20 represents the equivalent circuit of an amplifier. The dependent current source i_1 is equal to a constant, g_m,[6] multiplied by the voltage drop across r_π. Determine the voltage gain of the amplifier, v_{out}/v_{in}.

Figure 1.20

Solution We must compute V_{out} in terms of v_{in}, i.e., we must eliminate v_π from the equations. Writing a KVL in the "input loop," we have

$$v_{in} = v_\pi, \tag{1.7}$$

[6] What is the dimension of g_m?

and hence $g_m v_\pi = g_m v_{in}$. A KCL at the output node yields

$$g_m v_\pi + \frac{v_{out}}{R_L} = 0. \tag{1.8}$$

It follows that

$$\frac{v_{out}}{v_{in}} = -g_m R_L. \tag{1.9}$$

Note that the circuit amplifies the input if $g_m R_L > 1$. Unimportant in most cases, the negative sign simply means the circuit "inverts" the signal.

Exercise Repeat the above example if $r_\pi \to \infty$.

Example 1-6 Figure 1.21 shows another amplifier topology. Compute the gain.

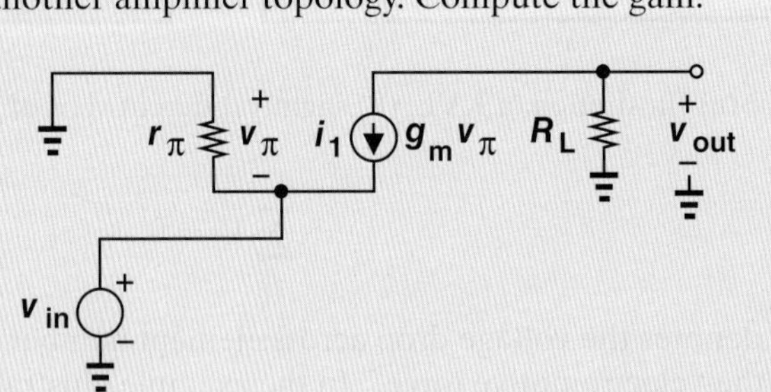

Figure 1.21

Solution Noting that r_π in fact appears in parallel with v_{in}, we write a KVL across these two components:

$$v_{in} = -v_\pi. \tag{1.10}$$

The KCL at the output node is similar to (1.8). Thus,

$$\frac{v_{out}}{v_{in}} = g_m R_L. \tag{1.11}$$

Interestingly, this type of amplifier does not invert the signal.

Exercise Repeat the above example if $r_\pi \to \infty$.

Example 1-7 A third amplifier topology is shown in Fig. 1.22. Determine the voltage gain.

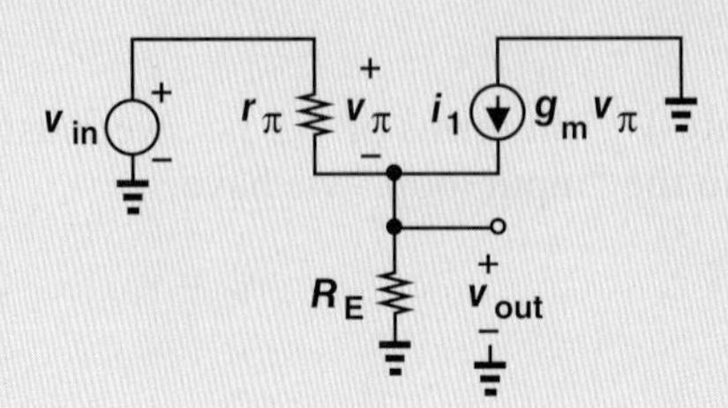

Figure 1.22

Solution We first write a KVL around the loop consisting of v_{in}, r_π, and R_E:

$$v_{in} = v_\pi + v_{out}. \tag{1.12}$$

That is, $v_\pi = v_{in} - v_{out}$. Next, noting that the currents v_π/r_π and $g_m v_\pi$ flow *into* the output node, and the current v_{out}/R_E flows *out* of it, we write a KCL:

$$\frac{v_\pi}{r_\pi} + g_m v_\pi = \frac{v_{out}}{R_E}. \tag{1.13}$$

Substituting $v_{in} - v_{out}$ for v_π gives

$$v_{in}\left(\frac{1}{r_\pi} + g_m\right) = v_{out}\left(\frac{1}{R_E} + \frac{1}{r_\pi} + g_m\right), \tag{1.14}$$

and hence

$$\frac{v_{out}}{v_{in}} = \frac{\dfrac{1}{r_\pi} + g_m}{\dfrac{1}{R_E} + \dfrac{1}{r_\pi} + g_m} \tag{1.15}$$

$$= \frac{(1 + g_m r_\pi)R_E}{r_\pi + (1 + g_m r_\pi)R_E}. \tag{1.16}$$

Note that the voltage gain always remains *below* unity. Would such an amplifier prove useful at all? In fact, this topology exhibits some important properties that make it a versatile building block.

Exercise Repeat the above example if $r_\pi \to \infty$.

The foregoing three examples relate to three amplifier topologies that are studied extensively in Chapter 5.

Thevenin and Norton Equivalents While Kirchoff's laws can always be utilized to solve any circuit, the Thevenin and Norton theorems can both simplify the algebra and, more importantly, provide additional insight into the operation of a circuit.

Thevenin's theorem states that a (linear) one-port network can be replaced with an equivalent circuit consisting of one voltage source in series with one impedance. Illustrated in Fig. 1.23(a), the term "port" refers to any two nodes whose voltage difference is of

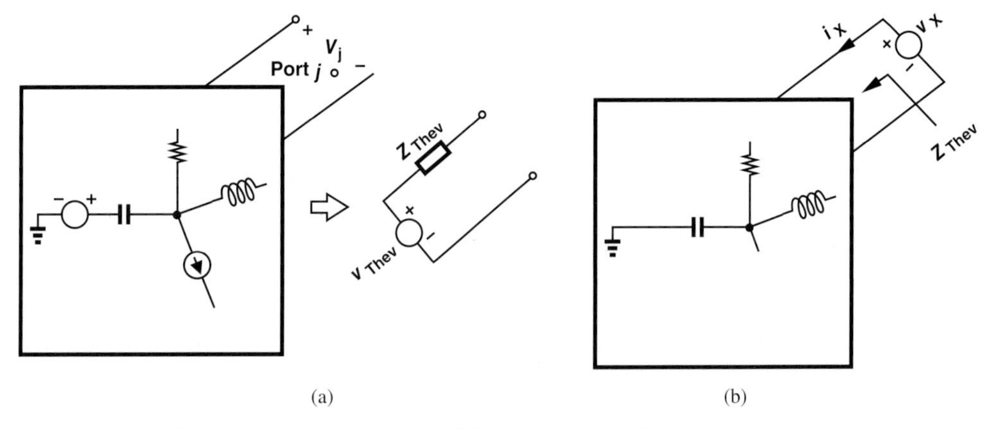

Figure 1.23 (a) Thevenin equivalent circuit, (b) computation of equivalent impedance.

interest. The equivalent voltage, v_{Thev}, is obtained by leaving the port *open* and computing the voltage created by the actual circuit at this port. The equivalent impedance, Z_{Thev}, is determined by setting all independent voltage and current sources in the circuit to zero and calculating the impedance between the two nodes. We also call Z_{Thev} the impedance "seen" when "looking" into the output port [Fig. 1.23(b)]. The impedance is computed by applying a voltage source across the port and obtaining the resulting current. A few examples illustrate these principles.

Example 1-8 Suppose the input voltage source and the amplifier shown in Fig. 1.20 are placed in a box and only the output port is of interest [Fig. 1.24(a)]. Determine the Thevenin equivalent of the circuit.

Solution We must compute the open-circuit output voltage and the impedance seen when looking into the output port. The Thevenin voltage is obtained from Fig. 1.24(a) and Eq. (1.9):

$$v_{\text{Thev}} = v_{out} \tag{1.17}$$

$$= -g_m R_L v_{in}. \tag{1.18}$$

To calculate Z_{Thev}, we set v_{in} to zero, apply a voltage source, v_X, across the output port, and determine the current drawn from the voltage source, i_X. As shown in Fig. 1.24(b), setting v_{in} to zero means replacing it with a *short circuit*. Also, note that the current source $g_m v_\pi$ remains in the circuit because it depends on the voltage across r_π, whose value is not known a priori.

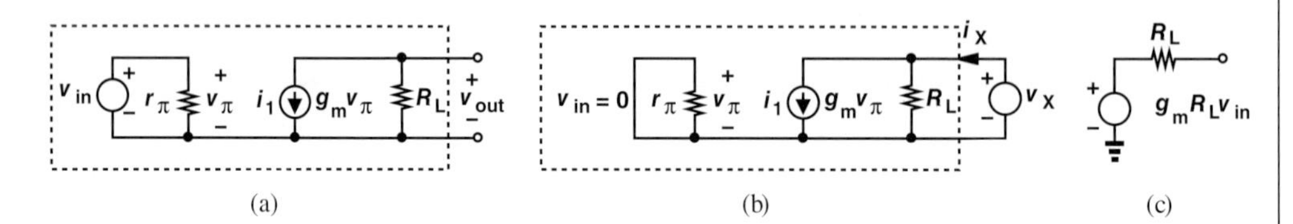

Figure 1.24

How do we solve the circuit of Fig. 1.24(b)? We must again eliminate v_π. Fortunately, since both terminals of r_π are tied to ground, $v_\pi = 0$ and $g_m v_\pi = 0$. The circuit thus reduces to R_L and

$$i_X = \frac{v_X}{R_L}. \tag{1.19}$$

That is,

$$R_{\text{Thev}} = R_L. \tag{1.20}$$

Figure 1.24(c) depicts the Thevenin equivalent of the input voltage source and the amplifier. In this case, we call $R_{\text{Thev}}(= R_L)$ the "output impedance" of the circuit.

Exercise Repeat the above example if $r_\pi \to \infty$.

With the Thevenin equivalent of a circuit available, we can readily analyze its behavior in the presence of a subsequent stage or "load."

Example 1-9 The amplifier of Fig. 1.20 must drive a speaker having an impedance of R_{sp}. Determine the voltage delivered to the speaker.

Solution Shown in Fig. 1.25(a) is the overall circuit arrangement that we must solve. Replacing the section in the dashed box with its Thevenin equivalent from Fig. 1.24(c), we greatly simplify the circuit [Fig. 1.25(b)], and write

$$v_{out} = -g_m R_L v_{in} \frac{R_{sp}}{R_{sp} + R_L} \tag{1.21}$$

$$= -g_m v_{in}(R_L || R_{sp}). \tag{1.22}$$

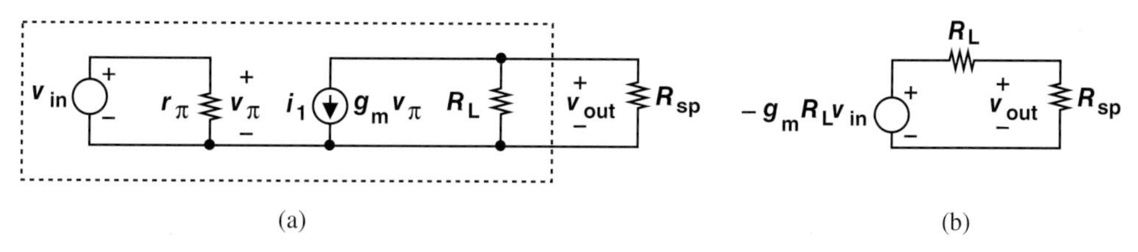

Figure 1.25

Exercise Repeat the above example if $r_\pi \to \infty$.

Example 1-10 Determine the Thevenin equivalent of the circuit shown in Fig. 1.22 if the output port is of interest.

Solution The open-circuit output voltage is simply obtained from (1.16):

$$v_{\text{Thev}} = \frac{(1 + g_m r_\pi)R_L}{r_\pi + (1 + g_m r_\pi)R_L} v_{in}. \tag{1.23}$$

To calculate the Thevenin impedance, we set v_{in} to zero and apply a voltage source across the output port as depicted in Fig. 1.26. To eliminate v_π, we recognize that the two terminals of r_π are tied to those of v_X and hence

$$v_\pi = -v_X. \tag{1.24}$$

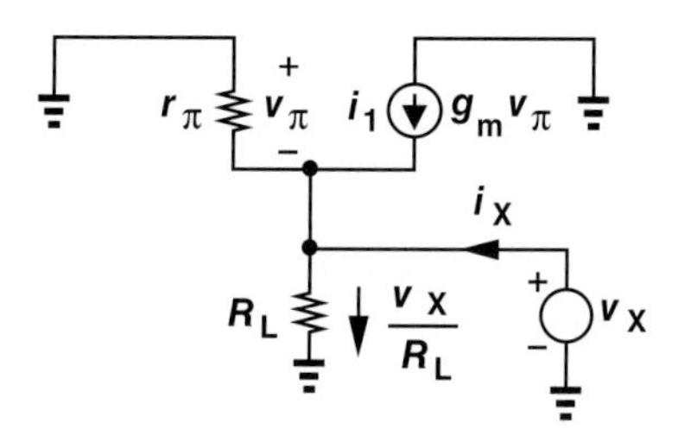

Figure 1.26

We now write a KCL at the output node. The currents v_π/r_π, $g_m v_\pi$, and i_X flow *into* this node and the current v_X/R_L flows out of it. Consequently,

$$\frac{v_\pi}{r_\pi} + g_m v_\pi + i_X = \frac{v_X}{R_L}, \tag{1.25}$$

or

$$\left(\frac{1}{r_\pi} + g_m\right)(-v_X) + i_X = \frac{v_X}{R_L}. \tag{1.26}$$

That is,

$$R_{\text{Thev}} = \frac{v_X}{i_X} \tag{1.27}$$

$$= \frac{r_\pi R_L}{r_\pi + (1 + g_m r_\pi)R_L}. \tag{1.28}$$

Exercise What happens if $R_L = \infty$?

Norton's theorem states that a (linear) one-port network can be represented by one current source in parallel with one impedance (Fig. 1.27). The equivalent current, i_{Nor}, is obtained by shorting the port of interest and computing the current that flows through it. The equivalent impedance, Z_{Nor}, is determined by setting all independent voltage and current sources in the circuit to zero and calculating the impedance seen at the port. Of course, $Z_{\text{Nor}} = Z_{\text{Thev}}$.

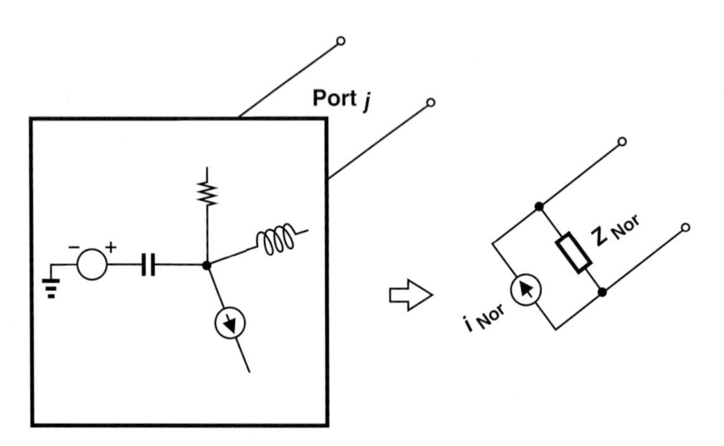

Figure 1.27 Norton's theorem.

Example 1-11 Determine the Norton equivalent of the circuit shown in Fig. 1.20 if the output port is of interest.

Solution As depicted in Fig. 1.28(a), we short the output port and seek the value of i_{Nor}. Since the voltage across R_L is now forced to zero, this resistor carries no current.

A KCL at the output node thus yields

$$i_{\text{Nor}} = -g_m v_\pi \tag{1.29}$$

$$= -g_m v_{in}. \tag{1.30}$$

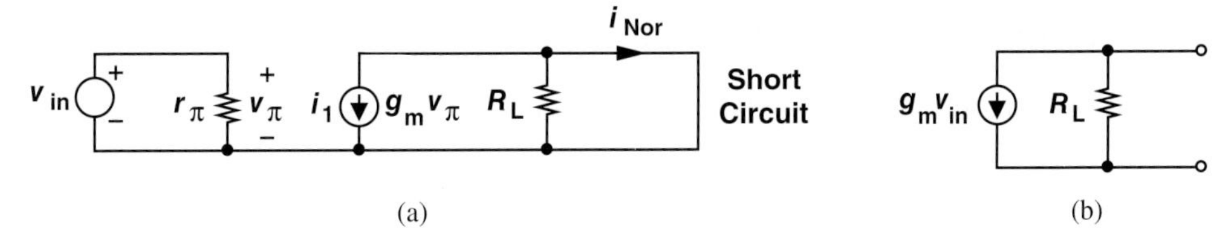

Figure 1.28

Also, from Example 1-8, R_{Nor} $(= R_{\text{Thev}}) = R_L$. The Norton equivalent therefore emerges as shown in Fig. 1.28(b). To check the validity of this model, we observe that the flow of i_{Nor} through R_L produces a voltage of $-g_m R_L v_{in}$, the same as the output voltage of the original circuit.

Exercise Repeat the above example if a resistor of value R_1 is added between the top terminal of v_{in} and the output node.

Example 1-12 Determine the Norton equivalent of the circuit shown in Fig. 1.22 if the output port is interest.

Solution Shorting the output port as illustrated in Fig. 1.29(a), we note that R_L carries no current. Thus,

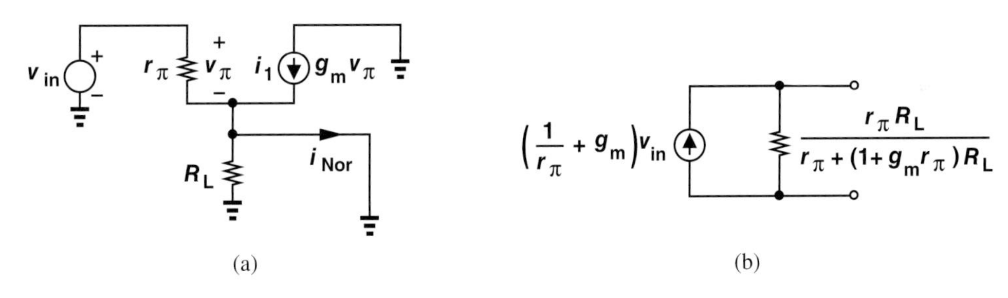

Figure 1.29

$$i_{\text{Nor}} = \frac{v_\pi}{r_\pi} + g_m v_\pi. \tag{1.31}$$

Also, $v_{in} = v_\pi$ (why?), yielding

$$i_{\text{Nor}} = \left(\frac{1}{r_\pi} + g_m\right) v_{in}. \tag{1.32}$$

With the aid of R_{Thev} found in Example 1-10, we construct the Norton equivalent depicted in Fig. 1.29(b).

Exercise What happens if $r_\pi = \infty$?

1.4 CHAPTER SUMMARY

- Electronic functions appear in many devices, including cellphones, digital cameras, laptop computers, etc.
- Amplification is an essential operation in many analog and digital systems.
- Analog circuits process signals that can assume various values at any time. By contrast, digital circuits deal with signals having only two levels and switching between these values at known points in time.
- Despite the "digital revolution," analog circuits find wide application in most of today's electronic systems.
- The voltage gain of an amplifier is defined as v_{out}/v_{in} and sometimes expressed in decibels (dB) as 20 $\log(v_{out}/v_{in})$.
- Kirchoff's current law (KCL) states that the sum of all currents flowing into any node is zero. Kirchoff's voltage law (KVL) states that the sum of all voltages around any loop is zero.
- Norton's theorem allows simplifying a one-port circuit to a current source in parallel with an impedance. Similarly, Thevenin's theorem reduces a one-port circuit to a voltage source in series with an impedance.

REFERENCE

1. H. Bharma, et al., "A subcutic millimeter wireless implantable intraocular pressure monitor microsystem," *IEEE Trans. Biomedical Circuits and Systems*, vol. 11, pp. 1204–1215, Dec. 2019.

2

Basic Physics of Semiconductors

Watch Companion YouTube Videos:

Razavi Electronics 1, Lec 1: Introduction, Charge Carriers, Doping
https://www.youtube.com/watch?v=yQDfVJzEyml

Razavi Electronics 1, Lec 2: Doping, Drift
https://www.youtube.com/watch?v=NWolpDgi6_Y

Razavi Electronics 1, Lec 3: Diffusion, Introduction to PN Junction
https://www.youtube.com/watch?v=mhtYm-USVD8

Razavi Electronics 1, Lec 4: PN Junction in Equilibrium & Reverse Bias
https://www.youtube.com/watch?v=6UNezo_GByE

Razavi Electronics 1, Lec 5: PN Junction in Forward Bias, Introduction to Diodes
https://www.youtube.com/watch?v=p9q5thwmlxY

Microelectronic circuits are based on complex semiconductor structures that have been under active research for the past six decades. While this book deals with the analysis and design of *circuits*, we should emphasize at the outset that a good understanding of *devices* is essential to our work. The situation is similar to many other engineering problems, e.g., one cannot design a high-performance automobile without a detailed knowledge of the engine and its limitations.

Nonetheless, we do face a dilemma. Our treatment of device physics must contain enough depth to provide adequate understanding, but must also be sufficiently brief to allow quick entry into circuits. This chapter accomplishes this task.

Our ultimate objective in this chapter is to study a fundamentally important and versatile device called the "diode." However, just as we need to eat our broccoli before having dessert, we must develop a basic understanding of "semiconductor" materials and their current conduction mechanisms before attacking diodes.

In this chapter, we begin with the concept of semiconductors and study the movement of charge (i.e., the flow of current) in them. Next, we deal with the "*pn* junction," which also serves as diode, and formulate its behavior. Our ultimate goal is to represent the device by a circuit model (consisting of resistors, voltage or current sources, capacitors, etc.), so that a circuit using such a device can be analyzed easily. The outline is shown below.

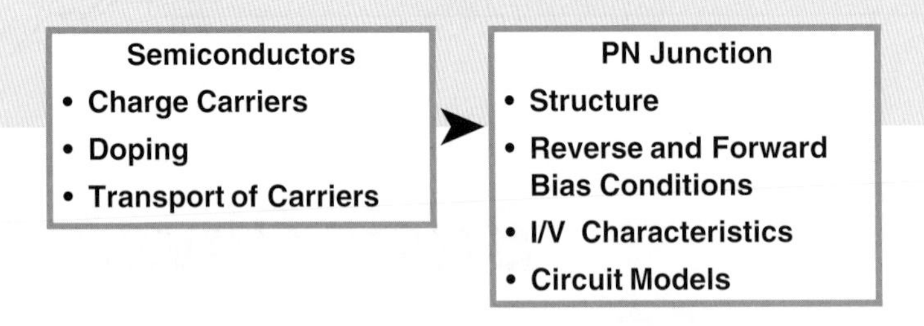

It is important to note that the task of developing accurate models proves critical for *all* microelectronic devices. The electronics industry continues to place greater demands on circuits, calling for aggressive designs that push semiconductor devices to their limits. Thus, a good understanding of the internal operation of devices is necessary.[1]

2.1 SEMICONDUCTOR MATERIALS AND THEIR PROPERTIES

Since this section introduces a multitude of concepts, it is useful to bear a general outline in mind (Fig. 2.1):

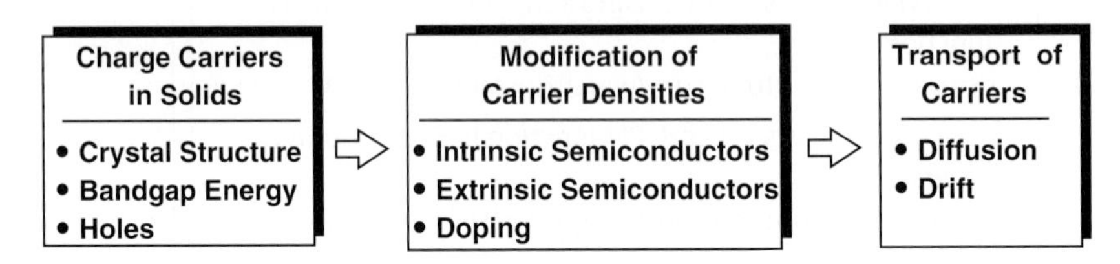

Figure 2.1 Outline of this section.

This outline represents a logical thought process: (a) we identify charge carriers in solids and formulate their role in current flow; (b) we examine means of modifying the density of charge carriers to create desired current flow properties; and (c) we determine current flow mechanisms. These steps naturally lead to the computation of the current/voltage (I/V) characteristics of actual diodes in the next section.

2.1.1 Charge Carriers in Solids

Recall from basic chemistry that the electrons in an atom orbit the nucleus in different "shells." The atom's chemical activity is determined by the electrons in the outermost shell, called "valence" electrons, and how complete this shell is. For example, neon exhibits a complete outermost shell (with eight electrons) and hence no tendency for chemical reactions. On the other hand, sodium has only one valence electron, ready to relinquish

[1] As design managers often say, "If you do not push the devices and circuits to their limit but your competitor does, then you lose to your competitor."

it, and chloride has seven valence electrons, eager to receive one more. Both elements are therefore highly reactive.

The above principles suggest that atoms having approximately four valence electrons fall somewhere between inert gases and highly volatile elements, possibly displaying interesting chemical and physical properties. Shown in Fig. 2.2 is a section of the periodic table containing a number of elements with three to five valence electrons. As the most popular material in microelectronics, silicon merits a detailed analysis.[2]

III	IV	V
Boron (B)	Carbon (C)	
Aluminum (Al)	Silicon (Si)	Phosphorus (P)
Galium (Ga)	Germanium (Ge)	Arsenic (As)

Figure 2.2 Section of the periodic table.

Covalent Bonds A silicon atom residing in isolation contains four valence electrons [Fig. 2.3(a)], requiring another four to complete its outermost shell. If processed properly, the silicon material can form a "crystal" wherein each atom is surrounded by exactly four others [Fig. 2.3(b)]. As a result, each atom *shares* one valence electron with its neighbors, thereby completing its own shell and those of the neighbors. The "bond" thus formed between atoms is called a "covalent bond" to emphasize the sharing of valence electrons.

The uniform crystal depicted in Fig. 2.3(b) plays a crucial role in semiconductor devices. But, does it carry current in response to a voltage? At temperatures near absolute zero, the valence electrons are confined to their respective covalent bonds, refusing to move freely. In other words, the silicon crystal behaves as an insulator for $T \rightarrow 0K$. However, at higher temperatures, electrons gain thermal energy, occasionally breaking away from the bonds and acting as free charge carriers [Fig. 2.3(c)] until they fall into another incomplete bond. We will hereafter use the term "electrons" to refer to free electrons.

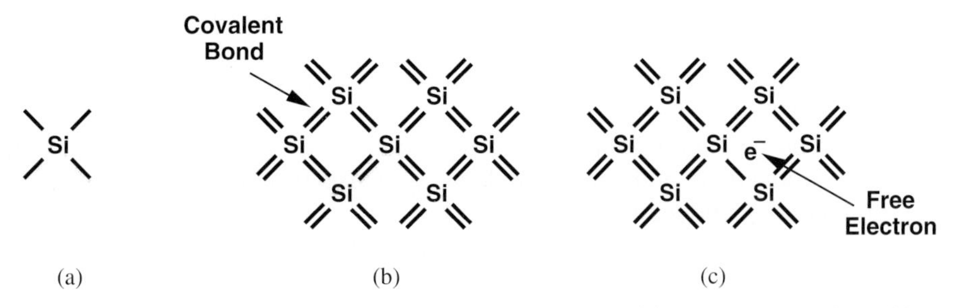

Figure 2.3 (a) Silicon atom, (b) covalent bonds between atoms, (c) free electron released by thermal energy.

[2]Silicon is obtained from sand after a great deal of processing.

Holes When freed from a covalent bond, an electron leaves a "void" behind because the bond is now incomplete. Called a "hole," such a void can readily absorb a free electron if one becomes available. Thus, we say an "electron-hole pair" is generated when an electron is freed, and an "electron-hole recombination" occurs when an electron "falls" into a hole.

Why do we bother with the concept of the hole? After all, it is the free electron that actually moves in the crystal. To appreciate the usefulness of holes, consider the time evolution illustrated in Fig. 2.4. Suppose covalent bond number 1 contains a hole after losing an electron some time before $t = t_1$. At $t = t_2$, an electron breaks away from bond number 2 and recombines with the hole in bond number 1. Similarly, at $t = t_3$, an electron leaves bond number 3 and falls into the hole in bond number 2. Looking at the three "snapshots," we can say one electron has traveled from right to left, or, alternatively, one hole has moved from left to right. This view of current flow by holes proves extremely useful in the analysis of semiconductor devices.

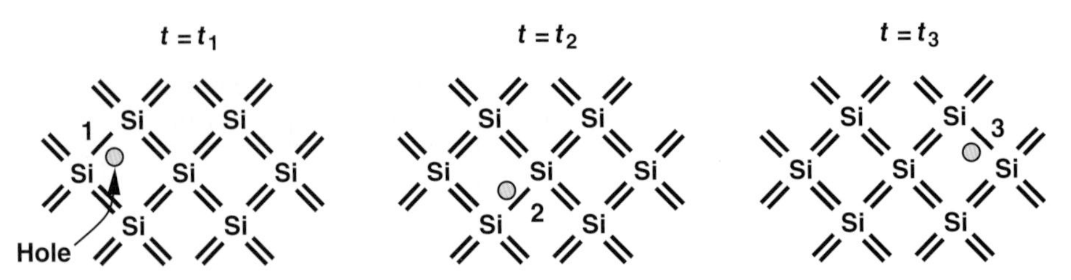

Figure 2.4 Movement of electron through crystal.

Bandgap Energy We must now answer two important questions. First, does *any* thermal energy create free electrons (and holes) in silicon? No, in fact, a minimum energy is required to dislodge an electron from a covalent bond. Called the "bandgap energy" and denoted by E_g, this minimum is a fundamental property of the material. For silicon, $E_g = 1.12$ eV.[3]

The second question relates to the conductivity of the material and is as follows. How *many* free electrons are created at a given temperature? From our observations thus far, we postulate that the number of electrons depends on both E_g and T: a greater E_g translates to fewer electrons, but a higher T yields more electrons. To simplify future derivations, we consider the *density* (or concentration) of electrons, i.e., the number of electrons per unit volume, n_i, and write for silicon:

$$n_i = 5.2 \times 10^{15} T^{3/2} \exp \frac{-E_g}{2kT} \text{ electrons/cm}^3 \tag{2.1}$$

where $k = 1.38 \times 10^{-23}$ J/K is called the Boltzmann constant. The derivation can be found in books on semiconductor physics, e.g., [1]. As expected, materials having a larger E_g exhibit a smaller n_i. Also, as $T \to 0$, so do $T^{3/2}$ and $\exp[-E_g/(2kT)]$, thereby bringing n_i toward zero.

The exponential dependence of n_i upon E_g reveals the effect of the bandgap energy on the conductivity of the material. Insulators display a high E_g; for example, $E_g = 2.5$ eV for diamond. Conductors, on the other hand, have a small bandgap. Finally, *semi*conductors exhibit a moderate E_g, typically ranging from 1 eV to 1.5 eV.

[3]The unit eV (electron volt) represents the energy necessary to move one electron across a potential difference of 1 V. Note that 1 eV $= 1.6 \times 10^{-19}$ J.

Example 2-1 Determine the density of electrons in silicon at $T = 300$ K (room temperature) and $T = 600$ K.

Solution Since $E_g = 1.12$ eV $= 1.792 \times 10^{-19}$ J, we have

$$n_i(T = 300\text{ K}) = 1.08 \times 10^{10}\text{ electrons/cm}^3, \tag{2.2}$$

$$n_i(T = 600\text{ K}) = 1.54 \times 10^{15}\text{ electrons/cm}^3. \tag{2.3}$$

Since for each free electron, a hole is left behind, the density of holes is also given by (2.2) and (2.3).

Exercise Repeat the above exercise for a material having a bandgap of 1.5 eV.

The n_i values obtained in the above example may appear quite high, but, noting that silicon has 5×10^{22} atoms/cm^3, we recognize that only one in 5×10^{12} atoms benefit from a free electron at room temperature. In other words, silicon still seems a very poor conductor. But, do not despair! We next introduce a means of making silicon more useful.

2.1.2 Modification of Carrier Densities

Intrinsic and Extrinsic Semiconductors

The "pure" type of silicon studied thus far is an example of "intrinsic semiconductors," suffering from a very high resistance. Fortunately, it is possible to modify the resistivity of silicon by replacing some of the atoms in the crystal with atoms of another material. In an intrinsic semiconductor, the electron density, $n(= n_i)$, is equal to the hole density, p. Thus,

$$np = n_i^2. \tag{2.4}$$

We return to this equation later.

Did you know?

The semiconductor industry manufactures microprocessors, memories, RF transceivers, imaging chips, and many other products, bringing in an annual revenue of 300 billion dollars. This means that, of the seven billion people in the world, each person spends an average of about \$40 on semiconductor chips every year. This starts when children buy their first video game device.

Recall from Fig. 2.2 that phosphorus (P) contains five valence electrons. What happens if some P atoms are introduced in a silicon crystal? As illustrated in Fig. 2.5, each P atom shares four electrons with the neighboring silicon atoms, leaving the fifth electron "unattached." This electron is free to move, serving as a charge carrier. Thus, if N phosphorus atoms are uniformly introduced in each cubic centimeter of a silicon crystal, then the density of free electrons rises by the same amount.

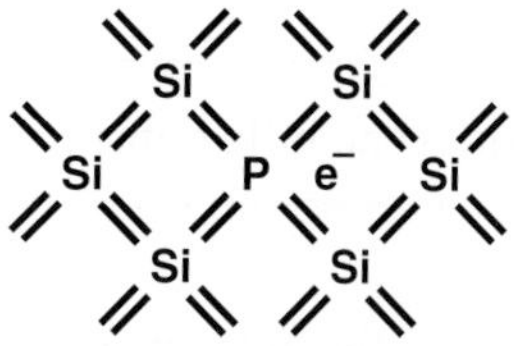

Figure 2.5 Loosely-attached electon with phosphorus doping.

The controlled addition of an "impurity" such as phosphorus to an intrinsic semiconductor is called "doping," and phosphorus itself a "dopant." Providing many more free electrons than in the intrinsic state, the doped silicon crystal is now called "extrinsic," more specifically, an "n-type" semiconductor to emphasize the abundance of free electrons.

As remarked earlier, the electron and hole densities in an intrinsic semiconductor are equal. But, how about these densities in a doped material? It can be proved that even in this case,

$$np = n_i^2, \tag{2.5}$$

where n and p respectively denote the electron and hole densities in the extrinsic semiconductor. The quantity n_i represents the densities in the intrinsic semiconductor (hence the subscript i) and is therefore independent of the doping level [e.g., Eq. (2.1) for silicon].

Example 2-2 The above result seems quite strange. How can np remain constant while we add more donor atoms and increase n?

Solution Equation (2.5) reveals that p must fall *below* its intrinsic level as more n-type dopants are added to the crystal. This occurs because many of the new electrons donated by the dopant "recombine" with the holes that were created in the intrinsic material.

Exercise Why can we not say that $n + p$ should remain constant?

Example 2-3 A piece of crystalline silicon is doped uniformly with phosphorus atoms. The doping density is 10^{16} atoms/cm^3. Determine the electron and hole densities in this material at the room temperature.

Solution The addition of 10^{16} P atoms introduces the same number of free electrons per cubic centimeter. Since this electron density exceeds that calculated in Example 2-1 by six orders of magnitude, we can assume

$$n = 10^{16} \text{ electrons/cm}^3. \tag{2.6}$$

It follows from (2.2) and (2.5) that

$$p = \frac{n_i^2}{n} \tag{2.7}$$

$$= 1.17 \times 10^4 \text{ holes/cm}^3. \tag{2.8}$$

Note that the hole density has dropped below the intrinsic level by six orders of magnitude. Thus, if a voltage is applied across this piece of silicon, the resulting current consists predominantly of electrons.

Exercise At what doping level does the hole density drop by three orders of magnitude?

This example justifies the reason for calling electrons the "majority carriers" and holes the "minority carriers" in an n-type semiconductor. We may naturally wonder if it is possible to construct a "p-type" semiconductor, thereby exchanging the roles of electrons and holes.

Indeed, if we can dope silicon with an atom that provides an *insufficient* number of electrons, then we may obtain many *incomplete* covalent bonds. For example, the table in Fig. 2.2 suggests that a boron (B) atom—with three valence electrons—can form only three complete covalent bonds in a silicon crystal (Fig. 2.6). As a result, the fourth bond contains a hole, ready to absorb a free electron. In other words, N boron atoms contribute N boron holes to the conduction of current in silicon. The structure in Fig. 2.6 therefore exemplifies a p-type semiconductor, providing holes as majority carriers. The boron atom is called an "acceptor" dopant.

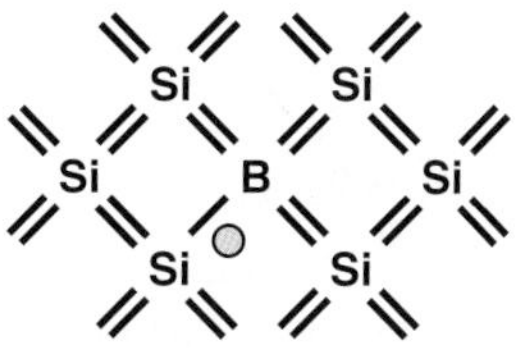

Figure 2.6 Available hole with boron doping.

Let us formulate our results thus far. If an intrinsic semiconductor is doped with a density of $N_D (\gg n_i)$ donor atoms per cubic centimeter, then the mobile charge densities are given by

$$\text{Majority Carriers: } n \approx N_D \tag{2.9}$$

$$\text{Minority Carriers: } p \approx \frac{n_i^2}{N_D}. \tag{2.10}$$

Similarly, for a density of $N_A (\gg n_i)$ acceptor atoms per cubic centimeter:

$$\text{Majority Carriers: } p \approx N_A \tag{2.11}$$

$$\text{Minority Carriers: } n \approx \frac{n_i^2}{N_A}. \tag{2.12}$$

Since typical doping densities fall in the range of 10^{15} to 10^{18} atoms/cm^3, the above expressions are quite accurate.

Example 2-4 Is it possible to use other elements of Fig. 2.2 as semiconductors and dopants?

Solution Yes, for example, some early diodes and transistors were based on germanium (Ge) rather than silicon. Also, arsenic (As) is another common dopant.

Exercise Can carbon be used for this purpose?

Figure 2.7 summarizes the concepts introduced in this section, illustrating the types of charge carriers and their densities in semiconductors.

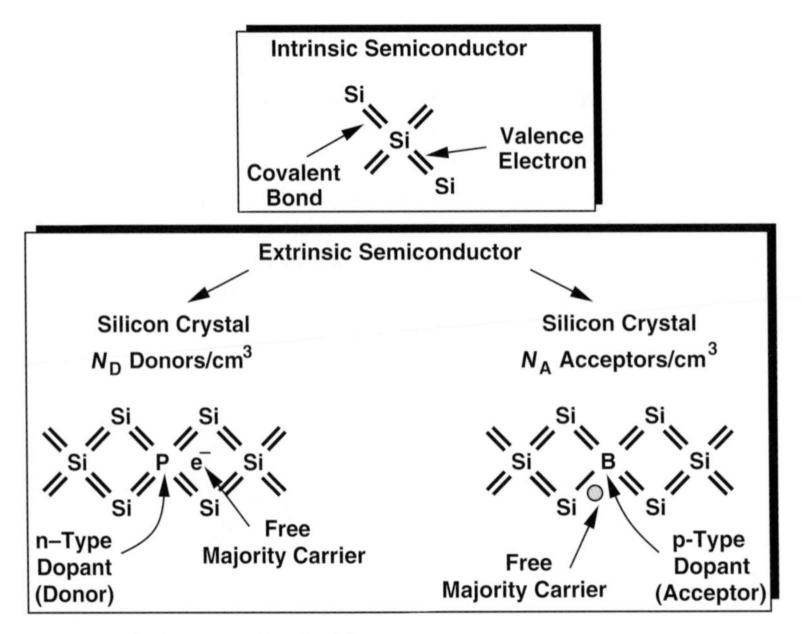

Figure 2.7 Summary of charge carriers in silicon.

2.1.3 Transport of Carriers

Having studied charge carriers and the concept of doping, we are ready to examine the *movement* of charge in semiconductors, i.e., the mechanisms leading to the flow of current.

Drift We know from basic physics and Ohm's law that a material can conduct current in response to a potential difference and hence an electric field.[4] The field accelerates the charge carriers in the material, forcing some to flow from one end to the other. Movement of charge carriers due to an electric field is called "drift."[5]

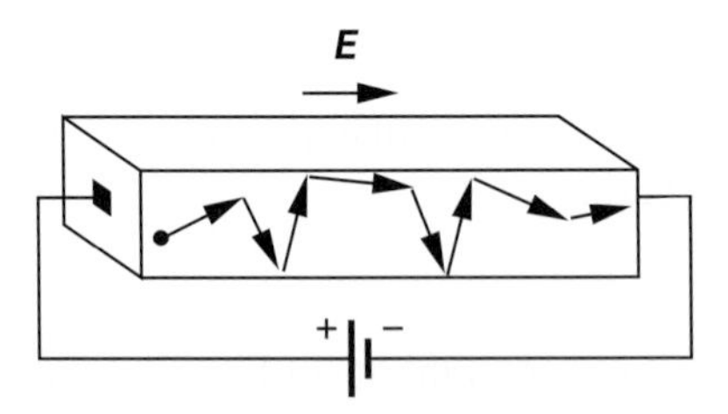

Figure 2.8 Drift in a semiconductor.

Semiconductors behave in a similar manner. As shown in Fig. 2.8, the charge carriers are accelerated by the field and accidentally collide with the atoms in the crystal, eventually

[4]Recall that the potential (voltage) difference, V, is equal to the negative integral of the electric field, E, with respect to distance: $V_{ab} = -\int_b^a E dx$.
[5]The convention for direction of current assumes flow of *positive* charge from a positive voltage to a negative voltage. Thus, if electrons flow from point A to point B, the current is considered to have a direction from B to A.

reaching the other end and flowing into the battery. The acceleration due to the field and the collision with the crystal counteract, leading to a *constant* velocity for the carriers.[6] We expect the velocity, v, to be proportional to the electric field strength, E:

$$v \propto E, \tag{2.13}$$

and hence

$$v = \mu E, \tag{2.14}$$

where μ is called the "mobility" and usually expressed in $\text{cm}^2/(\text{V} \cdot \text{s})$. For example, in silicon, the mobility of electrons, $\mu_n = 1350\ \text{cm}^2/(\text{V} \cdot \text{s})$, and that of holes, $\mu_p = 480\ \text{cm}^2/(\text{V} \cdot \text{s})$. Of course, since electrons move in a direction opposite to the electric field, we must express the velocity vector as

$$\vec{v_e} = -\mu_n \vec{E}. \tag{2.15}$$

For holes, on the other hand,

$$\vec{v_h} = \mu_p \vec{E}. \tag{2.16}$$

Example 2-5 A uniform piece of n-type of silicon that is 1 μm long senses a voltage of 1 V. Determine the velocity of the electrons.

Solution Since the material is uniform, we have $E = V/L$, where L is the length. Thus, $E = 10{,}000$ V/cm and hence $v = \mu_n E = 1.35 \times 10^7$ cm/s. In other words, electrons take $(1\ \mu\text{m})/(1.35 \times 10^7\ \text{cm/s}) = 7.4$ ps to cross the 1-μm length.

Exercise What happens if the mobility is halved?

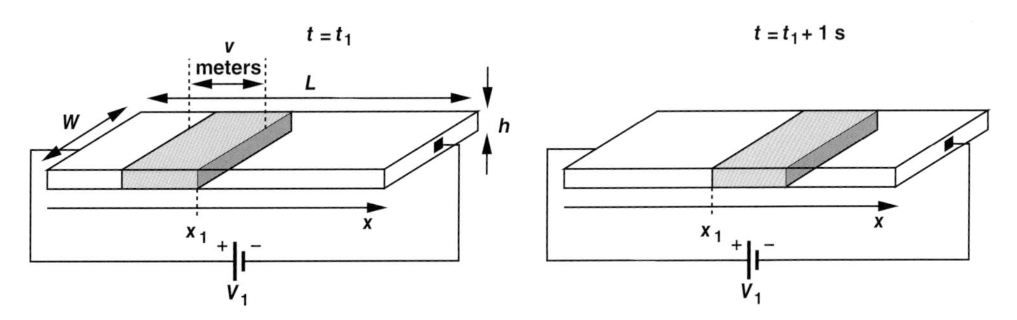

Figure 2.9 Current flow in terms of charge density.

With the velocity of carriers known, how is the current calculated? We first note that an electron carries a negative charge equal to $q = 1.6 \times 10^{-19}$ C. Equivalently, a hole carries a positive charge of the same value. Now suppose a voltage V_1 is applied across a uniform semiconductor bar having a free electron density of n (Fig. 2.9). Assuming the electrons move with a velocity of v m/s, considering a cross section of the bar at $x = x_1$ and taking two "snapshots" at $t = t_1$ and $t = t_1 + 1$ second, we note that the total charge in v meters

[6]This phenomenon is analogous to the "terminal velocity" that a sky diver with a parachute (hopefully, open) experiences.

passes the cross section in 1 second. In other words, the current is equal to the total charge enclosed in v meters of the bar's length. Since the bar has a width of W, we have:

$$I = -v \cdot W \cdot h \cdot n \cdot q, \tag{2.17}$$

where $v \cdot W \cdot h$ represents the volume, $n \cdot q$ denotes the charge density in coulombs, and the negative sign accounts for the fact that electrons carry negative charge.

Let us now reduce Eq. (2.13) to a more convenient form. Since for electrons, $v = -\mu_n E$, and since $W \cdot h$ is the cross section area of the bar, we write

$$J_n = \mu_n E \cdot n \cdot q, \tag{2.18}$$

where J_n denotes the "current density," i.e., the current passing through a *unit* cross section area, and is expressed in A/cm^2. We may loosely say, "the current is equal to the charge velocity times the charge density," with the understanding that "current" in fact refers to current density, and negative or positive signs are taken into account properly.

In the presence of both electrons and holes, Eq. (2.18) is modified to

$$J_{tot} = \mu_n E \cdot n \cdot q + \mu_p E \cdot p \cdot q \tag{2.19}$$

$$= q(\mu_n n + \mu_p p)E. \tag{2.20}$$

This equation gives the drift current density in response to an electric field E in a semiconductor having uniform electron and hole densities.

Example 2-6 In an experiment, it is desired to obtain equal electron and hole drift currents. How should the carrier densities be chosen?

Solution We must impose

$$\mu_n n = \mu_p p, \tag{2.21}$$

and hence

$$\frac{n}{p} = \frac{\mu_p}{\mu_n}. \tag{2.22}$$

We also recall that $np = n_i^2$. Thus,

$$p = \sqrt{\frac{\mu_n}{\mu_p}} n_i, \tag{2.23}$$

$$n = \sqrt{\frac{\mu_p}{\mu_n}} n_i. \tag{2.24}$$

For example, in silicon, $\mu_n/\mu_p = 1350/480 = 2.81$, yielding

$$p = 1.68 n_i, \tag{2.25}$$

$$n = 0.596 n_i. \tag{2.26}$$

Since p and n are of the same order as n_i, equal electron and hole drift currents can occur for only a very lightly doped material. This confirms our earlier notion of majority carriers in semiconductors having typical doping levels of 10^{15}–10^{18} atoms/cm^3.

Exercise How should the carrier densities be chosen so that the electron drift current is twice the hole drift current?

Velocity Saturation* We have thus far assumed that the mobility of carriers in semiconductors is *independent* of the electric field and the velocity rises linearly with E according to $v = \mu E$. In reality, if the electric field approaches sufficiently high levels, v no longer follows E linearly. This is because the carriers collide with the lattice so frequently and the time between the collisions is so short that they cannot accelerate much. As a result, v varies "sublinearly" at high electric fields, eventually reaching a saturated level, v_{sat} (Fig. 2.10). Called "velocity saturation," this effect manifests itself in some modern transistors, limiting the performance of circuits.

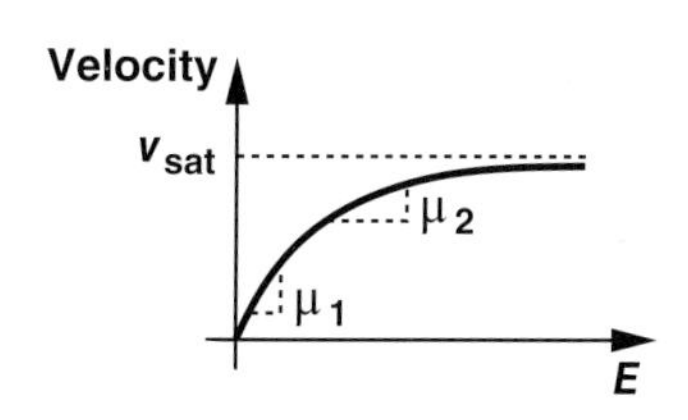

Figure 2.10 Velocity saturation.

In order to represent velocity saturation, we must modify $v = \mu E$ accordingly. A simple approach is to view the slope, μ, as a field-dependent parameter. The expression for μ must therefore gradually fall toward zero as E rises, but approach a constant value for small E; i.e.,

$$\mu = \frac{\mu_0}{1 + bE}, \tag{2.27}$$

where μ_0 is the "low-field" mobility and b a proportionality factor. We may consider μ as the "effective" mobility at an electric field E. Thus,

$$v = \frac{\mu_0}{1 + bE} E. \tag{2.28}$$

Since for $E \to \infty$, $v \to v_{sat}$, we have

$$v_{sat} = \frac{\mu_0}{b}, \tag{2.29}$$

and hence $b = \mu_0 / v_{sat}$. In other words,

$$v = \frac{\mu_0}{1 + \frac{\mu_0 E}{v_{sat}}} E. \tag{2.30}$$

Example 2-7 A uniform piece of semiconductor 0.2 μm long sustains a voltage of 1 V. If the low-field mobility is equal to 1350 $cm^2/(V \cdot s)$ and the saturation velocity of the carriers 10^7 cm/s, determine the effective mobility. Also, calculate the maximum allowable voltage such that the effective mobility is only 10% lower than μ_0.

Solution We have

$$E = \frac{V}{L} \tag{2.31}$$

$$= 50 \text{ kV/cm}. \tag{2.32}$$

*This section can be skipped in a first reading.

It follows that

$$\mu = \frac{\mu_0}{1 + \dfrac{\mu_0 E}{v_{sat}}} \tag{2.33}$$

$$= \frac{\mu_0}{7.75} \tag{2.34}$$

$$= 174\ \text{cm}^2/(\text{V} \cdot \text{s}). \tag{2.35}$$

If the mobility must remain within 10% of its low-field value, then

$$0.9\mu_0 = \frac{\mu_0}{1 + \dfrac{\mu_0 E}{v_{sat}}}, \tag{2.36}$$

and hence

$$E = \frac{1}{9}\frac{v_{sat}}{\mu_0} \tag{2.37}$$

$$= 823\ \text{V/cm}. \tag{2.38}$$

A device of length 0.2 μm experiences such a field if it sustains a voltage of (823 V/cm) × (0.2×10^{-4} cm) = 16.5 mV.

This example suggests that modern (submicron) devices incur substantial velocity saturation because they operate with voltages much greater than 16.5 mV.

Exercise At what voltage does the mobility fall by 20%?

Diffusion In addition to drift, another mechanism can lead to current flow. Suppose a drop of ink falls into a glass of water. Introducing a high local concentration of ink molecules, the drop begins to "diffuse," that is, the ink molecules tend to flow from a region of high concentration to regions of low concentration. This mechanism is called "diffusion."

A similar phenomenon occurs if charge carriers are "dropped" (injected) into a semiconductor so as to create a *nonuniform* density. Even in the absence of an electric field, the carriers move toward regions of low concentration, thereby carrying an electric current so long as the nonuniformity is sustained. Diffusion is therefore distinctly different from drift.

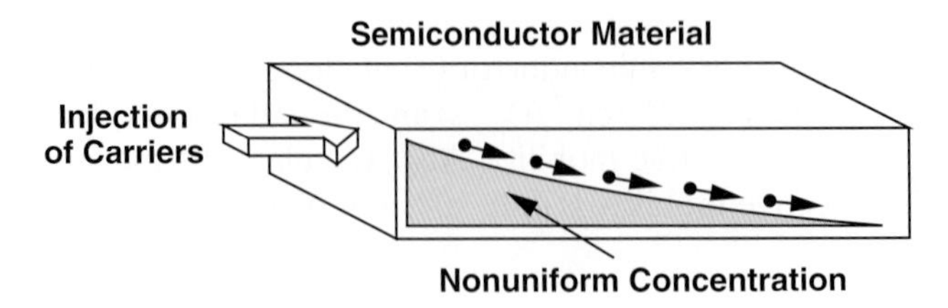

Figure 2.11 Diffusion in a semiconductor.

Figure 2.11 conceptually illustrates the process of diffusion. A source on the left continues to inject charge carriers into the semiconductor, a nonuniform charge profile is created along the x-axis, and the carriers continue to "roll down" the profile.

The reader may raise several questions at this point. What serves as the source of carriers in Fig. 2.11? Where do the charge carriers go after they roll down to the end of the profile at the far right? And, most importantly, why should we care?! Well, patience is a virtue and we will answer these questions in the next section.

Example 2-8 A source injects charge carriers into a semiconductor bar as shown in Fig. 2.12. Explain how the current flows.

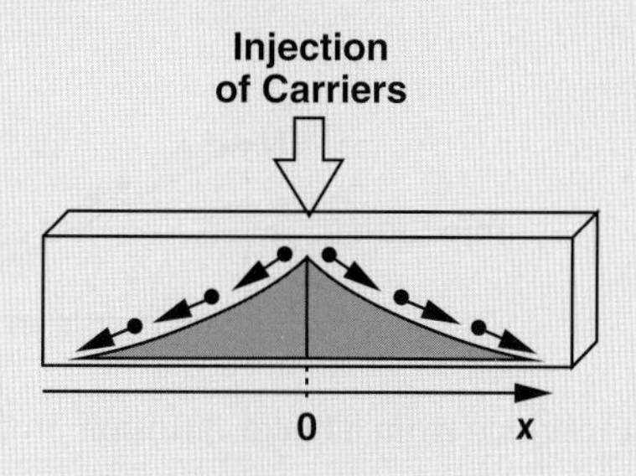

Figure 2.12 Injection of carriers into a semiconductor.

Solution In this case, two symmetric profiles may develop in both positive and negative directions along the x-axis, leading to current flow from the source toward the two ends of the bar.

Exercise Is KCL still satisfied at the point of injection?

Our qualitative study of diffusion suggests that the more nonuniform the concentration, the larger the current. More specifically, we can write:

$$I \propto \frac{dn}{dx}, \tag{2.39}$$

where n denotes the carrier concentration at a given point along the x-axis. We call dn/dx the concentration "gradient" with respect to x, assuming current flow only in the x direction. If each carrier has a charge equal to q, and the semiconductor has a cross section area of A, Eq. (2.39) can be written as

$$I \propto Aq\frac{dn}{dx}. \tag{2.40}$$

Thus,

$$I = AqD_n\frac{dn}{dx}, \tag{2.41}$$

where D_n is a proportionality factor called the "diffusion constant" and expressed in cm^2/s. For example, in intrinsic silicon, $D_n = 34\ \text{cm}^2/\text{s}$ (for electrons), and $D_p = 12\ \text{cm}^2/\text{s}$ (for holes).

As with the convention used for the drift current, we normalize the diffusion current to the cross section area, obtaining the current density as

$$J_n = qD_n\frac{dn}{dx}. \tag{2.42}$$

Similarly, a gradient in hole concentration yields:

$$J_p = -qD_p\frac{dp}{dx}. \tag{2.43}$$

With both electron and hole concentration gradients present, the total current density is given by

$$J_{tot} = q\left(D_n \frac{dn}{dx} - D_p \frac{dp}{dx}\right). \tag{2.44}$$

Example 2-9 Consider the scenario depicted in Fig. 2.11 again. Suppose the electron concentration is equal to N at $x = 0$ and falls linearly to zero at $x = L$ (Fig. 2.13). Determine the diffusion current.

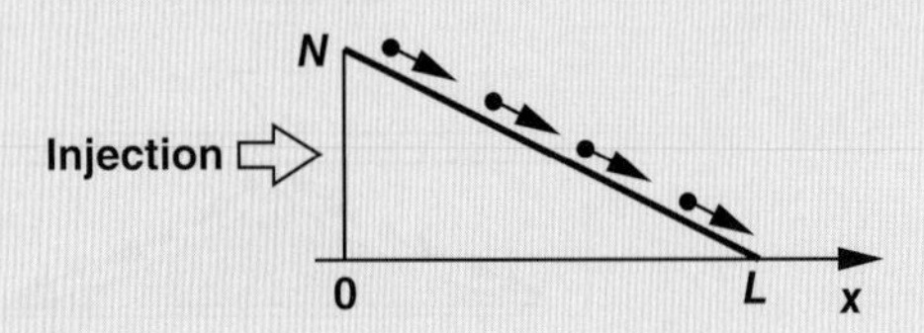

Figure 2.13 Current resulting from a linear diffusion profile.

Solution We have

$$J_n = qD_n \frac{dn}{dx} \tag{2.45}$$

$$= -qD_n \cdot \frac{N}{L}. \tag{2.46}$$

The current is constant along the x-axis; i.e., all of the electrons entering the material at $x = 0$ successfully reach the point at $x = L$. While obvious, this observation prepares us for the next example.

Exercise Repeat the above example for holes.

Example 2-10 Repeat the above example but assume an exponential gradient (Fig. 2.14):

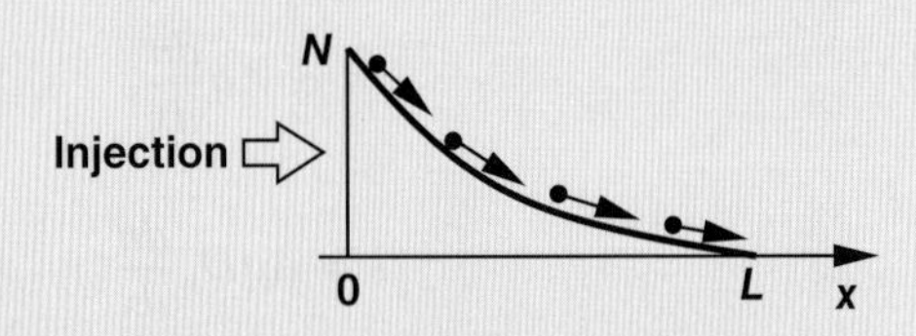

Figure 2.14 Current resulting from an exponential diffusion profile.

$$n(x) = N \exp \frac{-x}{L_d}, \tag{2.47}$$

where L_d is a constant.[7]

[7]The factor L_d is necessary to convert the exponent to a dimensionless quantity.

Solution We have

$$J_n = qD_n \frac{dn}{dx} \tag{2.48}$$

$$= \frac{-qD_n N}{L_d} \exp \frac{-x}{L_d}. \tag{2.49}$$

Interestingly, the current is *not* constant along the x-axis. That is, some electrons vanish while traveling from $x = 0$ to the right. What happens to these electrons? Does this example violate the law of conservation of charge? These are important questions and will be answered in the next section.

Exercise At what value of x does the current density drop to 1% of its maximum value?

Einstein Relation Our study of drift and diffusion has introduced a factor for each: μ_n (or μ_p) and D_n (or D_p), respectively. It can be proved that μ and D are related as:

$$\frac{D}{\mu} = \frac{kT}{q}. \tag{2.50}$$

Called the "Einstein Relation," this result is proved in semiconductor physics texts, e.g., [1]. Note that $kT/q \approx 26$ mV at $T = 300$ K.

Figure 2.15 summarizes the charge transport mechanisms studied in this section.

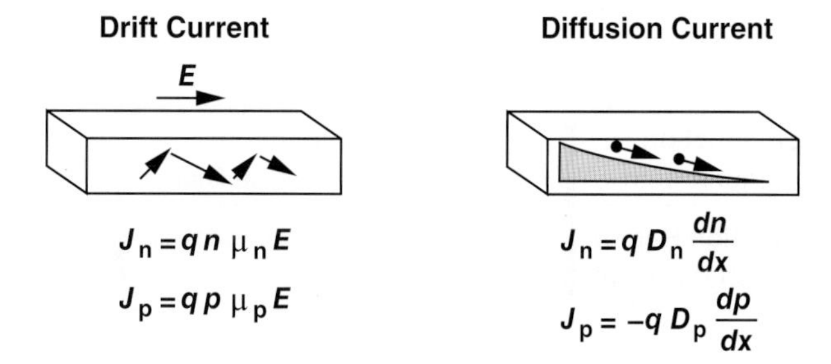

Figure 2.15 Summary of drift and diffusion mechanisms.

2.2 *pn* JUNCTION

We begin our study of semiconductor devices with the *pn* junction for three reasons. (1) The device finds application in many electronic systems, e.g., in adaptors that charge the batteries of cellphones. (2) The *pn* junction is among the simplest semiconductor devices, thus providing a good entry point into the study of the operation of such complex structures as transistors. (3) The *pn* junction also serves as part of transistors. We also use the term "diode" to refer to *pn* junctions.

Did you know?

The *pn* junction was inadvertantly invented by Russel Ohl of Bell Laboratories in 1940. He melted silicon in quartz tubes to achieve a high purity. During the cooling process, the *p*-type and *n*-type impurities redistributed themselves, creating a *pn* junction. Ohl even observed that the *pn* junction produced a current when it was exposed to light. One wonders if Ohl ever predicted that this property would eventually lead to the invention of the digital camera.

We have thus far seen that doping produces free electrons or holes in a semiconductor, and an electric field or a concentration gradient leads to the movement of these charge carriers. An interesting situation arises if we introduce *n*-type and *p*-type dopants into two adjacent sections of a piece of semiconductor. Depicted in Fig. 2.16 and called a "*pn* junction," this structure plays a fundamental role in many semiconductor devices. The *p* and *n* sides are called the "anode" and the "cathode," respectively.

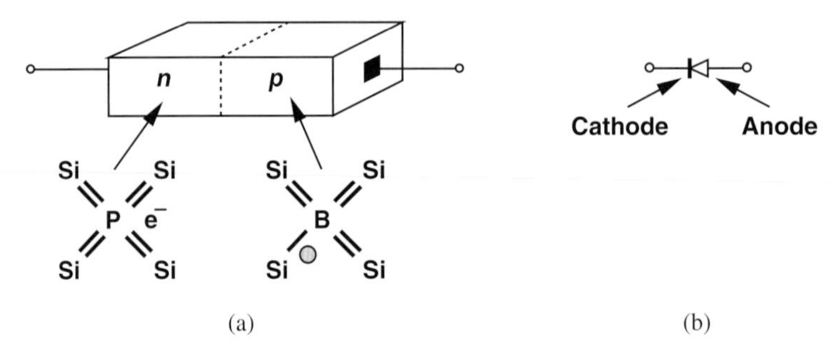

Figure 2.16 *pn* junction.

In this section, we study the properties and I/V characteristics of *pn* junctions. The following outline (Fig. 2.17) shows our thought process, indicating that our objective is to develop *circuit* models that can be used in analysis and design.

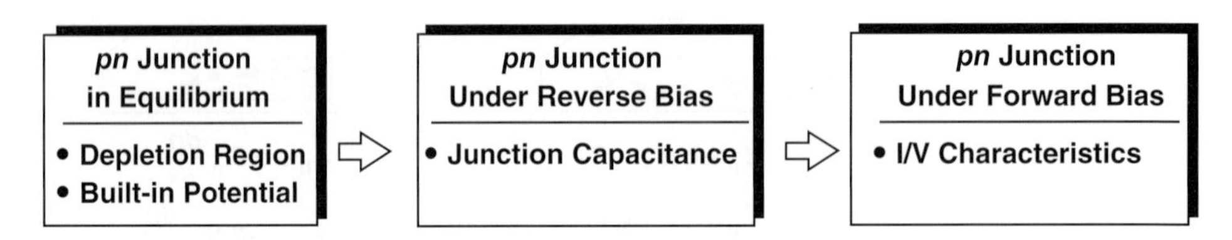

Figure 2.17 Outline of concepts to be studied.

2.2.1 *pn* Junction in Equilibrium

Let us first study the *pn* junction with no external connections, i.e., the terminals are open and no voltage is applied across the device. We say the junction is in "equilibrium." While seemingly of no practical value, this condition provides insights that prove useful in understanding the operation under nonequilibrium as well.

We begin by examining the interface between the *n* and *p* sections, recognizing that one side contains a large excess of holes and the other, a large excess of electrons. The sharp concentration gradient for both electrons and holes across the junction leads to two large diffusion currents: electrons flow from the *n* side to the *p* side, and holes flow in the opposite direction. Since we must deal with both electron and hole concentrations on each side of the junction, we introduce the notations shown in Fig. 2.18.

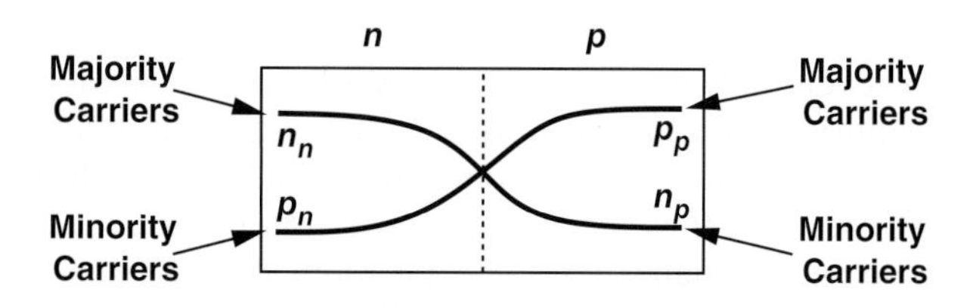

Figure 2.18

Example 2-11 A *pn* junction employs the following doping levels: $N_A = 10^{16}$ cm^{-3} and $N_D = 5 \times 10^{15}$ cm^{-3}. Determine the hole and electron concentrations on the two sides.

Solution From Eqs. (2.11) and (2.12), we express the concentrations of holes and electrons on the *p* side respectively as:

$$p_p \approx N_A \tag{2.51}$$

$$= 10^{16} \text{ cm}^{-3} \tag{2.52}$$

$$n_p \approx \frac{n_i^2}{N_A} \tag{2.53}$$

$$= \frac{(1.08 \times 10^{10} \text{ cm}^{-3})^2}{10^{16} \text{ cm}^{-3}} \tag{2.54}$$

$$\approx 1.1 \times 10^4 \text{ cm}^{-3}. \tag{2.55}$$

Similarly, the concentrations on the *n* side are given by

$$n_n \approx N_D \tag{2.56}$$

$$= 5 \times 10^{15} \text{ cm}^{-3} \tag{2.57}$$

$$p_n \approx \frac{n_i^2}{N_D} \tag{2.58}$$

$$= \frac{(1.08 \times 10^{10} \text{ cm}^{-3})^2}{5 \times 10^{15} \text{ cm}^{-3}} \tag{2.59}$$

$$= 2.3 \times 10^4 \text{ cm}^{-3}. \tag{2.60}$$

Note that the majority carrier concentration on each side is many orders of magnitude higher than the minority carrier concentration on either side.

Exercise Repeat the above example if N_D drops by a factor of four.

The diffusion currents transport a great deal of charge from each side to the other, but they must eventually decay to zero. This is because if the terminals are left open (equilibrium condition), the device cannot carry a net current indefinitely.

We must now answer an important question: what stops the diffusion currents? We may postulate that the currents stop after enough free carriers have moved across the junction so as to equalize the concentrations on the two sides. However, another effect dominates the situation and stops the diffusion currents well before this point is reached.

To understand this effect, we recognize that for every electron that departs from the n side, a *positive ion* is left behind, i.e., the junction evolves with time as conceptually shown in Fig. 2.19. In this illustration, the junction is suddenly formed at $t = 0$, and the diffusion currents continue to expose more ions as time progresses. Consequently, the immediate vicinity of the junction is depleted of free carriers and hence called the "depletion region."

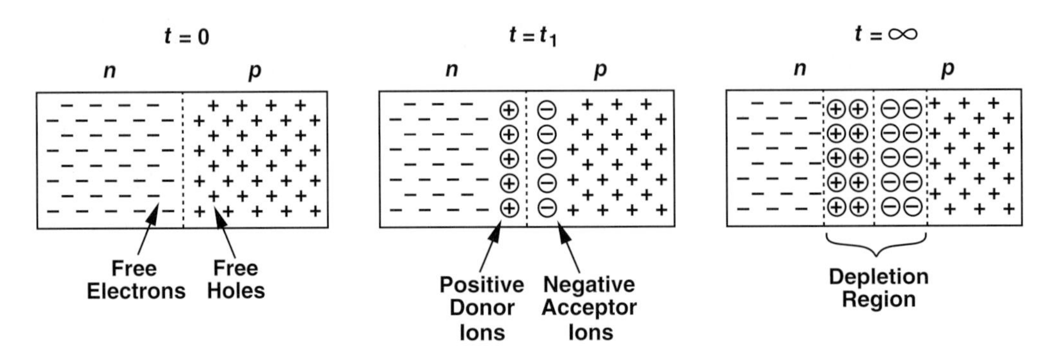

Figure 2.19 Evolution of charge concentrations in a pn junction.

Now recall from basic physics that a particle or object carrying a net (nonzero) charge creates an electric field around it. Thus, with the formation of the depletion region, an electric field emerges as shown in Fig. 2.20.[8] Interestingly, the field tends to force positive charge flow from left to right whereas the concentration gradients necessitate the flow of holes from right to left (and electrons from left to right). We therefore surmise that the junction reaches *equilibrium* once the electric field is strong enough to completely stop the diffusion currents. Alternatively, we can say, in equilibrium, the drift currents resulting from the electric field exactly cancel the diffusion currents due to the gradients.

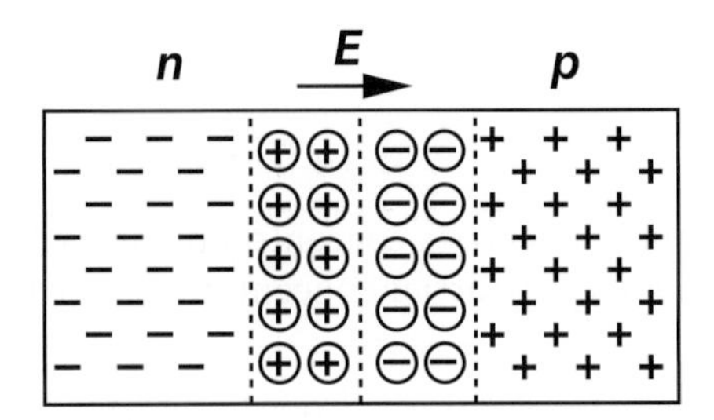

Figure 2.20 Electric field in a pn junction.

[8]The direction of the electric field is determined by placing a small positive test charge in the region and watching how it moves: away from positive charge and toward negative charge.

Example 2-12

In the junction shown in Fig. 2.21, the depletion region has a width of b on the n side and a on the p side. Sketch the electric field as a function of x.

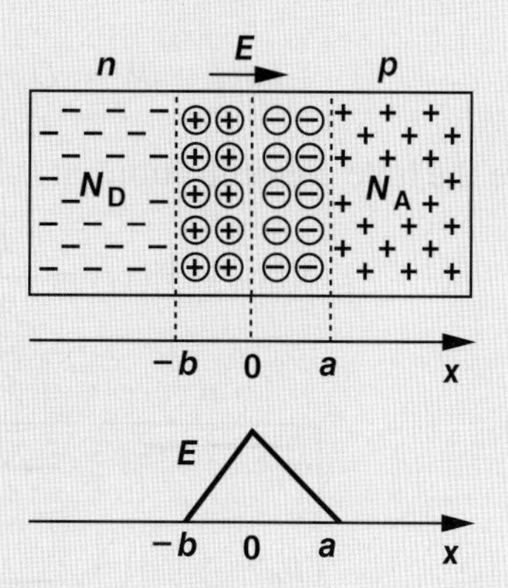

Figure 2.21 Electric field profile in a pn junction.

Solution Beginning at $x < -b$, we note that the absence of net charge yields $E = 0$. At $x > -b$, each positive donor ion contributes to the electric field, i.e., the magnitude of E rises as x approaches zero. As we pass $x = 0$, the negative acceptor atoms begin to contribute negatively to the field, i.e., E falls. At $x = a$, the negative and positive charge exactly cancel each other and $E = 0$.

Exercise Noting that potential voltage is negative integral of electric field with respect to distance, plot the potential as a function of x.

From our observation regarding the drift and diffusion currents under equilibrium, we may be tempted to write:

$$|I_{\text{drift},p} + I_{\text{drift},n}| = |I_{\text{diff},p} + I_{\text{diff},n}|, \tag{2.61}$$

where the subscripts p and n refer to holes and electrons, respectively, and each current term contains the proper polarity. This condition, however, allows an unrealistic phenomenon: if the number of the electrons flowing from the n side to the p side is equal to that of the holes going from the p side to the n side, then each side of this equation is zero while electrons continue to accumulate on the p side and holes on the n side. We must therefore impose the equilibrium condition on *each* carrier:

$$|I_{\text{drift},p}| = |I_{\text{diff},p}| \tag{2.62}$$

$$|I_{\text{drift},n}| = |I_{\text{diff},n}|. \tag{2.63}$$

Built-in Potential The existence of an electric field within the depletion region suggests that the junction may exhibit a "built-in potential." In fact, using (2.62) or (2.63), we can compute this potential. Since the electric field $E = -dV/dx$, and since (2.62) can be written as

$$q\mu_p pE = qD_p\frac{dp}{dx}, \tag{2.64}$$

we have

$$-\mu_p p \frac{dV}{dx} = D_p \frac{dp}{dx}. \tag{2.65}$$

Dividing both sides by p and taking the integral, we obtain

$$-\mu_p \int_{x_1}^{x_2} dV = D_p \int_{p_n}^{p_p} \frac{dp}{p}, \tag{2.66}$$

where p_n and p_p are the hole concentrations at x_1 and x_2, respectively (Fig. 2.22). Thus,

$$V(x_2) - V(x_1) = -\frac{D_p}{\mu_p} \ln \frac{p_p}{p_n}. \tag{2.67}$$

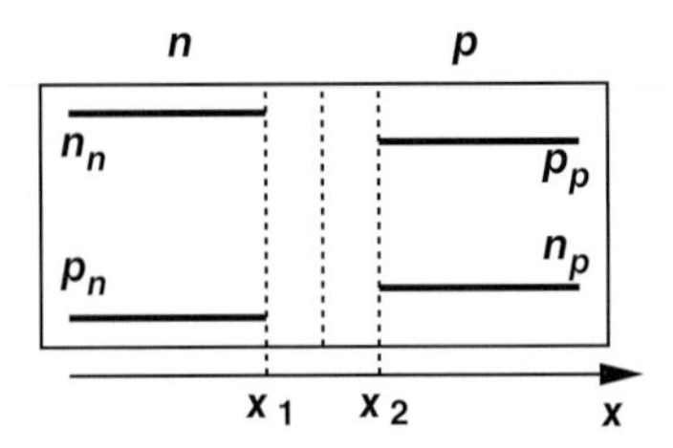

Figure 2.22 Carrier profiles in a *pn* junction.

The right side represents the voltage difference developed across the depletion region and will be denoted by V_0. Also, from Einstein's relation, Eq. (2.50), we can replace D_p/μ_p with kT/q:

$$|V_0| = \frac{kT}{q} \ln \frac{p_p}{p_n}. \tag{2.68}$$

Exercise Writing Eq. (2.64) for electron drift and diffusion currents, and carrying out the integration, derive an equation for V_0 in terms of n_n and n_p.

Finally, using (2.11) and (2.10) for p_p and p_n yields

$$V_0 = \frac{kT}{q} \ln \frac{N_A N_D}{n_i^2}. \tag{2.69}$$

Expressing the built-in potential in terms of junction parameters, this equation plays a central role in many semiconductor devices.

Example 2-13 A silicon *pn* junction employs $N_A = 2 \times 10^{16}$ cm^{-3} and $N_D = 4 \times 10^{16}$ cm^{-3}. Determine the built-in potential at room temperature ($T = 300$ K).

Solution Recall from Example 2-1 that $n_i(T = 300\text{ K}) = 1.08 \times 10^{10}$ cm^{-3}. Thus,

$$V_0 \approx (26\text{ mV}) \ln \frac{(2 \times 10^{16}) \times (4 \times 10^{16})}{(1.08 \times 10^{10})^2} \tag{2.70}$$

$$\approx 768\text{ mV}. \tag{2.71}$$

Exercise By what factor should N_D be changed to lower V_0 by 20 mV?

Example 2-14 Equation (2.69) reveals that V_0 is a weak function of the doping levels. How much does V_0 change if N_A or N_D is increased by one order of magnitude?

Solution We can write

$$\Delta V_0 = V_T \ln \frac{10 N_A \cdot N_D}{n_i^2} - V_T \ln \frac{N_A \cdot N_D}{n_i^2} \tag{2.72}$$

$$= V_T \ln 10 \tag{2.73}$$

$$\approx 60 \text{ mV (at } T = 300 \text{ K)}. \tag{2.74}$$

Exercise How much does V_0 change if N_A or N_D is increased by a factor of three?

An interesting question may arise at this point. The junction carries no net current (because its terminals remain open), but it sustains a voltage. How is that possible? We observe that the built-in potential is developed to *oppose* the flow of diffusion currents (and is, in fact, sometimes called the "potential barrier"). This phenomenon is in contrast to the behavior of a uniform conducting material, which exhibits no tendency for diffusion and hence no need to create a built-in voltage.

2.2.2 *pn* Junction Under Reverse Bias

Having analyzed the *pn* junction in equilibrium, we can now study its behavior under more interesting and useful conditions. Let us begin by applying an external voltage across the device as shown in Fig. 2.23, where the voltage source makes the *n* side more *positive* than the *p* side. We say the junction is under "reverse bias" to emphasize the connection of the positive voltage to the *n* terminal. Used as a noun or a verb, the term "bias" indicates operation under some "desirable" conditions. We will study the concept of biasing extensively in this and following chapters.

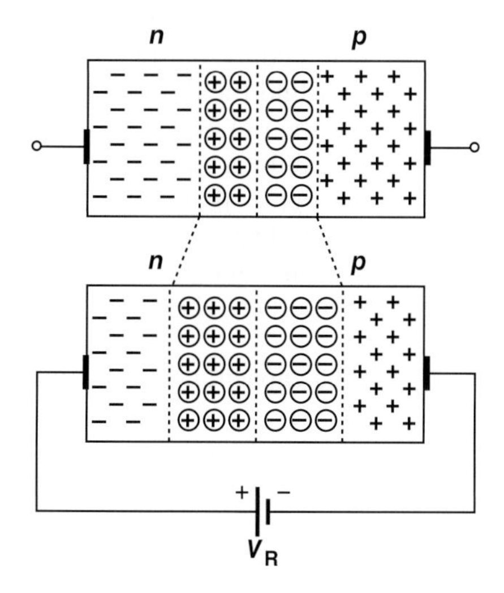

Figure 2.23 *pn* junction under reverse bias.

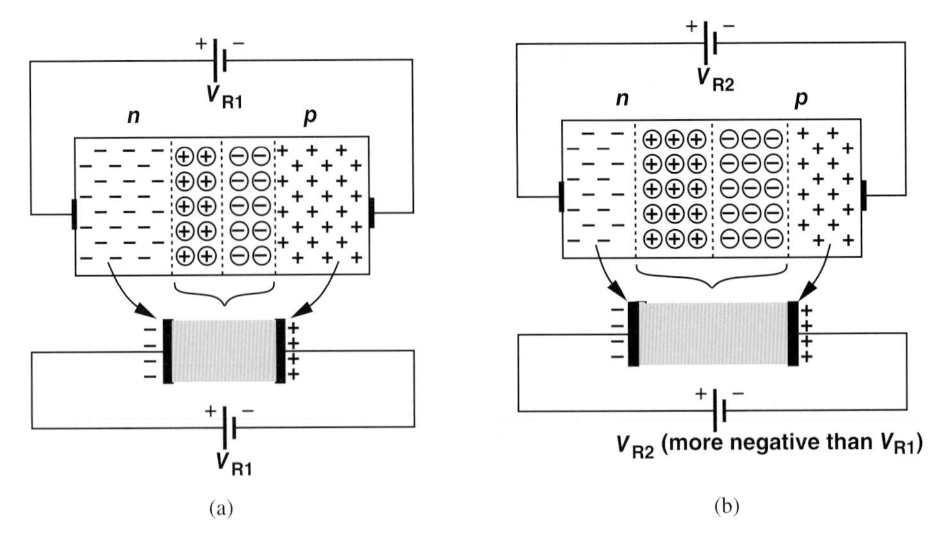

Figure 2.24 Reduction of junction capacitance with reverse bias.

We wish to reexamine the results obtained in equilibrium for the case of reverse bias. Let us first determine whether the external voltage *enhances* the built-in electric field or *opposes* it. Since under equilibrium, $\vec{E}$ is directed from the n side to the p side, V_R enhances the field. But, a higher electric field can be sustained only if a larger amount of fixed charge is provided, requiring that more acceptor and donor ions be exposed and, therefore, the depletion region be widened.

What happens to the diffusion and drift currents? Since the external voltage has strengthened the field, the barrier rises even higher than that in equilibrium, thus prohibiting the flow of current. In other words, the junction carries a negligible current under reverse bias.[9]

With no current conduction, a reverse-biased pn junction does not seem particularly useful. However, an important observation will prove otherwise. We note that in Fig. 2.23, as V_B increases, more positive charge appears on the n side and more negative charge on the p side. Thus, the device operates as a *capacitor* [Fig. 2.24(a)]. In essence, we can view the conductive n and p sections as the two plates of the capacitor. We also assume the charge in the depletion region equivalently resides on each plate.

The reader may still not find the device interesting. After all, since any two parallel plates can form a capacitor, the use of a pn junction for this purpose is not justified. But, reverse-biased pn junctions exhibit a unique property that becomes useful in circuit design. Returning to Fig. 2.23, we recognize that, as V_R increases, so does the width of the depletion region. That is, the conceptual diagram of Fig. 2.24(a) can be drawn as in Fig. 2.24(b) for increasing values of V_R, revealing that the capacitance of the structure *decreases* as the two plates move away from each other. The junction therefore displays a voltage-dependent capacitance.

It can be proved that the capacitance of the junction per unit area is equal to

$$C_j = \frac{C_{j0}}{\sqrt{1 - \frac{V_R}{V_0}}}, \tag{2.75}$$

[9]As explained in Section 2.2.3, the current is not exactly zero.

where C_{j0} denotes the capacitance corresponding to zero bias ($V_R = 0$) and V_0 is the built-in potential [Eq. (2.69)]. (This equation assumes V_R is negative for reverse bias.) The value of C_{j0} is in turn given by

$$C_{j0} = \sqrt{\frac{\epsilon_{si} q}{2} \frac{N_A N_D}{N_A + N_D} \frac{1}{V_0}}, \tag{2.76}$$

where ϵ_{si} represents the dielectric constant of silicon and is equal to $11.7 \times 8.85 \times 10^{-14}$ F/cm.[10] Plotted in Fig. 2.25, C_j indeed decreases as V_R increases.

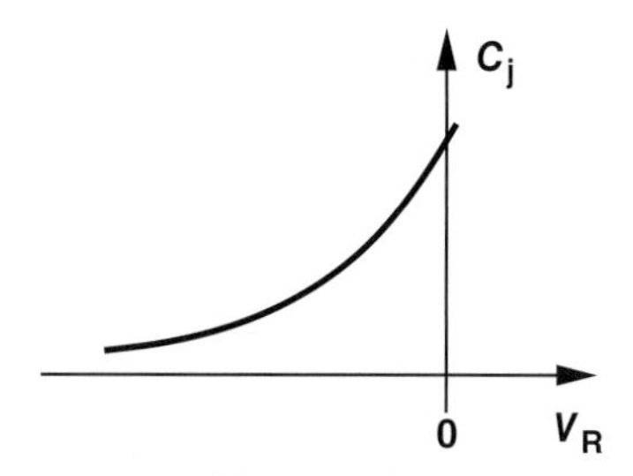

Figure 2.25 Junction capacitance under reverse bias.

Example 2-15 A *pn* junction is doped with $N_A = 2 \times 10^{16}$ cm^{-3} and $N_D = 9 \times 10^{15}$ cm^{-3}. Determine the capacitance of the device with (a) $V_R = 0$ and $V_R = 1$ V.

Solution We first obtain the built-in potential:

$$V_0 = V_T \ln \frac{N_A N_D}{n_i^2} \tag{2.77}$$

$$= 0.73 \text{ V}. \tag{2.78}$$

Thus, for $V_R = 0$ and $q = 1.6 \times 10^{-19}$ C, we have

$$C_{j0} = \sqrt{\frac{\epsilon_{si} q}{2} \frac{N_A N_D}{N_A + N_D} \cdot \frac{1}{V_0}} \tag{2.79}$$

$$= 2.65 \times 10^{-8} \text{ F/cm}^2. \tag{2.80}$$

In microelectronics, we deal with very small devices and may rewrite this result as

$$C_{j0} = 0.265 \text{ fF/}\mu\text{m}^2, \tag{2.81}$$

where 1 fF (femtofarad) $= 10^{-15}$ F. For $V_R = 1$ V,

$$C_j = \frac{C_{j0}}{\sqrt{1 + \frac{V_R}{V_0}}} \tag{2.82}$$

$$= 0.172 \text{ fF/}\mu\text{m}^2. \tag{2.83}$$

[10]The dielectric constant of materials is usually written in the form $\epsilon_r \epsilon_0$, where ϵ_r is the "relative" dielectric constant and a dimensionless factor (e.g., 11.7), and ϵ_0 the dielectric constant of vacuum (8.85×10^{-14} F/cm).

Exercise Repeat the above example if the donor concentration on the N side is doubled. Compare the results in the two cases.

The variation of the capacitance with the applied voltage makes the device a "nonlinear" capacitor because it does not satisfy $Q = CV$. Nonetheless, as demonstrated by the following example, a voltage-dependent capacitor leads to interesting circuit topologies.

Example 2-16 A cellphone incorporates a 2-GHz oscillator whose frequency is defined by the resonance frequency of an LC tank (Fig. 2.26). If the tank capacitance is realized as the *pn* junction of Example 2-15, calculate the change in the oscillation frequency while the reverse voltage goes from 0 to 2 V. Assume the circuit operates at 2 GHz at a reverse voltage of 0 V, and the junction area is 2000 μm².

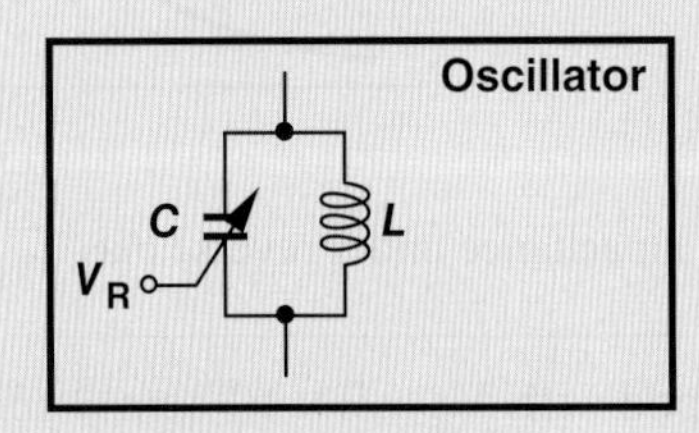

Figure 2.26 Variable capacitor used to tune an oscillator.

Solution Recall from basic circuit theory that the tank "resonates" if the impedances of the inductor and the capacitor are equal and opposite: $jL\omega_{res} = -(jC\omega_{res})^{-1}$. Thus, the resonance frequency is equal to

$$f_{res} = \frac{1}{2\pi}\frac{1}{\sqrt{LC}}. \tag{2.84}$$

At $V_R = 0$, $C_j = 0.265\ \text{fF}/\mu\text{m}^2$, yielding a total device capacitance of

$$C_{j,tot}(V_R = 0) = (0.265\ \text{fF}/\mu\text{m}^2) \times (2000\ \mu\text{m}^2) \tag{2.85}$$

$$= 530\ \text{fF}. \tag{2.86}$$

Setting f_{res} to 2 GHz, we obtain

$$L = 11.9\ \text{nH}. \tag{2.87}$$

If V_R goes to 2 V,

$$C_{j,tot}(V_R = 2\ \text{V}) = \frac{C_{j0}}{\sqrt{1 + \frac{2}{0.73}}} \times 2000\ \mu\text{m}^2 \tag{2.88}$$

$$= 274\ \text{fF}. \tag{2.89}$$

Using this value along with $L = 11.9$ nH in Eq. (2.84), we have

$$f_{res}(V_R = 2\ \text{V}) = 2.79\ \text{GHz}. \tag{2.90}$$

An oscillator whose frequency can be varied by an external voltage (V_R in this case) is called a "voltage-controlled oscillator" and used extensively in cellphones, microprocessors, personal computers, etc.

Exercise Some wireless systems operate at 5.2 GHz. Repeat the above example for this frequency, assuming the junction area is still 2000 μm^2 but the inductor value is scaled to reach 5.2 GHz.

In summary, a reverse-biased *pn* junction carries a negligible current but exhibits a voltage-dependent capacitance. Note that we have tacitly developed a circuit model for the device under this condition: a simple capacitance whose value is given by Eq. (2.75).

Another interesting application of reverse-biased diodes is in digital cameras (Chapter 1). If light of sufficient energy is applied to a *pn* junction, electrons are dislodged from their covalent bonds and hence electron-hole pairs are created. With a reverse bias, the electrons are attracted to the positive battery terminal and the holes to the negative battery terminal. As a result, a current flows through the diode that is proportional to the light intensity. We say the *pn* junction operates as a "photodiode."

Did you know?

Voltage-dependent capacitors are called "varactors" (or, in older literature, "varicaps"). The ability to "tune" a capacitor's value by a voltage proves essential in many systems. For example, your TV changes channels by changing the voltage applied to a varactor. By contrast, old TVs had a knob that mechanically switched different capacitors into the circuit. Imagine turning that knob by remote control!

Robotics Application: Infrared Transceivers

Some robots employ infrared (IR) transceivers to detect obstacles, control their distance with respect to other robots, or transmit and receive data. (The TV remote control also incorporates an IR transceiver.)

Obstacle detection in robotics proceeds in a manner similar to the operation of a radar: an IR light-emitting diode (LED) transmits a beam, the beam hits the obstacle and is reflected, and an IR sensor receives the reflected beam [Fig. 2.27(a)]. The roundtrip time is proportional to the distance.

The IR LED has the same structure as the diodes studied in this chapter, except that it is fabricated in GaAs or GaN technologies so as to produce light in the forward bias condition. (Silicon diodes do not "radiate.") The IR sensor is also a diode (a "photodiode") but operating in the reverse bias region. Upon receiving light, the sensor generates carriers in its depletion region, which are then swept out in the form of a current [Fig. 2.27(b)].

The situation depicted in Fig. 2.27(a) faces an interesting issue: the IR sensor can also pick up interference from the sunlight or indoor lighting, causing false alarms during obstacle detection. To avoid this difficulty, the transmitted light is "pulsed" periodically [Fig. 2.27(c)]. The beam received by the sensor thus generates a periodic current. This current is then applied to a filter that suppresses the interference and only passes the desired periodic waveform.

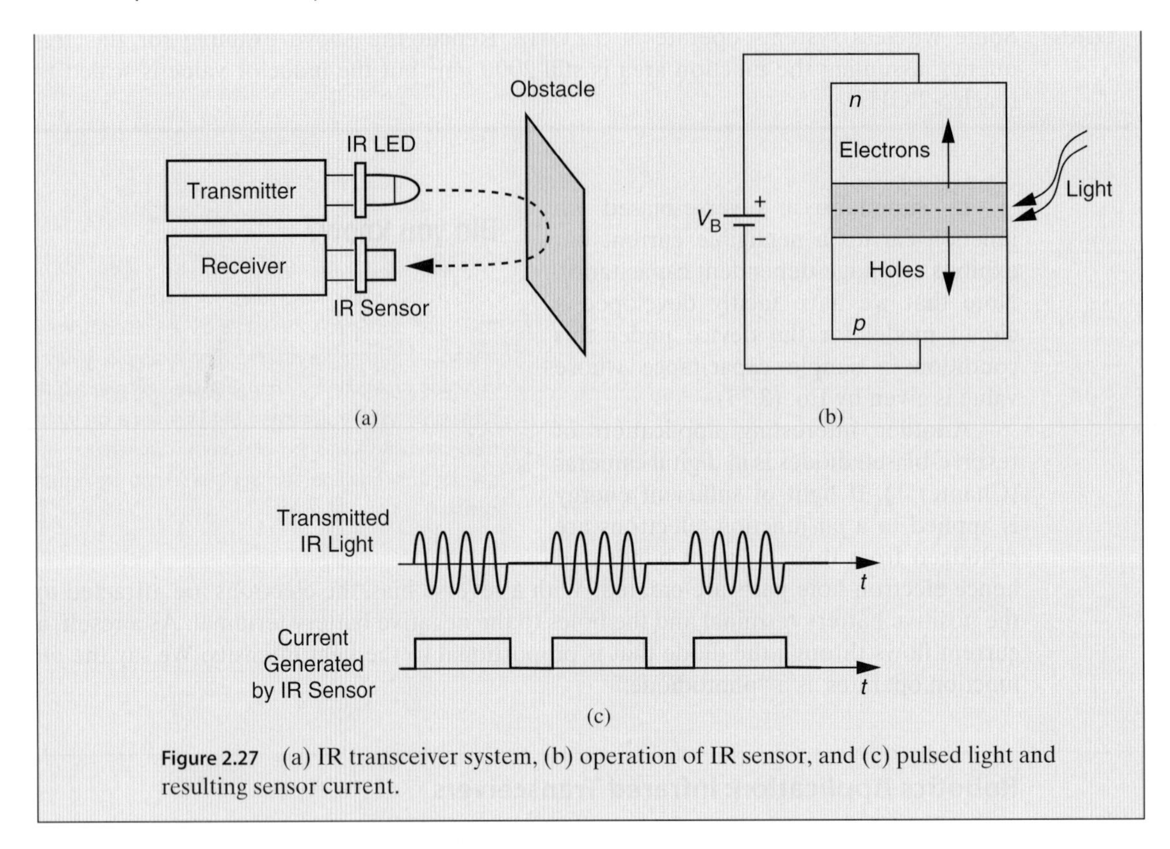

Figure 2.27 (a) IR transceiver system, (b) operation of IR sensor, and (c) pulsed light and resulting sensor current.

2.2.3 *pn* Junction Under Forward Bias

Our objective in this section is to show that the *pn* junction carries a current if the *p* side is raised to a more *positive* voltage than the *n* side (Fig. 2.28). This condition is called "forward bias." We also wish to compute the resulting current in terms of the applied voltage and the junction parameters, ultimately arriving at a circuit model.

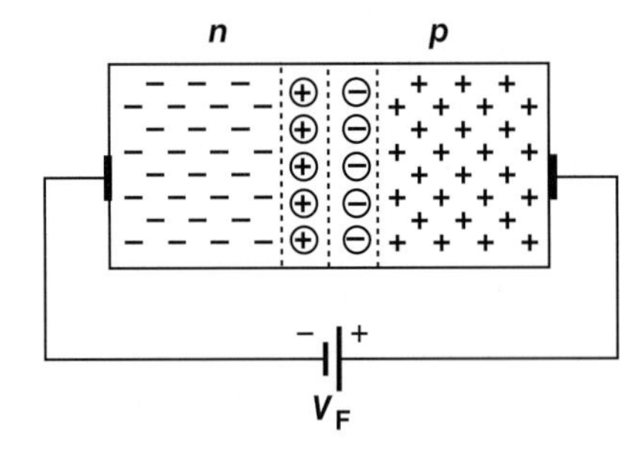

Figure 2.28 *pn* junction under forward bias.

From our study of the device in equilibrium and reverse bias, we note that the potential barrier developed in the depletion region determines the device's desire to conduct. In forward bias, the external voltage, V_F, tends to create a field directed from the *p* side toward the *n* side—opposite to the built-in field that was developed to stop the diffusion currents. We therefore surmise that V_F in fact *lowers* the potential barrier by weakening the field, thus allowing greater diffusion currents.

To derive the I/V characteristic in forward bias, we begin with Eq. (2.68) for the built-in voltage and rewrite it as

$$p_{n,e} = \frac{p_{p,e}}{\exp \dfrac{V_0}{V_T}}, \tag{2.91}$$

where the subscript e emphasizes equilibrium conditions [Fig. 2.29(a)] and $V_T = kT/q$ is called the "thermal voltage" ($\approx$26 mV at $T = 300$ K). In forward bias, the potential barrier is lowered by an amount equal to the applied voltage:

$$p_{n,f} = \frac{p_{p,f}}{\exp \dfrac{V_0 - V_F}{V_T}}. \tag{2.92}$$

where the subscript f denotes forward bias. Since the exponential denominator drops considerably, we expect $p_{n,f}$ to be much higher than $p_{n,e}$ (it can be proved that $p_{p,f} \approx p_{p,e} \approx N_A$). In other words, the *minority* carrier concentration on the p side rises rapidly with the forward bias voltage while the majority carrier concentration remains relatively constant. This statement applies to the n side as well.

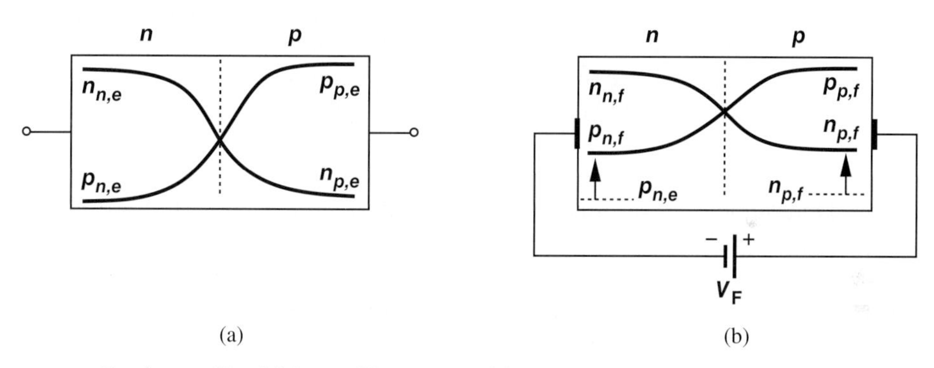

Figure 2.29 Carrier profiles (a) in equilibrium and (b) under forward bias.

Figure 2.29(b) illustrates the results of our analysis thus far. As the junction goes from equilibrium to forward bias, n_p and p_n increase dramatically, leading to a proportional change in the diffusion currents.[11] We can express the change in the hole concentration on the n side as:

$$\Delta p_n = p_{n,f} - p_{n,e} \tag{2.93}$$

$$= \frac{p_{p,f}}{\exp \dfrac{V_0 - V_F}{V_T}} - \frac{p_{p,e}}{\exp \dfrac{V_0}{V_T}} \tag{2.94}$$

$$\approx \frac{N_A}{\exp \dfrac{V_0}{V_T}} \left(\exp \frac{V_F}{V_T} - 1 \right). \tag{2.95}$$

[11] The width of the depletion region actually decreases in forward bias, but we neglect this effect here.

Similarly, for the electron concentration on the p side:

$$\Delta n_p \approx \frac{N_D}{\exp\frac{V_0}{V_T}}\left(\exp\frac{V_F}{V_T}-1\right). \tag{2.96}$$

Note that Eq. (2.69) indicates that $\exp(V_0/V_T) = N_A N_D / n_i^2$.

The increase in the minority carrier concentration suggests that the diffusion currents must rise by a proportional amount above their equilibrium value, i.e.,

$$I_{tot} \propto \frac{N_A}{\exp\frac{V_0}{V_T}}\left(\exp\frac{V_F}{V_T}-1\right)+\frac{N_D}{\exp\frac{V_0}{V_T}}\left(\exp\frac{V_F}{V_T}-1\right). \tag{2.97}$$

Indeed, it can be proved that [1]

$$I_{tot} = I_S\left(\exp\frac{V_F}{V_T}-1\right), \tag{2.98}$$

where I_S is called the "reverse saturation current" and given by

$$I_S = Aqn_i^2\left(\frac{D_n}{N_A L_n}+\frac{D_p}{N_D L_p}\right). \tag{2.99}$$

In this equation, A is the cross section area of the device, and L_n and L_p are electron and hole "diffusion lengths," respectively. Diffusion lengths are typically in the range of tens of micrometers. Note that the first and second terms in the parentheses correspond to the flow of electrons and holes, respectively.

Example 2-17 Determine I_S for the junction of Example 2-13 at $T = 300$ K if $A = 100\ \mu\text{m}^2$, $L_n = 20\ \mu\text{m}$, and $L_p = 30\ \mu\text{m}$.

Solution Using $q = 1.6\times10^{-19}$ C, $n_i = 1.08\times10^{10}$ electrons/cm^3 [Eq. (2.2)], $D_n = 34$ cm^2/s, and $D_p = 12$ cm^2/s, we have

$$I_S = 1.77\times10^{-17}\ \text{A}. \tag{2.100}$$

Since I_S is extremely small, the exponential term in Eq. (2.98) must assume very large values so as to yield a useful amount (e.g., 1 mA) for I_{tot}.

Exercise What junction area is necessary to raise I_S to 10^{-15} A?

An interesting question that arises here is: are the minority carrier concentrations *constant* along the x-axis? Depicted in Fig. 2.30(a), such a scenario would suggest that electrons continue to flow from the n side to the p side, but exhibit no tendency to go beyond $x = x_2$ because of the lack of a gradient. A similar situation exists for holes, implying that the charge carriers do not flow deep into the p and n sides and hence no net current results! Thus, the minority carrier concentrations must vary as shown in Fig. 2.30(b) so that diffusion can occur.

This observation reminds us of Example 2-10 and the question raised in conjunction with it: if the minority carrier concentration falls with x, what happens to the carriers and how can the current remain constant along the x-axis? Interestingly, as the electrons enter

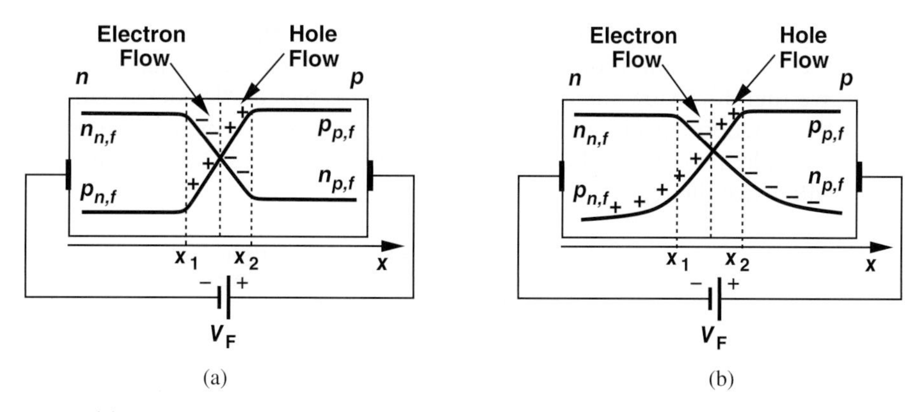

Figure 2.30 (a) Constant and (b) variable majority carrier profiles outside the depletion region.

the p side and roll down the gradient, they gradually *recombine* with the holes, which are abundant in this region. Similarly, the holes entering the n side recombine with the electrons. Thus, in the immediate vicinity of the depletion region, the current consists of mostly minority carriers, but towards the far contacts, it is primarily comprised of majority carriers (Fig. 2.31). At each point along the x-axis, the two components add up to I_{tot}.

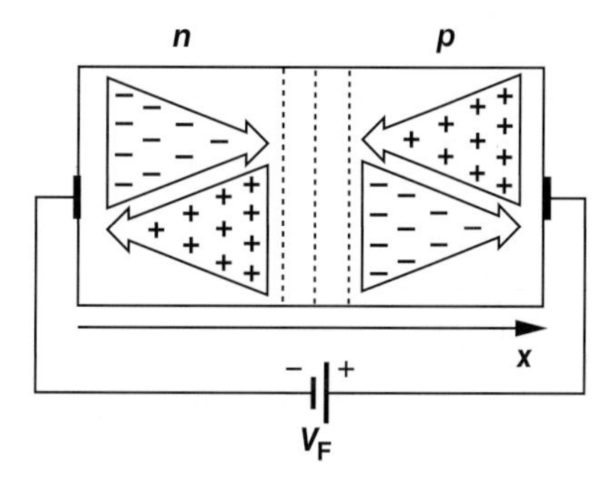

Figure 2.31 Minority and majority carrier currents.

2.2.4 I/V Characteristics

Let us summarize our thoughts thus far. In forward bias, the external voltage opposes the built-in potential, raising the diffusion currents substantially. In reverse bias, on the other hand, the applied voltage enhances the field, prohibiting current flow. We hereafter write the junction equation as:

$$I_D = I_S \left(\exp \frac{V_D}{V_T} - 1 \right), \tag{2.101}$$

where I_D and V_D denote the diode current and voltage, respectively. As expected, $V_D = 0$ yields $I_D = 0$. (Why is this expected?) As V_D becomes positive and exceeds several V_T, the exponential term grows rapidly and $I_D \approx I_S \exp(V_D/V_T)$. We hereafter assume $\exp(V_D/V_T) \gg 1$ in the forward bias region.

It can be proved that Eq. (2.101) also holds in reverse bias, i.e., for negative V_D. If $V_D < 0$ and $|V_D|$ reaches several V_T, then $\exp(V_D/V_T) \ll 1$ and

$$I_D \approx -I_S. \tag{2.102}$$

Figure 2.32 plots the overall I/V characteristic of the junction, revealing why I_S is called the "reverse saturation current." Example 2-17 indicates that I_S is typically very small. We therefore view the current under reverse bias as "leakage." Note that I_S and hence the junction current are proportional to the device cross section area [Eq. (2.99)]. For example, two identical devices placed in parallel (Fig. 2.33) behave as a single junction with twice the I_S.

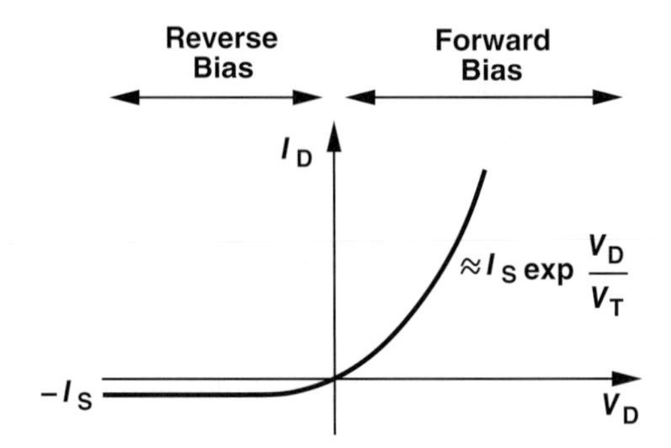

Figure 2.32 I/V characteristic of a *pn* junction.

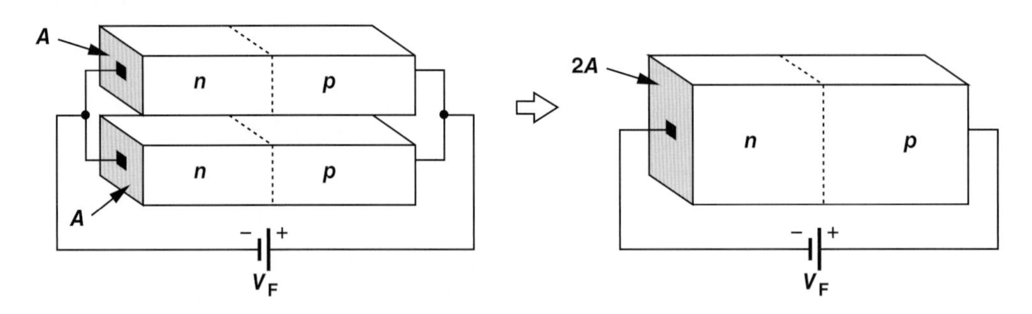

Figure 2.33 Equivalence of parallel devices to a larger device.

Example 2-18 Each junction in Fig. 2.33 employs the doping levels described in Example 2-13. Determine the forward bias current of the composite device for $V_D = 300$ mV and 800 mV at $T = 300$ K.

Solution From Example 2-17, $I_S = 1.77 \times 10^{-17}$ A for each junction. Thus, the total current is equal to

$$I_{D,tot}\ (V_D = 300\ \text{mV}) = 2I_S \left(\exp \frac{V_D}{V_T} - 1\right) \tag{2.103}$$

$$= 3.63\ \text{pA}. \tag{2.104}$$

Similarly, for $V_D = 800$ mV:

$$I_{D,tot}(V_D = 800\ \text{mV}) = 816\ \mu\text{A}. \tag{2.105}$$

Exercise How many of these diodes must be placed in parallel to obtain a current of 100 μA with a voltage of 750 mV?

Example 2-19 A diode operates in the forward bias region with a typical current level [i.e., $I_D \approx I_S \exp(V_D/V_T)$]. Suppose we wish to increase the current by a factor of 10. How much change in V_D is required?

Solution Let us first express the diode voltage as a function of its current:

$$V_D = V_T \ln \frac{I_D}{I_S}. \tag{2.106}$$

We define $I_1 = 10I_D$ and seek the corresponding voltage, V_{D1}:

$$V_{D1} = V_T \ln \frac{10I_D}{I_S} \tag{2.107}$$

$$= V_T \ln \frac{I_D}{I_S} + V_T \ln 10 \tag{2.108}$$

$$= V_D + V_T \ln 10. \tag{2.109}$$

Thus, the diode voltage must rise by $V_T \ln 10 \approx 60\,\text{mV}$ (at $T = 300\,\text{K}$) to accommodate a tenfold increase in the current. We say the device exhibits a 60-mV/decade characteristic, meaning V_D changes by 60 mV for a decade (tenfold) change in I_D. More generally, an n-fold change in I_D translates to a change of $V_T \ln n$ in V_D.

Exercise By what factor does the current change if the voltages changes by 120 mV?

Example 2-20 The cross section area of a diode operating in the forward bias region is increased by a factor of 10. (a) Determine the change in I_D if V_D is maintained constant. (b) Determine the change in V_D if I_D is maintained constant. Assume $I_D \approx I_S \exp(V_D/V_T)$.

Solution (a) Since $I_S \propto A$, the new current is given by

$$I_{D1} = 10I_S \exp \frac{V_D}{V_T} \tag{2.110}$$

$$= 10I_D. \tag{2.111}$$

(b) From the above example,

$$V_{D1} = V_T \ln \frac{I_D}{10I_S} \tag{2.112}$$

$$= V_T \ln \frac{I_D}{I_S} - V_T \ln 10. \tag{2.113}$$

Thus, a tenfold increase in the device area lowers the voltage by 60 mV if I_D remains constant.

Exercise A diode in forward bias with $I_D \approx I_S \exp(V_D/V_T)$ undergoes two simultaneous changes: the current is raised by a factor of m and the area is increased by a factor of n. Determine the change in the device voltage.

Constant-Voltage Model The exponential I/V characteristic of the diode results in nonlinear equations, making the analysis of circuits quite difficult. Fortunately, the above examples imply that the diode voltage is a relatively weak function of the device current and cross section area. With typical current levels and areas, V_D falls in the range of 700–800 mV. For this reason, we often approximate the forward bias voltage by a *constant* value of 800 mV (like an ideal battery), considering the device fully off if $V_D < 800$ mV. The resulting characteristic is illustrated in Fig. 2.34(a) with the turn-on voltage denoted by $V_{D,on}$. Note that the current goes to infinity as V_D tends to exceed $V_{D,on}$ because we assume the forward-biased diode operates as an ideal voltage source. Neglecting the leakage current in reverse bias, we derive the circuit model shown in Fig. 2.34(b). We say the junction operates as an open circuit if $V_D < V_{D,on}$ and as a constant voltage source if we attempt to increase V_D beyond $V_{D,on}$. While not essential, the voltage source placed in series with the switch in the off condition helps simplify the analysis of circuits: we can say that in the transition from off to on, only the switch turns on and the battery always resides in series with the switch.

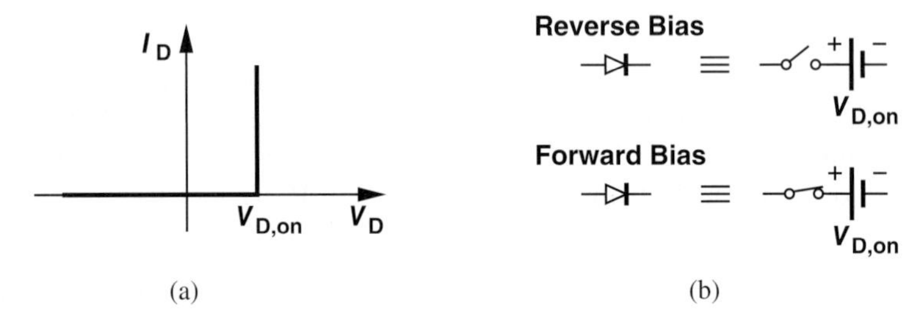

Figure 2.34 Constant-voltage diode model.

A number of questions may cross the reader's mind at this point. First, why do we subject the diode to such a seemingly inaccurate approximation? Second, if we indeed intend to use this simple approximation, why did we study the physics of semiconductors and *pn* junctions in such detail?

The developments in this chapter are representative of our treatment of *all* semiconductor devices: we carefully analyze the structure and physics of the device to understand its operation; we construct a "physics-based" circuit model; and we seek to approximate the resulting model, thus arriving at progressively simpler representations. Device models having different levels of complexity (and, inevitably, different levels of accuracy) prove essential to the analysis and design of circuits. Simple models allow a quick, intuitive understanding of the operation of a complex circuit, while more accurate models reveal the true performance.

Example 2-21

Consider the circuit of Fig. 2.35. Calculate I_X for $V_X = 3$ V and $V_X = 1$ V using (a) an exponential model with $I_S = 10^{-16}$ A and (b) a constant-voltage model with $V_{D,on} = 800$ mV.

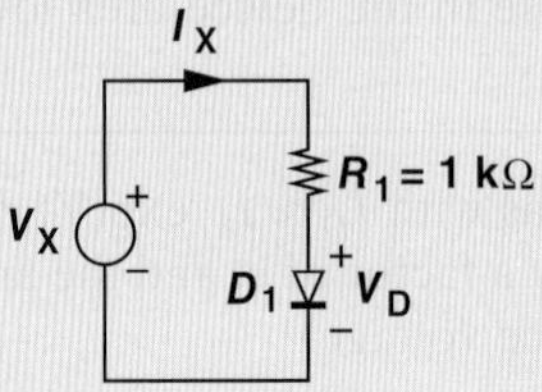

I_X, $R_1 = 1\ \text{k}\Omega$, V_X, D_1, V_D

Figure 2.35 Simple circuit using a diode.

Solution (a) Noting that $I_D = I_X$, we have

$$V_X = I_X R_1 + V_D \tag{2.114}$$

$$V_D = V_T \ln \frac{I_X}{I_S}. \tag{2.115}$$

This equation must be solved by iteration: we guess a value for V_D, compute the corresponding I_X from $I_X R_1 = V_X - V_D$, determine the new value of V_D from $V_D = V_T \ln (I_X/I_S)$ and iterate. Let us guess $V_D = 750$ mV and hence

$$I_X = \frac{V_X - V_D}{R_1} \tag{2.116}$$

$$= \frac{3\ \text{V} - 0.75\ \text{V}}{1\ \text{k}\Omega} \tag{2.117}$$

$$= 2.25\ \text{mA}. \tag{2.118}$$

Thus,

$$V_D = V_T \ln \frac{I_X}{I_S} \tag{2.119}$$

$$= 799\ \text{mV}. \tag{2.120}$$

With this new value of V_D, we can obtain a more accurate value for I_X:

$$I_X = \frac{3\ \text{V} - 0.799\ \text{V}}{1\ \text{k}\Omega} \tag{2.121}$$

$$= 2.201\ \text{mA}. \tag{2.122}$$

We note that the value of I_X rapidly converges. Following the same procedure for $V_X =$ 1 V, we have

$$I_X = \frac{1\ \text{V} - 0.75\ \text{V}}{1\ \text{k}\Omega} \tag{2.123}$$

$$= 0.25\ \text{mA}, \tag{2.124}$$

which yields $V_D = 0.742$ V and hence $I_X = 0.258$ mA. (b) A constant-voltage model readily gives

$$I_X = 2.2\ \text{mA for } V_X = 3\ \text{V} \tag{2.125}$$

$$I_X = 0.2\ \text{mA for } V_X = 1\ \text{V}. \tag{2.126}$$

The value of I_X incurs some error, but it is obtained with much less computational effort than that in part (a).

Exercise Repeat the above example if the cross section area of the diode is increased by a factor of 10.

Bioengineering Application: Cancer Detection

Certain cancerous cells can be detected through the use of optical techniques. The patient first receives an injection of "fluorophore," a substance that binds only to tumor cells in the body [2]. Next, a test sample of the patient's tissue is examined optically. As shown in Fig. 2.36, a "laser diode" driven by a current source generates light having a certain wavelength, λ_1. This light passes through the tissue sample, exciting the electrons in fluorophore to a higher energy level. Upon returning to their original level, these electrons emit light with a different wavelength, λ_2. The optical filter suppresses λ_1, allowing λ_2 to be received by a "photodiode," which converts the light to charge carriers. The resulting current, I_D, then represents the density of cancerous cells [2].

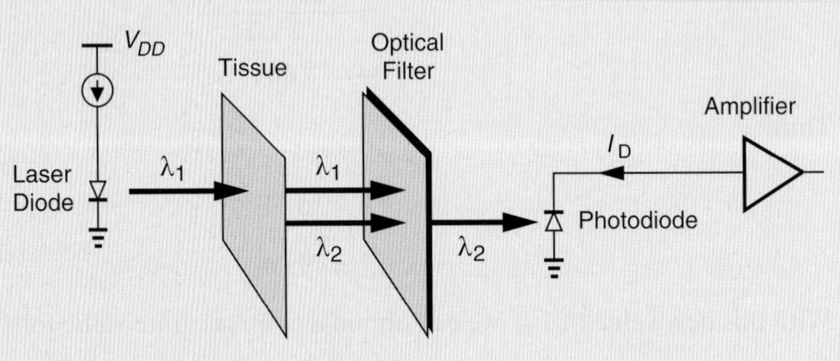

Figure 2.36 Optical method of detecting cancerous cells.

2.3 REVERSE BREAKDOWN*

Recall from Fig. 2.32 that the *pn* junction carries only a small, relatively constant current in reverse bias. However, as the reverse voltage across the device increases, eventually "breakdown" occurs and a sudden, enormous current is observed. Figure 2.37 plots the device I/V characteristic, displaying this effect.

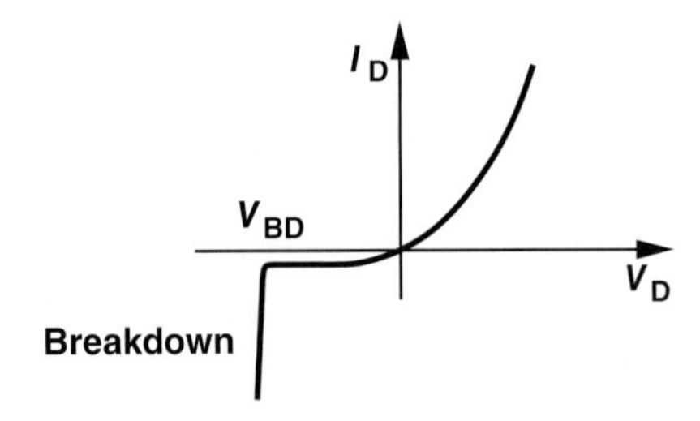

Figure 2.37 Reverse breakdown characteristic.

The breakdown resulting from a high voltage (and hence a high electric field) can occur in *any* material. A common example is lightning, in which case the electric field in the air reaches such a high level as to ionize the oxygen molecules, thus lowering the resistance of the air and creating a tremendous current.

*This section can be skipped in a first reading.

The breakdown phenomenon in *pn* junctions occurs by one of two possible mechanisms: "Zener effect" and "avalanche effect."

2.3.1 Zener Breakdown

The depletion region in a *pn* junction contains atoms that have lost an electron or a hole and, therefore, provide no loosely-connected carriers. However, a high electric field in this region may impart enough energy to the remaining covalent electrons to tear them from their bonds [Fig. 2.38(a)]. Once freed, the electrons are accelerated by the field and swept to the *n* side of the junction. This effect occurs at a field strength of about 10^6 V/cm (1 V/μm).

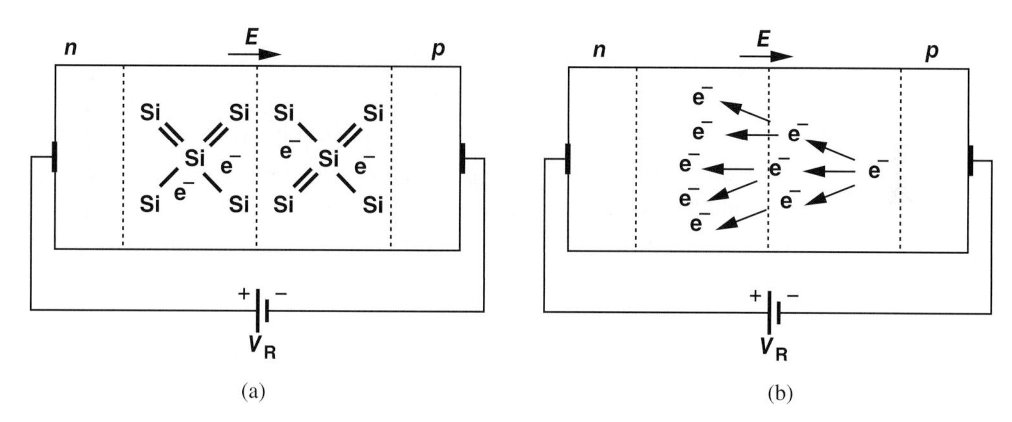

Figure 2.38 (a) Release of electrons due to high electric field, (b) avalanche effect.

In order to create such high fields with reasonable voltages, a *narrow* depletion region is required, which from Eq. (2.76) translates to high doping levels on both sides of the junction (why?). Called the "Zener effect," this type of breakdown appears for reverse bias voltages on the order of 3–8 V.

2.3.2 Avalanche Breakdown

Junctions with moderate or low doping levels ($<10^{15}$ cm^3) generally exhibit no Zener breakdown. But, as the reverse bias voltage across such devices increases, an avalanche effect takes place. Even though the leakage current is very small, each carrier entering the depletion region experiences a very high electric field and hence a large acceleration, thus gaining enough energy to break the electrons from their covalent bonds. Called "impact ionization," this phenomenon can lead to avalanche: each electron freed by the impact may itself speed up so much in the field as to collide with another atom with sufficient energy, thereby freeing one more covalent-bond electron. Now, these two electrons may again acquire energy and cause more ionizing collisions, rapidly raising the number of free carriers.

An interesting contrast between Zener and avalanche phenomena is that they display opposite temperature coefficients (TCs): V_{BD} has a negative TC for Zener effect and positive TC for avalanche effect. The two TCs cancel each other for $V_{BD} \approx 3.5$ V. For this reason, Zener diodes with 3.5-V rating find application in some voltage regulators.

The Zener and avalanche breakdown effects do not damage the diodes if the resulting current remains below a certain limit given by the doping levels and the geometry of the junction. Both the breakdown voltage and the maximum allowable reverse current are specified by diode manufacturers.

2.4 CHAPTER SUMMARY

- Silicon contains four atoms in its last orbital. It also contains a small number of free electrons at room temperature.
- When an electron is freed from a covalent bond, a "hole" is left behind.
- The bandgap energy is the minimum energy required to dislodge an electron from its covalent bond.
- To increase the number of free carriers, semiconductors are "doped" with certain impurities. For example, addition of phosphorus to silicon increases the number of free electrons because phosphorus contains five electrons in its last orbital.
- For doped or undoped semiconductors, $np = n_i^2$. For example, in an n-type material, $n \approx N_D$ and hence $p \approx n_i^2/N_D$.
- Charge carriers move in semiconductors via two mechanisms: drift and diffusion.
- The drift current density is proportional to the electric field and the mobility of the carriers and is given by $J_{tot} = q(\mu_n n + \mu_p p)E$.
- The diffusion current density is proportional to the gradient of the carrier concentration and given by $J_{tot} = q(D_n dn/dx - D_p dp/dx)$.
- A pn junction is a piece of semiconductor that receives n-type doping in one section and p-type doping in an adjacent section.
- The pn junction can be considered in three modes: equilibrium, reverse bias, and forward bias.
- Upon formation of the pn junction, sharp gradients of carrier densities across the junction result in a high current of electrons and holes. As the carriers cross, they leave ionized atoms behind, and a "depletion region" is formed. The electric field created in the depletion region eventually stops the current flow. This condition is called equilibrium.
- The electric field in the depletion results in a built-in potential across the region equal to $(kT/q)\ \ln\ (N_A N_D)/n_i^2$, typically in the range of 700 to 800 mV.
- Under reverse bias, the junction carries negligible current and operates as a capacitor. The capacitance itself is a function of the voltage applied across the device.
- Under forward bias, the junction carries a current that is an exponential function of the applied voltage: $I_S[\exp(V_F/V_T) - 1]$.
- Since the exponential model often makes the analysis of circuits difficult, a constant-voltage model may be used in some cases to estimate the circuit's response with less mathematical labor.
- Under a high reverse bias voltage, pn junctions break down, conducting a very high current. Depending on the structure and doping levels of the device, "Zener" or "avalanche" breakdown may occur.

PROBLEMS

Sec. 2.1 Semiconductor Materials and Their Properties

2.1. The intrinsic carrier concentration of germanium (GE) is expressed as

$$n_i = 1.66 \times 10^{15} T^{3/2} \exp \frac{-Eg}{2kT} \text{ cm}^{-3}, \qquad (2.127)$$

where $Eg = 0.66$ eV.

(a) Calculate n_i at 300 and 600 K and compare the results with those obtained in Example 2-1 for Si.

(b) Determine the electron and hole concentrations if Ge is doped with P at a density of 5×10^{16} cm^{-3}.

2.2. An n-type piece of silicon experiences an electric field equal to 0.1 V/μm.

(a) Calculate the velocity of electrons and holes in this material.

(b) What doping level is necessary to provide a current density of 1 mA/μm^2 under these conditions? Assume the hole current is negligible.

2.3. A n-type piece of silicon with a length of 0.1 μm and a cross section area of 0.05 μm × 0.05 μm sustains a voltage difference of 1 V.

(a) If the doping level is 10^{17} cm^{-3}, calculate the total current flowing through the device at $T = 300$ K.

(b) Repeat (a) for $T = 400$ K assuming for simplicity that mobility does not change with temperature. (This is not a good assumption.)

2.4. From the data in Problem 2.1, repeat Problem 2.3 for Ge. Assume $\mu_n = 3900$ cm^2/(V · s) and $\mu_p = 1900$ cm^2/(V · s).

SOL **2.5.** Figure 2.39 shows a p-type bar of silicon that is subjected to electron injection from the left and hole injection from the right. Determine the total current flowing through the device if the cross section area is equal to 1 μm × 1 μm.

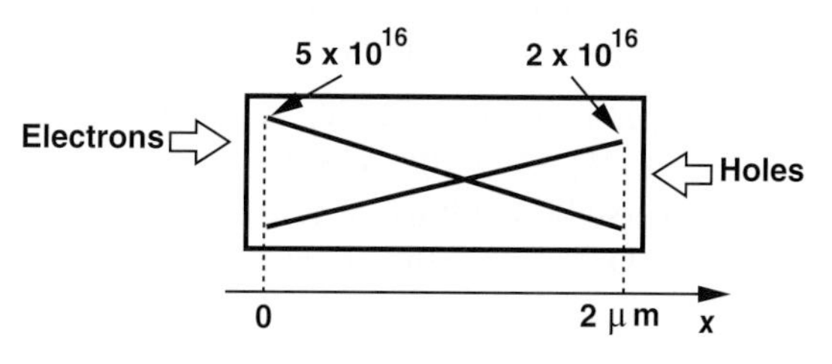

Figure 2.39

2.6. In Example 2-9, compute the total number of electrons "stored" in the material from $x = 0$ to $x = L$. Assume the cross section area of the bar is equal to a.

2.7. Repeat Problem 2.6 for Example 2-10 but for $x = 0$ to $x = \infty$. Compare the results for linear and exponential profiles.

***2.8.** Repeat Problem 2.7 if the electron and hole profiles are "sharp" exponentials, i.e., they fall to negligible values at $x = 2$ μm and $x = 0$, respectively (Fig. 2.40).

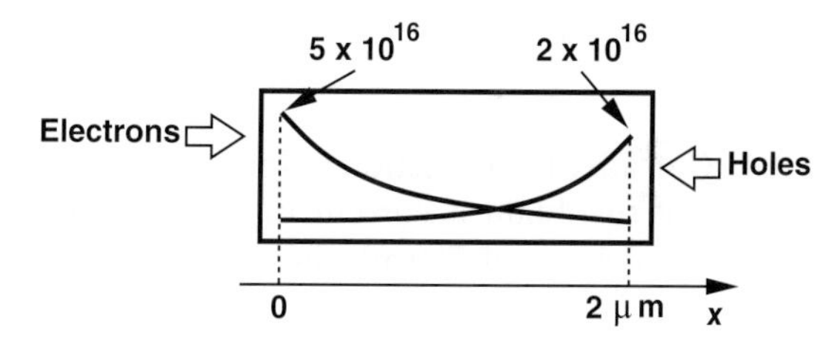

Figure 2.40

2.9. How do you explain the phenomenon of drift to a high school student? SOL

Sec. 2.2 *pn* Junctions

2.10. A junction employs $N_D = 5 \times 10^{17}$ cm^{-3} and $N_A = 4 \times 10^{16}$ cm^{-3}.

(a) Determine the majority and minority carrier concentrations on both sides.

(b) Calculate the built-in potential at $T =$ 250 K, 300 K, and 350 K. Explain the trend.

2.11. Due to a manufacturing error, the p-side of a pn junction has not been doped. If $N_D = 3 \times 10^{16}$ cm^{-3}, calculate the built-in potential at $T = 300$ K.

2.12. A pn junction with $N_D = 3 \times 10^{16}$ cm^{-3} and $N_A = 2 \times 10^{15}$ cm^{-3} experiences a reverse bias voltage of 1.6 V.

(a) Determine the junction capacitance per unit area.

(b) By what factor should N_A be increased to double the junction capacitance?

2.13. An oscillator application requires a variable capacitance with the characteristic shown in Fig. 2.41. Determine N_A and N_D.

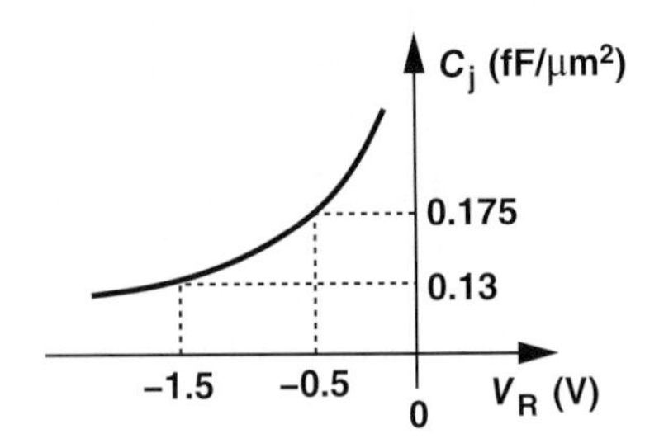

Figure 2.41

2.14. Consider a pn junction in forward bias.

(a) To obtain a current of 1 mA with a voltage of 750 mV, how should I_S be chosen?

(b) If the diode cross section area is now doubled, what voltage yields a current of 1 mA?

SOL **2.15.** Figure 2.42 shows two diodes with reverse saturation currents of I_{S1} and I_{S2} placed in parallel.

(a) Prove that the parallel combination operates as an exponential device.

(b) If the total current is I_{tot}, determine the current carried by each diode.

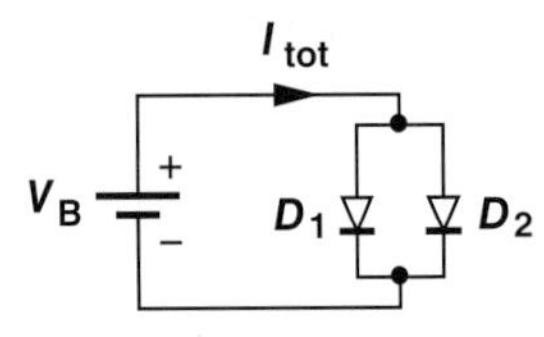

Figure 2.42

***2.16.** Two identical pn junctions are placed in series.

(a) Prove that this combination can be viewed as a single two-terminal device having an exponential characteristic.

(b) For a tenfold change in the current, how much voltage change does such a device require?

2.17. Figure 2.43 shows two diodes with reverse saturation currents of I_{S1} and I_{S2} placed in series. Calculate I_B, V_{D1}, and V_{D2} in terms of V_B, I_{S1}, and I_{S2}.

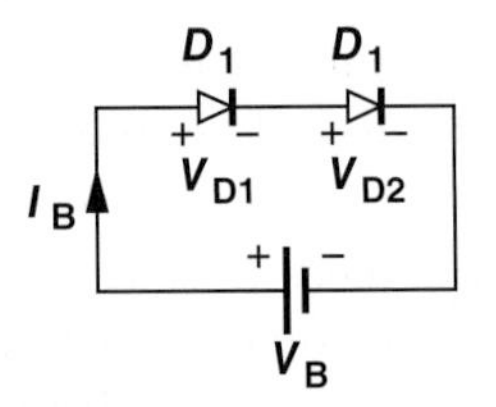

Figure 2.43

2.18. In the circuit of Problem 2.17, we wish to increase I_B by a factor of 10. What is the required change in V_B?

Sec. 2.2.5 I/V Characteristics

2.19. Consider the circuit shown in Fig. 2.44, where $I_S = 2 \times 10^{-15}$ A. Calculate V_{D1} and I_X for $V_X = 0.5$ V, 0.8 V, 1 V, and 1.2 V. Note that V_{D1} changes little for $V_X \geq$ 0.8 V.

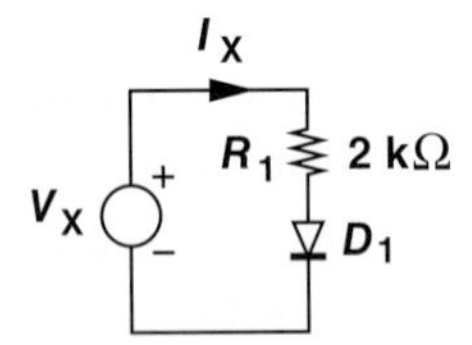

Figure 2.44

2.20. In the circuit of Fig. 2.44, the cross section area of D_1 is increased by a factor of 10. Determine V_{D1} and I_X for $V_X = 0.8$ V and 1.2 V. Compare the results with those obtained in Problem 2.19.

2.21. Suppose D_1 in Fig. 2.44 must sustain a voltage of 850 mV for $V_X = 2$ V. Calculate the required I_S. SOL

2.22. For what value of V_X in Fig. 2.44, does R_1 sustain a voltage equal to $V_X/2$? Assume $I_S = 2 \times 10^{-16}$ A.

2.23. We have received the circuit shown in Fig. 2.45 and wish to determine R_1 and I_S. We note that $V_X = 1\text{ V} \rightarrow I_X = 0.2$ mA and $V_X = 2\text{ V} \rightarrow I_X = 0.5$ mA. Calculate R_1 and I_S.

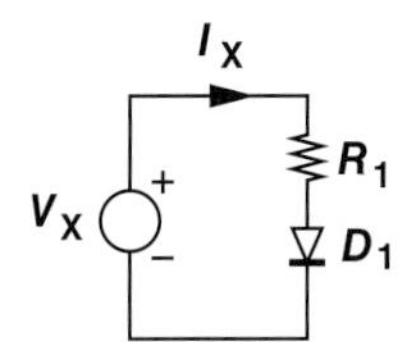

Figure 2.45

*__2.24.__ Figure 2.46 depicts a parallel resistor-diode combination. If $I_S = 3 \times 10^{-16}$ A, calculate V_{D1} for $I_X = 1$ mA, 2 mA, and 4 mA.

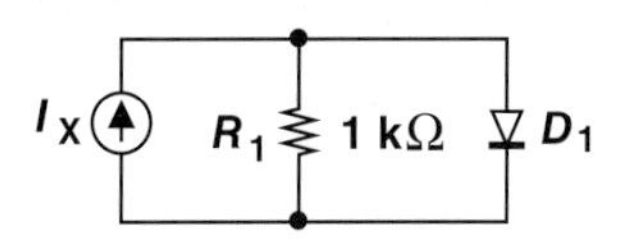

Figure 2.46

2.25. In the circuit of Fig. 2.46, we wish D_1 to carry a current of 0.5 mA for $I_X = 1.3$ mA. Determine the required I_S.

2.26. For what value of I_X in Fig. 2.46, does R_1 carry a current equal to $I_X/2$? Assume $I_S = 3 \times 10^{-16}$ A.

SOL *__2.27.__ We have received the circuit shown in Fig. 2.47 and wish to determine R_1 and I_S. Measurements indicate that $I_X = 1\text{ mA} \rightarrow V_X = 1.2$ V and $I_X = 2\text{ mA} \rightarrow V_X = 1.8$ V. Calculate R_1 and I_S.

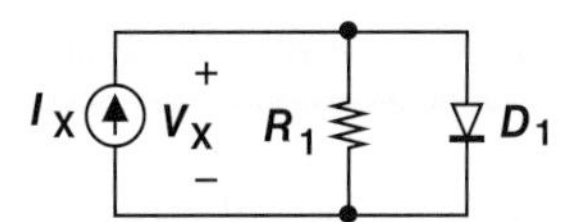

Figure 2.47

**__2.28.__ The circuit illustrated in Fig. 2.48 employs two identical diodes with $I_S = 5 \times 10^{-16}$ A. Calculate the voltage across R_1 for $I_X = 2$ mA.

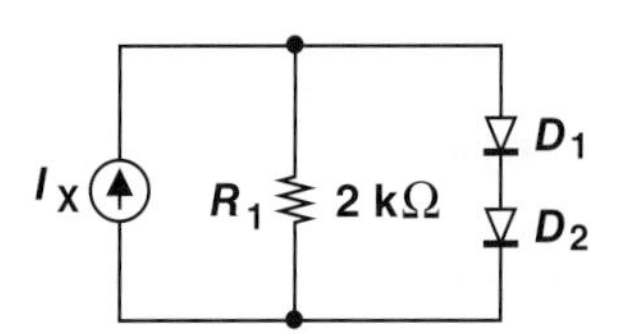

Figure 2.48

**__2.29.__ In the circuit of Fig. 2.49, determine the value of R_1 such that this resistor carries 0.5 mA. Assume $I_S = 5 \times 10^{-16}$ A for each diode.

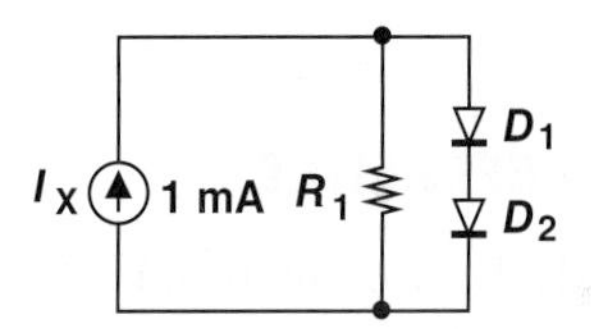

Figure 2.49

**__2.30.__ Sketch V_X as a function of I_X for the circuit shown in Fig. 2.50. Assume (a) a constant-voltage model, (b) an exponential model.

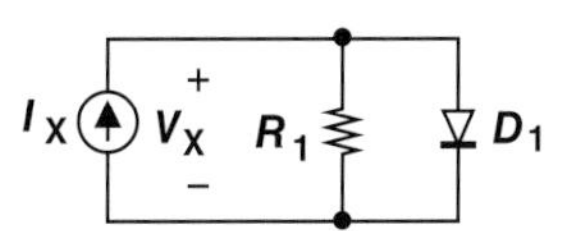

Figure 2.50

SPICE PROBLEMS

In the following problems, assume $I_S = 5 \times 10^{-16}$ A.

2.31. For the circuit shown in Fig. 2.51, plot V_{out} as a function of I_{in}. Assume I_{in} varies from 0 to 2 mA.

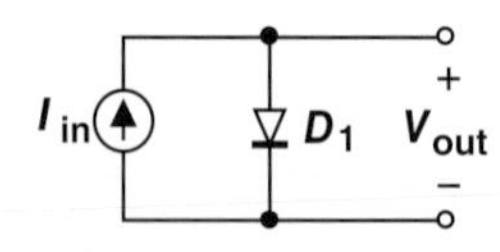

Figure 2.51

2.32. Repeat Problem 2.31 for the circuit depicted in Fig. 2.52, where $R_1 = 1\text{ k}\Omega$. At what value of I_{in} are the currents flowing through D_1 and R_1 equal?

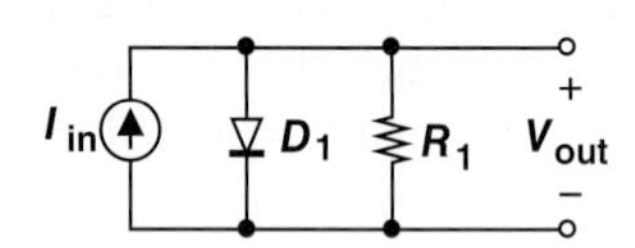

Figure 2.52

2.33. Using SPICE, determine the value of R_1 in Fig. 2.52 such that D_1 carries 1 mA if $I_{in} = 2$ mA.

2.34. In the circuit of Fig. 2.53, $R_1 = 500\ \Omega$. Plot V_{out} as a function of V_{in} if V_{in} varies from −2 V to +2 V. At what value of V_{in} are the voltage drops across R_1 and D_1 equal?

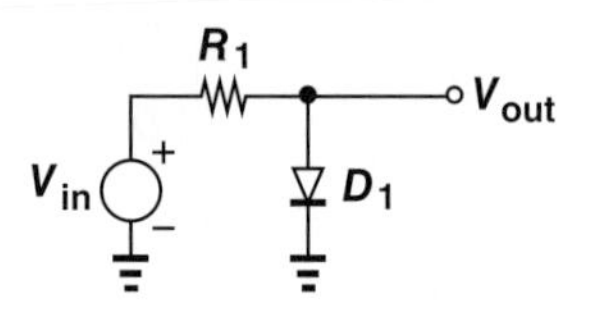

Figure 2.53

2.35. In the circuit of Fig. 2.53, use SPICE to select the value of R_1 such that $V_{out} < 0.7$ V for $V_{in} < 2$ V. We say the circuit "limits" the output.

REFERENCE

1. B. Streetman and S. Banerjee, *Solid-State Electronic Device*, fifth edition, Prentice-Hall, 1999.
2. E. P. Papageorgiou, et al., "Chip-scale angle-selective imager for in-vivo microscopic cancer detection," *IEEE Trans. Biomedical Circuits and Systems*, vol. 14, pp. 91–92, Feb. 2020.

3

Diode Models and Circuits

Watch Companion YouTube Videos:

Razavi Electronics 1, Lec 5: PN Junction in Forward Bias, Introduction to Diodes
https://www.youtube.com/watch?v=p9q5thwmlxY

Razavi Electronics 1, Lec 6: Diode Models
https://www.youtube.com/watch?v=EnA9m7TOaol

Razavi Electronics 1, Lec 7: Analysis of Diode Circuits I
https://www.youtube.com/watch?v=PoPp3Ea6KMs

Razavi Electronics 1, Lec 8: Analysis of Diode Circuits II
https://www.youtube.com/watch?v=5NSKxgWZFEE

Razavi Electronics 1, Lec 9: Other Examples of Diode Circuits, Half-Wave Rectifier
https://www.youtube.com/watch?v=4BoxbOs2gs0

Razavi Electronics 1, Lec 10: Half-Wave Rectifier with Different Loads
https://www.youtube.com/watch?v=cN-mNvzD6_Q

Razavi Electronics 1, Lec 11: Full-Wave Rectifier
https://www.youtube.com/watch?v=rtsamkkLcdY

Razavi Electronics 1, Lec 12: Limiters and Voltage Doublers
https://www.youtube.com/watch?v=uni-behlRN4

Having studied the physics of diodes in Chapter 2, we now rise to the next level of abstraction and deal with diodes as circuit elements, ultimately arriving at interesting and real-life applications. This chapter also prepares us for understanding transistors as circuit elements in subsequent chapters. We proceed as follows:

Diodes as Circuit Elements
- **Ideal Diode**
- **Circuit Characteristics**
- **Actual Diode**

➤

Applications
- **Regulators**
- **Rectifiers**
- **Limiting and Clamping Circuits**

3.1 IDEAL DIODE

3.1.1 Initial Thoughts

In order to appreciate the need for diodes, let us briefly study the design of a cellphone charger. The charger converts the line ac voltage at 110 V[1] and 60 Hz[2] to a dc voltage of 3.5 V. As shown in Fig. 3.1(a), this is accomplished by first stepping down the ac voltage by means of a transformer to about 4 V and subsequently converting the ac voltage to a dc quantity.[3] The same principle applies to adaptors that power other electronic devices.

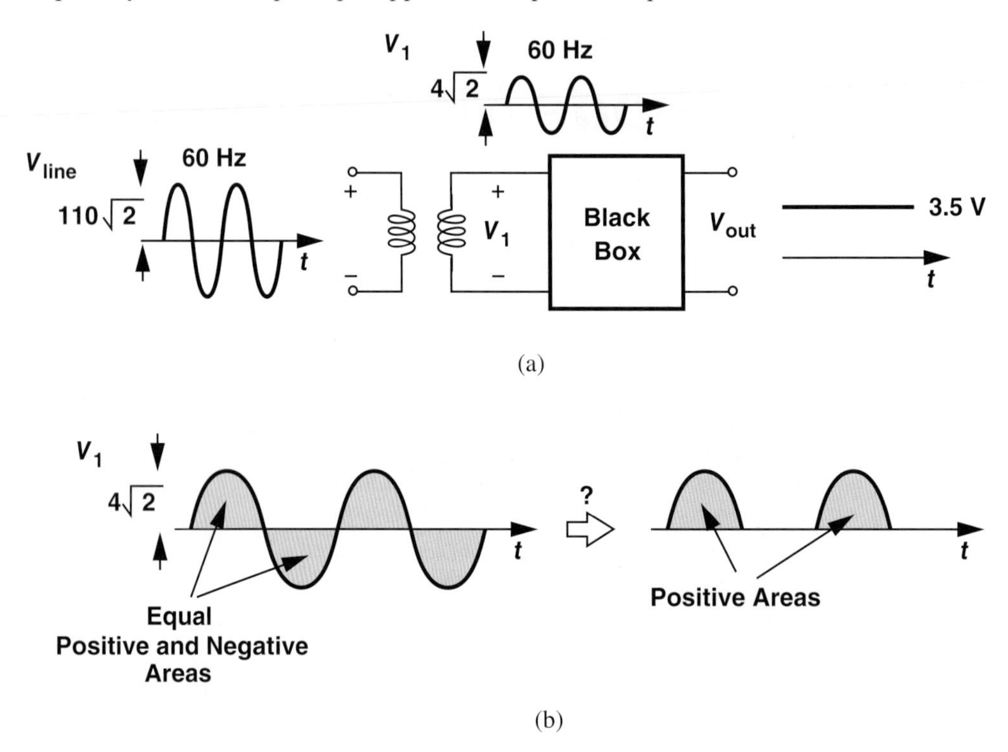

Figure 3.1 (a) Charger circuit, (b) elimination of negative half cycles.

Did you know?

The rectification effect was discovered by Carl Braun around 1875. He observed that a metal wire pressed against a sulphide carried more current in one direction than the other. In subsequent decades, researchers tried other structures, arriving at "cat's whisker" rectifiers, which served as detectors for the reception of wireless signals in early radios and radars. These were eventually supplanted by *pn* junctions.

How does the black box in Fig. 3.1(a) perform this conversion? As depicted in Fig. 3.1(b), the output of the transformer exhibits a zero dc content because the negative and positive half cycles enclose equal areas, leading to a zero average. Now suppose this waveform is applied to a mysterious device that passes the positive half cycles but blocks the negative ones. The result displays a positive average and some ac components, which can be removed by a low-pass filter (Section 3.5.1).

The waveform conversion in Fig. 3.1(b) points to the need for a device that *discriminates* between positive and negative voltages, passing only one and blocking the other.

[1]This value refers to the root-mean-square (rms) voltage. The peak value is therefore equal to $110\sqrt{2}$.

[2]The line ac voltage in most countries is at 220 V and 50 Hz.

[3]The actual operation of adaptors is somewhat different.

A simple resistor cannot serve in this role because it is *linear*. That is, Ohm's law, $V = IR$, implies that if the voltage across a resistor goes from positive to negative, so does the current through it. We must therefore seek a device that behaves as a short for positive voltages and as an open for negative voltages.

Figure 3.2 summarizes the result of our thought process thus far. The mysterious device generates an output equal to the input for positive half cycles and equal to zero for negative half cycles. Note that the device is nonlinear because it does not satisfy $y = \alpha x$; if $x \to -x$, $y \nrightarrow -y$.

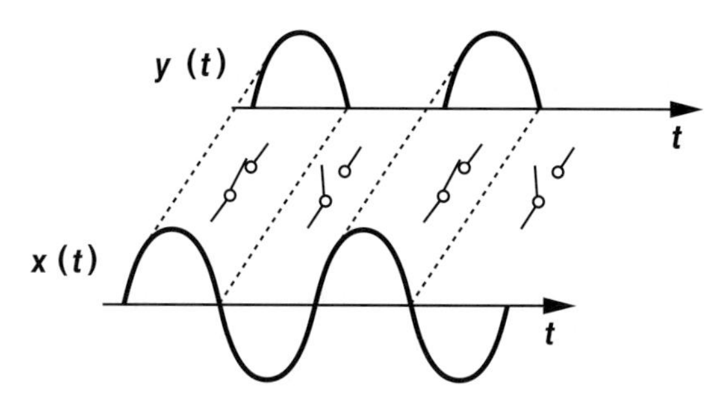

Figure 3.2 Conceptual operation of a diode.

3.1.2 Ideal Diode

The mysterious device mentioned above is called an "ideal diode." As shown in Fig. 3.3(a), the diode is a two-terminal device, with the triangular head denoting the allowable direction of current flow and the vertical bar representing the blocking behavior for currents in the opposite direction. The corresponding terminals are called the "anode" and the "cathode," respectively.

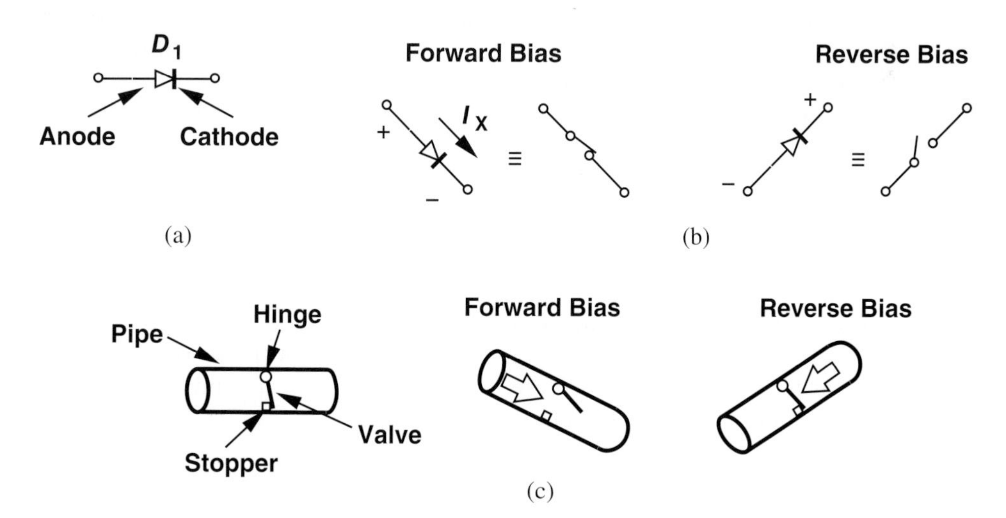

Figure 3.3 (a) Diode symbol, (b) equivalent circuit, (c) water pipe analogy.

Forward and Reverse Bias To serve as the mysterious device in the charger example of Fig. 3.3(a), the diode must turn "on" if $V_{anode} > V_{cathode}$ and "off" if $V_{anode} < V_{cathode}$ [Fig. 3.3(b)]. Defining $V_{anode} - V_{cathode} = V_D$, we say the diode is "forward-biased" if V_D tends to exceed zero and "reverse-biased" if $V_D < 0$.[4]

[4]In our drawings, we sometimes place more positive nodes higher to provide a visual picture of the circuit's operation. The diodes in Fig. 3.3(b) are drawn according to this convention.

A water pipe analogy proves useful here. Consider the pipe shown in Fig. 3.3(c), where a valve (a plate) is hinged on the top and faces a stopper on the bottom. If water pressure is applied from the left, the valve rises, allowing a current. On the other hand, if water pressure is applied from the right, the stopper keeps the valve shut.

Example 3-1 As with other two-terminal devices, diodes can be placed in series (or in parallel). Determine which one of the configurations in Fig. 3.4 can conduct current.

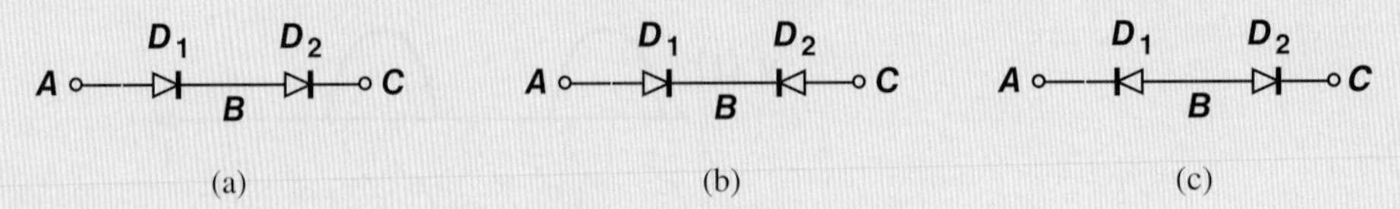

Figure 3.4 Series combinations of diodes.

Solution In Fig. 3.4(a), the anodes of D_1 and D_2 point to the same direction, allowing the flow of current from A to B to C but not in the reverse direction. In Fig. 3.4(b), D_1 stops current flow from B to A, and D_2, from B to C. Thus, no current can flow in either direction. By the same token, the topology of Fig. 3.4(c) behaves as an open for any voltage. Of course, none of these circuits appears particularly useful at this point, but they help us become comfortable with diodes.

Exercise Determine all possible series combinations of three diodes and study their conduction properties.

I/V Characteristics In studying electronic devices, it is often helpful to accompany equations with graphical visualizations. A common type of plot is that of the current/voltage (I/V) characteristic, i.e., the current that flows through the device as a function of the voltage across it.

Since an ideal diode behaves as a short or an open, we first construct the I/V characteristics for two special cases of Ohm's law:

$$R = 0 \Rightarrow I = \frac{V}{R} = \infty \tag{3.1}$$

$$R = \infty \Rightarrow I = \frac{V}{R} = 0. \tag{3.2}$$

The results are illustrated in Fig. 3.5(a). For an ideal diode, we combine the positive-voltage region of the first with the negative-voltage region of the second, arriving at the I_D/V_D characteristic in Fig. 3.5(b). Here, $V_D = V_{\text{anode}} - V_{\text{cathode}}$, and I_D is defined as the current flowing into the anode and out of the cathode.

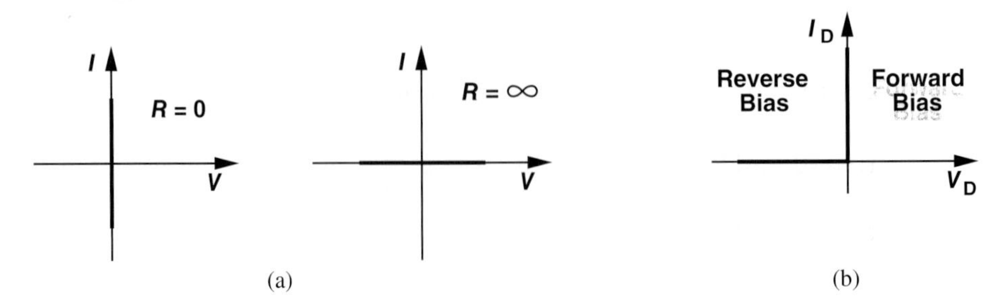

Figure 3.5 I/V characteristics of (a) zero and infinite resistors, (b) ideal diode.

Example 3-2 We said that an ideal diode turns on for positive anode-cathode voltages. But the characteristic in Fig. 3.5(b) does not appear to show any I_D values for $V_D > 0$. How do we interpret this plot?

Solution This characteristic indicates that as V_D exceeds zero by a very small amount, then the diode turns on and conducts infinite current *if* the circuit surrounding the diode can provide such a current. Thus, in circuits containing only finite currents, a forward-biased ideal diode sustains a zero voltage—similar to a short circuit.

Exercise How is the characteristic modified if we place a 1-Ω resistor in series with the diode?

Example 3-3 Plot the I/V characteristic for the "antiparallel" diodes shown in Fig. 3.6(a).

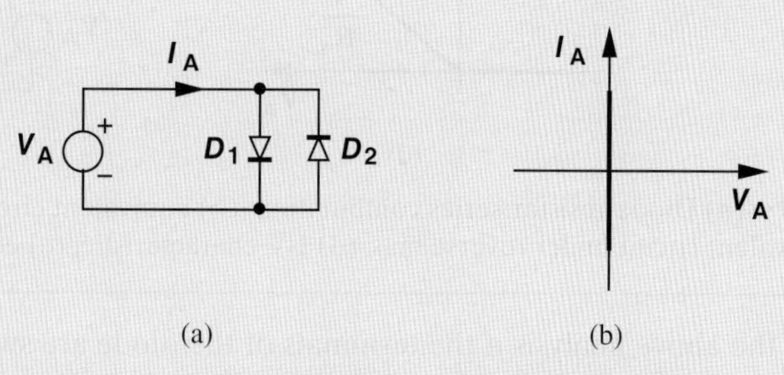

Figure 3.6 (a) Antiparallel diodes, (b) resulting I/V characteristic.

Solution If $V_A > 0$, D_1 is on and D_2 is off, yielding $I_A = \infty$. If $V_A < 0$, D_1 is off, but D_2 is on, again leading to $I_A = \infty$. The result is illustrated in Fig. 3.6(b). The antiparallel combination therefore acts as a short for all voltages. Seemingly a useless circuit, this topology becomes much more interesting with actual diodes (Section 3.5.3).

Exercise Repeat the above example if a 1-V battery is placed in series with the parallel combination of the diodes.

Example 3-4 Plot the I/V characteristic for the diode-resistor combination of Fig. 3.7(a).

Solution We surmise that, if $V_A > 0$, the diode is on [Fig. 3.7(b)] and $I_A = V_A/R_1$ because $V_{D1} = 0$ for an ideal diode. On the other hand, if $V_A < 0$, D_1 is probably off [Fig. 3.7(c)] and $I_D = 0$. Figure 3.7(d) plots the resulting I/V characteristic.

The above observations are based on guesswork. Let us study the circuit more rigorously. We begin with $V_A < 0$, postulating that the diode is off. To confirm the validity of this guess, let us assume D_1 is on and see if we reach a conflicting result. If D_1 is on, the circuit is reduced to that in Fig. 3.7(e), and if V_A is negative, so is I_A; i.e., the actual current flows from right to left. But this implies that D_1 carries a current from its cathode to its anode, violating the definition of the diode. Thus, for $V_A < 0$, D_1 remains off and $I_A = 0$.

As V_A rises above zero, it tends to forward bias the diode. Does D_1 turn on for any $V_A > 0$ or does R_1 shift the turn-on point? We again invoke proof by contradiction.

Suppose for some $V_A > 0$, D_1 is still off, behaving as an open circuit and yielding $I_A = 0$. The voltage drop across R_1 is therefore equal to zero, suggesting that $V_{D1} = V_A$ and hence $I_{D1} = \infty$ and contradicting the original assumption. In other words, D_1 turns on for any $V_A > 0$.

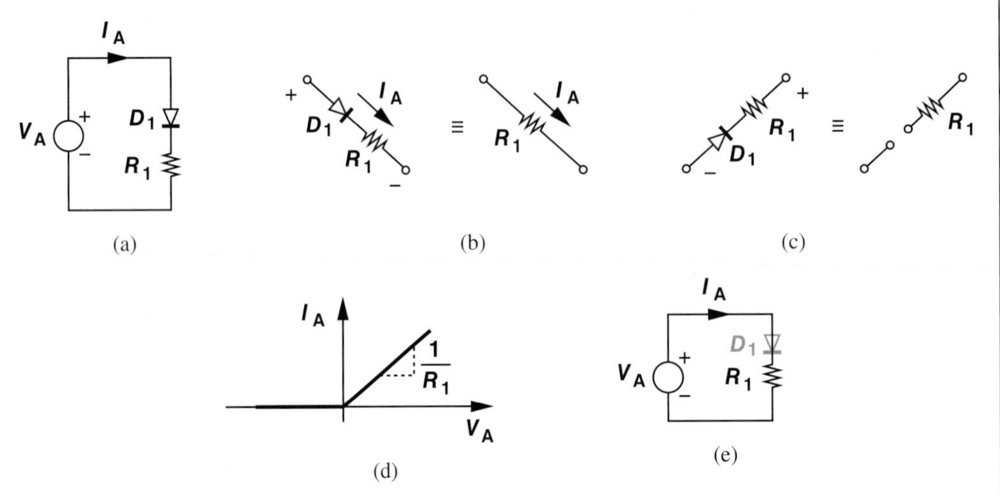

Figure 3.7 (a) Diode-resistor series combination, (b) equivalent circuit under forward bias, (c) equivalent circuit under reverse bias, (d) I/V characteristic, (e) equivalent circuit if D_1 is on.

Exercise Repeat the above analysis if the terminals of the diode are swapped.

The above example leads to two important points. First, the series combination of D_1 and R_1 acts as an open for negative voltages and as a resistor of value R_1 for positive voltages. Second, in the analysis of circuits, we can assume an arbitrary state (on or off) for each diode and proceed with the computation of voltages and currents; if the assumptions are incorrect, the final result contradicts the original assumptions. Of course, it is helpful to first examine the circuit carefully and make an intuitive guess.

Example 3-5 Why are we interested in I/V characteristics rather than V/I characteristics?

Solution In the analysis of circuits, we often prefer to consider the voltage to be the "cause" and the current, the "effect." This is because in typical circuits, voltage polarities can be predicted more readily and intuitively than current polarities. Also, devices such as transistors fundamentally produce current in response to voltage.

Exercise Plot the V/I characteristic of an ideal diode.

Example 3-6 In the circuit of Fig. 3.8, each input can assume a value of either zero or +3 V. Determine the response observed at the output.

Solution If $V_A = +3$ V, and $V_B = 0$, then we surmise that D_1 is forward-biased and D_2, reverse-biased. Thus, $V_{out} = V_A = +3$ V. If uncertain, we can assume both D_1 and D_2

are forward-biased, immediately facing a conflict: D_1 enforces a voltage of +3 V at the output whereas D_2 shorts V_{out} to $V_B = 0$. This assumption is therefore incorrect.

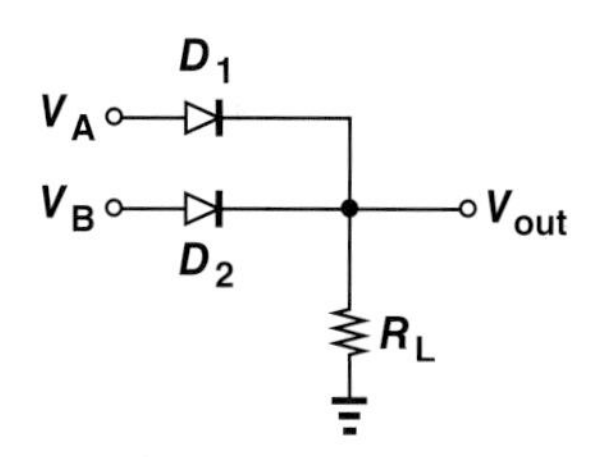

Figure 3.8 OR gate realized by diodes.

The symmetry of the circuit with respect to V_A and V_B suggests that $V_{out} = V_B =$ +3 V if $V_A = 0$ and $V_B = +3$ V. The circuit operates as a logical OR gate and was in fact used in early digital computers.

Exercise Construct a three-input OR gate.

Example 3-7 Is an ideal diode on or off if $V_D = 0$?

Solution An ideal diode experiencing a zero voltage must carry a zero current (why?). However, this does not mean it acts as an open circuit. After all, a piece of wire experiencing a zero voltage behaves similarly. Thus, the state of an ideal diode with $V_D = 0$ is somewhat arbitrary and ambiguous. In practice, we consider slightly positive or negative voltages to determine the response of a diode circuit.

Exercise Repeat the above example if a 1-Ω resistor is placed in series with the diode.

Input/Output Characteristics Electronic circuits process an input and generate a corresponding output. It is therefore instructive to construct the input/output characteristics of a circuit by varying the input across an allowable range and plotting the resulting output.

As an example, consider the circuit depicted in Fig. 3.9(a), where the output is defined as the voltage across D_1. If $V_{in} < 0$, D_1 is reverse biased, reducing the circuit to that in Fig. 3.9(b). Since no current flows through R_1, we have $V_{out} = V_{in}$. If $V_{in} > 0$, then D_1 is forward biased, shorting the output and forcing $V_{out} = 0$ [Fig. 3.9(c)]. Figure 3.9(d) illustrates the overall input/output characteristic.

3.1.3 Application Examples

Recall from Fig. 3.2 that we arrived at the concept of the ideal diode as a means of converting $x(t)$ to $y(t)$. Let us now design a circuit that performs this function. We may naturally construct the circuit as shown in Fig. 3.10(a). Unfortunately, however, the cathode of the diode is "floating," the output current is always equal to zero, and the state of the diode is ambiguous. We therefore modify the circuit as depicted in Fig. 3.10(b) and analyze its response to a sinusoidal input [Fig. 3.10(c)]. Since R_1 has a tendency to maintain the cathode of D_1 near zero, as V_{in} rises, D_1 is forward biased, shorting the output to the input.

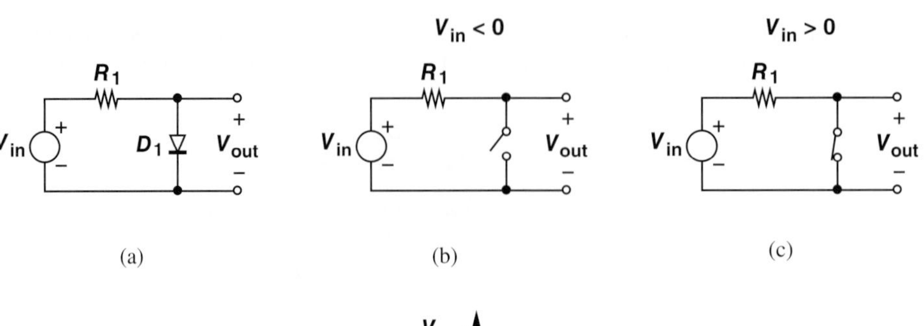

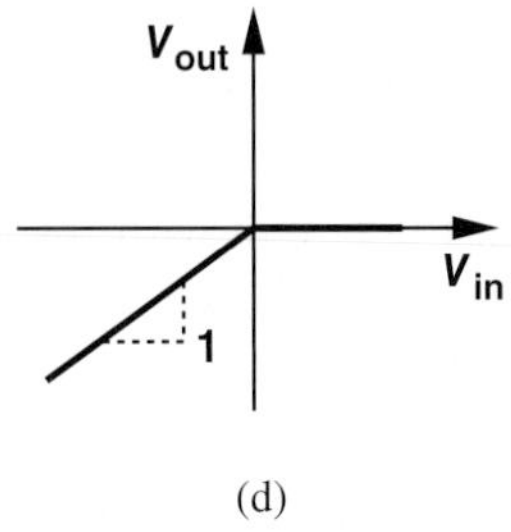

Figure 3.9 (a) Resistor-diode circuit, (b) equivalent circuit for negative input, (c) equivalent circuit for positive input, (d) input/output characteristic.

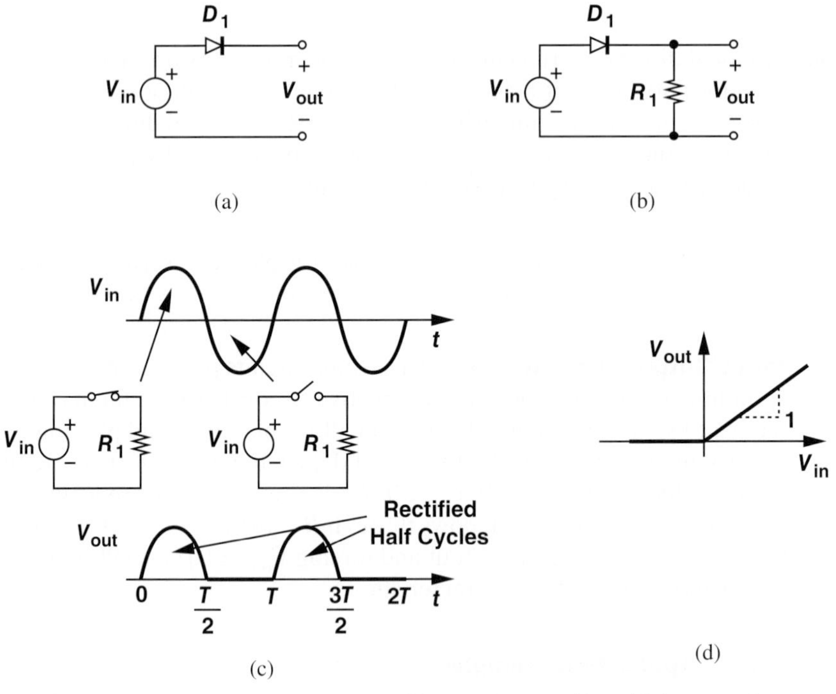

Figure 3.10 (a) A diode operating as a rectifier, (b) complete rectifier, (c) input and output waveforms, (d) input/output characteristic.

This state holds for the positive half cycle. When V_{in} falls below zero, D_1 turns off and R_1 ensures that $V_{out} = 0$ because $I_D R_1 = 0$.[5] The circuit of Fig. 3.10(b) is called a "rectifier."

It is instructive to plot the input/output characteristic of the circuit as well. Noting that if $V_{in} < 0$, D_1 is off and $V_{out} = 0$, and if $V_{in} > 0$, D_1 is on and $V_{out} = V_{in}$, we obtain the behavior shown in Fig. 3.10(d). The rectifier is a nonlinear circuit because if $V_{in} \rightarrow -V_{in}$ then $V_{out} \nrightarrow -V_{out}$.

Example 3-8 Is it a coincidence that the characteristics in Figs. 3.7(d) and 3.10(d) look similar?

Solution No, we recognize that the output voltage in Fig. 3.10(b) is simply equal to $I_A R_1$ in Fig. 3.7(a). Thus, the two plots differ by only a scaling factor equal to R_1.

Exercise Construct the characteristic if the terminals of D_1 are swapped.

We now determine the time average (dc value) of the output waveform in Fig. 3.10(c) to arrive at another interesting application. Suppose $V_{in} = V_p \sin \omega t$, where $\omega = 2\pi/T$ denotes the frequency in radians per second and T the period. Then, in the first cycle after $t = 0$, we have

$$V_{out} = V_p \sin \omega t \text{ for } 0 \leq t \leq \frac{T}{2} \tag{3.3}$$

$$= 0 \quad \text{for } \frac{T}{2} \leq t \leq T. \tag{3.4}$$

To compute the average, we obtain the area under V_{out} and normalize the result to the period:

$$V_{out,avg} = \frac{1}{T} \int_0^T V_{out}(t) dt \tag{3.5}$$

$$= \frac{1}{T} \int_0^{T/2} V_p \sin \omega t \, dt \tag{3.6}$$

$$= \frac{1}{T} \cdot \frac{V_p}{\omega} [-\cos \omega t]_0^{T/2} \tag{3.7}$$

$$= \frac{V_p}{\pi}. \tag{3.8}$$

Thus, the average is proportional to V_p, an expected result because a larger input amplitude yields a greater area under the rectified half cycles.

The above observation reveals that the average value of a rectified output can serve as a measure of the "strength" (amplitude) of the input. That is, a rectifier can operate as a "signal strength indicator." For example, since cellphones receive varying levels of signal depending on the user's location and environment, they require an indicator to determine how much the signal must be amplified.

[5] Note that without R_1, the output voltage is not defined because a floating node can assume any potential.

Example 3-9 A cellphone receives a 1.8-GHz signal with a peak amplitude ranging from 2 μV to 10 mV. If the signal is applied to a rectifier, what is the corresponding range of the output average?

Solution The rectified output exhibits an average value ranging from $2\ \mu V/(\pi) = 0.637\ \mu V$ to $10\ mV/(\pi) = 3.18\ mV$.

Exercise Do the above results change if a 1-Ω resistor is placed in series with the diode?

In our effort toward understanding the role of diodes, we examine another circuit that will eventually (in Section 3.5.3) lead to some important applications. First, consider the topology in Fig. 3.11(a), where a 1-V battery is placed in series with an ideal diode. How does this circuit behave? If $V_1 < 0$, the cathode voltage is higher than the anode voltage, placing D_1 in reverse bias. Even if V_1 is slightly greater than zero, e.g., equal to 0.9 V, the anode is not positive enough to forward bias D_1. Thus, V_1 must approach +1 V for D_1 to turn on. As shown in Fig. 3.11(a), the I/V characteristic of the diode-battery combination resembles that of a diode, but shifted to the right by 1 V.

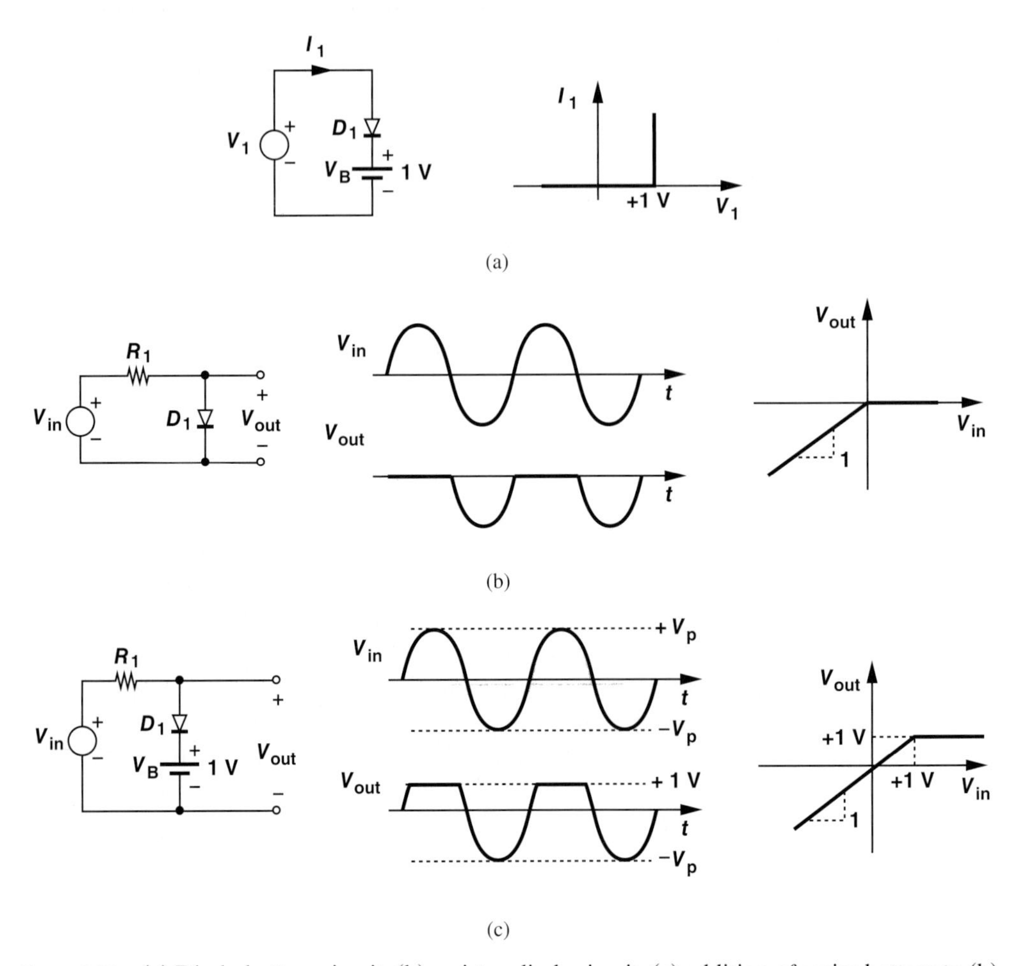

Figure 3.11 (a) Diode-battery circuit, (b) resistor-diode circuit, (c) addition of series battery to (b).

Now, let us examine the circuit in Fig. 3.11(b). Here, for $V_{in} < 0$, D_1 remains off, yielding $V_{out} = V_{in}$. For $V_{in} > 0$, D_1 acts a short, and $V_{out} = 0$. The circuit therefore does not allow the output to exceed zero, as illustrated in the output waveform and the input/output characteristic. But suppose we seek a circuit that must not allow the output to exceed +1 V (rather than zero). How should the circuit of Fig. 3.11(b) be modified? In this case, D_1 must turn on only when V_{out} approaches +1 V, implying that a 1-V battery must be inserted in series with the diode. Depicted in Fig. 3.11(c), the modification indeed guarantees $V_{out} \leq +1$ V for any input level. We say the circuit "clips" or "limits" at +1 V. "Limiters" prove useful in many applications and are described in Section 3.5.3.

Example 3-10 Sketch the time average of V_{out} in Fig. 3.11(c) for a sinusoidal input as the battery voltage, V_B, varies from $-\infty$ to $+\infty$.

Solution If V_B is very negative, D_1 is always on because $V_{in} \geq -V_p$. In this case, the output average is equal to V_B [Fig. 3.12(a)]. For $-V_p < V_B < 0$, D_1 turns off at some point in the negative half cycle and remains off in the positive half cycle, yielding an average greater than $-V_p$ but less than V_B. For $V_B = 0$, the average reaches $-V_p/(\pi)$. Finally, for $V_B \geq V_p$, no limiting occurs and the average is equal to zero. Figure 3.12(b) sketches this behavior.

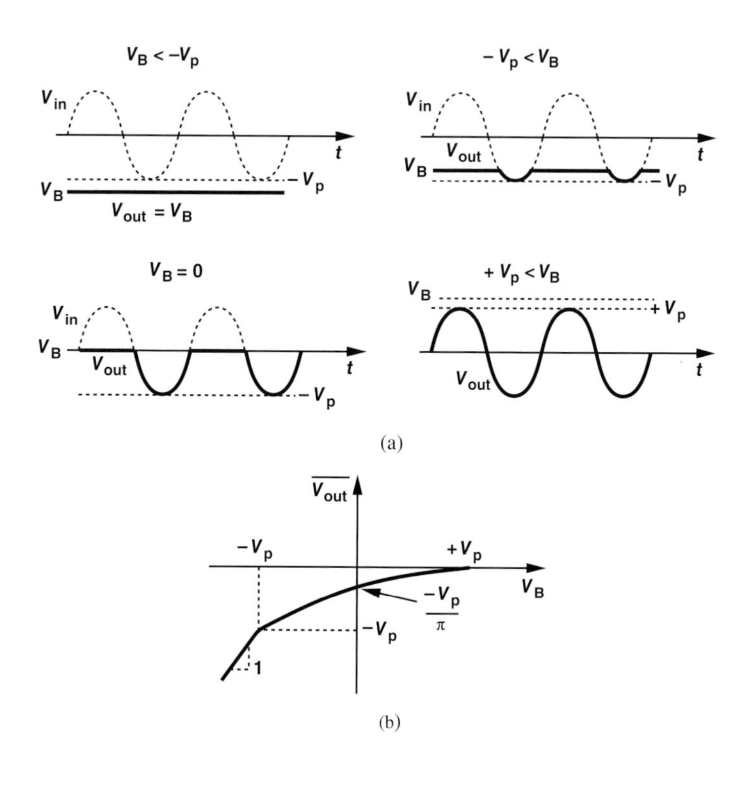

Figure 3.12

Exercise Repeat the above example if the terminals of the diode are swapped.

Example 3-11 Is the circuit of Fig. 3.11(b) a rectifier?

Solution Yes, indeed. The circuit passes only negative cycles to the output, producing a negative average.

Exercise How should the circuit of Fig. 3.11(b) be modified to pass only positive cycles to the output?

3.2 *pn* JUNCTION AS A DIODE

The operation of the ideal diode is somewhat reminiscent of the current conduction in *pn* junctions. In fact, the forward and reverse bias conditions depicted in Fig. 3.3(b) are quite similar to those studied for *pn* junctions in Chapter 2. Figures 3.13(a) and (b) plot the I/V characteristics of the ideal diode and the *pn* junction, respectively. The latter can serve as an approximation of the former by providing "unilateral" current conduction. Shown in Fig. 3.13 is the constant-voltage model developed in Chapter 2, providing a simple approximation of the exponential function and also resembling the characteristic plotted in Fig. 3.11(a).

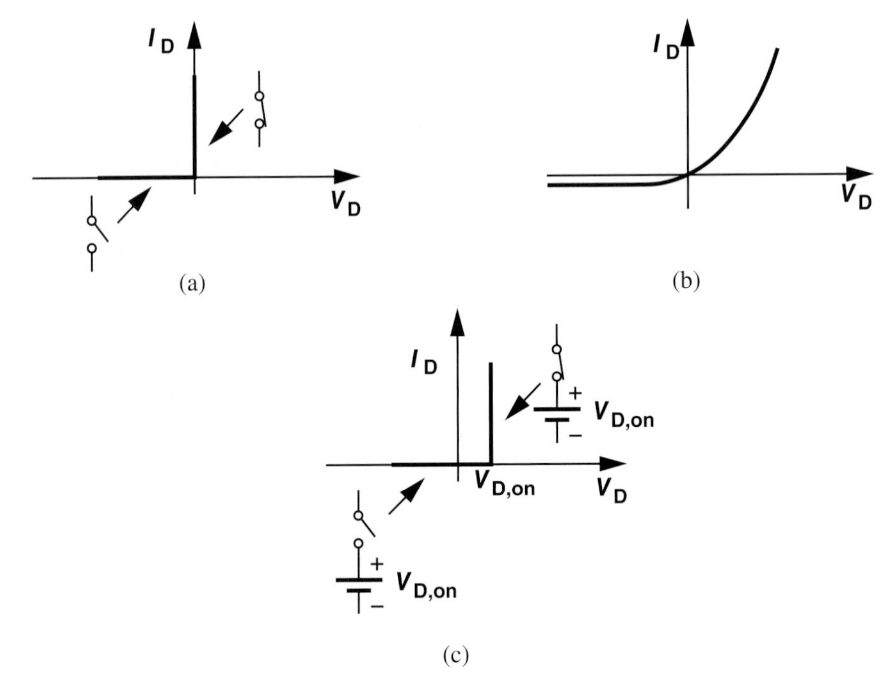

Figure 3.13 Diode characteristics: (a) ideal model, (b) exponential model, (c) constant-voltage model.

Given a circuit topology, how do we choose one of the above models for the diodes? We may utilize the ideal model so as to develop a quick, rough understanding of the circuit's operation. Upon performing this exercise, we may discover that this idealization is inadequate and hence employ the constant-voltage model. This model suffices in most cases, but we may need to resort to the exponential model for some circuits. The following examples illustrate these points.

Did you know?

Diodes are among few human-made devices that have a very wide range of sizes. The diodes in integrated circuits may have a cross-section area of 0.5 μm × 0.5 μm and carry a current of a few hundred microamperes. On the other hand, diodes used in industrial applications such as electroplating have a cross section of 10 cm × 10 cm and carry a current of several thousand amperes! Can you think of any other device that comes in such a wide range of sizes?

It is important to bear in mind two principles: (1) if a diode is at the edge of turning on or off, then $I_D \approx 0$ and $V_D \approx V_{D,on}$; (2) if a diode is on I_D *must* flow from anode to cathode.

Example 3-12 Plot the input/output characteristic of the circuit shown in Fig. 3.14(a) using (a) the ideal model and (b) the constant-voltage model.

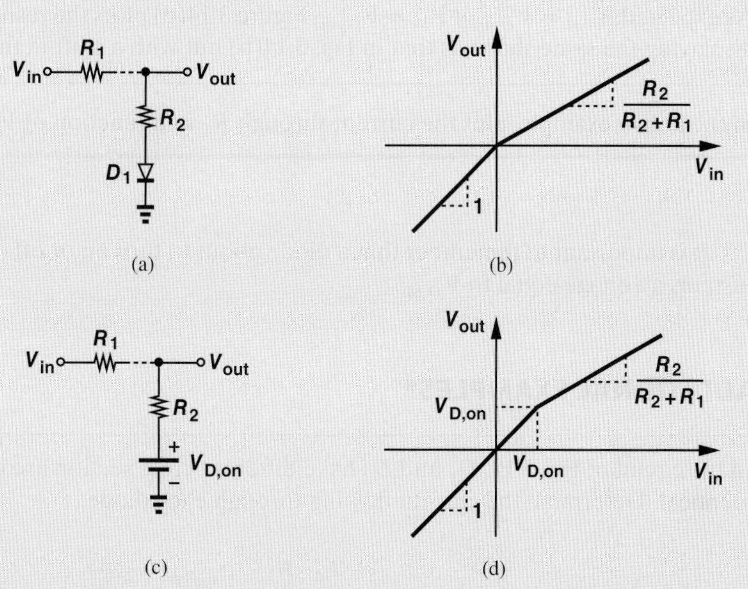

Figure 3.14 (a) Diode circuit, (b) input/output characteristic with ideal diode model, (c) input/output characteristic with constant-voltage diode model, and (d) circuit with ideal diode model.

Solution (a) We begin with $V_{in} = -\infty$, recognizing that D_1 is reverse biased. In fact, for $V_{in} < 0$, the diode remains off and no current flows through the circuit. Thus, the voltage drop across R_1 is zero and $V_{out} = V_{in}$.

As V_{in} exceeds zero, D_1 turns on, operating as a short and reducing the circuit to a voltage divider. That is,

$$V_{out} = \frac{R_2}{R_1 + R_2} V_{in} \text{ for } V_{in} > 0. \tag{3.9}$$

Figure 3.14(b) plots the overall characteristic, revealing a slope equal to unity for $V_{in} < 0$ and $R_2/(R_2 + R_1)$ for $V_{in} > 0$. In other words, the circuit operates as a voltage divider once the diode turns on and loads the output node with R_2.

(b) In this case, D_1 is reverse biased for $V_{in} < V_{D,on}$, yielding $V_{out} = V_{in}$. As V_{in} exceeds $V_{D,on}$, D_1 turns on, operating as a constant voltage source with a value $V_{D,on}$ [as illustrated in Fig. 3.13(c)]. Reducing the circuit to that in Fig. 3.14(c), we apply Kirchoff's current law to the output node:

$$\frac{V_{in} - V_{out}}{R_1} = \frac{V_{out} - V_{D,on}}{R_2}. \tag{3.10}$$

It follows that

$$V_{out} = \frac{\frac{R_2}{R_1} V_{in} + V_{D,on}}{1 + \frac{R_2}{R_1}}. \tag{3.11}$$

As expected, $V_{out} = V_{D,on}$ if $V_{in} = V_{D,on}$. Figure 3.14(d) plots the resulting characteristic, displaying the same shape as that in Fig. 3.14(b) but with a shift in the break point.

Exercise In the above example, plot the current through R_1 as a function of V_{in}.

It is important to remember that a diode about to turn on or off carries no current but sustains a voltage equal to $V_{D,on}$.

3.3 ADDITIONAL EXAMPLES*

Example 3-13 In the circuit of Fig. 3.15, D_1 and D_2 have different cross section areas but are otherwise identical. Determine the current flowing through each diode.

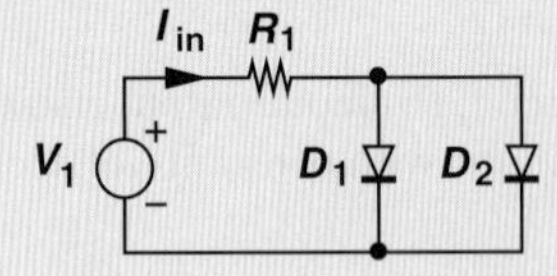

Figure 3.15 Diode circuit.

Solution In this case, we must resort to the exponential equation because the ideal and constant-voltage models do not include the device area. We have

$$I_{in} = I_{D1} + I_{D2}. \tag{3.12}$$

*This section can be skipped in a first reading.

We also equate the voltages across D_1 and D_2:

$$V_T \ln \frac{I_{D1}}{I_{S1}} = V_T \ln \frac{I_{D2}}{I_{S2}}; \tag{3.13}$$

that is,

$$\frac{I_{D1}}{I_{S1}} = \frac{I_{D2}}{I_{S2}}. \tag{3.14}$$

Solving (3.12) and (3.14) together yields

$$I_{D1} = \frac{I_{in}}{1 + \frac{I_{S2}}{I_{S1}}} \tag{3.15}$$

$$I_{D2} = \frac{I_{in}}{1 + \frac{I_{S1}}{I_{S2}}}. \tag{3.16}$$

As expected, $I_{D1} = I_{D2} = I_{in}/2$ if $I_{S1} = I_{S2}$.

Exercise For the circuit of Fig. 3.15, calculate V_D is terms of I_{in}, I_{S1}, and I_{S2}.

Example 3-14 Using the constant-voltage model, plot the input/output characteristics of the circuit depicted in Fig. 3.16(a). Note that a diode about to turn on carries *zero* current but sustains $V_{D,on}$.

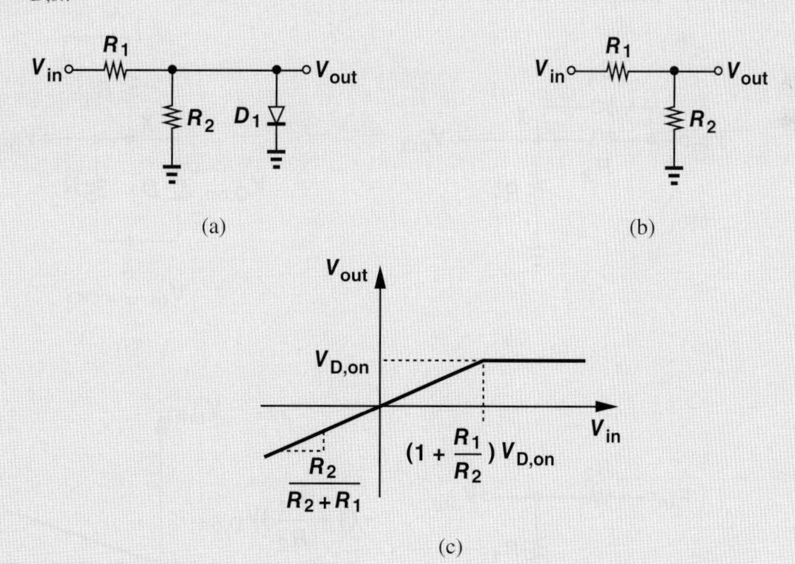

Figure 3.16 (a) Diode circuit, (b) equivalent circuit when D_1 is off, (c) input/output characteristic.

Solution In this case, the voltage across the diode happens to be equal to the output voltage. We note that if $V_{in} = -\infty$, D_1 is reverse biased and the circuit reduces to that in Fig. 3.16(b). Consequently,

$$v_{out} = \frac{R_2}{R_1 + R_2} V_{in}. \tag{3.17}$$

At what point does D_1 turn on? The diode voltage must reach $V_{D,on}$, requiring an input voltage given by:

$$\frac{R_2}{R_1+R_2}V_{in} = V_{D,on}, \tag{3.18}$$

and hence

$$V_{in} = \left(1+\frac{R_1}{R_2}\right)V_{D,on}. \tag{3.19}$$

The reader may question the validity of this result: if the diode is indeed on, it draws current and the diode voltage is no longer equal to $[R_2/(R_1+R_2)]V_{in}$. So why did we express the diode voltage as such in Eq. (3.18)? To determine the break point, we assume V_{in} gradually increases so that it places the diode at the *edge* of the turn-on, e.g., it creates $V_{out} \approx 799$ mV. The diode therefore still draws no current, but the voltage across it and hence the input voltage are almost sufficient to turn it on.

For $V_{in} > (1+R_1/R_2)V_{D,on}$, D_1 remains forward-biased, yielding $V_{out} = V_{D,on}$. Figure 3.16(c) plots the overall characteristic.

Exercise Repeat the above example but assume the terminals of D_1 are swapped, i.e., the anode is tied to ground and the cathode the output node.

Exercise For the above example, plot the current through R_1 as a function of V_{in}.

Example 3-15 Plot the input/output characteristic for the circuit shown in Fig. 3.17(a). Assume a constant-voltage model for the diode.

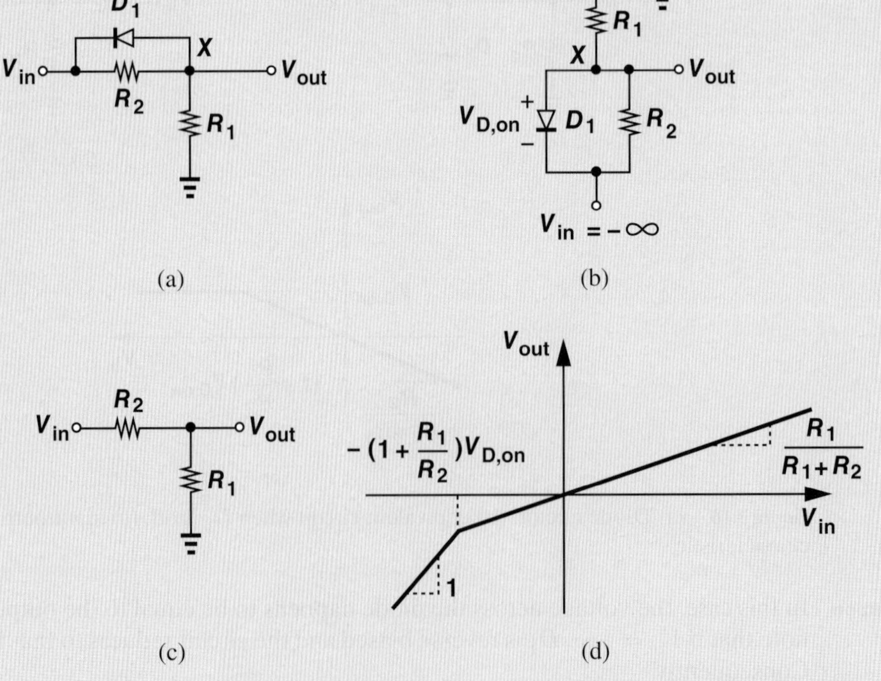

Figure 3.17 (a) Diode circuit, (b) illustration for very negative inputs, (c) equivalent circuit when D_1 is off, (d) input/output characteristic.

Solution We begin with $V_{in} = -\infty$, and redraw the circuit as depicted in Fig. 3.17(b), placing the more negative voltages on the bottom and the more positive voltages on the top. This diagram suggests that the diode operates in forward bias, establishing a voltage at node X equal to $V_{in} + V_{D,on}$. Note that in this regime, V_X is independent of R_2 because D_1 acts as a battery. Thus, so long as D_1 is on, we have

$$V_{out} = V_{in} + V_{D,on}. \tag{3.20}$$

We also compute the current flowing through R_2 and R_1:

$$I_{R2} = \frac{V_{D,on}}{R_2} \tag{3.21}$$

$$I_{R1} = \frac{0 - V_X}{R_1} \tag{3.22}$$

$$= \frac{-(V_{in} + V_{D,on})}{R_1}. \tag{3.23}$$

Thus, as V_{in} increases from $-\infty$, I_{R2} remains constant but $|I_{R1}|$ decreases; i.e., at some point $I_{R2} = I_{R1}$.

At what point does D_1 turn off? Interestingly, in this case it is simpler to seek the condition that results in a zero current through the diode rather than insufficient voltage across it. The observation that at some point, $I_{R2} = I_{R1}$ proves useful here as this condition also implies that D_1 carries no current (KCL at node X). In other words, D_1 turns off if V_{in} is chosen to yield $I_{R2} = I_{R1}$. From (3.21) and (3.23),

$$\frac{V_{D,on}}{R_2} = -\frac{V_{in} + V_{D,on}}{R_1} \tag{3.24}$$

and hence

$$V_{in} = -\left(1 + \frac{R_1}{R_2}\right) V_{D,on}. \tag{3.25}$$

As V_{in} exceeds this value, the circuit reduces to that shown in Fig. 3.17(c) and

$$V_{out} = \frac{R_1}{R_1 + R_2} V_{in}. \tag{3.26}$$

The overall characteristic is shown in Fig. 3.17(d).

The reader may find it interesting to recognize that the circuits of Figs. 3.16(a) and 3.17(a) are identical: in the former, the output is sensed across the diode whereas in the latter it is sensed across the series resistor.

Exercise Repeat the above example if the terminals of the diode are swapped.

As mentioned in Example 3-4, in more complex circuits, it may be difficult to correctly predict the region of operation of each diode by inspection. In such cases, we may simply make a guess, proceed with the analysis, and eventually determine if the final result agrees or conflicts with the original guess. Of course, we still apply intuition to minimize the guesswork. The following example illustrates this approach.

Example 3-16 Plot the input/output characteristic of the circuit shown in Fig. 3.18(a) using the constant-voltage diode model.

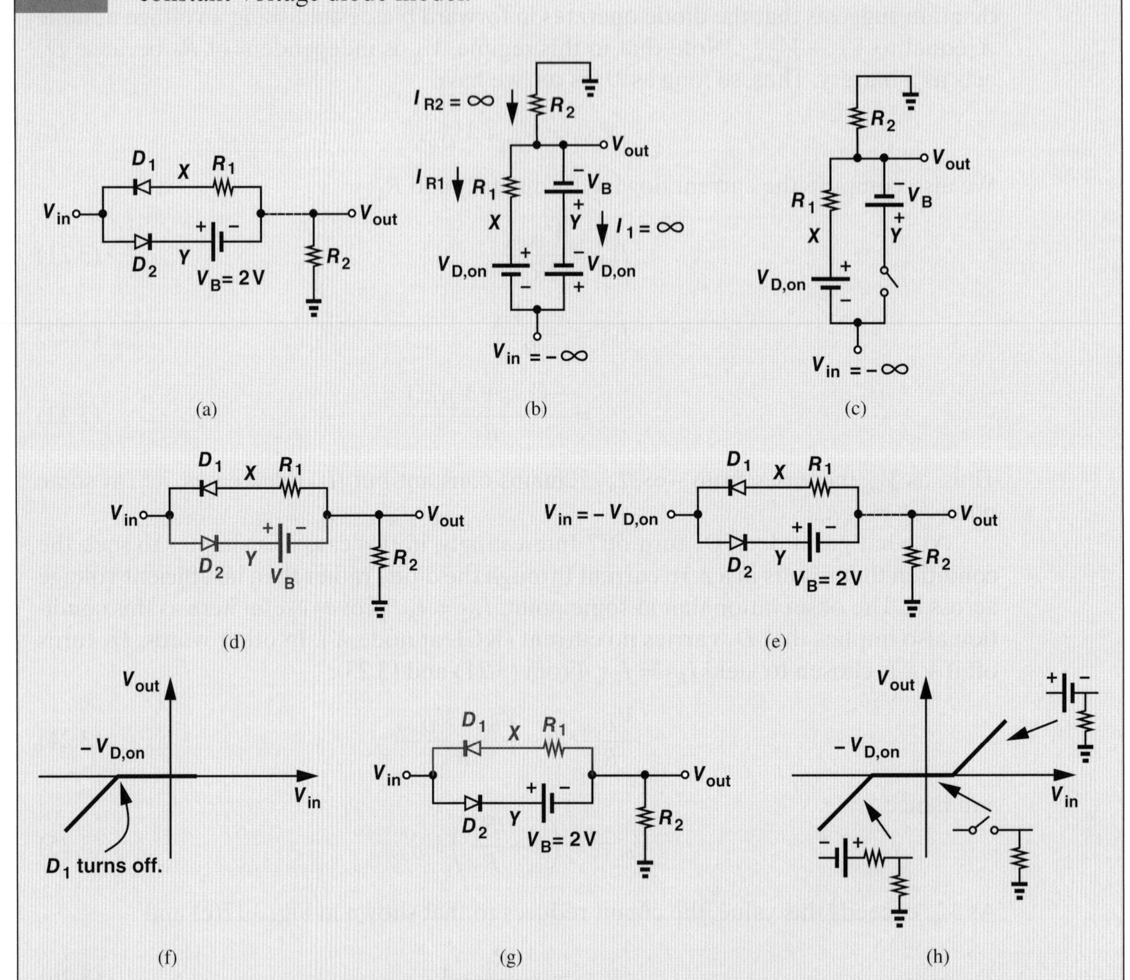

Figure 3.18 (a) Diode circuit, (b) possible equivalent circuit for very negative inputs, (c) simplified circuit, (d) equivalent circuit, (e) equivalent circuit for $V_{in} = -V_{D,on}$, (f) section of input/output characteristic, (g) equivalent circuit, (f) complete input/output characteristic.

Solution We begin with $V_{in} = -\infty$, predicting intuitively that D_1 is on. We also (blindly) assume that D_2 is on, thus reducing the circuit to that in Fig. 3.18(b). The path through $V_{D,on}$ and V_B creates a difference of $V_{D,on} + V_B$ between V_{in} and V_{out}, i.e., $V_{out} = V_{in} - (V_{D,on} + V_B)$. This voltage difference also appears across the branch consisting of R_1 and $V_{D,on}$, yielding

$$R_1 I_{R1} + V_{D,on} = -(V_B + V_{D,on}), \tag{3.27}$$

and hence

$$I_{R1} = \frac{-V_B - 2V_{D,on}}{R_1}. \tag{3.28}$$

That is, I_{R1} is independent of V_{in}. We must now analyze these results to determine whether they agree with our assumptions regarding the state of D_1 and D_2.

Consider the current flowing through R_2:

$$I_{R2} = -\frac{V_{out}}{R_2} \tag{3.29}$$

$$= -\frac{V_{in} - (V_{D,on} + V_B)}{R_2}, \tag{3.30}$$

which approaches $+\infty$ for $V_{in} = -\infty$. The large value of I_{R2} and the constant value of I_{R1} indicate that the branch consisting of V_B and D_2 carries a large current with the direction shown. That is, D_2 must conduct current from its cathode to its anode, which is not possible.

In summary, we have observed that the forward bias assumption for D_2 translates to a current in a prohibited direction. Thus, D_2 operates in reverse bias for $V_{in} = -\infty$. Redrawing the circuit as in Fig. 3.18(c) and noting that $V_X = V_{in} + V_{D,on}$, we have

$$V_{out} = (V_{in} + V_{D,on})\frac{R_2}{R_1 + R_2}. \tag{3.31}$$

We now raise V_{in} and determine the first break point, i.e., the point at which D_1 turns off or D_2 turns on. Which one occurs first? Let us assume D_1 turns off first and obtain the corresponding value of V_{in}. Since D_2 is assumed off, we draw the circuit as shown in Fig. 3.18(d). Assuming that D_1 is still slightly on, we recognize that at $V_{in} \approx -V_{D,on}$, $V_X = V_{in} + V_{D,on}$ approaches zero, yielding a zero current through R_1, R_2, and hence D_1. The diode therefore turns off at $V_{in} = -V_{D,on}$.

We must now verify the assumption that D_2 remains off. Since at this break point, $V_X = V_{out} = 0$, the voltage at node Y is equal to $+V_B$ whereas the cathode of D_2 is at $-V_{D,on}$ [Fig. 3.18(e)]. In other words, D_2 is indeed off. Fig. 3.18(f) plots the input/output characteristic to the extent computed thus far, revealing that $V_{out} = 0$ after the first break point because the current flowing through R_1 and R_2 is equal to zero.

At what point does D_2 turn on? The input voltage must exceed V_Y by $V_{D,on}$. Before D_2 turns on, $V_{out} = 0$, and $V_Y = V_B$; i.e., V_{in} must reach $V_B + V_{D,on}$, after which the circuit is configured as shown in Fig. 3.18(g). Consequently,

$$V_{out} = V_{in} - V_{D,on} - V_B. \tag{3.32}$$

Figure 3.18(h) plots the overall result, summarizing the regions of operation.

Exercise In the above example, assume D_2 turns on before D_1 turns off and show that the results conflict with the assumption.

Robotics Application: Relay Control

In order to control high-power motors, robots require switches that can pass tens of amperes of current and can be controlled electronically. Such switches are implemented as "relays," mechanical contacts that are controlled magnetically. As shown conceptually in Fig. 3.19(a), a relay consists of a main switch between nodes A and B and a bar that is magnetized when a current passes through the solenoid (the inductor) around it from X to Y. To turn on the relay and establish a path between A and B for conducting high currents, we pass a small current, I_r, through the solenoid.

Figure 3.19(b) depicts a typical environment, where a microcontroller turns the relay on or off to control a motor. But the solenoid introduces an interesting, yet fatal issue. Suppose the microcontroller decides to turn off the relay and drops I_r to zero. How does the solenoid respond to this sudden current change? Modeling the circuit as shown in Fig. 3.19(c) and writing

$$V_1 = L_1 \frac{dI_r}{dt}, \tag{3.33}$$

we recognize that, as I_r falls to zero, a *negative* voltage develops across L_1; this voltage can have a large magnitude if the current drops quickly. We now observe that this large, negative voltage appears at the output of the microcontroller, possibly *damaging* it.

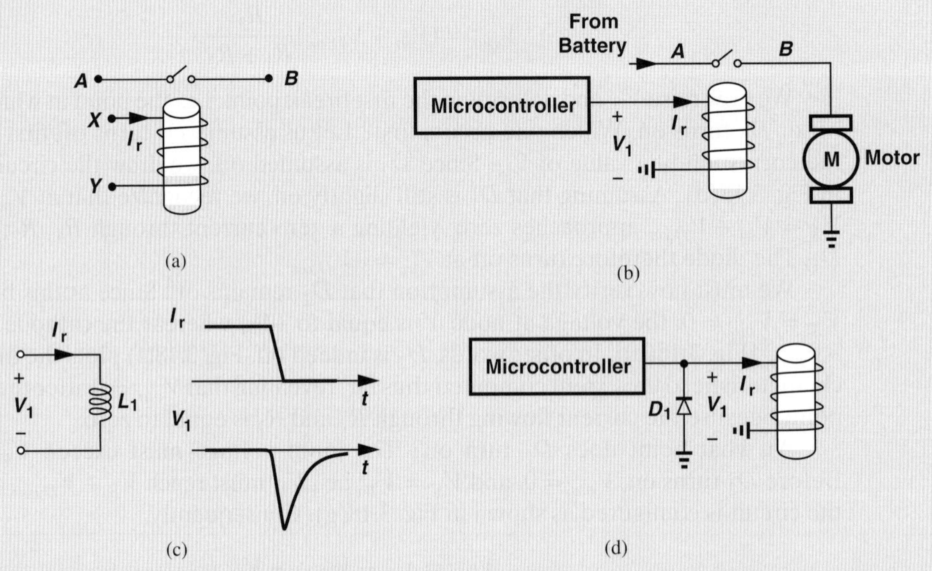

Figure 3.19 (a) Structure of a relay, (b) control of a motor by a relay, (c) problem of voltage surge, and (d) use of a diode to protect the microcontroller.

This issue is resolved by placing a diode in parallel with the solenoid, as illustrated in Fig. 3.19(d). When the microcontroller turns on the relay, $V_1 = L_1 dI_r/dt > 0$ (because I_r is rising) and D_1 is off. On the other hand, when the microcontroller stops the current and V_1 goes to negative values, D_1 turns on, limiting V_1 to about -0.7 V to -0.8 V, a value safe for the microcontroller.

3.4 LARGE-SIGNAL AND SMALL-SIGNAL OPERATION

Our treatment of diodes thus far has allowed arbitrarily large voltage and current changes, thereby requiring a "general" model such as the exponential I/V characteristic. We call this regime "large-signal operation" and the exponential characteristic the "large-signal model" to emphasize that the model can accommodate arbitrary signal levels. However, as seen in previous examples, this model often complicates the analysis, making it difficult to develop an intuitive understanding of the circuit's operation. Furthermore, as the number of nonlinear devices in the circuit increases, "manual" analysis eventually becomes impractical.

The ideal and constant-voltage diode models resolve the issues to some extent, but the sharp nonlinearity at the turn-on point still proves problematic. The following example illustrates the general difficulty.

Example 3-17

Having lost his 2.4-V cellphone charger, an electrical engineering student tries several stores but does not find adaptors with outputs less than 3 V. He then decides to put his knowledge of electronics to work and constructs the circuit shown in Fig. 3.20, where three identical diodes in forward bias produce a total voltage of $V_{out} = 3V_D \approx 2.4$ V and resistor R_1 sustains the remaining 600 mV. Neglect the current drawn by the cellphone.[6] (a) Determine the reverse saturation current, I_{S1} so that $V_{out} = 2.4$ V. (b) Compute V_{out} if the adaptor voltage is in fact 3.1 V.

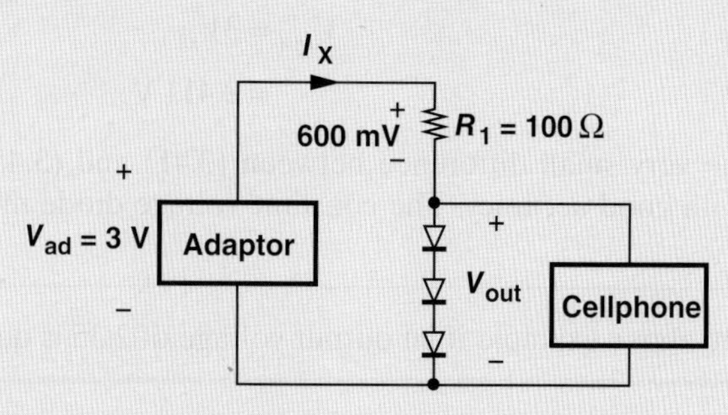

Figure 3.20 Adaptor feeding a cellphone.

Solution (a) With $V_{out} = 2.4$ V, the current flowing through R_1 is equal to

$$I_X = \frac{V_{ad} - V_{out}}{R_1} \tag{3.34}$$

$$= 6\ \text{mA}. \tag{3.35}$$

We note that each diode carries I_X and hence

$$I_X = I_S \exp\frac{V_D}{V_T}. \tag{3.36}$$

It follows that

$$6\ \text{mA} = I_S \exp\frac{800\ \text{mV}}{26\ \text{mV}} \tag{3.37}$$

and

$$I_S = 2.602 \times 10^{-16}\ \text{A}. \tag{3.38}$$

(b) If V_{ad} increases to 3.1 V, we expect that V_{out} increases only slightly. To understand why, first suppose V_{out} remains constant and equal to 2.4 V. Then, the additional

[6]Made for the sake of simplicity here, this assumption may not be valid.

0.1 V must drop across R_1, raising I_X to 7 mA. Since the voltage across each diode has a logarithmic dependence upon the current, the change from 6 mA to 7 mA indeed yields a small change in V_{out}.[7]

To examine the circuit quantitatively, we begin with $I_X = 7$ mA and iterate:

$$V_{out} = 3V_D \quad (3.39)$$

$$= 3V_T \ln \frac{I_X}{I_S} \quad (3.40)$$

$$= 2.412 \text{ V}. \quad (3.41)$$

This value of V_{out} gives a new value for I_X:

$$I_X = \frac{V_{ad} - V_{out}}{R_1} \quad (3.42)$$

$$= 6.88 \text{ mA}, \quad (3.43)$$

which translates to a new V_{out}:

$$V_{out} = 3V_D \quad (3.44)$$

$$= 2.411 \text{ V}. \quad (3.45)$$

Noting the very small difference between (3.41) and (3.45), we conclude that $V_{out} =$ 2.411 V with good accuracy. The constant-voltage diode model would not be useful in this case.

Exercise Repeat the above example if an output voltage of 2.35 is desired.

The situation[6] described above is an example of small "perturbations" in circuits. The change in V_{ad} from 3 V to 3.1 V results in a small change in the circuit's voltages and currents, motivating us to seek a simpler analysis method that can replace the nonlinear equations and the inevitable iterative procedure. Of course, since the above example does not present an overwhelmingly difficult problem, the reader may wonder if a simpler approach is really necessary. But, as seen in subsequent chapters, circuits containing complex devices such as transistors may indeed become impossible to analyze if the nonlinear equations are retained.

These thoughts lead us to the extremely important concept of "small-signal operation," whereby the circuit experiences only small changes in voltages and currents and can therefore be simplified through the use of "small-signal models" for nonlinear devices. The simplicity arises because such models are *linear*, allowing standard circuit analysis and obviating the need for iteration. The definition of "small" will become clear later.

To develop our understanding of small-signal operation, let us consider diode D_1 in Fig. 3.21(a), which sustains a voltage V_{D1} and carries a current I_{D1} [point A in Fig. 3.21(b)]. Now suppose a perturbation in the circuit changes the diode voltage by a small amount

[7] Recall from Eq. (2.109) that a tenfold change in a diode's current translates to a 60-mV change in its voltage.

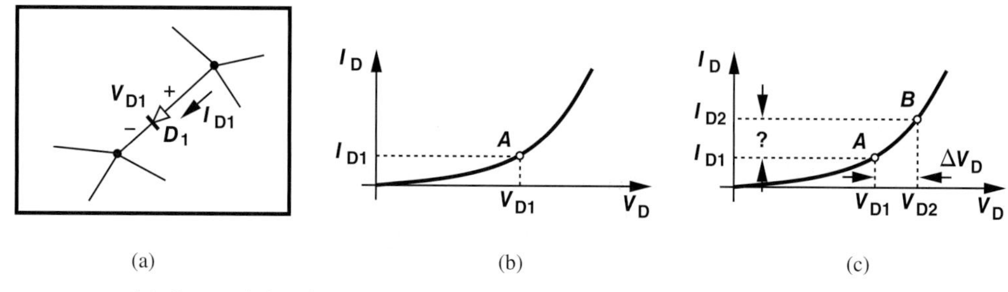

Figure 3.21 (a) General circuit containing a diode, (b) operating point of D_1, (c) change in I_D as a result of change in V_D.

ΔV_D [point B in Fig. 3.21(c)]. How do we predict the change in the diode current, ΔI_D? We can begin with the nonlinear characteristic:

$$I_{D2} = I_S \exp \frac{V_{D1} + \Delta V}{V_T} \tag{3.46}$$

$$= I_S \exp \frac{V_{D1}}{V_T} \exp \frac{\Delta V}{V_T}. \tag{3.47}$$

If $\Delta V \ll V_T$, then $\exp(\Delta V/V_T) \approx 1 + \Delta V/V_T$ and

$$I_{D2} = I_S \exp \frac{V_{D1}}{V_T} + \frac{\Delta V}{V_T} I_S \exp \frac{V_{D1}}{V_T} \tag{3.48}$$

$$= I_{D1} + \frac{\Delta V}{V_T} I_{D1}. \tag{3.49}$$

That is,

$$\Delta I_D = \frac{\Delta V}{V_T} I_{D1}. \tag{3.50}$$

The key observation here is that ΔI_D is a *linear* function of ΔV, with a proportionality factor equal to I_{D1}/V_T. (Note that larger values of I_{D1} lead to a greater ΔI_D for a given ΔV_D. The significance of this trend becomes clear later.)

The above result should not come as a surprise: if the change in V_D is small, the section of the characteristic in Fig. 3.21(c) between points A and B can be approximated by a straight line (Fig. 3.22), with a slope equal to the local slope of the characteristic.

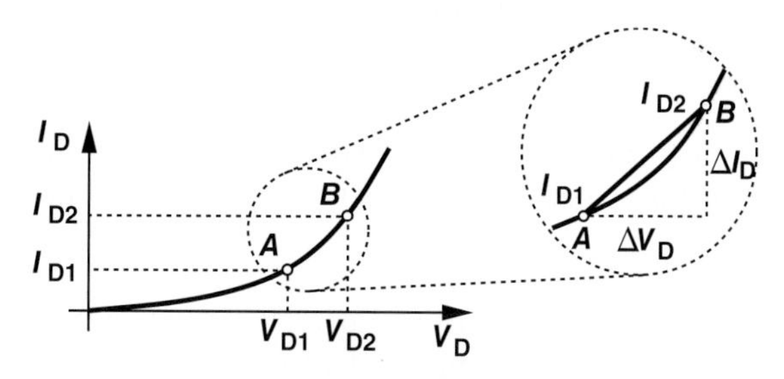

Figure 3.22 Approximation of characteristic by a straight line.

In other words,

$$\frac{\Delta I_D}{\Delta V_D} = \left.\frac{dI_D}{dV_D}\right|_{VD=VD1} \tag{3.51}$$

$$= \frac{I_S}{V_T} \exp \frac{V_{D1}}{V_T} \tag{3.52}$$

$$= \frac{I_{D1}}{V_T}, \tag{3.53}$$

which yields the same result as that in Eq. (3.50).[8]

Let us summarize our results thus far. If the voltage across a diode changes by a small amount (much less than V_T), then the change in the current is given by Eq. (3.50). Equivalently, for small-signal analysis, we can assume the operation is at a point such as A in Fig. 3.22 and, due to a small perturbation, it moves on a straight line to point B with a slope equal to the local slope of the characteristic (i.e., dI_D/dV_D calculated at $V_D = V_{D1}$ or $I_D = I_{D1}$). Point A is called the "bias" point, the "quiescent" point, or the "operating" point.

Example 3-18 A diode is biased at a current of 1 mA. (a) Determine the current change if V_D changes by 1 mV. (b) Determine the voltage change if I_D changes by 10%.

Solution (a) We have

$$\Delta I_D = \frac{I_D}{V_T}\Delta V_D \tag{3.54}$$

$$= 38.4\,\mu\text{A}. \tag{3.55}$$

(b) Using the same equation yields

$$\Delta V_D = \frac{V_T}{I_D}\Delta I_D \tag{3.56}$$

$$= \left(\frac{26\,\text{mV}}{1\,\text{mA}}\right) \times (0.1\,\text{mA}) \tag{3.57}$$

$$= 2.6\,\text{mV}. \tag{3.58}$$

Exercise In response to a current change of 1 mA, a diode exhibits a voltage change of 3 mV. Calculate the bias current of the diode.

Equation (3.59) in the above example reveals an interesting aspect of small-signal operation: as far as (small) changes in the diode current and voltage are concerned, the

[8]This is also to be expected. Writing Eq. (3.46) to obtain the change in I_D for a small change in V_D is in fact equivalent to taking the derivative.

device behaves as a linear resistor. In analogy with Ohm's law, we define the "small-signal resistance" of the diode as:

$$r_d = \frac{V_T}{I_D}. \tag{3.59}$$

This quantity is also called the "incremental" resistance to emphasize its validity for small changes. In the above example, $r_d = 26\,\Omega$.

Figure 3.23(a) summarizes the results of our derivations for a forward-biased diode. For bias calculations, the diode is replaced with an ideal voltage source of value $V_{D,on}$, and for small changes, with a resistance equal to r_d. For example, the circuit of Fig. 3.23(b) is transformed to that in Fig. 3.23(c) if only small changes in V_1 and V_{out} are of interest. Note that v_1 and v_{out} in Fig. 3.23(c) represent *changes* in voltage and are called small-signal quantities. In general, we denote small-signal voltages and currents by lowercase letters.

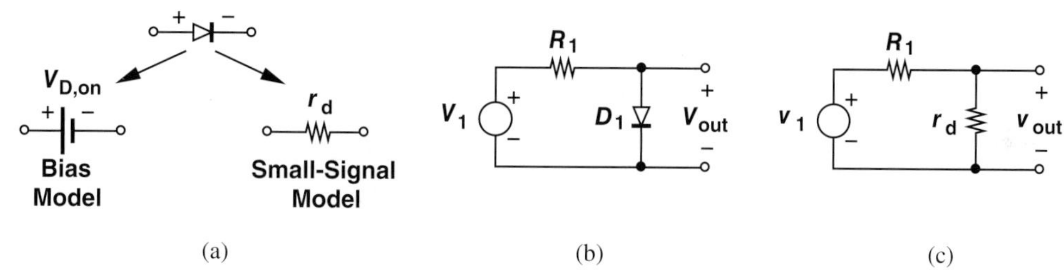

Figure 3.23 (a) Summary of diode models for bias and signal calculations, (b) circuit example, (c) small-signal model.

Example 3-19

A sinusoidal signal having a peak amplitude of V_p and a dc value of V_0 can be expressed as $V(t) = V_0 + V_p \cos \omega t$. If this signal is applied across a diode and $V_p \ll V_T$, determine the resulting diode current.

Solution The signal waveform is illustrated in Fig. 3.24(a). As shown in Fig. 3.24(b), we rotate this diagram by 90° so that its vertical axis is aligned with the voltage axis of the diode characteristic. With a signal swing much less than V_T, we can view V_0 and the corresponding current, I_0, as the bias point of the diode and V_p as a small perturbation. It follows that

$$I_0 = I_S \exp \frac{V_0}{V_T}, \tag{3.60}$$

and

$$r_d = \frac{V_T}{I_0}. \tag{3.61}$$

Thus, the peak current is simply equal to

$$I_p = V_p / r_d \tag{3.62}$$

$$= \frac{I_0}{V_T} V_p, \tag{3.63}$$

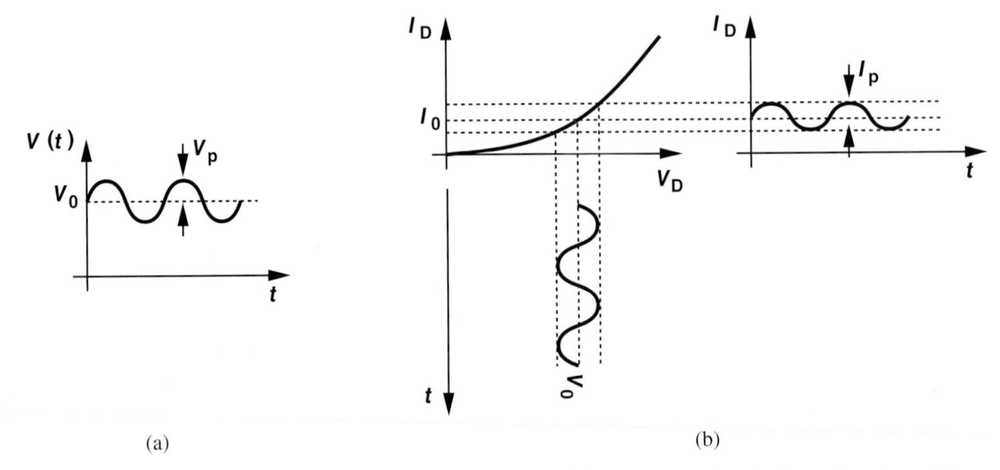

Figure 3.24 (a) Sinusoidal input along with a dc level, (b) response of a diode to the sinusoid.

yielding

$$I_D(t) = I_0 + I_p \cos \omega t \tag{3.64}$$

$$= I_S \exp \frac{V_0}{V_T} + \frac{I_0}{V_T} V_p \cos \omega t. \tag{3.65}$$

Exercise The diode in the above example produces a peak current of 0.1 mA in response to $V_0 = 800$ mV and $V_p = 1.5$ mV. Calculate I_S.

The above example demonstrates the utility of small-signal analysis. If V_p were large, we would need to solve the following equation:

$$I_D(t) = I_S \exp \frac{V_0 + V_p \cos \omega t}{V_T}, \tag{3.66}$$

a task much more difficult than the above linear calculations.[9]

Example 3-20 In the derivation leading to Eq. (3.50), we assumed a small change in V_D and obtained the resulting change in I_D. Beginning with $V_D = V_T \ln(I_D/I_s)$, investigate the reverse case, i.e., I_D changes by a small amount and we wish to compute the change in V_D.

Solution Denoting the change in V_D by ΔV_D, we have

$$V_{D1} + \Delta V_D = V_T \ln \frac{I_{D1} + \Delta I_D}{I_S} \tag{3.67}$$

$$= V_T \ln \left[\frac{I_{D1}}{I_S} \left(1 + \frac{\Delta I_D}{I_{D1}} \right) \right] \tag{3.68}$$

$$= V_T \ln \frac{I_{D1}}{I_S} + V_T \ln \left(1 + \frac{\Delta I_D}{I_{D1}} \right). \tag{3.69}$$

[9]The function exp(a sin bt) can be approximated by a Taylor expansion or Bessel functions.

For small-signal operation, we assume $\Delta I_D \ll I_{D1}$ and note that $\ln(1+\epsilon) \approx \epsilon$ if $\epsilon \ll 1$. Thus,

$$\Delta V_D = V_T \cdot \frac{\Delta I_D}{I_{D1}}, \tag{3.70}$$

which is the same as Eq. (3.50). Figure 3.25 illustrates the two cases, distinguishing between the cause and the effect.

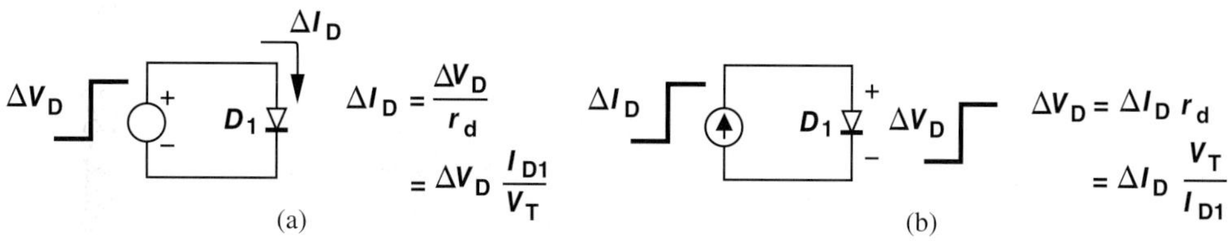

Figure 3.25 Change in diode current (voltage) due to a change in voltage (current).

Exercise Repeat the above example by taking the derivative of the diode voltage equation with respect to I_D.

With our understanding of small-signal operation, we now revisit Example 3-17.

Example 3-21 Repeat part (b) of Example 3-17 with the aid of a small-signal model for the diodes.

Solution Since each diode carries $I_{D1} = 6$ mA with an adaptor voltage of 3 V and $V_{D1} = 800$ mV, we can construct the small-signal model shown in Fig. 3.26, where $v_{ad} = 100$ mV and $r_d = (26\text{ mV})/(6\text{ mA}) = 4.33\ \Omega$. (As mentioned earlier, the voltages shown in this model denote small changes.) We can thus write:

$$v_{out} = \frac{3r_d}{R_1 + 3r_d} v_{ad} \tag{3.71}$$

$$= 11.5\text{ mV}. \tag{3.72}$$

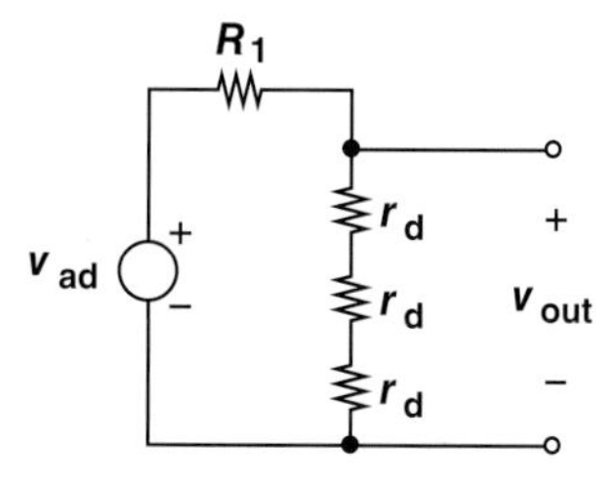

Figure 3.26 Small-signal model of adaptor.

That is, a 100-mV change in V_{ad} yields an 11.5-mV change in V_{out}. In Example 3-17, solution of nonlinear diode equations predicted an 11-mV change in V_{out}. The small-signal analysis therefore offers reasonable accuracy while requiring much less computational effort.

Exercise Repeat Examples (3-17) and (3-21) if the value of R_1 in Fig. 3.20 is changed to 200 Ω.

Considering the power of today's computer software tools, the reader may wonder if the small-signal model is really necessary. Indeed, we utilize sophisticated simulation tools in the design of integrated circuits today, but the intuition gained by hand analysis of a circuit proves invaluable in understanding fundamental limitations and various trade-offs that eventually lead to a compromise in the design. A good circuit designer analyzes and understands the circuit before giving it to the computer for a more accurate analysis. A bad circuit designer, on the other hand, allows the computer to think for him/her.

Example 3-22 In Examples 3-17 and 3-21, the current drawn by the cellphone is neglected. Now suppose, as shown in Fig. 3.27, the load pulls a current of 0.5 mA[10] and determine V_{out}.

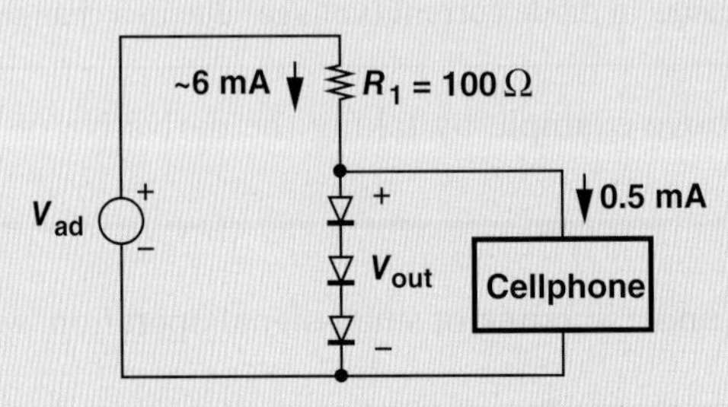

Figure 3.27 Adaptor feeding a cellphone.

Solution Since the current flowing through the diodes decreases by 0.5 mA and since this change is much less than the bias current (6 mA), we write the change in the output voltage as:

$$\Delta V_{out} = \Delta I_D \cdot (3r_d) \tag{3.73}$$

$$= 0.5\,\text{mA}(3 \times 4.33\,\Omega) \tag{3.74}$$

$$= 6.5\,\text{mV}. \tag{3.75}$$

Exercise Repeat the above example if R_1 is reduced to 80 Ω.

Did you know?

What would our life be like without diodes? Forget about cell phones, laptops, and digital cameras. We would not even have radios, TVs, GPS, radars, satellites, power plants, or long-distance telephone communication. And, of course, no Google or Facebook. In essence, we would return to the simple lifestyle of the early 1900s—which might not be that bad after all . . .

In summary, the analysis of circuits containing diodes (and other nonlinear devices such as transistors) proceeds in three steps: (1) determine—perhaps with the aid of the constant-voltage model—the initial values of voltages and currents (before an input change is applied); (2) develop the small-signal model for each diode (i.e., calculate r_d); (3) replace each diode with its small-signal model and compute the effect of the input change.

[10] A cellphone in reality draws a much higher current.

3.5 APPLICATIONS OF DIODES

The remainder of this chapter deals with circuit applications of diodes. A brief outline is shown in Fig. 3.28.

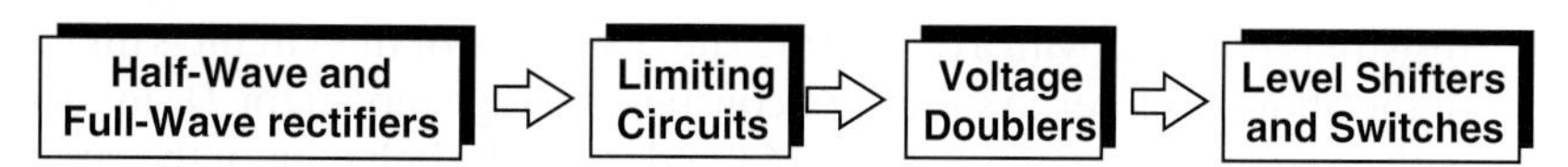

Figure 3.28 Applications of diodes.

3.5.1 Half-Wave and Full-Wave Rectifiers

Half-Wave Rectifier Let us return to the rectifier circuit of Fig. 3.10(b) and study it more closely. In particular, we no longer assume D_1 is ideal, but use a constant-voltage model. As illustrated in Fig. 3.29, V_{out} remains equal to zero until V_{in} exceeds $V_{D,on}$, at which point D_1 turns on and $V_{out} = V_{in} - V_{D,on}$. For $V_{in} < V_{D,on}$, D_1 is off[11] and $V_{out} = 0$. Thus, the circuit still operates as a rectifier but produces a slightly lower dc level.

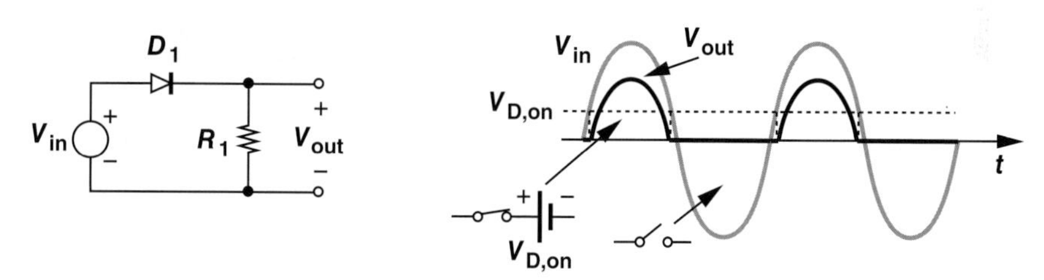

Figure 3.29 Simple rectifier.

Example 3-23 Prove that the circuit shown in Fig. 3.30(a) is also a rectifier.

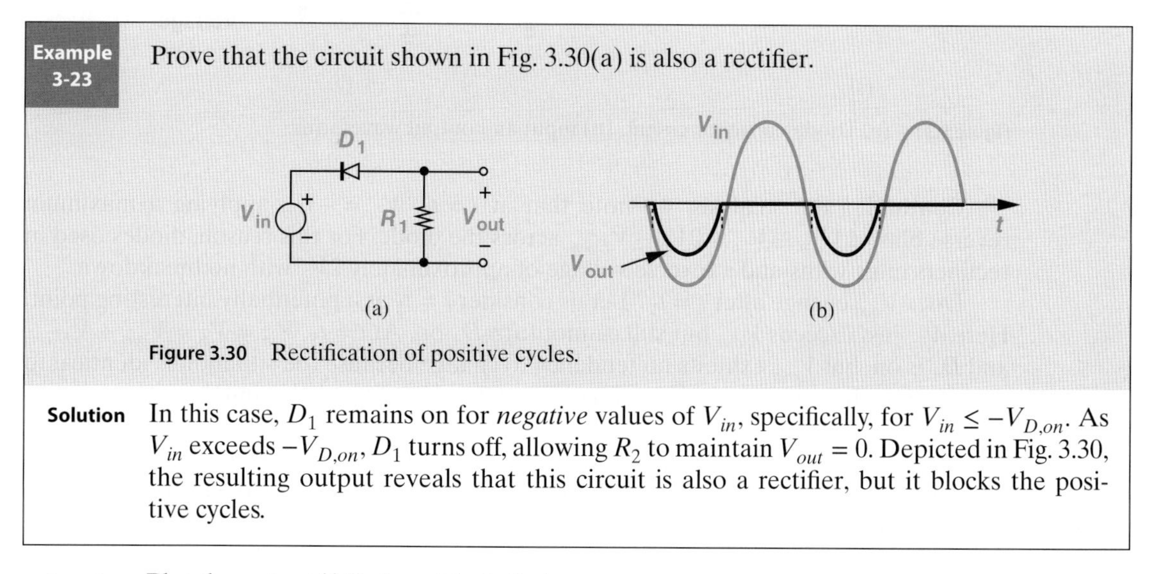

Figure 3.30 Rectification of positive cycles.

Solution In this case, D_1 remains on for *negative* values of V_{in}, specifically, for $V_{in} \leq -V_{D,on}$. As V_{in} exceeds $-V_{D,on}$, D_1 turns off, allowing R_2 to maintain $V_{out} = 0$. Depicted in Fig. 3.30, the resulting output reveals that this circuit is also a rectifier, but it blocks the positive cycles.

Exercise Plot the output if D_1 is an ideal diode.

[11] If $V_{in} < 0$, D_1 carries a small leakage current, but the effect is negligible.

Called a "half-wave rectifier," the circuit of Fig. 3.29 does not produce a useful output. Unlike a battery, the rectifier generates an output that *varies* considerably with time and cannot supply power to electronic devices. We must therefore attempt to create a *constant* output.

Fortunately, a simple modification solves the problem. As depicted in Fig. 3.31(a), the resistor is replaced with a capacitor. The operation of this circuit is quite different from that of the above rectifier. Assuming a constant-voltage model for D_1 in forward bias, we begin with a zero initial condition across C_1 and study the behavior of the circuit [Fig. 3.31(b)]. As V_{in} rises from zero, D_1 is off until $V_{in} > V_{D,on}$, at which point D_1 begins to act as a battery and $V_{out} = V_{in} - V_{D,on}$. Thus, V_{out} reaches a peak value of $V_p - V_{D,on}$. What happens as V_{in} passes its peak value? At $t = t_1$, we have $V_{in} = V_p$ and $V_{out} = V_p - V_{D,on}$. As V_{in} begins to fall, V_{out} must remain *constant*. This is because if V_{out} were to fall, then C_1 would need to be *discharged* by a current flowing from its top plate through the cathode of D_1, which is impossible.[12] The diode therefore turns off after t_1. At $t = t_2$, $V_{in} = V_p - V_{D,on} = V_{out}$, i.e., the diode sustains a zero voltage difference. At $t > t_2$, $V_{in} < V_{out}$ and the diode experiences a negative voltage.

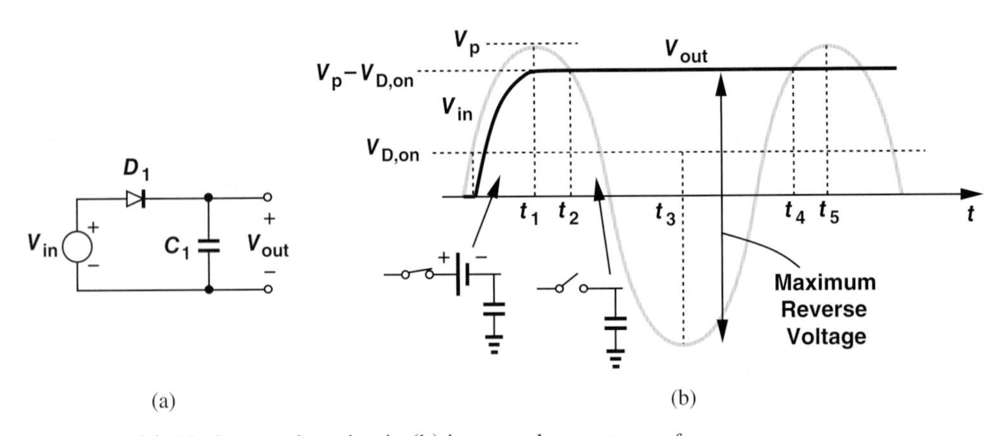

Figure 3.31 (a) Diode-capacitor circuit, (b) input and output waveforms.

Continuing our analysis, we note that at $t = t_3$, $V_{in} = -V_p$, applying a maximum reverse bias of $V_{out} - V_{in} = 2V_p - V_{D,on}$ across the diode. For this reason, diodes used in rectifiers must withstand a reverse voltage of approximately $2V_p$ with no breakdown.

Does V_{out} change after $t = t_1$? Let us consider $t = t_4$ as a potentially interesting point. Here, V_{in} just exceeds V_{out} but still cannot turn D_1 on. At $t = t_5$, $V_{in} = V_p = V_{out} + V_{D,on}$, and D_1 is on, but V_{out} exhibits no tendency to change because the situation is identical to that at $t = t_1$. In other words, V_{out} remains equal to $V_p - V_{D,on}$ indefinitely.

Example 3-24 Assuming an ideal diode model, (a) Repeat the above analysis. (b) Plot the voltage across D_1, V_{D1}, as a function of time.

Solution (a) With a zero initial condition across C_1, D_1 turns on as V_{in} exceeds zero and $V_{out} = V_{in}$. After $t = t_1$, V_{in} falls below V_{out}, turning D_1 off. Figure 3.32(a) shows the input and output waveforms.

[12]The water pipe analogy in Fig. 3.3(c) proves useful here.

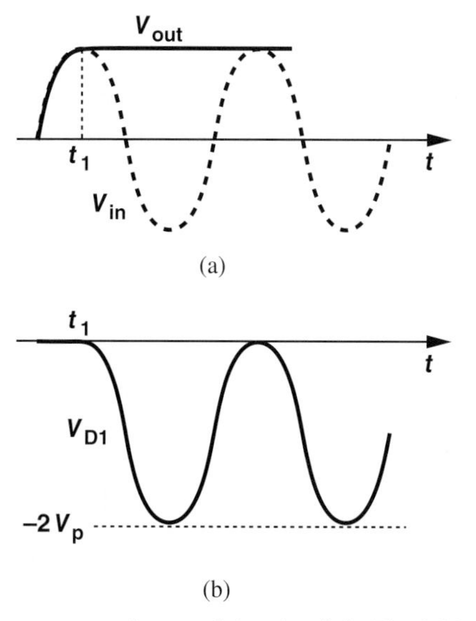

Figure 3.32 (a) Input and output waveforms of the circuit in Fig. 3.31 with an ideal diode, (b) voltage across the diode.

(b) The voltage across the diode is $V_{D1} = V_{in} - V_{out}$. Using the plots in Fig. 3.32(a), we readily arrive at the waveform in Fig. 3.32(b). Interestingly, V_{D1} is similar to V_{in} but with the average value shifted from zero to $-V_p$. We will exploit this result in the design of voltage doublers (Section 3.5.4).

Exercise Repeat the above example if the terminals of the diode are swapped.

The circuit of Fig. 3.31(a) achieves the properties required of an "ac-dc converter," generating a constant output equal to the peak value of the input sinusoid.[13] But how is the value of C_1 chosen? To answer this question, we consider a more realistic application where this circuit must provide a *current* to a load.

Example 3-25 A laptop computer consumes an average power of 25 W with a supply voltage of 3.3 V. Determine the average current drawn from the batteries or the adaptor.

Solution Since $P = V \cdot I$, we have $I \approx 7.58$ A. If the laptop is modeled by a resistor, R_L, then $R_L = V/I = 0.436\,\Omega$.

Exercise What power dissipation does a 1-Ω load represent for such a supply voltage?

As suggested by the above example, the load can be represented by a simple resistor in some cases [Fig. 3.33(a)]. We must therefore repeat our analysis with R_L present. From the waveforms in Fig. 3.33(b), we recognize that V_{out} behaves as before until $t = t_1$, still

[13]This circuit is also called a "peak detector."

exhibiting a value of $V_{in} - V_{D,on} = V_p - V_{D,on}$ if the diode voltage is assumed relatively constant. However, as V_{in} begins to fall after $t = t_1$, so does V_{out} because R_L provides a discharge path for C_1. Of course, since changes in V_{out} are undesirable, C_1 must be so large that the current drawn by R_L does not reduce V_{out} significantly. With such a choice of C_1, V_{out} falls slowly and D_1 remains reverse biased.

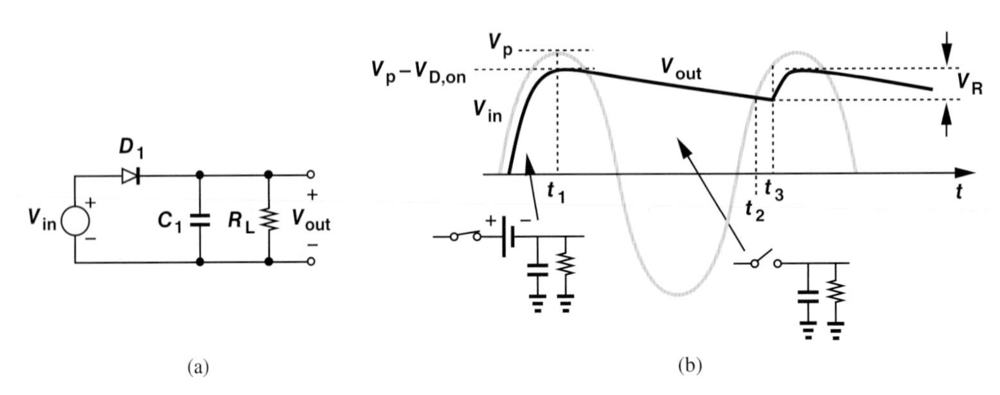

Figure 3.33 (a) Rectifier driving a resistive load, (b) input and output waveforms.

The output voltage continues to decrease while V_{in} goes through a negative excursion and returns to positive values. At some point, $t = t_2$, V_{in} and V_{out} become equal and slightly later, at $t = t_3$, V_{in} exceeds V_{out} by $V_{D,on}$, thereby turning D_1 on and forcing $V_{out} = V_{in} - V_{D,on}$. Hereafter, the circuit behaves as in the first cycle. The resulting variation in V_{out} is called the "ripple." Also, C_1 is called the "smoothing" or "filter" capacitor.

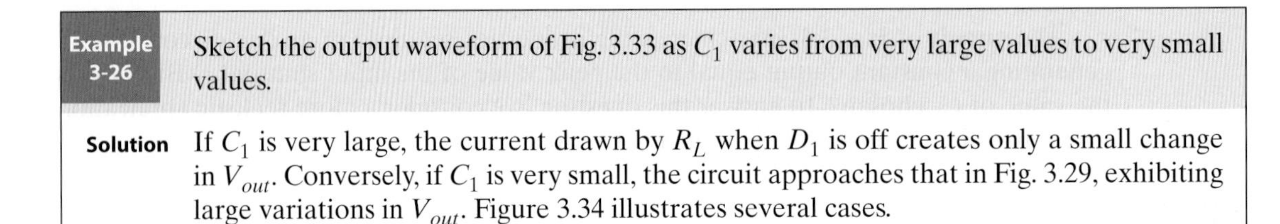

Example 3-26 Sketch the output waveform of Fig. 3.33 as C_1 varies from very large values to very small values.

Solution If C_1 is very large, the current drawn by R_L when D_1 is off creates only a small change in V_{out}. Conversely, if C_1 is very small, the circuit approaches that in Fig. 3.29, exhibiting large variations in V_{out}. Figure 3.34 illustrates several cases.

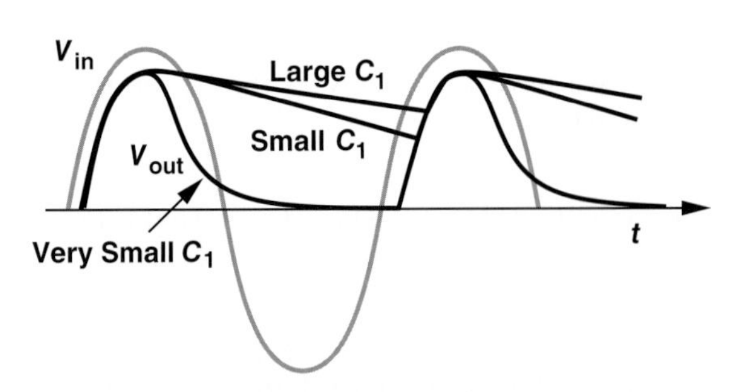

Figure 3.34 Output waveform of rectifier for different values of smoothing capacitor.

Exercise Repeat the above example for different values of R_L with C_1 constant.

Ripple Amplitude* In typical applications, the (peak-to-peak) amplitude of the ripple, V_R, in Fig. 3.33(b) must remain below 5% to 10% of the input peak voltage. If the maximum current drawn by the load is known, the value of C_1 is chosen large enough to yield an acceptable ripple. To this end, we must compute V_R analytically (Fig. 3.35). Since $V_{out} = V_p - V_{D,on}$ at $t = t_1$, the discharge of C_1 through R_L can be expressed as:

$$V_{out}(t) = (V_p - V_{D,on}) \exp \frac{-t}{R_L C_1} \quad 0 \le t \le t_3, \tag{3.76}$$

where we have chosen $t_1 = 0$ for simplicity. To ensure a small ripple, $R_L C_1$ must be much greater than $t_3 - t_1$; thus, noting that $\exp(-\epsilon) \approx 1 - \epsilon$ for $\epsilon \ll 1$,

$$V_{out}(t) \approx (V_p - V_{D,on}) \left(1 - \frac{t}{R_L C_1}\right) \tag{3.77}$$

$$\approx (V_p - V_{D,on}) - \frac{V_p - V_{D,on}}{R_L} \cdot \frac{t}{C_1}. \tag{3.78}$$

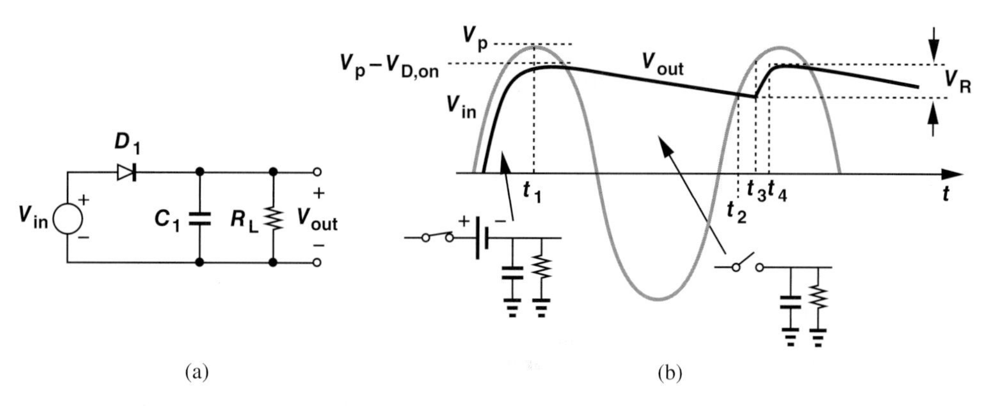

Figure 3.35 Ripple at output of a rectifier.

The first term on the right-hand side represents the initial condition across C_1 and the second term, a falling ramp—as if a constant current equal to $(V_p - V_{D,on})/R_L$ discharges C_1.[14] This result should not come as a surprise because the nearly constant voltage across R_L results in a relatively constant current equal to $(V_p - V_{D,on})/R_L$.

The peak-to-peak amplitude of the ripple is equal to the amount of discharge at $t = t_3$. Since $t_4 - t_1$ is equal to the input period, T_{in}, we write $t_3 - t_1 = T_{in} - \Delta T$, where $\Delta T (= t_4 - t_3)$ denotes the time during which D_1 is on. Thus,

$$V_R = \frac{V_p - V_{D,on}}{R_L} \frac{T_{in} - \Delta T}{C_1}. \tag{3.79}$$

Recognizing that if C_1 discharges by a small amount, then the diode turns on for only a brief period, we can assume $\Delta T \ll T_{in}$ and hence

$$V_R \approx \frac{V_p - V_{D,on}}{R_L} \cdot \frac{T_{in}}{C_1} \tag{3.80}$$

$$\approx \frac{V_p - V_{D,on}}{R_L C_1 f_{in}}, \tag{3.81}$$

where $f_{in} = T_{in}^{-1}$.

*This section can be skipped in a first reading.
[14]Recall that $I = C dV/dt$ and hence $dV = (I/C)dt$.

Example 3-27 A transformer converts the 110-V, 60-Hz line voltage to a peak-to-peak swing of 9 V. A half-wave rectifier follows the transformer to supply the power to the laptop computer of Example 3-25. Determine the minimum value of the filter capacitor that maintains the ripple below 0.1 V. Assume $V_{D,on} = 0.8$ V.

Solution We have $V_p = 4.5$ V, $R_L = 0.436\ \Omega$, and $T_{in} = 16.7$ ms. Thus,

$$C_1 = \frac{V_p - V_{D,on}}{V_R} \cdot \frac{T_{in}}{R_L} \tag{3.82}$$

$$= 1.417\ \text{F}. \tag{3.83}$$

This is a very large value. The designer must trade the ripple amplitude with the size, weight, and cost of the capacitor. In fact, limitations on size, weight, and cost of the adaptor may dictate a much greater ripple, e.g., 0.5 V, thereby demanding that the circuit following the rectifier tolerate such a large, periodic variation.

Exercise Repeat the above example for 220-V, 50-Hz line voltage, assuming the transformer still produces a peak-to-peak swing of 9 V. Which mains frequency gives a more desirable choice of C_1?

In many cases, the current drawn by the load is known. Repeating the above analysis with the load represented by a constant current source or simply viewing $(V_p - V_{D,on})/R_L$ in Eq. (3.81) as the load current, I_L, we can write

$$V_R = \frac{I_L}{C_1 f_{in}}. \tag{3.84}$$

Diode Peak Current* We noted in Fig. 3.31(b) that the diode must exhibit a reverse breakdown voltage of at least $2V_p$. Another important parameter of the diode is the maximum forward bias current that it must tolerate. For a given junction doping profile and geometry, if the current exceeds a certain limit, the power dissipated in the diode ($= V_D I_D$) may raise the junction temperature so much as to damage the device.

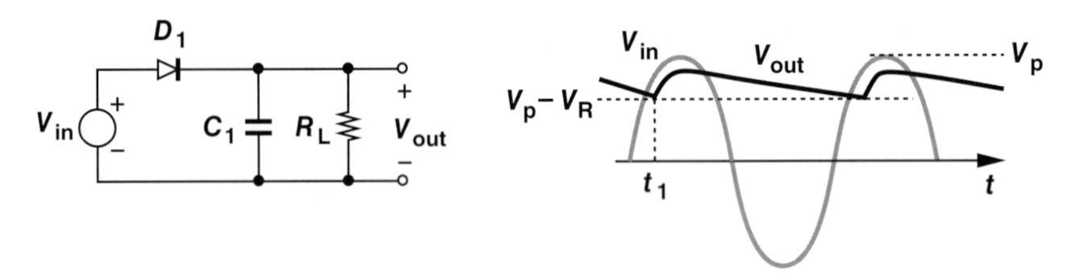

Figure 3.36 Rectifier circuit for calculation of I_D.

We recognize from Fig. 3.36, that the diode current in forward bias consists of two components: (1) the transient current drawn by C_1, $C_1 dV_{out}/dt$, and (2) the current supplied to R_L, approximately equal to $(V_p - V_{D,on})/R_L$. The peak diode current therefore occurs

*This section can be skipped in a first reading.

when the first component reaches a maximum, i.e., at the point D_1 turns on because the slope of the output waveform is maximum. Assuming $V_{D,on} \ll V_p$ for simplicity, we note that the point at which D_1 turns on is given by $V_{in}(t_1) = V_p - V_R$. Thus, for $V_{in}(t) = V_p \sin \omega_{in} t$,

$$V_p \sin \omega_{in} t_1 = V_p - V_R, \tag{3.85}$$

and hence

$$\sin \omega_{in} t_1 = 1 - \frac{V_R}{V_p}. \tag{3.86}$$

With $V_{D,on}$ neglected, we also have $V_{out}(t) \approx V_{in}(t)$, obtaining the diode current as

$$I_{D1}(t) = C_1 \frac{dV_{out}}{dt} + \frac{V_p}{R_L} \tag{3.87}$$

$$= C_1 \omega_{in} V_p \cos \omega_{in} t + \frac{V_p}{R_L}. \tag{3.88}$$

This current reaches a peak at $t = t_1$:

$$I_p = C_1 \omega_{in} V_p \cos \omega_{in} t_1 + \frac{V_p}{R_L}, \tag{3.89}$$

which, from (3.86), reduces to

$$I_p = C_1 \omega_{in} V_p \sqrt{1 - \left(1 - \frac{V_R}{V_p}\right)^2} + \frac{V_p}{R_L} \tag{3.90}$$

$$= C_1 \omega_{in} V_p \sqrt{\frac{2V_R}{V_p} - \frac{V_R^2}{V_p^2}} + \frac{V_p}{R_L}. \tag{3.91}$$

Since $V_R \ll V_p$, we neglect the second term under the square root:

$$I_p \approx C_1 \omega_{in} V_p \sqrt{\frac{2V_R}{V_p}} + \frac{V_p}{R_L} \tag{3.92}$$

$$\approx \frac{V_p}{R_L} \left(R_L C_1 \omega_{in} \sqrt{\frac{2V_R}{V_p}} + 1 \right). \tag{3.93}$$

Example 3-28 Determine the peak diode current in Example 3-27 assuming $V_{D,on} \approx 0$ and $C_1 = 1.417$ F.

Solution We have $V_p = 4.5$ V, $R_L = 0.436\ \Omega$, $\omega_{in} = 2\pi(60\text{ Hz})$, and $V_R = 0.1$ V. Thus,

$$I_p = 517\text{ A}. \tag{3.94}$$

This value is extremely large. Note that the current drawn by C_1 is much greater than that flowing through R_L.

Exercise Repeat the above example if $C_1 = 1000\ \mu$F.

Full-Wave Rectifier The half-wave rectifier studied above blocks the negative half cycles of the input, allowing the filter capacitor to be discharged by the load for almost the entire period. The circuit therefore suffers from a large ripple in the presence of a heavy load (a high current).

It is possible to reduce the ripple voltage by a factor of two through a simple modification. Illustrated in Fig. 3.37(a), the idea is to pass both positive and negative half cycles to the output, but with the negative half cycles *inverted* (i.e., multiplied by −1). We first implement a circuit that performs this function [called a "full-wave rectifier" (FWR)] and next prove that it indeed exhibits a smaller ripple. We begin with the assumption that the diodes are ideal to simplify the task of circuit synthesis. Figure 3.37(b) depicts the desired input/output characteristic of the full-wave rectifier.

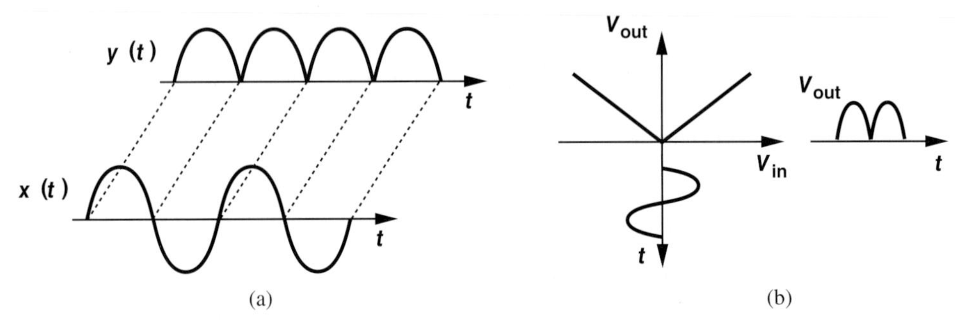

Figure 3.37 (a) Input and output waveforms and (b) input/output characteristic of a full-wave rectifier.

Consider the two half-wave rectifiers shown in Fig. 3.38(a), where one blocks negative half cycles and the other, positive half cycles. Can we combine these circuits to realize a full-wave rectifier? We may attempt the circuit in Fig. 3.38(b), but, unfortunately, the output contains both positive and negative half cycles, i.e., no rectification is performed because the negative half cycles are not inverted. Thus, the problem is reduced to that illustrated in Fig. 3.38(c): we must first design a half-wave rectifier that *inverts*. Shown in Fig. 3.38(d) is such a topology, which can also be redrawn as in Fig. 3.38(e) for simplicity. Note the polarity of V_{out} in the two diagrams. Here, if $V_{in} < 0$, both D_2 and D_1 are on and $V_{out} = -V_{in}$. Conversely, if $V_{in} > 0$, both diodes are off, yielding a zero current through

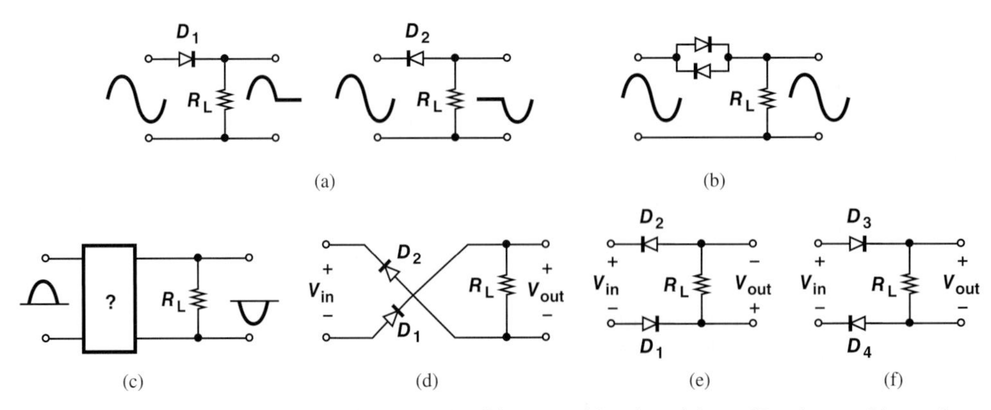

Figure 3.38 (a) Rectification of each half cycle, (b) no rectification, (c) rectification and inversion, (d) realization of (c), (e) path for negative half cycles, (f) path for positive half cycles.

R_L and hence $V_{out} = 0$. In analogy with this circuit, we also compose that in Fig. 3.38(f), which simply blocks the negative input half cycles; i.e., $V_{out} = 0$ for $V_{in} < 0$ and $V_{out} = V_{in}$ for $V_{in} > 0$.

With the foregoing developments, we can now combine the topologies of Figs. 3.38(d) and (f) to form a full-wave rectifier. Depicted in Fig. 3.39(a), the resulting circuit passes the negative half cycles through D_1 and D_2 with a sign reversal [as in Fig. 3.38(d)] and the positive half cycles through D_3 and D_4 with no sign reversal [as in Fig. 3.38(f)]. This configuration is usually drawn as in Fig. 3.39(b) and called a "bridge rectifier."

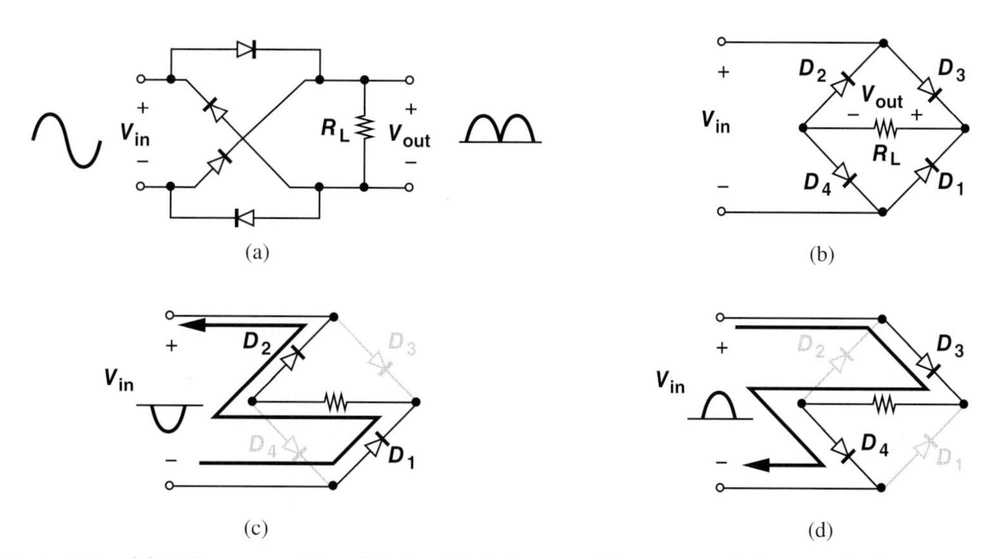

Figure 3.39 (a) Full-wave rectifier, (b) simplified diagram, (c) current path for negative input, (d) current path for positive input.

Let us summarize our thoughts with the aid of the circuit shown in Fig. 3.39(b). If $V_{in} < 0$, D_2 and D_1 are on and D_3 and D_4 are off, reducing the circuit to that shown in Fig. 3.39(c) and yielding $V_{out} = -V_{in}$. On the other hand, if $V_{in} > 0$, the bridge is simplified as shown in Fig. 3.39(d), and $V_{out} = V_{in}$.

How do these results change if the diodes are not ideal? Figures 3.39(c) and (d) reveal that the circuit introduces *two* forward-biased diodes in series with R_L, yielding $V_{out} = -V_{in} - 2V_{D,on}$ for $V_{in} < 0$. By contrast, the half-wave rectifier in Fig. 3.29 produces $V_{out} = V_{in} - V_{D,on}$. The drop of $2V_{D,on}$ may pose difficulty if V_p is relatively small and the output voltage must be close to V_p.

Example 3-29 Assuming a constant-voltage model for the diodes, plot the input/output characteristic of a full-wave rectifier.

Solution The output remains equal to zero for $|V_{in}| < 2V_{D,on}$ and "tracks" the input for $|V_{in}| > V_{D,on}$ with a slope of unity. Figure 3.40 plots the result.

Exercise What is the slope of the characteristic for $|V_{in}| > 2V_{D,on}$?

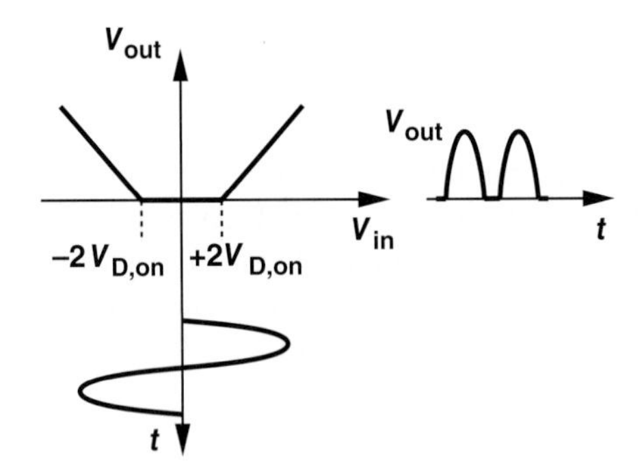

Figure 3.40 Input/output characteristic of full-wave rectifier with nonideal diodes.

We now redraw the bridge once more and add the smoothing capacitor to arrive at the complete design [Fig. 3.41(a)]. Since the capacitor discharge occurs for about half of the input cycle, the ripple is approximately equal to half of that in Eq. (3.81):

$$V_R \approx \frac{1}{2} \cdot \frac{V_p - 2V_{D,on}}{R_L C_1 f_{in}}, \tag{3.95}$$

where the numerator reflects the drop of $2V_{D,on}$ due to the bridge.

In addition to a lower ripple, the full-wave rectifier offers another important advantage: the maximum reverse bias voltage across each diode is approximately equal to V_p rather than $2V_p$. As illustrated in Fig. 3.41(b), when V_{in} is near V_p and D_3 is on, the voltage across D_2, V_{AB}, is simply equal to $V_{D,on} + V_{out} = V_p - V_{D,on}$. A similar argument applies to the other diodes.

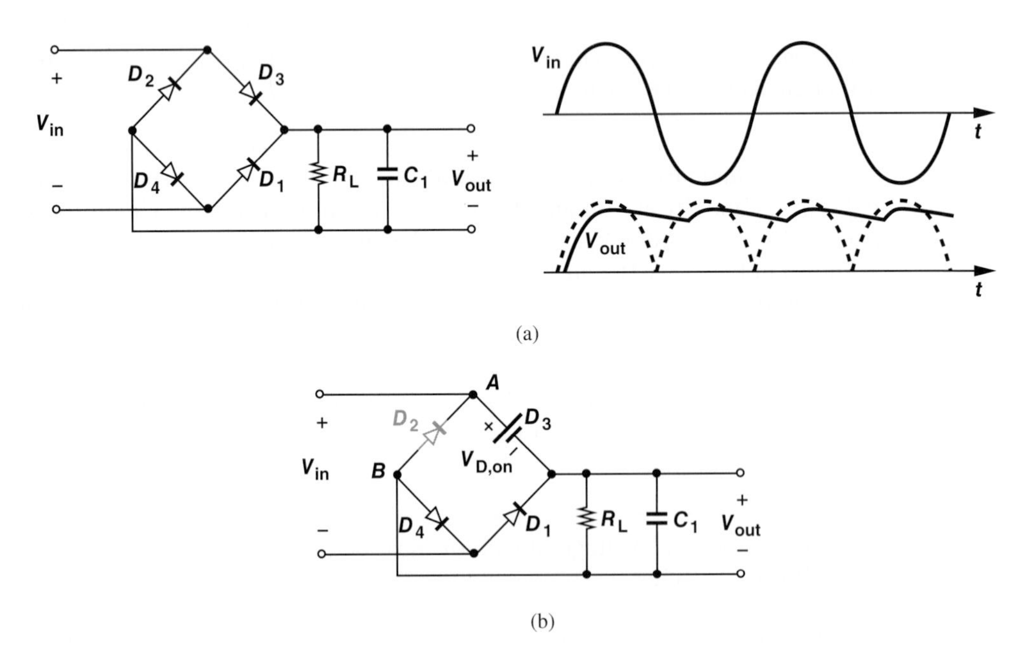

Figure 3.41 (a) Ripple in full-wave rectifier, (b) equivalent circuit.

Another point of contrast between half-wave and full-wave rectifiers is that the former has a common terminal between the input and output ports (node G in Fig. 3.29), whereas the latter does not. In Problem 3.38, we study the effect of shorting the input and output grounds of a full-wave rectifier and conclude that it disrupts the operation of the circuit.

Example 3-30 Plot the currents carried by each diode in a bridge rectifier as a function of time for a sinusoidal input. Assume no smoothing capacitor is connected to the output.

Solution From Figs. 3.39(c) and (d), we have $V_{out} = -V_{in} + 2V_{D,on}$ for $V_{in} < -2V_{D,on}$ and $V_{out} = V_{in} - 2V_{D,on}$ for $V_{in} > +2V_{D,on}$. In each half cycle, two of the diodes carry a current equal to V_{out}/R_L and the other two remain off. Thus, the diode currents appear as shown in Fig. 3.42.

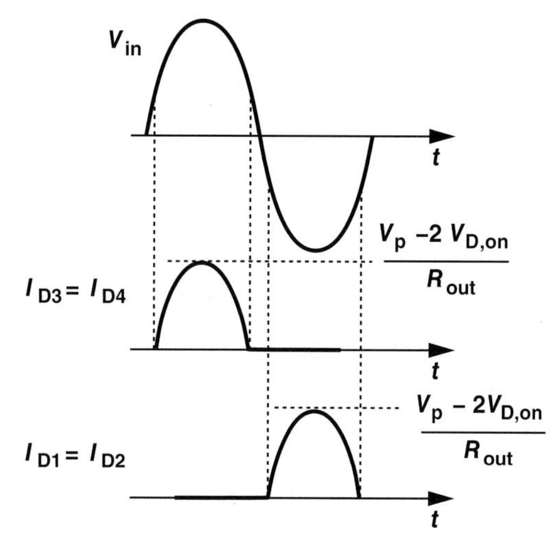

Figure 3.42 Currents carried by diodes in a full-wave rectifier.

Exercise Sketch the power consumed in each diode as a function of time.

The results of our study are summarized in Fig. 3.43. While using two more diodes, full-wave rectifiers exhibit a lower ripple and require only half the diode breakdown voltage, well justifying their use in adaptors and chargers.[15]

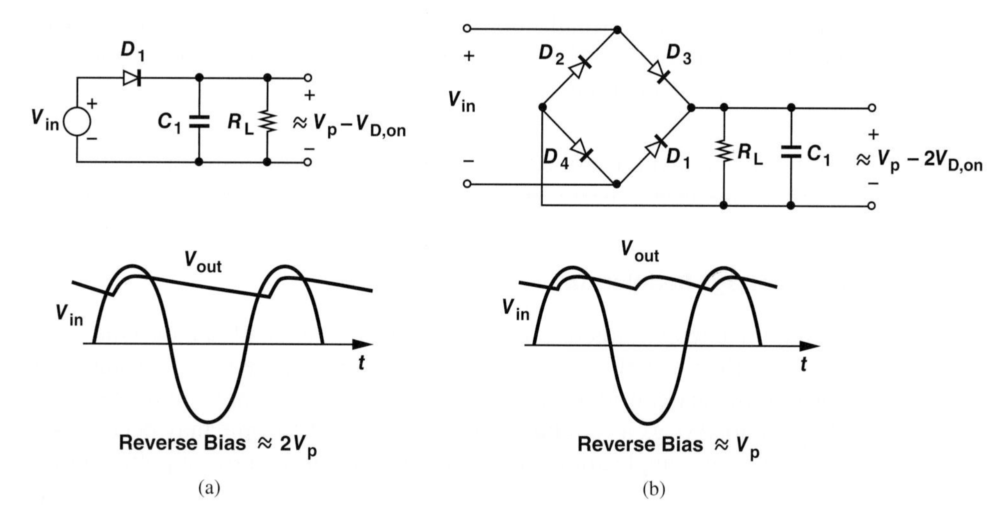

Figure 3.43 Summary of rectifier circuits.

[15]The four diodes are typically manufactured in a single package having four terminals.

Example 3-31 Design a full-wave rectifier to deliver an average power of 2 W to a cellphone with a voltage of 3.6 V and a ripple of 0.2 V.

Solution We begin with the required input swing. Since the output voltage is approximately equal to $V_p - 2V_{D,on}$, we have

$$V_{in,p} = 3.6\,\text{V} + 2V_{D,on} \tag{3.96}$$

$$\approx 5.2\,\text{V}. \tag{3.97}$$

Thus, the transformer preceding the rectifier must step the line voltage (110 V_{rms} or 220 V_{rms}) down to a peak value of 5.2 V.

Next, we determine the minimum value of the smoothing capacitor that ensures $V_R \leq 0.2$ V. Rewriting Eq. (3.84) for a full-wave rectifier gives

$$V_R = \frac{I_L}{2C_1 f_{in}} \tag{3.98}$$

$$= \frac{2\,\text{W}}{3.6\,\text{V}} \cdot \frac{1}{2C_1 f_{in}}. \tag{3.99}$$

For $V_R = 0.2$ V and $f_{in} = 60$ Hz,

$$C_1 = 23{,}000\,\mu\text{F}. \tag{3.100}$$

The diodes must withstand a reverse bias voltage of 5.2 V.

Exercise If cost and size limitations impose a maximum value of 1000 μF on the smoothing capacitor, what is the maximum allowable power drain in the above example?

Example 3-32 A radio frequency signal received and amplified by a cellphone exhibits a peak swing of 10 mV. We wish to generate a dc voltage representing the signal amplitude [Eq. (3.8)]. Is it possible to use the half-wave or full-wave rectifiers studied above?

Solution No, it is not. Owing to its small amplitude, the signal cannot turn actual diodes on and off, resulting in a zero output. For such signal levels, "precision rectification" is necessary, a subject studied in Chapter 8.

Exercise What if a constant voltage of 0.8 V is added to the desired signal?

3.5.2 Voltage Regulation*

The adaptor circuit studied above generally proves inadequate. Due to the significant variation of the line voltage, the peak amplitude produced by the transformer and hence the dc output vary considerably, possibly exceeding the maximum level that can be tolerated by the load (e.g., a cellphone). Furthermore, the ripple may become seriously objectionable in many applications. For example, if the adaptor supplies power to a stereo,

*This section can be skipped in a first reading.

the 120-Hz ripple can be heard from the speakers. Moreover, the finite output impedance of the transformer leads to changes in V_{out} if the current drawn by the load varies. For these reasons, the circuit of Fig. 3.41(a) is often followed by a "voltage regulator" so as to provide a constant output.

We have already encountered a voltage regulator without calling it such: the circuit studied in Example 3-17 provides a voltage of 2.4 V, incurring only an 11-mV change in the output for a 100-mV variation in the input. We may therefore arrive at the circuit shown in Fig. 3.44 as a more versatile adaptor having a nominal output of $3V_{D,on} \approx 2.4$ V. Unfortunately, as studied in Example 3-22, the output voltage varies with the load current.

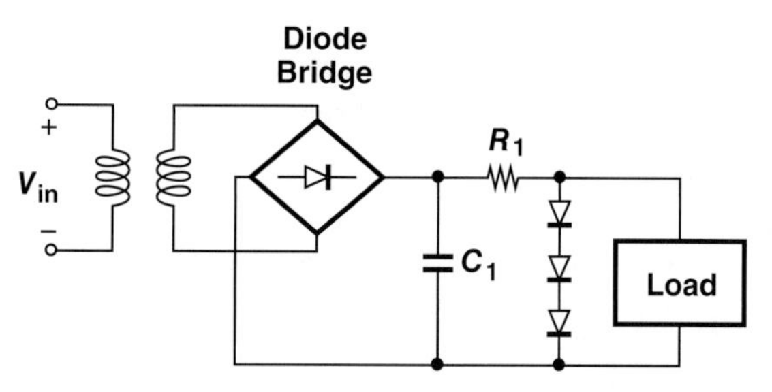

Figure 3.44 Voltage regulator block diagram.

Figure 3.45(a) shows another regulator circuit employing a Zener diode. Operating in the reverse breakdown region, D_1 exhibits a small-signal resistance, r_D, in the range of 1 to 10 Ω, thus providing a relatively constant output despite input variations if $r_D \ll R_1$. This can be seen from the small-signal model of Fig. 3.45(b):

$$V_{out} = \frac{r_D}{r_D + R_1} V_{in} \tag{3.101}$$

For example, if $r_D = 5\,\Omega$ and $R_1 = 1\,\text{k}\Omega$, then changes in V_{in} are attenuated by approximately a factor of 200 as they appear in V_{out}. The Zener regulator nonetheless has the same drawback as the circuit of Fig. 3.44, namely, poor stability if the load current varies significantly.

Our brief study of regulators thus far reveals two important aspects of their design: the stability of the output with respect to input variations, and the stability of the output with respect to load current variations. The former is quantified by "line regulation," defined as $\Delta V_{out}/\Delta V_{in}$, and the latter by "load regulation," defined as $\Delta V_{out}/\Delta I_L$.

Did you know?

The diodes used in power supplies may not seem to have a large carbon footprint, but if we add up the power consumption of all of the diodes in the world, we see a frightening picture.

As an example, consider Google's server farms. It is estimated that Google has about half a million servers. If one server draws 200 W from a 12-V supply, then each rectifier diode carries an average current of 200 W/12 V/2 ≈ 8.5 A. (We assume two diodes alternately turn on, each carrying half of the server's average current.) With a forward bias of about 0.7 V, two diodes consume 12 W, suggesting that the rectifier diodes in Google's servers dissipate a total power of roughly 6 MW. This is equivalent to the power generated by about 10 nuclear power plants! Indeed, a great deal of effort is presently expended on "green electronics," trying to reduce the power consumption of every circuit, including the power supply itself. (Our example is actually quite pessimistic as computers use "switching" power supplies to improve the efficiency.)

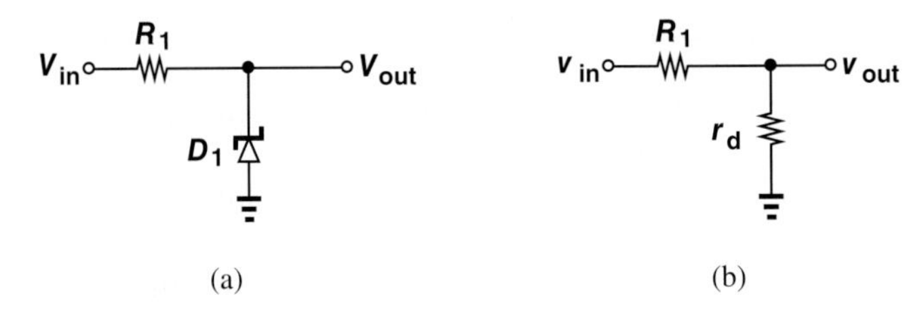

Figure 3.45 (a) Regulator using a Zener diode, (b) small-signal equivalent of (a).

Example 3-33

In the circuit of Fig. 3.46(a), V_{in} has a nominal value of 5 V, $R_1 = 100\,\Omega$, and D_2 has a reverse breakdown of 2.7 V and a small-signal resistance of 5 Ω. Assuming $V_{D,on} \approx 0.8$ V for D_1, determine the line and load regulation of the circuit.

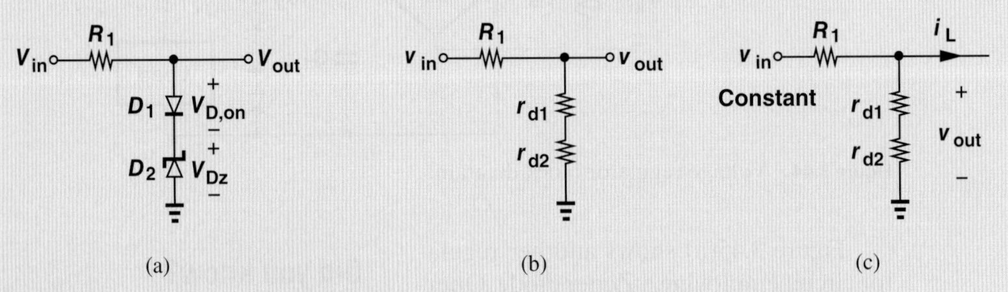

Figure 3.46 (a) Circuit using two diodes, (b) small-signal equivalent, (c) load regulation.

Solution We first determine the bias current of D_1 and hence its small-signal resistance:

$$I_{D1} = \frac{V_{in} - V_{D,on} - V_{D2}}{R_1} \tag{3.102}$$

$$= 15 \text{ mA}. \tag{3.103}$$

Thus,

$$r_{D1} = \frac{V_T}{I_{D1}} \tag{3.104}$$

$$= 1.73\,\Omega. \tag{3.105}$$

From the small-signal model of Fig. 3.45(b), we compute the line regulation as

$$\frac{v_{out}}{v_{in}} = \frac{r_{D1} + r_{D2}}{r_{D1} + r_{D2} + R_1} \tag{3.106}$$

$$= 0.063. \tag{3.107}$$

For load regulation, we assume the input is constant and study the effect of load current variations. Using the small-signal circuit shown in Fig. 3.46(c) (where $v_{in} = 0$ to represent a constant input), we have

$$\frac{v_{out}}{(r_{D1} + r_{D2}) \parallel R_1} = -i_L. \tag{3.108}$$

That is,

$$\left|\frac{v_{out}}{i_L}\right| = (r_{D1} + r_{D2}) \parallel R_1 \tag{3.109}$$

$$= 6.31\,\Omega. \tag{3.110}$$

This value indicates that a 1-mA change in the load current results in a 6.31-mV change in the output voltage.

Exercise Repeat the above example for $R_1 = 50\,\Omega$ and compare the results.

Figure 3.47 summarizes the results of our study in this section.

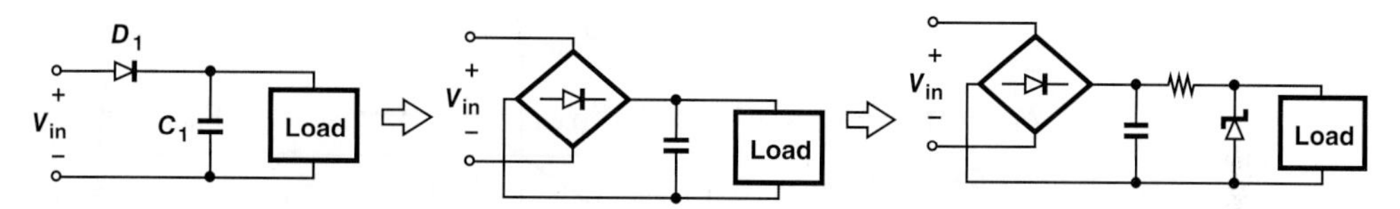

Figure 3.47 Summary of regulators.

3.5.3 Limiting Circuits

Consider the signal received by a cellphone as the user comes closer to a base station (Fig. 3.48). As the distance decreases from kilometers to hundreds of meters, the signal level may become large enough to "saturate" the circuits as it travels through the receiver chain. It is therefore desirable to "limit" the signal amplitude at a suitable point in the receiver.

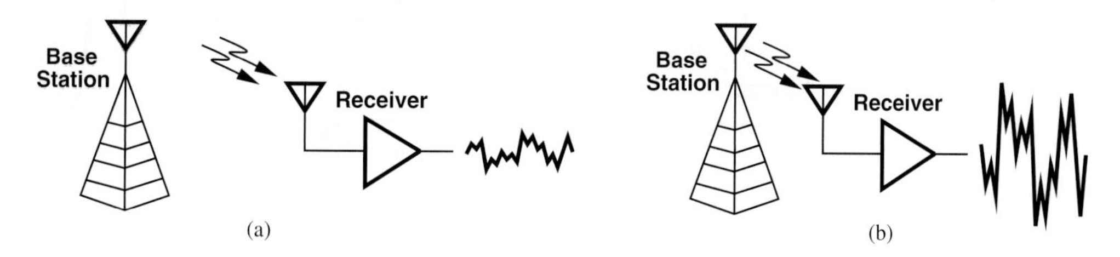

Figure 3.48 Signals received (a) far from or (b) near a base station.

How should a limiting circuit behave? For small input levels, the circuit must simply pass the input to the output, e.g., $V_{out} = V_{in}$, and as the input level exceeds a "threshold" or "limit," the output must remain constant. This behavior must hold for both positive and negative inputs, translating to the input/output characteristic shown in Fig. 3.49(a). As illustrated in Fig. 3.49(b), a signal applied to the input emerges at the output with its peak values "clipped" at $\pm V_L$.

We now implement a circuit that exhibits the above behavior. The nonlinear input/output characteristic suggests that one or more diodes must turn on or off as V_{in} approaches $\pm V_L$. In fact, we have already seen simple examples in Figs. 3.11(b) and (c), where the positive half cycles of the input are clipped at 0 V and +1 V, respectively.

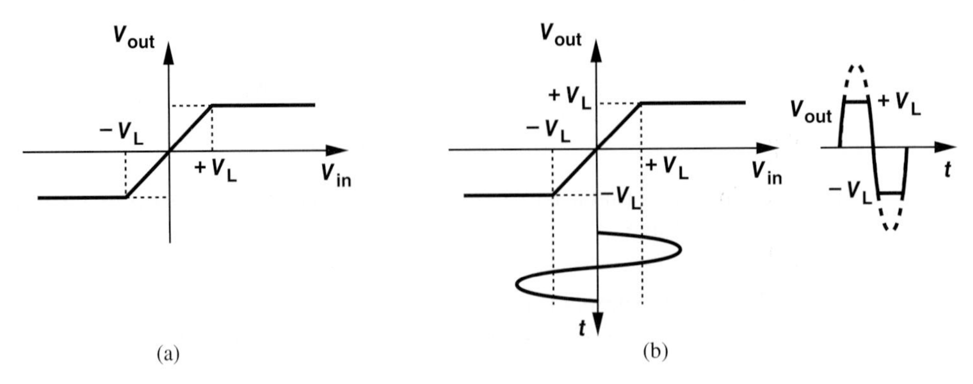

Figure 3.49 (a) Input/output characteristic of a limiting circuit, (b) response to a sinusoid.

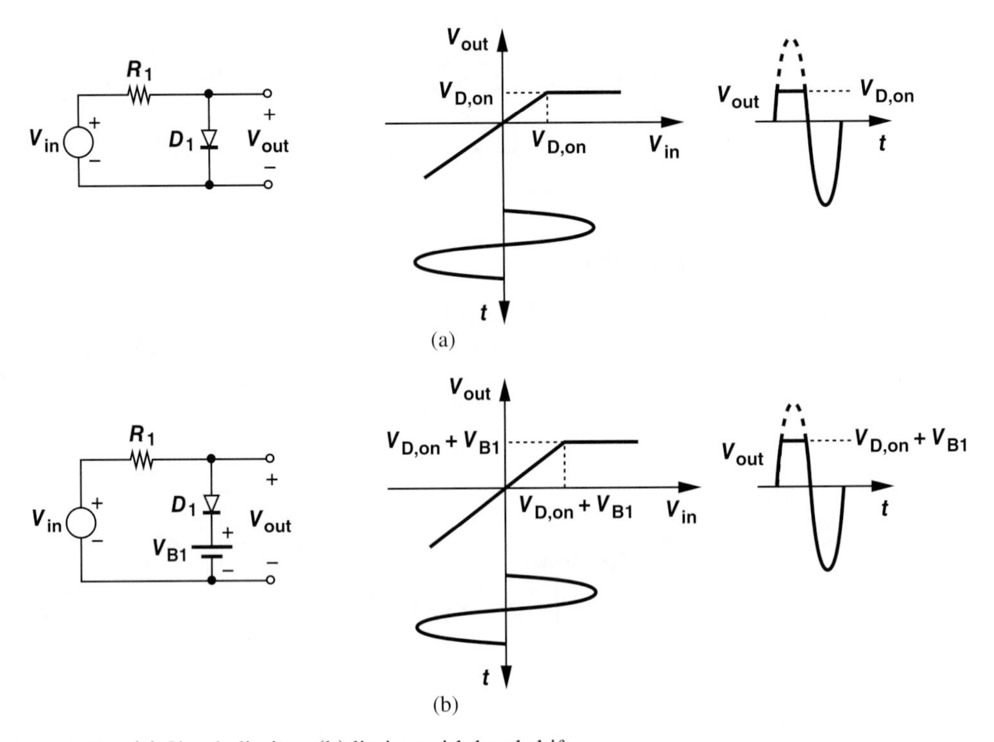

Figure 3.50 (a) Simple limiter, (b) limiter with level shift.

We reexamine the former assuming a more realistic diode, e.g., the constant-voltage model. As illustrated in Fig. 3.50(a), V_{out} is equal to V_{in} for $V_{in} < V_{D,on}$ and equal to $V_{D,on}$ thereafter.

To serve as a more general limiting circuit, the above topology must satisfy two other conditions. First, the limiting level, V_L, must be an arbitrary voltage and not necessarily equal to $V_{D,on}$. Inspired by the circuit of Fig. 3.11(c), we postulate that a constant voltage source in series with D_1 shifts the limiting point, accomplishing this objective. Depicted in Fig. 3.50(b), the resulting circuit limits at $V_L = V_{B1} + V_{D,on}$. Note that V_{B1} can be positive or negative to shift V_L to higher or lower values, respectively.

Second, the negative values of V_{in} must also experience limiting. Beginning with the circuit of Fig. 3.50(a), we recognize that if the anode and cathode of D_1 are swapped, then the circuit limits at $V_{in} = -V_{D,on}$ [Fig. 3.51(a)]. Thus, as shown in Fig. 3.51(b), two "antiparallel" diodes can create a characteristic that limits at $\pm V_{D,on}$. Finally, inserting constant voltage sources in series with the diodes shifts the limiting points to arbitrary levels (Fig. 3.52).

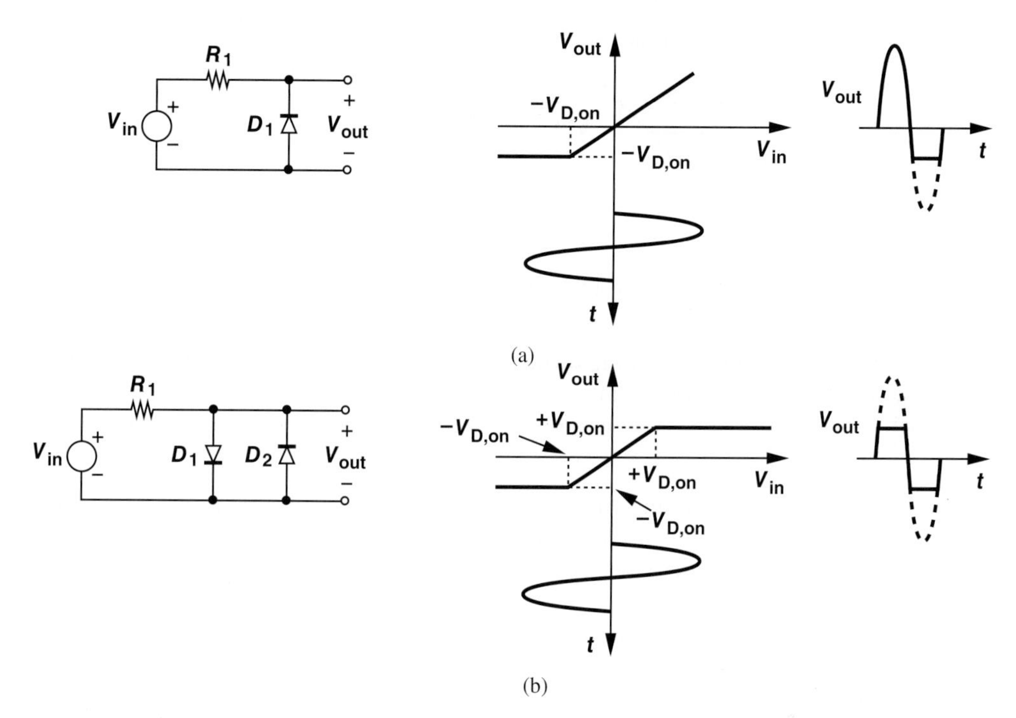

Figure 3.51 (a) Negative-cycle limiter, (b) limiter for both half cycles.

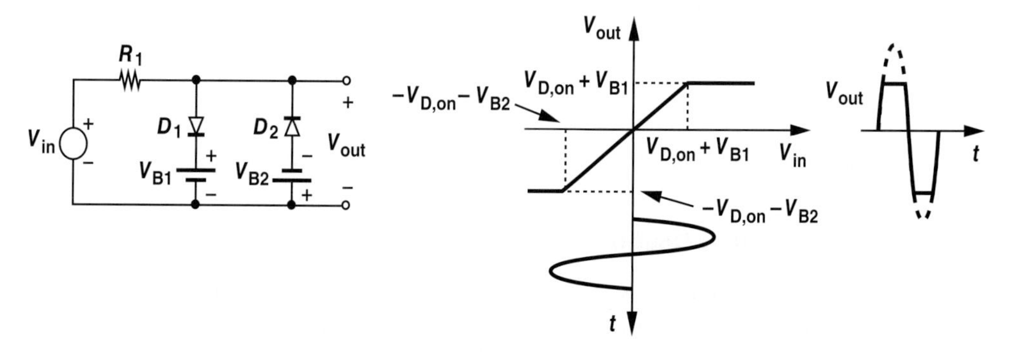

Figure 3.52 General limiter and its characteristic.

Example 3-34 A signal must be limited at ±100 mV. Assuming $V_{D,on} = 800$ mV, design the required limiting circuit.

Solution Figure 3.53(a) illustrates how the voltage sources must shift the break points. Since the positive limiting point must shift to the left, the voltage source in series with D_1 must be *negative* and equal to 700 mV. Similarly, the source in series with D_2 must be positive and equal to 700 mV. Figure 3.53(b) shows the result.

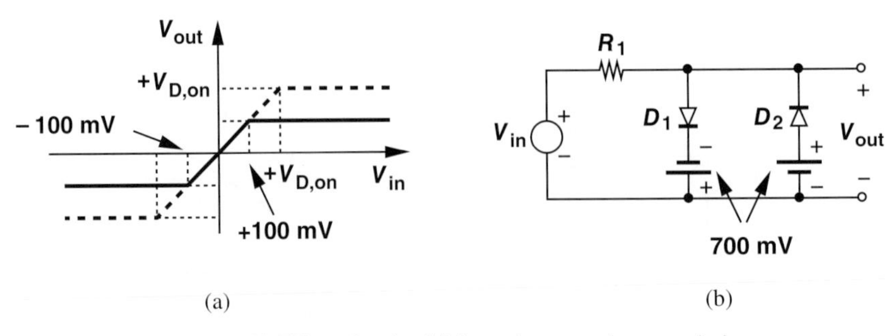

Figure 3.53 (a) Example of a limiting circuit, (b) input/output characteristic.

Exercise Repeat the above example if the positive values of the signal must be limited at +200 mV and the negative values at −1.1 V.

Before concluding this section, we make two observations. First, the circuits studied above actually display a nonzero slope in the limiting region (Fig. 3.54). This is because, as V_{in} increases, so does the current through the diode that is forward biased and hence the diode voltage.[16] Nonetheless, the 60-mV/decade rule expressed by Eq. (2.109) implies that this effect is typically negligible. Second, we have thus far assumed $V_{out} = V_{in}$ for $-V_L < V_{in} < +V_L$, but it is possible to realize a non-unity slope in the region: $V_{out} = \alpha V_{in}$.

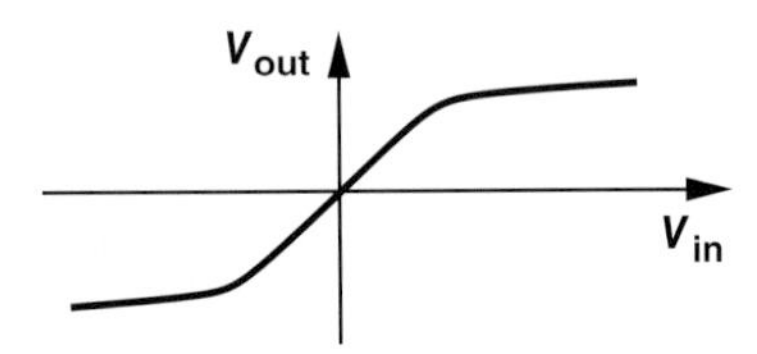

Figure 3.54 Effect of nonideal diodes on limiting characteristic.

3.5.4 Voltage Doublers*

Electronic systems typically employ a "global" supply voltage, e.g., 3 V, requiring that the discrete and integrated circuits operate with such a value. However, the design of some circuits in the system is greatly simplified if they run from a *higher* supply voltage, e.g., 6 V. "Voltage doublers" may serve this purpose.[17]

Before studying doublers, it is helpful to review some basic properties of capacitors. First, to charge one plate of a capacitor to $+Q$, the other plate *must* be charged to $-Q$. Thus, in the circuit of Fig. 3.55(a), the voltage across C_1 *cannot* change even if V_{in}

[16]Recall that $V_D = V_T \ln(I_D/I_S)$.
*This section can be skipped in a first reading.
[17]Voltage doublers are an example of "dc-dc converters."

changes because the right plate of C_1 cannot receive or release charge ($Q = CV$). Since V_{C1} remains constant, an input change ΔV_{in} appears directly at the output. This is an important observation.

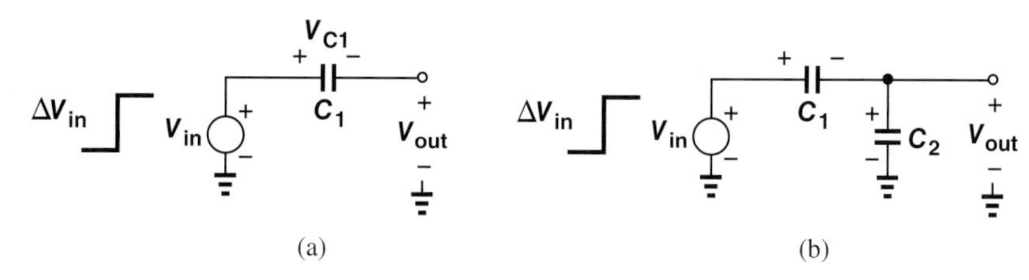

Figure 3.55 (a) Voltage change at one plate of a capacitor, (b) voltage division.

Second, a capacitive voltage divider such as that in Fig. 3.55(b) operates as follows. If V_{in} becomes more positive, the left plate of C_1 receives positive charge from V_{in}, thus requiring that the right plate absorb negative charge of the same magnitude from the top plate of C_2. Having lost negative charge, the top plate of C_2 equivalently holds more positive charge, and hence the bottom plate absorbs negative charge from ground. Note that all four plates receive or release equal amounts of charge because C_1 and C_2 are in series. To determine the change in V_{out}, ΔV_{out}, resulting from ΔV_{in}, we write the change in the charge on C_2 as $\Delta Q_2 = C_2 \cdot \Delta V_{out}$, which also holds for C_1: $\Delta Q_2 = \Delta Q_1$. Thus, the voltage change across C_1 is equal to $C_2 \cdot \Delta V_{out}/C_1$. Adding these two voltage changes and equating the result to ΔV_{in}, we have

$$\Delta V_{in} = \frac{C_2}{C_1}\Delta V_{out} + \Delta V_{out}. \tag{3.111}$$

That is,

$$\Delta V_{out} = \frac{C_1}{C_1 + C_2}\Delta V_{in}. \tag{3.112}$$

This result is similar to the voltage division expression for resistive dividers, except that C_1 (rather than C_2) appears in the numerator. Interestingly, the circuit of Fig. 3.55(a) is a special case of the capacitive divider with $C_2 = 0$ and hence $\Delta V_{out} = \Delta V_{in}$.

As our first step toward realizing a voltage doubler, recall the result illustrated in Fig. 3.32: the voltage across the diode in the peak detector exhibits an average value of $-V_p$ and, more importantly, a peak value of $-2V_p$ (with respect to zero). For further investigation, we redraw the circuit as shown in Fig. 3.56, where the diode and the capacitors are exchanged and the voltage across D_1 is labeled V_{out}. While V_{out} in this circuit behaves exactly the same as V_{D1} in Fig. 3.31(a), we derive the output waveform from a different perspective so as to gain more insight.

Assuming an ideal diode and a zero initial condition across C_1, we note that as V_{in} exceeds zero, the input tends to place positive charge on the left plate of C_1 and hence draw negative charge from D_1. Consequently, D_1 turns on, forcing $V_{out} = 0$.[18] As the input rises toward V_p, the voltage across C_1 remains equal to V_{in} because its right plate is "pinned" at zero by D_1. After $t = t_1$, V_{in} begins to fall and tends to discharge C_1, i.e., draw positive charge from the left plate and hence from D_1. The diode therefore turns off, reducing the circuit to that in Fig. 3.55(a). From this time, the output simply tracks the changes

[18]If we assume D_1 does *not* turn on, then the circuit resembles that in Fig. 3.55(a), requiring that V_{out} rise and D_1 turn on.

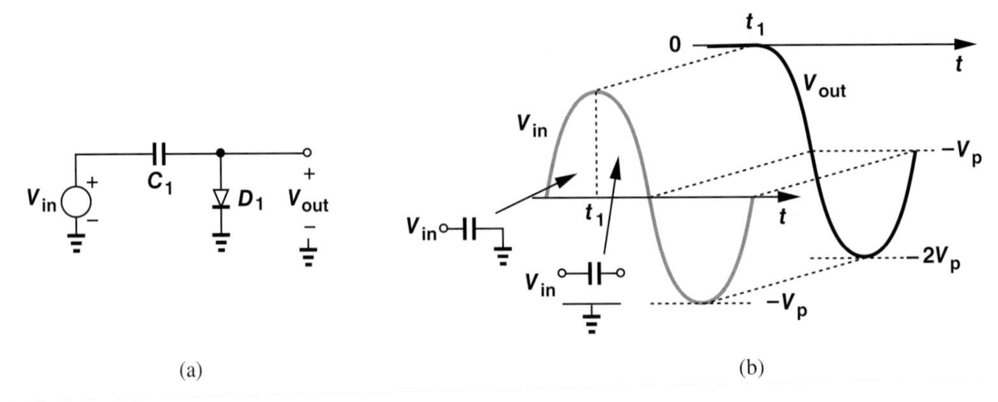

Figure 3.56 (a) Capacitor-diode circuit and (b) its waveforms.

in the input while C_1 sustains a constant voltage equal to V_p. In particular, as V_{in} varies from $+V_p$ to $-V_p$, the output goes from zero to $-2V_p$, and the cycle repeats indefinitely. The output waveform is thus identical to that obtained in Fig. 3.32(b).

Example 3-35 Plot the output waveform of the circuit shown in Fig. 3.57 if the initial condition across C_1 is zero.

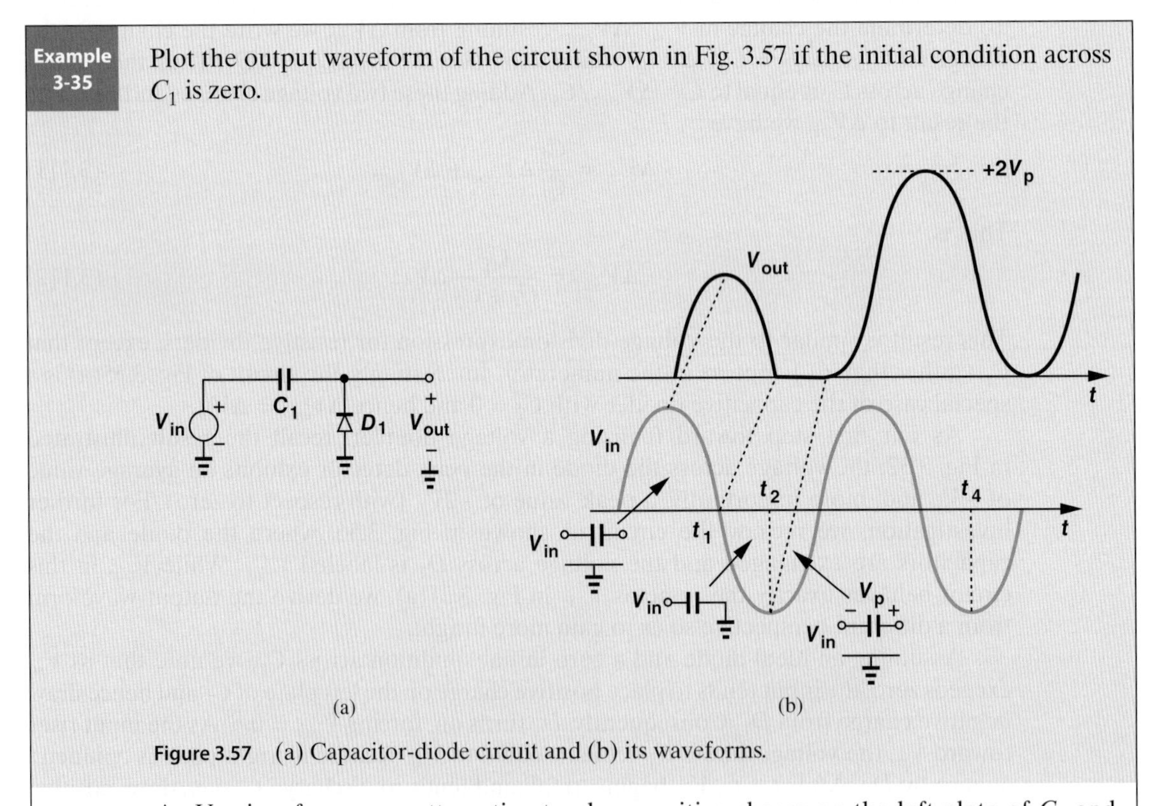

Figure 3.57 (a) Capacitor-diode circuit and (b) its waveforms.

Solution As V_{in} rises from zero, attempting to place positive charge on the left plate of C_1 and hence draw negative charge from D_1, the diode turns off. As a result, C_1 directly transfers the input change to the output for the entire positive half cycle. After $t = t_1$, the input tends to push negative charge into C_1, turning D_1 on and forcing $V_{out} = 0$. Thus, the voltage across C_1 remains equal to V_{in} until $t = t_2$, at which point the direction of the current through C_1 and D_1 must change, turning D_1 off. Now, C_1 carries a voltage equal to V_p and transfers the input change to the output; i.e., the output tracks the input but with a level shift of $+V_p$, reaching a peak value of $+2V_p$.

Exercise Repeat the above example if the right plate of C_1 is 1 V more positive than its left plate at $t = 0$.

We have thus far developed circuits that generate a periodic output with a peak value of $-2V_p$ or $+2V_p$ for an input sinusoid varying between $-V_p$ and $+V_p$. We surmise that if these circuits are followed by a *peak detector* [e.g., Fig. 3.31(a)], then a constant output equal to $-2V_p$ or $+2V_p$ may be produced. Figure 3.58 exemplifies this concept, combining the circuit of Fig. 3.57 with the peak detector of Fig. 3.31(a). Of course, since the peak detector "loads" the first stage when D_2 turns on, we must still analyze this circuit carefully and determine whether it indeed operates as a voltage doubler.

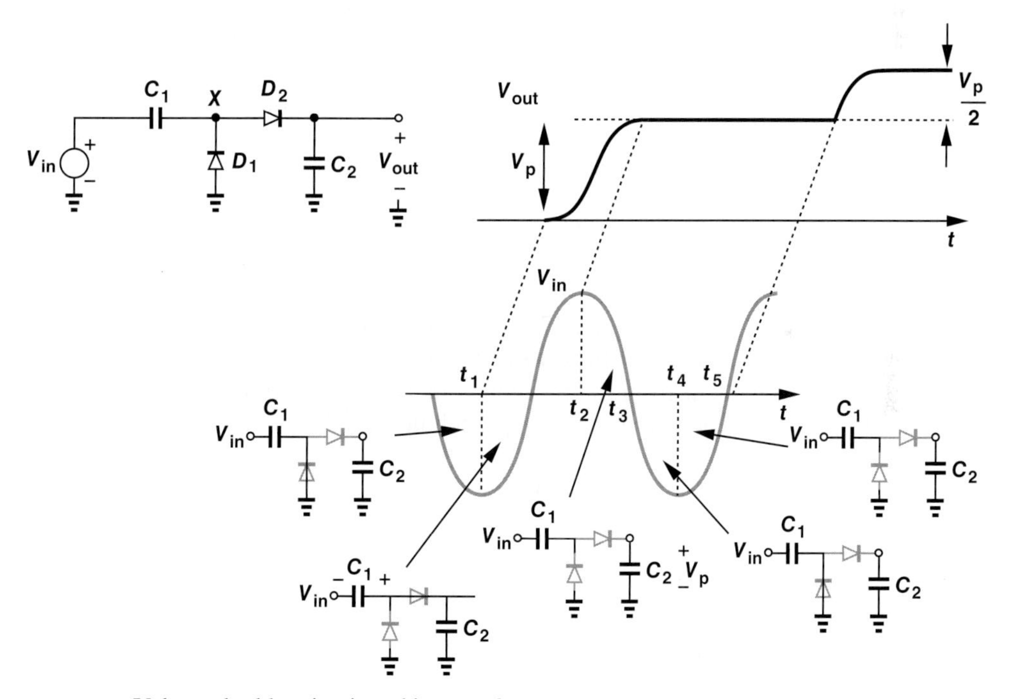

Figure 3.58 Voltage doubler circuit and its waveforms.

We assume ideal diodes, zero initial conditions across C_1 and C_2, and $C_1 = C_2$. In this case, the analysis is simplified if we begin with a negative cycle. As V_{in} falls below zero, D_1 turns on, pinning node X to zero.[19] Thus, for $t < t_1$, D_2 remains off and $V_{out} = 0$. At $t = t_1$, the voltage across C_1 reaches $-V_p$. For $t > t_1$, the input begins to rise and tends to deposit positive charge on the left plate of C_1, turning D_1 off and yielding the circuit shown in Fig. 3.58.

[19] As always, the reader is encouraged to assume otherwise (i.e., D_1 remains off) and arrive at a conflicting result.

How does D_2 behave in this regime? Since V_{in} is now rising, we postulate that V_X also tends to increase (from zero), turning D_2 on. (If D_2 remains off, then C_1 simply transfers the change in V_{in} to node X, raising V_X and hence turning D_2 on.) As a result, the circuit reduces to a simple capacitive divider that follows Eq. (3.112):

$$\Delta V_{out} = \frac{1}{2}\Delta V_{in}, \tag{3.113}$$

because $C_1 = C_2$. In other words, V_X and V_{out} begin from zero, remain equal, and vary sinusoidally but with an amplitude equal to $V_p/2$. Thus, from t_1 to t_2, a change of $2V_p$ in V_{in} appears as a change equal to V_p in V_X and V_{out}. Note at $t = t_2$, the voltage across C_1 is zero because both V_{in} and V_{out} are equal to $+V_p$.

What happens after $t = t_2$? Since V_{in} begins to fall and tends to draw charge from C_1, D_2 turns off, maintaining V_{out} at $+V_p$. The reader may wonder if something is wrong here; our objective was to generate an output equal to $2V_p$ rather than V_p. But again, patience is a virtue and we must continue the transient analysis. For $t > t_2$, both D_1 and D_2 are off, and each capacitor holds a constant voltage. Since the voltage across C_1 is zero, $V_X = V_{in}$, falling to zero at $t = t_3$. At this point, D_1 turns on again, allowing C_1 to charge to $-V_p$ at $t = t_4$. As V_{in} begins to rise again, D_1 turns off and D_2 remains off because $V_X = 0$ and $V_{out} = +V_p$. Now, with the right plate of C_1 floating, V_X tracks the change at the input, reaching $+V_p$ as V_{in} goes from $-V_p$ to 0. Thus, D_2 turns on at $t = t_5$, forming a capacitive divider again. After this time, the output change is equal to half of the input change, i.e., V_{out} increases from $+V_p$ to $+V_p + V_p/2$ as V_{in} goes from 0 to $+V_p$. The output has now reached $3V_p/2$.

Did you know?

Doublers are often used to generate larger dc voltages than the system's power supply can provide. For example, a "photomultiplier" requires a supply voltage of 1000 to 2000 V to accelerate electrons and, through a process called "secondary emission," amplify the number of photons. In this case, we begin with a line voltage of 110 V or 220 V and employ a cascade of voltage doublers to reach the necessary value. Photomultipliers find application in many fields, including astronomy, high-energy physics, and medicine. For example, blood cells are counted under low light level conditions with the aid of photomultipliers.

As is evident from the foregoing analysis, the output continues to rise by V_p, $V_p/2$, $V_p/4$, etc., in each input cycle, approaching a final value of

$$V_{out} = V_p\left(1 + \frac{1}{2} + \frac{1}{4} + \cdots\right) \tag{3.114}$$

$$= \frac{V_p}{1 - \frac{1}{2}} \tag{3.115}$$

$$= 2V_p. \tag{3.116}$$

The reader is encouraged to continue the analysis for a few more cycles and verify this trend.

Example 3-36 Sketch the current through D_1 in the doubler circuit as function of time.

Solution Using the diagram in Fig. 3.59(a), noting that D_1 and C_1 carry equal currents when D_1 is forward biased, and writing the current as $I_{D1} = -C_1 dV_{in}/dt$, we construct the plot shown in Fig. 3.59(b).[20] For $0 < t < t_1$, D_1 conducts and the peak current corresponds to

[20] As usual, I_{D1} denotes the current flowing from the anode to the cathode.

the maximum slope of V_{in}, i.e., immediately after $t = 0$. From $t = t_2$ to $t = t_3$, the diode remains off, repeating the same behavior in subsequent cycles.

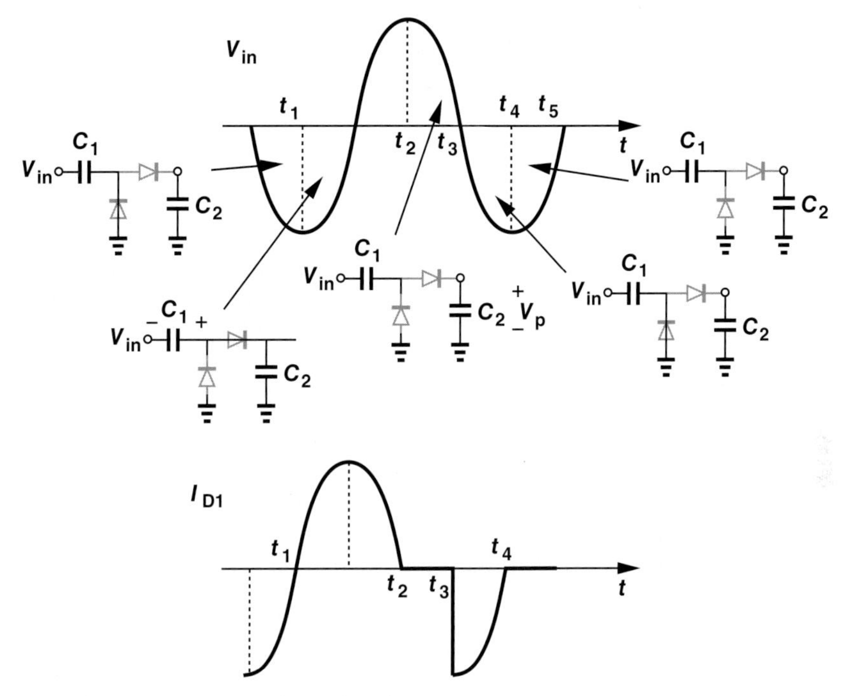

Figure 3.59 Diode current in a voltage doubler.

Exercise Plot the current through D_2 in the above example as a function of time.

Bioengineering Application: Energy Harvesting

Medical devices that are implanted within people's bodies must operate without wires. As such, they can be powered by rechargeable batteries or by "energy harvesting," wherein the energy from an external radio wave is collected by the device, stored on a capacitor, and used to power the electronics for a short time. Energy harvesting proves essential if the medical device must have small dimensions and hence cannot accommodate batteries.

An interesting example of such devices is an "intraocular" (within the eye) pressure monitor developed for glaucoma patients [1]. Owing to its small (less than 1 mm^3) volume, this device can be implanted in the patient's eye so as to measure the pressure. Figure 3.60 shows the simplified architecture [1]. The radio wave is received by a small loop antenna and converted to a dc voltage by means of a four-stage rectifier, which is somewhat similar to the voltage doubler described in this chapter. The dc voltage then acts as the supply for the pressure monitor circuit.

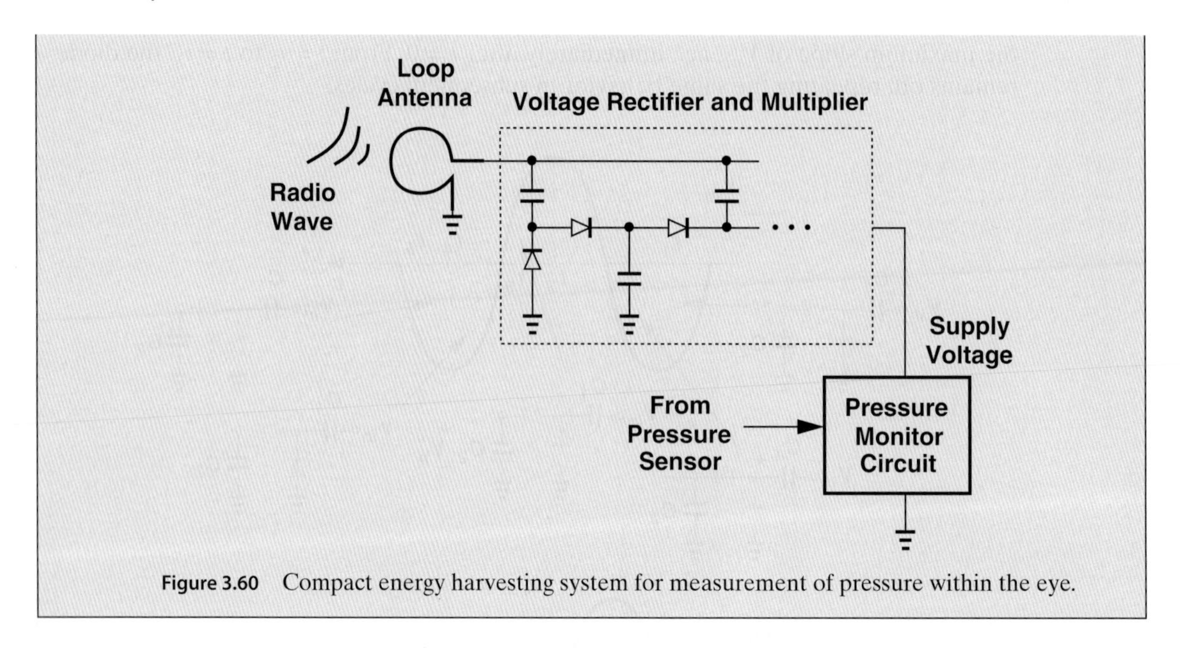

Figure 3.60 Compact energy harvesting system for measurement of pressure within the eye.

3.5.5 Diodes as Level Shifters and Switches*

In the design of electronic circuits, we may need to shift the average level of a signal up or down because the subsequent stage (e.g., an amplifier) may not operate properly with the present dc level.

Sustaining a relatively constant voltage in forward bias, a diode can be viewed as a battery and hence a device capable of shifting the signal level. In our first attempt, we consider the circuit shown in Fig. 3.61(a) as a candidate for shifting the level *down* by $V_{D,on}$. However, the diode current remains unknown and dependent on the next stage. To alleviate this issue we modify the circuit as depicted in Fig. 3.61(b), where I_1 draws a constant current, establishing $V_{D,on}$ across D_1.[21] If the current pulled by the next stage is negligible (or at least constant), V_{out} is simply lower than V_{in} by a constant amount, $V_{D,on}$.

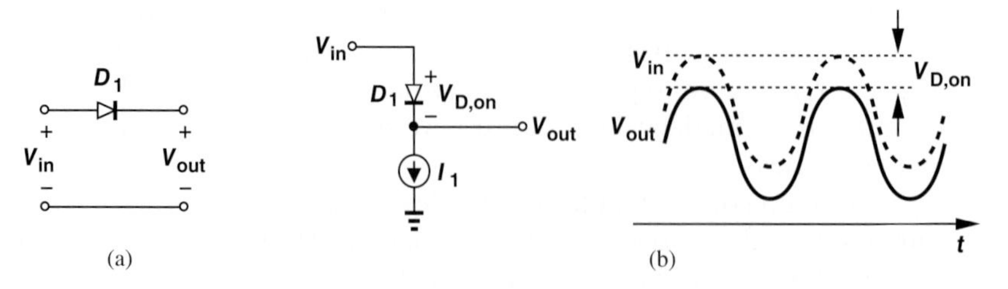

Figure 3.61 (a) Use of a diode for level shift, (b) practical implementation.

*This section can be skipped in a first reading.

[21]The diode is drawn vertically to emphasize that V_{out} is lower than V_{in}.

Example 3-37 Design a circuit that shifts up the dc level of a signal by $2V_{D,on}$.

Solution To shift the level *up*, we apply the input to the *cathode*. Also, to obtain a shift of $2V_{D,on}$, we place two diodes in series. Figure 3.62 shows the result.

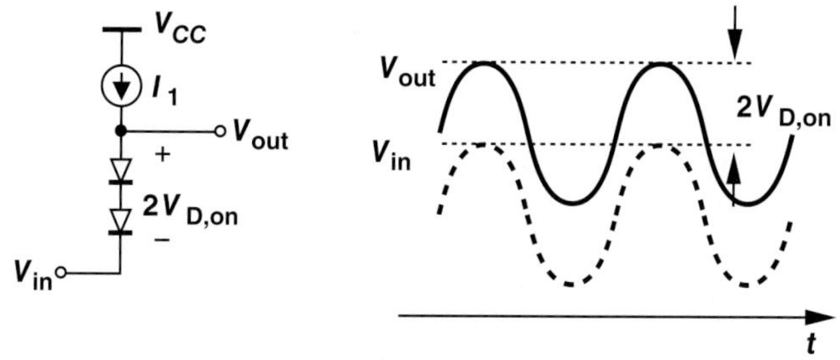

Figure 3.62 Positive voltage shift by two diodes.

Exercise What happens if I_1 is extremely small?

The level shift circuit of Fig. 3.61(b) can be transformed to an electronic switch. For example, many applications employ the topology shown in Fig. 3.63(a) to “sample” V_{in} across C_1 and “freeze” the value when S_1 turns off. Let us replace S_1 with the level shift circuit and allow I_1 to be turned on and off [Fig. 3.63(b)]. If I_1 is on, V_{out} tracks V_{in} except for a level shift equal to $V_{D,on}$. When I_1 turns off, so does D_1, evidently disconnecting C_1 from the input and freezing the voltage across C_1.

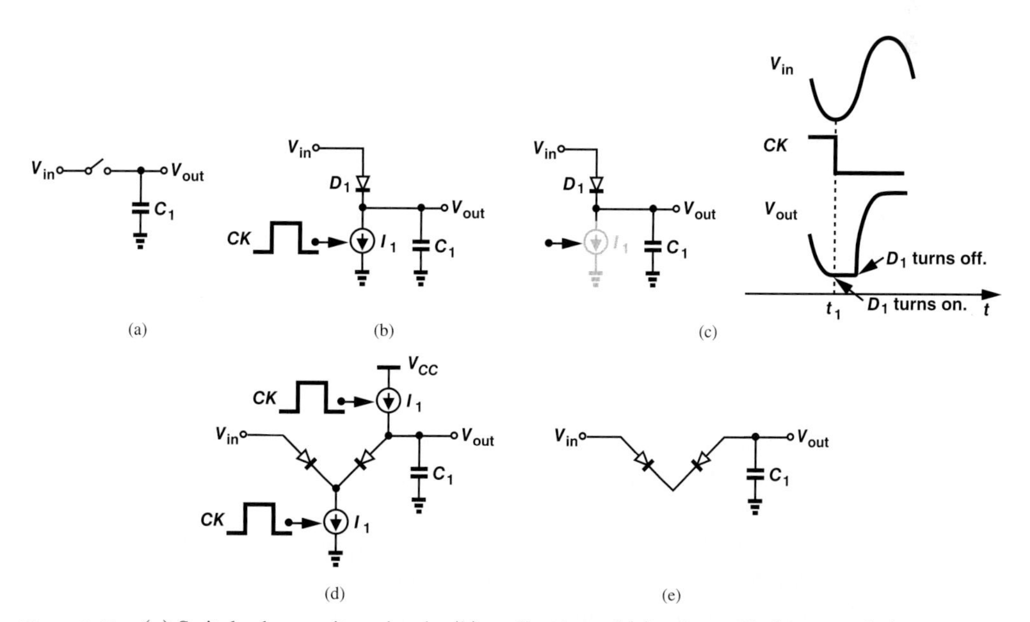

Figure 3.63 (a) Switched-capacitor circuit, (b) realization of (a) using a diode as a switch, (c) problem of diode conduction, (d) more complete circuit, (e) equivalent circuit when I_1 and I_2 are off.

We used the term "evidently" in the last sentence because the circuit's true behavior somewhat differs from the above description. The assumption that D_1 turns off holds only if C_1 draws no current from D_1, i.e., only if $V_{in} - V_{out}$ remains less than $V_{D,on}$. Now consider the case illustrated in Fig. 3.63(c), where I_1 turns off at $t = t_1$, allowing C_1 to store a value equal to $V_{in1} - V_{D,on}$. As the input waveform completes a negative excursion and exceeds V_{in1} at $t = t_2$, the diode is forward-biased again, charging C_1 with the input (in a manner similar to a peak detector). That is, even though I_1 is off, D_1 turns on for part of the cycle.

To resolve this issue, the circuit is modified as shown in Fig. 3.63(d), where D_2 is inserted between D_1 and C_1, and I_2 provides a bias current for D_2. With both I_1 and I_2 on, the diodes operate in forward bias, $V_X = V_{in} - V_{D1}$, and $V_{out} = V_X + V_{D2} = V_{in}$ if $V_{D1} = V_{D2}$. Thus, V_{out} tracks V_{in} with no level shift. When I_1 and I_2 turn off, the circuit reduces to that in Fig. 3.63(e), where the back-to-back diodes fail to conduct for any value of $V_{in} - V_{out}$, thereby isolating C_1 from the input. In other words, the two diodes and the two current sources form an electronic switch.

Example 3-38 Recall from Chapter 2 that diodes exhibit a junction capacitance in reverse bias. Study the effect of this capacitance on the operation of the above circuit.

Solution Figure 3.64 shows the equivalent circuit for the case where the diodes are off, suggesting that the conduction of the input through the junction capacitances disturbs the output. Specifically, invoking the capacitive divider of Fig. 3.55(b) and assuming $C_{j1} = C_{j2} = C_j$, we have

$$\Delta V_{out} = \frac{C_j/2}{C_j/2 + C_1}\Delta V_{in}. \tag{3.117}$$

To ensure this "feedthrough" is small, C_1 must be sufficiently large.

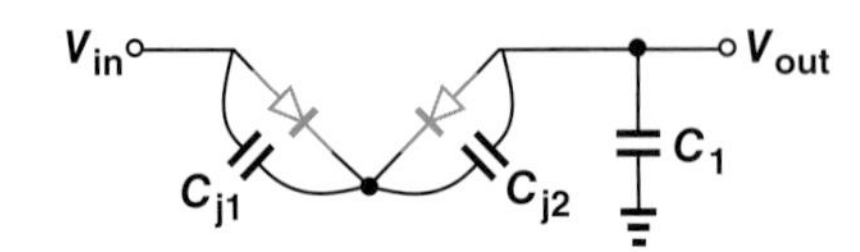

Figure 3.64 Feedthrough in the diode switch.

Exercise Calculate the change in the voltage at the left plate of C_{j1} (with respect to ground) in terms of ΔV_{in}.

3.6 CHAPTER SUMMARY

- In addition to the exponential and constant-voltage models, an "ideal" model is sometimes used to analyze diode circuits. The ideal model assumes the diode turns on with a very small forward bias voltage.
- For many electronic circuits, the "input/output characteristics" are studied to understand the response to various input levels, e.g., as the input level goes from $-\infty$ to $+\infty$.
- "Large-signal operation" occurs when a circuit or device experiences arbitrarily large voltage or current excursions. The exponential, constant-voltage, or ideal diode models are used in this case.

- If the *changes* in voltages and currents are sufficiently small, then nonlinear devices and circuits can be approximated by linear couterparts, greatly simplifying the analysis. This is called "small-signal operation."
- The small-signal model of a diode consists of an "incremental resistance" given by V_T/I_D.
- Diodes find application in many circuits, including rectifiers, limiting circuits, voltage doublers, and level shifters.
- Half-wave rectifiers pass the positive (negative) half cycles of the input wavefrom and block the negative (positive) half cycles. If followed by a capacitor, a rectifier can produce a dc level nearly equal to the peak of the input swing.
- A half-wave rectifier with a smoothing capacitor of value C_1 and load resistor R_L exhibits an output ripple equal to $(V_P - V_{D,on})/(R_L C_1 f_{in})$.
- Full-wave rectifiers convert both positive and negative input cycles to the same polarity at the output. If followed by a smoothing capacitor and a load resistor, these rectifiers exhibit an output ripple given by $0.5(V_P - 2V_{D,on})/(R_L C_1 f_{in})$.
- Diodes can operate as limiting devices, i.e., limit the output swing even if the input swing continues to increase.

PROBLEMS

In the following problems, assume $V_{D,on} =$ 800 mV for the constant-voltage diode model.

Section 3.2 *pn* Junction as a Diode

SOL **3.1.** Plot the I/V characteristic of the circuit shown in Fig. 3.65.

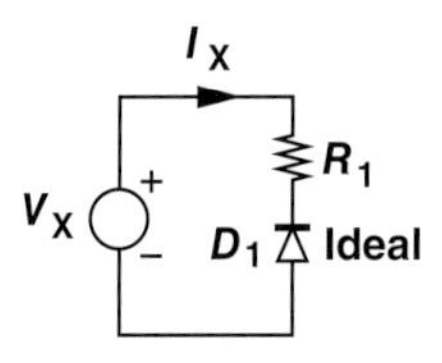

Figure 3.65

3.2. If the input in Fig. 3.65 is expressed as $V_X = V_0 \cos \omega t$, plot the current flowing through the circuit as a function of time.

3.3. Plot I_X as a function of V_X for the circuit shown in Fig. 3.66 for two cases: $V_B = -1$ V and $V_B = +1$ V.

3.4. If in Fig. 3.66, $V_X = V_0 \cos \omega t$, plot I_X as a function of time for two cases: $V_B = -1$ V and $V_B = +1$ V.

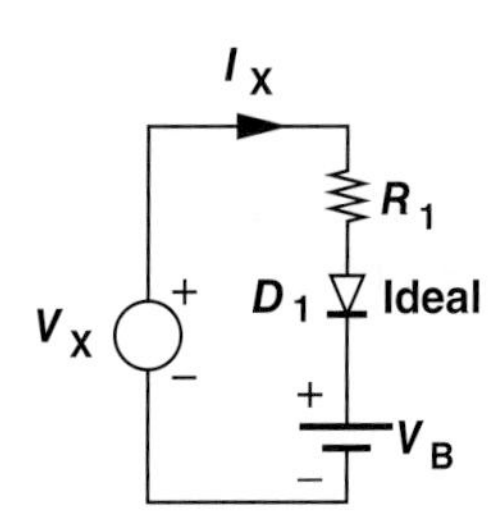

Figure 3.66

3.5. For the circuit depicted in Fig. 3.67, plot I_X as a function of V_X for two cases: $V_B = -1$ V and $V_B = +1$ V.

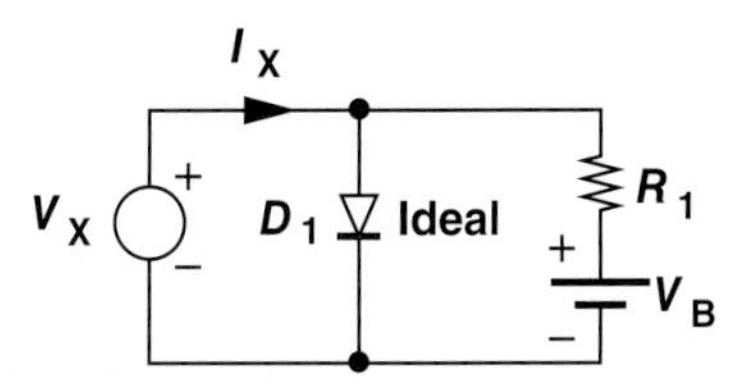

Figure 3.67

SOL **3.6.** Plot I_X and I_{D1} as a function of V_X for the circuit shown in Fig. 3.68. Assume $V_B > 0$.

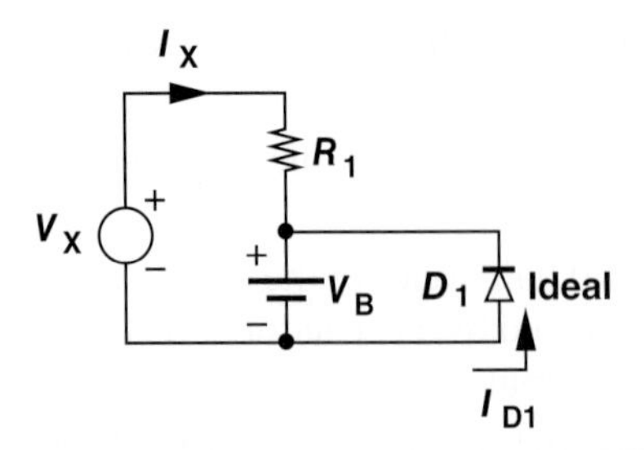

Figure 3.68

3.7. For the circuit depicted in Fig. 3.69, plot I_X and I_{R1} as a function of V_X for two cases: $V_B = -1$ V and $V_B = +1$ V.

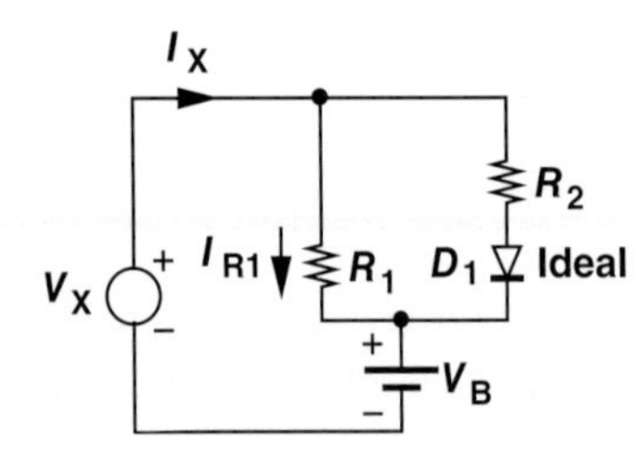

Figure 3.69

SOL **3.8.** In the circuit of Fig. 3.70, plot I_X and I_{R1} as a function of V_X for two cases: $V_B = -1$ V and $V_B = +1$ V.

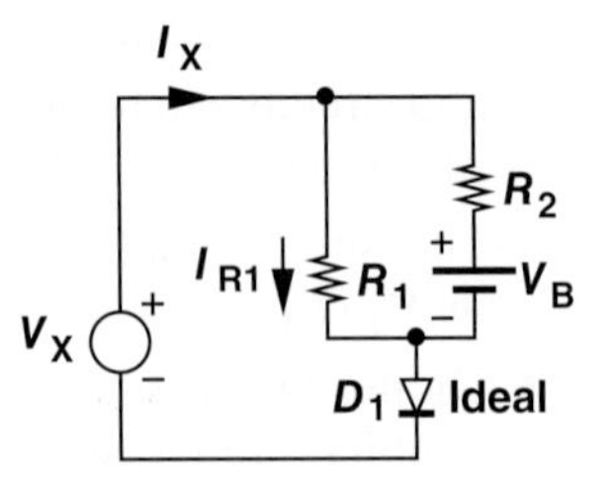

Figure 3.70

***3.9.** Plot the input/output characteristics of the circuits depicted in Fig. 3.71 using an ideal model for the diodes. Assume $V_B = 2$ V.

***3.10.** Repeat Problem 3.9 with a constant-voltage diode model.

***3.11.** If the input is given by $V_{in} = V_0 \cos \omega t$, plot the output of each circuit in Fig. 3.71 as a function of time. Assume an ideal diode model.

****3.12.** Plot the input/output characteristics of the circuits shown in Fig. 3.72 using an ideal model for the diodes.

****3.13.** Repeat Problem 3.12 with a constant-voltage diode model.

****3.14.** Assuming the input is expressed as $V_{in} = V_0 \cos \omega t$, plot the output of each circuit in Fig. 3.72 as a function of time. Use an ideal diode model.

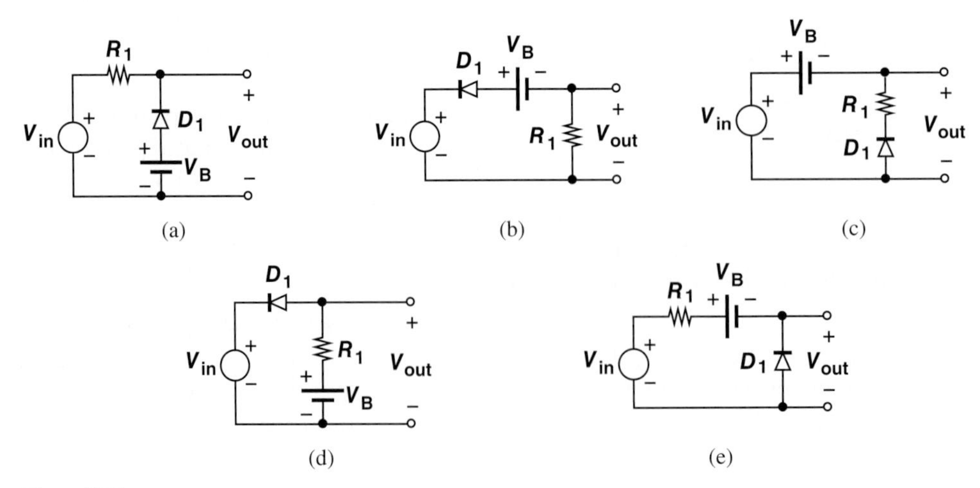

Figure 3.71

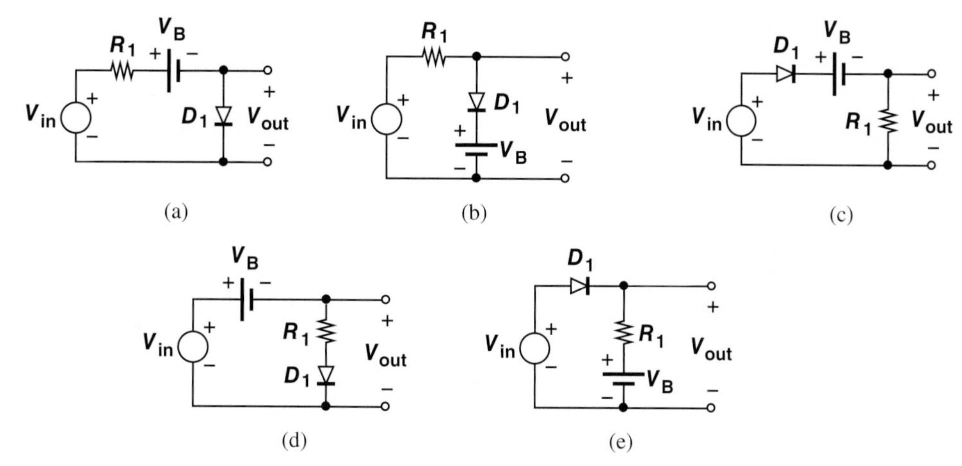

Figure 3.72

SOL **3.15. Assuming a constant-voltage diode model, plot V_{out} as a function of I_{in} for the circuits shown in Fig. 3.73.

**3.16. In the circuits of Fig. 3.73, plot the current flowing through R_1 as a function of V_{in}. Assume a constant-voltage diode model.

**3.17. For the circuits illustrated in Fig. 3.73, plot V_{out} as a function of time if $I_{in} = I_0 \cos \omega t$. Assume a constant-voltage model and a relatively large I_0.

**3.18. Plot V_{out} as a function of I_{in} for the circuits shown in Fig. 3.74. Assume a constant-voltage diode model.

*3.19. Plot the current flowing through R_1 in the circuits of Fig. 3.74 as a function of I_{in}. Assume a constant-voltage diode model.

*3.20. In the circuits depicted in Fig. 3.74, assume $I_{in} = I_0 \cos \omega t$, where I_0 is relatively large. Plot V_{out} as a function of time using a constant-voltage diode model. SOL

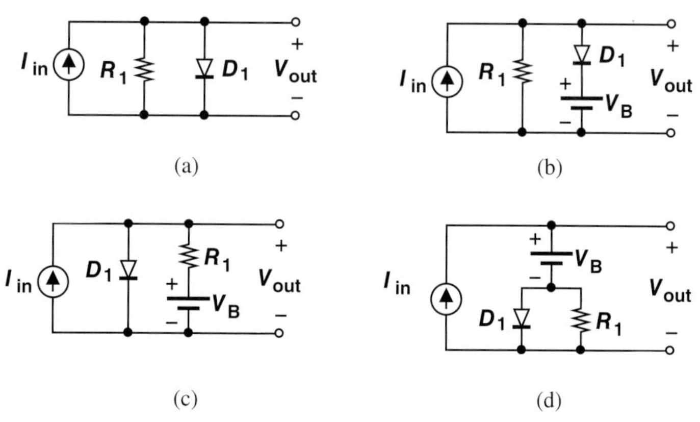

Figure 3.73

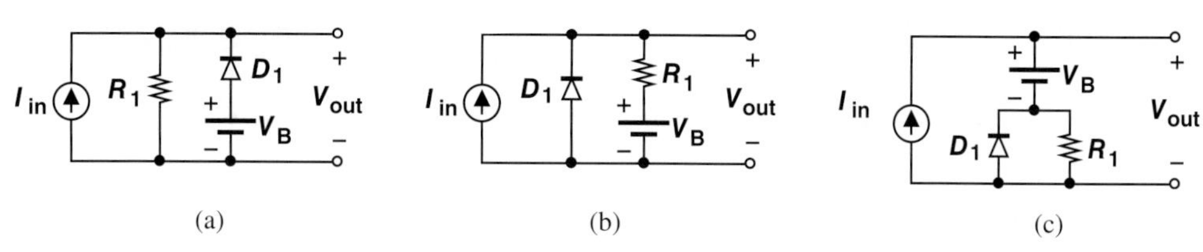

Figure 3.74

**3.21. For the circuits shown in Fig. 3.75, plot V_{out} as a function of I_{in} assuming a constant-voltage model for the diodes.

**3.22. Plot the current flowing through R_1 as a function of I_{in} for the circuits of Fig. 3.75. Assume a constant-voltage diode model.

*3.23. Plot the input/output characteristic of the circuits illustrated in Fig. 3.76 assuming a constant-voltage model.

SOL *3.24. Plot the currents flowing through R_1 and D_1 as a function of V_{in} for the circuits of Fig. 3.76. Assume a constant-voltage diode model.

**3.25. Plot the input/output characteristic of the circuits illustrated in Fig. 3.77 assuming a constant-voltage model.

**3.26. Plot the currents flowing through R_1 and D_1 as a function of V_{in} for the circuits of Fig. 3.77. Assume a constant-voltage diode model.

**3.27. Plot the input/output characteristic of the circuits illustrated in Fig. 3.78 assuming a constant-voltage model.

**3.28. Plot the input/output characteristic of the circuits illustrated in Fig. 3.79 assuming a constant-voltage model and $V_B = 2$ V.

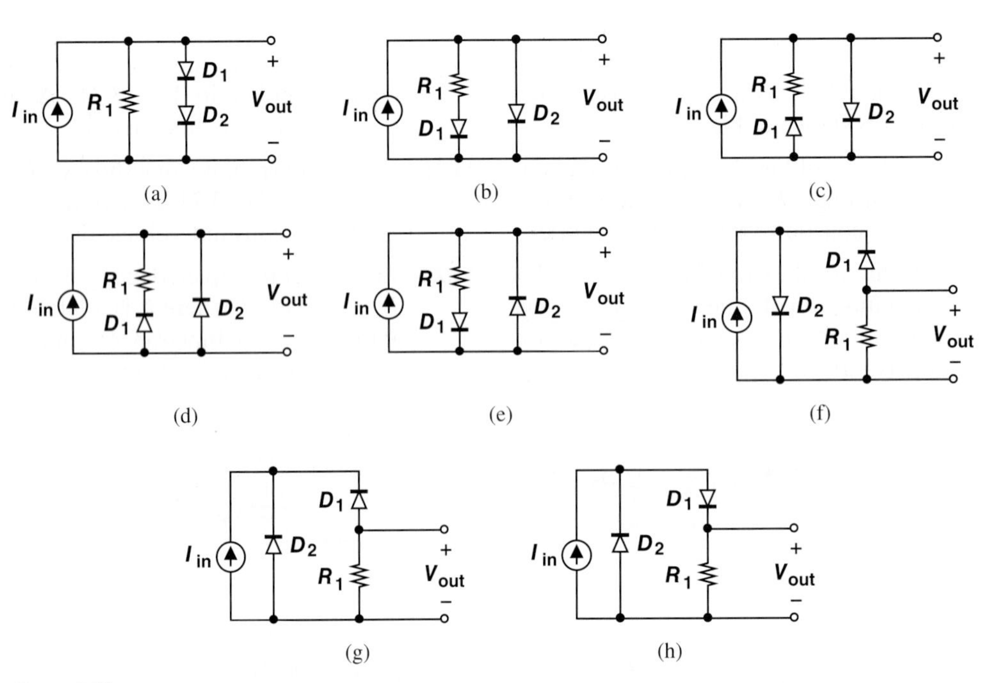

Figure 3.75

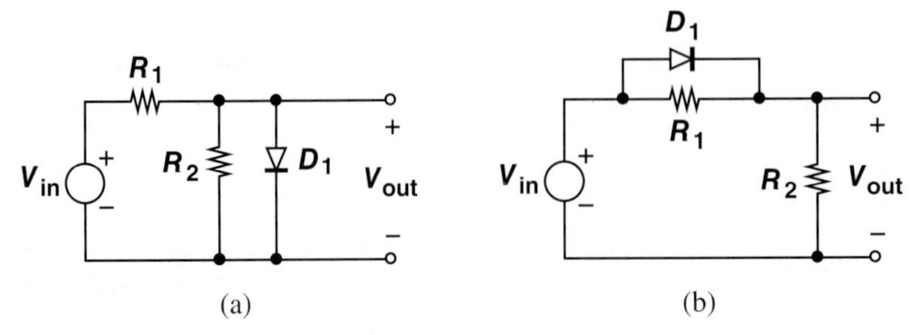

Figure 3.76

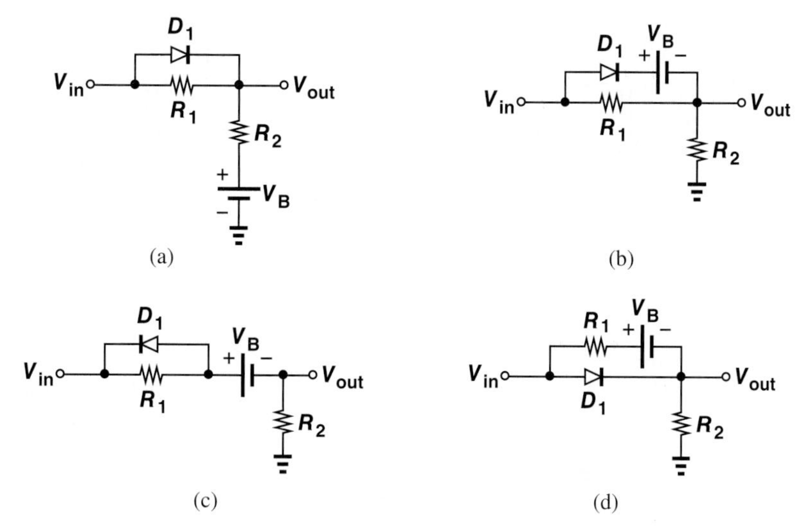

Figure 3.77

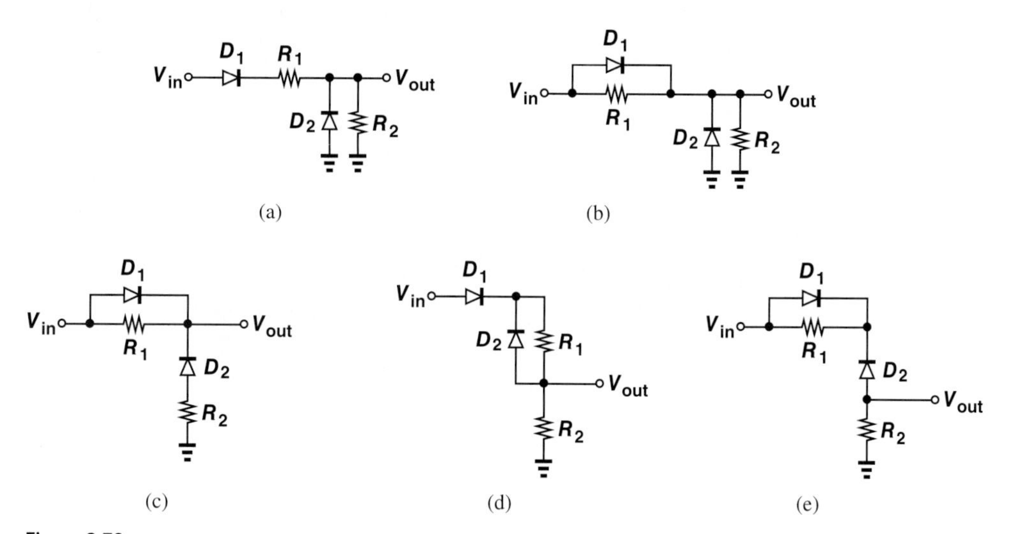

Figure 3.78

**3.29. Plot the currents flowing through R_1 and D_1 as a function of V_{in} for the circuits of Fig. 3.79. Assume a constant-voltage diode model and $V_B = 2$ V.

**3.30. Plot the currents flowing through R_1 and D_1 as a function of V_{in} for the circuits of Fig. 3.78. Assume constant-voltage diode model.

*3.31. Beginning with $V_{D,on} \approx 800$ mV for each diode, determine the change in V_{out} if V_{in} changes from +2.4 V to +2.5 V for the circuits shown in Fig. 3.80.

*3.32. Beginning with $V_{D,on} \approx 800$ mV for each diode, calculate the change in V_{out} if I_{in} changes from 3 mA to 3.1 mA in the circuits of Fig. 3.81.

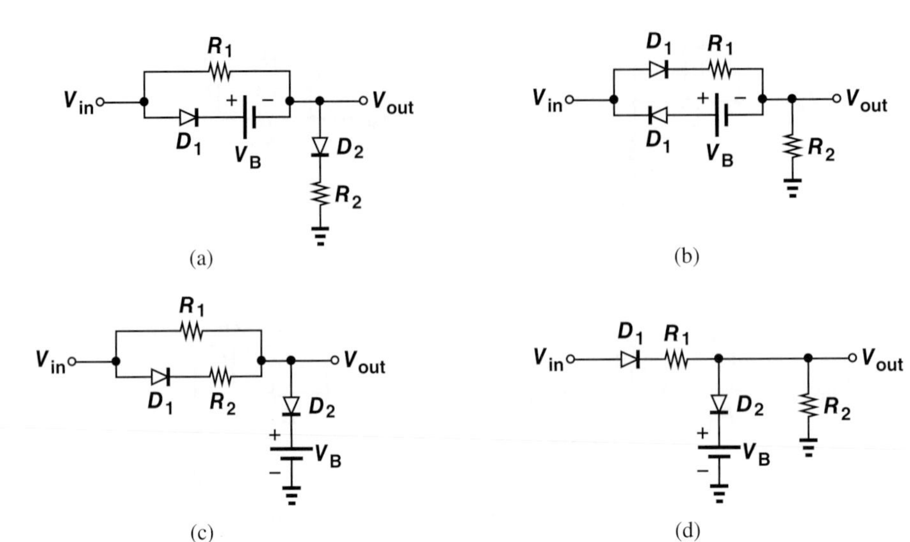

Figure 3.79

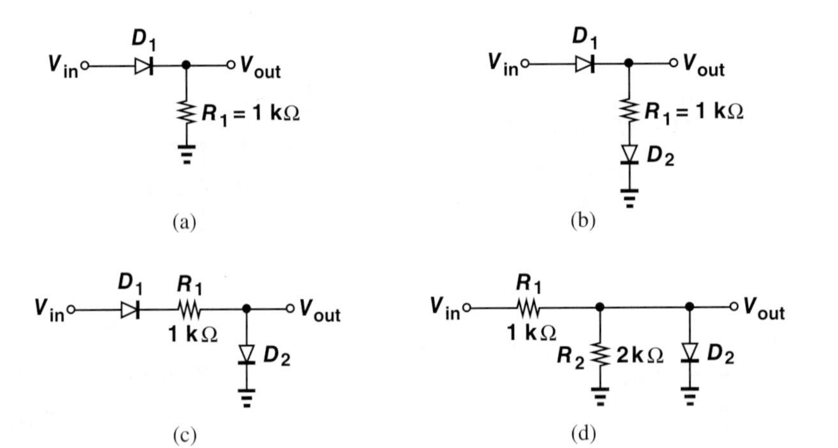

Figure 3.80

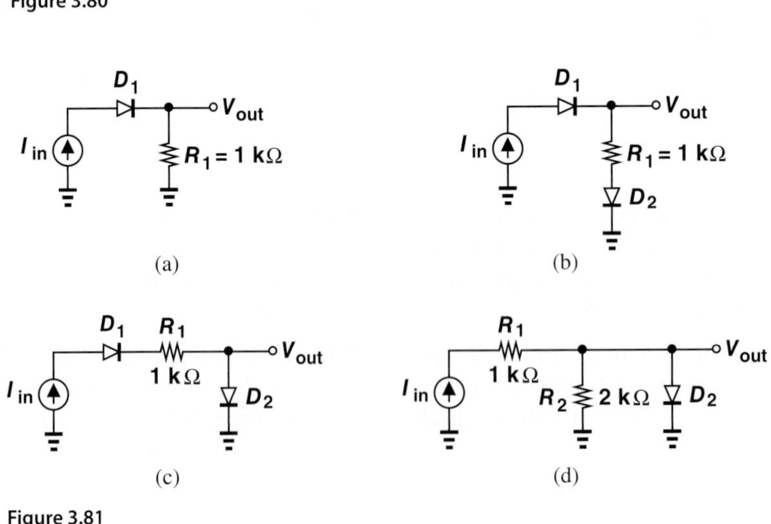

Figure 3.81

3.33. In Problem 3.32, determine the change in the current flowing through the 1-kΩ resistor in each circuit.

3.34. Assuming $V_{in} = V_p \sin \omega t$, plot the output waveform of the circuit depicted in Fig. 3.82 for an initial condition of +0.5 V across C_1. Assume $V_p = 5$ V.

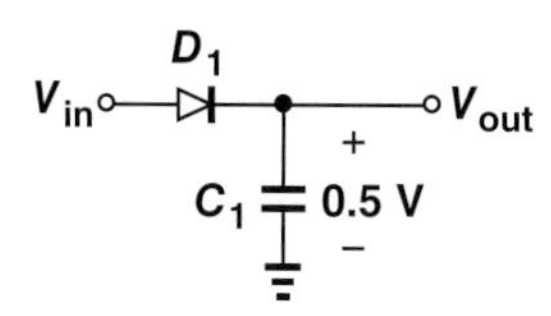

Figure 3.82

3.35. Repeat Problem 3.34 for the circuit shown in Fig. 3.83.

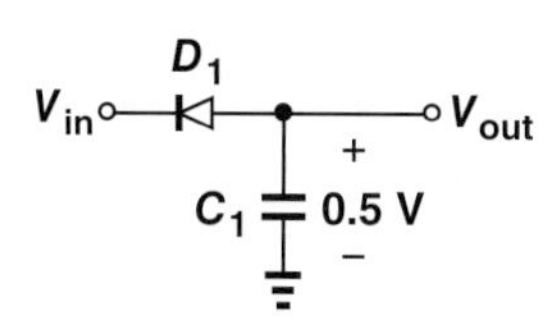

Figure 3.83

Section 3.5 Applications of Diodes

SOL **3.36.** Suppose the rectifier of Fig. 3.33 drives a 100-Ω load with a peak voltage of 3.5 V. For a 1000-μF smoothing capacitor, calculate the ripple amplitude if the frequency is 60 Hz.

3.37. A 3-V adaptor using a half-wave rectifier must supply a current of 0.5 A with a maximum ripple of 300 mV. For a frequency of 60 Hz, compute the minimum required smoothing capacitor.

3.38. Assume the input and output grounds in a full-wave rectifier are shorted together. Draw the output waveform with and without the load capacitor and explain why the circuit does not operate as a rectifier.

3.39. Plot the voltage across each diode in Fig. 3.39(b) as a function of time if $V_{in} = V_0 \cos \omega t$. Assume a constant-voltage diode model and $V_D > V_{D,on}$.

*__3.40.__ While constructing a full-wave rectifier, a student mistakenly has swapped the terminals of D_3 as depicted in Fig. 3.84. Explain what happens.

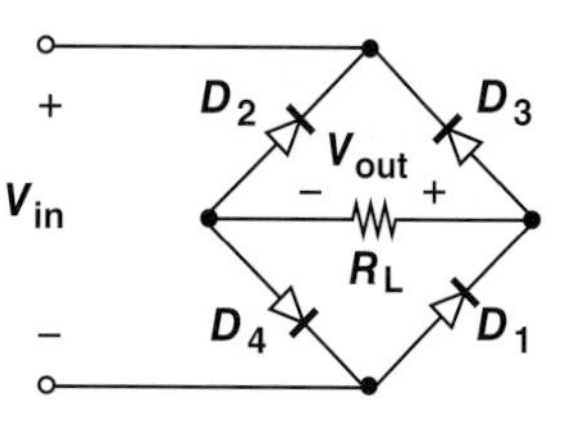

Figure 3.84

3.41. A full-wave rectifier is driven by a sinusoidal input $V_{in} = V_0 \cos \omega t$, where $V_0 = 3$ V and $\omega = 2\pi(60\text{ Hz})$. Assuming $V_{D,on} = 800$ mV, determine the ripple amplitude with a 1000-μF smoothing capacitor and a load resistance of 30 Ω. SOL

3.42. Suppose the negative terminals of V_{in} and V_{out} in Fig. 3.39(b) are shorted together. Plot the input-output characteristic assuming an ideal diode model and explaining why the circuit does not operate as a full-wave rectifier.

3.43. Suppose in Fig. 3.44, the diodes carry a current of 5 mA and the load, a current of 20 mA. If the load current increases to 21 mA, what is the change in the total voltage across the three diodes? Assume R_1 is much greater than $3r_d$.

3.44. In this problem, we estimate the ripple seen by the load in Fig. 3.44 so as to appreciate the regulation provided by the diodes. For simplicity, neglect the load. Also, $f_{in} = 60$ Hz, $C_1 = 100$ μF, $R_1 = 1000$ Ω, and the peak voltage produced by the transformer is equal to 5 V.

(a) Assuming R_1 carries a relatively constant current and $V_{D,on} \approx 800$ mV, estimate the ripple amplitude across C_1.

(b) Using the small-signal model of the diodes, determine the ripple amplitude across the load.

*__3.45.__ In the limiting circuit of Fig. 3.52, plot the currents flowing through D_1 and D_2 as a function of time if the input is given by $V_0 \cos \omega t$ and $V_0 > V_{D,on} + V_{B1}$ and $-V_0 > -V_{D,on} - V_{B2}$.

3.46. Design the limiting circuit of Fig. 3.52 for a negative threshold of -1.9 V and a positive threshold of $+2.2$ V. Assume the input peak voltage is equal to 5 V, the maximum allowable current through each diode is 2 mA, and $V_{D,on} \approx 800$ mV.

3.47. We wish to design a circuit that exhibits the input/output characteristic shown in Fig. 3.85. Using 1-kΩ resistors, ideal diodes, and other components, construct the circuit.

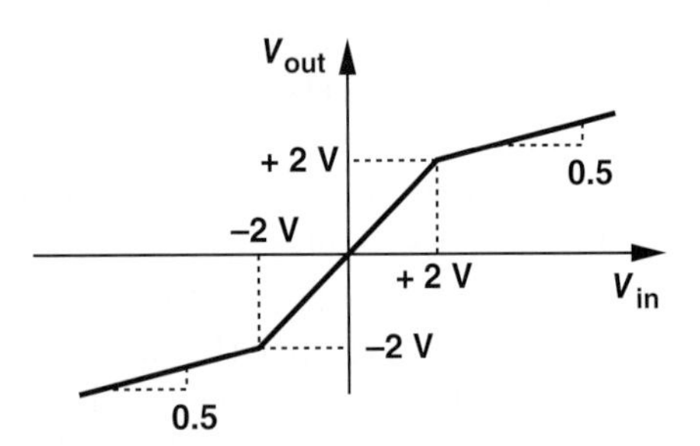

Figure 3.85

*3.48. "Wave-shaping" applications require the input/output characteristic illustrated in Fig. 3.86. Using ideal diodes and other components, construct a circuit that provides such a characteristic. (The value of resistors is not unique.)

**3.49. Suppose a triangular waveform is applied to the characteristic of Fig. 3.86 as shown in Fig. 3.87. Plot the output waveform and note that it is a rough approximation of a sinusoid. How should the input-output characteristic be modified so that the output becomes a better approximation of a sinusoid?

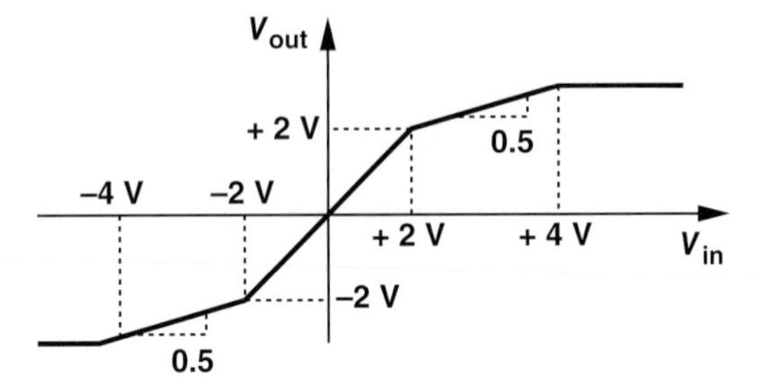

Figure 3.86

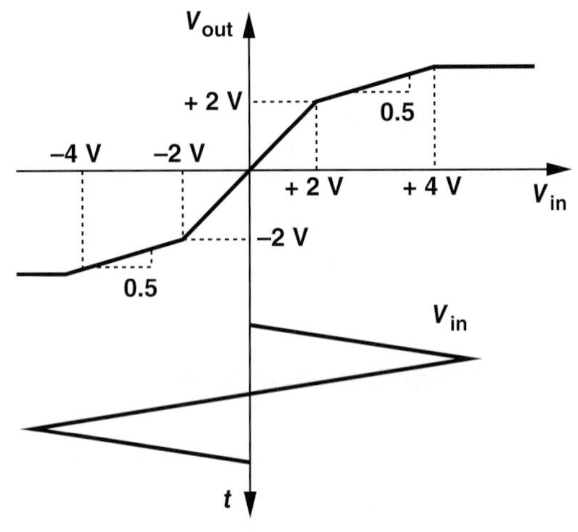

Figure 3.87

SPICE PROBLEMS

In the following problems, assume $I_S = 5 \times 10^{-16}$ A.

3.50. The half-wave rectifier of Fig. 3.88 must deliver a current of 5 mA to R_1 for a peak input level of 2 V.

(a) Using hand calculations, determine the required value of R_1.

(b) Verify the result by SPICE.

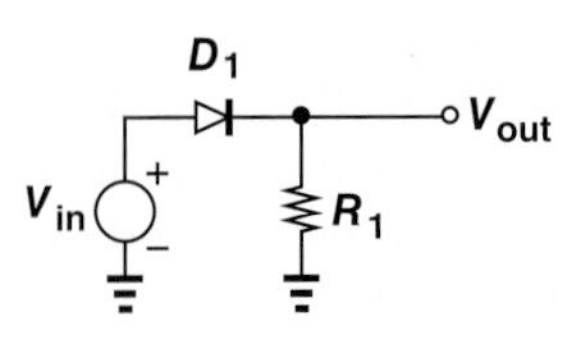

Figure 3.88

3.51. In the circuit of Fig. 3.89, $R_1 = 500\,\Omega$ and $R_2 = 1\,k\Omega$. Use SPICE to construct the input/output characteristic for $-2\text{ V} < V_{in} < +2\text{ V}$. Also, plot the current flowing through R_1 as a function of V_{in}.

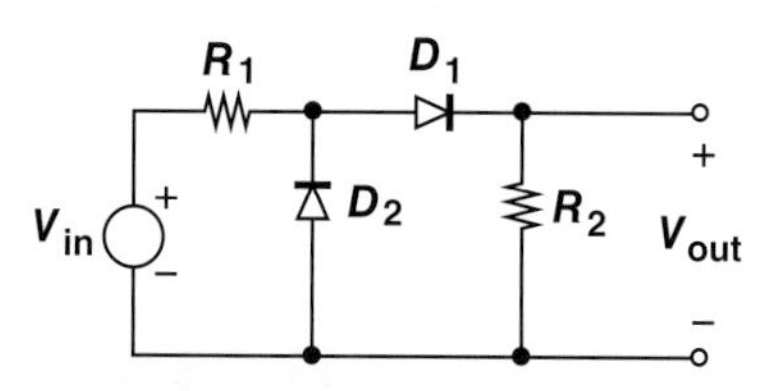

Figure 3.89

3.52. The rectifier shown in Fig. 3.90 is driven by a 60-Hz sinusoid input with a peak amplitude of 5 V. Using the transient analysis in SPICE,

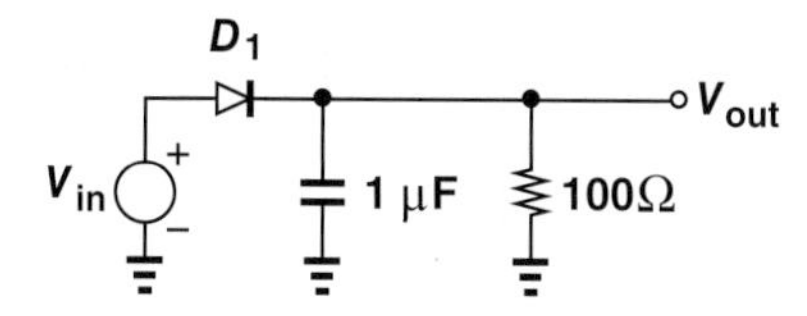

Figure 3.90

(a) Determine the peak-to-peak ripple at the output.
(b) Determine the peak current flowing through D_1.
(c) Compute the heaviest load (smallest R_L) that the circuit can drive while maintaining a ripple less than 200 mV_{pp}.

3.53. The circuit of Fig. 3.91 is used in some analog circuits. Plot the input/output characteristic for $-2\text{ V} < V_{in} < +2\text{ V}$ and determine the maximum input range across which $|V_{in} - V_{out}| < 5\text{ mV}$.

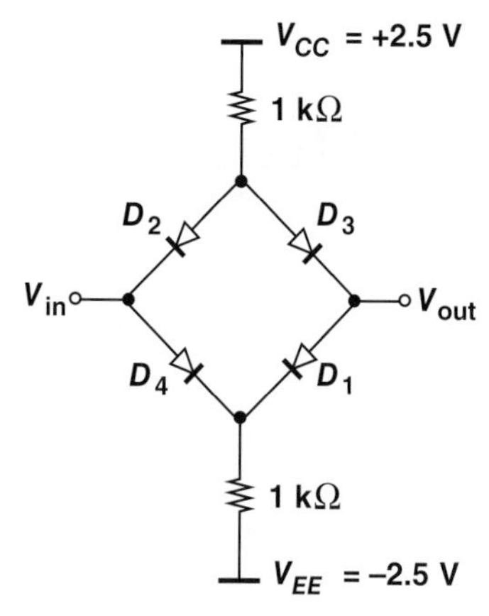

Figure 3.91

3.54. The circuit shown in Fig. 3.92 can provide an approximation of a sinusoid at the output in response to a triangular input waveform. Using the dc analysis in SPICE to plot the input/output characteristic for $0 < V_{in} < 4\text{ V}$, determine the values of V_{B1} and V_{B2} such that the characteristic closely resembles a sinusoid.

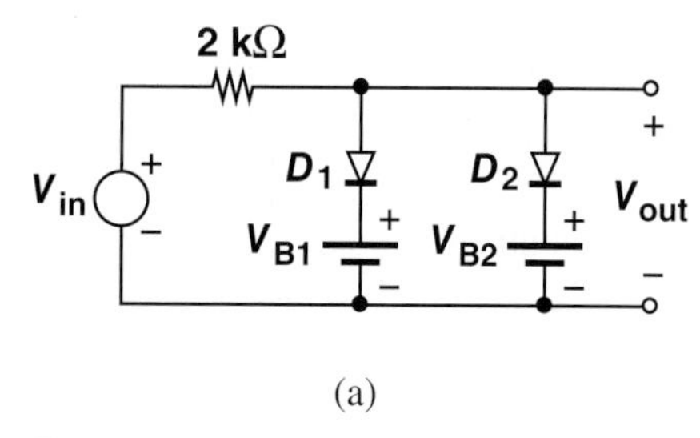

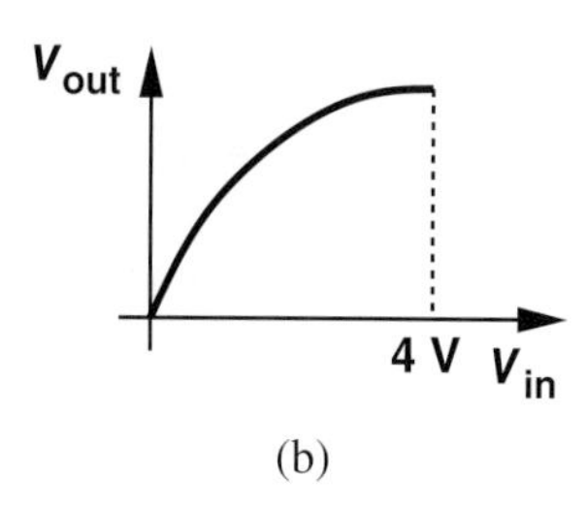

Figure 3.92

REFERENCE

1. H. Bharma, et al., "A subcutic millimeter wireless implantable intraocular pressure monitor microsystem," *IEEE Trans. Biomedical Circuits and Systems*, vol. 11, pp. 1204–1215, Dec. 2019.

4

Physics of Bipolar Transistors

Watch Companion YouTube Videos:

Razavi Electronics 1, Lec 13: Bipolar Transistor Structure & Operation
https://www.youtube.com/watch?v=j0rEXZYLksM

Razavi Electronics 1, Lec 14: Bipolar Transistor Characteristics, Introduction to Biasing
https://www.youtube.com/watch?v=_g6WdzfNvNU

Razavi Electronics 1, Lec 15: Transistor Biasing, Transconductance
https://www.youtube.com/watch?v=98dj7_rZvBA

Razavi Electronics 1, Lec 16: Large-Signal & Small-Signal Operation
https://www.youtube.com/watch?v=SueDJ13jLe8

Razavi Electronics 1, Lec 17: Bipolar Small-Signal Model, Early Effect
https://www.youtube.com/watch?v=XtRjv0EzSOQ

Razavi Electronics 1, Lec 18: PNP Transistor
https://www.youtube.com/watch?v=79ya97P-AkE

Razavi Electronics 1, Lec 19: PNP Transistor, Evolution of Amplifiers
https://www.youtube.com/watch?v=S4RgdPdBKY0

The bipolar transistor was invented in 1945 by Shockley, Brattain, and Bardeen at Bell Laboratories, subsequently replacing vacuum tubes in electronic systems and paving the way for integrated circuits.

In this chapter, we analyze the structure and operation of bipolar transistors, preparing ourselves for the study of circuits employing such devices. Following the same thought process as in Chapter 2 for *pn* junctions, we aim to understand the physics of the transistor, derive equations that represent its I/V characteristics, and develop an equivalent model that can be used in circuit analysis and design. The outline below illustrates the sequence of concepts introduced in this chapter.

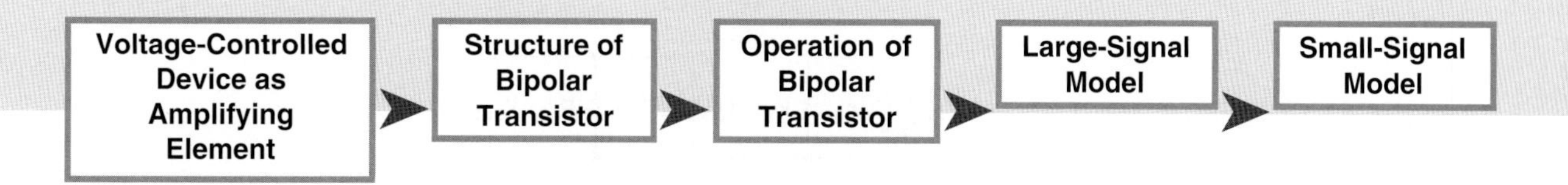

4.1 GENERAL CONSIDERATIONS

In its simplest form, the bipolar transistor can be viewed as a voltage-dependent current source. We first show how such a current source can form an amplifier and hence why bipolar devices are useful and interesting.

Consider the voltage-dependent current source depicted in Fig. 4.1(a), where I_1 is proportional to V_1: $I_1 = KV_1$. Note that K has a dimension of resistance^{-1}. For example, with $K = 0.001\ \Omega^{-1}$, an input voltage of 1 V yields an output current of 1 mA. Let us now construct the circuit shown in Fig. 4.1(b), where a voltage source V_{in} controls I_1 and the output current flows through a load resistor R_L, producing V_{out}. Our objective is to demonstrate that this circuit can operate as an amplifier, i.e., V_{out} is an amplified replica of V_{in}. Since $V_1 = V_{in}$ and $V_{out} = -R_L I_1$, we have

$$V_{out} = -KR_L V_{in}. \tag{4.1}$$

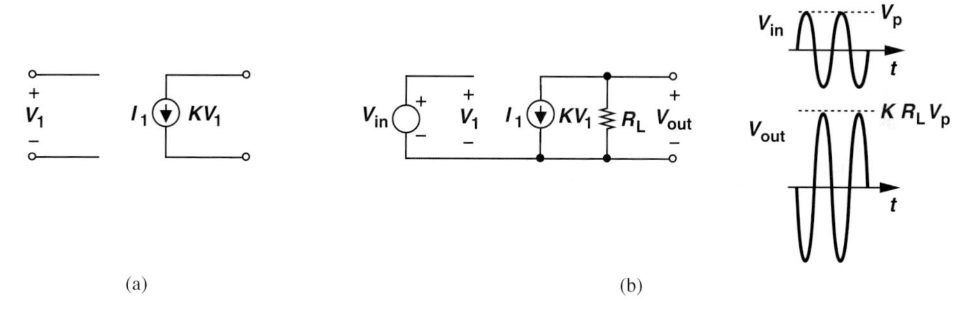

Figure 4.1 (a) Voltage-dependent current source, (b) simple amplifier.

Interestingly, if $KR_L > 1$, then the circuit amplifies the input. The negative sign indicates that the output is an "inverted" replica of the input circuit [Fig. 4.1(b)]. The amplification factor or "voltage gain" of the circuit, A_V, is defined as

$$A_V = \frac{V_{out}}{V_{in}} \tag{4.2}$$

$$= -KR_L, \tag{4.3}$$

and depends on both the characteristics of the controlled current source and the load resistor. Note that K signifies how strongly V_1 controls I_1, thus directly affecting the gain.

Example 4-1 Consider the circuit shown in Fig. 4.2, where the voltage-controlled current source exhibits an "internal" resistance of r_{in}. Determine the voltage gain of the circuit.

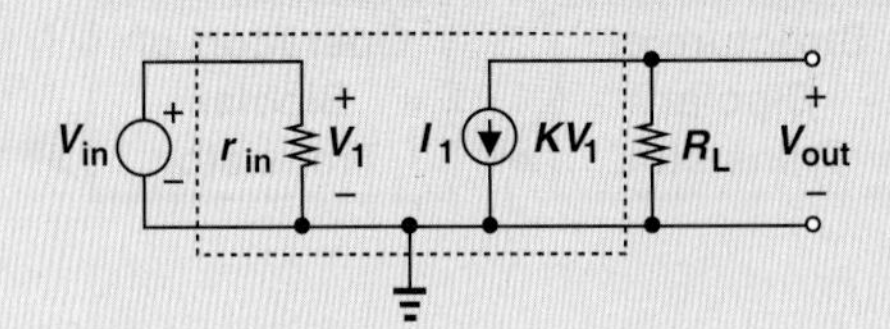

Figure 4.2 Voltage-dependent current source with an internal resistance r_{in}.

Solution Since V_1 is equal to V_{in} regardless of the value of r_{in}, the voltage gain remains unchanged. This point proves useful in our analyses later.

Exercise Repeat the above example if $r_{in} = \infty$.

The foregoing study reveals that a voltage-controlled current source can indeed provide signal amplification. Bipolar transistors are an example of such current sources and can ideally be modeled as shown in Fig. 4.3. Note that the device contains three terminals and its output current is an exponential function of V_1. We will see in Section 4.4.4 that under certain conditions, this model can be approximated by that in Fig. 4.1(a).

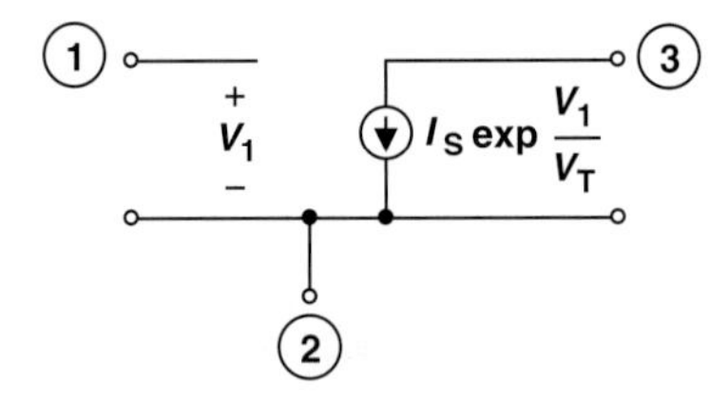

Figure 4.3 Exponential voltage-dependent current source.

As three-terminal devices, bipolar transistors make the analysis of circuits more difficult. Having dealt with two-terminal components such as resistors, capacitors, inductors, and diodes in elementary circuit analysis and the previous chapters of this book, we are accustomed to a one-to-one correspondence between the current through and the voltage across each device. With three-terminal elements, on the other hand, one may consider the current and voltage between every two terminals, arriving at a complex set of equations. Fortunately, as we develop our understanding of the transistor's operation, we discard some of these current and voltage combinations as irrelevant, thus obtaining a relatively simple model.

4.2 STRUCTURE OF BIPOLAR TRANSISTOR

The bipolar transistor consists of three doped regions forming a sandwich. Shown in Fig. 4.4(a) is an example comprising of a p layer sandwiched between two n regions and called an "*npn*" transistor. The three terminals are called the "base," the "emitter," and the "collector." As explained later, the emitter "emits" charge carriers and the collector "collects" them while the base controls the number of carriers that make this journey. The circuit symbol for the *npn* transistor is shown in Fig. 4.4(b). We denote the terminal voltages by V_E, V_B, and V_C, and the voltage differences by V_{BE}, V_{CB}, and V_{CE}. The transistor is labeled Q_1 here.

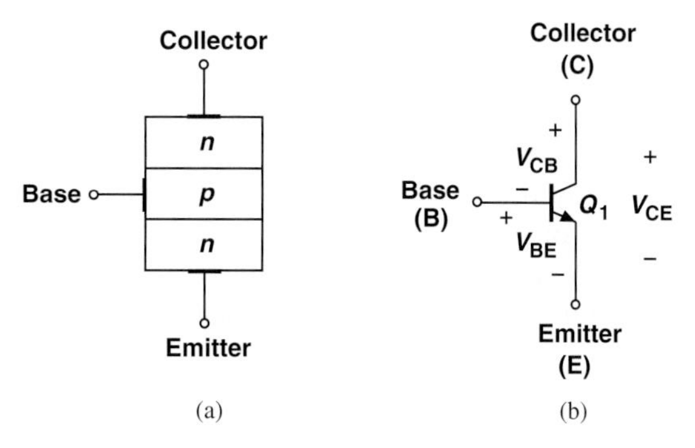

Figure 4.4 (a) Structure and (b) circuit symbol of bipolar transistor.

We readily note from Fig. 4.4(a) that the device contains two *pn* junction diodes: one between the base and the emitter and another between the base and the collector. For example, if the base is more positive than the emitter, $V_{BE} > 0$, then this junction is forward-biased. While this simple diagram may suggest that the device is symmetric with respect to the emitter and the collector, in reality, the dimensions and doping levels of these two regions are quite different. In other words, *E* and *C* cannot be interchanged. We will also see that proper operation requires a thin base region, e.g., about 100 Å in modern integrated bipolar transistors.

As mentioned in the previous section, the possible combinations of voltages and currents for a three-terminal device can prove overwhelming. For the device in Fig. 4.4(a), V_{BE}, V_{BC}, and V_{CE} can assume positive or negative values, leading to 2^3 possibilities for the terminal voltages of the transistor. Fortunately, only *one* of these eight combinations finds practical value and comes into our focus here.

Before continuing with the bipolar transistor, it is instructive to study an interesting effect in *pn* junctions. Consider the reverse-biased junction depicted in Fig. 4.5(a) and recall from Chapter 2 that the depletion region sustains a strong electric field. Now suppose an electron is somehow "injected" from outside into the right side of the depletion region. What happens to this electron? Serving as a minority carrier on the *p* side, the electron experiences the electric field and is rapidly swept away into the *n* side. The ability of a reverse-biased *pn* junction to efficiently "collect" externally-injected electrons proves essential to the operation of the bipolar transistor.

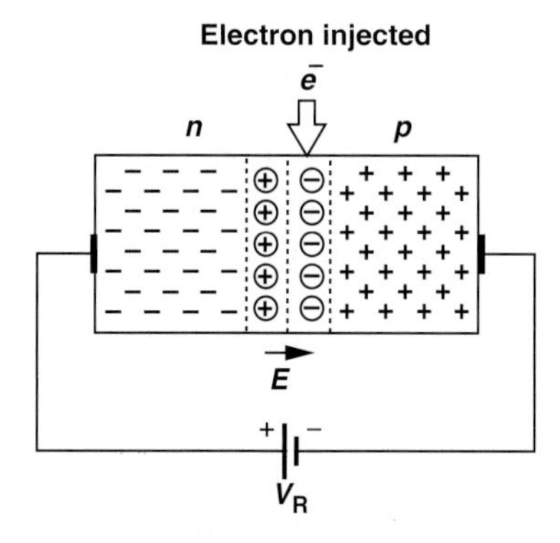

Figure 4.5 Injection of electrons into depletion region.

4.3 OPERATION OF BIPOLAR TRANSISTOR IN ACTIVE MODE

In this section, we analyze the operation of the transistor, aiming to prove that, under certain conditions, it indeed acts as a voltage-controlled current source. More specifically, we intend to show that (a) the current flow from the emitter to the collector can be viewed as a current source tied between these two terminals, and (b) this current is controlled by the voltage difference between the base and the emitter, V_{BE}.

We begin our study with the assumption that the base-emitter junction is forward-biased ($V_{BE} > 0$) and the base-collector junction is reverse-biased ($V_{BC} < 0$). Under these conditions, we say the device is biased in the "forward active region" or simply in the "active mode." For example, with the emitter connected to ground, the base voltage is set to about 0.8 V and the collector voltage to a *higher* value, e.g., 1 V [Fig. 4.6(a)]. The base-collector junction therefore experiences a reverse bias of 0.2 V.

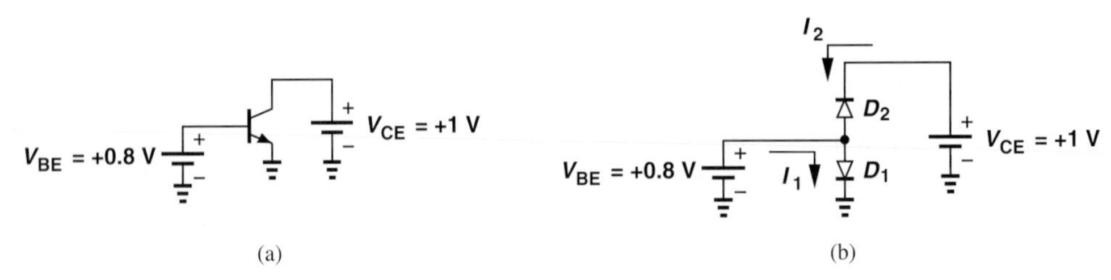

Figure 4.6 (a) Bipolar transistor with base and collector bias voltages, (b) simplistic view of bipolar transistor.

Let us now consider the operation of the transistor in the active mode. We may be tempted to simplify the example of Fig. 4.6(a) to the equivalent circuit shown in Fig. 4.6(b). After all, it appears that the bipolar transistor simply consists of two diodes sharing their anodes at the base terminal. This view implies that D_1 carries a current and D_2 does not; i.e., we should anticipate current flow from the base to the emitter but no current through the collector terminal. Were this true, the transistor would not operate as a voltage-controlled current source and would prove of little value.

To understand why the transistor cannot be modeled as merely two back-to-back diodes, we must examine the flow of charge inside the device, bearing in mind that the base region is very thin. Since the base-emitter junction is forward-biased, electrons flow from the emitter to the base and holes from the base to the emitter. For proper transistor operation, the former current component must be much greater than the latter, requiring that the emitter doping level be much greater than that of the base (Chapter 2). Thus, we denote the emitter region with n^+, where the superscript emphasizes the high doping level. Figure 4.7(a) summarizes our observations thus far, indicating that the emitter injects a large number of electrons into the base while receiving a small number of holes from it.

What happens to electrons as they enter the base? Since the base region is thin, most of the electrons reach the edge of the collector-base depletion region, beginning to experience the built-in electric field. Consequently, as illustrated in Fig. 4.5, the electrons are swept into the collector region (as in Fig. 4.5) and absorbed by the positive battery terminal. Figures 4.7(b) and (c) illustrate this effect in "slow motion." We therefore observe that the reverse-biased collector-base junction carries a current because minority carriers are "injected" into its depletion region.

Let us summarize our thoughts. In the active mode, an *npn* bipolar transistor carries a large number of electrons from the emitter, through the base, to the collector while drawing a small current of holes through the base terminal. We must now answer several questions. First, how do electrons travel through the base: by drift or diffusion? Second, how does the resulting current depend on the terminal voltages? Third, how large is the base current?

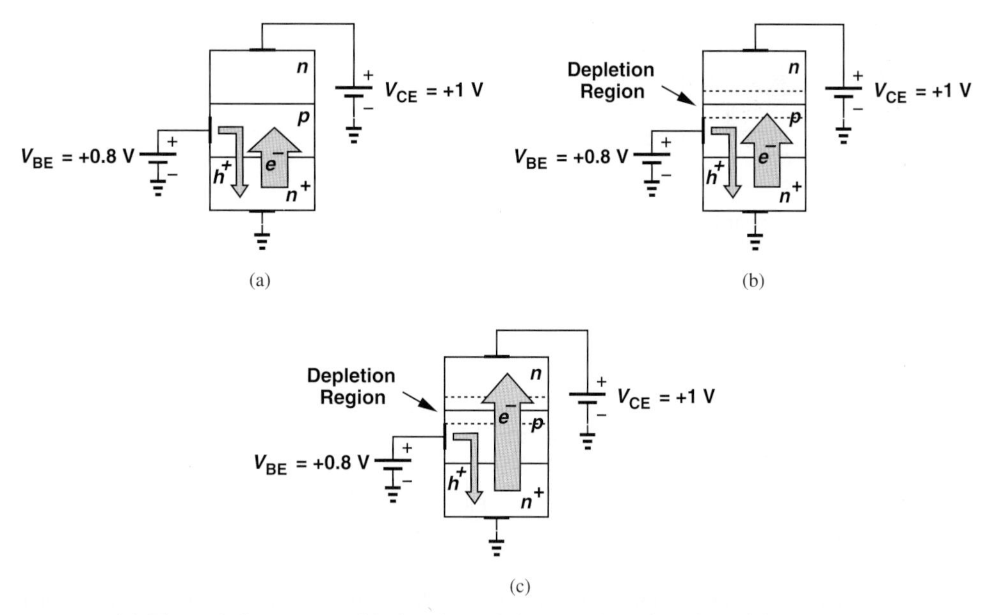

Figure 4.7 (a) Flow of electrons and holes through base-emitter junction, (b) electrons approaching collector junction, (c) electrons passing through collector junction.

Operating as a moderate conductor, the base region sustains but a small electric field, i.e., it allows most of the field to drop across the base-emitter depletion layer. Thus, as explained for *pn* junctions in Chapter 2, the drift current in the base is negligible,[1] leaving diffusion as the principal mechanism for the flow of electrons injected by the emitter. In fact, two observations directly lead to the necessity of diffusion: (1) redrawing the diagram of Fig. 2.29 for the emitter-base junction [Fig. 4.8(a)], we recognize that the density of electrons at $x = x_1$ is very high; (2) since any electron arriving at $x = x_2$ in Fig. 4.8(b) is swept away, the density of electrons falls to zero at this point. As a result, the electron density in the base assumes the profile depicted in Fig. 4.8(c), providing a gradient for the diffusion of electrons.

4.3.1 Collector Current

We now address the second question raised previously and compute the current flowing from the collector to the emitter.[2] As a forward-biased diode, the base-emitter junction exhibits a high concentration of electrons at $x = x_1$ in Fig. 4.8(c) given by Eq. (2.96):

$$\Delta n(x_1) = \frac{N_E}{\exp\dfrac{V_0}{V_T}}\left(\exp\frac{V_{BE}}{V_T} - 1\right) \tag{4.4}$$

$$= \frac{N_B}{n_i^2}\left(\exp\frac{V_{BE}}{V_T} - 1\right). \tag{4.5}$$

[1]This assumption simplifies the analysis here but may not hold in the general case.
[2]In an *npn* transistor, electrons go from the emitter to the collector. Thus, the conventional direction of the current is from the collector to the emitter.

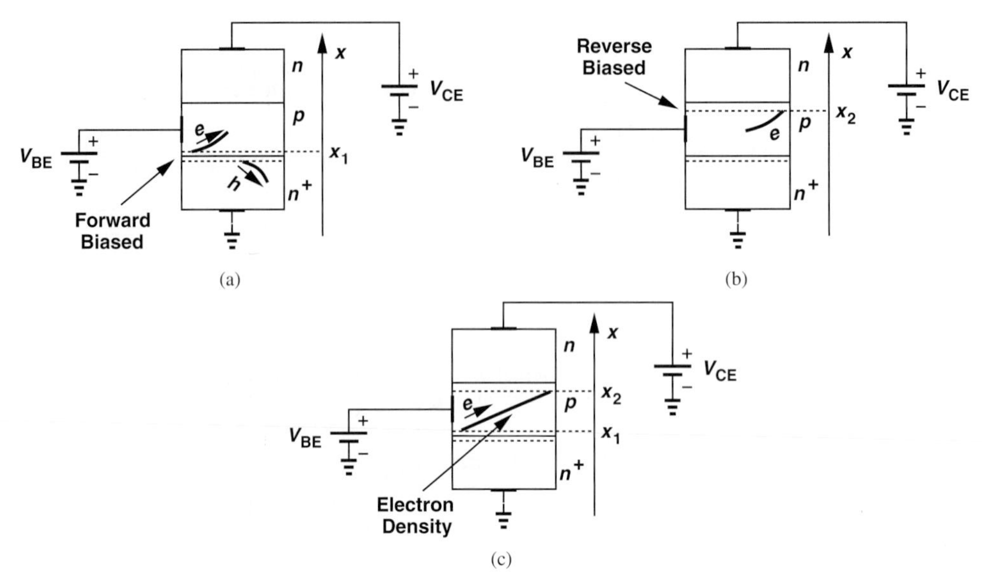

Figure 4.8 (a) Hole and electron profiles at base-emitter junction, (b) zero electron density near collector, (c) electron profile in base.

Here, N_E and N_B denote the doping levels in the emitter and the base, respectively, and we have utilized the relationship $\exp(V_0/V_T) = N_E N_B/n_i^2$. In this chapter, we assume $V_T = 26$ mV. Applying the law of diffusion [Eq. (2.42)], we determine the flow of electrons into the collector as

$$J_n = qD_n \frac{dn}{dx} \tag{4.6}$$

$$= qD_n \cdot \frac{0 - \Delta n(x_1)}{W_B}, \tag{4.7}$$

where W_B is the width of the base region. Multipling this quantity by the emitter cross section area, A_E, substituting for $\Delta n(x_1)$ from Eq. (4.5), and changing the sign to obtain the conventional current, we obtain

$$I_C = \frac{A_E q D_n n_i^2}{N_B W_B} \left(\exp \frac{V_{BE}}{V_T} - 1 \right). \tag{4.8}$$

In analogy with the diode current equation and assuming $\exp(V_{BE}/V_T) \gg 1$, we write

$$I_C = I_S \exp \frac{V_{BE}}{V_T}, \tag{4.9}$$

where

$$I_S = \frac{A_E q D_n n_i^2}{N_B W_B}. \tag{4.10}$$

Equation (4.9) implies that the bipolar transistor indeed operates as a voltage-controlled current source, proving a good candidate for amplification. We may alternatively say the transistor performs "voltage-to-current conversion."

Example 4-2 Determine the current I_X in Fig. 4.9(a) if Q_1 and Q_2 are identical and operate in the active mode and $V_1 = V_2$.

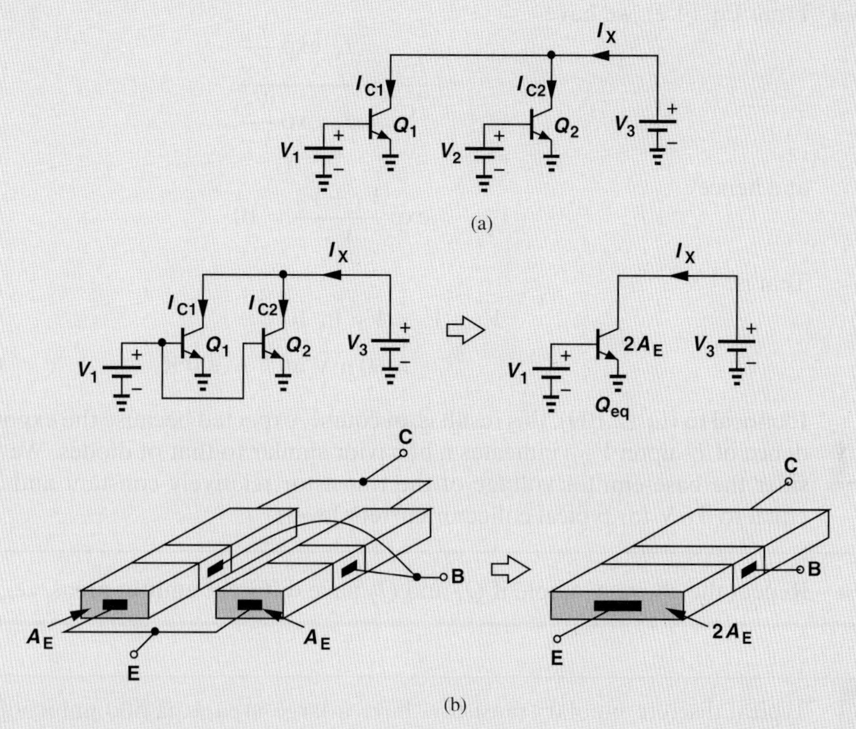

Figure 4.9 (a) Two identical transistors drawing current from V_3, (b) equivalence to a single transistor having twice the area.

Solution Since $I_X = I_{C1} + I_{C2}$, we have

$$I_X \approx 2\frac{A_E q D_n n_i^2}{N_B W_B} \exp \frac{V_1}{V_T}. \tag{4.11}$$

This result can also be viewed as the collector current of a *single* transistor having an emitter area of $2A_E$. In fact, redrawing the circuit as shown in Fig. 4.9(b) and noting that Q_1 and Q_2 experience identical voltages at their respective terminals, we say the two transistors are "in parallel," operating as a single transistor with twice the emitter area of each.

Exercise Repeat the above example if Q_1 has an emitter area of A_E and Q_2 an emitter area of $8A_E$.

Example 4-3 In the circuit of Fig. 4.9(a), Q_1 and Q_2 are identical and operate in the active mode. Determine $V_1 - V_2$ such that $I_{C1} = 10I_{C2}$.

Solution From Eq. (4.9), we have

$$\frac{I_{C1}}{I_{C2}} = \frac{I_S \exp \dfrac{V_1}{V_T}}{I_S \exp \dfrac{V_2}{V_T}}, \tag{4.12}$$

and hence

$$\exp \frac{V_1 - V_2}{V_T} = 10. \tag{4.13}$$

That is,

$$V_1 - V_2 = V_T \ln 10 \tag{4.14}$$

$$\approx 60 \text{ mV at } T = 300 \text{ K}. \tag{4.15}$$

Identical to Eq. (2.109), this result is, of course, expected because the exponential dependence of I_C upon V_{BE} indicates a behavior similar to that of diodes. We therefore consider the base-emitter voltage of the transistor relatively constant and approximately equal to 0.8 V for typical collector current levels.

Exercise Repeat the above example if Q_1 and Q_2 have different emitter areas, i.e., $A_{E1} = nA_{E2}$.

Example 4-4 Typical discrete bipolar transistors have a large area, e.g., 500 μm × 500 μm, whereas modern integrated devices may have an area as small as 0.5 μm × 0.2 μm. Assuming other device parameters are identical, determine the difference between the base-emitter voltage of two such transistors for equal collector currents.

Solution From Eq. (4.9), we have $V_{BE} = V_T \ln(I_C/I_S)$ and hence

$$V_{BEint} - V_{BEdis} = V_T \ln \frac{I_{S1}}{I_{S2}}, \tag{4.16}$$

where $V_{BEint} = V_T \ln(I_{C2}/I_{S2})$ and $V_{BEdis} = V_T \ln(I_{C1}/I_{S1})$ denote the base-emitter voltages of the integrated and discrete devices, respectively. Since $I_S \propto A_E$,

$$V_{BEint} - V_{BEdis} = V_T \ln \frac{A_{E2}}{A_{E1}}. \tag{4.17}$$

For this example, $A_{E2}/A_{E1} = 2.5 \times 10^6$, yielding

$$V_{BEint} - V_{BEdis} = 383 \text{ mV}. \tag{4.18}$$

In practice, however, $V_{BEint} - V_{BEdis}$ falls in the range of 100 to 150 mV because of differences in the base width and other parameters. The key point here is that $V_{BE} = 800$ mV is a reasonable approximation for integrated transistors and should be lowered to about 700 mV for discrete devices.

Exercise Repeat the above comparison for a very small integrated device with an emitter area of 0.15 μm × 0.15 μm.

Since many applications deal with *voltage* quantities, the collector current generated by a bipolar transistor typically flows through a resistor to produce a voltage.

Example 4-5 Determine the output voltage in Fig. 4.10 if $I_S = 5 \times 10^{-16}$ A.

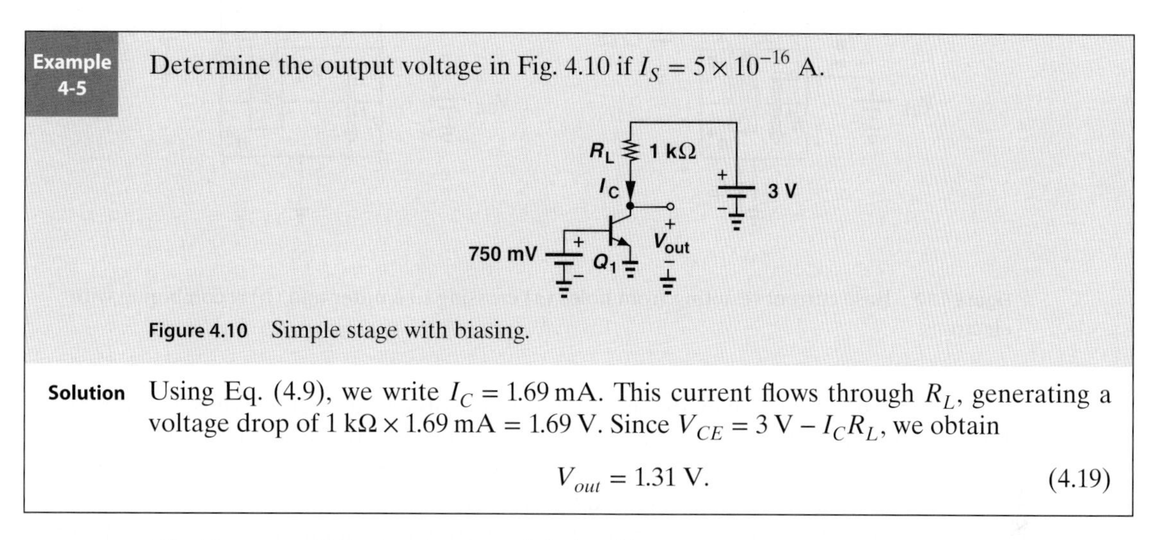

Figure 4.10 Simple stage with biasing.

Solution Using Eq. (4.9), we write $I_C = 1.69$ mA. This current flows through R_L, generating a voltage drop of 1 kΩ × 1.69 mA = 1.69 V. Since $V_{CE} = 3\text{ V} - I_C R_L$, we obtain

$$V_{out} = 1.31 \text{ V}. \tag{4.19}$$

Exercise What happens if the load resistor is halved?

Equation (4.9) reveals an interesting property of the bipolar transistor: the collector current does not depend on the collector voltage (so long as the device remains in the active mode). Thus, for a fixed base-emitter voltage, the device draws a constant current, acting as a current source [Fig. 4.11(a)]. Plotted in Fig. 4.11(b) is the current as a function of the collector-emitter voltage, exhibiting a constant value for $V_{CE} > V_1$.[3] Constant current sources find application in many electronic circuits and we will see numerous examples of their usage in this book. In Section 4.5, we study the behavior of the transistor for $V_{CE} < V_{BE}$.

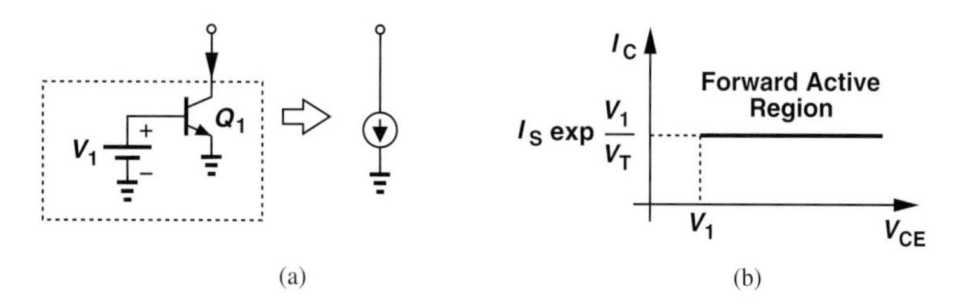

Figure 4.11 (a) Bipolar transistor as a current source, (b) I/V characteristic.

4.3.2 Base and Emitter Currents

Having determined the collector current, we now turn our attention to the base and emitter currents and their dependence on the terminal voltages. Since the bipolar transistor must satisfy Kirchoff's current law, calculation of the base current readily yields the emitter current as well.

In the *npn* transistor of Fig. 4.12(a), the base current, I_B, results from the flow of holes. Recall from Eq. (2.99) that the hole and electron currents in a forward-biased *pn* junction

[3]Recall that $V_{CE} > V_1$ is necessary to ensure the collector-base junction remains reverse biased.

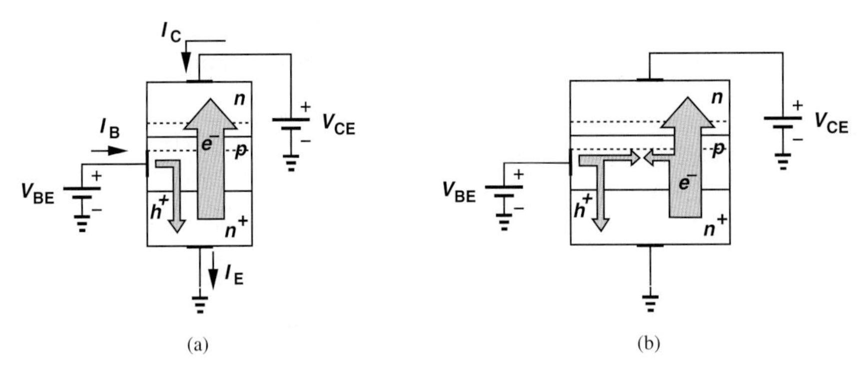

Figure 4.12 Base current resulting from holes (a) crossing to emitter and (b) recombining with electrons.

bear a *constant* ratio given by the doping levels and other parameters. Thus, the number of holes entering from the base to the emitter is a constant fraction of the number of electrons traveling from the emitter to the base. As an example, for every 200 electrons injected by the emitter, one hole must be supplied by the base.

In practice, the base current contains an additional component of holes. As the electrons injected by the emitter travel through the base, some may "recombine" with the holes [Fig. 4.12(b)]; inessence, some electrons and holes are "wasted" as a result of recombination. For example, on the average, out of every 200 electrons injected by the emitter, one recombines with a hole.

In summary, the base current must supply holes for both reverse injection into the emitter and recombination with the electrons traveling toward the collector. We can therefore view I_B as a constant fraction of I_E or a constant fraction of I_C. It is common to write

$$I_C = \beta I_B, \tag{4.20}$$

where β is called the "current gain" of the transistor because it shows how much the base current is "amplified." Depending on the device structure, the β of *npn* transistors typically ranges from 50 to 200.

In order to determine the emitter current, we apply the KCL to the transistor with the current directions depicted in Fig. 4.12(a):

$$I_E = I_C + I_B \tag{4.21}$$

$$= I_C\left(1 + \frac{1}{\beta}\right). \tag{4.22}$$

We can summarize our findings as follows:

$$I_C = I_S \exp\frac{V_{BE}}{V_T} \tag{4.23}$$

$$I_B = \frac{1}{\beta} I_S \exp\frac{V_{BE}}{V_T} \tag{4.24}$$

$$I_E = \frac{\beta + 1}{\beta} I_S \exp\frac{V_{BE}}{V_T}. \tag{4.25}$$

It is sometimes useful to write $I_C = [\beta/(\beta + 1)]I_E$ and denote $\beta/(\beta + 1)$ by α. For $\beta = 100$, $\alpha = 0.99$, suggesting that $\alpha \approx 1$ and $I_C \approx I_E$ are reasonable approximations. In this book, we assume that the collector and emitter currents are approximately equal.

Example 4-6

A bipolar transistor having $I_S = 5 \times 10^{-16}$ A is biased in the forward active region with $V_{BE} = 750$ mV. If the current gain varies from 50 to 200 due to manufacturing variations, calculate the minimum and maximum terminal currents of the device.

Solution For a given V_{BE}, the collector current remains independent of β:

$$I_C = I_S \exp \frac{V_{BE}}{V_T} \tag{4.26}$$

$$= 1.685 \text{ mA}. \tag{4.27}$$

The base current varies from $I_C/200$ to $I_C/50$:

$$8.43 \text{ µA} < I_B < 33.7 \text{ µA}. \tag{4.28}$$

On the other hand, the emitter current experiences only a small variation because $(\beta + 1)/\beta$ is near unity for large β:

$$1.005 I_C < I_E < 1.02 I_C \tag{4.29}$$

$$1.693 \text{ mA} < I_E < 1.719 \text{ mA}. \tag{4.30}$$

Exercise Repeat the above example if the area of the transistor is doubled.

4.4 BIPOLAR TRANSISTOR MODELS AND CHARACTERISTICS

4.4.1 Large-Signal Model

With our understanding of the transistor operation in the forward active region and the derivation of Eqs. (4.23)–(4.25), we can now construct a model that proves useful in the analysis and design of circuits—in a manner similar to the developments in Chapter 2 for the *pn* junction.

Did you know?

The first bipolar transistor introduced by Bell Labs in 1948 was implemented in germanium rather than silicon, had a base thickness of about 30 µm, and provided a maximum operation frequency (called the "transit frequency") of 10 MHz. By contrast, bipolar transistors realized in today's silicon integrated circuits have a base thickness less than 0.01 µm and a transit frequency of several hundred gigahertz.

Since the base-emitter junction is forward-biased in the active mode, we can place a diode between the base and emitter terminals. Moreover, since the current drawn from the collector and flowing into the emitter depends on only the base-emitter voltage, we add a voltage-controlled current source between the collector and the emitter, arriving at the model shown in Fig. 4.13. As illustrated in Fig. 4.11, this current remains independent of the collector-emitter voltage.

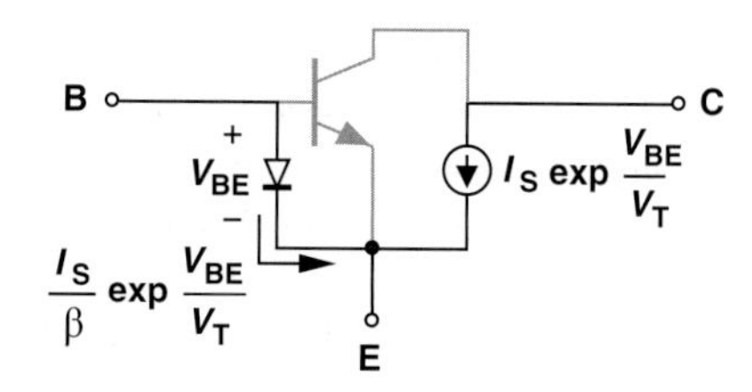

Figure 4.13 Large-signal model of bipolar transistor in active region.

But how do we ensure that the current flowing through the diode is equal to $1/\beta$ times the collector current? Equation (4.24) suggests that the base current is equal to that of a diode having a reverse saturation current of I_S/β. Thus, the base-emitter junction is modeled by a diode whose cross section area is $1/\beta$ times that of the actual emitter area.

With the interdependencies of currents and voltages in a bipolar transistor, the reader may wonder about the cause and effect relationships. We view the chain of dependencies as $V_{BE} \rightarrow I_C \rightarrow I_B \rightarrow I_E$; i.e., the base-emitter voltage generates a collector current, which requires a proportional base current, and the sum of the two flows through the emitter.

Example 4-7

Consider the circuit shown in Fig. 4.14(a), where $I_{S,Q1} = 5 \times 10^{-17}$ A and $V_{BE} = 800$ mV. Assume $\beta = 100$. (a) Determine the transistor terminal currents and voltages and verify that the device indeed operates in the active mode. (b) Determine the maximum value of R_C that permits operation in the active mode.

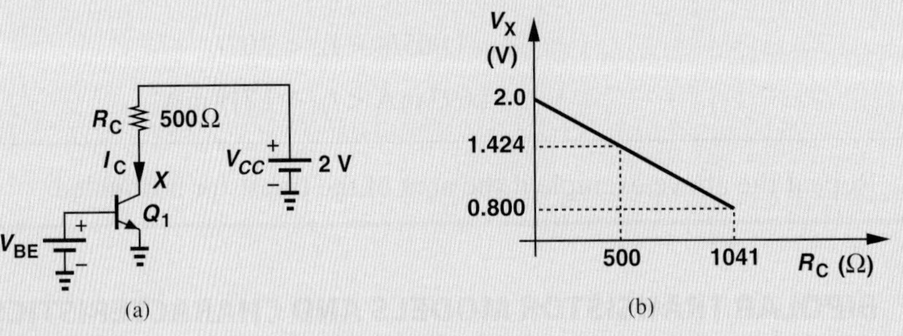

Figure 4.14 (a) Simple stage with biasing, (b) variation of collector voltage as a function of collector resistance.

Solution (a) Using Eqs. (4.23)–(4.25), we have

$$I_C = 1.153 \text{ mA} \tag{4.31}$$

$$I_B = 11.53\ \mu\text{A} \tag{4.32}$$

$$I_E = 1.165 \text{ mA}. \tag{4.33}$$

The base and emitter voltages are equal to +800 mV and zero, respectively. We must now calculate the collector voltage, V_X. Writing a KVL from the 2-V power supply and across R_C and Q_1, we obtain

$$V_{CC} = R_C I_C + V_X. \tag{4.34}$$

That is,

$$V_X = 1.424 \text{ V}. \tag{4.35}$$

Since the collector voltage is more positive than the base voltage, this junction is reverse-biased and the transistor operates in the active mode.

(b) What happens to the circuit as R_C increases? Since the voltage drop across the resistor, $R_C I_C$, increases while V_{CC} is constant, the voltage at node X *drops*. The device approaches the "edge" of the forward active region if the base-collector voltage falls to zero, i.e., as $V_X \rightarrow +800$ mV. Rewriting Eq. (4.33) yields:

$$R_C = \frac{V_{CC} - V_X}{I_C}, \tag{4.36}$$

which, for $V_X = +800\,\text{mV}$, reduces to

$$R_C = 1041\,\Omega. \tag{4.37}$$

Figure 4.14(b) plots V_X as a function of R_C.

This example implies that there exists a maximum allowable value of the collector resistance, R_C, in the circuit of Fig. 4.14(a). As we will see in Chapter 5, this limits the voltage gain that the circuit can provide.

Exercise In the above example, what is the minimum allowable value of V_{CC} for transistor operation in the active mode? Assume $R_C = 500\,\Omega$.

The reader may wonder why the equivalent circuit of Fig. 4.13 is called the "large-signal model." After all, the above example apparently contains *no* signals! This terminology emphasizes that the model can be used for *arbitrarily* large voltage and current changes in the transistor (so long as the device operates in the active mode). For example, if the base-emitter voltage varies from 800 to 300 mV, and hence the collector current by many *orders of magnitude*,[4] the model still applies. This is in contrast to the small-signal model, studied in Section 4.4.4.

4.4.2 I/V Characteristics

The large-signal model naturally leads to the I/V characteristics of the transistor. With three terminal currents and voltages, we may envision plotting different currents as a function of the potential difference between every two terminals—an elaborate task. However, as explained below, only a few of such characteristics prove useful.

The first characteristic to study is, of course, the exponential relationship inherent in the device. Figure 4.15(a) plots I_C versus V_{BE} with the assumption that the collector voltage is constant and no lower than the base voltage. As shown in Fig. 4.11, I_C is independent of V_{CE}; thus, different values of V_{CE} do not alter the characteristic.

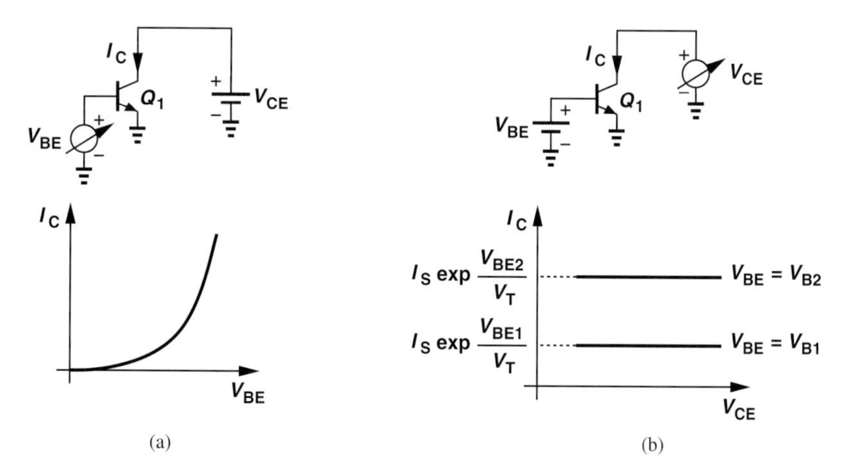

Figure 4.15 Collector current as a function of (a) base-emitter voltage and (b) collector-emitter voltage.

[4]A 500-mV change in V_{BE} leads to $500\,\text{mV}/60\,\text{mV} = 8.3$ decades of change in I_C.

Next, we examine I_C for a given V_{BE} but with V_{CE} varying. Illustrated in Fig. 4.15(b), the characteristic is a horizontal line because I_C is constant if the device remains in the active mode ($V_{CE} > V_{BE}$). On the other hand, if different values are chosen for V_{BE}, the characteristic moves up or down.

The two plots of Fig. 4.15 constitute the principal characteristics of interest in most analysis and design tasks. Equations (4.24) and (4.25) suggest that the base and emitter currents follow the same behavior.

Example 4-8 For a bipolar transistor, $I_S = 5 \times 10^{-17}$ A and $\beta = 100$. Construct the I_C-V_{BE}, I_C-V_{CE}, I_B-V_{BE}, and I_B-V_{CE} characteristics.

Solution We determine a few points along the I_C-V_{BE} characteristics, e.g.,

$$V_{BE1} = 700\,\text{mV} \Rightarrow I_{C1} = 24.6\,\mu\text{A} \tag{4.38}$$

$$V_{BE2} = 750\,\text{mV} \Rightarrow I_{C2} = 169\,\mu\text{A} \tag{4.39}$$

$$V_{BE3} = 800\,\text{mV} \Rightarrow I_{C3} = 1.153\,\text{mA}. \tag{4.40}$$

The characteristic is depicted in Fig. 4.16(a).

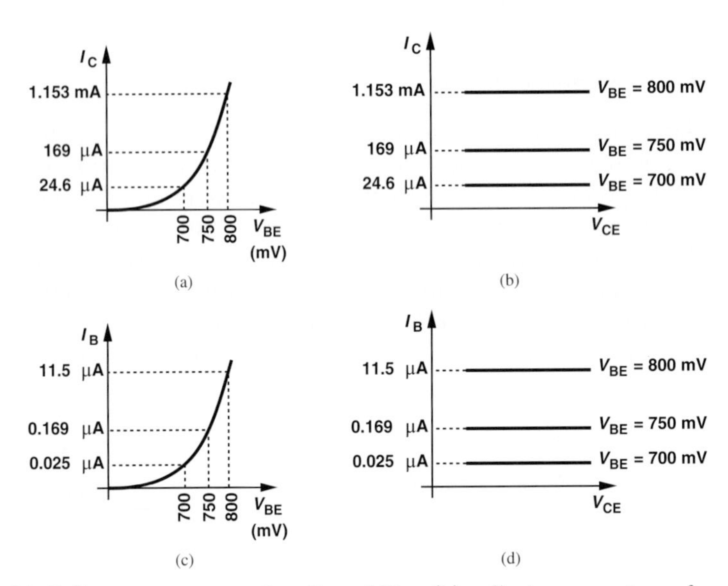

Figure 4.16 (a) Collector current as a function of V_{BE}, (b) collector current as a function of V_{CE}, (c) base current as a function of V_{BE}, (d) base current as a function of V_{CE}.

Using the values obtained above, we can also plot the I_C-V_{CE} characteristic as shown in Fig. 4.16(b), concluding that the transistor operates as a constant current source of, e.g., 169 μA if its base-emitter voltage is held at 750 mV. We also remark that, for equal increments in V_{BE}, I_C jumps by increasingly greater steps: 24.6 μA to 169 μA to 1.153 mA. We return to this property in Section 4.4.3.

For I_B characteristics, we simply divide the I_C values by 100 [Figs. 4.16(c) and (d)].

Exercise What change in V_{BE} doubles the base current?

The reader may wonder what exactly we learn from the I/V characteristics. After all, compared to Eqs. (4.23)–(4.25), the plots impart no additional information. However, as we will see throughout this book, the visualization of equations by means of such plots greatly enhances our understanding of the devices and the circuits employing them.

4.4.3 Concept of Transconductance

Our study thus far shows that the bipolar transistor acts as a voltage-dependent current source (when operating in the forward active region). An important question that arises here is, how is the *performance* of such a device quantified? In other words, what is the measure of the "goodness" of a voltage-dependent current source?

The example depicted in Fig. 4.1 suggests that the device becomes "stronger" as K increases because a given input voltage yields a larger output current. We must therefore concentrate on the voltage-to-current conversion property of the transistor, particularly as it relates to amplification of signals. More specifically, we ask, if a signal changes the base-emitter voltage of a transistor by a small amount (Fig. 4.17), how much *change* is produced in the collector current? Denoting the change in I_C by ΔI_C, we recognize that the "strength" of the device can be represented by $\Delta I_C/\Delta V_{BE}$. For example, if a base-emitter voltage change of 1 mV results in a ΔI_C of 0.1 mA in one transistor and 0.5 mA in another, we can view the latter as a better voltage-dependent current source or "voltage-to-current converter."

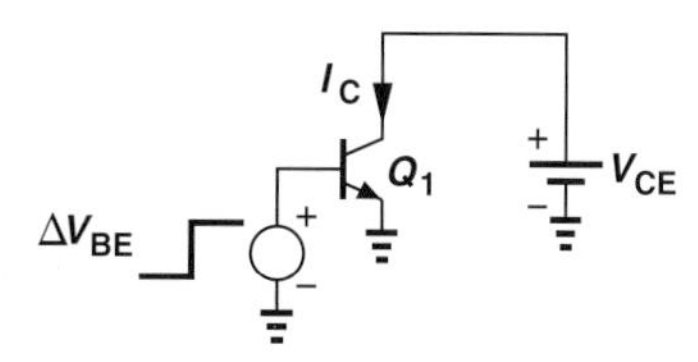

Figure 4.17 Test circuit for measurement of g_m.

The ratio $\Delta I_C/\Delta V_{BE}$ approaches dI_C/dV_{BE} for very small changes and, in the limit, is called the "transconductance," g_m*:

$$g_m = \frac{dI_C}{dV_{BE}}. \tag{4.41}$$

Note that this definition applies to any device that approximates a voltage-dependent current source (e.g., another type of transistor described in Chapter 6). For a bipolar transistor, Eq. (4.9) gives

$$g_m = \frac{d}{dV_{BE}}\left(I_S \exp\frac{V_{BE}}{V_T}\right) \tag{4.42}$$

$$= \frac{1}{V_T} I_S \exp\frac{V_{BE}}{V_T} \tag{4.43}$$

$$= \frac{I_C}{V_T}. \tag{4.44}$$

The close resemblance between this result and the small-signal resistance of diodes [Eq. (3.58)] is no coincidence and will become clearer in the next chapter.

*Note that V_{CE} is constant here.

Equation (4.44) reveals that, as I_C increases, the transistor becomes a better amplifying device by producing larger collector current excursions in response to a given signal level applied between the base and the emitter. The transconductance may be expressed in Ω^{-1} or "siemens," S. For example, if $I_C = 1$ mA, then with $V_T = 26$ mV, we have

$$g_m = 0.0385\ \Omega^{-1} \tag{4.45}$$

$$= 0.0385\ \text{S} \tag{4.46}$$

$$= 38.5\ \text{mS}. \tag{4.47}$$

However, as we will see throughout this book, it is often helpful to view g_m as the inverse of a resistance; e.g., for $I_C = 1$ mA, we may write

$$g_m = \frac{1}{26\ \Omega}. \tag{4.48}$$

The concept of transconductance can be visualized with the aid of the transistor I/V characteristics. As shown in Fig. 4.18, $g_m = dI_C/dV_{BE}$ simply represents the slope of I_C-V_{BE} characteristic at a given collector current, I_{C0}, and the corresponding base-emitter voltage, V_{BE0}. In other words, if V_{BE} experiences a small perturbation $\pm\Delta V$ around V_{BE0}, then the collector current displays a change of $\pm g_m\,\Delta V$ around I_{C0}, where $g_m = I_{C0}/V_T$. Thus, the value of I_{C0} must be chosen according to the required g_m and, ultimately, the required gain. We say the transistor is "biased" at a collector current of I_{C0}, meaning the device carries a bias (or "quiescent") current of I_{C0} in the absence of signals.[5]

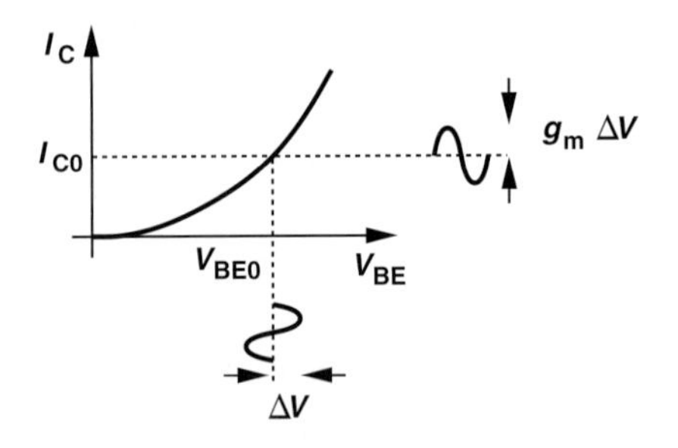

Figure 4.18 Illustration of transconductance.

Example 4-9

Consider the circuit shown in Fig. 4.19(a). What happens to the transconductance of Q_1 if the area of the device is increased by a factor of n?

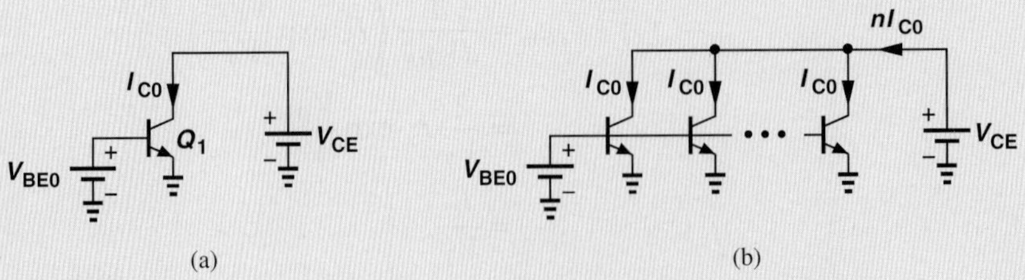

(a) (b)

Figure 4.19 (a) One transistor and (b) n transistors providing transconductance.

[5]Unless otherwise stated, we use the term "bias current" to refer to the collector bias current.

Solution Since $I_S \propto A_E$, I_S is multiplied by the same factor. Thus, $I_C = I_S \exp(V_{BE}/V_T)$ also rises by a factor of n because V_{BE} is constant. As a result, the transconductance increases by a factor of n. From another perspective, if n identical transistors, each carrying a collector current of I_{C0}, are placed in parallel, then the composite device exhibits a transconductance equal to n times that of each [Fig. 4.19(b)]. On the other hand, if the total collector current remains unchanged, then so does the transconductance.

Exercise Repeat the above example if V_{BE0} is reduced by $V_T \ln n$.

It is also possible to study the transconductance in the context of the I_C-V_{CE} characteristics of the transistor with V_{BE} as a parameter. Illustrated in Fig. 4.20 for two different bias currents I_{C1} and I_{C2}, the plots reveal that a change of ΔV in V_{BE} results in a greater change in I_C for operation around I_{C2} than around I_{C1} because $g_{m2} > g_{m1}$.

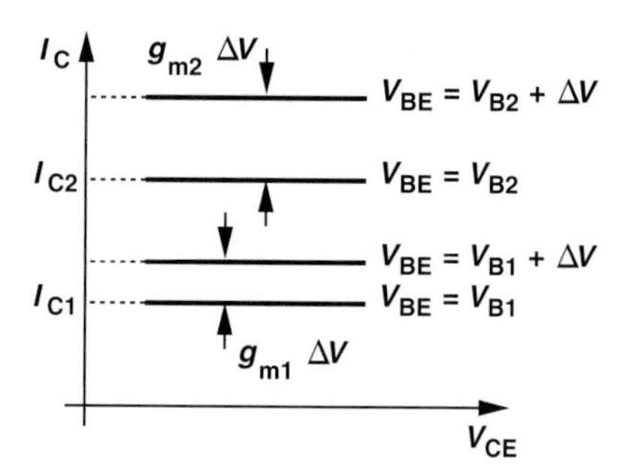

Figure 4.20 Transconductance for different collector bias currents.

The derivation of g_m in Eqs. (4.42)–(4.44) suggests that the transconductance is fundamentally a function of the collector current rather than the base current. For example, if I_C remains constant but β varies, then g_m does not change but I_B does. For this reason, the collector bias current plays a central role in the analysis and design, with the base current viewed as secondary, often undesirable effect.

As shown in Fig. 4.10, the current produced by the transistor may flow through a resistor to generate a proportional voltage. We exploit this concept in Chapter 5 to design amplifiers.

4.4.4 Small-Signal Model

Electronic circuits, e.g., amplifiers, may incorporate a large number of transistors, thus posing great difficulties in the analysis and design. Recall from Chapter 3 that diodes can be reduced to linear devices through the use of the small-signal model. A similar benefit accrues if a small-signal model can be developed for transistors.

The derivation of the small-signal model from the large-signal counterpart is relatively straightforward. We perturb the voltage difference between every two terminals (while the third terminal remains at a constant potential), determine the changes in the currents flowing through *all* terminals, and represent the results by proper circuit elements such as controlled current sources or resistors. Figure 4.21 depicts two conceptual examples where V_{BE} or V_{CE} is changed by ΔV and the changes in I_C, I_B, and I_E are examined.

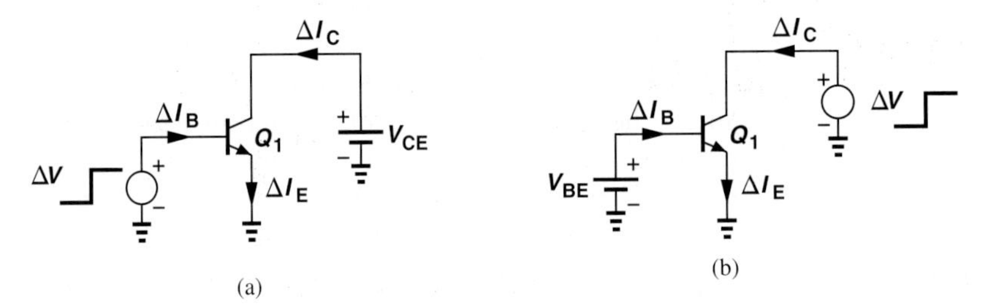

Figure 4.21 Excitation of bipolar transistor with small changes in (a) base-emitter and (b) collector-emitter voltage.

Let us begin with a change in V_{BE} while the collector voltage is constant (Fig. 4.22). We know from the definition of transconductance that

$$\Delta I_C = g_m \Delta V_{BE}, \tag{4.49}$$

concluding that a voltage-controlled current source must be connected between the collector and the emitter with a value equal to $g_m \Delta V$. For simplicity, we denote ΔV_{BE} by v_π and the change in the collector current by $g_m v_\pi$.

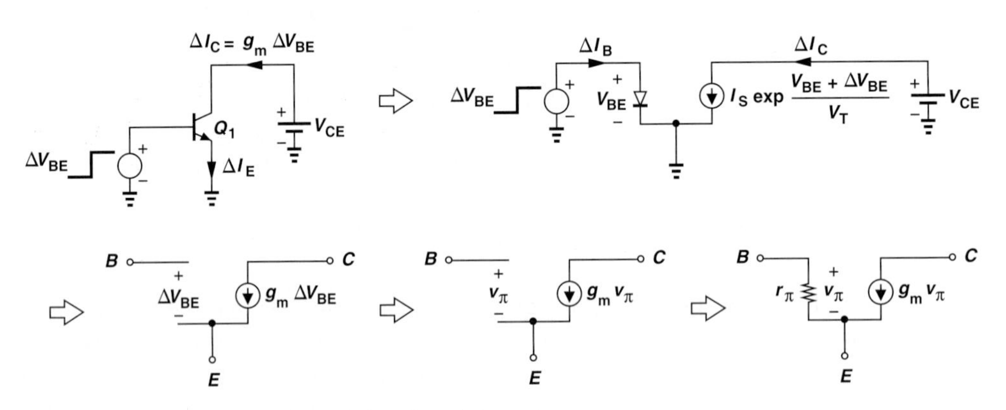

Figure 4.22 Development of small-signal model.

The change in V_{BE} creates another change as well:

$$\Delta I_B = \frac{\Delta I_C}{\beta} \tag{4.50}$$

$$= \frac{g_m}{\beta} \Delta V_{BE}. \tag{4.51}$$

That is, if the base-emitter voltage changes by ΔV_{BE}, the current flowing between these two terminals changes by $(g_m/\beta)\Delta V_{BE}$. Since the voltage and current correspond to the same two terminals, they can be related by Ohm's Law, i.e., by a resistor placed between the base and emitter having a value:

$$r_\pi = \frac{\Delta V_{BE}}{\Delta I_B} \tag{4.52}$$

$$= \frac{\beta}{g_m}. \tag{4.53}$$

Thus, the forward-biased diode between the base and the emitter is modeled by a small-signal resistance equal to β/g_m. This result is expected because the diode carries a bias current equal to I_C/β and, from Eq. (3.58), exhibits a small-signal resistance of $V_T/(I_C/\beta) = \beta(V_T/I_C) = \beta/g_m$.

We now turn our attention to the collector and apply a voltage change with respect to the emitter (Fig. 4.23). As illustrated in Fig. 4.11, for a constant V_{BE}, the collector voltage has no effect on I_C or I_B because $I_C = I_S \exp(V_{BE}/V_T)$ and $I_B = I_C/\beta$. Since ΔV_{CE} leads to no change in any of the terminal currents, the model developed in Fig. 4.22 need not be altered.

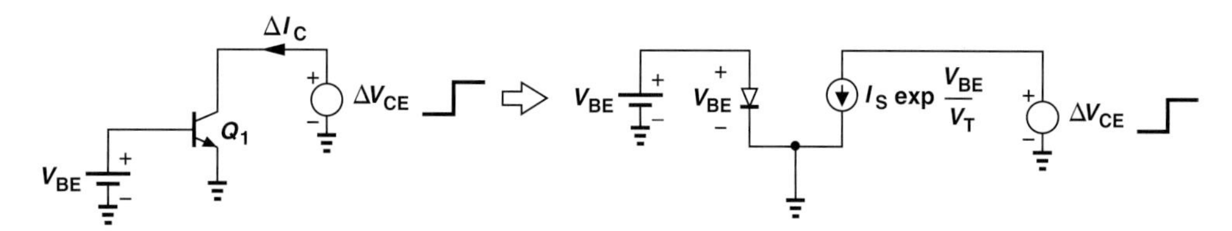

Figure 4.23 Response of bipolar transistor to small change in V_{CE}.

How about a change in the collector-base voltage? As studied in Problem 4.18, such a change also results in a zero change in the terminal currents.

The simple small-signal model developed in Fig. 4.22 serves as a powerful, versatile tool in the analysis and design of bipolar circuits. We should remark that both parameters of the model, g_m and r_π, depend on the bias current of the device. With a high collector bias current, a greater g_m is obtained, but the impedance between the base and emitter falls to lower values. Studied in Chapter 5, this trade-off proves undesirable in some cases.

Example 4-10

Consider the circuit shown in Fig. 4.24(a), where v_1 represents the signal generated by a microphone, $I_S = 3 \times 10^{-16}$ A, $\beta = 100$, and Q_1 operates in the active mode. (a) If $v_1 = 0$, determine the small-signal parameters of Q_1. (b) If the microphone generates a 1-mV signal, how much change is observed in the collector and base currents?

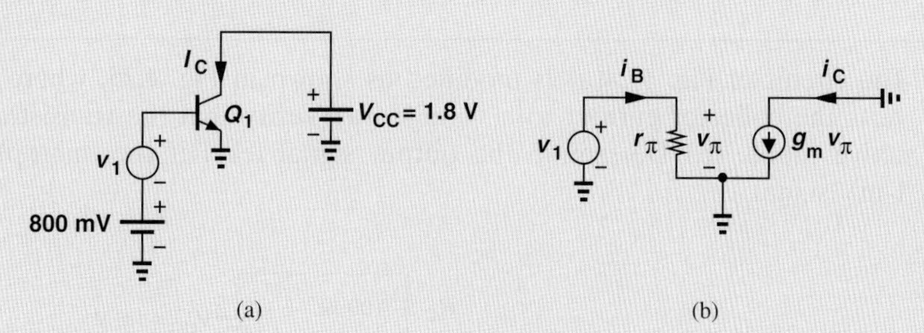

Figure 4.24 (a) Transistor with bias and small-signal excitation, (b) small-signal equivalent circuit.

Solution (a) Writing $I_C = I_S \exp(V_{BE}/V_T)$, we obtain a collector bias current of 6.92 mA for $V_{BE} = 800$ mV. Thus,

$$g_m = \frac{I_C}{V_T} \tag{4.54}$$

$$= \frac{1}{3.75\,\Omega}, \tag{4.55}$$

and

$$r_\pi = \frac{\beta}{g_m} \tag{4.56}$$

$$= 375\ \Omega. \tag{4.57}$$

(b) Drawing the small-signal equivalent of the circuit as shown in Fig. 4.24(b) and recognizing that $v_\pi = v_1$, we obtain the change in the collector current as:

$$\Delta I_C = g_m v_1 \tag{4.58}$$

$$= \frac{1\ \text{mV}}{3.75\ \Omega} \tag{4.59}$$

$$= 0.267\ \text{mA}. \tag{4.60}$$

The equivalent circuit also predicts the change in the base current as

$$\Delta I_B = \frac{v_1}{r_\pi} \tag{4.61}$$

$$= \frac{1\ \text{mV}}{375\ \Omega} \tag{4.62}$$

$$= 2.67\ \mu\text{A}, \tag{4.63}$$

which is, of course, equal to $\Delta I_C/\beta$.

Exercise Repeat the above example if I_S is halved.

The above example is not a useful circuit. The microphone signal produces a change in I_C, but the result flows through the 1.8-V battery. In other words, the circuit generates no output. On the other hand, if the collector current flows through a resistor, a useful output is provided.

Example 4-11 The circuit of Fig. 4.24(a) is modified as shown in Fig. 4.25, where resistor R_C converts the collector current to a voltage. (a) Verify that the transistor operates in the active mode. (b) Determine the output signal level if the microphone produces a 1-mV signal.

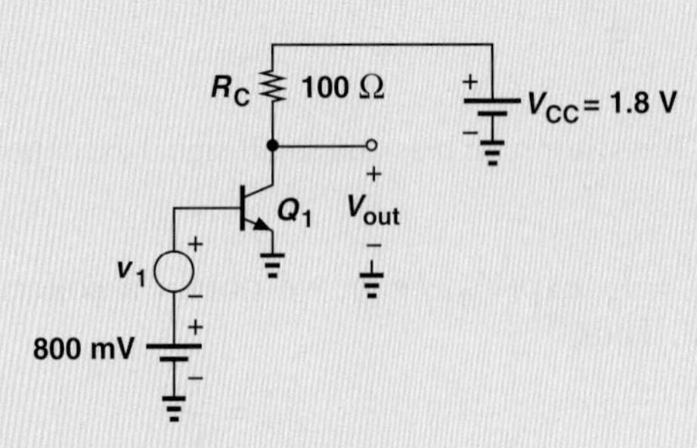

Figure 4.25 Simple stage with bias and small-signal excitation.

Solution (a) The collector bias current of 6.92 mA flows through R_C, leading to a potential drop of $I_C R_C = 692$ mV. The collector voltage, which is equal to V_{out}, is thus given by:

$$V_{out} = V_{CC} - R_C I_C \tag{4.64}$$

$$= 1.108 \text{ V}. \tag{4.65}$$

Since the collector voltage (with respect to ground) is more positive than the base voltage, the device operates in the active mode.

(b) As seen in the previous example, a 1-mV microphone signal leads to a 0.267-mA change in I_C. Upon flowing through R_C, this change yields a change of 0.267 mA × 100 Ω = 26.7 mV in V_{out}. The circuit therefore *amplifies* the input by a factor of 26.7.

Exercise What value of R_C results in a zero collector-base voltage?

The foregoing example demonstrates the amplification capability of the transistor. We will study and quantify the behavior of this and other amplifier topologies in the next chapter.

Did you know?

The first revolution afforded by the transistor was the concept of *portable* radios. Up to the 1940s, radios incorporated vacuum tubes, which required very high supply voltages (e.g., 60 V) and hence a bulky and heavy radio design. The transistor, on the other hand, could operate with a few batteries. The portable "transistor radio" was thus introduced by Regency and Texas Instruments in 1954. Interestingly, a Japanese company called Tsushin Kogyo had also been working on a transistor radio around that time and was eager to enter the American market. Since the company's name was difficult to pronounce for westerners, they picked the Latin word "sonus" for sound and called themselves Sony.

Small-Signal Model of Supply Voltage

We have seen that the use of the small-signal model of diodes and transistors can simplify the analysis considerably. In such an analysis, other components in the circuit must also be represented by a small-signal model. In particular, we must determine how the supply voltage, V_{CC}, behaves with respect to small changes in the currents and voltages of the circuit.

The key principle here is that the supply voltage (ideally) remains *constant* even though various voltages and currents within the circuit may change with time. Since the supply does not change and since the small-signal model of the circuit entails only changes in the quantities, we observe that V_{CC} must be replaced with a *zero* voltage to signify the zero change. Thus, we simply "ground" the supply voltage in small-signal analysis. Similarly, any other constant voltage in the circuit is replaced with a ground connection. To emphasize that such grounding holds for only signals, we sometimes say a node is an "ac ground."

4.4.5 Early Effect

Our treatment of the bipolar transistor has thus far concentrated on the fundamental principles, ignoring second-order effects in the device and their representation in the large-signal and small-signal models. However, some circuits require attention to such effects if meaningful results are to be obtained. The following example illustrates this point.

Example 4-12 Considering the circuit of Example 4-11, suppose we raise R_C to 200 Ω and V_{CC} to 3.6 V. Verify that the device operates in the active mode and compute the voltage gain.

Solution The voltage drop across R_C now increases to 6.92 mA × 200 Ω = 1.384 V, leading to a collector voltage of 3.6 V − 1.384 V = 2.216 V and guaranteeing operation in the active mode. Note that if V_{CC} is not doubled, then V_{out} = 1.8 V − 1.384 V = 0.416 V and the transistor is not in the forward active region.

Recall from part (b) of the above example that the change in the output voltage is equal to the change in the collector current multiplied by R_C. Since R_C is doubled, the voltage gain must also double, reaching a value of 53.4. This result is also obtained with the aid of the small-signal model. Illustrated in Fig. 4.26, the equivalent circuit yields $v_{out} = -g_m v_\pi R_C = -g_m v_1 R_C$ and hence $v_{out}/v_1 = -g_m R_C$. With $g_m = (3.75\ \Omega)^{-1}$ and $R_C = 200\ \Omega$, we have $v_{out}/v_1 = -53.4$.

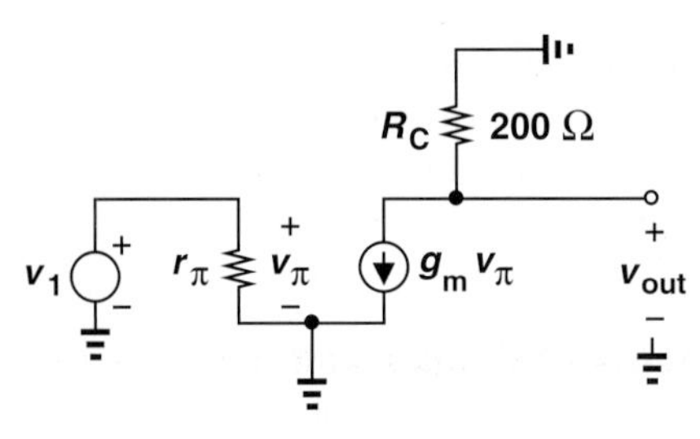

Figure 4.26 Small-signal equivalent circuit of the stage shown in Fig. 4.25.

Exercise What happens if $R_C = 250\ \Omega$?

This example points to an important trend: if R_C increases, so does the voltage gain of the circuit. Does this mean that, if $R_C \rightarrow \infty$, then the gain also grows indefinitely? Does another mechanism in the circuit, perhaps in the transistor, limit the maximum gain that can be achieved? Indeed, the "Early effect" translates to a nonideality in the device that can limit the gain of amplifiers.

To understand this effect, we return to the internal operation of the transistor and reexamine the claim shown in Fig. 4.11 that "the collector current does not depend on the collector voltage." Consider the device shown in Fig. 4.27(a), where the collector voltage is somewhat higher than the base voltage and the reverse bias across the junction creates a certain depletion region width. Now suppose V_{CE} is raised to V_{CE2} [Fig. 4.27(b)], thus increasing the reverse bias and widening the depletion region in the collector and base areas. Since the base charge profile must still fall to zero at the edge of depletion region, x_2', the *slope* of the profile increases. Equivalently, the effective base width, W_B, in Eq. (4.8) decreases, thereby increasing the collector current. Discovered by Early, this phenomenon poses interesting problems in amplifier design (Chapter 5).

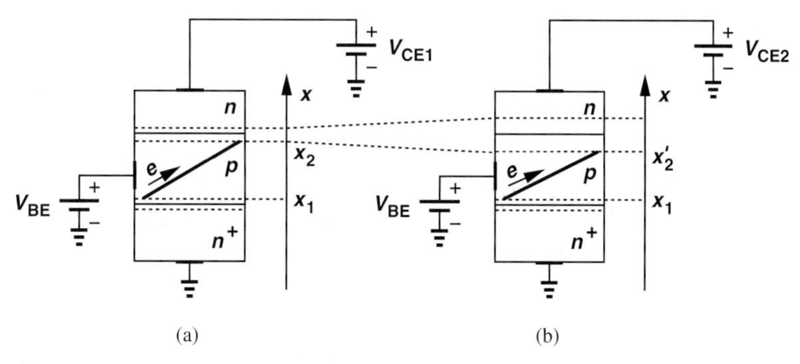

Figure 4.27 (a) Bipolar device with base and collector bias voltages, (b) effect of higher collector voltage.

How is the Early effect represented in the transistor model? We must first modify Eq. (4.9) to include this effect. It can be proved that the rise in the collector current with V_{CE} can be approximately expressed by a multiplicative factor:

$$I_C = \frac{A_E q D_n n_i^2}{N_B W_B}\left(\exp\frac{V_{BE}}{V_T} - 1\right)\left(1 + \frac{V_{CE}}{V_A}\right), \tag{4.66}$$

$$\approx \left(I_S \exp\frac{V_{BE}}{V_T}\right)\left(1 + \frac{V_{CE}}{V_A}\right). \tag{4.67}$$

where W_B is assumed constant and the second factor, $1 + V_{CE}/V_A$, models the Early effect. The quantity V_A is called the "Early voltage."

It is instructive to examine the I/V characteristics of Fig. 4.15 in the presence of the Early effect. For a constant V_{CE}, the dependence of I_C upon V_{BE} remains exponential but with a somewhat greater slope [Fig. 4.28(a)]. On the other hand, for a constant V_{BE}, the $I_C - V_{CE}$ characteristic displays a nonzero slope [Fig. 4.28(b)]. In fact, differentiation of Eq. (4.67) with respect to V_{CE} yields

$$\frac{\delta I_C}{\delta V_{CE}} = I_S\left(\exp\frac{V_{BE}}{V_T}\right)\left(\frac{1}{V_A}\right) \tag{4.68}$$

$$\approx \frac{I_C}{V_A}, \tag{4.69}$$

where it is assumed $V_{CE} \ll V_A$ and hence $I_C \approx I_S \exp(V_{BE}/V_T)$. This is a reasonable approximation in most cases.

The variation of I_C with V_{CE} in Fig. 4.28(b) reveals that the transistor in fact does not operate as an *ideal* current source, requiring modification of the perspective shown in

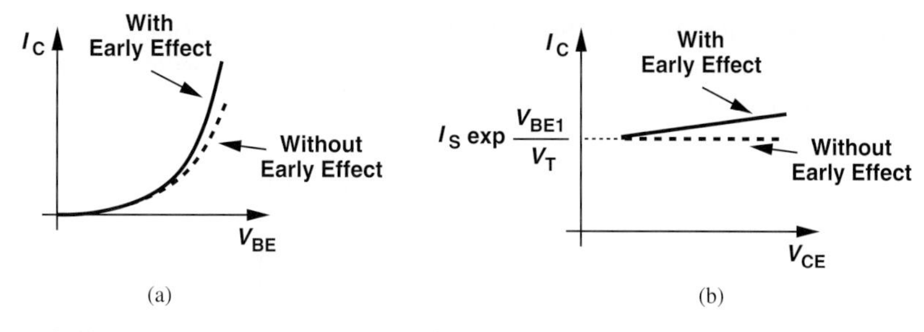

Figure 4.28 Collector current as a function of (a) V_{BE} and (b) V_{CE} with and without Early effect.

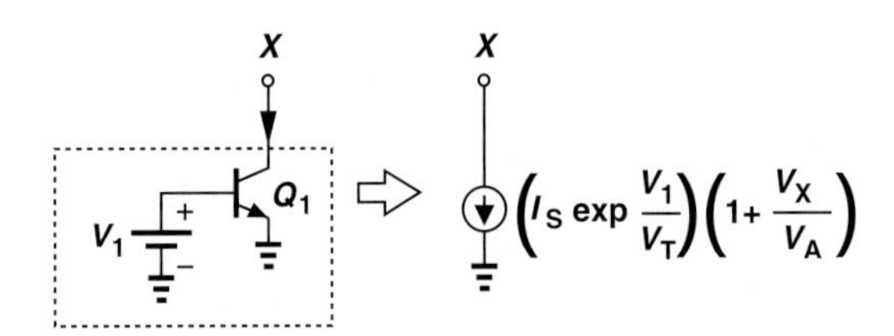

Figure 4.29 Realistic model of bipolar transistor as a current source.

Fig. 4.11(a). The transistor can still be viewed as a two-terminal device but with a current that varies to some extent with V_{CE} (Fig. 4.29).

Example 4-13 A bipolar transistor carries a collector current of 1 mA with $V_{CE} = 2$ V. Determine the required base-emitter voltage if $V_A = \infty$ or $V_A = 20$ V. Assume $I_S = 2 \times 10^{-16}$ A.

Solution With $V_A = \infty$, we have from Eq. (4.67)

$$V_{BE} = V_T \ln \frac{I_C}{I_S} \tag{4.70}$$

$$= 760.3\text{ mV}. \tag{4.71}$$

If $V_A = 20$ V, we rewrite Eq. (4.67) as

$$V_{BE} = V_T \ln \left(\frac{I_C}{I_S} \frac{1}{1 + \frac{V_{CE}}{V_A}} \right) \tag{4.72}$$

$$= 757.8\text{ mV}. \tag{4.73}$$

In fact, for $V_{CE} \ll V_A$, we have $(1 + V_{CE}/V_A)^{-1} \approx 1 - V_{CE}/V_A$

$$V_{BE} \approx V_T \ln \frac{I_C}{I_S} + V_T \ln \left(1 - \frac{V_{CE}}{V_A}\right) \tag{4.74}$$

$$\approx V_T \ln \frac{I_C}{I_S} - V_T \frac{V_{CE}}{V_A}, \tag{4.75}$$

where it is assumed $\ln(1 - \epsilon) \approx -\epsilon$ for $\epsilon \ll 1$.

Exercise Repeat the above example if two such transistors are placed in parallel.

Large-Signal and Small-Signal Models The presence of Early effect alters the transistor models developed in Sections 4.4.1 and 4.4.4. The large-signal model of Fig. 4.13 must now be modified to that in Fig. 4.30, where

$$I_C = \left(I_S \exp \frac{V_{BE}}{V_T}\right)\left(1 + \frac{V_{CE}}{V_A}\right) \tag{4.76}$$

$$I_B = \frac{1}{\beta}\left(I_S \exp \frac{V_{BE}}{V_T}\right) \tag{4.77}$$

$$I_E = I_C + I_B. \tag{4.78}$$

Note that I_B is independent of V_{CE} and still given by the base-emitter voltage.

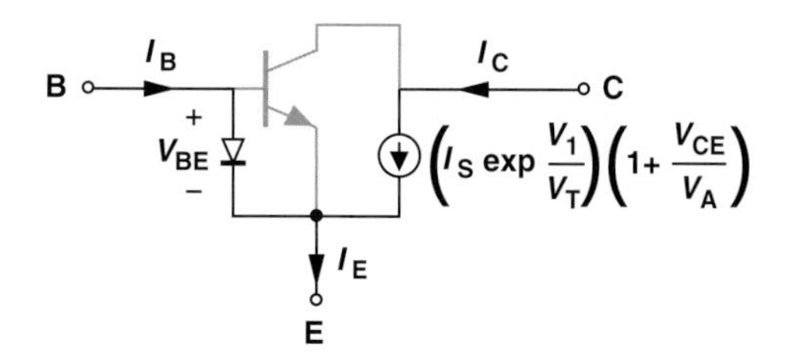

Figure 4.30 Large-signal model of bipolar transistor including Early effect.

For the small-signal model, we note that the controlled current source remains unchanged and g_m is expressed as

$$g_m = \frac{dI_C}{dV_{BE}} \tag{4.79}$$

$$= \frac{1}{V_T}\left(I_S \exp\frac{V_{BE}}{V_T}\right)\left(1+\frac{V_{CE}}{V_A}\right) \tag{4.80}$$

$$= \frac{I_C}{V_T}. \tag{4.81}$$

Similarly,

$$r_\pi = \frac{\beta}{g_m} \tag{4.82}$$

$$= \beta\frac{V_T}{I_C}. \tag{4.83}$$

Considering that the collector current does vary with V_{CE}, let us now apply a voltage change at the collector and measure the resulting current change [Fig. 4.31(a)]:

$$I_C + \Delta I_C = \left(I_S \exp\frac{V_{BE}}{V_T}\right)\left(1+\frac{V_{CE}+\Delta V_{CE}}{V_A}\right). \tag{4.84}$$

It follows that

$$\Delta I_C = \left(I_S \exp\frac{V_{BE}}{V_T}\right)\frac{\Delta V_{CE}}{V_A}, \tag{4.85}$$

which is consistent with Eq. (4.69). Since the voltage and current change correspond to the same two terminals, they satisfy Ohm's Law, yielding an equivalent resistor:

$$r_O = \frac{\Delta V_{CE}}{\Delta I_C} \tag{4.86}$$

$$= \frac{V_A}{I_S \exp\frac{V_{BE}}{V_T}} \tag{4.87}$$

$$\approx \frac{V_A}{I_C}. \tag{4.88}$$

Depicted in Fig. 4.31(b), the small-signal model contains only one extra element, r_O, to represent the Early effect. Called the "output resistance," r_O plays a critical role in high-gain amplifiers (Chapter 5). Note that both r_π and r_O are inversely proportionally to the bias current, I_C.

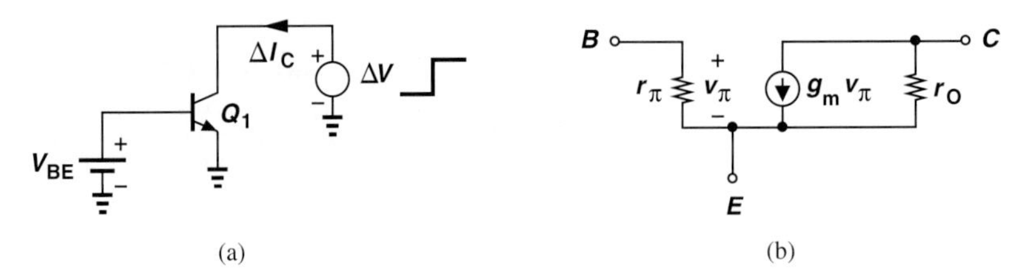

Figure 4.31 (a) Small change in V_{CE} and (b) small-signal model including Early effect.

Example 4-14 A transistor is biased at a collector current of 1 mA. Determine the small-signal model if $\beta = 100$ and $V_A = 15$ V.

Solution We have

$$g_m = \frac{I_C}{V_T} \tag{4.89}$$

$$= \frac{1}{26\ \Omega}, \tag{4.90}$$

and

$$r_\pi = \frac{\beta}{g_m} \tag{4.91}$$

$$= 2600\ \Omega. \tag{4.92}$$

Also,

$$r_O = \frac{V_A}{I_C} \tag{4.93}$$

$$= 15\ \text{k}\Omega. \tag{4.94}$$

Exercise What early voltage is required if the output resistance must reach 25 kΩ?

In the next chapter, we return to Example 4-12 and determine the gain of the amplifier in the presence of the Early effect. We will conclude that the gain is eventually limited by the transistor output resistance, r_O. Figure 4.32 summarizes the concepts studied in this section.

An important notion that has emerged from our study of the transistor is the concept of biasing. We must create proper dc voltages and currents at the device terminals to accomplish two goals: (1) guarantee operation in the active mode ($V_{BE} > 0$, $V_{CE} \geq 0$); e.g., the load resistance tied to the collector faces an upper limit for a given supply voltage (Example 4-7); (2) establish a collector current that yields the required values for the small-signal parameters g_m, r_O, and r_π. The analysis of amplifiers in the next chapter exercises these ideas extensively.

Finally, we should remark that the small-signal model of Fig. 4.31(b) does not reflect the high-frequency limitations of the transistor. For example, the base-emitter and base-collector junctions exhibit a depletion-region capacitance that impacts the speed. These properties are studied in Chapter 11.

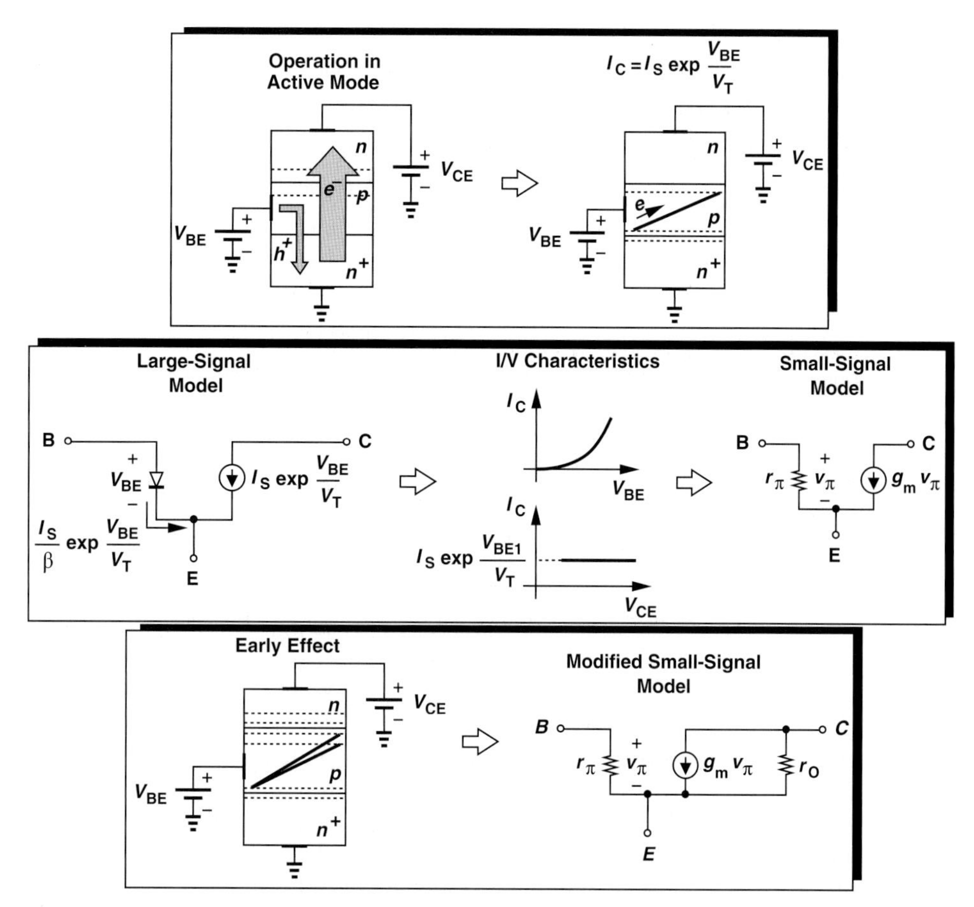

Figure 4.32 Summary of concepts studied thus far.

Robotics Application: Temperature Sensors

Robots use temperature sensors to decide on their actions. For example, a robot can detect elevated temperatures as it approaches a fire and avoid it to protect itself.

Temperature sensors provide an analog or digital output proportional to the ambient temperature. Bipolar transistors serve this purpose elegantly. Consider the branch shown in Fig. 4.33(a), where bias current I_0 flows through transistor Q_1. Neglecting the base current, we have $I_{C1} \approx I_0$ and, since $I_{C1} = I_{S1} \exp(V_{BE1}/V_T)$,

$$V_{BE1} = V_T \ln \frac{I_0}{I_{S1}}. \tag{4.95}$$

Note that $V_X = V_{CC} - V_{BE1}$.

Now, consider the circuit in Fig. 4.33(b). In addition to the branch on the left, we have another consisting of Q_2, R_1, and I_0. It follows that

$$V_{BE2} = V_T \ln \frac{I_0}{I_{S2}}. \tag{4.96}$$

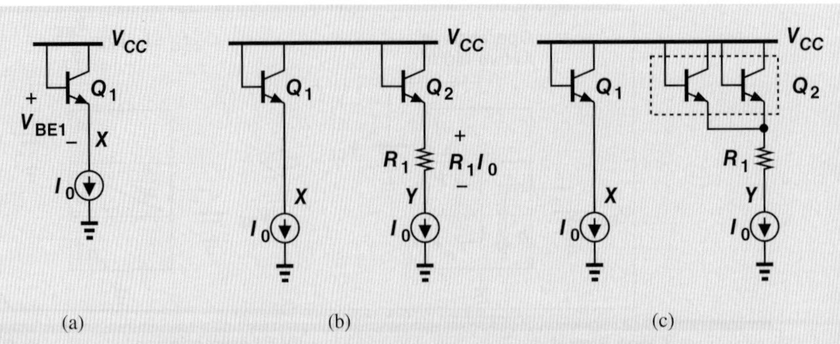

Figure 4.33 (a) A biased branch, (b) two branches providing $R_1I_0 \propto T$, and (c) a structure for $I_{S2} = 2I_{S1}$.

Also, $V_Y = V_{CC} - V_{BE2} - R_1I_0$. Next, we select I_0 so that $V_X = V_Y$:

$$V_{CC} - V_{BE1} = V_{CC} - V_{BE2} - R_1I_0. \tag{4.97}$$

That is,

$$R_1I_0 = V_{BE1} - V_{BE2} \tag{4.98}$$

$$= V_T \ln \frac{I_{S2}}{I_{S1}}. \tag{4.99}$$

Here, $V_T = kT/q$ and is proportional to the absolute temperature, and so is the voltage drop across R_1. This voltage serves as the output of the temperature sensor. Of course, we require that $I_{S1} \neq I_{S2}$. Figure 4.33(c) depicts an example where Q_2 comprises two unit transistors in parallel, each identical to Q_1. Thus, $I_{S2} = 2I_{S1}$. How do we guarantee that $V_X = V_Y$? This is explained in more advanced courses.

4.5 OPERATION OF BIPOLAR TRANSISTOR IN SATURATION MODE

As mentioned in the previous section, it is desirable to operate bipolar devices in the forward active region, where they act as voltage-controlled current sources. In this section, we study the behavior of the device outside this region and the resulting difficulties.

Let us set V_{BE} to a typical value, e.g., 750 mV, and vary the collector voltage from a high level to a low level [Fig. 4.34(a)]. As V_{CE} approaches V_{BE}, and V_{BC} goes from a negative value toward zero, the base-collector junction experiences less reverse bias. For $V_{CE} = V_{BE}$, the junction sustains a zero voltage difference, but its depletion region still absorbs most of the electrons injected by the emitter into the base. But what happens if $V_{CE} < V_{BE}$, i.e., $V_{BC} > 0$ and the B-C junction is forward biased? We say the transistor enters the "saturation region." Suppose $V_{CE} = 550$ mV and hence $V_{BC} = +200$ mV. We know from Chapter 2 that a typical diode sustaining 200 mV of forward bias carries an extremely small current.[6] Thus, even in this case the transistor continues to operate as in the active mode, and we say the device is in "soft saturation."

[6]About nine orders of magnitude less than one sustaining 750 mV: $(750\,\text{mV} - 200\,\text{mV})/(60\,\text{mV/dec}) \approx 9.2$.

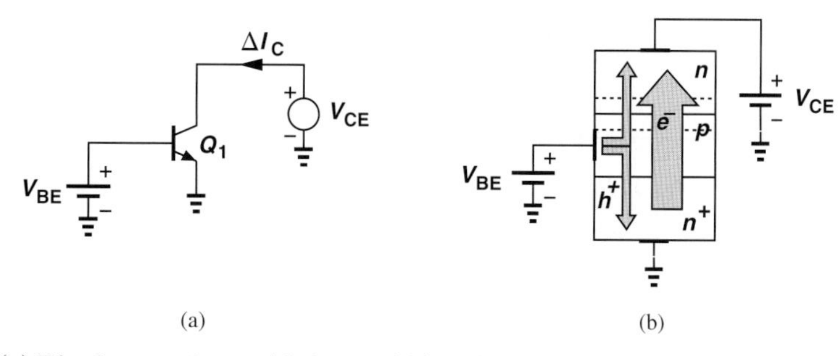

Figure 4.34 (a) Bipolar transistor with forward-biased base-collector junction, (b) flow of holes to collector.

If the collector voltage drops further, the B-C junction experiences greater forward bias, carrying a significant current [Fig. 4.34(b)]. Consequently, a large number of holes must be supplied to the base terminal—as if β is reduced. In other words, heavy saturation leads to a sharp rise in the base current and hence a rapid fall in β.

Example 4-15 A bipolar transistor is biased with $V_{BE} = 750$ mV and has a nominal β of 100. How much B-C forward bias can the device tolerate if β must not degrade by more than 10%? For simplicity, assume base-collector and base-emitter junctions have identical structures and doping levels.

Solution If the base-collector junction is forward-biased so much that it carries a current equal to one-tenth of the nominal base current, I_B, then the β degrades by 10%. Since $I_B = I_C/100$, the B-C junction must carry no more than $I_C/1000$. We therefore ask, what B-C voltage results in a current of $I_C/1000$ if $V_{BE} = 750$ mV gives a collector current of I_C? Assuming identical B-E and B-C junctions, we have

$$V_{BE} - V_{BC} = V_T \ln \frac{I_C}{I_S} - V_T \ln \frac{I_C/1000}{I_S} \tag{4.100}$$

$$= V_T \ln 1000 \tag{4.101}$$

$$\approx 180\,\text{mV}. \tag{4.102}$$

That is, $V_{BC} = 570$ mV.

Exercise Repeat the above example if $V_{BE} = 800$ mV.

It is instructive to study the transistor large-signal model and I-V characteristics in the saturation region. We construct the model as shown in Fig. 4.35(a), including the base-collector diode. Note that the net collector current *decreases* as the device enters saturation because part of the controlled current $I_{S1} \exp(V_{BE}/V_T)$ is provided by the B-C diode and need not flow from the collector terminal. In fact, as illustrated in Fig. 4.35(b), if the collector is left open, then D_{BC} is forward-biased so much that its current becomes equal to the controlled current.

The above observations lead to the I_C-V_{CE} characteristics depicted in Fig. 4.36, where I_C begins to fall for V_{CE} less than V_1, about a few hundred millivolts. The term "saturation" is used because increasing the base current in this region of operation leads to little change in the collector current.

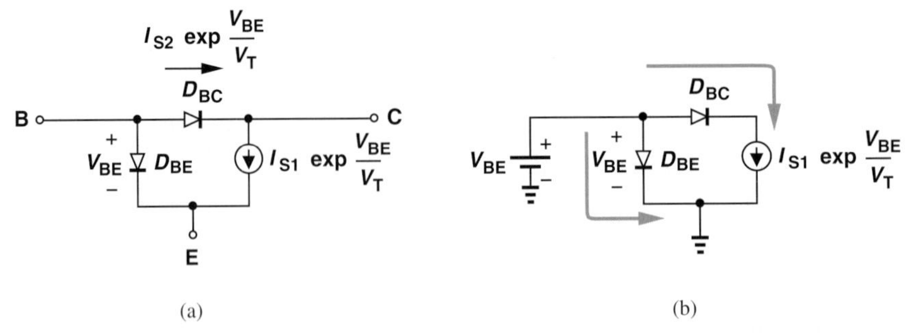

Figure 4.35 (a) Model of bipolar transistor including saturation effects, (b) case of open collector terminal.

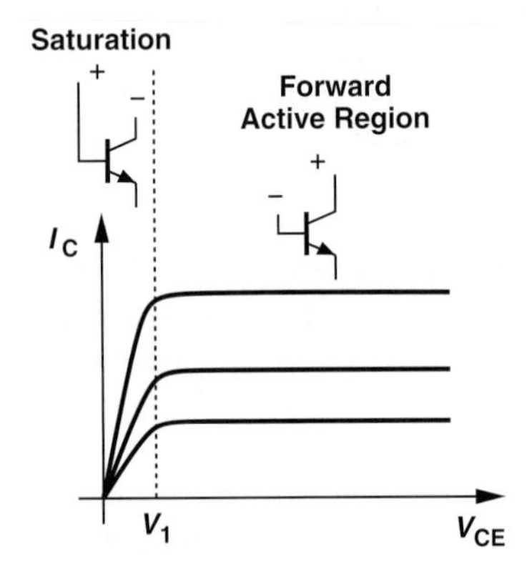

Figure 4.36 Transistor I/V characteristics in different regions of operation.

In addition to a drop in β, the *speed* of bipolar transistors also degrades in saturation (Chapter 11). Thus, electronic circuits rarely allow operation of bipolar devices in this mode. As a rule of thumb, we permit soft saturation with $V_{BC} < 400$ mV because the current in the B-C junction is negligible, provided that various tolerances in the component values do not drive the device into deep saturation.

It is important to recognize that the transistor simply draws a current from any component tied to its collector, e.g., a resistor. Thus, it is the external component that defines the collector voltage and hence the region of operation.

Example 4-16

For the circuit of Fig. 4.37, determine the relationship between R_C and V_{CC} that guarantees operation in soft saturation or active region.

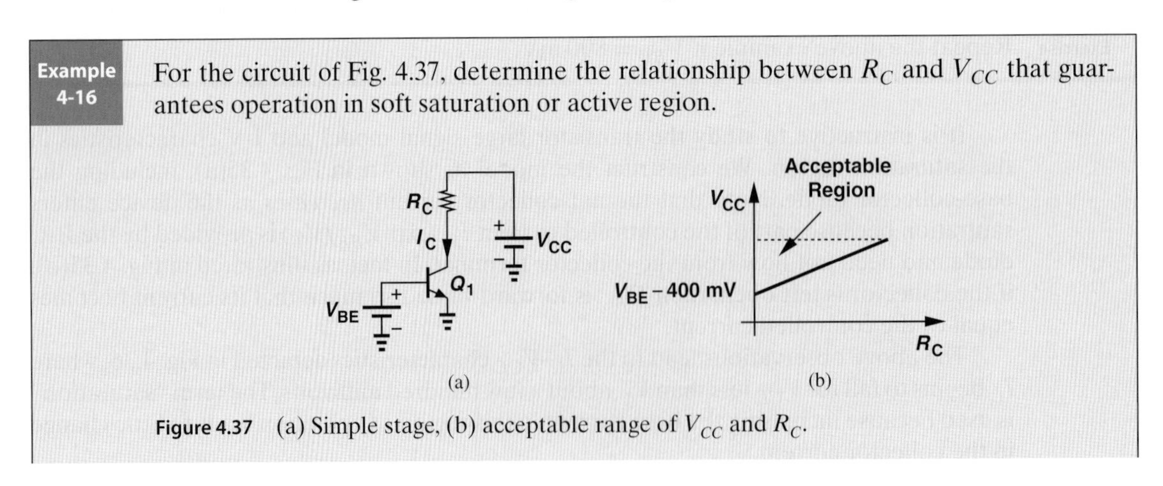

Figure 4.37 (a) Simple stage, (b) acceptable range of V_{CC} and R_C.

Solution In soft saturation, the collector current is still equal to $I_S \exp(V_{BE}/V_T)$. The collector voltage must not fall below the base voltage by more than 400 mV:

$$V_{CC} - R_C I_C \geq V_{BE} - 400\,\text{mV}. \tag{4.103}$$

Thus,

$$V_{CC} \geq I_C R_C + (V_{BE} - 400\,\text{mV}). \tag{4.104}$$

For a given value of R_C, V_{CC} must be sufficiently large so that $V_{CC} - I_C R_C$ still maintains a reasonable collector voltage.

Exercise Determine the maximum tolerable value of R_C.

In the deep saturation region, the collector-emitter voltage approaches a constant value called $V_{CE,sat}$ (about 200 mV). Under this condition, the transistor bears no resemblance to a controlled current source and can be modeled as shown in Fig. 4.38. (The battery tied between C and E indicates that V_{CE} is relatively constant in deep saturation.)

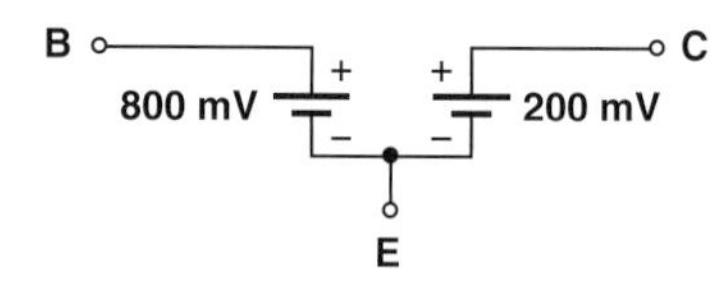

Figure 4.38 Transistor model in deep saturation.

4.6 THE *PNP* TRANSISTOR

We have thus far studied the structure and properties of the *npn* transistor, i.e., with the emitter and collector made of *n*-type materials and the base made of a *p*-type material. We may naturally wonder if the dopant polarities can be inverted in the three regions, forming a "*pnp*" device. More importantly, we may wonder why such a device would be useful.

4.6.1 Structure and Operation

Figure 4.39(a) shows the structure of a *pnp* transistor, emphasizing that the emitter is heavily doped. As with the *npn* counterpart, operation in the active region requires forward-biasing the base-emitter junction and reverse-biasing the collector junction. Thus, $V_{BE} < 0$ and $V_{BC} > 0$. Under this condition, majority carriers in the emitter (holes) are injected into the base and swept away into the collector. Also, a linear profile of holes is formed in the base region to allow diffusion. A small number of base majority carriers (electrons) are injected into the emitter or recombined with the holes in the base region, thus creating the base current. Figure 4.39(b) illustrates the flow of the carriers. All of the operation principles and equations described for *npn* transistors apply to *pnp* devices as well.

Figure 4.39(c) depicts the symbol of the *pnp* transistor along with constant voltage sources that bias the device in the active region. In contrast to the biasing of the *npn* transistor in Fig. 4.6, here the base and collector voltages are *lower* than the emitter voltage. Following our convention of placing more positive nodes on the top of the page, we redraw

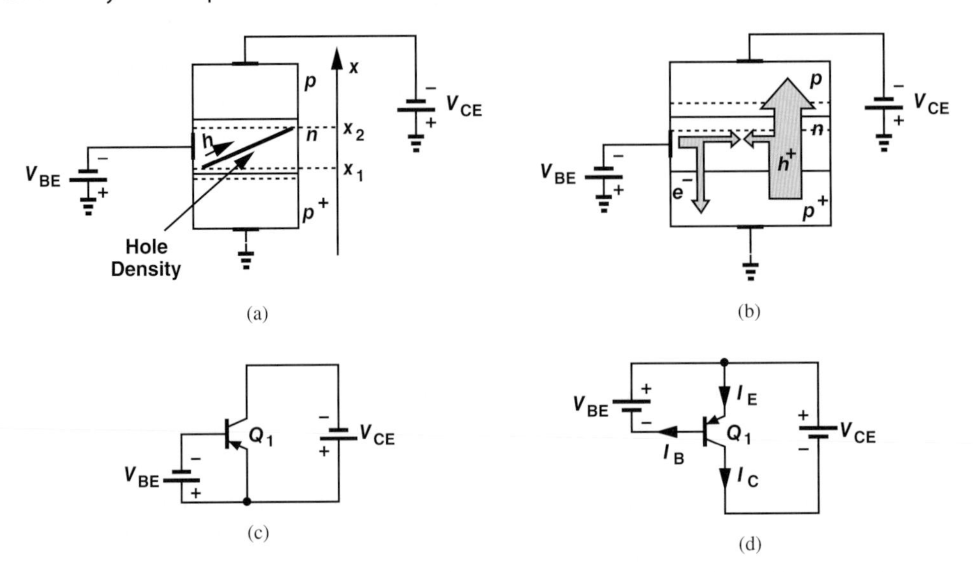

Figure 4.39 (a) Structure of *pnp* transistor, (b) current flow in *pnp* transistor, (c) proper biasing, (d) more intuitive view of (c).

the circuit as in Fig. 4.39(d) to emphasize $V_{EB} > 0$ and $V_{BC} > 0$ and to illustrate the actual direction of current flow into each terminal.

4.6.2 Large-Signal Model

The current and voltage polarities in *npn* and *pnp* transistors can be confusing. We address this issue by making the following observations. (1) The (conventional) current always flows from a positive supply (i.e., top of the page) toward a lower potential (i.e., bottom of the page). Figure 4.40(a) shows two branches employing *npn* and *pnp* transistors, illustrating that the (conventional) current flows from collector to emitter in *npn* devices and from emitter to collector in *pnp* counterparts. Since the base current must be included in the emitter current, we note that I_{B1} and I_{C1} add up to I_{E1}, whereas I_{E2} "loses" I_{B2} before emerging as I_{C2}. (2) The distinction between active and saturation regions is based on the B-C junction bias. The different cases are summarized in Fig. 4.40(b), where the relative position of the base and collector nodes signifies their potential difference. We note that an *npn* transistor is in the active mode if the collector (voltage) is *not* lower than the base (voltage). For the *pnp* device, on the other hand, the collector must not be *higher* than the base. (3) The *npn* current equations (4.23)–(4.25) must be modified as follows for the *pnp* device:

$$I_C = I_S \exp\frac{V_{EB}}{V_T} \tag{4.105}$$

$$I_B = \frac{I_S}{\beta}\exp\frac{V_{EB}}{V_T} \tag{4.106}$$

$$I_E = \frac{\beta+1}{\beta} I_S \exp\frac{V_{EB}}{V_T}, \tag{4.107}$$

where the current directions are defined in Fig. 4.41. The only difference between the *npn* and *pnp* equations relates to the base-emitter voltage that appears in the exponent, an

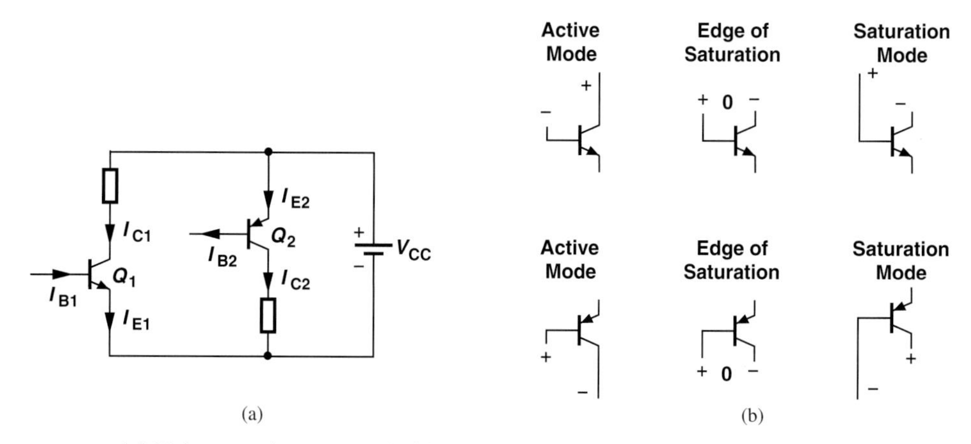

Figure 4.40 (a) Voltage and current polarities in *npn* and *pnp* transistors, (b) illustration of active and saturation regions.

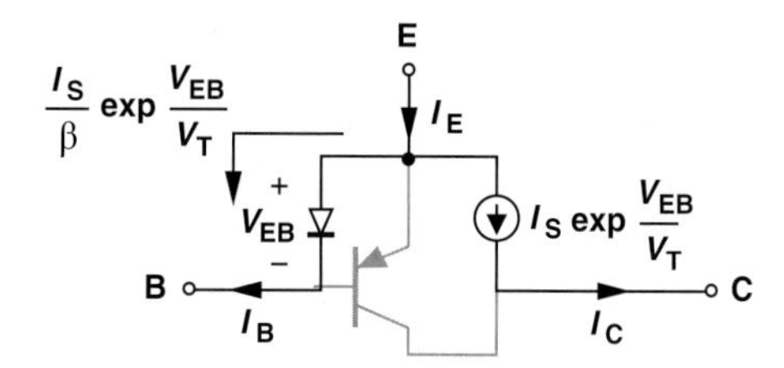

Figure 4.41 Large-signal model of *pnp* transistor.

expected result because $V_{BE} < 0$ for *pnp* devices and must be changed to V_{EB} to create a large exponential term. Also, the Early effect can be included as

$$I_C = \left(I_S \exp\frac{V_{EB}}{V_T}\right)\left(1 + \frac{V_{EC}}{V_A}\right). \tag{4.108}$$

Example 4-17

In the circuit shown in Fig. 4.42, determine the terminal currents of Q_1 and verify operation in the forward active region. Assume $I_S = 2 \times 10^{-16}$ A and $\beta = 50$, but $V_A = \infty$.

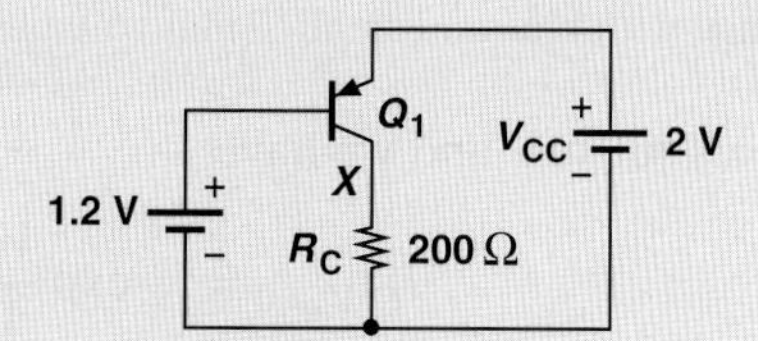

Figure 4.42 Simple stage using a *pnp* transistor.

Solution We have $V_{EB} = 2\text{ V} - 1.2\text{ V} = 0.8$ V and hence

$$I_C = I_S \exp\frac{V_{EB}}{V_T} \tag{4.109}$$

$$= 4.61\text{ mA}. \tag{4.110}$$

It follows that

$$I_B = 92.2\,\mu\text{A} \tag{4.111}$$

$$I_E = 4.70\,\text{mA}. \tag{4.112}$$

We must now compute the collector voltage and hence the bias across the B-C junction. Since R_C carries I_C,

$$V_X = R_C I_C \tag{4.113}$$

$$= 0.922\,\text{V}, \tag{4.114}$$

which is *lower* than the base voltage. Invoking the illustration in Fig. 4.40(b), we conclude that Q_1 operates in the active mode and the use of equations (4.105)–(4.107) is justified.

Exercise What is the maximum value of R_C if the transistor must remain in soft saturation?

We should mention that some books assume all of the transistor terminal currents flow into the device, thus requiring that the right-hand side of Eqs. (4.105) and (4.106) be multiplied by a negative sign. We nonetheless continue with our notation as it reflects the actual direction of currents and proves more efficient in the analysis of circuits containing many *npn* and *pnp* transistors.

Example 4-18 In the circuit of Fig. 4.43, V_{in} represents a signal generated by a microphone. Determine V_{out} for $V_{in} = 0$ and $V_{in} = +5\,\text{mV}$ if $I_S = 1.5 \times 10^{-16}\,\text{A}$.

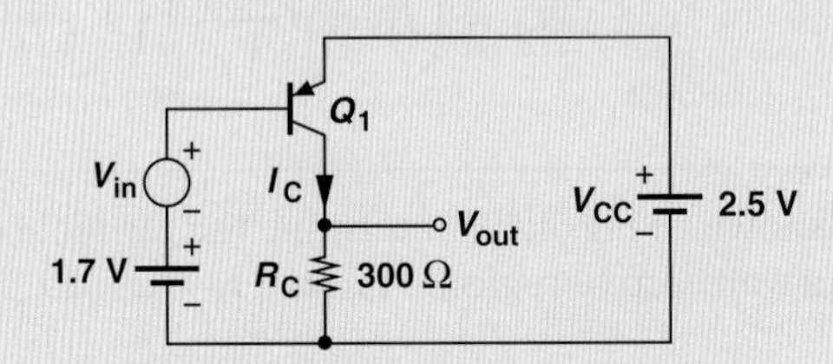

Figure 4.43 *PNP* stage with bias and small-signal voltages.

Solution For $V_{in} = 0$, $V_{EB} = +800\,\text{mV}$ and we have

$$I_C|_{V_{in}=0} = I_S \exp\frac{V_{EB}}{V_T} \tag{4.115}$$

$$= 3.46\,\text{mA}, \tag{4.116}$$

and hence

$$V_{out} = 1.038\,\text{V}. \tag{4.117}$$

If V_{in} increases to +5 mV, V_{EB} = +795 mV and

$$I_C|_{V_{in}=+5\text{ mV}} = 2.85\text{ mA}, \tag{4.118}$$

yielding

$$V_{out} = 0.856\text{ V}. \tag{4.119}$$

Note that as the base voltage *rises*, the collector voltage *falls*, a behavior similar to that of the *npn* counterparts in Fig. 4.25. Since a 5-mV change in V_{in} gives a 182-mV change in V_{out}, the voltage gain is equal to 36.4. These results are more readily obtained through the use of the small-signal model.

Exercise Determine V_{out} if V_{in} = −5 mV.

4.6.3 Small-Signal Model

Since the small-signal model represents *changes* in the voltages and currents, we expect *npn* and *pnp* transistors to have similar models. Depicted in Fig. 4.44(a), the small-signal model of the *pnp* transistor is indeed *identical* to that of the *npn* device. Following the convention in Fig. 4.39(d), we sometimes draw the model as shown in Fig. 4.44(b).

The reader may notice that the terminal currents in the small-signal model bear an opposite direction with respect to those in the large-signal model of Fig. 4.41. This is not an inconsistency and is studied in Problem 4.49.

Did you know?

Some of the early low-cost radios used only two germanium (rather than silicon) *pnp* transistors to form two amplifier stages. (Silicon transistors became manufacturable later.) However, these radios had poor performance and could receive only one or two stations. For this reason, many manufacturers would proudly print the number of the transistors inside a radio on its front panel along with the brand name, e.g., "Admiral—Eight Transistors."

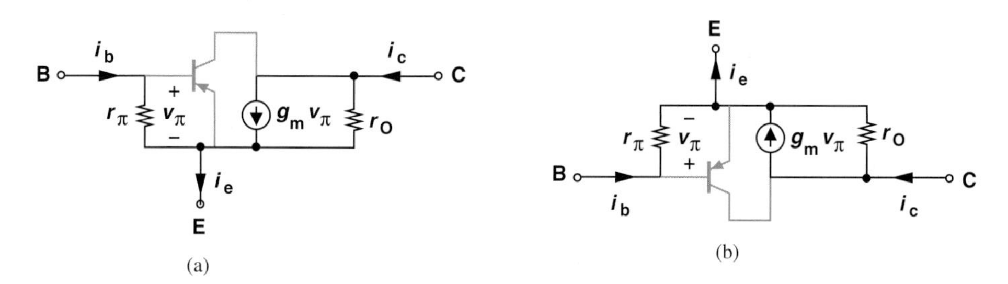

Figure 4.44 (a) Small-signal model of *pnp* transistor, (b) more intuitive view of (a).

The small-signal model of *pnp* transistors may cause confusion, especially if drawn as in Fig. 4.44(b). In analogy with *npn* transistors, one may automatically assume that the "top" terminal is the collector and hence the model in Fig. 4.44(b) is not identical to that in Fig. 4.31(b). We caution the reader about this confusion. A few examples prove helpful here.

Example 4-19

If the collector and base of a bipolar transistor are tied together, a two-terminal device results. Determine the small-signal impedance of the devices shown in Fig. 4.45(a). Assume $V_A = \infty$.

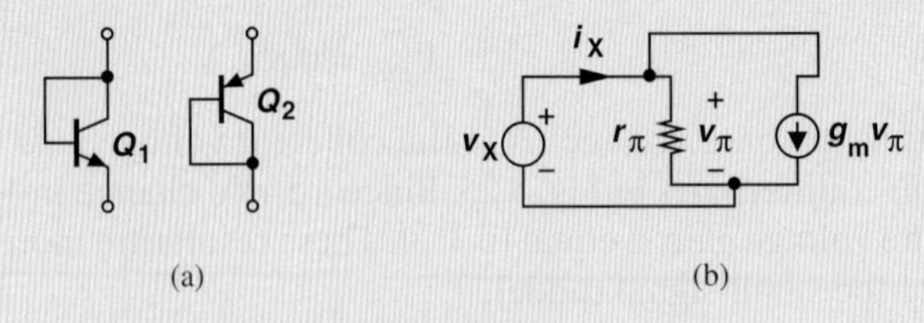

Figure 4.45

Solution We replace the bipolar transistor Q_1 with its small-signal model and apply a small-signal voltage across the device [Fig. 4.45(b)]. Noting that r_{pi} carries a current equal to v_X/r_π, we write a KCL at the input node:

$$\frac{v_X}{r_\pi} + g_m v_\pi = i_X. \tag{4.120}$$

Since $g_m r_\pi = \beta \gg 1$, we have

$$\frac{v_X}{i_X} = \frac{1}{g_m + r_\pi^{-1}} \tag{4.121}$$

$$\approx \frac{1}{g_m} \tag{4.122}$$

$$= \frac{V_T}{I_C}. \tag{4.123}$$

Interestingly, with a bias current of I_C, the device exhibits an impedance similar to that of a diode carrying the same bias current. We call this structure a "diode-connected transistor." The same results apply to the *pnp* configuration in Fig. 4.45(a).

Exercise What is the impedance of a diode-connected device operating at a current of 1 mA?

Example 4-20

Draw the small-signal equivalent circuits for the topologies shown in Figs. 4.46(a)–(c) and compare the results.

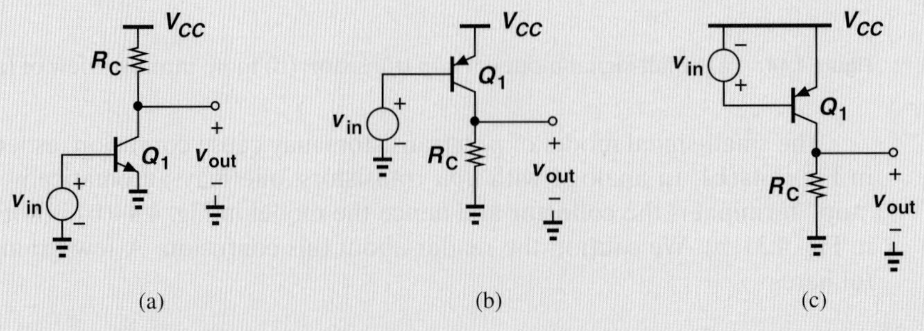

Figure 4.46 (a) Simple stage using an *npn* transistor, (b) simple stage using a *pnp* transistor, (c) another *pnp* stage, (d) small-signal equivalent of (a), (e) small-signal equivalent of (b), (f) small-signal equivalent of (f).

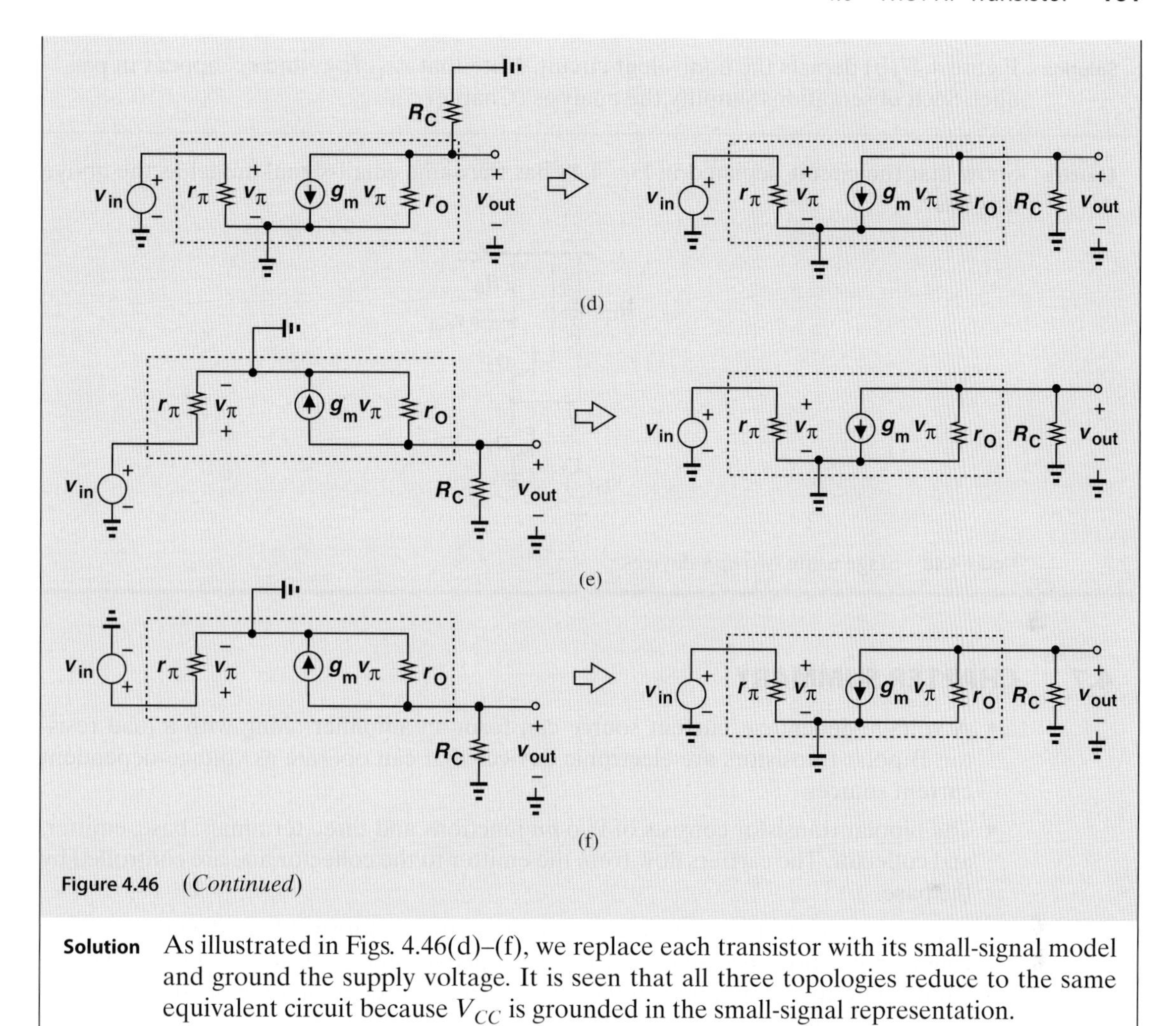

Figure 4.46 (*Continued*)

Solution As illustrated in Figs. 4.46(d)–(f), we replace each transistor with its small-signal model and ground the supply voltage. It is seen that all three topologies reduce to the same equivalent circuit because V_{CC} is grounded in the small-signal representation.

Exercise Repeat the preceding example if a resistor is placed between the collector and base of each transistor.

Example 4-21 Draw the small-signal equivalent circuit for the amplifier shown in Fig. 4.47(a).

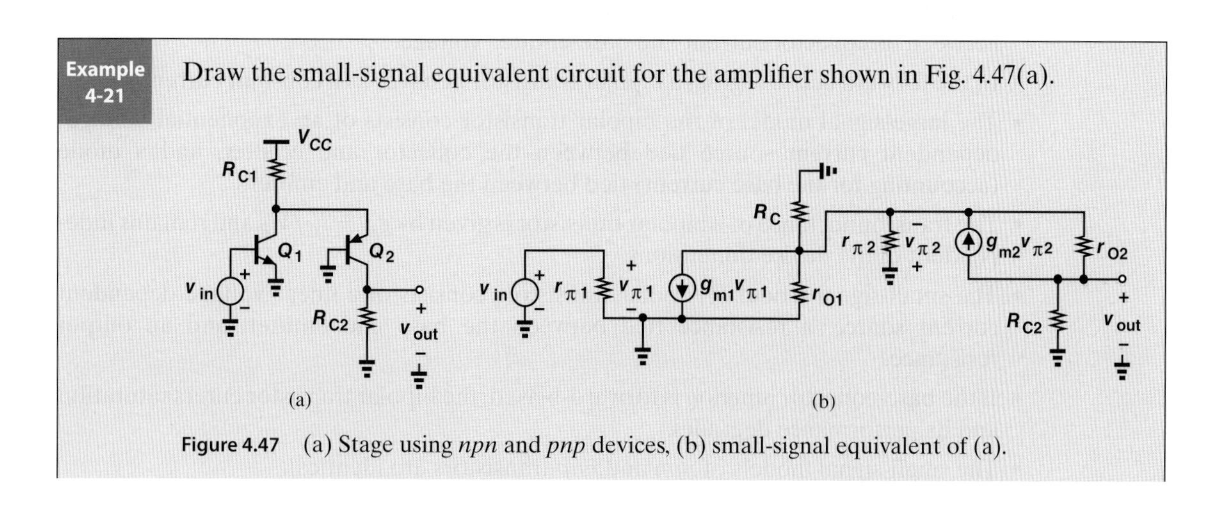

Figure 4.47 (a) Stage using *npn* and *pnp* devices, (b) small-signal equivalent of (a).

Solution Figure 4.47(b) depicts the equivalent circuit. Note that r_{O1}, R_{C1}, and $r_{\pi 2}$ appear in parallel. Such observations simplify the analysis (Chapter 5).

Exercise Show that the circuit depicted in Fig. 4.48 has the same small-signal model as the above amplifier.

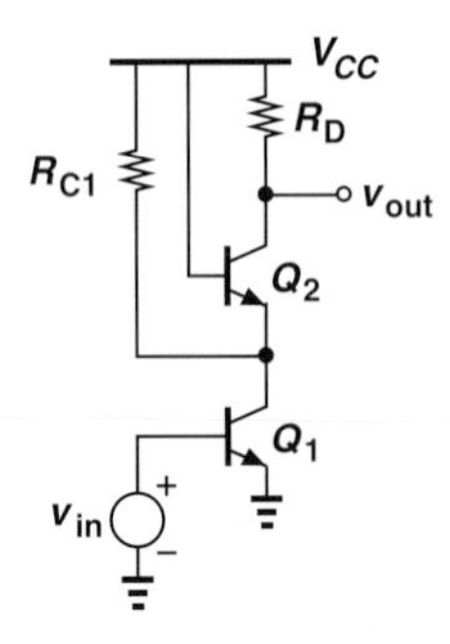

Figure 4.48 Stage using two *npn* devices.

4.7 CHAPTER SUMMARY

- A voltage-dependent current source can form an amplifier along with a load resistor. Bipolar transistors are electronic devices that can operate as voltage-dependent current sources.
- The bipolar transistor consists of two *pn* junctions and three terminals: base, emitter, and collector. The carriers flow from the emitter to the collector and are controlled by the base.
- For proper operation, the base-emitter junction is forward-biased and the base-collector junction reverse-biased (forward active region). Carriers injected by the emitter into the base approach the edge of collector depletion region and are swept away by the high electric field.
- The base terminal must provide a small flow of carriers, some of which go to the emitter and some others recombine in the base region. The ratio of collector current and base current is denoted by β.
- In the forward active region, the bipolar transistor exhibits an exponential relationship between its collector current and base-emitter voltage.
- In the forward active region, a bipolar transistor behaves as a constant current source.
- The large-signal model of the bipolar transistor consists of an exponential voltage-dependent current source tied between the collector and emitter, and a diode (accounting for the base current) tied between the base and emitter.
- The transconductance of a bipolar transistor is given by $g_m = I_C/V_T$ and remains independent of the device dimensions.
- The small-signal model of bipolar transistors consists of a linear voltage-dependent current source, a resistance tied between the base and emitter, and an output resistance.
- If the base-collector junction is forward-biased, the bipolar transistor enters saturation and its performance degrades.
- The small-signal models of *npn* and *pnp* transistors are identical.

PROBLEMS

In the following problems, unless otherwise stated, assume the bipolar transistors operate in the active mode.

Section 4.1 General Considerations

4.1. Suppose the voltage-dependent current source of Fig. 4.1(a) is constructed with $K = 20\,\text{mA/V}$. What value of load resistance in Fig. 4.1(b) is necessary to achieve a voltage gain of 15?

SOL **4.2.** A resistance of R_S is placed in series with the input voltage source in Fig. 4.2. Determine V_{out}/V_{in}.

4.3. Repeat Problem 4.2 but assuming that r_{in} and K are related: $r_{in} = a/x$ and $K = bx$. Plot the voltage gain as a function of x.

Section 4.3 Operation of Bipolar Transistor in Active Mode

4.4. Due to a manufacturing error, the base width of a bipolar transistor has increased by a factor of two. How does the collector current change?

4.5. In the circuit of Fig. 4.49, it is observed that the collector currents of Q_1 and Q_2 are equal if $V_{BE1} - V_{BE2} = 20\,\text{mV}$. Determine the ratio of transistor cross section areas if the other device parameters are identical.

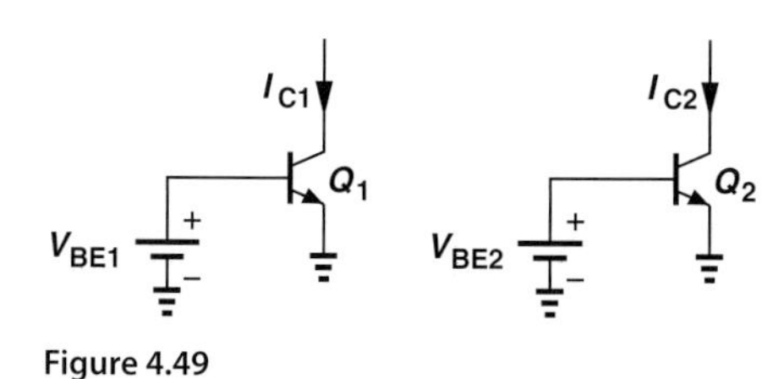

Figure 4.49

4.6. In the circuit of Fig. 4.50 $I_{S1} = I_{S2} = 3 \times 10^{-16}$ A.

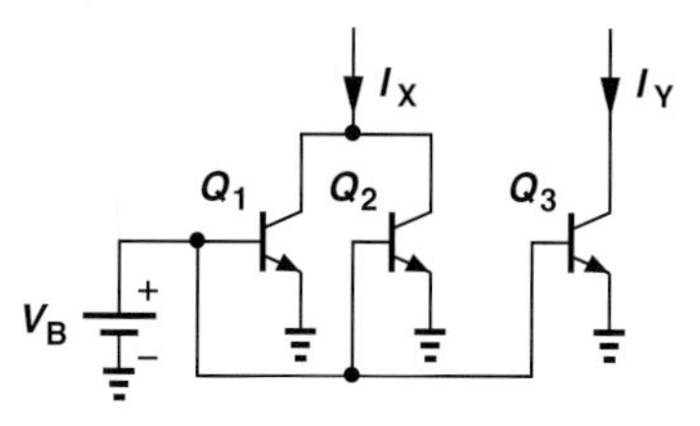

Figure 4.50

(a) Calculate V_B such that $I_X = 1\,\text{mA}$.
(b) With the value of V_B found in (a), choose I_{S3} such that $I_Y = 2.5\,\text{mA}$.

4.7. Consider the circuit shown in Fig. 4.51.

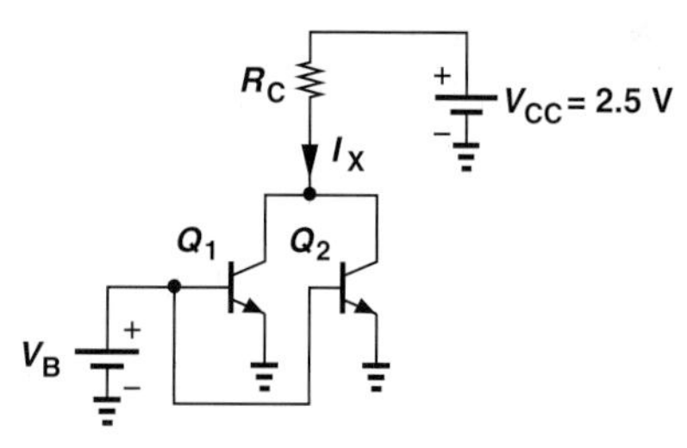

Figure 4.51

(a) If $I_{S1} = 2I_{S2} = 5 \times 10^{-16}$ A, determine V_B such that $I_X = 1.2\,\text{mA}$.
(b) What value of R_C places the transistors at the edge of the active mode?

4.8. Repeat Problem 4.7 if V_{CC} is lowered to 1.5 V. SOL

4.9. Consider the circuit shown in Fig. 4.52. Calculate the value of V_B that places Q_1 at the edge of the active region. Assume $I_S = 5 \times 10^{-16}$ A.

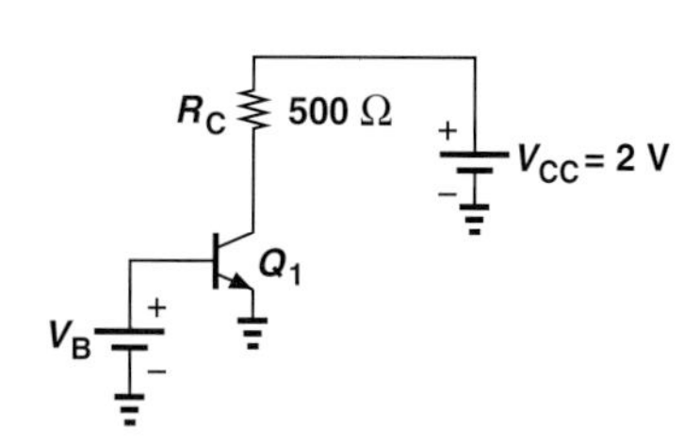

Figure 4.52

4.10. In the circuit of Fig. 4.53, determine the maximum value of V_{CC} that places Q_1 at the edge of saturation. Assume $I_S = 3 \times 10^{-16}$ A.

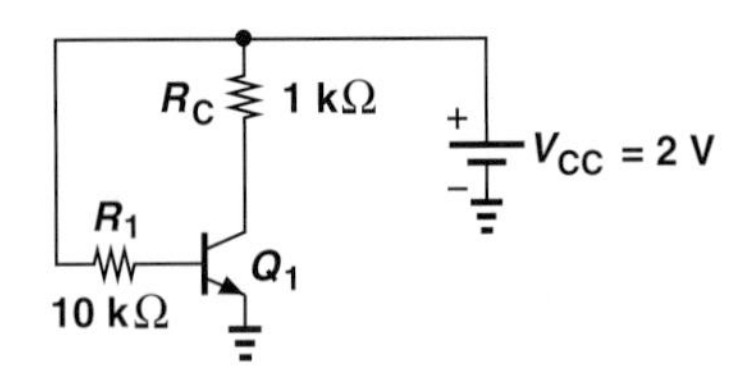

Figure 4.53

4.11. Calculate V_X in Fig. 4.54 if $I_S = 6 \times 10^{-16}$ A.

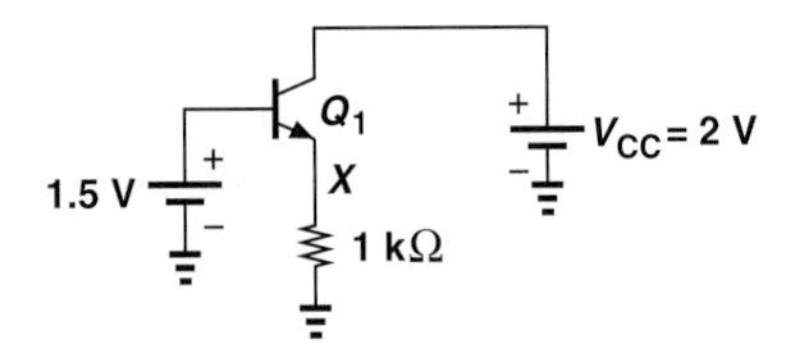

Figure 4.54

4.12. An integrated circuit requires two current sources: $I_1 = 1$ mA and $I_2 = 1.5$ mA. Assuming that only integer multiples of a unit bipolar transistor having $I_S = 3 \times 10^{-16}$ A can be placed in parallel, and only a single voltage source, V_B, is available (Fig. 4.55), construct the required circuit with minimum number of unit transistors.

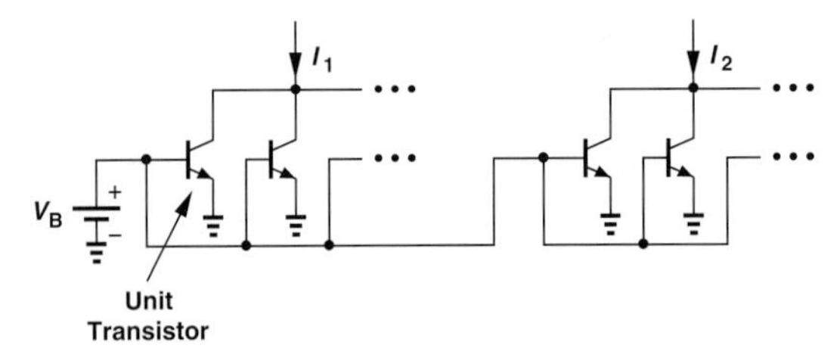

Figure 4.55

SOL **4.13.** Repeat Problem 4.12 for three current sources $I_1 = 0.2$ mA, $I_2 = 0.3$ mA, and $I_3 = 0.45$ mA.

4.14. Consider the circuit shown in Fig. 4.56, assuming $\beta = 100$ and $I_S = 7 \times 10^{-16}$ A. If $R_1 = 10$ kΩ, determine V_B such that $I_C = 1$ mA.

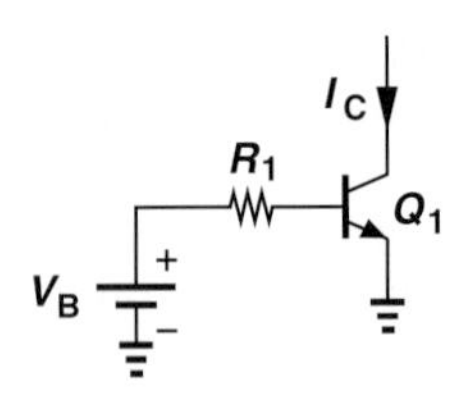

Figure 4.56

4.15. In the circuit of Fig. 4.56, $V_B = 800$ mV and $R_B = 10$ kΩ. Calculate the collector current.

4.16. In the circuit depicted in Fig. 4.57, $I_{S1} = 2I_{S2} = 4 \times 10^{-16}$ A. If $\beta_1 = \beta_2 = 100$ and $R_1 = 5$ kΩ, compute V_B such that $I_X = 1$ mA.

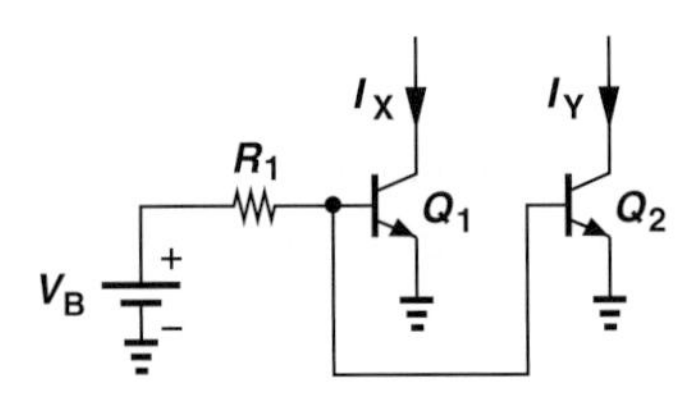

Figure 4.57

4.17. In the circuit of Fig. 4.57, $I_{S1} = 3 \times 10^{-16}$ A, $I_{S2} = 5 \times 10^{-16}$ A, $\beta_1 = \beta_2 = 100$, $R_1 = 5$ kΩ, and $V_B = 800$ mV. Calculate I_X and I_Y.

4.18. The base-emitter junction of a transistor is driven by a constant voltage. Suppose a voltage source is applied between the base and collector. If the device operates in the forward active region, prove that a change in base-collector voltage results in no change in the collector and base currents. (Neglect the Early effect.)

Section 4.4 Bipolar Transistor Models and Characteristics

4.19. A transistor with $I_S = 6 \times 10^{-16}$ A must provide a transconductance of $1/(13\ \Omega)$. What base-emitter voltage is required?

4.20. Most applications require that the transconductance of a transistor remain relatively constant as the signal level varies. Of course, since the signal changes SOL

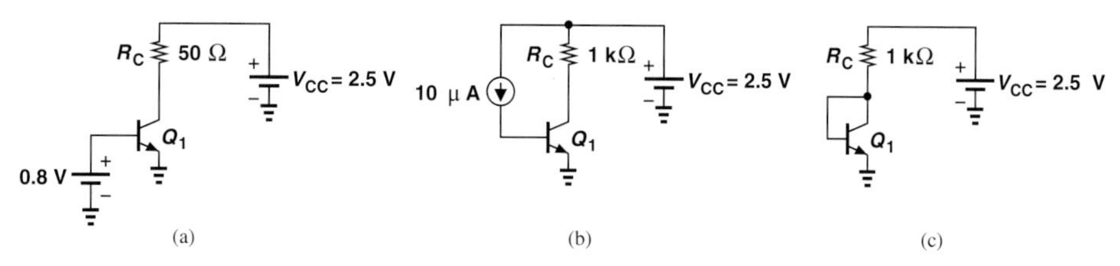

Figure 4.58

the collector current, $g_m = I_C/V_T$, does vary. Nonetheless, proper design ensures negligible variation, e.g., ±10%. If a bipolar device is biased at $I_C = 1$ mA, what is the largest change in V_{BE} that guarantees only ±10% variation in g_m?

4.21. Determine the operating point and the small-signal model of Q_1 for each of the circuits shown in Fig. 4.58. Assume $I_S = 8 \times 10^{-16}$ A, $\beta = 100$, and $V_A = \infty$.

4.22. Determine the operating point and the small-signal model of Q_1 for each of the circuits shown in Fig. 4.59. Assume $I_S = 8 \times 10^{-16}$ A, $\beta = 100$, and $V_A = \infty$.

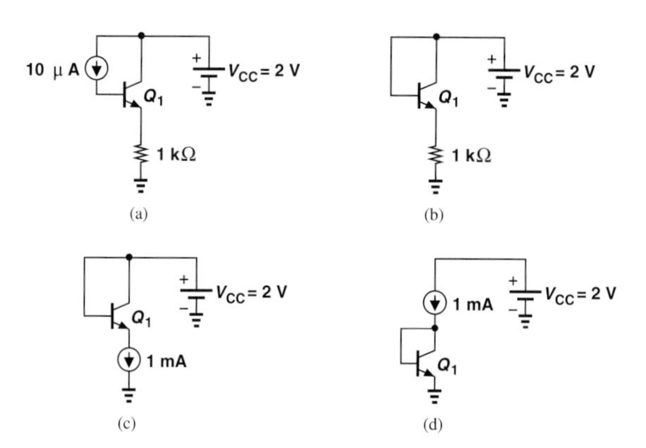

Figure 4.59

4.23. A fictitious bipolar transistor exhibits an I_C-V_{BE} characteristic given by

$$I_C = I_S \exp \frac{V_{BE}}{nV_T}, \qquad (4.124)$$

where n is a constant coefficient. Construct the small-signal model of the device if I_C is still equal to βI_B.

*__4.24.__ A fictitious bipolar transistor exhibits the following relationship between its base and collector currents:

$$I_C = aI_B^2, \qquad (4.125)$$

where a is a constant coefficient. Construct the small-signal model of the device if I_C is still equal to $I_S \exp(V_{BE}/V_T)$.

4.25. The collector voltage of a bipolar transistor varies from 1 V to 3 V while the base-emitter voltage remains constant. What Early voltage is necessary to ensure that the collector current changes by less than 5%?

*__4.26.__ In the circuit of Fig. 4.60, $I_S = 5 \times 10^{-17}$ A. Determine V_X for (a) $V_A = \infty$, and (b) $V_A = 5$ V.

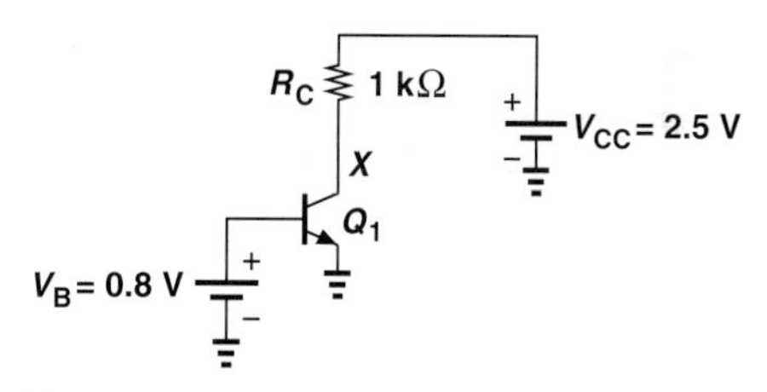

Figure 4.60

4.27. In the circuit of Fig. 4.61, V_{CC} changes from 2.5 to 3 V. Assuming $I_S = 1 \times 10^{-17}$ A and $V_A = 5$ V, determine the change in the collector current of Q_1.

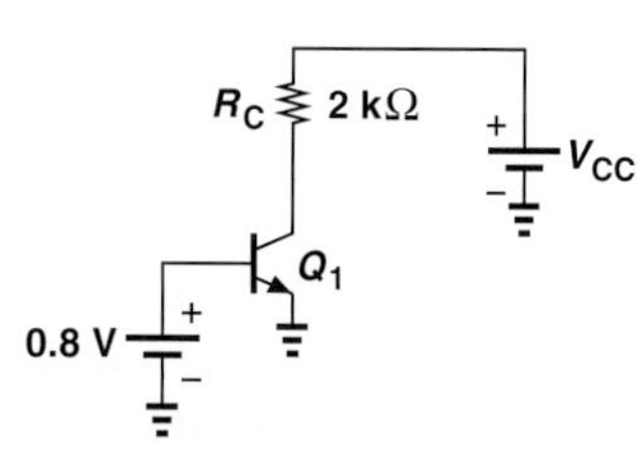

Figure 4.61

4.28. In Problem 4.27, we wish to decrease V_B to compensate for the change in I_C. Determine the new value of V_B.

*__4.29.__ Consider the circuit shown in Fig. 4.62, where I_1 is a 1-mA ideal current source and $I_S = 3 \times 10^{-17}$ A.

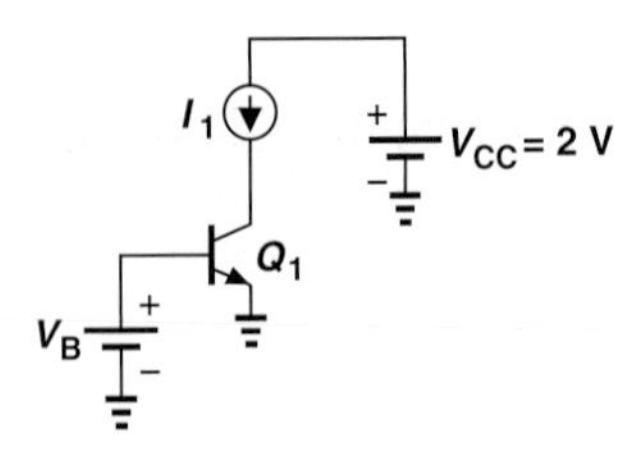

Figure 4.62

(a) Assuming $V_A = \infty$, determine V_B such that $I_C = 1$ mA.
(b) If $V_A = 5$ V, determine V_B such that $I_C = 1$ mA for a collector-emitter voltage of 1.5 V.

SOL **4.30.** A bipolar current source is designed for an output current of 2 mA. What value of V_A guarantees an output resistance greater than 10 kΩ?

4.31. In the circuit of Fig. 4.63, n identical transistors are placed in parallel. If $I_S = 5 \times 10^{-16}$ A and $V_A = 8$ V for each device, construct the small-signal model of the equivalent transistor.

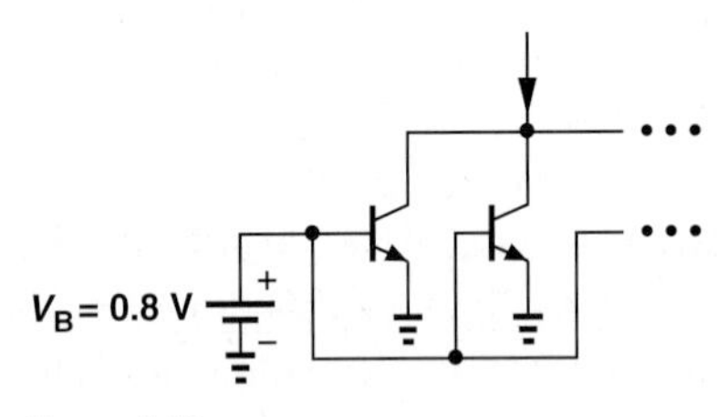

Figure 4.63

4.32. Consider the circuit shown in Fig. 4.64, where $I_S = 6 \times 10^{-16}$ A and $V_A = \infty$.
(a) Determine V_B such that Q_1 operates at the edge of the active region.
(b) If we allow soft saturation, e.g., a collector-base forward bias of 200 mV, by how much can V_B increase?

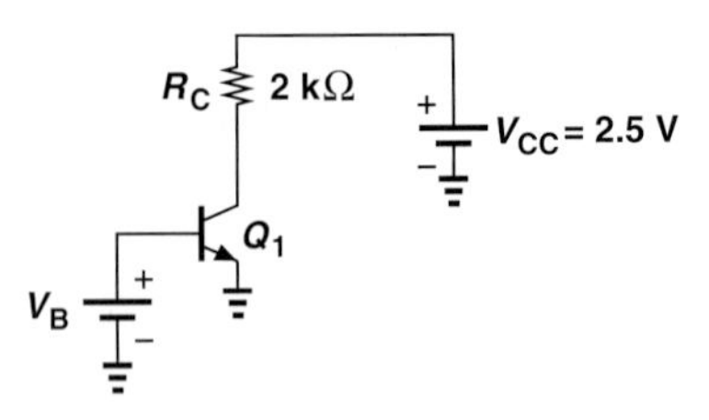

Figure 4.64

4.33. For the circuit depicted in Fig. 4.65, calculate the maximum value of V_{CC} that produces a collector-base forward bias of 200 mV. Assume $I_S = 7 \times 10^{-16}$ A and $V_A = \infty$.

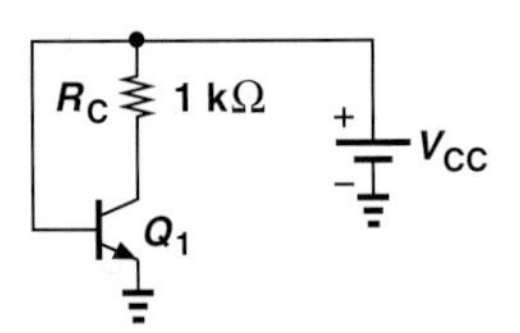

Figure 4.65

*__4.34.__ Assume $I_S = 2 \times 10^{-17}$ A, $V_A = \infty$, and $\beta = 100$ in Fig. 4.66. What is the maximum value of R_C if the collector-base must experience a forward bias of less than 200 mV?

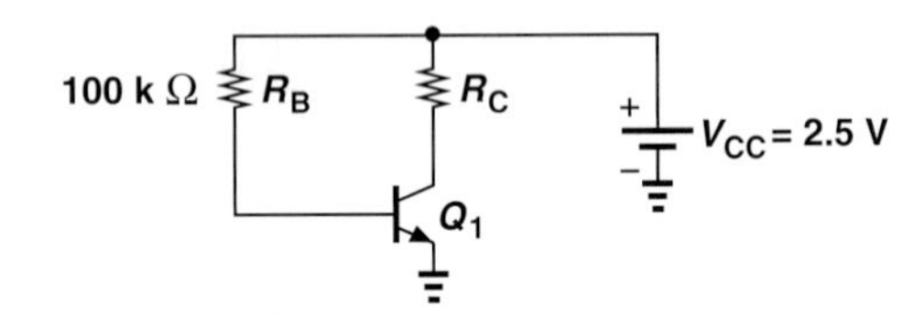

Figure 4.66

4.35. Consider the circuit shown in Fig. 4.67, where $I_S = 5 \times 10^{-16}$ A and $V_A = \infty$. If V_B is chosen to forward-bias the base-collector junction by 200 mV, what is the collector current?

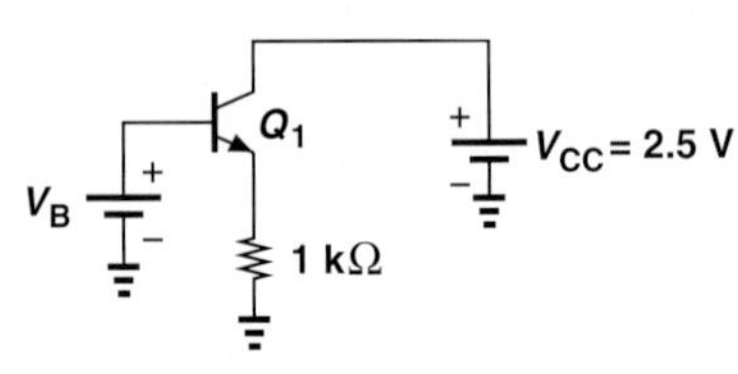

Figure 4.67

4.36. In the circuit of Fig. 4.68, $\beta = 100$ and $V_A = \infty$. Calculate the value of I_S such that the base-collector junction is forward-biased by 200 mV.

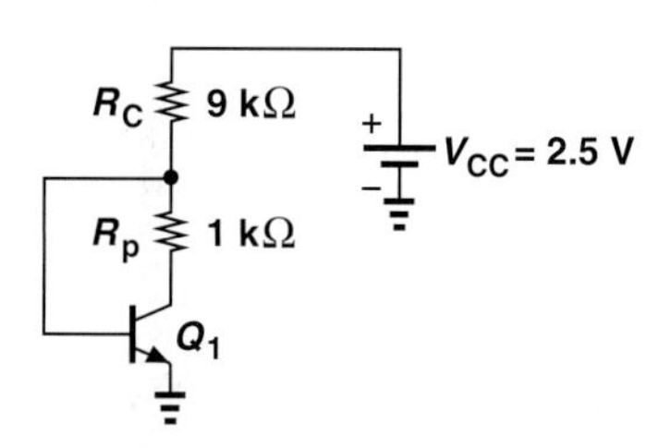

Figure 4.68

Section 4.6 The *PNP* Transistor

SOL **4.37.** If $I_{S1} = 3I_{S2} = 6 \times 10^{-16}$ A, calculate I_X in Fig. 4.69.

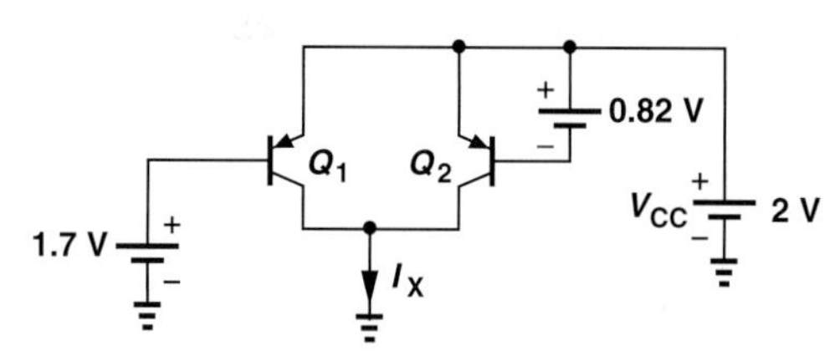

Figure 4.69

4.38. Determine the collector current of Q_1 in Fig. 4.70 if $I_S = 2 \times 10^{-17}$ A and $\beta = 100$.

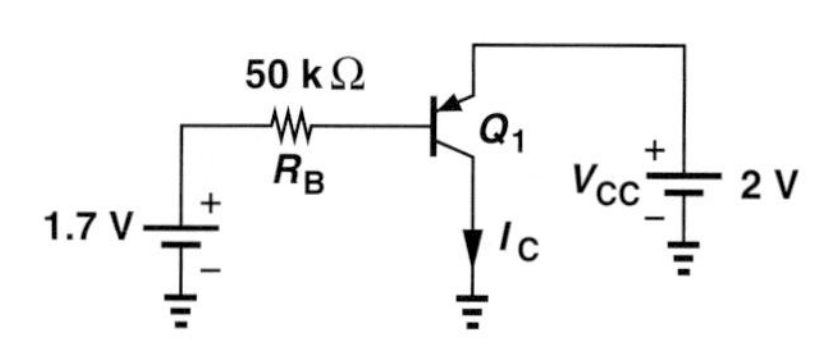

Figure 4.70

4.39. In the circuit of Fig. 4.71, it is observed that $I_C = 3$ mA. If $\beta = 100$, calculate I_S.

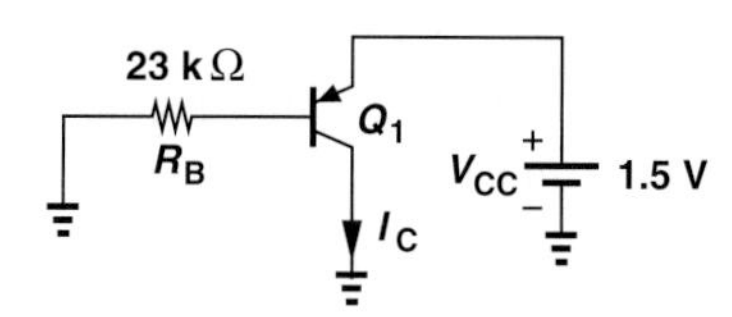

Figure 4.71

4.40. Determine the value of I_S in Fig. 4.72 such that Q_1 operates at the edge of the active mode.

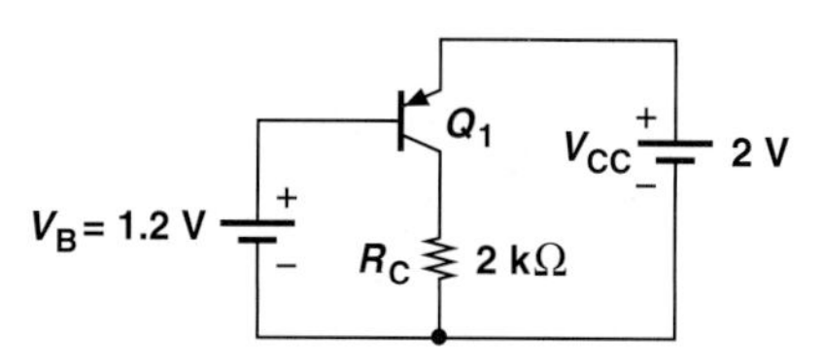

Figure 4.72

4.41. What is the value of β that places Q_1 at the edge of the active mode in Fig. 4.73? Assume $I_S = 8 \times 10^{-16}$ A.

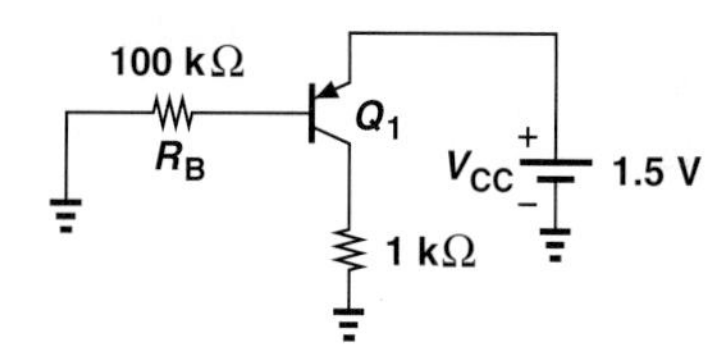

Figure 4.73

4.42. Calculate the collector current of Q_1 in Fig. 4.74 if $I_S = 3 \times 10^{-17}$ A.

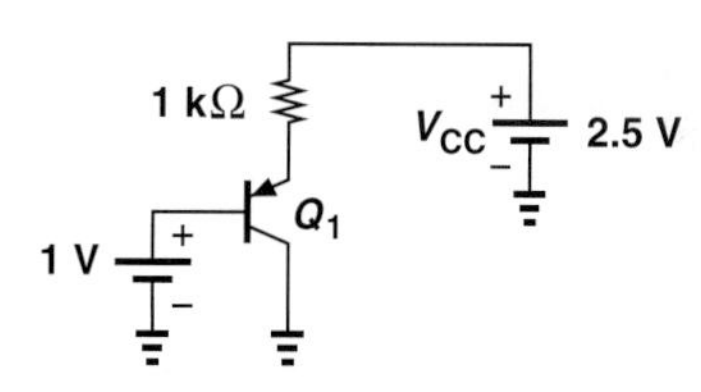

Figure 4.74

***4.43.** Determine the operating point and the small-signal model of Q_1 for each of the circuits shown in Fig. 4.75. Assume $I_S = 3 \times 10^{-17}$ A, $\beta = 100$, and $V_A = \infty$.

***4.44.** Determine the operating point and the small-signal model of Q_1 for each of the circuits shown in Fig. 4.76. Assume $I_S = 3 \times 10^{-17}$ A, $\beta = 100$, and $V_A = \infty$.

4.45. In the circuit of Fig. 4.77, $I_S = 5 \times 10^{-17}$ A. Calculate V_X for (a) $V_A = \infty$, and (b) $V_A = 6$ V.

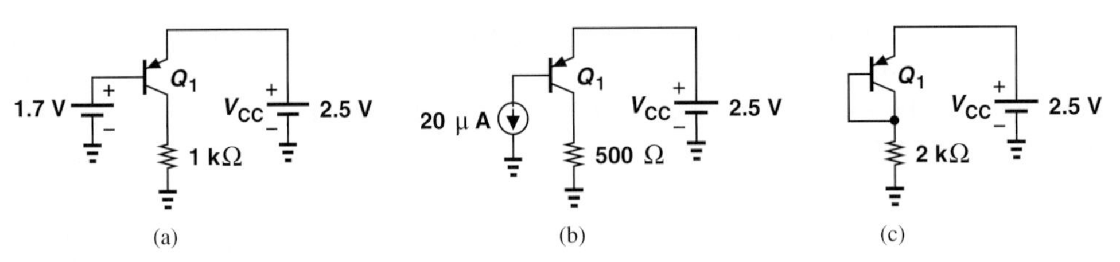

Figure 4.75

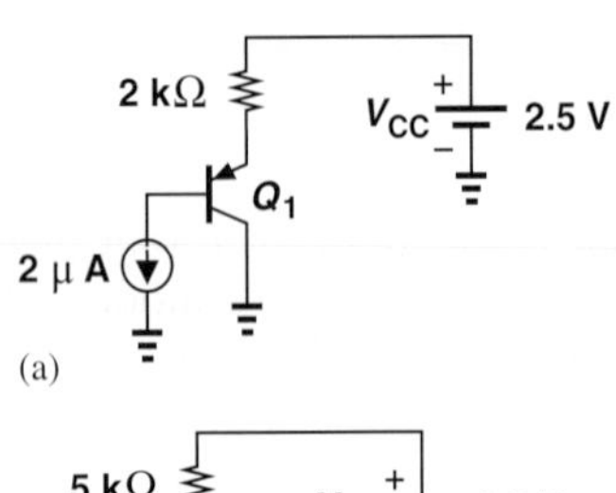

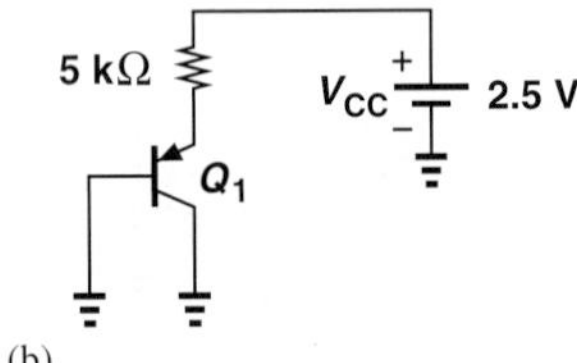

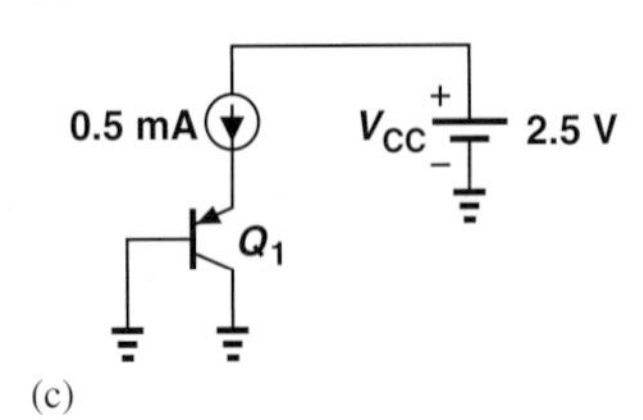

Figure 4.76

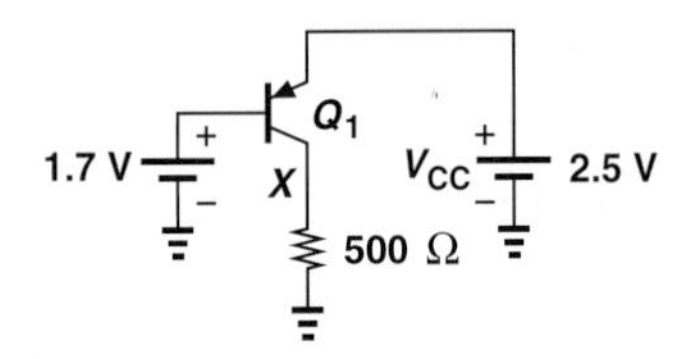

Figure 4.77

4.46. A *pnp* current source must provide an output current of 2 mA with an output resistance of 60 kΩ. What is the required Early voltage?

4.47. Repeat Problem 4.46 for a current of 1 mA and compare the results.

***4.48.** Suppose $V_A = 5\text{ V}$ in the circuit of Fig. 4.78.

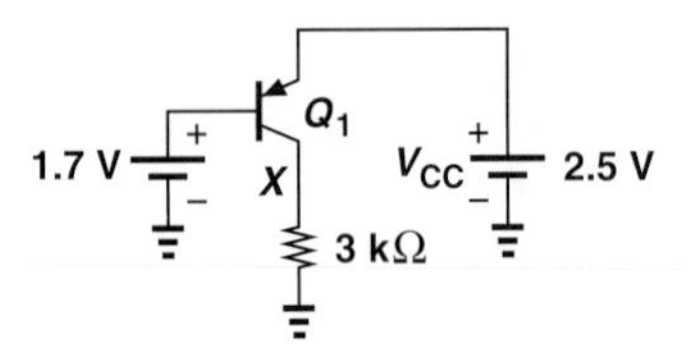

Figure 4.78

(a) What value of I_S places Q_1 at the edge of the active mode?

(b) How does the result in (a) change if $V_A = \infty$?

***4.49.** The terminal currents in the small-signal model of Fig. 4.44 do not seem to agree with those in the large-signal model of Fig. 4.41. Explain why this is not an inconsistency.

***4.50.** Consider the circuit depicted in Fig. 4.79, where $I_S = 6 \times 10^{-16}\text{ A}$, $V_A = 5\text{ V}$, and $I_1 = 2\text{ mA}$.

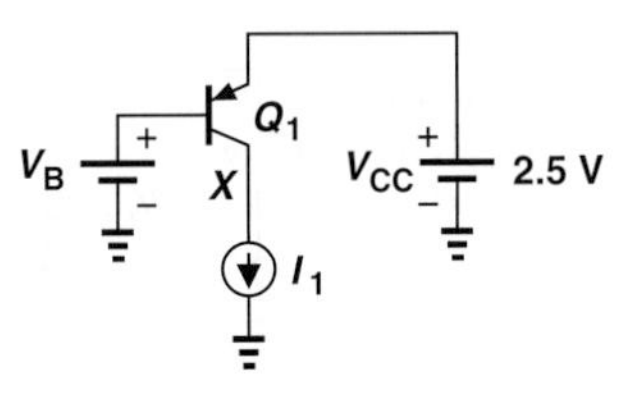

Figure 4.79

(a) What value of V_B yields $V_X = 1\text{ V}$?

(b) If V_B changes from the value found in (a) by 0.1 mV, what is the change in V_X?

(c) Construct the small-signal model of the transistor.

4.51. In the circuit of Fig. 4.80, $\beta = 100$ and $V_A = \infty$.

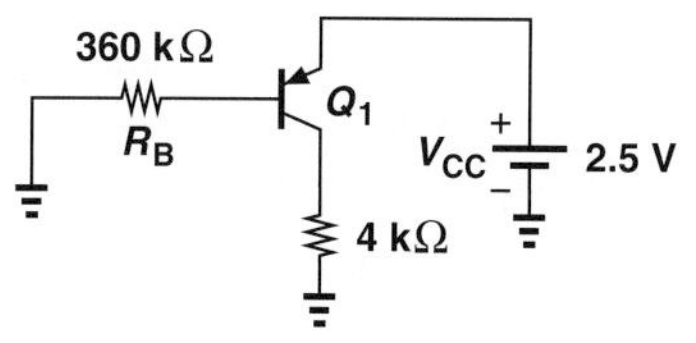

Figure 4.80

(a) Determine I_S such that Q_1 experiences a collector-base forward bias of 200 mV.

(b) Calculate the transconductance of the transistor.

**4.52. Determine the region of operation of Q_1 in each of the circuits shown in Fig. 4.81. Assume $I_S = 5 \times 10^{-16}$ A, $\beta = 100$, $V_A = \infty$.

**4.53. Consider the circuit shown in Fig. 4.82, where $I_{S1} = 3I_{S2} = 5 \times 10^{-16}$ A, $\beta_1 = 100$, $\beta_2 = 50$, $V_A = \infty$, and $R_C = 500\ \Omega$.

(a) We wish to forward-bias the collector-base junction of Q_2 by no more than 200 mV. What is the maximum allowable value of V_{in}?

(b) With the value found in (a), calculate the small-signal parameters of Q_1 and Q_2 and construct the equivalent circuit.

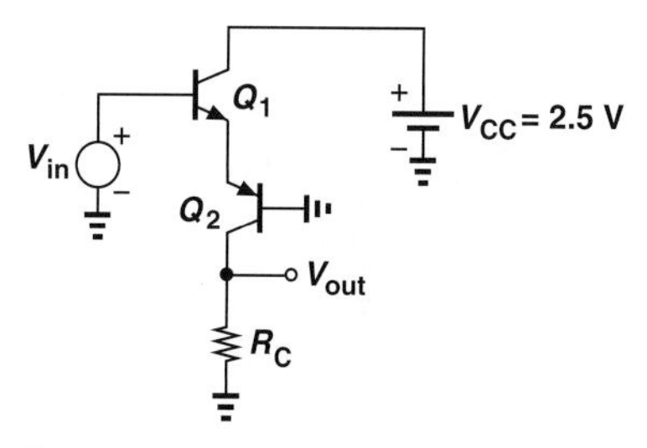

Figure 4.82

**4.54. Repeat Problem 4.53 for the circuit depicted in Fig. 4.83 but for part (a), determine the minimum allowable value of V_{in}. Verify that Q_1 operates in the active mode.

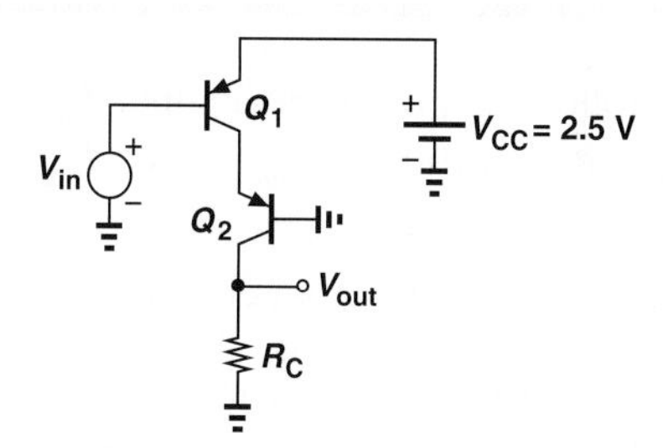

Figure 4.83

**4.55. Repeat Problem 4.53 for the circuit illustrated in Fig. 4.84.

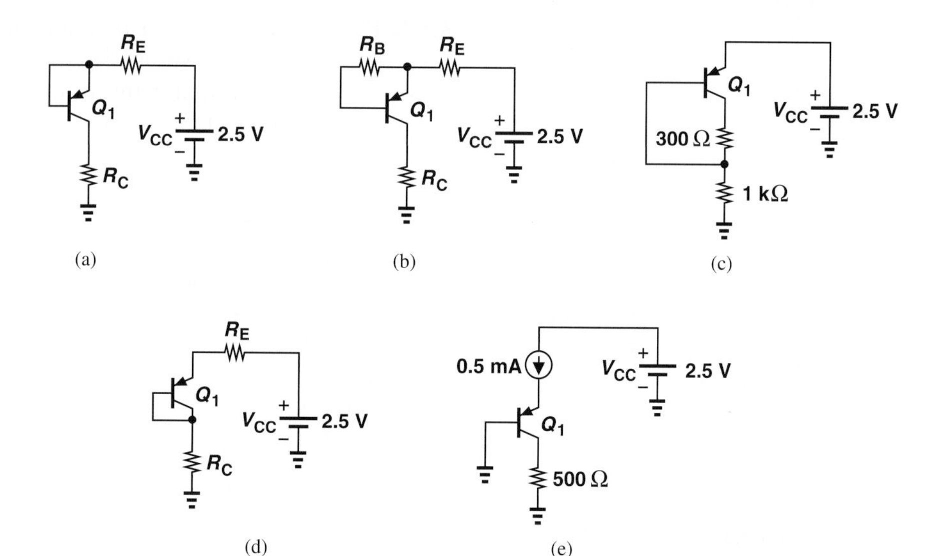

Figure 4.81

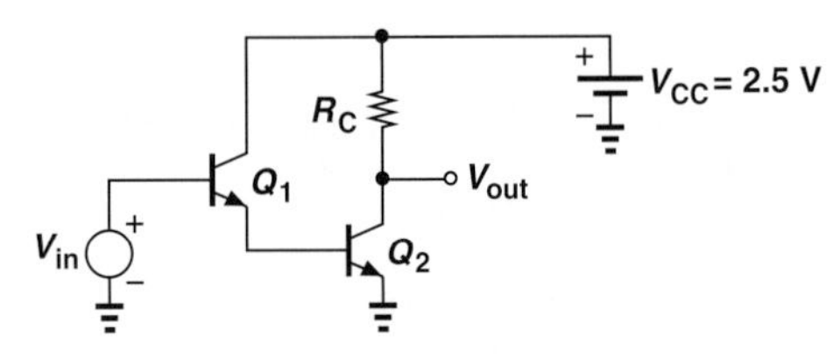

Figure 4.84

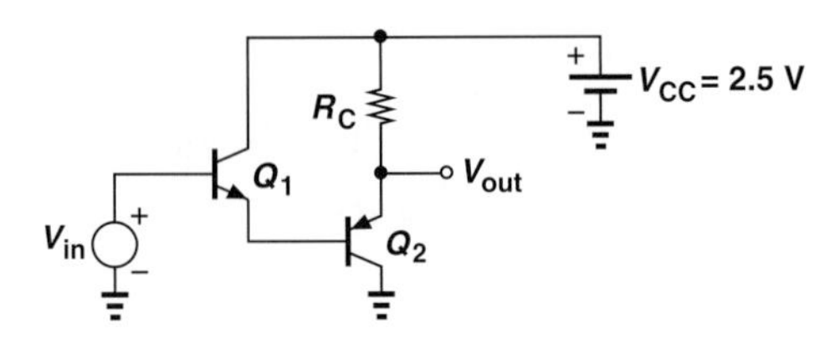

Figure 4.85

4.56. In the circuit of Fig. 4.85, $I_{S1} = 2I_{S2} = 6 \times 10^{-17}$ A, $\beta_1 = 80$ and $\beta_2 = 100$.
(a) What value of V_{in} yields a collector current of 2 mA for Q_2?
(b) With the value found in (a), calculate the small-signal parameters of Q_1 and Q_2 and construct the equivalent circuit.

SPICE PROBLEMS

In the following problems, assume $I_{S,npn} = 5 \times 10^{-16}$ A, $\beta_{npn} = 100$, $V_{A,npn} = 5$ V, $I_{S,pnp} = 8 \times 10^{-16}$ A, $\beta_{pnp} = 50$, $V_{A,pnp} = 3.5$ V.

4.57. Plot the input/output characteristic of the circuit shown in Fig. 4.86 for $0 < V_{in} < 2.5$ V. What value of V_{in} places the transistor at the edge of saturation?

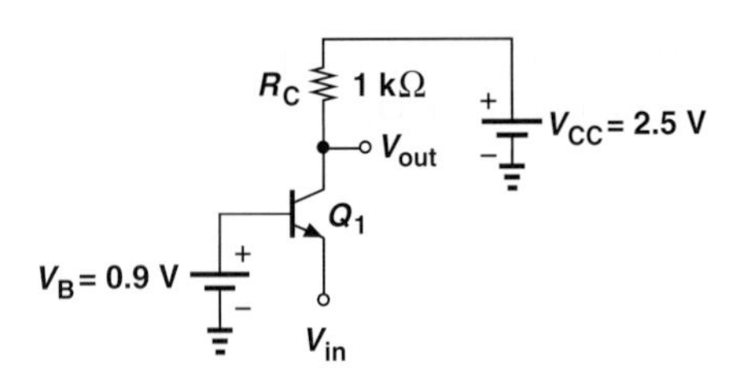

Figure 4.86

4.58. Repeat Problem 4.57 for the stage depicted in Fig. 4.87. At what value of V_{in} does Q_1 carry a collector current of 1 mA?

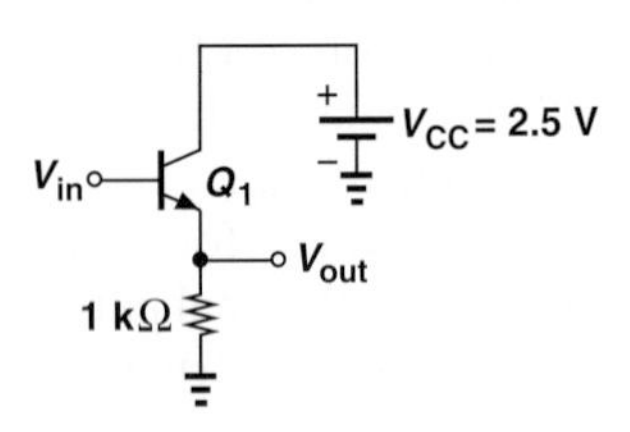

Figure 4.87

4.59. Plot I_{C1} and I_{C2} as a function of V_{in} for the circuits shown in Fig. 4.88 for $0 < V_{in} < 1.8$ V. Explain the dramatic difference between the two.

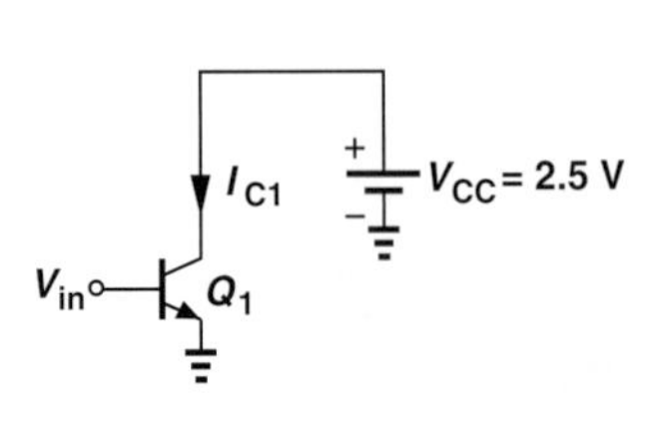

(a)

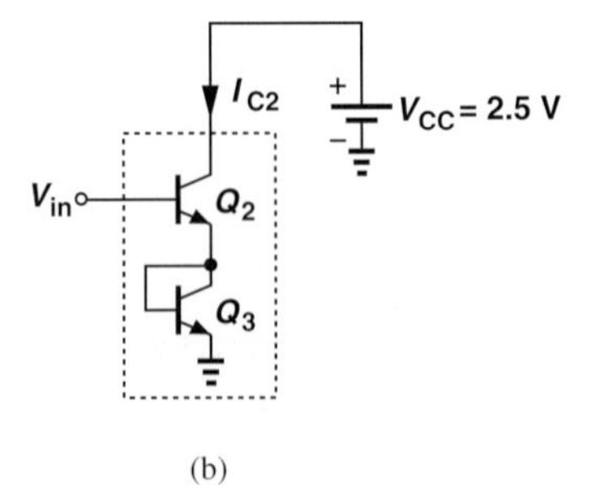

(b)

Figure 4.88

4.60. Plot the input/output characteristic of the circuit illustrated in Fig. 4.89 for $0 < V_{in} < 2$ V. What value of V_{in} yields a transconductance of $(50\ \Omega)^{-1}$ for Q_1?

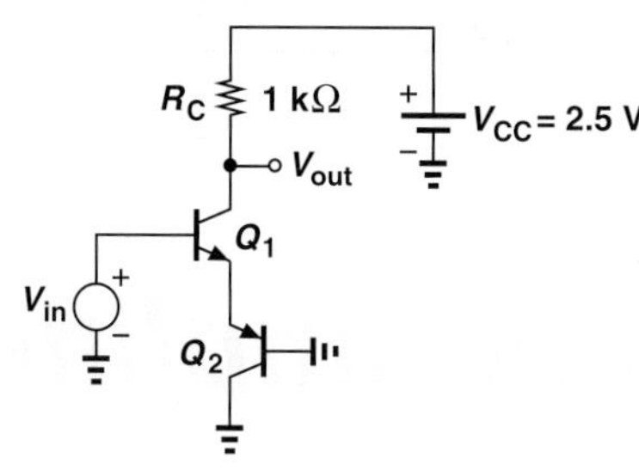

Figure 4.89

4.61. Plot the input/output characteristic of the stage shown in Fig. 4.90 for $0 < V_{in} < 2.5$ V. At what value of V_{in} do Q_1 and Q_2 carry equal collector currents? Can you explain this result intuitively?

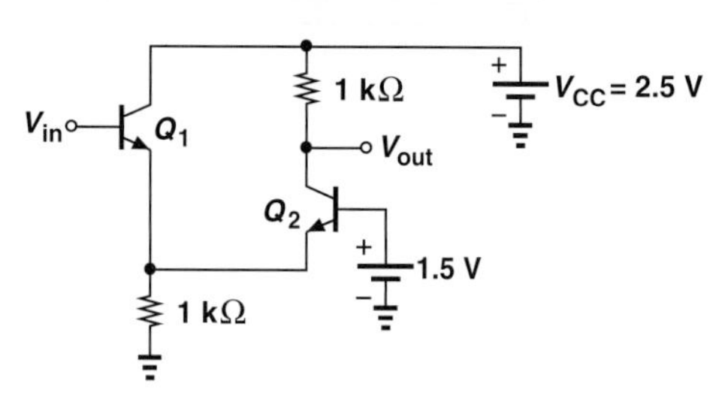

Figure 4.90

5

Bipolar Amplifiers

Watch Companion YouTube Videos:

Razavi Electronics 1, Lec 19: Evolution of Amplifiers
https://www.youtube.com/watch?v=S4RgdPdBKY0

Razavi Electronics 1, Lec 20: Common-Emitter Stage
https://www.youtube.com/watch?v=nl5ut-WpV9Y

Razavi Electronics 1, Lec 21: Input & Output Impedances
https://www.youtube.com/watch?v=HLJWeQYI9QI

Razavi Electronics 1, Lec 22: Common-Emitter Stage with Degeneration
https://www.youtube.com/watch?v=3J3r7o39aCM

Razavi Electronics 1, Lec 23: More on Emitter Degeneration
https://www.youtube.com/watch?v=jvUAUqlk0nA

Razavi Electronics 1, Lec 24: Biasing Techniques I
https://www.youtube.com/watch?v=2oxuxlZaq-Q

Razavi Electronics 1, Lec 25: Biasing Techniques II
https://www.youtube.com/watch?v=fo5hlmLZefY

Razavi Electronics 1, Lec 26: Common-Base Stage
https://www.youtube.com/watch?v=G7AkWcjtWD4

Razavi Electronics 1, Lec 27: Emitter Followers
https://www.youtube.com/watch?v=pQcH28h4vxk

Razavi Electronics 1, Lec 28: Emitter Followers & Summary
https://www.youtube.com/watch?v=NoExn4dkdSw

With the physics and operation of bipolar transistors described in Chapter 4, we now deal with amplifier circuits employing such devices. While the field of microelectronics involves much more than amplifiers, our study of cellphones and digital cameras in Chapter 1 indicates the extremely wide usage of amplification, motivating us to master the analysis and design of such building blocks. This chapter proceeds as follows.

General Concepts	Operating Point Analysis	Amplifier Topologies
• Input and Output Impedances • Biasing • DC and Small–Signal Analysis	• Simple Biasing • Emitter Degeneration • Self-Biasing • Biasing of *PNP* Devices	• Common–Emitter Stage • Common–Base Stage • Emitter Follower

Building the foundation for the remainder of this book, this chapter is quite long. Most of the concepts introduced here are invoked again in Chapter 7 (MOS amplifiers). The reader is therefore encouraged to take frequent breaks and absorb the material in small doses.

5.1 GENERAL CONSIDERATIONS

Recall from Chapter 4 that a voltage-controlled current source along with a load resistor can form an amplifier. In general, an amplifier produces an output (voltage or current) that is a magnified version of the input (voltage or current). Since most electronic circuits both sense and produce voltage quantities,[1] our discussion primarily centers around "voltage amplifiers" and the concept of "voltage gain," v_{out}/v_{in}.

What other aspects of an amplifier's performance are important? Three parameters that readily come to mind are (1) power dissipation (e.g., because it determines the battery lifetime in a cellphone or a digital camera); (2) speed (e.g., some amplifiers in a cellphone or analog-to-digital converters in a digital camera must operate at high frequencies); (3) noise (e.g., the front-end amplifier in a cellphone or a digital camera processes small signals and must introduce negligible noise of its own).

5.1.1 Input and Output Impedances

In addition to the above parameters, the input and output (I/O) impedances of an amplifier play a critical role in its capability to interface with preceding and following stages. To understand this concept, let us first determine the I/O impedances of an *ideal* voltage amplifier. At the input, the circuit must operate as a voltmeter, i.e., sense a voltage without disturbing (loading) the preceding stage. The ideal input impedance is therefore infinite. At the output, the circuit must behave as a voltage source, i.e., deliver a constant signal level to any load impedance. Thus, the ideal output impedance is equal to zero.

In reality, the I/O impedances of a voltage amplifier may considerably depart from the ideal values, requiring attention to the interface with other stages. The following example illustrates the issue.

Example 5-1

An amplifier with a voltage gain of 10 senses a signal generated by a microphone and applies the amplified output to a speaker [Fig. 5.1(a)]. Assume the microphone can be modeled with a voltage source having a 10-mV peak-to-peak signal and a series resistance of 200 Ω. Also assume the speaker can be represented by an 8-Ω resistor.

[1]Exceptions are described in Chapter 12.

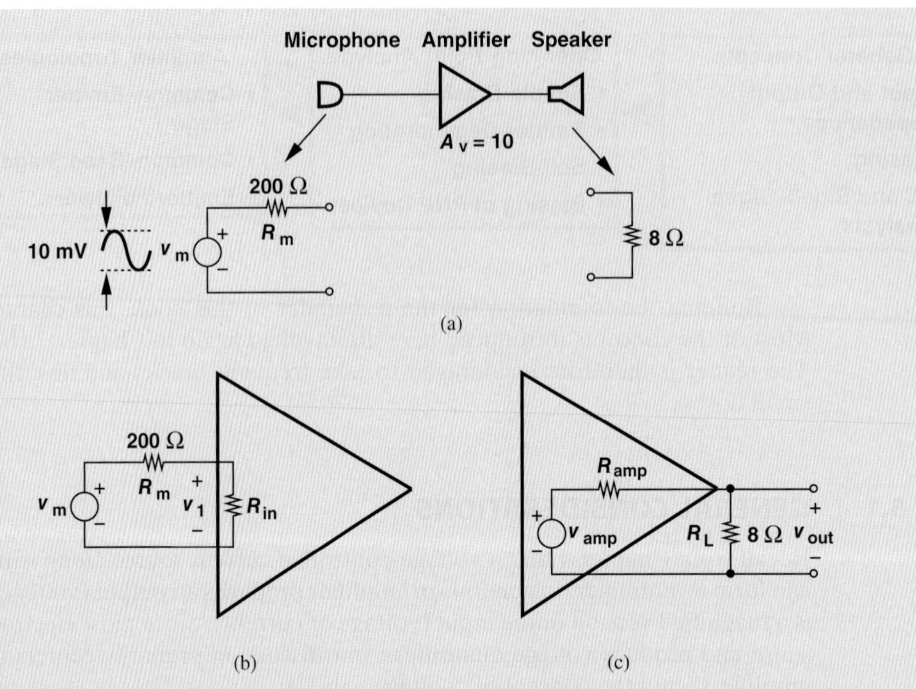

Figure 5.1 (a) Simple audio system, (b) signal loss due to amplifier input impedance, (c) signal loss due to amplifier output impedance.

(a) Determine the signal level sensed by the amplifier if the circuit has an input impedance of 2 kΩ or 500 Ω.

(b) Determine the signal level delivered to the speaker if the circuit has an output impedance of 10 Ω or 2 Ω.

Solution (a) Figure 5.1(b) shows the interface between the microphone and the amplifier. The voltage sensed by the amplifier is therefore given by

$$v_1 = \frac{R_{in}}{R_{in} + R_m} v_m. \tag{5.1}$$

For $R_{in} = 2\ \text{k}\Omega$,

$$v_1 = 0.91 v_m, \tag{5.2}$$

only 9% less than the microphone signal level. On the other hand, for $R_{in} = 500\ \Omega$,

$$v_1 = 0.71 v_m, \tag{5.3}$$

i.e., nearly 30% loss. It is therefore desirable to maximize the input impedance in this case.

(b) Drawing the interface between the amplifier and the speaker as in Fig. 5.1(c), we have

$$v_{out} = \frac{R_L}{R_L + R_{amp}} v_{amp}. \tag{5.4}$$

For $R_{amp} = 10\ \Omega$,

$$v_{out} = 0.44 v_{amp}, \tag{5.5}$$

a substantial attenuation. For $R_{amp} = 2\ \Omega$,

$$v_{out} = 0.8 v_{amp}. \tag{5.6}$$

Thus, the output impedance of the amplifier must be minimized.

Exercise If the signal delivered to the speaker is equal to $0.2v_m$, find the ratio of R_m and R_L.

The importance of I/O impedances encourages us to carefully prescribe the method of measuring them. As with the impedance of two-terminal devices such as resistors and capacitors, the input (output) impedance is measured between the input (output) nodes of the circuit while all independent sources in the circuit are set to zero.[2] Illustrated in Fig. 5.2, the method involves applying a voltage source to the two nodes (also called "ports") of interest, measuring the resulting current, and defining v_X/i_X as the impedance. Also shown are arrows to denote "looking into" the input or output port and the corresponding impedance.

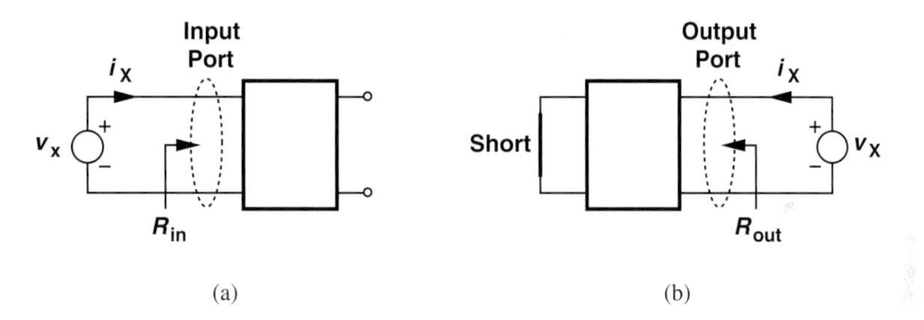

Figure 5.2 Measurement of (a) input and (b) output impedances.

The reader may wonder why the output port in Fig. 5.2(a) is left open whereas the input port in Fig. 5.2(b) is shorted. Since a voltage amplifier is driven by a voltage source during normal operation, and since all independent sources must be set to zero, the input port in Fig. 5.2(b) must be shorted to represent a zero voltage source. That is, the procedure for calculating the output impedance is identical to that used for obtaining the Thevenin impedance of a circuit (Chapter 1). In Fig. 5.2(a), on the other hand, the output remains open because it is not connected to any external sources.

Determining the transfer of signals from one stage to the next, the I/O impedances are usually regarded as small-signal quantities—with the tacit assumption that the signal levels are indeed small. For example, the input impedance is obtained by applying a small change in the input voltage and measuring the resulting change in the input current. The small-signal models of semiconductor devices therefore prove crucial here.

[2]Recall that a zero voltage source is replaced by a short and a zero current source by an open.

Example 5-2 Assuming that the transistor operates in the forward active region, determine the input impedance of the circuit shown in Fig. 5.3(a).

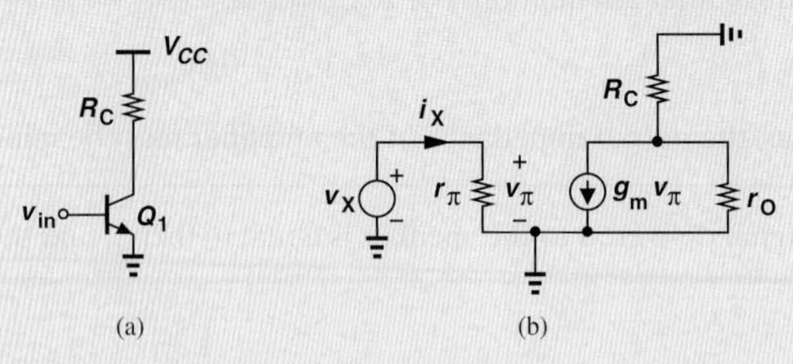

Figure 5.3 (a) Simple amplifier stage, (b) small-signal model.

Solution Constructing the small-signal equivalent circuit depicted in Fig. 5.3(b), we note that the input impedance is simply given by

$$\frac{v_x}{i_x} = r_\pi. \tag{5.7}$$

Since $r_\pi = \beta/g_m = \beta V_T/I_C$, we conclude that a higher β or lower I_C yield a higher input impedance.

Exercise What happens if R_C is doubled?

To simplify the notations and diagrams, we often refer to the impedance seen at a *node* rather than between two nodes (i.e., at a port). As illustrated in Fig. 5.4, such a convention simply assumes that the other node is the ground, i.e., the test voltage source is applied between the node of interest and ground.

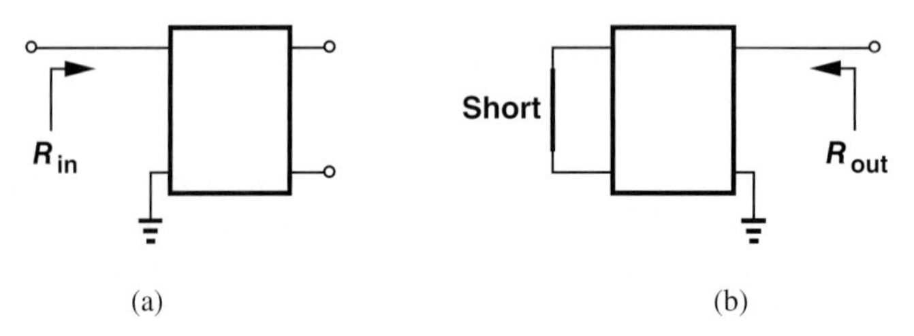

Figure 5.4 Concept of impedance seen at a node.

Example 5-3 Calculate the impedance seen looking into the collector of Q_1 in Fig. 5.5(a).

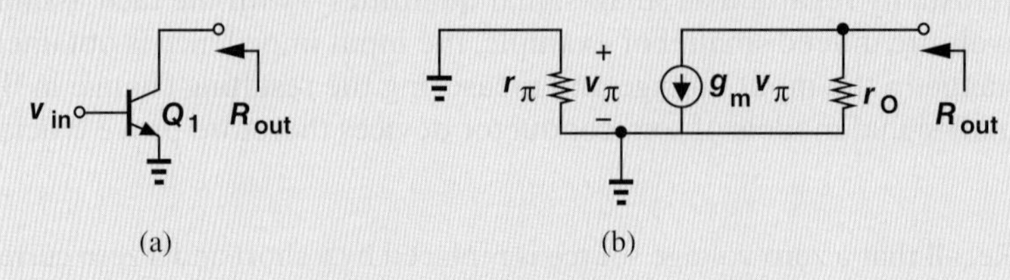

Figure 5.5 (a) Impedance seen at collector, (b) small-signal model.

Solution Setting the input voltage to zero and using the small-signal model in Fig. 5.5(b), we note that $v_\pi = 0$, $g_m v_\pi = 0$, and hence $R_{out} = r_O$.

Exercise What happens if a resistance of value R_1 is placed in series with the base?

Example 5-4 Calculate the impedance seen at the emitter of Q_1 in Fig. 5.6(a). Neglect the Early effect for simplicity.

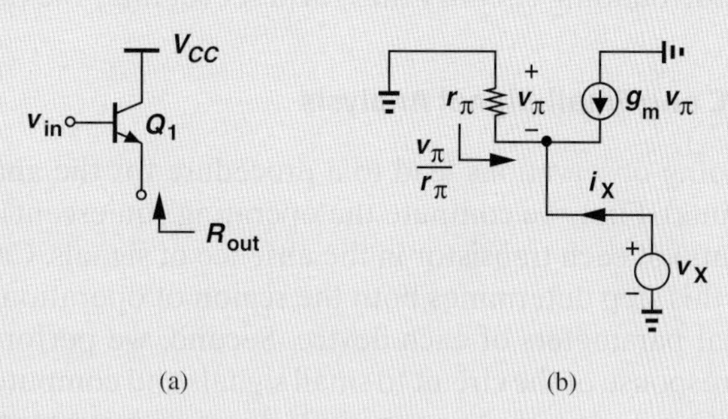

Figure 5.6 (a) Impedance seen at emitter, (b) small-signal model.

Solution Setting the input voltage to zero and replacing V_{CC} with ac ground, we arrive at the small-signal circuit shown in Fig. 5.6(b). Interestingly, $v_\pi = -v_X$ and

$$g_m v_\pi + \frac{v_\pi}{r_\pi} = -i_X. \tag{5.8}$$

That is,

$$\frac{v_X}{i_X} = \frac{1}{g_m + \frac{1}{r_\pi}}. \tag{5.9}$$

Since $r_\pi = \beta/g_m \gg 1/g_m$, we have $R_{out} \approx 1/g_m$.

Exercise What happens if a resistance of value R_1 is placed in series with the collector?

The above three examples provide three important rules that will be used throughout this book (Fig. 5.7): Looking into the base, we see r_π if the emitter is (ac) grounded. Looking into the collector, we see r_O if the emitter is (ac) grounded. Looking into the emitter,

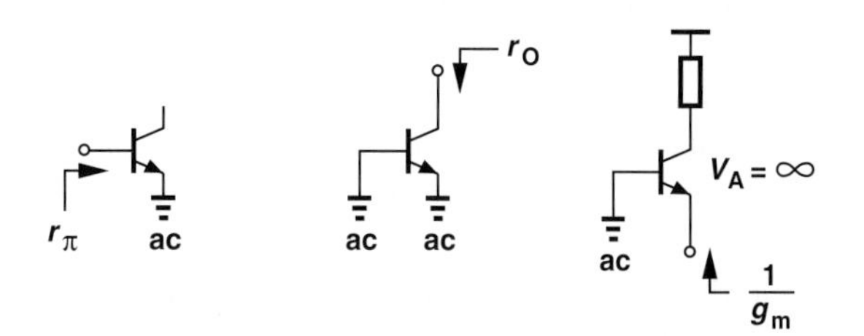

Figure 5.7 Summary of impedances seen at terminals of a transistor.

we see $1/g_m$ if the base is (ac) grounded and the Early effect is neglected. It is imperative that the reader master these rules and be able to apply them in more complex circuits.[3]

5.1.2 Biasing

Recall from Chapter 4 that a bipolar transistor operates as an amplifying device if it is biased in the active mode; that is, in the absence of signals, the environment surrounding the device must ensure that the base-emitter and base-collector junctions are forward- and reverse-biased, respectively. Moreover, as explained in Section 4.4, amplification properties of the transistor such as g_m, r_π, and r_O depend on the quiescent (bias) collector current. Thus, the surrounding circuitry must also set (define) the device bias currents properly.

5.1.3 DC and Small-Signal Analysis

The foregoing observations lead to a procedure for the analysis of amplifiers (and many other circuits). First, we compute the operating (quiescent) conditions (terminal voltages and currents) of each transistor in the absence of signals. Called the "dc analysis" or "bias analysis," this step determines both the region of operation (active or saturation) and the small-signal parameters of each device. Second, we perform "small-signal analysis," i.e., study the response of the circuit to small signals and compute quantities such as the voltage gain and I/O impedances. As an example, Fig. 5.8 illustrates the bias and signal components of a voltage and a current.

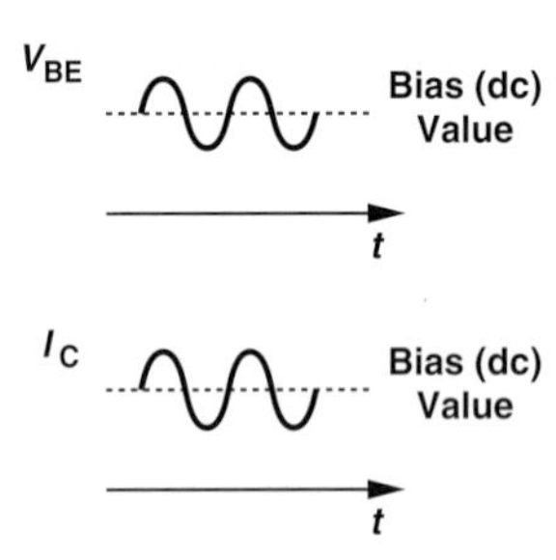

Figure 5.8 Bias and signal levels for a bipolar transistor.

It is important to bear in mind that small-signal analysis deals with only (small) *changes* in voltages and currents in a circuit around their quiescent values. Thus, as mentioned in Section 4.4.4, all *constant* sources, i.e., voltage and current sources that do not vary with time, must be set to zero for small-signal analysis. For example, the supply voltage is constant and, while establishing proper bias points, plays no role in the response to small signals. We therefore ground all constant voltage sources[4] and open all constant current sources while constructing the small-signal equivalent circuit. From another point of view, the two steps described above follow the superposition principle: first, we determine the effect of constant voltages and currents while signal sources are set to zero, and second, we analyze the response to signal sources while constant sources are set to zero. Figure 5.9 summarizes these concepts.

[3]While beyond the scope of this book, it can be shown that the impedance seen at the emitter is approximately equal to $1/g_m$ only if the collector is tied to a relatively low impedance.
[4]We say all constant voltage sources are replaced by an "ac ground."

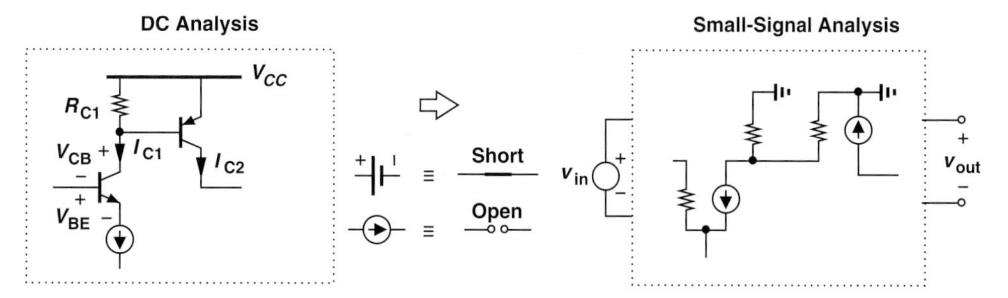

Figure 5.9 Steps in a general circuit analysis.

We should remark that the *design* of amplifiers follows a similar procedure. First, the circuitry around the transistor is designed to establish proper bias conditions and hence the necessary small-signal parameters. Second, the small-signal behavior of the circuit is studied to verify the required performance. Some iteration between the two steps may often be necessary so as to converge toward the desired behavior.

How do we differentiate between small-signal and large-signal operations? In other words, under what conditions can we represent the devices with their small-signal models? If the signal perturbs the bias point of the device only negligibly, we say the circuit operates in the small-signal regime. In Fig. 5.8, for example, the change in I_C due to the signal must remain small. This criterion is justified because the amplifying properties of the transistor such as g_m and r_π are considered *constant* in small-signal analysis even though they in fact vary as the signal perturbs I_C. That is, a *linear* representation of the transistor holds only if the small-signal parameters themselves vary negligibly. The definition of "negligibly" depends somewhat on the circuit and the application, but as a rule of thumb, we consider 10% variation in the collector current as the upper bound for small-signal operation.

In drawing circuit diagrams hereafter, we will employ some simplified notations and symbols. Illustrated in Fig. 5.10 is an example where the battery serving as the supply voltage is replaced with a horizontal bar labeled V_{CC}.[5] Also, the input voltage source is simplified to one node called V_{in}, with the understanding that the other node is ground.

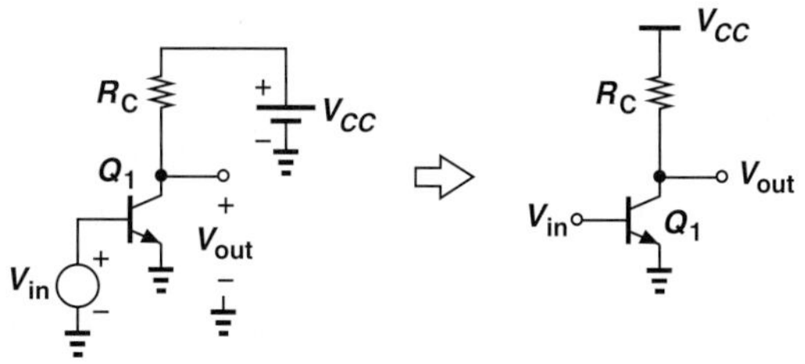

Figure 5.10 Notation for supply voltage.

In this chapter, we begin with the DC analysis and design of bipolar stages, developing skills to determine or create bias conditions. This phase of our study requires no knowledge of signals and hence the input and output ports of the circuit. Next, we introduce various amplifier topologies and examine their small-signal behavior.

[5]The subscript *CC* indicates supply voltage feeding the collector.

5.2 OPERATING POINT ANALYSIS AND DESIGN

It is instructive to begin our treatment of operating points with an example.

Example 5-5

A student familiar with bipolar devices constructs the circuit shown in Fig. 5.11 and attempts to amplify the signal produced by a microphone. If $I_S = 6 \times 10^{-16}$ A and the peak value of the microphone signal is 20 mV, determine the peak value of the output signal.

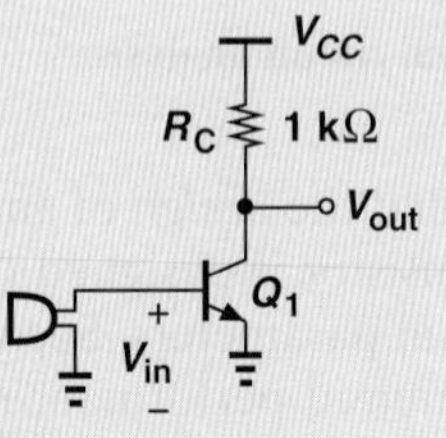

Figure 5.11 Amplifier driven directly by a microphone.

Solution Unfortunately, the student has forgotten to bias the transistor. (The microphone does not produce a dc output.) If V_{in} $(= V_{BE})$ reaches 20 mV, then

$$\Delta I_C = I_S \exp \frac{\Delta V_{BE}}{V_T} \tag{5.10}$$

$$= 1.29 \times 10^{-15} \text{ A}. \tag{5.11}$$

This change in the collector current yields a change in the output voltage equal to

$$R_C \, \Delta I_C = 1.29 \times 10^{-12} \text{ V}. \tag{5.12}$$

The circuit generates virtually no output because the bias current (in the absence of the microphone signal) is zero and so is the transconductance.

Exercise Repeat the above example if a constant voltage of 0.65 V is placed in series with the microphone.

As mentioned in Section 5.1.2, biasing seeks to fulfill two objectives: ensure operation in the forward active region, and set the collector current to the value required in the application. Let us return to the above example for a moment.

Example 5-6

Having realized the bias problem, the student in Example 5-5 modifies the circuit as shown in Fig. 5.12, connecting the base to V_{CC} to allow dc biasing for the base-emitter junction. Explain why the student needs to learn more about biasing.

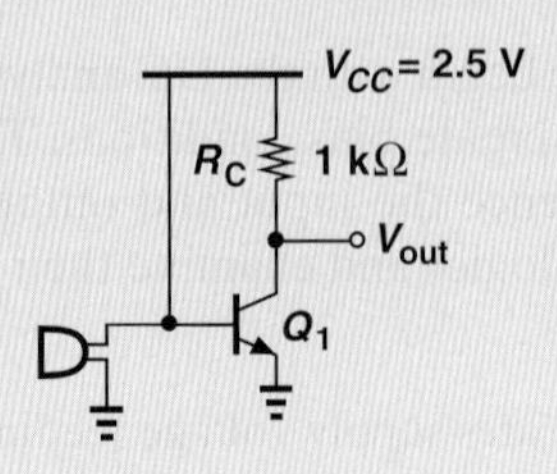

Figure 5.12 Amplifier with base tied to V_{CC}.

Solution The fundamental issue here is that the signal generated by the microphone is *shorted* to V_{CC}. Acting as an ideal voltage source, V_{CC} maintains the base voltage at a *constant* value, prohibiting any change introduced by the microphone. Since V_{BE} remains constant, so does V_{out}, leading to no amplification.

Another important issue relates to the value of V_{BE}: with $V_{BE} = V_{CC} = 2.5$ V, enormous currents flow into the transistor.

Exercise Does the circuit operate better if a resistor is placed in series with the emitter of Q_1?

5.2.1 Simple Biasing

Now consider the topology shown in Fig. 5.13, where the base is tied to V_{CC} through a relatively large resistor, R_B, so as to forward-bias the base-emitter junction. Our objective is to determine the terminal voltages and currents of Q_1 and obtain the conditions that ensure biasing in the active mode. How do we analyze this circuit? One can replace Q_1 with its large-signal model and apply KVL and KCL, but the resulting nonlinear equation(s) yield little intuition. Instead, we recall that the base-emitter voltage in most cases falls in the range of 700 to 800 mV and can be considered relatively constant. Since the voltage drop across R_B is equal to $R_B I_B$, we have

$$R_B I_B + V_{BE} = V_{CC} \tag{5.13}$$

and hence

$$I_B = \frac{V_{CC} - V_{BE}}{R_B}. \tag{5.14}$$

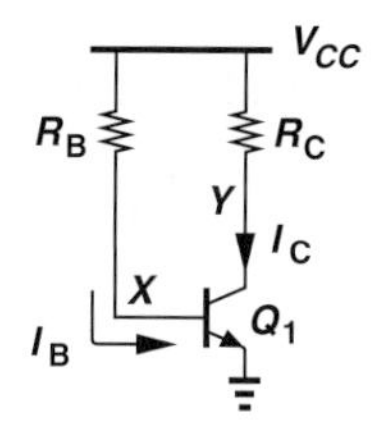

Figure 5.13 Use of base resistance for base current path.

With the base current known, we write

$$I_C = \beta \frac{V_{CC} - V_{BE}}{R_B}, \tag{5.15}$$

note that the voltage drop across R_C is equal to $R_C I_C$, and hence obtain V_{CE} as

$$V_{CE} = V_{CC} - R_C I_C \tag{5.16}$$

$$= V_{CC} - \beta \frac{V_{CC} - V_{BE}}{R_B} R_C. \tag{5.17}$$

Calculation of V_{CE} is necessary as it reveals whether the device operates in the active mode or not. For example, to avoid saturation completely, we require the collector voltage to remain above the base voltage:

$$V_{CC} - \beta \frac{V_{CC} - V_{BE}}{R_B} R_C > V_{BE}. \tag{5.18}$$

The circuit parameters can therefore be chosen so as to guarantee this condition.

In summary, using the sequence $I_B \to I_C \to V_{CE}$, we have computed the important terminal currents and voltages of Q_1. While not particularly interesting here, the emitter current is simply equal to $I_C + I_B$.

The reader may wonder about the error in the above calculations due to the assumption of a constant V_{BE} in the range of 700 to 800 mV. An example clarifies this issue.

Example 5-7 For the circuit shown in Fig. 5.14, determine the collector bias current. Assume $\beta = 100$ and $I_S = 10^{-17}$ A. Verify that Q_1 operates in the forward active region.

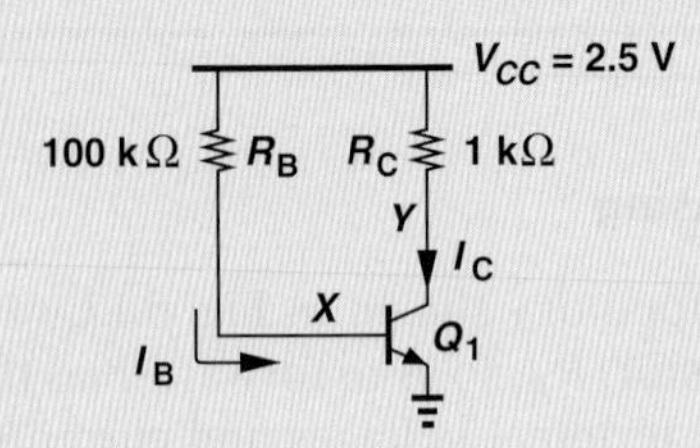

Figure 5.14 Simple biased stage.

Solution Since I_S is relatively small, we surmise that the base-emitter voltage required to carry typical current level is relatively large. Thus, we use $V_{BE} = 800$ mV as an initial guess and write Eq. (5.14) as

$$I_B = \frac{V_{CC} - V_{BE}}{R_B} \tag{5.19}$$

$$\approx 17\ \mu\text{A}. \tag{5.20}$$

It follows that

$$I_C = 1.7\ \text{mA}. \tag{5.21}$$

With this result for I_C, we calculate a new value for V_{BE}:

$$V_{BE} = V_T \ln \frac{I_C}{I_S} \tag{5.22}$$

$$= 852\ \text{mV}, \tag{5.23}$$

and iterate to obtain more accurate results. That is,

$$I_B = \frac{V_{CC} - V_{BE}}{R_B} \tag{5.24}$$

$$= 16.5\ \mu\text{A} \tag{5.25}$$

and hence

$$I_C = 1.65\ \text{mA}. \tag{5.26}$$

Since the values given by (5.21) and (5.26) are quite close, we consider $I_C = 1.65$ mA accurate enough and iterate no more.

Writing (5.16), we have

$$V_{CE} = V_{CC} - R_C I_C \tag{5.27}$$

$$= 0.85\ \text{V}, \tag{5.28}$$

a value nearly equal to V_{BE}. The transistor therefore operates near the edge of active and saturation modes.

Exercise What value of R_B provides a reverse bias of 200 mV across the base-collector junction?

The biasing scheme of Fig. 5.13 merits a few remarks. First, the effect of V_{BE} "uncertainty" becomes more pronounced at low values of V_{CC} because $V_{CC} - V_{BE}$ determines the base current. Thus, in low-voltage design—an increasingly common paradigm in modern electronic systems—the bias is more sensitive to V_{BE} variations among transistors or with temperature. Second, we recognize from Eq. (5.15) that I_C heavily depends on β, a parameter that may change considerably. In the above example, if β increases from 100 to 120, then I_C rises to 1.98 mA and V_{CE} falls to 0.52, driving the transistor toward heavy saturation. For these reasons, the topology of Fig. 5.13 is rarely used in practice.

Did you know?

Even in the first transistor radio introduced by Regency and Texas Instruments, it was recognized that the simple biasing technique would be a poor choice for production. The four transistors in this radio would need to maintain a relatively constant bias despite manufacturing variations and extreme temperatures in different places around the globe. For this reason, each stage incorporated "emitter degeneration" to stabilize the bias point.

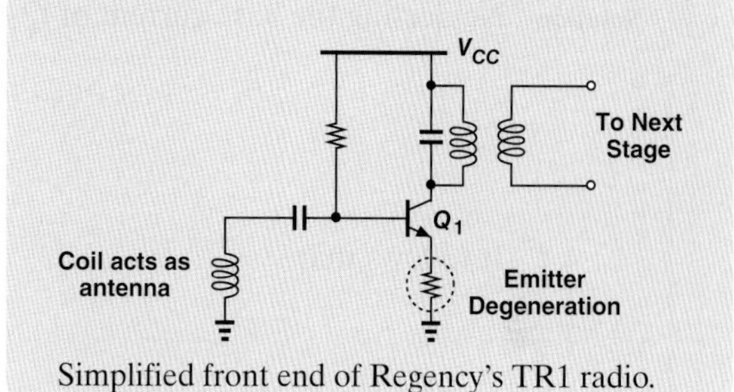

Simplified front end of Regency's TR1 radio.

5.2.2 Resistive Divider Biasing

In order to suppress the dependence of I_C upon β, we return to the fundamental relationship $I_C = I_S \exp(V_{BE}/V_T)$ and postulate that I_C must be set by applying a well-defined V_{BE}. Figure 5.15 depicts an example, where R_1 and R_2 act as a voltage divider, providing a base-emitter voltage equal to

$$V_X = \frac{R_2}{R_1 + R_2} V_{CC}, \tag{5.29}$$

if the base current is negligible. Thus,

$$I_C = I_S \exp\left(\frac{R_2}{R_1 + R_2} \cdot \frac{V_{CC}}{V_T}\right), \tag{5.30}$$

a quantity independent of β. Nonetheless, the design must ensure that the base current remains negligible.

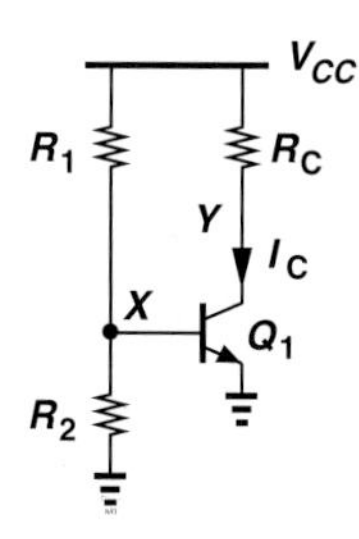

Figure 5.15 Use of resistive divider to define V_{BE}.

Example 5-8

Determine the collector current of Q_1 in Fig. 5.16 if $I_S = 10^{-17}$ A and $\beta = 100$. Verify that the base current is negligible and the transistor operates in the active mode.

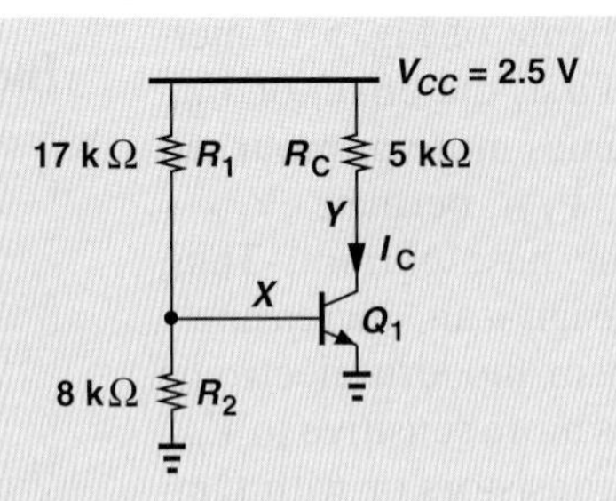

Figure 5.16 Example of biased stage.

Solution Neglecting the base current of Q_1, we have

$$V_X = \frac{R_2}{R_1 + R_2} V_{CC} \tag{5.31}$$

$$= 800 \text{ mV}. \tag{5.32}$$

It follows that

$$I_C = I_S \exp \frac{V_{BE}}{V_T} \tag{5.33}$$

$$= 231 \text{ μA} \tag{5.34}$$

and

$$I_B = 2.31 \text{ μA}. \tag{5.35}$$

Is the base current negligible? With which quantity should this value be compared? Provided by the resistive divider, I_B must be negligible with respect to the current flowing through R_1 and R_2:

$$I_B \stackrel{?}{\ll} \frac{V_{CC}}{R_1 + R_2}. \tag{5.36}$$

This condition indeed holds in this example because $V_{CC}/(R_1 + R_2) = 100$ μA $\approx 43 I_B$.

We also note that

$$V_{CE} = 1.345 \text{ V}, \tag{5.37}$$

and hence Q_1 operates in the active region.

Exercise What is the maximum value of R_C if Q_1 must remain in soft saturation?

The analysis approach taken in the above example assumes a negligible base current, requiring verification at the end. But what if the end result indicates that I_B is *not* negligible? We now analyze the circuit without this assumption. Let us replace the voltage divider with a Thevenin equivalent (Fig. 5.17), noting that V_{Thev} is equal to the open-circuit output voltage (V_X when the amplifier is disconnected):

$$V_{Thev} = \frac{R_2}{R_1 + R_2} V_{CC}. \tag{5.38}$$

Moreover, R_{Thev} is given by the output resistance of the network if V_{CC} is set to zero:

$$R_{Thev} = R_1 || R_2. \tag{5.39}$$

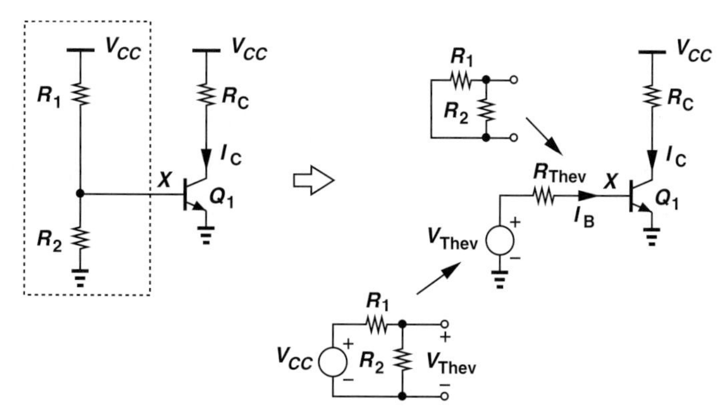

Figure 5.17 Use of Thevenin equivalent to calculate bias.

The simplified circuit yields:

$$V_X = V_{Thev} - I_B R_{Thev} \tag{5.40}$$

and

$$I_C = I_S \exp \frac{V_{Thev} - I_B R_{Thev}}{V_T}. \tag{5.41}$$

This result along with $I_C = \beta I_B$ forms the system of equations leading to the values of I_C and I_B. As in the previous examples, iterations prove useful here, but the exponential dependence in Eq. (5.41) gives rise to wide fluctuations in the intermediate solutions. For this reason, we rewrite Eq. (5.41) as

$$I_B = \left(V_{Thev} - V_T \ln \frac{I_C}{I_S} \right) \cdot \frac{1}{R_{Thev}}, \tag{5.42}$$

and begin with a guess for $V_{BE} = V_T \ln(I_C/I_S)$. The iteration then follows the sequence $V_{BE} \rightarrow I_B \rightarrow I_C \rightarrow V_{BE} \rightarrow \cdots$.

Example 5-9 Calculate the collector current of Q_1 in Fig. 5.18(a). Assume $\beta = 100$ and $I_S = 10^{-17}$ A.

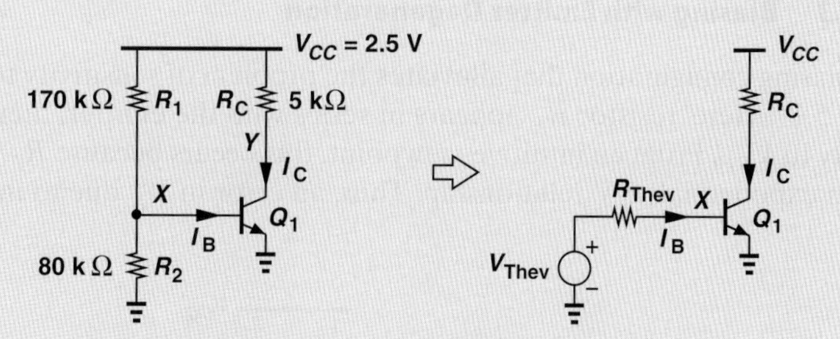

Figure 5.18 (a) Stage with resistive divider bias, (b) stage with Thevenin equivalent for the resistive divider and V_{CC}.

Solution Constructing the equivalent circuit shown in Fig. 5.18(b), we note that

$$V_{Thev} = \frac{R_2}{R_1 + R_2} V_{CC} \tag{5.43}$$

$$= 800 \text{ mV} \tag{5.44}$$

and

$$R_{Thev} = R_1||R_2 \tag{5.45}$$

$$= 54.4\ \text{k}\Omega. \tag{5.46}$$

We begin the iteration with an initial guess $V_{BE} = 750$ mV (because we know that the voltage drop across R_{Thev} makes V_{BE} *less* than V_{Thev}), thereby arriving at the base current:

$$I_B = \frac{V_{Thev} - V_{BE}}{R_{Thev}} \tag{5.47}$$

$$= 0.919\ \mu\text{A}. \tag{5.48}$$

Thus, $I_C = \beta I_B = 91.9\ \mu$A and

$$V_{BE} = V_T \ln \frac{I_C}{I_S} \tag{5.49}$$

$$= 776\ \text{mV}. \tag{5.50}$$

It follows that $I_B = 0.441\ \mu$A and hence $I_C = 44.1\ \mu$A, still a large fluctuation with respect to the first value from above. Continuing the iteration, we obtain $V_{BE} = 757$ mV, $I_B = 0.79\ \mu$A and $I_C = 79.0\ \mu$A. After many iterations, $V_{BE} \approx 766$ mV and $I_C = 63\ \mu$A.

Exercise How much can R_2 be increased if Q_1 must remain in soft saturation?

While proper choice of R_1 and R_2 in the topology of Fig. 5.15 makes the bias relatively insensitive to β, the exponential dependence of I_C upon the voltage generated by the resistive divider still leads to substantial bias variations. For example, if R_2 is 1% higher than its nominal value, so is V_X, thus multiplying the collector current by $\exp(0.01V_{BE}/V_T) \approx 1.36$ (for $V_{BE} = 800$ mV). In other words, a 1% error in one resistor value introduces a 36% error in the collector current. The circuit is therefore still of little practical value.

5.2.3 Biasing with Emitter Degeneration

A biasing configuration that alleviates the problem of sensitivity to β and V_{BE} is shown in Fig. 5.19. Here, resistor R_E appears in series with the emitter, thereby lowering the sensitivity to V_{BE}. From an intuitive viewpoint, this occurs because R_E exhibits a *linear* (rather than exponential) I-V relationship. Thus, an error in V_X due to inaccuracies in R_1, R_2, or

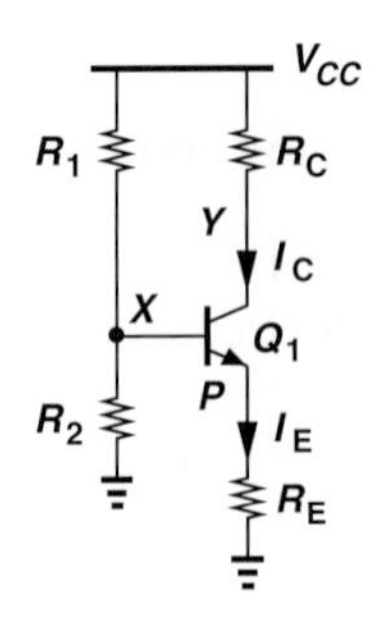

Figure 5.19 Addition of degeneration resistor to stabilize bias point.

V_{CC} is partly "absorbed" by R_E, introducing a smaller error in V_{BE} and hence I_C. Called "emitter degeneration," the addition of R_E in series with the emitter alters many attributes of the circuit, as described later in this chapter.

To understand the above property, let us determine the bias currents of the transistor. Neglecting the base current, we have $V_X = V_{CC}R_2/(R_1 + R_2)$. Also, $V_P = V_X - V_{BE}$, yielding

$$I_E = \frac{V_P}{R_E} \tag{5.51}$$

$$= \frac{1}{R_E}\left(V_{CC}\frac{R_2}{R_1 + R_2} - V_{BE}\right) \tag{5.52}$$

$$\approx I_C, \tag{5.53}$$

if $\beta \gg 1$. How can this result be made less sensitive to V_X or V_{BE} variations? If the voltage drop across R_E, i.e., the difference between $V_{CC}R_2/(R_1 + R_2)$ and V_{BE}, is large enough to absorb and swamp such variations, then I_E and I_C remain relatively constant. An example illustrates this point.

Example 5-10

Calculate the bias currents in the circuit of Fig. 5.20 and verify that Q_1 operates in the forward active region. Assume $\beta = 100$ and $I_S = 5 \times 10^{-17}$ A. How much does the collector current change if R_2 is 1% higher than its nominal value?

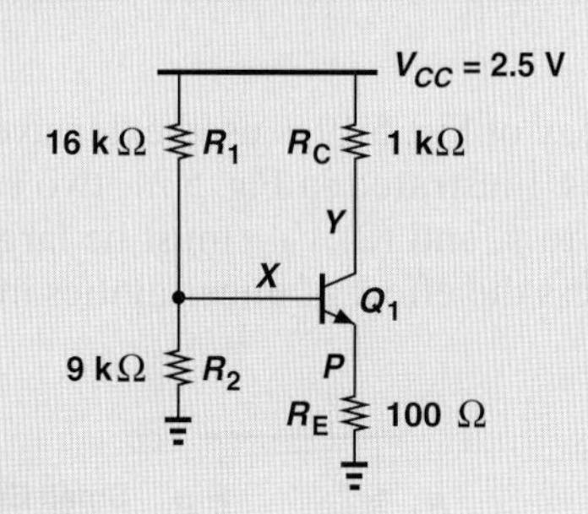

Figure 5.20 Example of biased stage.

Solution We neglect the base current and write

$$V_X = V_{CC}\frac{R_2}{R_1 + R_2} \tag{5.54}$$

$$= 900 \text{ mV}. \tag{5.55}$$

Using $V_{BE} = 800$ mV as an initial guess, we have

$$V_P = V_X - V_{BE} \tag{5.56}$$

$$= 100 \text{ mV}, \tag{5.57}$$

and hence

$$I_E \approx I_C \approx 1 \text{ mA}. \tag{5.58}$$

With this result, we must reexamine the assumption of $V_{BE} = 800$ mV. Since

$$V_{BE} = V_T \ln \frac{I_C}{I_S} \tag{5.59}$$

$$= 796 \text{ mV}, \tag{5.60}$$

we conclude that the initial guess is reasonable. Furthermore, Eq. (5.57) suggests that a 4-mV error in V_{BE} leads to a 4% error in V_P and hence I_E, indicating a good approximation.

Let us now determine if Q_1 operates in the active mode. The collector voltage is given by

$$V_Y = V_{CC} - I_C R_C \tag{5.61}$$

$$= 1.5\ \text{V}. \tag{5.62}$$

With the base voltage at 0.9 V, the device is indeed in the active region.

Is the assumption of negligible base current valid? With $I_C \approx 1$ mA, $I_B \approx 10$ μA whereas the current flowing through R_1 and R_2 is equal to 100 μA. The assumption is therefore reasonable. For greater accuracy, an iterative procedure similar to that in Example 5-9 can be followed.

If R_2 is 1.6% higher than its nominal value, then Eq. (5.54) indicates that V_X rises to approximately 909 mV. We may assume that the 9-mV change directly appears across R_E, raising the emitter current by 9 mV/100 Ω = 90 μA. From Eq. (5.56), we note that this assumption is equivalent to considering V_{BE} constant, which is reasonable because the emitter and collector currents have changed by only 9%.

Exercise What value of R_2 places Q_1 at the edge of saturation?

The bias topology of Fig. 5.19 is used extensively in discrete circuits and occasionally in integrated circuits. Illustrated in Fig. 5.21, two rules are typically followed: (1) $I_1 \gg I_B$ to lower sensitivity to β, and (2) V_{RE} must be large enough (100 mV to several hundred millivolts) to suppress the effect of uncertainties in V_X and V_{BE}.

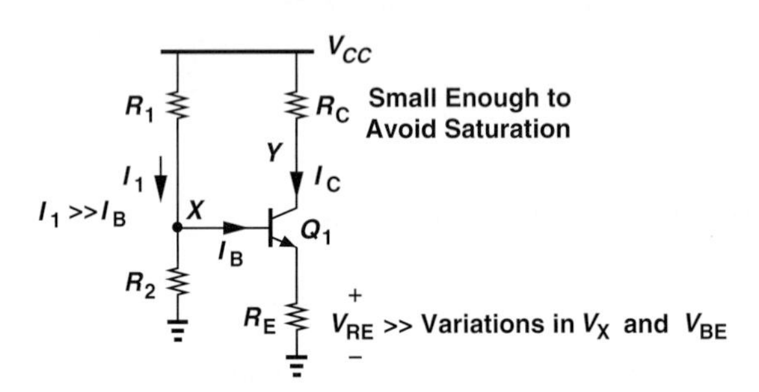

Figure 5.21 Summary of robust bias conditions.

Design Procedure It is possible to prescribe a design procedure for the bias topology of Fig. 5.21 that serves most applications: (1) decide on a collector bias current that yields proper small-signal parameters such as g_m and r_π; (2) based on the expected variations of R_1, R_2, and V_{BE}, choose a value for $V_{RE} \approx I_C R_E$, e.g., 200 mV; (3) calculate $V_X = V_{BE} + I_C R_E$ with $V_{BE} = V_T \ln(I_C/I_S)$; (4) choose R_1 and R_2 so as to provide the necessary value of V_X *and* establish $I_1 \gg I_B$. Determined by small-signal gain requirements, the value of R_C is bounded by a maximum that places Q_1 at the edge of saturation. The following example illustrates these concepts.

Example 5-11 Design the circuit of Fig. 5.21 so as to provide a transconductance of $1/(52\ \Omega)$ for Q_1. Assume $V_{CC} = 2.5$ V, $\beta = 100$, and $I_S = 5 \times 10^{-17}$ A. What is the maximum tolerable value of R_C?

Solution A g_m of $(52\ \Omega)^{-1}$ translates to a collector current of 0.5 mA and a V_{BE} of 778 mV. Assuming $R_E I_C = 200$ mV, we obtain $R_E = 400\ \Omega$. To establish $V_X = V_{BE} + R_E I_C = 978$ mV, we must have

$$\frac{R_2}{R_1 + R_2} V_{CC} = V_{BE} + R_E I_C, \tag{5.63}$$

where the base current is neglected. For the base current $I_B = 5\ \mu$A to be negligible,

$$\frac{V_{CC}}{R_1 + R_2} \gg I_B, \tag{5.64}$$

e.g., by a factor of 10. Thus, $R_1 + R_2 = 50$ kΩ, which in conjunction with Eq. (5.63) yields

$$R_1 = 30.45\ \text{k}\Omega \tag{5.65}$$

$$R_2 = 19.55\ \text{k}\Omega. \tag{5.66}$$

How large can R_C be? Since the collector voltage is equal to $V_{CC} - R_C I_C$, we pose the following constraint to ensure active mode operation:

$$V_{CC} - R_C I_C > V_X; \tag{5.67}$$

that is,

$$R_C I_C < 1.522\ \text{V}. \tag{5.68}$$

Consequently,

$$R_C < 3.044\ \text{k}\Omega. \tag{5.69}$$

If R_C exceeds this value, the collector voltage falls below the base voltage. As mentioned in Chapter 4, the transistor can tolerate soft saturation, i.e., up to about 400 mV of base-collector forward bias. Thus, in low-voltage applications, we may allow $V_Y \approx V_X - 400$ mV and hence a greater value for R_C.

Exercise Repeat the above example if the power budget is only 1 mW and the transconductance of Q_1 is not given.

The two rules depicted in Fig. 5.21 to lower sensitivities do impose some trade-offs. Specifically, an overly conservative design faces the following issues: (1) if we wish I_1 to be much much greater than I_B, then $R_1 + R_2$ and hence R_1 and R_2 are quite small, leading to a low *input impedance*; (2) if we choose a very large V_{RE}, then $V_X\ (= V_{BE} + V_{RE})$ must be high, thereby limiting the minimum value of the collector voltage to avoid saturation. Let us return to the above example and study these issues.

Example 5-12 Repeat Example 5-11 but assuming $V_{RE} = 500$ mV and $I_1 \geq 100I_B$.

Solution The collector current and base-emitter voltage remain unchanged. The value of R_E is now given by $500\text{ mV}/0.5\text{ mA} = 1\text{ k}\Omega$. Also, $V_X = V_{BE} + R_E I_C = 1.278$ V and Eq. (5.63) still holds. We rewrite Eq. (5.64) as

$$\frac{V_{CC}}{R_1 + R_2} \geq 100I_B, \tag{5.70}$$

obtaining $R_1 + R_2 = 5\text{ k}\Omega$. It follows that

$$R_1 = 1.45\text{ k}\Omega \tag{5.71}$$

$$R_2 = 3.55\text{ k}\Omega. \tag{5.72}$$

Since the base voltage has risen to 1.278 V, the collector voltage must exceed this value to avoid saturation, leading to

$$R_C < \frac{V_{CC} - V_X}{I_C} \tag{5.73}$$

$$< 1.044\text{ k}\Omega. \tag{5.74}$$

As seen in Section 5.3.1, the reduction in R_C translates to a lower voltage gain. Also, the much smaller values of R_1 and R_2 here than in Example 5-11 introduce a low input impedance, loading the preceding stage. We compute the exact input impedance of this circuit in Section 5.3.1.

Exercise Repeat the above example if V_{RE} is limited to 100 mV.

5.2.4 Self-Biased Stage

Another biasing scheme commonly used in discrete and integrated circuits is shown in Fig. 5.22. Called "self-biased" because the base current and voltage are provided from the collector, this stage exhibits many interesting and useful attributes.

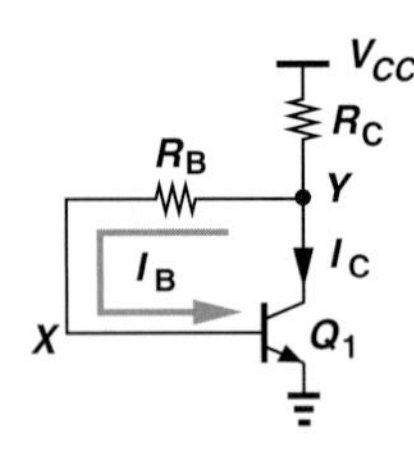

Figure 5.22 Self-biased stage.

Let us begin the analysis of the circuit with the observation that the base voltage is always *lower* than the collector voltage: $V_X = V_Y - I_B R_B$. A result of self-biasing, this important property guarantees that Q_1 operates in the active mode regardless of device and circuit parameters. For example, if R_C increases indefinitely, Q_1 remains in the active region, a critical advantage over the circuit of Fig. 5.21.

We now determine the collector bias current by assuming $I_B \ll I_C$; i.e., R_C carries a current equal to I_C, thereby yielding

$$V_Y = V_{CC} - R_C I_C. \tag{5.75}$$

Also,

$$V_Y = R_B I_B + V_{BE} \tag{5.76}$$

$$= \frac{R_B I_C}{\beta} + V_{BE}. \tag{5.77}$$

Equating the right-hand sides of Eqs. (5.75) and (5.77) gives

$$I_C = \frac{V_{CC} - V_{BE}}{R_C + \dfrac{R_B}{\beta}}. \tag{5.78}$$

As usual, we begin with an initial guess for V_{BE}, compute I_C, and utilize $V_{BE} = V_T \ln(I_C/I_S)$ to improve the accuracy of our calculations.

Example 5-13 Determine the collector current and voltage of Q_1 in Fig. 5.22 if $R_C = 1\ \text{k}\Omega$, $R_B = 10\ \text{k}\Omega$, $V_{CC} = 2.5$ V, $I_S = 5 \times 10^{-17}$ A, and $\beta = 100$. Repeat the calculations for $R_C = 2\ \text{k}\Omega$.

Solution Assuming $V_{BE} = 0.8$ V, we have from Eq. (5.78):

$$I_C = 1.545\ \text{mA}, \tag{5.79}$$

and hence $V_{BE} = V_T \ln(I_C/I_S) = 807.6$ mV, concluding that the initial guess for V_{BE} and the value of I_C given by it are reasonably accurate. We also note that $R_B I_B = 154.5$ mV and $V_Y = R_B I_B + V_{BE} \approx 0.955$ V.

If $R_C = 2\ \text{k}\Omega$, then with $V_{BE} = 0.8$ V, Eq. (5.78) gives

$$I_C = 0.810\ \text{mA}. \tag{5.80}$$

To check the validity of the initial guess, we write $V_{BE} = V_T \ln(I_C/I_S) = 791$ mV. Compared with $V_{CC} - V_{BE}$ in the numerator of Eq. (5.78), the 9-mV error is negligible and the value of I_C in Eq. (5.80) is acceptable. Since $R_B I_B = 81$ mV, $V_Y \approx 0.881$ V.

Exercise What happens if the base resistance is doubled?

Equation (5.78) and the preceding example suggest two important guidelines for the design of the self-biased stage: (1) $V_{CC} - V_{BE}$ must be much greater than the uncertainties in the value of V_{BE}; (2) R_C must be much greater than R_B/β to lower sensitivity to β. In fact, if $R_C \gg R_B/\beta$, then

$$I_C \approx \frac{V_{CC} - V_{BE}}{R_C}, \tag{5.81}$$

and $V_Y = V_{CC} - I_C R_C \approx V_{BE}$. This result serves as a quick estimate of the transistor bias conditions.

Design Procedure Equation (5.78) together with the condition $R_C \gg R_B/\beta$ provides the basic expressions for the design of the circuit. With the required value of I_C known from small-signal considerations, we choose $R_C = 10R_B/\beta$ and rewrite Eq. (5.78) as

$$I_C = \frac{V_{CC} - V_{BE}}{1.1R_C}, \tag{5.82}$$

where $V_{BE} = V_T \ln(I_C/I_S)$. That is,

$$R_C = \frac{V_{CC} - V_{BE}}{1.1I_C} \tag{5.83}$$

$$R_B = \frac{\beta R_C}{10}. \tag{5.84}$$

The choice of R_B also depends on small-signal requirements and may deviate from this value, but it must remain substantially lower than βR_C.

Example 5-14 Design the self-biased stage of Fig. 5.22 for $g_m = 1/(13\ \Omega)$ and $V_{CC} = 1.8$ V. Assume $I_S = 5 \times 10^{-16}$ A and $\beta = 100$.

Solution Since $g_m = I_C/V_T = 1/(13\ \Omega)$, we have $I_C = 2$ mA, $V_{BE} = 754$ mV, and

$$R_C \approx \frac{V_{CC} - V_{BE}}{1.1I_C} \tag{5.85}$$

$$\approx 475\ \Omega. \tag{5.86}$$

Also,

$$R_B = \frac{\beta R_C}{10} \tag{5.87}$$

$$= 4.75\ \text{k}\Omega. \tag{5.88}$$

Note that $R_B I_B = 95$ mV, yielding a collector voltage of 754 mV + 95 mV = 849 mV.

Exercise Repeat the above design with a supply voltage of 2.5 V.

Did you know?

The self-biased stage incorporates "negative feedback" to stabilize the bias. For example, if the collector voltage tends to increase due to a temperature change, then resistor R_B passes this voltage increase to the base, raising the V_{BE} and hence the collector current. The collector voltage thus falls back to nearly its original value.

The use of negative feedback also improves the *speed* of the circuit. In fact, optical communication links used in the backbone of the Internet often employ this type of amplifier topology so as to accommodate data rates as high as 40 Gb/s. Next time you send an email, bear in mind that your data may go through a self-biased stage.

Figure 5.23 summarizes the biasing principles studied in this section.

5.2.5 Biasing of *PNP* Transistors

The dc bias topologies studied thus far incorporate *npn* transistors. Circuits using *pnp* devices follow the same analysis and design procedures while requiring attention to voltage and current polarities. We illustrate these points with the aid of some examples.

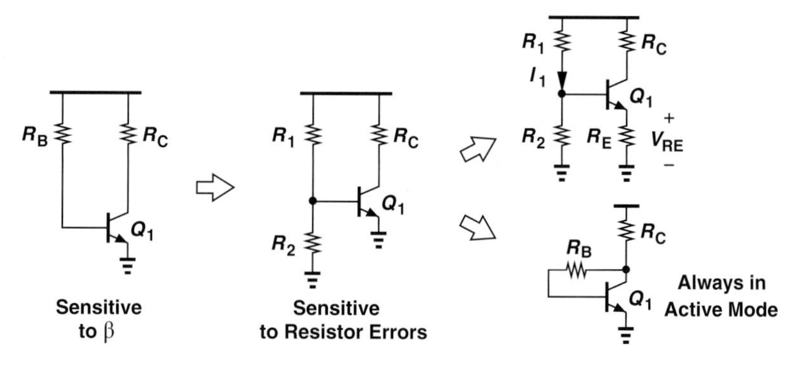

Figure 5.23 Summary of biasing techniques.

Example 5-15 Calculate the collector and voltage of Q_1 in the circuit of Fig. 5.24 and determine the maximum allowable value of R_C for operation in the active mode.

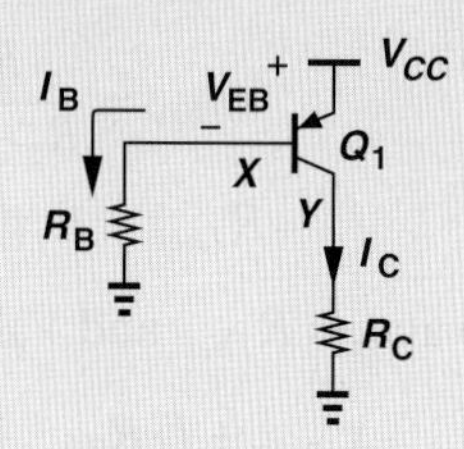

Figure 5.24 Simple biasing of *pnp* stage.

Solution The topology is the same as that in Fig. 5.13 and we have

$$I_B R_B + V_{EB} = V_{CC}. \tag{5.89}$$

That is,

$$I_B = \frac{V_{CC} - V_{EB}}{R_B} \tag{5.90}$$

and

$$I_C = \beta \frac{V_{CC} - V_{EB}}{R_B}. \tag{5.91}$$

The circuit suffers from sensitivity to β.

If R_C is increased, V_Y rises, thus approaching V_X $(= V_{CC} - V_{EB})$ and bringing Q_1 closer to saturation. The transistor enters saturation at $V_Y = V_X$, i.e.,

$$I_C R_{C,max} = V_{CC} - V_{EB} \tag{5.92}$$

and hence

$$R_{C,max} = \frac{V_{CC} - V_{EB}}{I_C} \tag{5.93}$$

$$= \frac{R_B}{\beta}. \tag{5.94}$$

From another perspective, since $V_X = I_B R_B$ and $V_Y = I_C R_C$, we have $I_B R_B = I_C R_{C,max}$ as the condition for edge of saturation, obtaining $R_B = \beta R_{C,max}$.

Exercise For a given R_C, what value of R_B places the device at the edge of saturation?

Example 5-16 Determine the collector current and voltage of Q_1 in the circuit of Fig. 5.25(a).

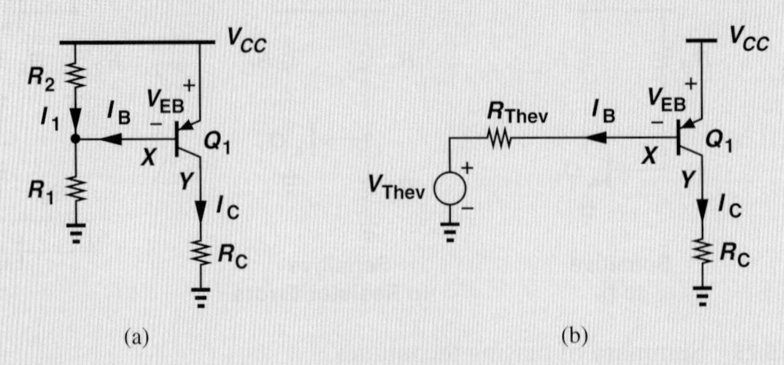

Figure 5.25 (a) *PNP* stage with resistive divider biasing, (b) Thevenin equivalent of divider and V_{CC}.

Solution As a general case, we assume I_B is significant and construct the Thevenin equivalent of the voltage divider as depicted in Fig. 5.25(b):

$$V_{Thev} = \frac{R_1}{R_1 + R_2} V_{CC} \tag{5.95}$$

$$R_{Thev} = R_1 || R_2. \tag{5.96}$$

Adding the voltage drop across R_{Thev} and V_{EB} to V_{Thev} yields

$$V_{Thev} + I_B R_{Thev} + V_{EB} = V_{CC}; \tag{5.97}$$

that is,

$$I_B = \frac{V_{CC} - V_{Thev} - V_{EB}}{R_{Thev}} \tag{5.98}$$

$$= \frac{\frac{R_2}{R_1 + R_2} V_{CC} - V_{EB}}{R_{Thev}}. \tag{5.99}$$

It follows that

$$I_C = \beta \frac{\frac{R_2}{R_1 + R_2} V_{CC} - V_{EB}}{R_{Thev}}. \tag{5.100}$$

As in Example 5-9, some iteration between I_C and V_{EB} may be necessary.

Equation (5.100) indicates that if I_B is significant, then the transistor bias heavily depends on β. On the other hand, if $I_B \ll I_1$, we equate the voltage drop across R_2 to V_{EB}, thereby obtaining the collector current:

$$\frac{R_2}{R_1 + R_2} V_{CC} = V_{EB} \tag{5.101}$$

$$I_C = I_S \exp\left(\frac{R_2}{R_1 + R_2} \frac{V_{CC}}{V_T}\right). \tag{5.102}$$

Note that this result is identical to Eq. (5.30).

Exercise What is the maximum value of R_C if Q_1 must remain in soft saturation?

Example 5-17 Assuming a negligible base current, calculate the collector current and voltage of Q_1 in the circuit of Fig. 5.26. What is the maximum allowable value of R_C for Q_1 to operate in the forward active region?

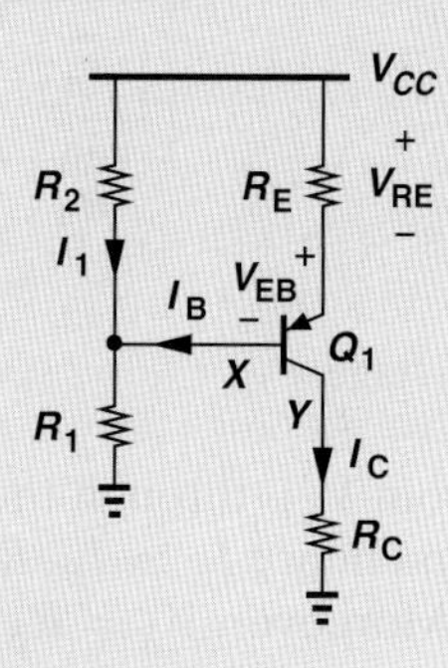

Figure 5.26 *PNP* stage with degeneration resistor.

Solution With $I_B \ll I_1$, we have $V_X = V_{CC}R_1/(R_1 + R_2)$. Adding to V_X the emitter-base voltage and the drop across R_E, we obtain

$$V_X + V_{EB} + R_E I_E = V_{CC} \tag{5.103}$$

and hence

$$I_E = \frac{1}{R_E}\left(\frac{R_2}{R_1 + R_2}V_{CC} - V_{EB}\right). \tag{5.104}$$

Using $I_C \approx I_E$, we can compute a new value for V_{EB} and iterate if necessary. Also, with $I_B = I_C/\beta$, we can verify the assumption $I_B \ll I_1$.

In arriving at Eq. (5.104), we have written a KVL from V_{CC} to ground, Eq. (5.103). But a more straightforward approach is to recognize that the voltage drop across R_2 is equal to $V_{EB} + I_E R_E$, i.e.,

$$V_{CC}\frac{R_2}{R_1 + R_2} = V_{EB} + I_E R_E, \tag{5.105}$$

which yields the same result as in Eq. (5.104).

The maximum allowable value of R_C is obtained by equating the base and collector voltages:

$$V_{CC}\frac{R_1}{R_1 + R_2} = R_{C,max} I_C \tag{5.106}$$

$$\approx \frac{R_{C,max}}{R_E}\left(\frac{R_2}{R_1 + R_2}V_{CC} - V_{EB}\right). \tag{5.107}$$

It follows that

$$R_{C,max} = R_E V_{CC}\frac{R_1}{R_1 + R_2} \cdot \frac{1}{\dfrac{R_2}{R_1 + R_2}V_{CC} - V_{EB}}. \tag{5.108}$$

Exercise Repeat the above example if $R_2 = \infty$.

Example 5-18 Determine the collector current and voltage of Q_1 in the self-biased circuit of Fig. 5.27.

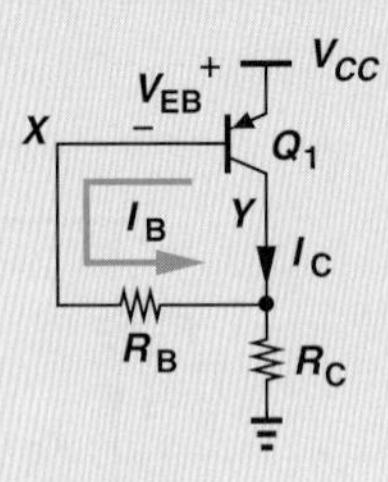

Figure 5.27 Self-biased *pnp* stage.

Solution We must write a KVL from V_{CC} through the emitter-base junction of Q_1, R_B, and R_C to ground. Since $\beta \gg 1$ and hence $I_C \gg I_B$, R_C carries a current approximately equal to I_C, creating $V_Y = R_C I_C$. Moreover, $V_X = R_B I_B + V_Y = R_B I_B + R_C I_C$, yielding

$$V_{CC} = V_{EB} + V_X \tag{5.109}$$

$$= V_{EB} + R_B I_B + I_C R_C \tag{5.110}$$

$$= V_{EB} + \left(\frac{R_B}{\beta} + R_C\right) I_C. \tag{5.111}$$

Thus,

$$I_C = \frac{V_{CC} - V_{EB}}{\dfrac{R_B}{\beta} + R_C}, \tag{5.112}$$

a result similar to Eq. (5.78). As usual, we begin with a guess for V_{EB}, compute I_C, and determine a new value for V_{EB}, etc. Note that, since the base is *higher* than the collector voltage, Q_1 always remains in the active mode.

Exercise How far is Q_1 from saturation?

5.3 BIPOLAR AMPLIFIER TOPOLOGIES

Following our detailed study of biasing, we can now delve into different amplifier topologies and examine their small-signal properties.[6]

Since the bipolar transistor contains three terminals, we may surmise that three possibilities exist for applying the input signal to the device, as conceptually illustrated in Figs. 5.28(a)–(c). Similarly, the output signal can be sensed from any of the terminals (with respect to ground) [Figs. 5.28(d)–(f)], leading to nine possible combinations of input and output networks and hence nine amplifier topologies.

However, as seen in Chapter 4, bipolar transistors operating in the active mode respond to base-emitter voltage variations by varying their collector current. This property rules out the input connection shown in Fig. 5.28(c) because here V_{in} does not affect the base or emitter voltages. Also, the topology in Fig. 5.28(f) proves of no value as V_{out} is not a function of the collector current. The number of possibilities therefore falls to

[6]While beyond the scope of this book, the large-signal behavior of amplifiers also becomes important in many applications.

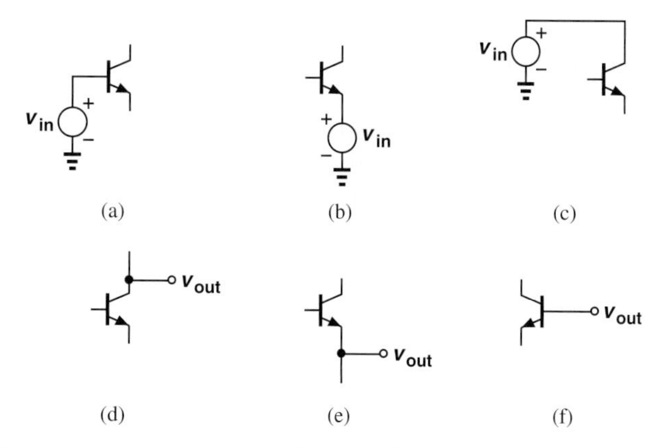

Figure 5.28 Possible input and output connections to a bipolar transistor.

four. But we note that the input and output connections in Figs. 5.28(b) and (e) remain incompatible because V_{out} would be sensed at the *input* node (the emitter) and the circuit would provide no function.

The foregoing observations reveal three possible amplifier topologies. We study each carefully, seeking to compute its gain and input and output impedances. In all cases, the bipolar transistors operate in the active mode. The reader is encouraged to review Examples (5-2)–(5-4) and the three resulting rules illustrated in Fig. 5.7 before proceeding further.

5.3.1 Common-Emitter Topology

Our initial thoughts in Section 4.1 pointed to the circuit of Fig. 4.1(b) and hence the topology of Fig. 4.25 as an amplifier. If the input signal is applied to the base [Fig. 5.28(a)] and the output signal is sensed at the collector [Fig. 5.28(d)], the circuit is called a "common-emitter" (CE) stage (Fig. 5.29). We have encountered and analyzed this circuit in different contexts without giving it a name. The term "common-emitter" is used because the emitter terminal is grounded and hence appears *in common* to the input and output ports. Nevertheless, we identify the stage based on the input and output connections (to the base and from the collector, respectively) so as to avoid confusion in more complex topologies.

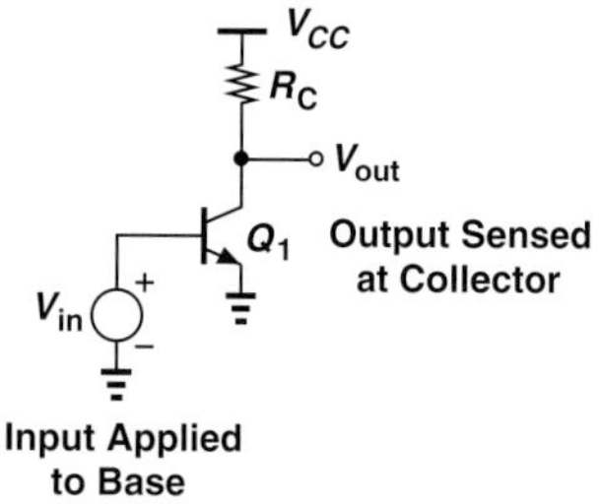

Figure 5.29 Common-emitter stage.

We deal with the CE amplifier in two phases: (a) analysis of the CE core to understand its fundamental properties, and (b) analysis of the CE stage including the bias circuitry as a more realistic case.

Analysis of CE Core Recall from the definition of transconductance in Section 4.4.3 that a small increment of ΔV applied to the base of Q_1 in Fig. 5.29 increases the collector current by $g_m \Delta V$ and hence the voltage drop across R_C by $g_m \Delta V R_C$. In order to examine the amplifying properties of the CE stage, we construct the small-signal equivalent of the circuit, shown in Fig. 5.30. As explained in Chapter 4, the supply voltage node, V_{CC}, acts as an ac ground because its value remains constant with time. We neglect the Early effect for now.

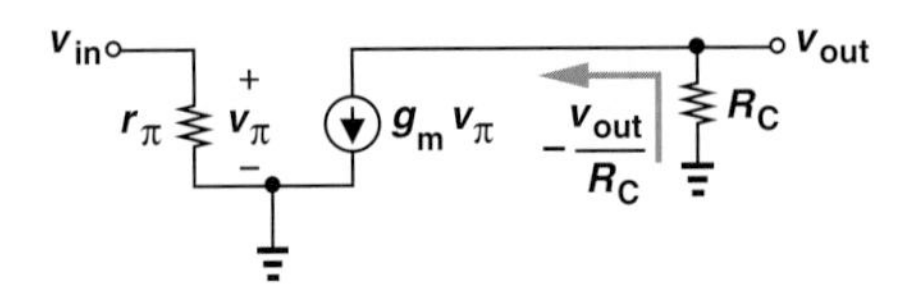

Figure 5.30 Small-signal model of CE stage.

Let us first compute the small-signal voltage gain $A_v = v_{out}/v_{in}$. Beginning from the output port and writing a KCL at the collector node, we have

$$-\frac{v_{out}}{R_C} = g_m v_\pi, \tag{5.113}$$

and $v_\pi = v_{in}$. It follows that

$$A_v = -g_m R_C. \tag{5.114}$$

Equation (5.114) embodies two interesting and important properties of the CE stage. First, the small-signal gain is *negative* because raising the base voltage and hence the collector current in Fig. 5.29 *lowers* V_{out}. Second, A_v is proportional to g_m (i.e., the collector bias current) and the collector resistor, R_C.

Interestingly, the voltage gain of the stage is limited by the supply voltage. A higher collector bias current or a larger R_C demands a greater voltage drop across R_C, but this drop cannot exceed V_{CC}. In fact, denoting the dc drop across R_C with V_{RC} and writing $g_m = I_C/V_T$, we express Eq. (5.113) as

$$|A_v| = \frac{I_C R_C}{V_T} \tag{5.115}$$

$$= \frac{V_{RC}}{V_T}. \tag{5.116}$$

Since $V_{RC} < V_{CC}$,

$$|A_v| < \frac{V_{CC}}{V_T}. \tag{5.117}$$

Furthermore, the transistor itself requires a minimum collector-emitter voltage of about V_{BE} to remain in the active region, lowering the limit to

$$|A_v| < \frac{V_{CC} - V_{BE}}{V_T}. \tag{5.118}$$

Example 5-19 Design a CE core with $V_{CC} = 1.8$ V and a power budget, P, of 1 mW while achieving maximum voltage gain.

Solution Since $P = I_C \cdot V_{CC} = 1$ mW, we have $I_C = 0.556$ mA. The value of R_C that places Q_1 at the edge of saturation is given by

$$V_{CC} - R_C I_C = V_{BE}, \tag{5.119}$$

which, along with $V_{BE} \approx 800$ mV, yields

$$R_C \leq \frac{V_{CC} - V_{BE}}{I_C} \tag{5.120}$$

$$\leq 1.8 \text{ k}\Omega. \tag{5.121}$$

The voltage gain is therefore equal to

$$A_v = -g_m R_C \tag{5.122}$$

$$= -38.5. \tag{5.123}$$

Under this condition, an input signal drives the transistor into saturation. As illustrated in Fig. 5.31(a), a 2-mV$_{pp}$ input results in a 77-mV$_{pp}$ output, forward-biasing the base-collector junction for half of each cycle. Nevertheless, so long as Q_1 remains in soft saturation ($V_{BC} > 400$ mV), the circuit amplifies properly.

A more aggressive design may allow Q_1 to operate in soft saturation, e.g., $V_{CE} \approx 400$ mV and hence

$$R_C \leq \frac{V_{CC} - 400 \text{ mV}}{I_C} \tag{5.124}$$

$$\leq 2.52 \text{ k}\Omega. \tag{5.125}$$

In this case, the maximum voltage gain is given by

$$A_v = -53.9. \tag{5.126}$$

Of course, the circuit can now tolerate only very small voltage swings at the output. For example, a 2-mV$_{pp}$ input signal gives rise to a 107.8-mV$_{pp}$ output, driving Q_1 into heavy saturation [Fig. 5.31(b)]. We say the circuit suffers from a trade-off between voltage gain and voltage "headroom."

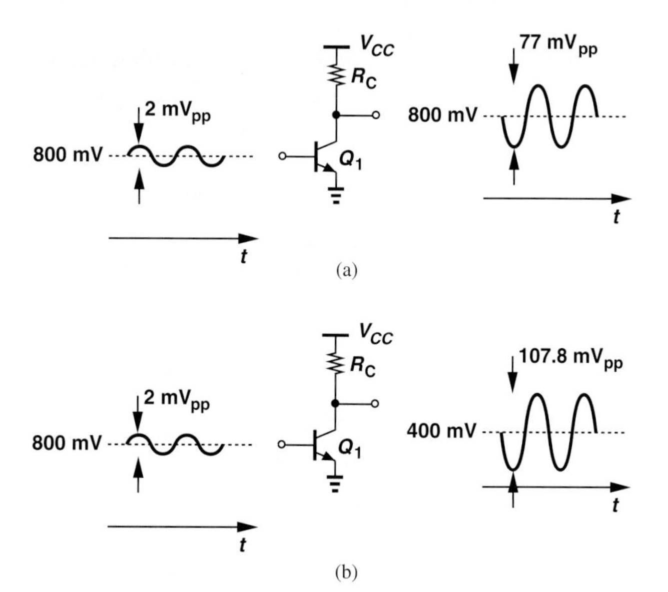

Figure 5.31 CE stage (a) with some signal levels, (b) in saturation.

Exercise Repeat the above example if $V_{CC} = 2.5$ V and compare the results.

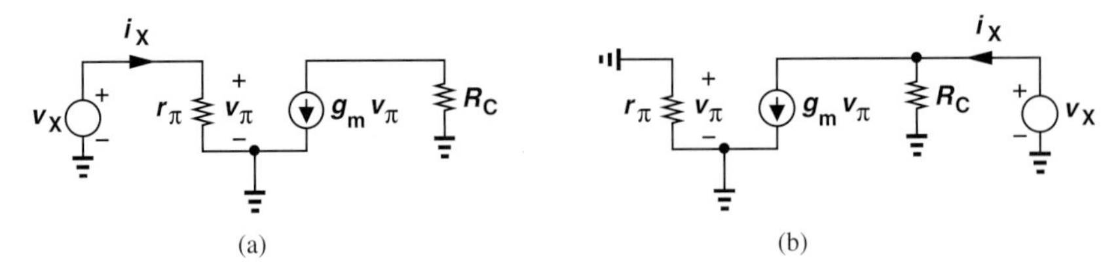

Figure 5.32 (a) Input and (b) output impedance calculation of CE stage.

Let us now calculate the I/O impedances of the CE stage. Using the equivalent circuit depicted in Fig. 5.32(a), we write

$$R_{in} = \frac{v_X}{i_X} \tag{5.127}$$

$$= r_\pi. \tag{5.128}$$

Thus, the input impedance is simply equal to $\beta/g_m = \beta V_T/I_C$ and decreases as the collector bias increases.

The output impedance is obtained from Fig. 5.32(b), where the input voltage source is set to zero (replaced with a short). Since $v_\pi = 0$, the dependent current source also vanishes, leaving R_C as the only component seen by v_X. In other words,

$$R_{out} = \frac{v_X}{i_X} \tag{5.129}$$

$$= R_C. \tag{5.130}$$

The output impedance therefore trades with the voltage gain, $-g_m R_C$.

Figure 5.33 summarizes the trade-offs in the performance of the CE topology along with the parameters that create such trade-offs. For example, for a given value of output impedance, R_C is fixed and the voltage gain can be increased by increasing I_C, thereby lowering both the voltage headroom and the input impedance.

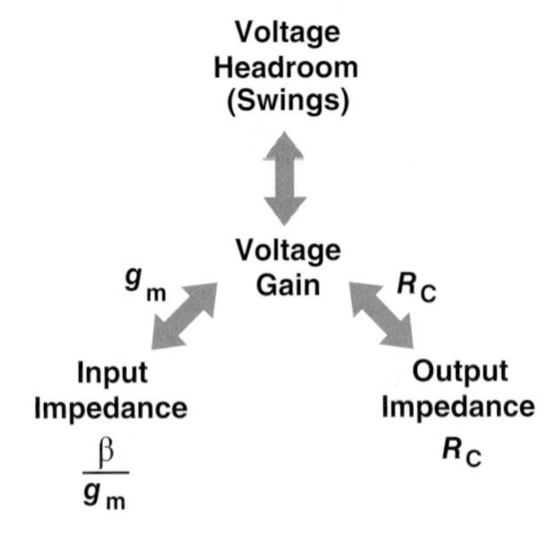

Figure 5.33 CE stage trade-offs.

Example 5-20 A CE stage must achieve an input impedance of R_{in} and an output impedance of R_{out}. What is the voltage gain of the circuit?

Solution Since $R_{in} = r_\pi = \beta/g_m$ and $R_{out} = R_C$, we have

$$A_v = -g_m R_C \tag{5.131}$$

$$= -\beta \frac{R_{out}}{R_{in}}. \tag{5.132}$$

Interestingly, if the I/O impedances are specified, then the voltage gain is automatically set. We will develop other circuits in this book that avoid this "coupling" of design specifications.

Exercise What happens to this result if the supply voltage is halved?

Inclusion of Early Effect Equation (5.114) suggests that the voltage gain of the CE stage can be increased indefinitely if $R_C \to \infty$ while g_m remains constant. Mentioned in Section 4.4.5, this trend appears valid if V_{CC} is also raised to ensure the transistor remains in the active mode. From an intuitive point of view, a given change in the input voltage and hence the collector current gives rise to an increasingly larger output swing as R_C increases.

In reality, however, the Early effect limits the voltage gain even if R_C approaches infinity. Since achieving a high gain proves critical in circuits such as operational amplifiers, we must reexamine the above derivations in the presence of the Early effect.

Figure 5.34 depicts the small-signal equivalent circuit of the CE stage including the transistor output resistance. Note that r_O appears in parallel with R_C, allowing us to rewrite Eq. (5.114) as

$$A_v = -g_m(R_C || r_O). \tag{5.133}$$

We also recognize that the input impedance remains equal to r_π whereas the output impedance falls to

$$R_{out} = R_C || r_O. \tag{5.134}$$

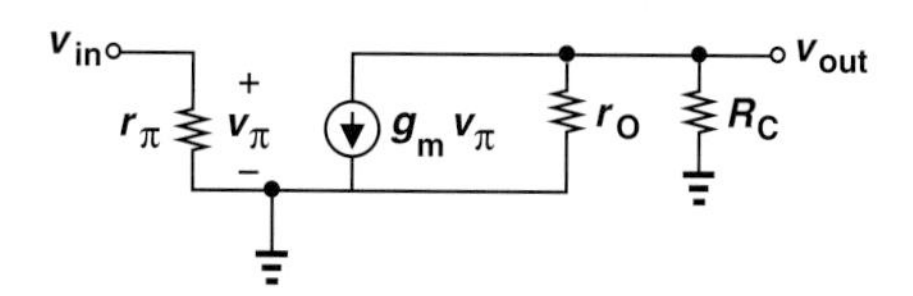

Figure 5.34 CE stage including Early effect.

Example 5-21 The circuit of Fig. 5.29 is biased with a collector current of 1 mA and $R_C = 1$ kΩ. If $\beta = 100$ and $V_A = 10$ V, determine the small-signal voltage gain and the I/O impedances.

Solution We have

$$g_m = \frac{I_C}{V_T} \tag{5.135}$$

$$= (26\ \Omega)^{-1} \tag{5.136}$$

and

$$r_O = \frac{V_A}{I_C} \tag{5.137}$$

$$= 10\ \text{k}\Omega. \tag{5.138}$$

Thus,

$$A_v = -g_m(R_C||r_O) \tag{5.139}$$

$$\approx 35. \tag{5.140}$$

(As a comparison, if $V_A = \infty$, then $A_v \approx 38$.) For the I/O impedances, we write

$$R_{in} = r_\pi \tag{5.141}$$

$$= \frac{\beta}{g_m} \tag{5.142}$$

$$= 2.6\ \text{k}\Omega \tag{5.143}$$

and

$$R_{out} = R_C||r_O \tag{5.144}$$

$$= 0.91\ \text{k}\Omega. \tag{5.145}$$

Exercise Calculate the gain if $V_A = 5$ V.

Let us determine the gain of a CE stage as $R_C \to \infty$. Equation (5.132) gives

$$A_v = -g_m r_O. \tag{5.146}$$

Called the "intrinsic gain" of the transistor to emphasize that no external device loads the circuit, $g_m r_O$ represents the *maximum* voltage gain provided by a single transistor, playing a fundamental role in high-gain amplifiers.

We now substitute $g_m = I_C/V_T$ and $r_O = V_A/I_C$ in Eq. (5.133), thereby arriving at

$$|A_v| = \frac{V_A}{V_T}. \tag{5.147}$$

Interestingly, the intrinsic gain of a bipolar transistor is independent of the bias current. In modern integrated bipolar transistors, V_A falls in the vicinity of 5 V, yielding a gain of nearly 200.[7] In this book, we assume $g_m r_O \gg 1$ (and hence $r_O \gg 1/g_m$) for all transistors.

Another parameter of the CE stage that may prove relevant in some applications is the "current gain," defined as

$$A_I = \frac{i_{out}}{i_{in}}, \tag{5.148}$$

[7] But other second-order effects limit the actual gain to about 50.

where i_{out} denotes the current delivered to the load and i_{in} the current flowing to the input. We rarely deal with this parameter for voltage amplifiers, but note that $A_I = \beta$ for the stage shown in Fig. 5.29 because the entire collector current is delivered to R_C.

CE Stage with Emitter Degeneration In many applications, the CE core of Fig. 5.29 is modified as shown in Fig. 5.35(a), where a resistor R_E appears in series with the emitter. Called "emitter degeneration," this technique improves the "linearity" of the circuit and provides many other interesting properties that are studied in more advanced courses.

Did you know?

The Early effect was discovered by James Early, who worked under Shockley at Bell Laboratories in the 1950s. Early was also involved in the development of transistors and solar cells for satellites. Early was the first to recognize that the speed of a bipolar transistor can be improved if the base region is made much thinner than the collector region.

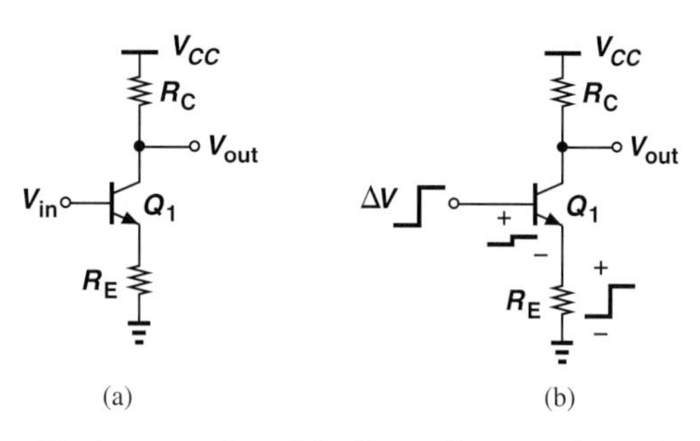

Figure 5.35 (a) CE stage with degeneration, (b) effect of input voltage change.

As with the CE core, we intend to determine the voltage gain and I/O impedances of the circuit, assuming Q_1 is biased properly. Before delving into a detailed analysis, it is instructive to make some qualitative observations. Suppose the input signal raises the base voltage by ΔV [Fig. 5.35(b)]. If R_E were zero, then the base-emitter voltage would also increase by ΔV, producing a collector current change of $g_m\,\Delta V$. But with $R_E \neq 0$, some fraction of ΔV appears across R_E, thus leaving a voltage change across the BE junction that is *less* than ΔV. Consequently, the collector current change is also less than $g_m\,\Delta V$. We therefore expect that the voltage gain of the degenerated stage is *lower* than that of the CE core with no degeneration. While undesirable, the reduction in gain is incurred to improve other aspects of the performance.

How about the input impedance? Since the collector current change is less than $g_m\,\Delta V$, the base current also changes by less than $g_m\,\Delta V/\beta$, yielding an input impedance *greater* than $\beta/g_m = r_\pi$. Thus, emitter degeneration *increases* the input impedance of the CE stage, a desirable property. A common mistake is to conclude that $R_{in} = r_\pi + R_E$, but as explained below, $R_{in} = r_\pi + (\beta+1)R_E$.

We now quantify the foregoing observations by analyzing the small-signal behavior of the circuit. Depicted in Fig. 5.36 is the small-signal equivalent circuit, where V_{CC} is replaced with an ac ground and the Early effect is neglected. Note that v_π appears across r_π and *not* from the base to ground. To determine v_{out}/v_{in}, we first write a KCL at the output node,

$$g_m v_\pi = -\frac{v_{out}}{R_C}, \tag{5.149}$$

obtaining

$$v_\pi = -\frac{v_{out}}{g_m R_C}. \tag{5.150}$$

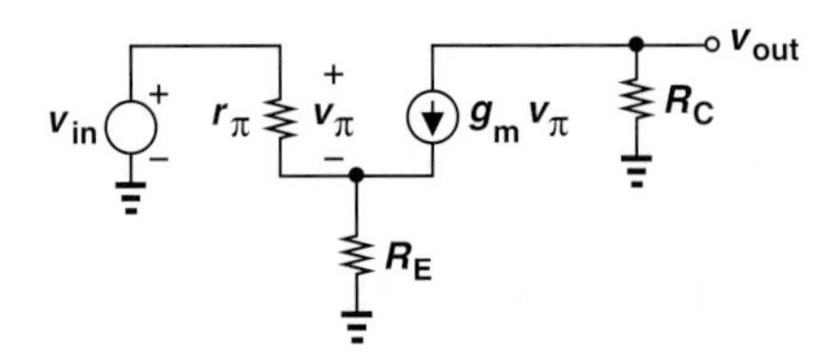

Figure 5.36 Small-signal model of CE stage with emitter degeneration.

We also recognize that two currents flow through R_E: one originating from r_π equal to v_π/r_π and another equal to $g_m v_\pi$. Thus, the voltage drop across R_E is given by

$$v_{RE} = \left(\frac{v_\pi}{r_\pi} + g_m v_\pi\right) R_E. \tag{5.151}$$

Since the voltage drop across r_π and R_E must add up to v_{in}, we have

$$v_{in} = v_\pi + v_{RE} \tag{5.152}$$

$$= v_\pi + \left(\frac{v_\pi}{r_\pi} + g_m v_\pi\right) R_E \tag{5.153}$$

$$= v_\pi \left[1 + \left(\frac{1}{r_\pi} + g_m\right) R_E\right]. \tag{5.154}$$

Substituting for v_π from Eq. (5.150) and rearranging the terms, we arrive at

$$\frac{v_{out}}{v_{in}} = -\frac{g_m R_C}{1 + \left(\frac{1}{r_\pi} + g_m\right) R_E}. \tag{5.155}$$

As predicted earlier, the magnitude of the voltage gain is lower than $g_m R_C$ for $R_E \neq 0$. With $\beta \gg 1$, we can assume $g_m \gg 1/r_\pi$ and hence

$$A_v = -\frac{g_m R_C}{1 + g_m R_E}. \tag{5.156}$$

Thus, the gain falls by a factor of $1 + g_m R_E$.

To arrive at an interesting interpretation of Eq. (5.156), we divide the numerator and denominator by g_m,

$$A_v = -\frac{R_C}{\frac{1}{g_m} + R_E}. \tag{5.157}$$

It is helpful to memorize this result as "the gain of the degenerated CE stage is equal to the total load resistance seen at the collector (to ground) divided by $1/g_m$ plus the total resistance placed in series with the emitter." (In verbal descriptions, we often ignore the negative sign in the gain, with the understanding that it must be included.) This and similar interpretations throughout this book greatly simplify the analysis of amplifiers—often obviating the need for drawing small-signal circuits.

Example 5-22 Determine the voltage gain of the stage shown in Fig. 5.37(a).

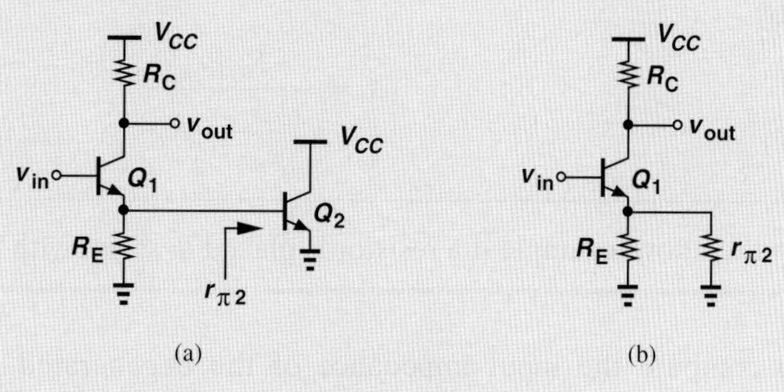

Figure 5.37 (a) CE stage example, (b) simplified circuit.

Solution We identify the circuit as a CE stage because the input is applied to the base of Q_1 and the output is sensed at its collector. This transistor is degenerated by two devices: R_E and the base-emitter junction of Q_2. The latter exhibits an impedance of $r_{\pi 2}$ (as illustrated in Fig. 5.7), leading to the simplified model depicted in Fig. 5.37(b). The total resistance placed in series with the emitter is therefore equal to $R_E||r_{\pi 2}$, yielding

$$A_v = -\frac{R_C}{\frac{1}{g_{m1}} + R_E||r_{\pi 2}}. \tag{5.158}$$

Without the above observations, we would need to draw the small-signal model of both Q_1 and Q_2 and solve a system of several equations.

Exercise Repeat the above example if a resistor is placed in series with the emitter of Q_2.

Example 5-23 Calculate the voltage gain of the circuit in Fig. 5.38(a).

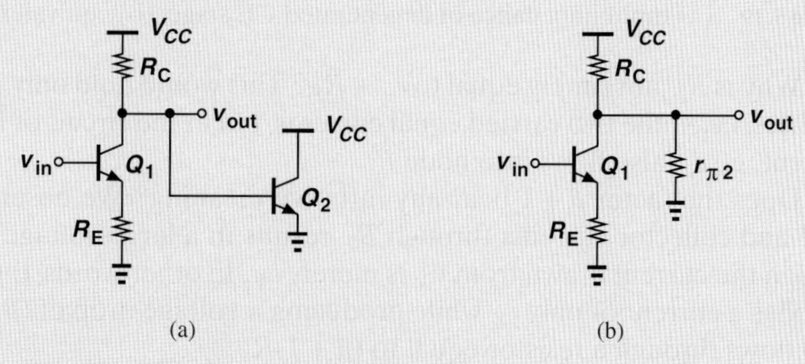

Figure 5.38 (a) CE stage example, (b) simplified circuit.

Solution The topology is a CE stage degenerated by R_E, but the load resistance between the collector of Q_1 and ac ground consists of R_C and the base-emitter junction of Q_2. Modeling the latter by $r_{\pi 2}$, we reduce the circuit to that shown in Fig. 5.38(b), where the total load resistance seen at the collector of Q_1 is equal to $R_C||r_{\pi 2}$. The voltage gain is thus given by

$$A_v = -\frac{R_C||r_{\pi 2}}{\frac{1}{g_{m1}} + R_E}. \tag{5.159}$$

Exercise Repeat the above example if a resistor is placed in series with the emitter of Q_2.

To compute the input impedance of the degenerated CE stage, we redraw the small-signal model as in Fig. 5.39(a) and calculate v_X/i_X. Since $v_\pi = r_\pi i_X$, the current flowing through R_E is equal to $i_X + g_m r_\pi i_X = (1+\beta)i_X$, creating a voltage drop of $R_E(1+\beta)i_X$. Summing v_π and v_{RE} and equating the result to v_X, we have

$$v_X = r_\pi i_X + R_E(1+\beta)i_X, \tag{5.160}$$

and hence

$$R_{in} = \frac{v_X}{i_X} \tag{5.161}$$

$$= r_\pi + (\beta + 1)R_E. \tag{5.162}$$

As predicted by our qualitative reasoning, emitter degeneration increases the input impedance [Fig. 5.39(b)].

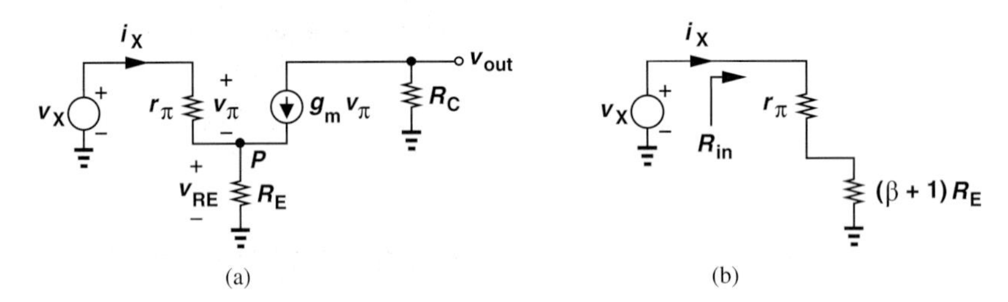

Figure 5.39 (a) Input impedance of degenerated CE stage, (b) equivalent circuit.

Why is R_{in} not simply equal to $r_\pi + R_E$? This would hold only if r_π and R_E were exactly in series, i.e., if the two carried equal currents, but in the circuit of Fig. 5.39(a), the collector current, $g_m v_\pi$, also flows into node P.

Does the factor $\beta + 1$ bear any intuitive meaning? We observe that the flow of both base and collector currents through R_E results in a large voltage drop, $(\beta+1)i_X R_E$, even though the current drawn from v_X is merely i_X. In other words, the test voltage source, v_X, supplies a current of only i_X while producing a voltage drop of $(\beta+1)i_X R_E$ across R_E—as if i_X flows through a resistor equal to $(\beta+1)R_E$.

The above observation is articulated as follows: any impedance tied between the emitter and ground is multiplied by $\beta + 1$ when "seen from the base." The expression "seen from the base" means the impedance measured between the base and ground.

We also calculate the output impedance of the stage with the aid of the equivalent shown in Fig. 5.40, where the input voltage is set to zero. Equation (5.153) applies to this circuit as well:

$$v_{in} = 0 = v_\pi + \left(\frac{v_\pi}{r_\pi} + g_m v_\pi\right) R_E, \tag{5.163}$$

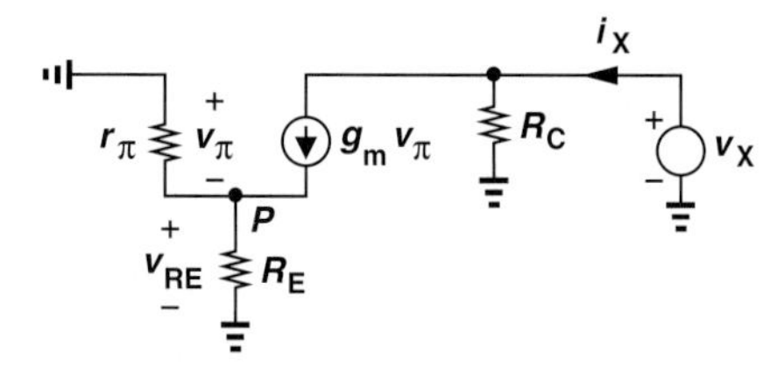

Figure 5.40 Output impedance of degenerated stage.

yielding $v_\pi = 0$ and hence $g_m v_\pi = 0$. Thus, all of i_X flows through R_C, and

$$R_{out} = \frac{v_X}{i_X} \tag{5.164}$$

$$= R_C, \tag{5.165}$$

revealing that emitter degeneration does not alter the output impedance if the Early effect is neglected.

Example 5-24 A CE stage is biased at a collector current of 1 mA. If the circuit provides a voltage gain of 20 with no emitter degeneration and 10 with degeneration, determine R_C, R_E, and the I/O impedances. Assume $\beta = 100$.

Solution For $A_v = 20$ in the absence of degeneration, we require

$$g_m R_C = 20, \tag{5.166}$$

which, together with $g_m = I_C/V_T = (26\ \Omega)^{-1}$, yields

$$R_C = 520\ \Omega. \tag{5.167}$$

Since degeneration lowers the gain by a factor of two,

$$1 + g_m R_E = 2, \tag{5.168}$$

i.e.,

$$R_E = \frac{1}{g_m} \tag{5.169}$$

$$= 26\ \Omega. \tag{5.170}$$

The input impedance is given by

$$R_{in} = r_\pi + (\beta + 1)R_E \tag{5.171}$$

$$= \frac{\beta}{g_m} + (\beta + 1)R_E \tag{5.172}$$

$$\approx 2r_\pi \tag{5.173}$$

because $\beta \gg 1$ and $R_E = 1/g_m$ in this example. Thus, $R_{in} = 5200\ \Omega$. Finally,

$$R_{out} = R_C \tag{5.174}$$

$$= 520\ \Omega. \tag{5.175}$$

Exercise What bias current would result in a gain of 5 with such emitter and collector resistor values?

Example 5-25 Compute the voltage gain and I/O impedances of the circuit depicted in Fig. 5.41. Assume a very large value for C_1.

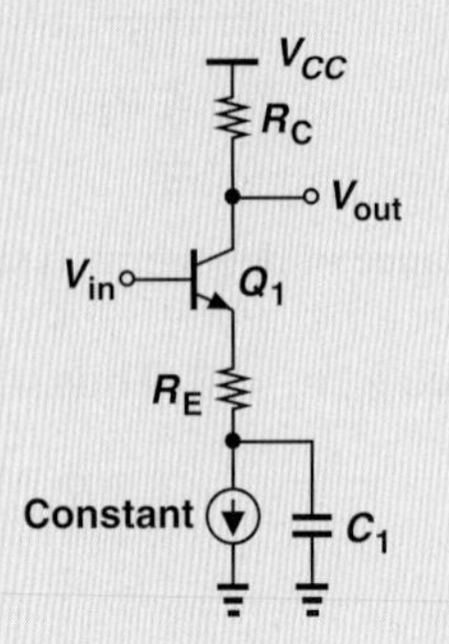

Figure 5.41 CE stage example.

Solution If C_1 is very large, it acts as a short circuit for the signal frequencies of interest. Also, the constant current source is replaced with an open circuit in the small-signal equivalent circuit. Thus, the stage reduces to that in Fig. 5.35(a) and Eqs. (5.157), (5.162), (5.165) apply.

Exercise Repeat the above example if we tie another capacitor from the base to ground.

The degenerated CE stage can be analyzed from a different perspective to provide more insight. Let us place the transistor and the emitter resistor in a black box having still three terminals [Fig. 5.42(a)]. For small-signal operation, we can view the box as a new transistor (or "active" device) and model its behavior by new values of transconductance and impedances. Denoted by G_m to avoid confusion with g_m of Q_1, the equivalent transconductance is obtained from Fig. 5.42(b). Since Eq. (5.154) still holds, we have

$$i_{out} = g_m v_\pi \tag{5.176}$$

$$= g_m \frac{v_{in}}{1 + (r_\pi^{-1} + g_m)R_E}, \tag{5.177}$$

and hence

$$G_m = \frac{i_{out}}{v_{in}} \tag{5.178}$$

$$\approx \frac{g_m}{1 + g_m R_E}. \tag{5.179}$$

For example, the voltage gain of the stage with a load resistance of R_D is given by $-G_m R_D$.

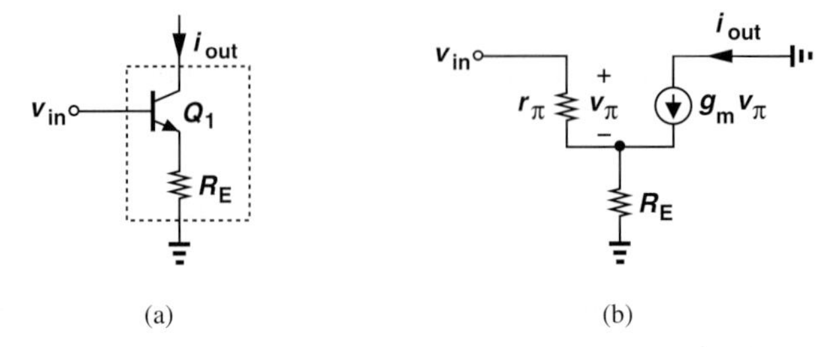

Figure 5.42 (a) Degenerated bipolar transistor viewed as a black box, (b) small-signal equivalent.

An interesting property of the degenerated CE stage is that its voltage gain becomes relatively independent of the transistor transconductance and hence bias current if $g_m R_E \gg 1$. From Eq. (5.157), we note that $A_v \rightarrow -R_C/R_E$ under this condition. As studied in Problem 5.40, this trend in fact represents the "linearizing" effect of emitter degeneration.

As a more general case, we now consider a degenerated CE stage containing a resistance in series with the base [Fig. 5.43(a)]. As seen below, R_B only *degrades* the performance of the circuit, but often proves inevitable. For example, R_B may represent the output resistance of a microphone connected to the input of the amplifier.

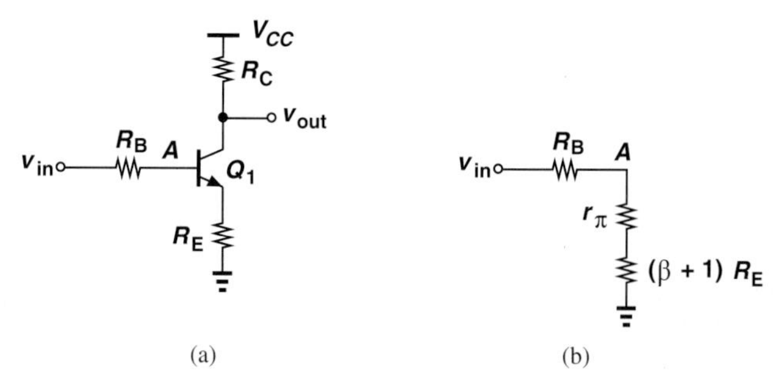

Figure 5.43 (a) CE stage with base resistance, (b) equivalent circuit.

To analyze the small-signal behavior of this stage, we can adopt one of two approaches: (a) draw the small-signal model of the entire circuit and solve the resulting equations, or (b) recognize that the signal at node A is simply an attenuated version of v_{in} and write

$$\frac{v_{out}}{v_{in}} = \frac{v_A}{v_{in}} \cdot \frac{v_{out}}{v_A}. \tag{5.180}$$

Here, v_A/v_{in} denotes the effect of voltage division between R_B and the impedance seen at the base of Q_1, and v_{out}/v_A represents the voltage gain from the base of Q_1 to the output, as already obtained in Eqs. (5.155) and (5.157). We leave the former approach for Problem 5.44 and continue with the latter here.

Let us first compute v_A/v_{in} with the aid of Eq. (5.162) and the model depicted in Fig. 5.39(b), as illustrated in Fig. 5.43(b). The resulting voltage divider yields

$$\frac{v_A}{v_{in}} = \frac{r_\pi + (\beta+1)R_E}{r_\pi + (\beta+1)R_E + R_B}. \tag{5.181}$$

Combining Eqs. (5.155) and (5.157), we arrive at the overall gain as

$$\frac{v_{out}}{v_{in}} = \frac{r_\pi + (\beta+1)R_E}{r_\pi + (\beta+1)R_E + R_B} \cdot \frac{-g_m R_C}{1 + \left(\frac{1}{r_\pi} + g_m\right) R_E} \tag{5.182}$$

$$= \frac{r_\pi + (\beta+1)R_E}{r_\pi + (\beta+1)R_E + R_B} \cdot \frac{-g_m r_\pi R_C}{r_\pi + (1+\beta)R_E} \tag{5.183}$$

$$= \frac{-\beta R_C}{r_\pi + (\beta+1)R_E + R_B}. \tag{5.184}$$

To obtain a more intuitive expression, we divide the numerator and the denominator by β:

$$A_v \approx \frac{-R_C}{\frac{1}{g_m} + R_E + \frac{R_B}{\beta + 1}}. \tag{5.185}$$

Compared to Eq. (5.157), this result contains only one additional term in the denominator equal to the base resistance divided by $\beta + 1$.

The above results reveal that resistances in series with the emitter and the base have similar effects on the voltage gain, but R_B is scaled down by $\beta + 1$. The significance of this observation becomes clear later.

For the stage of Fig. 5.43(a), we can define two different input impedances, one seen at the base of Q_1 and another at the left terminal of R_B (Fig. 5.44). The former is equal to

$$R_{in1} = r_\pi + (\beta + 1)R_E \tag{5.186}$$

and the latter,

$$R_{in2} = R_B + r_\pi + (\beta + 1)R_E. \tag{5.187}$$

In practice, R_{in1} proves more relevant and useful. We also note that the output impedance of the circuit remains equal to

$$R_{out} = R_C \tag{5.188}$$

even with $R_B \neq 0$. This is studied in Problem 5.45.

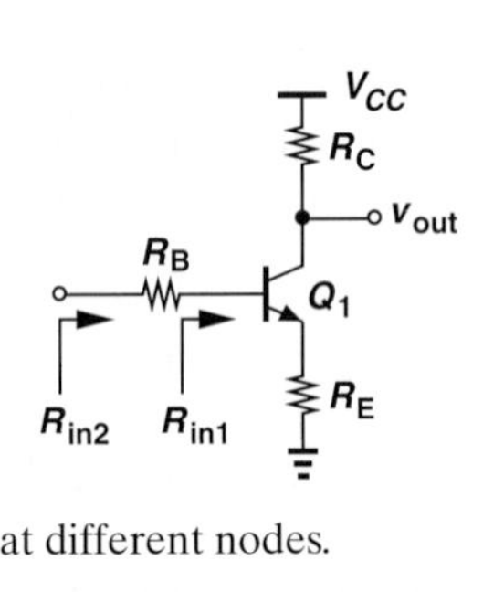

Figure 5.44 Input impedances seen at different nodes.

Example 5-26 A microphone having an output resistance of 1 kΩ generates a peak signal level of 2 mV. Design a CE stage with a bias current of 1 mA that amplifies this signal to 40 mV. Assume $R_E = 4/g_m$ and $\beta = 100$.

Solution The following quantities are obtained: $R_B = 1$ kΩ, $g_m = (26\ \Omega)^{-1}$, $|A_v| = 20$, and $R_E = 104\ \Omega$. From Eq. (5.185),

$$R_C = |A_v| \left(\frac{1}{g_m} + R_E + \frac{R_B}{\beta + 1} \right) \tag{5.189}$$

$$\approx 2.8\ \text{k}\Omega. \tag{5.190}$$

Exercise Repeat the above example if the microphone output resistance is doubled.

Example 5-27 Determine the voltage gain and I/O impedances of the circuit shown in Fig. 5.45(a). Assume a very large value for C_1 and neglect the Early effect.

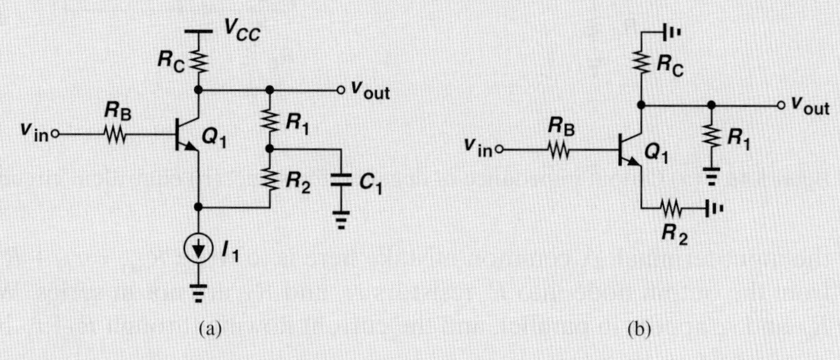

Figure 5.45 (a) CE stage example, (b) simplified circuit.

Solution Replacing C_1 with a short circuit, I_1 with an open circuit, and V_{CC} with ac ground, we arrive at the simplified model in Fig. 5.45(b), where R_1 and R_C appear in parallel and R_2 acts as an emitter degeneration resistor. Equations (5.185)–(5.188) are therefore written respectively as

$$A_v = \frac{-(R_C||R_1)}{\frac{1}{g_m} + R_2 + \frac{R_B}{\beta + 1}} \tag{5.191}$$

$$R_{in} = R_B + r_\pi + (\beta + 1)R_2 \tag{5.192}$$

$$R_{out} = R_C||R_1. \tag{5.193}$$

Exercise What happens if a very large capacitor is tied from the emitter of Q_1 to ground?

Effect of Transistor Output Resistance The analysis of the degenerated CE stage has thus far neglected the Early effect. Somewhat beyond the scope of this book, the derivation of the circuit properties in the presence of this effect is outlined in Problem 5.48 for the interested reader. We nonetheless explore one aspect of the circuit, namely, the output resistance, as it provides the foundation for many other topologies studied later.

Our objective is to determine the output impedance seen looking into the collector of a degenerated transistor [Fig. 5.46(a)]. Recall from Fig. 5.7 that $R_{out} = r_O$ if $R_E = 0$. Also, $R_{out} = \infty$ if $V_A = \infty$ (why?). To include the Early effect, we draw the small-signal equivalent circuit as in Fig. 5.46(b), grounding

Did you know?

Our extensive treatment of emitter degeneration is justified. As mentioned previously, even in early transistor radios, emitter degeneration was used to stabilize the bias point and gain of the amplifier stages against transistor and temperature variations. Today, we often employ this technique to improve the *linearity* of circuits. Without degeneration, the circuit's voltage gain is equal to $g_m R_C = (I_C/V_T)R_C$. If the input *signal* goes up and down substantially, so do I_C and the gain, causing a nonlinear characteristic. (If the gain depends on the input signal level, then the circuit is nonlinear.) With degeneration, on the other hand, the gain is given by $g_m R_C/(1 + g_m R_E)$, varying to a lesser extent. In fact, if $g_m R_E \gg 1$, $|A_v \approx R_C/R_E$, a quantity relatively independent of the input signal level. Linearity is a critical parameter in most analog circuits, e.g., audio and video amplifiers, RF circuits, etc.

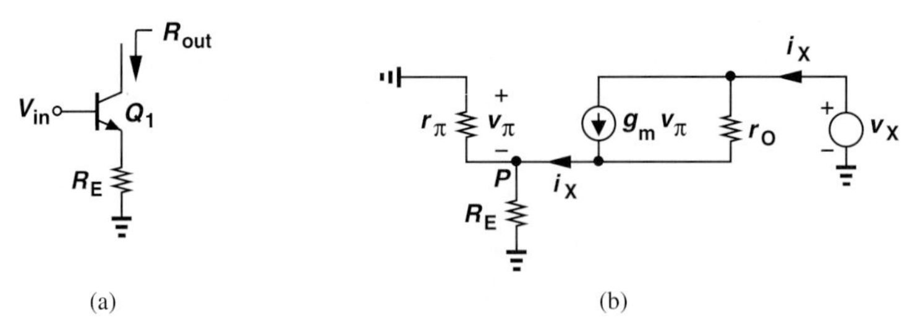

Figure 5.46 (a) Output impedance of degenerated stage, (b) equivalent circuit.

the input terminal. A common mistake here is to write $R_{out} = r_O + R_E$. Since $g_m v_\pi$ flows from the output node into P, resistors r_O and R_E are not in series. We readily note that R_E and r_π appear in parallel, and the current flowing through $R_E||r_\pi$ is equal to i_X. Thus,

$$v_\pi = -i_X(R_E||r_\pi), \tag{5.194}$$

where the negative sign arises because the positive side of v_π is at ground. We also recognize that r_O carries a current of $i_X - g_m v_\pi$ and hence sustains a voltage of $(i_X - g_m v_\pi)r_O$. Adding this voltage to that across R_E $(= -v_\pi)$ and equating the result to v_X, we obtain

$$v_X = (i_X - g_m v_\pi)r_O - v_\pi \tag{5.195}$$

$$= [i_X + g_m i_X(R_E||r_\pi)]r_O + i_X(R_E||r_\pi). \tag{5.196}$$

It follows that

$$R_{out} = [1 + g_m(R_E||r_\pi)]r_O + R_E||r_\pi \tag{5.197}$$

$$= r_O + (g_m r_O + 1)(R_E||r_\pi). \tag{5.198}$$

Recall from Eq. (5.146) that the intrinsic gain of the transistor, $g_m r_O \gg 1$, and hence

$$R_{out} \approx r_O + g_m r_O(R_E||r_\pi) \tag{5.199}$$

$$\approx r_O[1 + g_m(R_E||r_\pi)]. \tag{5.200}$$

Interestingly, emitter degeneration *raises* the output impedance from r_O to the above value, i.e., by a factor of $1 + g_m(R_E||r_\pi)$.

The reader may wonder if the increase in the output resistance is desirable or undesirable. The "boosting" of output resistance as a result of degeneration proves extremely useful in circuit design, producing amplifiers with a higher gain as well as creating more ideal current sources. These concepts are studied in Chapter 9.

It is instructive to examine Eq. (5.200) for two special cases $R_E \gg r_\pi$ and $R_E \ll r_\pi$. For $R_E \gg r_\pi$, we have $R_E||r_\pi \to r_\pi$ and

$$R_{out} \approx r_O(1 + g_m r_\pi) \tag{5.201}$$

$$\approx \beta r_O, \tag{5.202}$$

because $\beta \gg 1$. Thus, the maximum resistance seen at the collector of a bipolar transistor is equal to βr_O—if the degeneration impedance becomes much larger than r_π.

For $R_E \ll r_\pi$, we have $R_E||r_\pi \to R_E$ and

$$R_{out} \approx (1 + g_m R_E)r_O. \tag{5.203}$$

Thus, the output resistance is boosted by a factor of $1 + g_m R_E$.

In the analysis of circuits, we sometimes draw the transistor output resistance explicitly to emphasize its significance (Fig. 5.47). This representation, of course, assumes Q_1 itself does not contain another r_O.

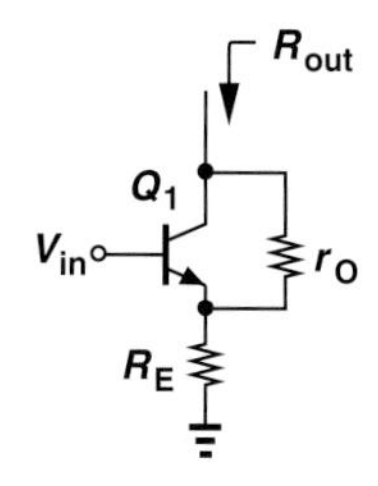

Figure 5.47 Stage with explicit depiction of r_O.

Example 5-28 We wish to design a current source having a value of 1 mA and an output resistance of 20 kΩ. The available bipolar transistor exhibits $\beta = 100$ and $V_A = 10$ V. Determine the minimum required value of emitter degeneration resistance.

Solution Since $r_O = V_A/I_C = 10$ kΩ, degeneration must raise the output resistance by a factor of two. We postulate that the condition $R_E \ll r_\pi$ holds and write

$$1 + g_m R_E = 2. \tag{5.204}$$

That is,

$$R_E = \frac{1}{g_m} \tag{5.205}$$

$$= 26\ \Omega. \tag{5.206}$$

Note that indeed $r_\pi = \beta/g_m \gg R_E$.

Exercise What is the output impedance if R_E is doubled?

Example 5-29 Calculate the output resistance of the circuit shown in Fig. 5.48(a) if C_1 is very large.

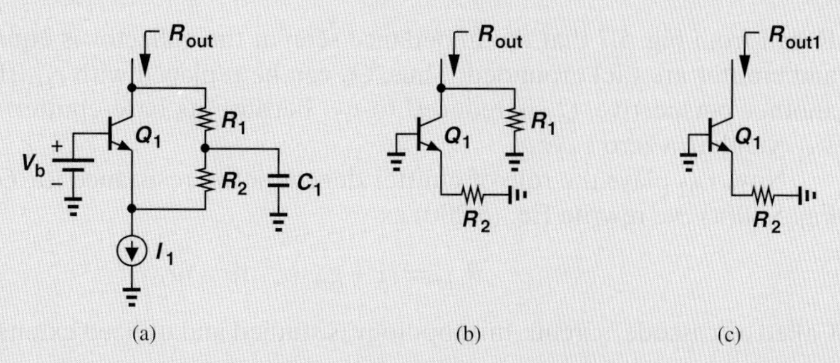

Figure 5.48 (a) CE stage example, (b) simplified circuit, (c) resistance seen at the collector.

Solution Replacing V_b and C_1 with an ac ground and I_1 with an open circuit, we arrive at the simplified model in Fig. 5.48(b). Since R_1 appears in parallel with the resistance seen looking into the collector of Q_1, we ignore R_1 for the moment, reducing the circuit to that in Fig. 5.48(c). In analogy with Fig. 5.40, we rewrite Eq. (5.200) as

$$R_{out1} = [1 + g_m(R_2||r_\pi)]r_O. \quad (5.207)$$

Returning to Fig. 5.48(b), we have

$$R_{out} = R_{out1}||R_1 \quad (5.208)$$

$$= \{[1 + g_m(R_2||r_\pi)]r_O\}||R_1. \quad (5.209)$$

Exercise What is the output resistance if a very large capacitor is tied between the emitter of Q_1 and ground?

The procedure of progressively simplifying a circuit until it resembles a known topology proves extremely critical in our work. Called "analysis by inspection," this method obviates the need for complex small-signal models and lengthy calculations. The reader is encouraged to attempt the above example using the small-signal model of the overall circuit to appreciate the efficiency and insight provided by our intuitive approach.

Example 5-30 Determine the output resistance of the stage shown in Fig. 5.49(a).

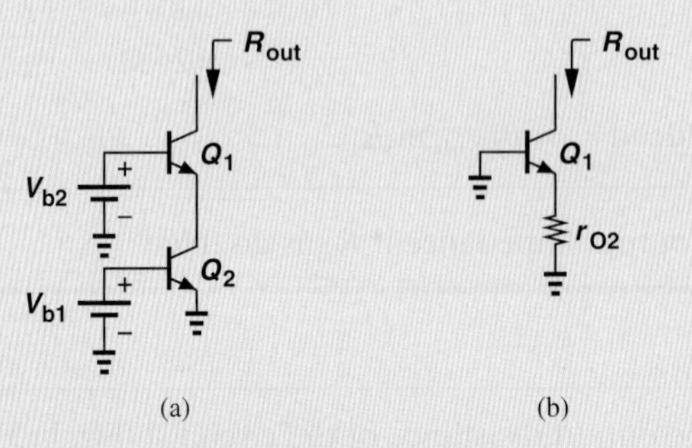

Figure 5.49 (a) CE stage example, (b) simplified circuit.

Solution Recall from Fig. 5.7 that the impedance seen at the collector is equal to r_O if the base and emitter are (ac) grounded. Thus, Q_2 can be replaced with r_{O2} [Fig. 5.49(b)]. From another perspective, Q_2 is reduced to r_{O2} because its base-emitter voltage is fixed by V_{b1}, yielding a zero $g_{m2}v_{\pi 2}$.

Now, r_{O2} plays the role of emitter degeneration resistance for Q_1. In analogy with Fig. 5.40(a), we rewrite Eq. (5.200) as

$$R_{out} = [1 + g_{m1}(r_{O2}||r_{\pi 1})]r_{O1}. \quad (5.210)$$

Called a "cascode" circuit, this topology is studied and utilized extensively in Chapter 9.

Exercise Repeat the above example for a "stack" of three transistors.

CE Stage with Biasing Having learned the small-signal properties of the common-emitter amplifier and its variants, we now study a more general case wherein the circuit contains a bias network as well. We begin with simple biasing schemes described in Section 5.2 and progressively add complexity (and more robust performance) to the circuit. Let us begin with an example.

Example 5-31

A student familiar with the CE stage and basic biasing constructs the circuit shown in Fig. 5.50 to amplify the signal produced by a microphone. Unfortunately, Q_1 carries no current, failing to amplify. Explain the cause of this problem.

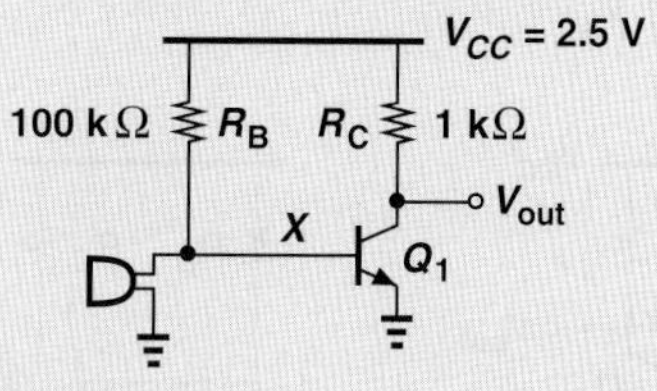

Figure 5.50 Microphone amplifier.

Solution Many microphones exhibit a small low-frequency resistance (e.g., $< 100\ \Omega$). If used in this circuit, such a microphone creates a low resistance from the base of Q_1 to ground, forming a voltage divider with R_B and providing a very low base voltage. For example, a microphone resistance of 100 Ω yields

$$V_X = \frac{100\ \Omega}{100\ \text{k}\Omega + 100\ \Omega} \times 2.5\ \text{V} \tag{5.211}$$

$$\approx 2.5\ \text{mV}. \tag{5.212}$$

Thus, the microphone low-frequency resistance disrupts the bias of the amplifier.

Exercise Does the circuit operate better if R_B is halved?

How should the circuit of Fig. 5.50 be fixed? Since only the *signal* generated by the microphone is of interest, a series capacitor can be inserted as depicted in Fig. 5.51 so as to isolate the dc biasing of the amplifier from the microphone. That is, the bias point of Q_1 remains independent of the resistance of the microphone because C_1 carries no bias current. The value of C_1 is chosen so that it provides a relatively low impedance (almost a short circuit) for the frequencies of interest. We say C_1 is a "coupling" capacitor and the input of this stage is "ac-coupled" or "capacitively coupled." Many circuits employ capacitors to isolate the bias conditions from "undesirable" effects. More examples clarify this point later.

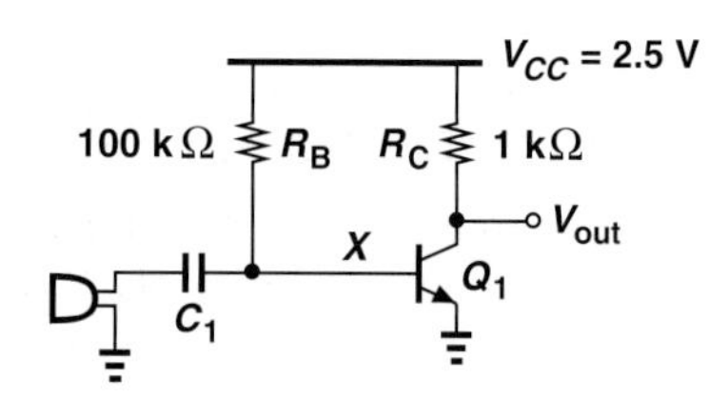

Figure 5.51 Capacitive coupling at the input of microphone amplifier.

The foregoing observation suggests that the methodology illustrated in Fig. 5.9 must include an additional rule: replace all capacitors with an open circuit for dc analysis and a short circuit for small-signal analysis.

Let us begin with the stage depicted in Fig. 5.52(a). For bias calculations, the signal source is set to zero and C_1 is opened, leading to Fig. 5.52(b). From Section 5.2.1, we have

$$I_C = \beta \frac{V_{CC} - V_{BE}}{R_B}, \tag{5.213}$$

$$V_Y = V_{CC} - \beta R_C \frac{V_{CC} - V_{BE}}{R_B}. \tag{5.214}$$

To avoid saturation, $V_Y \geq V_{BE}$.

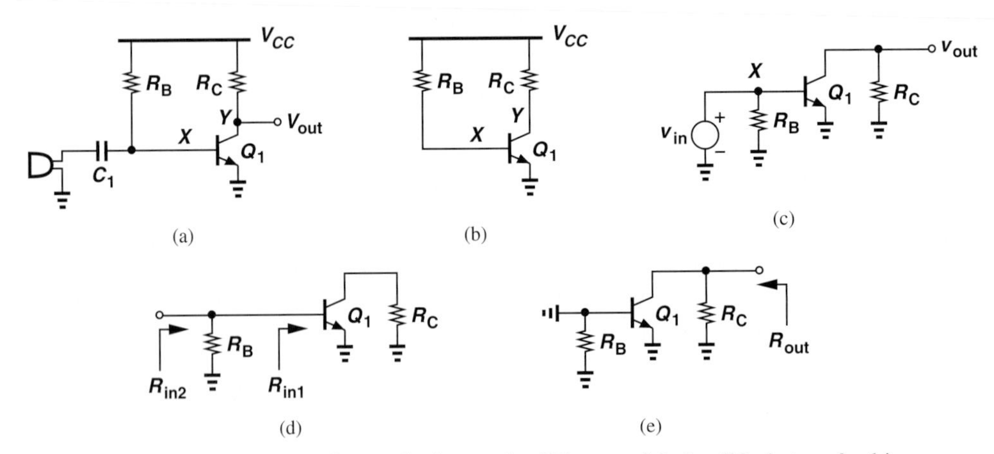

Figure 5.52 (a) Capacitive coupling at the input of a CE stage, (b) simplified stage for bias calculation, (c) simplified stage for small-signal calculation, (d) simplified circuit for input impedance calculation, (e) simplified circuit for output impedance calculation.

With the bias current known, the small-signal parameters g_m, r_π, and r_O can be calculated. We now turn our attention to small-signal analysis, considering the simplified circuit of Fig. 5.52(c). Here, C_1 is replaced with a short and V_{CC} with ac ground, but Q_1 is maintained as a symbol. We attempt to solve the circuit by inspection: if unsuccessful, we will resort to using a small-signal model for Q_1 and writing KVLs and KCLs.

The circuit of Fig. 5.52(c) resembles the CE core illustrated in Fig. 5.29, except for R_B. Interestingly, R_B has no effect on the voltage at node X so long as v_{in} remains an ideal voltage source; i.e., $v_X = v_{in}$ regardless of the value of R_B. Since the voltage gain from the base to the collector is given by $v_{out}/v_X = -g_m R_C$, we have

$$\frac{v_{out}}{v_{in}} = -g_m R_C. \tag{5.215}$$

If $V_A < \infty$, then

$$\frac{v_{out}}{v_{in}} = -g_m(R_C || r_O). \tag{5.216}$$

However, the input impedance is affected by R_B [Fig. 5.52(d)]. Recall from Fig. 5.7 that the impedance seen looking into the base, R_{in1}, is equal to r_π if the emitter is grounded. Here, R_B simply appears in parallel with R_{in1}, yielding

$$R_{in2} = r_\pi || R_B. \tag{5.217}$$

Thus, the bias resistor lowers the input impedance. Nevertheless, as shown in Problem 5.51, this effect is usually negligible.

To determine the output impedance, we set the input source to zero [Fig. 5.52(e)]. Comparing this circuit with that in Fig. 5.32(b), we recognize that R_{out} remains unchanged:

$$R_{out} = R_C || r_O. \tag{5.218}$$

because both terminals of R_B are shorted to ground.

In summary, the bias resistor, R_B, negligibly impacts the performance of the stage shown in Fig. 5.52(a).

Example 5-32 Having learned about ac coupling, the student in Example 5-31 modifies the design to that shown in Fig. 5.53 and attempts to drive a speaker. Unfortunately, the circuit still fails. Explain why.

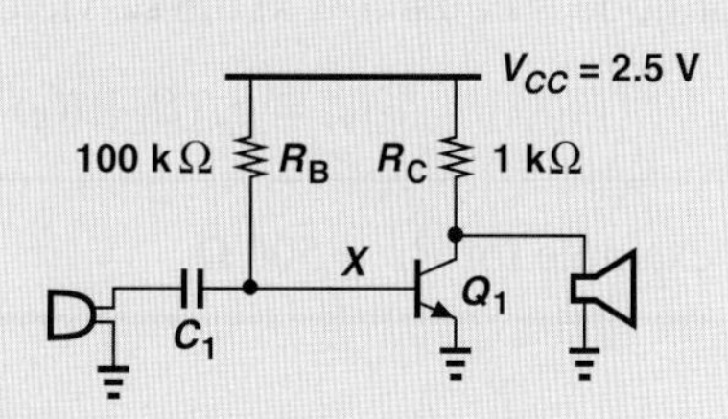

Figure 5.53 Amplifier with direct connection of speaker.

Solution Typical speakers incorporate a solenoid (inductor) to actuate a membrane. The solenoid exhibits a very low dc resistance, e.g., less than 1 Ω. Thus, the speaker in Fig. 5.53 shorts the collector to ground, driving Q_1 into deep saturation.

Exercise Does the circuit operate better if the speaker is tied between the output node and V_{CC}?

Example 5-33 The student applies ac coupling to the output as well [Fig. 5.54(a)] and measures the quiescent points to ensure proper biasing. The collector bias voltage is 1.5 V, indicating that Q_1 operates in the active region. However, the student still observes no gain in the circuit. (a) If $I_S = 5 \times 10^{-17}$ A and $V_A = \infty$, compute the β of the transistor. (b) Explain why the circuit provides no gain.

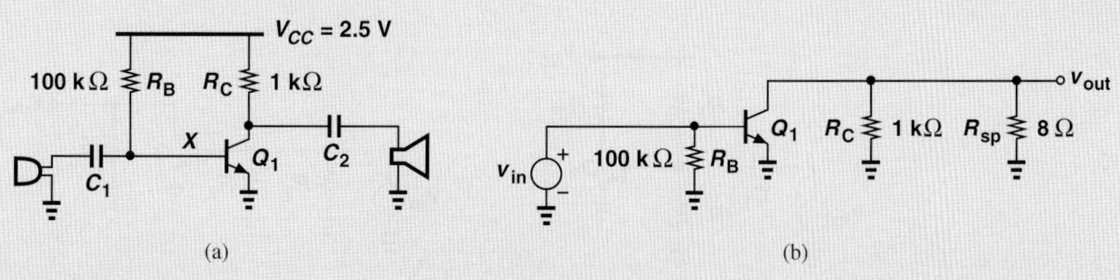

Figure 5.54 (a) Amplifier with capacitive coupling at the input and output, (b) simplified small-signal model.

Solution (a) A collector voltage of 1.5 V translates to a voltage drop of 1 V across R_C and hence a collector current of 1 mA. Thus,

$$V_{BE} = V_T \ln \frac{I_C}{I_S} \tag{5.219}$$

$$= 796 \text{ mV}. \tag{5.220}$$

It follows that

$$I_B = \frac{V_{CC} - V_{BE}}{R_B} \tag{5.221}$$

$$= 17 \text{ µA}, \tag{5.222}$$

and $\beta = I_C/I_B = 58.8$.

(b) Speakers typically exhibit a low impedance in the audio frequency range, e.g., 8 Ω. Drawing the ac equivalent as in Fig. 5.54(b), we note that the total resistance seen at the collector node is equal to 1 kΩ||8 Ω, yielding a gain of

$$|A_v| = g_m(R_C||R_S) = 0.31. \tag{5.223}$$

Exercise Repeat the above example for $R_C = 500$ Ω.

The design in Fig. 5.54(a) exemplifies an improper interface between an amplifier and a load: the output impedance is so much higher than the load impedance that the connection of the load to the amplifier drops the gain drastically.

How can we remedy the problem of loading here? Since the voltage gain is proportional to g_m, we can bias Q_1 at a much higher current to raise the gain. This is studied in Problem 5.53. Alternatively, we can interpose a "buffer" stage between the CE amplifier and the speaker (Section 5.3.3).

Let us now consider the biasing scheme shown in Fig. 5.15 and repeated in Fig. 5.55(a). To determine the bias conditions, we set the signal source to zero and open the capacitor(s). Equations (5.38)–(5.41) can then be used. For small-signal analysis, the simplified circuit in Fig. 5.55(b) reveals a resemblance to that in Fig. 5.52(b), except that both R_1 and R_2 appear in parallel with the input. Thus, the voltage gain is still equal to $-g_m(R_C||r_O)$ and the input impedance is given by

$$R_{in} = r_\pi||R_1||R_2. \tag{5.224}$$

The output resistance is equal to $R_C||r_O$.

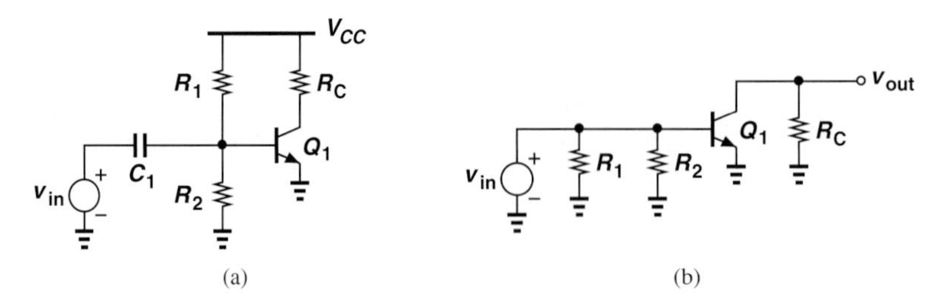

Figure 5.55 (a) Biased stage with capacitive coupling, (b) simplified circuit.

We next study the more robust biasing scheme of Fig. 5.19, repeated in Fig. 5.56(a) along with an input coupling capacitor. The bias point is determined by opening C_1 and following Eqs. (5.52) and (5.53). With the collector current known, the small-signal parameters of Q_1 can be computed. We also construct the simplified ac circuit shown in Fig. 5.56(b), noting that the voltage gain is not affected by R_1 and R_2 and remains equal to

$$A_v = \frac{-R_C}{\frac{1}{g_m} + R_E}, \tag{5.225}$$

where Early effect is neglected. On the other hand, the input impedance is lowered to:

$$R_{in} = [r_\pi + (\beta + 1)R_E]||R_1||R_2, \tag{5.226}$$

whereas the output impedance remains equal to R_C if $V_A = \infty$.

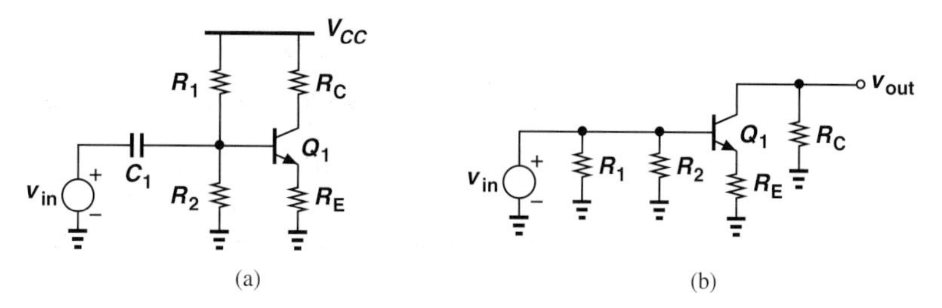

Figure 5.56 (a) Degenerated stage with capacitive coupling, (b) simplified circuit.

As explained in Section 5.2.3, the use of emitter degeneration can effectively stabilize the bias point despite variations in β and I_S. However, as evident from Eq. (5.225), degeneration also lowers the gain. Is it possible to apply degeneration to biasing but *not* to the signal? Illustrated in Fig. 5.57 is such a topology, where C_2 is large enough to act as a short circuit for signal frequencies of interest. We can therefore write

$$A_v = -g_m R_C \tag{5.227}$$

and

$$R_{in} = r_\pi ||R_1||R_2 \tag{5.228}$$

$$R_{out} = R_C. \tag{5.229}$$

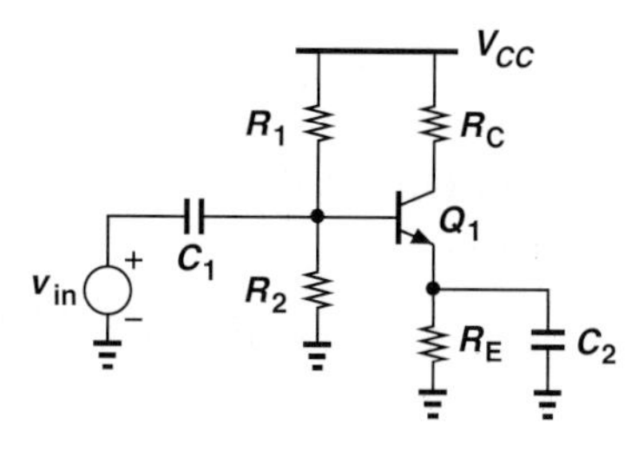

Figure 5.57 Use of capacitor to eliminate degeneration.

Example 5-34 Design the stage of Fig. 5.57 to satisfy the following conditions: $I_C = 1$ mA, voltage drop across $R_E = 400$ mV, voltage gain = 20 in the audio frequency range (20 Hz to 20 kHz), input impedance > 2 kΩ. Assume $\beta = 100$, $I_S = 5 \times 10^{-16}$, and $V_{CC} = 2.5$ V.

Solution With $I_C = 1\ \text{mA} \approx I_E$, the value of R_E is equal to 400 Ω. For the voltage gain to remain unaffected by degeneration, the maximum impedance of C_1 must be much smaller than $1/g_m = 26\ \Omega$.[8] Occurring at 20 Hz, the maximum impedance must remain below roughly $0.1 \times (1/g_m) = 2.6\ \Omega$:

$$\frac{1}{C_2\omega} \leq \frac{1}{10} \cdot \frac{1}{g_m} \text{ for } \omega = 2\pi \times 20 \text{ Hz}. \tag{5.230}$$

Thus,

$$C_2 \geq 30{,}607\ \mu\text{F}. \tag{5.231}$$

(This value is unrealistically large, requiring modification of the design.) We also have

$$|A_v| = g_m R_C = 20, \tag{5.232}$$

obtaining

$$R_C = 520\ \Omega. \tag{5.233}$$

Since the voltage across R_E is equal to 400 mV and $V_{BE} = V_T \ln(I_C/I_S) = 736$ mV, we have $V_X = 1.14$ V. Also, with a base current of 10 μA, the current flowing through R_1 and R_2 must exceed 100 μA to lower sensitivity to β:

$$\frac{V_{CC}}{R_1 + R_2} > 10I_B \tag{5.234}$$

and hence

$$R_1 + R_2 < 25\ \text{k}\Omega. \tag{5.235}$$

Under this condition,

$$V_X \approx \frac{R_2}{R_1 + R_2} V_{CC} = 1.14\ \text{V}, \tag{5.236}$$

yielding

$$R_2 = 11.4\ \text{k}\Omega \tag{5.237}$$

$$R_1 = 13.6\ \text{k}\Omega. \tag{5.238}$$

We must now check to verify that this choice of R_1 and R_2 satisfies the condition $R_{in} >$ 2 kΩ. That is,

$$R_{in} = r_\pi || R_1 || R_2 \tag{5.239}$$

$$= 1.83\ \text{k}\Omega. \tag{5.240}$$

Unfortunately, R_1 and R_2 lower the input impedance excessively. To remedy the problem, we can allow a smaller current through R_1 and R_2 than $10I_B$, at the cost of creating more sensitivity to β. For example, if this current is set to $5I_B = 50$ μA and we still neglect I_B in the calculation of V_X,

$$\frac{V_{CC}}{R_1 + R_2} > 5I_B \tag{5.241}$$

[8] A common mistake here is to make the impedance of C_1 much less than R_E.

and

$$R_1 + R_2 < 50\ \text{k}\Omega. \tag{5.242}$$

Consequently,

$$R_2 = 22.8\ \text{k}\Omega \tag{5.243}$$

$$R_1 = 27.2\ \text{k}\Omega, \tag{5.244}$$

giving

$$R_{in} = 2.15\ \text{k}\Omega. \tag{5.245}$$

Exercise Redesign the above stage for a gain of 10 and compare the results.

We conclude our study of the CE stage with a brief look at the more general case depicted in Fig. 5.58(a), where the input signal source exhibits a finite resistance and the output is tied to a load R_L. The biasing remains identical to that of Fig. 5.56(a), but R_S and R_L lower the voltage gain v_{out}/v_{in}. The simplified ac circuit of Fig. 5.58(b) reveals V_{in} is attenuated by the voltage division between R_S and the impedance seen at node X, $R_1||R_2||[r_\pi + (\beta + 1)R_E]$, i.e.,

$$\frac{v_X}{v_{in}} = \frac{R_1||R_2||[r_\pi + (\beta + 1)R_E]}{R_1||R_2||[r_\pi + (\beta + 1)R_E] + R_S}. \tag{5.246}$$

The voltage gain from v_{in} to the output is given by

$$\frac{v_{out}}{v_{in}} = \frac{v_X}{v_{in}} \cdot \frac{v_{out}}{v_X} \tag{5.247}$$

$$= -\frac{R_1||R_2||[r_\pi + (\beta + 1)R_E]}{R_1||R_2||[r_\pi + (\beta + 1)R_E] + R_S}\frac{R_C||R_L}{\frac{1}{g_m} + R_E}. \tag{5.248}$$

As expected, lower values of R_1 and R_2 reduce the gain.

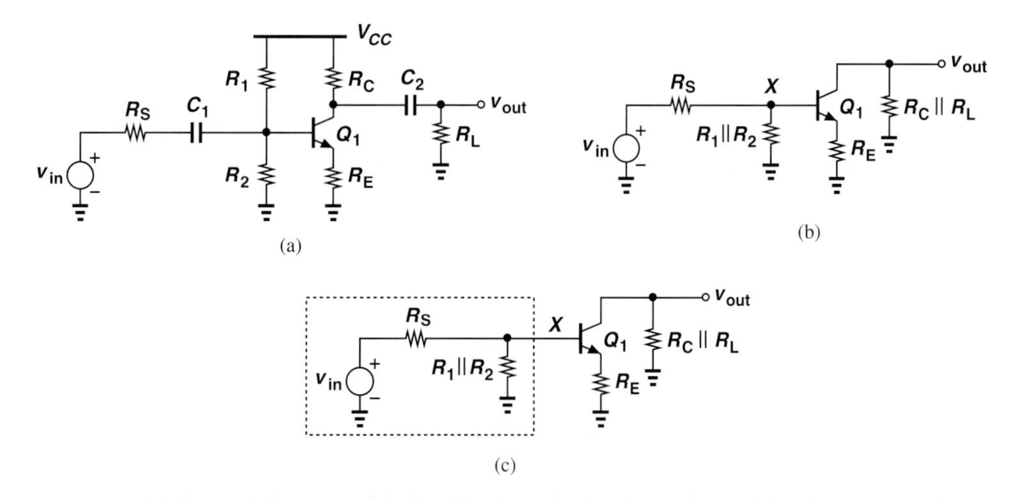

Figure 5.58 (a) General CE stage, (b) simplified circuit, (c) Thevenin model of input network.

Did you know?

What we have learned so far readily allows us to construct interesting circuits. One of the first gadgets that you can build are radio receivers and transmitters. Shown below is an FM transmitter operating in the vicinity of 100 MHz. A self-biased CE stage amplifies the microphone signal, applying the result to an oscillator. The audio signal "modulates" (varies) the frequency of the oscillator, creating an FM output.

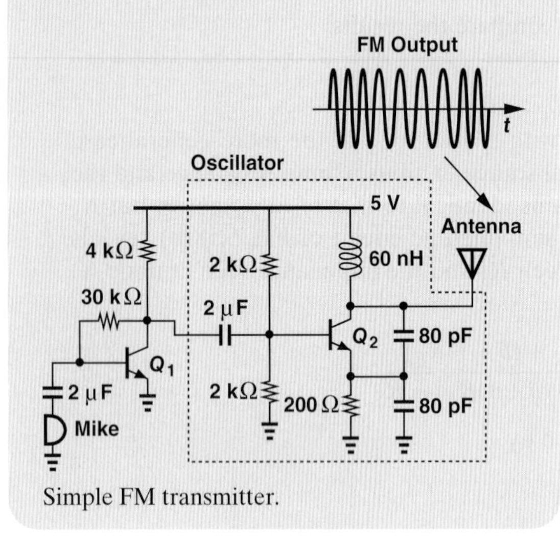

Simple FM transmitter.

The above computation views the input network as a voltage divider. Alternatively, we can utilize a Thevenin equivalent to include the effect of R_S, R_1, and R_2 on the voltage gain. Illustrated in Fig. 5.58(c), the idea is to replace v_{in}, R_S and $R_1||R_2$ with v_{Thev} and R_{Thev}:

$$v_{Thev} = \frac{R_1||R_2}{R_1||R_2 + R_S} v_{in} \tag{5.249}$$

$$R_{Thev} = R_S||R_1||R_2. \tag{5.250}$$

The resulting circuit resembles that in Fig. 5.43(a) and follows Eq. (5.185):

$$A_v = -\frac{R_C||R_L}{\dfrac{1}{g_m} + R_E + \dfrac{R_{Thev}}{\beta + 1}} \cdot \frac{R_1||R_2}{R_1||R_2 + R_S}, \tag{5.251}$$

where the second fraction on the right accounts for the voltage attenuation given by Eq. (5.249). The reader is encouraged to prove that Eqs. (5.248) and (5.251) are identical.

The two approaches described above exemplify analysis techniques used to solve circuits and gain insight. Neither requires drawing the small-signal model of the transistor because the reduced circuits can be "mapped" into known topologies.

Figure 5.59 summarizes the concepts studied in this section.

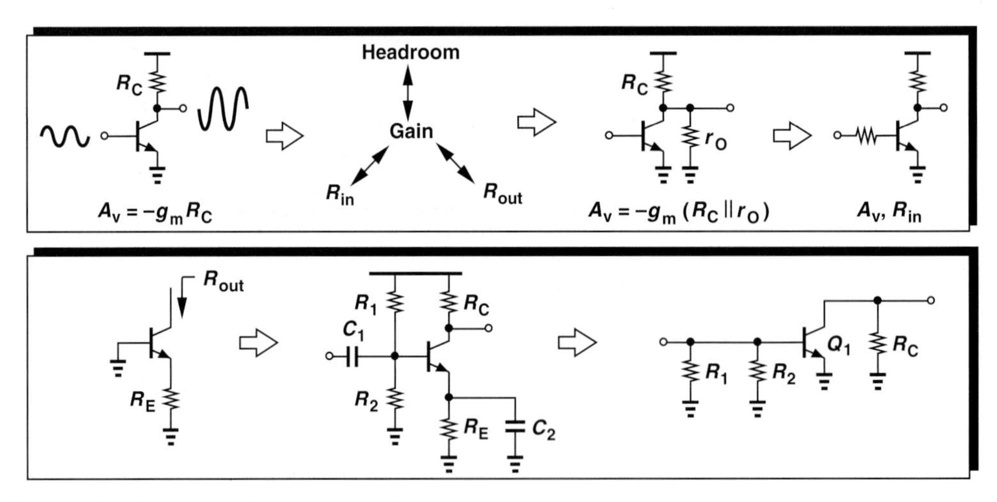

Figure 5.59 Summary of concepts studied thus far.

Robotics Application: Digital-to-Analog Converters

It is common in robotics to use a microcontroller (or a microprocessor) to perform computations in the digital domain. Nonetheless, the results of these computations must often be delivered in analog form to various devices. For example, a robot's arm moves continuously and must be driven by an analog signal. As the arm moves to pick up an object, it must slow down when it comes closer. Based on measurements performed by a camera or another type of sensor, the microcontroller determines that the motor controlling the arm must gradually receive less current rather than simply turn off. The digital output of the microcontroller must therefore be transformed to an analog quantity by a digital-to-analog converter (DAC). Figure 5.60 conceptually depicts a possible implementation. The DAC receives a binary input, $D_n \cdots D_1$, from the microcontroller and consists of n current sources that flow from the motor if their corresponding bit is high. The current sources scale up by a factor of 2 from the least-significant bit (LSB) to the most significant bit (MSB). For example, if only D_n is high, then $I_{out} = 2^{n-1}I_0$. The minimum change in the output current is equal to I_0. We say the DAC has a "resolution" of n bits. Each current source can be realized by a bipolar transistor operating in the forward active region. As explained in more advanced courses, the current sources employ resistive degeneration to achieve a greater accuracy. Note that the LSB current, I_0, determines how smoothly the robotic arm moves.

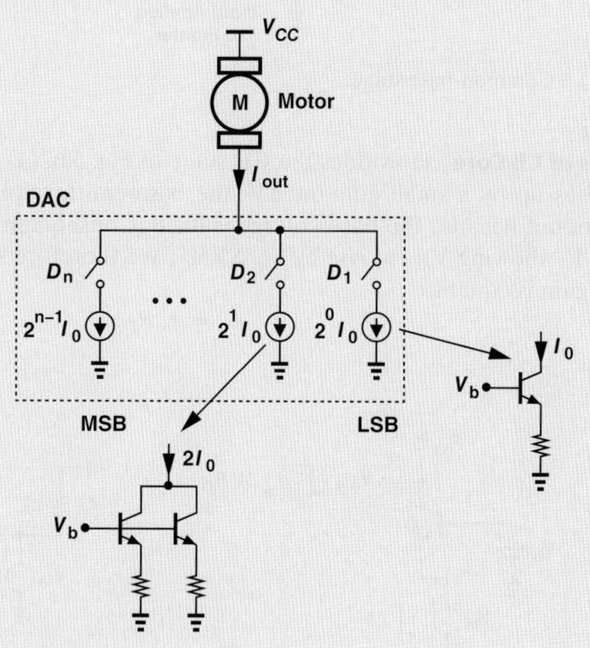

Figure 5.60 Control of a motor by a DAC.

5.3.2 Common-Base Topology

Following our extensive study of the CE stage, we now turn our attention to the "common-base" (CB) topology. Nearly all of the concepts described for the CE configuration apply here as well. We therefore follow the same train of thought, but at a slightly faster pace.

Given the amplification capabilities of the CE stage, the reader may wonder why we study other amplifier topologies. As we will see, other configurations provide different circuit properties that are preferable to those of the CE stage in some applications. The reader is encouraged to review Examples 5-2–5-4, their resulting rules illustrated in Fig. 5.7, and the possible topologies in Fig. 5.28 before proceeding further.

Figure 5.61 shows the CB stage. The input is applied to the emitter and the output is sensed at the collector. Biased at a proper voltage, the base acts as ac ground and hence as a node "common" to the input and output ports. As with the CE stage, we first study the core and subsequently add the biasing elements.

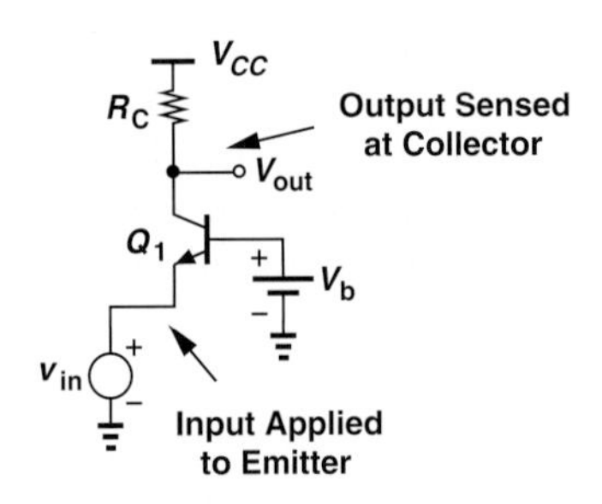

Figure 5.61 Common-base stage.

Analysis of CB Core How does the CB stage of Fig. 5.62(a) respond to an input signal?[9] If V_{in} goes up by a small amount ΔV, the base-emitter voltage of Q_1 *decreases* by the same amount because the base voltage is fixed. Consequently, the collector current falls by $g_m\,\Delta V$, allowing V_{out} to *rise* by $g_m\,\Delta V R_C$. We therefore surmise that the small-signal voltage gain is equal to

$$A_v = g_m R_C. \tag{5.252}$$

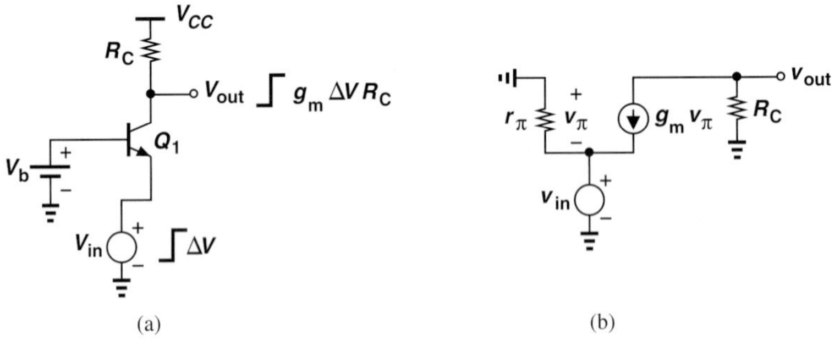

(a) (b)

Figure 5.62 (a) Response of CB stage to small input change, (b) small-signal model.

[9]Note that the topologies of Figs. 5.61 and 5.62(a) are identical even though Q_1 is drawn differently.

Interestingly, this expression is identical to the gain of the CE topology. Unlike the CE stage, however, this circuit exhibits a *positive* gain because an increase in V_{in} leads to an increase in V_{out}.

Let us confirm these results with the aid of the small-signal equivalent depicted in Fig. 5.62(b), where the Early effect is neglected. Beginning with the output node, we equate the current flowing through R_C to $g_m v_\pi$:

$$-\frac{v_{out}}{R_C} = g_m v_\pi, \tag{5.253}$$

obtaining $v_\pi = -v_{out}/(g_m R_C)$. Considering the input node next, we recognize that $v_\pi = -v_{in}$. It follows that

$$\frac{v_{out}}{v_{in}} = g_m R_C. \tag{5.254}$$

As with the CE stage, the CB topology suffers from trade-offs among the gain, the voltage headroom, and the I/O impedances. We first examine the circuit's headroom limitations. How should the base voltage, V_b, in Fig. 5.62(a) be chosen? Recall that the operation in the active region requires $V_{BE} > 0$ and $V_{BC} \leq 0$ (for *npn* devices). Thus, V_b must remain *higher* than the input by about 800 mV, and the output must remain higher than or equal to V_b. For example, if the dc level of the input is zero (Fig. 5.63), then the output must not fall below approximately 800 mV, i.e., the voltage drop across R_C cannot exceed $V_{CC} - V_{BE}$. Similar to the CE stage limitation, this condition translates to

$$A_v = \frac{I_C}{V_T} \cdot R_C \tag{5.255}$$

$$= \frac{V_{CC} - V_{BE}}{V_T}. \tag{5.256}$$

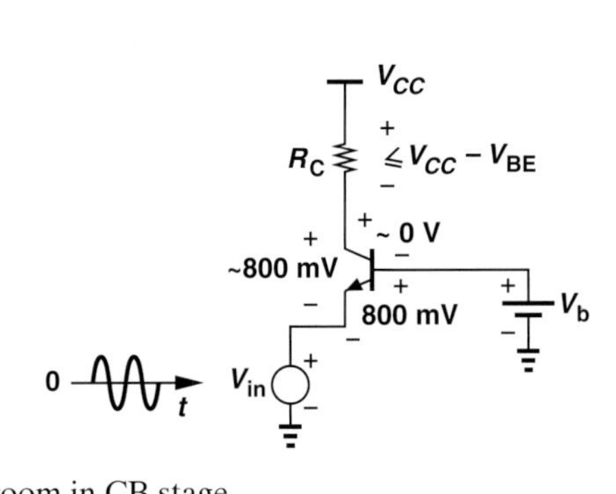

Figure 5.63 Voltage headroom in CB stage.

Example 5-35 The voltage produced by an electronic thermometer is equal to 600 mV at room temperature. Design a CB stage to sense the thermometer voltage and amplify the change with maximum gain. Assume $V_{CC} = 1.8$ V, $I_C = 0.2$ mA, $I_S = 5 \times 10^{-17}$ A, and $\beta = 100$.

Solution Illustrated in Fig. 5.64(a), the circuit must operate properly with an input level of 600 mV. Thus, $V_b = V_{BE} + 600\text{ mV} = V_T \ln(I_C/I_S) + 600\text{ mV} = 1.354$ V. To avoid saturation, the collector voltage must not fall below the base voltage, thereby allowing a maximum voltage drop across R_C equal to 1.8 V − 1.354 V = 0.446 V. Thus $R_C = 2.23\ \text{k}\Omega$. We can then write

$$A_v = g_m R_C \tag{5.257}$$

$$= \frac{I_C R_C}{V_T} \tag{5.258}$$

$$= 17.2. \tag{5.259}$$

The reader is encouraged to repeat the problem with $I_C = 0.4$ mA to verify that the maximum gain remains relatively independent of the bias current.[10]

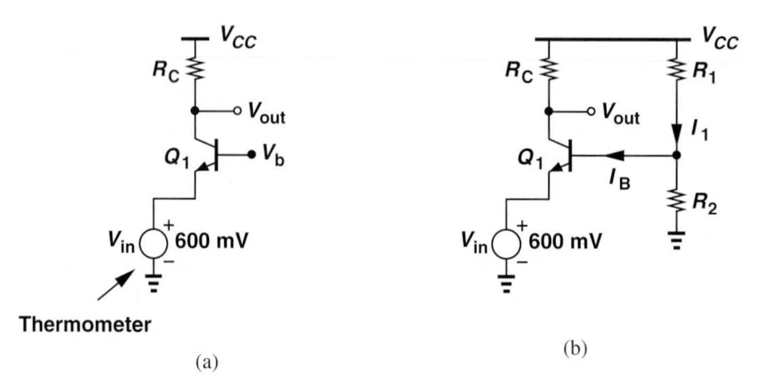

Figure 5.64 (a) CB stage sensing an input, (b) bias network for base.

We must now generate V_b. A simple approach is to employ a resistive divider as depicted in Fig. 5.64(b). To lower sensitivity to β, we choose $I_1 \approx 10 I_B \approx 20\ \mu\text{A} \approx V_{CC}/(R_1 + R_2)$. Thus, $R_1 + R_2 = 90\ \text{k}\Omega$. Also,

$$V_b \approx \frac{R_2}{R_1 + R_2} V_{CC} \tag{5.260}$$

and hence

$$R_2 = 67.7\ \text{k}\Omega \tag{5.261}$$

$$R_1 = 22.3\ \text{k}\Omega. \tag{5.262}$$

Exercise Repeat the above example if the thermometer voltage is 300 mV.

Let us now compute the I/O impedances of the CB topology so as to understand its capabilities in interfacing with preceding and following stages. The rules illustrated in Fig. 5.7 prove extremely useful here, obviating the need for small-signal equivalent circuits. As shown in Fig. 5.65(a), the simplified ac circuit reveals that R_{in} is simply the impedance seen looking into the emitter with the base at ac ground. From the rules in Fig. 5.7, we have

$$R_{in} = \frac{1}{g_m} \tag{5.263}$$

[10]This example serves only as an illustration of the CB stage. A CE stage may prove more suited to sensing a thermometer voltage.

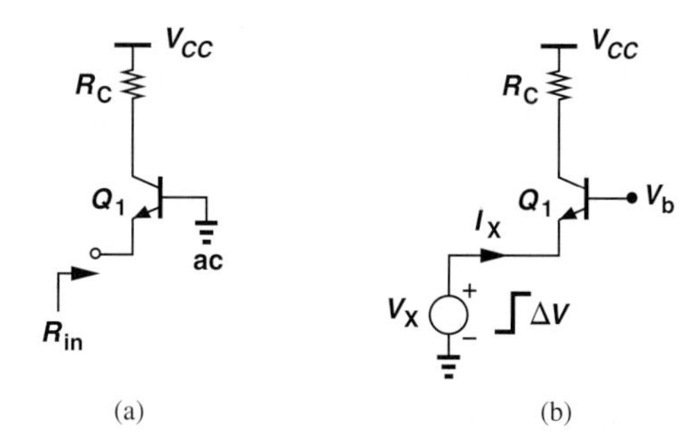

Figure 5.65 (a) Input impedance of CB stage, (b) response to a small change in input.

if $V_A = \infty$. The input impedance of the CB stage is therefore relatively *low*, e.g., 26 Ω for $I_C = 1$ mA (in sharp contrast to the corresponding value for a CE stage, β/g_m).

The input impedance of the CB stage can also be determined intuitively [Fig. 5.65(b)]. Suppose a voltage source V_X tied to the emitter of Q_1 changes by a small amount ΔV. The base-emitter voltage therefore changes by the same amount, leading to a change in the collector current equal to $g_m\ \Delta V$. Since the collector current flows through the input source, the current supplied by V_X also changes by $g_m\ \Delta V$. Consequently, $R_{in} = \Delta V_X/\Delta I_X = 1/g_m$.

Does an amplifier with a low input impedance find any practical use? Yes, indeed. For example, many stand-alone high-frequency amplifiers are designed with an input resistance of 50 Ω to provide "impedance matching" between modules in a cascade and the transmission lines (traces on a printed-circuit board) connecting the modules (Fig. 5.66).[11]

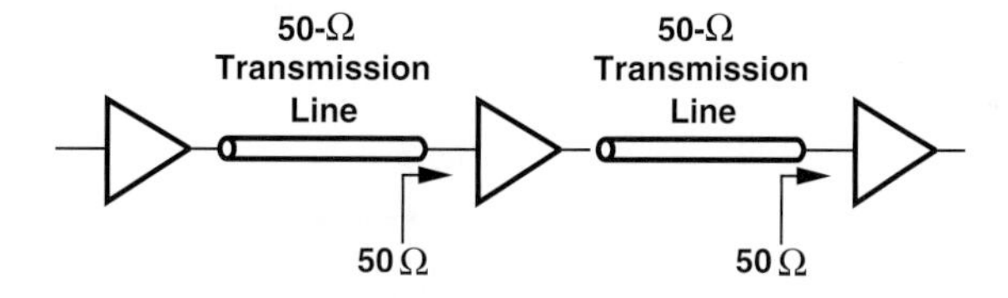

Figure 5.66 System using transmission lines.

The output impedance of the CB stage is computed with the aid of Fig. 5.67, where the input voltage source is set to zero. We note that $R_{out} = R_{out1}||R_C$, where R_{out1} is the impedance seen at the collector with the emitter grounded. From the rules of Fig. 5.7, we have $R_{out1} = r_O$ and hence

$$R_{out} = r_O||R_C \tag{5.264}$$

or

$$R_{out} = R_C \text{ if } V_A = \infty. \tag{5.265}$$

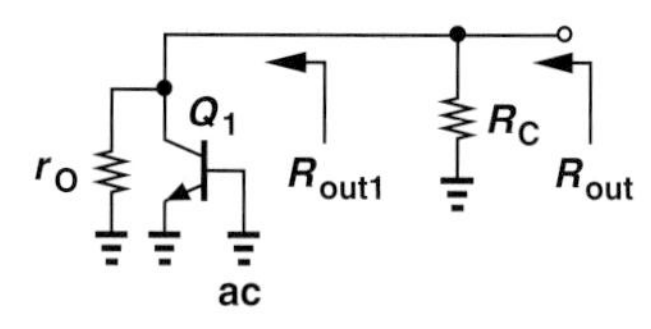

Figure 5.67 Output impedance of CB stage.

[11]If the input impedance of each stage is not matched to the characteristic impedance of the preceding transmission line, then "reflections" occur, corrupting the signal or at least creating dependence on the length of the lines.

Example 5-36 A common-base amplifier is designed for an input impedance of R_{in} and an output impedance of R_{out}. Neglecting the Early effect, determine the voltage gain of the circuit.

Solution Since $R_{in} = 1/g_m$ and $R_{out} = R_C$, we have

$$A_v = \frac{R_{out}}{R_{in}}. \tag{5.266}$$

Exercise Compare this value with that obtained for the CE stage.

From Eqs. (5.256) and (5.266), we conclude that the CB stage exhibits a set of trade-offs similar to those depicted in Fig. 5.33 for the CE amplifier.

It is instructive to study the behavior of the CB topology in the presence of a finite source resistance. Shown in Fig. 5.68, such a circuit suffers from signal attenuation from the input to node X, thereby providing a smaller voltage gain. More specifically, since the impedance seen looking into the emitter of Q_1 (with the base grounded) is equal to $1/g_m$ (for $V_A = \infty$), we have

$$v_X = \frac{\frac{1}{g_m}}{R_S + \frac{1}{g_m}} v_{in} \tag{5.267}$$

$$= \frac{1}{1 + g_m R_S} v_{in}. \tag{5.268}$$

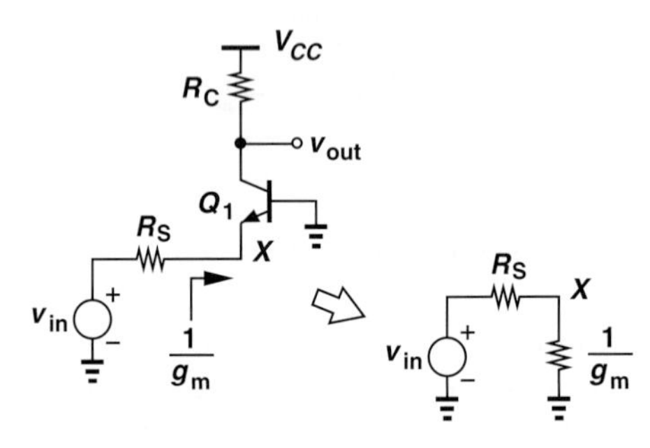

Figure 5.68 CB stage with source resistance.

We also recall from Eq. (5.254) that the gain from the emitter to the output is given by

$$\frac{v_{out}}{v_X} = g_m R_C. \tag{5.269}$$

It follows that

$$\frac{v_{out}}{v_{in}} = \frac{g_m R_C}{1 + g_m R_S} \tag{5.270}$$

$$= \frac{R_C}{\frac{1}{g_m} + R_S}, \tag{5.271}$$

a result identical to that of the CE stage (except for a negative sign) if R_S is viewed as an emitter degeneration resistor.

Example 5-37 A common-base stage is designed to amplify an RF signal received by a 50-Ω antenna. Determine the required bias current if the input impedance of the amplifier must "match" the impedance of the antenna. What is the voltage gain if the CB stage also *drives* a 50-Ω load? Assume $V_A = \infty$.

Solution Figure 5.69 depicts the amplifier[12] and the equivalent circuit with the antenna modeled by a voltage source, v_{in}, and a resistance, $R_S = 50\ \Omega$. For impedance matching, it is necessary that the input impedance of the CB core, $1/g_m$, be equal to R_S, and hence

$$I_C = g_m V_T \tag{5.272}$$

$$= 0.52\ \text{mA}. \tag{5.273}$$

If R_C itself is replaced by a 50-Ω load, then Eq. (5.271) reveals that

$$A_v = \frac{R_C}{\frac{1}{g_m} + R_S} \tag{5.274}$$

$$= \frac{1}{2}. \tag{5.275}$$

The circuit is therefore not suited to driving a 50-Ω load directly.

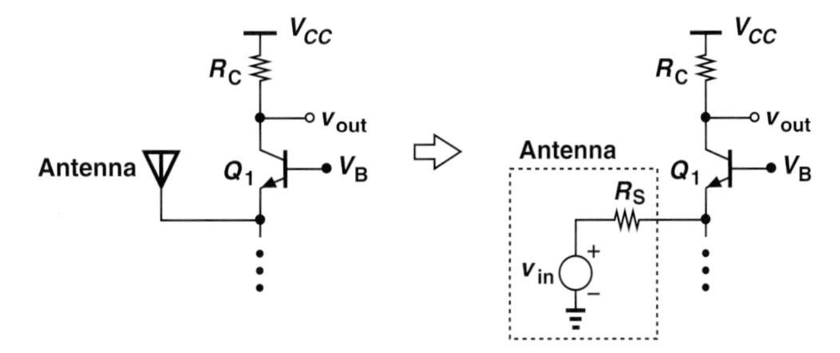

Figure 5.69 (a) CB stage sensing a signal received by an antenna, (b) equivalent circuit.

Exercise What is the voltage gain if a 50-Ω resistor is also tied from the emitter of Q_1 to ground?

Another interesting point of contrast between the CE and CB stages relates to their current gains. The CB stage displays a current gain of *unity* because the current flowing into the emitter simply emerges from the collector (if the base current is neglected). On the other hand, as mentioned in Section 5.3.1, $A_I = \beta$ for the CE stage. In fact, in the preceding example, $i_{in} = v_{in}/(R_S + 1/g_m)$, which upon flowing through R_C, yields $v_{out} = R_C v_{in}/(R_S + 1/g_m)$. It is thus not surprising that the voltage gain does not exceed 0.5 if $R_C \le R_S$.

As with the CE stage, we may desire to analyze the CB topology in the general case: with emitter degeneration, $V_A < \infty$, and a resistance in series with the base [Fig. 5.70(a)].

[12]The dots denote the need for biasing circuitry, as described later in this section.

Outlined in Problem 5.64, this analysis is somewhat beyond the scope of this book. Nevertheless, it is instructive to consider a special case where $R_B = 0$ but $V_A < \infty$, and we wish to compute the output impedance. As illustrated in Fig. 5.70(b), R_{out} is equal to R_C in parallel with the impedance seen looking into the collector, R_{out1}. But R_{out1} is identical to the output resistance of an emitter-degenerated *common emitter* stage, i.e., Fig. 5.46, and hence given by Eq. (5.197):

$$R_{out1} = [1 + g_m(R_E||r_\pi)]r_O + (R_E||r_\pi). \tag{5.276}$$

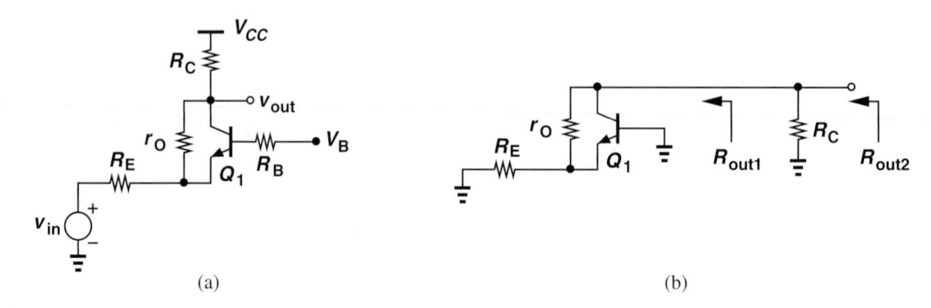

Figure 5.70 (a) General CB stage, (b) output impedance seen at different nodes.

It follows that

$$R_{out} = R_C||\{[1 + g_m(R_E||r_\pi)]r_O + (R_E||r_\pi)\}. \tag{5.277}$$

The reader may have recognized that the output impedance of the CB stage is equal to that of the CE stage. Is this true in general? Recall that the output impedance is determined by setting the input source to zero. In other words, when calculating R_{out}, we have no knowledge of the input terminal of the circuit, as illustrated in Fig. 5.71 for CE and CB stages. It is therefore no coincidence that the output impedances are identical *if* the same assumptions are made for both circuits (e.g., identical values of V_A and emitter degeneration).

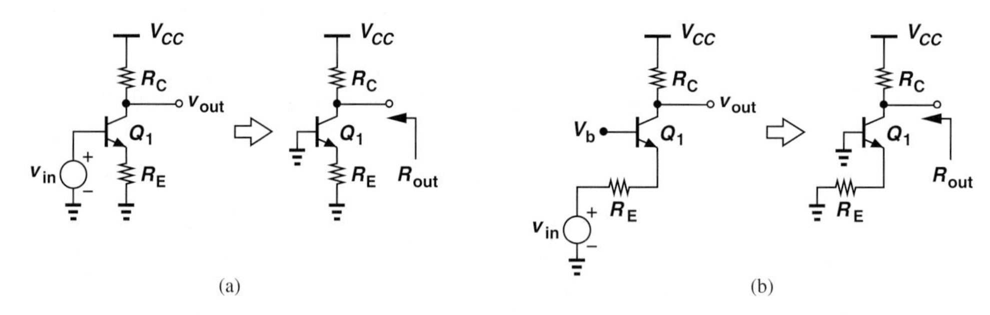

Figure 5.71 (a) CE stage and (b) CB stage simplified for output impedance calculation.

Example 5-38

Old wisdom says "the output impedance of the CB stage is substantially higher than that of the CE stage." This claim is justified by the tests illustrated in Fig. 5.72. If a constant current is injected into the base while the collector voltage is varied, I_C exhibits a slope equal to r_O^{-1} [Fig. 5.72(a)]. On the other hand, if a constant current is drawn from the emitter, I_C displays much less dependence on the collector voltage. Explain why these tests do not represent practical situations.

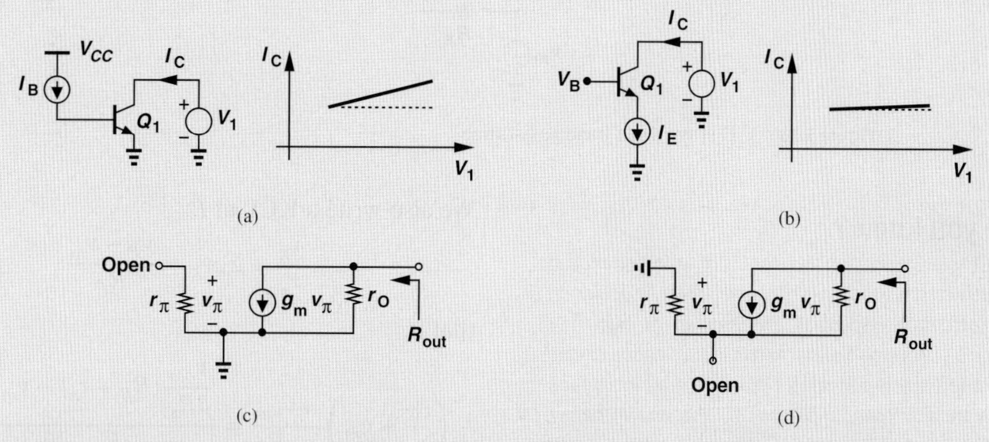

Figure 5.72 (a) Resistance seen at collector with emitter grounded, (b) resistance seen at collector with an ideal current source in emitter, (c) small-signal model of (a), (d) small-signal model of (b).

Solution The principal issue in these tests relates to the use of *current* sources to drive each stage. From a small-signal point of view, the two circuits reduce to those depicted in Figs. 5.72(c) and (d), with current sources I_B and I_E replaced with open circuits because they are constant. In Fig. 5.72(c), the current through r_π is zero, yielding $g_m v_\pi = 0$ and hence $R_{out} = r_O$. On the other hand, Fig. 5.72(d) resembles an emitter-degenerated stage (Fig. 5.46) with an infinite emitter resistance, exhibiting an output resistance of

$$R_{out} = [1 + g_m(R_E||r_\pi)]r_O + (R_E||r_\pi) \tag{5.278}$$

$$= (1 + g_m r_\pi)r_O + r_\pi \tag{5.279}$$

$$\approx \beta r_O + r_\pi, \tag{5.280}$$

which is, of course, much greater than r_O. In practice, however, each stage may be driven by a *voltage* source having a finite impedance, making the above comparison irrelevant.

Exercise Repeat the above example if a resistor of value R_1 is inserted in series with the emitter.

Another special case of the topology shown in Fig. 5.70(a) occurs if $V_A = \infty$ but $R_B > 0$. Since this case does not reduce to any of the configurations studied earlier, we employ the small-signal model shown in Fig. 5.73 to study its behavior. As usual, we write $g_m v_\pi = -v_{out}/R_C$ and hence $v_\pi = -v_{out}/(g_m R_C)$. The current flowing through r_π (and R_B) is then equal to $v_\pi/r_\pi = -v_{out}/(g_m r_\pi R_C) = -v_{out}/(\beta R_C)$. Multiplying this current by $R_B + r_\pi$, we obtain the voltage at node P:

$$v_P = -\frac{-v_{out}}{\beta R_C}(R_B + r_\pi) \tag{5.281}$$

$$= \frac{v_{out}}{\beta R_C}(R_B + r_\pi). \tag{5.282}$$

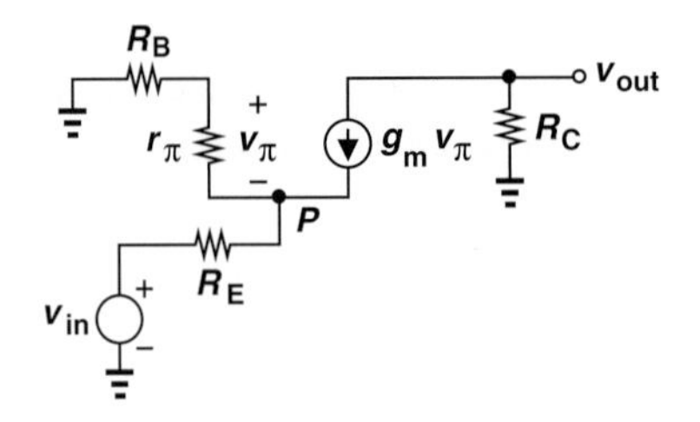

Figure 5.73 CB stage with base resistance.

Did you know?

The CB gain reduction as a result of an impedance in series with the base has stumbled many a designer. This effect is especially pronounced at high frequencies because the parasitic inductances of the package (which contains the chip) may introduce a significant impedance ($= L\omega$). As shown below, the "bond wires" connecting the chip to the package behave like inductors, thus degrading the performance. For example, a 5-mm bond wire has an inductance of about 5 nH, exhibiting an impedance of 157 Ω at 5 GHz.

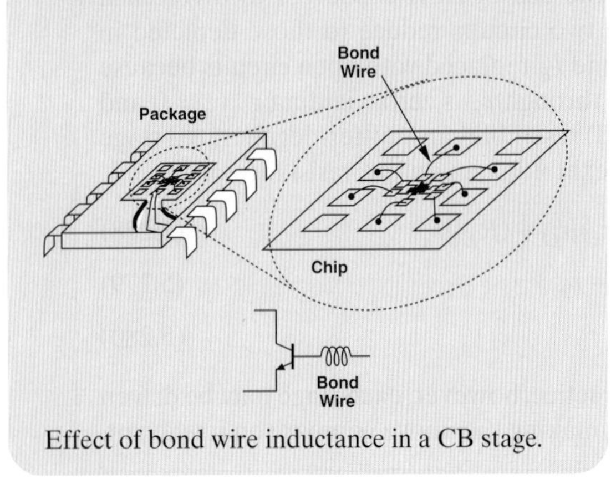

Effect of bond wire inductance in a CB stage.

We also write a KCL at P:

$$\frac{v_\pi}{r_\pi} + g_m v_\pi = \frac{v_P - v_{in}}{R_E}; \tag{5.283}$$

that is,

$$\left(\frac{1}{r_\pi} + g_m\right)\frac{-v_{out}}{g_m R_C} = \frac{\dfrac{v_{out}}{\beta R_C}(R_B + r_\pi) - v_{in}}{R_E}. \tag{5.284}$$

It follows that

$$\frac{v_{out}}{v_{in}} = \frac{\beta R_C}{(\beta+1)R_E + R_B + r_\pi}. \tag{5.285}$$

Dividing the numerator and denominator by $\beta + 1$, we have

$$\frac{v_{out}}{v_{in}} \approx \frac{R_C}{R_E + \dfrac{R_B}{\beta+1} + \dfrac{1}{g_m}}. \tag{5.286}$$

As expected, the gain is positive. Furthermore, this expression is identical to that in Eq. (5.185) for the CE stage. Figure 5.74 illustrates the results, revealing that, except for a negative sign, the two stages exhibit equal gains. Note that R_B degrades the gain and is not added to the circuit deliberately. As explained later in this section, R_B may arise from the biasing network.

Figure 5.74 Comparison of CE and CB stages with base resistance.

Let us now determine the input impedance of the CB stage in the presence of a resistance in series with the base, still assuming $V_A = \infty$. From the small-signal equivalent circuit shown in Fig. 5.75, we recognize that

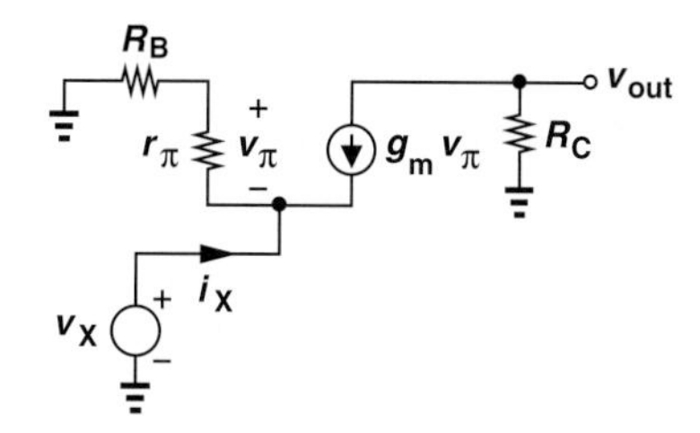

Figure 5.75 Input impedance of CB stage with base resistance.

r_π and R_B form a voltage divider, thereby producing[13]

$$v_\pi = -\frac{r_\pi}{r_\pi + R_B} v_X. \tag{5.287}$$

Moreover, KCL at the input node gives

$$\frac{v_\pi}{r_\pi} + g_m v_\pi = -i_X. \tag{5.288}$$

Thus,

$$\left(\frac{1}{r_\pi} + g_m\right) \frac{-r_\pi}{r_\pi + R_B} v_X = -i_X \tag{5.289}$$

and

$$\frac{v_X}{i_X} = \frac{r_\pi + R_B}{\beta + 1} \tag{5.290}$$

$$\approx \frac{1}{g_m} + \frac{R_B}{\beta + 1}. \tag{5.291}$$

Note that $R_{in} = 1/g_m$ if $R_B = 0$, an expected result from the rules illustrated in Fig. 5.7. Interestingly, the base resistance is divided by $\beta + 1$ when "seen" from the emitter. This is in contrast to the case of emitter degeneration, where the emitter resistance is *multiplied* by $\beta + 1$ when seen from the base. Figure 5.76 summarizes the two cases. Note that these results remain independent of R_C if $V_A = \infty$.

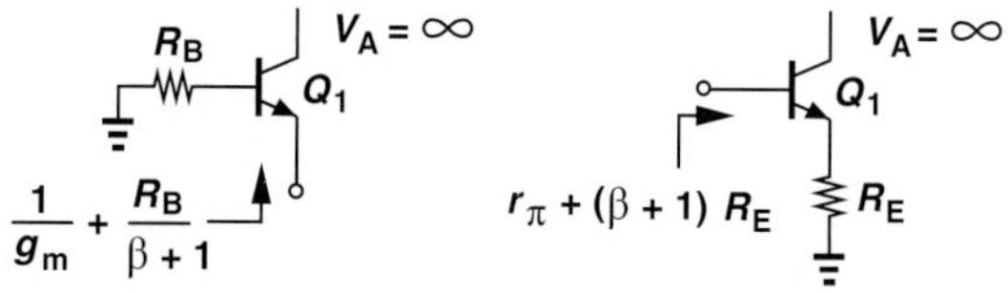

Figure 5.76 Impedance seen at the emitter or base of a transistor.

[13]Alternatively, the current through $r_\pi + R_B$ is equal to $v_X/(r_\pi + R_B)$, yielding a voltage of $-r_\pi v_X/(r_\pi + R_B)$ across r_π.

Example 5-39 Determine the impedance seen at the emitter of Q_2 in Fig. 5.77(a) if the two transistors are identical and $V_A = \infty$.

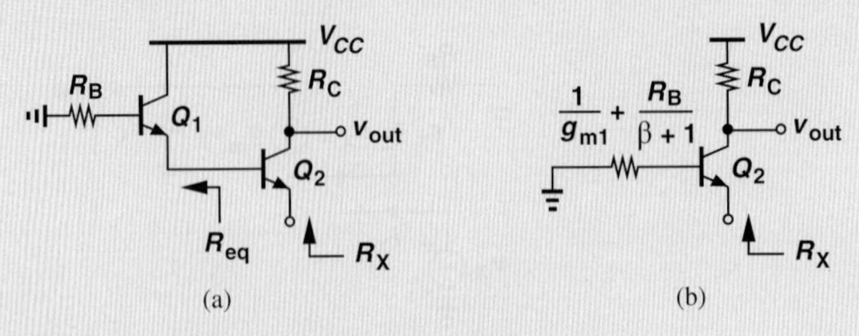

Figure 5.77 (a) Example of CB stage, (b) simplified circuit.

Solution The circuit employs Q_2 as a common-base device, but with its base tied to a finite series resistance equal to that seen at the emitter of Q_1. Thus, we must first obtain the equivalent resistance R_{eq}, which from Eq. (5.291) is simply equal to

$$R_{eq} = \frac{1}{g_{m1}} + \frac{R_B}{\beta + 1}. \tag{5.292}$$

Reducing the circuit to that shown in Fig. 5.77(b), we have

$$R_X = \frac{1}{g_{m2}} + \frac{R_{eq}}{\beta + 1} \tag{5.293}$$

$$= \frac{1}{g_{m2}} + \frac{1}{\beta + 1}\left(\frac{1}{g_{m1}} + \frac{R_B}{\beta + 1}\right). \tag{5.294}$$

Exercise What happens if a resistor of value R_1 is placed in series with the collector of Q_1?

CB Stage with Biasing Having learned the small-signal properties of the CB core, we now extend our analysis to the circuit including biasing. An example proves instructive at this point.

Example 5-40 The student in Example 5-31 decides to incorporate ac coupling at the input of a CB stage to ensure the bias is not affected by the signal source, drawing the design as shown in Fig. 5.78. Explain why this circuit does not work.

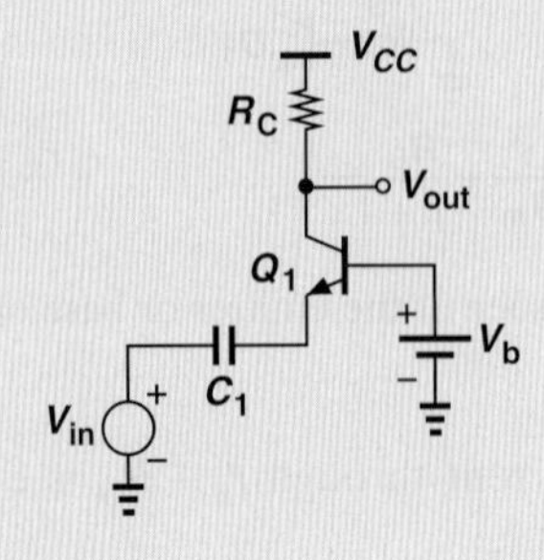

Figure 5.78 CB stage lacking bias current.

Solution Unfortunately, the design provides no dc path for the emitter current of Q_1, forcing a zero bias current and hence a zero transconductance. The situation is similar to the CE counterpart in Example 5-5, where no base current can be supported.

Exercise In what region does Q_1 operate if $V_b = V_{CC}$?

Example 5-41 Somewhat embarrassed, the student quickly connects the emitter to ground so that $V_{BE} = V_b$ and a reasonable collector current can be established (Fig. 5.79). Explain why "haste makes waste."

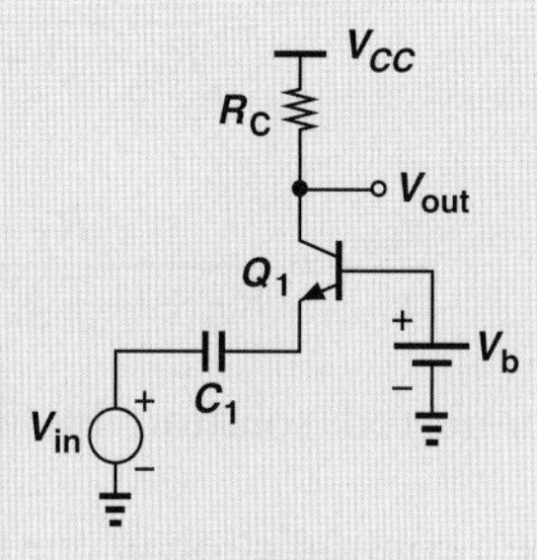

Figure 5.79 CB stage with emitter shorted to ground.

Solution As in Example 5-6, the student has shorted the *signal* to ac ground. That is, the emitter voltage is equal to zero regardless of the value of v_{in}, yielding $v_{out} = 0$.

Exercise Does the circuit operate better if V_b is raised?

The above examples imply that the emitter can remain neither open nor shorted to ground, thereby requiring some bias element. Shown in Fig. 5.80(a) is an example in which R_E provides a path for the bias current at the cost of lowering the input impedance. We recognize that R_{in} now consists of two *parallel* components: (1) $1/g_m$, seen looking "up" into the emitter (with the base at ac ground) and (2) R_E, seen looking "down." Thus,

$$R_{in} = \frac{1}{g_m} || R_E. \quad (5.295)$$

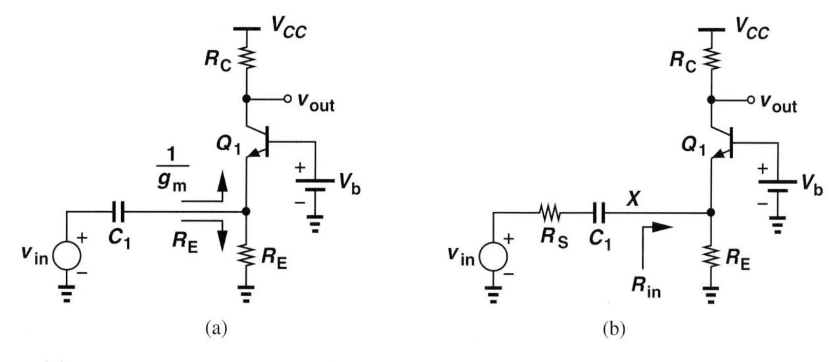

Figure 5.80 (a) CB stage with biasing, (b) inclusion of source resistance.

As with the input biasing network in the CE stage (Fig. 5.58), the reduction in R_{in} manifests itself if the source voltage exhibits a finite output resistance. Depicted in Fig. 5.80(b), such a circuit attenuates the signal, lowering the overall voltage gain.

Following the analysis illustrated in Fig. 5.68, we can write

$$\frac{v_X}{v_{in}} = \frac{R_{in}}{R_{in} + R_S} \tag{5.296}$$

$$= \frac{\frac{1}{g_m}||R_E}{\frac{1}{g_m}||R_E + R_S} \tag{5.297}$$

$$= \frac{R_E}{R_E + (1 + g_m R_E)R_S}. \tag{5.298}$$

Since $v_{out}/v_X = g_m R_C$,

$$\frac{v_{out}}{v_{in}} = \frac{R_E}{R_E + (1 + g_m R_E)R_S} \cdot g_m R_C. \tag{5.299}$$

As usual, we have preferred solution by inspection over drawing the small-signal equivalent.

The reader may see a contradiction in our thoughts: on the one hand, we view the low input impedance of the CB stage as a *useful* property; on the other hand, we consider the reduction of the input impedance due to R_E *undesirable*. To resolve this apparent contradiction, we must distinguish between the two components $1/g_m$ and R_E, noting that the latter shunts the input source current to ground, thus "wasting" the signal. As shown in Fig. 5.81, i_{in} splits two ways, with only i_2 reaching R_C and contributing to the output signal. If R_E decreases while $1/g_m$ remains constant, then i_2 also falls.[14] Thus, reduction of R_{in} due to R_E is undesirable. By contrast, if $1/g_m$ decreases while R_E remains constant, then i_2 rises. For R_E to affect the input impedance negligibly, we must have

$$R_E \gg \frac{1}{g_m} \tag{5.300}$$

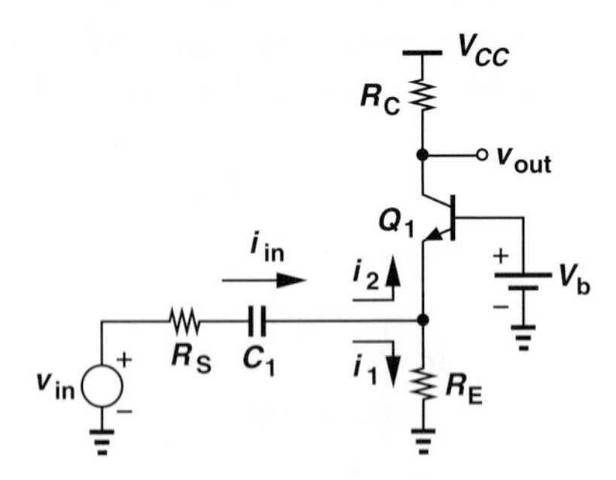

Figure 5.81 Small-signal input current components in a CB stage.

and hence

$$I_C R_E \gg V_T. \tag{5.301}$$

That is, the dc voltage drop across R_E muts be much greater than V_T.

[14]In the extreme case, $R_E = 0$ (Example 5-41) and $i_2 = 0$.

How is the base voltage, V_b, generated? We can employ a resistive divider similar to that used in the CE stage. As shown in Fig. 5.82(a), such a topology must ensure $I_1 \gg I_B$ to minimize sensitivity to β, yielding

$$V_b \approx \frac{R_2}{R_1 + R_2} V_{CC}. \tag{5.302}$$

However, recall from Eq. (5.286) that a resistance in series with the base *reduces* the voltage gain of the CB stage. Substituting a Thevenin equivalent for R_1 and R_2 as depicted in Fig. 5.82(b), we recognize that a resistance of $R_{Thev} = R_1||R_2$ now appears in series with the base. For this reason, a "bypass capacitor" is often tied from the base to ground, acting as a short circuit at frequencies of interest [Fig. 5.82(c)].

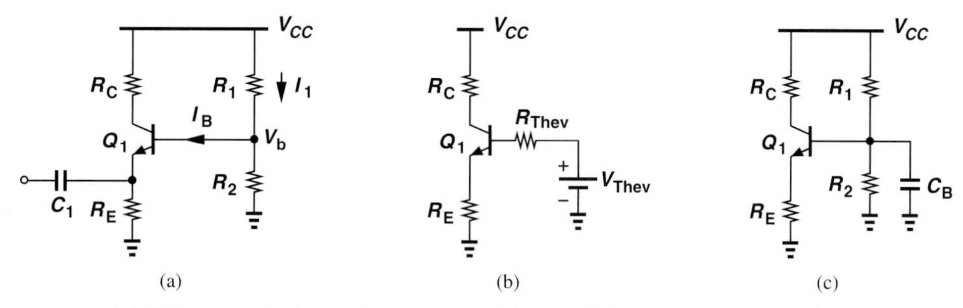

Figure 5.82 (a) CB stage with base bias network, (b) use of Thevenin equivalent, (c) effect of bypass capacitor.

Example 5-42

Design a CB stage (Fig. 5.83) for a voltage gain of 10 and an input impedance of 50 Ω. Assume $I_S = 5 \times 10^{-16}$ A, $V_A = \infty$, $\beta = 100$, and $V_{CC} = 2.5$ V.

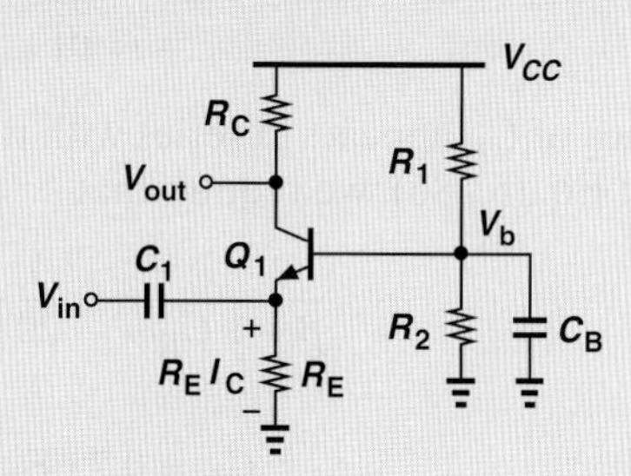

Figure 5.83 Example of CB stage with biasing.

Solution

We begin by selecting $R_E \gg 1/g_m$, e.g., $R_E = 500$ Ω, to minimize the undesirable effect of R_E. Thus,

$$R_{in} \approx \frac{1}{g_m} = 50\ \Omega \tag{5.303}$$

and hence

$$I_C = 0.52\ \text{mA}. \tag{5.304}$$

If the base is bypassed to ground

$$A_v = g_m R_C, \tag{5.305}$$

yielding

$$R_C = 500\ \Omega. \tag{5.306}$$

We now determine the base bias resistors. Since the voltage drop across R_E is equal to $500\ \Omega \times 0.52$ mA = 260 mV and $V_{BE} = V_T \ln(I_C/I_S) = 899$ mV, we have

$$V_b = I_E R_E + V_{BE} \tag{5.307}$$

$$= 1.16\ \text{V}. \tag{5.308}$$

Selecting the current through R_1 and R_2 to be $10I_B = 52\ \mu\text{A}$, we write

$$V_b \approx \frac{R_2}{R_1 + R_2} V_{CC}. \tag{5.309}$$

$$\frac{V_{CC}}{R_1 + R_2} = 52\ \mu\text{A}. \tag{5.310}$$

It follows that

$$R_1 = 25.8\ \text{k}\Omega \tag{5.311}$$

$$R_2 = 22.3\ \text{k}\Omega. \tag{5.312}$$

The last step in the design is to compute the required values of C_1 and C_B according to the signal frequency. For example, if the amplifier is used at the receiver front end of a 900-MHz cellphone, the impedances of C_1 and C_B must be sufficiently small at this frequency. Appearing in series with the emitter of Q_1, C_1 plays a role similar to R_S in Fig. 5.68 and Eq. (5.271). Thus, its impedance, $|C_1\omega|^{-1}$, must remain much less than $1/g_m = 50\ \Omega$. In high-performance applications such as cellphones, we may choose $|C_1\omega|^{-1} = (1/g_m)/20$ to ensure negligible gain degradation. Consequently, for $\omega = 2\pi \times$ (900 MHz):

$$C_1 = \frac{20g_m}{\omega} \tag{5.313}$$

$$= 71\ \text{pF}. \tag{5.314}$$

Since the impedance of C_B appears in series with the base and plays a role similar to the term $R_B/(\beta + 1)$ in Eq. (5.286), we require that

$$\frac{1}{\beta + 1}\frac{1}{C_B\omega} = \frac{1}{20}\frac{1}{g_m} \tag{5.315}$$

and hence

$$C_B = 0.7\ \text{pF}. \tag{5.316}$$

(A common mistake is to make the impedance of C_B negligible with respect to $R_1||R_2$ rather than with respect to $1/g_m$.)

Exercise Design the above circuit for an input impedance of $100\ \Omega$.

5.3.3 Emitter Follower

Another important circuit topology is the emitter follower (also called the "common-collector" stage). The reader is encouraged to review Examples 5-2–5-3, rules illustrated in Fig. 5.7, and the possible topologies in Fig. 5.28 before proceeding further. For the sake of brevity, we may also use the term "follower" to refer to emitter followers in this chapter.

Shown in Fig. 5.84, the emitter follower senses the input at the base of the transistor and produces the output at the emitter. The collector is tied to V_{CC} and hence ac ground. We first study the core and subsequently add the biasing elements.

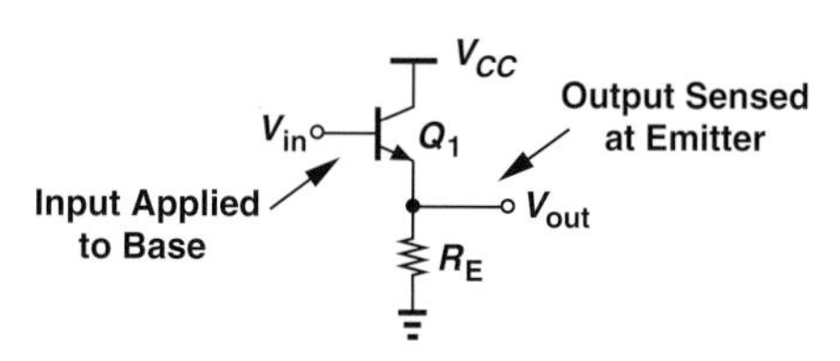

Figure 5.84 Emitter follower.

Emitter Follower Core How does the follower in Fig. 5.85(a) respond to a change in V_{in}? If V_{in} rises by a small amount ΔV_{in}, the base-emitter voltage of Q_1 tends to increase, raising the collector and emitter currents. The higher emitter current translates to a greater drop across R_E and hence a *higher* V_{out}. From another perspective, if we assume, for example, V_{out} is constant, then V_{BE} must rise and so must I_E, requiring that V_{out} go up. Since V_{out} changes in the same direction as V_{in}, we expect the voltage gain to be positive. Note that V_{out} is always lower than V_{in} by an amount equal to V_{BE}, and the circuit is said to provide "level shift."

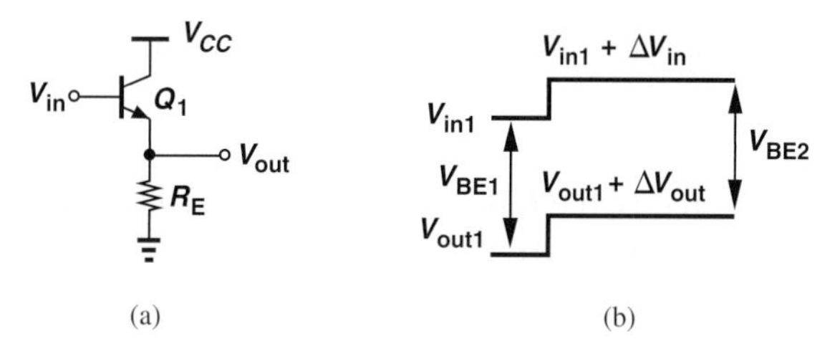

Figure 5.85 (a) Emitter follower sensing an input change, (b) response of the circuit.

Another interesting and important observation here is that the change in V_{out} cannot be larger than the change in V_{in}. Suppose V_{in} increases from V_{in1} to $V_{in1} + \Delta V_{in}$ and V_{out} from V_{out1} to $V_{out1} + \Delta V_{out}$ [Fig. 5.85(b)]. If the output changes by a *greater* amount than the input, $\Delta V_{out} > \Delta V_{in}$, then V_{BE2} must be *less* than V_{BE1}. But this means the emitter current also decreases and so does $I_E R_E = V_{out}$, contradicting the assumption that V_{out} has increased. Thus, $\Delta V_{out} < \Delta V_{in}$, implying that the follower exhibits a voltage gain less than unity.[15]

The reader may wonder if an amplifier with a subunity gain has any practical value. As explained later, the input and output impedances of the emitter follower make it a particularly useful circuit for some applications.

Let us now derive the small-signal properties of the follower, first assuming $V_A = \infty$. Shown in Fig. 5.86, the equivalent circuit yields

$$\frac{v_\pi}{r_\pi} + g_m v_\pi = \frac{v_{out}}{R_E} \tag{5.317}$$

[15]In an extreme case described in Example 5-43, the gain becomes equal to unity.

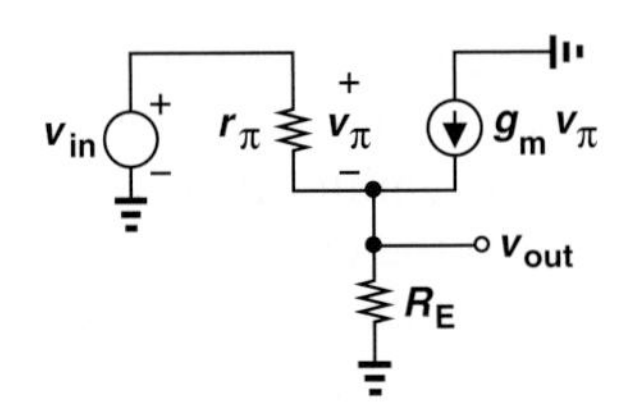

Figure 5.86 Small-signal model of emitter follower.

and hence

$$v_\pi = \frac{r_\pi}{\beta + 1} \cdot \frac{v_{out}}{R_E}. \tag{5.318}$$

We also have

$$v_{in} = v_\pi + v_{out}. \tag{5.319}$$

Substituting for v_π from (5.318), we obtain

$$\frac{v_{out}}{v_{in}} = \frac{1}{1 + \frac{r_\pi}{\beta + 1} \cdot \frac{1}{R_E}} \tag{5.320}$$

$$\approx \frac{R_E}{R_E + \frac{1}{g_m}}. \tag{5.321}$$

The voltage gain is therefore positive and less than unity.

Example 5-43 In integrated circuits, the follower is typically realized as shown in Fig. 5.87. Determine the voltage gain if the current source is ideal and $V_A = \infty$.

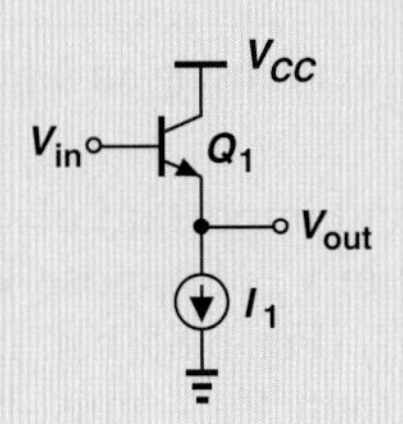

Figure 5.87 Follower with current source.

Solution Since the emitter resistor is replaced with an ideal current source, the value of R_E in Eq. (5.321) must tend to infinity, yielding

$$A_v = 1. \tag{5.322}$$

This result can also be derived intuitively. A constant current source flowing through Q_1 requires that $V_{BE} = V_T \ln(I_C/I_S)$ remain constant. Writing $V_{out} = V_{in} - V_{BE}$, we recognize that V_{out} exactly *follows* V_{in} if V_{BE} is constant.

Exercise Repeat the above example if a resistor of value R_1 is placed in series with the collector.

Equation (5.321) suggests that the emitter follower acts as a voltage divider, a perspective that can be reinforced by an alternative analysis. Suppose, as shown in Fig. 5.88(a), we wish to model v_{in} and Q_1 by a Thevenin equivalent. The Thevenin voltage is given by the open-circuit output voltage produced by Q_1 [Fig. 5.88(b)], as if Q_1 operates with $R_E = \infty$ (Example 5-43). Thus, $v_{Thev} = v_{in}$. The Thevenin resistance is obtained by setting the input to zero [Fig. 5.88(c)] and is equal to $1/g_m$. The circuit of Fig. 5.88(a) therefore reduces to that shown in Fig. 5.88(d), confirming operation as a voltage divider.

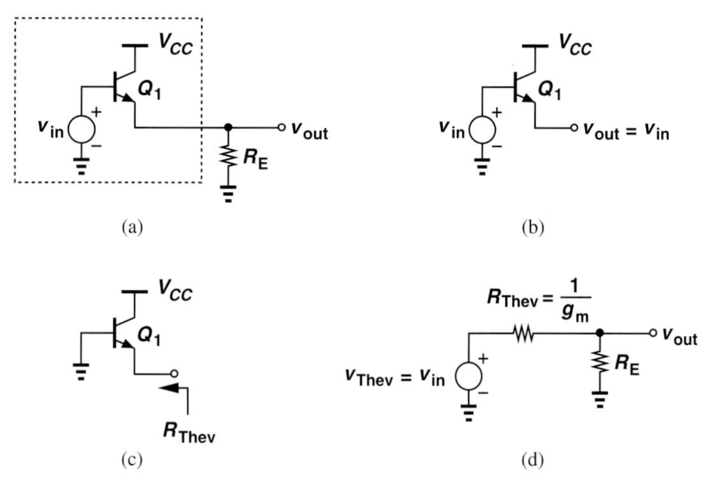

Figure 5.88 (a) Emitter follower stage, (b) Thevenin voltage, (c) Thevenin resistance, (d) simplified circuit.

Example 5-44

Determine the voltage gain of a follower driven by a finite source impedance of R_S [Fig. 5.89(a)] if $V_A = \infty$.

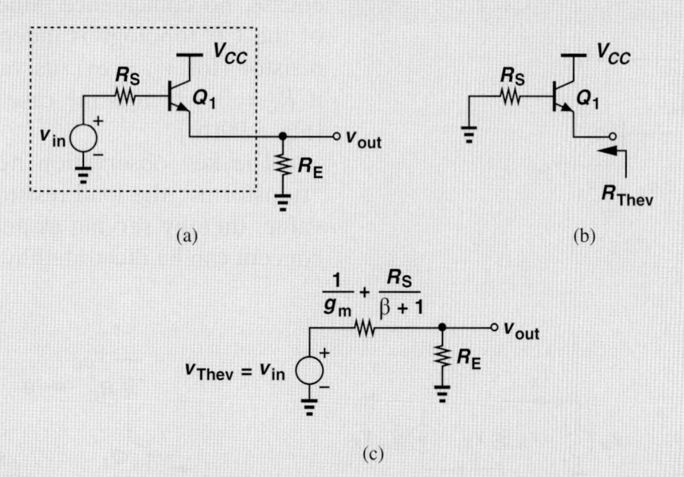

Figure 5.89 (a) Follower with source impedance, (b) Thevenin resistance seen at emitter, (c) simplified circuit.

Solution We model v_{in}, R_S, and Q_1 by a Thevenin equivalent. The reader can show that the open-circuit voltage is equal to v_{in}. Furthermore, the Thevenin resistance [Fig. 5.89(b)] is given by Eq. (5.291) as $R_S/(\beta+1)+1/g_m$. Figure 5.89(c) depicts the equivalent circuit, revealing that

$$\frac{v_{out}}{v_{in}} = \frac{R_E}{R_E + \dfrac{R_S}{\beta+1} + \dfrac{1}{g_m}}. \tag{5.323}$$

This result can also be obtained by solving the small-signal equivalent circuit of the follower.

Exercise What happens if $R_E = \infty$?

Did you know?

Shortly after the invention of the bipolar transistor, a Bell Labs engineer named Sidney Darlington borrowed a few transistors from his boss and took them home for the weekend. It was over that weekend that he came up with what is known as the "Darlington Pair" [Fig. (a)]. Here, Q_1 and Q_2 can be viewed as two emitter followers in a cascade. The circuit provides a high input impedance but since I_{C1} is small ($= I_{B2}$), Q_1 is quite slow. For this reason, the modified topologies in Figs. (b) and (c) are preferred.

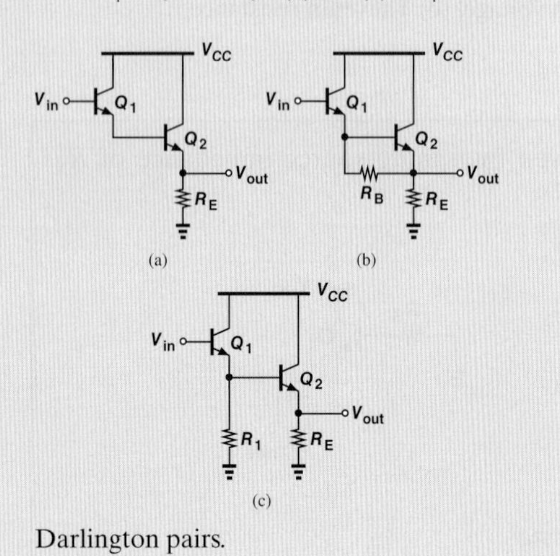

Darlington pairs.

In order to appreciate the usefulness of emitter followers, let us compute their input and output impedances. In the equivalent circuit of Fig. 5.90(a), we have $i_X r_\pi = v_\pi$. Also, the current i_X and $g_m v_\pi$ flow through R_E, producing a voltage drop equal to $(i_X + g_m v_\pi)R_E$. Adding the voltages across r_π and R_E and equating the result to v_X, we have

$$v_X = v_\pi + (i_X + g_m v_\pi)R_E \tag{5.324}$$

$$= i_X r_\pi + (i_X + g_m i_X r_\pi)R_E, \tag{5.325}$$

and hence

$$\frac{v_X}{i_X} = r_\pi + (1+\beta)R_E. \tag{5.326}$$

This expression is identical to that in Eq. (5.162) derived for a degenerated CE stage. This is, of course, no coincidence. Since the input impedance of the CE topology is independent of the collector resistor (for $V_A = \infty$), its value remains unchanged if $R_C = 0$, which is the case for an emitter follower [Fig. 5.90(b)].

The key observation here is that the follower "transforms" the load resistor, R_E, to a much larger value, thereby serving as an efficient "buffer." This concept can be illustrated by an example.

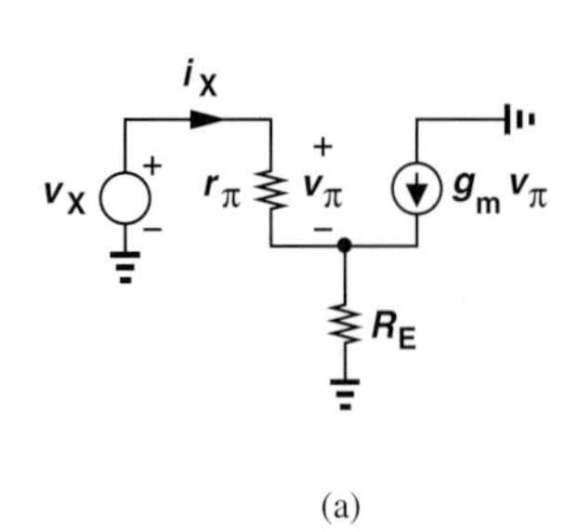

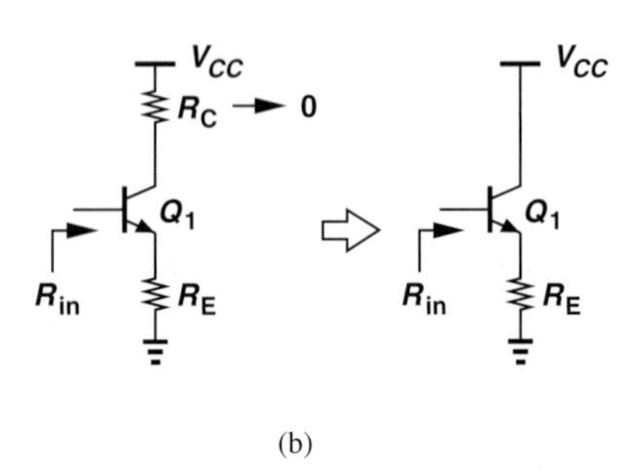

(a) (b)

Figure 5.90 (a) Input impedance of emitter follower, (b) equivalence of CE and follower stages.

Example 5-45 A CE stage exhibits a voltage gain of 20 and an output resistance of 1 kΩ. Determine the voltage gain of the CE amplifier if

(a) The stage drives an 8-Ω speaker directly.

(b) An emitter follower biased at a current of 5 mA is interposed between the CE stage and the speaker. Assume $\beta = 100$, $V_A = \infty$, and the follower is biased with an ideal current source.

Solution (a) As depicted in Fig. 5.91(a), the equivalent resistance seen at the collector is now given by the parallel combination of R_C and the speaker impedance, R_{sp}, reducing the gain from 20 to $20 \times (R_C||8\ \Omega)/R_C = 0.159$. The voltage gain therefore degrades drastically.

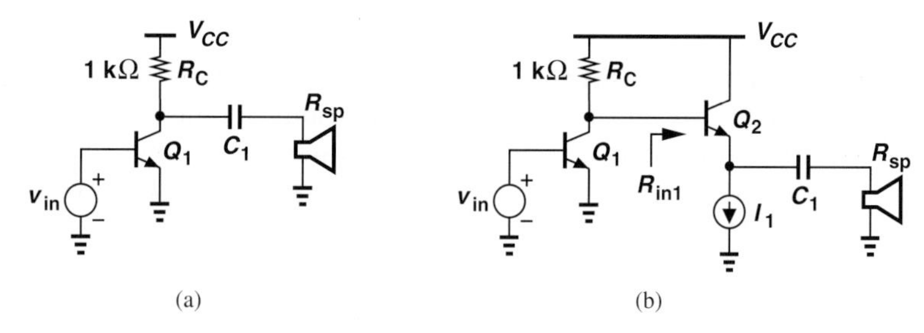

Figure 5.91 (a) CE stage and (b) two-stage circuit driving a speaker.

(b) From the arrangement in Fig. 5.91(b), we note that

$$R_{in1} = r_{\pi 2} + (\beta + 1)R_{sp} \tag{5.327}$$

$$= 1328\ \Omega. \tag{5.328}$$

Thus, the voltage gain of the CE stage drops from 20 to $20 \times (R_C||R_{in1})/R_C = 11.4$, a substantial improvement over case (a).

Exercise Repeat the above example if the emitter follower is biased at a current of 10 mA.

We now calculate the output impedance of the follower, assuming the circuit is driven by a source impedance R_S [Fig. 5.92(a)]. Interestingly, we need not resort to a small-signal model here as R_{out} can be obtained by inspection. As illustrated in Fig. 5.92(b), the output resistance can be viewed as the parallel combination of two components: one seen looking "up" into the emitter and another looking "down" into R_E. From Fig. 5.89, the former is equal to $R_S/(\beta + 1) + 1/g_m$, and hence

$$R_{out} = \left(\frac{R_S}{\beta + 1} + \frac{1}{g_m}\right) ||R_E. \tag{5.329}$$

This result can also be derived from the Thevenin equivalent shown in Fig. 5.89(c) by setting v_{in} to zero.

Equation (5.329) reveals another important attribute of the follower: the circuit transforms the source impedance, R_S, to a much lower value, thereby providing higher "driving" capability. We say the follower operates as a good "voltage buffer" because it

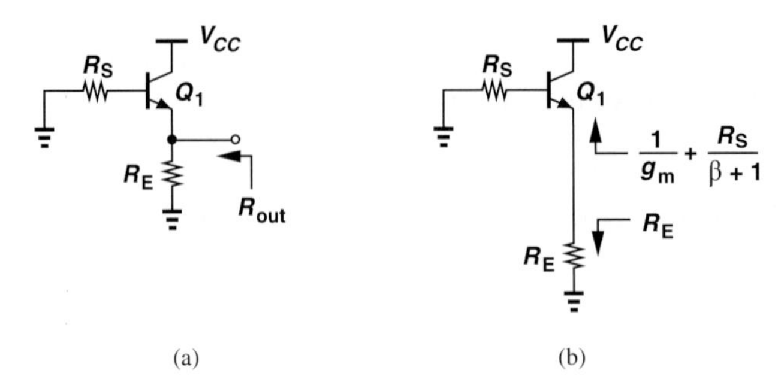

Figure 5.92 (a) Output impedance of a follower, (b) components of output resistance.

displays a high input impedance (like a voltmeter) and a low output impedance (like a voltage source).

Effect of Transistor Output Resistance Our analysis of the follower has thus far neglected the Early effect. Fortunately, the results obtained above can be readily modified to reflect this nonideality. Figure 5.93 illustrates a key point that facilitates the analysis: in small-signal operation, r_O appears in parallel with R_E. We can therefore rewrite Eqs. (5.323), (5.326), and (5.329) as

$$A_v = \frac{R_E||r_O}{R_E||r_O + \frac{R_S}{\beta + 1} + \frac{1}{g_m}} \tag{5.330}$$

$$R_{in} = r_\pi + (\beta + 1)(R_E||r_O) \tag{5.331}$$

$$R_{out} = \left(\frac{R_S}{\beta + 1} + \frac{1}{g_m}\right)||R_E||r_O. \tag{5.332}$$

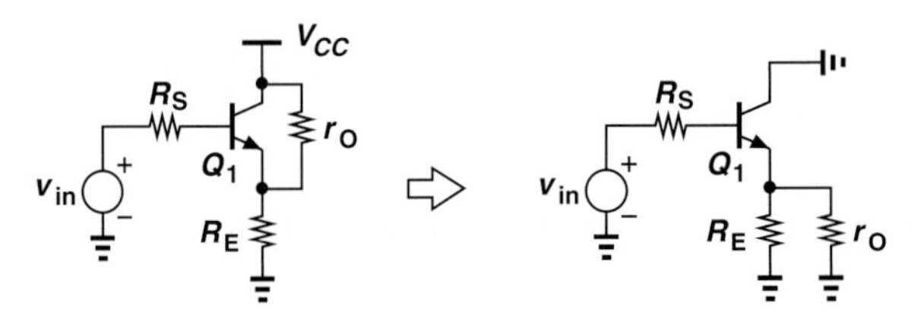

Figure 5.93 Follower including transistor output resistance.

Example 5-46 Determine the small-signal properties of an emitter follower using an ideal current source (as in Example 5-43) but with a finite source impedance R_S.

Solution Since $R_E = \infty$, we have

$$A_v = \frac{r_O}{r_O + \frac{R_S}{\beta + 1} + \frac{1}{g_m}} \tag{5.333}$$

$$R_{in} = r_\pi + (\beta + 1)r_O \tag{5.334}$$

$$R_{out} = \left(\frac{R_S}{\beta + 1} + \frac{1}{g_m}\right)||r_O. \tag{5.335}$$

Also, $g_m r_O \gg 1$, and hence

$$A_v \approx \frac{r_O}{r_O + \frac{R_S}{\beta+1}} \tag{5.336}$$

$$R_{in} \approx (\beta+1) r_O. \tag{5.337}$$

We note that A_v approaches unity if $R_S \ll (\beta+1) r_O$, a condition typically valid.

Exercise How are the results modified if $R_E < \infty$?

The buffering capability of followers is sometimes attributed to their "current gain." Since a base current i_B results in an emitter current of $(\beta+1)i_B$, we can say that for a current i_L delivered to the load, the follower draws only $i_L/(\beta+1)$ from the source voltage (Fig. 5.94). Thus, v_X sees the load impedance multiplied by $(\beta+1)$.

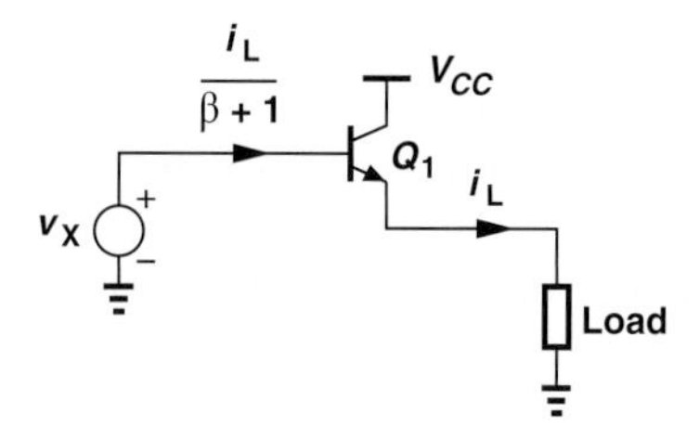

Figure 5.94 Current amplification in a follower.

Emitter Follower with Biasing The biasing of emitter followers entails defining both the base voltage and the collector (emitter) current. Figure 5.95(a) depicts an example similar to the scheme illustrated in Fig. 5.19 for the CE stage. As usual, the current flowing through R_1 and R_2 is chosen to be much greater than the base current.

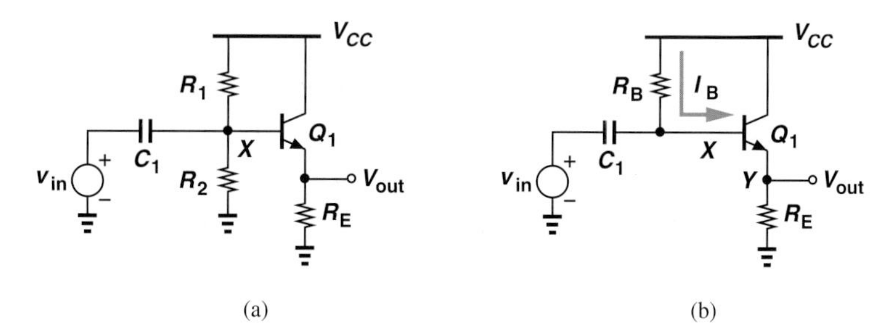

Figure 5.95 Biasing a follower by means of (a) resistive divider, (b) single base resistor.

It is interesting to note that, unlike the CE topology, the emitter follower can operate with a base voltage near V_{CC}. This is because the collector is tied to V_{CC}, allowing the same voltage for the base without driving Q_1 into saturation. For this reason, followers are often biased as shown in Fig. 5.95(b), where $R_B I_B$ is chosen much less than the voltage drop across R_E, thus lowering the sensitivity to β. The following example illustrates this point.

Example 5-47 The follower of Fig. 5.95(b) employs $R_B = 10\ \text{k}\Omega$ and $R_E = 1\ \text{k}\Omega$. Calculate the bias current and voltages if $I_S = 5 \times 10^{-16}$ A, $\beta = 100$, and $V_{CC} = 2.5$ V. What happens if β drops to 50?

Solution To determine the bias current, we follow the iterative procedure described in Section 5.2.3. Writing a KVL through R_B, the base-emitter junction, and R_E gives

$$\frac{R_B I_C}{\beta} + V_{BE} + R_E I_C = V_{CC}, \tag{5.338}$$

which, with $V_{BE} \approx 800$ mV, leads to

$$I_C = 1.545\ \text{mA}. \tag{5.339}$$

It follows that $V_{BE} = V_T \ln(I_C/I_S) = 748$ mV. Using this value in Eq. (5.338), we have

$$I_C = 1.593\ \text{mA}, \tag{5.340}$$

a value close to that in Eq. (5.339) and hence relatively accurate. Under this condition, $I_B R_B = 159$ mV whereas $R_E I_C = 1.593$ V.

Since $I_B R_B \ll R_E I_C$, we expect that variation of β and hence $I_B R_B$ negligibly affects the voltage drop across R_E and hence the emitter and collector currents. As a rough estimate, for $\beta = 50$, $I_B R_B$ is doubled (≈ 318 mV), reducing the drop across R_E by 159 mV. That is, $I_E = (1.593\ \text{V} - 0.159\ \text{V})/1\ \text{k}\Omega = 1.434$ mA, implying that a twofold change in β leads to a 10% change in the collector current. The reader is encouraged to repeat the above iterations with $\beta = 50$ and determine the exact current.

Exercise If R_B is doubled, is the circuit more or less sensitive to the variation in β?

As manifested by Eq. (5.338), the topologies of Fig. 5.95 suffer from supply-dependent biasing. In integrated circuits, this issue is resolved by replacing the emitter resistor with a constant current source (Fig. 5.96). Now, since I_{EE} is constant, so are V_{BE} and $R_B I_B$. Thus, if V_{CC} rises, so do V_X and V_Y, but the bias current remains constant.

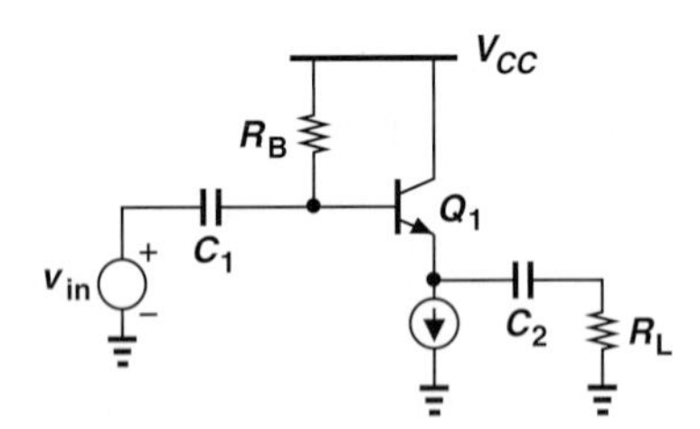

Figure 5.96 Capacitive coupling at input and output of a follower.

5.4 SUMMARY AND ADDITIONAL EXAMPLES

This chapter has created a foundation for amplifier design, emphasizing that a proper bias point must be established to define the small-signal properties of each circuit. Depicted in Fig. 5.97, the three amplifier topologies studied here exhibit different gains and I/O impedances, each serving a specific application. CE and CB stages can provide a voltage gain greater than unity and their input and output impedances are independent of the load and source impedances, respectively (if $V_A = \infty$). On the other hand, followers display a voltage gain of at most unity but their terminal impedances depend on the load and source impedances.

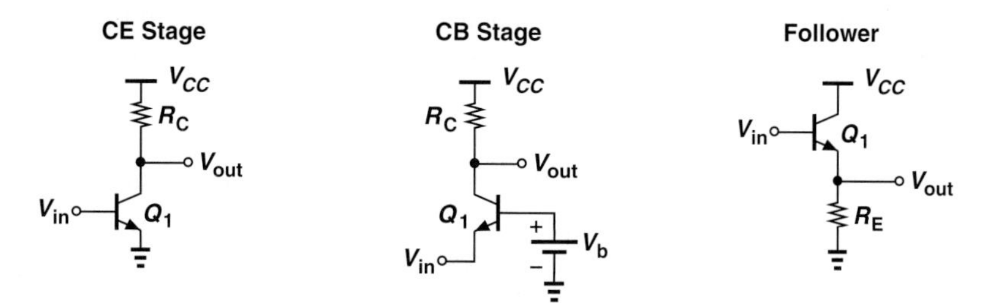

Figure 5.97 Summary of bipolar amplifier topologies.

In this section, we consider a number of challenging examples, seeking to improve our circuit analysis techniques. As usual, our emphasis is on solution by inspection and hence intuitive understanding of the circuits. We assume various capacitors used in each circuit have a negligible impedance at the signal frequencies of interest.

Example 5-48

Assuming $V_A = \infty$, determine the voltage gain of the circuit shown in Fig. 5.98(a).

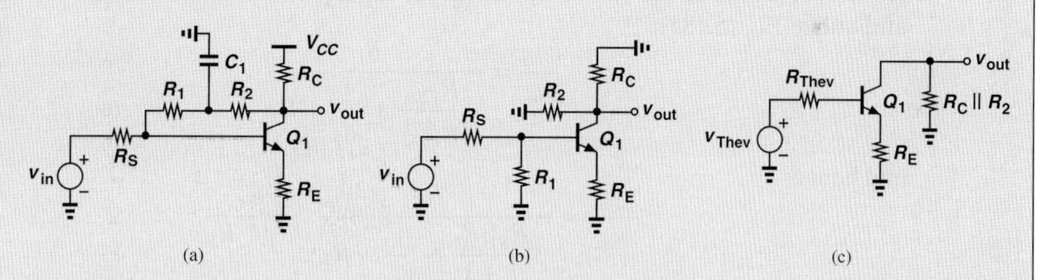

Figure 5.98 (a) Example of CE stage, (b) equivalent circuit with C_1 shorted, (c) simplified circuit.

Solution The simplified ac model is depicted in Fig. 5.98(b), revealing that R_1 appears between base and ground, and R_2 between collector and ground. Replacing v_{in}, R_S, and R_1 with a Thevenin equivalent [Fig. 5.98(c)], we have

$$v_{Thev} = \frac{R_1}{R_1 + R_S} v_{in} \tag{5.341}$$

$$R_{Thev} = R_1 || R_S. \tag{5.342}$$

The resulting circuit resembles that in Fig. 5.43(a) and satisfies Eq. (5.185):

$$\frac{v_{out}}{v_{Thev}} = -\frac{R_2 || R_C}{\dfrac{R_{Thev}}{\beta + 1} + \dfrac{1}{g_m} + R_E}. \tag{5.343}$$

Substituting for v_{Thev} and R_{Thev} gives

$$\frac{v_{out}}{v_{in}} = -\frac{R_2||R_C}{\frac{R_1||R_S}{\beta+1} + \frac{1}{g_m} + R_E} \cdot \frac{R_1}{R_1 + R_S}. \tag{5.344}$$

Exercise What happens if a very large capacitor is added from the emitter of Q_1 to ground?

Example 5-49 Assuming $V_A = \infty$, compute the voltage gain of the circuit shown in Fig. 5.99(a).

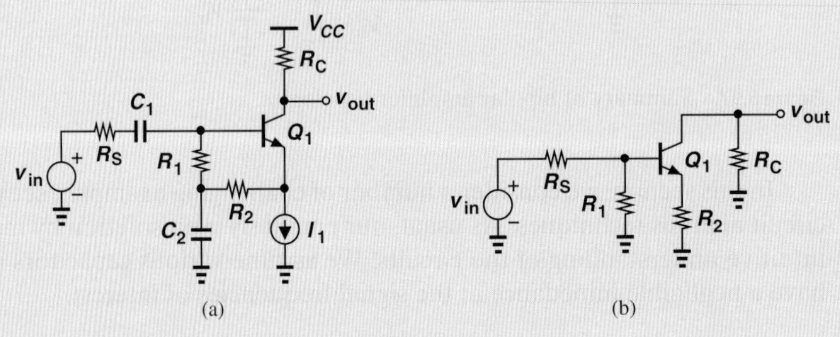

Figure 5.99 (a) Example of CE stage, (b) simplified circuit.

Solution As shown in the simplified diagram of Fig. 5.99(b), R_2 appears as an emitter degeneration resistor. As in the above example, we replace v_{in}, R_S, and R_1 with a Thevenin equivalent and utilize Eq. (5.185):

$$\frac{v_{out}}{v_{in}} = -\frac{R_C}{\frac{R_{Thev}}{\beta+1} + \frac{1}{g_m} + R_2} \tag{5.345}$$

and hence

$$\frac{v_{out}}{v_{in}} = -\frac{R_C}{\frac{R_S||R_1}{\beta+1} + \frac{1}{g_m} + R_2} \cdot \frac{R_1}{R_1 + R_S}. \tag{5.346}$$

Exercise What happens if C_2 is tied from the emitter of Q_1 to ground?

Example 5-50 Assuming $V_A = \infty$, compute the voltage gain and input impedance of the circuit shown in Fig. 5.100(a).

Solution The circuit resembles a CE stage (why?) degenerated by the impedance seen at the emitter of Q_2, R_{eq}. Recall from Fig. 5.76 that

$$R_{eq} = \frac{R_1}{\beta+1} + \frac{1}{g_{m2}}. \tag{5.347}$$

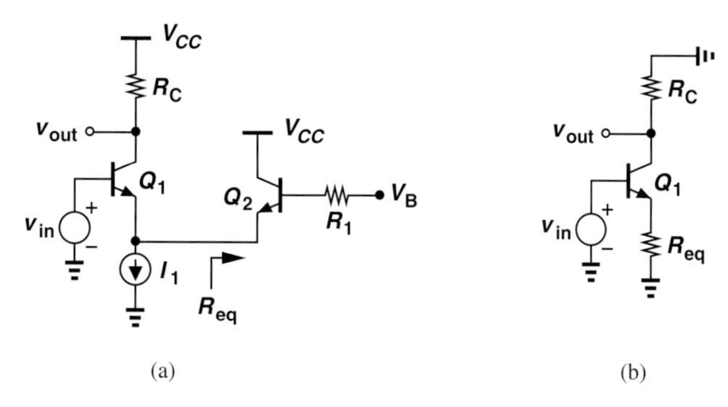

Figure 5.100 (a) Example of CE stage, (b) simplified circuit.

The simplified model in Fig. 5.100(b) thus yields

$$A_v = \frac{-R_C}{\frac{1}{g_{m1}} + R_{eq}} \tag{5.348}$$

$$= \frac{-R_C}{\frac{1}{g_{m1}} + \frac{R_1}{\beta + 1} + \frac{1}{g_{m2}}}. \tag{5.349}$$

The input impedance is also obtained from Fig. 5.76:

$$R_{in} = r_{\pi 1} + (\beta + 1)R_{eq} \tag{5.350}$$

$$= r_{\pi 1} + R_1 + r_{\pi 2}. \tag{5.351}$$

Exercise Repeat the above example if R_1 is placed in series with the emitter of Q_2.

Example 5-51 Calculate the voltage gain of the circuit in Fig. 5.101(a) if $V_A = \infty$.

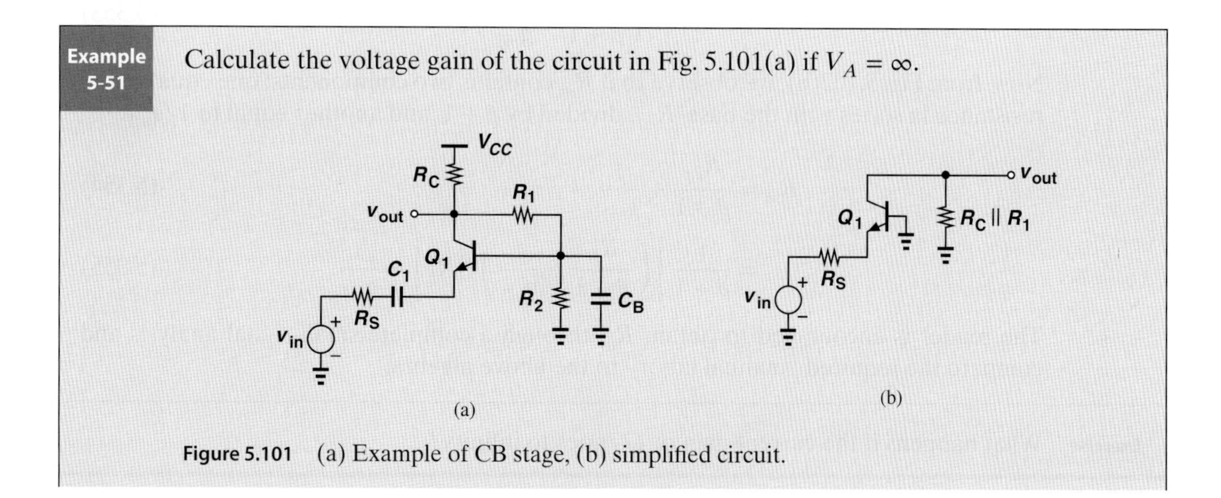

Figure 5.101 (a) Example of CB stage, (b) simplified circuit.

Solution Since the base is at ac ground, R_1 appears in parallel with R_C and R_2 is shorted to ground on both ends [Fig. 5.101(b)]. The voltage gain is given by Eq. (5.271) but with R_C replaced by $R_C||R_1$:

$$A_v = \frac{R_C||R_1}{R_S + \dfrac{1}{g_m}}. \tag{5.352}$$

Exercise What happens if R_C is replaced by an ideal current source?

Example 5-52 Determine the input impedance of the circuit shown in Fig. 5.102(a) if $V_A = \infty$.

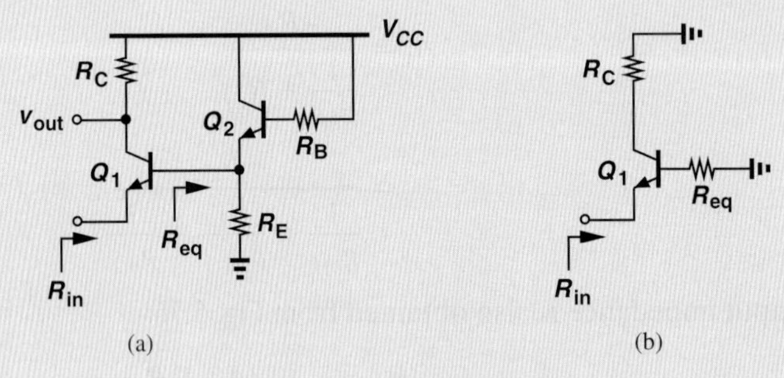

Figure 5.102 (a) Example of CB stage, (b) simplified circuit.

Solution In this circuit, Q_1 operates as a common-base device (why?) but with a resistance R_{eq} in series with its base [Fig. 5.102(b)]. To obtain R_{eq}, we recognize that Q_2 resembles an emitter follower, e.g., the topology in Fig. 5.92(a), concluding that R_{eq} can be viewed as the output resistance of such a stage, as given by Eq. (5.329):

$$R_{eq} = \left(\frac{R_B}{\beta + 1} + \frac{1}{g_{m2}}\right)||R_E. \tag{5.353}$$

Now, from Fig. 5.102(b), we observe that R_{in} contains two components: one equal to the resistance in series with the base, R_{eq}, divided by $\beta + 1$, and another equal to $1/g_{m1}$:

$$R_{in} = \frac{R_{eq}}{\beta + 1} + \frac{1}{g_{m1}} \tag{5.354}$$

$$= \frac{1}{\beta + 1}\left[\left(\frac{R_B}{\beta + 1} + \frac{1}{g_{m2}}\right)||R_E\right] + \frac{1}{g_{m1}}. \tag{5.355}$$

The reader is encouraged to obtain R_{in} through a complete small-signal analysis and compare the required "manual labor" to the above algebra.

Exercise What happens if the current gain of Q_2 goes to infinity?

Example 5-53

Compute the voltage gain and the output impedance of the circuit depicted in Fig. 5.103(a) with $V_A < \infty$.

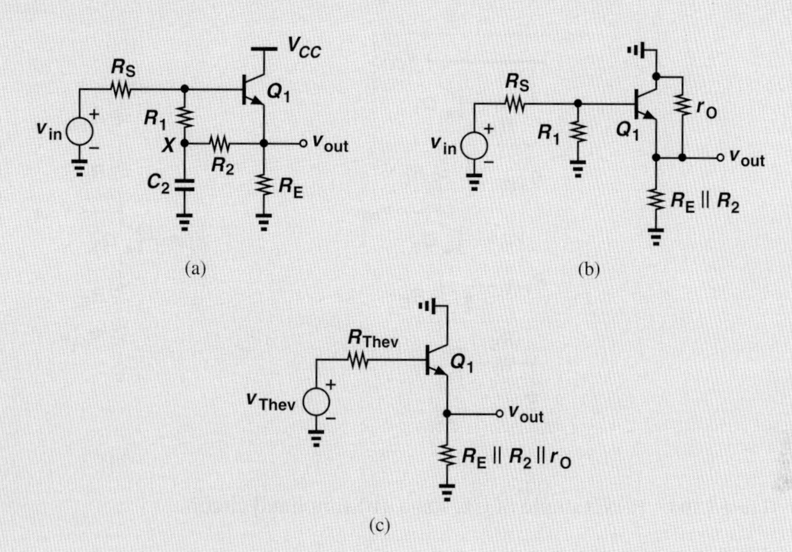

Figure 5.103 (a) Example of emitter follower, (b) circuit with C_1 shorted, (c) simplified circuit.

Solution Noting that X is at ac ground, we construct the simplified circuit shown in Fig. 5.103(b), where the output resistance of Q_1 is explicitly drawn. Replacing v_{in}, R_S, and R_1 with their Thevenin equivalent and recognizing that R_E, R_2, and r_O appear in parallel [Fig. 5.103(c)], we employ Eq. (5.330) and write

$$\frac{v_{out}}{v_{Thev}} = \frac{R_E||R_2||r_O}{R_E||R_2||r_O + \dfrac{1}{g_m} + \dfrac{R_{Thev}}{\beta + 1}} \tag{5.356}$$

and hence

$$\frac{v_{out}}{v_{in}} = \frac{R_E||R_2||r_O}{R_E||R_2||r_O + \dfrac{1}{g_m} + \dfrac{R_S||R_1}{\beta + 1}} \cdot \frac{R_1}{R_1 + R_S}. \tag{5.357}$$

For the output resistance, we refer to Eq. (5.332):

$$R_{out} = \left(\frac{R_{Thev}}{\beta + 1} + \frac{1}{g_m}\right) ||(R_E||R_2||r_O) \tag{5.358}$$

$$= \left(\frac{R_S||R_1}{\beta + 1} + \frac{1}{g_m}\right) ||R_E||R_2||r_O. \tag{5.359}$$

Exercise What happens if $R_S = 0$?

Example 5-54 Determine the voltage gain and I/O impedances of the topology shown in Fig. 5.104(a). Assume $V_A = \infty$ and equal βs for *npn* and *pnp* transistors.

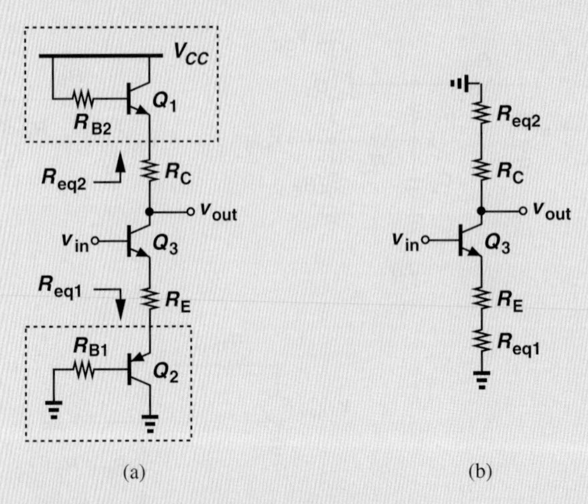

Figure 5.104 (a) Example of CE stage, (b) simplified circuit.

Solution We identify the stage as a CE amplifier with emitter degeneration and a composite collector load. As the first step, we represent the role of Q_2 and Q_3 by the impedances that they create at their emitter. Since R_{eq1} denotes the impedance seen looking into the emitter of Q_2 with a base resistance of R_{B1}, we have from Fig. 5.76

$$R_{eq1} = \frac{R_{B1}}{\beta + 1} + \frac{1}{g_{m2}}. \tag{5.360}$$

Similarly,

$$R_{eq2} = \frac{R_{B2}}{\beta + 1} + \frac{1}{g_{m1}}, \tag{5.361}$$

leading to the simplified circuit shown in Fig. 5.104(b). It follows that

$$A_v = -\frac{R_C + R_{eq2}}{R_{eq1} + \dfrac{1}{g_{m3}} + R_E} \tag{5.362}$$

$$= -\frac{R_C + \dfrac{R_{B2}}{\beta + 1} + \dfrac{1}{g_{m1}}}{\dfrac{R_{B1}}{\beta + 1} + \dfrac{1}{g_{m2}} + \dfrac{1}{g_{m3}} + R_E}. \tag{5.363}$$

Also,

$$R_{in} = r_{\pi 3} + (\beta + 1)(R_E + R_{eq1}) \tag{5.364}$$

$$= r_{\pi 3} + (\beta + 1)\left(R_E + \frac{R_{B1}}{\beta + 1} + \frac{1}{g_{m2}}\right), \tag{5.365}$$

and

$$R_{out} = R_C + R_{eq2} \tag{5.366}$$

$$= R_C + \frac{R_{B2}}{\beta + 1} + \frac{1}{g_{m1}}. \tag{5.367}$$

Exercise What happens if $R_{B2} \to \infty$?

5.5 CHAPTER SUMMARY

- In addition to gain, the input and output impedances of amplifiers determine the ease with which various stages can be cascaded.
- Voltage amplifiers must ideally provide a high input impedance (so that they can sense a voltage without disturbing the node) and a low output impedance (so that they can drive a load without reduction in gain).
- The impedances seen looking into the base, collector, and emitter of a bipolar transistor are equal to r_π (with emitter grounded), r_O (with emitter grounded), and $1/g_m$ (with base grounded), respectively.
- In order to obtain the required small-signal bipolar device parameters such as g_m, r_π, and r_O, the transistor must be "biased," i.e., carry a certain collector current and operate in the active region. Signals simply perturb these conditions.
- Biasing techniques establish the required base-emitter and base-collector voltages while providing the base current.
- With a single bipolar transistor, only three amplifier topologies are possible: common-emitter and common-base stages and emitter followers.
- The CE stage provides a moderate voltage gain, a moderate input impedance, and a moderate output impedance.
- Emitter degeneration improves the linearity but lowers the voltage gain.
- Emitter degeneration raises the output impedance of CE stages considerably.
- The CB stage provides a moderate voltage gain, a low input impedance, and a moderate output impedance.
- The voltage gain expressions for CE and CB stages are similar but for a sign.
- The emitter follower provides a voltage gain less than unity, a high input impedance, and a low output impedance, serving as a good voltage buffer.

PROBLEMS

Section 5.1.1 Input and Output Impedances

5.1. An antenna can be modeled as a Thevenin equivalent having a sinusoidal voltage source $V_0 \cos \omega t$ and an output resistance R_{out}. Determine the average power delivered to a load resistance R_L and plot the result as a function of R_L.

5.2. Determine the small-signal input resistance of the circuits shown in Fig. 5.105. Assume all diodes are forward-biased. (Recall from Chapter 3 that each diode behaves as a linear resistance if the voltage and current changes are small.)

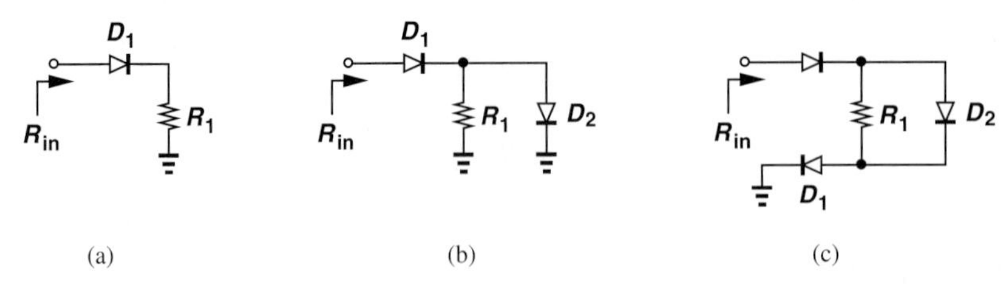

Figure 5.105

5.3. Compute the input resistance of the circuits depicted in Fig. 5.106. Assume $V_A = \infty$.

5.4. Compute the output resistance of the circuits depicted in Fig. 5.107.

5.5. Determine the input impedance of the circuits depicted in Fig. 5.108. Assume $V_A = \infty$.

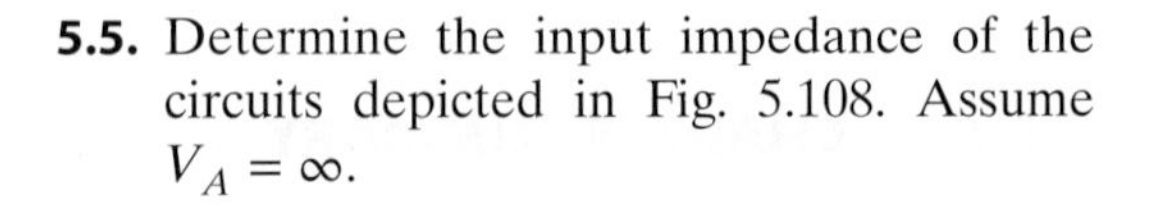

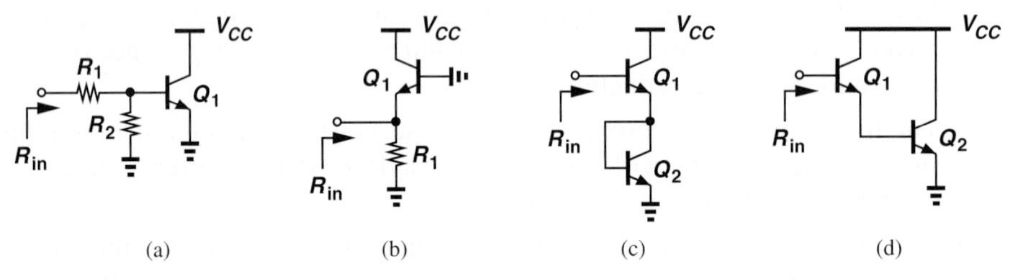

Figure 5.106

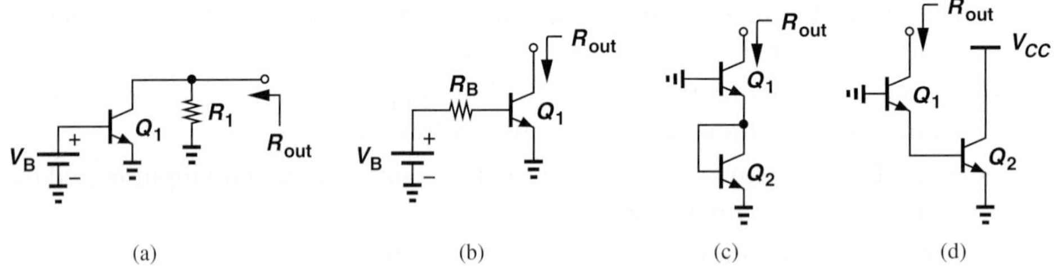

Figure 5.107

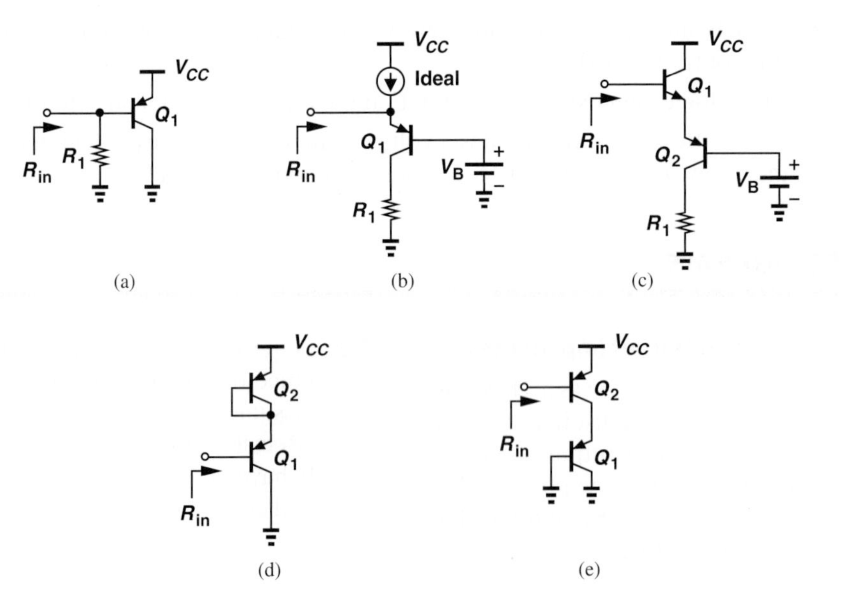

Figure 5.108

5.6. Compute the output impedance of the circuits shown in Fig. 5.109.

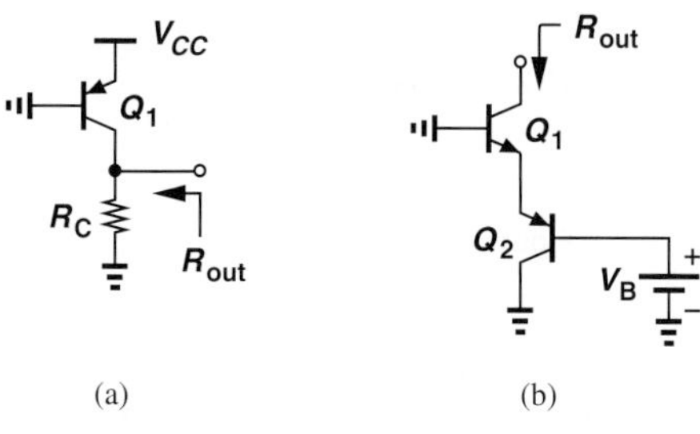

Figure 5.109

Sections 5.2.1 and 5.2.2 Simple and Resistive Divider Biasing

5.7. Compute the bias point of the circuits depicted in Fig. 5.110. Assume $\beta = 100$, $I_S = 6 \times 10^{-16}$ A, and $V_A = \infty$.

5.8. Construct the small-signal equivalent of each of the circuits in Problem 5.7.

***5.9.** Calculate the bias point of the circuits shown in Fig. 5.111. Assume $\beta = 100$, $I_S = 5 \times 10^{-16}$ A, and $V_A = \infty$.

5.10. Construct the small-signal equivalent of each of the circuits in Problem 5.9.

5.11. Consider the circuit shown in Fig. 5.112, where $\beta = 100$, $I_S = 6 \times 10^{-16}$ A, and $V_A = \infty$.

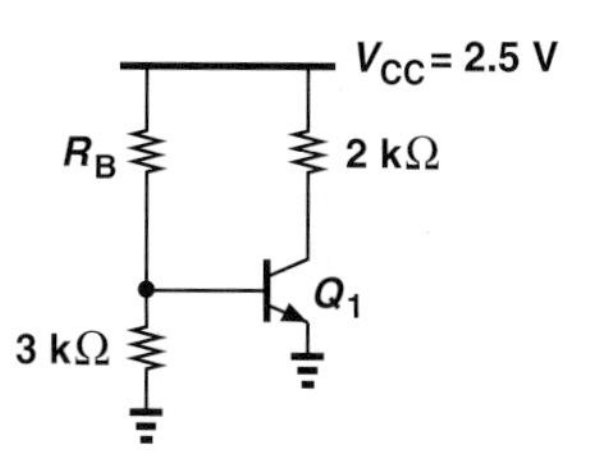

Figure 5.112

(a) What is the minimum value of R_B that guarantees operation in the active mode?

(b) With the value found in R_B, how much base-collector forward bias is sustained if β rises to 200?

5.12. In the circuit of Fig. 5.113, $\beta = 100$ and $V_A = \infty$. SOL

(a) If the collector current of Q_1 is equal to 0.5 mA, calculate the value of I_S.

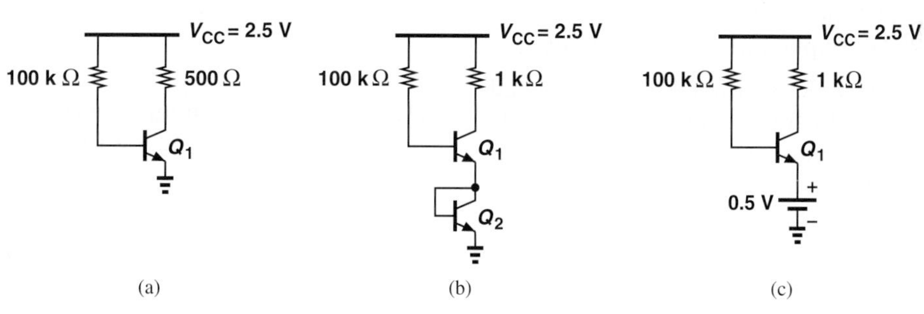

Figure 5.110

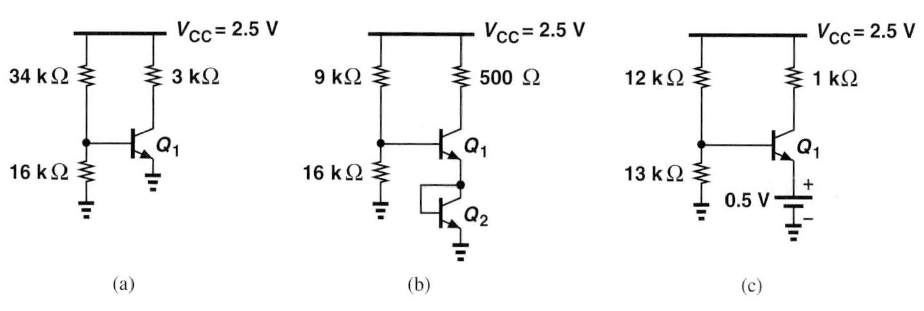

Figure 5.111

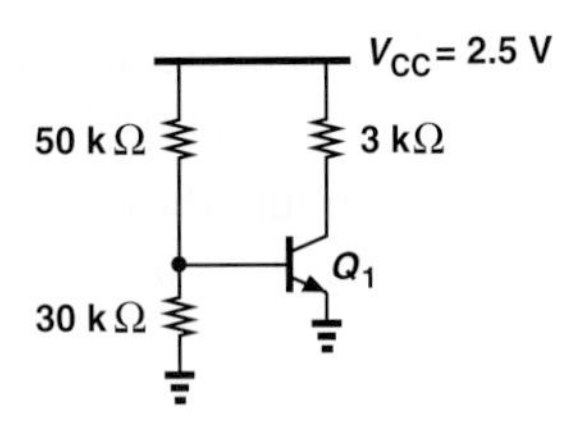

Figure 5.113

(b) If Q_1 is biased at the edge of saturation, calculate the value of I_S.

5.13. The circuit of Fig. 5.114 must be designed for an input impedance of greater than 10 kΩ and a g_m of at least 1/(260 Ω). If $\beta = 100$, $I_S = 2 \times 10^{-17}$ A, and $V_A = \infty$, determine the minimum allowable values of R_1 and R_2.

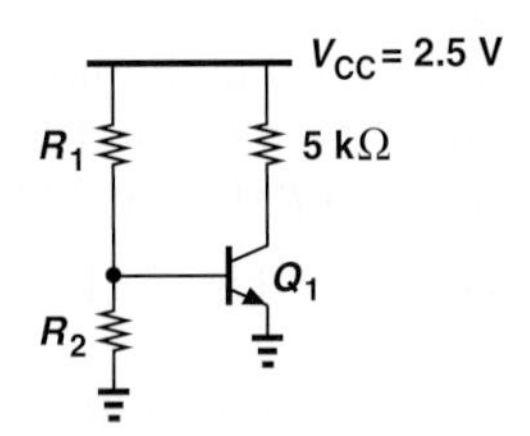

Figure 5.114

SOL *__5.14.__ Repeat Problem 5.13 for a g_m of at least 1/(26 Ω). Explain why no solution exists.

***5.15.** We wish to design the CE stage depicted in Fig. 5.115 for a gain ($= g_m R_C$) of A_0 with an output impedance of R_0. What is the maximum achievable input impedance here? Assume $V_A = \infty$.

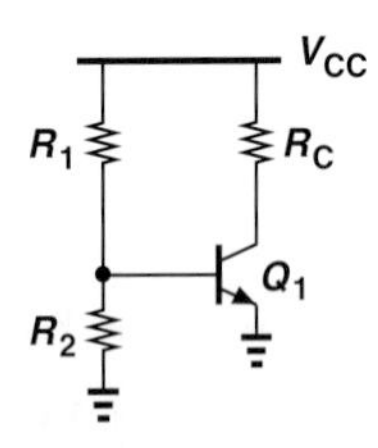

Figure 5.115

Section 5.2.3 Biasing with Emitter Degeneration

5.16. The circuit of Fig. 5.116 is designed for a collector current of 0.25 mA. Assume $I_S = 6 \times 10^{-16}$ A, $\beta = 100$, and $V_A = \infty$.

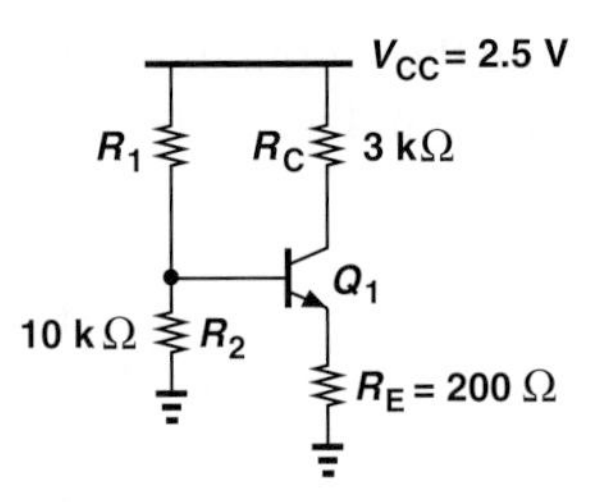

Figure 5.116

(a) Determine the required value of R_1.
(b) What is the error in I_C if R_E deviates from its nominal value by 5%?

5.17. In the circuit of Fig. 5.117, determine the maximum value of R_2 that guarantees operation of Q_1 in the active mode. Assume $\beta = 100$, $I_S = 10^{-17}$ A, and $V_A = \infty$.

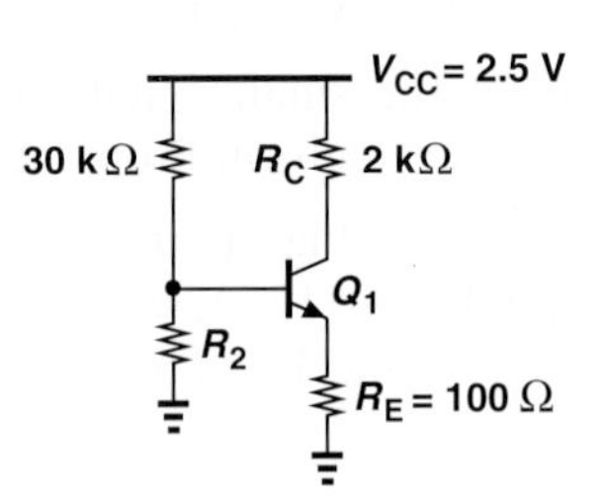

Figure 5.117

***5.18.** Consider the circuit shown in Fig. 5.118, where $I_{S1} = 2I_{S2} = 5 \times 10^{-16}$ A, $\beta_1 = \beta_2 = 100$, and $V_A = \infty$.

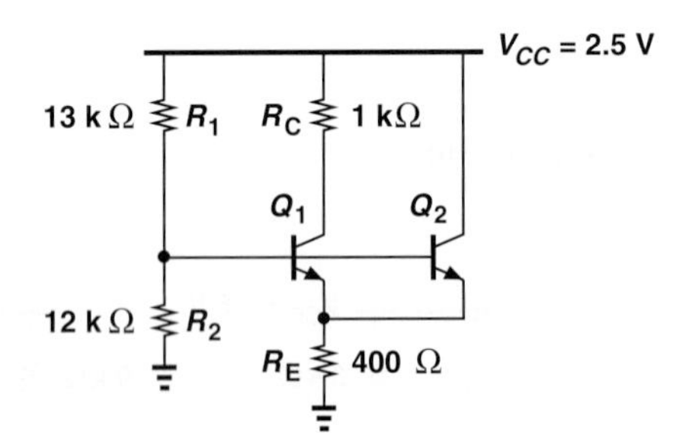

Figure 5.118

(a) Determine the collector currents of Q_1 and Q_2.
(b) Construct the small-signal equivalent circuit.

5.19. In the circuit depicted in Fig. 5.119, $I_{S1} = I_{S2} = 4 \times 10^{-16}$ A, $\beta_1 = \beta_2 = 100$, and $V_A = \infty$.

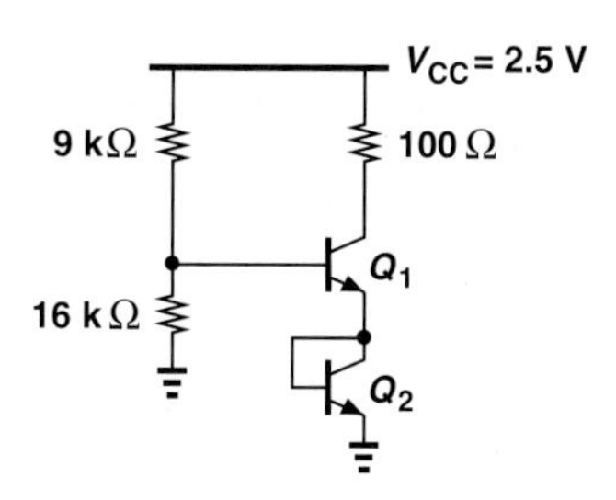

Figure 5.119

(a) Determine the operating point of the transistor.
(b) Draw the small-signal equivalent circuit.

Section 5.2.4 Self-Biased Stage

SOL **5.20.** The circuit of Fig. 5.120 must be biased with a collector current of 1 mA. Compute the required value of R_B if $I_S = 3 \times 10^{-16}$ A, $\beta = 100$, and $V_A = \infty$.

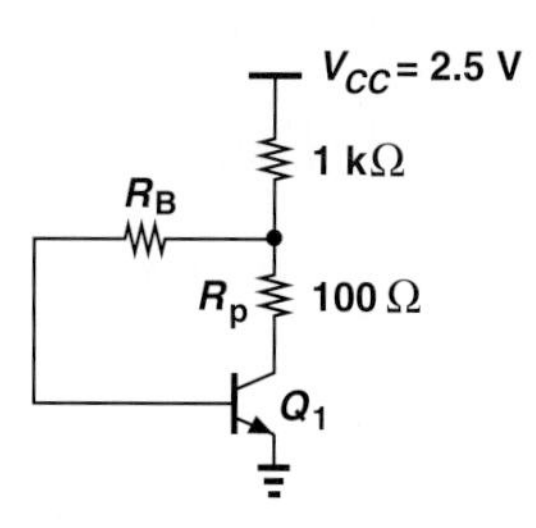

Figure 5.120

5.21. In the circuit of Fig. 5.121, $V_X = 1.1$ V. If $\beta = 100$ and $V_A = \infty$, what is the value of I_S?

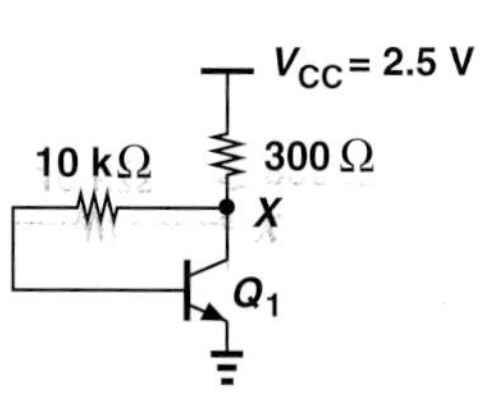

Figure 5.121

5.22. Consider the circuit shown in Fig. 5.122, where $I_S = 6 \times 10^{-16}$ A, $\beta = 100$, and $V_A = \infty$. Calculate the operating point of Q_1.

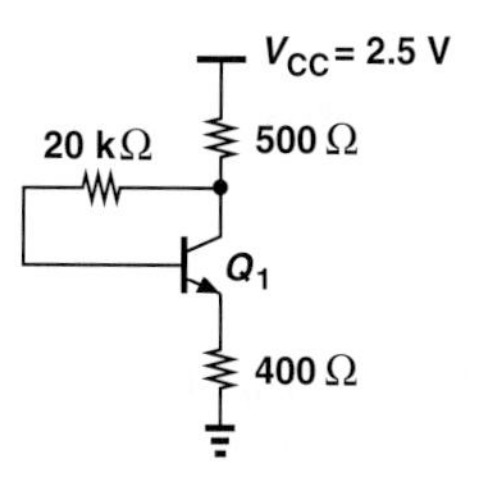

Figure 5.122

5.23. Due to a manufacturing error, a parasitic resistor, R_P, has appeared in series with the collector of Q_1 in Fig. 5.123. What is the minimum allowable value of R_B if the base-collector forward bias must not exceed 200 mV? Assume $I_S = 3 \times 10^{-16}$ A, $\beta = 100$, and $V_A = \infty$.

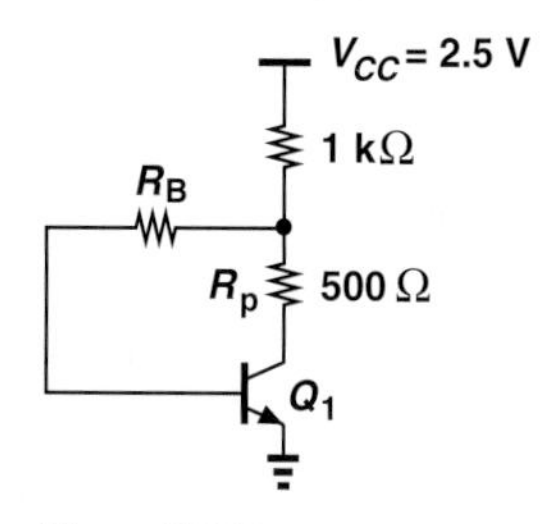

Figure 5.123

*__5.24.__ In the circuit of Fig. 5.124, $I_S = 8 \times 10^{-16}$ A, $\beta = 100$, and $V_A = \infty$.

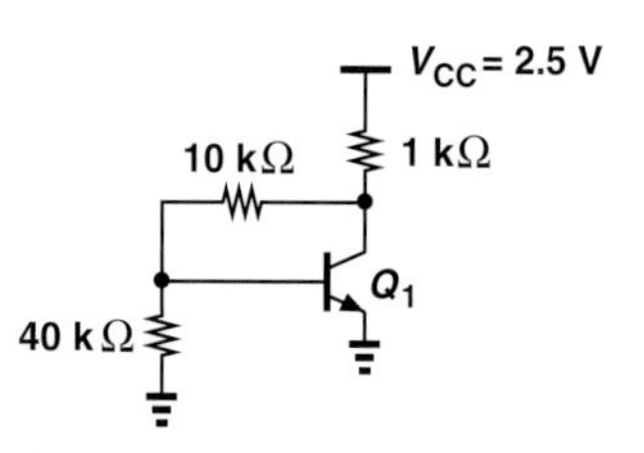

Figure 5.124

(a) Determine the operating point of Q_1.
(b) Draw the small-signal equivalent circuit.

**__5.25.__ In the circuit of Fig. 5.125, $I_{S1} = I_{S2} = 3 \times 10^{-16}$ A, $\beta = 100$, and $V_A = \infty$.

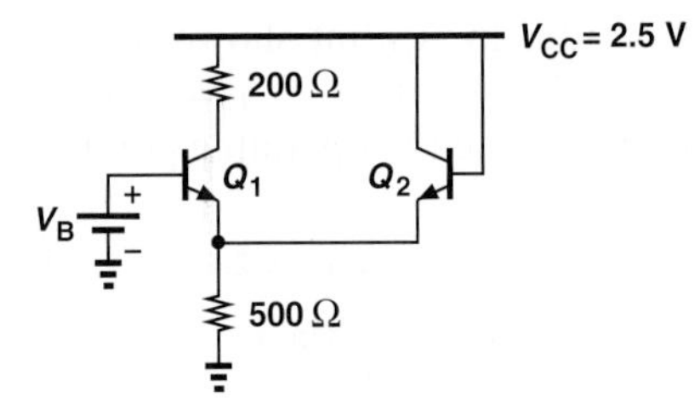

Figure 5.125

(a) Calculate V_B such that Q_1 carries a collector current of 1 mA.
(b) Construct the small-signal equivalent circuit.

Section 5.2.5 Biasing of *PNP* Transistors

5.26. Determine the bias point of each circuit shown in Fig. 5.126. Assume $\beta_{npn} = 2\beta_{pnp} = 100$, $I_S = 9 \times 10^{-16}$ A, and $V_A = \infty$.

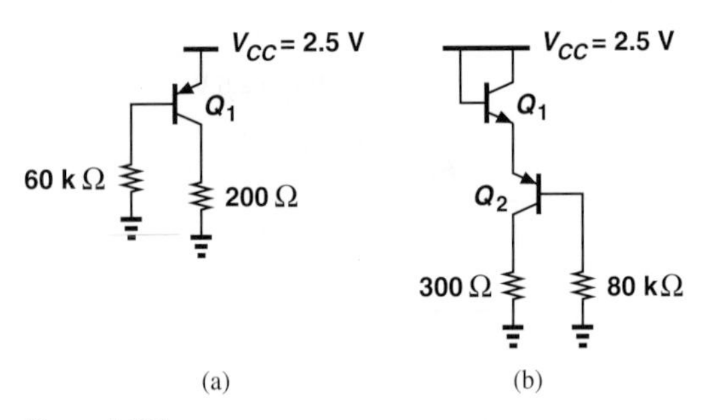

Figure 5.126

5.27. Construct the small-signal model of the circuits in Problem 5.26.

5.28. Calculate the bias point of the circuits shown in Fig. 5.127. Assume $\beta_{npn} = 2\beta_{pnp} = 100$, $I_S = 9 \times 10^{-16}$ A, and $V_A = \infty$.

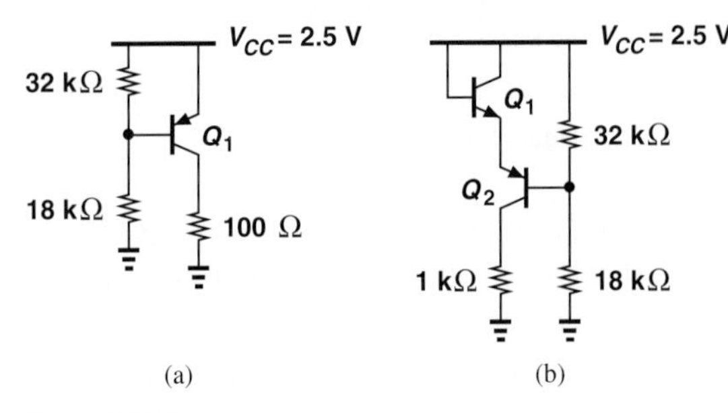

Figure 5.127

5.29. Draw the small-signal model of the circuits in Problem 5.28.

5.30. We have chosen R_B in Fig. 5.128 to place Q_1 at the edge of saturation. But the actual value of this resistor can vary by $\pm 5\%$. Determine the forward- or reverse-bias across the base-collector junction at these two extremes. Assume $\beta = 50$, $I_S = 8 \times 10^{-16}$ A, and $V_A = \infty$.

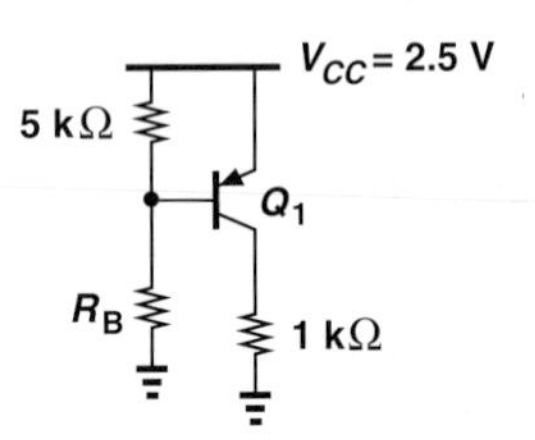

Figure 5.128

5.31. Calculate the value of R_E in Fig. 5.129 such that Q_1 sustains a reverse bias of 300 mV across its base-collector junction. Assume $\beta = 50$, $I_S = 8 \times 10^{-16}$ A, and $V_A = \infty$. What happens if the value of R_E is halved?

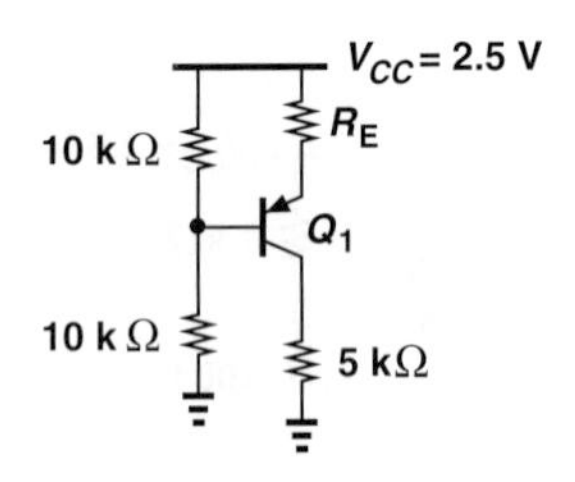

Figure 5.129

5.32. If $\beta = 80$ and $V_A = \infty$, what value of I_S yields a collector current of 1 mA in Fig. 5.130?

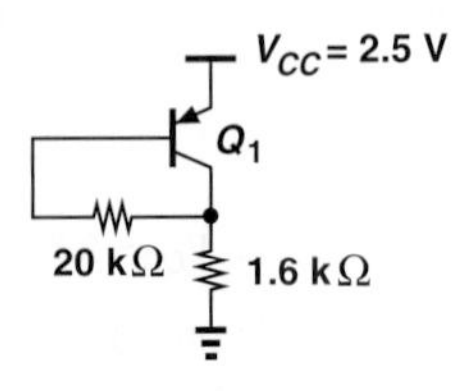

Figure 5.130

***5.33.** The topology depicted in Fig. 5.131(a) is called a "V_{BE} multiplier." (The *npn* counterpart has a similar topology.) Constructing the circuit shown in Fig. 5.131(b), determine the collector-emitter voltage of Q_1 if the base current is negligible. (The *npn* counterpart can also be used.)

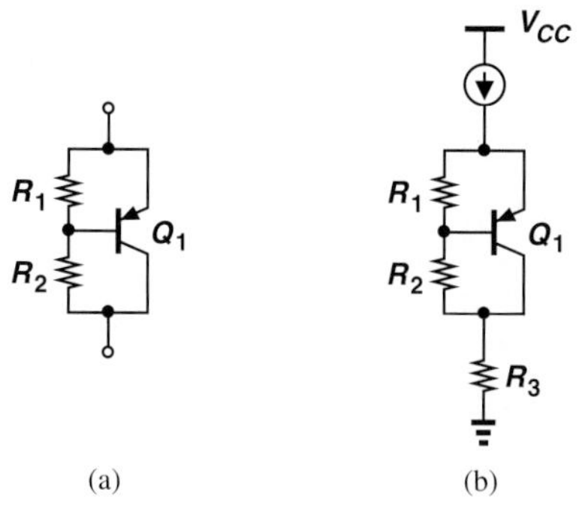

Figure 5.131

Section 5.3.1 Common-Emitter Topology

SOL **5.34.** We wish to design the CE stage of Fig. 5.132 for a voltage gain of 20. What is the minimum allowable supply voltage if Q_1 must remain in the active mode? Assume $V_A = \infty$ and $V_{BE} = 0.8$ V.

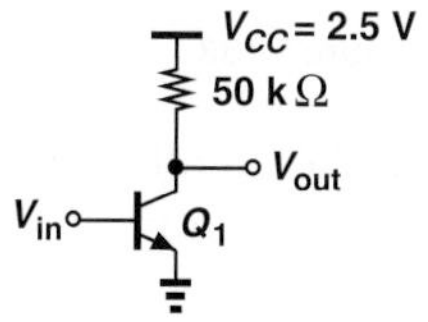

Figure 5.132

5.35. The circuit of Fig. 5.133 must be designed for maximum voltage gain while maintaining Q_1 in the active mode. If $V_A = 10$ V and $V_{BE} = 0.8$ V, calculate the required bias current.

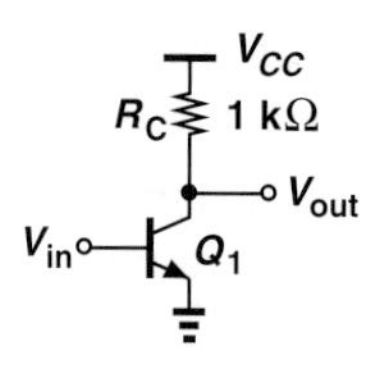

Figure 5.133

5.36. The CE stage of Fig. 5.134 employs an ideal current source as the load. If the voltage gain is equal to 50 and the output impedance equal to 10 kΩ, determine the bias current of the transistor. SOL

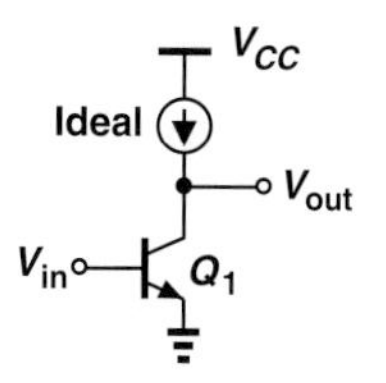

Figure 5.134

5.37. Suppose the bipolar transistor in Fig. 5.135 exhibits the following hypothetical characteristic:

$$I_C = I_S \exp \frac{V_{BE}}{2V_T}, \qquad (5.368)$$

and no Early effect. Compute the voltage gain for a bias current of 1 mA.

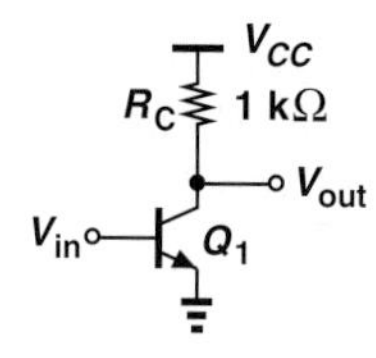

Figure 5.135

****5.38.** Determine the voltage gain and I/O impedances of the circuits shown in Fig. 5.136. Assume $V_A = \infty$. Transistor Q_2 in Figs. 5.136(d) and (e) operates in soft saturation.

****5.39.** Repeat Problem 5.38 with $V_A < \infty$.

5.40. Consider Eq. (5.157) for the gain of a degenerated CE stage. Writing $g_m = I_C/V_T$, we note that g_m and hence the voltage gain vary if I_C changes with the signal level. For the following two cases, determine the relative change in the gain if I_C varies by 10%: (a) $g_m R_E$ is nominally equal to 3; (b) $g_m R_E$ is nominally equal to 7. The more constant gain in the second case translates to greater circuit linearity.

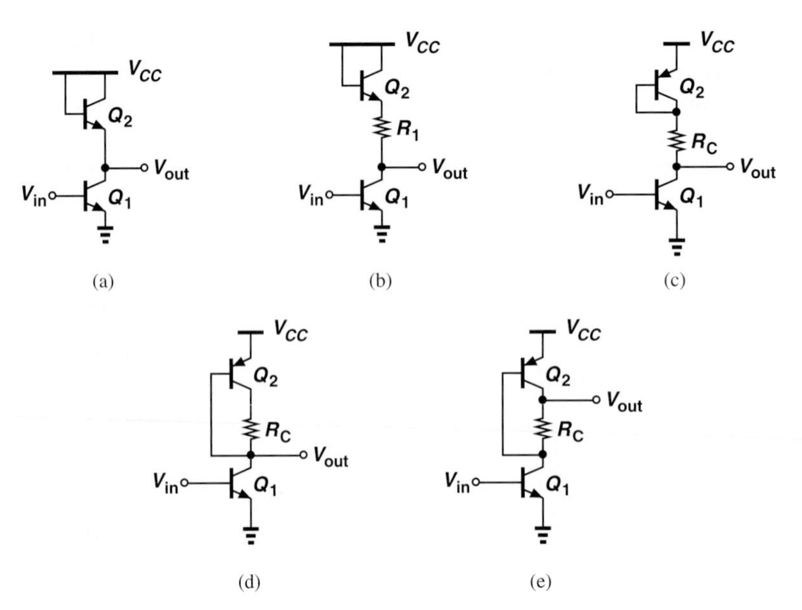

Figure 5.136

SOL **5.41.** Express the voltage gain of the stage depicted in Fig. 5.137 in terms of the collector bias current, I_C, and V_T. If $V_A = \infty$, what is the gain if the dc voltage drops across R_C and R_E are equal to $20V_T$ and $5V_T$, respectively?

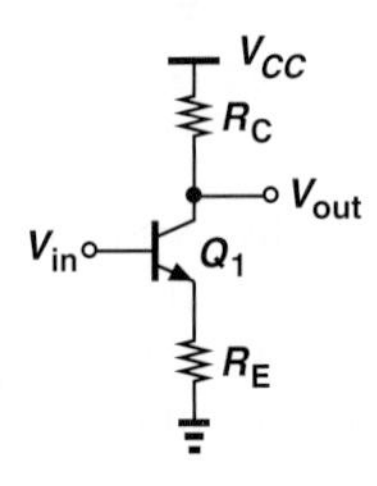

Figure 5.137

5.42. We wish to design the degenerated stage of Fig. 5.138 for a voltage gain of 10 with Q_1 operating at the edge of saturation. Calculate the bias current and the value of R_C if $\beta = 100$, $I_S = 5 \times 10^{-16}$A, and $V_A = \infty$. Calculate the input impedance of the circuit.

*__5.43.__ Repeat Problem 5.42 for a voltage gain of 100. Explain why no solution exists. What is the maximum gain that can be achieved in this stage?

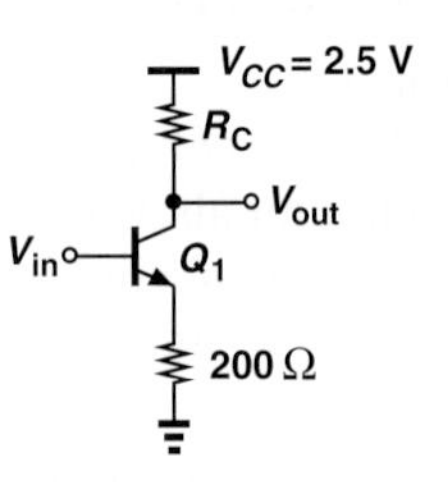

Figure 5.138

5.44. Construct the small-signal model of the CE stage shown in Fig. 5.43(a) and calculate the voltage gain. Assume $V_A = \infty$.

5.45. Construct the small-signal model of the CE stage shown in Fig. 5.43(a) and prove that the output impedance is equal to R_C if the Early effect is neglected.

**__5.46.__ Determine the voltage gain and I/O impedances of the circuits shown in Fig. 5.139. Assume $V_A = \infty$.

**__5.47.__ Compute the voltage gain the I/O impedances of the circuits depicted in Fig. 5.140. Assume $V_A = \infty$.

*__5.48.__ Using a small-signal equivalent circuit, compute the output impedance of a degenerated CE stage with $V_A < \infty$. Assume $\beta \gg 1$.

*__5.49.__ Calculate the output impedance of the circuits shown in Fig. 5.141. Assume $\beta \gg 1$.

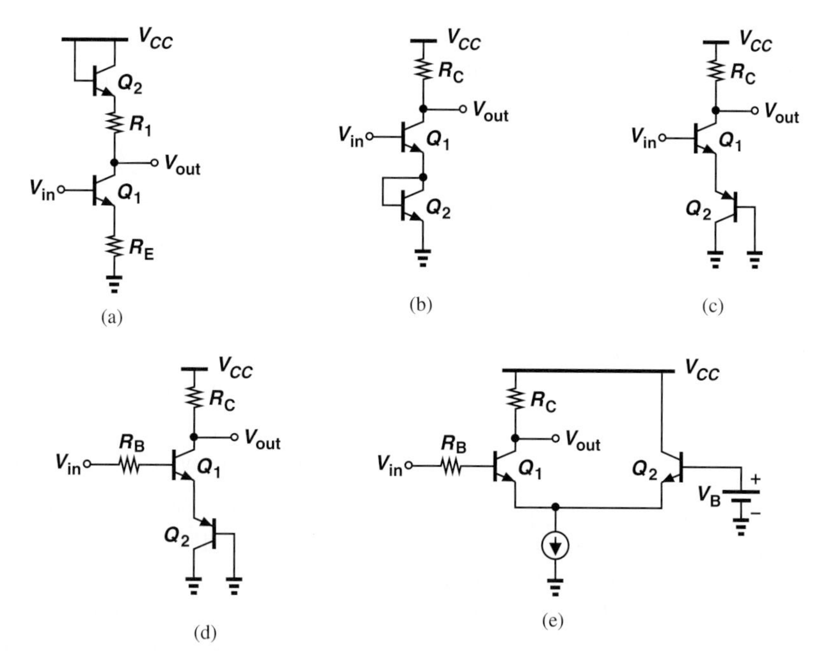

Figure 5.139

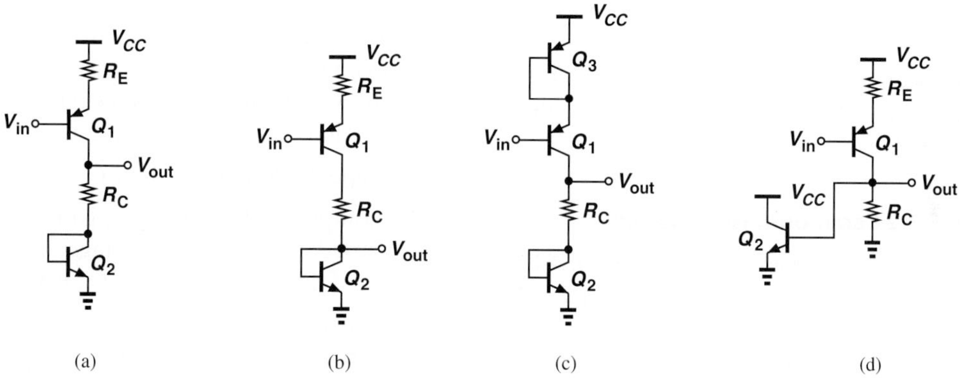

Figure 5.140

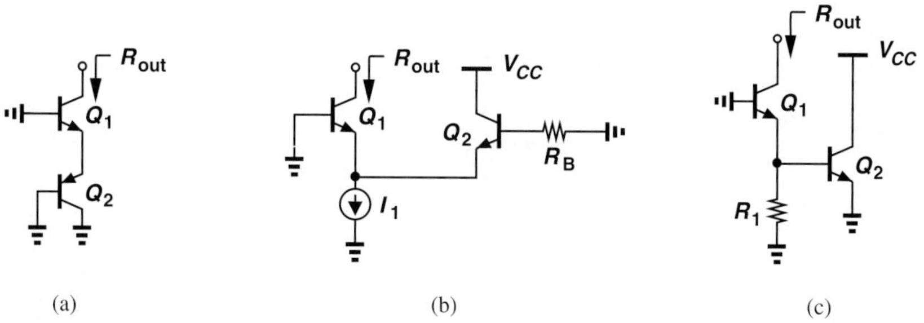

Figure 5.141

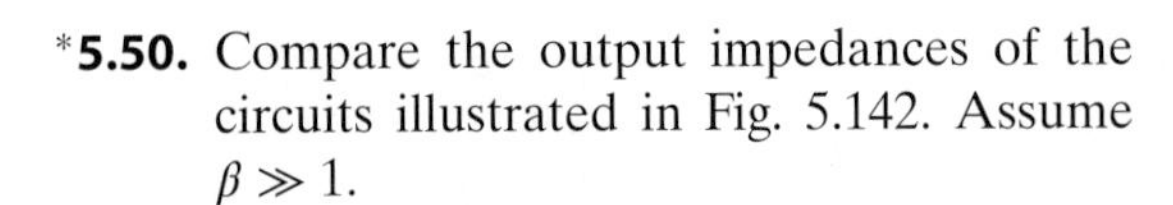

*__5.50.__ Compare the output impedances of the circuits illustrated in Fig. 5.142. Assume $\beta \gg 1$.

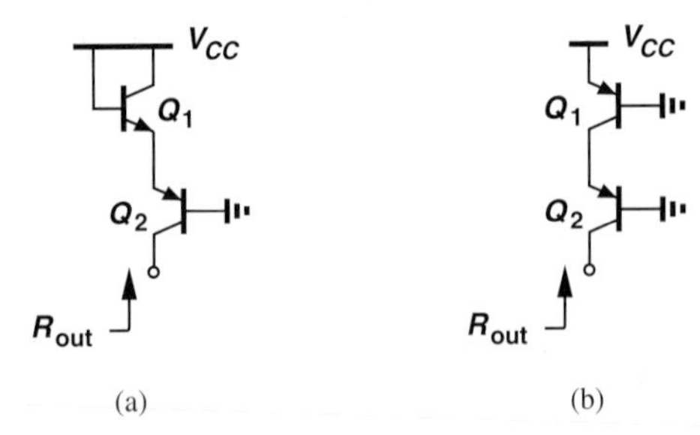

Figure 5.142

*__5.51.__ Writing $r_\pi = \beta V_T/I_C$, expand Eq. (5.217) and prove that the result remains close to r_π if $I_B R_B \gg V_T$ (which is valid because V_{CC} and V_{BE} typically differ by about 0.5 V or higher).

5.52. Calculate v_{out}/v_{in} for each of the circuits depicted in Fig. 5.143. Assume $I_S = 8 \times 10^{-16}$ A, $\beta = 100$, and $V_A = \infty$. Also, assume the capacitors are very large.

5.53. Repeat Example 5-33 with $R_B = 25k\Omega$ and $R_C = 250\ \Omega$. Is the gain greater than unity?

Section 5.3.2 Common-Base Topology

SOL **5.54.** The common-base stage of Fig. 5.144 is biased with a collector current of 2 mA. Assume $V_A = \infty$.

(a) Calculate the voltage gain and I/O impedances of the circuit.

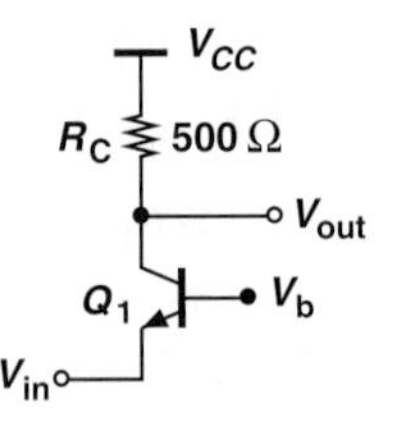

Figure 5.144

(b) How should V_B and R_C be chosen to maximize the voltage gain with a bias current of 2 mA?

5.55. Determine the voltage gain of the circuits shown in Fig. 5.145. Assume $V_A = \infty$.

*__5.56.__ Compute the input impedance of the stages depicted in Fig. 5.146. Assume $V_A = \infty$.

5.57. Calculate the voltage gain and I/O impedances of the CB stage shown in Fig. 5.147. Assume $V_A < \infty$.

5.58. Consider the CB stage depicted in Fig. 5.148, where $\beta = 100$, $I_S = 8 \times 10^{-16}$ A, $V_A = \infty$, and C_B is very large.

(a) Determine the operating point of Q_1.

(b) Calculate the voltage gain and I/O impedances of the circuit.

5.59. Repeat Problem 5.58 for $C_B = 0$.

5.60. Compute the voltage gain and I/O impedances of the stage shown in Fig. 5.149 if $V_A = \infty$ and C_B is very large. SOL

*__5.61.__ Calculate the voltage gain and the I/O impedances of the stage depicted in Fig. 5.150 if $V_A = \infty$ and C_B is very large.

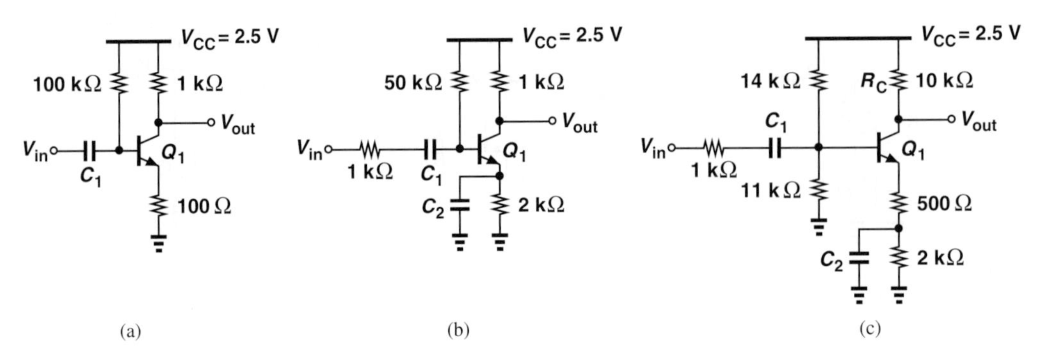

Figure 5.143

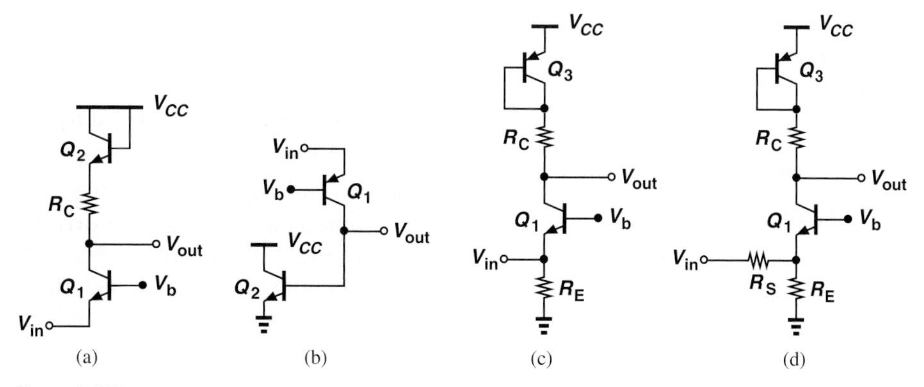

Figure 5.145

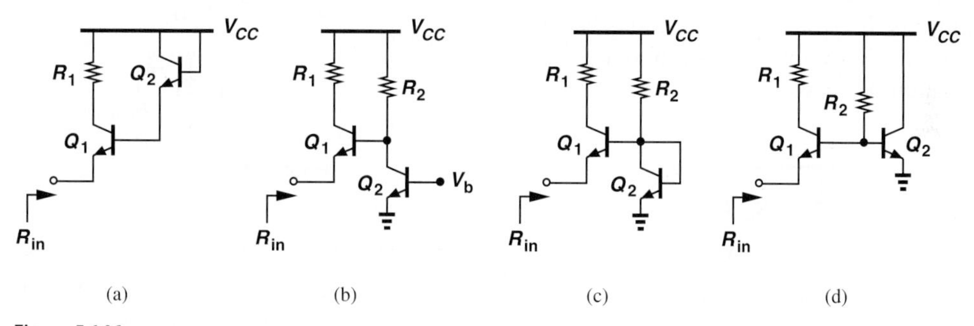

Figure 5.146

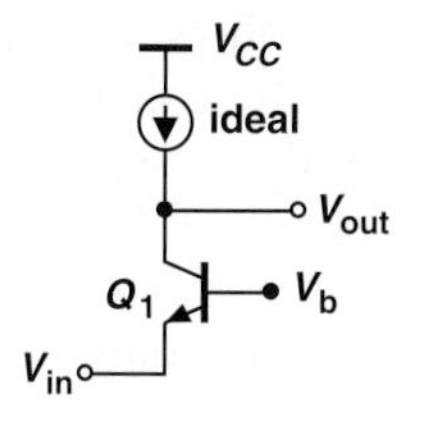

Figure 5.147

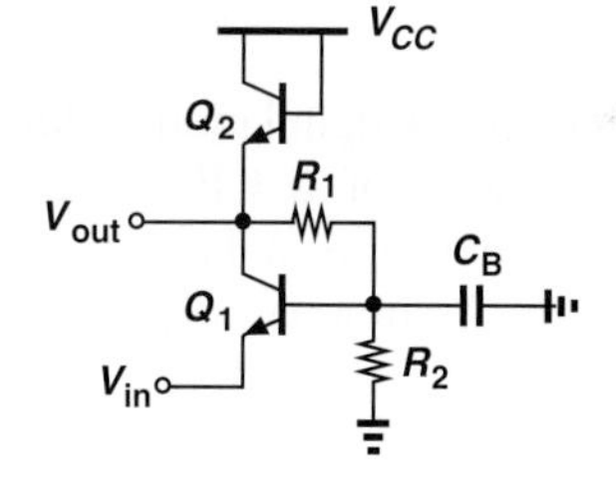

Figure 5.149

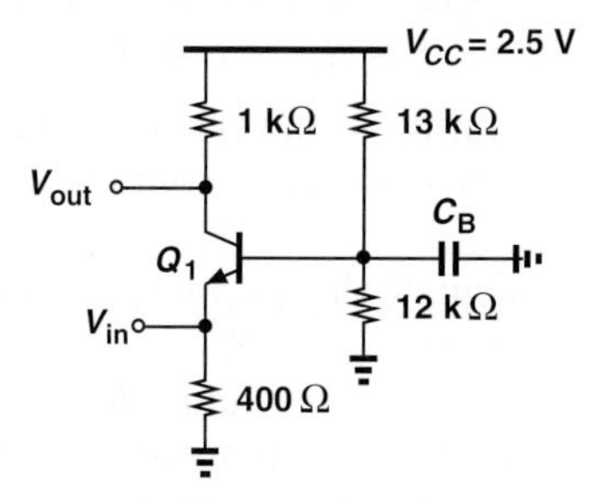

Figure 5.148

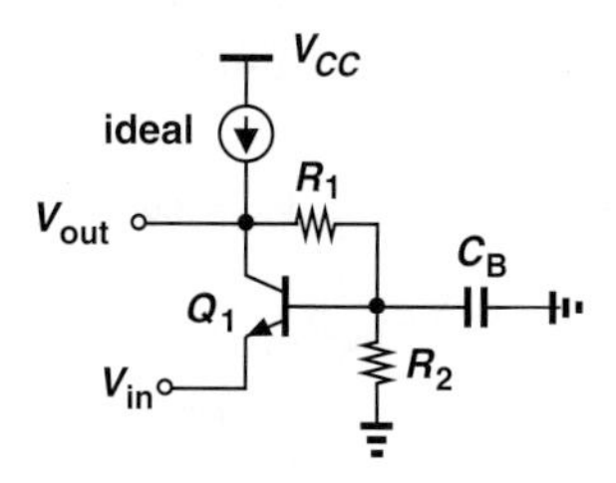

Figure 5.150

*5.62. Calculate the voltage gain of the circuit shown in Fig. 5.151 if $V_A < \infty$.

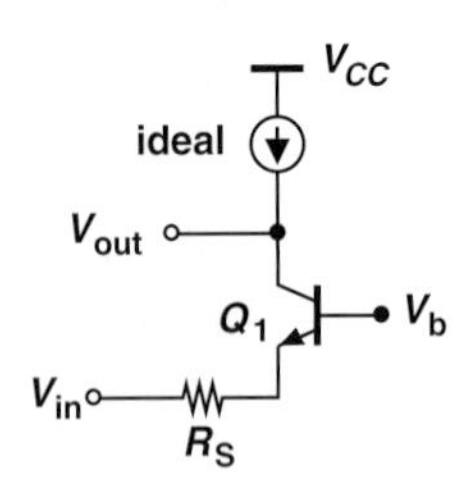

Figure 5.151

5.63. The circuit of Fig. 5.152 provides two outputs. If $I_{S1} = 2I_{S2}$, determine the relationship between v_{out1}/v_{in} and v_{out2}/v_{in}. Assume $V_A = \infty$.

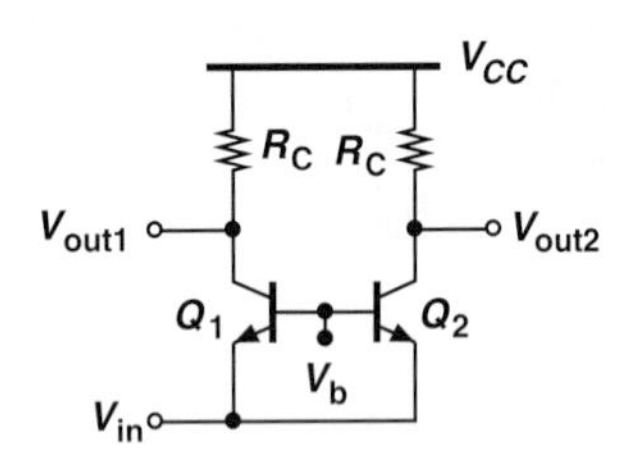

Figure 5.152

**5.64. Using a small-signal model, determine the voltage gain of a CB stage with emitter degeneration, a base resistance, and $V_A < \infty$. Assume $\beta \gg 1$.

Section 5.3.3 Emitter Follower

5.65. For $R_E = 100\ \Omega$ in Fig. 5.153, determine the bias current of Q_1 such that the gain is equal to 0.8. Assume $V_A = \infty$.

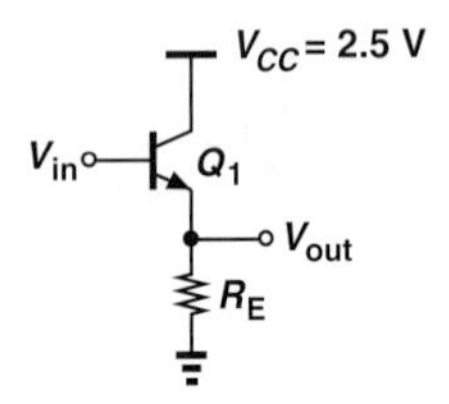

Figure 5.153

5.66. The circuit of Fig. 5.153 must provide an input impedance of greater than 10 kΩ with a minimum gain of 0.9. Calculate the required bias current and R_E. Assume $\beta = 100$ and $V_A = \infty$.

5.67. A microphone having an output impedance $R_S = 200\ \Omega$ drives an emitter follower as shown in Fig. 5.154. Determine the bias current such that the output impedance does not exceed 5 Ω. Assume $\beta = 100$ and $V_A = \infty$.

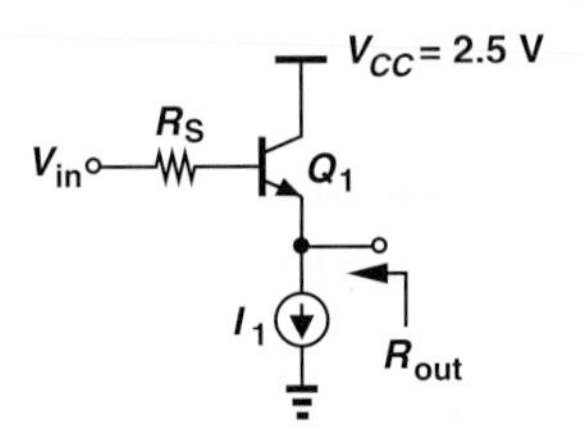

Figure 5.154

5.68. Compute the voltage gain and I/O impedances of the circuits shown in Fig. 5.155. Assume $V_A = \infty$.

*5.69. Figure 5.156 depicts a "Darlington pair," where Q_1 plays a role somewhat similar to an emitter follower driving Q_2. Assume $V_A = \infty$ and the collectors of Q_1 and Q_2 are tied to V_{CC}. Note that $I_{E1} (\approx I_{C1}) = I_{B2} = I_{C2}/\beta$.

(a) If the emitter of Q_2 is grounded, determine the impedance seen at the base of Q_1.

(b) If the base of Q_1 is grounded, calculate the impedance seen at the emitter of Q_2.

(c) Compute the current gain of the pair, defined as $(I_{C1} + I_{C2})/I_{B1}$.

*5.70. In the emitter follower shown in Fig. 5.157, Q_2 serves as a current source for the input device Q_1. SOL

(a) Calculate the output impedance of the current source, R_{CS}.

(b) Replace Q_2 and R_E with the impedance obtained in (a) and compute the voltage gain and I/O impedances of the circuit.

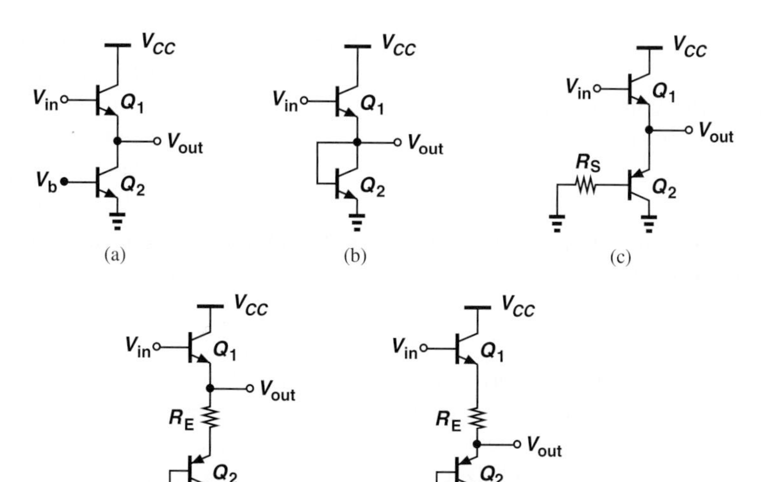

Figure 5.155

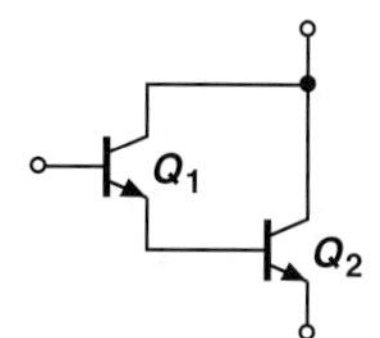

Figure 5.156

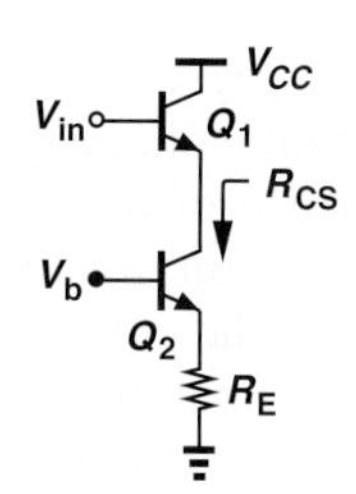

Figure 5.157

5.71. Determine the voltage gain of the follower depicted in Fig. 5.158. Assume $I_S = 7 \times 10^{-16}$ A, $\beta = 100$, and $V_A = 5$ V. (But for bias calculations, assume $V_A = \infty$.) Also, assume the capacitors are very large.

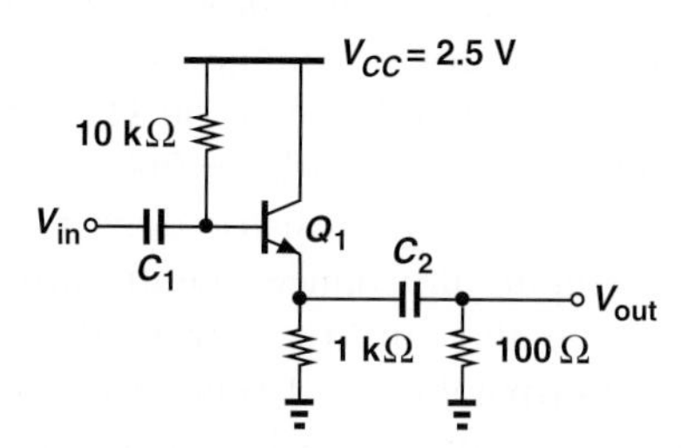

Figure 5.158

5.72. Figure 5.159 illustrates a cascade of an emitter follower and a common-emitter stage. Assume $V_A < \infty$.

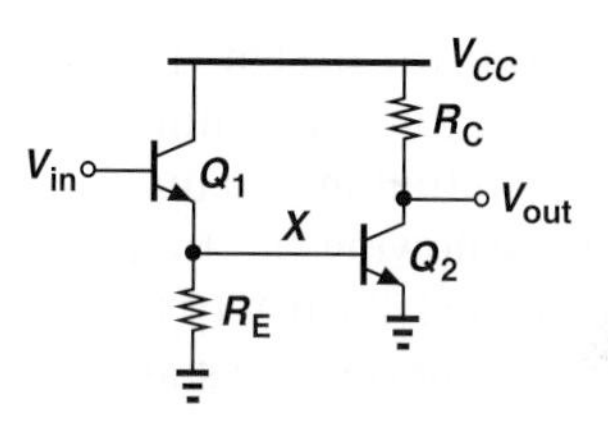

Figure 5.159

(a) Calculate the input and output impedances of the circuit.

(b) Determine the voltage gain, $v_{out}/v_{in} = (v_X/v_{in})(v_{out}/v_X)$.

***5.73.** Figure 5.160 shows a cascade of an emitter follower and a common-base stage. Assume $V_A = \infty$.

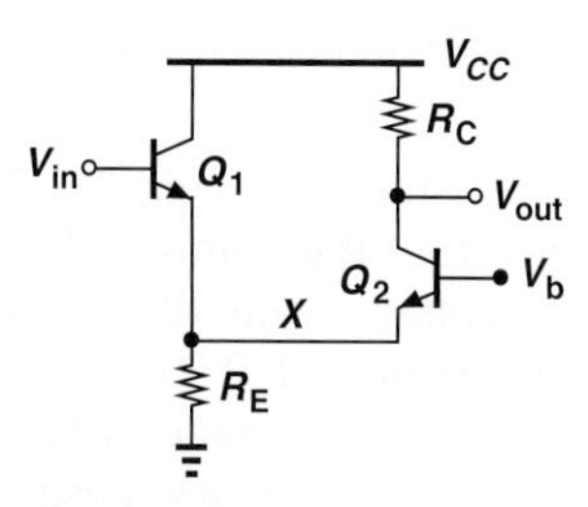

Figure 5.160

(a) Calculate the I/O impedances of the circuit.

(b) Calculate the voltage gain, $v_{out}/v_{in} = (v_X/v_{in})(v_{out}/v_X)$.

Design Problems

In the following problems, unless otherwise stated, assume $\beta = 100$, $I_S = 6 \times 10^{-16}$A, and $V_A = \infty$.

5.74. Design the CE stage shown in Fig. 5.161 for a voltage gain of 10, and input impedance of greater than 5 kΩ, and an output impedance of 1 kΩ. If the lowest signal frequency of interest is 200 Hz, estimate the minimum allowable value of C_B.

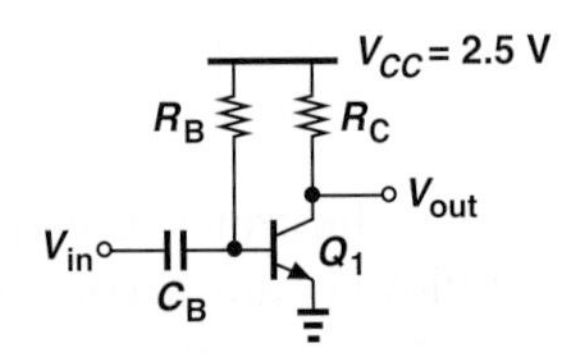

Figure 5.161

5.75. We wish to design the CE stage of Fig. 5.162 for maximum voltage gain but with an output impedance no greater than 500 Ω. Allowing the transistor to experience at most 400 mV of base-collector forward bias, design the stage.

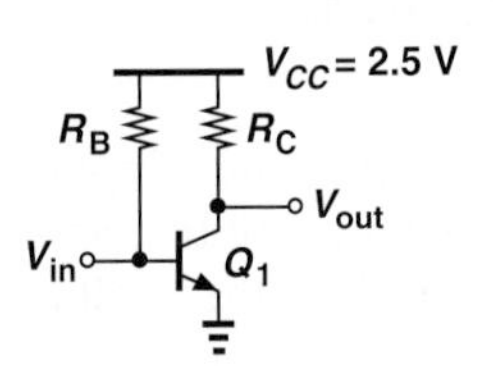

Figure 5.162

5.76. The stage depicted in Fig. 5.162 must achieve maximum input impedance but with a voltage gain of at least 20 and an output impedance of 1 kΩ. Design the stage.

5.77. The CE stage of Fig. 5.162 must be designed for minimum supply voltage but with a voltage gain of 15 and an output impedance of 2 kΩ. If the transistor is allowed to sustain a base-collector forward bias of 400 mV, design the stage and calculate the required supply voltage.

5.78. We wish to design the CE stage of Fig. 5.162 for minimum power dissipation. If the voltage gain must be equal to A_0, determine the trade-off between the power dissipation and the output impedance of the circuit.

5.79. Design the CE stage of Fig. 5.162 for a power budget of 1 mW and a voltage gain of 20. SOL

5.80. Design the degenerated CE stage of Fig. 5.163 for a voltage gain of 5 and an output impedance of 500 Ω. Assume R_E sustains a voltage drop of 300 mV and the current flowing through R_1 is approximately 10 times the base current.

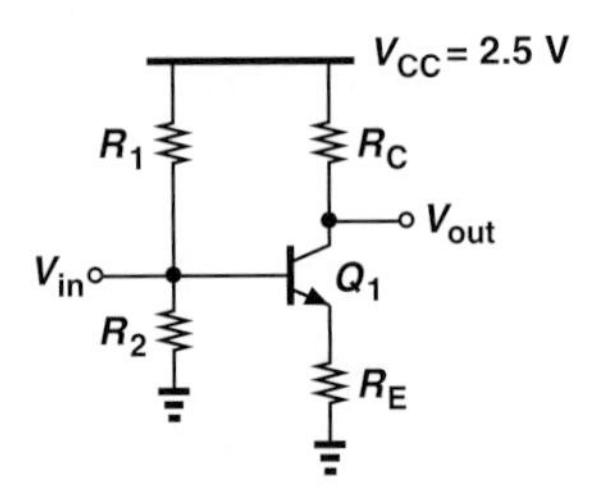

Figure 5.163

5.81. The stage of Fig. 5.163 must be designed for maximum voltage gain but an output impedance of no greater than 1 kΩ. Design the circuit, assuming that R_E sustains 200 mV, the current flowing through R_1 is approximately 10 times the base current, and Q_1 experiences a maximum base-collector forward bias of 400 mV.

5.82. Design the stage of Fig. 5.163 for a power budget of 5 mW, a voltage gain of 5, and a voltage drop of 200 mV across R_E. Assume the current flowing through R_1 is approximately 10 times the base current.

5.83. Design the common-base stage shown in Fig. 5.164 for a voltage gain of 20 and an input impedance of 50 Ω. Assume a voltage drop of $10V_T = 260$ mV across R_E so that this resistor does not affect the input impedance significantly. Also, assume the current flowing through R_1 is approximately 10 times the base current, and the lowest frequency of interest is 200 Hz.

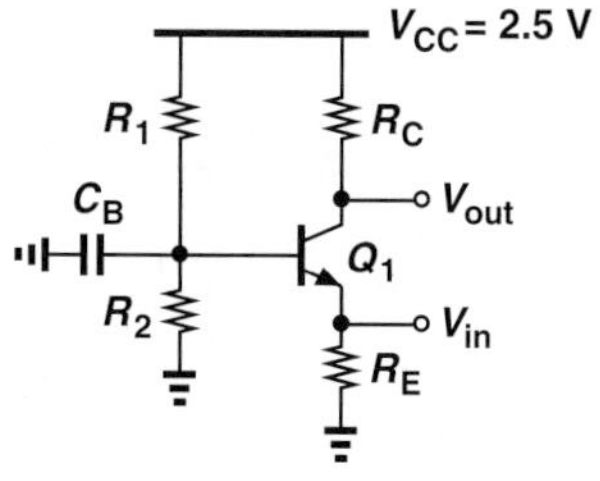

Figure 5.164

5.84. The CB amplifier of Fig. 5.164 must achieve a voltage gain of 8 with an output impedance of 500 Ω. Design the circuit with the same assumptions as those in Problem 5.83.

5.85. We wish to design the CB stage of Fig. 5.164 for an output impedance of 200 Ω and a voltage gain of 20. What is the minimum required power dissipation? Make the same assumptions as those in Problem 5.83.

5.86. Design the CB amplifier of Fig. 5.164 for a power budget of 5 mW and a voltage gain of 10. Make the same assumptions as those in Problem 5.83.

5.87. Design the CB stage of Fig. 5.164 for the minimum supply voltage if an input impedance of 50 Ω and a voltage gain of 20 are required. Make the same assumptions as those in Problem 5.83.

5.88. Design the emitter follower shown in Fig. 5.165 for a voltage gain of 0.85 and an input impedance of greater than 10 kΩ. Assume $R_L = 200\ \Omega$.

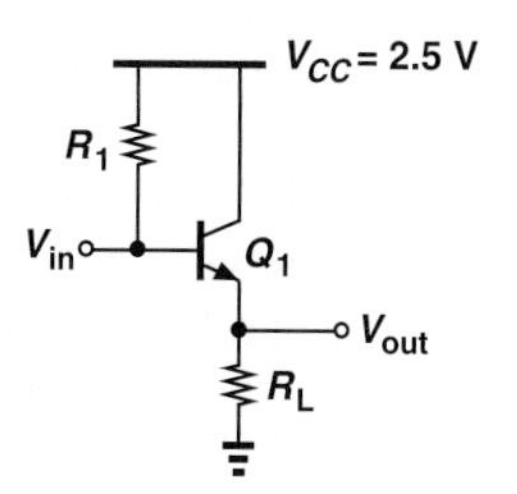

Figure 5.165

5.89. The follower of Fig. 5.165 must consume 5 mW of power while achieving a voltage gain of 0.9. What is the minimum load resistance, R_L, that it can drive?

5.90. The follower shown in Fig. 5.166 must drive a load resistance, $R_L = 50\ \Omega$, with a voltage gain of 0.8. Design the circuit assuming that the lowest frequency of interest is 100 MHz. (Hint: Select the voltage drop across R_E to be much greater than V_T so that this resistor does not affect the voltage gain significantly.) SOL

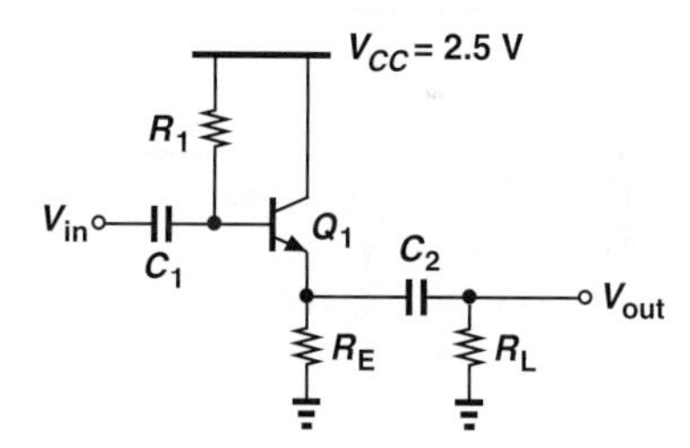

Figure 5.166

SPICE PROBLEMS

In the following problems, assume $I_{S,npn} = 5 \times 10^{-16}$ A, $\beta_{npn} = 100$, $V_{A,npn} = 5$ V, $I_{S,pnp} = 8 \times 10^{-16}$ A, $\beta_{pnp} = 50$, $V_{A,pnp} = 3.5$ V.

5.91. The common-emitter shown in Fig. 5.167 must amplify signals in the range of 1 MHz to 100 MHz.

(a) Using the .op command, determine the bias conditions of Q_1 and verify that it operates in the active region.

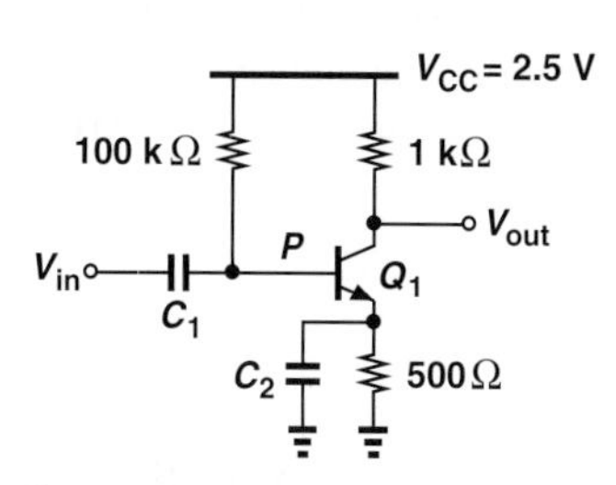

Figure 5.167

(b) Running an ac analysis, choose the value of C_1 such that $|V_P/V_{in}| \approx 0.99$ at 1 MHz. This ensures that C_1 acts as a short circuit at all frequencies of interest.

(c) Plot $|V_{out}/V_{in}|$ as a function of frequency for several values of C_2, e.g., 1 μF, 1 nF, and 1 pF. Determine the value of C_2 such that the gain of the circuit at 10 MHz is only 2% below its maximum (i.e., for $C_2 = 1$ μF).

(d) With the proper value of C_2 found in (c), determine the input impedance of the circuit at 10 MHz. (One approach is to insert a resistor in series with V_{in} and adjust its value until V_P/V_{in} or V_{out}/V_{in} drops by a factor of two.)

5.92. Predicting an output impedance of about 1 kΩ for the stage shown in Fig. 5.167, a student constructs the circuit depicted in Fig. 5.168, where V_X represents an ac source with zero dc value. Unfortunately, V_N/V_X is far from 0.5. Explain why.

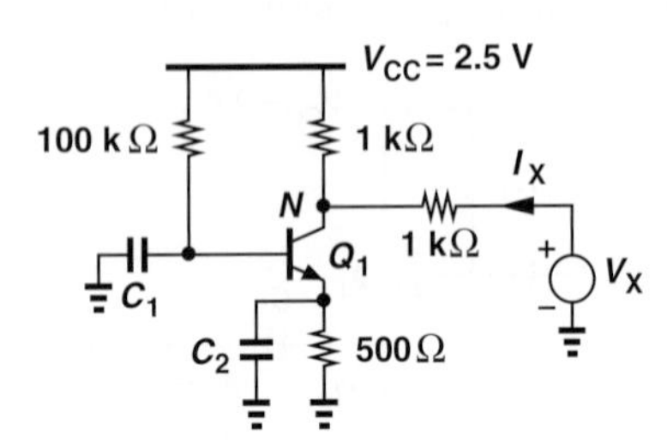

Figure 5.168

5.93. Consider the self-biased stage shown in Fig. 5.169.

(a) Determine the bias conditions of Q_1.

(b) Select the value of C_1 such that it operates as nearly a short circuit (e.g., $|V_P/V_{in}| \approx 0.99$) at 10 MHz.

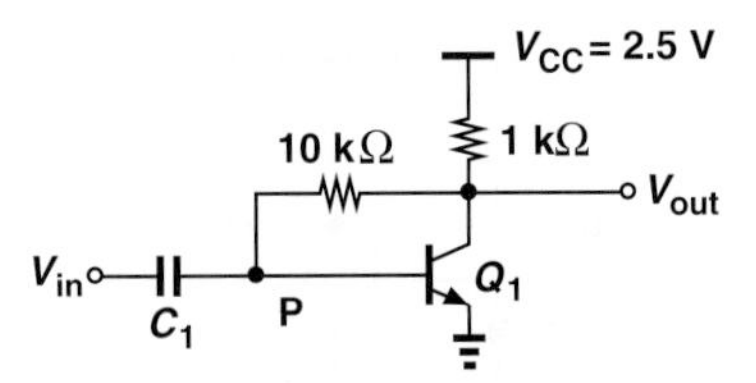

Figure 5.169

(c) Compute the voltage gain of the circuit at 10 MHz.

(d) Determine the input impedance of the circuit at 10 MHz.

(e) Suppose the supply voltage is provided by an aging battery. How much can V_{CC} fall while the gain of the circuit degrades by only 5%?

5.94. Repeat Problem 5.93 for the stage illustrated in Fig. 5.170. Which one of the two circuits is less sensitive to supply variations?

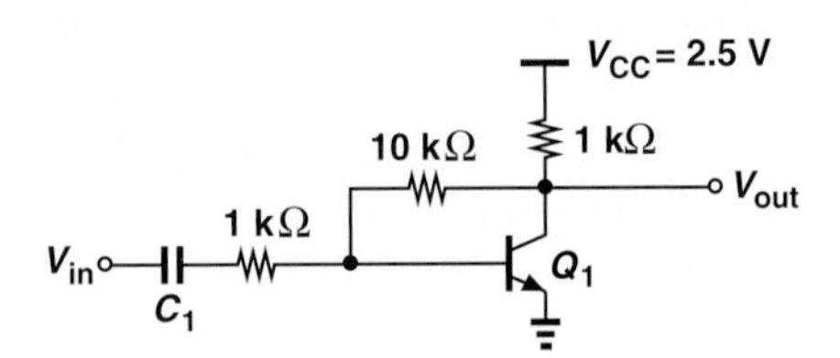

Figure 5.170

5.95. The amplifier shown in Fig. 5.171 employs an emitter follower to drive a 50-Ω load at a frequency of 100 MHz.

(a) Determine the value of R_{E1} such that Q_2 carries a bias current of 2 mA.

(b) Determine the minimum acceptable value of C_1, C_2, and C_3 if each one is to degrade the gain by less than 1%.

(c) What is the signal attenuation of the emitter follower? Does the overall gain increase if R_{G2} is reduced to 100 Ω? Why?

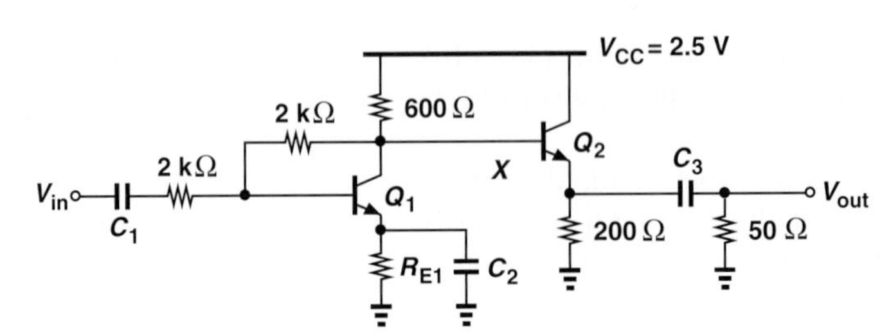

Figure 5.171

6

Physics of MOS Transistors

Watch Companion YouTube Videos:

Razavi Electronics 1, Lec 29: Introduction to MOSFETs
https://www.youtube.com/watch?v=dlOlxAcfBo4

Razavi Electronics 1, Lec 30: MOS Characteristics I
https://www.youtube.com/watch?v=0OH9d72ZX7s

Razavi Electronics 1, Lec 31: MOS Characteristics II
https://www.youtube.com/watch?v=GmhFnm6mdxY

Razavi Electronics 1, Lec 32: Biasing, Transconductance
https://www.youtube.com/watch?v=jfizMWm3Q0I

Razavi Electronics 1, Lec 33: Large-Signal & Small-Signal Operation
https://www.youtube.com/watch?v=LvE0curDf_Q

Razavi Electronics 1, Lec 34: MOS Small-Signal Model, PMOS Device
https://www.youtube.com/watch?v=HviPXCkrdXg

Razavi Electronics 1, Lec 35: CMOS Technology, Introduction to Amplifiers
https://www.youtube.com/watch?v=UTo4IGXxzlk

Today's field of microelectronics is dominated by a type of device called the metal-oxide-semiconductor field-effect transistor (MOSFET). Conceived in the 1930s but first realized in the 1960s, MOSFETs (also called MOS devices) offer unique properties that have led to the revolution of the semiconductor industry. This revolution has culminated in microprocessors having 100 million transistors, memory chips containing billions of transistors, and sophisticated communication circuits providing tremendous signal processing capability.

Our treatment of MOS devices and circuits follows the same procedure as that taken in Chapters 2 and 3 for *pn* junctions. In this chapter, we analyze the structure and operation of MOSFETs, seeking models that prove useful in circuit design. In Chapter 7, we utilize the models to study MOS amplifier topologies. The outline below illustrates the sequence of concepts covered in this chapter.

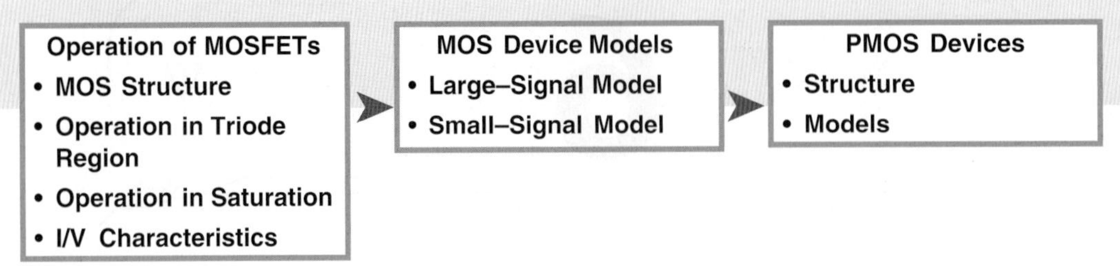

6.1 STRUCTURE OF MOSFET

Recall from Chapter 5 that any voltage-controlled current source can provide signal amplification. MOSFETs also behave as such controlled sources but their characteristics are different from those of bipolar transistors.

In order to arrive at the structure of the MOSFET, we begin with a simple geometry consisting of a conductive (e.g., metal) plate, an insulator ("dielectric"), and a doped piece of silicon. Illustrated in Fig. 6.1(a), such a structure operates as a capacitor because the p-type silicon is somewhat conductive, "mirroring" any charge deposited on the top plate.

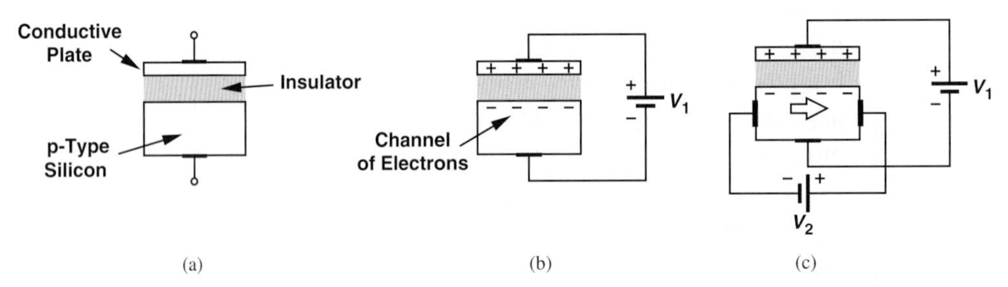

Figure 6.1 (a) Hypothetical semiconductor device, (b) operation as a capacitor, (c) current flow as a result of potential difference.

What happens if a potential difference is applied as shown in Fig. 6.1(b)? As positive charge is placed on the top plate, it attracts negative charge, e.g., electrons, from the piece of silicon. (Even though doped with acceptors, the p-type silicon does contain a small number of electrons.) We therefore observe that a "channel" of *free* electrons may be created at the interface between the insulator and the piece of silicon, potentially serving as a good conductive path if the electron density is sufficiently high. The key point here is that the density of electrons in the channel *varies* with V_1, as evident from $Q = CV$, where C denotes the capacitance between the two plates.

The dependence of the electron density upon V_1 leads to an interesting property: if, as depicted in Fig. 6.1(c), we allow a current to flow from left to right through the silicon material, V_1 can *control* the current by adjusting the resistivity of the channel. (Note that the current prefers to take the path of least resistance, thus flowing primarily through the channel rather than through the entire body of silicon.) This will serve our objective of building a voltage-controlled current source.

Equation $Q = CV$ suggests that, to achieve a strong control of Q by V, the value of C must be maximized, for example, by *reducing* the thickness of the dielectric layer separating the two plates.[1] The ability of silicon fabrication technology to produce

[1]The capacitance between two plates is given by $\epsilon A/t$, where ϵ is the "dielectric constant" (also called the "permitivity"), A is the area of each plate, and t is the dielectric thickness.

extremely thin but uniform dielectric layers (with thicknesses below 20 Å today) has proven essential to the rapid advancement of microelectronic devices.

The foregoing thoughts lead to the MOSFET structure shown in Fig. 6.2(a) as a candidate for an amplifying device. Called the "gate" (G), the top conductive plate resides on a thin dielectric (insulator) layer, which itself is deposited on the underlying *p*-type silicon "substrate." To allow current flow through the silicon material, two contacts are attached to the substrate through two heavily-doped *n*-type regions because direct connection of metal to the substrate would not produce a good "ohmic" contact.[2] These two terminals are called "source" (S) and "drain" (D) to indicate that the former can *provide* charge carriers and the latter can *absorb* them. Figure 6.2(a) reveals that the device is symmetric with respect to S and D; i.e., depending on the voltages applied to the device, either of these two terminals can drain the charge carriers from the other. As explained in Section 6.2, with *n*-type source/drain and *p*-type substrate, this transistor operates with electrons rather than holes and is therefore called an *n*-type MOS (NMOS) device. (The *p*-type counterpart is studied in Section 6.4.) We draw the device as shown in Fig. 6.2(b) for simplicity. Figure 6.2(c) depicts the circuit symbol for an NMOS transistor, wherein the arrow signifies the source terminal.

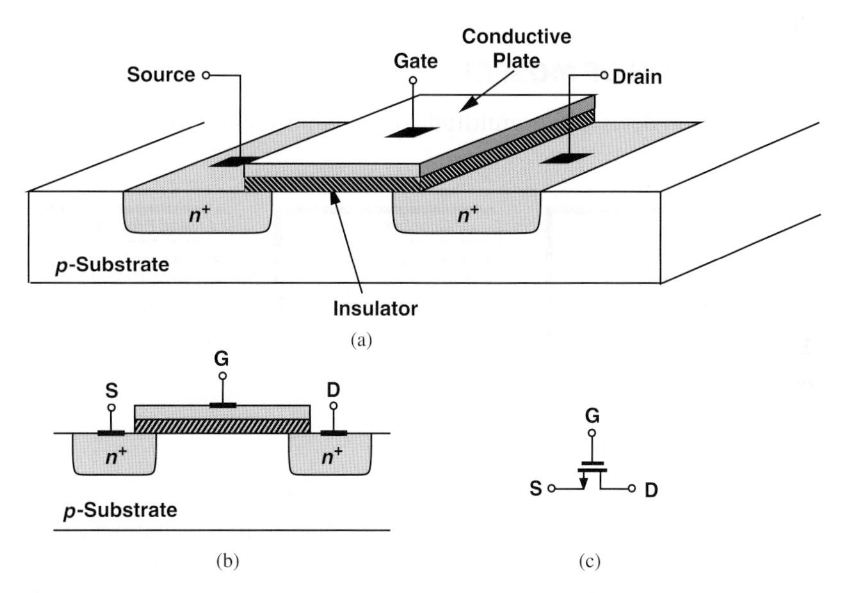

Figure 6.2 (a) Structure of MOSFET, (b) side view, (c) circuit symbol.

Before delving into the operation of the MOSFET, let us consider the types of materials used in the device. The gate plate must serve as a good conductor and was in fact realized by metal (aluminum) in the early generations of MOS technology. However, it was discovered that noncrystalline silicon ("polysilicon" or simply "poly") with heavy doping (for low resistivity) exhibits better fabrication and physical properties. Thus, today's MOSFETs employ polysilicon gates.

The dielectric layer sandwiched between the gate and the substrate plays a critical role in the performance of transistors and is created by growing silicon dioxide (or simply "oxide") on top of the silicon area. The n^+ regions are sometimes called source/drain "diffusion," referring to a fabrication method used in early days of microelectronics. We should

[2]Used to distinguish it from other types of contacts such as diodes, the term "ohmic" contact emphasizes bi-directional current flow—as in a resistor.

also remark that these regions in fact form *diodes* with the p-type substrate (Fig. 6.3). As explained later, proper operation of the transistor requires that these junctions remain reverse-biased. Thus, only the depletion region capacitance associated with the two diodes must be taken into account. Figure 6.3 shows some of the device dimensions in today's state-of-the-art MOS technologies. The oxide thickness is denoted by t_{ox}.

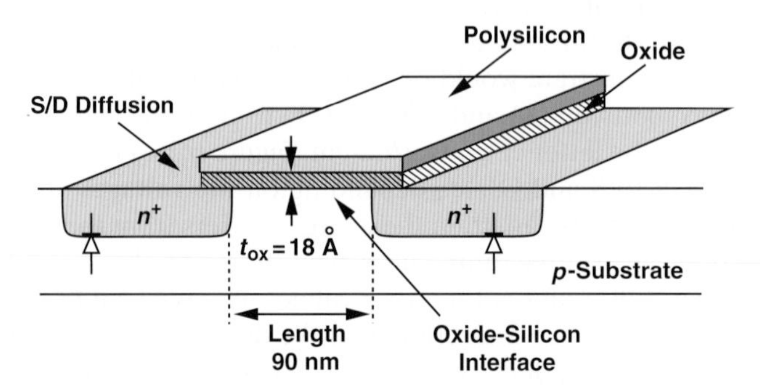

Figure 6.3 Typical dimensions of today's MOSFETs.

6.2 OPERATION OF MOSFET

This section deals with a multitude of concepts related to MOSFETs. The outline is shown in Fig. 6.4.

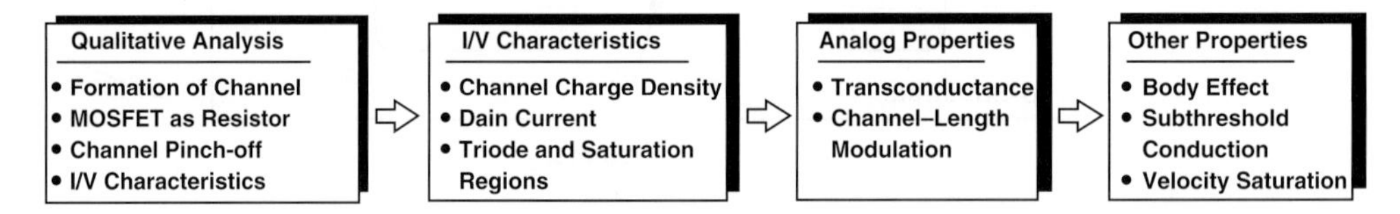

Figure 6.4 Outline of concepts to be studied.

6.2.1 Qualitative Analysis

Our study of the simple structures shown in Figs. 6.1 and 6.2 suggests that the MOSFET may conduct current between the source and drain if a channel of electrons is created by making the gate voltage sufficiently positive. Moreover, we expect that the magnitude of the current can be controlled by the gate voltage. Our analysis will indeed confirm these conjectures while revealing other subtle effects in the device. Note that the gate terminal draws no (low-frequency) current as it is insulated from the channel by the oxide.

Since the MOSFET contains three terminals,[3] we may face many combinations of terminal voltages and currents. Fortunately, with the (low-frequency) gate current being zero, the only current

Did you know?

The concept of MOSFET was proposed by Julins Edgar Lilienfeld in 1925, decades before the invention of the bipolar transistor. But why did it take until the 1960s for the MOS transistor to be successfully fabricated? The critical issue was the oxide-silicon interface. In initial attempts, this interface contained many "surface states," trapping the charge carriers and leading to poor conduction. As semiconductor technology advanced and "clean rooms" were invented for fabrication, the oxide could be grown on silicon with almost no surface states, yielding a high transconductance.

[3]The substrate acts as a fourth terminal, but we ignore that for now.

of interest is that flowing between the source and the drain. We must study the dependence of this current upon the gate voltage (e.g., for a constant drain voltage) and upon the drain voltage (e.g., for a constant gate voltage). These concepts become clearer below.

Let us first consider the arrangement shown in Fig. 6.5(a), where the source and drain are grounded and the gate voltage is varied. This circuit does not appear particularly useful but it gives us a great deal of insight. Recall from Fig. 6.1(b) that, as V_G rises, the positive charge on the gate must be mirrored by negative charge in the substrate. While we stated in Section 6.1 that electrons are attracted to the interface, in reality, another phenomenon precedes the formation of the channel. As V_G increases from zero, the positive charge on the gate *repels* the holes in the substrate, thereby exposing negative ions and creating a depletion region [Fig. 6.5(b)].[4] Note that the device still acts as a capacitor—positive charge on the gate is mirrored by negative charge in the substrate—but no channel of *mobile* charge is created yet. Thus, no current can flow from the source to the drain. We say the MOSFET is off.

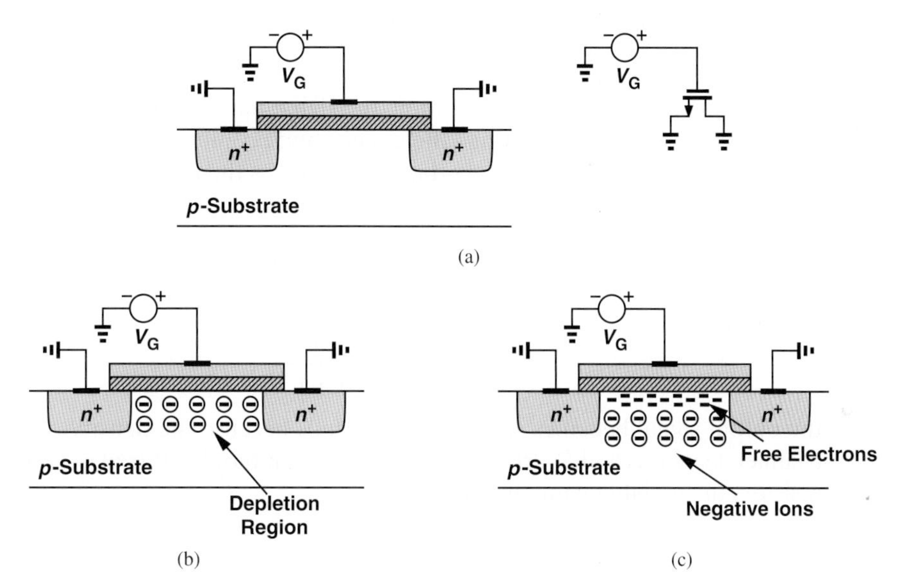

Figure 6.5 (a) MOSFET with gate voltage, (b) formation of depletion region, (c) formation of channel.

Can the source-substrate and drain-substrate junctions carry current in this mode? To avoid this effect, the substrate itself is also tied to zero, ensuring that these diodes are not forward-biased. For simplicity, we do not show this connection in the diagrams.

What happens as V_G increases? To mirror the charge on the gate, more negative ions are exposed and the depletion region under the oxide becomes deeper. Does this mean the transistor never turns on?! Fortunately, if V_G becomes sufficiently positive, free electrons are attracted to the oxide-silicon interface, forming a conductive channel [Fig. 6.5(c)]. We say the MOSFET is on. The gate potential at which the channel begins to appear is called the "threshold voltage," V_{TH}, and falls in the range of 300 mV to 500 mV. Note that the electrons are readily provided by the n^+ source and drain regions, and need not be supplied by the substrate.

[4]Note that this depletion region contains only one immobile charge polarity, whereas the depletion region in a *pn* junction consists of two areas of negative and positive ions on the two sides of the junction.

It is interesting to recognize that the gate terminal of the MOSFET draws no (low-frequency) current. Resting on top of the oxide, the gate remains insulated from other terminals and simply operates as a plate of a capacitor.

MOSFET as a Variable Resistor The conductive channel between S and D can be viewed as a resistor. Furthermore, since the density of electrons in the channel must increase as V_G becomes more positive (why?), the value of this resistor *changes* with the gate voltage. Conceptually illustrated in Fig. 6.6, such a voltage-dependent resistor proves extremely useful in analog and digital circuits.

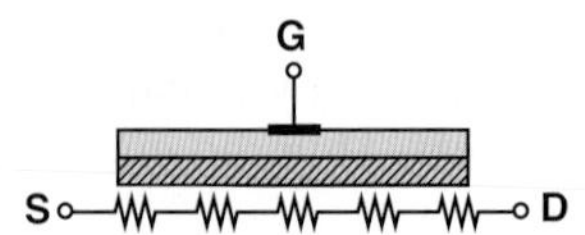

Figure 6.6 MOSFET viewed as a voltage-dependent resistor.

Example 6-1 In the vicinity of a wireless base station, the signal received by a cellphone may become very strong, possibly "saturating" the circuits and prohibiting proper operation. Devise a variable-gain circuit that lowers the signal level as the cellphone approaches the base station.

Solution A MOSFET can form a voltage-controlled attenuator along with a resistor as shown in Fig. 6.7.

Since

$$\frac{v_{out}}{v_{in}} = \frac{R_1}{R_M + R_1}, \tag{6.1}$$

the output signal becomes smaller as V_{cont} falls because the density of electrons in the channel decreases and R_M rises. MOSFETs are commonly utilized as voltage-dependent resistors in "variable-gain amplifiers."

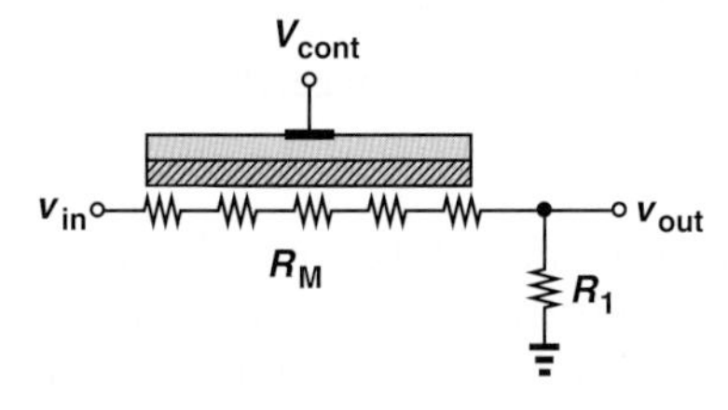

Figure 6.7 Use of MOSFET to adjust signal levels.

Exercise What happens to R_M if the channel length is doubled?

In the arrangement of Fig. 6.5(c), no current flows between S and D because the two terminals are at the same potential. We now raise the drain voltage as shown in Fig. 6.8(a) and examine the drain current (= source current). If $V_G < V_{TH}$, no channel exists, the device is off, and $I_D = 0$ regardless of the value of V_D. On the other hand, if $V_G > V_{TH}$, then $I_D > 0$ [Fig. 6.8(b)]. In fact, the source-drain path may act as a simple resistor, yielding

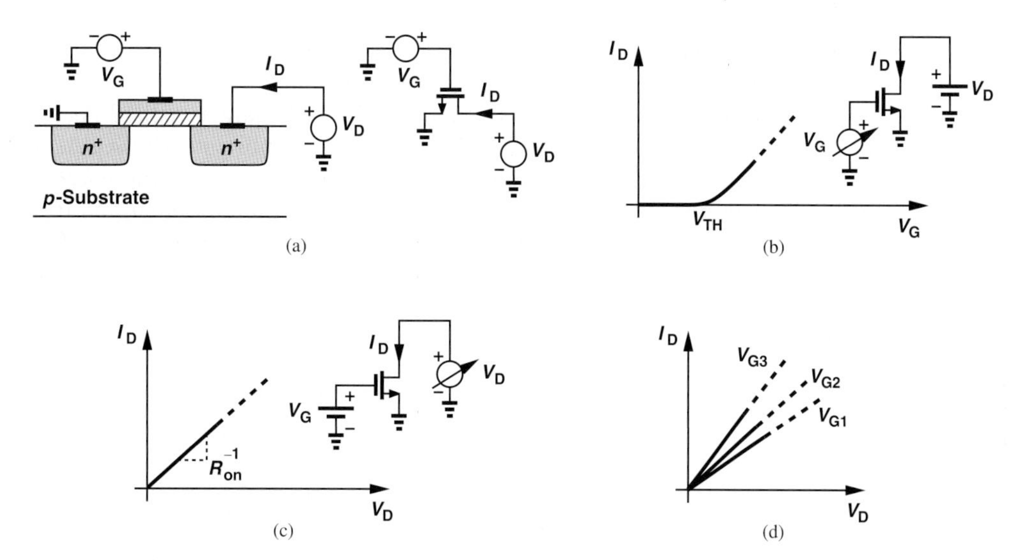

Figure 6.8 (a) MOSFET with gate and drain voltages, (b) I_D-V_G characteristic, (c) I_D-V_D characteristic, (d) I_D-V_D characteristics for various gate voltages.

the I_D-V_D characteristic shown in Fig. 6.8(c). The slope of the characteristic is equal to $1/R_{on}$, where R_{on} denotes the "on-resistance" of the transistor.[5]

Our brief treatment of the MOS I-V characteristics thus far points to two different views of the operation: in Fig. 6.8(b), V_G is varied while V_D remains constant whereas in Fig. 6.8(c), V_D is varied while V_G remains constant. Each view provides valuable insight into the operation of the transistor.

How does the characteristic of Fig. 6.8(c) change if V_G increases? The higher density of electrons in the channel lowers the on-resistance, yielding a *greater* slope. Depicted in Fig. 6.8(d), the resulting characteristics strengthen the notion of voltage-dependent resistance.

Recall from Chapter 2 that charge flow in semiconductors occurs by diffusion or drift. How about the transport mechanism in a MOSFET? Since the voltage source tied to the drain creates an electric field along the channel, the current results from the *drift* of charge.

The I_D-V_G and I_D-V_D characteristics shown in Figs. 6.8(b) and (c), respectively, play a central role in our understanding of MOS devices. The following example reinforces the concepts studied thus far.

Example 6-2 Sketch the I_D-V_G and I_D-V_D characteristics for (a) different channel lengths, and (b) different oxide thicknesses.

Solution As the channel length increases, so does the on-resistance.[6] Thus, for $V_G > V_{TH}$, the drain current begins with lesser values as the channel length increases [Fig. 6.9(a)]. Similarly, I_D exhibits a smaller slope as a function of V_D [Fig. 6.9(b)]. It is therefore

[5]The term "on-resistance" always refers to that between the source and drain, as no resistance exists between the gate and other terminals.

[6]Recall that the resistance of a conductor is proportional to the length.

desirable to *minimize* the channel length so as to achieve large drain currents—an important trend in the MOS technology development.

How does the oxide thickness, t_{ox}, affect the I-V characteristics? As t_{ox} increases, the capacitance between the gate and the silicon substrate *decreases*. Thus, from $Q = CV$, we note that a given voltage results in *less* charge on the gate and hence a lower electron density in the channel. Consequently, the device suffers from a *higher* on-resistance, producing less drain current for a given gate voltage [Fig. 6.9(c)] or drain voltage [Fig. 6.9(d)]. For this reason, the semiconductor industry has continued to reduce the gate oxide thickness.

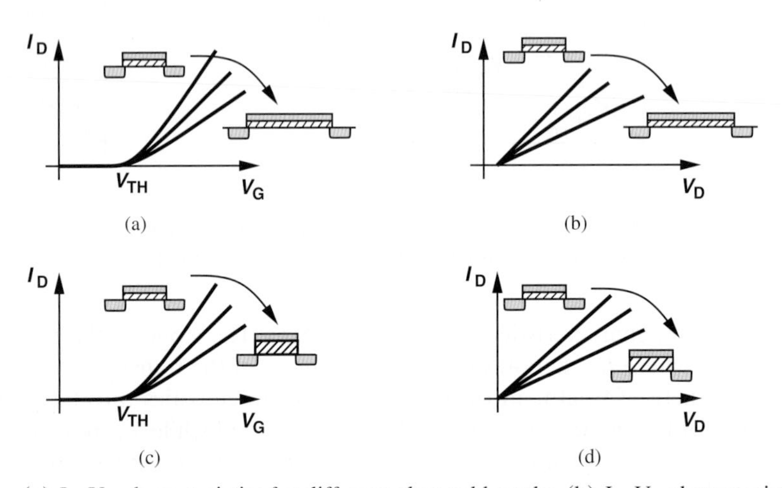

Figure 6.9 (a) I_D-V_G characteristics for different channel lengths, (b) I_D-V_D characteristics for different channel lengths, (c) I_D-V_G characteristics for different oxide thicknesses, (d) I_D-V_D characteristics for different oxide thicknesses.

Exercise The current conduction in the channel is in the form of drift. If the mobility falls at high temperatures, what can we say about the on-resistance as the temperature goes up?

While both the length and the oxide thickness affect the performance of MOSFETs, only the former is under the circuit designer's control, i.e., it can be specified in the "layout" of the transistor. The latter, on the other hand, is defined during fabrication and remains constant for all transistors in a given generation of the technology.

Another MOS parameter controlled by circuit designers is the *width* of the transistor, the dimension perpendicular to the length [Fig. 6.10(a)]. We therefore observe that "lateral" dimensions such as L and W can be chosen by circuit designers whereas "vertical" dimensions such as t_{ox} cannot.

How does the gate width impact the I-V characteristics? As W increases, so does the width of the channel, thus *lowering* the resistance between the source and the drain[7] and yielding the trends depicted in Fig. 6.10(b). From another perspective, a wider device can be viewed as two narrower transistors *in parallel*, producing a high drain current [Fig. 6.10(c)]. We may then surmise that W must be maximized, but we must also note

[7]Recall that the resistance of a conductor is inversely proportional to the cross section area, which itself is equal to the product of the width and thickness of the conductor.

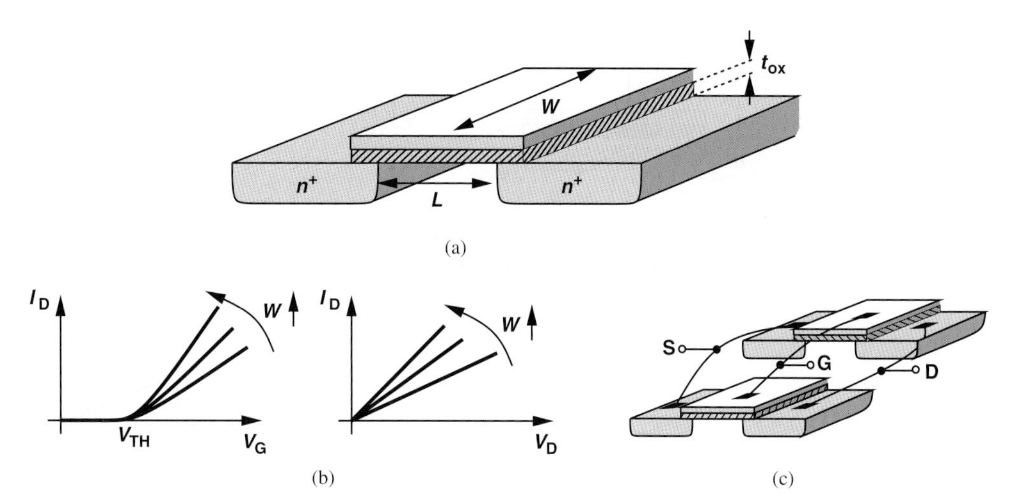

Figure 6.10 (a) Dimensions of a MOSFET (W and L are under circuit designer's control), (b) I_D characteristics for different values of W, (c) equivalence to devices in parallel.

that the total gate capacitance increases with W, possibly limiting the speed of the circuit. Thus, the width of each device in the circuit must be chosen carefully.

Channel Pinch-Off Our qualitative study of the MOSFET thus far implies that the device acts as a voltage-dependent resistor if the gate voltage exceeds V_{TH}. In reality, however, the transistor operates as a *current source* if the drain voltage is sufficiently positive. To understand this effect, we make two observations: (1) to form a channel, the potential difference between the gate and the oxide-silicon interface must exceed V_{TH}; (2) if the drain voltage remains higher than the source voltage, then the voltage at each point along the channel with respect to ground increases as we go from the source towards the drain. Illustrated in Fig. 6.11(a), this effect arises from the gradual voltage drop along the channel resistance. Since the gate voltage is constant (because the gate is conductive but carries no current in any direction), and since the potential at the oxide-silicon interface rises from the source to the drain, the potential difference *between* the gate and the oxide-silicon interface *decreases* along the x-axis [Fig. 6.11(b)]. The density of electrons in the channel follows the same trend, falling to a minimum at $x = L$.

From these observations, we conclude that, if the drain voltage is high enough to produce $V_G - V_D \leq V_{TH}$, then the channel ceases to exist near the drain. We say the gate-substrate potential difference is not sufficient at $x = L$ to attract electrons and the channel is "pinched off" [Fig. 6.12(a)].

What happens if V_D rises even higher than $V_G - V_{TH}$? Since $V(x)$ now goes from 0 at $x = 0$ to $V_D > V_G - V_{TH}$ at $x = L$, the voltage difference between the gate and the substrate falls to V_{TH} at some point $L_1 < L$ [Fig. 6.12(b)]. The device therefore contains no channel between L_1 and L. Does this mean the transistor cannot conduct current? No, the device still conducts: as illustrated in Fig. 6.12(c), once the electrons reach the end of the channel, they experience the high electric field in the depletion region surrounding the drain junction and are rapidly swept to the drain terminal. Nonetheless, as shown in the next section, the drain voltage no longer affects the current significantly, and the MOSFET acts as a constant current source—similar to a bipolar transistor in the forward active region. Note that the source-substrate and drain-substrate junctions carry no current.

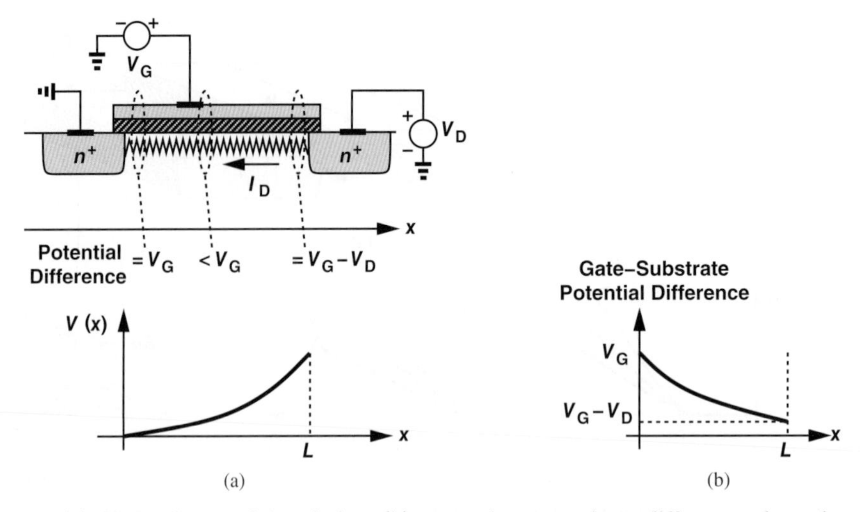

Figure 6.11 (a) Channel potential variation, (b) gate-substrate voltage difference along the channel.

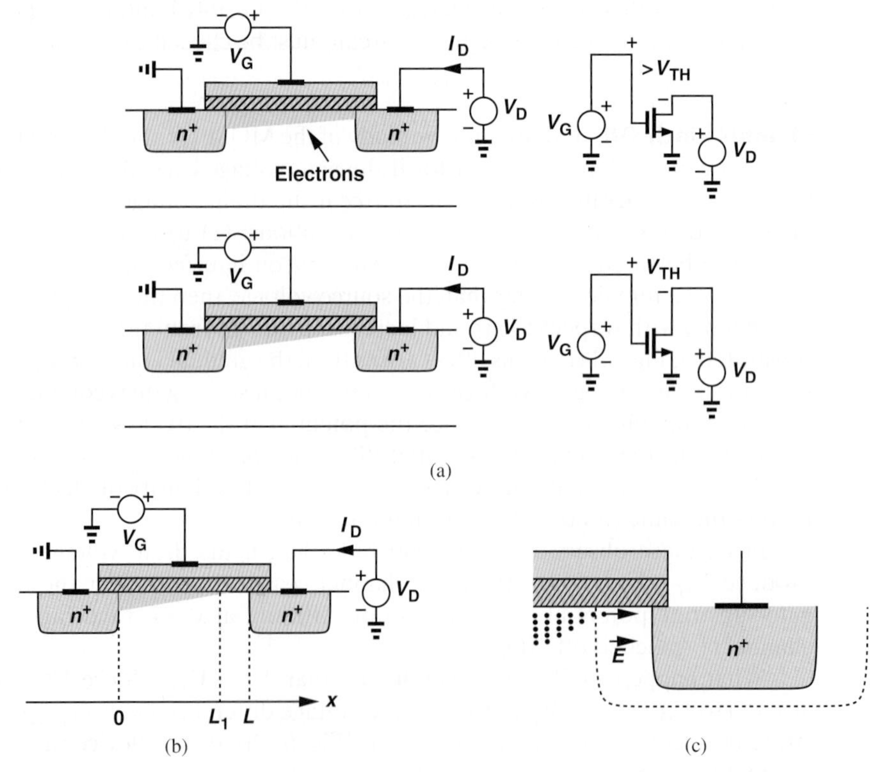

Figure 6.12 (a) Pinchoff, (b) variation of length with drain voltage, (c) detailed operation near the drain.

6.2.2 Derivation of I-V Characteristics

With the foregoing qualitative study, we can now formulate the behavior of MOSFETs in terms of their terminal voltages.

Channel Charge Density Our derivations require an expression for the channel charge (i.e., free electrons) per unit *length*, also called the "charge density." From $Q = CV$, we note that if C is the gate capacitance per unit length and V the voltage difference between the gate and the channel, then Q is the desired charge density. Denoting the gate capacitance per unit *area* by C_{ox} (expressed in F/m^2 or fF/μm^2), we write $C = WC_{ox}$ to account for the width of the transistor (Fig. 6.13). Moreover, we have $V = V_{GS} - V_{TH}$ because no mobile charge exists for $V_{GS} < V_{TH}$. (Hereafter, we denote both the gate and drain voltages with respect to the source.) It follows that

$$Q = WC_{ox}(V_{GS} - V_{TH}). \tag{6.2}$$

Note that Q is expressed in coulomb/meter. Now recall from Fig. 6.11(a) that the channel voltage varies along the length of the transistor, and the charge density falls as we go from the source to the drain. Thus, Eq. (6.2) is valid only near the source terminal, where the channel potential remains close to zero. As shown in Fig. 6.14, we denote the channel potential at x by $V(x)$ and write

$$Q(x) = WC_{ox}[V_{GS} - V(x) - V_{TH}], \tag{6.3}$$

noting that $V(x)$ goes from zero to V_D if the channel is not pinched off.

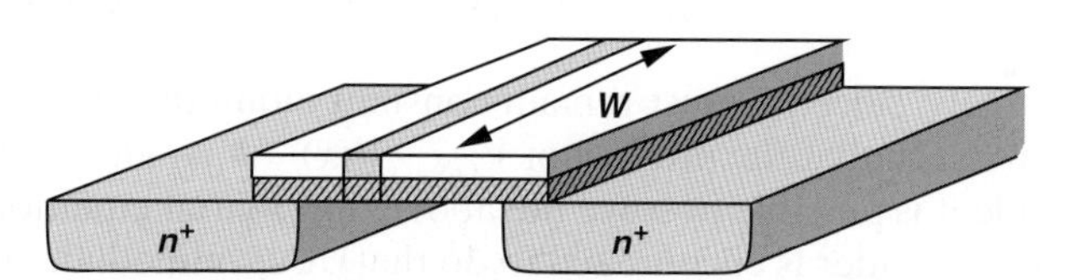

Figure 6.13 Illustration of capacitance per unit length.

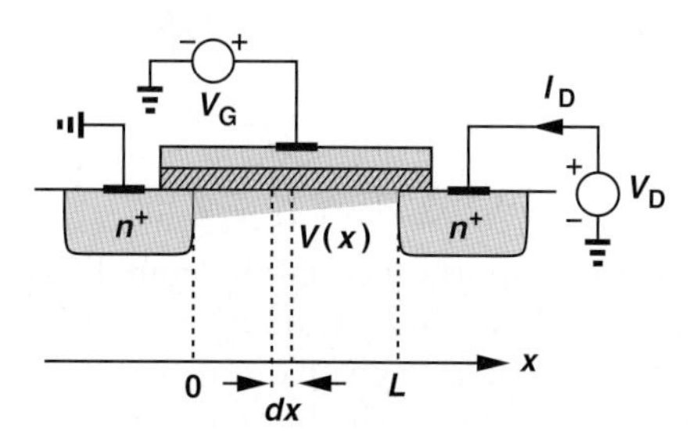

Figure 6.14 Device illustration for calculation of drain current.

Drain Current What is the relationship between the mobile charge density and the current? Consider a bar of semiconductor having a uniform charge density (per unit length) equal to Q and carrying a current I (Fig. 6.15). Note from Chapter 2 that (1) I is given by the total charge that passes through the cross section of the bar in one second, and (2) if the carriers move with a velocity of v m/s, then the charge enclosed in v meters along the bar passes through the cross section in one second. Since the charge enclosed in v meters is equal to $Q \cdot v$, we have

$$I = Q \cdot v. \tag{6.4}$$

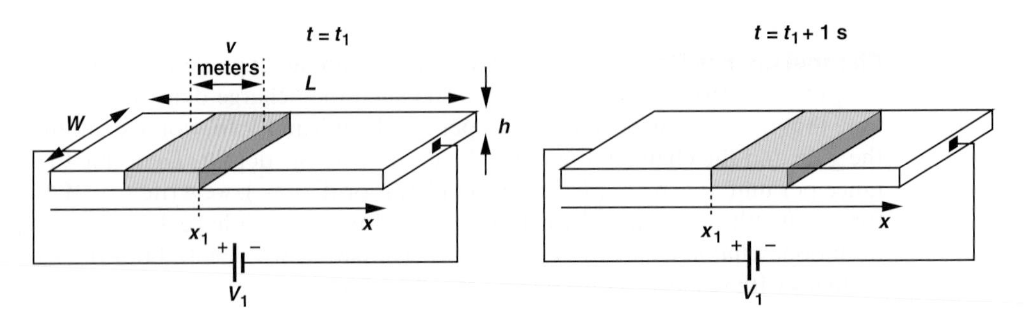

Figure 6.15 Relationship between charge velocity and current.

As explained in Chapter 2,

$$v = -\mu_n E, \tag{6.5}$$

$$= +\mu_n \frac{dV}{dx}, \tag{6.6}$$

where dV/dx denotes the derivative of the voltage at a given point. Combining Eqs. (6.3), (6.4), and (6.6), we obtain

$$I_D = WC_{ox}[V_{GS} - V(x) - V_{TH}]\mu_n \frac{dV(x)}{dx}. \tag{6.7}$$

Interestingly, since I_D must remain constant along the channel (why?), $V(x)$ and dV/dx must vary such that the product of $V_{GS} - V(x) - V_{TH}$ and dV/dx is independent of x.

While it is possible to solve the above differential equation to obtain $V(x)$ in terms of I_D (and the reader is encouraged to do that), our immediate need is to find an expression for I_D in terms of the terminal voltages. To this end, we write

$$\int_{x=0}^{x=L} I_D \, dx = \int_{V(x)=0}^{V(x)=V_{DS}} \mu_n C_{ox} W[V_{GS} - V(x) - V_{TH}] \, dV. \tag{6.8}$$

That is,

$$I_D = \frac{1}{2}\mu_n C_{ox} \frac{W}{L} \left[2(V_{GS} - V_{TH})V_{DS} - V_{DS}^2\right]. \tag{6.9}$$

We now examine this important equation from different perspectives to gain more insight. First, the linear dependence of I_D upon μ_n, C_{ox}, and W/L is to be expected: a higher mobility yields a greater current for a given drain-source voltage; a higher gate oxide capacitance leads to a larger electron density in the channel for a given gate-source voltage; and a larger W/L (called the device "aspect ratio") is equivalent to placing more transistors in parallel [Fig. 6.10(c)]. Second, for a constant V_{GS}, I_D varies *parabolically* with V_{DS} (Fig. 6.16), reaching a maximum of

$$I_{D,max} = \frac{1}{2}\mu_n C_{ox} \frac{W}{L}(V_{GS} - V_{TH})^2 \tag{6.10}$$

at $V_{DS} = V_{GS} - V_{TH}$. It is common to write W/L as the ratio of two values, e.g., 5 μm/0.18 μm (rather than 27.8) to emphasize the choice of W and L. While only the ratio appears in many MOS equations, the individual values of W and L also become critical in most cases. For example, if both W and L are doubled, the ratio remains unchanged but the gate capacitance increases.

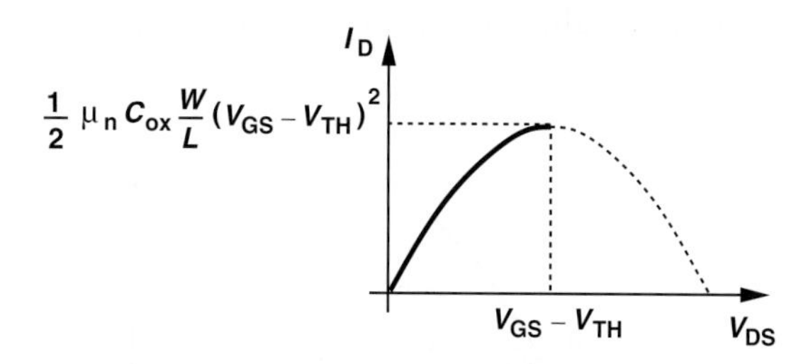

Figure 6.16 Parabolic I_D-V_{DS} characteristic.

Example 6-3 Plot the I_D-V_{DS} characteristics for different values of V_{GS}.

Solution As V_{GS} increases, so do $I_{D,max}$ and $V_{GS} - V_{TH}$. Illustrated in Fig. 6.17, the characteristics exhibit maxima that follow a parabolic shape themselves because $I_{D,max} \propto (V_{GS} - V_{TH})^2$.

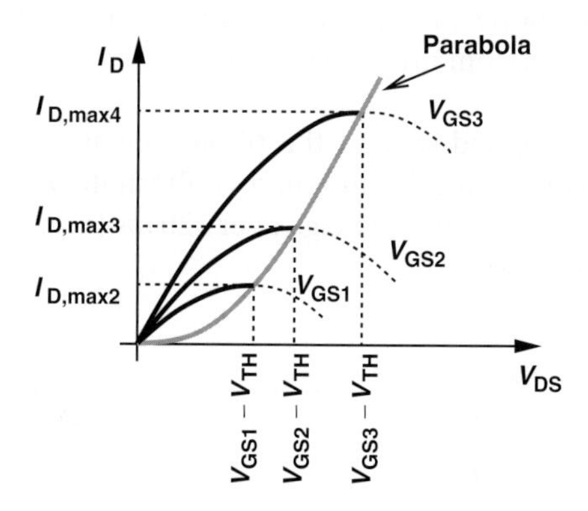

Figure 6.17 MOS characteristics for different gate-source voltages.

Exercise What happens to the above plots if t_{ox} is halved?

The nonlinear relationship between I_D and V_{DS} reveals that the transistor *cannot* generally be modeled as a simple linear resistor. However, if $V_{DS} \ll 2(V_{GS} - V_{TH})$, Eq. (6.9) reduces to:

$$I_D \approx \mu_n C_{ox} \frac{W}{L}(V_{GS} - V_{TH})V_{DS}, \tag{6.11}$$

exhibiting a linear I_D-V_{DS} behavior for a given V_{GS}. In fact, the equivalent on-resistance is given by V_{DS}/I_D:

$$R_{on} = \frac{1}{\mu_n C_{ox}\frac{W}{L}(V_{GS} - V_{TH})}. \tag{6.12}$$

From another perspective, at small V_{DS} (near the origin), the parabolas in Fig. 6.17 can be approximated by straight lines having different slopes (Fig. 6.18).

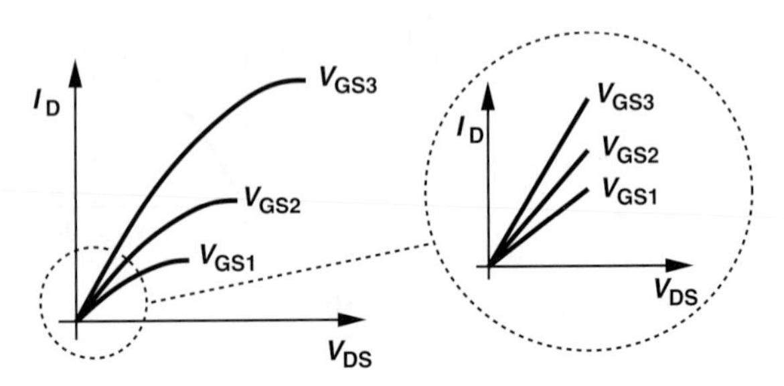

Figure 6.18 Detailed characteristics for small V_{DS}.

As predicted in Section 6.2.1, Eq. (6.12) suggests that the on-resistance can be controlled by the gate-source voltage. In particular, for $V_{GS} = V_{TH}$, $R_{on} = \infty$, i.e., the device can operate as an electronic switch.

Example 6-4 A cordless telephone incorporates a single antenna for reception and transmission. Explain how the system must be configured.

Solution The system is designed so that the phone receives for half of the time and transmits for the other half. Thus, the antenna is alternately connected to the receiver and the transmitter in regular intervals, e.g., every 20 ms (Fig. 6.19). An electronic antenna switch is therefore necessary here.[8]

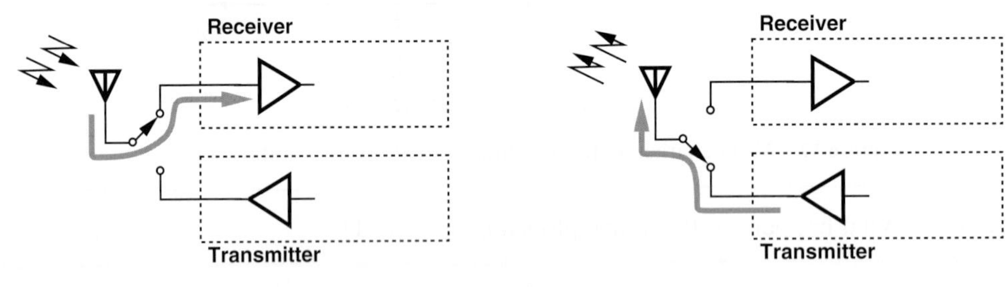

Figure 6.19 Role of antenna switch in a cordless phone.

Exercise Some systems employ two antennas, each of which receives and transmits signals. How many switches are needed?

[8]Some cellphones operate in the same manner.

In most applications, it is desirable to achieve a low on-resistance for MOS switches. The circuit designer must therefore maximize W/L and V_{GS}. The following example illustrates this point.

Example 6-5 In the cordless phone of Example 6-4, the switch connecting the transmitter to the antenna must negligibly attenuate the signal, e.g., by no more than 10%. If $V_{DD} = 1.8$ V, $\mu_n C_{ox} = 100\ \mu\text{A/V}^2$, and $V_{TH} = 0.4$ V, determine the minimum required aspect ratio of the switch. Assume the antenna can be modeled as a 50-Ω resistor.

Solution As depicted in Fig. 6.20, we wish to ensure

$$\frac{V_{out}}{V_{in}} \geq 0.9 \tag{6.13}$$

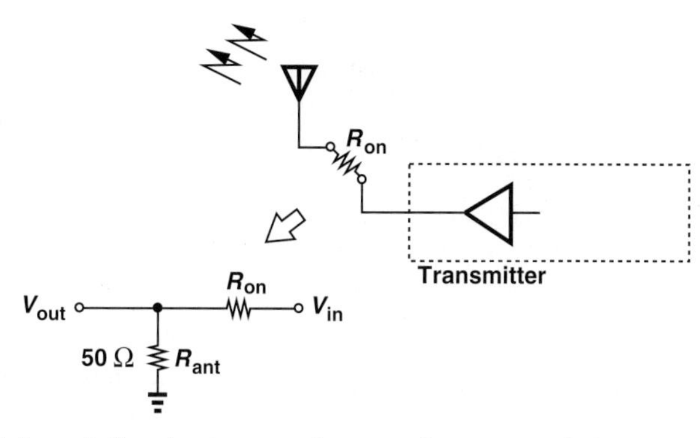

Figure 6.20 Signal degradation due to on-resistance of antenna switch.

and hence

$$R_{on} \leq 5.6\ \Omega. \tag{6.14}$$

Setting V_{GS} to the maximum value, V_{DD}, we obtain from Eq. (6.12),

$$\frac{W}{L} \geq 1276. \tag{6.15}$$

(Since wide transistors introduce substantial capacitance in the signal path, this choice of W/L may still attenuate high-frequency signals.)

Exercise What W/L is necessary if V_{DD} drops to 1.2 V?

Triode and Saturation Regions Equation (6.9) expresses the drain current in terms of the device terminal voltages, implying that the current begins to *fall* for $V_{DS} > V_{GS} - V_{TH}$. We say the device operates in the "triode region" (also called the "linear region") if $V_{DS} < V_{GS} - V_{TH}$ (the rising section of the parabola). We also use the term "deep triode region" for $V_{DS} \ll 2(V_{GS} - V_{TH})$, where the transistor operates as a resistor.

In reality, the drain current reaches "saturation," that is, becomes *constant* for $V_{DS} > V_{GS} - V_{TH}$ (Fig. 6.21). To understand why, recall from Fig. 6.12 that the channel experiences pinch-off if $V_{DS} = V_{GS} - V_{TH}$. Thus, further increase in V_{DS} simply shifts the

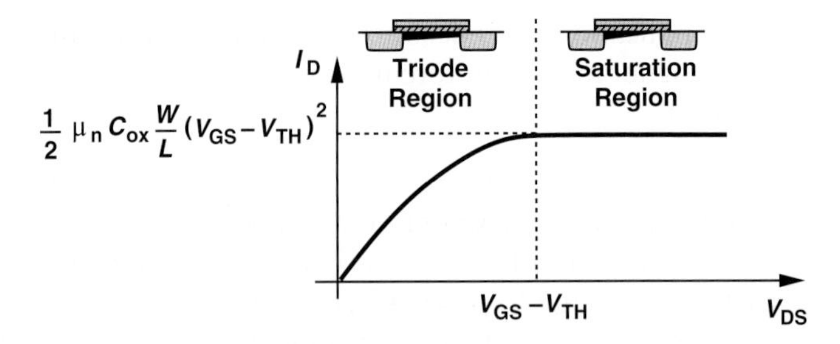

Figure 6.21 Overall MOS characteristic.

Did you know?

The explosive growth of MOS technology is attributed to two factors: the ability to shrink the dimensions of the MOS device (W, L, t_{ox}, etc.) and the ability to integrate a greater number of MOS devices on a chip each year. The latter trend was predicted by one of Intel's founders, Gordon Moore, in 1965. He observed that the number of transistors per chip doubled every two years. Indeed, starting with 50 devices per chip in 1965, we now have reached tens of billions on memory chips and several billion on microprocessor chips. Can you think of any other product in human history that has grown so much so fast (except for Bill Gates' wealth)?

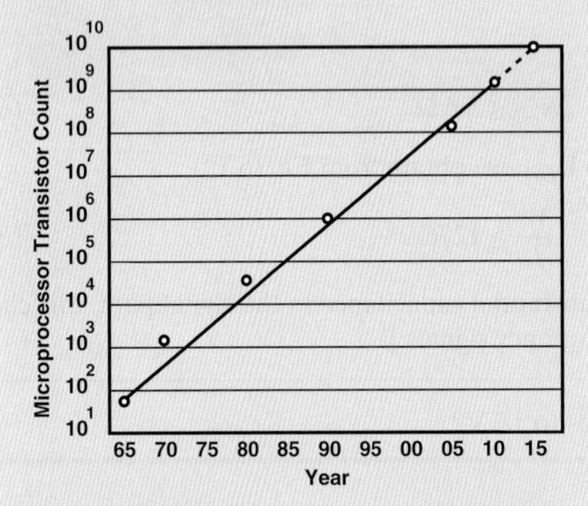

Moore's Law: transistor count per chip throughout the years.

pinch-off point slightly toward the source. Also, recall that Eqs. (6.7) and (6.8) are valid only where channel charge exists. It follows that the integration in Eq. (6.8) must encompass only the channel, i.e., from $x = 0$ to $x = L_1$ in Fig. 6.12(b), and be modified to

$$\int_{x=0}^{x=L_1} I_D\,dx = \int_{V(x)=0}^{V(x)=V_{GS}-V_{TH}} \mu_n C_{ox} W[V_{GS} - V(x) - V_{TH}]\,dV. \quad (6.16)$$

Note that the upper limits correspond to the channel pinch-off point. In particular, the integral on the right-hand side is evaluated up to $V_{GS} - V_{TH}$ rather than V_{DS}. Consequently,

$$I_D = \frac{1}{2}\mu_n C_{ox}\frac{W}{L_1}(V_{GS} - V_{TH})^2, \quad (6.17)$$

a result independent of V_{DS} and identical to $I_{D,max}$ in Eq. (6.10) if we assume $L_1 \approx L$. Called the "overdrive voltage," the quantity $V_{GS} - V_{TH}$ plays a key role in MOS circuits. MOSFETs are sometimes called "square-law" devices to emphasize the relationship between I_D and the overdrive. For the sake of brevity, we hereafter denote L_1 with L.

The I-V characteristic of Fig. 6.21 resembles that of bipolar devices, with the triode and saturation regions in MOSFETs appearing similar to saturation and forward active regions in bipolar transistors, respectively. It is unfortunate that the term "saturation" refers to completely different regions in MOS and bipolar I-V characteristics.

We employ the conceptual illustration in Fig. 6.22 to determine the region of operation. Note that the gate-drain potential difference suits this purpose and we need not compute the gate-source and gate-drain voltages separately.

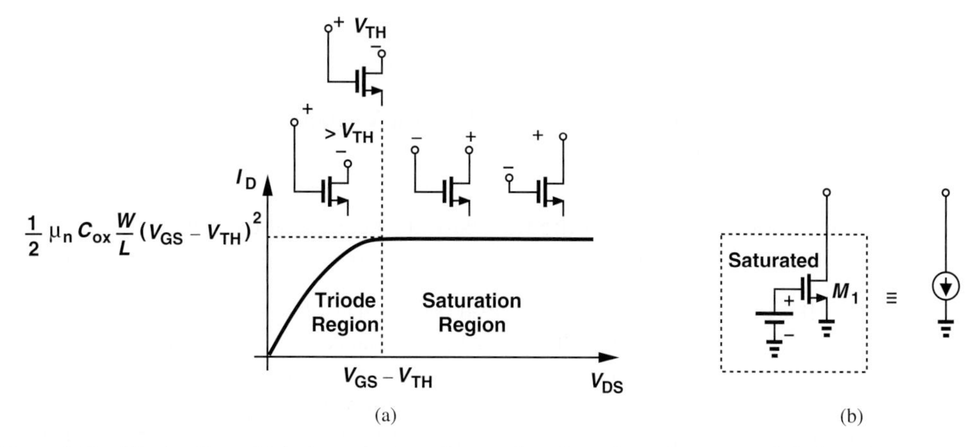

Figure 6.22 Illustration of triode and saturation regions based on the gate and drain voltages.

Exhibiting a "flat" current in the saturation region, a MOSFET can operate as a current source having a value given by Eq. (6.17). Furthermore, the square-law dependence of I_D upon $V_{GS} - V_{TH}$ suggests that the device can act as a voltage-controlled current source.

Example 6-6 Calculate the bias current of M_1 in Fig. 6.23. Assume $\mu_n C_{ox} = 100\ \mu\text{A/V}^2$ and $V_{TH} = 0.4$ V. If the gate voltage increases by 10 mV, what is the change in the drain voltage?

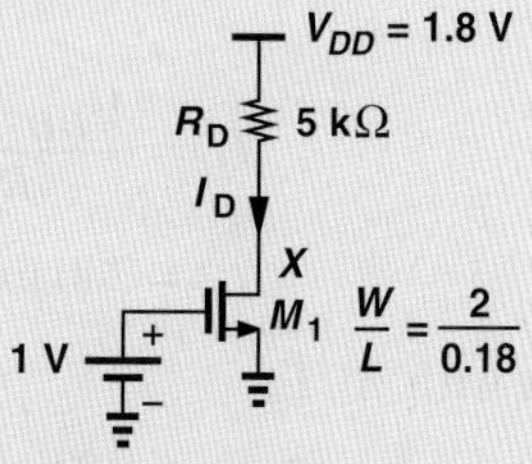

Figure 6.23 Simple MOS circuit.

Solution It is unclear a priori in which region M_1 operates. Let us assume M_1 is saturated and proceed. Since $V_{GS} = 1$ V,

$$I_D = \frac{1}{2}\mu_n C_{ox}\frac{W}{L}(V_{GS} - V_{TH})^2 \tag{6.18}$$

$$= 200\ \mu\text{A}. \tag{6.19}$$

We must check our assumption by calculating the drain potential:

$$V_X = V_{DD} - R_D I_D \tag{6.20}$$

$$= 0.8\ \text{V}. \tag{6.21}$$

The drain voltage is lower than the gate voltage, but by less than V_{TH}. The illustration in Fig. 6.22 therefore indicates that M_1 indeed operates in saturation.

If the gate voltage increases to 1.01 V, then

$$I_D = 206.7\ \mu\text{A}, \tag{6.22}$$

lowering V_X to

$$V_X = 0.766\ \text{V}. \tag{6.23}$$

Fortunately, M_1 is still saturated. The 34-mV change in V_X reveals that the circuit can *amplify* the input.

Exercise What choice of R_D places the transistor at the edge of the triode region?

It is instructive to identify several points of contrast between bipolar and MOS devices. (1) A bipolar transistor with $V_{BE} = V_{CE}$ resides at the edge of the active region whereas a MOSFET approaches the edge of saturation if its drain voltage falls below its gate voltage by V_{TH}. (2) Bipolar devices exhibit an exponential I_C-V_{BE} characteristic while MOSFETs display a square-law dependence. That is, the former provide a greater transconductance than the latter (for a given bias current). (3) In bipolar circuits, most transistors have the same dimensions and hence the same I_S, whereas in MOS circuits, the aspect ratio of each device may be chosen differently to satisfy the design requirements. (4) The gate of MOSFETs draws no bias current.[9]

Example 6-7 Determine the value of W/L in Fig. 6.23 that places M_1 at the edge of saturation and calculate the drain voltage change for a 1-mV change at the gate. Assume $V_{TH} = 0.4$ V.

Solution With $V_{GS} = +1$ V, the drain voltage must fall to $V_{GS} - V_{TH} = 0.6$ V for M_1 to enter the triode region. That is,

$$I_D = \frac{V_{DD} - V_{DS}}{R_D} \tag{6.24}$$

$$= 240\ \mu\text{A}. \tag{6.25}$$

Since I_D scales linearly with W/L,

$$\left.\frac{W}{L}\right|_{max} = \frac{240\ \mu\text{A}}{200\ \mu\text{A}} \cdot \frac{2}{0.18} \tag{6.26}$$

$$= \frac{2.4}{0.18}. \tag{6.27}$$

If V_{GS} increases by 1 mV,

$$I_D = 248.04\ \mu\text{A}, \tag{6.28}$$

changing V_X by

$$\Delta V_X = \Delta I_D \cdot R_D \tag{6.29}$$

$$= 4.02\ \text{mV}. \tag{6.30}$$

The voltage gain is thus equal to 4.02 in this case.

Exercise Repeat the above example if R_D is doubled.

[9]New generations of MOSFETs suffer from gate "leakage" current, but we neglect this effect here.

Example 6-8

Calculate the maximum allowable gate voltage in Fig. 6.24 if M_1 must remain saturated.

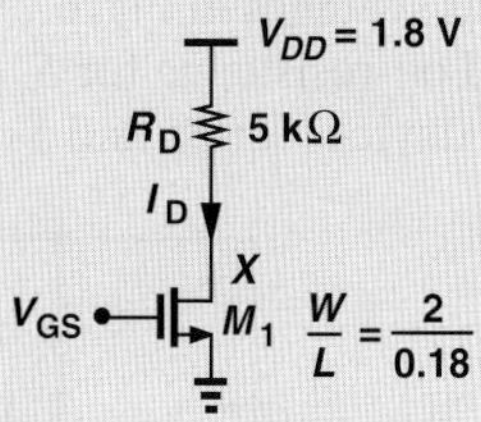

Figure 6.24 Simple MOS circuit.

Solution At the edge of saturation, $V_{GS} - V_{TH} = V_{DS} = V_{DD} - R_D I_D$. Substituting for I_D from Eq. (6.17) gives

$$V_{GS} - V_{TH} = V_{DD} - \frac{R_D}{2}\mu_n C_{ox}\frac{W}{L}(V_{GS} - V_{TH})^2, \tag{6.31}$$

and hence

$$V_{GS} - V_{TH} = \frac{-1 + \sqrt{1 + 2R_D V_{DD}\mu_n C_{ox}\frac{W}{L}}}{R_D \mu_n C_{ox}\frac{W}{L}}. \tag{6.32}$$

Thus,

$$V_{GS} = \frac{-1 + \sqrt{1 + 2R_D V_{DD}\mu_n C_{ox}\frac{W}{L}}}{R_D \mu_n C_{ox}\frac{W}{L}} + V_{TH}. \tag{6.33}$$

Exercise Calculate the value of V_{GS} if $\mu_n C_{ox} = 100\ \mu A/V^2$ and $V_{TH} = 0.4$.

Robotics Application: Binary Motor Control

Mobile robots use motors for motion. Depending on the polarity of the voltage or current applied to a motor, it rotates in the clockwise or counterclockwise direction. Thus, for the robot to go forward or backward, we must swap the positive and negative connections to the motors. Shown in Fig. 6.25 is an arrangement serving this purpose. We observe that, if S_1 and S_4 are on, the A and B terminals of the motor are tied to the positive and negative terminals of the battery, respectively. On the other hand, if S_2 and S_3 are on, the motor is driven with the opposite polarity. This configuration is called an "H-bridge" as it resembles the letter H. We call this action "binary control" because the motor has only two states.

Owing to the on-resistance of S_1-S_4, the voltage delivered to the motor is less than V_B. If S_1 and S_4 are on and the motor draws a current of I_M, we have

$$V_{AB} = V_{DD} - I_M R_1 - I_M R_4, \tag{6.34}$$

where R_1 and R_4 denote the resistances of the corresponding switches. We wish to minimize the voltage drops across the switches by minimizing their on-resistance.

MOSFETs can act as switches in Fig. 6.25. With proper choice of W/L and V_{GS}, we ensure a relatively small on-resistance for S_1-S_4. For reasons that are beyond our scope here, we choose PMOS devices for S_1 and S_2 and NMOS devices for S_3 and S_4.

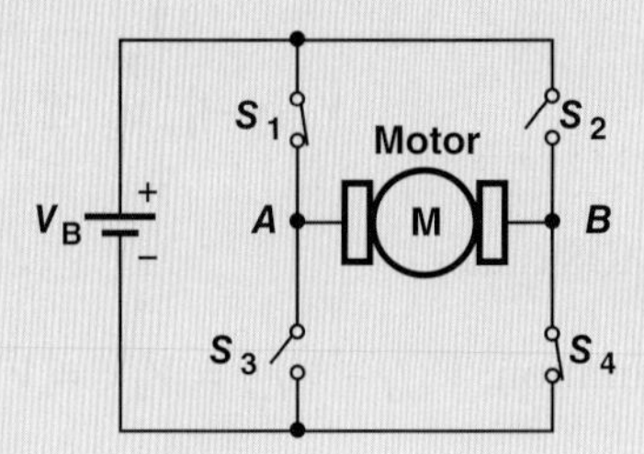

Figure 6.25 Control of a motor by an H-bridge.

6.2.3 Channel-Length Modulation

In our study of the pinch-off effect, we observed that the point at which the channel vanishes in fact moves toward the source as the drain voltage increases. In other words, the value of L_1 in Fig. 6.12(b) varies with V_{DS} to some extent. Called "channel-length modulation" and illustrated in Fig. 6.26, this phenomenon yields a larger drain current as V_{DS} increases because $I_D \propto 1/L_1$ in Eq. (6.17). Similar to the Early effect in bipolar devices, channel-length modulation results in a finite output impedance given by the inverse of the I_D-V_{DS} slope in Fig. 6.26.

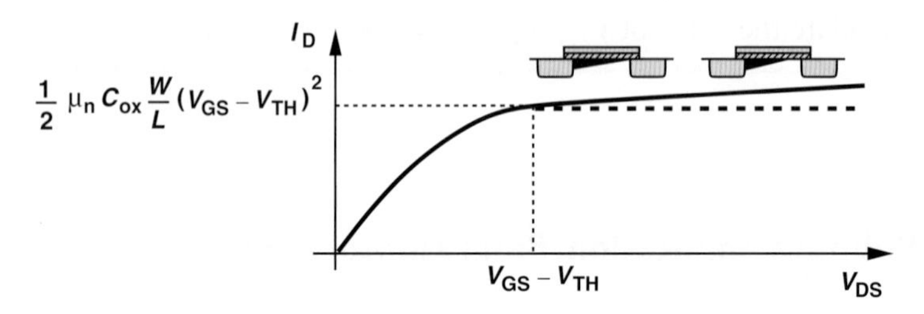

Figure 6.26 Variation of I_D in saturation region.

To account for channel-length modulation, we assume L is constant, but multiply the right-hand side of Eq. (6.17) by a corrective term:

$$I_D = \frac{1}{2}\mu_n C_{ox}\frac{W}{L}(V_{GS} - V_{TH})^2(1 + \lambda V_{DS}), \tag{6.35}$$

where λ is called the "channel-length modulation coefficient." While only an approximation, this linear dependence of I_D upon V_{DS} still provides a great deal of insight into the circuit design implications of channel-length modulation.

Unlike the Early effect in bipolar devices (Chapter 4), the amount of channel-length modulation is under the circuit designer's control. This is because λ is inversely proportional to L: for a longer channel, the *relative* change in L (and hence in I_D) for a given change in V_{DS} is smaller (Fig. 6.27).[10] (By contrast, the base width of bipolar devices cannot be adjusted by the circuit designer, yielding a constant Early voltage for all transistors in a given technology.)

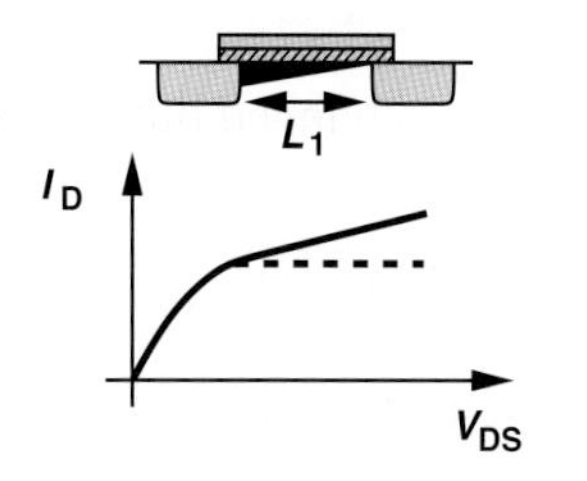

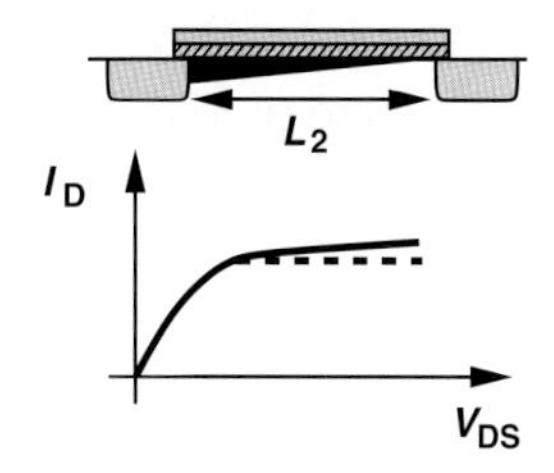

Figure 6.27 Channel-length modulation.

Example 6-9 A MOSFET carries a drain current of 1 mA with $V_{DS} = 0.5$ V in saturation. Determine the change in I_D if V_{DS} rises to 1 V and $\lambda = 0.1\ \text{V}^{-1}$. What is the device output impedance?

Solution We write

$$I_{D1} = \frac{1}{2}\mu_n C_{ox}\frac{W}{L}(V_{GS} - V_{TH})^2(1 + \lambda V_{DS1}) \tag{6.36}$$

$$I_{D2} = \frac{1}{2}\mu_n C_{ox}\frac{W}{L}(V_{GS} - V_{TH})^2(1 + \lambda V_{DS2}) \tag{6.37}$$

and hence

$$I_{D2} = I_{D1}\frac{1 + \lambda V_{DS2}}{1 + \lambda V_{DS1}}. \tag{6.38}$$

With $I_{D1} = 1$ mA, $V_{DS1} = 0.5$ V, $V_{DS2} = 1$ V, and $\lambda = 0.1\ \text{V}^{-1}$,

$$I_{D2} = 1.048\ \text{mA}. \tag{6.39}$$

The change in I_D is therefore equal to 48 μA, yielding an output impedance of

$$r_O = \frac{\Delta V_{DS}}{\Delta I_D} \tag{6.40}$$

$$= 10.42\ \text{k}\Omega. \tag{6.41}$$

Exercise Does W affect the above results?

[10]Since different MOSFETs in a circuit may be sized for different λ's, we do not define a quantity similar to the Early voltage here.

The above example reveals that channel-length modulation limits the output impedance of MOS current sources. The same effect was observed for bipolar current sources in Chapters 4 and 5.

Example 6-10 Assuming $\lambda \propto 1/L$, calculate ΔI_D and r_O in Example 6-9 if both W and L are doubled.

Solution In Eqs. (6.36) and (6.37), W/L remains unchanged but λ drops to 0.05 V^{-1}. Thus,

$$I_{D2} = I_{D1}\frac{1 + \lambda V_{DS2}}{1 + \lambda V_{DS1}} \tag{6.42}$$

$$= 1.024 \text{ mA}. \tag{6.43}$$

That is, $\Delta I_D = 24\ \mu\text{A}$ and

$$r_O = 20.84 \text{ k}\Omega. \tag{6.44}$$

Exercise What output impedance is achieved if W and L are quadrupled and I_D is halved?

6.2.4 MOS Transconductance

As a voltage-controlled current source, a MOS transistor can be characterized by its transconductance:

$$g_m = \frac{\partial I_D}{\partial V_{GS}}. \tag{6.45}$$

This quantity serves as a measure of the "strength" of the device: a higher value corresponds to a greater change in the drain current for a given change in V_{GS}. Using Eq. (6.17) for the saturation region, we have

$$g_m = \mu_n C_{ox}\frac{W}{L}(V_{GS} - V_{TH}), \tag{6.46}$$

concluding that (1) g_m is linearly proportional to W/L for a given $V_{GS} - V_{TH}$, and (2) g_m is linearly proportional to $V_{GS} - V_{TH}$ for a given W/L. Also, substituting for $V_{GS} - V_{TH}$ from Eq. (6.17), we obtain

$$g_m = \sqrt{2\mu_n C_{ox}\frac{W}{L}I_D}. \tag{6.47}$$

That is, (1) g_m is proportional to $\sqrt{W/L}$ for a given I_D, and (2) g_m is proportional to $\sqrt{I_D}$ for a given W/L. Moreover, dividing Eq. (6.46) by (6.17) gives

$$g_m = \frac{2I_D}{V_{GS} - V_{TH}}, \tag{6.48}$$

revealing that (1) g_m is linearly proportional to I_D for a given $V_{GS} - V_{TH}$, and (2) g_m is inversely proportional to $V_{GS} - V_{TH}$ for a given I_D. Summarized in Table 6.1, these

TABLE 6.1 Various dependencies of g_m.

$\frac{W}{L}$ Constant $V_{GS} - V_{TH}$ Variable	$\frac{W}{L}$ Variable $V_{GS} - V_{TH}$ Constant	$\frac{W}{L}$ Variable I_D Constant
$g_m \propto \sqrt{I_D}$	$g_m \propto I_D$	$g_m \propto \sqrt{\frac{W}{L}}$
$g_m \propto V_{GS} - V_{TH}$	$g_m \propto \frac{W}{L}$	$g_m \propto \frac{1}{V_{GS} - V_{TH}}$

dependencies prove critical in understanding performance trends of MOS devices and have no counterpart in bipolar transistors.[11] Among these three expressions for g_m, Eq. (6.47) is more frequently used because I_D may be predetermined by power dissipation requirements.

Example 6-11 For a MOSFET operating in saturation, how do g_m and $V_{GS} - V_{TH}$ change if both W/L and I_D are doubled?

Solution Equation (6.47) indicates that g_m is also doubled. Moreover, Eq. (6.17) suggests that the overdrive remains constant. These results can be understood intuitively if we view the doubling of W/L and I_D as shown in Fig. 6.28. Indeed, if V_{GS} remains constant and the width of the device is doubled, it is as if two transistors carrying equal currents are placed in parallel, thereby doubling the transconductance. The reader can show that this trend applies to any type of transistor.

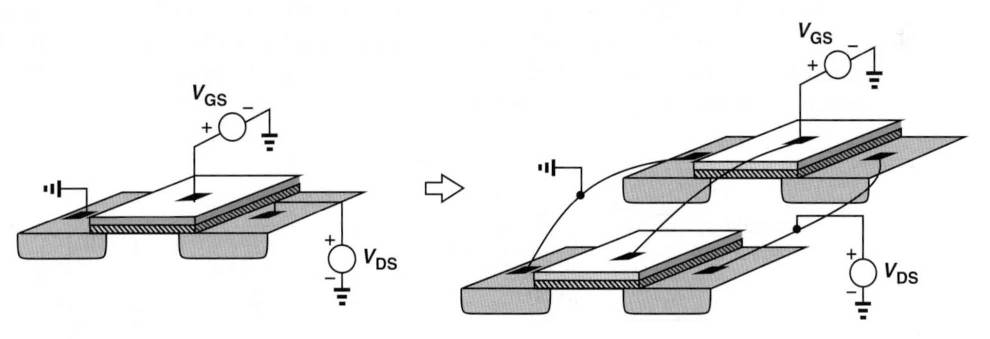

Figure 6.28 Equivalence of a wide MOSFET to two in parallel.

Exercise How do g_m and $V_{GS} - V_{TH}$ change if only W and I_D are doubled?

[11]There is some resemblance between the second column and the behavior of $g_m = I_C/V_T$. If the bipolar transistor width is increased while V_{BE} remains constant, then both I_C and g_m increase linearly.

6.2.5 Velocity Saturation*

Recall from Section 2.1.3 that at high electric fields, carrier mobility degrades, eventually leading to a *constant* velocity. Owing to their very short channels (e.g., 0.1 μm), modern MOS devices experience velocity saturation even with drain-source voltages as low as 1 V. As a result, the I/V characteristics no longer follow the square-law behavior.

Let us examine the derivations in Section 6.2.2 under velocity saturation conditions. Denoting the saturated velocity by v_{sat}, we have

$$I_D = v_{sat} \cdot Q \quad (6.49)$$

$$= v_{sat} \cdot WC_{ox}(V_{GS} - V_{TH}). \quad (6.50)$$

Interestingly, I_D now exhibits a *linear* dependence on $V_{GS} - V_{TH}$ and no dependence on L.[12] We also recognize that

$$g_m = \frac{\partial I_D}{\partial V_{GS}} \quad (6.51)$$

$$= v_{sat} WC_{ox}, \quad (6.52)$$

a quantity independent of L and I_D.

6.2.6 Other Second-Order Effects

Body Effect In our study of MOSFETs, we have assumed that both the source and the substrate (also called the "bulk" or the "body") are tied to ground. However, this condition need not hold in all circuits. For example, if the source terminal rises to a positive voltage while the substrate is at zero, then the source-substrate junction remains reverse-biased and the device still operates properly.

Figure 6.29 illustrates this case. The source terminal is tied to a potential V_S with respect to ground while the substrate is grounded through a p^+ contact.[13] The dashed line added to the transistor symbol indicates the substrate terminal. We denote the voltage difference between the source and the substrate (the bulk) by V_{SB}.

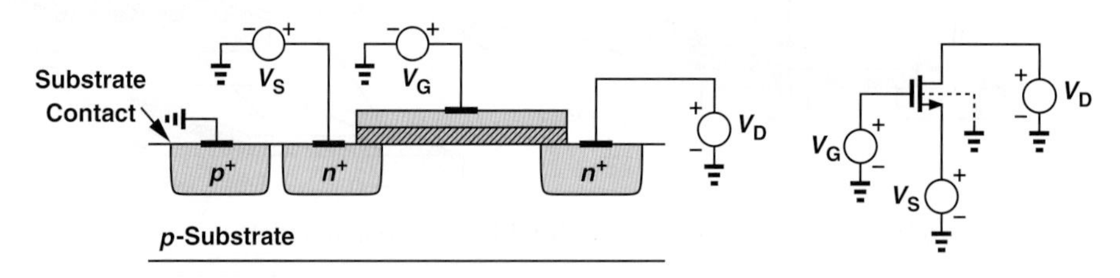

Figure 6.29 Body effect.

An interesting phenomenon occurs as the source-substrate potential difference departs from zero: the threshold voltage of the device *changes*. In particular, as the source

*This section can be skipped in a first reading.
[12]Of course, if L is increased substantially, while V_{DS} remains constant, then the device experiences less velocity saturation and Eq. (6.50) is not accurate.
[13]The p^+ island is necessary to achieve an "ohmic" contact with low resistance.

becomes more positive with respect to the substrate, V_{TH} *increases*. Called "body effect," this phenomenon is formulated as

$$V_{TH} = V_{TH0} + \gamma\,(\sqrt{|2\phi_F + V_{SB}|} - \sqrt{|2\phi_F|}), \tag{6.53}$$

where V_{TH0} denotes the threshold voltage with $V_{SB} = 0$ (as studied earlier), and γ and ϕ_F are technology-dependent parameters having typical values of $0.4\ \sqrt{\text{V}}$ and 0.4 V, respectively.

Example 6-12 In the circuit of Fig. 6.29, assume $V_S = 0.5$ V, $V_G = V_D = 1.4$ V, $\mu_n C_{ox} = 100\ \mu\text{A/V}^2$, $W/L = 50$, and $V_{TH0} = 0.6$ V. Determine the drain current if $\lambda = 0$.

Solution Since the source-body voltage, $V_{SB} = 0.5$ V, Eq. (6.53) and the typical values for γ and ϕ_F yield

$$V_{TH} = 0.698\ \text{V}. \tag{6.54}$$

Also, with $V_G = V_D$, the device operates in saturation (why?) and hence

$$I_D = \frac{1}{2}\mu_n C_{ox}\frac{W}{L}(V_G - V_S - V_{TH})^2 \tag{6.55}$$

$$= 102\ \mu\text{A}. \tag{6.56}$$

Exercise Sketch the drain current as a function of V_S as V_S goes from zero to 1 V.

Body effect manifests itself in some analog and digital circuits and is studied in more advanced texts. We neglect body effect in this book.

Subthreshold Conduction The derivation of the MOS I-V characteristic has assumed that the transistor abruptly turns on as V_{GS} reaches V_{TH}. In reality, formation of the channel is a gradual effect, and the device conducts a small current even for $V_{GS} < V_{TH}$. Called "subthreshold conduction," this effect has become a critical issue in modern MOS devices and is studied in more advanced texts.

6.3 MOS DEVICE MODELS

With our study of MOS I-V characteristics in the previous section, we now develop models that can be used in circuit analysis and design.

6.3.1 Large-Signal Model

For arbitrary voltage and current levels, we must resort to Eqs. (6.9) and (6.35) to express the device behavior:

$$I_D = \frac{1}{2}\mu_n C_{ox}\frac{W}{L}\left[2(V_{GS} - V_{TH})V_{DS} - V_{DS}^2\right] \quad \text{Triode Region} \tag{6.57}$$

$$I_D = \frac{1}{2}\mu_n C_{ox}\frac{W}{L}(V_{GS} - V_{TH})^2(1 + \lambda V_{DS}) \quad \text{Saturation Region} \tag{6.58}$$

In the saturation region, the transistor acts as a voltage-controlled current source, lending itself to the model shown in Fig. 6.30(a). Note that I_D does depend on V_{DS} and is therefore not an ideal current source. For $V_{DS} < V_{GS} - V_{TH}$, the model must reflect the triode region, but it can still incorporate a voltage-controlled current source, as depicted in Fig. 6.30(b). Finally, if $V_{DS} \ll 2(V_{GS} - V_{TH})$, the transistor can be viewed as a voltage-controlled resistor [Fig. 6.30(c)]. In all three cases, the gate remains an open circuit to represent the zero gate current.

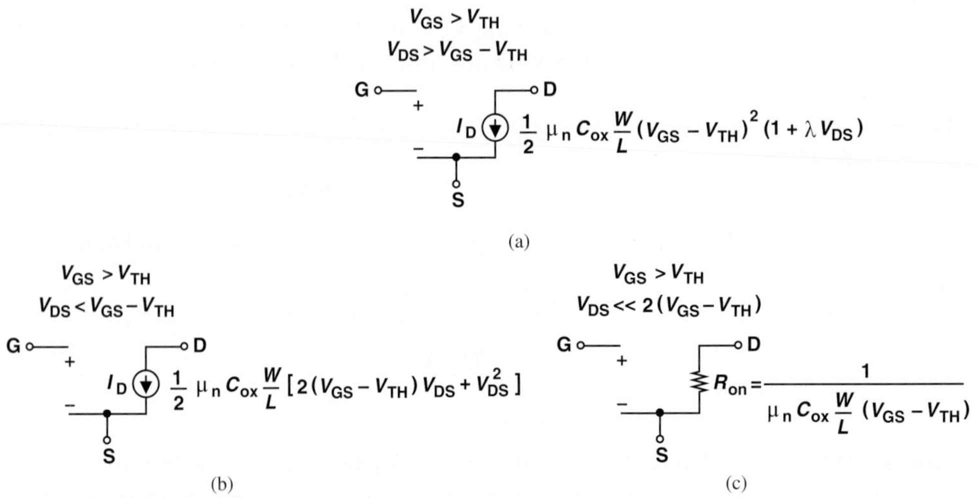

Figure 6.30 MOS models for (a) saturation region, (b) triode region, (c) deep triode region.

Example 6-13 Sketch the drain current of M_1 in Fig. 6.31(a) versus V_1 as V_1 varies from zero to V_{DD}. Assume $\lambda = 0$.

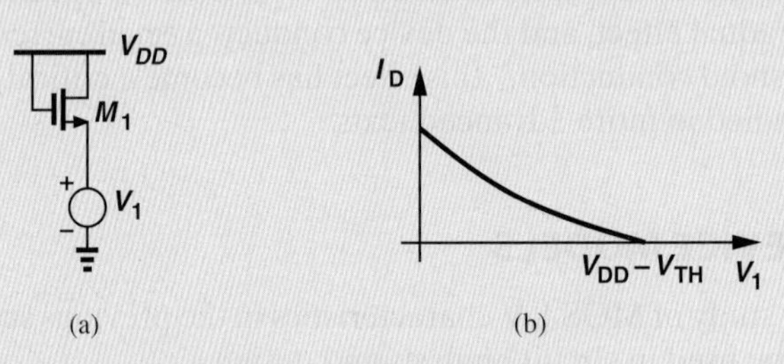

Figure 6.31 (a) Simple MOS circuit, (b) variation of I_D with V_1.

Solution Noting that the device operates in saturation (why?), we write

$$I_D = \frac{1}{2}\mu_n C_{ox}\frac{W}{L}(V_{GS} - V_{TH})^2 \tag{6.59}$$

$$= \frac{1}{2}\mu_n C_{ox}\frac{W}{L}(V_{DD} - V_1 - V_{TH})^2. \tag{6.60}$$

At $V_1 = 0$, $V_{GS} = V_{DD}$ and the device carries maximum current. As V_1 rises, V_{GS} falls and so does I_D. If V_1 reaches $V_{DD} - V_{TH}$, V_{GS} drops to V_{TH}, turning the transistor off. The drain current thus varies as illustrated in Fig. 6.31(b). Note that, owing to body effect, V_{TH} *varies* with V_1 if the substrate is tied to ground.

Exercise Repeat the above example if the gate of M_1 is tied to a voltage equal to 1.5 V and $V_{DD} =$ 2 V.

6.3.2 Small-Signal Model

If the bias currents and voltages of a MOSFET are only slightly disturbed by signals, the nonlinear, large-signal models can be reduced to linear, small-signal representations. The development of the model proceeds in a manner similar to that in Chapter 4 for bipolar devices. Of particular interest to us in this book is the small-signal model for the saturation region.

Viewing the transistor as a voltage-controlled current source, we draw the basic model as in Fig. 6.32(a), where $i_D = g_m v_{GS}$ and the gate remains open. To represent channel-length modulation, i.e., variation of i_D with v_{DS}, we add a resistor as in Fig. 6.32(b):

$$r_O = \left(\frac{\partial I_D}{\partial V_{DS}}\right)^{-1} \tag{6.61}$$

$$= \frac{1}{\frac{1}{2}\mu_n C_{ox}\frac{W}{L}(V_{GS} - V_{TH})^2 \cdot \lambda}. \tag{6.62}$$

Since channel-length modulation is relatively small, the denominator of Eq. (6.62) can be approximated as $I_D \cdot \lambda$, yielding

$$r_O \approx \frac{1}{\lambda I_D}. \tag{6.63}$$

Did you know?

In addition to integrating a larger number of transistors per chip, MOS technology has also benefited tremendously from "scaling," i.e., the reduction of the transistors' dimensions. The minimum channel length has fallen from about 10 μm to about 25 nm today and the *speed* of MOSFETs has improved by more than 4 orders of magnitude. For example, the clock frequency of Intel's microprocessors has risen from 100 kHz to 4 GHz. But have analog circuits taken advantage of the scaling as well? Yes, indeed. Plotted below is the frequency of MOS oscillators as a function of time over the past three decades.

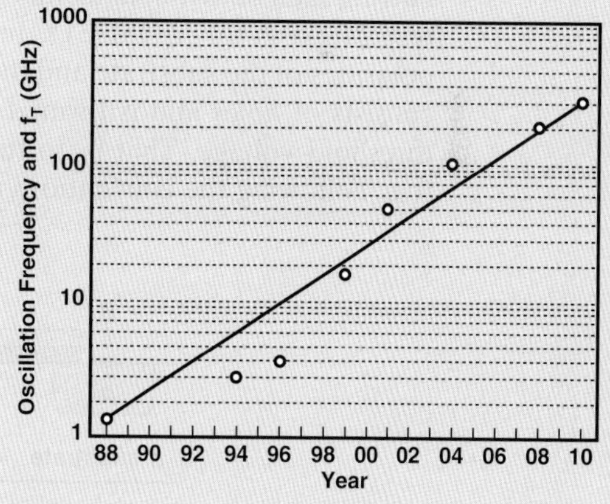

MOS oscillator frequency as a function of time.

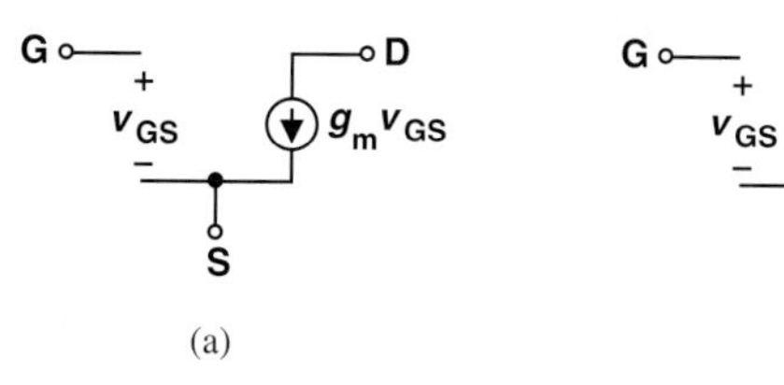

Figure 6.32 (a) Small-signal model of MOSFET, (b) inclusion of channel-length modulation.

Example 6-14 A MOSFET is biased at a drain current of 0.5 mA. If $\mu_n C_{ox} = 100\ \mu\text{A/V}^2$, $W/L = 10$, and $\lambda = 0.1\ \text{V}^{-1}$, calculate its small-signal parameters.

Solution We have

$$g_m = \sqrt{2\mu_n C_{ox}\frac{W}{L}I_D} \tag{6.64}$$

$$= \frac{1}{1\ \text{k}\Omega}. \tag{6.65}$$

Also,

$$r_O = \frac{1}{\lambda I_D} \tag{6.66}$$

$$= 20\ \text{k}\Omega. \tag{6.67}$$

This means that the intrinsic gain, $g_m r_O$ (Chapter 4), is equal to 20 for this choice of device dimensions and bias current.

Exercise Repeat the above example if W/L is doubled.

6.4 PMOS TRANSISTOR

Having seen both *npn* and *pnp* bipolar transistors, the reader may wonder if a *p*-type counterpart exists for MOSFETs. Indeed, as illustrated in Fig. 6.33(a), changing the doping polarities of the substrate and the S/D areas results in a "PMOS" device. The channel now consists of *holes* and is formed if the gate voltage is *below* the source potential by one threshold voltage. That is, to turn the device on, $V_{GS} < V_{TH}$, where V_{TH} itself is negative. Following the conventions used for bipolar devices, we draw the PMOS device as in

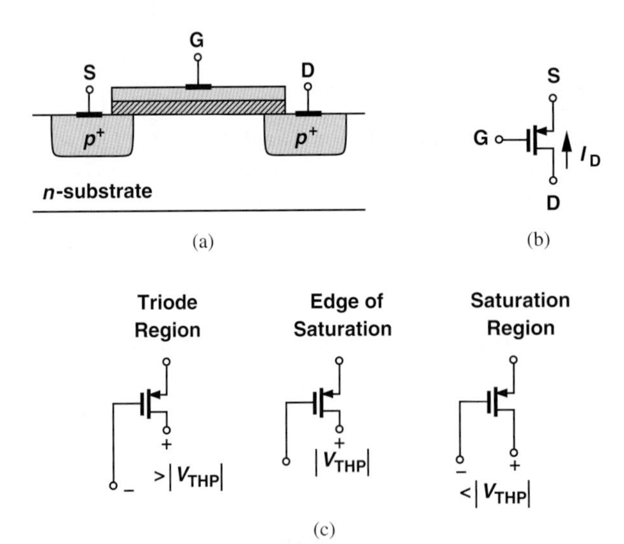

Figure 6.33 (a) Structure of PMOS device, (b) PMOS circuit symbol, (c) illustration of triode and saturation regions based on gate and drain voltages.

Fig. 6.33(b), with the source terminal identified by the arrow and placed on top to emphasize its higher potential. The transistor operates in the triode region if the drain voltage is near the source potential, approaching saturation as V_D falls to $V_G - V_{TH} = V_G + |V_{TH}|$. Figure 6.33(c) conceptually illustrates the gate-drain voltages required for each region of operation. We say that if V_{DS} of a PMOS (NMOS) device is sufficiently negative (positive), then it is in saturation.

Example 6-15

In the circuit of Fig. 6.34, determine the region of operation of M_1 as V_1 goes from V_{DD} to zero. Assume $V_{DD} = 2.5$ V and $|V_{TH}| = 0.5$ V.

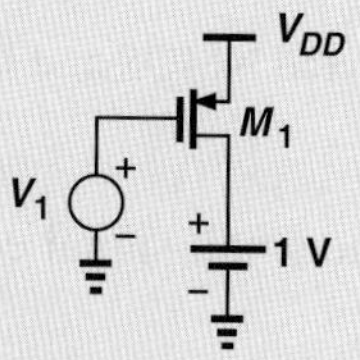

Figure 6.34 Simple PMOS circuit.

Solution For $V_1 = V_{DD}$, $V_{GS} = 0$ and M_1 is off. As V_1 falls and approaches $V_{DD} - |V_{TH}|$, the gate-source potential is negative enough to form a channel of holes, turning the device on. At this point, $V_G = V_{DD} - |V_{TH}| = +2$ V while $V_D = +1$ V; i.e., M_1 is saturated [Fig. 6.33(c)]. As V_1 falls further, V_{GS} becomes more negative and the transistor current rises. For $V_1 = +1\text{ V} - |V_{TH}| = 0.5$ V, M_1 is at the edge of the triode region. As V_1 goes below 0.5 V, the transistor enters the triode region further.

The voltage and current polarities in PMOS devices can prove confusing. Using the current direction shown in Fig. 6.33(b), we express I_D in the saturation region as

$$I_{D,sat} = -\frac{1}{2}\mu_p C_{ox}\frac{W}{L}(V_{GS} - V_{TH})^2(1 - \lambda V_{DS}), \tag{6.68}$$

where λ is multiplied by a negative sign.[14] In the triode region,

$$I_{D,tri} = -\frac{1}{2}\mu_p C_{ox}\frac{W}{L}\left[2(V_{GS} - V_{TH})V_{DS} - V_{DS}^2\right]. \tag{6.69}$$

Alternatively, both equations can be expressed in terms of absolute values:

$$|I_{D,sat}| = \frac{1}{2}\mu_p C_{ox}\frac{W}{L}(|V_{GS}| - |V_{TH}|)^2(1 + \lambda|V_{DS}|) \tag{6.70}$$

$$|I_{D,tri}| = \frac{1}{2}\mu_p C_{ox}\frac{W}{L}\left[2(|V_{GS}| - |V_{TH}|)|V_{DS}| - V_{DS}^2\right]. \tag{6.71}$$

The small-signal model of PMOS transistor is identical to that of NMOS devices (Fig. 6.32). The following example illustrates this point.

[14]To make this equation more consistent with that of NMOS devices [Eq. (6.35)], we can define λ itself to be negative and express I_D as $(1/2)\mu_p C_{ox}(W/L)(V_{GS} - V_{TH})^2(1 + \lambda V_{DS})$. But a negative λ carries little physical meaning.

Example 6-16 For the configurations shown in Fig. 6.35(a), determine the small-signal resistances R_X and R_Y. Assume $\lambda \neq 0$.

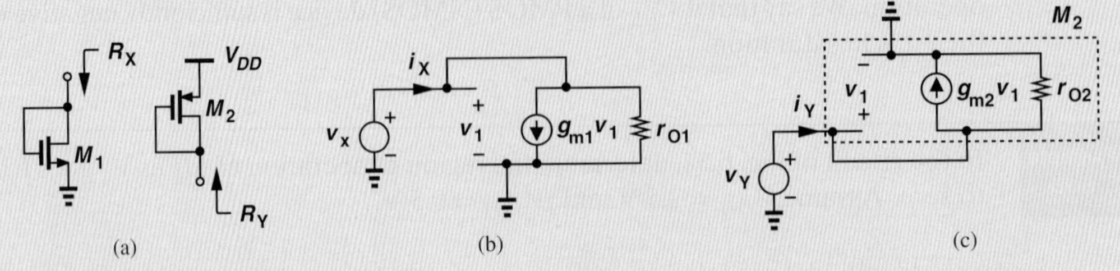

Figure 6.35 (a) Diode-connected NMOS and PMOS devices, (b) small-signal model of (a), (c) small-signal model of (b).

Solution For the NMOS version, the small-signal equivalent appears as depicted in Fig. 6.35(b), yielding

$$R_X = \frac{v_X}{i_X} \tag{6.72}$$

$$= \left(g_{m1}v_X + \frac{v_X}{r_{O1}}\right)^{-1} v_x \tag{6.73}$$

$$= \frac{1}{g_{m1}} \| r_{O1}. \tag{6.74}$$

For the PMOS version, we draw the equivalent as shown in Fig. 6.35(c) and write

$$R_Y = \frac{v_Y}{i_Y} \tag{6.75}$$

$$= \left(g_{m2}v_Y + \frac{v_Y}{r_{O1}}\right)^{-1} v_x \tag{6.76}$$

$$= \frac{1}{g_{m2}} \| r_{O2}. \tag{6.77}$$

In both cases, the small-signal resistance is equal to $1/g_m$ if $\lambda \to 0$.

In analogy with their bipolar counterparts [Fig. 4.45(a)], the structures shown in Fig. 6.35(a) are called "diode-connected" devices and act as two-terminal components: we will encounter many applications of diode-connected devices in Chapters 9 and 10.

Owing to the lower mobility of holes (Chapter 2), PMOS devices exhibit a poorer performance than NMOS transistors. For example, Eq. (6.47) indicates that the transconductance of a PMOS device is lower for a given drain current. We therefore prefer to use NMOS transistors wherever possible.

Bioengineering Application: Cancer Detection

Some cancerous cells produce a substance called "vascular endothelial growth factor" (VEGF), which is released into the patient's blood stream. VEGF alters the permittivity (ϵ) of the environment and hence the capacitance between two conductors placed in such an environment. The work in [1] inserts two microneedles into the patient's blood stream and measures the capacitance change [Fig. 6.36(a)]. As shown in Fig. 6.36(b), when S_3 is off, I_1 flows through the capacitance, causing V_X to rise linearly. Then, S_3 turns on and discharges C_B rapidly. This cycle repeats periodically, yielding a sawtooth waveform for V_X. The amplitude of V_X is given by I_1T_1/C_B and measured with high precision so as to detect small changes in C_B [1]. MOS current sources and switches prove essential here.

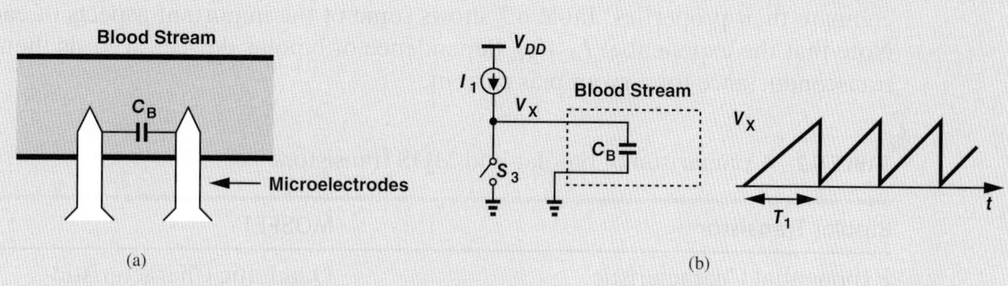

Figure 6.36 (a) Detection of cancer by measuring the capacitance between two microneedles and (b) a circuit for measuring capacitance change.

6.5 CMOS TECHNOLOGY

Is it possible to build both NMOS and PMOS devices on the same wafer? Figures 6.2(a) and 6.33(a) reveal that the two require *different* types of substrate. Fortunately, a *local* n-type substrate can be created in a p-type substrate, thereby accommodating PMOS transistors. As illustrated in Fig. 6.37, an "n-well" encloses a PMOS device while the NMOS transistor resides in the p-substrate.

Called "complementary MOS" (CMOS) technology, the above structure requires more complex processing than simple NMOS or PMOS devices. In fact, the first few generations of MOS technology contained only NMOS transistors,[15] and the higher cost of CMOS processes seemed prohibitive. However, many significant advantages of complementary devices eventually made CMOS technology dominant and NMOS technology obsolete.

[15]The first Intel microprocessor, the 4004, was realized in NMOS technology.

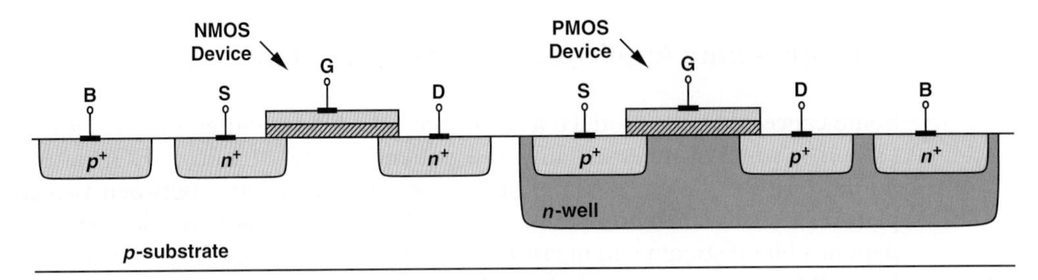

Figure 6.37 CMOS technology.

6.6 COMPARISON OF BIPOLAR AND MOS DEVICES

Having studied the physics and operation of bipolar and MOS transistors, we can now compare their properties. Table 6.2 shows some of the important aspects of each device. Note that the exponential I_C-V_{BE} dependence of bipolar devices accords them a higher transconductance for a given bias current.

TABLE 6.2 Comparison of bipolar and MOS transistors.

Bipolar Transistor	MOSFET
Exponential Characteristic	Quadratic Characteristic
Active: $V_{CB} > 0$	Saturation: $V_{DS} > V_{GS} - V_{TH}$ (NMOS)
Saturation: $V_{CB} < 0$	Triode: $V_{DS} < V_{GS} - V_{TH}$ (NMOS)
Finite Base Current	Zero Gate Current
Early Effect	Channel-Length Modulation
Diffusion Current	Drift Current
—	Voltage-Dependent Resistor

6.7 CHAPTER SUMMARY

- A voltage-dependent current source can form an amplifier along with a load resistor. MOSFETs are electronic devices that can operate as voltage-dependent current sources.
- A MOSFET consists of a conductive plate (the "gate") atop a semiconductor substrate and two junctions ("source" and "drain") in the substrate. The gate controls the current flow from the source to the drain. The gate draws nearly zero current because an insulating layer separates it from the substrate.
- As the gate voltage rises, a depletion region is formed in the substrate under the gate area. Beyond a certain gate-source voltage (the "threshold voltage"), mobile carriers are attracted to the oxide-silicon interface and a channel is formed.
- If the drain-source voltage is small, the device operates a voltage-dependent resistor.

- As the drain voltage rises, the charge density near the drain falls. If the drain voltage reaches one threshold below the gate voltage, the channel ceases to exist near the drain, leading to "pinch-off."
- MOSFETs operate in the "triode" region if the drain voltage is more than one threshold below the gate voltage. In this region, the drain current is a function of V_{GS} and V_{DS}. The current is also proportional to the device aspect ratio, W/L.
- MOSFETs enter the "saturation region" if channel pinch-off occurs, i.e., the drain voltage is less than one threshold below the gate voltage. In this region, the drain current is proportional to $(V_{GS} - V_{TH})^2$.
- MOSFETs operating in the saturation region behave as current sources and find wide application in microelectronic circuits.
- As the drain voltage exceeds $V_{GS} - V_{TH}$ and pinch-off occurs, the drain end of the channel begins to move toward the source, reducing the effective length of the device. Called "channel-length modulation," this effect leads to variation of drain current in the saturation region. That is, the device is not an ideal current source.
- A measure of the small-signal performance of voltage-dependent current sources is the "transconductance," defined as the change in the output current divided by the change in the input voltage. The transconductance of MOSFETs can be expressed by one of three equations in terms of the bias voltages and currents.
- Operation across different regions and/or with large swings exemplifies "large-signal behavior." If the signal swings are sufficiently small, the MOSFET can be represented by a small-signal model consisting of a *linear* voltage-dependent current source and an output resistance.
- The small-signal model is derived by making a small change in the voltage difference between two terminals while the other voltages remain constant.
- The small-signal models of NMOS and PMOS devices are identical.
- NMOS and PMOS transistors are fabricated on the same substrate to create CMOS technology.

PROBLEMS

In the following problems, unless otherwise stated, assume $\mu_n C_{ox} = 200\ \mu A/V^2$, $\mu_p C_{ox} = 100\ \mu A/V^2$, and $V_{TH} = 0.4$ V for NMOS devices and -0.4 V for PMOS devices.

Sec. 6.2 Operation of MOSFET

***6.1.** Two identical MOSFETs are placed in series as shown in Fig. 6.38. If both devices operate as resistors, explain intuitively why this combination is equivalent to a single transistor, M_{eq}. What are the width and length of M_{eq}?

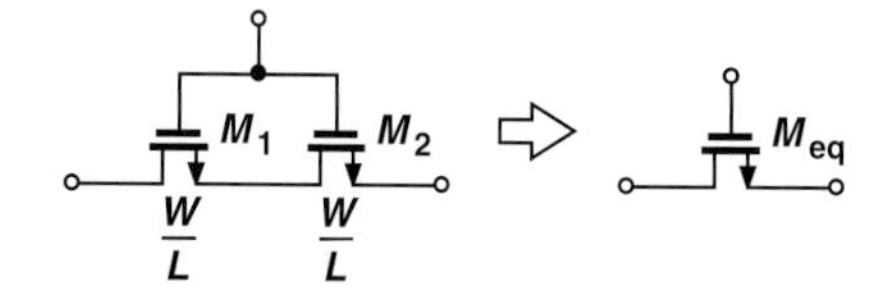

Figure 6.38

***6.2.** Consider a MOSFET experiencing pinch-off near the drain. Equation (6.4) indicates that the charge density and carrier velocity must change in opposite directions if the current remains constant. How can this relationship be interpreted at the pinch-off point, where the charge density approaches zero?

SOL **6.3.** Calculate the total charge stored in the channel of an NMOS device if $C_{ox} = 10\ \text{fF}/\mu\text{m}^2$, $W = 5\ \mu\text{m}$, $L = 0.1\ \mu\text{m}$, and $V_{GS} - V_{TH} = 1$ V. Assume $V_{DS} = 0$.

6.4. Referring to Fig. 6.11 and assuming that $V_D > 0$,
(a) Sketch the electron density in the channel as a function of x.
(b) Sketch the local resistance of the channel (per unit length) as a function of x.

6.5. Assuming I_D is constant, solve Eq. (6.7) to obtain an expression for $V(x)$. Plot both $V(x)$ and dV/dx as a function of x for different values of W or V_{TH}.

***6.6.** The drain current of a MOSFET in the triode region is expressed as

$$I_D = \mu_n C_{ox}\frac{W}{L}\left[(V_{GS} - V_{TH})V_{DS} - \frac{1}{2}V_{DS}^2\right]. \tag{6.78}$$

Suppose the values of $\mu_n C_{ox}$ and W/L are unknown. Is it possible to determine these quantities by applying different values of $V_{GS} - V_{TH}$ and V_{DS} and measuring I_D?

6.7. An NMOS device carries 1 mA with $V_{GS} - V_{TH} = 0.6$ V and 1.6 mA with $V_{GS} - V_{TH} = 0.8$ V. If the device operates in the triode region, calculate V_{DS} and W/L.

***6.8.** Compute the transconductance of a MOSFET operating in the triode region. Define $g_m = \partial I_D/\partial V_{GS}$ for a constant V_{DS}. Explain why $g_m = 0$ for $V_{DS} = 0$.

SOL **6.9.** An NMOS device operating with a small drain-source voltage serves as a resistor. If the supply voltage is 1.8 V, what is the minimum on-resistance that can be achieved with $W/L = 20$?

6.10. We wish to use an NMOS transistor as a variable resistor with $R_{on} = 500\ \Omega$ at $V_{GS} = 1$ V and $R_{on} = 400\ \Omega$ at $V_{GS} = 1.5$ V. Explain why this is not possible.

6.11. For a MOS transistor biased in the triode region, we can define an incremental drain-source resistance as

$$r_{DS,tri} = \left(\frac{\partial I_D}{\partial V_{DS}}\right)^{-1}. \tag{6.79}$$

Derive an expression for this quantity.

6.12. It is possible to define an "intrinsic time constant" for a MOSFET operating as a resistor:

$$\tau = R_{on}C_{GS}, \tag{6.80}$$

where $C_{GS} = WLC_{ox}$. Obtain an expression for τ and explain what the circuit designer must do to minimize the time constant.

6.13. In the circuit of Fig. 6.39, M_1 serves as an electronic switch. If $V_{in} \approx 0$, determine W/L such that the circuit attenuates the signal by only 5%. Assume $V_G = 1.8$ V and $R_L = 100\ \Omega$.

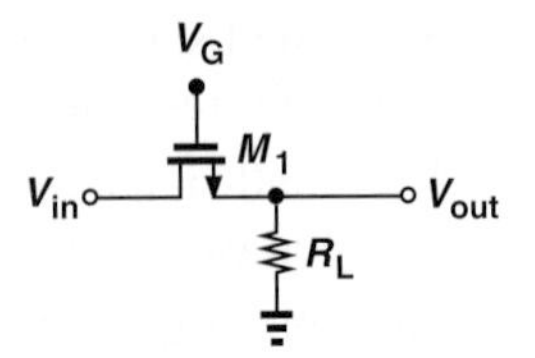

Figure 6.39

6.14. In the circuit of Fig. 6.39, the input is a small sinusoid superimposed on a dc level: $V_{in} = V_0 \cos\omega t + V_1$, where V_0 is on the order of a few millivolts.
(a) For $V_1 = 0$, obtain W/L in terms of R_L and other parameters so that $V_{out} = 0.95V_{in}$.
(b) Repeat part (a) for $V_1 = 0.5$ V. Compare the results.

6.15. For an NMOS device, plot I_D as a function of V_{GS} for different values of V_{DS}.

6.16. In Fig. 6.17, explain why the peaks of the parabolas lie on a parabola themselves. SOL

6.17. Advanced MOS devices do not follow the square-law behavior expressed by Eq. (6.17). A somewhat better approximation is:

$$I_D = \frac{1}{2}\mu_n C_{ox}\frac{W}{L}(V_{GS} - V_{TH})^{\alpha}, \tag{6.81}$$

where α is less than 2. Determine the transconductance of such a device.

6.18. For MOS devices with very short channel lengths, the square-law behavior is not valid, and we may instead write:

$$I_D = WC_{ox}(V_{GS} - V_{TH})v_{sat}, \qquad (6.82)$$

where v_{sat} is a relatively *constant* velocity. Determine the transconductance of such a device.

***6.19.** Determine the region of operation of M_1 in each of the circuits shown in Fig. 6.40.

***6.20.** Determine the region of operation of M_1 in each of the circuits shown in Fig. 6.41.

6.21. Two current sources realized by identical MOSFETs (Fig. 6.42) match to within 1%, i.e., $0.99I_{D2} < I_{D1} < 1.01I_{D2}$. If $V_{DS1} = 0.5$ V and $V_{DS2} = 1$ V, what is the maximum tolerable value of λ?

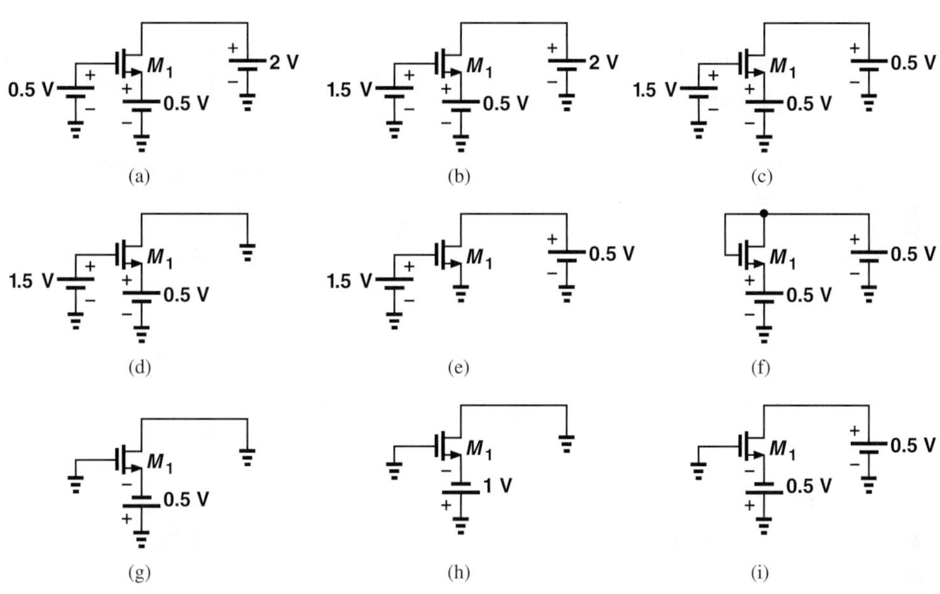

Figure 6.40

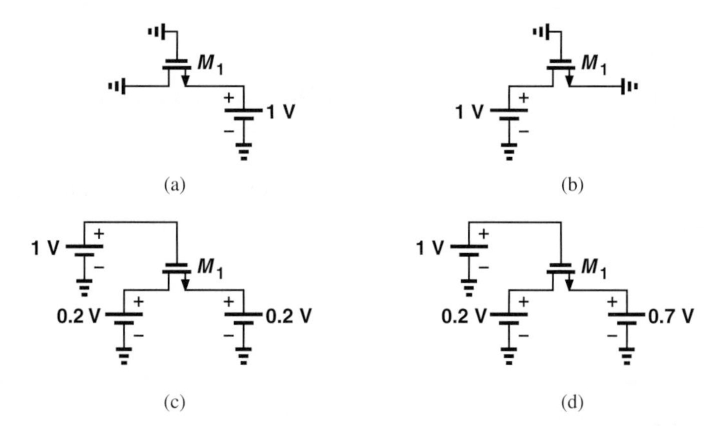

Figure 6.41

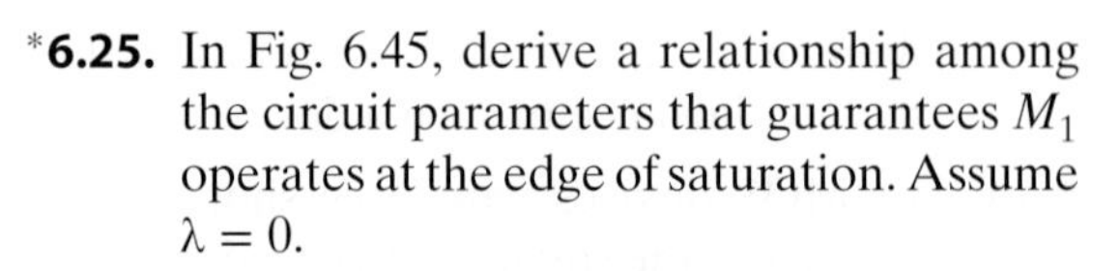

Figure 6.42

6.22. Assume $\lambda = 0$, compute W/L of M_1 in Fig. 6.43 such that the device operates at the edge of saturation.

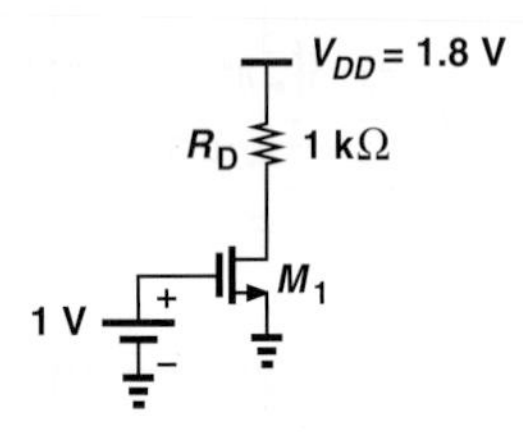

Figure 6.43

SOL **6.23.** Using the value of W/L found in Problem 6.22, explain what happens if the gate oxide thickness is doubled due to a manufacturing error.

6.24. In the Fig. 6.44, what is the minimum allowable value of V_{DD} if M_1 must not enter the triode region? Assume $\lambda = 0$.

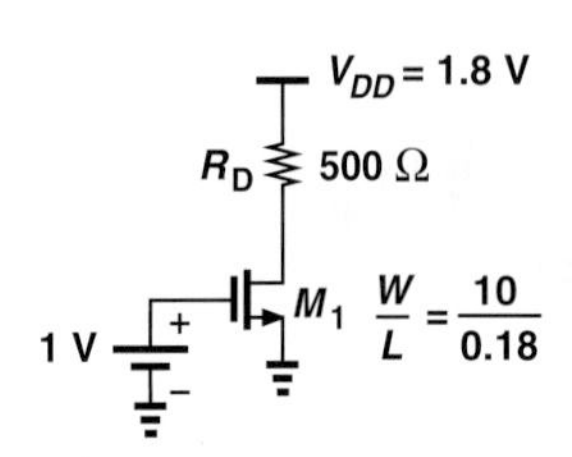

Figure 6.44

*__6.25.__ In Fig. 6.45, derive a relationship among the circuit parameters that guarantees M_1 operates at the edge of saturation. Assume $\lambda = 0$.

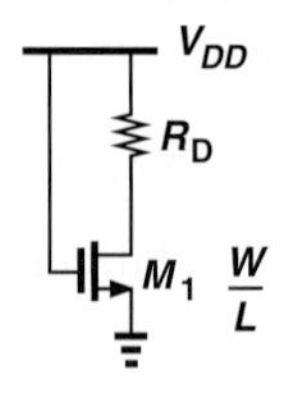

Figure 6.45

6.26. Compute the value of W/L for M_1 in Fig. 6.46 for a bias current of I_1. Assume $\lambda = 0$.

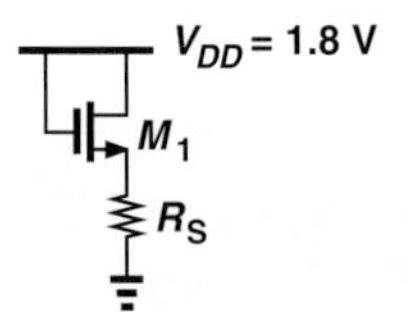

Figure 6.46

6.27. Calculate the bias current of M_1 in Fig. 6.47 if $\lambda = 0$.

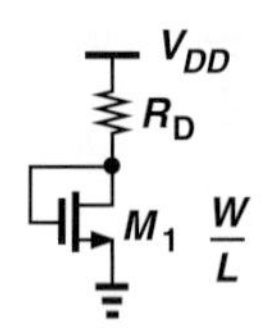

Figure 6.47

**__6.28.__ Sketch I_X as a function of V_X for the circuits shown in Fig. 6.48. Assume V_X goes from 0 to $V_{DD} = 1.8$ V. Also, $\lambda = 0$. Determine at what value of V_X the device changes its region of operation.

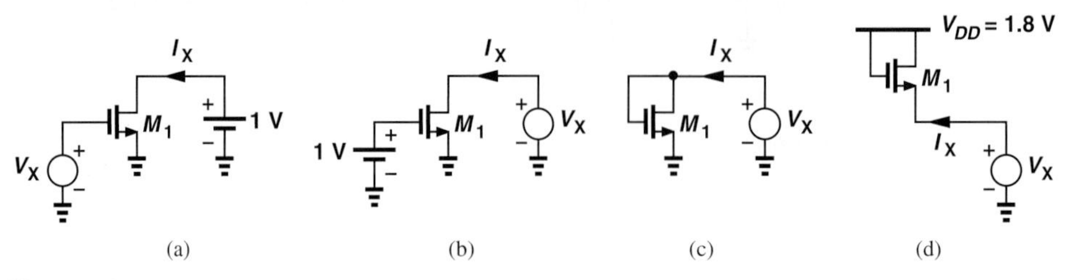

Figure 6.48

6.29. Assuming $W/L = 10/0.18$ $\lambda = 0.1\ \text{V}^{-1}$, and $V_{DD} = 1.8$ V, calculate the drain current of M_1 in Fig. 6.49.

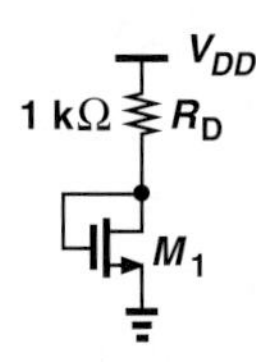

Figure 6.49

6.30. In the circuit of Fig. 6.50, $W/L = 20/0.18$ and $\lambda = 0.1\ \text{V}^{-1}$. What value of V_B places the transistor at the edge of saturation?

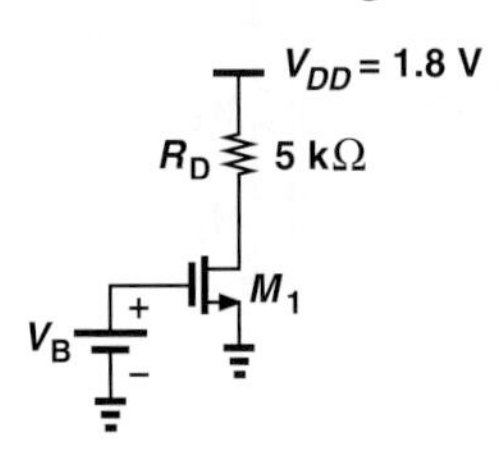

Figure 6.50

SOL **6.31.** An NMOS device operating in saturation with $\lambda = 0$ must provide a transconductance of $1/(50\ \Omega)$.

(a) Determine W/L if $I_D = 0.5$ mA.

(b) Determine W/L if $V_{GS} - V_{TH} = 0.5$ V.

(c) Determine I_D if $V_{GS} - V_{TH} = 0.5$ V.

****6.32.** Determine how the transconductance of a MOSFET (operating in saturation) changes if

(a) W/L is doubled but I_D remains constant.

(b) $V_{GS} - V_{TH}$ is doubled but I_D remains constant.

(c) I_D is doubled but W/L remains constant.

(d) I_D is doubled but $V_{GS} - V_{TH}$ remains constant.

6.33. If $\lambda = 0.1\ \text{V}^{-1}$ and $W/L = 20/0.18$, construct the small-signal model of each of the circuits shown in Fig. 6.51.

6.34. The "intrinsic gain" of a MOSFET operating in saturation is defined as $g_m r_O$. Derive an expression for $g_m r_O$ and plot the result as a function of I_D. Assume V_{DS} is constant.

***6.35.** Assuming a constant V_{DS}, plot the intrinsic gain, $g_m r_O$, of a MOSFET

(a) as a function of $V_{GS} - V_{TH}$ if I_D is constant.

(b) as a function of I_D if $V_{GS} - V_{TH}$ is constant.

6.36. An NMOS device with $\lambda = 0.1\ \text{V}^{-1}$ must provide a $g_m r_O$ of 20 with $V_{DS} = 1.5$ V. Determine the required value of W/L if $I_D = 0.5$ mA.

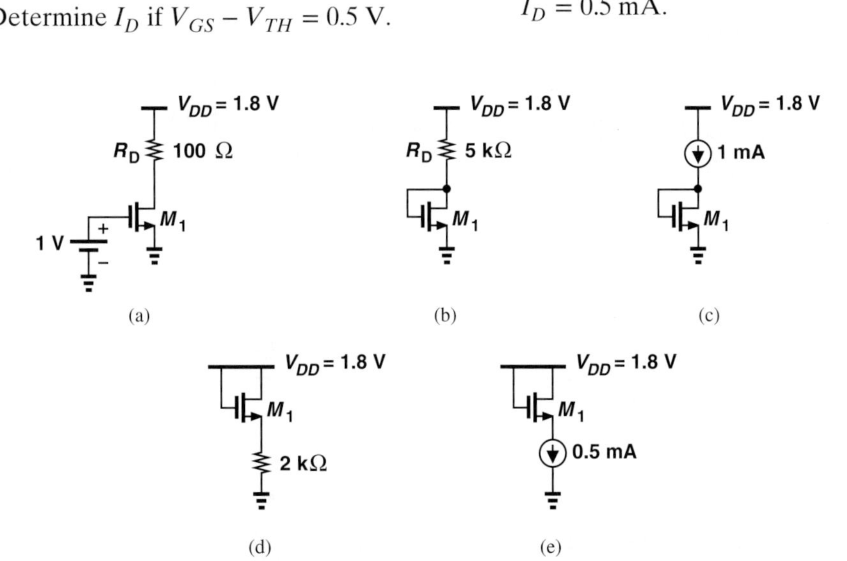

Figure 6.51

SOL **6.37.** Repeat Problem 6.36 for $\lambda = 0.2\ V^{-1}$.

6.38. Construct the small-signal model of the circuits depicted in Fig. 6.52. Assume all transistors operate in saturation and $\lambda \neq 0$.

Sec. 6.4 PMOS Transistor

***6.39.** Determine the region of operation of M_1 in each circuit shown in Fig. 6.53.

***6.40.** Determine the region of operation of M_1 in each circuit shown in Fig. 6.54.

6.41. If $\lambda = 0$, what value of W/L places M_1 at the edge of saturation in Fig. 6.55?

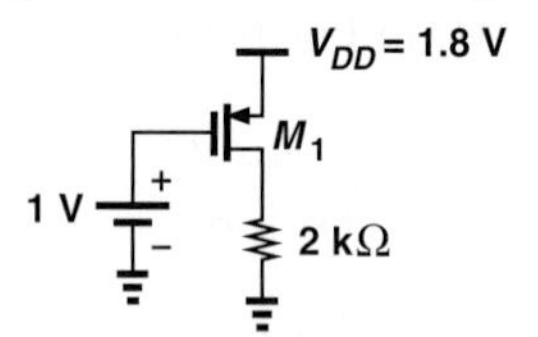

Figure 6.55

6.42. With the value of W/L obtained in Problem 6.41, what happens if V_B changes to +0.8 V?

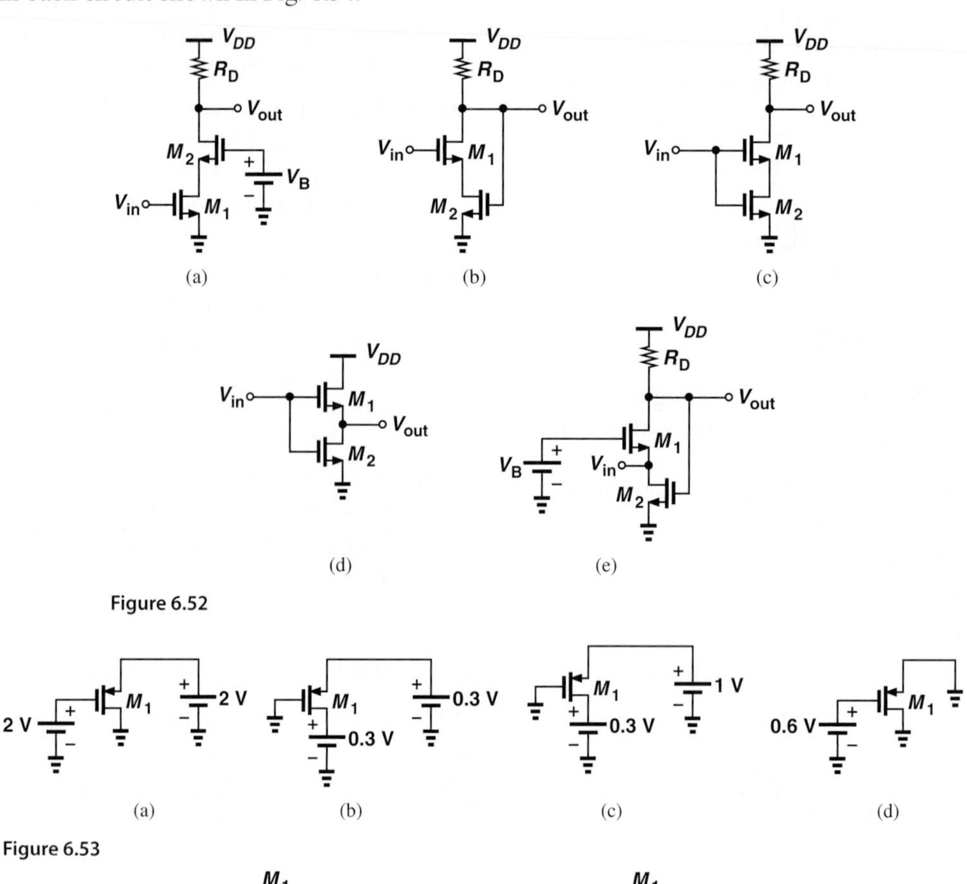

Figure 6.52

Figure 6.53

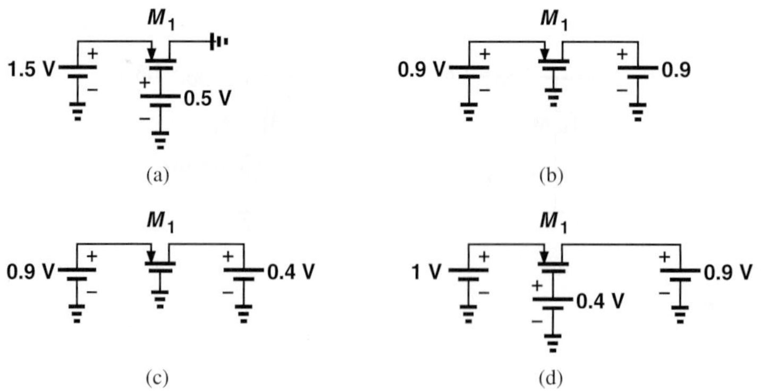

M_1

1.5 V

0.5 V

(a)

M_1

0.9 V

0.9

(b)

M_1

0.9 V

0.4 V

(c)

M_1

1 V

0.4 V

0.9 V

(d)

Figure 6.54

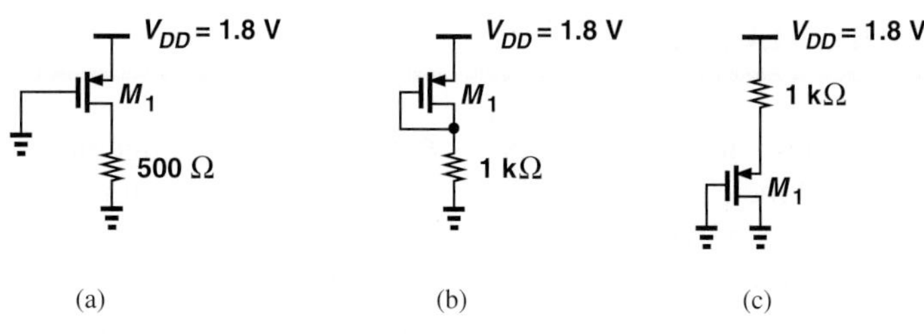

Figure 6.56

6.43. If $W/L = 10/0.18$ and $\lambda = 0$, determine the operating point of M_1 in each circuit depicted in Fig. 6.56.

**6.44. Sketch I_X as a function of V_X for the circuits shown in Fig. 6.57. Assume V_X goes from 0 to $V_{DD} = 1.8$ V. Also, $\lambda = 0$. Determine at what value of V_X the device changes its region of operation.

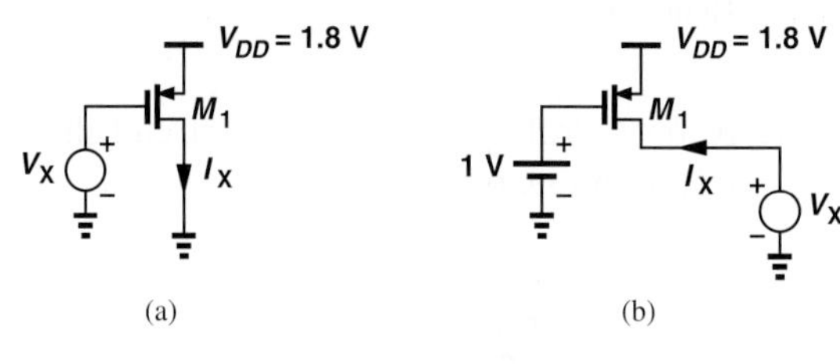

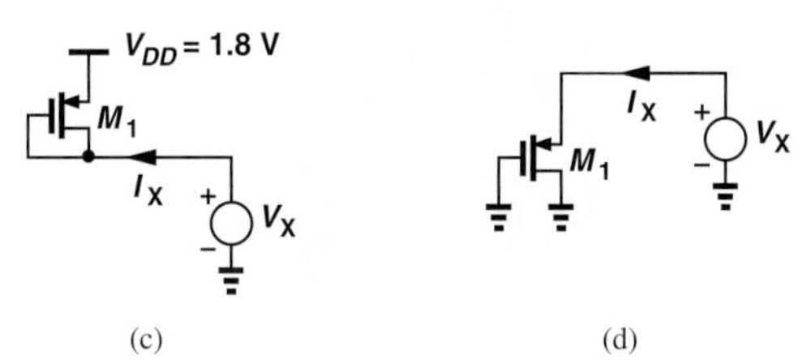

Figure 6.57

6.45. Construct the small-signal model of each circuit shown in Fig. 6.58 if all of the transistors operate in saturation and $\lambda \neq 0$.

**6.46. Consider the circuit depicted in Fig. 6.59, where M_1 and M_2 operate in saturation and exhibit channel-length modulation coefficients λ_n and λ_p, respectively.

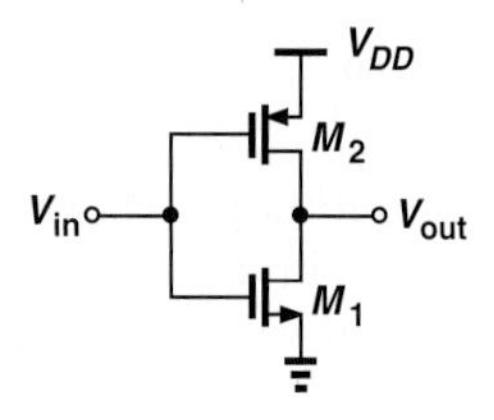

Figure 6.59

(a) Construct the small-signal equivalent circuit and explain why M_1 and M_2 appear in "parallel."

(b) Determine the small-signal voltage gain of the circuit.

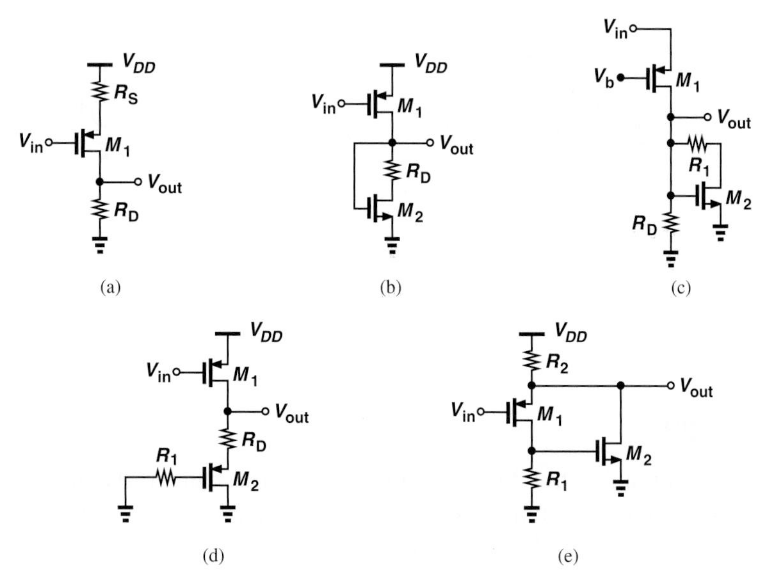

Figure 6.58

SPICE PROBLEMS

In the following problems, use the MOS models and source/drain dimensions given in Appendix A. Assume the substrates of NMOS and PMOS devices are tied to ground and V_{DD}, respectively.

6.47. For the circuit shown in Fig. 6.60, plot V_X as a function of I_X for $0 < I_X < 3$ mA. Explain the sharp change in V_X as I_X exceeds a certain value.

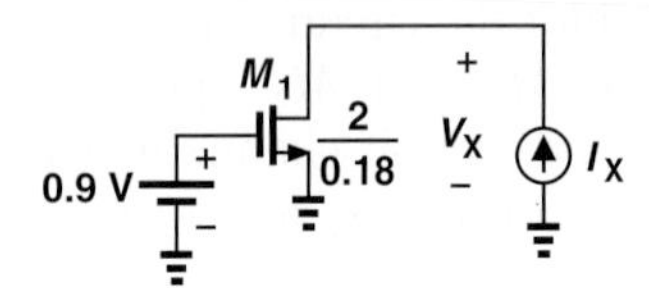

Figure 6.60

6.48. Plot the input/output characteristic of the stage shown in Fig. 6.61 for $0 < V_{in} <$ 1.8 V. At what value of V_{in} does the slope (gain) reach a maximum?

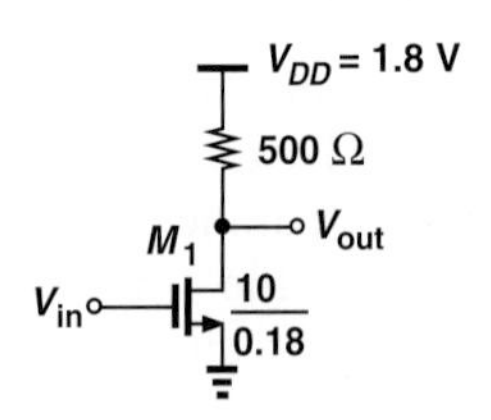

Figure 6.61

6.49. For the arrangements shown in Fig. 6.62, plot I_D as a function of V_X as V_X varies from 0 to 1.8 V. Can we say these two arrangements are equivalent?

6.50. Plot I_X as a function of V_X for the arrangement depicted in Fig. 6.63 as V_X varies from 0 to 1.8 V. Can you explain the behavior of the circuit?

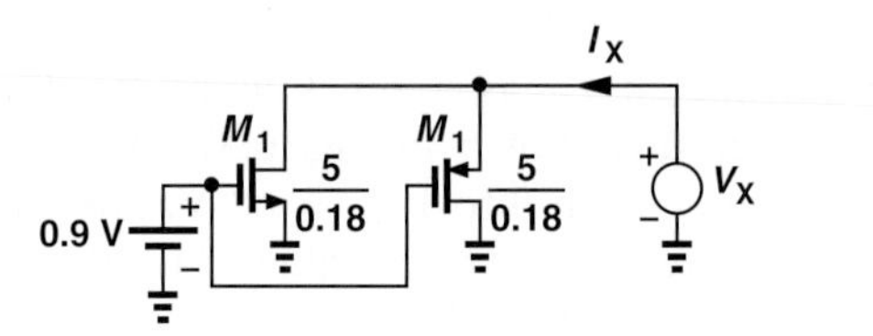

Figure 6.63

6.51. Repeat Problem 6.50 for the circuit illustrated in Fig. 6.64.

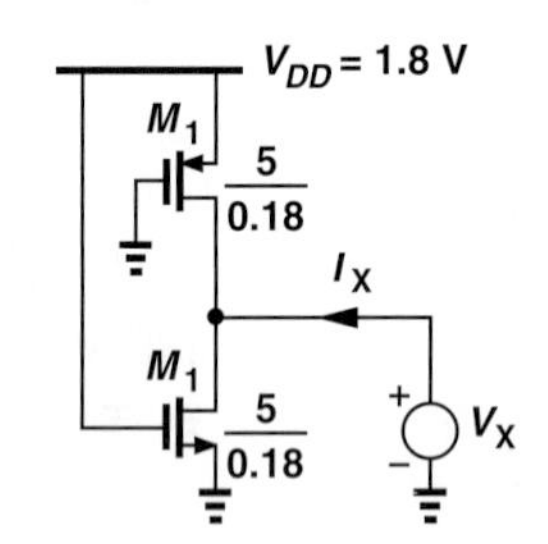

Figure 6.64

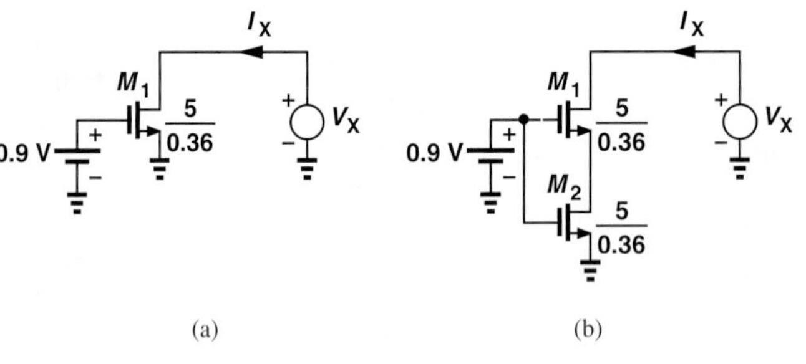

Figure 6.62

REFERENCE

1. S. Song, et al., "A CMOS VEGF sensor for cancer diagnosis using a peptide aptamer-based functionalized microneedle," *IEEE Trans. Biomedical Circuits and Systems*, vol. 13, pp. 1288–1299, Nov. 2019.

7

CMOS Amplifiers

Watch Companion YouTube Videos:

Razavi Electronics 1, Lec 35: Introduction to Amplifiers, Common-Source Stage I
https://www.youtube.com/watch?v=UTo4lGXxzlk

Razavi Electronics 1, Lec 36: Common-Source Stage II
https://www.youtube.com/watch?v=tdjabkYv-Dc

Razavi Electronics 1, Lec 37: Common-Source Variants
https://www.youtube.com/watch?v=UeXtyA42ElU

Razavi Electronics 1, Lec 38: Common-Source Stage with Degeneration
https://www.youtube.com/watch?v=qc6pDZJqFRs

Razavi Electronics 1, Lec 39: Biasing Techniques, Introduction to Common-Gate Stage
https://www.youtube.com/watch?v=qNNk1cmnYCo

Razavi Electronics 1, Lec 40: Common-Gate Stage
https://www.youtube.com/watch?v=-Gm7BskTC1A

Razavi Electronics 1, Lec 41: Source Followers & Summary
https://www.youtube.com/watch?v=hol6CUOoMmU

Most CMOS amplifiers have identical bipolar counterparts and can therefore be analyzed in the same fashion. Our study in this chapter parallels the developments in Chapter 5, identifying both similarities and differences between CMOS and bipolar circuit topologies. It is recommended that the reader review Chapter 5, specifically, Section 5.1. We assume the reader is familiar with concepts such as I/O impedances, biasing, and dc and small-signal analysis. The outline of the chapter is shown below.

General Concepts
- **Biasing of MOS Stages**
- **Realization of Current Sources**

MOS Amplifiers
- **Common–Source Stage**
- **Common–Gate Stage**
- **Source Follower**

7.1 GENERAL CONSIDERATIONS

7.1.1 MOS Amplifier Topologies

Recall from Section 5.3 that the nine possible circuit topologies using a bipolar transistor in fact reduce to three useful configurations. The similarity of bipolar and MOS small-signal models (i.e., a voltage-controlled current source) suggests that the same must hold for MOS amplifiers. In other words, we expect three basic CMOS amplifiers: the "common-source" (CS) stage, the "common-gate" (CG) stage, and the "source follower."

7.1.2 Biasing

Depending on the application, MOS circuits may incorporate biasing techniques that are quite different from those described in Chapter 5 for bipolar stages. Most of these techniques are beyond the scope of this book and some methods are studied in Chapter 5. Nonetheless, it is still instructive to apply some of the biasing concepts of Chapter 5 to MOS stages.

Consider the circuit shown in Fig. 7.1, where the gate voltage is defined by R_1 and R_2. We assume M_1 operates in saturation. Also, in most bias calculations, we can neglect channel-length modulation. Noting that the gate current is zero, we have

$$V_X = \frac{R_2}{R_1 + R_2} V_{DD}. \tag{7.1}$$

Since $V_X = V_{GS} + I_D R_S$,

$$\frac{R_2}{R_1 + R_2} V_{DD} = V_{GS} + I_D R_S. \tag{7.2}$$

Also,

$$I_D = \frac{1}{2}\mu_n C_{ox}\frac{W}{L}(V_{GS} - V_{TH})^2. \tag{7.3}$$

Equations (7.2) and (7.3) can be solved to obtain I_D and V_{GS}, either by iteration or by finding I_D from Eq. (7.2) and replacing for it in Eq. (7.3):

$$\left(\frac{R_2}{R_1 + R_2} V_{DD} - V_{GS}\right)\frac{1}{R_S} = \frac{1}{2}\mu_n C_{ox}\frac{W}{L}(V_{GS} - V_{TH})^2. \tag{7.4}$$

That is,

$$V_{GS} = -(V_1 - V_{TH}) + \sqrt{(V_1 - V_{TH})^2 - V_{TH}^2 + \frac{2R_2}{R_1 + R_2}V_1 V_{DD}}, \tag{7.5}$$

$$= -(V_1 - V_{TH}) + \sqrt{V_1^2 + 2V_1\left(\frac{R_2 V_{DD}}{R_1 + R_2} - V_{TH}\right)}, \tag{7.6}$$

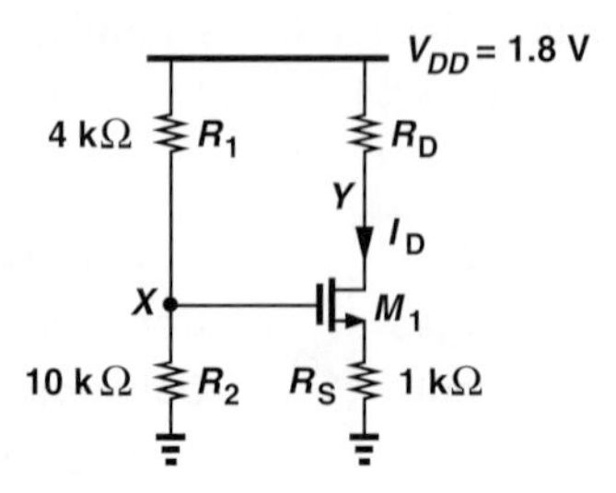

Figure 7.1 MOS stage with biasing.

where

$$V_1 = \frac{1}{\mu_n C_{ox} \frac{W}{L} R_S}. \tag{7.7}$$

This value of V_{GS} can then be substituted in Eq. (7.2) to obtain I_D. Of course, V_Y must exceed $V_X - V_{TH}$ to ensure operation in the saturation region.

Example 7-1 Determine the bias current of M_1 in Fig. 7.1 assuming $V_{TH} = 0.5$ V, $\mu_n C_{ox} = 100\ \mu\text{A/V}^2$, $W/L = 5/0.18$, and $\lambda = 0$. What is the maximum allowable value of R_D for M_1 to remain in saturation?

Solution We have

$$V_X = \frac{R_2}{R_1 + R_2} V_{DD} \tag{7.8}$$

$$= 1.286\ \text{V}. \tag{7.9}$$

With an initial guess $V_{GS} = 1$ V, the voltage drop across R_S can be expressed as $V_X - V_{GS} = 286$ mV, yielding a drain current of 286 μA. Substituting for I_D in Eq. (7.3) gives the new value of V_{GS} as

$$V_{GS} = V_{TH} + \sqrt{\frac{2I_D}{\mu_n C_{ox} \frac{W}{L}}} \tag{7.10}$$

$$= 0.954\ \text{V}. \tag{7.11}$$

Consequently,

$$I_D = \frac{V_X - V_{GS}}{R_S} \tag{7.12}$$

$$= 332\ \mu\text{A}, \tag{7.13}$$

and hence

$$V_{GS} = 0.989\ \text{V}. \tag{7.14}$$

This gives $I_D = 297$ μA.

As seen from the iterations, the solutions converge more slowly than those encountered in Chapter 5 for bipolar circuits. This is due to the quadratic (rather than exponential) I_D-V_{GS} dependence. We may therefore utilize the exact result in Eq. (7.6) to avoid lengthy calculations. Since $V_1 = 0.36$ V,

$$V_{GS} = 0.974\ \text{V} \tag{7.15}$$

and

$$I_D = \frac{V_X - V_{GS}}{R_S} \tag{7.16}$$

$$= 312\ \mu\text{A}. \tag{7.17}$$

The maximum allowable value of R_D is obtained if $V_Y = V_X - V_{TH} = 0.786$ V. That is,

$$R_D = \frac{V_{DD} - V_Y}{I_D} \tag{7.18}$$

$$= 3.25\ \text{k}\Omega. \tag{7.19}$$

Exercise What is the value of R_2 that places M_1 at the edge of saturation?

Example 7-2 In the circuit of Example 7-1, assume M_1 is in saturation and $R_D = 2.5\ \text{k}\Omega$ and compute (a) the maximum allowable value of W/L and (b) the minimum allowable value of R_S (with $W/L = 5/0.18$). Assume $\lambda = 0$.

Solution (a) As W/L becomes larger, M_1 can carry a larger current for a given V_{GS}. With $R_D = 2.5\ \text{k}\Omega$ and $V_X = 1.286$ V, the maximum allowable value of I_D is given by

$$I_D = \frac{V_{DD} - V_Y}{R_D} \tag{7.20}$$

$$= 406\ \mu\text{A}. \tag{7.21}$$

The voltage drop across R_S is then equal to 406 mV, yielding $V_{GS} = 1.286\ \text{V} - 0.406\ \text{V} = 0.88$ V. In other words, M_1 must carry a current of 406 μA with $V_{GS} = 0.88$ V:

$$I_D = \frac{1}{2}\mu_n C_{ox}\frac{W}{L}(V_{GS} - V_{TH})^2 \tag{7.22}$$

$$406\ \mu\text{A} = (50\ \mu\text{A/V}^2)\frac{W}{L}(0.38\ \text{V})^2; \tag{7.23}$$

thus,

$$\frac{W}{L} = 56.2. \tag{7.24}$$

(b) With $W/L = 5/0.18$, the minimum allowable value of R_S gives a drain current of 406 μA. Since

$$V_{GS} = V_{TH} + \sqrt{\frac{2I_D}{\mu_n C_{ox}\frac{W}{L}}} \tag{7.25}$$

$$= 1.041\ \text{V}, \tag{7.26}$$

the voltage drop across R_S is equal to $V_X - V_{GS} = 245$ mV. It follows that

$$R_S = \frac{V_X - V_{GS}}{I_D} \tag{7.27}$$

$$= 604\ \Omega. \tag{7.28}$$

Exercise Repeat the above example if $V_{TH} = 0.35$ V.

The self-biasing technique of Fig. 5.22 can also be applied to MOS amplifiers. As depicted in Fig. 7.2, the circuit can be analyzed by noting that M_1 is in saturation (why?) and the voltage drop across R_G is zero. Thus,

$$I_D R_D + V_{GS} + R_S I_D = V_{DD}. \tag{7.29}$$

Finding V_{GS} from this equation and substituting it in Eq. (7.3), we have

$$I_D = \frac{1}{2}\mu_n C_{ox}\frac{W}{L}[V_{DD} - (R_S + R_D)I_D - V_{TH}]^2, \tag{7.30}$$

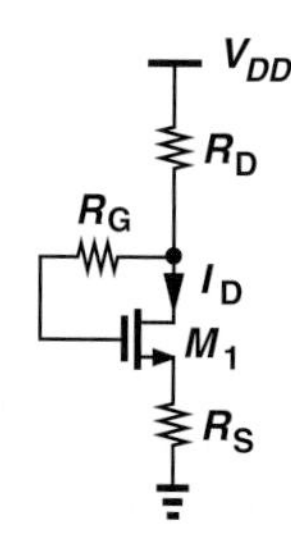

Figure 7.2 Self-biased MOS stage.

where channel-length modulation is neglected. It follows that

$$(R_S + R_D)^2 I_D^2 - 2\left[(V_{DD} - V_{TH})(R_S + R_D) + \frac{1}{\mu_n C_{ox}\frac{W}{L}}\right] I_D + (V_{DD} - V_{TH})^2 = 0. \quad (7.31)$$

Example 7-3

Calculate the drain current of M_1 in Fig. 7.3 if $\mu_n C_{ox} = 100\ \mu\text{A/V}^2$, $V_{TH} = 0.5$ V, and $\lambda = 0$. What value of R_D is necessary to reduce I_D by a factor of two?

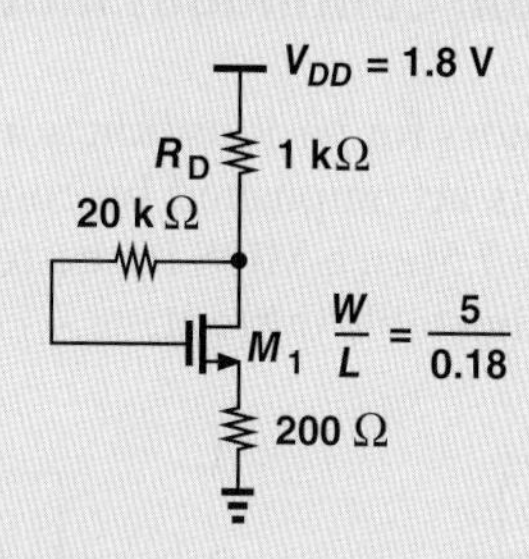

Figure 7.3 Example of self-biased MOS stage.

Solution Equation (7.31) gives

$$I_D = 556\ \mu\text{A}. \quad (7.32)$$

To reduce I_D to 278 μA, we solve Eq. (7.31) for R_D:

$$R_D = 2.867\ \text{k}\Omega. \quad (7.33)$$

Exercise Repeat the above example if V_{DD} drops to 1.2 V.

7.1.3 Realization of Current Sources

MOS transistors operating in saturation can act as current sources. As illustrated in Fig. 7.4(a), an NMOS device serves as a current source with one terminal tied to ground, i.e., it draws current from node X to ground. On the other hand, a PMOS transistor [Fig. 7.4(b)] draws current from V_{DD} to node Y. If $\lambda = 0$, these currents remain independent of V_X or V_Y (so long as the transistors are in saturation).

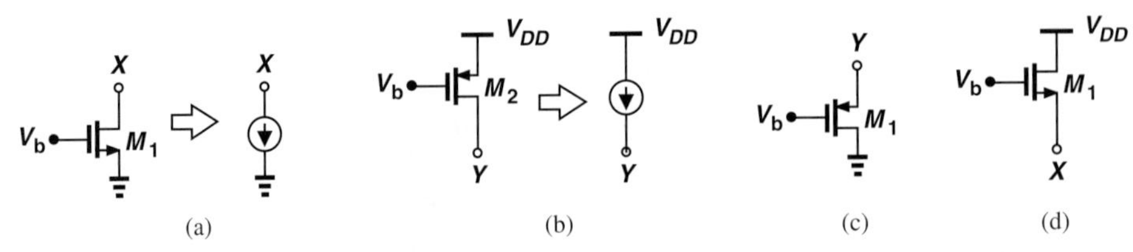

Figure 7.4 (a) NMOS device operating as a current source, (b) PMOS device operating as a current source, (c) PMOS topology not operating as a current source, (d) NMOS topology not operating as a current source.

It is important to understand that only the *drain* terminal of a MOSFET can draw a dc current and still present a high impedance. Specifically, NMOS or PMOS devices configured as shown in Figs. 7.4(c) and (d) do *not* operate as current sources because variation of V_X or V_Y directly changes the gate-source voltage of each transistor, thus changing the drain current considerably. From another perspective, the small-signal model of these two structures is identical to that of the diode-connected devices in Fig. 6.34, revealing a small-signal impedance of only $1/g_m$ (if $\lambda = 0$) rather than infinity.

Robotics Application: Continuous Motor Control

Robots incorporate motors for various functions. The motors must typically run with an adjustable speed. This is accomplished by controlling the current flowing through the motor. As shown in Fig. 7.5(a), a MOS transistor can act as a voltage-dependent current source, delivering an adjustable current.

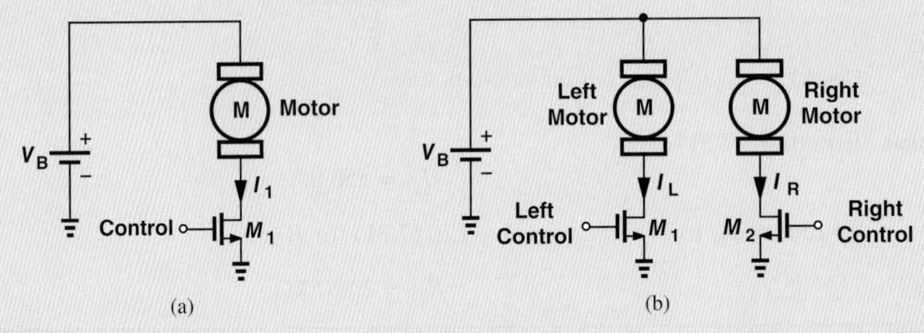

Figure 7.5 (a) Continuous control of a motor by a MOS current source, and (b) control of left and right motors in a robot.

Mobile robots employ motors for the wheels on the left and on the right [Fig. 7.5(b)]. If the robot is to travel on a straight line, the two motors run at the same speed. For making left or right turns, one motor is slowed down by reducing its current. For example, if the motor on the left runs more slowly, then the robot turns left.

It is interesting to note that, even if $I_L = I_R$ in Fig. 7.5(b), the robot may not follow a straight line because the two motors generally suffer from "mismatches" due to manufacturing tolerances. Similarly, even if the motors exhibit no mismatches, M_1 and M_2 may. Thus, the system must be initially calibrated to correct for the mismatches.

7.2 COMMON-SOURCE STAGE

7.2.1 CS Core

As shown in Fig. 7.6(a), the basic CS stage is similar to the common-emitter topology, with the input applied to the gate and the output sensed at the drain. For small signals, M_1 converts the input voltage variations to proportional drain current changes, and R_D transforms the drain currents to the output voltage. If channel-length modulation is neglected, the small-signal model in Fig. 7.6(b) yields $v_{in} = v_1$ and $v_{out} = -g_m v_1 R_D$. That is,

$$\frac{v_{out}}{v_{in}} = -g_m R_D, \tag{7.34}$$

a result similar to that obtained for the common emitter stage in Chapter 5.

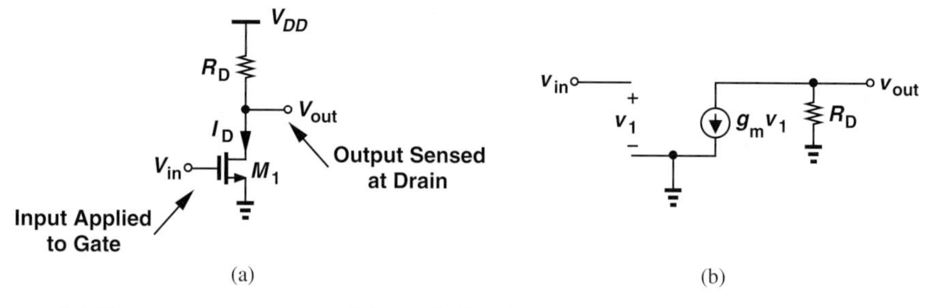

Figure 7.6 (a) Common-source stage, (b) small-signal mode.

The voltage gain of the CS stage is also limited by the supply voltage. Since $g_m = \sqrt{2\mu_n C_{ox}(W/L)I_D}$, we have

$$A_v = -\sqrt{2\mu_n C_{ox}\frac{W}{L}I_D}R_D, \tag{7.35}$$

concluding that if I_D or R_D is increased, so is the voltage drop across R_D ($= I_D R_D$).[1] For M_1 to remain in saturation,

$$V_{DD} - R_D I_D > V_{GS} - V_{TH}, \tag{7.36}$$

that is,

$$R_D I_D < V_{DD} - (V_{GS} - V_{TH}). \tag{7.37}$$

Example 7-4 Calculate the small-signal voltage gain of the CS stage shown in Fig. 7.7 if $I_D = 1$ mA, $\mu_n C_{ox} = 100\ \mu A/V^2$, $V_{TH} = 0.5$ V, and $\lambda = 0$. Verify that M_1 operates in saturation.

Solution We have

$$g_m = \sqrt{2\mu_n C_{ox}\frac{W}{L}I_D} \tag{7.38}$$

$$= \frac{1}{300\ \Omega}. \tag{7.39}$$

[1] It is possible to raise the gain to some extent by increasing W, but "subthreshold conduction" eventually limits the transconductance. This concept is beyond the scope of this book.

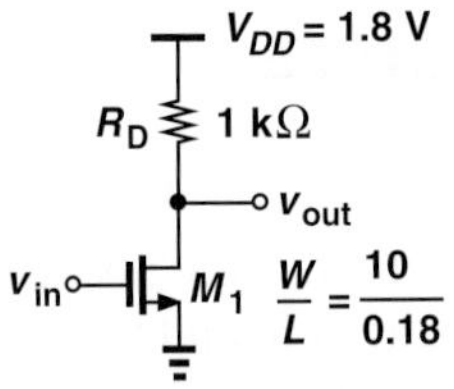

Figure 7.7 Example of CS stage.

Thus,

$$A_v = -g_m R_D \tag{7.40}$$

$$= 3.33. \tag{7.41}$$

To check the operation region, we first determine the gate-source voltage:

$$V_{GS} = V_{TH} + \sqrt{\frac{2I_D}{\mu_n C_{ox}\frac{W}{L}}} \tag{7.42}$$

$$= 1.1 \text{ V}. \tag{7.43}$$

The drain voltage is equal to $V_{DD} - R_D I_D = 0.8$ V. Since $V_{GS} - V_{TH} = 0.6$ V, the device indeed operates in saturation and has a margin of 0.2 V with respect to the triode region. For example, if R_D is doubled with the intention of doubling A_v, then M_1 enters the triode region and its transconductance drops.

Exercise What value of V_{TH} places M_1 at the edge of saturation?

Since the gate terminal of MOSFETs draws a zero current (at very low frequencies), we say the CS amplifier provides a current gain of infinity. By contrast, the current gain of a common-emitter stage is equal to β.

Let us now compute the I/O impedances of the CS amplifier. Since the gate current is zero (at low frequencies),

$$R_{in} = \infty, \tag{7.44}$$

a point of contrast to the CE stage (whose R_{in} is equal to r_π). The high input impedance of the CS topology plays a critical role in many analog circuits.

The similarity between the small-signal equivalents of CE and CS stages indicates that the output impedance of the CS amplifier is simply equal to

$$R_{out} = R_D. \tag{7.45}$$

This is also seen from Fig. 7.8.

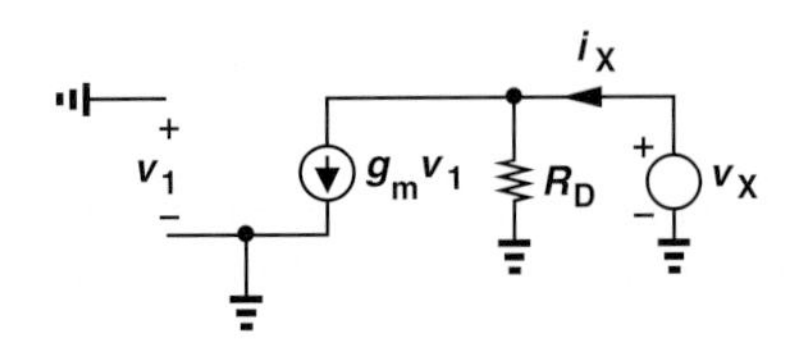

Figure 7.8 Output impedance of CS stage.

In practice, channel-length modulation may not be negligible, especially if R_D is large. The small-signal model of CS topology is therefore modified as shown in Fig. 7.9, revealing that

$$A_v = -g_m(R_D||r_O) \tag{7.46}$$

$$R_{in} = \infty \tag{7.47}$$

$$R_{out} = R_D||r_O. \tag{7.48}$$

In other words, channel-length modulation and the Early effect impact the CS and CE stages, respectively, in a similar manner.

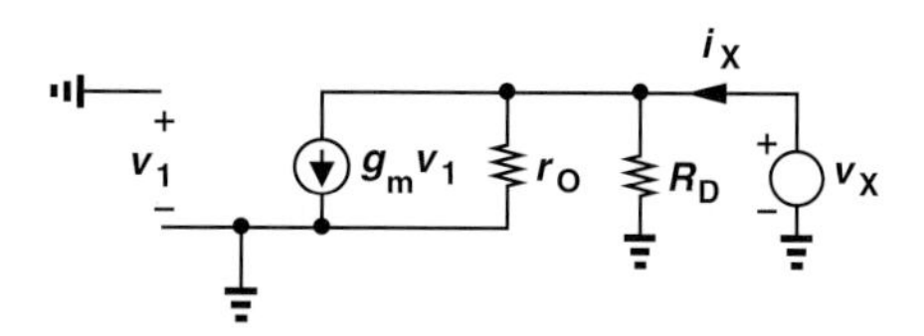

Figure 7.9 Effect of channel-length modulation on CS stage.

Example 7-5 Assuming M_1 operates in saturation, determine the voltage gain of the circuit depicted in Fig. 7.10(a) and plot the result as a function of the transistor channel length while other parameters remain constant.

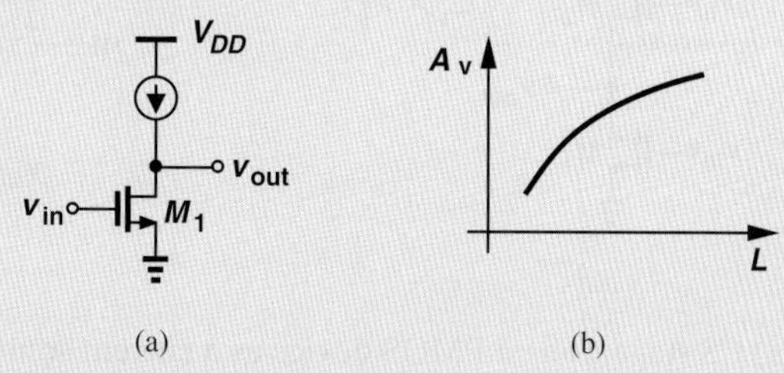

Figure 7.10 (a) CS stage with ideal current source as a load, (b) gain as a function of device channel length.

Solution The ideal current source presents an infinite small-signal resistance, allowing the use of Eq. (7.46) with $R_D = \infty$:

$$A_v = -g_m r_O. \tag{7.49}$$

This is the highest voltage gain that a single transistor can provide. Writing $g_m = \sqrt{2\mu_n C_{ox}(W/L)I_D}$ and $r_O = (\lambda I_D)^{-1}$, we have

$$|A_v| = \frac{\sqrt{2\mu_n C_{ox}\frac{W}{L}}}{\lambda\sqrt{I_D}}. \tag{7.50}$$

This result may imply that $|A_v|$ falls as L increases, but recall from Chapter 6 that $\lambda \propto L^{-1}$:

$$|A_v| \propto \sqrt{\frac{2\mu_n C_{ox} WL}{I_D}}. \tag{7.51}$$

Consequently, $|A_v|$ increases with L [Fig. 7.10(b)].

Exercise Repeat the above example if a resistor of value R_1 is tied between the gate and drain of M_1.

Did you know?

The intrinsic gain, $g_m r_O$, of MOSFETs has fallen with technology scaling, i.e., as the minimum channel length has gone from about 10 μm in the 1960s to 25 nm today. Due to severe channel-length modulation, the intrinsic gain of these short-channel devices is on the order of 5 to 10, making it difficult to achieve a high voltage gain in many analog circuits. This issue has prompted extensive research on analog design using low-gain building blocks. For example, the analog-to-digital converter that digitizes the image in your camera may need an op amp with a gain of 4,000 but must now be designed with a gain of only 20.

7.2.2 CS Stage with Current-Source Load

As seen in the above example, the trade-off between the voltage gain and the voltage headroom can be relaxed by replacing the load resistor with a current source. The observations made in relation to Fig. 7.4(b) therefore suggest the use of a PMOS device as the load of an NMOS CS amplifier [Fig. 7.11(a)].

Let us determine the small-signal gain and output impedance of the circuit. Having a constant gate-source voltage, M_2 simply behaves as a resistor equal to its output impedance [Fig. 7.11(b)] because $v_1 = 0$ and hence $g_{m2}v_1 = 0$. Thus, the drain node of M_1 sees both r_{O1} and r_{O2} to ac ground. Equations (7.46) and (7.48) give

$$A_v = -g_{m1}(r_{O1}||r_{O2}) \tag{7.52}$$

$$R_{out} = r_{O1}||r_{O2}. \tag{7.53}$$

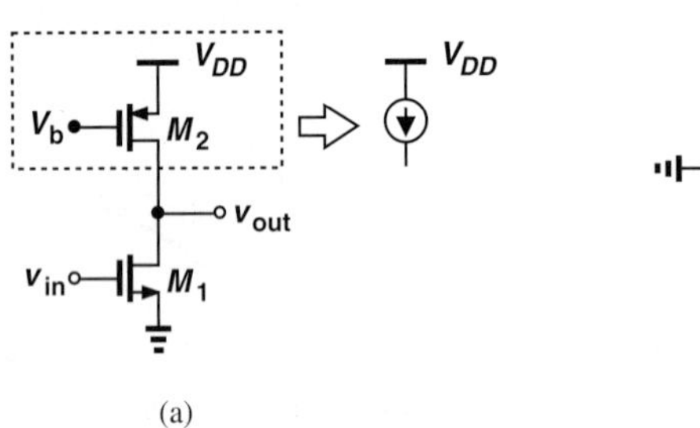

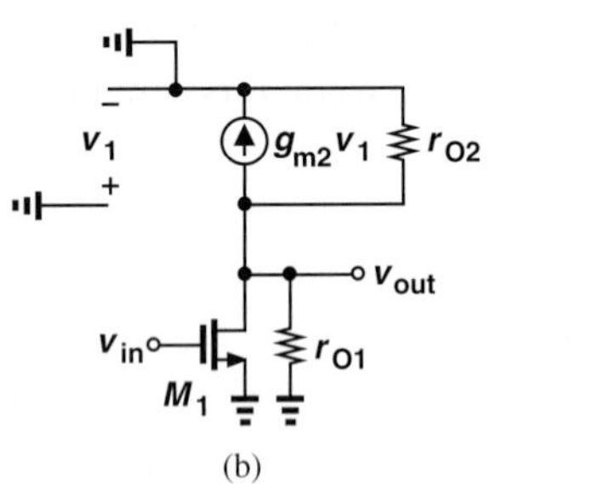

Figure 7.11 (a) CS stage using a PMOS device as a current source, (b) small-signal model.

Example 7-6 Figure 7.12 shows a PMOS CS stage using an NMOS current source load. Compute the voltage gain of the circuit.

Figure 7.12 CS stage using an NMOS device as current source.

Solution Transistor M_2 generates a small-signal current equal to $g_{m2}v_{in}$, which then flows through $r_{O1}||r_{O2}$, producing $v_{out} = -g_{m2}v_{in}(r_{O1}||r_{O2})$. Thus,

$$A_v = -g_{m2}(r_{O1}||r_{O2}). \tag{7.54}$$

Exercise Calculate the gain if the circuit drives a loads resistance equal to R_L.

7.2.3 CS Stage with Diode-Connected Load

In some applications, we may use a diode-connected MOSFET as the drain load. Illustrated in Fig. 7.13(a), such a topology exhibits only a moderate gain due to the relatively low impedance of the diode-connected device (Section 7.1.3). With $\lambda = 0$, M_2 acts as a small-signal resistance equal to $1/g_{m2}$, and Eq. (7.34) yields

$$A_v = -g_{m1} \cdot \frac{1}{g_{m2}} \tag{7.55}$$

$$= -\frac{\sqrt{2\mu_n C_{ox}(W/L)_1 I_D}}{\sqrt{2\mu_n C_{ox}(W/L)_2 I_D}} \tag{7.56}$$

$$= -\sqrt{\frac{(W/L)_1}{(W/L)_2}}. \tag{7.57}$$

Interestingly, the gain is given by the dimensions of M_1 and M_2 and remains independent of process parameters μ_n and C_{ox} and the drain current, I_D.

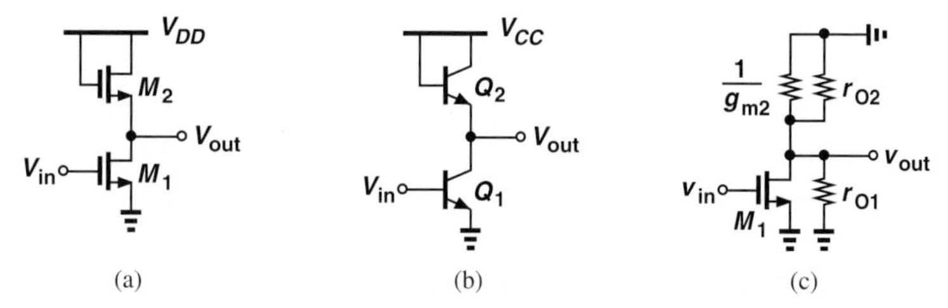

Figure 7.13 (a) MOS stage using a diode-connected load, (b) bipolar counterpart, (c) simplified circuit of (a).

The reader may wonder why we did not consider a common-emitter stage with a diode-connected load in Chapter 5. As shown in Fig. 7.13(b), such a circuit is not used because it provides a voltage gain of only unity:

$$A_v = -g_{m1} \cdot \frac{1}{g_{m2}} \tag{7.58}$$

$$= -\frac{I_{C1}}{V_T} \cdot \frac{1}{I_{C2}/V_T} \tag{7.59}$$

$$\approx -1. \tag{7.60}$$

The contrast between Eqs. (7.57) and (7.60) arises from a fundamental difference between MOS and bipolar devices: transconductance of the former depends on device dimensions whereas that of the latter does not.

A more accurate expression for the gain of the stage in Fig. 7.13(a) must take channel-length modulation into account. As depicted in Fig. 7.13(c), the resistance seen at the drain is now equal to $(1/g_{m2})||r_{O2}||r_{O1}$, and hence

$$A_v = -g_{m1}\left(\frac{1}{g_{m2}}||r_{O2}||r_{O1}\right). \tag{7.61}$$

Similarly, the output resistance of the stage is given by

$$R_{out} = \frac{1}{g_{m2}} || r_{O2} || r_{O1}. \tag{7.62}$$

Example 7-7 Determine the voltage gain of the circuit shown in Fig. 7.14(a) if $\lambda \neq 0$.

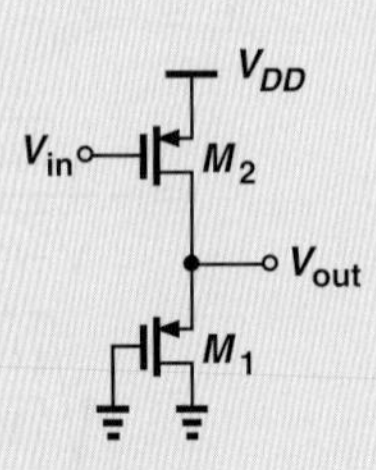

Figure 7.14 CS stage with diode-connected PMOS device.

Solution This stage is similar to that in Fig. 7.13(a), but with NMOS devices changed to PMOS transistors: M_1 serves as a common-source device and M_2 as a diode-connected load. Thus,

$$A_v = -g_{m2}\left(\frac{1}{g_{m1}} || r_{O1} || r_{O2}\right). \tag{7.63}$$

Exercise Repeat the above example if the gate of M_1 is tied to a constant voltage equal to 0.5 V.

7.2.4 CS Stage with Degeneration

Recall from Chapter 5 that a resistor placed in series with the emitter of a bipolar transistor alters characteristics such as gain, I/O impedances, and linearity. We expect similar results for a degenerated CS amplifier.

Figure 7.15 depicts the stage along with its small-signal equivalent (if $\lambda = 0$). As with the bipolar counterpart, the degeneration resistor sustains a fraction of the input voltage change. From Fig. 7.15(b), we have

$$v_{in} = v_1 + g_m v_1 R_S \tag{7.64}$$

and hence

$$v_1 = \frac{v_{in}}{1 + g_m R_S}. \tag{7.65}$$

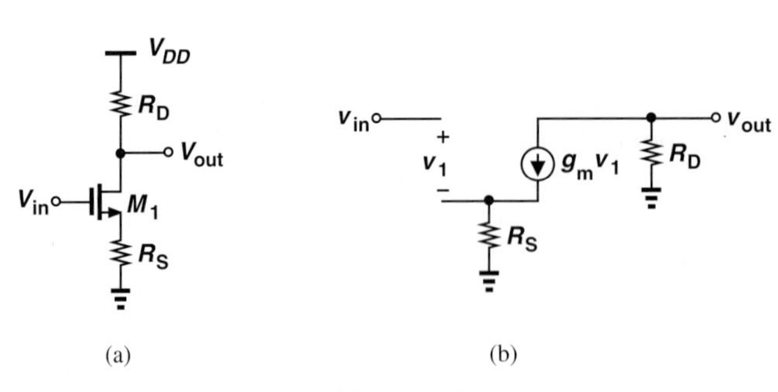

Figure 7.15 (a) CS stage with degeneration, (b) small-signal model.

Since $g_m v_1$ flows through R_D, $v_{out} = -g_m v_1 R_D$ and

$$\frac{v_{out}}{v_{in}} = -\frac{g_m R_D}{1 + g_m R_S} \tag{7.66}$$

$$= -\frac{R_D}{\dfrac{1}{g_m} + R_S}, \tag{7.67}$$

a result identical to that expressed by Eq. (5.157) for the bipolar counterpart.

Example 7-8 Compute the voltage gain of the circuit shown in Fig. 7.16(a) if $\lambda = 0$.

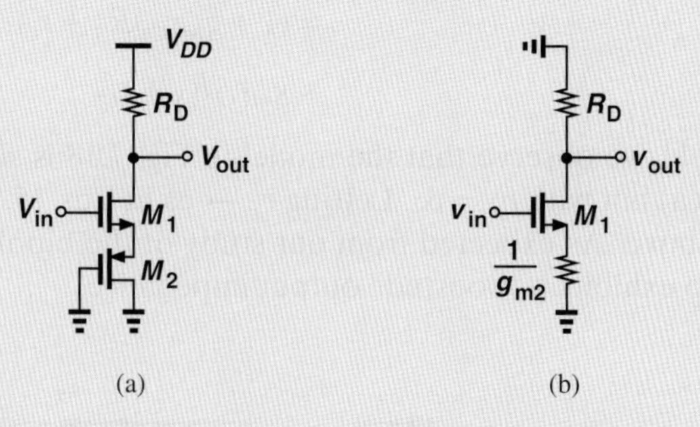

Figure 7.16 (a) Example of CS stage with degeneration, (b) simplified circuit.

Solution Transistor M_2 serves as a diode-connected device, presenting an impedance of $1/g_{m2}$ [Fig. 7.16(b)]. The gain is therefore given by Eq. (7.67) if R_S is replaced with $1/g_{m2}$:

$$A_v = -\frac{R_D}{\dfrac{1}{g_{m1}} + \dfrac{1}{g_{m2}}}. \tag{7.68}$$

Exercise What happens if $\lambda \neq 0$ for M_2?

In parallel with the developments in Chapter 5, we may study the effect of a resistor appearing in series with the gate (Fig. 7.17). However, since the gate current is zero (at low frequencies), R_G sustains no voltage drop and does not affect the voltage gain or the I/O impedances.

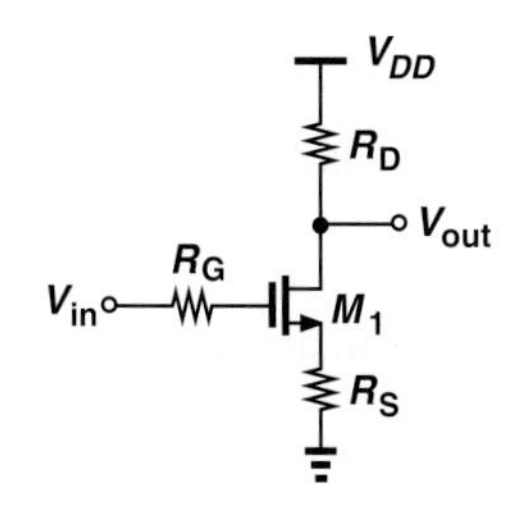

Figure 7.17 CS stage with gate resistance.

Effect of Transistor Output Impedance As with the bipolar counterparts, the inclusion of the transistor output impedance complicates the analysis and is studied in Problem 7.32. Nonetheless, the output impedance of the degenerated CS stage plays a critical role in analog design and is worth studying here.

Figure 7.18 shows the small-signal equivalent of the circuit. Since R_S carries a current equal to i_X (why?), we have $v_1 = -i_X R_S$. Also, the current through r_O is equal to $i_X - g_m v_1 = i_X - g_m(-i_X R_S) = i_X + g_m i_X R_S$. Adding the voltage drops across r_O and R_S and equating the result to v_X, we have

$$r_O(i_X + g_m i_X R_S) + i_X R_S = v_X, \tag{7.69}$$

and hence

$$\frac{v_X}{i_X} = r_O(1 + g_m R_S) + R_S \tag{7.70}$$

$$= (1 + g_m r_O)R_S + r_O \tag{7.71}$$

$$\approx g_m r_O R_S + r_O. \tag{7.72}$$

Alternatively, we observe that the model in Fig. 7.18 is similar to its bipolar counterpart in Fig. 5.46(a) but with $r_\pi = \infty$. Letting $r_\pi \to \infty$ in Eqs. (5.196) and (5.197) yields the same results as above. As expected from our study of the bipolar degenerated stage, the MOS version also exhibits a "boosted" output impedance.

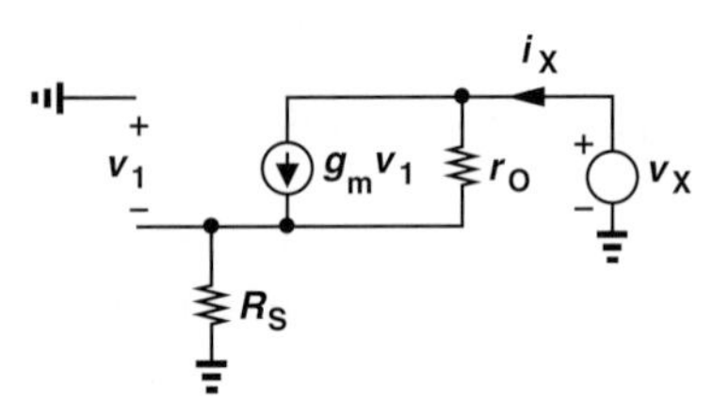

Figure 7.18 Output impedance of CS stage with degeneration.

Example 7-9 Compute the output resistance of the circuit in Fig. 7.19(a) if M_1 and M_2 are identical.

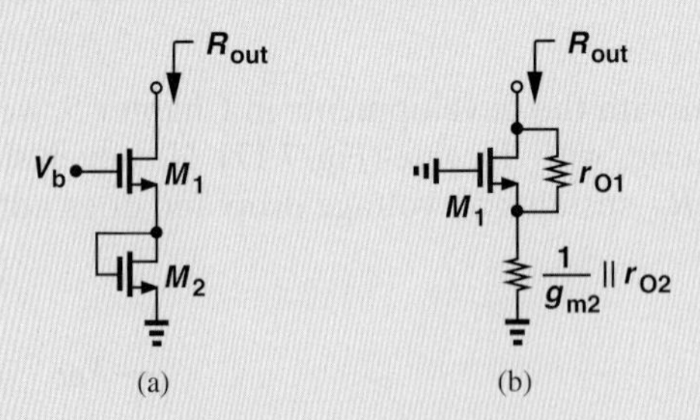

Figure 7.19 (a) Example of CS stage with degeneration, (b) simplified circuit.

Solution The diode-connected device M_2 can be represented by a small-signal resistance of $(1/g_{m2})||r_{O2} \approx 1/g_{m2}$. Transistor M_1 is degenerated by this resistance, and from Eq. (7.70):

$$R_{out} = r_{O1}\left(1 + g_{m1}\frac{1}{g_{m2}}\right) + \frac{1}{g_{m2}} \tag{7.73}$$

which, since $g_{m1} = g_{m2} = g_m$, reduces to

$$R_{out} = 2r_{O1} + \frac{1}{g_m} \tag{7.74}$$

$$\approx 2r_{O1}. \tag{7.75}$$

Exercise Do the results remain unchanged if M_2 is replaced with a diode-connected PMOS device?

Example 7-10 Determine the output resistance of the circuit in Fig. 7.20(a) and compare the result with that in the above example. Assume M_1 and M_2 are in saturation.

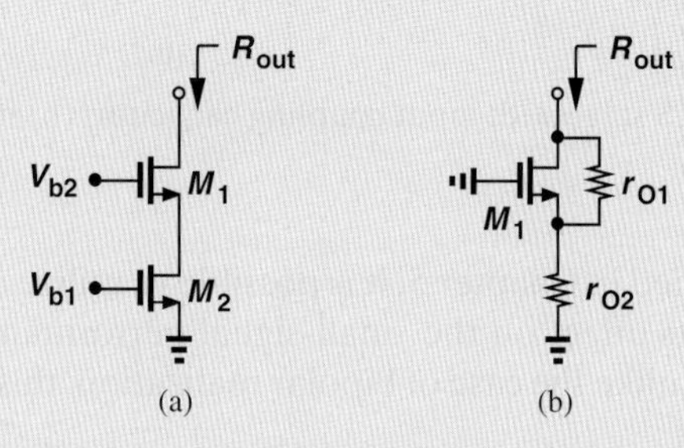

Figure 7.20 (a) Example of CS stage with degeneration, (b) simplified circuit.

Solution With its gate-source voltage fixed, transistor M_2 operates as a current source, introducing a resistance of r_{O2} from the source of M_1 to ground [Fig. 7.20(b)].

Equation (7.71) can therefore be written as

$$R_{out} = (1 + g_{m1}r_{O1})r_{O2} + r_{O1} \tag{7.76}$$

$$\approx g_{m1}r_{O1}r_{O2} + r_{O1}. \tag{7.77}$$

Assuming $g_{m1}r_{O2} \gg 1$ (which is valid in practice), we have

$$R_{out} \approx g_{m1}r_{O1}r_{O2}. \tag{7.78}$$

We observe that this value is quite higher than that in Eq. (7.75).

Exercise Repeat the above example for the PMOS counterpart of the circuit.

7.2.5 CS Core with Biasing

The effect of the simple biasing network shown in Fig. 7.1 is similar to that analyzed for the bipolar stage in Chapter 5. Depicted in Fig. 7.21(a) along with an input coupling capacitor (assumed a short circuit), such a circuit no longer exhibits an infinite input impedance:

$$R_{in} = R_1||R_2. \tag{7.79}$$

Thus, if the circuit is driven by a finite source impedance [Fig. 7.21(b)], the voltage gain falls to

$$A_v = \frac{R_1||R_2}{R_G + R_1||R_2} \cdot \frac{-R_D}{\dfrac{1}{g_m} + R_S}, \tag{7.80}$$

where λ is assumed to be zero.

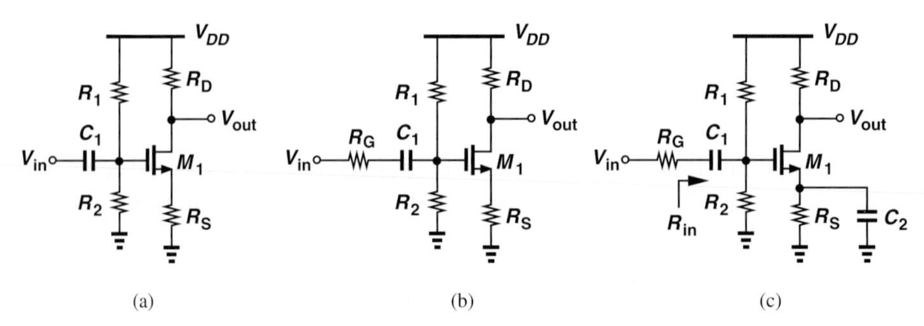

Figure 7.21 (a) CS stage with input coupling capacitor, (b) inclusion of gate resistance, (c) use of bypass capacitor.

As mentioned in Chapter 5, it is possible to utilize degeneration for bias point stability but eliminate its effect on the small-signal performance by means of a bypass capacitor [Fig. 7.21(c)]. Unlike the case of bipolar realization, this does not alter the input impedance of the CS stage:

$$R_{in} = R_1||R_2, \tag{7.81}$$

but raises the voltage gain:

$$A_v = -\frac{R_1||R_2}{R_G + R_1||R_2} g_m R_D. \tag{7.82}$$

Example 7-11

Design the CS stage of Fig. 7.21(c) for a voltage gain of 5, an input impedance of 50 kΩ, and a power budget of 5 mW. Assume $\mu_n C_{ox} = 100\ \mu\text{A/V}^2$, $V_{TH} = 0.5$ V, $\lambda = 0$, and $V_{DD} = 1.8$ V. Also, assume a voltage drop of 400 mV across R_S.

Solution

The power budget along with $V_{DD} = 1.8$ V implies a maximum supply current of 2.78 mA. As an initial guess, we allocate 2.7 mA to M_1 and the remaining 80 μA to R_1 and R_2. It follows that

$$R_S = 148\ \Omega. \tag{7.83}$$

As with typical design problems, the choice of g_m and R_D is somewhat flexible so long as $g_m R_D = 5$. However, with I_D known, we must ensure a reasonable value for V_{GS}, e.g., $V_{GS} = 1$ V. This choice yields

$$g_m = \frac{2I_D}{V_{GS} - V_{TH}} \tag{7.84}$$

$$= \frac{1}{92.6\ \Omega}, \tag{7.85}$$

and hence

$$R_D = 463\ \Omega. \tag{7.86}$$

Writing

$$I_D = \frac{1}{2}\mu_n C_{ox} \frac{W}{L}(V_{GS} - V_{TH})^2 \tag{7.87}$$

gives

$$\frac{W}{L} = 216. \tag{7.88}$$

With $V_{GS} = 1$ V and a 400-mV drop across R_S, the gate voltage reaches 1.4 V, requiring that

$$\frac{R_2}{R_1 + R_2} V_{DD} = 1.4\ \text{V}, \tag{7.89}$$

which, along with $R_{in} = R_1 || R_2 = 50\ \text{k}\Omega$, yields

$$R_1 = 64.3\ \text{k}\Omega \tag{7.90}$$

$$R_2 = 225\ \text{k}\Omega. \tag{7.91}$$

We must now check to verify that M_1 indeed operates in saturation. The drain voltage is given by $V_{DD} - I_D R_D = 1.8\ \text{V} - 1.25\ \text{V} = 0.55\ \text{V}$. Since the gate voltage is equal to 1.4 V, the gate-drain voltage difference exceeds V_{TH}, driving M_1 into the triode region!

How did our design procedure lead to this result? For the given I_D, we have chosen an excessively large R_D, i.e., an excessively small g_m (because $g_m R_D = 5$), even though V_{GS} is reasonable. We must therefore increase g_m so as to allow a lower value for R_D. For example, suppose we halve R_D and double g_m by increasing W/L by a factor of four:

$$\frac{W}{L} = 864 \tag{7.92}$$

$$g_m = \frac{1}{46.3\ \Omega}. \tag{7.93}$$

The corresponding gate-source overdrive voltage is obtained from (7.84):

$$V_{GS} - V_{TH} = 250\ \text{mV}, \tag{7.94}$$

yielding a gate voltage of 1.17 V.

Is M_1 in saturation? The drain voltage is equal to $V_{DD} - R_D I_D = 1.17$ V, a value higher than the gate voltage minus V_{TH}. Thus, M_1 operates in saturation.

Exercise Repeat the above example for a power budget of 3 mW and $V_{DD} = 1.2$ V.

7.3 COMMON-GATE STAGE

As shown in Fig. 7.22, the CG topology resembles the common-base stage studied in Chapter 5. Here, if the input rises by a small value, ΔV, then the gate-source voltage of M_1 *decreases* by the same amount, thereby lowering the drain current by $g_m \Delta V$ and *raising* V_{out} by $g_m \Delta V R_D$. That is, the voltage gain is positive and equal to

$$A_v = g_m R_D. \tag{7.95}$$

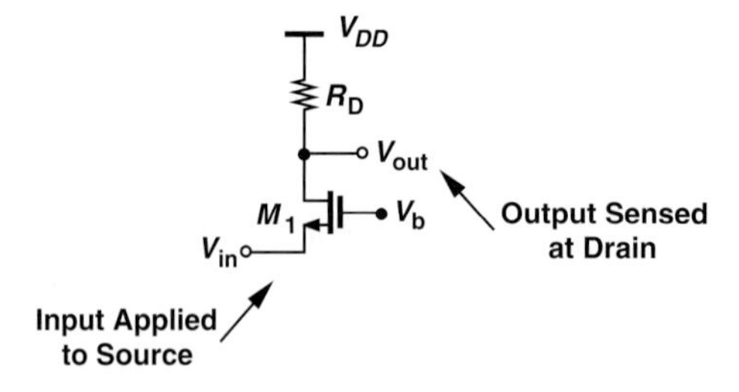

Figure 7.22 Common-gate stage.

The CG stage suffers from voltage headroom-gain trade-offs similar to those of the CB topology. In particular, to achieve a high gain, a high I_D or R_D is necessary, but the drain voltage, $V_{DD} - I_D R_D$, must remain above $V_b - V_{TH}$ to ensure M_1 is saturated.

Example 7-12 A microphone having a dc level of zero drives a CG stage biased at $I_D = 0.5$ mA. If $W/L = 50$, $\mu_n C_{ox} = 100\ \mu\text{A/V}^2$, $V_{TH} = 0.5$ V, and $V_{DD} = 1.8$ V, determine the maximum allowable value of R_D and hence the maximum voltage gain. Neglect channel-length modulation.

Solution With W/L known, the gate-source voltage can be determined from

$$I_D = \frac{1}{2}\mu_n C_{ox}\frac{W}{L}(V_{GS} - V_{TH})^2 \tag{7.96}$$

as

$$V_{GS} = 0.947\ \text{V}. \tag{7.97}$$

For M_1 to remain in saturation,

$$V_{DD} - I_D R_D > V_b - V_{TH} \tag{7.98}$$

and hence

$$R_D < 2.71\ \text{k}\Omega. \tag{7.99}$$

Also, the above value of W/L and I_D yield $g_m = (447\ \Omega)^{-1}$ and

$$A_v \leq 6.06. \tag{7.100}$$

Figure 7.23 summarizes the allowable signal levels in this design. The gate voltage can be generated using a resistive divider similar to that in Fig. 7.21(a).

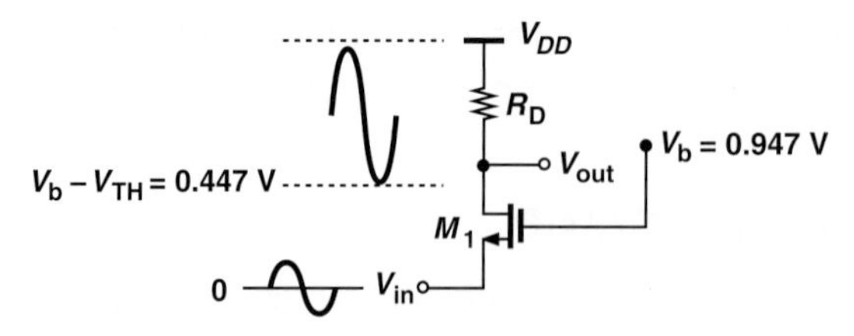

Figure 7.23 Signal levels in CG stage.

Exercise If a gain of 10 is required, what value should be chosen for W/L?

We now compute the I/O impedances of the CG stage, expecting to obtain results similar to those of the CB topology. Neglecting channel-length modulation for now, we have from Fig. 7.24(a) $v_1 = -v_X$ and

$$i_X = -g_m v_1 \tag{7.101}$$

$$= g_m v_X. \tag{7.102}$$

That is,

$$R_{in} = \frac{1}{g_m}, \tag{7.103}$$

a relatively *low* value. Also, from Fig. 7.24(b), $v_1 = 0$ and hence

$$R_{out} = R_D, \tag{7.104}$$

an expected result because the circuits of Figs. 7.24(b) and 7.8 are identical.

Did you know?

The common-gate stage is sometimes used as a low-noise RF amplifier, e.g., at the input of your WiFi receiver. This topology is attractive because its low input impedance allows simple "impedance matching" with the antenna. However, with the reduction of the intrinsic gain, $g_m r_O$, as a result of scaling, the input impedance, R_{in}, is now too high! It can be shown that with channel-length modulation

$$R_{in} = \frac{R_D + r_O}{1 + g_m r_O},$$

which reduces to $1/g_m$ if $g_m r_O \gg 1$ and $R_D \ll r_O$. Since neither of these conditions holds anymore, the CG stage presents new challenges to RF designers.

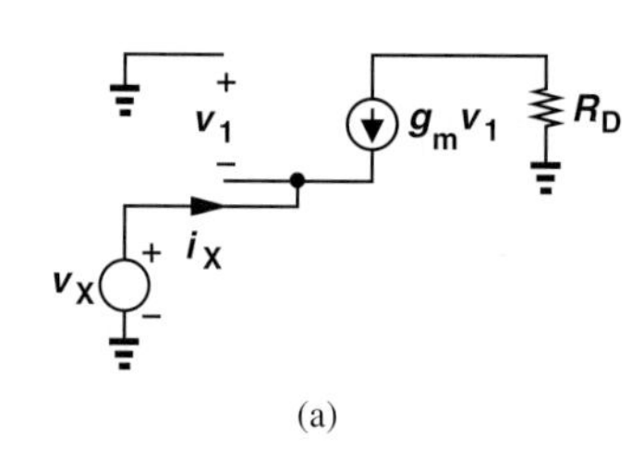

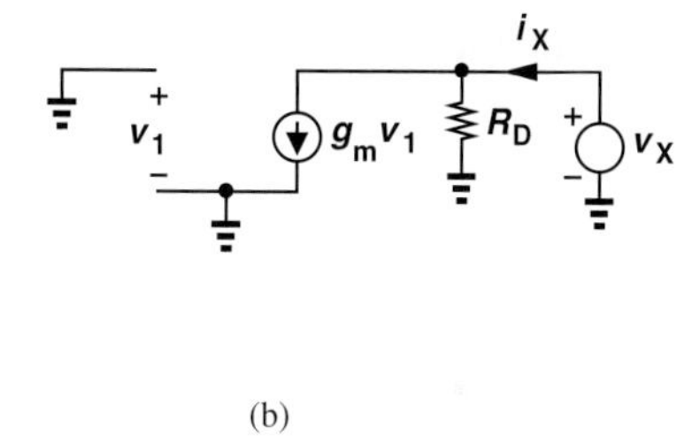

Figure 7.24 (a) Input and (b) output impedances of CG stage.

Let us study the behavior of the CG stage in the presence of a finite source impedance (Fig. 7.25) but still with $\lambda = 0$. In a manner similar to that depicted in Chapter 5 for the CB topology, we write

$$v_X = \frac{\frac{1}{g_m}}{\frac{1}{g_m} + R_S} v_{in} \tag{7.105}$$

$$= \frac{1}{1 + g_m R_S} v_{in}. \tag{7.106}$$

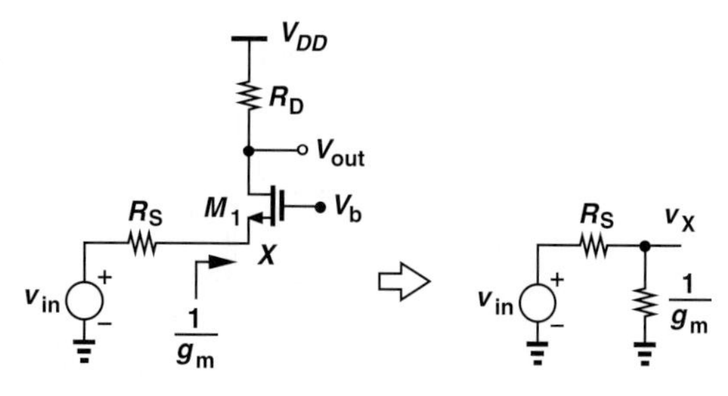

Figure 7.25 Simplification of CG stage with signal source resistance.

Thus,

$$\frac{v_{out}}{v_{in}} = \frac{v_{out}}{v_X} \cdot \frac{v_X}{v_{in}} \tag{7.107}$$

$$= \frac{g_m R_D}{1 + g_m R_S} \tag{7.108}$$

$$= \frac{R_D}{\frac{1}{g_m} + R_S}. \tag{7.109}$$

The gain is therefore equal to that of the degenerated CS stage except for a negative sign.

In contrast to the common-source stage, the CG amplifier exhibits a current gain of unity: the current provided by the input voltage source simply flows through the channel and emerges from the drain node.

The analysis of the common-gate stage in the general case, i.e., including both channel-length modulation and a finite source impedance, is beyond the scope of this book (Problem 7.42). However, we can make two observations. First, a resistance appearing in series with the gate terminal [Fig. 7.26(a)] does not alter the gain or I/O impedances (at low frequencies) because it sustains a zero potential drop—as if its value were zero. Second, the output resistance of the CG stage in the general case [Fig. 7.26(b)] is identical to that of the degenerated CS topology:

$$R_{out} = (1 + g_m r_O) R_S + r_O. \tag{7.110}$$

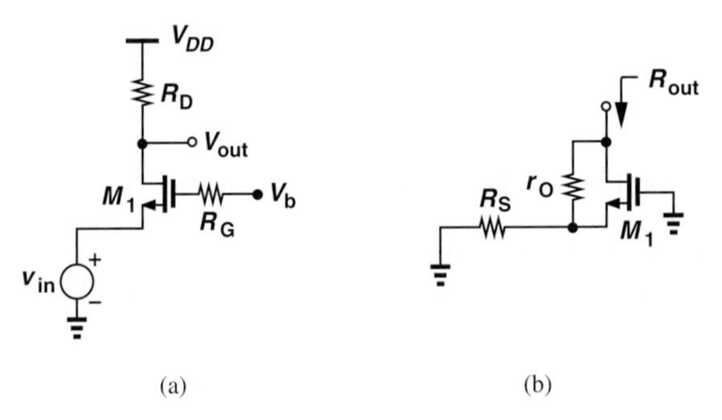

Figure 7.26 (a) CG stage with gate resistance, (b) output resistance of CG stage.

Example 7-13 For the circuit shown in Fig. 7.27(a), calculate the voltage gain if $\lambda = 0$ and the output impedance if $\lambda > 0$.

Solution We first compute v_X/v_{in} with the aid of the equivalent circuit depicted in Fig. 7.27(b):

$$\frac{v_X}{v_{in}} = \frac{\frac{1}{g_{m2}} \| \frac{1}{g_{m1}}}{\frac{1}{g_{m2}} \| \frac{1}{g_{m1}} + R_S} \tag{7.111}$$

$$= \frac{1}{1 + (g_{m1} + g_{m2}) R_S}. \tag{7.112}$$

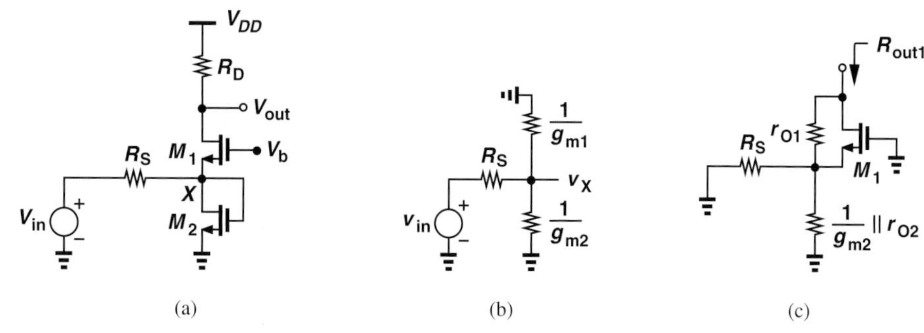

Figure 7.27 (a) Example of CG stage, (b) equivalent input network, (c) calculation of output resistance.

Noting that $v_{out}/v_X = g_{m1}R_D$, we have

$$\frac{v_{out}}{v_{in}} = \frac{g_{m1}R_D}{1 + (g_{m1} + g_{m2})R_S}. \tag{7.113}$$

To compute the output impedance, we first consider R_{out1}, as shown in Fig. 7.27(c), which from Eq. (7.110) is equal to

$$R_{out1} = (1 + g_{m1}r_{O1})\left(\frac{1}{g_{m2}}||r_{O2}||R_S\right) + r_{O1} \tag{7.114}$$

$$\approx g_{m1}r_{O1}\left(\frac{1}{g_{m2}}||R_S\right) + r_{O1}. \tag{7.115}$$

The overall output impedance is then given by

$$R_{out} = R_{out1}||R_D \tag{7.116}$$

$$\approx \left[g_{m1}r_{O1}\left(\frac{1}{g_{m2}}||R_S\right) + r_{O1}\right]\Bigg|\Bigg| R_D. \tag{7.117}$$

Exercise Calculate the output impedance if the gate of M_2 is tied to a constant voltage.

7.3.1 CG Stage with Biasing

Following our study of the CB biasing in Chapter 5, we surmise the CG amplifier can be biased as shown in Fig. 7.28. Providing a path for the bias current to ground, resistor R_3 lowers the input impedance—and hence the voltage gain—if the signal source exhibits a finite output impedance, R_S.

Since the impedance seen to the right of node X is equal to $R_3||(1/g_m)$, we have

$$\frac{v_{out}}{v_{in}} = \frac{v_X}{v_{in}} \cdot \frac{v_{out}}{v_X} \tag{7.118}$$

$$= \frac{R_3||(1/g_m)}{R_3||(1/g_m) + R_S} \cdot g_mR_D, \tag{7.119}$$

where channel-length modulation is neglected. As mentioned earlier, the voltage divider consisting of R_1 and R_2 does not affect the small-signal behavior of the circuit (at low frequencies).

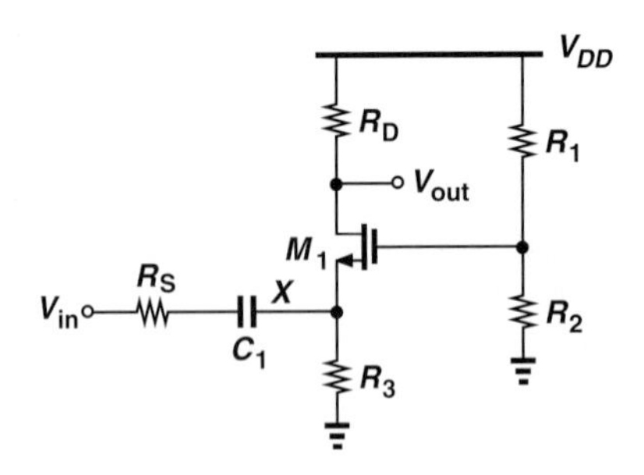

Figure 7.28 CG stage with biasing.

Example 7-14 Design the common-gate stage of Fig. 7.28 for the following parameters: $v_{out}/v_{in} = 5$, $R_S = 0$, $R_3 = 500\,\Omega$, $1/g_m = 50\,\Omega$, power budget = 2 mW, $V_{DD} = 1.8$ V. Assume $\mu_n C_{ox} = 100\,\mu\text{A/V}^2$, $V_{TH} = 0.5$ V, and $\lambda = 0$.

Solution From the power budget, we obtain a total supply current of 1.11 mA. Allocating 10 μA to the voltage divider, R_1 and R_2, we leave 1.1 mA for the drain current of M_1. Thus, the voltage drop across R_3 is equal to 550 mV.

We must now compute two interrelated parameters: W/L and R_D. A larger value of W/L yields a greater g_m, allowing a lower value of R_D. As in Example 7-11, we choose an initial value for V_{GS} to arrive at a reasonable guess for W/L. For example, if $V_{GS} = 0.8$ V, then $W/L = 244$, and $g_m = 2I_D/(V_{GS} - V_{TH}) = (136.4\,\Omega)^{-1}$, dictating $R_D = 682\,\Omega$ for $v_{out}/v_{in} = 5$.

Let us determine whether M_1 operates in saturation. The gate voltage is equal to V_{GS} plus the drop across R_3, amounting to 1.35 V. On the other hand, the drain voltage is given by $V_{DD} - I_D R_D = 1.05$ V. Since the drain voltage exceeds $V_G - V_{TH}$, M_1 is indeed in saturation.

The resistive divider consisting of R_1 and R_2 must establish a gate voltage equal to 1.35 V while drawing 10 μA:

$$\frac{V_{DD}}{R_1 + R_2} = 10\,\mu\text{A} \tag{7.120}$$

$$\frac{R_2}{R_1 + R_2} V_{DD} = 1.35\text{ V}. \tag{7.121}$$

It follows that $R_1 = 45\text{ k}\Omega$ and $R_2 = 135\text{ k}\Omega$.

Exercise If W/L cannot exceed 100, what voltage gain can be achieved?

Example 7-15 Suppose in Example 7-14, we wish to minimize W/L (and hence transistor capacitances). What is the minimum acceptable value of W/L?

Solution For a given I_D, as W/L decreases, $V_{GS} - V_{TH}$ increases. Thus, we must first compute the maximum allowable V_{GS}. We impose the condition for saturation as

$$V_{DD} - I_D R_D > V_{GS} + V_{R3} - V_{TH}, \tag{7.122}$$

where V_{R3} denotes the voltage drop across R_3, and set $g_m R_D$ to the required gain:

$$\frac{2I_D}{V_{GS} - V_{TH}} R_D = A_v. \tag{7.123}$$

Eliminating R_D from Eqs. (7.122) and (7.123) gives:

$$V_{DD} - \frac{A_v}{2}(V_{GS} - V_{TH}) > V_{GS} - V_{TH} + V_{R3} \tag{7.124}$$

and hence

$$V_{GS} - V_{TH} < \frac{V_{DD} - V_{R3}}{\frac{A_v}{2} + 1}. \tag{7.125}$$

In other words,

$$W/L > \frac{2I_D}{\mu_n C_{ox}\left(2\frac{V_{DD} - V_{R3}}{A_v + 2}\right)^2}. \tag{7.126}$$

It follows that

$$W/L > 172.5. \tag{7.127}$$

Exercise Repeat the above example for $A_v = 10$.

7.4 SOURCE FOLLOWER

The MOS counterpart of the emitter follower is called the "source follower" (or the "common-drain" stage) and shown in Fig. 7.29. The amplifier senses the input at the gate and produces the output at the source, with the drain tied to V_{DD}. The circuit's behavior is similar to that of the bipolar counterpart.

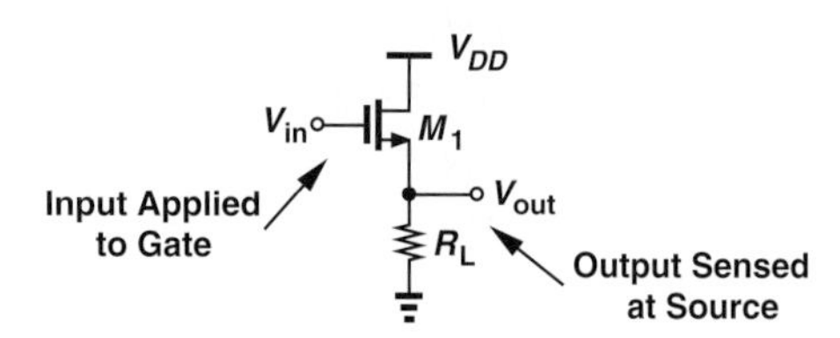

Figure 7.29 Source follower.

7.4.1 Source Follower Core

If the gate voltage of M_1 in Fig. 7.29 is raised by a small amount, ΔV_{in}, the gate-source voltage tends to increase, thereby raising the source current and hence the output voltage. Thus, V_{out} "follows" V_{in}. Since the dc level of V_{out} is lower than that of V_{in} by V_{GS}, we say the follower can serve as a "level shift" circuit. From our analysis of emitter followers in Chapter 5, we expect this topology to exhibit a subunity gain, too.

Figure 7.30(a) depicts the small-signal equivalent of the source follower, including channel-length modulation. Recognizing that r_O appears in parallel with R_L, we have

$$g_m v_1 (r_O || R_L) = v_{out}. \tag{7.128}$$

Also,

$$v_{in} = v_1 + v_{out}. \tag{7.129}$$

It follows that

$$\frac{v_{out}}{v_{in}} = \frac{g_m(r_O||R_L)}{1 + g_m(r_O||R_L)} \tag{7.130}$$

$$= \frac{r_O||R_L}{\frac{1}{g_m} + r_O||R_L}. \tag{7.131}$$

The voltage gain is therefore positive and less than unity. It is desirable to maximize R_L (and r_O).

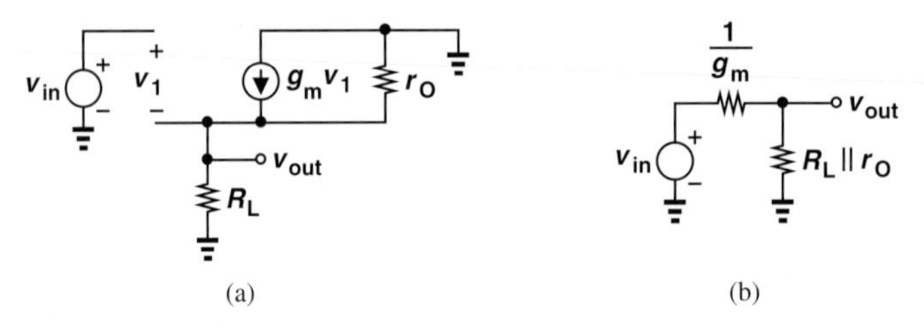

Figure 7.30 (a) Small-signal equivalent of source follower, (b) simplified circuit.

As with emitter followers, we can view the above result as voltage division between a resistance equal to $1/g_m$ and another equal to $r_O||R_L$ [Fig. 7.30(b)]. Note, however, that a resistance placed in series with the gate does not affect Eq. (7.131) (at low frequencies) because it sustains a zero drop.

Example 7-16 A source follower is realized as shown in Fig. 7.31(a), where M_2 serves as a current source. Calculate the voltage gain of the circuit.

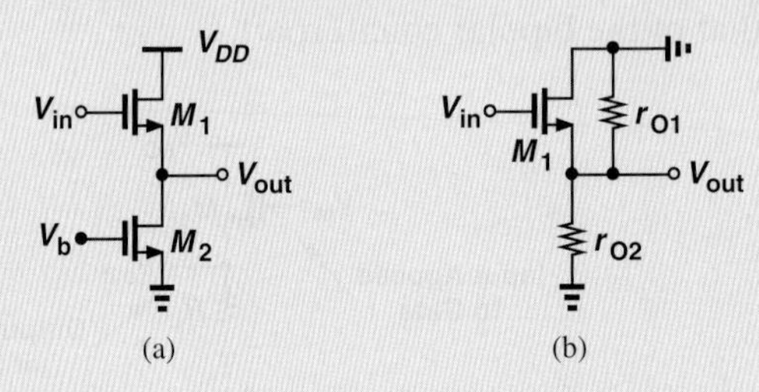

Figure 7.31 (a) Follower with ideal current source, (b) simplified circuit.

Solution Since M_2 simply presents an impedance of r_{O2} from the output node to ac ground [Fig. 7.31(b)], we substitute $R_L = r_{O2}$ in Eq. (7.131):

$$A_v = \frac{r_{O1}||r_{O2}}{\frac{1}{g_{m1}} + r_{O1}||r_{O2}}. \tag{7.132}$$

If $r_{O1}||r_{O2} \gg 1/g_{m1}$, then $A_v \approx 1$.

Exercise Repeat the above example if a resistance of value R_S is placed in series with the source of M_2.

Example 7-17 Design a source follower to drive a 50-Ω load with a voltage gain of 0.5 and a power budget of 10 mW. Assume $\mu_n C_{ox} = 100\ \mu\text{A/V}^2$, $V_{TH} = 0.5$ V, $\lambda = 0$, and $V_{DD} = 1.8$ V.

Solution With $R_L = 50\ \Omega$ and $r_O = \infty$ in Fig. 7.29, we have

$$A_v = \frac{R_L}{\frac{1}{g_m} + R_L} \tag{7.133}$$

and hence

$$g_m = \frac{1}{50\ \Omega}. \tag{7.134}$$

The power budget and supply voltage yield a maximum supply current of 5.56 mA. Using this value for I_D in $g_m = \sqrt{2\mu_n C_{ox}(W/L)I_D}$ gives

$$W/L = 360. \tag{7.135}$$

Exercise What voltage gain can be achieved if the power budget is raised to 15 mW?

It is instructive to compute the output impedance of the source follower.[2] As illustrated in Fig. 7.32, R_{out} consists of the resistance seen looking up into the source in parallel with that seen looking down into R_L. With $\lambda \neq 0$, the former is equal to $(1/g_m)||r_O$, yielding

$$R_{out} = \frac{1}{g_m}||r_O||R_L \tag{7.136}$$

$$\approx \frac{1}{g_m}||R_L. \tag{7.137}$$

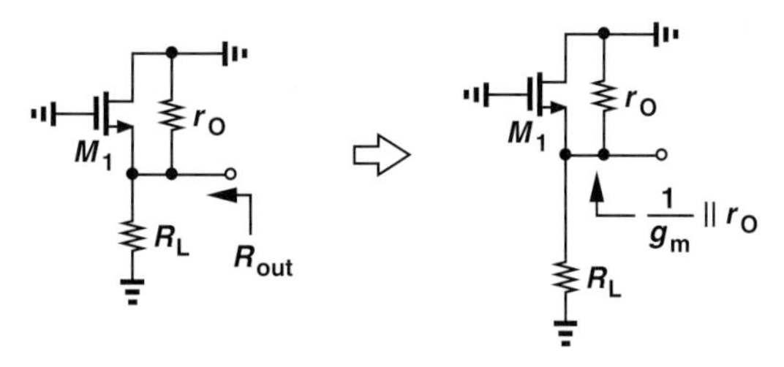

Figure 7.32 Output resistance of source follower.

In summary, the source follower exhibits a very high input impedance and a relatively low output impedance, thereby providing buffering capability.

7.4.2 Source Follower with Biasing

The biasing of source followers is similar to that of emitter followers (Chapter 5). Figure 7.33 depicts an example where R_G establishes a dc voltage equal to V_{DD} at the

[2]The input impedance is infinite at low frequencies.

gate of M_1 (why?) and R_S sets the drain bias current. Note that M_1 operates in saturation because the gate and drain voltages are equal. Also, the input impedance of the circuit has dropped from infinity to R_G.

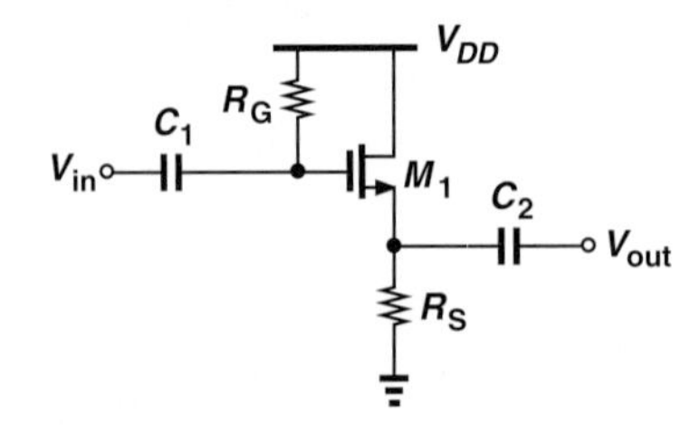

Figure 7.33 Source follower with input and output coupling capacitors.

Let us compute the bias current of the circuit. With a zero voltage drop across R_G, we have

$$V_{GS} + I_D R_S = V_{DD}. \tag{7.138}$$

Neglecting channel-length modulation, we write

$$I_D = \frac{1}{2}\mu_n C_{ox}\frac{W}{L}(V_{GS} - V_{TH})^2 \tag{7.139}$$

$$= \frac{1}{2}\mu_n C_{ox}\frac{W}{L}(V_{DD} - I_D R_S - V_{TH})^2. \tag{7.140}$$

The resulting quadratic equation can be solved to obtain I_D.

Example 7-18

Design the source follower of Fig. 7.33 for a drain current of 1 mA and a voltage gain of 0.8. Assume $\mu_n C_{ox} = 100\ \mu\text{A/V}^2$, $V_{TH} = 0.5$ V, $\lambda = 0$, $V_{DD} = 1.8$ V, and $R_G = 50\ \text{k}\Omega$.

Solution The unknowns in this problem are V_{GS}, W/L, and R_S. The following three equations can be formed:

$$I_D = \frac{1}{2}\mu_n C_{ox}\frac{W}{L}(V_{GS} - V_{TH})^2 \tag{7.141}$$

$$I_D R_S + V_{GS} = V_{DD} \tag{7.142}$$

$$A_v = \frac{R_S}{\frac{1}{g_m} + R_S}. \tag{7.143}$$

If g_m is written as $2I_D/(V_{GS} - V_{TH})$, then Eqs. (7.142) and (7.143) do not contain W/L and can be solved to determine V_{GS} and R_S. With the aid of Eq. (7.142), we write Eq. (7.143) as

$$A_v = \frac{R_S}{\frac{V_{GS} - V_{TH}}{2I_D} + R_S} \tag{7.144}$$

$$= \frac{2I_D R_S}{V_{GS} - V_{TH} + 2I_D R_S} \tag{7.145}$$

$$= \frac{2I_D R_S}{V_{DD} - V_{TH} + I_D R_S}. \tag{7.146}$$

Thus,

$$R_S = \frac{V_{DD} - V_{TH}}{I_D} \frac{A_v}{2 - A_v} \tag{7.147}$$

$$= 867\ \Omega. \tag{7.148}$$

and

$$V_{GS} = V_{DD} - I_D R_S \tag{7.149}$$

$$= V_{DD} - (V_{DD} - V_{TH}) \frac{A_v}{2 - A_v} \tag{7.150}$$

$$= 0.933\ \text{V}. \tag{7.151}$$

It follows from Eq. (7.141) that

$$\frac{W}{L} = 107. \tag{7.152}$$

Exercise What voltage gain can be achieved if W/L cannot exceed 50?

Equation (7.140) reveals that the bias current of the source follower varies with the supply voltage. To avoid this effect, integrated circuits bias the follower by means of a current source (Fig. 7.34).

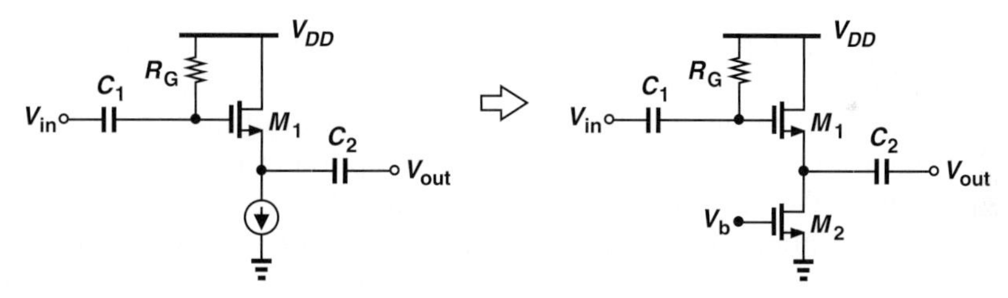

Figure 7.34 Source follower with biasing.

Robotics Application: CMOS Cameras

Vision-based robots rely on cameras to "see" the environment and act accordingly. Examples include robots that must perform inspection of parts in a factory or navigate a terrain while avoiding collision with other objects.

As explained in Chapter 1, a digital camera consists of a matrix of pixels, each of which receives light and, by means of a reverse-biased photodiode, converts it to current. How do we process this result? CMOS devices prove essential here. As shown in Fig. 7.35, we first set the voltage across the photodiode, D_1, to V_{DD} when S_0 is on. Next, we turn off S_0 and allow the carriers generated within D_1 by light to discharge C_1. This voltage change is sensed by a source follower and made available to a column that connects to other pixels. When S_j is turned on, the voltage provided by pixel j is measured, digitized, and stored by the column processing circuit. This switch then turns off and the next pixel is sensed.

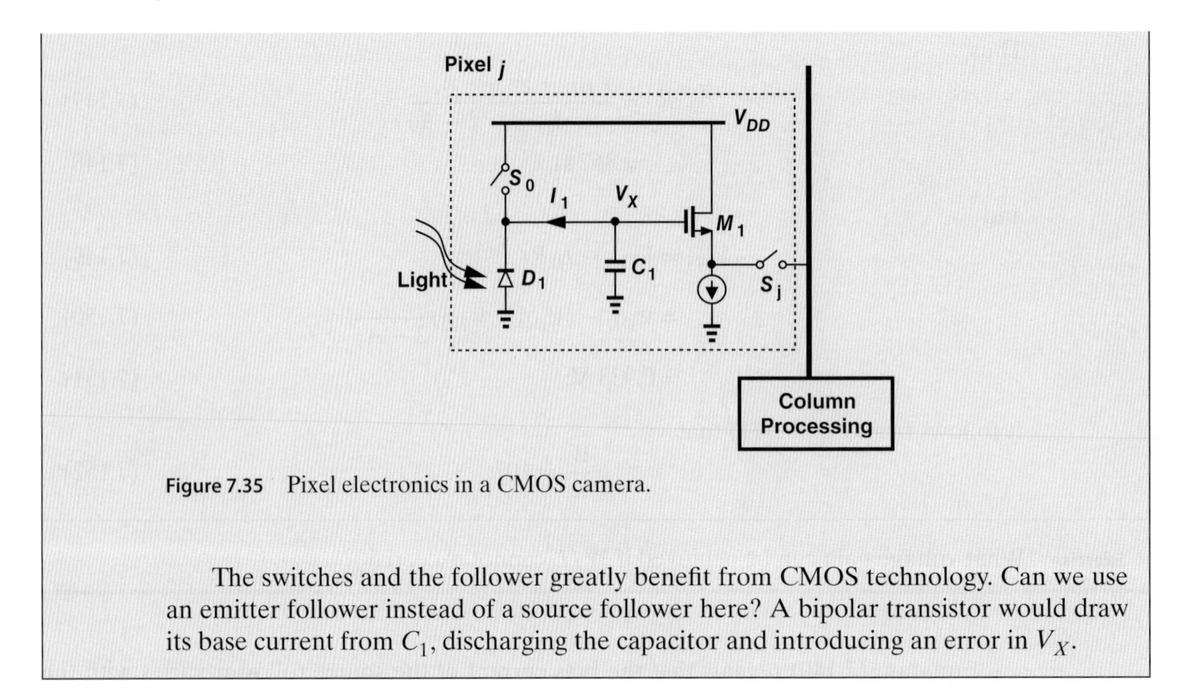

Figure 7.35 Pixel electronics in a CMOS camera.

The switches and the follower greatly benefit from CMOS technology. Can we use an emitter follower instead of a source follower here? A bipolar transistor would draw its base current from C_1, discharging the capacitor and introducing an error in V_X.

7.5 SUMMARY AND ADDITIONAL EXAMPLES

In this chapter, we have studied three basic CMOS building blocks, namely, the common-source stage, the common-gate stage, and the source follower. As observed throughout the chapter, the small-signal behavior of these circuits is quite similar to that of their bipolar counterparts, with the exception of the high impedance seen at the gate terminal. We have noted that the biasing schemes are also similar, with the quadratic I_D-V_{GS} relationship supplanting the exponential I_C-V_{BE} characteristic.

In this section, we consider a number of additional examples to solidify the concepts introduced in this chapter, emphasizing analysis by inspection.

Example 7-19 Calculate the voltage gain and output impedance of the circuit shown in Fig. 7.36(a).

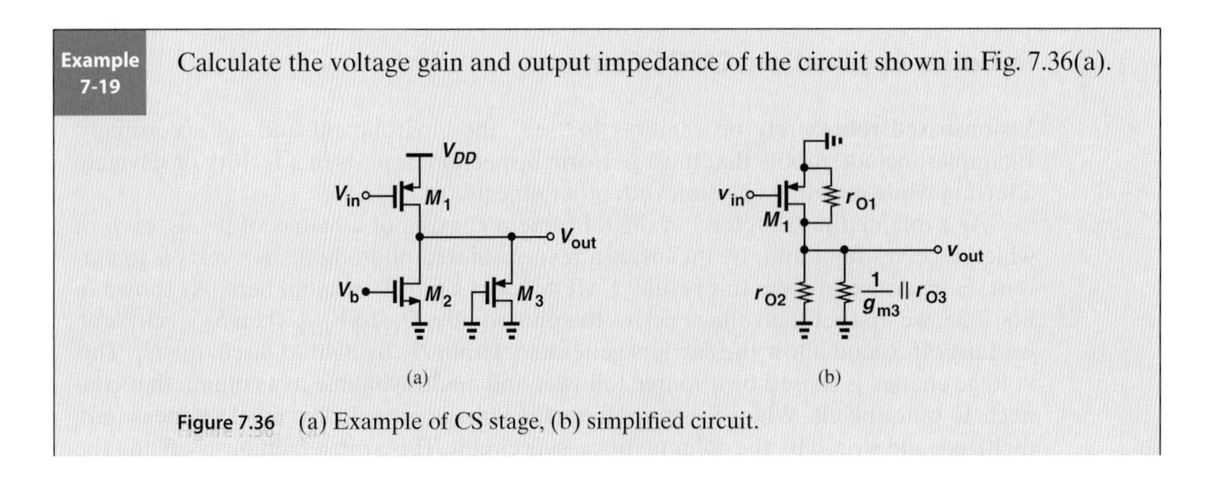

Figure 7.36 (a) Example of CS stage, (b) simplified circuit.

Solution We identify M_1 as a common-source device because it senses the input at its gate and generates the output at its drain. Transistors M_2 and M_3 therefore act as the load, with the former serving as a current source and the latter as a diode-connected device. Thus, M_2 can be replaced with a small-signal resistance equal to r_{O2}, and M_3 with another equal to $(1/g_{m3})||r_{O3}$. The circuit now reduces to that depicted in Fig. 7.36(b), yielding

$$A_v = -g_{m1}\left(\frac{1}{g_{m3}}||r_{O1}||r_{O2}||r_{O3}\right) \tag{7.153}$$

and

$$R_{out} = \frac{1}{g_{m3}}||r_{O1}||r_{O2}||r_{O3}. \tag{7.154}$$

Note that $1/g_{m3}$ is dominant in both expressions.

Exercise Repeat the above example if M_2 is converted to a diode-connected device.

Example 7-20 Compute the voltage gain of the circuit shown in Fig. 7.37(a). Neglect channel-length modulation in M_1.

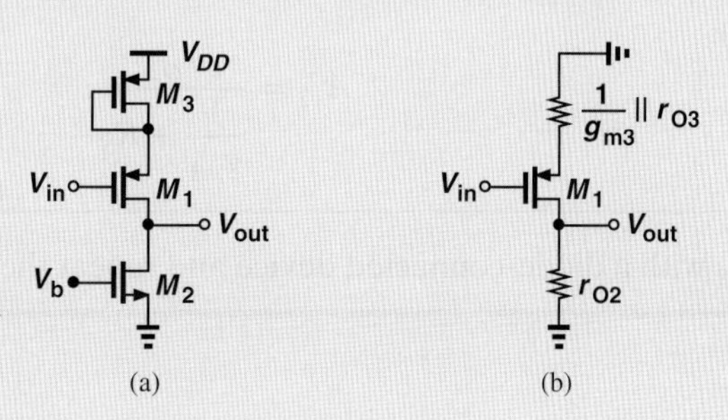

Figure 7.37 (a) Example of CS stage, (b) simplified circuit.

Solution Operating as a CS stage and degenerated by the diode-connected device M_3, transistor M_1 drives the current-source load, M_2. Simplifying the amplifier to that in Fig. 7.37(b), we have

$$A_v = -\frac{r_{O2}}{\dfrac{1}{g_{m1}} + \dfrac{1}{g_{m3}}||r_{O3}}. \tag{7.155}$$

Exercise Repeat the above example if the gate of M_3 is tied to a constant voltage.

Example 7-21 Determine the voltage gain of the amplifiers illustrated in Fig. 7.38. For simplicity, assume $r_{O1} = \infty$ in Fig. 7.38(b).

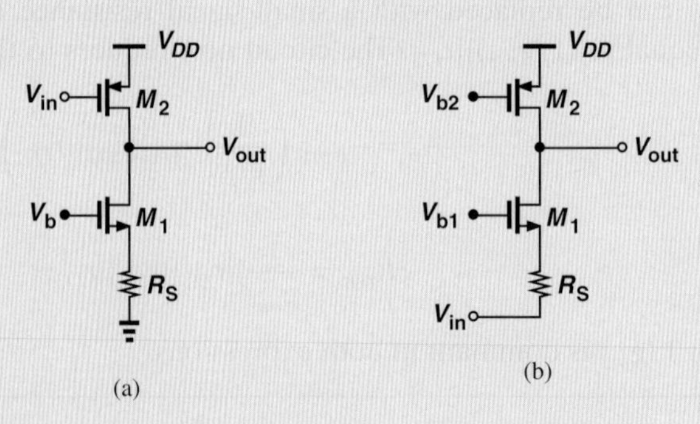

Figure 7.38 Examples of (a) CS and (b) CG stages.

Solution Degenerated by R_S, transistor M_1 in Fig. 7.38(a) presents an impedance of $(1 + g_{m1}r_{O1})R_S + r_{O1}$ to the drain of M_2. Thus the total impedance seen at the drain is equal to $[(1 + g_{m1}r_{O1})R_S + r_{O1}]||r_{O2}$, giving a voltage gain of

$$A_v = -g_{m2}\{[(1 + g_{m1}r_{O1})R_S + r_{O1}]||r_{O2}\}. \tag{7.156}$$

In Fig. 7.38(b), M_1 operates as a common-gate stage and M_2 as the load, obtaining Eq. (7.109):

$$A_{v2} = \frac{r_{O2}}{\dfrac{1}{g_{m1}} + R_S}. \tag{7.157}$$

Exercise Replace R_S with a diode-connected device and repeat the analysis.

Example 7-22 Calculate the voltage gain of the circuit shown in Fig. 7.39(a) if $\lambda = 0$.

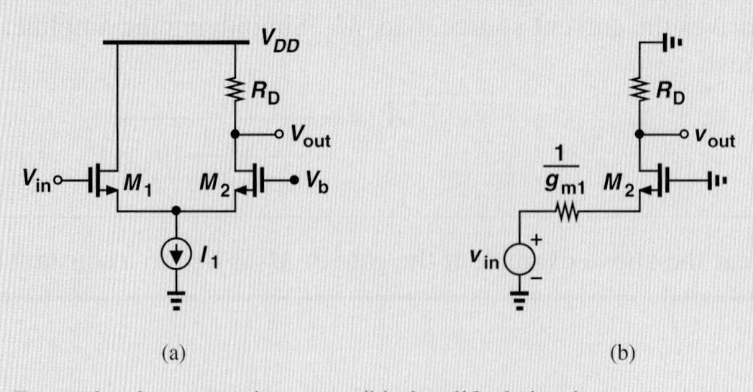

Figure 7.39 (a) Example of a composite stage, (b) simplified circuit.

Solution In this circuit, M_1 operates as a source follower and M_2 as a CG stage (why?). A simple method of analyzing the circuit is to replace v_{in} and M_1 with a Thevenin equivalent. From Fig. 7.30(b), we derive the model depicted in Fig. 7.39(b). Thus,

$$A_v = \frac{R_D}{\frac{1}{g_{m1}} + \frac{1}{g_{m2}}}. \tag{7.158}$$

Exercise What happens if a resistance of value R_1 is placed in series with the drain of M_1?

Example 7-23 The circuit of Fig. 7.40 produces two outputs. Calculate the voltage gain from the input to Y and to X. Assume $\lambda = 0$ for M_1.

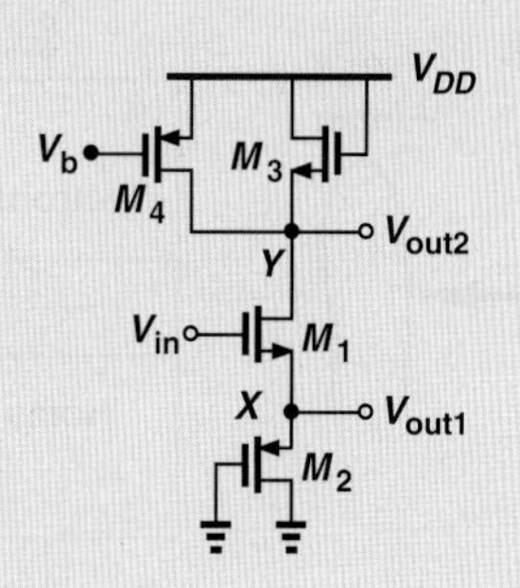

Figure 7.40 Example of composite stage.

Solution For V_{out1}, the circuit serves as a source follower. The reader can show that if $r_{O1} = \infty$, then M_3 and M_4 do not affect the source follower operation. Exhibiting a small-signal impedance of $(1/g_{m2})||r_{O2}$, transistor M_2 acts as a load for the follower, yielding from Eq. (7.131)

$$\frac{v_{out1}}{v_{in}} = \frac{\frac{1}{g_{m2}}||r_{O2}}{\frac{1}{g_{m2}}||r_{O2} + \frac{1}{g_{m1}}}. \tag{7.159}$$

For V_{out2}, M_1 operates as a degenerated CS stage with a drain load consisting of the diode-connected device M_3 and the current source M_4. This load impedance is equal to $(1/g_{m3})||r_{O3}||r_{O4}$, resulting in

$$\frac{v_{out2}}{v_{in}} = -\frac{\frac{1}{g_{m3}}||r_{O3}||r_{O4}}{\frac{1}{g_{m1}} + \frac{1}{g_{m2}}||r_{O2}}. \tag{7.160}$$

Exercise Which one of the two gains is higher? Explain intuitively why.

Bioengineering Application: Health Monitoring

Various devices have been introduced that can continuously monitor and record a user's vital signs. Such gathering of data proves useful for any individual: a doctor looking over several *years* of a patient's health history can more easily make a diagnosis than by simply asking the patient some questions.

It is possible to monitor a user's heart rate, blood pressure, and oxygen saturation by examining the blood flow in his/her finger. As illustrated in Fig. 7.41, an LED transmits light through the finger, and an array of photodiodes collects the light on the other side [1]. The intensity of this light has a large, relatively constant component (which we can view as a "bias" value) and a small, time-varying one that results from blood flow in the vessels.

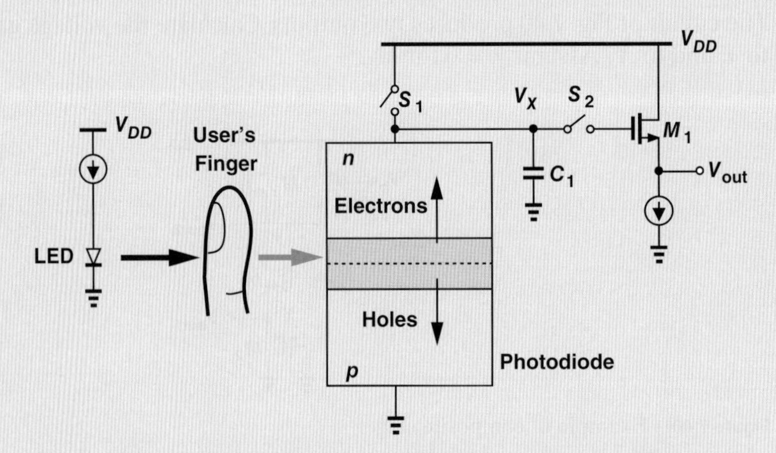

Figure 7.41 Monitoring vital signs by examining blood flow in the user's finger.

The circuit operates as follows [1]. First, S_1 is on, precharging the photodiode capacitance to V_{DD}. Next, S_1 turns off, S_2 turns on, and the photodiode current discharges the capacitance. The change in V_X is proportional to the light intensity and is sensed by the source follower. Tens of these photodiodes and their associated circuits are necessary to capture a sufficiently large area of the finger.

7.6 CHAPTER SUMMARY

- The impedances seen looking into the gate, drain, and source of a MOSFET are equal to infinity, r_O (with source grounded), and $1/g_m$ (with gate grounded), respectively.
- In order to obtain the required small-signal MOS parameters such as g_m and r_O, the transistor must be "biased," i.e., carry a certain drain current and sustain certain gate-source and drain-source voltages. Signals simply perturb these conditions.
- Biasing techniques establish the required gate voltage by means of a resistive path to the supply rails or the output node (self-biasing).
- With a single transistor, only three amplifier topologies are possible: common-source and common-gate stages and source followers.

- The CS stage provides a moderate voltage gain, a high input impedance, and a moderate output impedance.
- Source degeneration improves the linearity but lowers the voltage gain.
- Source degeneration raises the output impedance of CS stages considerably.
- The CG stage provides a moderate voltage gain, a low input impedance, and a moderate output impedance.
- The voltage gain expressions for CS and CG stages are similar but for a sign.
- The source follower provides a voltage gain less than unity, a high input impedance, and a low output impedance, serving as a good voltage buffer.

PROBLEMS

In the following problems, unless otherwise stated, assume $\mu_n C_{ox} = 200\ \mu\text{A/V}^2$, $\mu_p C_{ox} = 100\ \mu\text{A/V}^2$, $\lambda = 0$, and $V_{TH} = 0.4$ V for NMOS devices and -0.4 V for PMOS devices.

Sec. 7.1.2 Biasing

7.1. In the circuit of Fig. 7.42, determine the maximum allowable value of W/L if M_1 must remain in saturation. Assume $\lambda = 0$.

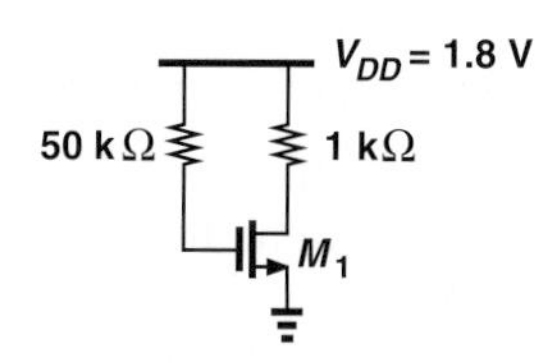

Figure 7.42

7.2. We wish to design the circuit of Fig. 7.43 for a drain current of 1 mA. If $W/L = 20/0.18$, compute R_1 and R_2 such that the input impedance is at least 20 kΩ.

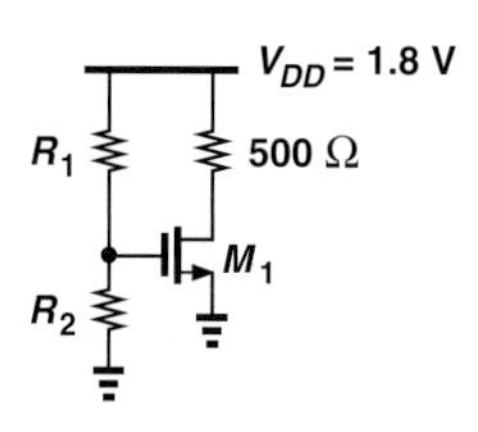

Figure 7.43

7.3. Consider the circuit shown in Fig. 7.44. Calculate the maximum transconductance that M_1 can provide (without going into the triode region.)

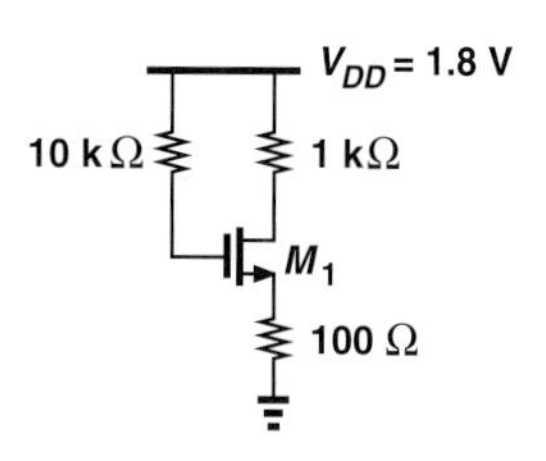

Figure 7.44

7.4. The circuit of Fig. 7.45 must be designed for a voltage drop of 200 mV across R_S.

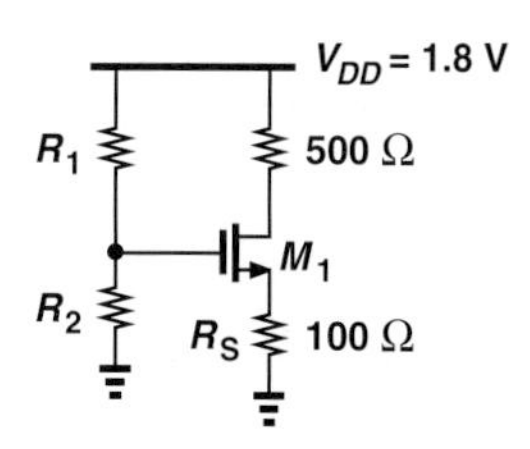

Figure 7.45

(a) Calculate the minimum allowable value of W/L if M_1 must remain in saturation.

(b) What are the required values of R_1 and R_2 if the input impedance must be at least 30 kΩ?

7.5. SOL Consider the circuit depicted in Fig. 7.46, where $W/L = 20/0.18$. Assuming the current flowing through R_2 is one-tenth of

I_{D1}, calculate the values of R_1 and R_2 so that $I_{D1} = 0.5$ mA.

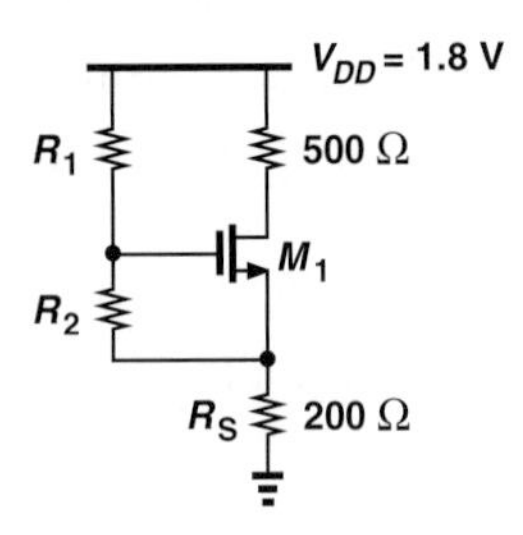

Figure 7.46

7.6. The self-biased stage of Fig. 7.47 must be designed for a drain current of 1 mA. If M_1 is to provide a transconductance of 1/(100 Ω), calculate the required value of R_D.

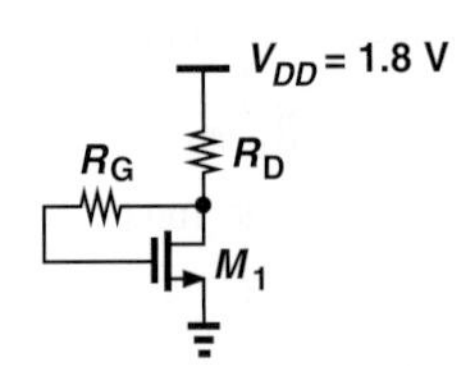

Figure 7.47

7.7. We wish to design the stage in Fig. 7.48 for a drain current of 0.5 mA. If $W/L = 50/0.18$, calculate the values of R_1 and R_2 such that these resistors carry a current equal to one-tenth of I_{D1}.

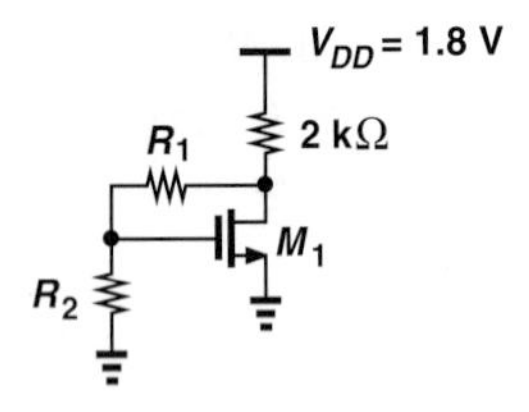

Figure 7.48

***7.8.** Due to a manufacturing error, a parasitic resistor, R_P has appeared in the circuit of Fig. 7.49. We know that circuit samples free from this error exhibit $V_{GS} = V_{DS} +$ 100 mV whereas defective samples exhibit $V_{GS} = V_{DS} + 50$ mV. Determine the values of W/L and R_P.

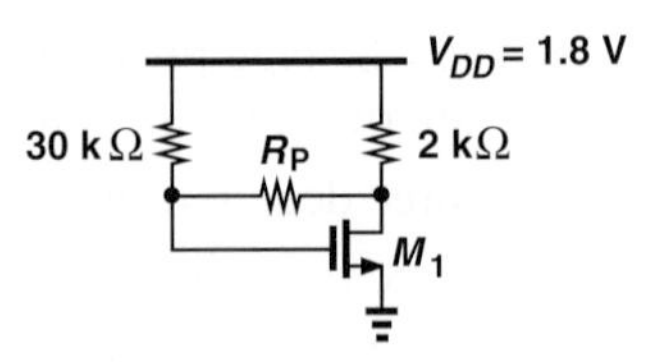

Figure 7.49

***7.9.** Due to a manufacturing error, a parasitic resistor, R_P has appeared in the circuit of Fig. 7.50. We know that circuit samples free from this error exhibit $V_{GS} = V_{DS}$ whereas defective samples exhibit $V_{GS} = V_{DS} + V_{TH}$. Determine the values of W/L and R_P if the drain current is 1 mA without R_P.

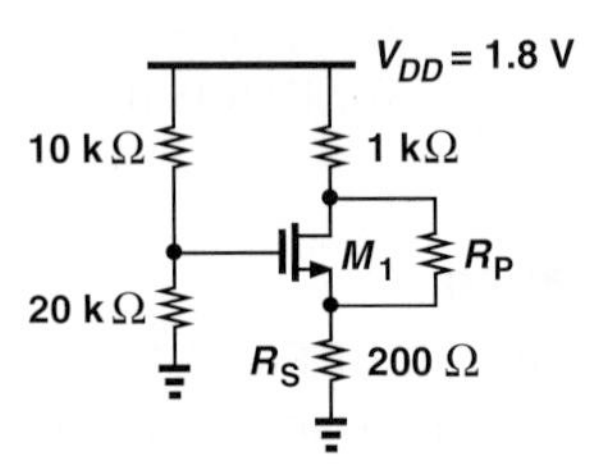

Figure 7.50

Sec. 7.1.3 Realization of Current Sources

7.10. In the circuit of Fig. 7.51, M_1 and M_2 have lengths equal to 0.25 μm and $\lambda = 0.1\ \text{V}^{-1}$. Determine W_1 and W_2 such that $I_X = 2I_Y = 1$ mA. Assume $V_{DS1} = V_{DS2} = V_B = 0.8$ V. What is the output resistance of each current source?

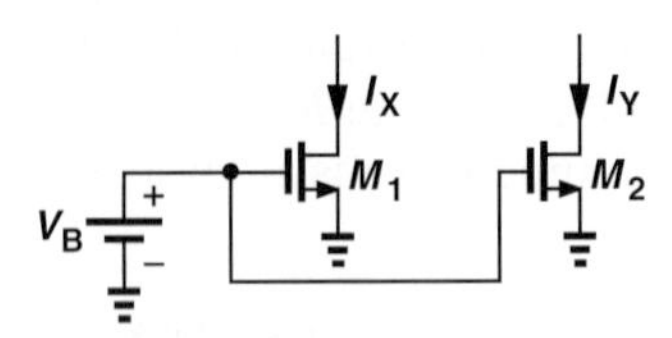

Figure 7.51

7.11. An NMOS current source must be designed for an output resistance of 20 kΩ and an output current of 0.5 mA. What is the maximum tolerable value of λ? SOL

7.12. The two current sources in Fig. 7.52 must be designed for $I_X = I_Y = 0.5$ mA. If $V_{B1} = 1$ V, $V_{B2} = 1.2$ V, $\lambda = 0.1\ \text{V}^{-1}$, and

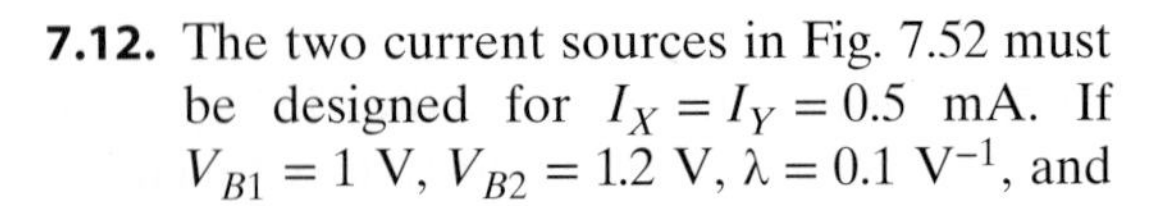

$L_1 = L_2 = 0.25$ μm, calculate W_1 and W_2. Compare the output resistances of the two current sources.

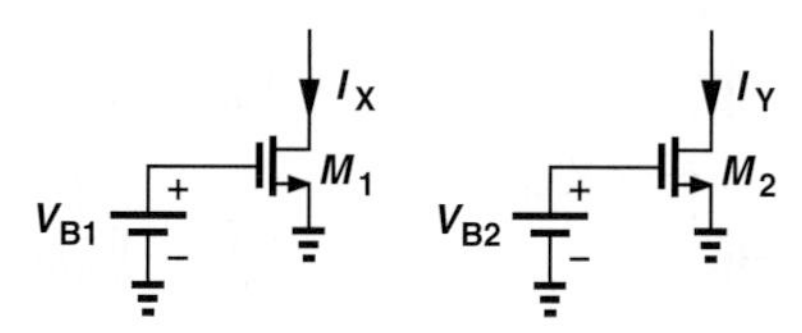

Figure 7.52

7.13. Consider the circuit shown in Fig. 7.53, where $(W/L)_1 = 10/0.18$ and $(W/L)_2 = 30/0.18$. If $\lambda = 0.1$ V^{-1}, calculate V_B such that $V_X = 0.9$ V.

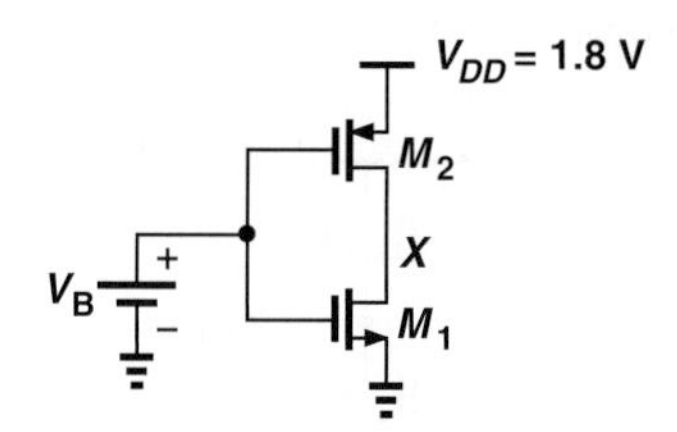

Figure 7.53

7.14. In the circuit of Fig. 7.54, M_1 and M_2 serve as current sources. Calculate I_X and I_Y if $V_B = 1$ V and $W/L = 20/0.25$. How are the output resistances of M_1 and M_2 related?

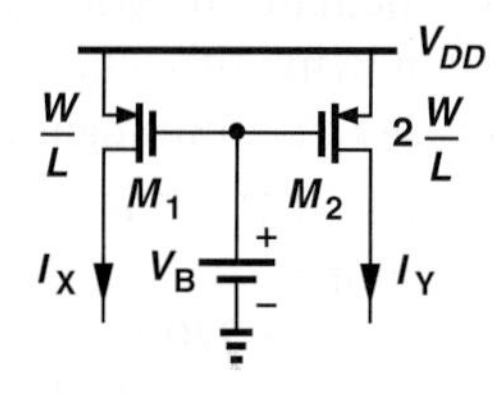

Figure 7.54

*__7.15.__ A student mistakenly uses the circuit of Fig. 7.55 as a current source. If $W/L = 10/0.25$, $\lambda = 0.1$ V^{-1}, $V_{B1} = 0.2$ V, and V_X has a dc level of 1.2 V, calculate the impedance seen at the source of M_1.

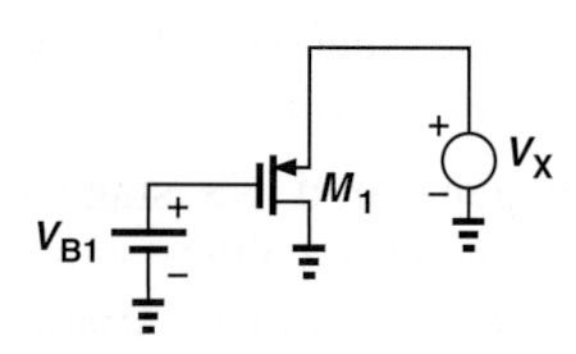

Figure 7.55

****7.16.** In the circuit of Fig. 7.56, $(W/L)_1 = 5/0.18$, $(W/L)_2 = 10/0.18$, $\lambda_1 = 0.1$ V^{-1}, and $\lambda_2 = 0.15$ V^{-1}.

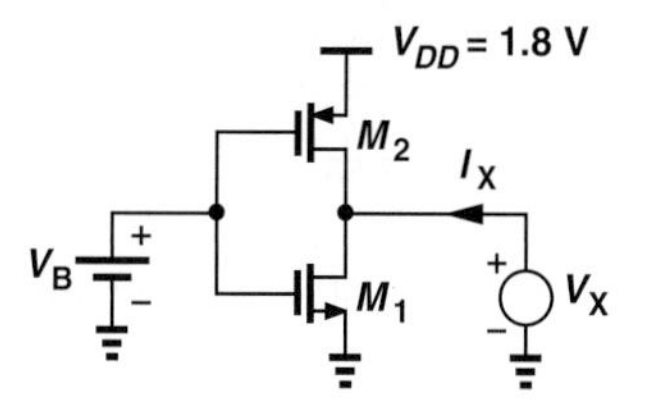

Figure 7.56

(a) Determine V_B such that $I_{D1} = |I_{D2}| = 0.5$ mA for $V_X = 0.9$ V.
(b) Now sketch I_X as a function of V_X as V_X goes from 0 to V_{DD}.

Sec. 7.2 Common-Source Stage

7.17. In the common-source stage of Fig. 7.57, $W/L = 30/0.18$ and $\lambda = 0$.

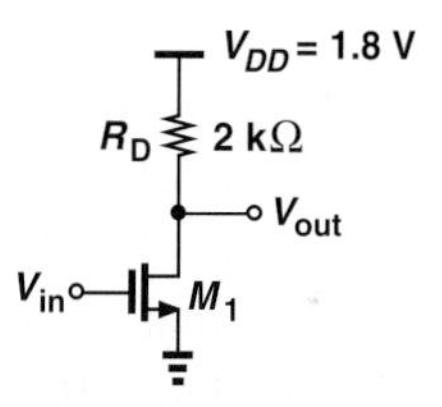

Figure 7.57

(a) What gate voltage yields a drain current of 0.5 mA? (Verify that M_1 operates in saturation.)
(b) With such a drain bias current, calculate the voltage gain of the stage.

7.18. The circuit of Fig. 7.57 is designed with $W/L = 20/0.18$, $\lambda = 0$, and $I_D = 0.25$ mA.
(a) Compute the required gate bias voltage.
(b) With such a gate voltage, how much can W/L be increased while M_1 remains in saturation? What is the maximum voltage gain that can be achieved as W/L increases?

7.19. We wish to design the stage of Fig. 7.58 for a voltage gain of 5 with $W/L \le 20/0.18$. Determine the required value of R_D if the power dissipation must not exceed 1 mW. SOL

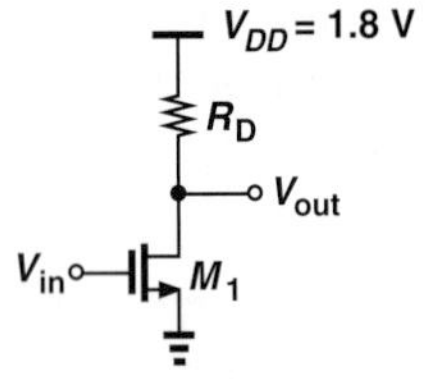

Figure 7.58

7.20. The CS stage of Fig. 7.59 must provide a voltage gain of 10 with a bias current of 0.5 mA. Assume $\lambda_1 = 0.1\ \text{V}^{-1}$, and $\lambda_2 = 0.15\ \text{V}^{-1}$.

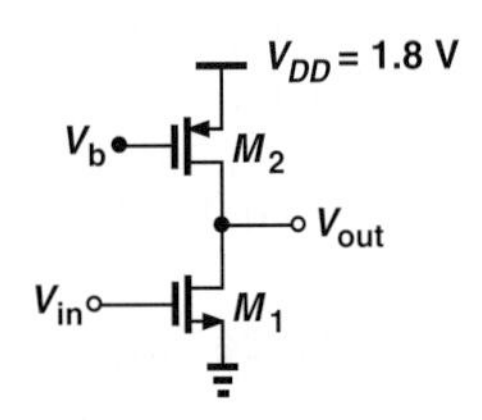

Figure 7.59

(a) Compute the required value of $(W/L)_1$.

(b) If $(W/L)_2 = 20/0.18$, calculate the required value of V_B.

7.21. In the stage of Fig. 7.59, M_2 has a long length so that $\lambda_2 \ll \lambda_1$. Calculate the voltage gain if $\lambda_1 = 0.1\ \text{V}^{-1}$, $(W/L)_1 = 20/0.18$, and $I_D = 1$ mA.

****7.22.** The circuit of Fig. 7.59 is designed for a bias current of I_1 with certain dimensions for M_1 and M_2. If the width and the length of both transistors are doubled, how does the voltage gain change? Consider two cases: (a) the bias current remains constant, or (b) the bias current is doubled.

7.23. The CS stage depicted in Fig. 7.60 must achieve a voltage gain of 15 at a bias current of 0.5 mA. If $\lambda_1 = 0.15\ \text{V}^{-1}$ and $\lambda_2 = 0.05\ \text{V}^{-1}$, determine the required value of $(W/L)_2$.

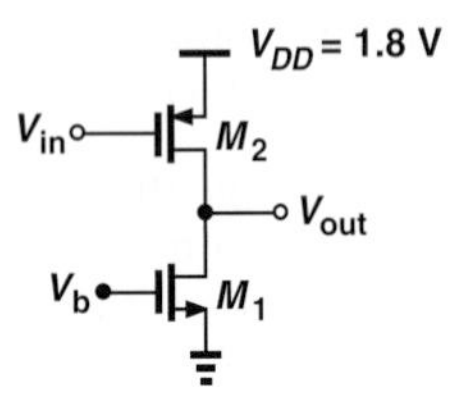

Figure 7.60

7.24. Explain which one of the topologies shown in Fig. 7.61 is preferred.

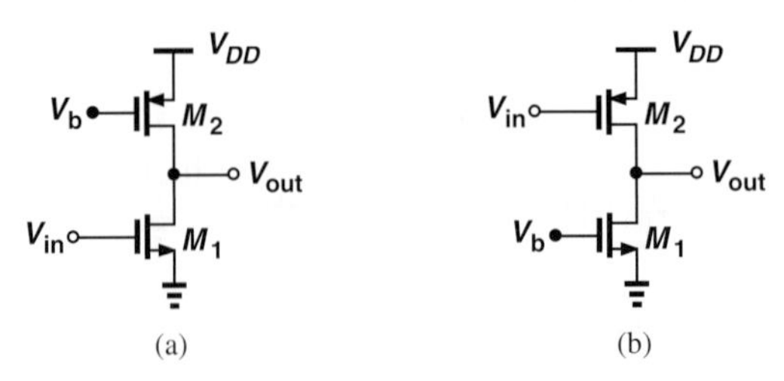

Figure 7.61

7.25. We wish to design the circuit shown in Fig. 7.62 for a voltage gain of 3. If $(W/L)_1 = 20/0.18$, determine $(W/L)_2$. Assume $\lambda = 0$. SOL

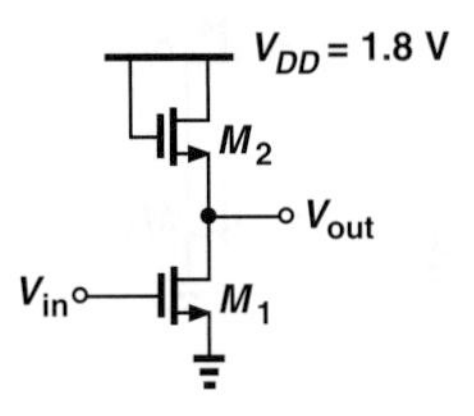

Figure 7.62

7.26. In the circuit of Fig. 7.62, $(W/L)_1 = 10/0.18$ and $I_{D1} = 0.5$ mA.

(a) If $\lambda = 0$, determine $(W/L)_2$ such that M_1 operates at the edge of saturation.

(b) Now calculate the voltage gain.

(c) Explain why this choice of $(W/L)_2$ yields the maximum gain.

7.27. The CS stage of Fig. 7.62 must achieve a voltage gain of 5.

(a) If $(W/L)_2 = 2/0.18$, compute the required value of $(W/L)_1$.

(b) What is the maximum allowable bias current if M_1 must operate in saturation?

****7.28.** If $\lambda \neq 0$, determine the voltage gain of the stages shown in Fig. 7.63.

***7.29.** In the circuit of Fig. 7.64, determine the gate voltage such that M_1 operates at the edge of saturation. Assume $\lambda = 0$.

***7.30.** The degenerated CS stage of Fig. 7.64 must provide a voltage gain of 4 with a bias current of 1 mA. Assume a drop of 200 mV across R_S and $\lambda = 0$.

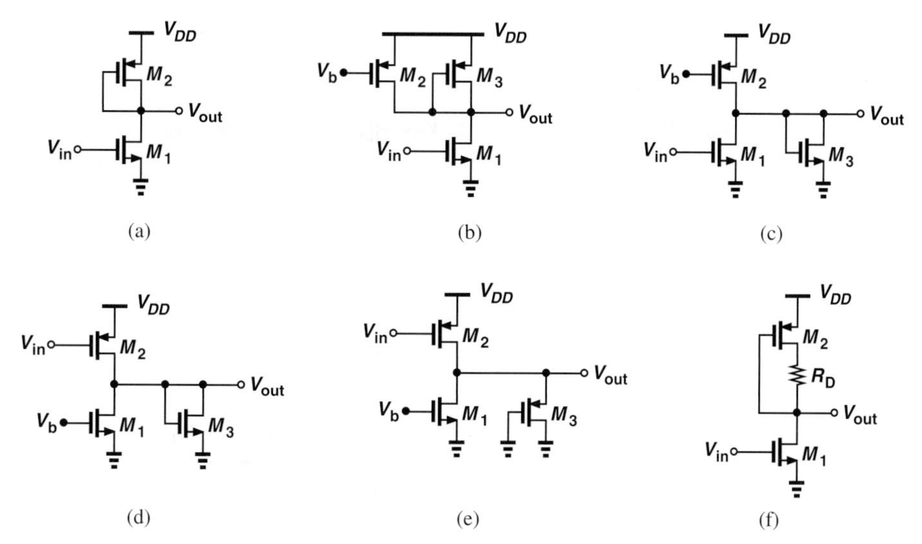

Figure 7.63

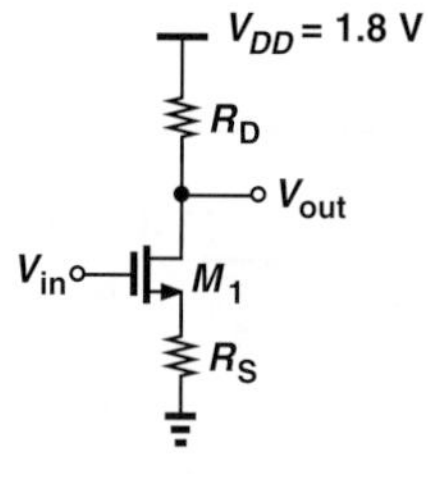

Figure 7.64

(a) If $R_D = 1\ \text{k}\Omega$, determine the required value of W/L. Does the transistor operate in saturation for this choice of W/L?

(b) If $W/L = 50/0.18$, determine the required value of R_D. Does the transistor operate in saturation for this choice of R_D?

7.31. Calculate the voltage gain of the circuits depicted in Fig. 7.65. Assume $\lambda = 0$.

7.32. Consider a degenerated CS stage with $\lambda > 0$. Assuming $g_m r_O \gg 1$, calculate the voltage gain of the circuit.

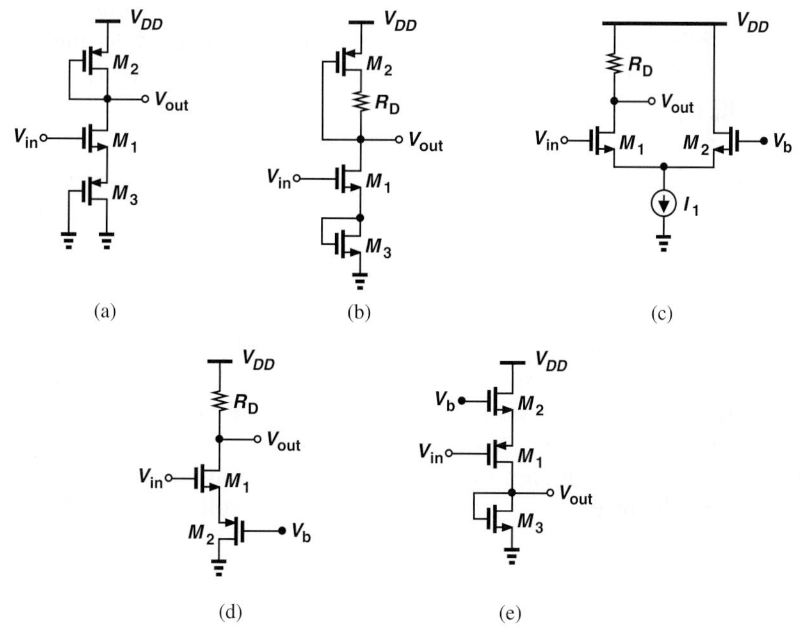

Figure 7.65

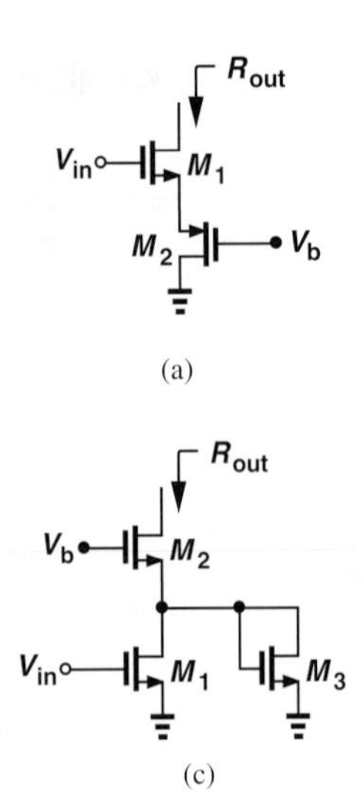
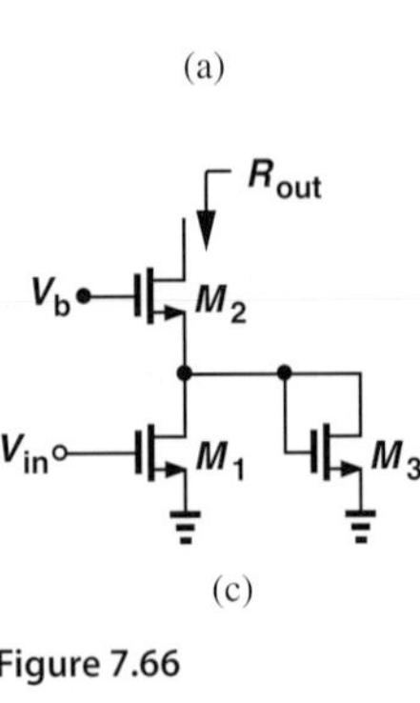
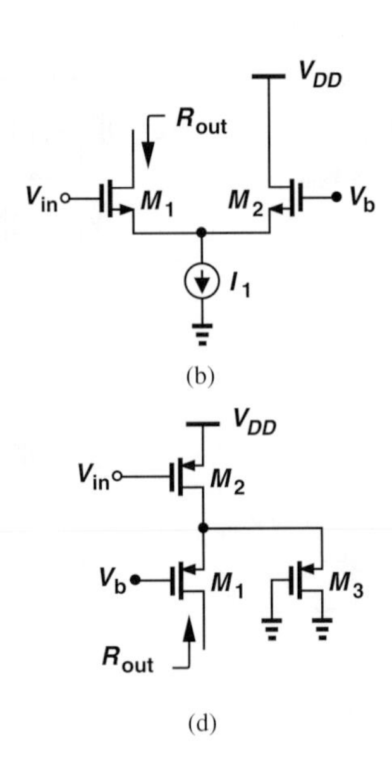

Figure 7.66

SOL *7.33. Determine the output impedance of each circuit shown in Fig. 7.66. Assume $\lambda \neq 0$.

7.34. The CS stage of Fig. 7.67 carries a bias current of 1 mA. If $R_D = 1\ \text{k}\Omega$ and $\lambda = 0.1\ \text{V}^{-1}$, compute the required value of W/L for a gate voltage of 1 V. What is the voltage gain of the circuit?

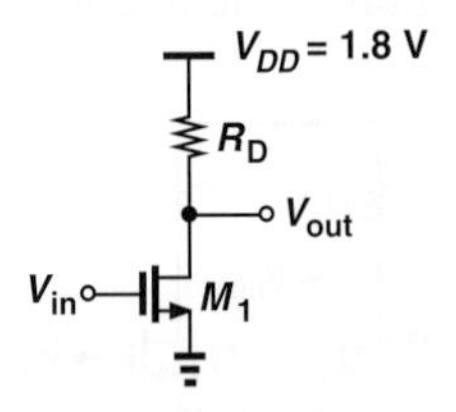

Figure 7.67

7.35. Repeat Problem 7.34 with $\lambda = 0$ and compare the results.

7.36. An adventurous student decides to try a new circuit topology wherein the input is applied to the drain and the output is sensed at the source (Fig. 7.68). Assume $\lambda \neq 0$, determine the voltage gain of the circuit and discuss the result.

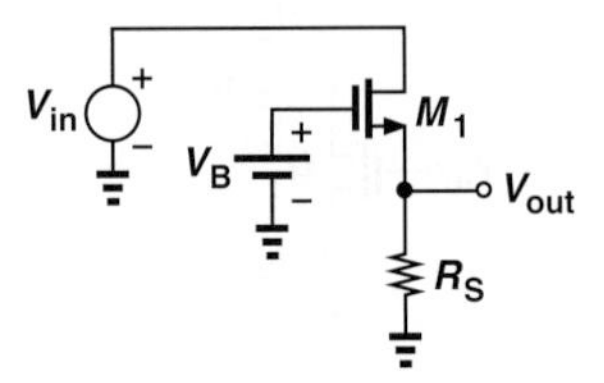

Figure 7.68

7.37. In the common-source stage depicted in Fig. 7.69, the drain current of M_1 is defined by the ideal current source I_1 and remains independent of R_1 and R_2 (why?). Suppose $I_1 = 1$ mA, $R_D = 500\ \Omega$, $\lambda = 0$, and C_1 is very large.

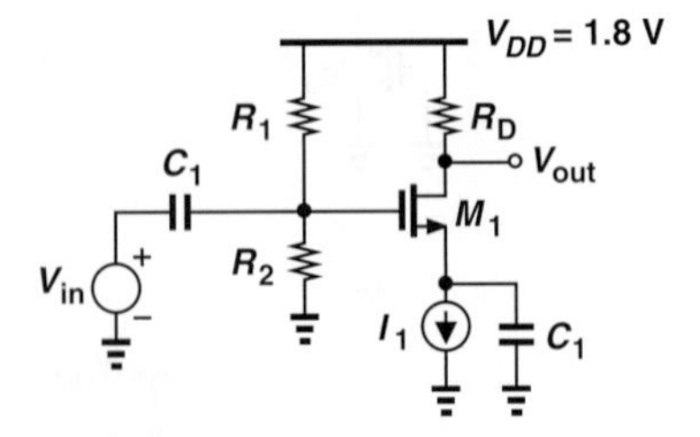

Figure 7.69

(a) Compute the value of W/L to obtain a voltage gain of 5.

(b) Choose the values of R_1 and R_2 to place the transistor 200 mV away from the triode region while $R_1 + R_2$ draws no more than 0.1 mA from the supply.

(c) With the values found in (b), what happens if W/L is twice that found in (a)? Consider both the bias conditions (e.g., whether M_1 comes closer to the triode region) and the voltage gain.

7.38. Consider the CS stage shown in Fig. 7.70, where I_1 defines the bias current of M_1 and C_1 is very large.

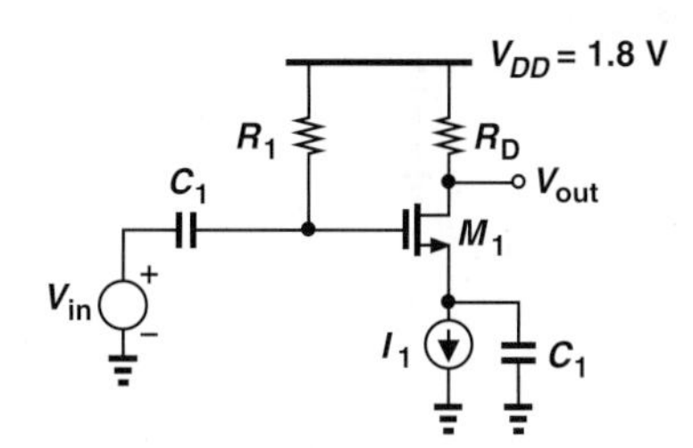

Figure 7.70

(a) If $\lambda = 0$ and $I_1 = 1$ mA, what is the maximum allowable value of R_D for M_1 to remain in saturation?

(b) With the value found in (a), determine W/L to obtain a voltage gain of 5.

Sec. 7.3 Common-Gate Stage

7.39. The common-gate stage shown in Fig. 7.71 must provide a voltage gain of 4 and an input impedance of 50 Ω. If $I_D = 0.5$ mA, and $\lambda = 0$, determine the values of R_D and W/L.

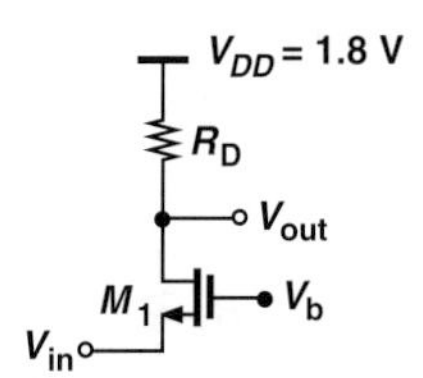

Figure 7.71

7.40. Suppose in Fig. 7.71, $I_D = 0.5$ mA, $\lambda = 0$, and $V_b = 1$ V. Determine the values of W/L and R_D for an input impedance of 50 Ω and maximum voltage gain (while M_1 remains in saturation).

7.41. The CG stage depicted in Fig. 7.72 must provide an input impedance of 50 Ω and an output impedance of 500 Ω. Assume $\lambda = 0$.

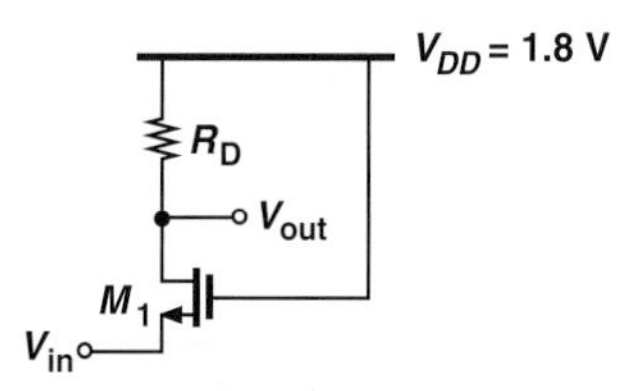

Figure 7.72

(a) What is the maximum allowable value of I_D?

(b) With the value obtained in (a), calculate the required value of W/L.

(c) Compute the voltage gain.

***7.42.** A CG stage with a source resistance of R_S employs a MOSFET with $\lambda > 0$. Assuming $g_m r_O \gg 1$, calculate the voltage gain of the circuit.

7.43. The CG amplifier shown in Fig. 7.73 is biased by means of $I_1 = 1$ mA. Assume $\lambda = 0$ and C_1 is very large.

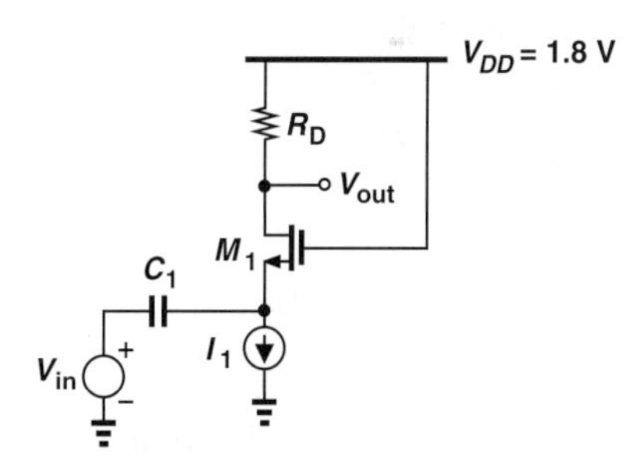

Figure 7.73

(a) What value of R_D places the transistor M_1 100 mV away from the triode region?

(b) What is the required W/L if the circuit must provide a voltage gain of 5 with the value of R_D obtained in (a)?

***7.44.** Determine the voltage gain of each stage depicted in Fig. 7.74. Assume $\lambda = 0$.

7.45. Consider the circuit of Fig. 7.75, where a common-source stage (M_1 and R_{D1}) is followed by a common-gate stage (M_2 and R_{D2}).

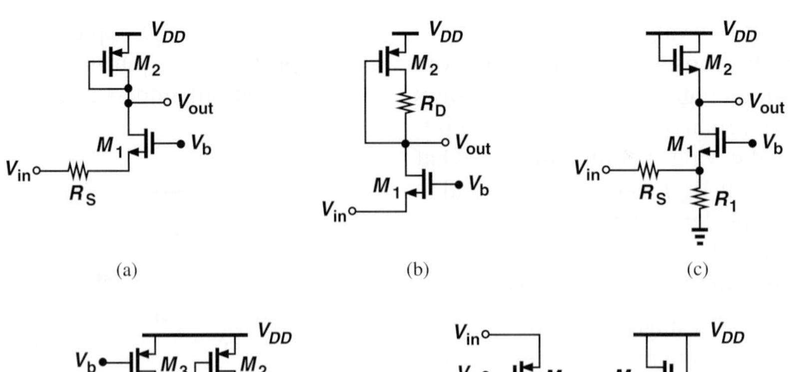

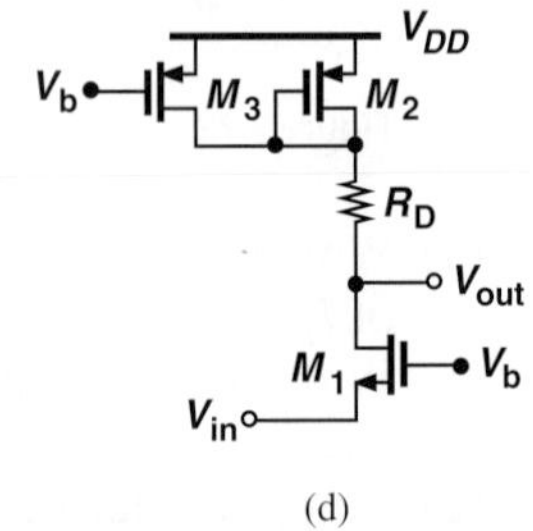

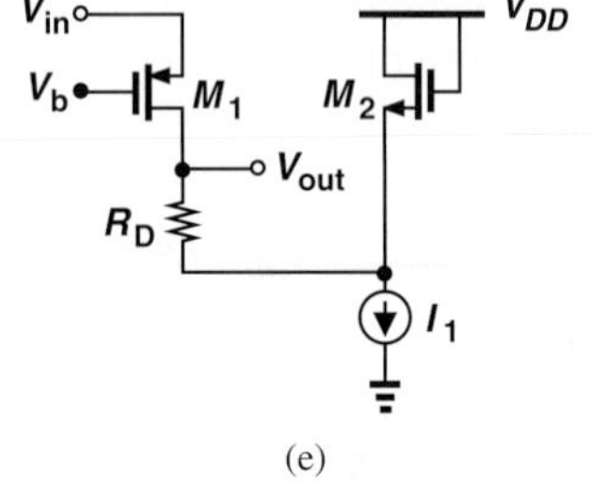

Figure 7.74

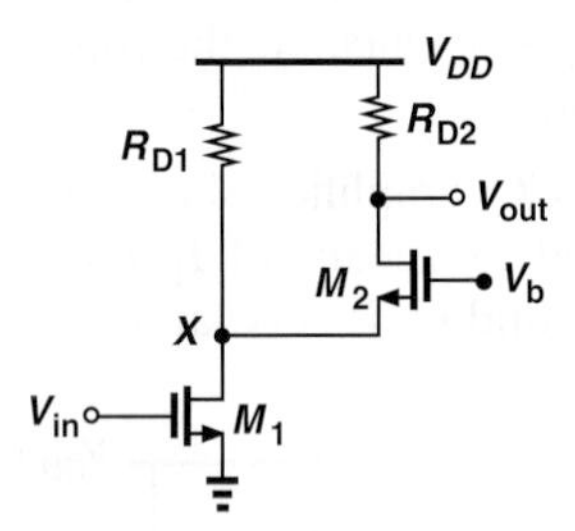

Figure 7.75

(a) Writing $v_{out}/v_{in} = (v_X/v_{in})(v_{out}/v_X)$ and assuming $\lambda = 0$, compute the overall voltage gain.

(b) Simplify the result obtained in (a) if $R_{D1} \to \infty$. Explain why this result is to be expected.

SOL **7.46.** Repeat Problem 7.45 for the circuit shown in Fig. 7.76.

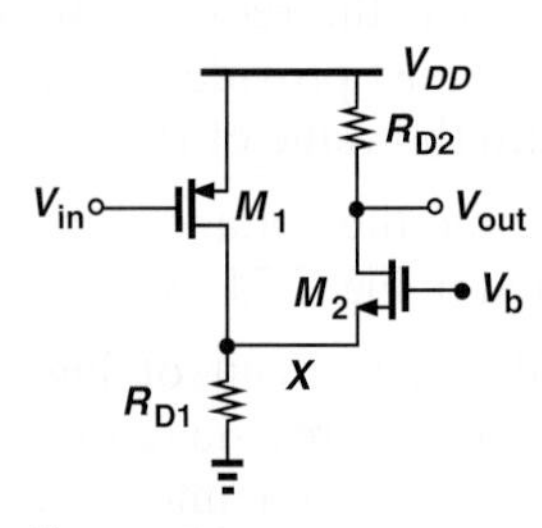

Figure 7.76

7.47. Assuming $\lambda = 0$, calculate the voltage gain of the circuit shown in Fig. 7.77. Explain why this stage is *not* a common-gate amplifier.

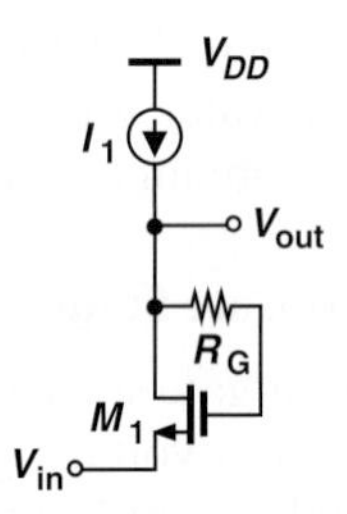

Figure 7.77

7.48. Calculate the voltage gain of the stage depicted in Fig. 7.78. Assume $\lambda = 0$ and the capacitors are very large.

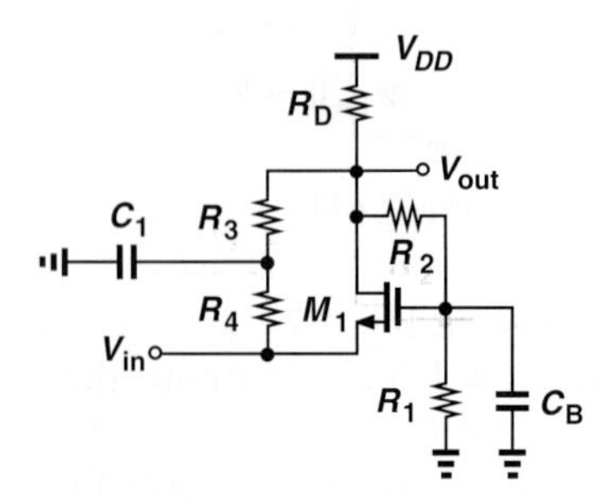

Figure 7.78

Sec. 7.4 Source Follower

7.49. The source follower shown in Fig. 7.79 is biased through R_G. Calculate the voltage gain if $W/L = 20/0.18$ and $\lambda = 0.1\ \text{V}^{-1}$.

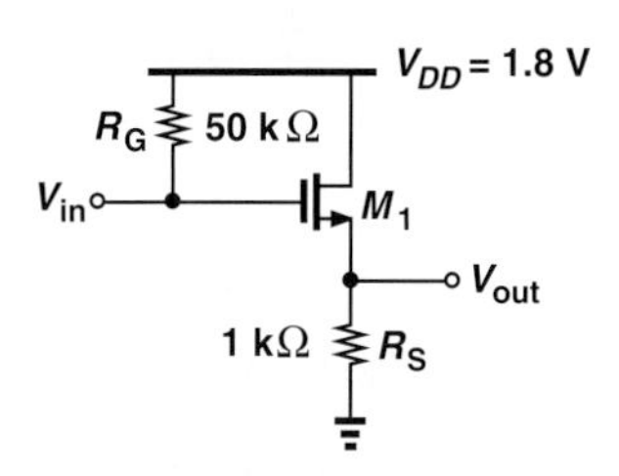

Figure 7.79

7.50. We wish to design the source follower shown in Fig. 7.80 for a voltage gain of 0.8. If $W/L = 30/0.18$ and $\lambda = 0$, determine the required gate bias voltage.

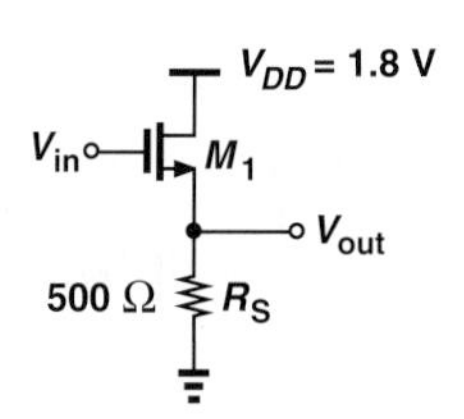

Figure 7.80

7.51. The source follower of Fig. 7.80 is to be designed with a maximum bias gate voltage of 1.8 V. Compute the required value of W/L for a voltage gain of 0.8 if $\lambda = 0$.

7.52. The source follower depicted in Fig. 7.81 employs a current source. Determine the values of I_1 and W/L if the circuit must provide an output impedance less than 100 Ω with $V_{GS} = 0.9$ V. Assume $\lambda = 0$.

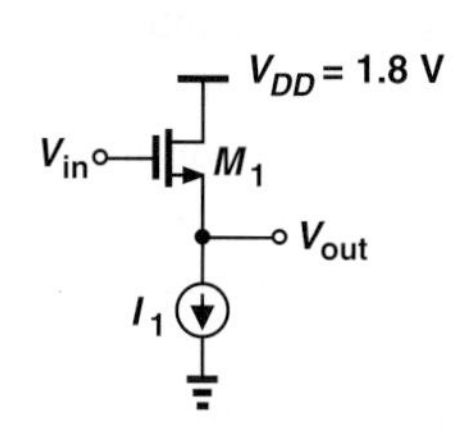

Figure 7.81

SOL **7.53.** The circuit of Fig. 7.81 must exhibit an output impedance of less than 50 Ω with a power budget of 2 mW. Determine the required value of W/L. Assume $\lambda = 0$.

7.54. We wish to design the source follower of Fig. 7.82 for a voltage gain of 0.8 with a power budget of 3 mW. Compute the required value of W/L. Assume C_1 is very large and $\lambda = 0$.

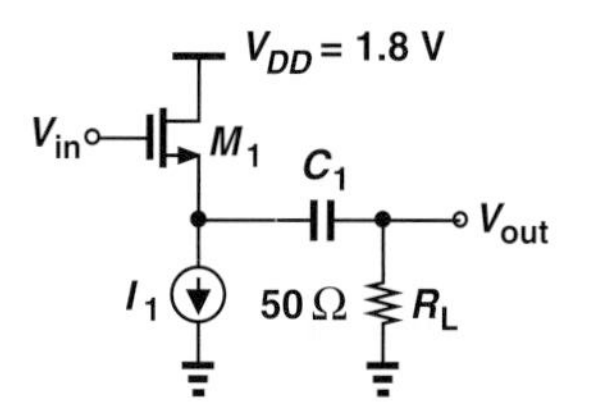

Figure 7.82

***7.55.** Determine the voltage gain of the stages shown in Fig. 7.83. Assume $\lambda \neq 0$.

***7.56.** Consider the circuit shown in Fig. 7.84, where a source follower (M_1 and I_1) precedes a common-gate stage (M_2 and R_D).
(a) Writing $v_{out}/v_{in} = (v_X/v_{in})(v_{out}/v_X)$, compute the overall voltage gain.
(b) Simplify the result obtained in (a) if $g_{m1} = g_{m2}$.

Design Problems

In the following problems, unless otherwise stated, assume $\lambda = 0$.

7.57. Design the CS stage shown in Fig. 7.85 for a voltage gain of 5 and an output impedance of 1 kΩ. Bias the transistor so that it operates 100 mV away from the triode region. Assume the capacitors are very large and $R_D = 10$ kΩ.

7.58. The CS amplifier of Fig. 7.85 must be designed for a voltage gain of 5 with a power budget of 2 mW. If $R_D I_D = 1$ V, determine the required value of W/L. Make the same assumptions as those in Problem 7.57.

7.59. We wish to design the CS stage of Fig. 7.85 for maximum voltage gain but with $W/L \leq 50/0.18$ and a maximum output impedance of 500 Ω. Determine the required current. Make the same assumptions as those in Problem 7.57.

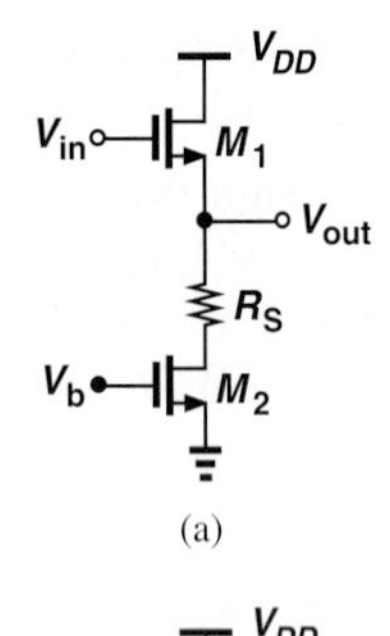

(a)

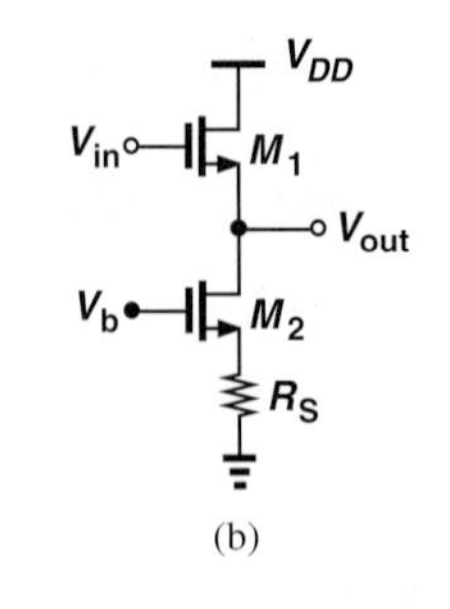

(b)

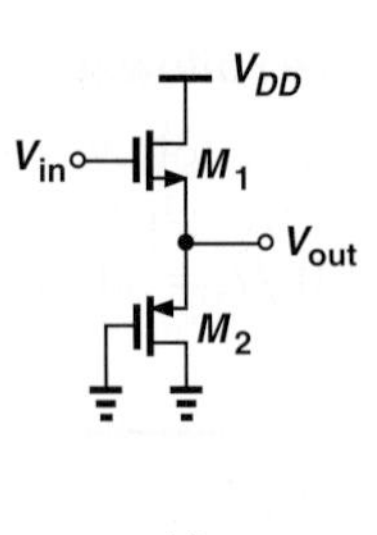

(c)

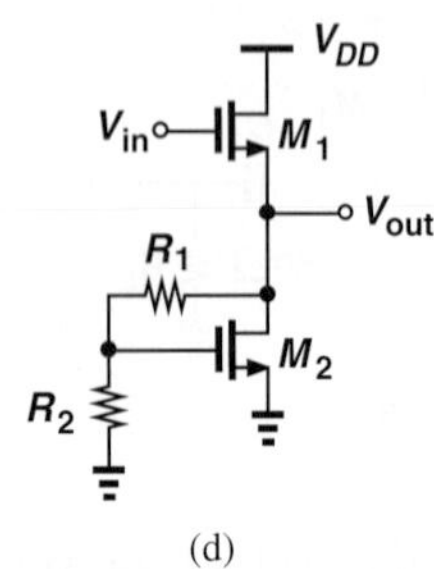

(d)

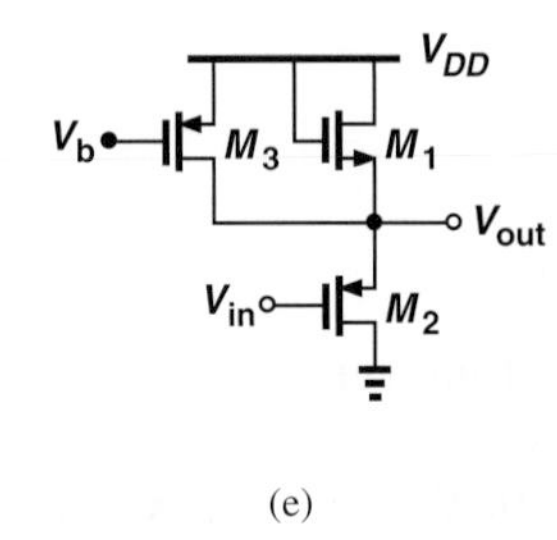

(e)

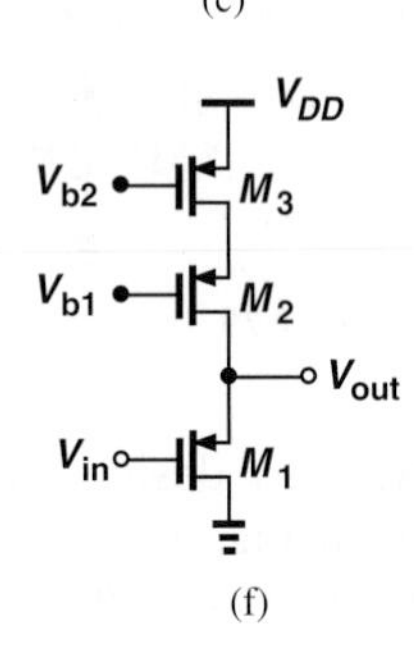

(f)

Figure 7.83

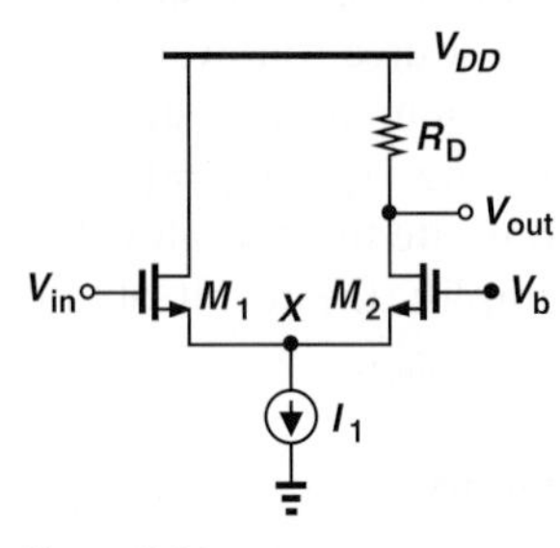

Figure 7.84

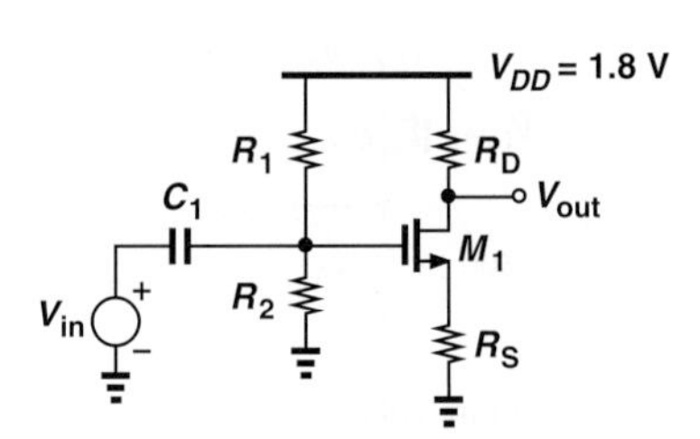

Figure 7.86

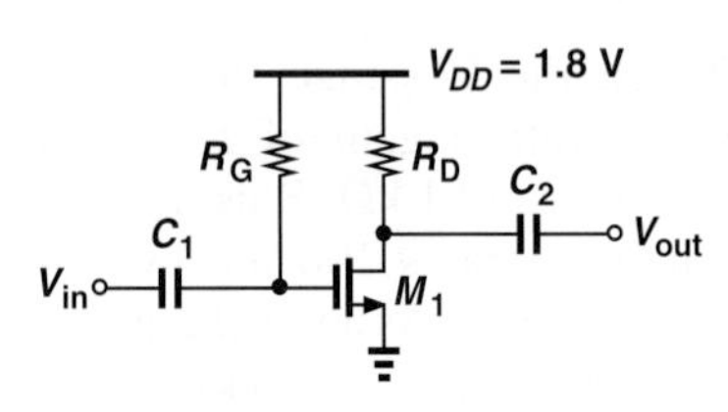

Figure 7.85

7.60. The degenerated stage depicted in Fig. 7.86 must provide a voltage gain of 4 with a power budget of 2 mW while the voltage drop across R_S is equal to 200 mV. If the overdrive voltage of the transistor must not exceed 300 mV and $R_1 + R_2$ must consume less than 5% of the allocated power, design the circuit. Make the same assumptions as those in Problem 7.57.

7.61. Design the circuit of Fig. 7.86 for a voltage gain of 5 and a power budget of 6 mW. Assume the voltage drop across R_S is equal to the overdrive voltage of the transistor and $R_D = 200\ \Omega$.

7.62. The circuit shown in Fig. 7.87 must provide a voltage gain of 6, with C_S serving as a low impedance at the frequencies of interest. Assuming a power budget of 2 mW and an input impedance of 20 kΩ, design the circuit such that M_1 operates 200 mV away from the triode region. Select the values of C_1 and C_S so that their impedance is negligible at 1 MHz.

7.63. In the circuit of Fig. 7.88, M_2 serves as a current source. Design the stage for a voltage gain of 20 and a power budget of 2 mW. Assume $\lambda = 0.1\ \text{V}^{-1}$ for both transistors and the maximum allowable level

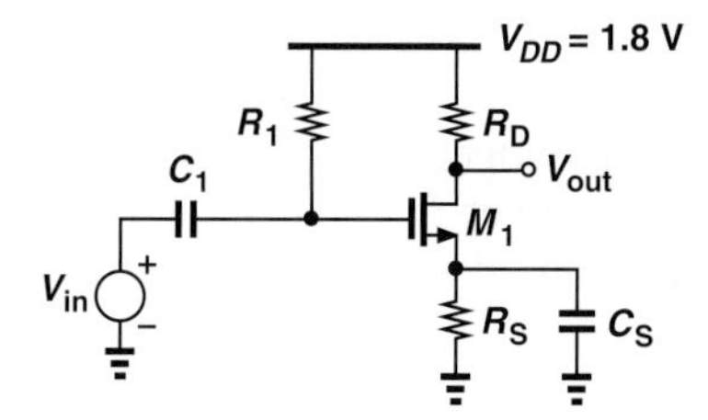

Figure 7.87

at the output is 1.5 V (i.e., M_2 must remain in saturation if $V_{out} \le 1.5$ V).

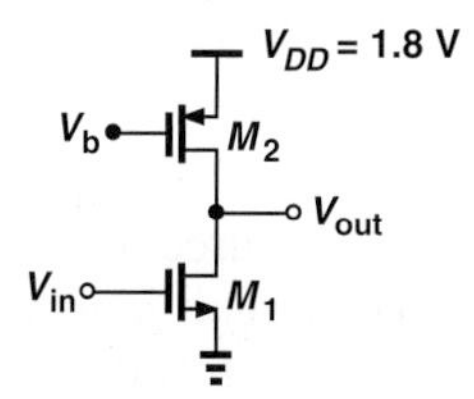

Figure 7.88

7.64. Consider the circuit shown in Fig. 7.89, where C_B is very large and $\lambda_n = 0.5\lambda_p = 0.1\ \text{V}^{-1}$.

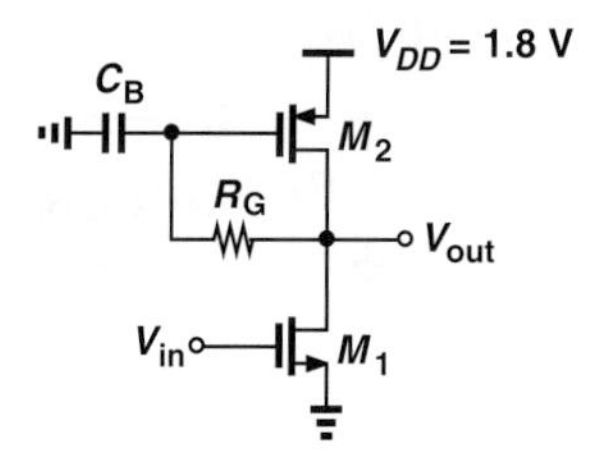

Figure 7.89

(a) Calculate the voltage gain.
(b) Design the circuit for a voltage gain of 15 and a power budget of 3 mW. Assume $R_G \approx 10(r_{O1}||r_{O2})$ and the dc level of the output must be equal to $V_{DD}/2$.

7.65. The CS stage of Fig. 7.90 incorporates a degenerated PMOS current source. The degeneration must raise the output impedance of the current source to about $10r_{O1}$ such that the voltage gain remains nearly equal to the intrinsic gain of M_1. Assume $\lambda = 0.1\ \text{V}^{-1}$ for both transistors and a power budget of 2 mW.

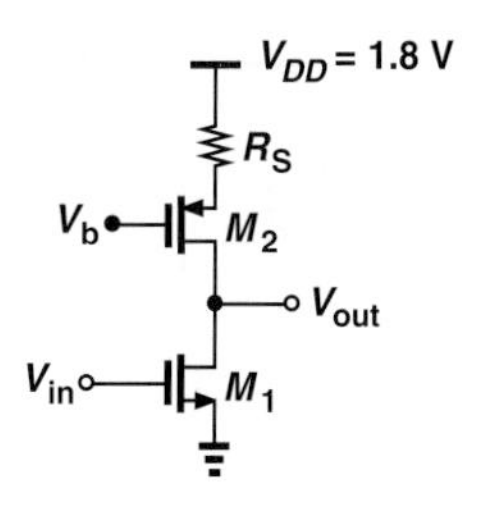

Figure 7.90

(a) If $V_B = 1$ V, determine the values of $(W/L)_2$ and R_S so that the impedance seen looking into the drain of M_2 is equal to $10r_{O1}$.
(b) Determine $(W/L)_1$ to achieve a voltage gain of 30.

7.66. Assuming a power budget of 1 mW and an overdrive of 200 mV for M_1, design the circuit shown in Fig. 7.91 for a voltage gain of 4. SOL

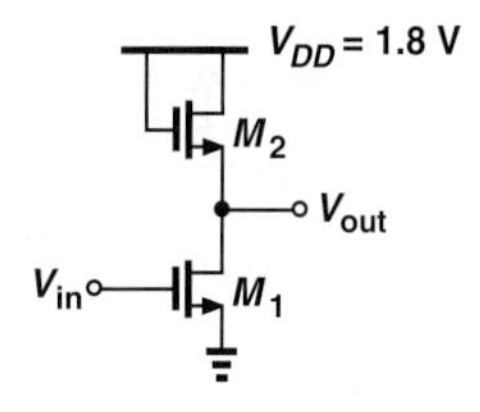

Figure 7.91

7.67. Design the common-gate stage depicted in Fig. 7.92 for an input impedance of 50 Ω and a voltage gain of 5. Assume a power budget of 3 mW.

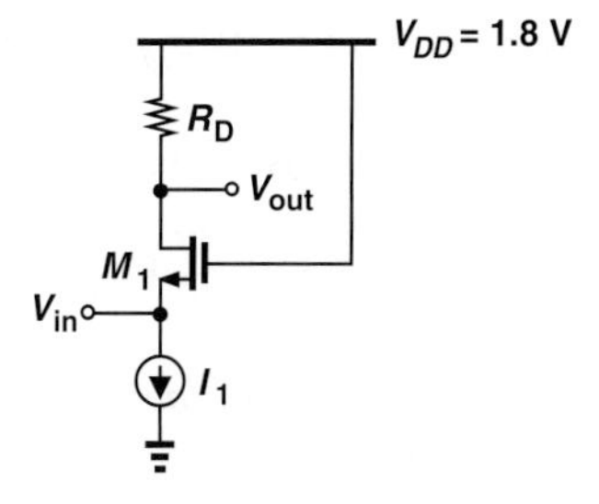

Figure 7.92

7.68. Design the circuit of Fig. 7.93 such that M_1 operates 100 mV away from the triode region while providing a voltage gain of 4. Assume a power budget of 2 mW.

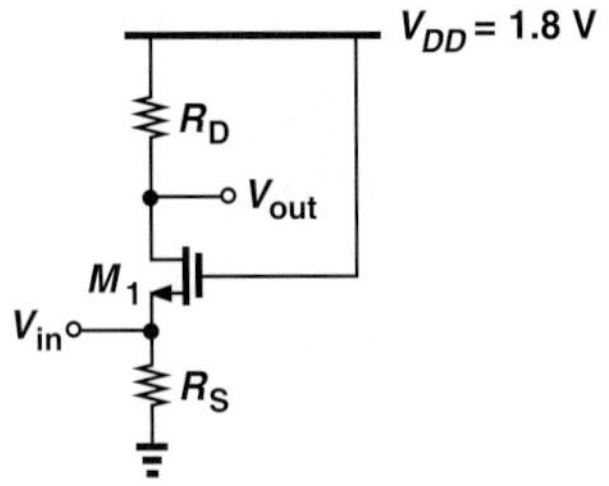

Figure 7.93

7.69. Figure 7.94 shows a self-biased common-gate stage, where $R_G \approx 10R_D$ and C_G serves as a low impedance so that the voltage gain is still given by $g_m R_D$. Design the circuit for a power budget of 5 mW and a voltage gain of 5. Assume $R_S \approx 10/g_m$ so that the input impedance remains approximately equal to $1/g_m$.

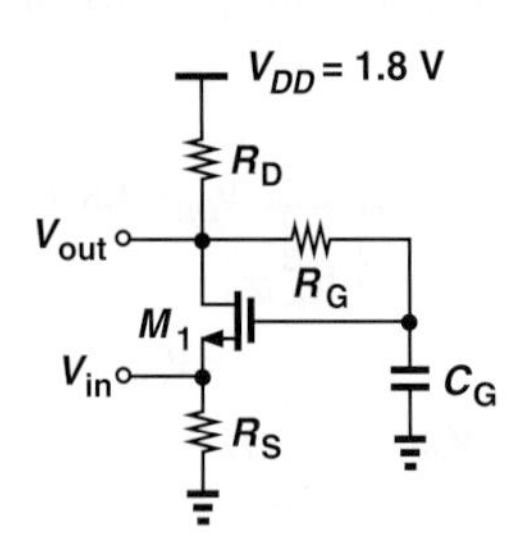

Figure 7.94

7.70. Design the CG stage shown in Fig. 7.95 such that it can accommodate an output swing of 500 mV_{pp}, i.e., V_{out} can fall below its bias value by 250 mV without driving M_1 into the triode region. Assume a voltage gain of 4 and an input impedance of 50 Ω. Select $R_S \approx 10/g_m$ and $R_1 + R_2 =$ 20 kΩ.
(Hint: Since M_1 is biased 250 mV away from the triode region, we have $R_S I_D + V_{GS} - V_{TH} + 250\text{ mV} = V_{DD} - I_D R_D$.)

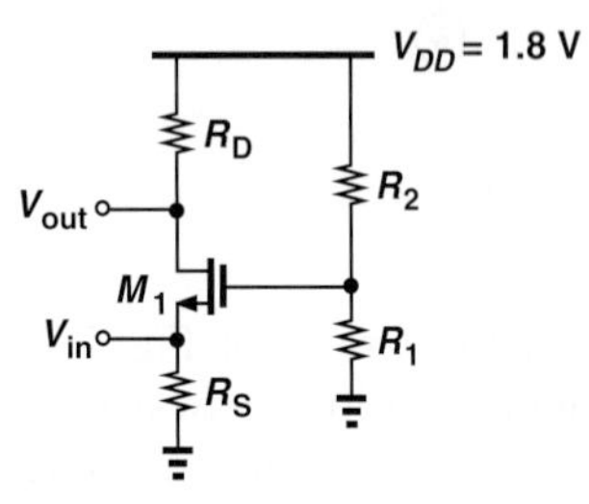

Figure 7.95

7.71. Design the source follower depicted in Fig. 7.96 for a voltage gain of 0.8 and a power budget of 2 mW. Assume the output dc level is equal to $V_{DD}/2$ and the input impedance exceeds 10 kΩ.

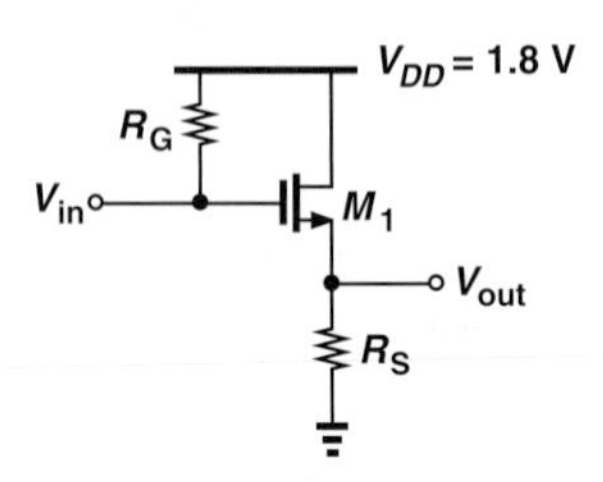

Figure 7.96

7.72. Consider the source follower shown in Fig. 7.97. The circuit must provide a voltage gain of 0.6 at 100 MHz. Design the circuit such that the dc voltage at node X is equal to $V_{DD}/2$. Assume the input impedance exceeds 20 kΩ.

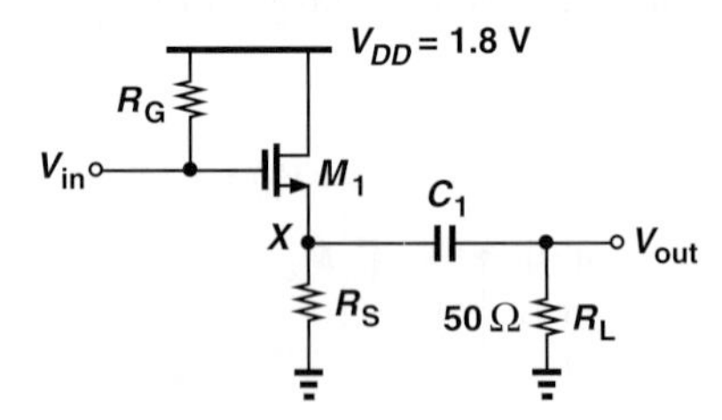

Figure 7.97

7.73. In the source follower of Fig. 7.98, M_2 serves as a current source. The circuit must operate with a power budget of 3 mW, a voltage gain of 0.9, and a minimum allowable output of 0.3 V (i.e., M_2 must remain in saturation if $V_{DS2} \geq 0.3$ V). Assuming $\lambda = 0.1\text{ V}^{-1}$ for both transistors, design the circuit.

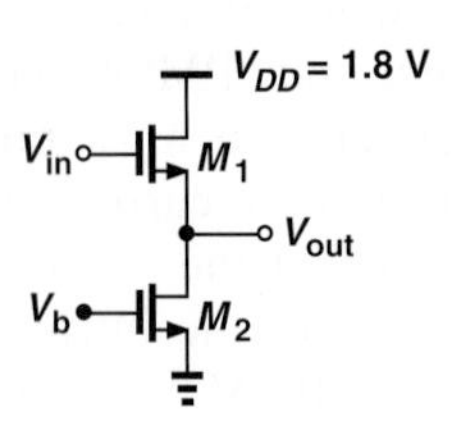

Figure 7.98

SPICE PROBLEMS

In the following problems, use the MOS models and source/drain dimensions given in Appendix A. Assume the substrates of NMOS and PMOS devices are tied to ground and V_{DD}, respectively.

7.74. In the circuit of Fig. 7.99, I_1 is an ideal current source equal to 1 mA.

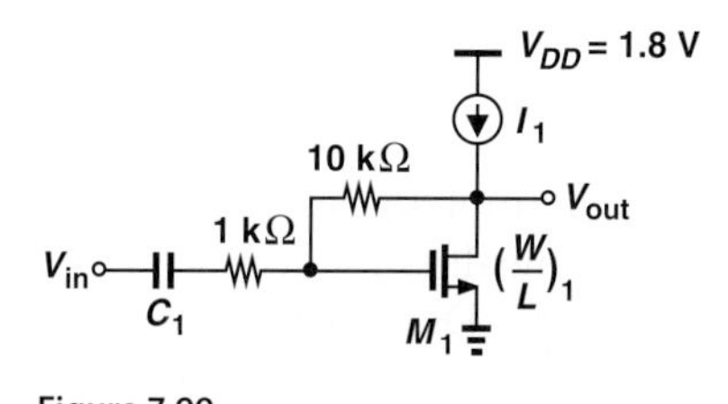

Figure 7.99

(a) Using hand calculations, determine $(W/L)_1$ such that $g_{m1} = (100\ \Omega)^{-1}$.

(b) Select C_1 for an impedance of $\approx 100\ \Omega\ (\ll 1\ \text{k}\Omega)$ at 50 MHz.

(c) Simulate the circuit and obtain the voltage gain and output impedance at 50 MHz.

(d) What is the change in the gain if I_1 varies by ±20%?

7.75. The source follower of Fig. 7.100 employs a bias current source, M_2.

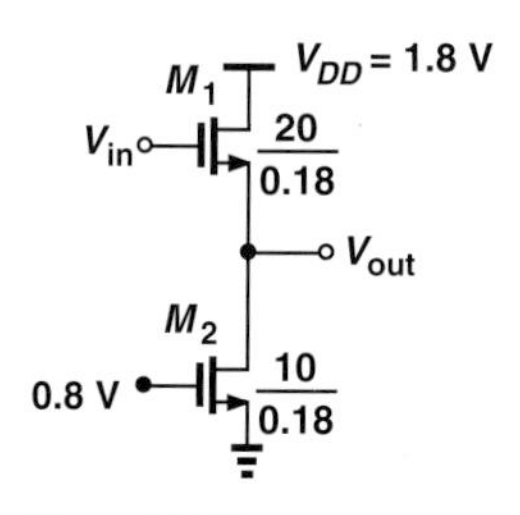

Figure 7.100

(a) What value of V_{in} places M_2 at the edge of saturation?

(b) What value of V_{in} places M_1 at the edge of saturation?

(c) Determine the voltage gain if V_{in} has a dc value of 1.5 V.

(d) What is the change in the gain if V_b changes by ±50 mV?

7.76. Figure 7.101 depicts a cascade of a source follower and a common-gate stage. Assume $V_b = 1.2$ V and $(W/L)_1 = (W/L)_2 = 10\ \mu\text{m}/0.18\ \mu\text{m}$.

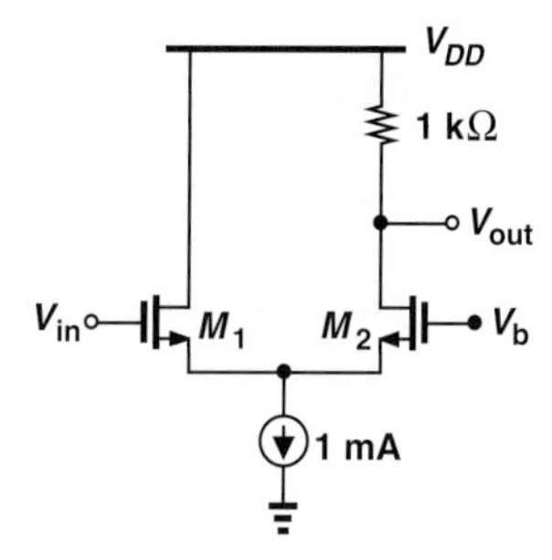

Figure 7.101

(a) Determine the voltage gain if V_{in} has a dc value of 1.2 V.

(b) Verify that the gain drops if the dc value of V_{in} is higher or lower than 1.2 V.

(c) What dc value at the input reduces the gain by 10% with respect to that obtained in (a)?

7.77. Consider the CS stage shown in Fig. 7.102, where M_2 operates as a resistor.

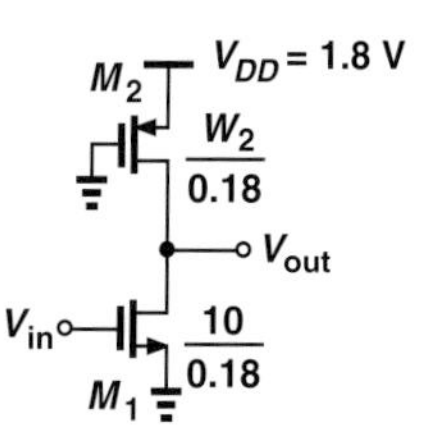

Figure 7.102

(a) Determine W_2 such that an input dc level of 0.8 V yields an output dc level of 1 V. What is the voltage gain under these conditions?

(b) What is the change in the gain if the mobility of the NMOS device varies by ±10%? Can you explain this result using the expressions derived in Chapter 6 for the transconductance?

7.78. Repeat Problem 7.77 for the circuit illustrated in Fig. 7.103 and compare the sensitivities to the mobility.

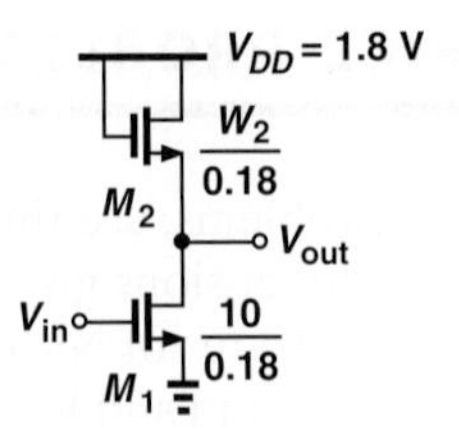

Figure 7.103

REFERENCE

1. A. Caizzone, et al., "A 2.6-μW monolithic CMOS photoplethysmographic (PPG) sensor operating with 2-μW LED power for continuous health monitoring," *IEEE Trans. Biomedical Circuits and Systems*, vol. 13, pp. 1243–1253, Dec. 2019.

8

Operational Amplifier as a Black Box

Watch Companion YouTube Videos:

Razavi Electronics 1, Lec 42: Op Amp Circuits I
https://www.youtube.com/watch?v=WzdmaSUCQGM

Razavi Electronics 1, Lec 43: Op Amp Circuits II
https://www.youtube.com/watch?v=oWMBjDfRacc

Razavi Electronics 1, Lec 44: Nonlinear Op Amp Circuits, Op Amp Nonidealities I
https://www.youtube.com/watch?v=FWBVjEgPx0U

Razavi Electronics 1, Lec 45: Op Amp Nonidealities II
https://www.youtube.com/watch?v=VN8SeVA8LnU

The term "operational amplifier" (op amp) was coined in the 1940s, well before the invention of the transistor and the integrated circuit. Op amps realized by vacuum tubes[1] served as the core of electronic "integrators," "differentiators," etc., thus forming systems whose behavior followed a given differential equation. Called "analog computers," such circuits were used to study the stability of differential equations that arose in fields such as control or power systems. Since each op amp implemented a mathematical *operation* (e.g., integration), the term "operational amplifier" was born.

Op amps find wide application in today's discrete and integrated electronics. In the cellphone studied in Chapter 1, for example, integrated op amps serve as building blocks in (active) filters. Similarly, the analog-to-digital converter(s) used in digital cameras often employ op amps.

In this chapter, we study the operational amplifier as a black box, developing op-amp-based circuits that perform interesting and useful functions. The outline is shown below.

[1]Vacuum tubes were amplifying devices consisting of a filament that released electrons, a plate that collected them, and another that controlled the flow—somewhat similar to MOSFETs.

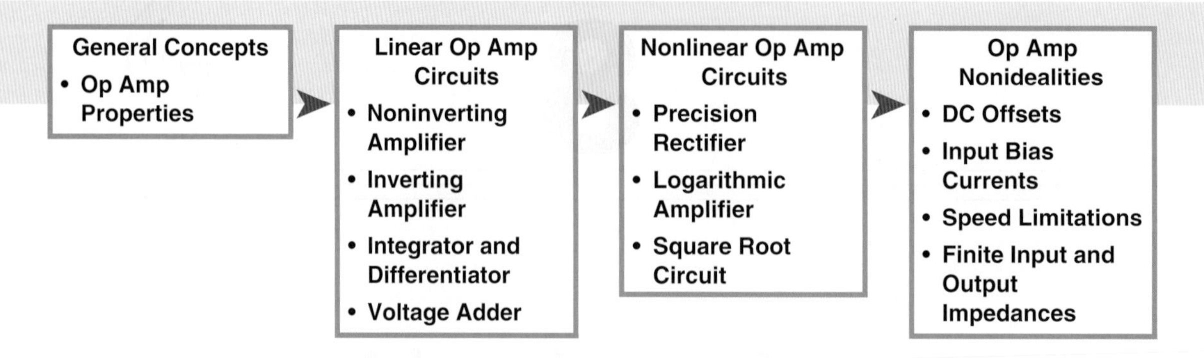

8.1 GENERAL CONSIDERATIONS

The operational amplifier can be abstracted as a black box having two inputs and one output.[2] Shown in Fig. 8.1(a), the op amp symbol distinguishes between the two inputs by the plus and minus sign; V_{in1} and V_{in2} are called the "noninverting" and "inverting" inputs, respectively. We view the op amp as a circuit that amplifies the *difference* between the two inputs, arriving at the equivalent circuit depicted in Fig. 8.1(b). The voltage gain is denoted by A_0:

$$V_{out} = A_0(V_{in1} - V_{in2}). \tag{8.1}$$

We call A_0 the "open-loop" gain.

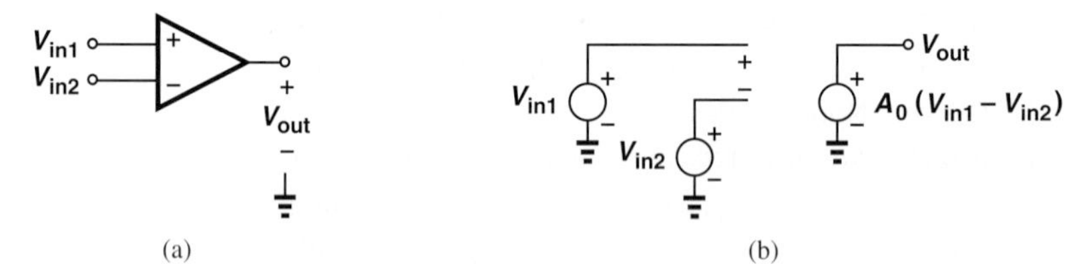

Figure 8.1 (a) Op amp symbol, (b) equivalent circuit.

It is instructive to plot V_{out} as a function of one input while the other remains at zero. With $V_{in2} = 0$, we have $V_{out} = A_0 V_{in1}$, obtaining the behavior shown in Fig. 8.2(a). The positive slope (gain) is consistent with the label "noninverting" given to V_{in1}. On the other hand, if $V_{in1} = 0$, $V_{out} = -A_0 V_{in2}$ [Fig. 8.2(b)], revealing a negative slope and hence an "inverting" behavior.

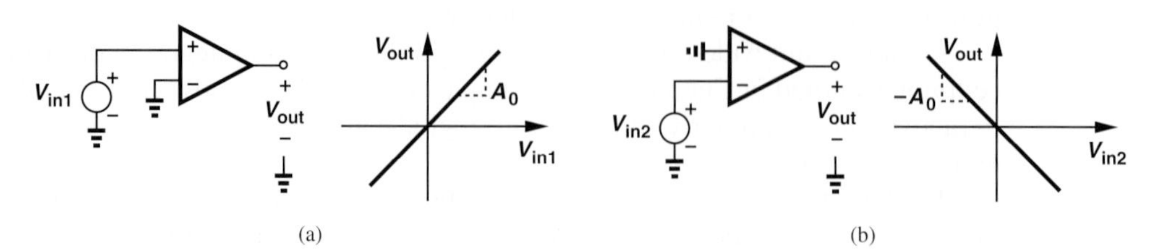

Figure 8.2 Op amp characteristics from (a) noninverting and (b) inverting inputs to output.

[2] In modern integrated circuits, op amps typically have two outputs that vary by equal and opposite amounts.

The reader may wonder why the op amp has *two* inputs. After all, the amplifier stages studied in Chapters 5 and 7 have only one input node (i.e., they sense the input voltage with respect to ground). As seen throughout this chapter, the principal property of the op amp, $V_{out} = A_0(V_{in1} - V_{in2})$, forms the foundation for many circuit topologies that would be difficult to realize using an amplifier having $V_{out} = AV_{in}$. Amplifier circuits having two inputs are studied in Chapter 10.

How does the "ideal" op amp behave? Such an op amp would provide an *infinite* voltage gain, an infinite input impedance, a zero output impedance, and infinite speed. In fact, the first-order analysis of an op-amp-based circuit typically begins with this idealization, quickly revealing the basic function of the circuit. We can then consider the effect of the op amp "nonidealities" on the performance.

The very high gain of the op amp leads to an important observation. Since realistic circuits produce finite output swings, e.g., 2 V, the difference between V_{in1} and V_{in2} in Fig. 8.1(a) is always small:

$$V_{in1} - V_{in2} = \frac{V_{out}}{A_0}. \tag{8.2}$$

In other words, the op amp, along with the circuitry around it, brings V_{in1} and V_{in2} close to each other. Following the above idealization, we may say $V_{in1} = V_{in2}$ if $A_0 = \infty$.

A common mistake is to interpret $V_{in1} = V_{in2}$ as if the two terminals V_{in1} and V_{in2} are *shorted* together. It must be borne in mind that $V_{in1} - V_{in2}$ becomes only *infinitesimally* small as $A_0 \to \infty$ but cannot be assumed *exactly* equal to zero.

Example 8-1

The circuit shown in Fig. 8.3 is called a "unity-gain" buffer. Note that the output is tied to the inverting input. Determine the output voltage if $V_{in1} = +1$ V and $A_0 = 1000$.

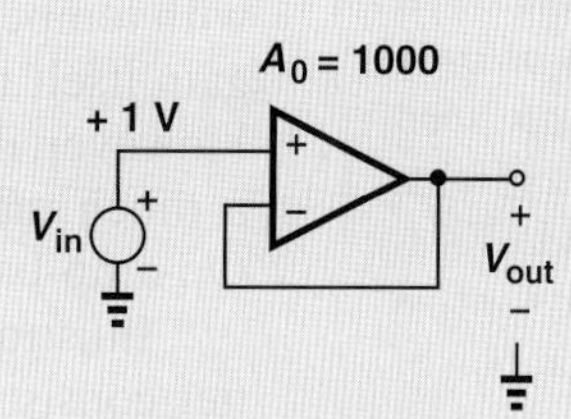

Figure 8.3 Unity-gain buffer.

Solution If the voltage gain of the op amp were infinite, the difference between the two inputs would be zero and $V_{out} = V_{in}$; hence the term "unity-gain buffer." For a finite gain, we write

$$V_{out} = A_0(V_{in1} - V_{in2}) \tag{8.3}$$

$$= A_0(V_{in} - V_{out}). \tag{8.4}$$

That is,

$$\frac{V_{out}}{V_{in}} = \frac{A_0}{1 + A_0}. \tag{8.5}$$

As expected, the gain approaches unity as A_0 becomes large. In this example, $A_0 = 1000$, $V_{in} = 1$ V, and $V_{out} = 0.999$ V. Indeed, $V_{in1} - V_{in2}$ is small compared to V_{in} and V_{out}.

Exercise What value of A_0 is necessary so that the output voltage is equal to 0.9999?

Op amps are sometimes represented as shown in Fig. 8.4 to indicate explicitly the supply voltages, V_{EE} and V_{CC}. For example, an op amp may operate between ground and a positive supply, in which case $V_{EE} = 0$.

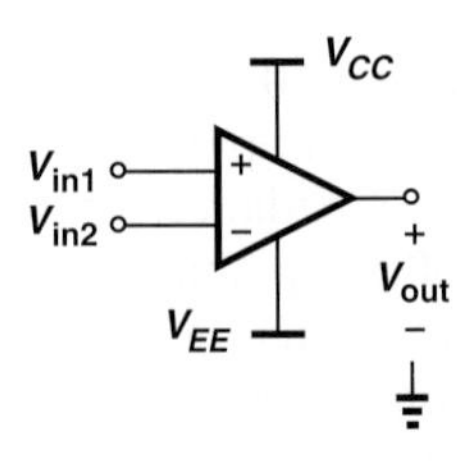

Figure 8.4 Op amp with supply rails.

8.2 OP-AMP-BASED CIRCUITS

In this section, we study a number of circuits that utilize op amps to process analog signals. In each case, we first assume an ideal op amp to understand the underlying principles and subsequently examine the effect of the finite gain on the performance.

8.2.1 Noninverting Amplifier

Did you know?

Early op amps were implemented using discrete bipolar transistors and other components and packaged as a "module" with dimensions of 5 to 10 cm. More importantly, each of these op amps cost \$100 to \$200 (in the 1960s). By contrast, today's off-the-shelf op amp ICs, e.g., the 741, cost less than a dollar. Op amps realized within larger integrated circuits cost a small fraction of a cent.

Recall from Chapters 5 and 7 that the voltage gain of amplifiers typically depends on the load resistor and other parameters that may vary considerably with temperature or process.[3] As a result, the voltage gain itself may suffer from a variation of, say, ±20%. However, in some applications (e.g., A/D converters), a much more precise gain (e.g., 2.000) is required. Op-amp-based circuits can provide such precision.

Illustrated in Fig. 8.5, the noninverting amplifier consists of an op amp and a voltage divider that returns a fraction of the output voltage to the inverting input:

$$V_{in2} = \frac{R_2}{R_1 + R_2} V_{out}. \tag{8.6}$$

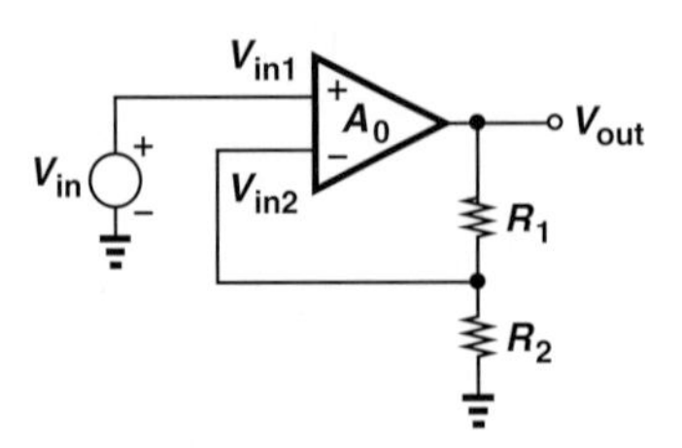

Figure 8.5 Noninverting amplifier.

[3]Variation with process means that circuits fabricated in different "batches" exhibit somewhat different characteristics.

Since a high op amp gain translates to a small difference between V_{in1} and V_{in2}, we have

$$V_{in1} \approx V_{in2} \tag{8.7}$$

$$\approx \frac{R_2}{R_1 + R_2} V_{out}; \tag{8.8}$$

and hence

$$\frac{V_{out}}{V_{in}} \approx 1 + \frac{R_1}{R_2}. \tag{8.9}$$

Due to the positive gain, the circuit is called a "noninverting amplifier." We call this result the "closed-loop" gain of the circuit.

Interestingly, the voltage gain depends on only the *ratio* of the resistors; if R_1 and R_2 increase by 20%, R_1/R_2 remains constant. The idea of creating dependence on only the ratio of quantities that have the same dimension plays a central role in circuit design.

Example 8-2 Study the noninverting amplifier for two extreme cases: $R_1/R_2 = \infty$ and $R_1/R_2 = 0$.

Solution If $R_1/R_2 \rightarrow \infty$, e.g., if R_2 approaches zero, we note that $V_{out}/V_{in} \rightarrow \infty$. Of course, as depicted in Fig. 8.6(a), this occurs because the circuit reduces to the op amp itself, with no fraction of the output fed back to the input. Resistor R_1 simply loads the output node, with no effect on the gain if the op amp is ideal.

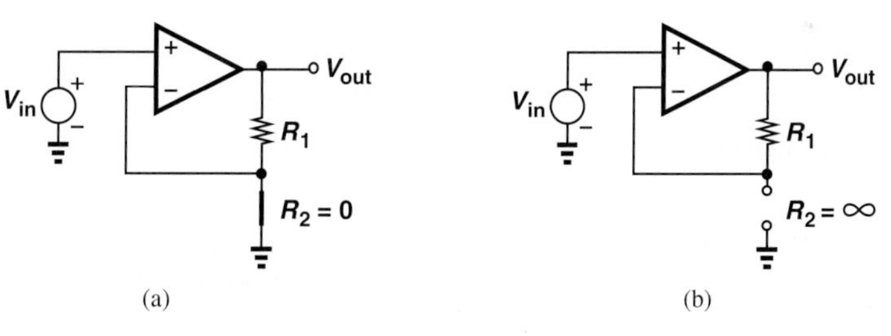

Figure 8.6 Noninverting amplifier with (a) zero and (b) infinite value for R_2.

If $R_1/R_2 \rightarrow 0$, e.g., if R_2 approaches infinity, we have $V_{out}/V_{in} \rightarrow 1$. Shown in Fig. 8.6(b), this case in fact reduces to the unity-gain buffer of Fig. 8.3 because the ideal op amp draws no current at its inputs, yielding a zero drop across R_1 and hence $V_{in2} = V_{out}$.

Exercise Suppose the circuit is designed for a nominal gain of 2.00 but the R_1 and R_2 suffer from a mismatch of 5% (i.e., $R_1 = (1 \pm 0.05)R_2$). What is the actual voltage gain?

Let us now take into account the finite gain of the op amp. Based on the model shown in Fig. 8.1(b), we write

$$(V_{in1} - V_{in2})A_0 = V_{out}, \tag{8.10}$$

and substitute for V_{in2} from Eq. (8.6):

$$\frac{V_{out}}{V_{in}} = \frac{A_0}{1 + \dfrac{R_2}{R_1 + R_2} A_0}. \tag{8.11}$$

As expected, this result reduces to Eq. (8.9) if $A_0 R_2/(R_1 + R_2) \gg 1$. To avoid confusion between the gain of the op amp, A_0, and the gain of the overall amplifier, V_{out}/V_{in}, we call the former the "open-loop" gain and the latter the "closed-loop" gain.

Equation (8.11) indicates that the finite gain of the op amp creates a small error in the value of V_{out}/V_{in}. If much greater than unity, the term $A_0 R_2/(R_1 + R_2)$ can be factored from the denominator to permit the approximation $(1 + \epsilon)^{-1} \approx 1 - \epsilon$ for $\epsilon \ll 1$:

$$\frac{V_{out}}{V_{in}} \approx \left(1 + \frac{R_1}{R_2}\right) \left[1 - \left(1 + \frac{R_1}{R_2}\right) \frac{1}{A_0}\right]. \tag{8.12}$$

Called the "gain error," the term $(1 + R_1/R_2)/A_0$ must be minimized according to each application's requirements.

Example 8-3 A noninverting amplifier incorporates an op amp having a gain of 1000. Determine the gain error if the circuit is to provide a nominal gain of (a) 5, or (b) 50.

Solution For a nominal gain of 5, we have $1 + R_1/R_2 = 5$, obtaining a gain error of:

$$\left(1 + \frac{R_1}{R_2}\right) \frac{1}{A_0} = 0.5\%. \tag{8.13}$$

On the other hand, if $1 + R_1/R_2 = 50$, then

$$\left(1 + \frac{R_1}{R_2}\right) \frac{1}{A_0} = 5\%. \tag{8.14}$$

In other words, a higher closed-loop gain inevitably suffers from less accuracy.

Exercise Repeat the above example if the op amp has a gain of 500.

With an ideal op amp, the noninverting amplifier exhibits an infinite input impedance and a zero output impedance. For a nonideal op amp, the I/O impedances are derived in Problem 8.6.

8.2.2 Inverting Amplifier

Depicted in Fig. 8.7(a), the "inverting amplifier" incorporates an op amp along with resistors R_1 and R_2 while the noninverting input is grounded. Recall from Section 8.1 that if the op amp gain is infinite, then a finite output swing translates to $V_{in1} - V_{in2} \rightarrow 0$; i.e., node X bears a zero potential even though it is *not* shorted to ground. For this reason, node X is called a "virtual ground." Under this condition, the entire input voltage appears across R_2, producing a current of V_{in}/R_2, which must then flow through R_1 if the op amp input draws no current [Fig. 8.7(b)]. Since the left terminal of R_1 remains at zero and the right terminal at V_{out},

$$\frac{0 - V_{out}}{R_1} = \frac{V_{in}}{R_2} \tag{8.15}$$

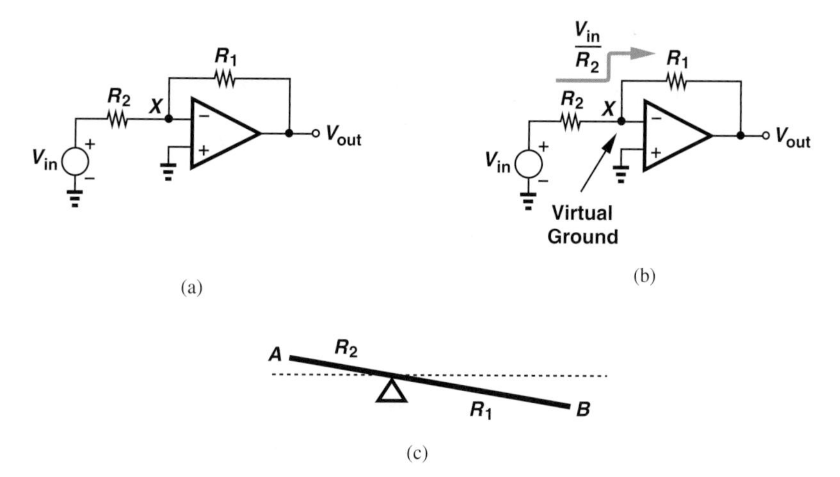

Figure 8.7 (a) Inverting amplifier, (b) currents flowing in resistors, (c) analogy with a seesaw.

yielding

$$\frac{V_{out}}{V_{in}} = \frac{-R_1}{R_2}. \tag{8.16}$$

Due to the negative gain, the circuit is called the "inverting amplifier." As with its noninverting counterpart, the gain of this circuit is given by the *ratio* of the two resistors, thereby experiencing only small variations with temperature and process.

It is important to understand the role of the virtual ground in this circuit. If the inverting input of the op amp were not near zero potential, then neither V_{in}/R_2 nor V_{out}/R_1 would accurately represent the currents flowing through R_2 and R_1, respectively. This behavior is similar to a seesaw [Fig. 8.7(c)], where the point between the two arms is "pinned" (e.g., does not move), allowing displacement of point A to be "amplified" (and "inverted") at point B.

The above development also reveals why the virtual ground cannot be *shorted* to the actual ground. Such a short in Fig. 8.7(b) would force to ground all of the current flowing through R_2, yielding $V_{out} = 0$. It is interesting to note that the inverting amplifier can also be drawn as shown in Fig. 8.8, displaying a similarity with the noninverting circuit but with the input applied at a different point.

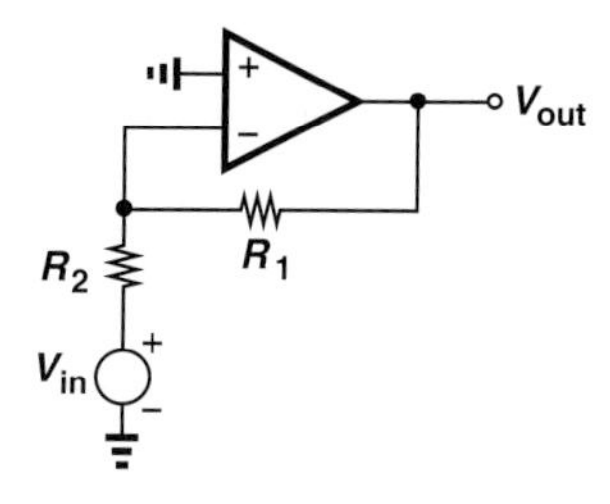

Figure 8.8 Inverting amplifier.

In contrast to the noninverting amplifier, the topology of Fig. 8.7(a) exhibits an input impedance equal to R_2—as can be seen from the input current, V_{in}/R_2, in Fig. 8.7(b). That is, a lower R_2 results in a greater gain but a smaller input impedance. This trade-off sometimes makes this amplifier less attractive than its noninverting counterpart.

Let us now compute the closed-loop gain of the inverting amplifier with a finite op amp gain. We note from Fig. 8.7(a) that the currents flowing through R_2 and R_1 are given by $(V_{in} - V_X)/R_2$ and $(V_X - V_{out})/R_1$, respectively. Moreover,

$$V_{out} = A_0(V_{in1} - V_{in2}) \tag{8.17}$$

$$= -A_0 V_X. \tag{8.18}$$

Equating the currents through R_2 and R_1 and substituting $-V_{out}/A_0$ for V_X, we obtain

$$\frac{V_{out}}{V_{in}} = -\frac{1}{\dfrac{1}{A_0} + \dfrac{R_2}{R_1}\left(\dfrac{1}{A_0} + 1\right)} \tag{8.19}$$

$$= -\frac{1}{\dfrac{R_2}{R_1} + \dfrac{1}{A_0}\left(1 + \dfrac{R_2}{R_1}\right)}. \tag{8.20}$$

Factoring R_2/R_1 from the denominator and assuming $(1 + R_1/R_2)/A_0 \ll 1$, we have

$$\frac{V_{out}}{V_{in}} \approx -\frac{R_1}{R_2}\left[1 - \frac{1}{A_0}\left(1 + \frac{R_1}{R_2}\right)\right]. \tag{8.21}$$

As expected, a higher closed-loop gain ($\approx -R_1/R_2$) is accompanied with a greater gain error. Note that the gain error expression is the same for noninverting and inverting amplifiers.

Example 8-4 Design the inverting amplifier of Fig. 8.7(a) for a nominal gain of 4, a gain error of 0.1%, and an input impedance of at least 10 kΩ.

Solution Since both the nominal gain and the gain error are given, we must first determine the minimum op amp gain. We have

$$\frac{R_1}{R_2} = 4 \tag{8.22}$$

$$\frac{1}{A_0}\left(1 + \frac{R_1}{R_2}\right) = 0.1\%. \tag{8.23}$$

Thus,

$$A_0 = 5000. \tag{8.24}$$

Since the input impedance is approximately equal to R_2, we choose:

$$R_2 = 10\ \text{k}\Omega \tag{8.25}$$

$$R_1 = 40\ \text{k}\Omega. \tag{8.26}$$

Exercise Repeat the above example for a gain error of 1% and compare the results.

In the above example, we assumed the input impedance is approximately equal to R_2. How accurate is this assumption? With $A_0 = 5000$, the virtual ground experiences a voltage of $-V_{out}/5000 \approx -4V_{in}/5000$, yielding an input current of $(V_{in} + 4V_{in}/5000)/R_1$. That is, our assumption leads to an error of about 0.08%—an acceptable value in most applications.

Bioengineering Application: Measuring Biosignals

Many functions in our body generate voltages that can be measured on the skin. Examples include brain and heart activities.

Suppose we use an inverting amplifier to sense a voltage on a patient's body [Fig. 8.9(a)]. In this case, the amplifier's low input impedance draws some current from the patient's skin. This is a serious issue because the skin, especially when it is dry, exhibits a high output impedance, leading to considerable attenuation of the signal [Fig. 8.9(b)]. Here, we have

$$V_{out} = -\frac{R_2}{R_1 + R_{skin}} V_{out}. \tag{8.27}$$

For this reason, we prefer a non-inverting amplifier as it presents a high input impedance [Fig. 8.9(c)].

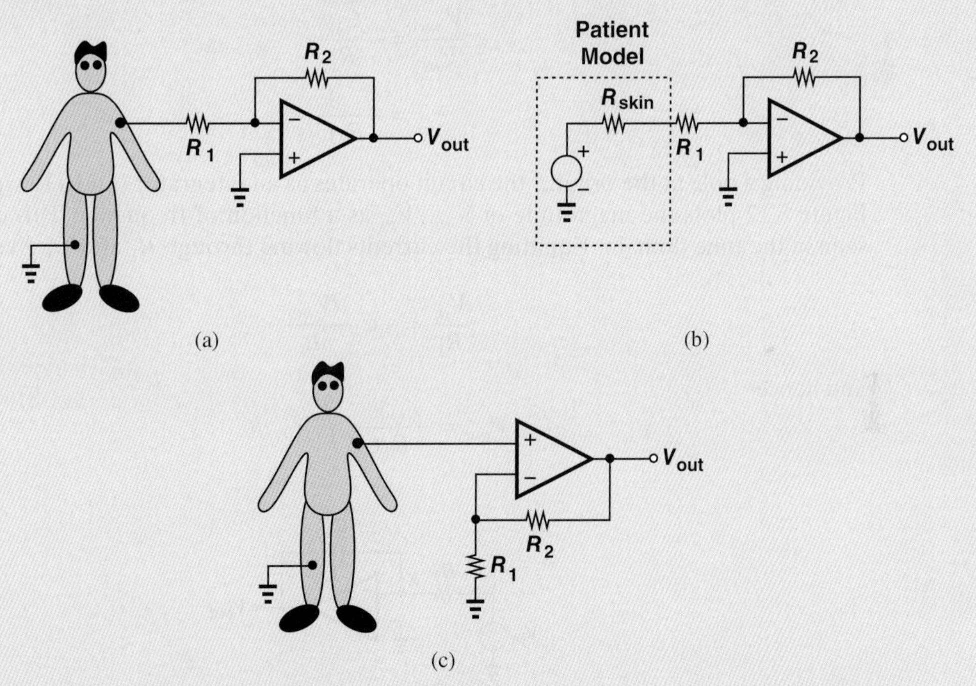

Figure 8.9 (a) Measuring a voltage on the body by an inverting amplifier, (b) skin model illustrating heavy voltage attenuation, and (c) use of a noninverting amplifier to avoid signal attenuation.

8.2.3 Integrator and Differentiator

Our study of the inverting topology in previous sections has assumed a resistive network around the op amp. In general, it is possible to employ complex impedances instead (Fig. 8.10). In analogy with Eq. (8.16), we can write

$$\frac{V_{out}}{V_{in}} \approx -\frac{Z_1}{Z_2}, \tag{8.28}$$

where the gain of the op amp is assumed large. If Z_1 or Z_2 is a capacitor, two interesting functions result.

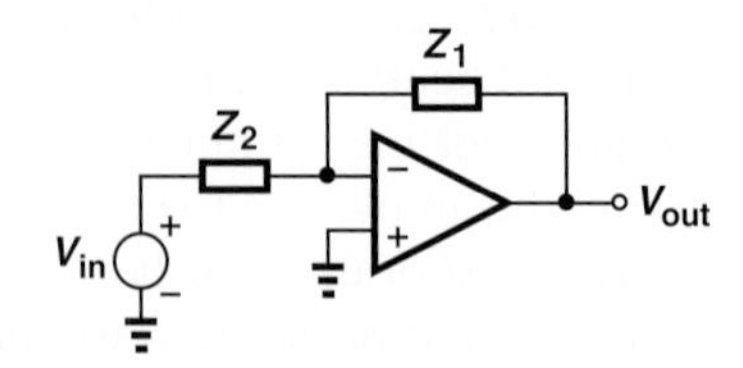

Figure 8.10 Circuit with general impedances around the op amp.

Integrator Suppose in Fig. 8.10, Z_1 is a capacitor and Z_2 a resistor (Fig. 8.11). That is, $Z_1 = (C_1 s)^{-1}$ and $Z_2 = R_1$. With an ideal op amp, we have

$$\frac{V_{out}}{V_{in}} = -\frac{\frac{1}{C_1 s}}{R_1} \tag{8.29}$$

$$= -\frac{1}{R_1 C_1 s}. \tag{8.30}$$

Providing a pole at the origin,[4] the circuit operates as an integrator (and a low-pass filter). Figure 8.12 plots the magnitude of V_{out}/V_{in} as a function of frequency. This can also be seen in the time domain. Equating the currents flowing through R_1 and C_1 gives

$$\frac{V_{in}}{R_1} = -C_1 \frac{dV_{out}}{dt} \tag{8.31}$$

and hence

$$V_{out} = -\frac{1}{R_1 C_1} \int V_{in}\, dt. \tag{8.32}$$

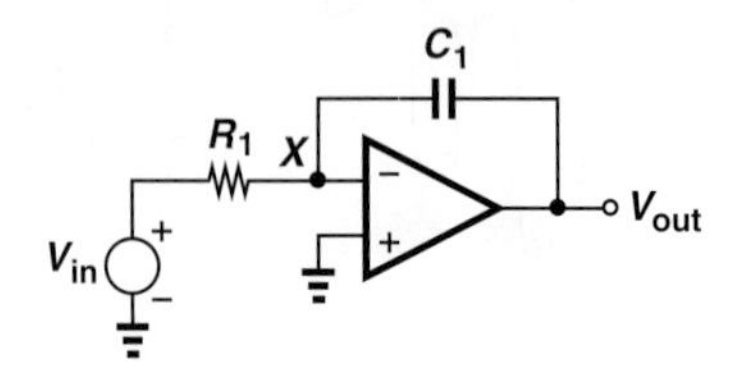

Figure 8.11 Integrator.

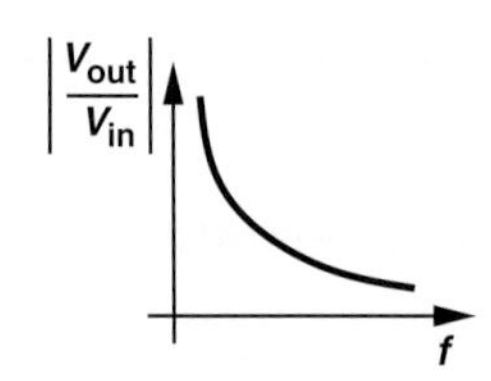

Figure 8.12 Frequency response of integrator.

[4]Pole frequencies are obtained by setting the denominator of the transfer function to zero.

Equation (8.30) indicates that V_{out}/V_{in} approaches infinity as the input frequency goes to zero. This is to be expected: the capacitor impedance becomes very large at low frequencies, approaching an open circuit and reducing the circuit to the open-loop op amp.

As mentioned at the beginning of the chapter, integrators originally appeared in analog computers to simulate differential equations. Today, electronic integrators find usage in analog filters, control systems, and many other applications.

Example 8-5 Plot the output waveform of the circuit shown in Fig. 8.13(a). Assume a zero initial condition across C_1 and an ideal op amp.

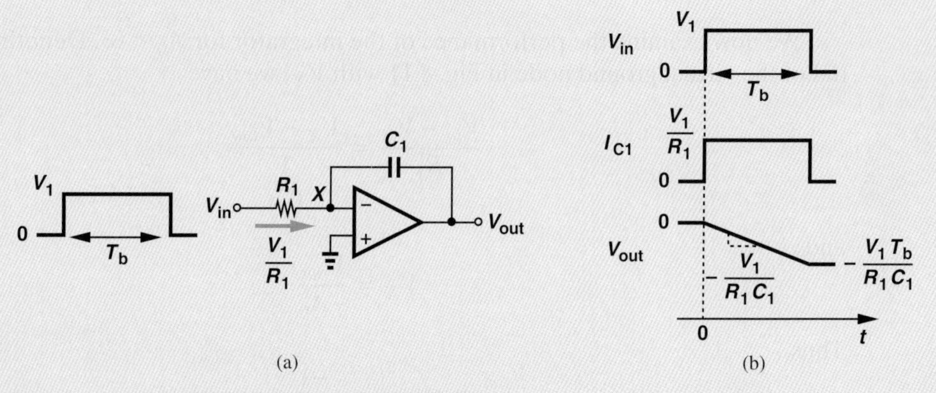

Figure 8.13 (a) Integrator with pulse input, (b) input and output waveforms.

Solution When the input jumps from 0 to V_1, a constant current equal to V_1/R_1 begins to flow through the resistor and hence the capacitor, forcing the right plate voltage of C_1 to fall linearly with time while its left plate is pinned at zero [Fig. 8.13(b)]:

$$V_{out} = -\frac{1}{R_1 C_1} \int V_{in}\, dt \tag{8.33}$$

$$= -\frac{V_1}{R_1 C_1} t \qquad 0 < t < T_b. \tag{8.34}$$

(Note that the output waveform becomes "sharper" as R_1C_1 decreases.) When V_{in} returns to zero, so do the currents through R_1 and C_1. Thus, the voltage across the capacitor and hence V_{out} remain equal to $-V_1T_b/(R_1C_1)$ (proportional to the area under the input pulse) thereafter.

Exercise Repeat the previous example if V_1 is negative.

The previous example demonstrates the role of the virtual ground in the integrator. The ideal integration expressed by Eq. (8.33) occurs because the left plate of C_1 is pinned at zero. To gain more insight, let us compare the integrator with a first-order RC filter in terms of their step response. As illustrated in Fig. 8.14, the integrator forces a *constant* current (equal to V_1/R_1) through the capacitor. On the other hand, the RC filter creates a current equal to $(V_{in} - V_{out})/R_1$, which *decreases* as V_{out} rises, leading to an increasingly slower voltage variation across C_1. We may therefore consider the RC filter as a "passive" approximation of the integrator. In fact, for a large R_1C_1 product, the exponential response of Fig. 8.14(b) becomes slow enough to be approximated as a ramp.

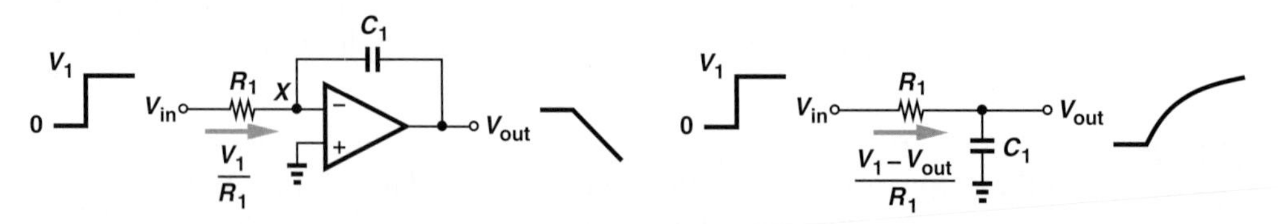

Figure 8.14 Comparison of integrator with and RC circuit.

We now examine the performance of the integrator for $A_0 < \infty$. Denoting the potential of the virtual ground node in Fig. 8.11 with V_X, we have

$$\frac{V_{in} - V_X}{R_1} = \frac{V_X - V_{out}}{\frac{1}{C_1 s}} \tag{8.35}$$

and

$$V_X = \frac{V_{out}}{-A_0}. \tag{8.36}$$

Thus,

$$\frac{V_{out}}{V_{in}} = \frac{-1}{\frac{1}{A_0} + \left(1 + \frac{1}{A_0}\right) R_1 C_1 s}, \tag{8.37}$$

revealing that the gain at $s = 0$ is limited to A_0 (rather than infinity) and the pole frequency has moved from zero to

$$s_p = \frac{-1}{(A_0 + 1)R_1C_1}. \tag{8.38}$$

Such a circuit is sometimes called a "lossy" integrator to emphasize the nonideal gain and pole position.

Example 8-6 Recall from basic circuit theory that the RC filter shown in Fig. 8.15 contains a pole at $-1/(R_X C_X)$. Determine R_X and C_X such that this circuit exhibits the same pole as that of the above integrator.

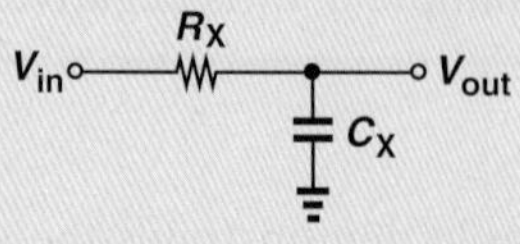

Figure 8.15 Simple low-pass filter.

Solution From Eq. (8.38),

$$R_X C_X = (A_0 + 1)R_1C_1. \tag{8.39}$$

The choice of R_X and C_X is arbitrary so long as their product satisfies Eq. (8.39). An interesting choice is

$$R_X = R_1 \tag{8.40}$$

$$C_X = (A_0 + 1)C_1. \tag{8.41}$$

It is as if the op amp "boosts" the value of C_1 by a factor of $A_0 + 1$.

Exercise What value of R_X is necessary if $C_X = C_1$?

Did you know?

In addition to the circuits described in this chapter, many other functions employ op amps. Examples include filters, voltage references, analog multipliers, oscillators, and regulated power supplies. An interesting application is in accelerometers used in automobile airbags, etc. An accelerometer is formed as two microscopic pillars, one of which bends in proportion to the acceleration [Fig. (a)]. As a result, the capacitance between the pillars varies by a small amount. In order to measure this change, an oscillator whose frequency is defined by this capacitance can be constructed. Shown in Fig. (b) is an oscillator consisting of two integrators and an inverting amplifier. The change in the oscillation frequency, f_{out}, can be measured precisely to determine the acceleration.

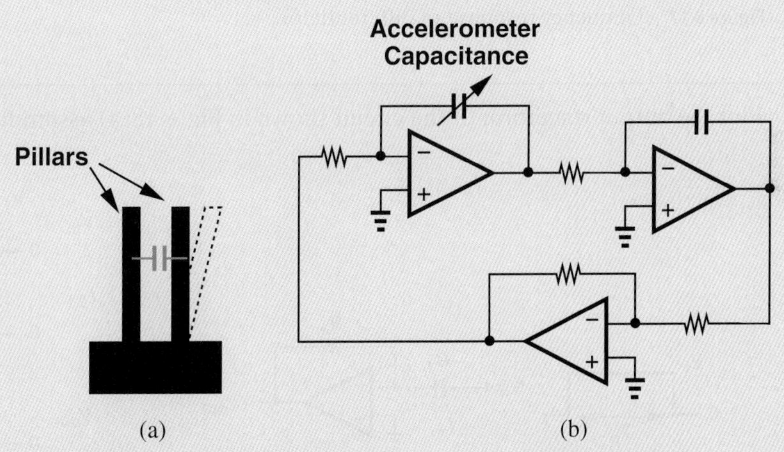

(a) Conceptual diagram of an accelerometer, (b) oscillator using an accelerometer to produce a frequency proportional to acceleration.

Differentiator If in the general topology of Fig. 8.10, Z_1 is a resistor and Z_2 a capacitor (Fig. 8.16), we have

$$\frac{V_{out}}{V_{in}} = -\frac{R_1}{\dfrac{1}{C_1 s}} \tag{8.42}$$

$$= -R_1 C_1 s. \tag{8.43}$$

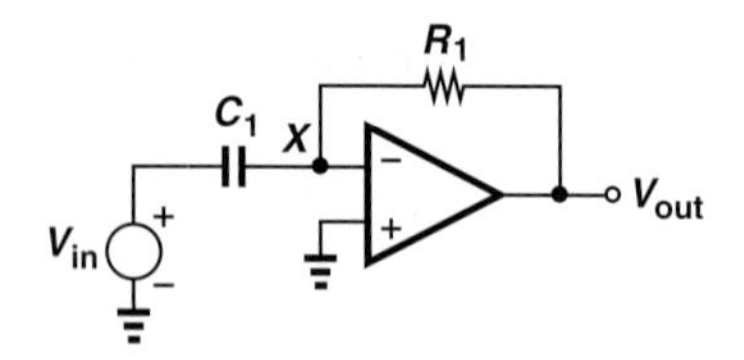

Figure 8.16 Differentiator.

Exhibiting a zero at the origin, the circuit acts as a differentiator (and a high-pass filter). Figure 8.17 plots the magnitude of V_{out}/V_{in} as a function of frequency. From a time-domain perspective, we can equate the currents flowing through C_1 and R_1:

$$C_1 \frac{dV_{in}}{dt} = -\frac{V_{out}}{R_1}, \tag{8.44}$$

arriving at

$$V_{out} = -R_1 C_1 \frac{dV_{in}}{dt}. \tag{8.45}$$

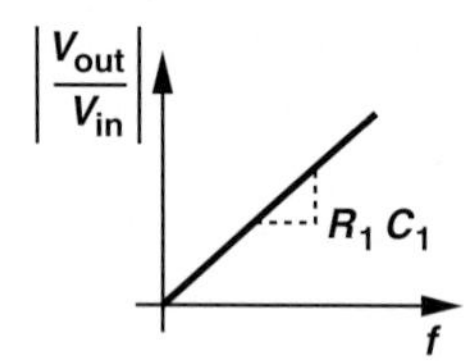

Figure 8.17 Frequency response of differentiator.

Example 8-7 Plot the output waveform of the circuit shown in Fig. 8.18(a) assuming an ideal op amp.

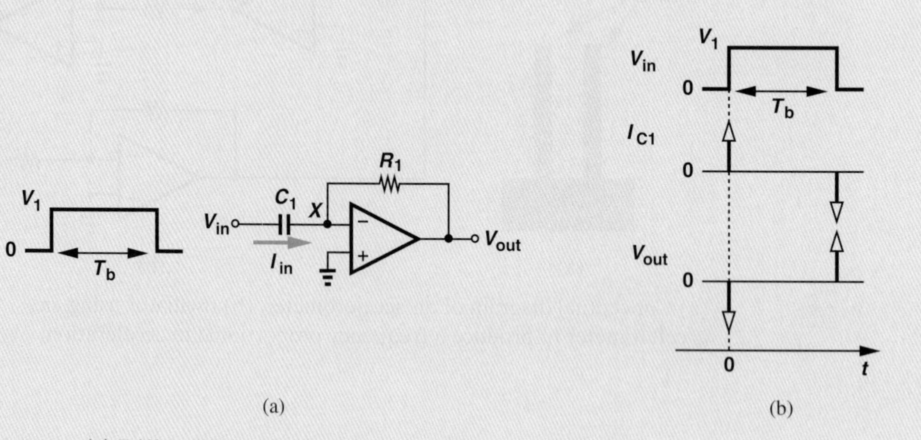

Figure 8.18 (a) Differentiator with pulse input, (b) input and output waveforms.

Solution At $t = 0^-$, $V_{in} = 0$ and $V_{out} = 0$ (why?). When V_{in} jumps to V_1, an *impulse* of current flows through C_1 because the op amp maintains V_X constant:

$$I_{in} = C_1 \frac{dV_{in}}{dt} \tag{8.46}$$

$$= C_1 V_1 \delta(t). \tag{8.47}$$

The current flows through R_1, generating an output given by

$$V_{out} = -I_{in}R_1 \tag{8.48}$$

$$= -R_1 C_1 V_1 \delta(t). \tag{8.49}$$

Figure 8.18(b) depicts the result. At $t = T_b$, V_{in} returns to zero, again creating an impulse of current in C_1:

$$I_{in} = C_1 \frac{dV_{in}}{dt} \tag{8.50}$$

$$= C_1 V_1 \delta(t). \tag{8.51}$$

It follows that

$$V_{out} = -I_{in}R_1 \tag{8.52}$$

$$= R_1 C_1 V_1 \delta(t). \tag{8.53}$$

We can therefore say that the circuit generates an impulse of current $[\pm C_1 V_1 \delta(t)]$ and "amplifies" it by R_1 to produce V_{out}. In reality, of course, the output exhibits neither an infinite height (limited by the supply voltage) nor a zero width (limited by the op amp nonidealities).

Exercise Plot the output if V_1 is negative.

It is instructive to compare the operation of the differentiator with that of its "passive" counterpart (Fig. 8.19). In the ideal differentiator, the virtual ground node permits the input to change the voltage across C_1 instantaneously. In the RC filter, on the other hand, node X is not "pinned," thereby following the input change at $t = 0$ and limiting the initial current in the circuit to V_1/R_1. If the decay time constant, R_1C_1, is sufficiently small, the passive circuit can be viewed as an approximation of the ideal differentiator.

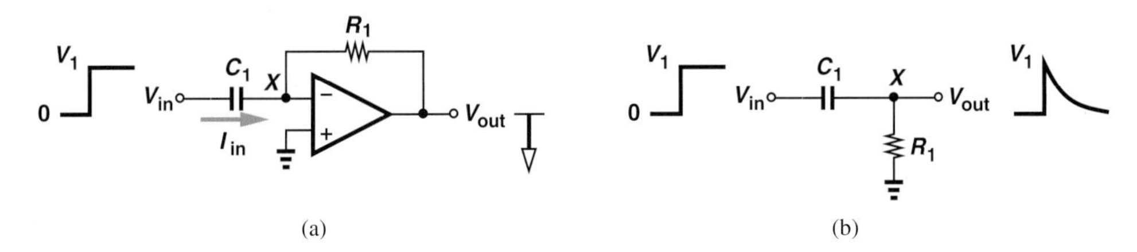

Figure 8.19 Comparison of differentiator and RC circuit.

Let us now study the differentiator with a finite op amp gain. Equating the capacitor and resistor currents in Fig. 8.16 gives

$$\frac{V_{in} - V_X}{\frac{1}{C_1 s}} = \frac{V_X - V_{out}}{R_1}. \tag{8.54}$$

Substituting $-V_{out}/A_0$ for V_X, we have

$$\frac{V_{out}}{V_{in}} = \frac{-R_1 C_1 s}{1 + \frac{1}{A_0} + \frac{R_1 C_1 s}{A_0}}. \tag{8.55}$$

In contrast to the ideal differentiator, the circuit contains a pole at

$$s_p = -\frac{A_0 + 1}{R_1 C_1}. \tag{8.56}$$

Example 8-8 Determine the transfer function of the high-pass filter shown in Fig. 8.20 and choose R_X and C_X such that the pole of this circuit coincides with Eq. (8.56).

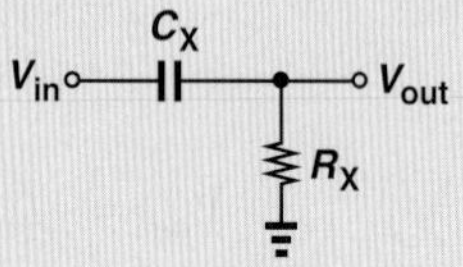

Figure 8.20 Simple high-pass filter.

Solution The capacitor and resistor operate as a voltage divider:

$$\frac{V_{out}}{V_{in}} = \frac{R_X}{R_X + \dfrac{1}{C_X s}} \tag{8.57}$$

$$= \frac{R_X C_X s}{R_X C_X s + 1}. \tag{8.58}$$

The circuit therefore exhibits a zero at the origin ($s = 0$) and a pole at $-1/(R_X C_X)$. For this pole to be equal to Eq. (8.56), we require

$$\frac{1}{R_X C_X} = \frac{A_0 + 1}{R_1 C_1}. \tag{8.59}$$

One choice of R_X and C_X is

$$R_X = \frac{R_1}{A_0 + 1} \tag{8.60}$$

$$C_X = C_1, \tag{8.61}$$

Exercise What is the necessary value of C_X if $R_X = R_1$?

An important drawback of differentiators stems from the amplification of high-frequency *noise*. As suggested by Eq. (8.43) and Fig. 8.17, the increasingly larger gain of the circuit at high frequencies tends to boost noise in the circuit.

The general topology of Fig. 8.10 and its integrator and differentiator descendants operate as *inverting* circuits. The reader may wonder if it is possible to employ a configuration similar to the noninverting amplifier of Fig. 8.5 to avoid the sign reversal. As shown in Fig. 8.21, such a circuit provides the following transfer function:

$$\frac{V_{out}}{V_{in}} = 1 + \frac{Z_1}{Z_2}, \tag{8.62}$$

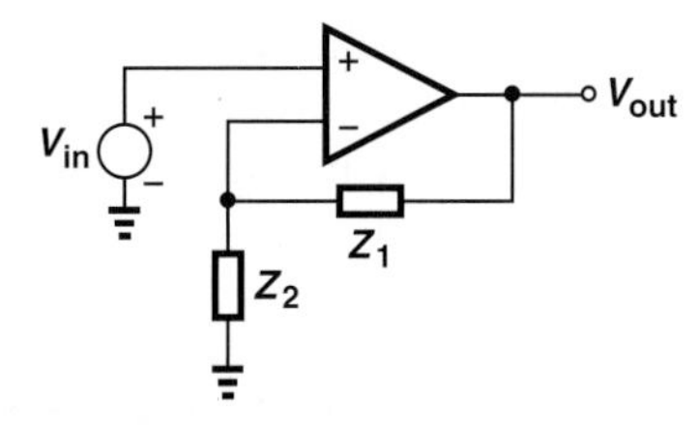

Figure 8.21 Op amp with general network.

if the op amp is ideal. Unfortunately, this function does not translate to ideal integration or differentiation. For example, $Z_1 = R_1$ and $Z_2 = 1/(C_2 s)$ yield a nonideal differentiator (why?).

8.2.4 Voltage Adder

The need for adding voltages arises in many applications. In audio recording, for example, a number of microphones may convert the sounds of various musical instruments to voltages, and these voltages must then be added to create the overall musical piece. This operation is called "mixing" in the audio industry.[5] For example, in "noise cancelling" headphones, the environmental noise is applied to an inverting amplifier and subsequently added to the signal so as to cancel itself.

Figure 8.22 depicts a voltage adder ("summer") incorporating an op amp. With an ideal op amp, $V_X = 0$, and R_1 and R_2 carry currents proportional to V_1 and V_2, respectively. The two currents *add* at the virtual ground node and flow through R_F:

$$\frac{V_1}{R_1} + \frac{V_2}{R_2} = \frac{-V_{out}}{R_F}. \tag{8.63}$$

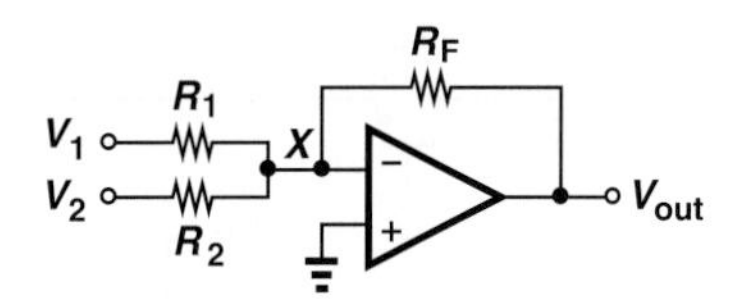

Figure 8.22 Voltage adder.

That is,

$$V_{out} = -R_F \left(\frac{V_1}{R_1} + \frac{V_2}{R_2} \right). \tag{8.64}$$

For example, if $R_1 = R_2 = R$, then

$$V_{out} = \frac{-R_F}{R}(V_1 + V_2). \tag{8.65}$$

This circuit can therefore add and amplify voltages. Extension to more than two voltages is straightforward.

Equation (8.64) indicates that V_1 and V_2 can be added with different *weightings*: R_F/R_1 and R_F/R_2, respectively. This property also proves useful in many applications.

[5]The term "mixing" bears a completely different meaning in the RF and wireless industry.

For example, in audio recording it may be necessary to lower the "volume" of one musical instrument for part of the piece, a task possible by varying R_1 and R_2.

The behavior of the circuit in the presence of a finite op amp gain is studied in Problem 8.31.

Robotics Application: Voltage Regulators

Mobile robots run from a battery but they employ different devices that require different supply voltages. For example, one must generate a 3-V supply for a microcontroller from a 6-V battery. This is accomplished by "voltage regulators."

Figure 8.23 shows a typical regulator configuration. It consists of a bipolar (or MOS) transistor that delivers the voltage or current necessary for the load. The output voltage is divided by R_1 and R_2 and the result is applied to the inverting input of op amp A_0. The other input senses an accurate reference, V_{REF}. The op amp controls the base voltage of Q_1. Recall that in a well-designed circuit, the difference between the input voltages of an op amp is small. Thus, $V_X \approx V_{REF}$. Since $V_X = [R_2/(R_1 + R_2)]V_{out}$, we have

$$V_{out} = \left(1 + \frac{R_1}{R_2}\right) V_{REF}. \tag{8.66}$$

To see how the circuit "regulates" the output voltage, suppose the load decides to draw a larger current. As a result, V_{out} tends to drop and so does V_X. The output voltage of the op amp thus rises (why?). How does Q_1 react to this change? We recognize that Q_1 is configured as an emitter follower because the op amp output signal is applied to the base of Q_1 and the output, V_{out}, is sensed at its emitter. Consequently, as the base voltage increases, so does V_{out}, returning to its original value even though the load demands a greater current.

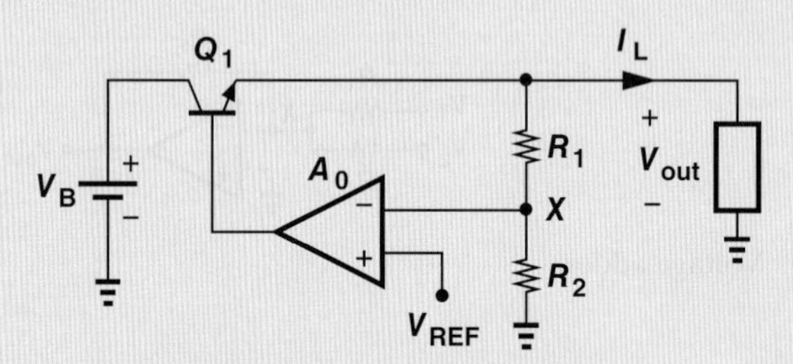

Figure 8.23 Voltage regulator topology.

Why do we divide V_{out} before subtracting it from V_{REF}? It is possible that the accurate reference that is available is not equal to the desired output voltage (but it is typically *less*). We thus resort to Eq. (8.66) to generate V_{out} from V_{REF}.

8.3 NONLINEAR FUNCTIONS

It is possible to implement useful nonlinear functions through the use of op amps and nonlinear devices such as transistors. The virtual ground property plays an essential role here as well.

8.3.1 Precision Rectifier

The rectifier circuits described in Chapter 3 suffer from a "dead zone" due to the finite voltage required to turn on the diodes. That is, if the input signal amplitude is less than approximately 0.7 V, the diodes remain off and the output voltage remains at zero. This drawback prohibits the use of the circuit in high-precision applications, e.g., if a small signal received by a cellphone must be rectified to determine its amplitude.

It is possible to place a diode around an op amp to form a "precision rectifier," i.e., a circuit that rectifies even very small signals. Let us begin with a unity-gain buffer tied to a resistive load [Fig. 8.24(a)]. We note that the high gain of the op amp ensures that node X tracks V_{in} (for both positive and negative cycles). Now suppose we wish to hold X at zero during negative cycles, i.e., "break" the connection between the output of the op amp and its inverting input. This can be accomplished as depicted in Fig. 8.24(b), where D_1 is inserted in the feedback loop. Note that V_{out} is sensed at X rather than at the output of the op amp.

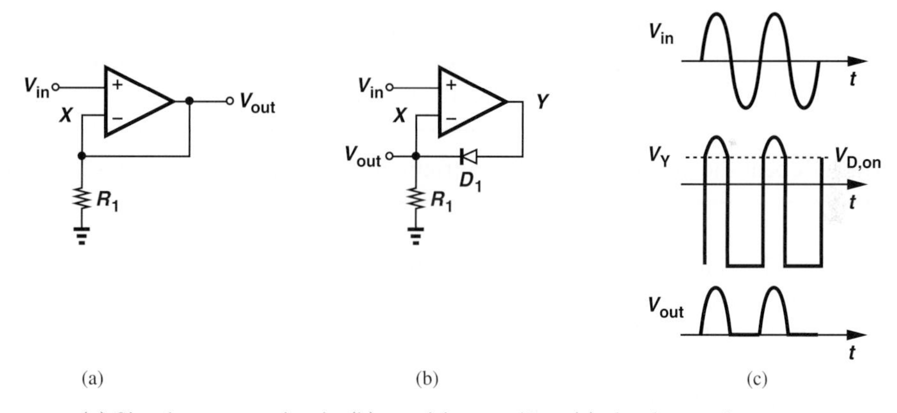

Figure 8.24 (a) Simple op amp circuit, (b) precision rectifier, (c) circuit waveforms.

To analyze the operation of this circuit, let us first assume that $V_{in} = 0$. In its attempt to minimize the voltage difference between the noninverting and the inverting inputs, the op amp raises V_Y to approximately $V_{D1,on}$, turning D_1 barely on but with little current so that $V_X \approx 0$. Now if V_{in} becomes slightly positive, V_Y rises further so that the current flowing through D_1 and R_1 yields $V_{out} \approx V_{in}$. That is, even small positive levels at the input appear at the output.

What happens if V_{in} becomes slightly negative? For V_{out} to assume a negative value, D_1 must carry a current from X to Y, which is not possible. Thus, D_1 turns off and the op amp produces a very large negative output (near the negative supply rail) because its non-inverting input falls below its inverting input. Figure 8.24(c) plots the circuit's waveforms in response to an input sinusoid.

Example 8-9 Plot the waveforms in the circuit of Fig. 8.25(a) in response to an input sinusoid.

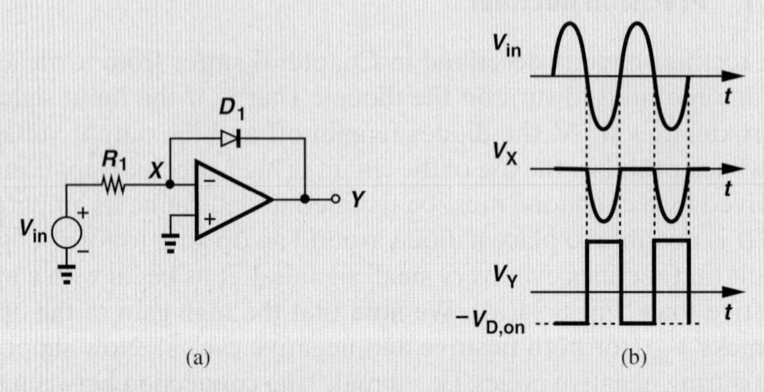

Figure 8.25 (a) Inverting precision rectifier, (b) circuit waveforms.

Solution For $V_{in} = 0$, the op amp creates $V_Y \approx -V_{D,on}$ so that D_1 is barely on, R_1 carries little current, and X is a virtual ground. As V_{in} becomes positive, thus raising the current through R_1, V_Y only slightly decreases to allow D_1 to carry the higher current. That is, $V_X \approx 0$ and $V_Y \approx -V_{D,on}$ for positive input cycles.

For $V_{in} < 0$, D_1 turns off (why?), leading to $V_X = V_{in}$ and driving V_Y to a very positive value. Figure 8.25(b) shows the resulting waveforms.

Exercise Repeat the above example for a triangular input that goes from −2 V to +2 V.

The large swings at the output of the op amp in Figs. 8.24(b) and 8.25(a) lower the speed of the circuit as the op amp must "recover" from a saturated value before it can turn D_1 on again. Additional techniques can resolve this issue (Problem 8.39).

8.3.2 Logarithmic Amplifier

Consider the circuit of Fig. 8.26, where a bipolar transistor is placed around the op amp. With an ideal op amp, R_1 carries a current equal to V_{in}/R_1 and so does Q_1. Thus,

$$V_{BE} = V_T \ln \frac{V_{in}/R_1}{I_S}. \tag{8.67}$$

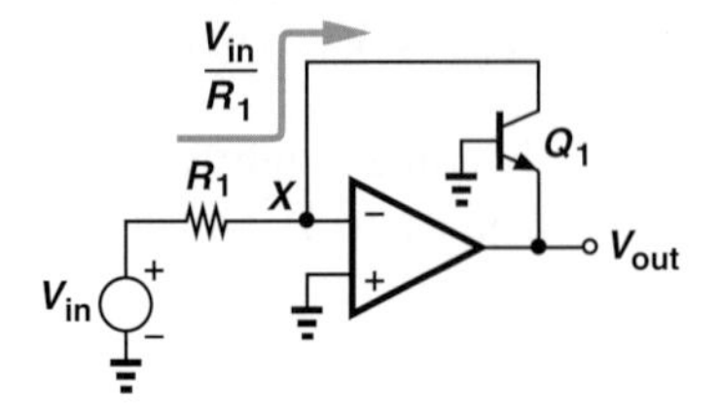

Figure 8.26 Logarithmic amplifier.

Also, $V_{out} = -V_{BE}$ and hence

$$V_{out} = -V_T \ln \frac{V_{in}}{R_1 I_S}. \tag{8.68}$$

The output is therefore proportional to the natural logarithm of V_{in}. As with previous linear and nonlinear circuits, the virtual ground plays an essential role here as it guarantees the current flowing through Q_1 is exactly proportional to V_{in}.

Logarithmic amplifiers ("logamps") prove useful in applications where the input signal level may vary by a large factor. It may be desirable in such cases to amplify weak signals and attenuate ("compress") strong signals hence a logarithmic dependence.

The negative sign in Eq. (8.68) is to be expected: if V_{in} rises, so do the currents flowing through R_1 and Q_1, requiring an *increase* in V_{BE}. Since the base is at zero, the emitter voltage must fall *below* zero to provide a greater collector current. Note that Q_1 operates in the active region because both the base and the collector remain at zero. The effect of finite op amp gain is studied in Problem 8.41.

The reader may wonder what happens if V_{in} becomes negative. Equation (8.68) predicts that V_{out} is not defined. In the actual circuit, Q_1 cannot carry a "negative" current, the loop around the op amp is broken, and V_{out} approaches the positive supply rail. It is therefore necessary to ensure V_{in} remains positive.

8.3.3 Square-Root Amplifier

Recognizing that the logarithmic amplifier of Fig. 8.26 in fact implements the *inverse* function of the exponential characteristic, we surmise that replacing the bipolar transistor with a MOSFET leads to a "square-root" amplifier. Illustrated in Fig. 8.27, such a circuit requires that M_1 carry a current equal to V_{in}/R_1:

$$\frac{V_{in}}{R_1} = \frac{1}{2}\mu_n C_{ox} \frac{W}{L}(V_{GS} - V_{TH})^2. \tag{8.69}$$

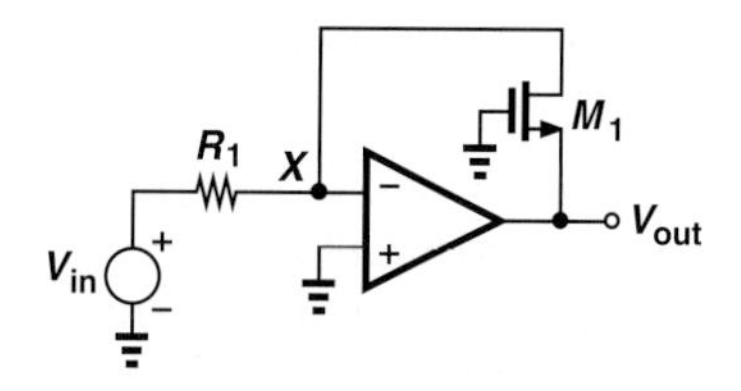

Figure 8.27 Square-root circuit.

(Channel-length modulation is neglected here.) Since $V_{GS} = -V_{out}$,

$$V_{out} = -\sqrt{\frac{2V_{in}}{\mu_n C_{ox} \frac{W}{L} R_1}} - V_{TH}. \tag{8.70}$$

If V_{in} is near zero, then V_{out} remains at $-V_{TH}$, placing M_1 at the edge of conduction. As V_{in} becomes more positive, V_{out} falls to allow M_1 to carry a greater current. With its gate and drain at zero, M_1 operates in saturation.

8.4 OP AMP NONIDEALITIES

Our study in previous sections has dealt with a relatively idealized op amp model—except for the finite gain—so as to establish insight. In practice, however, op amps suffer from other imperfections that may affect the performance significantly. In this section, we deal with such nonidealities.

8.4.1 DC Offsets

The op amp characteristics shown in Fig. 8.2 imply that $V_{out} = 0$ if $V_{in1} = V_{in2}$. In reality, a zero input difference may not give a zero output difference! Illustrated in Fig. 8.28(a), the characteristic is "offset" to the right or to the left; i.e., for $V_{out} = 0$, the input difference must be raised to a certain value, V_{os}, called the input "offset voltage."

What causes offset? The internal circuit of the op amp experiences random asymmetries ("mismatches") during fabrication and packaging. For example, as shown conceptually in Fig. 8.28(b), the bipolar transistors sensing the two inputs may display slightly different base-emitter voltages. The same effect occurs for MOSFETs. We model the offset by a single voltage source placed in series with one of the inputs [Fig. 8.28(c)]. Since offsets are random and hence can be positive or negative, V_{os} can appear at either input with arbitrary polarity.

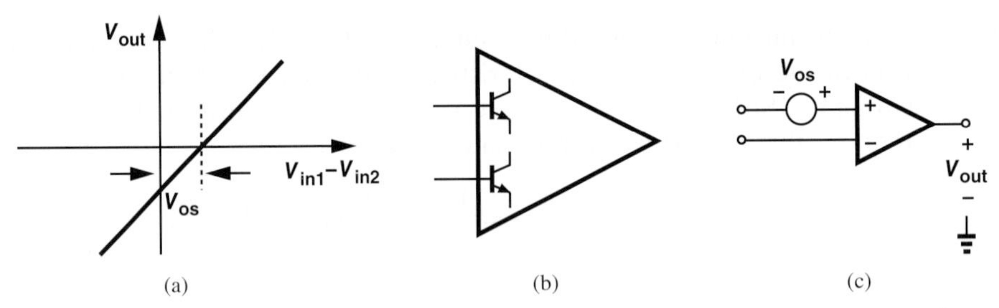

Figure 8.28 (a) Offset in an op amp, (b) mismatch between input devices, (c) representation of offset.

Why are DC offsets important? Let us reexamine some of the circuit topologies studied in Section 8.2 in the presence of op amp offsets. Depicted in Fig. 8.29, the noninverting amplifier now sees a total input of $V_{in} + V_{os}$, thereby generating

$$V_{out} = \left(1 + \frac{R_1}{R_2}\right)(V_{in} + V_{os}). \tag{8.71}$$

In other words, the circuit *amplifies* the offset as well as the signal, thus incurring accuracy limitations.[6]

Example 8-10

A truck weighing station employs an electronic pressure meter whose output is amplified by the circuit of Fig. 8.29. If the pressure meter generates 20 mV for every 100 kg of load and if the op amp offset is 2 mV, what is the accuracy of the weighing station?

[6]The reader can show that placing V_{os} in series with the inverting input of the op amp yields the same result.

Solution An offset of 2 mV corresponds to a load of 10 kg. We therefore say the station has an error of ±10 kg in its measurements.

Exercise What offset voltage is required for an accuracy of ±1 kg?

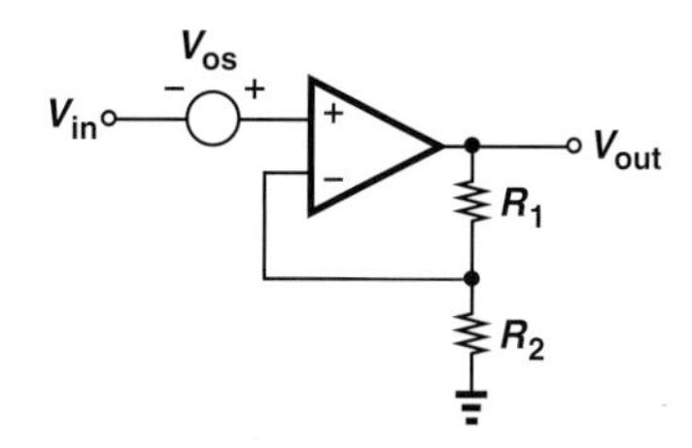

Figure 8.29 Offset in noninverting amplifier.

DC offsets may also cause “saturation” in amplifiers. The following example illustrates this point.

Example 8-11 An electrical engineering student constructs the circuit shown in Fig. 8.30 to amplify the signal produced by a microphone. The targeted gain is 10^4 so that very low level sounds (i.e., microvolt signals) can be detected. Explain what happens if op amp A_1 exhibits an offset of 2 mV.

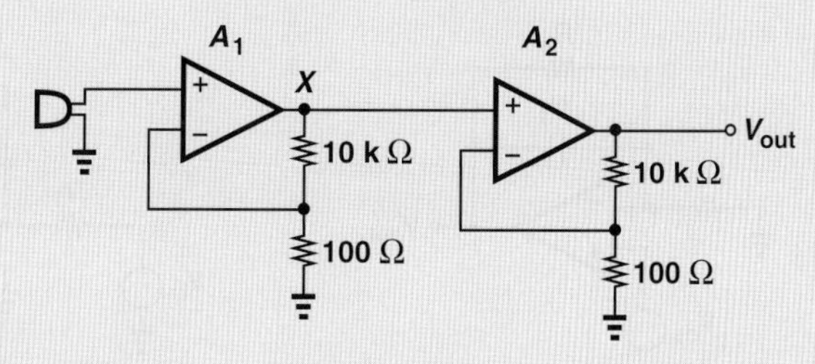

Figure 8.30 Two-stage amplifier.

Solution From Fig. 8.29, we recognize that the first stage amplifies the offset by a factor of 100, generating a dc level of 200 mV at node X (if the microphone produces a zero dc output). The second stage now amplifies V_X by another factor of 100, thereby attempting to generate $V_{out} = 20$ V. If A_2 operates with a supply voltage of, say, 3 V, the output cannot exceed this value, the second op amp drives its transistors into saturation (for bipolar devices) or triode region (for MOSFETs), and its gain falls to a small value. We say the second stage is saturated. (The problem of offset amplification in cascaded stages can be resolved through ac coupling.)

Exercise Repeat the above example if the second stage has a voltage gain of 10.

DC offsets impact the inverting amplifier of Fig. 8.7(a) in a similar manner. This is studied in Problem 8.48.

We now examine the effect of offset on the integrator of Fig. 8.11. Suppose the input is set to zero and V_{os} is referred to the noninverting input [Fig. 8.31(a)]. What happens at the output? Recall from Fig. 8.21 and Eq. (8.62) that the response to this "input" consists of the input itself [the unity term in Eq. (8.62)] and the integral of the input [the second term in Eq. (8.62)]. We can therefore express V_{out} in the time domain as

$$V_{out} = V_{os} + \frac{1}{R_1 C_1} \int_0^t V_{os}\, dt \tag{8.72}$$

$$= V_{os} + \frac{V_{os}}{R_1 C_1} t, \tag{8.73}$$

where the initial condition across C_1 is assumed zero. In other words, the circuit integrates the op amp offset, generating an output that tends to $+\infty$ or $-\infty$ depending on the sign of V_{os}. Of course, as V_{out} approaches the positive or negative supply voltages, the transistors in the op amp fail to provide gain and the output saturates [Fig. 8.31(b)].

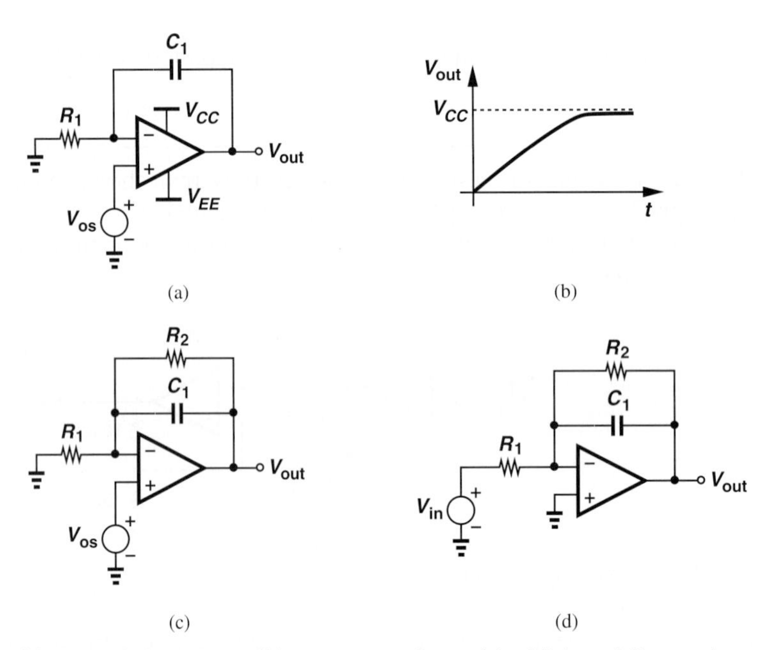

Figure 8.31 (a) Offset in integrator, (b) output waveform, (c) addition of R_2 to reduce effect of offset, (d) determination of transfer function.

The problem of offsets proves quite serious in integrators. Even in the presence of an input signal, the circuit of Fig. 8.31(a) integrates the offset and reaches saturation. Figure 8.31(c) depicts a modification where resistor R_2 is placed in parallel with C_1. Now the effect of V_{os} at the output is given by (8.9) because the circuits of Figs. 8.5 and 8.31(c) are similar at low frequencies:

$$V_{out} = V_{os}\left(1 + \frac{R_2}{R_1}\right). \tag{8.74}$$

For example, if $V_{os} = 2$ mV and $R_2/R_1 = 100$, then V_{out} contains a dc error of 202 mV, but at least remains away from saturation.

How does R_2 affect the integration function? Disregarding V_{os}, viewing the circuit as shown in Fig. 8.31(d), and using (8.28), we have

$$\frac{V_{out}}{V_{in}} = -\frac{R_2}{R_1}\frac{1}{R_2C_1s+1}. \tag{8.75}$$

Thus, the circuit now contains a pole at $-1/(R_2C_1)$ rather than at the origin. If the input signal frequencies of interest lie well above this value, then $R_2C_1s \gg 1$ and

$$\frac{V_{out}}{V_{in}} = -\frac{1}{R_1C_1s}. \tag{8.76}$$

That is, the integration function holds for input frequencies much higher than $1/(R_2C_1)$. Thus, R_2/R_1 must be sufficiently small so as to minimize the amplified offset given by Eq. (8.74) whereas R_2C_1 must be sufficiently large so as to negligibly impact the signal frequencies of interest.

8.4.2 Input Bias Current

Op amps implemented in bipolar technology draw a base current from each input. While relatively small ($\approx$ 0.1–1 μA), the input bias currents may create inaccuracies in some circuits. As shown in Fig. 8.32, each bias current is modeled by a current source tied between the corresponding input and ground. Nominally, $I_{B1} = I_{B2}$.

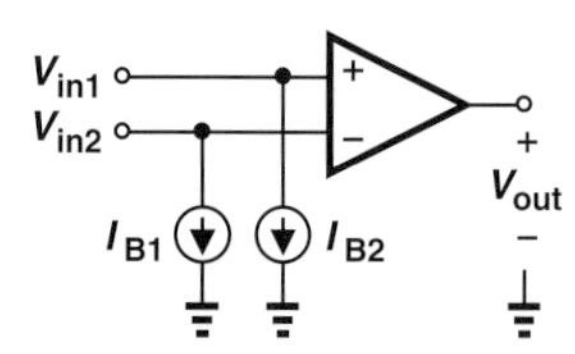

Figure 8.32 Input bias currents.

Let us study the effect of the input currents on the noninverting amplifier. As depicted in Fig. 8.33(a), I_{B1} has no effect on the circuit because it flows through a voltage source. The current I_{B2}, on the other hand, flows through R_1 and R_2, introducing an error. Using superposition and setting V_{in} to zero, we arrive at the circuit in Fig. 8.33(b), which can be transformed to that in Fig. 8.33(c) if I_{B2} and R_2 are replaced with their Thevenin equivalent. Interestingly, the circuit now resembles the inverting amplifier of Fig. 8.7(a), thereby yielding

$$V_{out} = -R_2I_{B2}\left(-\frac{R_1}{R_2}\right) \tag{8.77}$$

$$= R_1I_{B2} \tag{8.78}$$

if the op amp gain is infinite. This expression suggests that I_{B2} flows only through R_1, an expected result because the virtual ground at X in Fig. 8.33(b) forces a zero voltage across R_2 and hence a zero current through it.

The error due to the input bias current appears similar to the DC offset effects illustrated in Fig. 8.29, corrupting the output. However, unlike DC offsets, this phenomenon is *not* random; for a given bias current in the bipolar transistors used in the op amp, the base currents drawn from the inverting and noninverting inputs remain approximately equal.

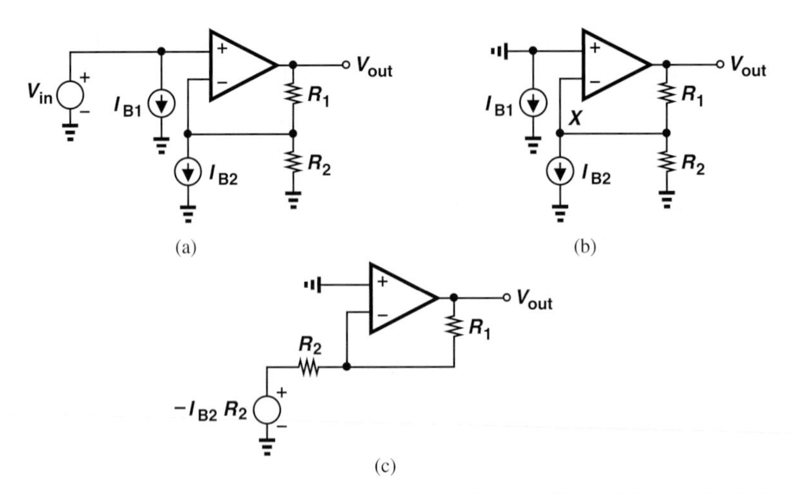

Figure 8.33 (a) Effect of input bias currents on noninverting amplifier, (b) simplified circuit, (c) Thevenin equivalent.

We may therefore seek a method of canceling this error. For example, we can insert a corrective voltage in series with the noninverting input so as to drive V_{out} to zero (Fig. 8.34). Since V_{corr} "sees" a noninverting amplifier, we have

$$V_{out} = V_{corr}\left(1 + \frac{R_1}{R_2}\right) + I_{B2}R_1. \tag{8.79}$$

For $V_{out} = 0$,

$$V_{corr} = -I_{B2}(R_1||R_2). \tag{8.80}$$

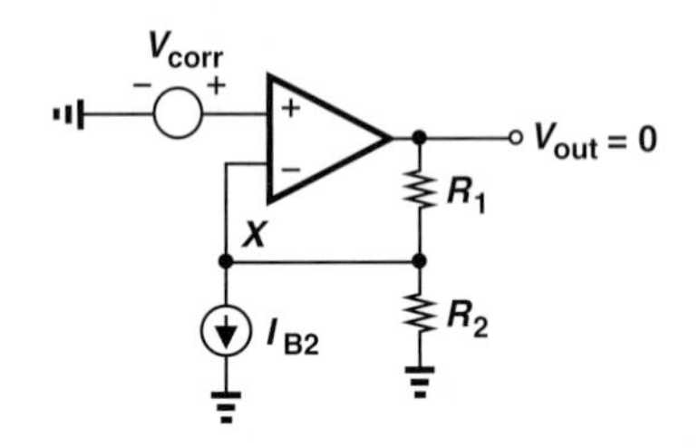

Figure 8.34 Addition of voltage source to correct for input bias currents.

Example 8-12 A bipolar op amp employs a collector current of 1 mA in each of the input devices. If $\beta = 100$ and the circuit of Fig. 8.34 incorporates $R_2 = 1\ \text{k}\Omega$, $R_1 = 10\ \text{k}\Omega$, determine the output error and the required value of V_{corr}.

Solution We have $I_B = 10\ \mu\text{A}$ and hence

$$V_{out} = 0.1\ \text{mV}. \tag{8.81}$$

Thus, V_{corr} is chosen as

$$V_{corr} = -9.1\ \mu\text{V}. \tag{8.82}$$

Exercise Determine the correction voltage if $\beta = 200$.

Equation (8.80) implies that V_{corr} depends on I_{B2} and hence the current gain of transistors. Since β varies with process and temperature, V_{corr} cannot remain at a *fixed* value and must "track" β. Fortunately, (8.80) also reveals that V_{corr} can be obtained by passing a base current through a resistor equal to $R_1||R_2$, leading to the topology shown in Fig. 8.35. Here, if $I_{B1} = I_{B2}$, then $V_{out} = 0$ for $V_{in} = 0$. The reader is encouraged to take the finite gain of the op amp into account and prove that V_{out} is still near zero.

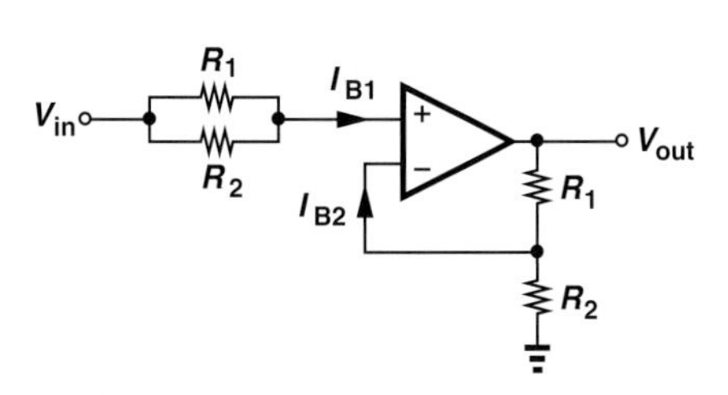

Figure 8.35 Correction for variation of beta.

From the drawing in Fig. 8.33(b), we observe that the input bias currents have an identical effect on the inverting amplifier. Thus, the correction technique shown in Fig. 8.35 applies to this circuit as well.

In reality, asymmetries in the op amp's internal circuitry introduce a slight (random) mismatch between I_{B1} and I_{B2}. Problem 8.53 studies the effect of this mismatch on the output in Fig. 8.35.

We now consider the effect of the input bias currents on the performance of integrators. Illustrated in Fig. 8.36(a) with $V_{in} = 0$ and I_{B1} omitted (why?), the circuit forces I_{B2} to flow through C_1 because R_1 sustains a zero voltage drop. In fact, the Thevenin equivalent of R_1 and I_{B2} [Fig. 8.36(b)] yields

$$V_{out} = -\frac{1}{R_1C_1}\int V_{in}\,dt \qquad (8.83)$$

$$= +\frac{1}{R_1C_1}I_{B2}R_1\,dt \qquad (8.84)$$

$$= \frac{I_{B2}}{C_1}\,dt. \qquad (8.85)$$

(Of course, the flow of I_{B2} through C_1 leads to the same result.) In other words, the circuit integrates the input bias current, thereby forcing V_{out} to eventually saturate near the positive or negative supply rails.

Can we apply the correction technique of Fig. 8.35 to the integrator? The model in Fig. 8.36(b) suggests that a resistor equal to R_1 placed in series with the noninverting input can cancel the effect. The result is depicted in Fig. 8.37.

Did you know?

Op amp design has always faced interesting challenges and elicited new ideas. In fact, many of the techniques invented for use within op amps have quickly found their way to other circuits as well. Early op amps strove for a high gain but encountered instability; that is, they would sometimes *oscillate*. Subsequent generations were designed for a high input impedance, a low output impedance, or low supply voltages. Many others were introduced for low-noise or high-speed operation. Most of the issues persist in today's op amp design as well. Interestingly, the number of transistors used in op amps has not changed much over five decades: the 741 contained about 20 transistors and so do modern op amps.

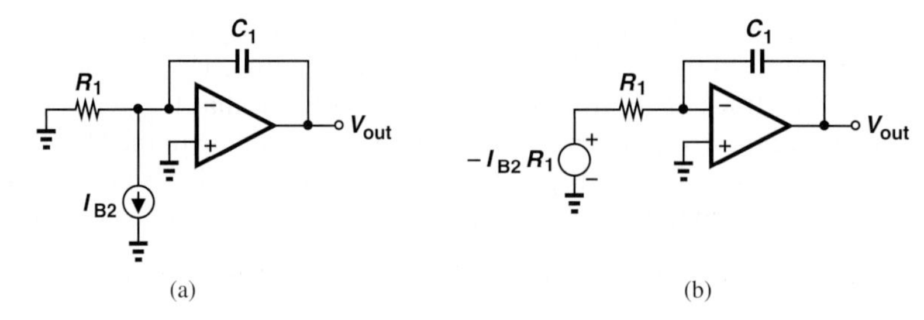

Figure 8.36 (a) Effect of input bias currents on integrator, (b) Thevenin equivalent.

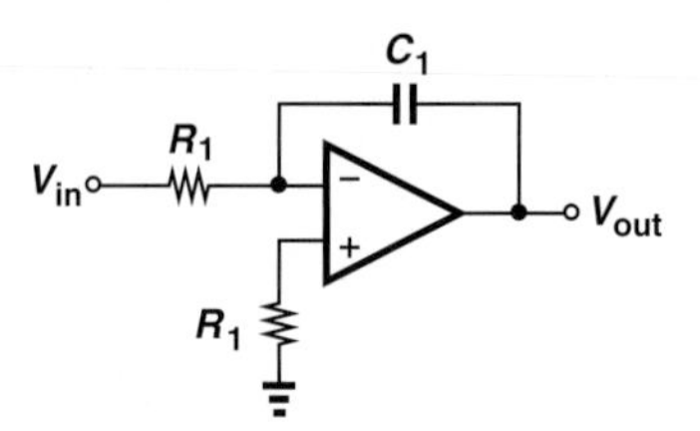

Figure 8.37 Correction for input currents in an integrator.

Example 8-13 An electrical engineering student attempts the topology of Fig. 8.37 in the laboratory but observes that the output still saturates. Give three possible reasons for this effect.

Solution First, the DC offset voltage of the op amp itself is still integrated (Section 8.4.1). Second, the two input bias currents always suffer from a slight mismatch, thus causing incomplete cancellation. Third, the two resistors in Fig. 8.37 also exhibit mismatches, creating an additional error.

Exercise Is resistor R_1 necessary if the internal circuitry of the op amp uses MOS devices?

The problem of input bias current mismatch requires a modification similar to that in Fig. 8.31(c). The mismatch current then flows through R_2 rather than through C_1 (why?).

8.4.3 Speed Limitations

Finite Bandwidth Our study of op amps has thus far assumed no speed limitations. In reality, the internal capacitances of the op amp degrade the performance at high frequencies. For example, as illustrated in Fig. 8.38, the gain begins to fall as the frequency of operation exceeds f_1. In this chapter, we provide a simple analysis of such effects, deferring a more detailed study to Chapter 11.

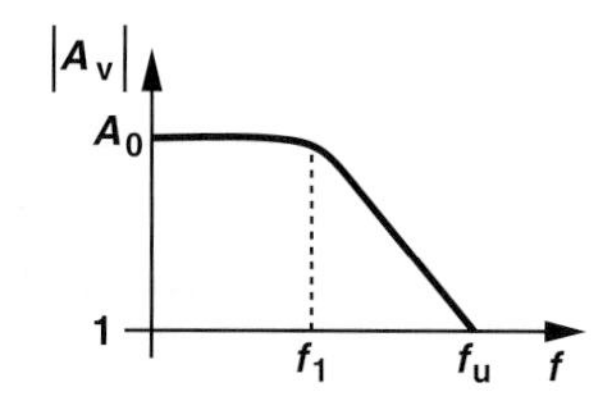

Figure 8.38 Frequency response of an op amp.

To represent the gain roll-off shown in Fig. 8.38, we must modify the op amp model offered in Fig. 8.1. As a simple approximation, the internal circuitry of the op amp can be modeled by a first-order (one-pole) system having the following transfer function:

$$\frac{V_{out}}{V_{in1} - V_{in2}}(s) = \frac{A_0}{1 + \dfrac{s}{\omega_1}}, \tag{8.86}$$

where $\omega_1 = 2\pi f_1$. Note that at frequencies well below ω_1, $s/\omega_1 \ll 1$ and the gain is equal to A_0. At very high frequencies, $s/\omega_1 \gg 1$, and the gain of the op amp falls to *unity* at $\omega_u = A_0\omega_1$. This frequency is called the "unity-gain bandwidth" of the op amp. Using this model, we can reexamine the performance of the circuits studied in the previous sections.

Consider the noninverting amplifier of Fig. 8.5. We utilize Eq. (8.11) but replace A_0 with the above transfer function:

$$\frac{V_{out}}{V_{in}}(s) = \frac{\dfrac{A_0}{1 + \dfrac{s}{\omega_1}}}{1 + \dfrac{R_2}{R_1 + R_2}\dfrac{A_0}{1 + \dfrac{s}{\omega_1}}}. \tag{8.87}$$

Multiplying the numerator and the denominator by $(1 + s/\omega_1)$ gives

$$\frac{V_{out}}{V_{in}}(s) = \frac{A_0}{\dfrac{s}{\omega_1} + \dfrac{R_2}{R_1 + R_2}A_0 + 1}. \tag{8.88}$$

The system is still of first order and the pole of the closed-loop transfer function is given by

$$|\omega_{p,closed}| = \left(1 + \frac{R_2}{R_1 + R_2}A_0\right)\omega_1. \tag{8.89}$$

As depicted in Fig. 8.39, the bandwidth of the closed-loop circuit is substantially higher than that of the op amp itself. This improvement, of course, accrues at the cost of a proportional reduction in gain—from A_0 to $1 + R_2A_0/(R_1 + R_2)$.

Example 8-14

A noninverting amplifier incorporates an op amp having an open-loop gain of 100 and bandwidth of 1 MHz. If the circuit is designed for a closed-loop gain of 16, determine the resulting bandwidth and time constant.

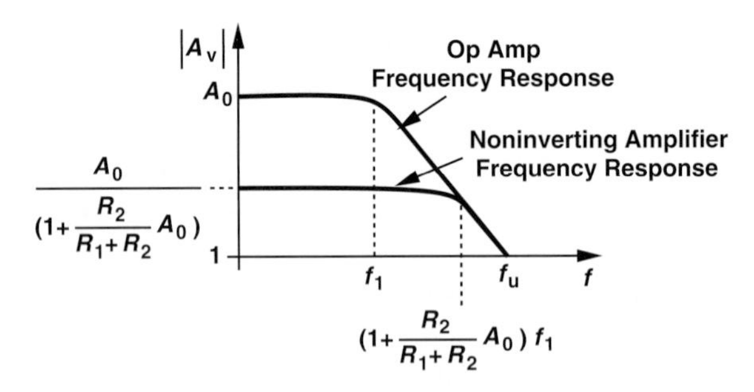

Figure 8.39 Frequency response of open-loop op amp and closed-loop circuit.

Solution For a closed-loop gain of 16, we require that $1 + R_1/R_2 = 16$ and hence

$$|\omega_{p,closed}| = \left(1 + \frac{R_2}{R_1 + R_2} A_0\right) \omega_1 \tag{8.90}$$

$$= \left(1 + \frac{1}{\frac{R_1}{R_2} + 1} A_0\right) \omega_1 \tag{8.91}$$

$$= 2\pi \times (7.25 \text{ MHz}). \tag{8.92}$$

Given by $|\omega_{p,closed}|^{-1}$, the time constant of the circuit is equal to 2.51 ns.

Exercise Repeat the above example if the op amp gain is 500.

The above analysis can be repeated for the inverting amplifier as well. The reader can prove that the result is similar to Eq. (8.89).

The finite bandwidth of the op amp may considerably degrade the performance of integrators. The analysis is beyond the scope of this book, but it is outlined in Problem 8.57 for the interested reader.

Another critical issue in the use of op amps is *stability*; if placed in the topologies seen above, some op amps may *oscillate*. Arising from the internal circuitry of the op amp, this phenomenon often requires internal or external *stabilization*, also called "frequency compensation." These concepts are studied in Chapter 12.

Slew Rate In addition to bandwidth and stability problems, another interesting effect is observed in op amps that relates to their response to large signals. Consider the noninverting configuration shown in Fig. 8.40(a), where the closed-loop transfer function is given by Eq. (8.88). A small step of ΔV at the input thus results in an amplified output waveform having a time constant equal to $|\omega_{p,closed}|^{-1}$ [Fig. 8.40(b)]. If the input step is raised to $2\Delta V$, each point on the output waveform also rises by a factor of two.[7] In other words, doubling the input amplitude doubles both the output amplitude and the output *slope*.

[7]Recall that in a linear system, if $x(t) \rightarrow y(t)$, then $2x(t) \rightarrow 2y(t)$.

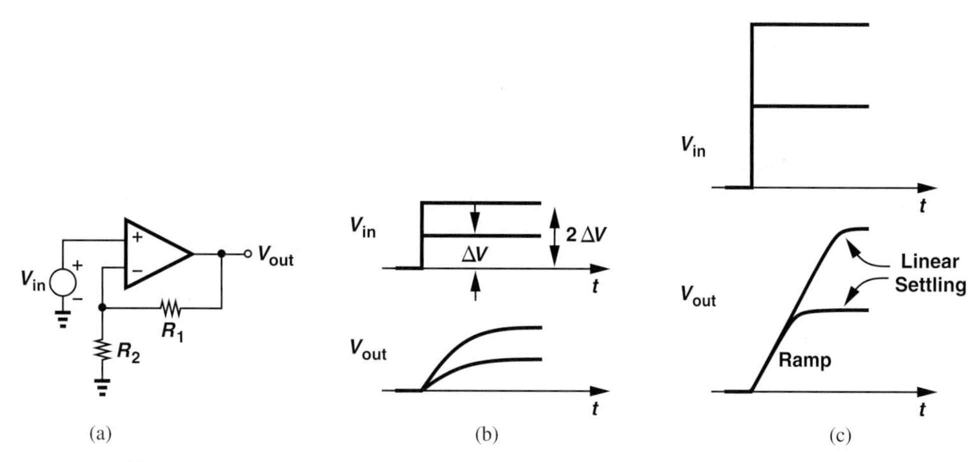

Figure 8.40 (a) Noninverting amplifier, (b) input and output waveforms in linear regime, (c) input and output waveforms in slewing regime.

In reality, op amps do not exhibit the above behavior if the signal amplitudes are large. As illustrated in Fig. 8.40(c), the output first rises with a *constant* slope (i.e., as a ramp) and eventually settles as in the linear case of Fig. 8.40(b). The ramp section of the waveform arises because, with a large input step, the internal circuitry of the op amp reduces to a constant current source charging a capacitor. We say the op amp "slews" during this time. The slope of the ramp is called the "slew rate" (SR).

Slewing further limits the speed of op amps. While for small-signal steps, the output response is determined by the closed-loop time constant, large-signal steps must face slewing prior to linear settling. Figure 8.41 compares the response of a non-slewing circuit with that of a slewing op amp, revealing the longer settling time in the latter case.

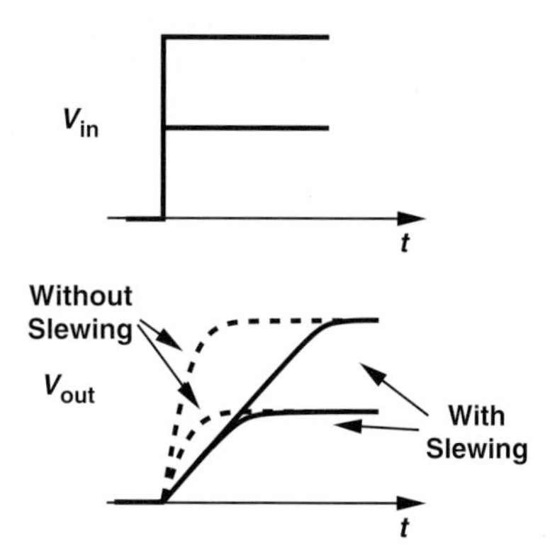

Figure 8.41 Output settling speed with and without slewing.

It is important to understand that slewing is a *nonlinear* phenomenon. As suggested by the waveforms in Fig. 8.40(c), the points on the ramp section do not follow linear scaling (if $x \rightarrow y$, then $2x \nrightarrow 2y$). The nonlinearity can also be observed by applying a large-signal sine wave to the circuit of Fig. 8.40(a) and gradually increasing the frequency (Fig. 8.42). At low frequencies, the op amp output "tracks" the sine wave because the maximum slope of the sine wave remains less than the op amp slew rate [Fig. 8.42(a)]. Writing $V_{in}(t) = V_0 \sin \omega t$

and $V_{out}(t) = V_0(1 + R_1/R_2)\sin\omega t$, we observe that

$$\frac{dV_{out}}{dt} = V_0\left(1 + \frac{R_1}{R_2}\right)\omega\cos\omega t. \tag{8.93}$$

The output therefore exhibits a maximum slope of $V_0\omega(1 + R_1/R_2)$ (at its zero crossing points), and the op amp slew rate must exceed this value to avoid slewing.

What happens if the op amp slew rate is insufficient? The output then fails to follow the sinusoidal shape while passing through zero, exhibiting the distorted behavior shown in Fig. 8.42(b). Note that the output *tracks* the input so long as the slope of the waveform does not exceed the op amp slew rate, e.g., between t_1 and t_2.

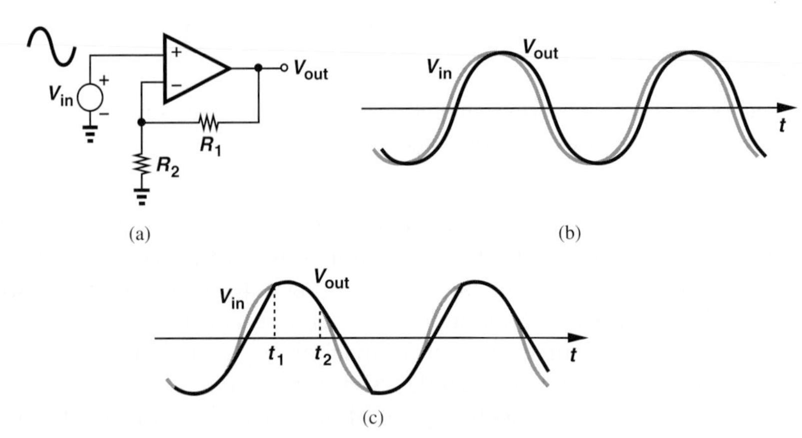

Figure 8.42 (a) Simple noninverting amplifier, (b) input and output waveforms without slewing, (c) input and output waveforms with slewing.

Example 8-15 The internal circuitry of an op amp can be simplified to a 1-mA current source charging a 5-pF capacitor during large-signal operation. If an amplifier using this op amp produces a sinusoid with a peak amplitude of 0.5 V, determine the maximum frequency of operation that avoids slewing.

Solution The slew rate is given by $I/C = 0.2$ V/ns. For an output given by $V_{out} = V_p\sin\omega t$, where $V_p = 0.5$ V, the maximum slope is equal to

$$\left.\frac{dV_{out}}{dt}\right|_{max} = V_p\omega. \tag{8.94}$$

Equating this to the slew rate, we have

$$\omega = 2\pi(63.7\text{ MHz}). \tag{8.95}$$

That is, for frequencies above 63.7 MHz, the zero crossings of the output experience slewing.

Exercise Plot the output waveform if the input frequency is 200 MHz.

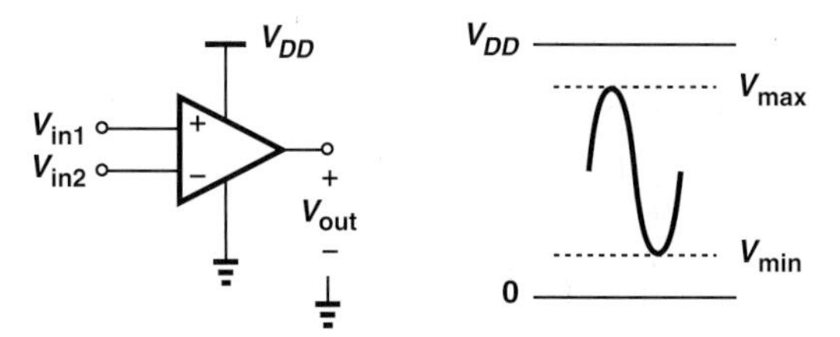

Figure 8.43 Maximum op amp output swings.

Equation (8.93) indicates that the onset of slewing depends on the closed-loop gain, $1 + R_1/R_2$. To define the maximum sinusoidal frequency that remains free from slewing, it is common to assume the worst case, namely, when the op amp produces its maximum allowable voltage swing without saturation. As exemplified by Fig. 8.43, the largest sinusoid permitted at the output is given by

$$V_{out} = \frac{V_{max} - V_{min}}{2} \sin \omega t + \frac{V_{max} + V_{min}}{2}, \tag{8.96}$$

where V_{max} and V_{min} denote the bounds on the output level without saturation. If the op amp provides a slew rate of SR, then the maximum frequency of the above sinusoid can be obtained by writing

$$\left.\frac{dV_{out}}{dt}\right|_{max} = SR \tag{8.97}$$

and hence

$$\omega_{FP} = \frac{SR}{\frac{V_{max} - V_{min}}{2}}. \tag{8.98}$$

Called the "full-power bandwidth," ω_{FP} serves as a measure of the useful large-signal speed of the op amp.

8.4.4 Finite Input and Output Impedances

Actual op amps do not provide an infinite input impedance[8] or a zero output impedance—the latter often creating limitations in the design. We analyze the effect of this nonideality on one circuit here.

Consider the inverting amplifier shown in Fig. 8.44(a), assuming the op amp suffers from an output resistance, R_{out}. How should the circuit be analyzed? We return to the model in Fig. 8.1 and place R_{out} in series with the output voltage source [Fig. 8.44(b)]. We must solve the circuit in the presence of R_{out}. Recognizing that the current flowing through R_{out} is equal to $(-A_0 v_X - v_{out})/R_{out}$, we write a KVL from v_{in} to v_{out} through R_2 and R_1:

$$v_{in} + (R_1 + R_2)\frac{-A_0 v_X - v_{out}}{R_{out}} = v_{out}. \tag{8.99}$$

[8]Op amps employing MOS transistors at their input exhibit a very high input impedance at low frequencies.

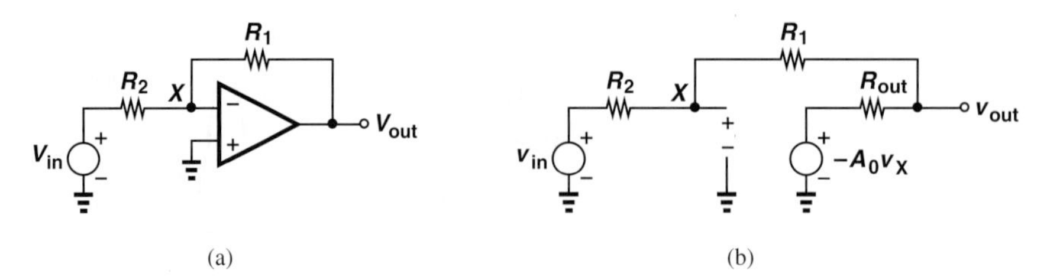

Figure 8.44 (a) Inverting amplifier, (b) effect of finite output resistance of op amp.

To construct another equation for v_X, we view R_1 and R_2 as a voltage divider:

$$v_X = \frac{R_2}{R_1 + R_2}(v_{out} - v_{in}) + v_{in}. \tag{8.100}$$

Substituting for v_X in Eq. (8.99) thus yields

$$\frac{v_{out}}{v_{in}} = -\frac{R_1}{R_2}\frac{A_0 - \dfrac{R_{out}}{R_1}}{1 + \dfrac{R_{out}}{R_2} + A_0 + \dfrac{R_1}{R_2}}. \tag{8.101}$$

The additional terms $-R_{out}/R_1$ in the numerator and R_{out}/R_2 in the denominator increase the gain error of the circuit.

Example 8-16 An electrical engineering student purchases an op amp with $A_0 = 10{,}000$ and $R_{out} = 1\ \Omega$ and constructs the amplifier of Fig. 8.44(a) using $R_1 = 50\ \Omega$ and $R_2 = 10\ \Omega$. Unfortunately, the circuit fails to provide large voltage swings at the output even though R_{out}/R_1 and R_{out}/R_2 remain much less than A_0 in Eq. (8.101). Explain why.

Solution For an output swing of, say, 2 V, the op amp may need to deliver a current as high as 40 mA to R_1 (why?). Many op amps can provide only a small output current even though their small-signal output impedance is very low.

Exercise If the op amp can deliver a current of 5 mA, what value of R_1 is acceptable for output voltages as high as 1 V?

8.5 DESIGN EXAMPLES

Following our study of op amp applications in the previous sections, we now consider several examples of the design procedure for op amp circuits. We begin with simple examples and gradually proceed to more challenging problems.

Example 8-17 Design an inverting amplifier with a nominal gain of 4, a gain error of 0.1%, and an input impedance of at least 10 kΩ. Determine the minimum op amp gain required here.

Solution For an input impedance of 10 kΩ, we choose the same value of R_2 in Fig. 8.7(a), arriving at $R_1 = 40$ kΩ for a nominal gain of 4. Under these conditions, Eq. (8.21) demands that

$$\frac{1}{A_0}\left(1+\frac{R_1}{R_2}\right) < 0.1\% \tag{8.102}$$

and hence

$$A_0 > 5000. \tag{8.103}$$

Exercise Repeat the above example for a nominal gain of 8 and compare the results.

Example 8-18 Design a noninverting amplifier for the following specifications: closed-loop gain = 5, gain error = 1%, closed-loop bandwidth = 50 MHz. Determine the required open-loop gain and bandwidth of the op amp. Assume the op amp has an input bias current of 0.2 μA.

Solution From Fig. 8.5 and Eq. (8.9), we have

$$\frac{R_1}{R_2} = 4. \tag{8.104}$$

The choice of R_1 and R_2 themselves depends on the "driving capability" (output resistance) of the op amp. For example, we may select $R_1 = 4$ kΩ and $R_2 = 1$ kΩ and check the gain error from Eq. (8.101) at the end. For a gain error of 1%,

$$\frac{1}{A_0}\left(1+\frac{R_1}{R_2}\right) < 1\% \tag{8.105}$$

and hence

$$A_0 > 500. \tag{8.106}$$

Also, from Eq. (8.89), the open-loop bandwidth is given by

$$\omega_1 > \frac{\omega_{p,closed}}{1+\dfrac{R_2}{R_1+R_2}A_0} \tag{8.107}$$

$$\omega_1 > \frac{\omega_{p,closed}}{1+\left(1+\dfrac{R_1}{R_2}\right)^{-1}A_0} \tag{8.108}$$

$$> \frac{2\pi(50\text{ MHz})}{100}. \tag{8.109}$$

Thus, the op amp must provide an open-loop bandwidth of at least 500 kHz.

Exercise Repeat the above example for a gain error of 2% and compare the results.

Example 8-19 Design an integrator for a unity-gain frequency of 10 MHz and an input impedance of 20 kΩ. If the op amp provides a slew rate of 0.1 V/ns, what is the largest peak-to-peak sinusoidal swing at the input at 1 MHz that produces an output free from slewing?

Solution From Eq. (8.30), we have

$$\frac{1}{R_1C_1(2\pi \times 10\text{ MHz})} = 1 \tag{8.110}$$

and, with $R_1 = 20\text{ k}\Omega$,

$$C_1 = 0.796\text{ pF}. \tag{8.111}$$

(In discrete design, such a small capacitor value may prove impractical.)

For an input given by $V_{in} = V_p \cos \omega t$,

$$V_{out} = \frac{-1}{R_1C_1}\frac{V_p}{\omega}\sin \omega t, \tag{8.112}$$

with a maximum slope of

$$\left.\frac{dV_{out}}{dt}\right|_{max} = \frac{1}{R_1C_1}V_p. \tag{8.113}$$

Equating this result to 0.1 V/ns gives

$$V_p = 1.59\text{ V}. \tag{8.114}$$

In other words, the input peak-to-peak swing at 1 MHz must remain below 3.18 V for the output to be free from slewing.

Exercise How do the above results change if the op amp provides a slew rate of 0.5 V/ns?

8.6 CHAPTER SUMMARY

- An op amp is a circuit that provides a high voltage gain and an output proportional to the *difference* between two inputs.
- Due to its high voltage gain, an op amp producing a moderate output swing requires only a very small input difference.
- The noninverting amplifier topology exhibits a nominal gain equal to one plus the ratio of two resistors. The circuit also suffers from a gain error that is inversely proportional to the gain of the op amp.
- The inverting amplifier configuration provides a nominal gain equal to the ratio of two resistors. Its gain error is the same as that of the noninverting configuration. With the noninverting input of the op amp tied to ground, the inverting input also remains close to the ground potential and is thus called a "virtual ground."
- If the feedback resistor in an inverting configuration is replaced with a capacitor, the circuit operates as an integrator. Integrator find wide application in analog filters and analog-to-digital converters.

- If the input resistor in an inverting configuration is replaced with a capacitor, the circuit acts as a differentiator. Due to their higher noise, differentiators are less common than integrators.
- An inverting configuration using multiple input resistors tied to the virtual ground node can serve as a voltage adder.
- Placing a diode around an op amp leads to a precision rectifier, i.e., a circuit that can rectify very small input swings.
- Placing a bipolar device around an op amp provides a logarithmic function.
- Op amps suffer from various imperfections, including dc offsets and input bias currents. These effects impact the performance of various circuits, most notably integrators.
- The speed of op amp circuits is limited by the bandwidth of the op amps. Also, for large signals, the op amp suffers from a finite slew rate, distorting the output waveform.

PROBLEMS

Sec. 8.1 General Considerations

8.1. Actual op amps exhibit "nonlinear" characteristics. For example, the voltage gain (the slope) may be equal to 1000 for $-1\text{ V} < V_{out} < +1\text{ V}$, 500 for $1\text{ V} < |V_{out}| < 2\text{ V}$, and close to zero for $|V_{out}| > 2\text{ V}$.

(a) Plot the input/output characteristic of this op amp.

(b) What is the largest input swing that the op amp can sense without producing "distortion" (i.e., nonlinearity)?

8.2. An op amp exhibits the following nonlinear characteristic:

$$V_{out} = \alpha \tanh[\beta(V_{in1} - V_{in2})]. \quad (8.115)$$

Sketch this characteristic and determine the small-signal gain of the op amp in the vicinity of $V_{in1} - V_{in2} \approx 0$.

Sec. 8.2.1 Noninverting Amplifier

SOL **8.3.** A noninverting amplifier employs an op amp having a nominal gain of 2000 to achieve a nominal closed-loop gain of 8. Determine the gain error.

8.4. A noninverting amplifier must provide a nominal gain of 4 with a gain error of 0.1%. Compute the minimum required op amp gain.

8.5. Looking at Equation (8.11), an adventurous student decides that it is possible to achieve a *zero* gain error with a finite A_0 if $R_2/(R_1 + R_2)$ is slightly adjusted from its nominal value.

(a) Suppose a nominal closed-loop gain of α_1 is required. How should $R_2/(R_1 + R_2)$ be chosen?

(b) With the value obtained in (a), determine the gain error if A_0 drops to $0.6A_0$.

*__8.6.__ A noninverting amplifier employs an op amp with a finite output impedance, R_{out}. Representing the op amp as depicted in Fig. 8.45, compute the closed-loop gain and output impedance. What happens if $A_0 \to \infty$?

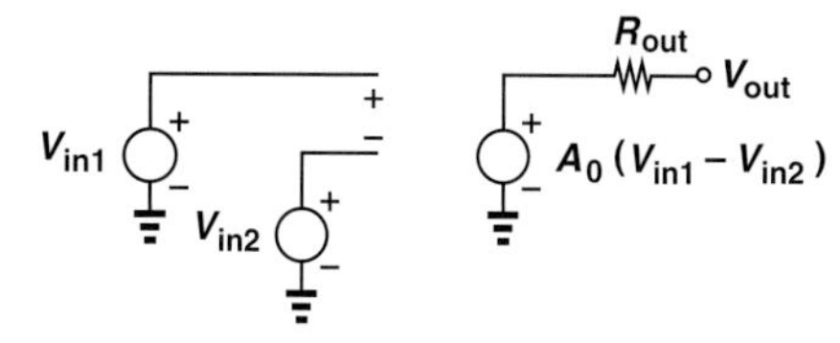

Figure 8.45

*__8.7.__ A noninverting amplifier incorporates an op amp having an input impedance of R_{in}. Modeling the op amp as shown in Fig. 8.46, determine the closed-loop gain and input impedance. What happens if $A_0 \to \infty$?

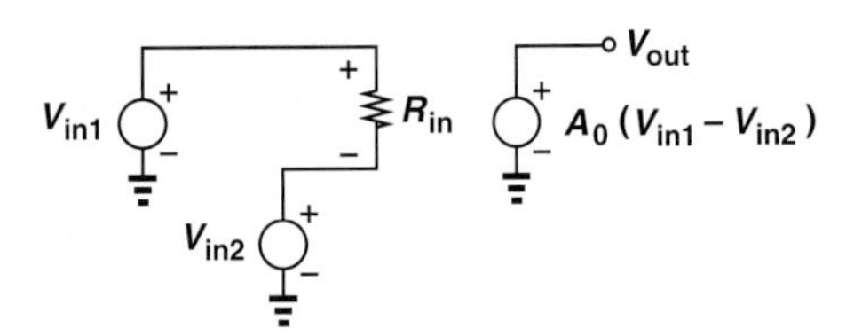

Figure 8.46

8.8. In the noninverting amplifier shown in Fig. 8.47, resistor R_2 deviates from its nominal value by ΔR. Calculate the gain error of the circuit if $\Delta R/R_2 \ll 1$.

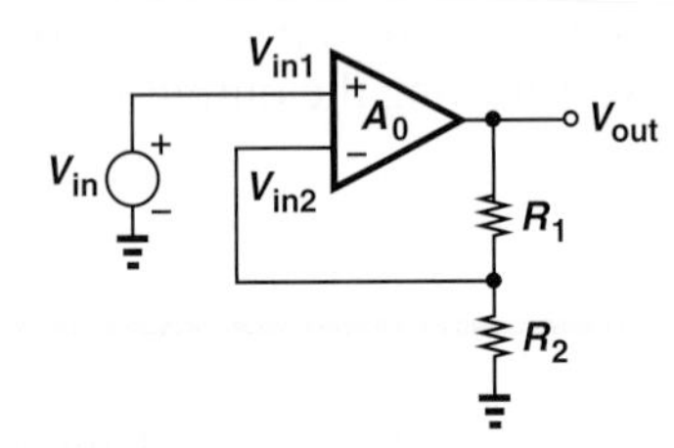

Figure 8.47

*__8.9.__ The input/output characteristic of an op amp can be approximated by the piecewise-linear behavior illustrated in Fig. 8.48, where the gain drops from A_0 to $0.8A_0$ and eventually to zero as $|V_{in1} - V_{in2}|$ increases. Suppose this op amp is used in a noninverting amplifier with a nominal gain of 5. Plot the closed-loop input/output characteristic of the circuit. (Note that the closed-loop gain experiences much less variation; i.e., the closed-loop circuit is much more linear.)

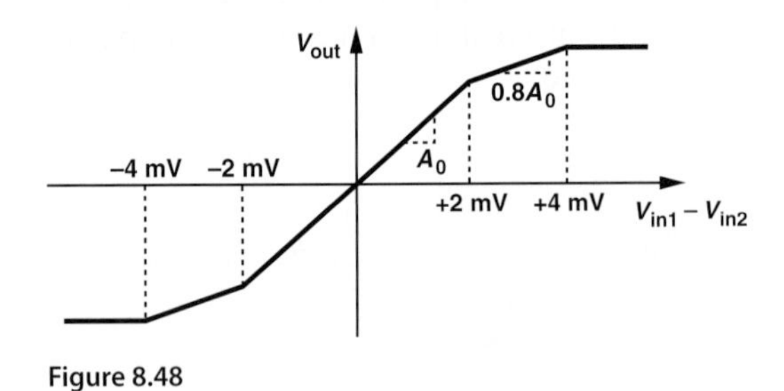

Figure 8.48

SOL **8.10.** A truck weighing station incorporates a sensor whose resistance varies linearly with the weight: $R_S = R_0 + \alpha W$. Here R_0 is a constant value, α a proportionality factor, and W the weight of each truck. Suppose R_S plays the role of R_2 in the noninverting amplifier (Fig. 8.49). Also, $V_{in} = 1$ V. Determine the gain of the system, defined as the change in V_{out} divided by the change in W.

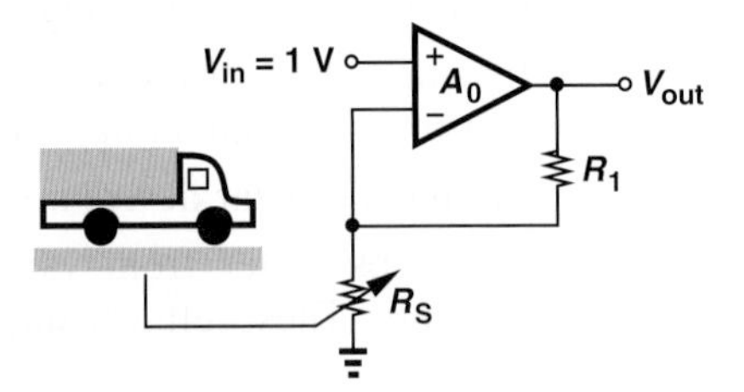

Figure 8.49

*__8.11.__ Calculate the closed-loop gain of the noninverting amplifier shown in Fig. 8.50 if $A_0 = \infty$. Verify that the result reduces to expected values if $R_1 \to 0$ or $R_3 \to 0$.

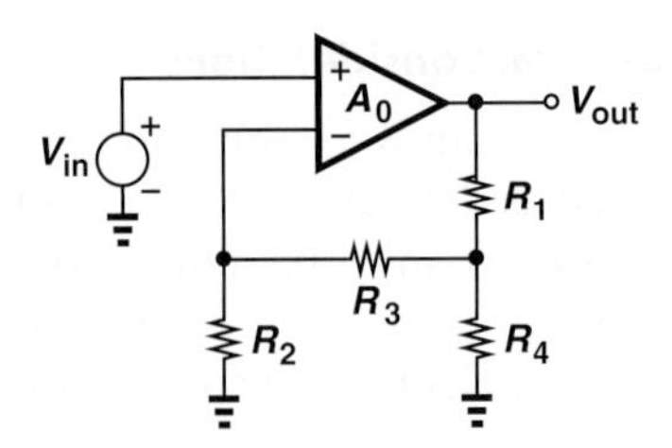

Figure 8.50

Sec. 8.2.2 Inverting Amplifier

8.12. An inverting amplifier must provide a nominal gain of 8 with a gain error of 0.2%. Determine the minimum required op amp gain.

8.13. The op amp used in an inverting amplifier exhibits a finite input impedance, R_{in}. Modeling the op amp as shown in Fig. 8.45, determine the closed-loop gain and input impedance.

8.14. An inverting amplifier employs an op amp having an output impedance of R_{out}. Modeling the op amp as depicted in Fig. 8.46, compute the closed-loop gain and output impedance.

8.15. An inverting amplifier must provide an input impedance of approximately 10 kΩ and a nominal gain of 4. If the op amp exhibits an open-loop gain of 1000 and an output impedance of 1 kΩ, determine the gain error.

8.16. Assuming $A_0 = \infty$, compute the closed-loop gain of the inverting amplifier shown in Fig. 8.51. Verify that the result reduces to expected values if $R_1 \to 0$ or $R_3 \to 0$.

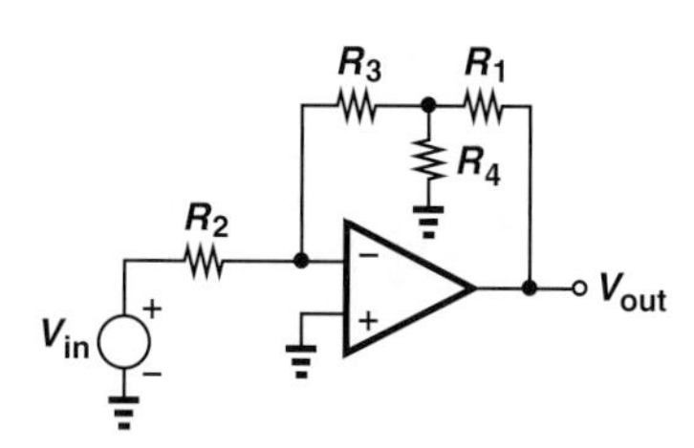

Figure 8.51

8.17. An inverting amplifier is designed for a nominal gain of 8 and a gain error of 0.1% using an op amp that exhibits an output impedance of 2 kΩ. If the input impedance of the circuit must be equal to approximately 1 kΩ, calculate the required open-loop gain of the op amp.

****8.18.** Determine the closed-loop gain of the circuit depicted in Fig. 8.52 if $A_0 = \infty$.

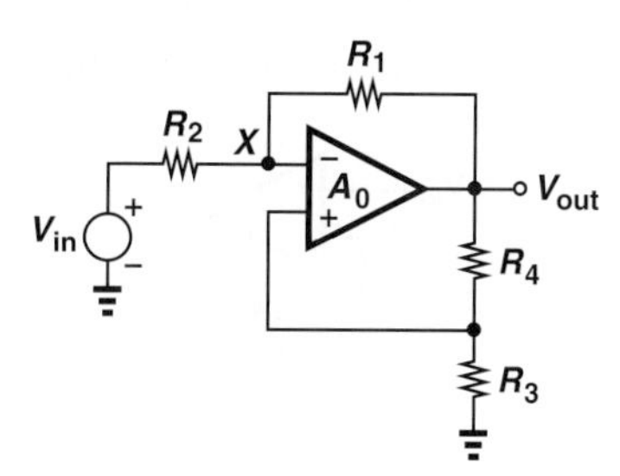

Figure 8.52

Sec. 8.2.3 Integrator and Differentiator

SOL **8.19.** The integrator of Fig. 8.53 senses an input signal given by $V_{in} = V_0 \sin \omega t$. Determine the output signal amplitude if $A_0 = \infty$.

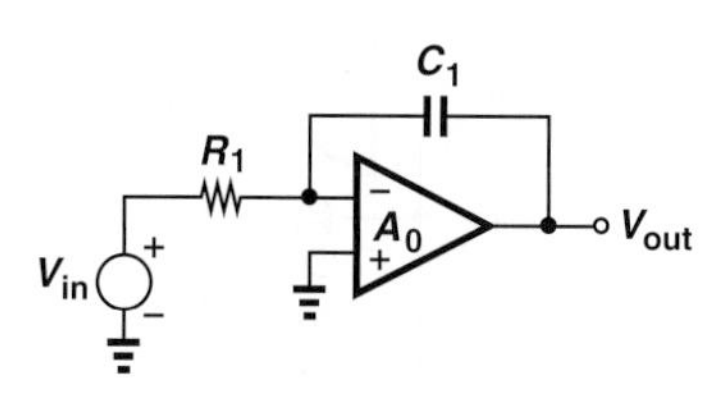

Figure 8.53

8.20. The integrator of Fig. 8.53 is used to amplify a sinusoidal input by a factor of 10. If $A_0 = \infty$ and $R_1C_1 = 10$ ns, compute the frequency of the sinusoid.

8.21. The integrator of Fig. 8.53 must provide a pole at no higher than 1 Hz. If the values of R_1 and C_1 are limited to 10 kΩ and 1 nF, respectively, determine the required gain of the op amp.

***8.22.** Consider the integrator shown in Fig. 8.53 and suppose the op amp is modeled as shown in Fig. 8.45. Determine the transfer function V_{out}/V_{in} and compare the location of the pole with that given by Eq. (8.38).

***8.23.** The op amp used in the integrator of Fig. 8.53 exhibits a finite output impedance and is modeled as depicted in Fig. 8.46. Compute the transfer function V_{out}/V_{in} and compare the location of the pole with that given by Eq. (8.38).

8.24. The differentiator of Fig. 8.54 is used to amplify a sinusoidal input at a frequency of 1 MHz by a factor of 5. If $A_0 = \infty$, determine the value of R_1C_1.

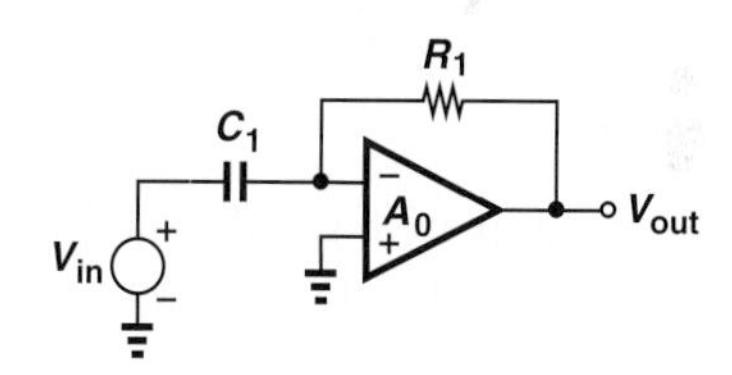

Figure 8.54

8.25. We wish to design the differentiator of Fig. 8.54 for a pole frequency of 100 MHz. If the values of R_1 and C_1 cannot be lower than 1 kΩ and 1 nF, respectively, compute the required gain of the op amp.

***8.26.** Suppose the op amp in Fig. 8.54 exhibits a finite input impedance and is modeled as shown in Fig. 8.45. Determine the transfer function V_{out}/V_{in} and compare the result with Eq. (8.43).

***8.27.** The op amp used in the differentiator of Fig. 8.54 suffers from a finite output impedance and is modeled as depicted in Fig. 8.46. Compute the transfer function and compare the result with Eq. (8.43).

8.28. Calculate the transfer function of the circuit shown in Fig. 8.55 if $A_0 = \infty$. What choice of component values reduces $|V_{out}/V_{in}|$ to unity at all frequencies?

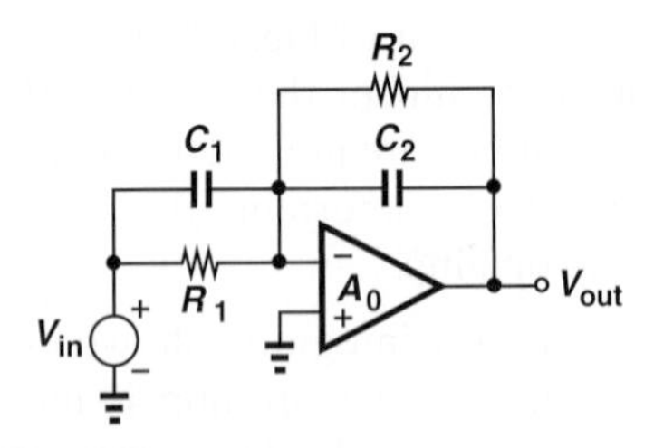

Figure 8.55

8.29. Repeat Problem 8.28 if $A_0 < \infty$. Can the resistors and capacitors be chosen so as to reduce $|V_{out}/V_{in}|$ to approximately unity?

Sec. 8.2.4 Voltage Adder

SOL **8.30.** Consider the voltage adder shown in Fig. 8.56. Plot V_{out} as a function of time if $V_1 = V_0 \sin \omega t$ and $V_2 = V_0 \sin(3\omega t)$. Assume $R_1 = R_2$ and $A_0 = \infty$.

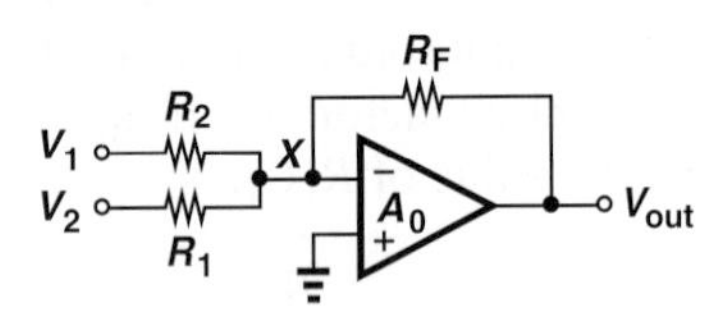

Figure 8.56

8.31. The op amp in Fig. 8.56 suffers from a finite gain. Calculate V_{out} in terms of V_1 and V_2.

8.32. The voltage adder of Fig. 8.56 employs an op amp having a finite output impedance, R_{out}. Using the op amp model depicted in Fig. 8.46, compute V_{out} in terms of V_1 and V_2.

8.33. Due to a manufacturing error, a parasitic resistance R_P has appeared in the adder of Fig. 8.57. Calculate V_{out} in terms of V_1 and V_2 for $A_0 = \infty$ and $A_0 < \infty$. (Note that R_P can also represent the input impedance of the op amp.)

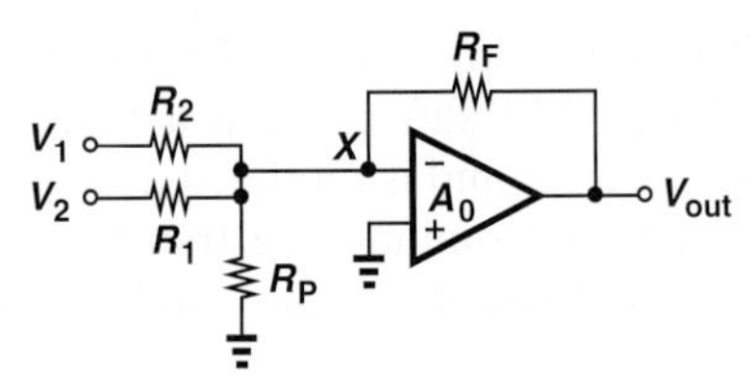

Figure 8.57

8.34. Consider the voltage adder illustrated in Fig. 8.58, where R_P is a parasitic resistance and the op amp exhibits a finite input impedance. With the aid of the op amp model shown in Fig. 8.45, determine V_{out} in terms of V_1 and V_2.

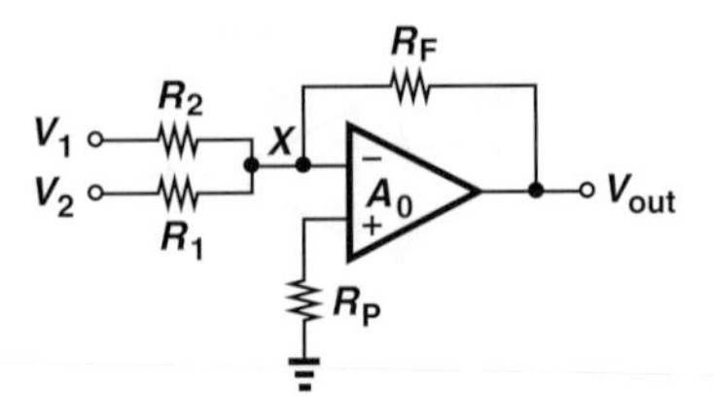

Figure 8.58

Sec. 8.3 Nonlinear Functions

8.35. Plot the current flowing through D_1 in the precision rectifier of Fig. 8.24(b) as a function of time for a sinusoidal input.

8.36. Plot the current flowing through D_1 in the precision rectifier of Fig. 8.25(a) as a function of time for a sinusoidal input.

8.37. Figure 8.59 shows a precision rectifier producing negative cycles. Plot V_Y, V_{out}, and the current flowing through D_1 as a function of time for a sinusoidal input.

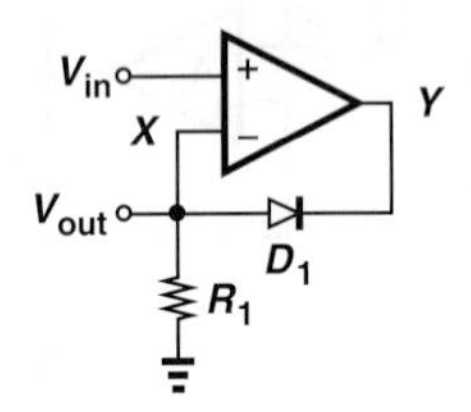

Figure 8.59

8.38. Consider the precision rectifier depicted in Fig. 8.60, where a parasitic resistor R_P has appeared in parallel with D_1. Plot V_X and V_Y as a function of time in response to a sinusoidal input. Use a constant-voltage model for the diode.

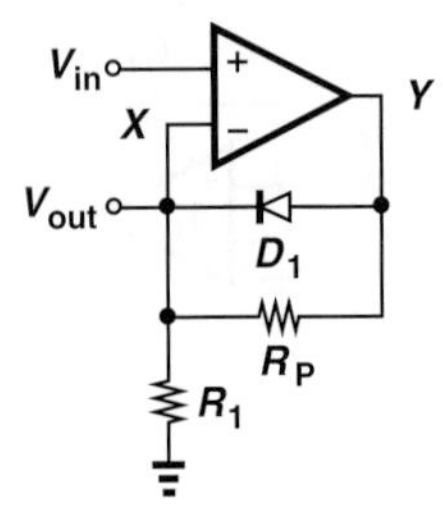

Figure 8.60

8.39. We wish to improve the speed of the rectifier shown in Fig. 8.24(b) by connecting a diode from node Y to ground. Explain how this can be accomplished.

SOL **8.40.** Suppose V_{in} in Fig. 8.26 varies from −1 V to +1 V. Sketch V_{out} and V_X as a function of V_{in} if the op amp is ideal.

8.41. Suppose the gain of the op amp in Fig. 8.26 is finite. Determine the input/output characteristic of the circuit.

8.42. Determine the small-signal voltage gain of the logarithmic amplifier depicted in Fig. 8.26 by differentiating both sides of Eq. (8.68) with respect to V_{in}. Plot the magnitude of the gain as a function of V_{in} and explain why the circuit is said to provide a "compressive" characteristic.

***8.43.** A student attempts to construct a *noninverting* logarithmic amplifier as illustrated in Fig. 8.61. Describe the operation of this circuit.

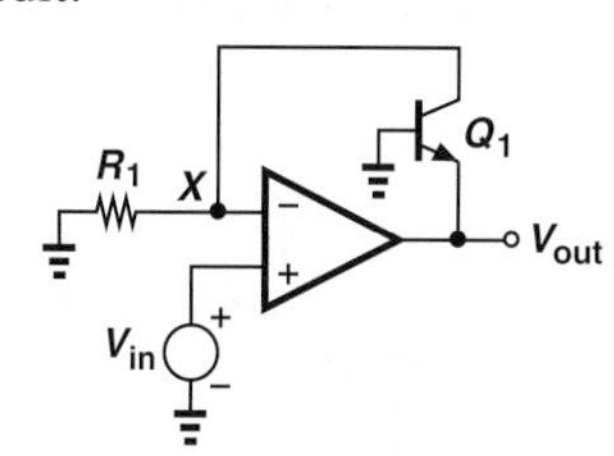

Figure 8.61

8.44. The logarithmic amplifier of Fig. 8.26 must "map" an input voltage of 1 V to an output voltage of −50 mV.

(a) Determine the required value of $I_S R_1$.

(b) Calculate the small-signal voltage gain at this input level.

8.45. The circuit illustrated in Fig. 8.62 can be considered a "true" square-root amplifier. Determine V_{out} in terms of V_{in} and compute the small-signal gain by differentiating the result with respect to V_{in}.

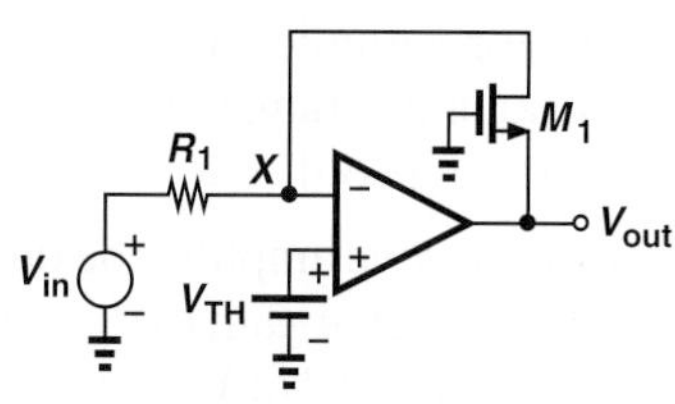

Figure 8.62

***8.46.** Calculate V_{out} in terms of V_{in} for the circuit shown in Fig. 8.63.

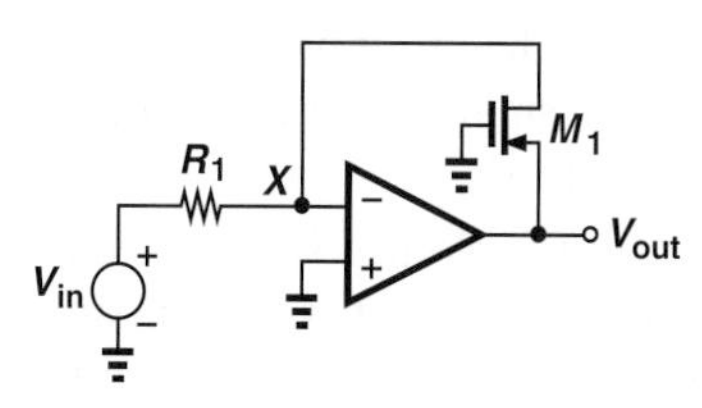

Figure 8.63

Sec. 8.4 Op Amp Nonidealities

8.47. In the noninverting amplifier of Fig. 8.64, the op amp offset is represented by a voltage source in series with the inverting input. Calculate V_{out}.

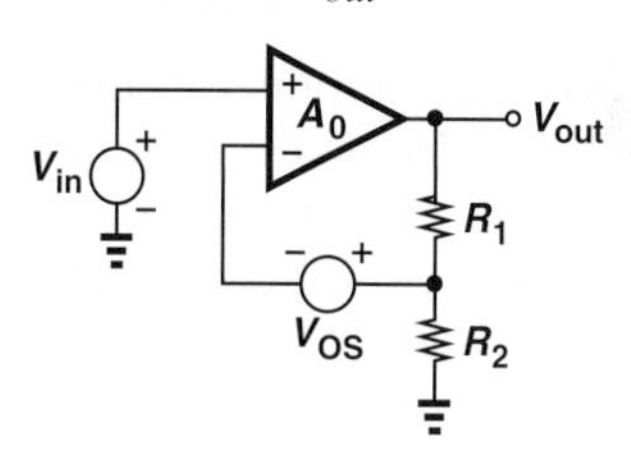

Figure 8.64

8.48. For the inverting amplifier illustrated in Fig. 8.65, calculate V_{out} if the op amp exhibits an input offset of V_{os}. Assume $A_0 = \infty$.

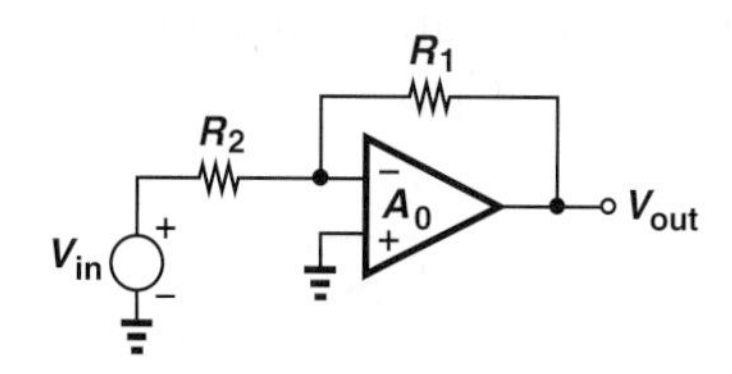

Figure 8.65

8.49. Suppose each op amp in Fig. 8.30 suffers from an input offset of 3 mV. Determine the maximum offset error in V_{out} if each amplifier is designed for a gain of 10.

8.50. The integrator of Fig. 8.31(c) must operate with frequencies as low as 1 kHz while providing an output offset of less than 20 mV with an op amp offset of 3 mV. Determine the required values of R_1 and R_2 if $C_1 \le 100$ pF.

8.51. Explain why dc offsets are not considered a serious issue in differentiators.

SOL **8.52.** Explain the effect of op amp offset on the output of a logarithmic amplifier.

8.53. Suppose the input bias currents in Fig. 8.33 incur a small offset, i.e., $I_{B1} = I_{B2} + \Delta I$. Calculate V_{out}.

8.54. Repeat Problem 8.55 for the circuit shown in Fig. 8.35. What is the maximum allowable value of $R_1||R_2$ if the output error due to this mismatch must remain below a certain value, ΔV?

8.55. A noninverting amplifier must provide a bandwidth of 100 MHz with a nominal gain of 4. Determine which one of the following op amp specifications is adequate:
(a) $A_0 = 1000, f_1 = 50$ Hz.
(b) $A_0 = 500, f_1 = 1$ MHz.

8.56. An inverting amplifier incorporates an op amp whose frequency response is given by Eq. (8.86). Determine the transfer function of the closed-loop circuit and compute the bandwidth.

** **8.57.** Figure 8.66 shows an integrator employing an op amp whose frequency response is given by

$$A(s) = \frac{A_0}{1 + \frac{s}{\omega_0}}. \qquad (8.116)$$

Determine the transfer function of the overall integrator. Simplify the result if $\omega_0 \gg 1/(R_1 C_1)$.

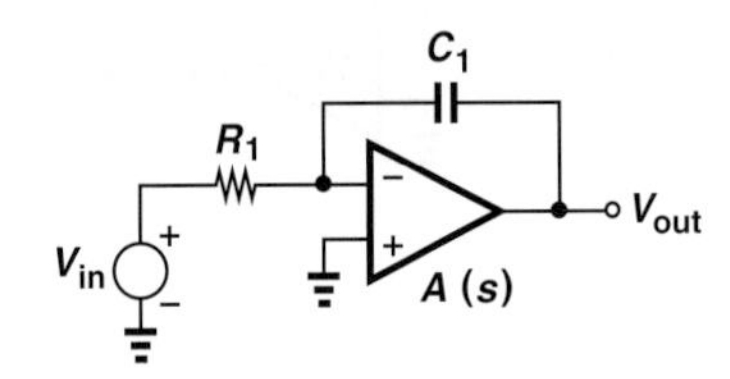

Figure 8.66

8.58. The unity-gain buffer of Fig. 8.3 must be designed to drive a 100 Ω load with a gain error of 0.5%. Determine the required op amp gain if the op amp has an output resistance of 1 kΩ.

SOL **8.59.** A noninverting amplifier with a nominal gain of 4 senses a sinusoid having a peak amplitude of 0.5 V. If the op amp provides a slew rate of 1 V/ns, what is the highest input frequency for which no slewing occurs?

Design Problems

8.60. Design a noninverting amplifier with a nominal gain of 4, a gain error of 0.2%, and a total resistance, $R_1 + R_2$, of 20 kΩ. Assume the op amp has a finite gain but is otherwise ideal.

8.61. Design the inverting amplifier of Fig. 8.7(a) for a nominal gain of 8 and a gain error of 0.1%. Assume $R_{out} = 100$ Ω.

8.62. Design an integrator that attenuates input frequencies above 100 kHz and exhibits a pole at 100 Hz. Assume the largest available capacitor is 50 pF.

8.63. With a finite op amp gain, the step response of an integrator is a slow exponential rather than an ideal ramp. Design an integrator whose step response approximates $V(t) = \alpha t$ with an error less than 0.1% for the range $0 < V(t) < V_0$ (Fig. 8.67). Assume $\alpha = 10$ V/μs, $V_0 =$ 1 V, and the capacitor must remain below 20 pF.

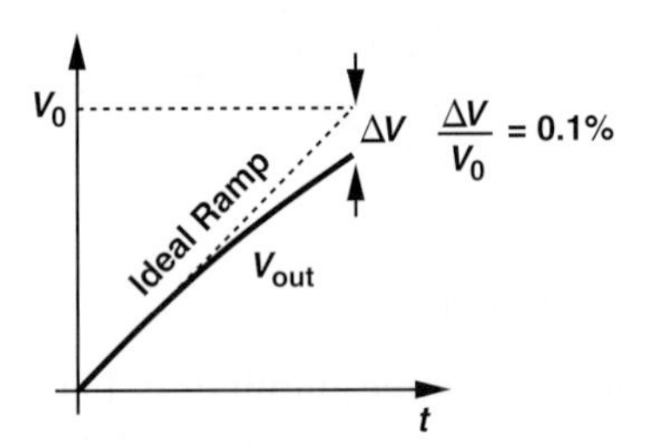

Figure 8.67

8.64. A voltage adder must realize the following function: $V_{out} = \alpha_1 V_1 + \alpha_2 V_2$, where $\alpha_1 = -0.5$ and $\alpha_2 = -1.5$. Design the circuit if the worst-case error in α_1 or α_2 must remain below 0.5% and the input impedance seen by V_1 or V_2 must exceed 10 kΩ.

8.65. Design a logarithmic amplifier that "compresses" an input range of [0.1 V 2 V] to an output range of [−0.5 V −1 V].

8.66. Can a logarithmic amplifier be designed to have a small-signal gain (dV_{out}/dV_{in}) of 2 at $V_{in} = 1$ V and 0.2 at $V_{in} = 2$ V? Assume the gain of the op amp is sufficiently high.

SPICE PROBLEMS

8.67. Assuming an op amp gain of 1000 and $I_S = 10^{-17}$ A for D_1, plot the input/output characteristic of the precision rectifier shown in Fig. 8.68.

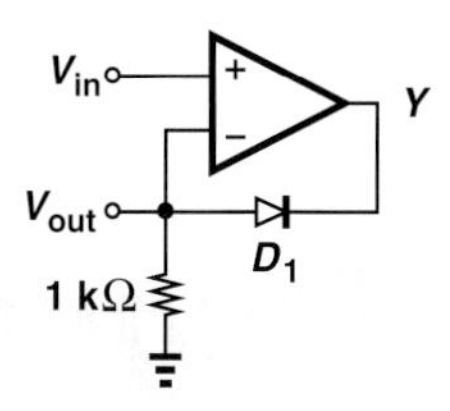

Figure 8.68

8.68. Repeat Problem 8.67 but assuming that the op amp suffers from an output resistance of 1 kΩ.

8.69. In the circuit of Fig. 8.69, each op amp provides a gain of 500. Apply a 10-MHz sinusoid at the input and plot the output as a function of time. What is the error in the output amplitude with respect to the input amplitude?

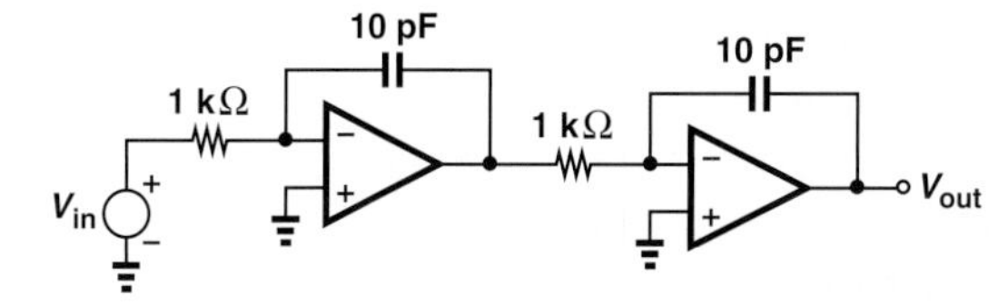

Figure 8.69

8.70. Using ac analysis in SPICE, plot the frequency response of the circuit depicted in Fig. 8.70.

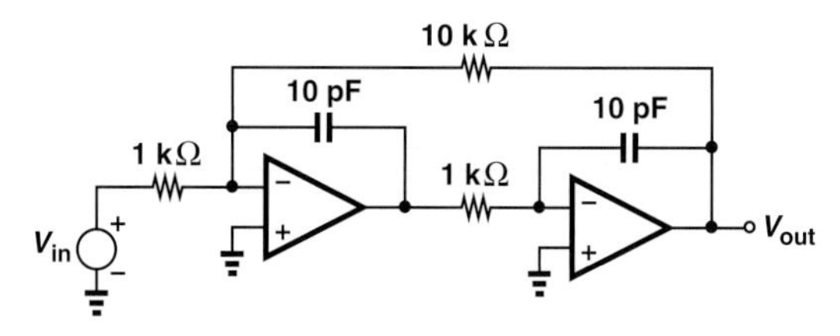

Figure 8.70

8.71. The arrangement shown in Fig. 8.71 incorporates an op amp to "linearize" a common-emitter stage. Assume $I_{S,Q1} = 5 \times 10^{-16}$ A, and $\beta = 100$.

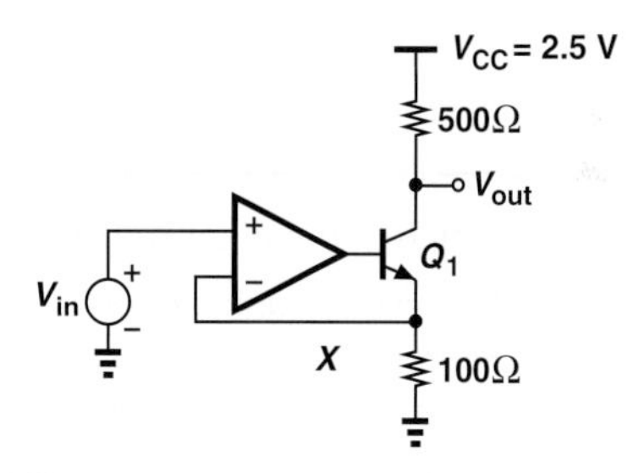

Figure 8.71

(a) Explain why the small-signal gain of the circuit approaches R_C/R_E if the gain of the op amp is very high. (Hint: $V_X \approx V_{in}$.)

(b) Plot the input/output characteristic of the circuit for $0.1\text{ V} < V_{in} < 0.2\text{ V}$ and an op amp gain of 100.

(c) Subtract $V_{out} = 5\,V_{in}$ (e.g., using a voltage-dependent voltage source) from the input/output characteristic and determine the maximum error.

9

Cascode Stages and Current Mirrors

Watch Companion YouTube Videos:

Razavi Electronics 2, Lec 1: Introduction, Cascode Current Sources
https://www.youtube.com/watch?v=pK2elUcXWzs

Razavi Electronics 2, Lec 2: MOS and Bipolar Cascode Current Sources, Introduction to Cascode Amplifiers
https://www.youtube.com/watch?v=GAvZLyii3aw

Razavi Electronics 2, Lec 3: MOS and Bipolar Cascode Amplifiers
https://www.youtube.com/watch?v=icrn9cYIPUA

Razavi Electronics 2, Lec 4: Additional Cascode Examples, Cascode Amp with PMOS Input
https://www.youtube.com/watch?v=ElbiWDyYb0Q

Razavi Electronics 2, Lec 5: Problem of Biasing; Introduction to Current Mirrors
https://www.youtube.com/watch?v=aDteBKenpWk

Razavi Electronics 2, Lec 6: Current Mirror Examples, Proper Scaling, Bipolar Current Mirrors
https://www.youtube.com/watch?v=PxsY1U8aMso

Following our study of basic bipolar and MOS amplifiers in previous chapters, we deal with two other important building blocks in this chapter. The "cascode"[1] stage is a modified version of common-emitter or common-source topologies and proves useful in high-performance circuit design, and the "current mirror" is an interesting and versatile technique employed extensively in integrated circuits. Our study includes both bipolar and MOS implementations of each building block. Shown below is the outline of the chapter.

[1]Coined in the vacuum-tube era, the term "cascode" is believed to be an abbreviation of "cascaded triodes."

Cascode Stages
- Cascode as Current Source
- Cascode as Amplifier

Current Mirrors
- Bipolar Mirrors
- MOS Mirrors

9.1 CASCODE STAGE

9.1.1 Cascode as a Current Source

Recall from Chapters 5 and 7 that the use of current-source loads can markedly increase the voltage gain of amplifiers. We also know that a single transistor can operate as a current source but its output impedance is limited due to the Early effect (in bipolar devices) or channel-length modulation (in MOSFETs).

How can we increase the output impedance of a transistor that acts as a current source? An important observation made in Chapters 5 and 7 forms the foundation for our study here: emitter or source degeneration "boosts" the impedance seen looking into the collector or drain, respectively. For the circuits shown in Fig. 9.1, we have

$$R_{out1} = [1 + g_m(R_E||r_\pi)]r_O + R_E||r_\pi \tag{9.1}$$

$$= (1 + g_m r_O)(R_E||r_\pi) + r_O \tag{9.2}$$

$$R_{out2} = (1 + g_m R_S)r_O + R_S \tag{9.3}$$

$$= (1 + g_m r_O)R_S + r_O, \tag{9.4}$$

observing that R_E or R_S can be increased to raise the output resistance. Unfortunately, however, the voltage drop across the degeneration resistor also increases proportionally, consuming voltage headroom and ultimately limiting the voltage swings provided by the circuit using such a current source. For example, if R_E sustains 300 mV and Q_1 requires a minimum collector-emitter voltage of 500 mV, then the degenerated current source "consumes" a headroom of 800 mV.

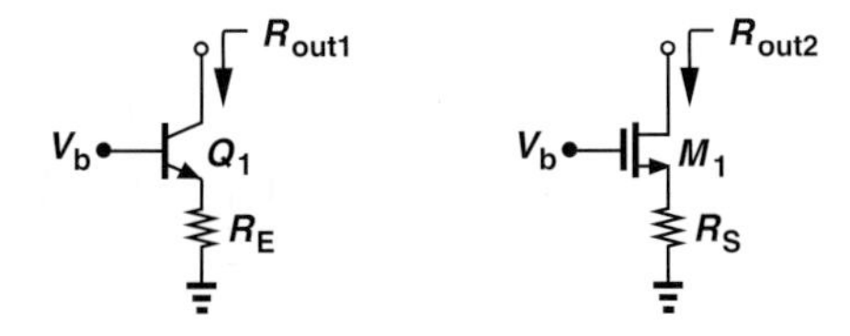

Figure 9.1 Output impedance of degenerated bipolar and MOS devices.

Bipolar Cascode In order to relax the trade-off between the output impedance and the voltage headroom, we can replace the degeneration resistor with a transistor. Depicted in Fig. 9.2(a) for the bipolar version, the idea is to introduce a high small-signal resistance ($= r_{O2}$) in the emitter of Q_1 while consuming a headroom *independent* of the current. In this case, Q_2 requires a headroom of approximately 0.4 V to remain in soft saturation. This configuration is called the "cascode" stage.[2] To emphasize that Q_1 and Q_2 play distinctly

[2]Or simply the "cascode."

different roles here, we call Q_1 the cascode transistor and Q_2 the degeneration transistor. Note that $I_{C1} \approx I_{C2}$ if $\beta_1 \gg 1$.

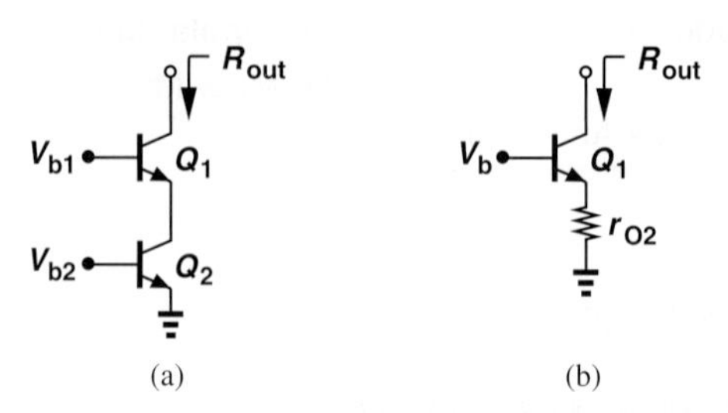

Figure 9.2 (a) Cascode bipolar current source, (b) equivalent circuit.

Let us compute the output impedance of the bipolar cascode of Fig. 9.2(a). Since the base-emitter voltage of Q_2 is constant, this transistor simply operates as a small-signal resistance equal to r_{O2} [Fig. 9.2(b)]. In analogy with the resistively-degenerated counterpart in Fig. 9.1, we have

$$R_{out} = [1 + g_{m1}(r_{O2}||r_{\pi 1})]r_{O1} + r_{O2}||r_{\pi 1}. \tag{9.5}$$

Since typically $g_{m1}(r_{O2}||r_{\pi 1}) \gg 1$,

$$R_{out} \approx (1 + g_{m1}r_{O1})(r_{O2}||r_{\pi 1}) \tag{9.6}$$

$$\approx g_{m1}r_{O1}(r_{O2}||r_{\pi 1}). \tag{9.7}$$

Note, however, that r_O cannot generally be assumed much greater than r_π.

Example 9-1

If Q_1 and Q_2 in Fig. 9.2(a) are biased at a collector current of 1 mA, determine the output resistance. Assume $\beta = 100$ and $V_A = 5$ V for both transistors.

Solution Since Q_1 and Q_2 are identical and biased at the same current level, Eq. (9.7) can be simplified by noting that $g_m = I_C/V_T$, $r_O = V_A/I_C$, and $r_\pi = \beta V_T/I_C$:

$$R_{out} \approx \frac{I_{C1}}{V_T} \cdot \frac{V_{A1}}{I_{C1}} \cdot \frac{\dfrac{V_{A2}}{I_{C2}} \cdot \dfrac{\beta V_T}{I_{C1}}}{\dfrac{V_{A2}}{I_{C2}} + \dfrac{\beta V_T}{I_{C1}}} \tag{9.8}$$

$$\approx \frac{1}{I_{C1}} \cdot \frac{V_A}{V_T} \cdot \frac{\beta V_A V_T}{V_A + \beta V_T}, \tag{9.9}$$

where $I_C = I_{C1} = I_{C2}$ and $V_A = V_{A1} = V_{A2}$. At room temperature, $V_T \approx 26$ mV and hence

$$R_{out} \approx 328.9 \text{ k}\Omega. \tag{9.10}$$

By comparison, the output resistance of Q_1 with no degeneration would be equal to $r_{O1} = 5$ kΩ; i.e., "cascoding" has boosted R_{out} by a factor of 66 here. Note that r_{O2} and $r_{\pi 1}$ are comparable in this example.

Exercise What Early voltage is required for an output resistance of 500 kΩ?

It is interesting to note that if r_{O2} becomes much greater than $r_{\pi 1}$, then R_{out1} approaches

$$R_{out,max} \approx g_{m1} r_{O1} r_{\pi 1} \tag{9.11}$$

$$\approx \beta_1 r_{O1}. \tag{9.12}$$

This is the maximum output impedance provided by a bipolar cascode. After all, even with $r_{O2} = \infty$ (Fig. 9.3) [or $R_E = \infty$ in Eq. (9.1)], $r_{\pi 1}$ still appears from the emitter of Q_1 to ac ground, thereby limiting R_{out} to $\beta_1 r_{O1}$.

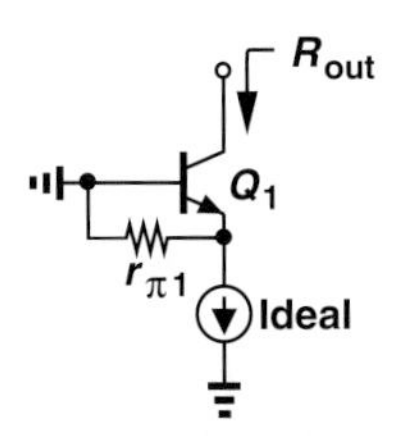

Figure 9.3 Cascode topology using an ideal current source.

Example 9-2 Suppose in Example 9-1, the Early voltage of Q_2 is equal to 50 V.[3] Compare the resulting output impedance of the cascode with the upper bound given by Eq. (9.12).

Solution Since $g_{m1} = (26\ \Omega)^{-1}$, $r_{\pi 1} = 2.6\ \text{k}\Omega$, $r_{O1} = 5\ \text{k}\Omega$, and $r_{O2} = 50\ \text{k}\Omega$, we have

$$R_{out} \approx g_{m1} r_{O1} (r_{O2} || r_{\pi 1}) \tag{9.13}$$

$$\approx 475\ \text{k}\Omega. \tag{9.14}$$

The upper bound is equal to 500 kΩ, about 5% higher.

Exercise Repeat the above example if the Early voltage of Q_1 is 10 V.

Example 9-3 We wish to increase the output resistance of the bipolar cascode of Fig. 9.2(a) by a factor of two through the use of resistive degeneration in the emitter of Q_2. Determine the required value of the degeneration resistor if Q_1 and Q_2 are identical.

Solution As illustrated in Fig. 9.4, we replace Q_2 and R_E with their equivalent resistance from Eq. (9.1):

$$R_{outA} = [1 + g_{m2}(R_E || r_{\pi 2})] r_{O2} + R_E || r_{\pi 2}. \tag{9.15}$$

It follows from Eq. (9.7) that

$$R_{out} \approx g_{m1} r_{O1} (R_{outA} || r_{\pi 1}). \tag{9.16}$$

We wish this value to be twice that given by Eq. (9.7):

$$R_{outA} || r_{\pi 1} = 2(r_{O2} || r_{\pi 1}). \tag{9.17}$$

[3] In integrated circuits, all bipolar transistors fabricated on the same wafer exhibit the same Early voltage. This example applies to discrete implementations.

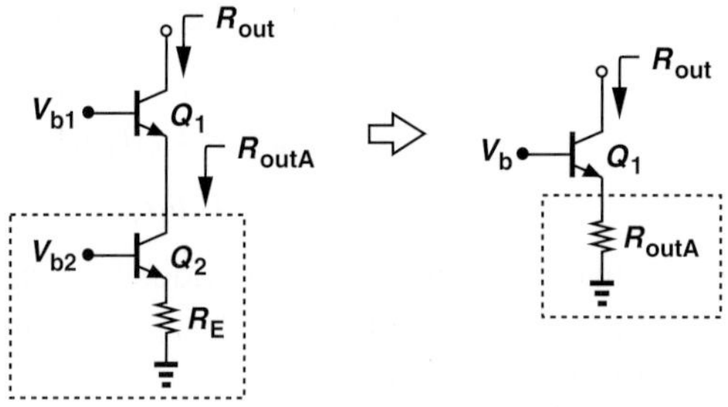

Figure 9.4

That is,

$$R_{outA} = \frac{2r_{O2}r_{\pi1}}{r_{\pi1} - r_{O2}}. \tag{9.18}$$

In practice, $r_{\pi1}$ is typically *less* than r_{O2}, and no positive value of R_{outA} exists! In other words, it is impossible to double the output impedance of the cascode by emitter degeneration.

Exercise Is there a solution if the output impedance must increase by a factor of 1.5?

What does the above result mean? Comparing the output resistances obtained in Examples 9-1 and 9-2, we recognize that even identical transistors yield an R_{out} (= 328.9 kΩ) that is not far from the upper bound (= 500 kΩ). More specifically, the ratio of (9.7) and (9.12) is equal to $r_{O2}/(r_{O2} + r_{\pi1})$, a value greater than 0.5 if $r_{O2} > r_{\pi1}$.

For completeness, Fig. 9.5 shows a *pnp* cascode, where Q_1 serves as the cascode device and Q_2 as the degeneration device. The output impedance is given by Eq. (9.5).

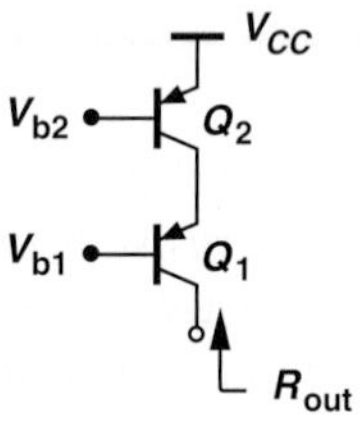

Figure 9.5 *PNP* cascode current source.

While we have arrived at the cascode as an extreme case of emitter degeneration, it is also possible to view the evolution as illustrated in Fig. 9.6. That is, since Q_2 provides only an output impedance of r_{O2}, we "stack" Q_1 *on top of it* to raise R_{out}.

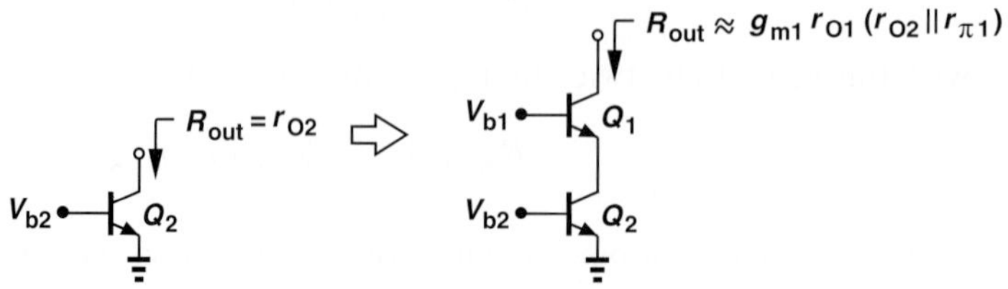

Figure 9.6 Evolution of cascode topology viewed as stacking Q_1 atop Q_2.

Example 9-4 Explain why the topologies depicted in Fig. 9.7 are *not* cascodes.

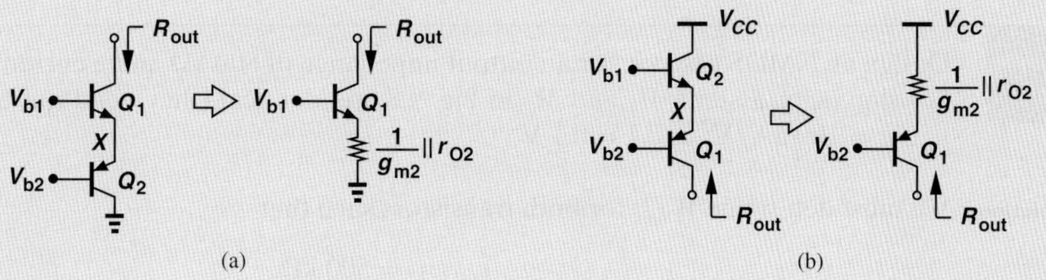

(a) (b)

Figure 9.7

Solution Unlike the cascode of Fig. 9.2(a), the circuits of Fig. 9.7 connect the emitter of Q_1 to the *emitter* of Q_2. Transistor Q_2 now operates as a diode-connected device (rather than a current source), thereby presenting an impedance of $(1/g_{m2})||r_{O2}$ (rather than r_{O2}) at node X. Given by Eq. (9.1), the output impedance, R_{out}, is therefore considerably lower:

$$R_{out} = \left[1 + g_{m1}\left(\frac{1}{g_{m2}}||r_{O2}||r_{\pi 1}\right)\right] r_{O1} + \frac{1}{g_{m2}}||r_{O2}||r_{\pi 1}. \tag{9.19}$$

In fact, since $1/g_{m2} \ll r_{O2}, r_{\pi 1}$ and since $g_{m1} \approx g_{m2}$ (why?),

$$R_{out} \approx \left(1 + \frac{g_{m1}}{g_{m2}}\right) r_{O1} + \frac{1}{g_{m2}} \tag{9.20}$$

$$\approx 2r_{O1}. \tag{9.21}$$

The same observations apply to the topology of Fig. 9.7(b).

Exercise Estimate the output impedance for a collector bias current of 1 mA and $V_A = 8$ V.

MOS Cascodes The similarity of Eqs. (9.1) and (9.3) for degenerated stages suggests that cascoding can also be realized with MOSFETs so as to increase the output impedance of a current source. As illustrated in Fig. 9.8, the idea is to replace the degeneration resistor with a MOS current source, thus presenting a small-signal resistance of r_{O2} from X to ground. Equation (9.3) can now be written as

$$R_{out} = (1 + g_{m1}r_{O2})r_{O1} + r_{O2} \tag{9.22}$$

$$\approx g_{m1}r_{O1}r_{O2}, \tag{9.23}$$

where it is assumed $g_{m1}r_{O1}r_{O2} \gg r_{O1}, r_{O2}$.

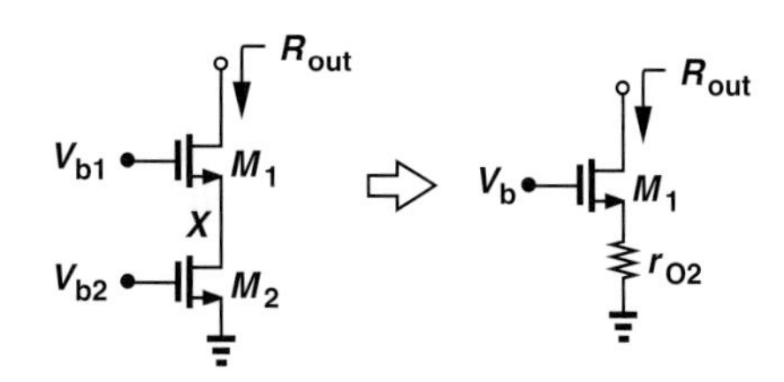

Figure 9.8 MOS cascode current source and its equivalent.

Equation (9.23) is an extremely important result, implying that the output impedance is proportional to the intrinsic gain of the cascode device.

Example 9-5 Design an NMOS cascode for an output impedance of 500 kΩ and a current of 0.5 mA. For simplicity, assume M_1 and M_2 in Fig. 9.8 are identical (they need not be). Assume $\mu_n C_{ox} = 100\ \mu A/V^2$ and $\lambda = 0.1\ V^{-1}$.

Solution We must determine W/L for both transistors such that

$$g_{m1} r_{O1} r_{O2} = 500\ k\Omega. \tag{9.24}$$

Since $r_{O1} = r_{O2} = (\lambda I_D)^{-1} = 20\ k\Omega$, we require that $g_{m1} = (800\ \Omega)^{-1}$ and hence

$$\sqrt{2\mu_n C_{ox} \frac{W}{L} I_D} = \frac{1}{800\ \Omega}. \tag{9.25}$$

It follows that

$$\frac{W}{L} = 15.6. \tag{9.26}$$

We should also note that $g_{m1} r_{O1} = 25 \gg 1$.

Exercise What is the output resistance if $W/L = 32$?

Invoking the alternative view depicted in Fig. 9.6 for the MOS counterpart (Fig. 9.9), we recognize that stacking a MOSFET on top of a current source "boosts" the impedance by a factor of $g_{m2} r_{O2}$ (the intrinsic gain of the cascode transistor). This observation reveals an interesting point of contrast between bipolar and MOS cascodes: in the former, raising r_{O2} eventually leads to $R_{out,bip} = \beta r_{O1}$, whereas in the latter, $R_{out,MOS} = g_{m1} r_{O1} r_{O2}$ increases with no bound.[4] This is because in MOS devices, β and r_π are infinite (at low frequencies).

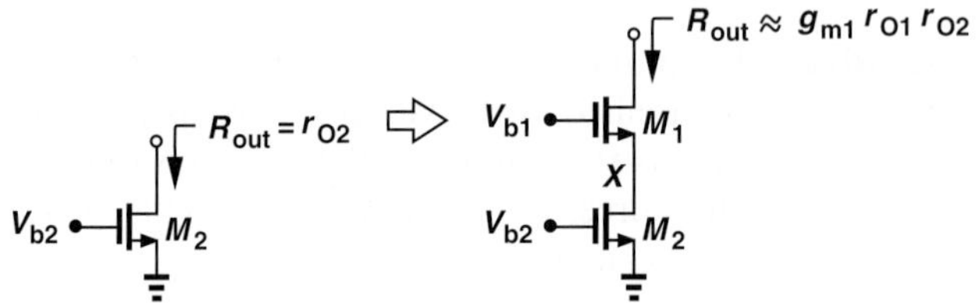

Figure 9.9 MOS cascode viewed as stack of M_1 atop M_2.

Figure 9.10 illustrates a PMOS cascode. The output resistance is given by Eq. (9.22).

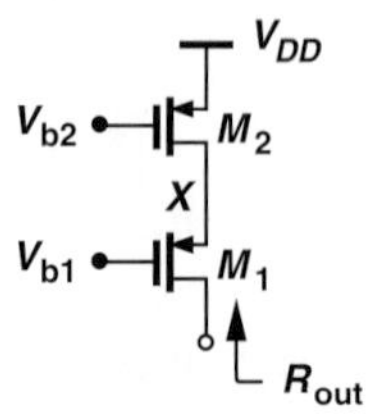

Figure 9.10 PMOS cascode current source.

[4]In reality, other second-order effects limit the output impedance of MOS cascodes.

Example 9-6

During manufacturing, a large parasitic resistor, R_P, has appeared in a cascode as shown in Fig. 9.11. Determine the output resistance.

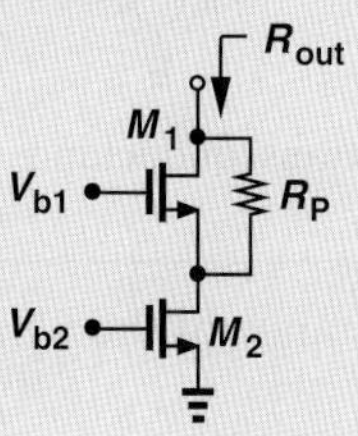

Figure 9.11

Solution We observe that R_P is in parallel with r_{O1}. It is therefore possible to rewrite Eq. (9.23) as

$$R_{out} = g_{m1}(r_{O1}||R_P)r_{O2}. \tag{9.27}$$

If $g_{m1}(r_{O1}||R_P)$ is not much greater than unity, we return to the original equation, (9.22), substituting $r_{O1}||R_P$ for r_{O1}:

$$R_{out} = (1 + g_{m1}r_{O2})(r_{O1}||R_P) + r_{O2}. \tag{9.28}$$

Exercise What value of R_P degrades the output impedance by a factor of two?

9.1.2 Cascode as an Amplifier

In addition to providing a high output impedance as a current source, the cascode topology can also serve as a high-gain amplifier. In fact, the output impedance and the gain of amplifiers are closely related.

For our study below, we need to understand the concept of the transconductance for circuits. In Chapters 4 and 6, we defined the transconductance of a *transistor* as the change in the collector or drain current divided by the change in the base-emitter or gate-source voltage. This concept can be generalized to circuits as well. As illustrated in Fig. 9.12, the output voltage is set to zero by shorting the output node to ground, and the "short-circuit transconductance" of the circuit is defined as

$$G_m = \left.\frac{i_{out}}{v_{in}}\right|_{vout=0}. \tag{9.29}$$

The transconductance signifies the "strength" of a circuit in converting the input voltage to a current.[5] Note the direction of i_{out} in Fig. 9.12.

Did you know?

The cascode can be viewed as a two-transistor circuit, i.e., as a CE/CB cascade. An interesting question is, how many meaningful two-transistor circuit topologies can we realize? Shown below are some: which ones look familiar? Can you think of any others? How about *npn-pnp* combinations?

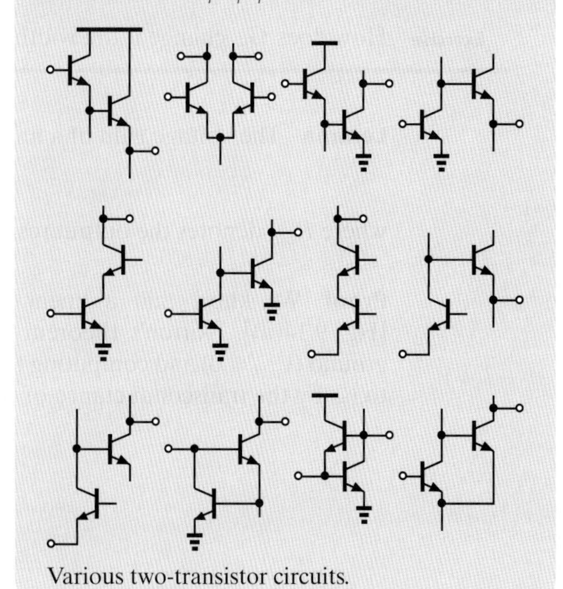

Various two-transistor circuits.

[5]While omitted for simplicity in Chapters 4 and 6, the condition $v_{out} = 0$ is also required for the transconductance of transistors. That is, the collector or drain must by shorted to ac ground.

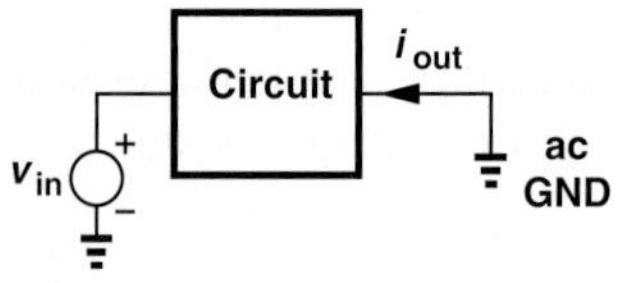

Figure 9.12 Computation of transconductance for a circuit.

Example 9-7 Calculate the transconductance of the CS stage shown in Fig. 9.13(a).

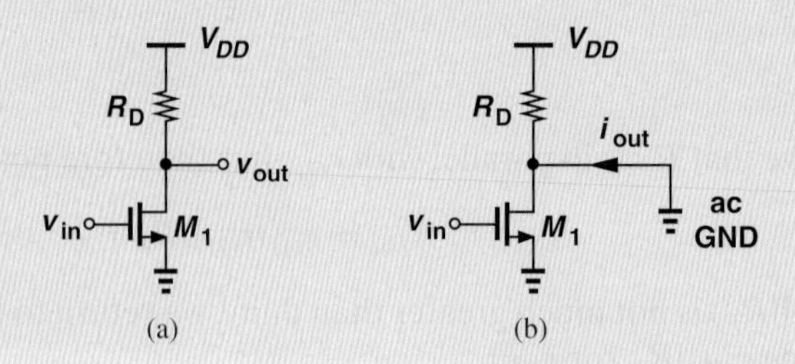

Figure 9.13

Solution As depicted in Fig. 9.13(b), we short the output node to ac ground and, noting that R_D carries no current (why?), write

$$G_m = \frac{i_{out}}{v_{in}} \tag{9.30}$$

$$= \frac{i_{D1}}{v_{GS1}} \tag{9.31}$$

$$= g_{m1}. \tag{9.32}$$

Thus, in this case, the transconductance of the circuit is equal to that of the transistor.

Exercise How does G_m change if the width and bias current of the transistor are doubled?

Lemma The voltage gain of a linear circuit can be expressed as

$$A_v = -G_m R_{out}, \tag{9.33}$$

where R_{out} denotes the output resistance of the circuit (with the input voltage set to zero).

Proof We know that a linear circuit can be replaced with its Norton equivalent [Fig. 9.14(a)]. Norton's theorem states that i_{out} is obtained by shorting the output to ground ($v_{out} = 0$) and computing the short-circuit current [Fig. 9.14(b)]. We also relate i_{out} to v_{in} by the transconductance of the circuit, $G_m = i_{out}/v_{in}$. Thus, in Fig. 9.14(a),

$$v_{out} = -i_{out} R_{out} \tag{9.34}$$

$$= -G_m v_{in} R_{out} \tag{9.35}$$

and hence

$$\frac{v_{out}}{v_{in}} = -G_m R_{out}. \tag{9.36}$$

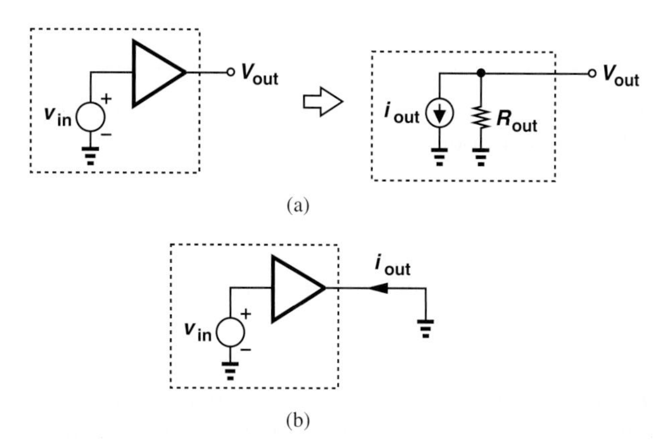

Figure 9.14 (a) Norton equivalent of a circuit, (b) computation of short-circuit output current.

Example 9-8

Determine the voltage gain of the common-emitter stage shown in Fig. 9.15(a).

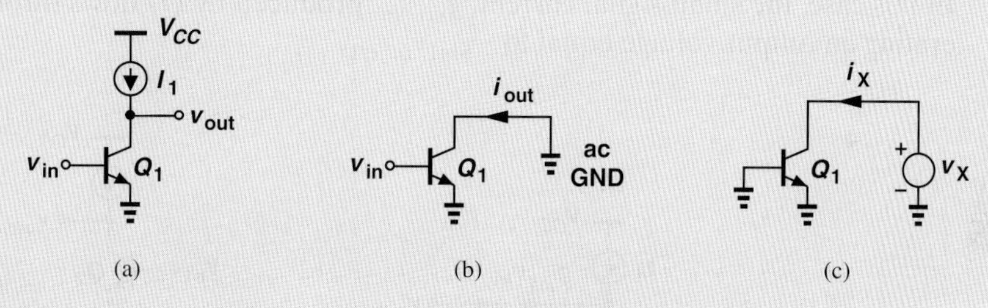

Figure 9.15

Solution To calculate the short-circuit transconductance of the circuit, we place an ac short from the output to ground and find the current through it [Fig. 9.15(b)]. In this case, i_{out} is simply equal to the collector current of Q_1, $g_{m1}v_{in}$, i.e.,

$$G_m = \frac{i_{out}}{v_{in}} \tag{9.37}$$

$$= g_{m1}. \tag{9.38}$$

Note that r_O does not carry a current in this test (why?). Next, we obtain the output resistance as depicted in Fig. 9.15(c):

$$R_{out} = \frac{v_X}{i_X} \tag{9.39}$$

$$= r_{O1}. \tag{9.40}$$

It follows that

$$A_v = -G_m R_{out} \tag{9.41}$$

$$= -g_{m1} r_{O1}. \tag{9.42}$$

Exercise Suppose the transistor is degenerated by an emitter resistor equal to R_E. The transconductance falls but the ouput resistance rises. Does the voltage gain increase or decrease?

The above lemma serves as an alternative method of gain calculation. It also indicates that the voltage gain of a circuit can be increased by raising the *output impedance*, as in cascodes.

Bipolar Cascode Amplifier Recall from Chapter 4 that to maximize the voltage gain of a common-emitter stage, the collector load impedance must be maximized. In the limit, an ideal current source serving as the load [Fig. 9.16(a)] yields a voltage gain of

$$A_v = -g_{m1} r_{O1} \tag{9.43}$$

$$= -\frac{V_A}{V_T}. \tag{9.44}$$

In this case, the small-signal current, $g_{m1} v_{in}$, produced by Q_1 flows through r_{O1}, thus generating an output voltage equal to $-g_{m1} v_{in} r_{O1}$.

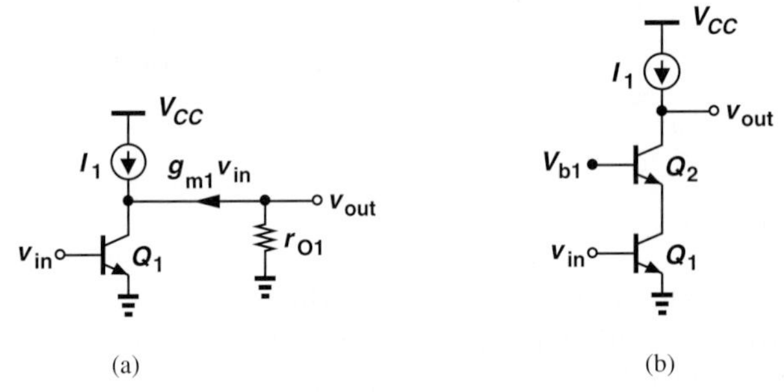

Figure 9.16 (a) Flow of output current generated by a CE stage through r_{O1}, (b) use of cascode to increase the output impedance.

Now, suppose we stack a transistor on top of Q_1 as shown in Fig. 9.16(b). We know from Section 9.1.1 that the circuit achieves a high output impedance and, from the above lemma, a voltage gain higher than that of a CE stage.

Let us determine the voltage gain of the bipolar cascode with the aid of the above lemma. As shown in Fig. 9.17(a), the short-circuit transconductance is equal to i_{out}/v_{in}. As a common-emitter stage, Q_1 still produces a collector current of $g_{m1} v_{in}$, which subsequently flows through Q_2 and hence through the output short:

$$i_{out} = g_{m1} v_{in}. \tag{9.45}$$

That is,

$$G_m = g_{m1}. \tag{9.46}$$

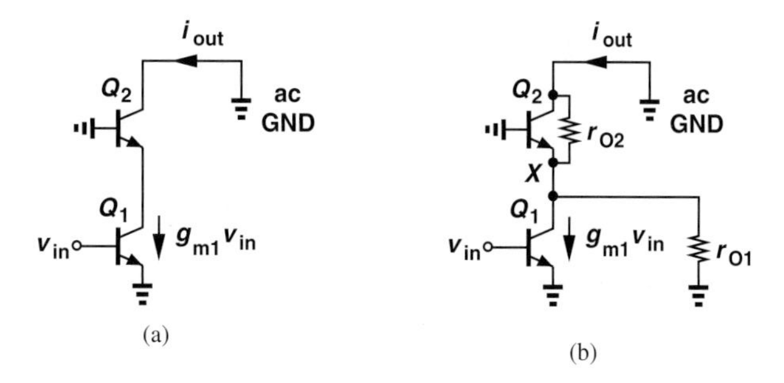

Figure 9.17 (a) Short-circuit output current of a cascode, (b) detailed view of (a).

The reader may view Eq. (9.46) dubiously. After all, as shown in Fig. 9.17(b), the collector current of Q_1 must split between r_{O1} and the impedance seen looking into the emitter of Q_2. We must therefore verify that only a negligible fraction of $g_{m1}v_{in}$ is "lost" in r_{O1}. Since the base and collector voltages of Q_2 are equal, this transistor can be viewed as a diode-connected device having an impedance of $(1/g_{m2})||r_{O2}$. Dividing $g_{m1}v_{in}$ between this impedance and r_{O1}, we have

$$i_{out} = g_{m1}v_{in}\frac{r_{O1}||r_{\pi 2}}{r_{O1}||r_{\pi 2} + \frac{1}{g_{m2}}||r_{O2}}. \tag{9.47}$$

For typical transistors, $1/g_{m2} \ll r_{O2}, r_{O1}$, and hence

$$i_{out} \approx g_{m1}v_{in}. \tag{9.48}$$

That is, the approximation $G_m = g_{m1}$ is reasonable.

To obtain the overall voltage gain, we write from Eqs. (9.33) and (9.5),

$$A_v = -G_m R_{out} \tag{9.49}$$

$$= -g_{m1}\{[1 + g_{m2}(r_{O1}||r_{\pi 2})]r_{O2} + r_{O1}||r_{\pi 2}\} \tag{9.50}$$

$$\approx -g_{m1}[g_{m2}(r_{O1}||r_{\pi 2})r_{O2} + r_{O1}||r_{\pi 2}]. \tag{9.51}$$

Also, since Q_1 and Q_2 carry approximately equal bias currents, $g_{m1} \approx g_{m2}$ and $r_{O1} \approx r_{O2}$:

$$A_v = -g_{m1}r_{O1}[g_{m1}(r_{O1}||r_{\pi 2}) + 1] \tag{9.52}$$

$$\approx -g_{m1}r_{O1}g_{m1}(r_{O1}||r_{\pi 2}). \tag{9.53}$$

Compared to the simple CE stage of Fig. 9.16(a), the cascode amplifier exhibits a gain that is higher by a factor of $g_{m1}(r_{O1}||r_{\pi 2})$—a relatively large value because r_{O1} and $r_{\pi 2}$ are much greater than $1/g_{m1}$.

Example 9-9

The bipolar cascode of Fig. 9.16(b) is biased at a current of 1 mA. If $V_A = 5$ V and $\beta = 100$ for both transistors, determine the voltage gain. Assume the load is an ideal current source.

Solution We have $g_{m1} = (26\ \Omega)^{-1}$, $r_{\pi 1} \approx r_{\pi 2} \approx 2600\ \Omega$, $r_{O1} \approx r_{O2} = 5\ \text{k}\Omega$. Thus,

$$g_{m1}(r_{O1}||r_{\pi 2}) = 65.8 \tag{9.54}$$

and from Eq. (9.53),

$$|A_v| = 12{,}654. \tag{9.55}$$

Cascoding thus raises the voltage gain by a factor of 65.8.

Exercise What Early voltage gives a gain of 5,000?

It is possible to view the cascode amplifier as a common-emitter stage followed by a common-base stage. As illustrated in Fig. 9.18, the idea is to consider the cascode device, Q_2, as a common-base transistor that senses the small-signal current produced by Q_1. This perspective may prove useful in some cases.

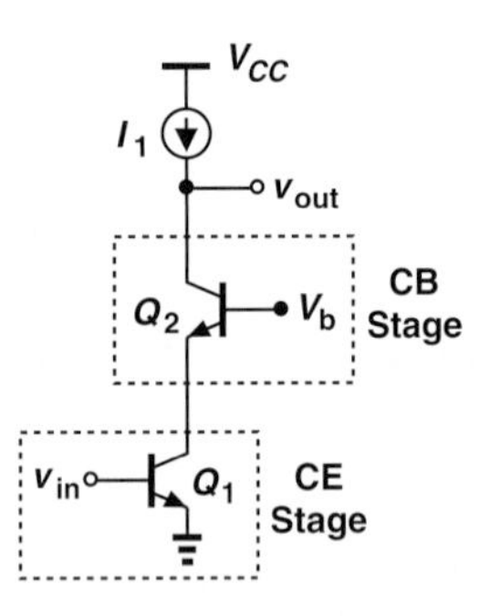

Figure 9.18 Cascode amplifier as a cascade of a CE stage and a CB stage.

The high voltage gain of the cascode topology makes it attractive for many applications. But, in the circuit of Fig. 9.16(b), the load is assumed to be an ideal current source. An actual current source lowers the impedance seen at the output node and hence the voltage gain. For example, the circuit illustrated in Fig. 9.19(a) suffers from a low gain because the *pnp* current source introduces an impedance of only r_{O3} from the output node to ac ground, dropping the output impedance to

$$R_{out} = r_{O3}||\{[1 + g_{m2}(r_{O1}||r_{\pi 2})]r_{O2} + r_{O1}||r_{\pi 2}\} \tag{9.56}$$

$$\approx r_{O3}||[g_{m2}r_{O2}(r_{O1}||r_{\pi 2}) + r_{O1}||r_{\pi 2}]. \tag{9.57}$$

How should we realize the load current source to maintain a high gain? We know from Section 9.1.1 that cascoding also raises the output impedance of current sources, postulating that the circuit of Fig. 9.5 is a good candidate and arriving at the stage depicted in Fig. 9.19(b). The output impedance is now given by the parallel combination of those of the *npn* and *pnp* cascodes, R_{on} and R_{op}, respectively. Using Eq. (9.7), we have

$$R_{on} \approx g_{m2}r_{O2}(r_{O1}||r_{\pi 2}) \tag{9.58}$$

$$R_{op} \approx g_{m3}r_{O3}(r_{O4}||r_{\pi 3}). \tag{9.59}$$

Note that, since *npn* and *pnp* devices may display different Early voltages, r_{O1} $(= r_{O2})$ may not be equal to r_{O3} $(= r_{O4})$.

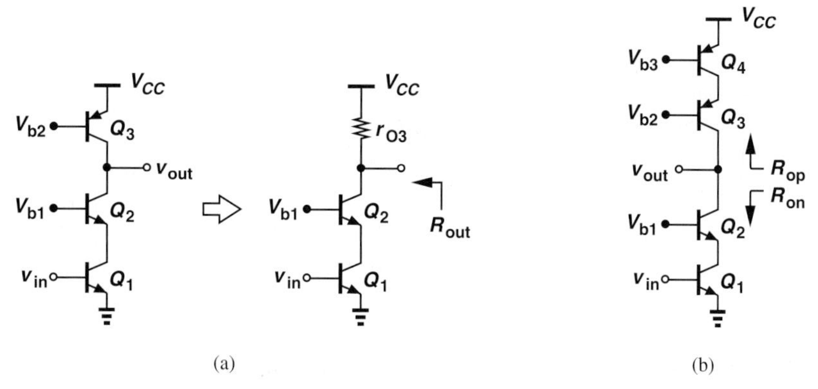

Figure 9.19 (a) Cascode with a simple current-source load, (b) use of cascode in the load to raise the voltage gain.

Recognizing that the short-circuit transconductance, G_m, of the stage is still approximately equal to g_{m1} (why?), we express the voltage gain as

$$A_v = -g_{m1}(R_{on}||R_{op}) \tag{9.60}$$

$$\approx -g_{m1}\{[g_{m2}r_{O2}(r_{O1}||r_{\pi 2})]||[g_{m3}r_{O3}(r_{O4}||r_{\pi 3})]\}. \tag{9.61}$$

This result represents the highest voltage gain that can be obtained in a cascode stage. For comparable values of R_{on} and R_{op}, this gain is about half of that expressed by Eq. (9.53).

Example 9-10 Suppose the circuit of Example 9-9 incorporates a cascode load using *pnp* transistors with $V_A = 4$ V and $\beta = 50$. What is the voltage gain?

Solution The load transistors carry a collector current of approximately 1 mA. Thus,

$$R_{op} = g_{m3}r_{O3}(r_{O4}||r_{\pi 3}) \tag{9.62}$$

$$= 151 \text{ k}\Omega \tag{9.63}$$

and

$$R_{on} = 329 \text{ k}\Omega. \tag{9.64}$$

It follows that

$$|A_v| = g_{m1}(R_{on}||R_{op}) \tag{9.65}$$

$$= 3{,}981. \tag{9.66}$$

Compared to the ideal current source case, the gain has fallen by approximately a factor of 3 because the *pnp* devices suffer from a lower Early voltage and β.

Exercise Repeat the above example for a collector bias current of 0.5 mA.

Did you know?

The cascode stage was originally invented to suppress an undesirable high-frequency phenomenon called the "Miller effect" (Chapter 11). In fact, many discrete bipolar amplifiers still employ cascoding for the same reason. However, it was recognized in early days of CMOS technology that cascodes could also provide a high output impedance and a high voltage gain, hence their widespread use.

It is important to take a step back and appreciate our analysis techniques. The cascode of Fig. 9.19(b) proves quite formidable if we attempt to replace each transistor with its small-signal model and solve the resulting circuit. Our gradual approach to constructing this stage reveals the role of each device, allowing straightforward calculation of the output impedance. Moreover, the lemma illustrated in Fig. 9.14 utilizes our knowledge of the output impedance to quickly provide the voltage gain of the stage.

CMOS Cascode Amplifier The foregoing analysis of the bipolar cascode amplifier can readily be extended to the CMOS counterpart. Depicted in Fig. 9.20(a) with an ideal current-source load, this stage also provides a short-circuit transconductance $G_m \approx g_{m1}$ if $1/g_{m2} \ll r_{O1}$. The output resistance is given by Eq. (9.22), yielding a voltage gain of

$$A_v = -G_m R_{out} \tag{9.67}$$

$$\approx -g_{m1}[(1 + g_{m2}r_{O2})r_{O1} + r_{O2}] \tag{9.68}$$

$$\approx -g_{m1}r_{O1}g_{m2}r_{O2}. \tag{9.69}$$

In other words, compared to a simple common-source stage, the voltage gain has risen by a factor of $g_{m2}r_{O2}$ (the intrinsic gain of the cascode device). Since β and r_π are infinite for MOS devices (at low frequencies), we can also utilize Eq. (9.53) to arrive at Eq. (9.69). Note, however, that M_1 and M_2 need not exhibit equal transconductances or output resistances (their widths and lengths need not be the same) even though they carry equal currents (why?).

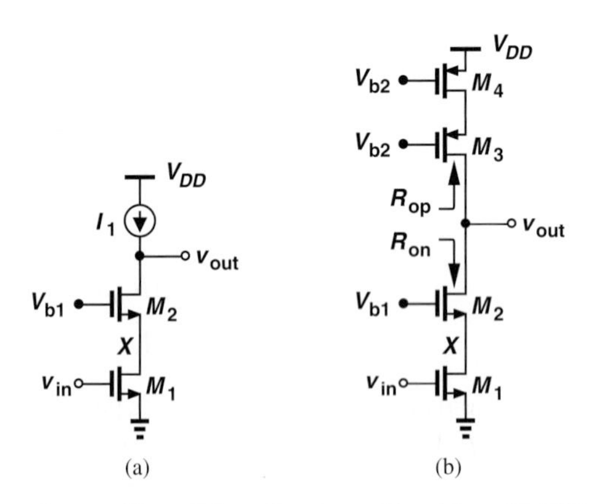

Figure 9.20 (a) MOS cascode amplifier, (b) realization of load by a PMOS cascode.

As with the bipolar counterpart, the MOS cascode amplifier must incorporate a cascode PMOS current source so as to maintain a high voltage gain. As illustrated in Fig. 9.20(b), the circuit exhibits the following output impedance components:

$$R_{on} \approx g_{m2} r_{O2} r_{O1} \tag{9.70}$$

$$R_{op} \approx g_{m3} r_{O3} r_{O4}. \tag{9.71}$$

The voltage gain is therefore equal to

$$A_v \approx -g_{m1}[(g_{m2} r_{O2} r_{O1}) || (g_{m3} r_{O3} r_{O4})]. \tag{9.72}$$

Example 9-11 The cascode amplifier of Fig. 9.20(b) incorporates the following device parameters: $(W/L)_{1,2} = 30$, $(W/L)_{3,4} = 40$, $I_{D1} = \cdots = I_{D4} = 0.5$ mA. If $\mu_n C_{ox} = 100\ \mu\text{A/V}^2$, $\mu_p C_{ox} = 50\ \mu\text{A/V}^2$, $\lambda_n = 0.1\ \text{V}^{-1}$, and $\lambda_p = 0.15\ \text{V}^{-1}$, determine the voltage gain.

Solution With the particular choice of device parameters here, $g_{m1} = g_{m2}$, $r_{O1} = r_{O2}$, $g_{m3} = g_{m4}$, and $r_{O3} = r_{O4}$. We have

$$g_{m1,2} = \sqrt{2\mu_n C_{ox} \left(\frac{W}{L}\right)_{1,2} I_{D1,2}} \tag{9.73}$$

$$= (577\ \Omega)^{-1} \tag{9.74}$$

and

$$g_{m3,4} = (707\ \Omega)^{-1}. \tag{9.75}$$

Also,

$$r_{O1,2} = \frac{1}{\lambda_n I_{D1,2}} \tag{9.76}$$

$$= 20\ \text{k}\Omega \tag{9.77}$$

and

$$r_{O3,4} = 13.3\ \text{k}\Omega. \tag{9.78}$$

Equations (9.70) and (9.71) thus respectively give

$$R_{on} \approx 693\ \text{k}\Omega \tag{9.79}$$

$$R_{op} \approx 250\ \text{k}\Omega \tag{9.80}$$

and

$$A_v = -g_{m1}(R_{on} || R_{op}) \tag{9.81}$$

$$\approx -318. \tag{9.82}$$

Exercise Explain why a lower bias current results in a higher output impedance in the above example. Calculate the output impedance for a drain current of 0.25 mA.

Robotics Application: the GPS

The Global Positioning System (GPS) is gradually finding its way into robotics. If used in outdoor applications, e.g., in agriculture, robots must incorporate GPS for navigation. An interesting case is where drones fly over a farm to detect plants damaged by pests and then robots are deployed to reach those plants and apply pesticide to them.

The GPS consists of three satellites located in the earth's atmosphere. The GPS receiver (RX) in a robot picks up three signals from the satellites and, based on their respective delays or phases, computes the location. Owing to the vast distance across which they travel, the satellite signals are heavily attenuated as they reach the receiver. To detect such weak signals, the RX must add minimal "noise," hence the need for a "low-noise amplifier" (LNA) [Fig. 9.21(a)].

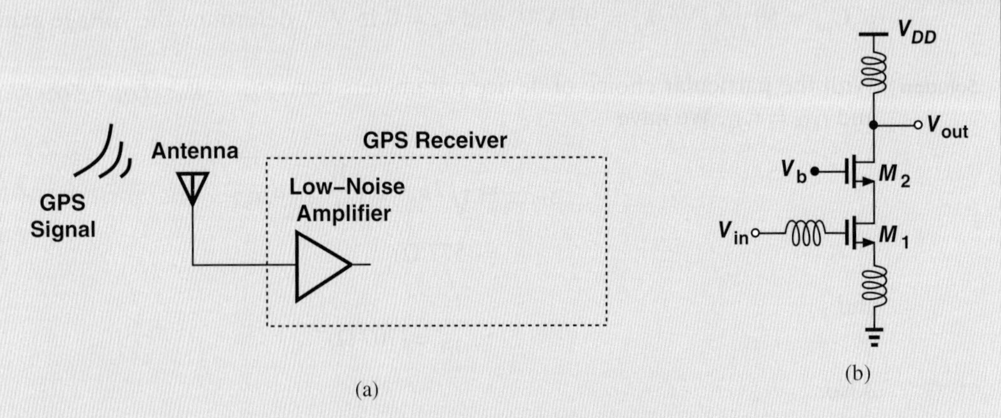

Figure 9.21 (a) GPS receiver and (b) cascode low-noise amplifier.

A common LNA topology employs the cascode structure, as shown in Fig. 9.21(b). The role of the inductors is explained in more advanced courses. Our study in this chapter has arrived at the cascode by seeking a higher output impedance, but in this LNA, we rely on M_2 to *stabilize* the circuit, i.e., to avoid oscillations.

9.2 CURRENT MIRRORS

9.2.1 Initial Thoughts

The biasing techniques studied for bipolar and MOS amplifiers in Chapters 4 and 6 prove inadequate for high-performance microelectronic circuits. For example, the bias current of CE and CS stages is a function of the supply voltage—a serious issue because in practice, this voltage experiences some variation. The rechargeable battery in a cellphone or laptop computer, for example, gradually loses voltage as it is discharged, thereby mandating that the circuits operate properly across a *range* of supply voltages.

Another critical issue in biasing relates to ambient temperature variations. A cellphone must maintain its performance at –20°C in Finland and +50°C in Saudi Arabia. To understand how temperature affects the biasing, consider the bipolar current source shown in Fig. 9.22(a), where R_1 and R_2 divide V_{CC} down to the required V_{BE}. That is, for a desired current I_1, we have

$$\frac{R_2}{R_1 + R_2} V_{CC} = V_T \ln \frac{I_1}{I_S}, \tag{9.83}$$

where the base current is neglected. But, what happens if the temperature varies? The left-hand side remains constant if the resistors are made of the same material and hence vary by the same percentage. The right-hand side, however, contains two temperature-dependent parameters: $V_T = kT/q$ and I_S. Thus, even if the base-emitter voltage remains constant with temperature, I_1 does not.

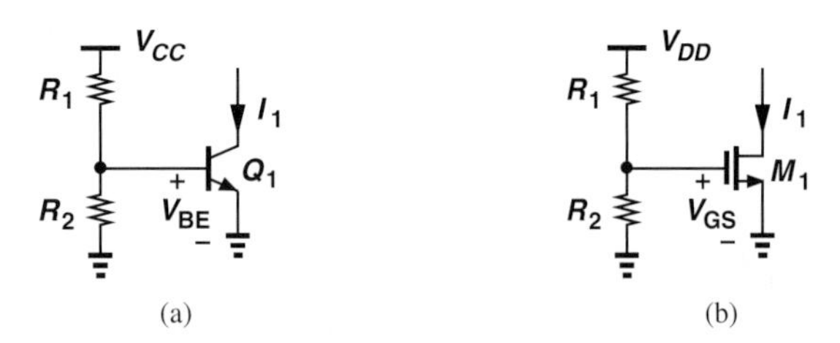

Figure 9.22 Impractical biasing of (a) bipolar and (b) MOS current sources.

A similar situation arises in CMOS circuits. As illustrated in Fig. 9.22(b), a MOS current source biased by means of a resistive divider suffers from dependence on V_{DD} and temperature. Here, we can write

$$I_1 = \frac{1}{2}\mu_n C_{ox}\frac{W}{L}(V_{GS} - V_{TH})^2 \tag{9.84}$$

$$= \frac{1}{2}\mu_n C_{ox}\frac{W}{L}\left(\frac{R_2}{R_1 + R_2}V_{DD} - V_{TH}\right)^2. \tag{9.85}$$

Since both the mobility and the threshold voltage vary with temperature, I_1 is not constant even if V_{GS} is.

In summary, the typical biasing schemes introduced in Chapters 4 and 6 fail to establish a constant collector or drain current if the supply voltage or the ambient temperature are subject to change. Fortunately, an elegant method of creating supply- and temperature-independent voltages and currents exists and appears in almost all microelectronic systems. Called the "bandgap reference circuit" and employing several tens of devices, this scheme is studied in more advanced books [1].

The bandgap circuit by itself does not solve all of our problems! An integrated circuit may incorporate hundreds of current sources, e.g., as the load impedance of CE or CS stages to achieve a high gain. Unfortunately, the complexity of the bandgap prohibits its use for each current source in a large integrated circuit.

Let us summarize our thoughts thus far. In order to avoid supply and temperature dependence, a bandgap reference can provide a "golden current" while requiring a few tens of devices. We must therefore seek a method of "copying" the golden current without duplicating the entire bandgap circuitry. Current mirrors serve this purpose.

Figure 9.23 conceptually illustrates our goal here. The golden current generated by a bandgap reference is "read" by the current mirror and a copy having the same characteristics as those of I_{REF} is produced. For example, $I_{copy} = I_{REF}$ or $2I_{REF}$.

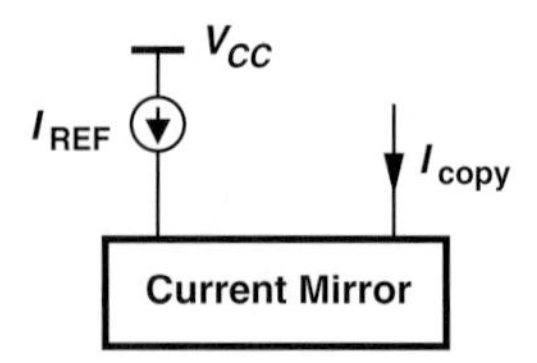

Figure 9.23 Concept of current mirror.

9.2.2 Bipolar Current Mirror

Since the current source generating I_{copy} in Fig. 9.23 must be implemented as a bipolar or MOS transistor, we surmise that the current mirror resembles the topology shown in Fig. 9.24(a), where Q_1 operates in the forward active region and the black box guarantees $I_{copy} = I_{REF}$ regardless of temperature or transistor characteristics. (The MOS counterpart is similar.)

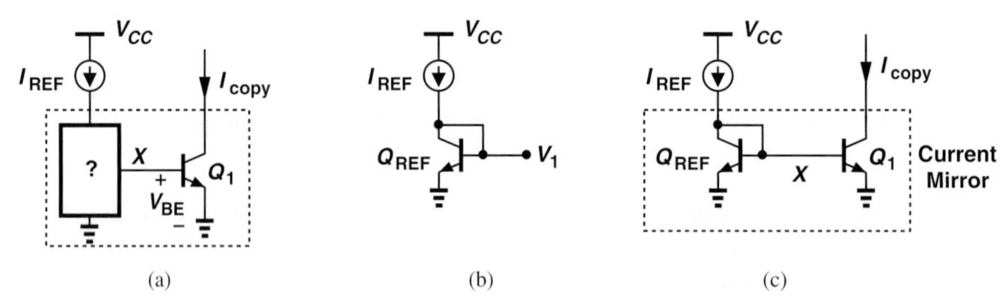

Figure 9.24 (a) Conceptual illustration of current copying, (b) voltage proportional to natural logarithm of current, (c) bipolar current mirror.

How should the black box of Fig. 9.24(a) be realized? The black box generates an output voltage, $V_X(= V_{BE})$, such that Q_1 carries a current equal to I_{REF}:

$$I_{S1} \exp \frac{V_X}{V_T} = I_{REF}, \tag{9.86}$$

where the Early effect is neglected. Thus, the black box satisfies the following relationship:

$$V_X = V_T \ln \frac{I_{REF}}{I_{S1}}. \tag{9.87}$$

We must therefore seek a circuit whose output voltage is proportional to the natural logarithm of its input, i.e., the inverse function of bipolar transistor characteristics. Fortunately, a single diode-connected device satisfies Eq. (9.87). Neglecting the base-current in Fig. 9.24(b), we have

$$V_1 = V_T \ln \frac{I_{REF}}{I_{S,REF}}, \tag{9.88}$$

where $I_{S,REF}$ denotes the reverse saturation current of Q_{REF}. In other words, $V_1 = V_X$ if $I_{S,REF} = I_{S1}$, i.e., if Q_{REF} is identical to Q_1.

Figure 9.24(c) consolidates our thoughts, displaying the current mirror circuitry. We say Q_1 "mirrors" or copies the current flowing through Q_{REF}. For now, we neglect the base currents. From one perspective, Q_{REF} takes the natural logarithm of I_{REF} and Q_1 takes the exponential of V_X, thereby yielding $I_{copy} = I_{REF}$. From another perspective, since Q_{REF} and Q_1 have equal base-emitter voltages, we can write

$$I_{REF} = I_{S,REF} \exp \frac{V_X}{V_T} \tag{9.89}$$

$$I_{copy} = I_{S1} \exp \frac{V_X}{V_T} \tag{9.90}$$

and hence

$$I_{copy} = \frac{I_{S1}}{I_{S,REF}} I_{REF}, \tag{9.91}$$

which reduces to $I_{copy} = I_{REF}$ if Q_{REF} and Q_1 are identical. This holds even though V_T and I_S vary with temperature. Note that V_X does vary with temperature but in such a way that I_{copy} does not.

Example 9-12 An electrical engineering student who is excited by the concept of the current mirror constructs the circuit but forgets to tie the base of Q_{REF} to its collector (Fig. 9.25). Explain what happens.

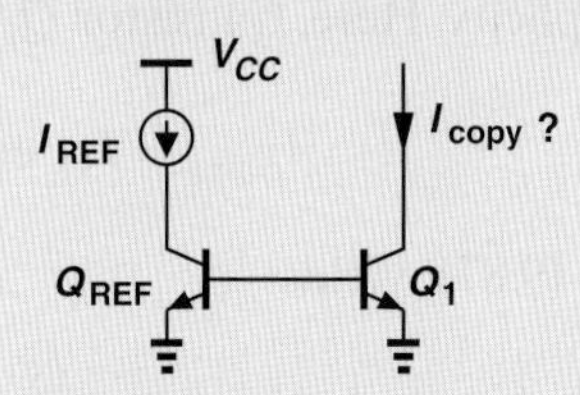

Figure 9.25

Solution The circuit provides no path for the base currents of the transistors. More fundamentally, the base-emitter voltage of the devices is not defined. The lack of the base currents translates to $I_{copy} = 0$.

Exercise What is the region of operation of Q_{REF}?

Example 9-13 Realizing the mistake in the above circuit, the student makes the modification shown in Fig. 9.26, hoping that the battery V_X provides the base currents and defines the base-emitter voltage of Q_{REF} and Q_1. Explain what happens.

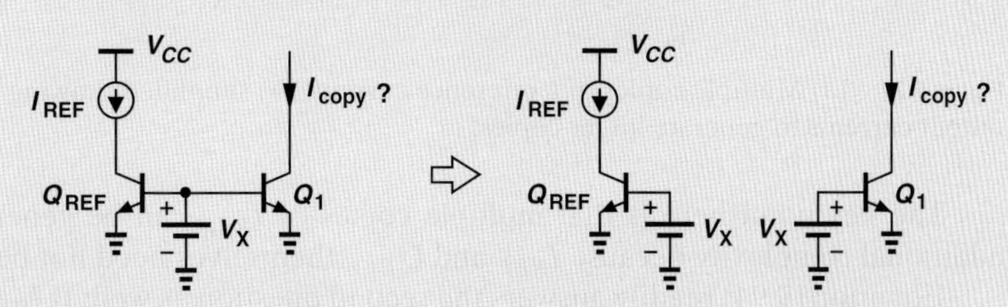

Figure 9.26

Solution While Q_1 now carries a finite current, the biasing of Q_1 is no different from that in Fig. 9.22; i.e.,

$$I_{copy} = I_{S1} \exp \frac{V_X}{V_T}, \tag{9.92}$$

which is a function of temperature if V_X is constant. The student has forgotten that a diode-connected device is necessary here to ensure that V_X remains proportional to $\ln(I_{REF}/I_{S,REF})$.

Exercise Suppose V_X is slightly greater than the necessary value, $V_T \ln(I_{REF}/I_{S,REF})$. In what region does Q_{REF} operate?

We must now address two important questions. First, how do we make additional copies of I_{REF} to feed different parts of an integrated circuit? Second, how do we obtain different values for these copies, e.g., $2I_{REF}$, $5I_{REF}$, etc.? Considering the topology in Fig. 9.23(c), we recognize that V_X can serve as the base-emitter voltage of multiple transistors, thus arriving at the circuit shown in Fig. 9.27(a). The circuit is often drawn as in Fig. 9.27(b) for simplicity. Here, transistor Q_j carries a current $I_{copy,j}$, given by

$$I_{copy,j} = I_{S,j} \exp \frac{V_X}{V_T}, \tag{9.93}$$

which, along with Eq. (9.87), yields

$$I_{copy,j} = \frac{I_{S,j}}{I_{S,REF}} I_{REF}. \tag{9.94}$$

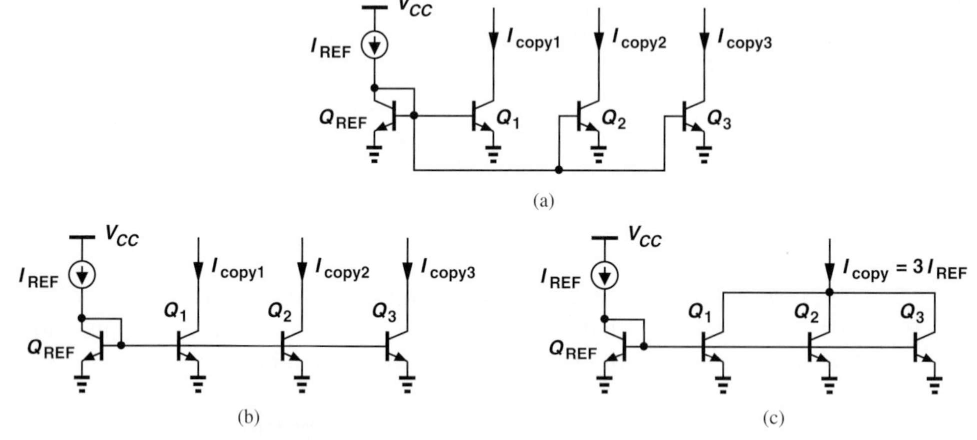

Figure 9.27 (a) Multiple copies of a reference current, (b) simplified drawing of (a), (c) combining output currents to generate larger copies.

The key point here is that multiple copies of I_{REF} can be generated with minimal additional complexity because I_{REF} and Q_{REF} themselves need not be duplicated.

Equation (9.94) readily answers the second question as well: If $I_{S,j}$ ($\propto$ the emitter area of Q_j) is chosen to be n times $I_{S,REF}$ ($\propto$ the emitter area of Q_{REF}), then $I_{copy,j} = nI_{REF}$. We say the copies are "scaled" with respect to I_{REF}. Recall from Chapter 4 that this is equivalent to placing n unit transistors in parallel. Figure 9.27(c) depicts an example where Q_1-Q_3 are identical to Q_{REF}, providing $I_{copy} = 3I_{REF}$.

Example 9-14 A multistage amplifier incorporates two current sources of values 0.75 mA and 0.5 mA. Using a bandgap reference current of 0.25 mA, design the required current sources. Neglect the effect of the base current for now.

Solution Figure 9.28 illustrates the circuit. Here, all transistors are identical to ensure proper scaling of I_{REF}.

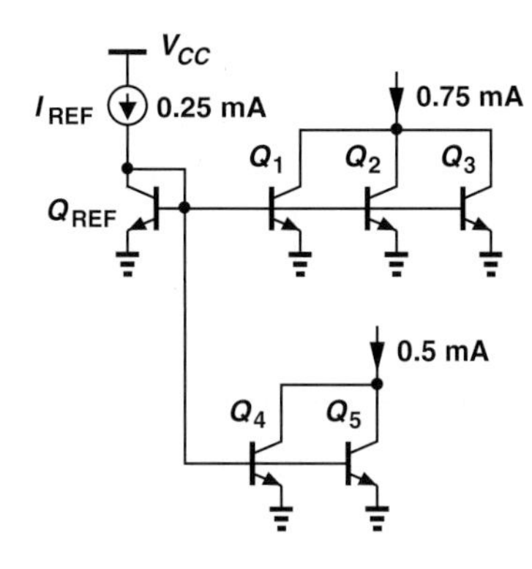

Figure 9.28

Exercise Repeat the above example if the bandgap reference current is 0.1 mA.

The use of multiple transistors in parallel provides an accurate means of scaling the reference in current mirrors. But, how do we create *fractions* of I_{REF}? This is accomplished by realizing Q_{REF} itself as multiple parallel transistors. Exemplified by the circuit in Fig. 9.29, the idea is to begin with a larger $I_{S,REF}$ ($= 3I_S$ here) so that a unit transistor, Q_1, can generate a smaller current. Repeating the expressions in Eqs. (9.89) and (9.90), we have

$$I_{REF} = 3I_S \exp \frac{V_X}{V_T} \tag{9.95}$$

$$I_{copy} = I_S \exp \frac{V_X}{V_T} \tag{9.96}$$

and hence

$$I_{copy} = \frac{1}{3} I_{REF}. \tag{9.97}$$

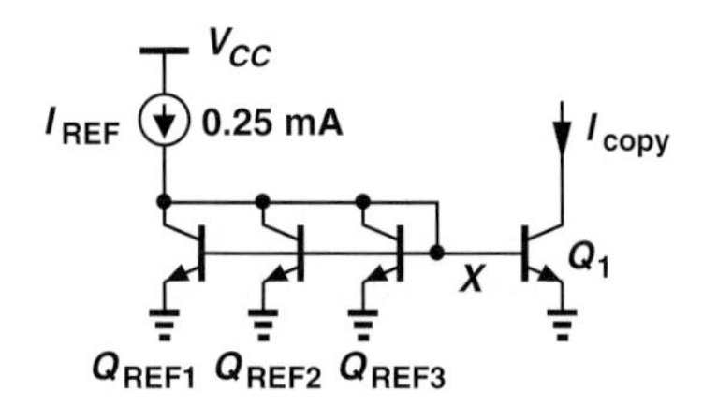

Figure 9.29 Copying a fraction of a reference current.

Example 9-15 It is desired to generate two currents equal to 50 μA and 500 μA from a reference of 200 μA. Design the current mirror circuit.

Solution To produce the smaller current, we must employ four unit transistors for Q_{REF} such that each carries 50 μA. A unit transistor thus generates 50 μA (Fig. 9.30). The current of 500 μA requires 10 unit transistors, denoted by $10A_E$ for simplicity.

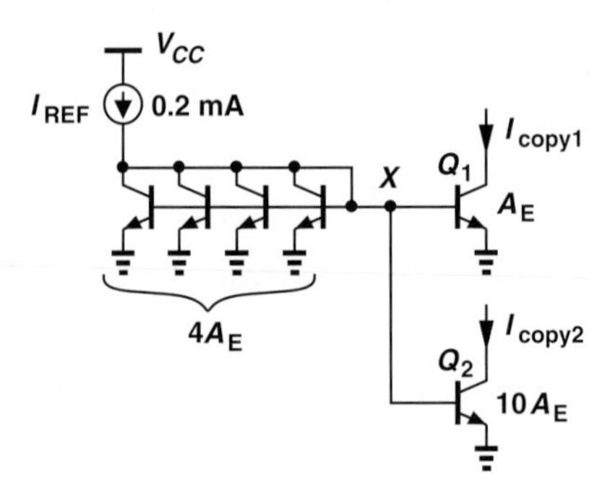

Figure 9.30

Exercise Repeat the above example for a reference current of 150 μA.

Effect of Base Current We have thus far neglected the base current drawn from node X in Fig. 9.27(a) by all transistors, an effect leading to a significant error as the number of copies (i.e., the total copied current) increases. The error arises because a fraction of I_{REF} flows through the bases rather than through the collector of Q_{REF}. We analyze the error with the aid of the diagram shown in Fig. 9.31, where A_E and nA_E denote one unit transistor and n unit transistors, respectively. Our objective is to calculate I_{copy}, recognizing that Q_{REF} and Q_1 still have equal base-emitter voltages and hence carry currents with a ratio of n. Thus, the base currents of Q_1 and Q_{REF} can be expressed as

$$I_{B1} = \frac{I_{copy}}{\beta} \tag{9.98}$$

$$I_{B,REF} = \frac{I_{copy}}{\beta} \cdot \frac{1}{n}. \tag{9.99}$$

Writing a KCL at X therefore yields

$$I_{REF} = I_{C,REF} + \frac{I_{copy}}{\beta} \cdot \frac{1}{n} + \frac{I_{copy}}{\beta}, \tag{9.100}$$

which, since $I_{C,REF} = I_{copy}/n$, leads to

$$I_{copy} = \frac{nI_{REF}}{1 + \frac{1}{\beta}(n+1)}. \tag{9.101}$$

For a large β and moderate n, the second term in the denominator is much less than unity and $I_{copy} \approx nI_{REF}$. However, as the copied current ($\propto n$) increases, so does the error in I_{copy}.

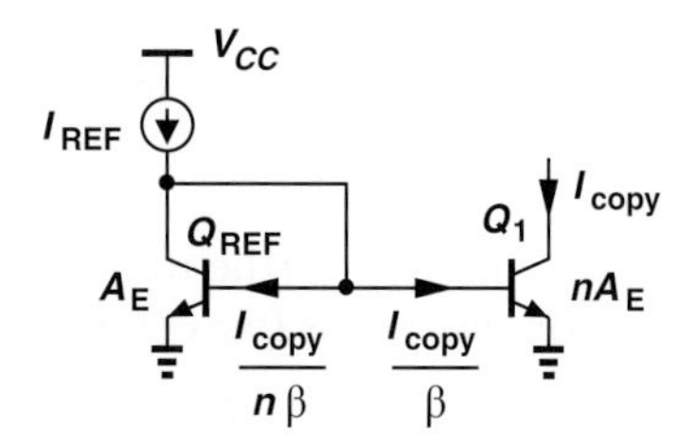

Figure 9.31 Error due to base currents.

To suppress the above error, the bipolar current mirror can be modified as illustrated in Fig. 9.32. Here, emitter follower Q_F is interposed between the collector of Q_{REF} and node X, thereby reducing the effect of the base currents by a factor of β. More specifically, assuming $I_{C,F} \approx I_{E,F}$, we can repeat the above analysis by writing a KCL at X:

$$I_{C,F} = \frac{I_{copy}}{\beta} + \frac{I_{copy}}{\beta} \cdot \frac{1}{n}, \tag{9.102}$$

obtaining the base current of Q_F as

$$I_{B,F} = \frac{I_{copy}}{\beta^2}\left(1 + \frac{1}{n}\right). \tag{9.103}$$

Another KCL at node P gives

$$I_{REF} = I_{B,F} + I_{C,REF} \tag{9.104}$$

$$= \frac{I_{copy}}{\beta^2}\left(1 + \frac{1}{n}\right) + \frac{I_{copy}}{n} \tag{9.105}$$

and hence

$$I_{copy} = \frac{nI_{REF}}{1 + \frac{1}{\beta^2}(n+1)}. \tag{9.106}$$

That is, the error is lowered by a factor of β.*

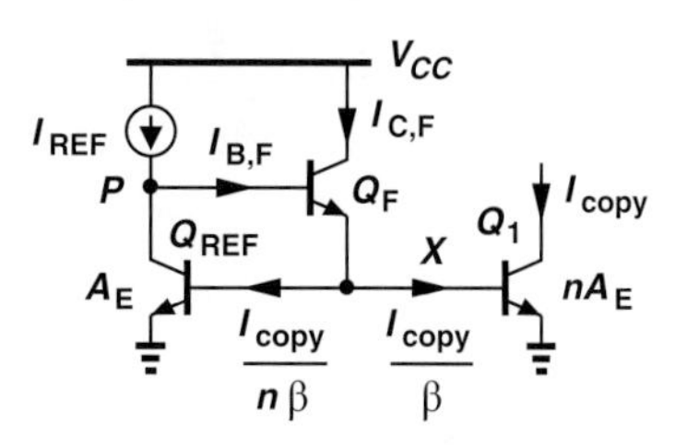

Figure 9.32 Addition of emitter follower to reduce error due to base currents.

*In more advanced designs, a constant current is drawn from X.

Example 9-16 Compute the error in I_{copy1} and I_{copy2} in Fig. 9.30 before and after adding a follower.

Solution Noting that I_{copy1}, I_{copy2}, and $I_{C,REF}$ (the total current flowing through four unit transistors) still retain their nominal ratios (why?), we write a KCL at X:

$$I_{REF} = I_{C,REF} + \frac{I_{copy1}}{\beta} + \frac{I_{copy2}}{\beta} + \frac{I_{C,REF}}{\beta} \tag{9.107}$$

$$= 4I_{copy1} + \frac{I_{copy1}}{\beta} + \frac{10I_{copy1}}{\beta} + \frac{I_{C,REF}}{\beta}. \tag{9.108}$$

Thus,

$$I_{copy1} = \frac{I_{REF}}{4 + \frac{15}{\beta}} \tag{9.109}$$

$$I_{copy2} = \frac{10I_{REF}}{4 + \frac{15}{\beta}}. \tag{9.110}$$

With the addition of emitter follower (Fig. 9.33), we have at X:

$$I_{C,F} = \frac{I_{C,REF}}{\beta} + \frac{I_{copy1}}{\beta} + \frac{I_{copy2}}{\beta} \tag{9.111}$$

$$= \frac{4I_{copy1}}{\beta} + \frac{I_{copy1}}{\beta} + \frac{10I_{copy1}}{\beta} \tag{9.112}$$

$$= \frac{15I_{copy1}}{\beta}. \tag{9.113}$$

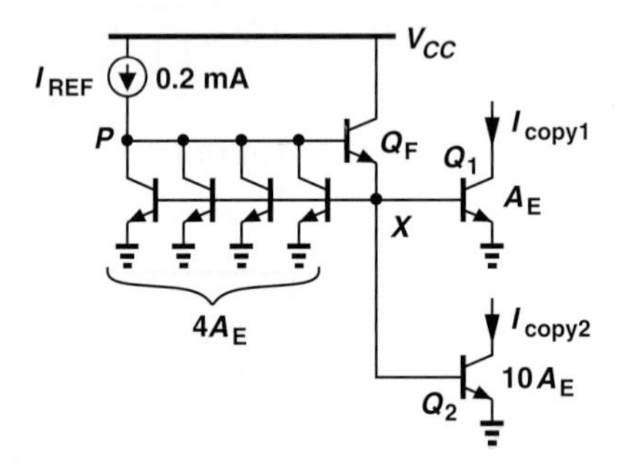

Figure 9.33

A KCL at P therefore yields

$$I_{REF} = \frac{15I_{copy1}}{\beta^2} + I_{C,REF} \tag{9.114}$$

$$= \frac{15I_{copy1}}{\beta^2} + 4I_{copy1}, \tag{9.115}$$

and hence

$$I_{copy1} = \frac{I_{REF}}{4 + \frac{15}{\beta^2}} \tag{9.116}$$

$$I_{copy2} = \frac{10I_{REF}}{4 + \frac{15}{\beta^2}}. \tag{9.117}$$

Exercise Calculate I_{copy1} if one of the four unit transistors is omitted, i.e., the reference transistor has an area of $3A_E$.

***PNP* Mirrors** Consider the common-emitter stage shown in Fig. 9.34(a), where a current source serves as a load to achieve a high voltage gain. The current source can be realized as a *pnp* transistor operating in the active region [Fig. 9.34(b)]. We must therefore define the bias current of Q_2 properly. In analogy with the *npn* counterpart of Fig. 9.24(c), we form the *pnp* current mirror depicted in Fig. 9.34(c). For example, if Q_{REF} and Q_2 are identical and the base currents negligible, then Q_2 carries a current equal to I_{REF}.

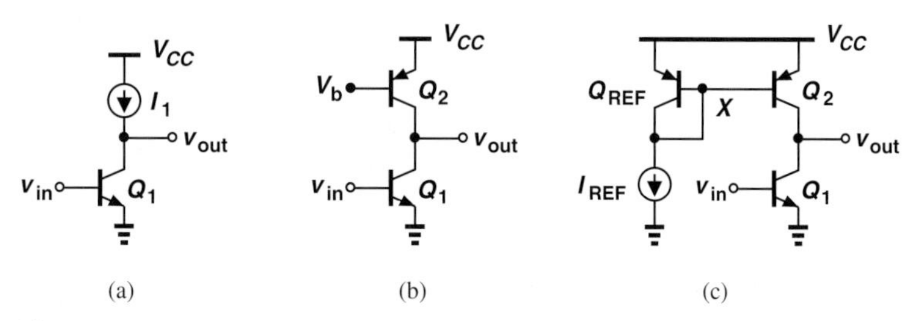

Figure 9.34 (a) CE stage with current-source load, (b) realization of current source by a *pnp* device, (c) proper biasing of Q_2.

Example 9-17 Design the circuit of Fig. 9.34(c) for a voltage gain of 100 and a power budget of 2 mW. Assume $V_{A,npn} = 5$ V, $V_{A,pnp} = 4$ V, $I_{REF} = 100$ μA, and $V_{CC} = 2.5$ V.

Solution From the power budget and $V_{CC} = 2.5$ V, we obtain a total supply current of 800 μA, of which 100 μA is dedicated to I_{REF} and Q_{REF}. Thus, Q_1 and Q_2 are biased at a current of 700 μA, requiring that the (emitter) area of Q_2 be 7 times that of Q_{REF}. (For example, Q_{REF} incorporates one unit device and Q_1 seven unit devices.)

The voltage gain can be written as

$$A_v = -g_{m1}(r_{O1}||r_{O2}) \tag{9.118}$$

$$= -\frac{1}{V_T} \cdot \frac{V_{A,npn}V_{A,pnp}}{V_{A,npn} + V_{A,pnp}} \tag{9.119}$$

$$= -85.5. \tag{9.120}$$

What happened here?! We sought a gain of 100 but inevitably obtained a value of 85.5! This is because the gain of the stage is simply given by the Early voltages and V_T, a fundamental constant of the technology and independent of the bias current. Thus, with the above choice of Early voltages, the circuit's gain cannot reach 100.

Exercise What Early voltage is necessary for a voltage gain of 100?

We must now address an interesting problem. In the mirror of Fig. 9.24(c), it is assumed that the golden current flows from V_{CC} to node X, whereas in Fig. 9.34(c) it flows from X to ground. How do we generate the latter from the former? It is possible to combine the *npn* and *pnp* mirrors for this purpose, as illustrated in Fig. 9.35. Assuming for simplicity that Q_{REF1}, Q_M, Q_{REF2}, and Q_2 are identical and neglecting the base currents, we observe that Q_M draws a current of I_{REF} from Q_{REF2}, thereby forcing the same current through Q_2 and Q_1. We can also create various scaling scenarios between Q_{REF1} and Q_M and between Q_{REF2} and Q_2. Note that the base currents introduce a cumulative error as I_{REF} is copied onto $I_{C,M}$, and $I_{C,M}$ onto I_{C2}.

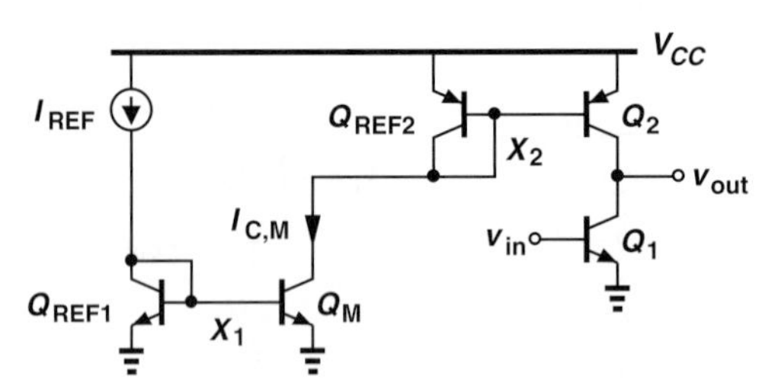

Figure 9.35 Generation of current for *pnp* devices.

Example 9-18 We wish to bias Q_1 and Q_2 in Fig. 9.35 at a collector current of 1 mA while $I_{REF} = 25\ \mu A$. Choose the scaling factors in the circuit so as to minimize the number of unit transistors.

Solution For an overall scaling factor of $1\ mA/25\ \mu A = 40$, we can choose either

$$I_{C,M} = 8I_{REF} \tag{9.121}$$

$$|I_{C2}| = 5I_{C,M} \tag{9.122}$$

or

$$I_{C,M} = 10I_{REF} \tag{9.123}$$

$$|I_{C2}| = 4I_{C,M}. \tag{9.124}$$

(In each case, the *npn* and *pnp* scaling factors can be swapped.) In the former case, the four transistors in the current mirror circuitry require 15 units, and in the latter case, 16 units. Note that we have implicitly dismissed the case $I_{C,M} = 40I_{C,REF1}$ and $I_{C2} = I_{C,REF2}$ as it would necessitate 43 units.

Exercise Calculate the exact value of I_{C2} if $\beta = 50$ for all transistors.

Example 9-19 An electrical engineering student purchases two nominally identical discrete bipolar transistors and constructs the current mirror shown in Fig. 9.24(c). Unfortunately, I_{copy} is 30% higher than I_{REF}. Explain why.

Solution It is possible that the two transistors were fabricated in different batches and hence underwent slightly different processing. Random variations during manufacturing may lead to changes in the device parameters and even the emitter area. As a result, the two transistors suffer from significant I_S mismatch. This is why current mirrors are rarely used in discrete design.

Exercise How much I_S mismatch results in a 30% collector current mismatch?

9.2.3 MOS Current Mirror

The developments in Section 9.2.2 can be applied to MOS current mirrors as well. In particular, drawing the MOS counterpart of Fig. 9.24(a) as in Fig. 9.36(a), we recognize that the black box must generate V_X such that

$$\frac{1}{2}\mu_n C_{ox}\left(\frac{W}{L}\right)_1 (V_X - V_{TH1})^2 = I_{REF}, \tag{9.125}$$

where channel-length modulation is neglected. Thus, the black box must satisfy the following input (current)/output (voltage) characteristic:

$$V_X = \sqrt{\frac{2I_{REF}}{\mu_n C_{ox}\left(\frac{W}{L}\right)_1}} + V_{TH1}. \tag{9.126}$$

That is, it must operate as a "square-root" circuit. From Chapter 6, we recall that a diode-connected MOSFET provides such a characteristic [Fig. 9.36(b)], thus arriving at the NMOS current mirror depicted in Fig. 9.36(c). As with the bipolar version, we can view the circuit's operation from two perspectives: (1) M_{REF} takes the square root of I_{REF} and M_1 squares the result; or (2) the drain currents of the two transistors can be expressed as

$$I_{D,REF} = \frac{1}{2}\mu_n C_{ox}\left(\frac{W}{L}\right)_{REF} (V_X - V_{TH})^2 \tag{9.127}$$

$$I_{copy} = \frac{1}{2}\mu_n C_{ox}\left(\frac{W}{L}\right)_1 (V_X - V_{TH})^2, \tag{9.128}$$

Did you know?

A simple current mirror may also be considered a two-transistor circuit topology. We can even implement a *current amplifier* using such an arrangement. As shown below, if the reference (diode-connected) device is smaller than the current-source transistor, then $I_{out}/I_{in} > 1$. For example, if I_{in} changes by 1 μA, then I_{out} changes by 1 μA × *n*. Current amplifiers are occasionally used in analog systems. In fact, some analog designers promoted the notion of "current-mode circuits" in the 1990s, but the movement did not catch on.

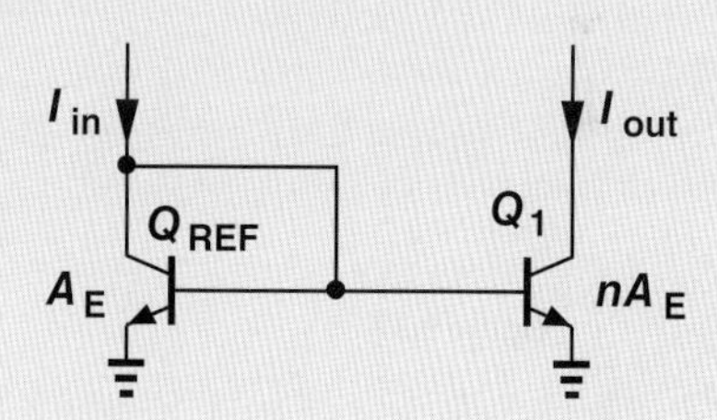

Current mirror as an amplifier.

where the threshold voltages are assumed equal. It follows that

$$I_{copy} = \frac{\left(\frac{W}{L}\right)_1}{\left(\frac{W}{L}\right)_{REF}} I_{REF}, \tag{9.129}$$

which reduces to $I_{copy} = I_{REF}$ if the two transistors are identical.

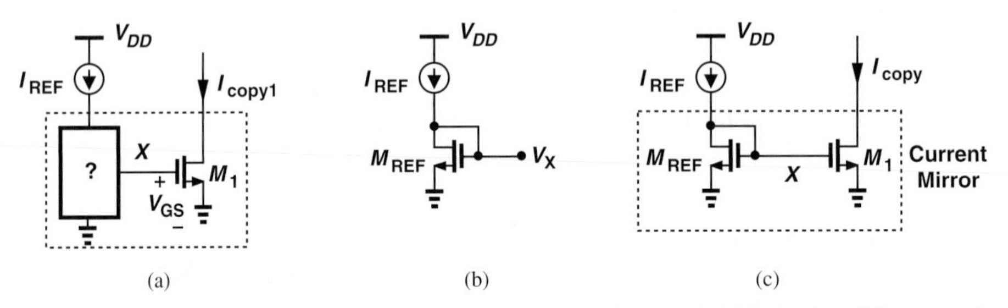

Figure 9.36 (a) Conceptual illustration of copying a current by an NMOS device, (b) generation of a voltage proportional to square root of current, (c) MOS current mirror.

Example 9-20

The student working on the circuits in Examples 9-12 and 9-13 decides to try the MOS counterpart, thinking that the gate current is zero and hence leaving the gates floating (Fig. 9.37). Explain what happens.

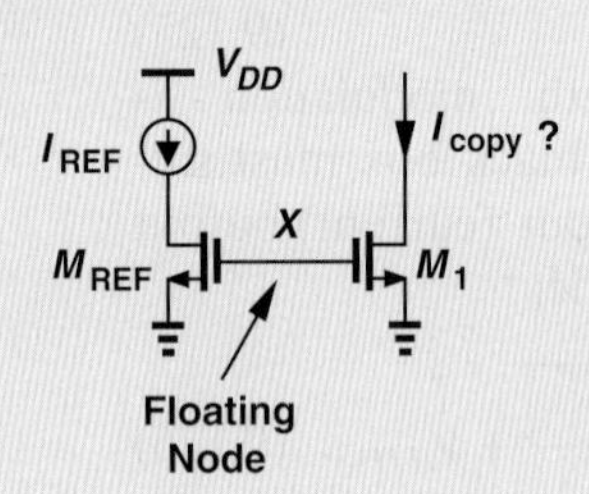

Figure 9.37

Solution This circuit is not a current mirror because only a diode-connected device can establish Eq. (9.129) and hence a copy current independent of device parameters and temperature. Since the gates of M_{REF} and M_1 are floating, they can assume any voltage, e.g., an initial condition created at node X when the power supply is turned on. In other words, I_{copy} is very poorly defined.

Exercise Is M_{REF} always off in this circuit?

Generation of additional copies of I_{REF} with different scaling factors also follows the principles shown in Fig. 9.27. The following example illustrates these concepts.

Example 9-21 An integrated circuit employs the source follower and the common-source stage shown in Fig. 9.38(a). Design a current mirror that produces I_1 and I_2 from a 0.3-mA reference.

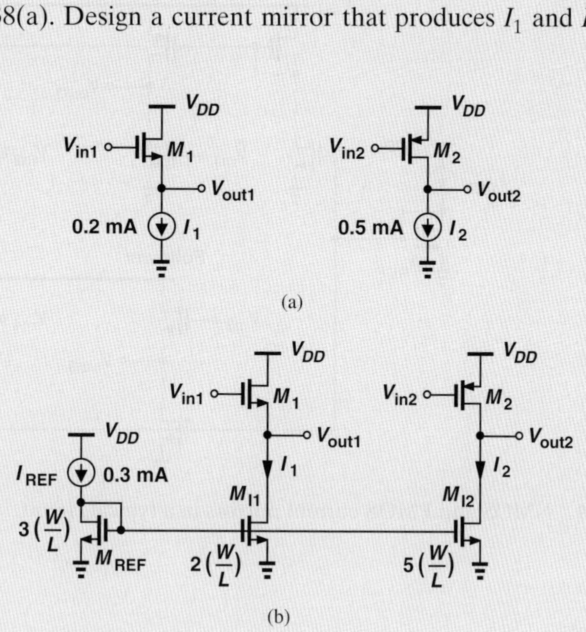

Figure 9.38

Solution Following the methods depicted in Figs. 9.29 and 9.30, we select an aspect ratio of $3(W/L)$ for the diode-connected device, $2(W/L)$ for M_{I1}, and $5(W/L)$ for M_{I2}. Figure 9.38(b) shows the overall circuit.

Exercise Repeat the above example if $I_{REF} = 0.8$ mA.

Since MOS devices draw a negligible gate current,[6] MOS mirrors need not resort to the technique shown in Fig. 9.32. On the other hand, channel-length modulation in the current-source transistors does lead to additional errors. Investigated in Problem 9.53, this effect mandates circuit modifications that are described in more advanced texts [1].

The idea of combining NMOS and PMOS current mirrors follows the bipolar counterpart depicted in Fig. 9.35. The circuit of Fig. 9.39 exemplifies these ideas.

[6]In deep-submicron CMOS technologies, the gate oxide thickness is reduced to less than 30 Å, leading to "tunneling" and hence noticeable gate current. This effect is beyond the scope of this book.

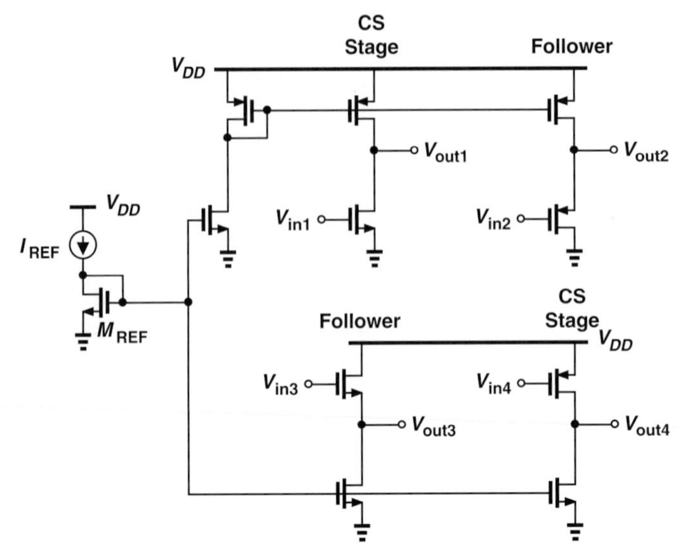

Figure 9.39 NMOS and PMOS current mirrors in a typical circuit.

Bioengineering Application: Electronic Skin

People who have lost an arm or a leg can benefit from prosthetic replacements. These mechanical body parts provide some ability for the person similar to those of the original limb. However, they lack the sense of *touch*. Recent work on the "electronic skin" addresses this issue. As illustrated in Fig. 9.40(a), a prosthetic hand, for example, is covered with pressure sensors that send their signals to a "stimulator." The stimulator applies corresponding electric signals to the person's nerves, causing the brain to feel a sensation. Interestingly, even with only three sensors, the user can distinguish objects with different curvatures or feel pain when touching sharp objects [2].

A large electronic skin requires a wide array of sensors and their associated circuits. Piezoelectric force sensors act as a variable resistance whose value must be measured accurately. Shown in Fig. 9.40(b) is an example [2], where one sensor is connected to a virtual ground at a given time, thereby generating an output voltage inversely proportional to the sensor resistance. The op amp's internal circuit is shown in Fig. 9.40(c). We recognize M_3 and M_4 as a current mirror and M_5 as a common-source stage (why?). The topology consisting of M_1-M_4 is studied in the next chapter.

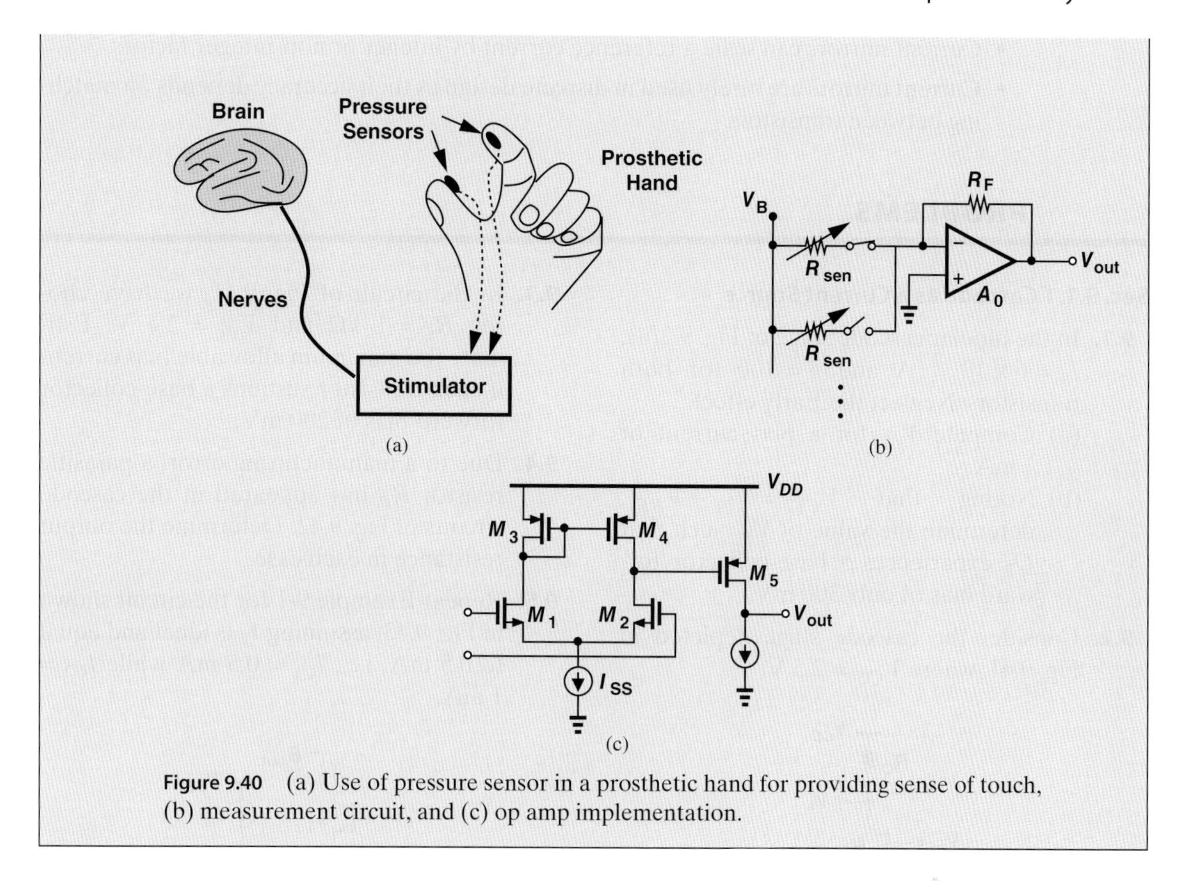

Figure 9.40 (a) Use of pressure sensor in a prosthetic hand for providing sense of touch, (b) measurement circuit, and (c) op amp implementation.

9.3 CHAPTER SUMMARY

- Stacking a transistor atop another forms a cascode structure, resulting in a high output impedance.
- The cascode topology can also be considered an extreme case of source or emitter degeneration.
- The voltage gain of an amplifier can be expressed as $-G_m R_{out}$, where G_m denotes the short-circuit transconductance of the amplifier. This relationship indicates that the gain of amplifiers can be maximized by maximing their output impedance.
- With its high output impedance, a cascode stage can operate as a high-gain amplifier.
- The load of a cascode stage is also realized as a cascode circuit so as to approach an ideal current source.
- Setting the bias currents of analog circuits to well-defined values is difficult. For example, resistive dividers tied to the base or gate of transistors result in supply- and temperature-dependent currents.
- If V_{BE} or V_{GS} are well-defined, then I_C or I_D are not.
- Current mirrors can "copy" a well-defined reference current numerous times for various blocks in an analog system.

- Current mirrors can scale a reference current by integer or non-integer factors.
- Current mirrors are rarely used in disrcete design as their accuracy depends on matching between transistors.

PROBLEMS

Sec. 9.1.1 Cascode as a Current Source

9.1. In the bipolar cascode stage of Fig. 9.2(a), $I_S = 6 \times 10^{-17}$ A and $\beta = 100$ for both transistors. Neglect the Early effect.
(a) Compute V_{b2} for a bias current of 1 mA.
(b) Noting that $V_{CE2} = V_{b1} - V_{BE1}$, determine the value of V_{b1} such that Q_2 experiences a base-collector forward bias of only 300 mV.

9.2. Consider the cascode stage depicted in Fig. 9.41, where $V_{CC} = 2.5$ V.

Figure 9.41

(a) Repeat Problem 9.1 for this circuit, assuming a bias current of 0.5 mA.
(b) With the minimum allowable value of V_{b1}, compute the maximum allowable value of R_C such that Q_1 experiences a base-collector forward bias of no more than 300 mV.

9.3. In the circuit of Fig. 9.41, we have chosen $R_C = 1$ kΩ and $V_{CC} = 2.5$ V. Estimate the maximum allowable bias current if each transistor sustains a base-collector forward bias of 200 mV. SOL

***9.4.** Due to a manufacturing error, a parasitic resistor R_P has appeared in the cascode circuits of Fig. 9.42. Determine the output resistance in each case.

9.5. Repeat Example 9-1 for the circuit shown in Fig. 9.43, assuming I_1 is ideal and equal to 0.5 mA, i.e., $I_{C1} = 0.5$ mA while $I_{C2} =$ 1 mA.

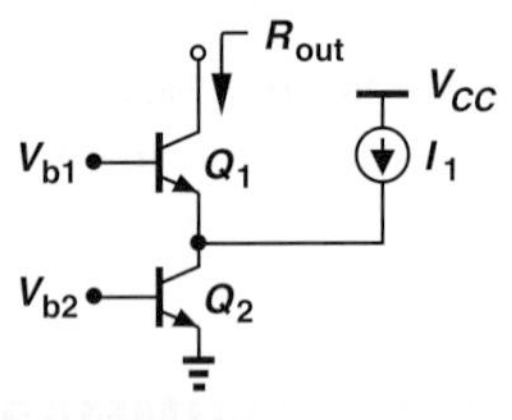

Figure 9.43

9.6. Suppose the circuit of Fig. 9.43 is realized as shown in Fig. 9.44, where Q_3 plays the role of I_1. Assuming $V_{A1} = V_{A2} = V_{A,n}$ and $V_{A3} = V_{A,p}$, determine the output impedance of the circuit.

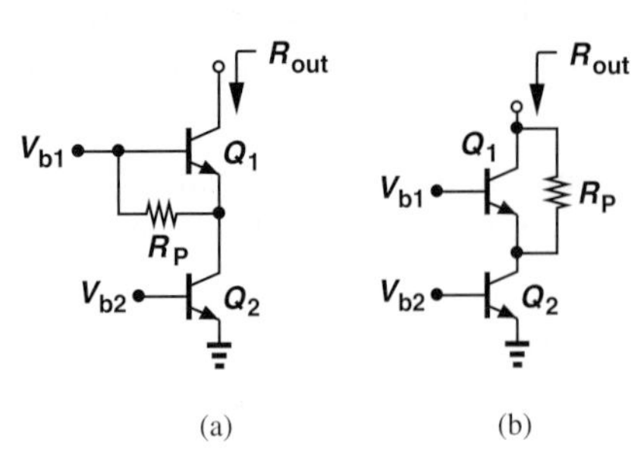

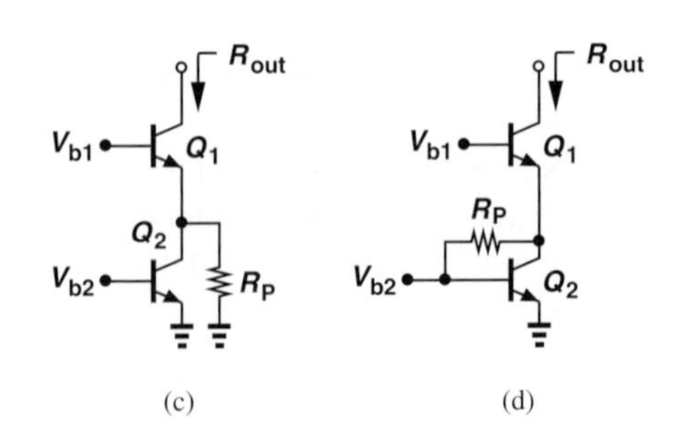

Figure 9.42

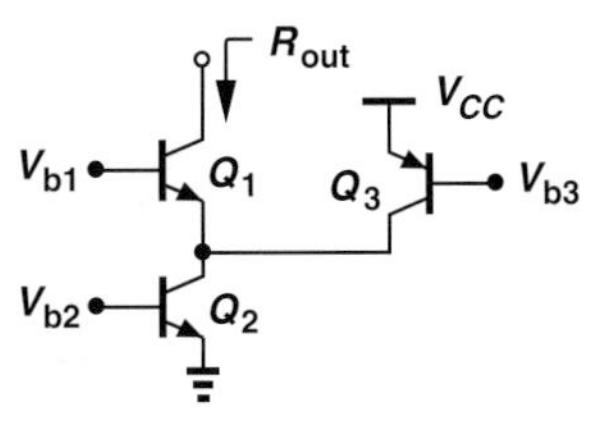

Figure 9.44

*9.7. Excited by the output impedance "boosting" capability of cascodes, a student decides to extend the idea as illustrated in Fig. 9.45. What is the maximum output impedance that the student can achieve? Assume the transistors are identical.

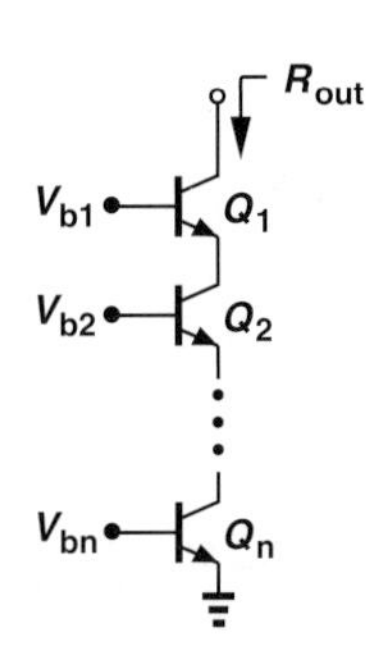

Figure 9.45

*9.8. While constructing a cascode stage, a student adventurously swaps the collector and base terminals of the degeneration transistor, arriving at the circuit shown in Fig. 9.46.

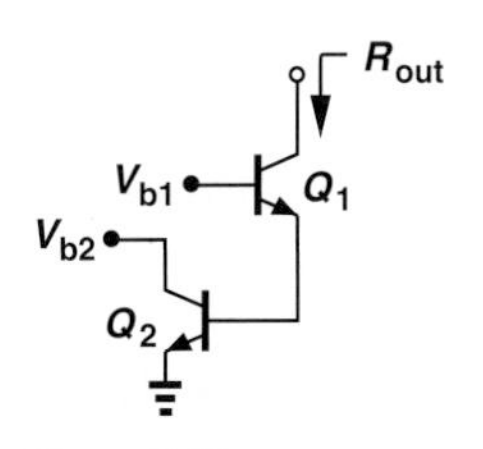

Figure 9.46

(a) Assuming both transistors operate in the active region, determine the output impedance of the circuit.
(b) Compare the result with that of a cascode stage for a *given bias current* (I_{C1}) and explain why this is generally not a good idea.

9.9. For discrete bipolar transistors, the Early voltage reaches tens of volts, allowing the approximation $V_A \gg \beta V_T$ if $\beta < 100$. Using this approximation, simplify Eq. (9.9) and explain why the result resembles that in Eq. (9.12). SOL

*9.10. The *pnp* cascode depicted in Fig. 9.47 must provide a bias current of 0.5 mA to a circuit. If $I_S = 10^{-16}$ and $\beta = 100$,

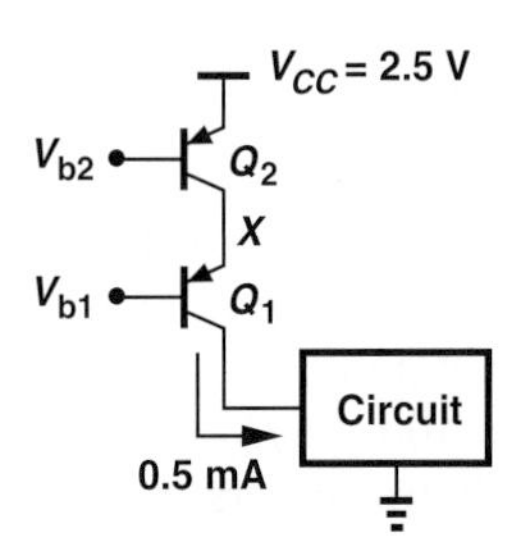

Figure 9.47

(a) Calculate the required value of V_{b2}.
(b) Noting that $V_X = V_{b1} + |V_{BE1}|$, determine the maximum allowable value of V_{b1} such that Q_2 experiences a base-collector forward bias of only 200 mV.

**9.11. Determine the output impedance of each circuit shown in Fig. 9.48. Assume $\beta \gg 1$. Explain which ones are considered cascode stages.

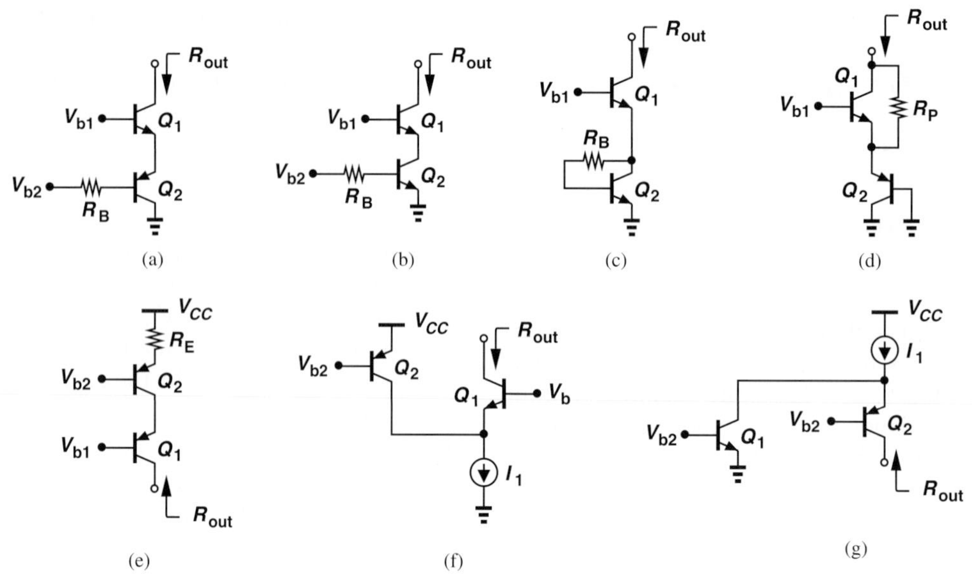

Figure 9.48

9.12. The MOS cascode of Fig. 9.49 must provide a bias current of 0.5 mA with an output impedance of at least 50 kΩ. If $\mu_n C_{ox} = 100\ \mu A/V^2$ and $W/L = 20/0.18$ for both transistors, compute the maximum tolerable value of λ.

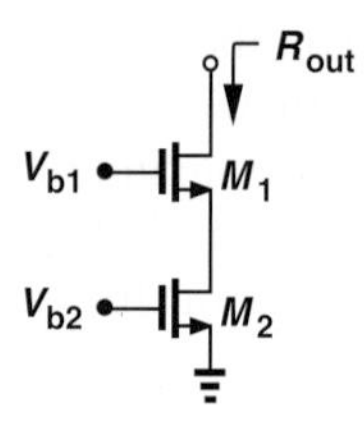

Figure 9.49

9.13. (a) Writing $G_m = \sqrt{2\mu_n C_{ox}(W/L)I_D}$, express Eq. (9.23) in terms of I_D and plot the result as a function of I_D.

(b) Compare this expression with that in Eq. (9.9) for the bipolar counterpart. Which one is a stronger function of the bias current?

9.14. The cascode current source shown in Fig. 9.50 must be designed for a bias current of 0.5 mA. Assume $\mu_n C_{ox} = 100\ \mu A/V^2$ and $V_{TH} = 0.4$ V.

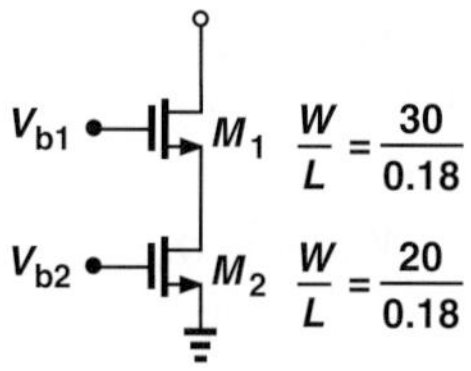

Figure 9.50

(a) Neglecting channel-length modulation, compute the required value of V_{b2}. What is the minimum tolerable value of V_{b1} if M_2 must remain in saturation?

(b) Assuming $\lambda = 0.1\ V^{-1}$, calculate the output impedance of the circuit.

9.15. Consider the circuit shown in Fig. 9.51, where $V_{DD} = 1.8$ V, $(W/L)_1 = 20/0.18$, and $(W/L)_2 = 40/0.18$. Assume $\mu_n C_{ox} = 100\ \mu A/V^2$ and $V_{TH} = 0.4$ V.

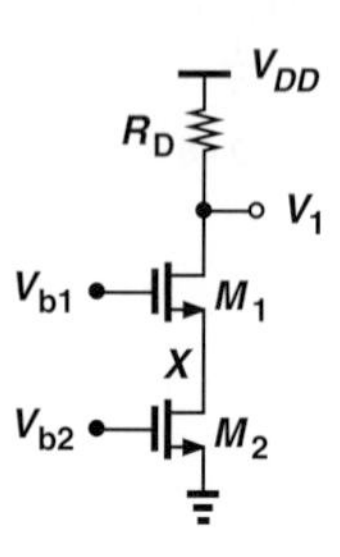

Figure 9.51

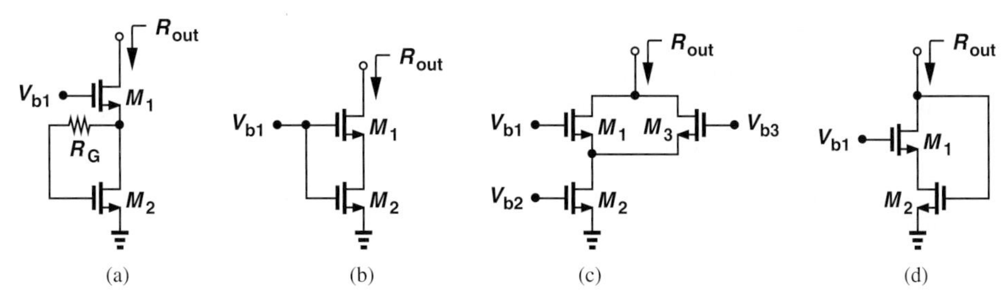

Figure 9.52

(a) If we require a bias current of 1 mA and $R_D = 500\ \Omega$, what is the highest allowable value of V_{b1}?

(b) With such a value chosen for V_{b1}, what is the value of V_X?

**9.16. Compute the output resistance of the circuits depicted in Fig. 9.52. Assume all of the transistors operate in saturation and $g_m r_O \gg 1$.

9.17. The PMOS cascode of Fig. 9.53 must provide a bias current of 0.5 mA with an output impedance of 40 kΩ. If $\mu_p C_{ox} = 50\ \mu\text{A/V}^2$ and $\lambda = 0.2\ \text{V}^{-1}$, determine the required value of $(W/L)_1 = (W/L)_2$.

SOL ***9.18.** The PMOS cascode of Fig. 9.53 is designed for a given output impedance, R_{out}. Using Eq. (9.23), explain what happens if the widths of both transistors are increased by a factor of N while the transistor lengths and bias currents remain unchanged. Assume $\lambda \propto L^{-1}$.

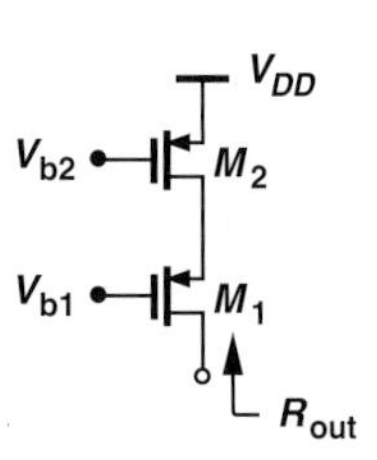

Figure 9.53

**9.19. Determine the output impedance of the stages shown in Fig. 9.54. Assume all of the transistors operate in saturation and $g_m r_O \gg 1$.

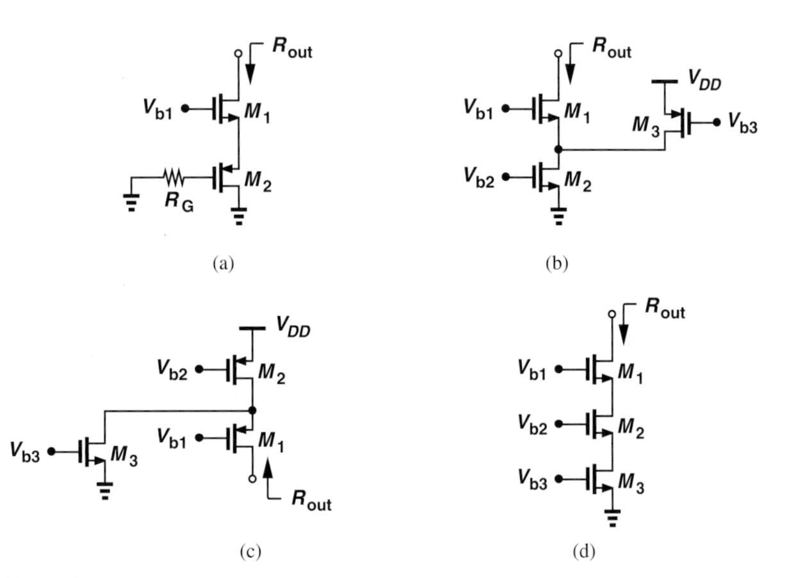

Figure 9.54

Sec. 9.1.2 Cascode as an Amplifier

9.20. Compute the short-circuit transconductance and the voltage gain of each of the stages in Fig. 9.55. Assume $\lambda > 0$ and $V_A < \infty$.

*__9.21.__ Prove that Eq. (9.53) reduces to

$$A_v \approx \frac{-\beta V_A^2}{V_T(V_A + \beta V_T)}, \tag{9.130}$$

a quantity independent of the bias current.

9.22. The cascode stage of Fig. 9.16(b) must be designed for a voltage gain of 500. If $\beta_1 = \beta_2 = 100$, determine the minimum required value of $V_{A1} = V_{A2}$. Assume $I_1 = 1$ mA.

**__9.23.__ Having learned about the high voltage gain of the cascode stage, a student adventurously constructs the circuit depicted in Fig. 9.56, where the input is applied to the base of Q_2 rather than to the base of Q_1.

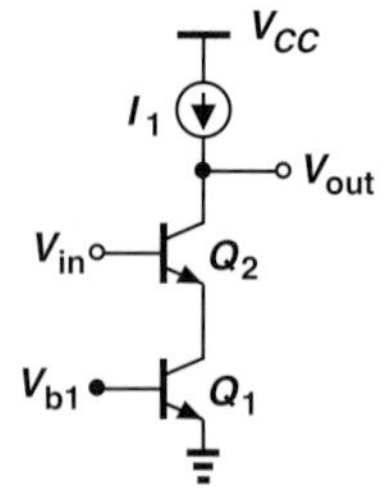

Figure 9.56

(a) Replacing Q_1 with r_{O1}, explain intuitively why the voltage gain of this stage cannot be as high as that of the cascode.

(b) Assuming $g_m r_O \gg 1$, compute the short-circuit transconductance and the voltage gain.

*__9.24.__ Determine the short-circuit transconductance and the voltage gain of the circuit shown in Fig. 9.57.

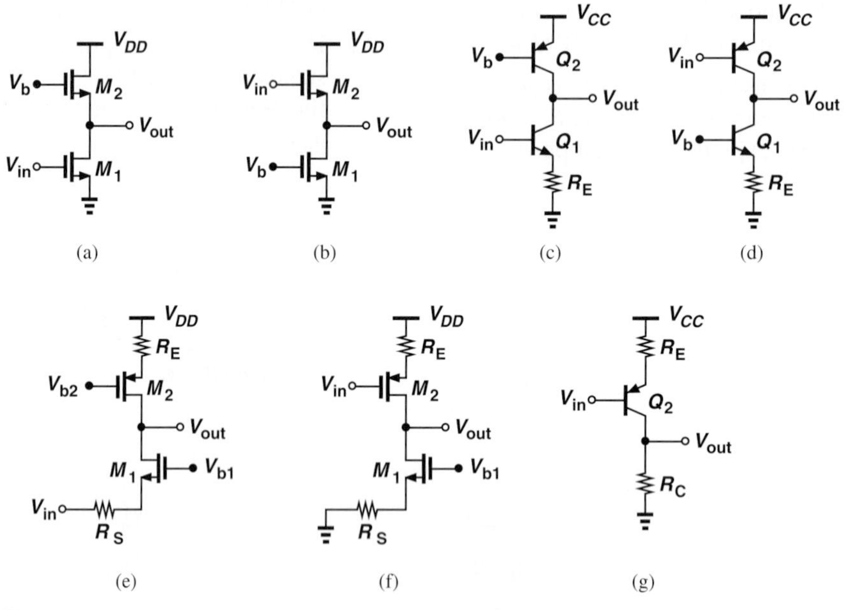

(a) (b) (c) (d) (e) (f) (g)

Figure 9.55

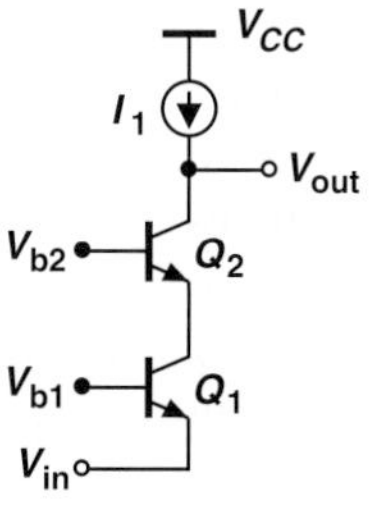

Figure 9.57

9.25. ** Calculate the voltage gain of each stage illustrated in Fig. 9.58.

9.26. Consider the cascode amplifier of Fig. 9.19 and assume $\beta_1 = \beta_2 = \beta_N$, $V_{A1} = V_{A2} = V_{A,N}$, $\beta_3 = \beta_4 = \beta_P$, $V_{31} = V_{A4} = V_{A,P}$. Express Eq. (9.61) in terms of these quantities. Does the result depend on the bias current?

9.27. * Due to a manufacturing error, a bipolar cascode amplifier has been configured as shown in Fig. 9.59. Determine the voltage gain of the circuit.

9.28. Writing $g_m = \sqrt{2\mu_n C_{ox}(W/L)I_D}$ and $r_O = 1/(\lambda I_D)$, express Eq. (9.72) in terms of the device parameters and plot the result as a function of I_D. SOL

9.29. The MOS cascode of Fig. 9.20(a) must provide a voltage gain of 200 with a bias current of 1 mA. If $\mu_n C_{ox} = 100\ \mu A/V^2$ and $\lambda = 0.1\ V^{-1}$ for both transistors, determine the required value of $(W/L)_1 = (W/L)_2$.

9.30. * The MOS cascode of Fig. 9.20(a) is designed for a given voltage gain, A_v. Using Eq. (9.79) and the result obtained in Problem 9.28, explain what happens if the widths of the transistors are increased by a factor of N while the transistor lengths and bias currents remain unchanged.

9.31. * Repeat Problem 9.30 if the lengths of both transistors are increased by a factor of N while the transistor widths and bias currents remain unchanged.

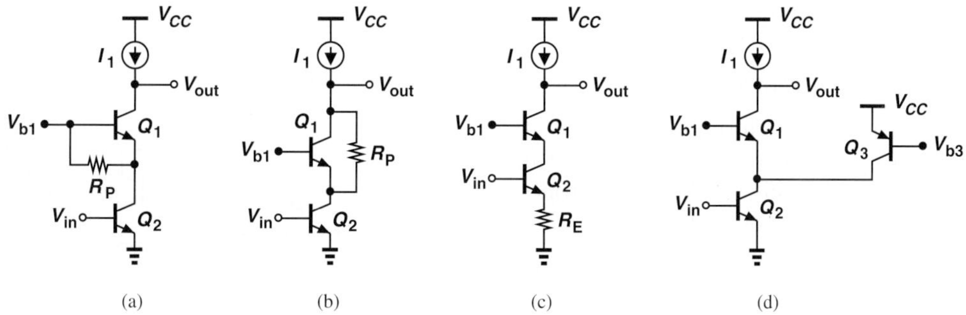

Figure 9.58

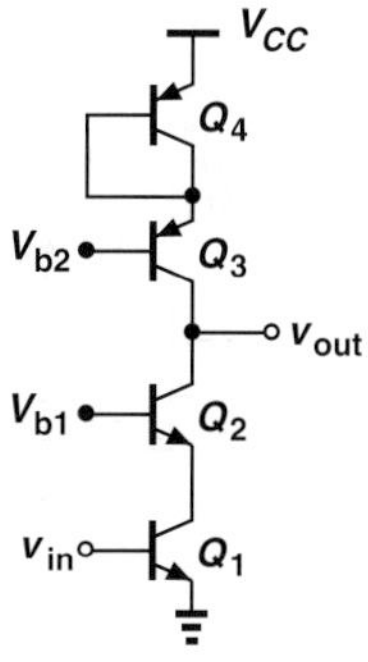

Figure 9.59

9.32. Due to a manufacturing error, a CMOS cascode amplifier has been configured as shown in Fig. 9.60. Calculate the voltage gain of the circuit.

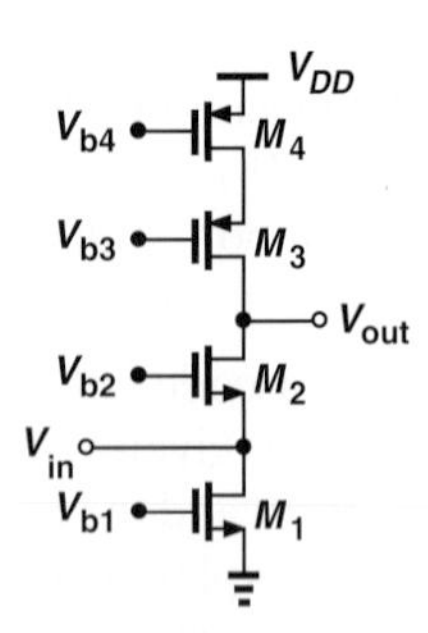

Figure 9.60

9.33. In the cascode stage of Fig. 9.20(b), $(W/L)_1 = \cdots = (W/L)_4 = 20/0.18$. If $\mu_n C_{ox} = 100\ \mu\text{A/V}^2$, and $\mu_p C_{ox} = 50\ \mu\text{A/V}^2$, $\lambda_n = 0.1\ \text{V}^{-1}$, and $\lambda_p = 0.15\ \text{V}^{-1}$, calculate the bias current such that the circuit achieves a voltage gain of 500.

****9.34.** Determine the voltage gain of each circuit in Fig. 9.61. Assume $g_m r_O \gg 1$.

Sec. 9.2.2 Bipolar Current Mirror

***9.35.** From Eq. (9.83), determine the sensitivity of I_1 to V_{CC}, defined as $\partial I_1/\partial V_{CC}$. Explain intuitively why this sensitivity is proportional to the transconductance of Q_1.

SOL **9.36.** Repeat Problem 9.35 for Eq. (9.85) (in terms of V_{DD}).

9.37. The parameters $\mu_n C_{ox}$ and V_{TH} in Eq. (9.85) also vary with the fabrication process. (Integrated circuits fabricated in different batches exhibit slightly different parameters.) Determine the sensitivity of I_1 to V_{TH} and explain why this issue becomes more serious at low supply voltages.

9.38. Having learned about the logarithmic function of the circuit in Fig. 9.24(b), a student remembers the logarithmic amplifier studied in Chapter 8 and constructs the circuit depicted in Fig. 9.62. Explain what happens.

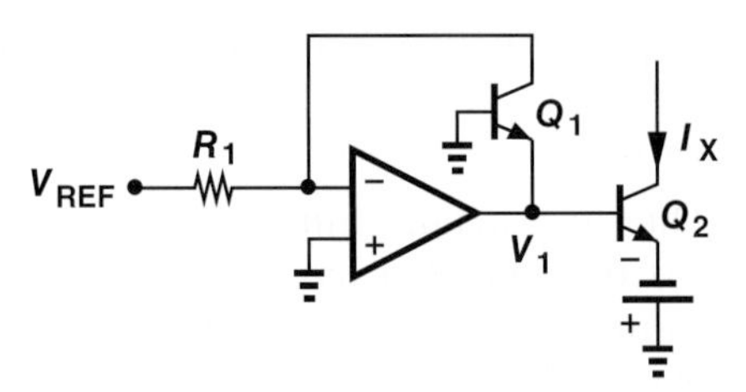

Figure 9.62

9.39. Repeat Problem 9.38 for the topology shown in Fig. 9.63.

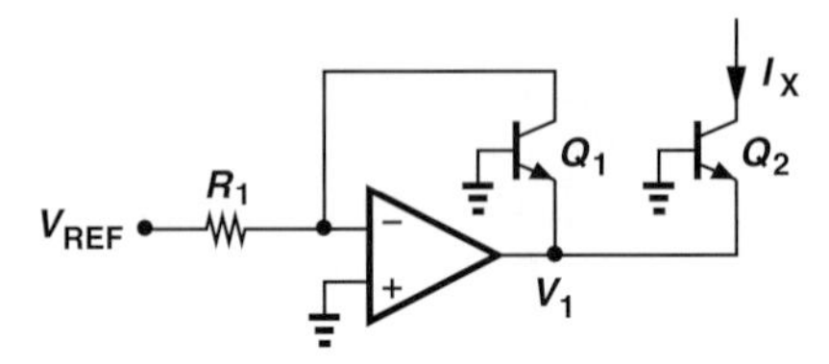

Figure 9.63

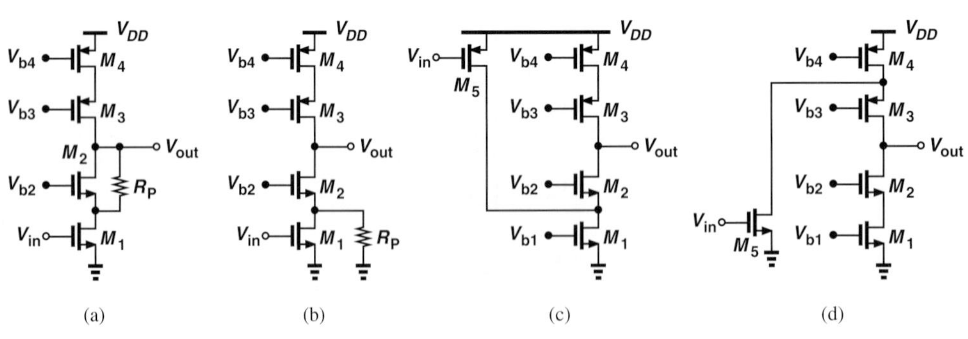

Figure 9.61

*9.40. Due to a manufacturing error, resistor R_P has appeared in series with the emitter of Q_1 in Fig. 9.64. If I_1 is half of its nominal value, express the value of R_P in terms of other circuit parameters. Assume Q_{REF} and Q_1 are identical and $\beta \gg 1$.

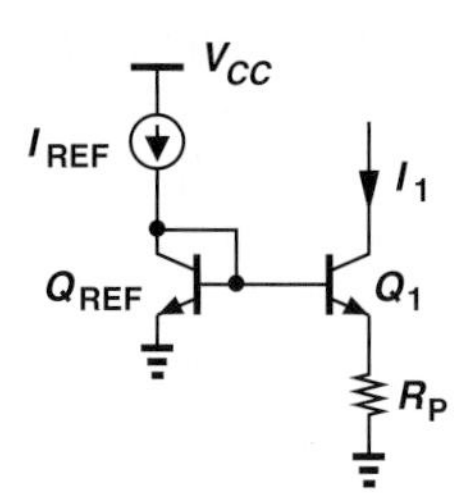

Figure 9.64

*9.41. Repeat Problem 9.40 for the circuit shown in Fig. 9.65, but assuming that I_1 is twice its nominal value.

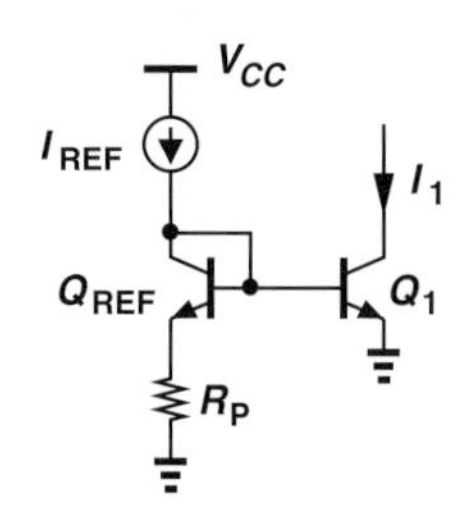

Figure 9.65

9.42. We wish to generate two currents equal to 50 μA and 230 μA from a reference of 130 μA. Design an *npn* current mirror for this purpose. Neglect the base currents.

9.43. Repeat Example 9-15 if the reference current is equal to 180 μA.

**9.44. Due to a manufacturing error, resistor R_P has appeared in series with the base of Q_{REF} in Fig. 9.66. If I_1 is 10% greater than its nominal value, express the value of R_P in terms of other circuit parameters. Assume Q_{REF} and Q_1 are identical.

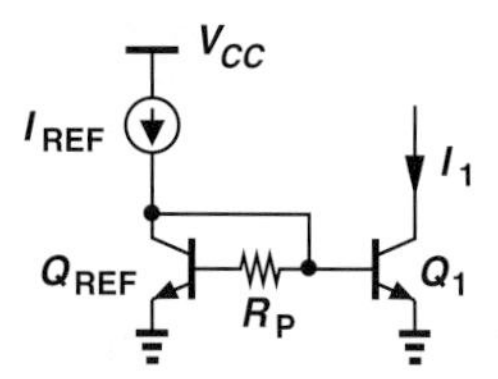

Figure 9.66

*9.45. Repeat Problem 9.44 for the circuit shown in Fig. 9.67, but assuming I_1 is 10% less than its nominal value.

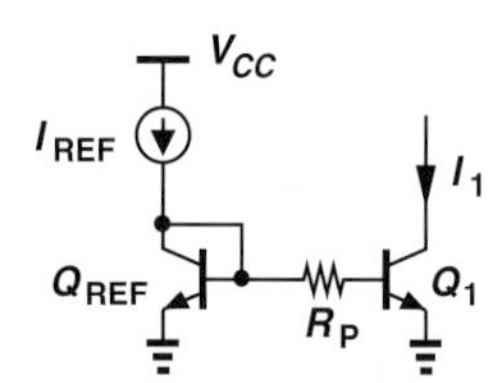

Figure 9.67

9.46. Taking base currents into account, determine the value of I_{copy} in each circuit depicted in Fig. 9.68. Normalize the error to the nominal value of I_{copy}.

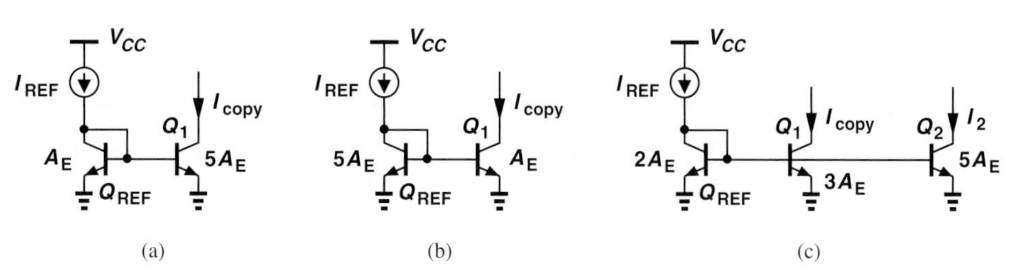

Figure 9.68

*__9.47.__ Calculate the error in I_{copy} for the circuit shown in Fig. 9.69.

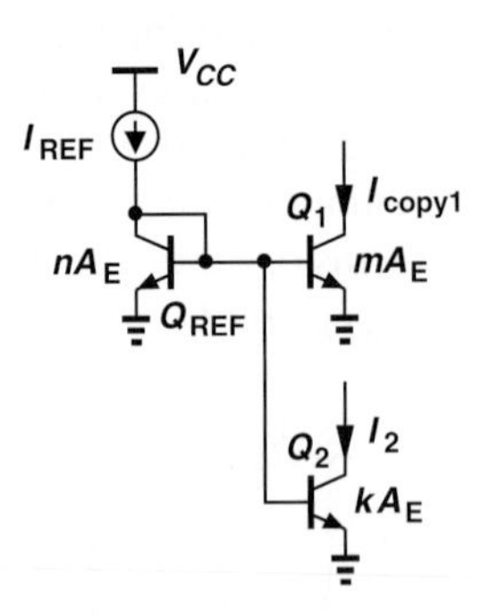

Figure 9.69

9.48. Taking base currents into account, compute the error in I_{copy} for each of the circuits illustrated in Fig. 9.70.

Sec. 9.2.3 MOS Current Mirror

9.49. Determine the value of R_P in the circuit of Fig. 9.71 such that $I_1 = I_{REF}/2$. With this choice of R_P, does I_1 change if the threshold voltage of both transistors increases by ΔV?

9.50. Determine the value of R_P in the circuit of Fig. 9.72 such that $I_1 = 2I_{REF}$. With this choice of R_P, does I_1 change if the threshold voltage of both transistors increases by ΔV?

9.51. Repeat Example 9-21 if the reference current is 0.35 mA. SOL

9.52. Calculate I_{copy} in each of the circuits shown in Fig. 9.73. Assume all of the transistors operate in saturation.

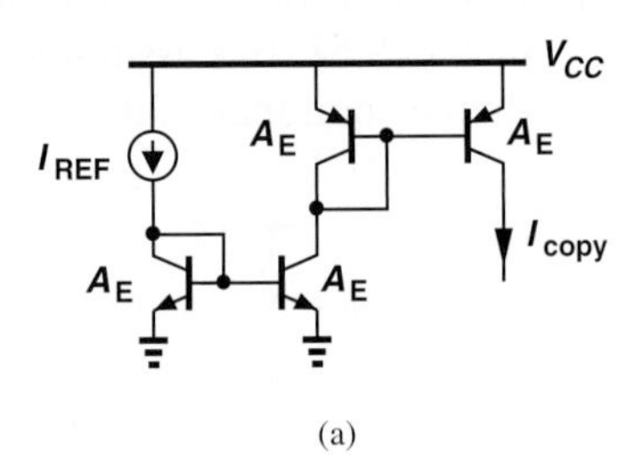

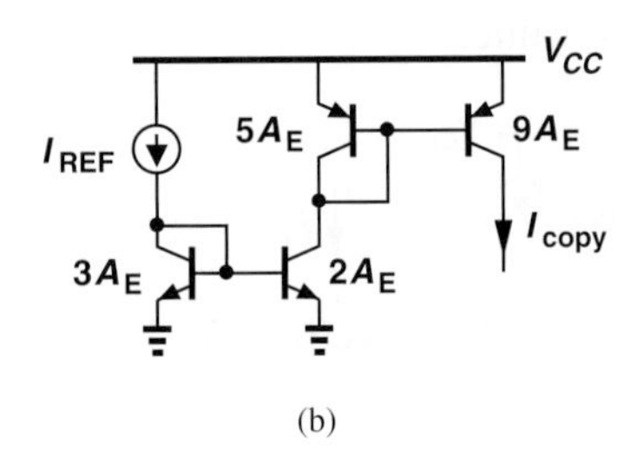

Figure 9.70

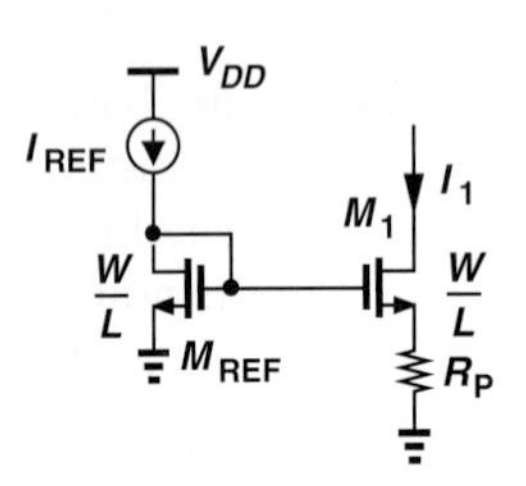

Figure 9.71

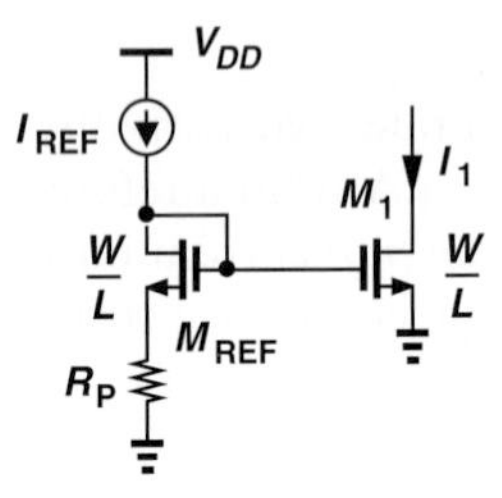

Figure 9.72

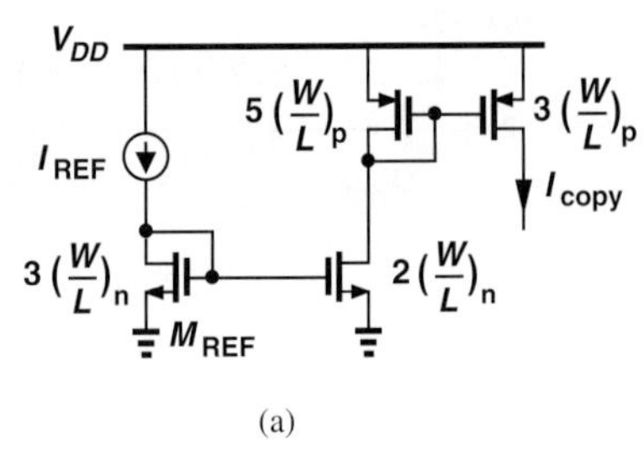

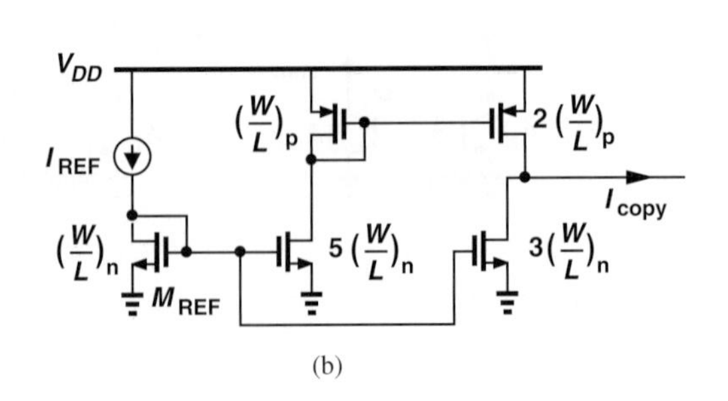

Figure 9.73

****9.53.** Consider the MOS current mirror shown in Fig. 9.36(c) and assume M_1 and M_2 are identical but $\lambda \neq 0$.

(a) How should V_{DS1} be chosen so that I_{copy1} is exactly equal to I_{REF}?

(b) Determine the error in I_{copy1} with respect to I_{REF} if V_{DS1} is equal to $V_{GS} - V_{TH}$ (so that M_1 resides at the edge of saturation).

Design Problems

In the following problems, unless otherwise stated, assume $I_{S,n} = I_{S,p} = 6 \times 10^{-16}$ A, $V_{A,n} = V_{A,p} = 5$ V, $\beta_n = 100, \beta_p = 50, \mu_n C_{ox} = 100\ \mu A/V^2$, $\mu_p C_{ox} = 50\ \mu A/V^2$, $V_{TH,n} = 0.4$ V, and $V_{TH,p} = -0.5$ V, where the subscripts n and p refer to n-type (npn or NMOS) and p-type (pnp or PMOS) devices, respectively.

9.54. Assuming a bias current of 1 mA, design the degenerated current source of Fig. 9.74(a) such that R_E sustains a voltage approximately equal to the minimum required collector-emitter voltage of Q_2 in Fig. 9.74(b) ($\approx$ 0.5 V). Compare the output impedances of the two circuits.

Rout
Vb
Q1
RE
Rout
Vb1
Q1
Vb2
Q2

(a) (b)

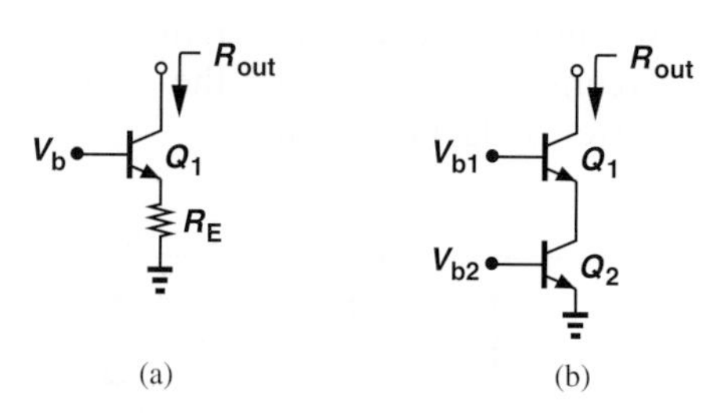

Figure 9.74

9.55. Design the cascode current source of Fig. 9.74(b) for an output impedance of 50 kΩ. Select V_{b1} such that Q_2 experiences a base-collector forward bias of only 100 mV. Assume a bias current of 1 mA.

9.56. We wish to design the MOS cascode of Fig. 9.75 for an output impedance of 200 kΩ and a bias current of 0.5 mA.

Rout
Vb1
M1
Vb2
M2

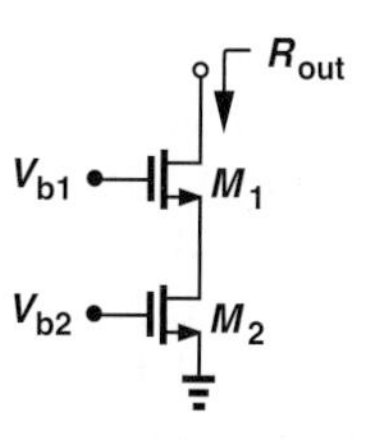

Figure 9.75

(a) Determine $(W/L)_1 = (W/L)_2$ if $\lambda = 0.1\ V^{-1}$.

(b) Calculate the required value of V_{b2}.

9.57. The bipolar cascode amplifier of Fig. 9.76 must be designed for a voltage gain of 500. Use Eq. (9.53) and assume $\beta = 100$.

(a) What is the minimum required value of V_A?

(b) For a bias current of 0.5 mA, calculate the required bias component in V_{in}.

(c) Compute the value of V_{b1} such that Q_1 sustains a collector-emitter voltage of 500 mV.

VCC
I1
Vout
Vb1
Q2
Vin
Q1

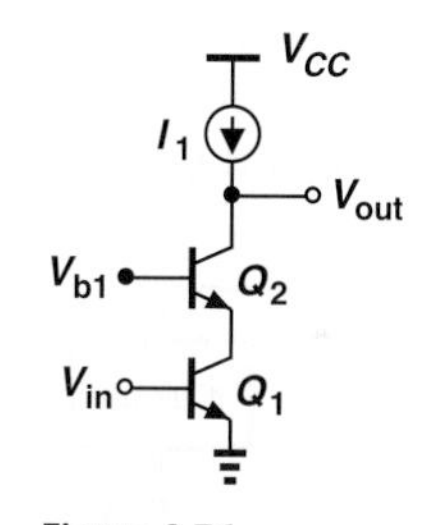

Figure 9.76

9.58. Design the cascode amplifier shown in Fig. 9.77 for a power budget of 2 mW. Select V_{b1} and V_{b2} such that Q_1 and Q_4 sustain a base-collector forward bias of 200 mV. What voltage gain is achieved?

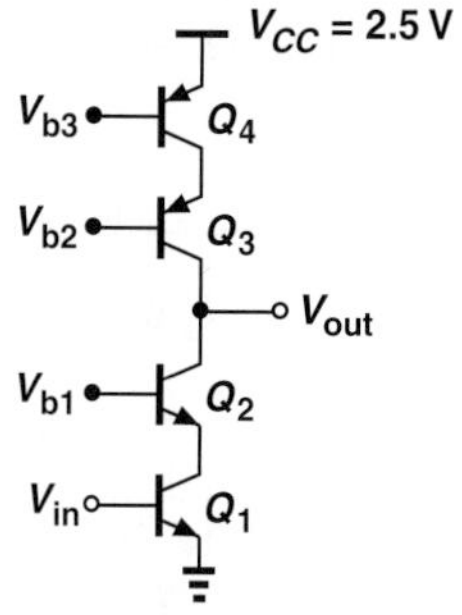

VCC = 2.5 V
Vb3
Q4
Vb2
Q3
Vout
Vb1
Q2
Vin
Q1

Figure 9.77

9.59. Design the CMOS cascode amplifier of Fig. 9.78 for a voltage gain of 200 and a power budget of 2 mW with $V_{DD} =$ 1.8 V. Assume $(W/L)_1 = \cdots = (W/L)_4 =$ 20/0.18 and $\lambda_p = 2\lambda_n = 0.2\ \text{V}^{-1}$. Determine the required dc levels of V_{in} and V_{b3}. For simplicity, assume $V_{b1} = V_{b2} = 0.9$ V.

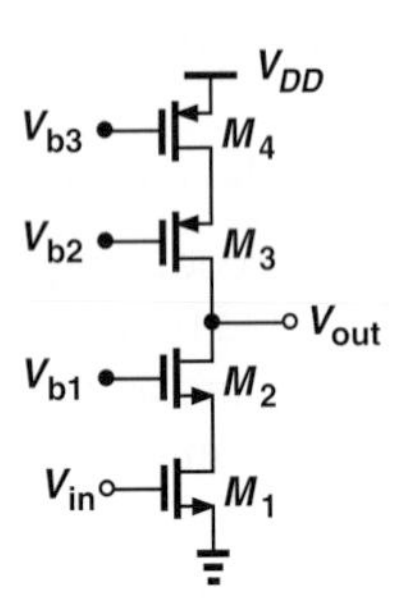

Figure 9.78

SOL **9.60.** The current mirror shown in Fig. 9.79 must deliver $I_1 = 0.5$ mA to a circuit with a total power budget of 2 mW. Assuming $V_A = \infty$ and $\beta \gg 1$, determine the required value of I_{REF} and the relative sizes of Q_{REF} and Q_1.

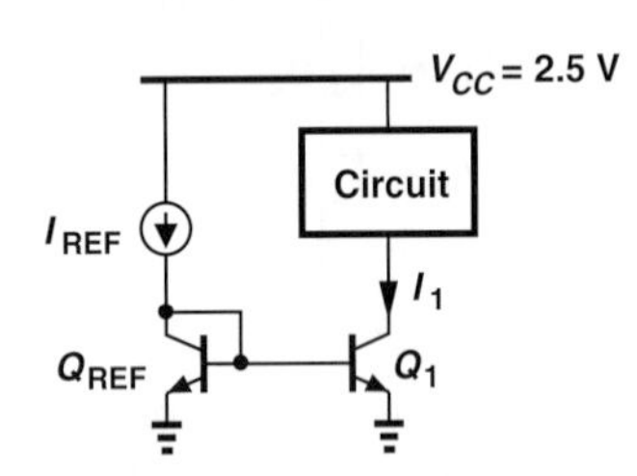

Figure 9.79

9.61. In the circuit of Fig. 9.80, Q_2 operates as an emitter follower. Design the circuit for a power budget of 3 mW and an output impedance of 50 Ω. Assume $V_A = \infty$ and $\beta \gg 1$.

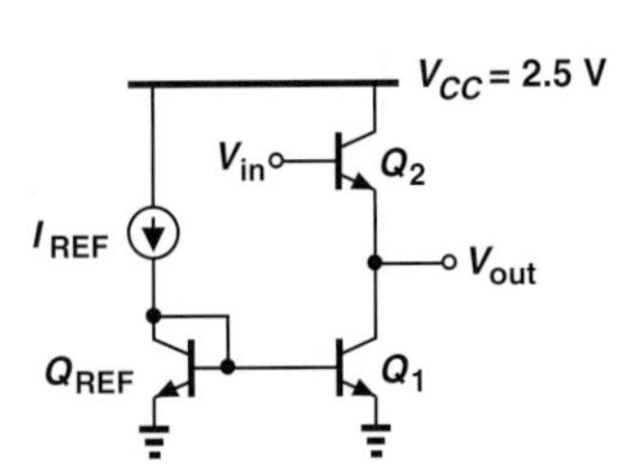

Figure 9.80

9.62. In the circuit of Fig. 9.81, Q_2 operates as a common-base stage. Design the circuit for an output impedance of 500 Ω, a voltage gain of 20, and a power budget of 3 mW. Assume $V_A = \infty$ and $\beta \gg 1$.

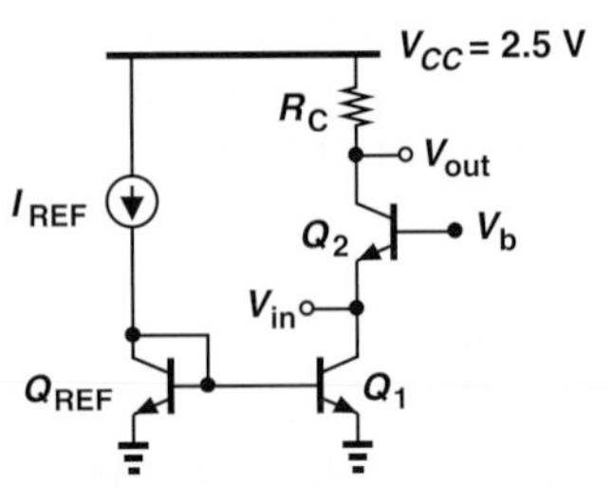

Figure 9.81

9.63. Design the circuit of Fig. 9.31 for $I_{copy} =$ 0.5 mA and an error of less than 1% with respect to the nominal value. Explain the trade-off between accuracy and power dissipation in this circuit. Assume $V_{CC} =$ 2.5 V.

9.64. Design the circuit of Fig. 9.35 such that the bias current of Q_2 is 1 mA and the error in I_{C1} with respect to its nominal value is less than 10%. Is the solution unique?

9.65. Figure 9.82 shows an arrangement where M_1 and M_2 serve as current sources for circuits 1 and 2. Design the circuit for a power budget of 3 mW.

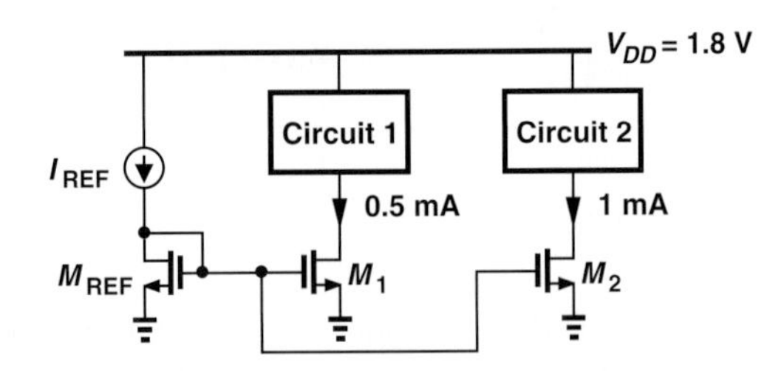

Figure 9.82

9.66. The common-source stage depicted in Fig. 9.83 must be designed for a voltage gain of 20 and a power budget of 2 mW. Assuming $(W/L)_1 = 20/0.18$, $\lambda_n =$ 0.1 V^{-1}, and $\lambda_p = 0.2\ \text{V}^{-1}$, design the circuit.

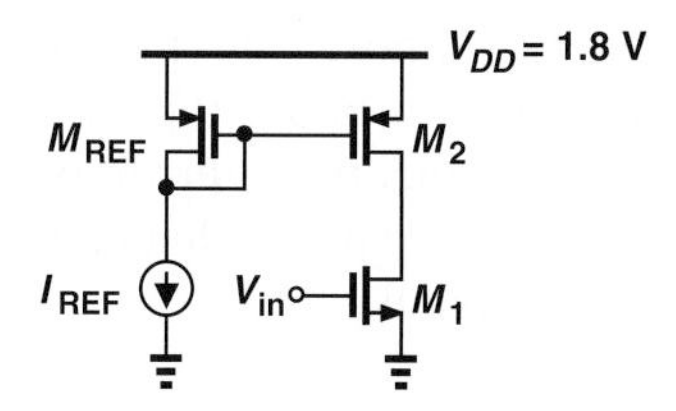

Figure 9.83

9.67. The source follower of Fig. 9.84 must achieve a voltage gain of 0.85 and an output impedance of 100 Ω. Assuming $(W/L)_2 = 10/0.18$, $\lambda_n = 0.1\ \text{V}^{-1}$, and $\lambda_p = 0.2\ \text{V}^{-1}$, design the circuit.

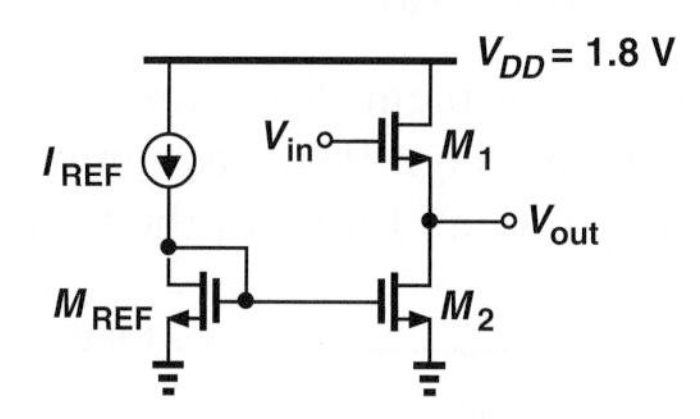

Figure 9.84

9.68. The common-gate stage of Fig. 9.85 employs the current source M_3 as the load to achieve a high voltage gain. For simplicity, neglect channel-length modulation in M_1. Assuming $(W/L)_3 = 40/0.18$, $\lambda_n = 0.1\ \text{V}^{-1}$, and $\lambda_p = 0.2\ \text{V}^{-1}$, design the circuit for a voltage gain of 20, an input impedance of 50 Ω, and a power budget of 13 mW. (You may not need all of the power budget.)

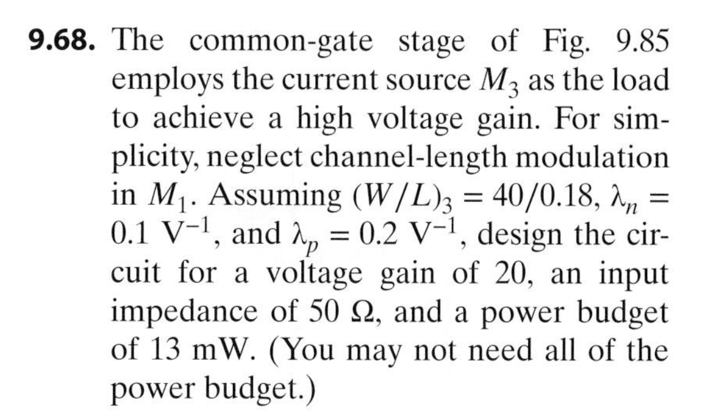

Figure 9.85

SPICE PROBLEMS

In the following problems, use the MOS device models given in Appendix A. For bipolar transistors, assume $I_{S,npn} = 5 \times 10^{-16}$ A, $\beta_{npn} = 100$, $V_{A,npn} = 5$ V, $I_{S,pnp} = 8 \times 10^{-16}$ A, $\beta_{pnp} = 50$, $V_{A,pnp} = 3.5$ V.

9.69. In the circuit of Fig. 9.86, we wish to suppress the error due to the base currents by means of resistor R_P.

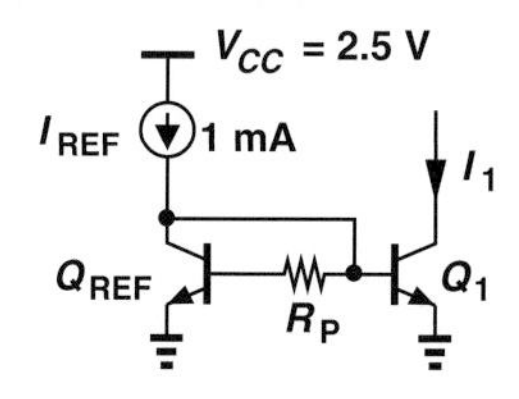

Figure 9.86

(a) Tying the collector of Q_2 to V_{CC}, select the value of R_P so as to minimize the error between I_1 and I_{REF}.

(b) What is the change in the error if the β of both transistors varies by ±3%?

(c) What is the change in the error if R_P changes by ±10%?

9.70. Repeat Problem 9.69 for the circuit shown in Fig. 9.87. Which circuit exhibits less sensitivity to variations in β and R_P?

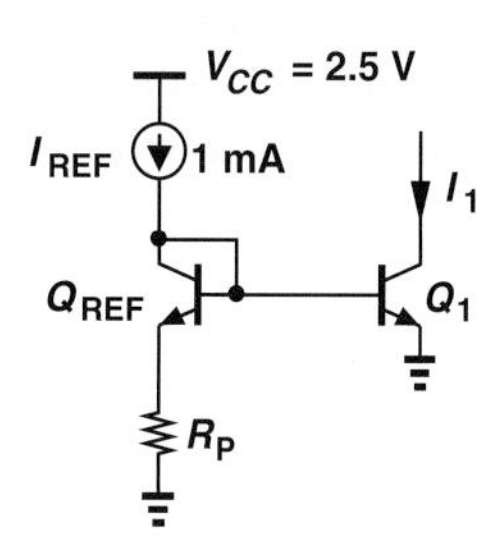

Figure 9.87

9.71. Figure 9.88 depicts a cascode current source whose value is defined by the mirror arrangement, M_1-M_2. Assume $W/L = 5\ \mu m/0.18\ \mu m$ for M_1-M_3.

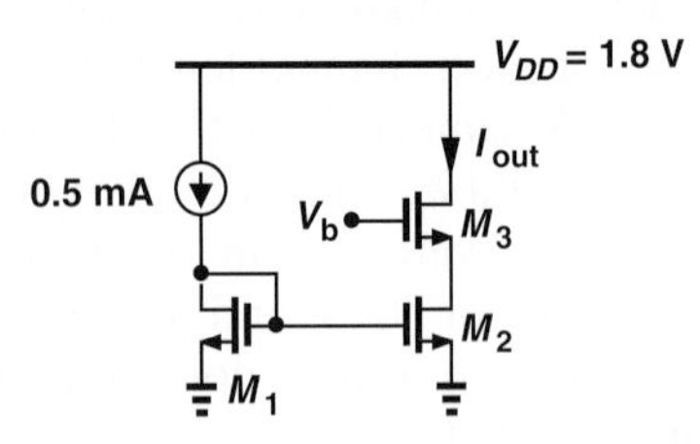

Figure 9.88

(a) Select the value of V_b so that I_{out} is precisely equal to 0.5 mA.
(b) Determine the change in I_{out} if V_b varies by ±100 mV. Explain the cause of this change.
(c) Using both hand analysis and SPICE simulations, determine the output impedance of the cascode and compare the results.

9.72. We wish to study the problem of biasing in a high-gain cascode stage, Fig. 9.89. Assume $(W/L)_{1,2} = 10\ \mu m/0.18\ \mu m$, $V_b = 0.9$ V, and $I_1 = 1$ mA is an ideal current source.

(a) Plot the input/output characteristic and determine the value of V_{in} at which the slope (small-signal gain) reaches a maximum.
(b) Now, suppose the biasing circuitry that must produce the above dc value for V_{in} incurs an error of ±20 mV. From (a), explain what happens to the small-signal gain.

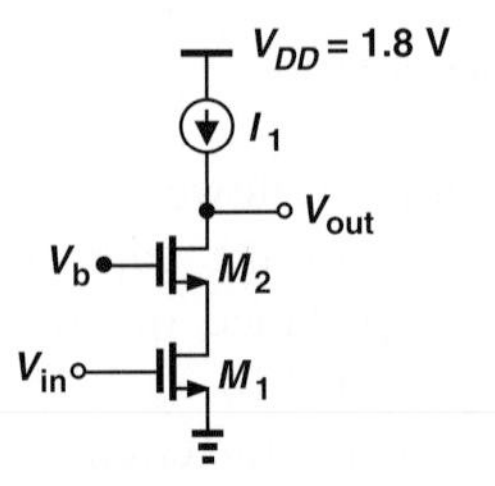

Figure 9.89

9.73. Repeat Problem 9.72 for the cascode shown in Fig. 9.90, assuming $W/L = 10\ \mu m/0.18\ \mu m$ for all of the transistors.

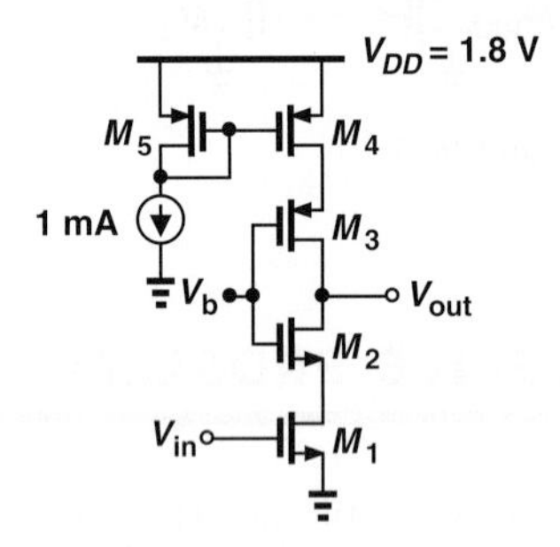

Figure 9.90

REFERENCES

1. B. Razavi, *Design of Analog CMOS Integrated Circuits*, McGraw-Hill, 2001.
2. L. E. Aygun, et al., "Hybrid LAE-CMOS force-sensing system employing TFT-based compressed sensing for scalability of tactile sensing," *IEEE Trans. Biomedical Circuits and Systems*, vol. 13, pp. 1264–1276, Dec. 2019.

10

Differential Amplifiers

Watch Companion YouTube Videos:

Razavi Electronics 2, Lec 7: Problem of Noise Coupling, Introduction to Differential Pair
https://www.youtube.com/watch?v=GXCsDRKJPag

Razavi Electronics 2, Lec 8: Intuitive Study of Bipolar Diff. Pair, CM and Diff. Characs.
https://www.youtube.com/watch?v=xRBHZRDYPqE

Razavi Electronics 2, Lec 9: Large-Signal Analysis of Bipolar Differential Pair
https://www.youtube.com/watch?v=CW6yT69n9BA

Razavi Electronics 2, Lec 10: Small-Signal Analysis of Bipolar Differential Pair
https://www.youtube.com/watch?v=4GGfQvvvcNE

Razavi Electronics 2, Lec 11: Additional Examples of Bipolar Diff. Pairs
https://www.youtube.com/watch?v=gUgOsZfyhOE

Razavi Electronics 2, Lec 12: Large-Signal Analysis of MOS Differential Pair
https://www.youtube.com/watch?v=RgpH9_sy5iI

Razavi Electronics 2, Lec 13: Additional Analysis of MOS Differential Pair
https://www.youtube.com/watch?v=5MWQWe0S9mo

Razavi Electronics 2, Lec 14: Small-Signal Analysis of MOS Differential Pair
https://www.youtube.com/watch?v=cBcjMVxBJ4A

Razavi Electronics 2, Lec 15: High-Gain Differential Pairs, Introduction to Diff. Pair with Active Load
https://www.youtube.com/watch?v=gjKTAx9FQHo

Razavi Electronics 2, Lec 16: Small-Signal Behavior of Diff. Pair with Active Load
https://www.youtube.com/watch?v=cst7wjPd6R4

The elegant concept of "differential" signals and amplifiers was invented in the 1940s and first utilized in vacuum-tube circuits. Since then, differential circuits have found increasingly wider usage in microelectronics and serve as a robust, high-performance design paradigm in many of today's systems. This chapter describes bipolar and MOS differential amplifiers and formulates their large-signal and small-signal properties. The concepts are outlined below.

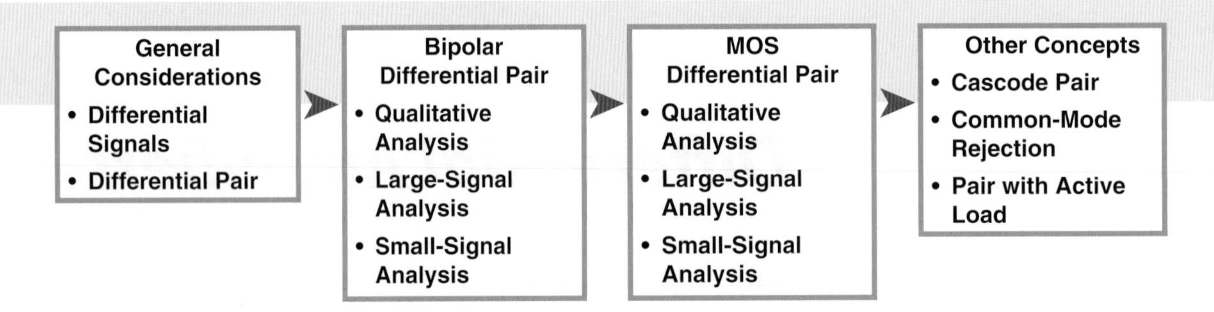

10.1 GENERAL CONSIDERATIONS

10.1.1 Initial Thoughts

We have already seen that op amps have *two* inputs, a point of contrast to the amplifiers studied in previous chapters. In order to further understand the need for differential circuits, let us first consider an example.

Example 10-1

Having learned the design of rectifiers and basic amplifier stages, an electrical engineering student constructs the circuit shown in Fig. 10.1(a) to amplify the signal produced by a microphone. Unfortunately, upon applying the result to a speaker, the student observes that the amplifier output contains a strong "humming" noise, i.e., a steady low-frequency component. Explain what happens.

Figure 10.1 (a) CE stage powered by a rectifier, (b) ripple on supply voltage, (c) effect at output, (d) ripple and signal paths to output.

Solution Recall from Chapter 3 that the current drawn from the rectified output creates a ripple waveform at twice the ac line frequency (50 or 60 Hz) [Fig. 10.1(b)]. Examining the output of the common-emitter stage, we can identify two components: (1) the amplified version of the microphone signal and (2) the ripple waveform present on V_{CC}. For the latter, we can write

$$V_{out} = V_{CC} - R_C I_C, \tag{10.1}$$

noting that V_{out} simply "tracks" V_{CC} and hence contains the ripple in its entirety. The "hum" originates from the ripple. Figure 10.1(c) depicts the overall output in the presence of both the signal and the ripple. As illustrated in Fig. 10.1(d), this phenomenon is summarized as the "supply noise goes to the output with a gain of unity." (A MOS implementation would suffer from the same problem.)

Exercise What is the hum frequency for a full-wave rectifier or a half-wave rectifier?

How should we suppress the hum in the above example? We can increase C_1, thus lowering the ripple amplitude, but the required capacitor value may become prohibitively large if many circuits draw current from the rectifier. Alternatively, we can modify the amplifier topology such that the output is insensitive to V_{CC}. How is that possible? Equation (10.1) implies that a change in V_{CC} directly appears in V_{out}, fundamentally because both V_{out} and V_{CC} are measured with respect to ground and differ by $R_C I_C$. But what if V_{out} is *not* "referenced" to ground?! More specifically, what if V_{out} is measured with respect to another point that itself experiences the supply ripple to the same extent? It is thus possible to eliminate the ripple from the "net" output.

While rather abstract, the above conjecture can be readily implemented. Figure 10.2(a) illustrates the core concept. The CE stage is duplicated on the right, and the output is now measured *between* nodes X and Y rather than from X to ground. What happens if V_{CC} contains ripple? Both V_X and V_Y rise and fall by the same amount and hence the *difference* between V_X and V_Y remains free from the ripple.

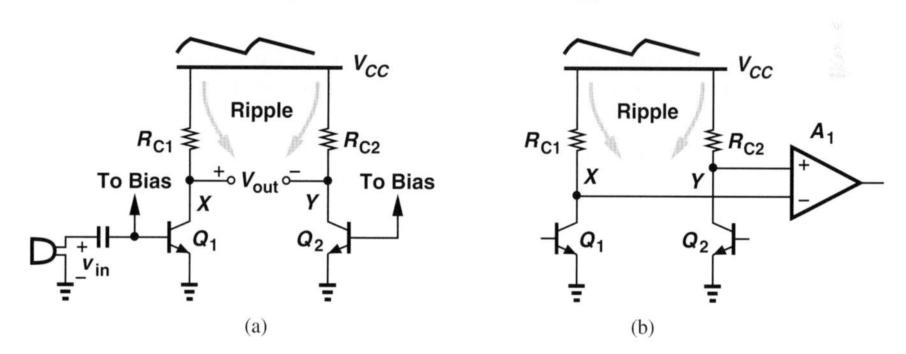

Figure 10.2 Use of two CE stages to remove effect of ripple.

In fact, denoting the ripple by v_r, we express the small-signal voltages at these nodes as

$$v_X = A_v v_{in} + v_r \tag{10.2}$$

$$v_Y = v_r. \tag{10.3}$$

That is,

$$v_X - v_Y = A_v v_{in}. \tag{10.4}$$

Note that Q_2 carries no signal, simply serving as a constant current source.

The above development serves as the foundation for differential amplifiers: the symmetric CE stages provide *two* output nodes whose voltage difference remains free from the supply ripple.

10.1.2 Differential Signals

Let us return to the circuit of Fig. 10.2(a) and recall that the duplicate stage consisting of Q_2 and R_{C2} remains "idle," thereby "wasting" current. We may therefore wonder if this stage can provide signal amplification in addition to establishing a reference point for V_{out}. In our first attempt, we directly apply the input signal to the base of Q_2 [Fig. 10.3(a)]. Unfortunately, the signal components at X and Y are in phase, canceling each other as they appear in $v_X - v_Y$:

$$v_X = A_v v_{in} + v_r \tag{10.5}$$

$$v_Y = A_v v_{in} + v_r \tag{10.6}$$

$$\Rightarrow v_X - v_Y = 0. \tag{10.7}$$

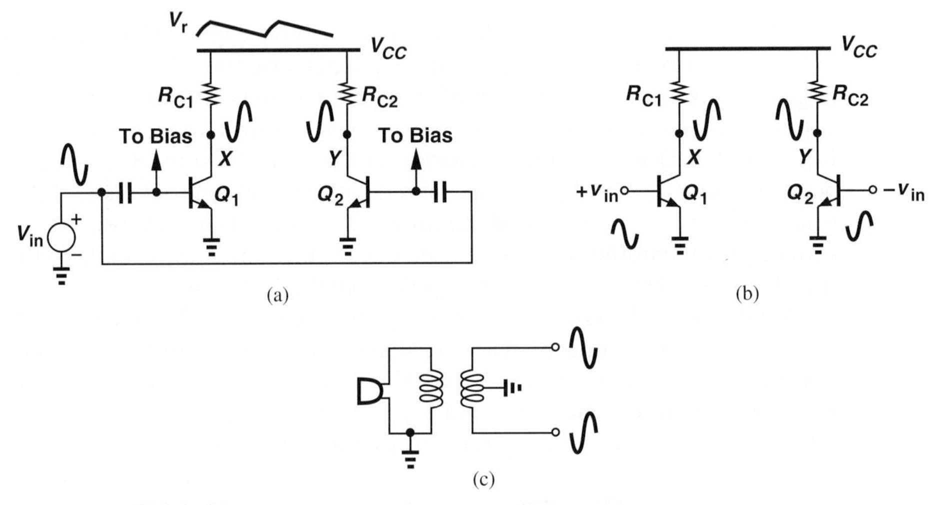

Figure 10.3 (a) Application of one input signal to two CE stages, (b) use of differential input signals, (c) generation of differential phases from one signal.

For the signal components to *enhance* each other at the output, we can *invert* one of the input phases as shown in Fig. 10.3(b), obtaining

$$v_X = A_v v_{in} + v_r \tag{10.8}$$

$$v_Y = -A_v v_{in} + v_r \tag{10.9}$$

and hence

$$v_X - v_Y = 2A_v v_{in}. \tag{10.10}$$

Compared to the circuit of Fig. 10.2(a), this topology provides twice the output swing by exploiting the amplification capability of the duplicate stage.

The reader may wonder how $-v_{in}$ can be generated. As illustrated in Fig. 10.3(c), a simple approach is to utilize a transformer to convert the microphone signal to two components bearing a phase difference of 180°.

Our thought process has led us to the specific waveforms in Fig. 10.3(b): the circuit senses two inputs that vary by equal and opposite amounts and generates two outputs that behave in a similar fashion. These waveforms are examples of "differential" signals and stand in contrast to "single-ended" signals—the type to which we are accustomed from basic circuits and previous chapters of this book. More specifically, a single-ended signal is one measured with respect to the common ground [Fig. 10.4(a)] and "carried by one line," whereas a differential signal is measured between two nodes that have equal and opposite swings [Fig. 10.4(b)] and is thus "carried by two lines."

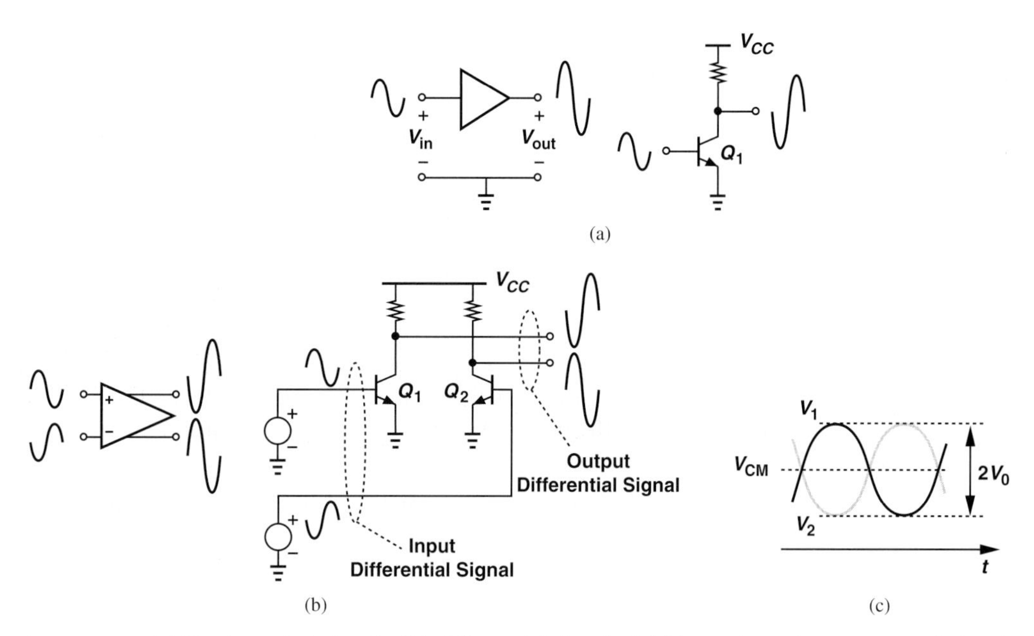

Figure 10.4 (a) Single-ended signals, (b) differential signals, (c) illustration of common-mode level.

Figure 10.4(c) summarizes the foregoing development. Here, V_1 and V_2 vary by equal and opposite amounts *and* have the same average (dc) level, V_{CM}, with respect to ground:

$$V_1 = V_0 \sin \omega t + V_{CM} \tag{10.11}$$

$$V_2 = -V_0 \sin \omega t + V_{CM}. \tag{10.12}$$

Since each of V_1 and V_2 has a peak-to-peak swing of $2V_0$, we say the "differential swing" is $4V_0$. We may also say V_1 and V_2 are differential signals to emphasize that they vary by equal and opposite amounts around a fixed level, V_{CM}.

The dc voltage that is common to both V_1 and V_2 [V_{CM} in Fig. 10.4(c)] is called the "common-mode (CM) level." That is, in the absence of differential signals, the two nodes remain at a potential equal to V_{CM} with respect to the global ground. For example, in the transformer of Fig. 10.3(c), $+v_{in}$ and $-v_{in}$ display a CM level of zero because the center tap of the transformer is grounded.

Example 10-2 How can the transformer of Fig. 10.3(c) produce an output CM level equal to +2 V?

Solution The center tap can simply be tied to a voltage equal to +2 V (Fig. 10.5).

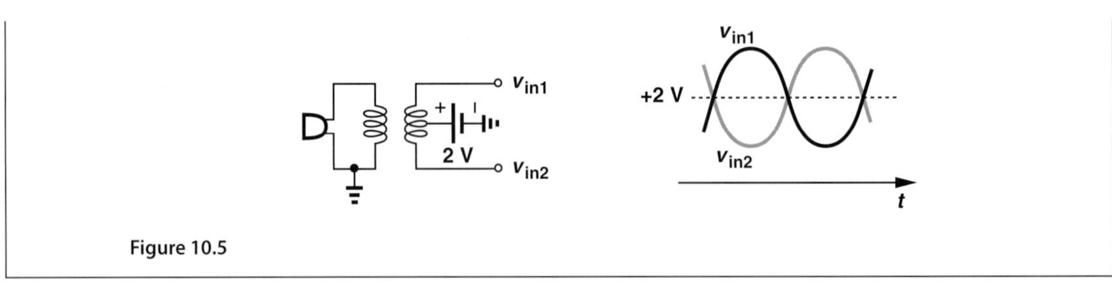

Figure 10.5

Exercise Does the CM level change if the inputs of the amplifier draw a bias current?

Example 10-3 Determine the common-mode level at the output of the circuit shown in Fig. 10.3(b).

Solution In the absence of signals, $V_X = V_Y = V_{CC} - R_C I_C$ (with respect to ground), where $R_C = R_{C1} = R_{C2}$ and I_C denotes the bias current of Q_1 and Q_2. Thus, $V_{CM} = V_{CC} - R_C I_C$. Interestingly, the ripple affects V_{CM} but not the differential output.

Exercise If a resistor of value R_1 is inserted between V_{CC} and the top terminals of R_{C1} and R_{C2}, what is the output CM level?

Our observations regarding supply ripple and the use of the "duplicate stage" provide sufficient justification for studying differential signals. But, how about the common-mode level? What is the significance of $V_{CM} = V_{CC} - R_C I_C$ in the above example? Why is it interesting that the ripple appears in V_{CM} but not in the differential output? We will answer these important questions in the following sections.

Bioengineering Application: Electrocardiography

Electrocardiography deals with the measurement of our heart activity. Every time our heart beats, it generates a voltage difference between the left side and right side of our body. An "electrocardiogram" (ECG) employs two electrodes to sense this difference [Fig. 10.6(a)]. We can view the result as simply V_1, or we can say nodes X and Y have certain voltages with respect to the ground and we are only interested in $V_X - V_Y = V_1$. The voltage waveform has an amplitude in the millivolt range, carries many details about our heart condition, and must be sensed with minimal corruption.

A great challenge in ECG electronics is that the wires attached to the electrodes pick up noise from the line voltage (also called the "mains" voltage) (110 V at 60 Hz or 220 V at 50 Hz) [Fig. 10.6(b)]. We can see that, if the wires are symmetrically packed, the noise coupling onto X is equal to that onto Y. That is, the coupling of 60 Hz or 50 Hz becomes a common-mode effect. As shown in Fig. 10.6(c), V_X and V_Y vary differentially (we have not proved this point) but also experience a common-mode fluctuation as well. Nonetheless, $V_X - V_Y$ is free from the noise.

The differential and CM components can be better understood if we consider two cases: (1) the 60-Hz disturbance is absent [Fig. 10.6(d)], or (2) the patient dies and has no heart activity [Fig. 10.6(e)]. In the latter case, $V_X - V_Y = 0$, and we say the patient has "flatlined."

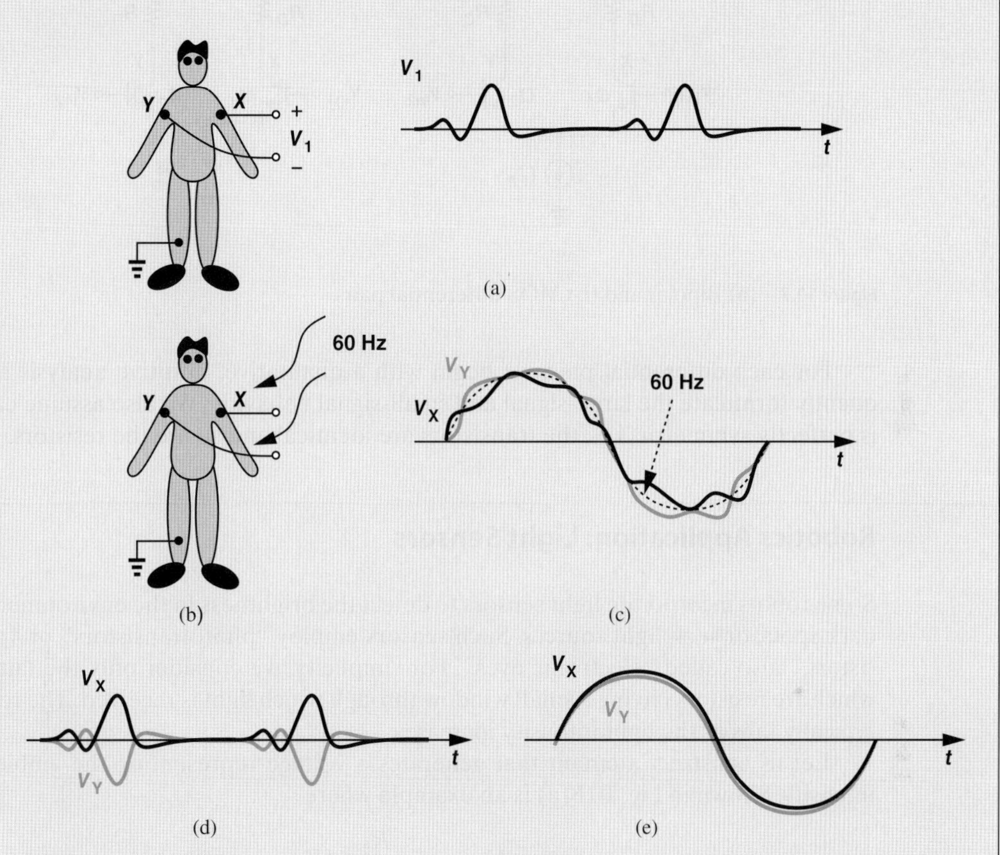

Figure 10.6 (a) Sensing heart activity by voltage difference on the body, (b) problem of 60-Hz noise, (c) effect of 60-Hz noise on V_X and V_Y, (d) behavior of V_X and V_Y with no 60-Hz noise, and (e) behavior of V_X and V_Y for a patient who has expired.

10.1.3 Differential Pair

Before formally introducing the differential pair, we must recognize that the circuit of Fig. 10.4(b) senses *two* inputs and can therefore serve as A_1 in Fig. 10.2(b). This observation leads to the differential pair.

While sensing and producing differential signals, the circuit of Fig. 10.4(b) suffers from some drawbacks. Fortunately, a simple modification yields an elegant, versatile topology. As illustrated in Fig. 10.7(a), the (bipolar) "differential pair"[1] is similar to the circuit of Fig. 10.4(b), except that the emitters of Q_1 and Q_2 are tied to a constant current source rather than to ground. We call I_{EE} the "tail current source." The MOS counterpart is shown in Fig. 10.7(b). In both cases, the sum of the transistor currents is equal to the tail current.

[1]Also called the "emitter-coupled pair" or the "long-tailed pair."

Our objective is to analyze the large-signal and small-signal behavior of these circuits and demonstrate their advantages over the "single-ended" stages studied in previous chapters.

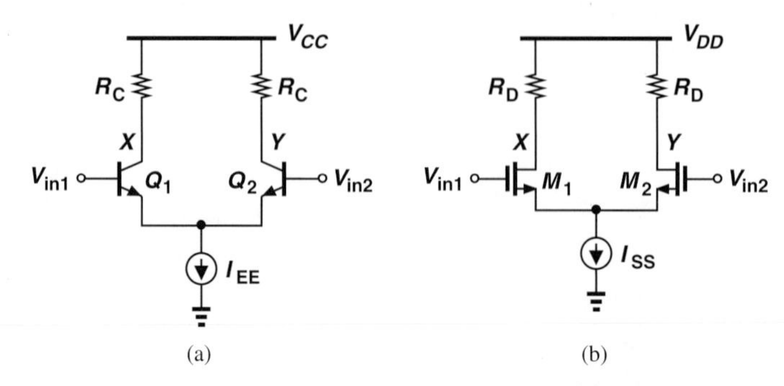

Figure 10.7 (a) Bipolar and (b) MOS differential pairs.

For each differential pair, we begin with a qualitative, intuitive analysis and subsequently formulate the large-signal and small-signal behavior. We also assume each circuit is perfectly symmetric, i.e., the transistors are identical and so are the resistors.

Robotics Application: Light Sensors

Some robots incorporate light sensors to detect the brightness in the environment, avoid dark spots, or seek light sources. Such sensors employ "phototransistors" or "photoresistors," also called "electronic eyes." For simplicity, we consider only the latter here, which are resistors having a small window through which light is received. The resistance typically falls as the light intensity increases.

Let us construct a circuit that generates a voltage representing the ambient light intensity. Shown in Fig. 10.8(a) is an example where

$$V_{out} = \frac{R_1}{R_1 + R_{p1}} V_B. \tag{10.13}$$

Thus, V_{out} rises as the intensity rises. A typical situation is that the robot must stop if the intensity is below a certain threshold. This means that as R_{p1} and hence V_{out} *exceed* a certain value, the robot must stop. For example, if, in typical ambient light, $R_{p1} = 1\ \text{k}\Omega$, we also select $R_1 = 1\ \text{k}\Omega$, obtaining $V_{out} = V_B/2$, which we consider the threshold. As the robot reaches a dark spot, $V_{out} < V_B/2$, commanding the robot to stop.

The circuit of Fig. 10.8(a) faces a critical issue. In a robotics environment, the operation of various motors and actuators introduces considerable "noise" on the supply voltage. The exact mechanisms of this noise generation are beyond the scope of this book, but we note that the seemingly quiet V_B in fact suffers from fluctuations [Fig. 10.8(b)]. This noise occasionally causes V_{out} in Fig. 10.8(a) to fall below the threshold, forcing the robot to stop even though it has not entered a dark spot. We say the "single-ended" output is prone to noise.

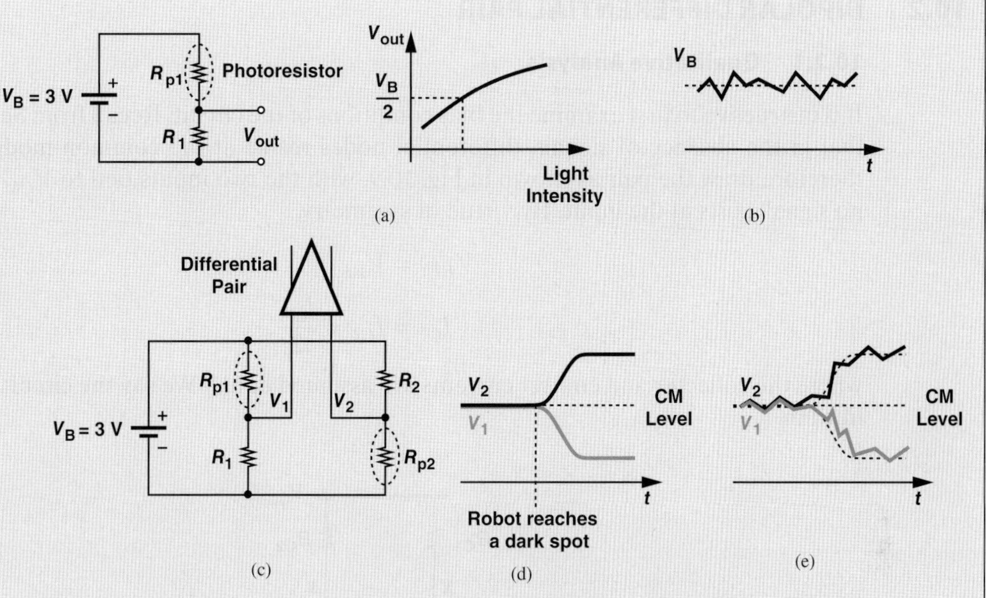

Figure 10.8 (a) Simple circuit for measuring light intensity, (b) noise on V_B, (c) differential circuit for measuring light intensity, (d) the V_1 and V_2 waveforms, and (e) effect of CM noise.

We now develop an improved design, one whose output is much less sensitive to supply noise. Consider the arrangement shown in Fig. 10.8(c), where two photoresistors R_{p1} and R_{p2} are employed in two voltage dividers. We assume that, in typical light, $R_{p1} = R_{p2} = R_1 = R_2$, and view the *difference* between V_1 and V_2 as the output:

$$V_1 - V_2 = \frac{R_1}{R_1 + R_{p1}} V_B - \frac{R_{p2}}{R_{p2} + R_2} V_B. \tag{10.14}$$

What happens as the robot approaches a dark spot? Both R_{p1} and R_{p2} rise by a small amount. We have

$$\frac{dV_1}{dR_{p1}} = \frac{-R_1}{(R_1 + R_{p1})^2} V_B \tag{10.15}$$

$$\frac{dV_2}{dR_{p2}} = \frac{R_2}{(R_2 + R_{p2})^2} V_B. \tag{10.16}$$

Thus, V_1 and V_2 change by equal and opposite amounts [Fig. 10.8(d)]. We say V_1 and V_2 begin from a common-mode level and change differentially. We can also prove that noise on V_B affects $V_1 - V_2$ only by a small amount. Since

$$V_1 - V_2 = \frac{R_1 R_2 - R_{p1} R_{p2}}{(R_1 + R_{p2})(R_{p2} + R_2)} V_B, \tag{10.17}$$

and since $R_1 R_2 \approx R_{p1} R_{p2}$ in nominal light conditions, fluctuations in V_B have little effect on $V_1 - V_2$. Note that V_1 and V_2 do suffer from significant common-mode noise in this case [Fig. 10.8(e)].

10.2 BIPOLAR DIFFERENTIAL PAIR

10.2.1 Qualitative Analysis

It is instructive to first examine the bias conditions of the circuit. Recall from Section 10.1.2 that in the absence of signals, differential nodes reside at the common-mode level. We therefore draw the pair as shown in Fig. 10.9, with the two inputs tied to V_{CM} to indicate no signal exists at the input. By virtue of symmetry,

$$V_{BE1} = V_{BE2} \tag{10.18}$$

$$I_{C1} = I_{C2} = \frac{I_{EE}}{2}, \tag{10.19}$$

where the collector and emitter currents are assumed equal. We say the circuit is in "equilibrium."

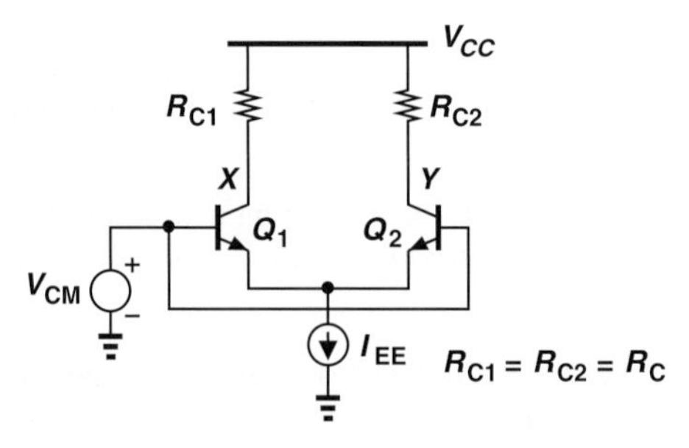

Figure 10.9 Response of differential pair to input CM change.

Did you know?

The amplifiers studied in previous chapters have *one* input with respect to ground [Fig. (a)] whereas the differential pair has two distinct inputs. An important application of the differential pair is at the input of op amps, where inverting and noninverting input terminals are necessary [Fig. (b)]. Without this second input, many op-amp-based functions would be difficult to realize. For example, the non-inverting amplifier and the precision rectifier utilize both inputs.

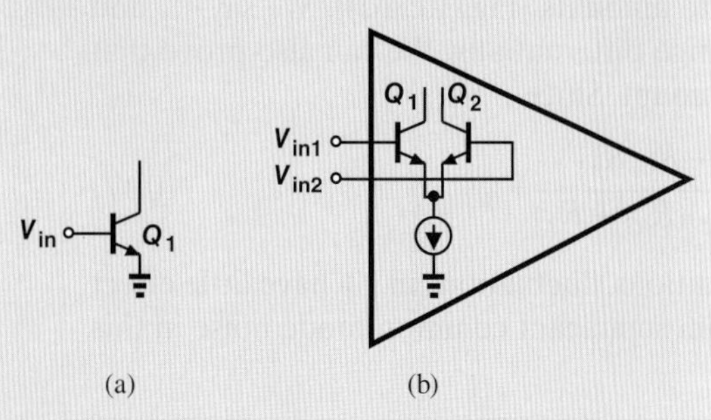

Amplifiers with single-ended and differential inputs.

Thus, the voltage drop across each load resistor is equal to $R_C I_{EE}/2$ and hence

$$V_X = V_Y = V_{CC} - R_C \frac{I_{EE}}{2}. \tag{10.20}$$

In other words, if the two input voltages are equal, so are the two outputs. We say a zero differential input produces a zero differential output. The circuit also "rejects" the effect of supply ripple: if V_{CC} experiences a change, the differential output, $V_X - V_Y$, does not.

Are Q_1 and Q_2 in the active region? To avoid saturation, the collector voltages must not fall below the base voltages:

$$V_{CC} - R_C \frac{I_{EE}}{2} \geq V_{CM}, \tag{10.21}$$

revealing that V_{CM} cannot be arbitrarily high.

Example 10-4 A bipolar differential pair employs a load resistance of 1 kΩ and a tail current of 1 mA. How close to V_{CC} can V_{CM} be chosen?

Solution Equation 10.21 gives

$$V_{CC} - V_{CM} \geq R_C \frac{I_{EE}}{2} \tag{10.22}$$

$$\geq 0.5\ \text{V}. \tag{10.23}$$

That is, V_{CM} must remain below V_{CC} by at least 0.5 V.

Exercise What value of R_C allows the input CM level to approach V_{CC} if the transistors can tolerate a base-collector forward bias of 400 mV?

Now, let us vary V_{CM} in Fig. 10.9 by a small amount and determine the circuit's response. Interestingly, Eqs. (10.18)–(10.20) remain unchanged, thereby suggesting that neither the collector current nor the collector voltage of the transistors is affected. We say the circuit does not respond to changes in the input common-mode level, or the circuit "rejects" input CM variations. Figure 10.10 summarizes these results.

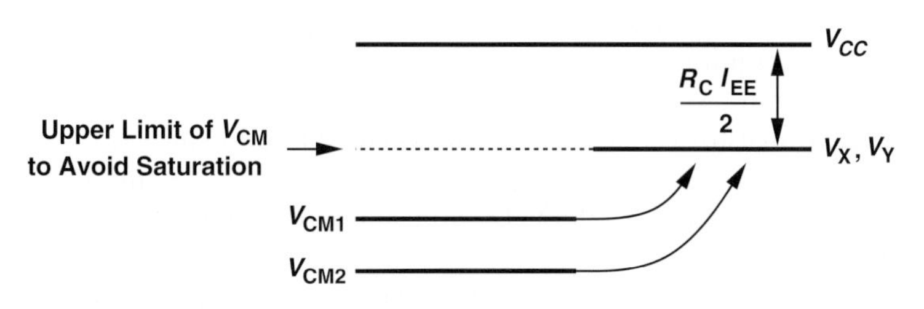

Figure 10.10 Effect of V_{CM1} and V_{CM2} at output.

The "common-mode rejection" capability of the differential pair distinctly sets it apart from our original circuit in Fig. 10.4(b). In the latter, if the base voltage of Q_1 and Q_2 changes, so do their collector currents and voltages (why?). The reader may recognize that it is the tail current source in the differential pair that guarantees constant collector currents and hence rejection of the input CM level.

With our treatment of the common-mode response, we now turn to the more interesting case of *differential* response. We hold one input constant, vary the other, and examine the currents flowing in the two transistors. While not exactly differential, such input signals provide a simple, intuitive starting point. Recall that $I_{C1} + I_{C2} = I_{EE}$.

Consider the circuit shown in Fig. 10.11(a), where the two transistors are drawn with a vertical offset to emphasize that Q_1 senses a more positive base voltage. Since the difference between the base voltages of Q_1 and Q_2 is so large, we postulate that Q_1 "hogs" all of the tail current, thereby turning Q_2 off. That is,

$$I_{C1} = I_{EE} \tag{10.24}$$

$$I_{C2} = 0, \tag{10.25}$$

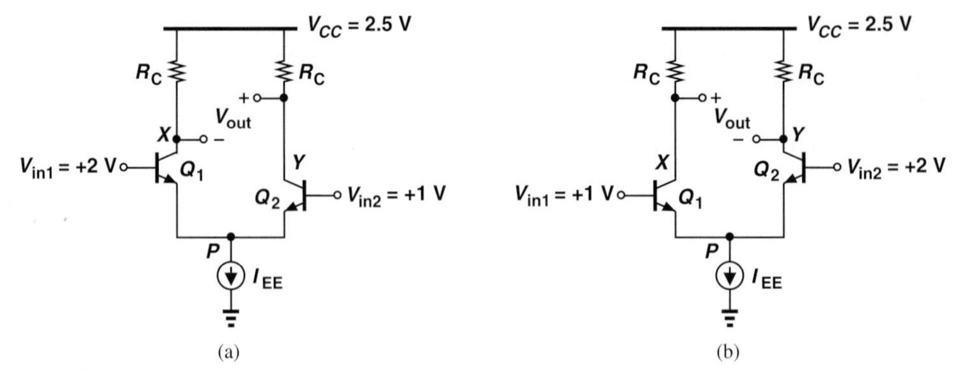

Figure 10.11 Response of bipolar differential pair to (a) large positive input difference and (b) large negative input difference.

and hence

$$V_X = V_{CC} - R_C I_{EE} \tag{10.26}$$

$$V_Y = V_{CC}. \tag{10.27}$$

But, how can we *prove* that Q_1 indeed absorbs all of I_{EE}? Let us assume that it is not so; i.e., $I_{C1} < I_{EE}$ and $I_{C2} \neq 0$. If Q_2 carries an appreciable current, then its base-emitter voltage must reach a typical value of, say, 0.8 V. With its base held at +1 V, the device therefore requires an emitter voltage of $V_P \approx 0.2$ V. However, this means that Q_1 sustains a base-emitter voltage of $V_{in1} - V_P = +2\text{ V} - 0.2\text{ V} = 1.8$ V!! Since with $V_{BE} = 1.8$ V, a typical transistor carries an enormous current, and since I_{C1} cannot exceed I_{EE}, we conclude that the conditions $V_{BE1} = 1.8$ V and $V_P \approx 0.2$ V cannot occur. In fact, with a typical base-emitter voltage of 0.8 V, Q_1 holds node P at approximately +1.2 V, ensuring that Q_2 remains off.

Symmetry of the circuit implies that swapping the base voltages of Q_1 and Q_2 reverses the situation [Fig. 10.11(b)], giving

$$I_{C2} = I_{EE} \tag{10.28}$$

$$I_{C1} = 0, \tag{10.29}$$

and hence

$$V_Y = V_{CC} - R_C I_{EE} \tag{10.30}$$

$$V_X = V_{CC}. \tag{10.31}$$

The above experiments reveal that, as the difference between the two inputs departs from zero, the differential pair "steers" the tail current from one transistor to the other. In fact, based on Eqs. (10.19), (10.24), and (10.28), we can sketch the collector currents of Q_1 and Q_2 as a function of the input difference [Fig. 10.12(a)]. We have not yet formulated these characteristics but we do observe that the collector current of each transistor goes from 0 to I_{EE} if $|V_{in1} - V_{in2}|$ becomes sufficiently large.

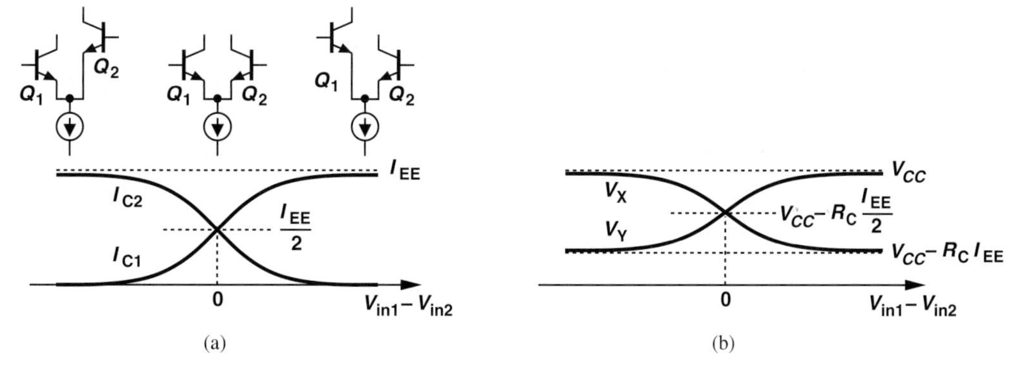

Figure 10.12 Variation of (a) collector currents and (b) output voltages as a function of input.

It is also important to note that V_X and V_Y vary differentially in response to $V_{in1} - V_{in2}$. From Eqs. (10.20), (10.26), and (10.30), we can sketch the input/output characteristics of the circuit as shown in Fig. 10.12(b). That is, a nonzero differential input yields a nonzero differential output—a behavior in sharp contrast to the CM response. Since V_X and V_Y are differential, we can define a common-mode level for them. Given by $V_{CC} - R_C I_{EE}/2$, this quantity is called the "output CM level."

Example 10-5 A bipolar differential pair employs a tail current of 0.5 mA and a collector resistance of 1 kΩ. What is the maximum allowable base voltage if the differential input is large enough to completely steer the tail current? Assume $V_{CC} = 2.5$ V.

Solution If I_{EE} is completely steered, the transistor carrying the current lowers its collector voltage to $V_{CC} - R_C I_{EE} = 2$ V. Thus, the base voltage must remain below this value so as to avoid saturation.

Exercise Repeat the above example if the tail current is raised to 1 mA.

In the last step of our qualitative analysis, we "zoom in" around $V_{in1} - V_{in2} = 0$ (the equilibrium condition) and study the circuit's behavior for a *small* input difference. As illustrated in Fig. 10.13(a), the base voltage of Q_1 is raised from V_{CM} by ΔV while that of Q_2 is lowered from V_{CM} by the same amount. We surmise that I_{C1} increases slightly and, since $I_{C1} + I_{C2} = I_{EE}$, I_{C2} decreases by the same amount:

$$I_{C1} = \frac{I_{EE}}{2} + \Delta I \tag{10.32}$$

$$I_{C2} = \frac{I_{EE}}{2} - \Delta I. \tag{10.33}$$

How is ΔI related to ΔV? If the emitters of Q_1 and Q_2 were directly tied to ground, then ΔI would simply be equal to $g_m \Delta V$. In the differential pair, however, node P is free to rise or fall. We must therefore compute the change in V_P.

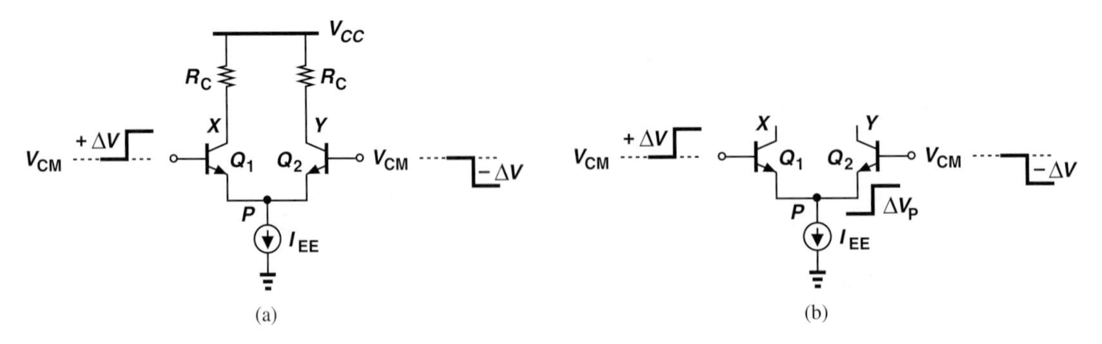

Figure 10.13 (a) Differential pair sensing small, differential input changes, (b) hypothetical change at P.

Suppose, as shown in Fig. 10.13(b), V_P rises by ΔV_P. As a result, the net increase in V_{BE1} is equal to $\Delta V - \Delta V_P$ and hence

$$\Delta I_{C1} = g_m(\Delta V - \Delta V_P). \tag{10.34}$$

Similarly, the net decrease in V_{BE2} is equal to $\Delta V + \Delta V_P$, yielding

$$\Delta I_{C2} = -g_m(\Delta V + \Delta V_P). \tag{10.35}$$

But recall from Eqs. (10.32) and (10.33) that ΔI_{C1} must be equal to $-\Delta I_{C2}$, dictating that

$$g_m(\Delta V - \Delta V_P) = g_m(\Delta V + \Delta V_P) \tag{10.36}$$

and hence

$$\Delta V_P = 0. \tag{10.37}$$

Interestingly, the tail voltage remains constant if the two inputs vary differentially and by a small amount—an observation critical to the small-signal analysis of the circuit.

The reader may wonder why Eq. (10.37) does not hold if ΔV is large. Which one of the above equations is violated? For a large differential input, Q_1 and Q_2 carry significantly *different* currents, thus exhibiting unequal transconductances and prohibiting the omission of g_m's from the two sides of Eq. (10.36).

With $\Delta V_P = 0$ in Fig. 10.13(a), we can rewrite Eqs. (10.34) and (10.35) respectively as

$$\Delta I_{C1} = g_m \Delta V \tag{10.38}$$

$$\Delta I_{C2} = -g_m \Delta V \tag{10.39}$$

and

$$\Delta V_X = -g_m \Delta V R_C \tag{10.40}$$

$$\Delta V_Y = g_m \Delta V R_C. \tag{10.41}$$

The differential output therefore goes from 0 to

$$\Delta V_X - \Delta V_Y = -2g_m \Delta V R_C. \tag{10.42}$$

We define the small-signal differential gain of the circuit as

$$A_v = \frac{\text{Change in Differential Output}}{\text{Change in Differential Input}} \tag{10.43}$$

$$= \frac{-2g_m \Delta V R_C}{2\Delta V} \tag{10.44}$$

$$= -g_m R_C. \tag{10.45}$$

(Note that the change in the differential input is equal to $2\Delta V$.) This expression is similar to that of the common-emitter stage.

Example 10-6 Design a bipolar differential pair for a gain of 10 and a power budget of 1 mW with a supply voltage of 2 V.

Solution With $V_{CC} = 2$ V, the power budget translates to a tail current of 0.5 mA. Each transistor thus carries a current of 0.25 mA near equilibrium, providing a transconductance of $0.25 \text{ mA}/26 \text{ mV} = (104\ \Omega)^{-1}$. It follows that

$$R_C = \frac{|A_v|}{g_m} \tag{10.46}$$

$$= 1040\ \Omega. \tag{10.47}$$

Exercise Redesign the circuit for a power budget of 0.5 mW and compare the results.

Example 10-7 Compare the power dissipation of a bipolar differential pair with that of a CE stage if both circuits are designed for equal voltage gains, collector resistances, and supply voltages.

Solution The gain of the differential pair is written from Eq. (10.45) as

$$|A_{V,\text{diff}}| = g_{m1,2} R_C, \tag{10.48}$$

where $g_{m1,2}$ denotes the transconductance of each of the two transistors. For a CE stage

$$|A_{V,CE}| = g_m R_C. \tag{10.49}$$

Thus,

$$g_{m1,2} R_C = g_m R_C \tag{10.50}$$

and hence

$$\frac{I_{EE}}{2V_T} = \frac{I_C}{V_T}, \tag{10.51}$$

where $I_{EE}/2$ is the bias current of each transistor in the differential pair, and I_C represents the bias current of the CE stage. In other words,

$$I_{EE} = 2I_C, \tag{10.52}$$

indicating that the differential pair consumes twice as much power. This is one of the drawbacks of differential circuits.

Exercise If both circuits are designed for the same power budget, equal collector resistances, and equal supply voltages, compare their voltage gains.

10.2.2 Large-Signal Analysis

Having obtained insight into the operation of the bipolar differential pair, we now quantify its large-signal behavior, aiming to formulate the input/output characteristic of the circuit (the sketches in Fig. 10.12). Not having seen any large-signal analysis in the previous chapters, the reader may naturally wonder why we are suddenly interested in this aspect of the differential pair. Our interest arises from (a) the need to understand the circuit's limitations in serving as a *linear* amplifier, and (b) the application of the differential pair as a (nonlinear) current-steering circuit.

In order to derive the relationship between the differential input and output of the circuit, we first note from Fig. 10.14 that

$$V_{out1} = V_{CC} - R_C I_{C1} \tag{10.53}$$

$$V_{out2} = V_{CC} - R_C I_{C2} \tag{10.54}$$

and hence

$$V_{out} = V_{out1} - V_{out2} \tag{10.55}$$

$$= -R_C(I_{C1} - I_{C2}). \tag{10.56}$$

We must therefore compute I_{C1} and I_{C2} in terms of the input difference. Assuming $\alpha = 1$ and $V_A = \infty$, and recalling from Chapter 4 that $V_{BE} = V_T \ln (I_C/I_S)$, we write a KVL around the input network,

$$V_{in1} - V_{BE1} = V_P = V_{in2} - V_{BE2}, \tag{10.57}$$

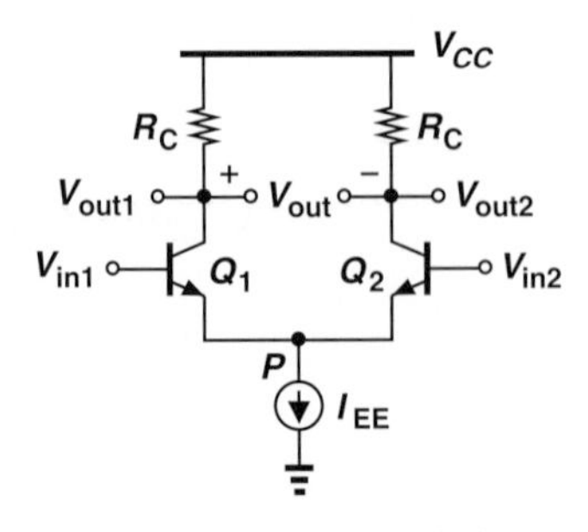

Figure 10.14 Bipolar differential pair for large-signal analysis.

obtaining

$$V_{in1} - V_{in2} = V_{BE1} - V_{BE2} \tag{10.58}$$

$$= V_T \ln \frac{I_{C1}}{I_{S1}} - V_T \ln \frac{I_{C2}}{I_{S2}} \tag{10.59}$$

$$= V_T \ln \frac{I_{C1}}{I_{C2}}. \tag{10.60}$$

Also, a KCL at node P gives

$$I_{C1} + I_{C2} = I_{EE}. \tag{10.61}$$

Equations (10.60) and (10.61) contain two unknowns. Substituting for I_{C1} from Eq. (10.60) in Eq. (10.61) yields

$$I_{C2} \exp \frac{V_{in1} - V_{in2}}{V_T} + I_{C2} = I_{EE} \tag{10.62}$$

and, therefore,

$$I_{C2} = \frac{I_{EE}}{1 + \exp \dfrac{V_{in1} - V_{in2}}{V_T}}. \tag{10.63}$$

The symmetry of the circuit with respect to V_{in1} and V_{in2} and with respect to I_{C1} and I_{C2} suggests that I_{C1} exhibits the same behavior as Eq. (10.63) but with the roles of V_{in1} and V_{in2} exchanged:

$$I_{C1} = \frac{I_{EE}}{1 + \exp \dfrac{V_{in2} - V_{in1}}{V_T}} \tag{10.64}$$

$$= \frac{I_{EE} \exp \dfrac{V_{in1} - V_{in2}}{V_T}}{1 + \exp \dfrac{V_{in1} - V_{in2}}{V_T}}. \tag{10.65}$$

Alternatively, the reader can substitute for I_{C2} from Eq. (10.63) in Eq. (10.61) to obtain I_{C1}.

Equations (10.63) and (10.65) play a crucial role in our quantitative understanding of the differential pair's operation. In particular, if $V_{in1} - V_{in2}$ is very *negative*, then $\exp(V_{in1} - V_{in2})/V_T \rightarrow 0$ and

$$I_{C1} \rightarrow 0 \tag{10.66}$$

$$I_{C2} \rightarrow I_{EE}, \tag{10.67}$$

as predicted by our qualitative analysis [Fig. 10.11(b)]. Similarly, if $V_{in1} - V_{in2}$ is very *positive*, $\exp(V_{in1} - V_{in2})/V_T \rightarrow \infty$ and

$$I_{C1} \rightarrow I_{EE} \tag{10.68}$$

$$I_{C2} \rightarrow 0. \tag{10.69}$$

What is meant by "very" negative or positive? For example, can we say $I_{C1} \approx 0$ and $I_{C2} \approx I_{EE}$ if $V_{in1} - V_{in2} = -10V_T$? Since $\exp(-10) \approx 4.54 \times 10^{-5}$,

$$I_{C1} \approx \frac{I_{EE} \times 4.54 \times 10^{-5}}{1 + 4.54 \times 10^{-5}} \tag{10.70}$$

$$\approx 4.54 \times 10^{-5} I_{EE} \tag{10.71}$$

and

$$I_{C2} \approx \frac{I_{EE}}{1 + 4.54 \times 10^{-5}} \tag{10.72}$$

$$\approx I_{EE}(1 - 4.54 \times 10^{-5}). \tag{10.73}$$

In other words, Q_1 carries only 0.0045% of the tail current; and I_{EE} can be considered steered completely to Q_2.

Example 10-8 Determine the differential input voltage that steers 98% of the tail current to one transistor.

Solution We require that

$$I_{C1} = 0.02 I_{EE} \tag{10.74}$$

$$\approx I_{EE} \exp \frac{V_{in1} - V_{in2}}{V_T} \tag{10.75}$$

and hence

$$V_{in1} - V_{in2} \approx -3.91 V_T. \tag{10.76}$$

We often say a differential input of $4V_T$ is sufficient to turn one side of the bipolar pair nearly off. Note that this value remains independent of I_{EE} and I_S.

Exercise What differential input is necessary to steer 90% of the tail current?

For the output voltages in Fig. 10.14, we have

$$V_{out1} = V_{CC} - R_C I_{C1} \tag{10.77}$$

$$= V_{CC} - R_C \frac{I_{EE} \exp \frac{V_{in1} - V_{in2}}{V_T}}{1 + \exp \frac{V_{in1} - V_{in2}}{V_T}} \tag{10.78}$$

and

$$V_{out2} = V_{CC} - R_C I_{C2} \tag{10.79}$$

$$= V_{CC} - R_C \frac{I_{EE}}{1 + \exp \frac{V_{in1} - V_{in2}}{V_T}}. \tag{10.80}$$

Of particular importance is the output *differential* voltage:

$$V_{out1} - V_{out2} = -R_C(I_{C1} - I_{C2}) \tag{10.81}$$

$$= R_C I_{EE} \frac{1 - \exp\dfrac{V_{in1} - V_{in2}}{V_T}}{1 + \exp\dfrac{V_{in1} - V_{in2}}{V_T}} \tag{10.82}$$

$$= -R_C I_{EE} \tanh \frac{V_{in1} - V_{in2}}{2V_T}. \tag{10.83}$$

Figure 10.15 summarizes the results, indicating that the differential output voltage begins from a "saturated" value of $+R_C I_{EE}$ for a very negative differential input, gradually becomes a linear function of $V_{in1} - V_{in2}$ for relatively small values of $|V_{in1} - V_{in2}|$, and reaches a saturated level of $-R_C I_{EE}$ as $V_{in1} - V_{in2}$ becomes very positive. From Example 10-8, we recognize that even a differential input of $4V_T \approx 104$ mV "switches" the differential pair, thereby concluding that $|V_{in1} - V_{in2}|$ must remain well below this value for linear operation.

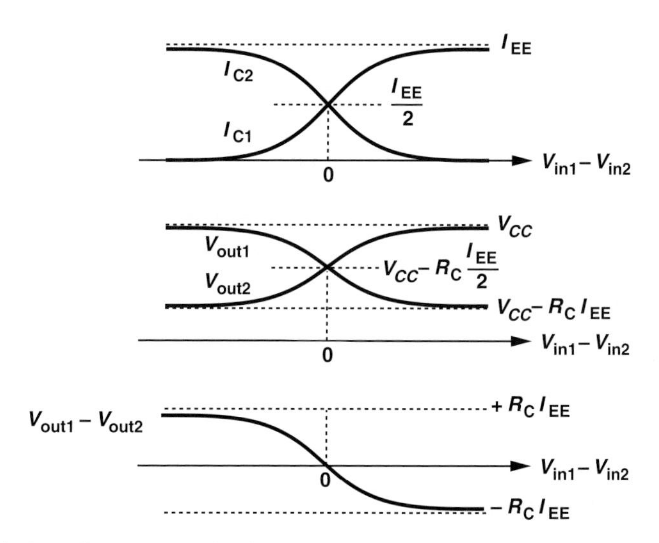

Figure 10.15 Variation of currents and voltages as a function of input.

Example 10-9 Sketch the output waveforms of the bipolar differential pair in Fig. 10.16(a) in response to the sinusoidal inputs shown in Figs. 10.16(b) and (c). Assume Q_1 and Q_2 remain in the forward active region.

Solution For the sinusoids depicted in Fig. 10.16(b), the circuit operates linearly because the maximum differential input is equal to ± 2 mV. The outputs are therefore sinusoids having a peak amplitude of 1 mV $\times g_m R_C$ [Fig. 10.16(d)]. On the other hand, the sinusoids in Fig. 10.16(c) force a maximum input difference of ± 200 mV, turning Q_1 or Q_2

off. For example, as V_{in1} approaches 50 mV above V_{CM} and V_{in2} reaches 50 mV below V_{CM} (at $t = t_1$), Q_1 absorbs most of the tail current, thus producing

$$V_{out1} \approx V_{CC} - R_C I_{EE} \tag{10.84}$$

$$V_{out2} \approx V_{CC}. \tag{10.85}$$

Thereafter, the outputs remain saturated until $|V_{in1} - V_{in2}|$ falls to less than 100 mV. The result is sketched in Fig. 10.16(e). We say the circuit operates as a "limiter" in this case, playing a role similar to the diode limiters studied in Chapter 3.

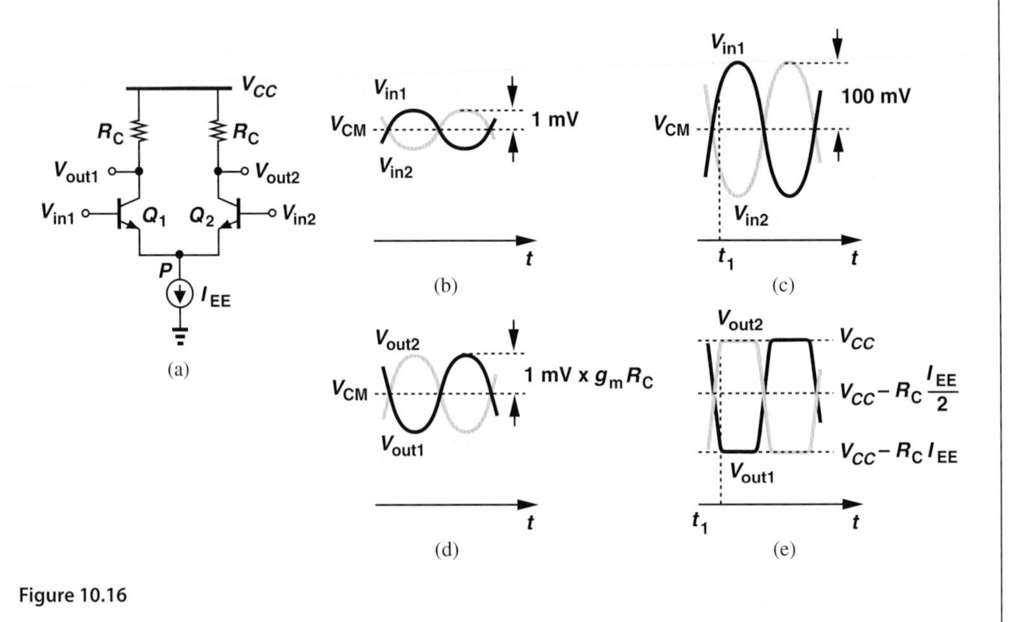

Figure 10.16

Exercise What happens to the above results if the tail current is halved?

Did you know?

Differential pairs can operate as limiters if their input difference is large. In this case, they "clip" the top and bottom of the signal waveform. Of course, limiting a signal such as your voice creates distortion and is undesirable. However, limiting proves useful in some applications. A common example is in FM radios, where the RF waveform ideally contains only frequency modulation [Fig. (a)], but in practice incurs some noise [Fig. (b)]. A limiter removes the noise on the amplitude, improving the signal quality.

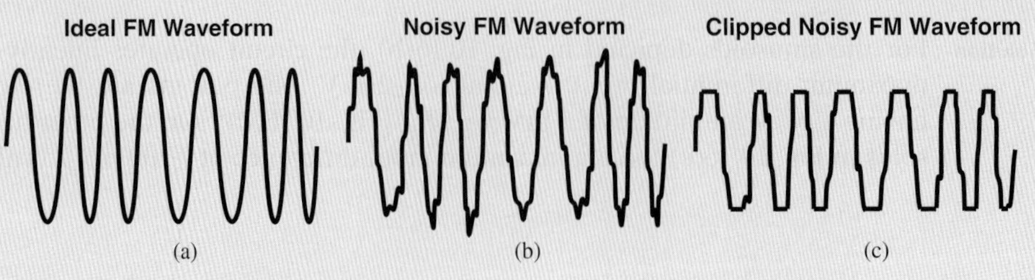

Effect of clipping on FM waveform.

10.2.3 Small-Signal Analysis

Our brief investigation of the differential pair in Fig. 10.13 revealed that, for small differential inputs, the tail node maintains a constant voltage (and hence is called a "virtual ground"). We also obtained a voltage gain equal to $g_m R_C$. We now study the small-signal behavior of the circuit in greater detail. As explained in previous chapters, the definition of "small signals" is somewhat arbitrary, but the requirement is that the input signals not influence the bias currents of Q_1 and Q_2 appreciably. In other words, the two transistors must exhibit approximately equal transconductances—the same condition required for node P to appear as virtual ground. In practice, an input difference of less than 10 mV is considered "small" for most applications.

Assuming perfect symmetry, an ideal tail current source, and $V_A = \infty$, we construct the small-signal model of the circuit as shown in Fig. 10.17(a). Here, v_{in1} and v_{in2} represent small *changes* in each input and must satisfy $v_{in1} = -v_{in2}$ for differential operation. Note that the tail current source is replaced with an open circuit. As with the foregoing large-signal analysis, let us write a KVL around the input network and a KCL at node P:

$$v_{in1} - v_{\pi 1} = v_P = v_{in2} - v_{\pi 2} \tag{10.86}$$

$$\frac{v_{\pi 1}}{r_{\pi 1}} + g_{m1} v_{\pi 1} + \frac{v_{\pi 2}}{r_{\pi 2}} + g_{m2} v_{\pi 2} = 0. \tag{10.87}$$

With $r_{\pi 1} = r_{\pi 2}$ and $g_{m1} = g_{m2}$, Eq. (10.87) yields

$$v_{\pi 1} = -v_{\pi 2} \tag{10.88}$$

and since $v_{in1} = -v_{in2}$, Eq. (10.86) translates to

$$2v_{in1} = 2v_{\pi 1}. \tag{10.89}$$

That is,

$$v_P = v_{in1} - v_{\pi 1} \tag{10.90}$$

$$= 0. \tag{10.91}$$

Thus, the small-signal model confirms the prediction made by Eq. (10.37). In Problem 10.28, we prove that this property holds in the presence of the Early effect as well.

The virtual-ground nature of node P for differential small-signal inputs simplifies the analysis considerably. Since $v_P = 0$, this node can be shorted to ac ground, reducing the differential pair of Fig. 10.17(a) to two "half circuits" [Fig. 10.17(b)]. With each half resembling a common-emitter stage, we can write

$$v_{out1} = -g_m R_C v_{in1} \tag{10.92}$$

$$v_{out2} = -g_m R_C v_{in2}. \tag{10.93}$$

It follows that the differential voltage gain of the differential pair is equal to

$$\frac{v_{out1} - v_{out2}}{v_{in1} - v_{in2}} = -g_m R_C, \tag{10.94}$$

the same as that expressed by Eq. (10.45). For simplicity, we may draw the two half circuits as in Fig. 10.17(c), with the understanding that the incremental inputs are small and differential. Also, since the two halves are identical, we may draw only one half.

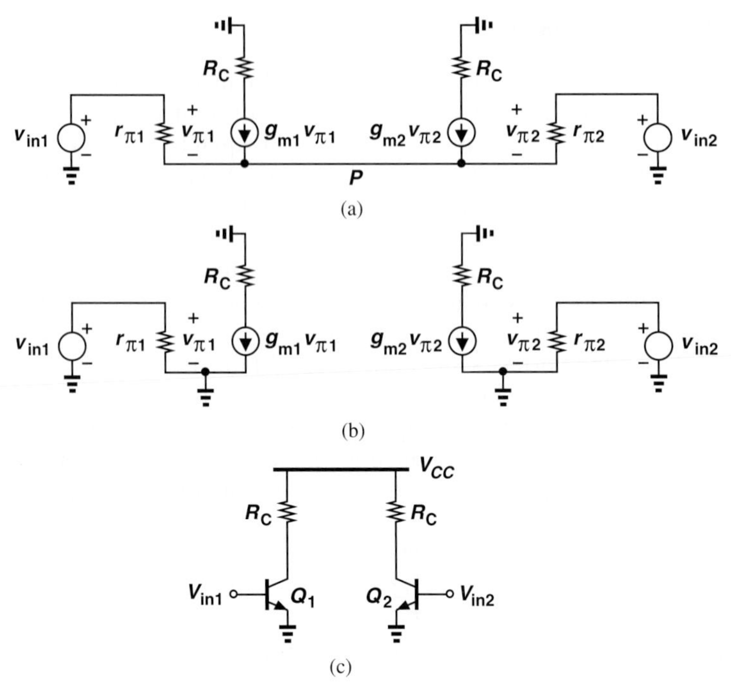

Figure 10.17 (a) Small-signal model of bipolar pair, (b) simplified small-signal model, (c) simplified diagram.

Example 10-10

Compute the differential gain of the circuit shown in Fig. 10.18(a), where ideal current sources are used as loads to maximize the gain.

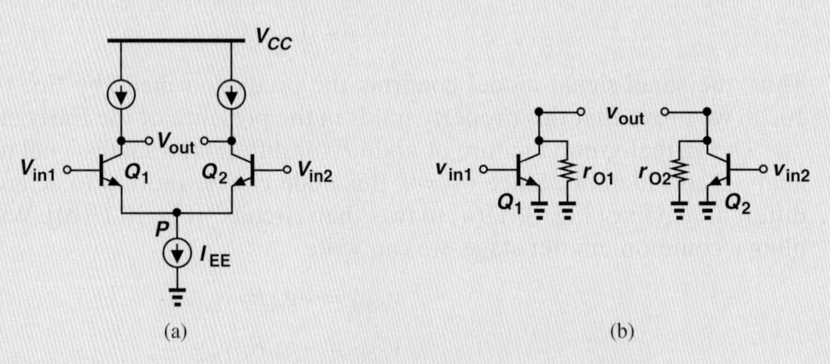

Figure 10.18

Solution With ideal current sources, the Early effect in Q_1 and Q_2 cannot be neglected, and the half circuits must be visualized as depicted in Fig. 10.18(b). Thus,

$$v_{out1} = -g_m r_O v_{in1} \tag{10.95}$$

$$v_{out2} = -g_m r_O v_{in2} \tag{10.96}$$

and hence

$$\frac{v_{out1} - v_{out2}}{v_{in1} - v_{in2}} = -g_m r_O. \tag{10.97}$$

Exercise Calculate the gain for $V_A = 5$ V.

Example 10-11 Figure 10.19(a) illustrates an implementation of the topology shown in Fig. 10.18(a). Calculate the differential voltage gain.

Solution Noting that each *pnp* device introduces a resistance of r_{OP} at the output nodes and drawing the half circuit as in Fig. 10.19(b), we have

$$\frac{v_{out1} - v_{out2}}{v_{in1} - v_{in2}} = -g_m(r_{ON}||r_{OP}), \tag{10.98}$$

where r_{ON} denotes the output impedance of the *npn* transistors.

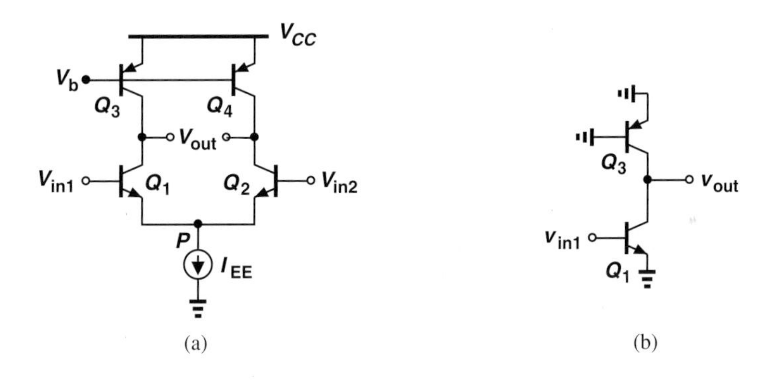

Figure 10.19

Exercise Calculate the gain if Q_3 and Q_4 are configured as diode-connected devices.

We must emphasize that the differential voltage gain is defined as the difference between the outputs divided by the difference between the inputs. As such, this gain is equal to the single-ended gain of each half circuit.

We now make an observation that proves useful in the analysis of differential circuits. As noted above, the symmetry of the circuit ($g_{m1} = g_{m2}$) establishes a virtual ground at node P in Fig. 10.14 if the incremental inputs are small and differential. This property holds for any other node that appears on the axis of symmetry. For example, the two resistors shown in Fig. 10.20 create a virtual ground at X if (1) $R_1 = R_2$ and (2) nodes A and B vary by equal and opposite amounts.[2] Additional examples make this concept clearer. We assume perfect symmetry in each case.

[2]Since the resistors are linear, the signals need not be small in this case.

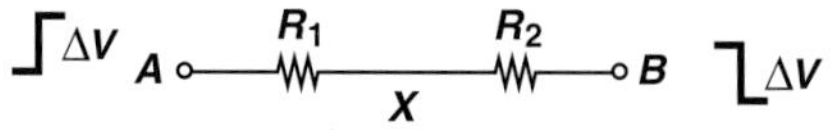

Figure 10.20

Example 10-12 Determine the differential gain of the circuit in Fig. 10.21(a) if $V_A < \infty$ and the circuit is symmetric.

Solution Drawing one of the half circuits as shown in Fig. 10.21(b), we express the total resistance seen at the collector of Q_1 as

$$R_{out} = r_{O1}||r_{O3}||R_1. \tag{10.99}$$

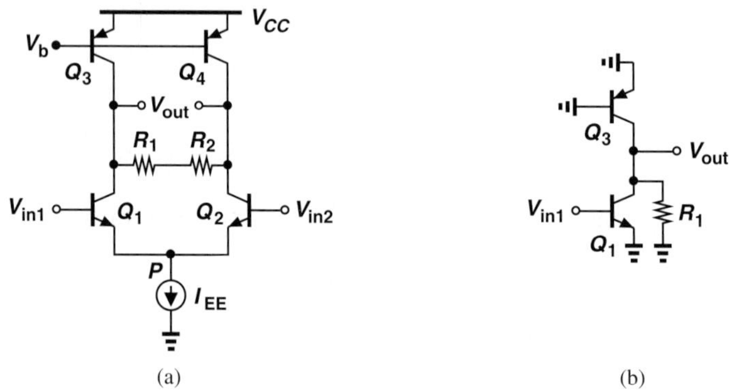

(a) (b)

Figure 10.21

Thus, the voltage gain is equal to

$$A_v = -g_{m1}(r_{O1}||r_{O3}||R_1). \tag{10.100}$$

Exercise Repeat the above example if $R_1 \neq R_2$.

Example 10-13 Calculate the differential gain of the circuit illustrated in Fig. 10.22(a) if $V_A < \infty$.

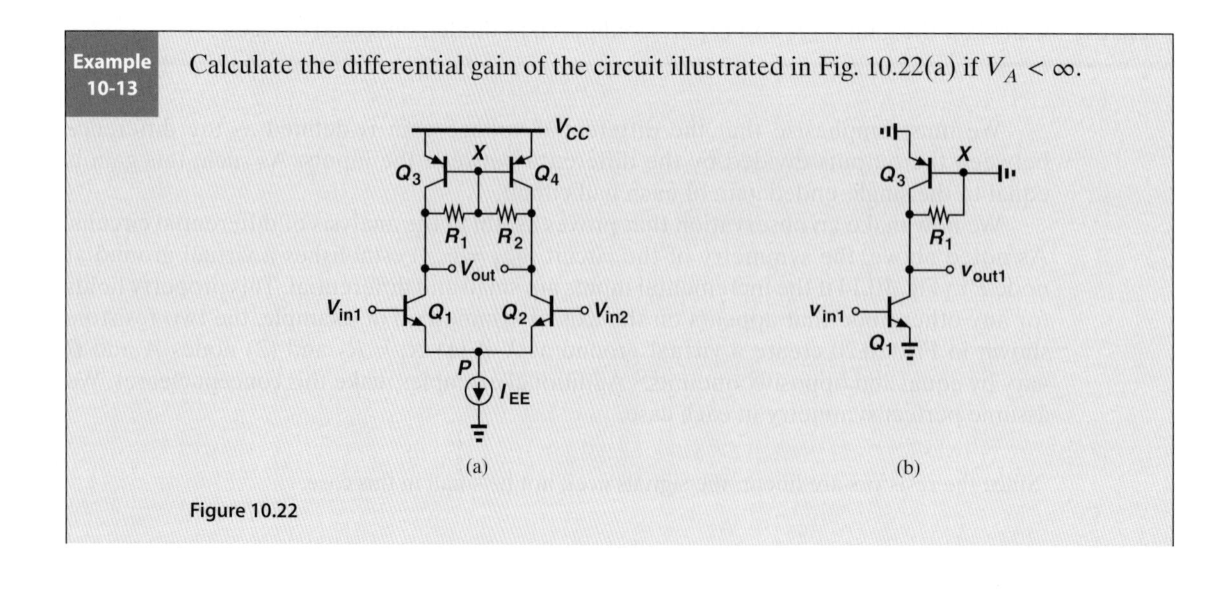

Figure 10.22

Solution For small differential inputs and outputs, V_X remains constant, leading to the conceptual half circuit shown in Fig. 10.22(b)—the same as that in the above example. This is because Q_3 and Q_4 experience a *constant* base-emitter voltage in both cases, thereby serving as current sources and exhibiting only an output resistance. It follows that

$$A_v = -g_{m1}(r_{O1}||r_{O3}||R_1). \tag{10.101}$$

Exercise Calculate the gain if $V_A = 4$ V for all transistors, $R_1 = R_2 = 10$ kΩ, and $I_{EE} = 1$ mA.

Example 10-14 Determine the gain of the degenerated differential pairs shown in Figs. 10.23(a) and (b). Assume $V_A = \infty$.

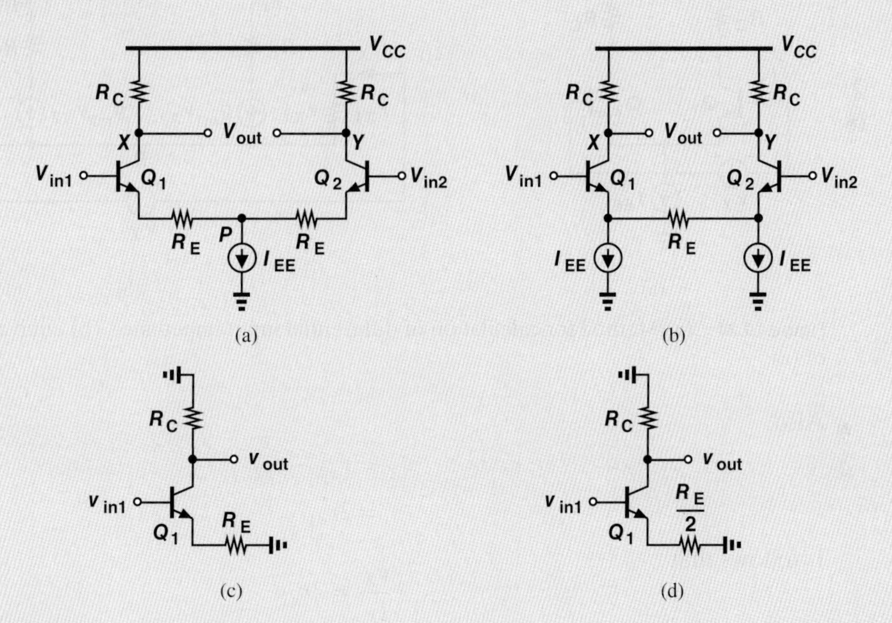

Figure 10.23

Solution In the topology of Fig. 10.23(a), node P is a virtual ground, yielding the half circuit depicted in Fig. 10.23(c). From Chapter 5, we have

$$A_v = -\frac{R_C}{R_E + \dfrac{1}{g_m}}. \tag{10.102}$$

In the circuit of Fig. 10.23(b), the line of symmetry passes through the "midpoint" of R_E. In other words, if R_E is regarded as two $R_E/2$ units in series, then the node between the units acts as a virtual ground [Fig. 10.23(d)]. It follows that

$$A_v = -\frac{R_C}{\dfrac{R_E}{2} + \dfrac{1}{g_m}}. \tag{10.103}$$

The two circuits provide equal gains if the pair in Fig. 10.23(b) incorporates a total degeneration resistance of $2R_E$.

Exercise Design each circuit for a gain of 5 and power consumption of 2 mW. Assume $V_{CC} = 2.5$ V, $V_A = \infty$, and $R_E = 2/g_m$.

I/O Impedances For a differential pair, we can define the input impedance as illustrated in Fig. 10.24(a). From the equivalent circuit in Fig. 10.24(b), we have

$$\frac{v_{\pi 1}}{r_{\pi 1}} = i_X = -\frac{v_{\pi 2}}{r_{\pi 2}}. \tag{10.104}$$

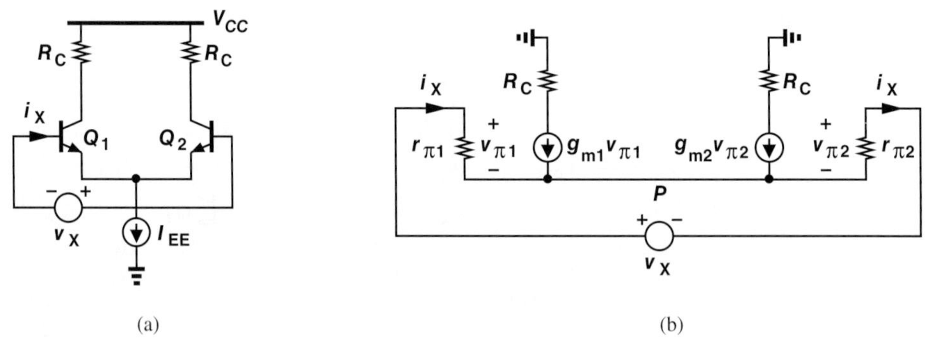

Figure 10.24 (a) Method for calculation of differential input impedance, (b) equivalent circuit of (a).

Also,

$$v_X = v_{\pi 1} - v_{\pi 2} \tag{10.105}$$

$$= 2r_{\pi 1} i_X. \tag{10.106}$$

It follows that

$$\frac{v_X}{i_X} = 2r_{\pi 1}, \tag{10.107}$$

as if the two base-emitter junctions appear in series.

The above quantity is called the "differential input impedance" of the circuit. It is also possible to define a "single-ended input impedance" with the aid of a half circuit (Fig. 10.25), obtaining

$$\frac{v_X}{i_X} = r_{\pi 1}. \tag{10.108}$$

This result provides no new information with respect to that in Eq. (10.107) but proves useful in some calculations.

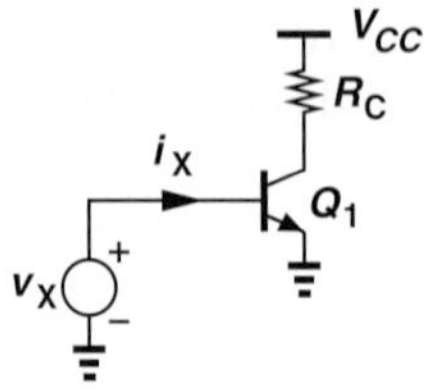

Figure 10.25 Calculation of single-ended input impedance.

In a manner similar to the foregoing development, the reader can show that the differential and single-ended output impedances are equal to $2R_C$ and R_C, respectively.

10.3 MOS DIFFERENTIAL PAIR

Most of the principles studied in the previous section for the bipolar differential pair apply directly to the MOS counterpart as well. For this reason, our treatment of the MOS circuit in this section is more concise. We continue to assume perfect symmetry.

10.3.1 Qualitative Analysis

Figure 10.26(a) depicts the MOS pair with the two inputs tied to V_{CM}, yielding

$$I_{D1} = I_{D2} = \frac{I_{SS}}{2} \tag{10.109}$$

and

$$V_X = V_Y = V_{DD} - R_D \frac{I_{SS}}{2}. \tag{10.110}$$

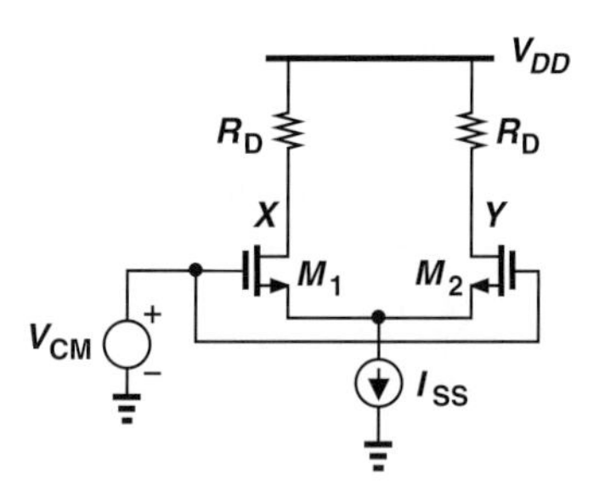

Figure 10.26 Response of MOS differential pair to input CM variation.

That is, a zero differential input gives a zero differential output. Note that the output CM level is equal to $V_{DD} - R_D I_{SS}/2$.

For our subsequent derivations, it is useful to compute the "equilibrium overdrive voltage" of M_1 and M_2, $(V_{GS} - V_{TH})_{equil.}$. We assume $\lambda = 0$ and hence $I_D = (1/2)\mu_n C_{ox}(W/L)(V_{GS} - V_{TH})^2$. Carrying a current of $I_{SS}/2$, each device exhibits an overdrive of

$$(V_{GS} - V_{TH})_{equil.} = \sqrt{\frac{I_{SS}}{\mu_n C_{ox} \frac{W}{L}}}. \tag{10.111}$$

As expected, a greater tail current or a smaller W/L translates to a larger equilibrium overdrive.

To guarantee that M_1 and M_2 operate in saturation, we require that their drain voltages not fall below $V_{CM} - V_{TH}$:

$$V_{DD} - R_D \frac{I_{SS}}{2} > V_{CM} - V_{TH}. \tag{10.112}$$

It can also be observed that a change in V_{CM} cannot alter $I_{D1} = I_{D2} = I_{SS}/2$, leaving V_X and V_Y undisturbed. The circuit thus rejects input CM variations.

Example 10-15 A MOS differential pair is driven with an input CM level of 1.6 V. If $I_{SS} = 0.5$ mA, $V_{TH} = 0.5$ V, and $V_{DD} = 1.8$ V, what is the maximum allowable load resistance?

Solution From Eq. (10.112), we have

$$R_D < 2\frac{V_{DD} - V_{CM} + V_{TH}}{I_{SS}} \tag{10.113}$$

$$< 2.8\ \text{k}\Omega. \tag{10.114}$$

We may suspect that this limitation in turn constrains the voltage gain of the circuit, as explained later.

Exercise What is the maximum tail current if the load resistance is 5 kΩ?

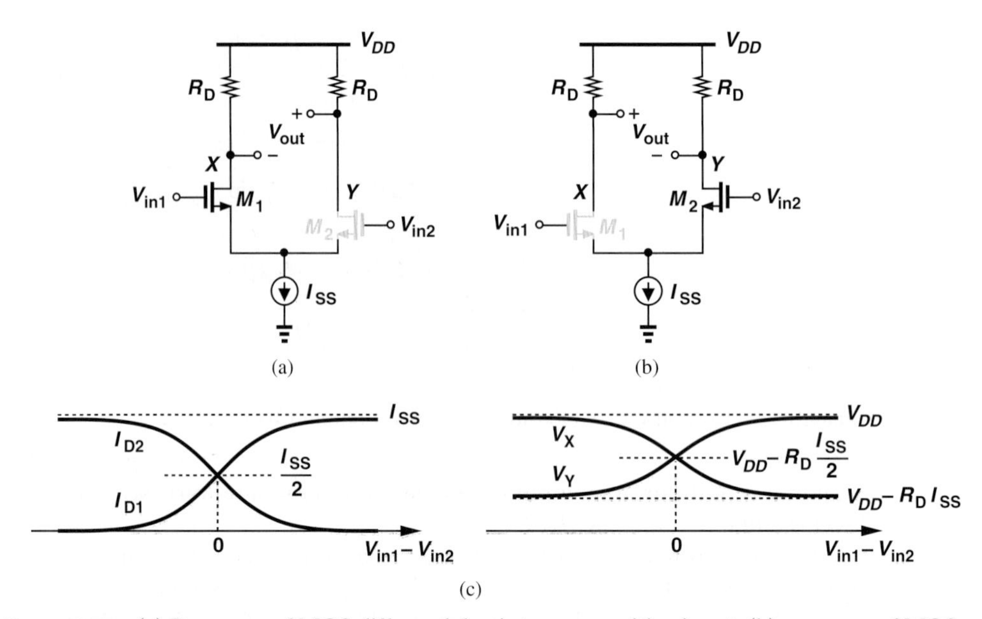

Figure 10.27 (a) Response of MOS differential pair to very positive input, (b) response of MOS differential pair to very negative input, (c) qualitative plots of currents and voltages.

Figure 10.27 illustrates the response of the MOS pair to large differential inputs. If V_{in1} is well above V_{in2} [Fig. 10.27(a)], then M_1 carries the entire tail current, generating

$$V_X = V_{DD} - R_D I_{SS} \tag{10.115}$$

$$V_Y = V_{DD}. \tag{10.116}$$

Similarly, if V_{in2} is well above V_{in1} [Fig. 10.27(b)], then

$$V_X = V_{DD} \tag{10.117}$$

$$V_Y = V_{DD} - R_D I_{SS}. \tag{10.118}$$

The circuit therefore steers the tail current from one side to the other, producing a differential output in response to a differential input. Figure 10.27(c) sketches the characteristics of the circuit.

Let us now examine the circuit's behavior for a *small* input difference. Depicted in Fig. 10.28(a), such a scenario maintains V_P constant because Eqs. (10.32)–(10.37) apply to this case equally well. It follows that

$$\Delta I_{D1} = g_m \Delta V \tag{10.119}$$

$$\Delta I_{D2} = -g_m \Delta V \tag{10.120}$$

and

$$\Delta V_X - \Delta V_Y = -2g_m R_D \Delta V. \tag{10.121}$$

As expected, the differential voltage gain is given by

$$A_v = -g_m R_D, \tag{10.122}$$

similar to that of a common-source stage.

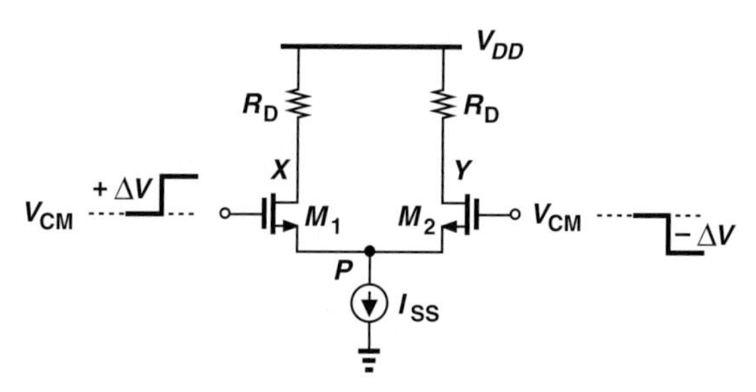

Figure 10.28 Response of MOS pair to small differential inputs.

Example 10-16

Design an NMOS differential pair for a voltage gain of 5 and a power budget of 2 mW subject to the condition that the stage following the differential pair requires an input CM level of at least 1.6 V. Assume $\mu_n C_{ox} = 100\ \mu A/V^2$, $\lambda = 0$, and $V_{DD} = 1.8$ V.

Solution From the power budget and the supply voltage, we have

$$I_{SS} = 1.11\ \text{mA}. \tag{10.123}$$

The output CM level (in the absence of signals) is equal to

$$V_{CM,out} = V_{DD} - R_D \frac{I_{SS}}{2}. \tag{10.124}$$

For $V_{CM,out} = 1.6$ V, each resistor must sustain a voltage drop of no more than 200 mV, thereby assuming a maximum value of

$$R_D = 360\ \Omega. \tag{10.125}$$

Setting $g_m R_D = 5$, we must choose the transistor dimensions such that $g_m = 5/(360\ \Omega)$. Since each transistor carries a drain current of $I_{SS}/2$,

$$g_m = \sqrt{2\mu_n C_{ox} \frac{W}{L} \frac{I_{SS}}{2}}, \tag{10.126}$$

and hence

$$\frac{W}{L} = 1738. \tag{10.127}$$

The large aspect ratio arises from the small drop allowed across the load resistors.

Exercise If the aspect ratio must remain below 200, what voltage gain can be achieved?

Example 10-17 What is the maximum allowable input CM level in the previous example if $V_{TH} = 0.4$ V?

Solution We rewrite Eq. (10.112) as

$$V_{CM,in} < V_{DD} - R_D\frac{I_{SS}}{2} + V_{TH} \tag{10.128}$$

$$< V_{CM,out} + V_{TH}. \tag{10.129}$$

This is conceptually illustrated in Fig. 10.29. Thus,

$$V_{CM,in} < 2 \text{ V}. \tag{10.130}$$

Interestingly, the input CM level can comfortably remain at V_{DD}. In contrast to Example 10-5, the constraint on the load resistor in this case arises from the *output* CM level requirement.

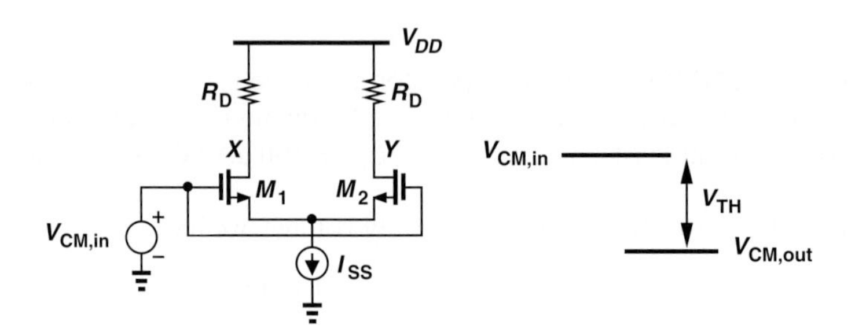

Figure 10.29

Exercise Does the above result hold if $V_{TH} = 0.2$ V?

Example 10-18 The common-source stage and the differential pair shown in Fig. 10.30 incorporate equal load resistors. If the two circuits are designed for the same voltage gain and the same supply voltage, discuss the choice of (a) transistor dimensions for a given power budget, (b) power dissipation for given transistor dimensions.

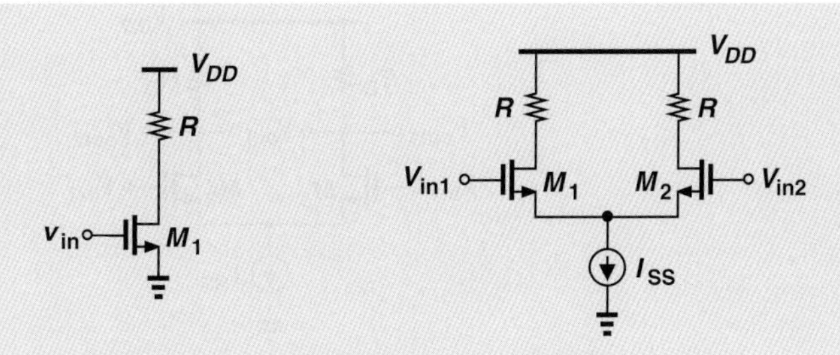

Figure 10.30

Solution (a) For the two circuits to consume the same amount of power $I_{D1} = I_{SS} = 2I_{D2} = 2I_{D3}$; i.e., each transistor in the differential pair carries a current equal to half of the drain current of the CS transistor. Equation (10.126) therefore requires that the differential pair transistors be twice as wide as the CS device to obtain the same voltage gain. (b) If the transistors in both circuits have the same dimensions, then the tail current of the differential pair must be twice the bias current of the CS stage for M_1-M_3 to have the same transconductance, doubling the power consumption.

Exercise Discuss the above results if the CS stage and the differential pair incorporate equal source degeneration resistors.

10.3.2 Large-Signal Analysis

As with the large-signal analysis of the bipolar pair, our objective here is to derive the input/output characteristics of the MOS pair as the differential input varies from very negative to very positive values. From Fig. 10.31, we have

$$V_{out} = V_{out1} - V_{out2} \tag{10.131}$$

$$= -R_D(I_{D1} - I_{D2}). \tag{10.132}$$

To obtain $I_{D1} - I_{D2}$, we neglect channel-length modulation and write a KVL around the input network and a KCL at the tail node:

$$V_{in1} - V_{GS1} = V_{in2} - V_{GS2} \tag{10.133}$$

$$I_{D1} + I_{D2} = I_{SS}. \tag{10.134}$$

Since $I_D = (1/2)\mu_n C_{ox}(W/L)(V_{GS} - V_{TH})^2$,

$$V_{GS} = V_{TH} + \sqrt{\frac{2I_D}{\mu_n C_{ox}\frac{W}{L}}}. \tag{10.135}$$

Did you know?

The MOS differential pair was a natural extension of its bipolar counterpart, offering an important advantage: very little input current. But the need for high-input-impedance op amps had in fact arisen well before MOS op amps became manufacturable. The challenge was to integrate field-effect transistors (which generally have a small gate current) on the same chip as the other bipolar devices. The FETs were poorly controlled and suffered from a large offset. It was not until 1970s that high-input-impedance op amps became common. MOS op amps also began to play a profound role in analog integrated circuits, serving as the core of "switched-capacitor" filters and analog-to-digital converters.

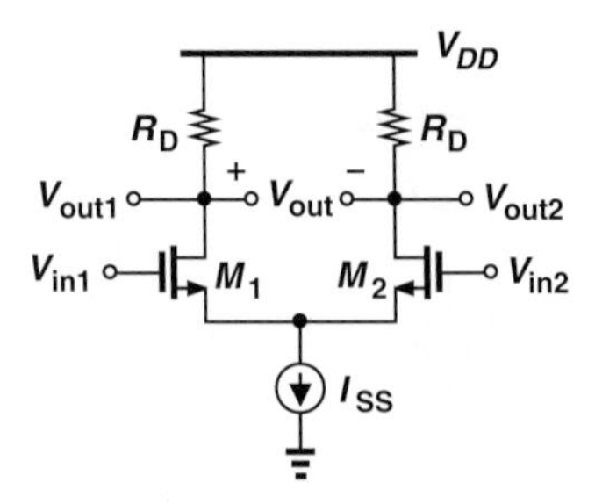

Figure 10.31 MOS differential pair for large-signal analysis.

Substituting for V_{GS1} and V_{GS2} in Eq. (10.133), we have

$$V_{in1} - V_{in2} = V_{GS1} - V_{GS2} \tag{10.136}$$

$$= \sqrt{\frac{2}{\mu_n C_{ox}\frac{W}{L}}}(\sqrt{I_{D1}} - \sqrt{I_{D2}}). \tag{10.137}$$

Squaring both sides yields

$$(V_{in1} - V_{in2})^2 = \frac{2}{\mu_n C_{ox}\frac{W}{L}}(I_{D1} + I_{D2} - 2\sqrt{I_{D1}I_{D2}}) \tag{10.138}$$

$$= \frac{2}{\mu_n C_{ox}\frac{W}{L}}(I_{SS} - 2\sqrt{I_{D1}I_{D2}}). \tag{10.139}$$

We now find $\sqrt{I_{D1}I_{D2}}$,

$$4\sqrt{I_{D1}I_{D2}} = 2I_{SS} - \mu_n C_{ox}\frac{W}{L}(V_{in1} - V_{in2})^2, \tag{10.140}$$

square the result again,

$$16 I_{D1} I_{D2} = \left[2I_{SS} - \mu_n C_{ox}\frac{W}{L}(V_{in1} - V_{in2})^2\right]^2, \tag{10.141}$$

and substitute $I_{SS} - I_{D1}$ for I_{D2},

$$16 I_{D1}(I_{SS} - I_{D1}) = \left[2I_{SS} - \mu_n C_{ox}\frac{W}{L}(V_{in1} - V_{in2})^2\right]^2. \tag{10.142}$$

It follows that

$$16 I_{D1}^2 - 16 I_{SS} I_{D1} + \left[2I_{SS} - \mu_n C_{ox}\frac{W}{L}(V_{in1} - V_{in2})^2\right]^2 = 0 \tag{10.143}$$

and hence

$$I_{D1} = \frac{I_{SS}}{2} \pm \frac{1}{4}\sqrt{4I_{SS}^2 - \left[\mu_n C_{ox}\frac{W}{L}(V_{in1} - V_{in2})^2 - 2I_{SS}\right]^2}. \tag{10.144}$$

In Problem 10.44, we show that only the solution with the *sum* of the two terms is acceptable:

$$I_{D1} = \frac{I_{SS}}{2} + \frac{V_{in1} - V_{in2}}{4}\sqrt{\mu_n C_{ox}\frac{W}{L}\left[4I_{SS} - \mu_n C_{ox}\frac{W}{L}(V_{in1} - V_{in2})^2\right]}. \tag{10.145}$$

The symmetry of the circuit also implies that

$$I_{D2} = \frac{I_{SS}}{2} + \frac{V_{in2} - V_{in1}}{4}\sqrt{\mu_n C_{ox}\frac{W}{L}\left[4I_{SS} - \mu_n C_{ox}\frac{W}{L}(V_{in2} - V_{in1})^2\right]}. \tag{10.146}$$

That is,

$$I_{D1} - I_{D2} = \frac{1}{2}\mu_n C_{ox}\frac{W}{L}(V_{in1} - V_{in2})\sqrt{\frac{4I_{SS}}{\mu_n C_{ox}\frac{W}{L}} - (V_{in1} - V_{in2})^2}. \tag{10.147}$$

Equations (10.145)–(10.147) form the foundation of our understanding of the MOS differential pair.

Let us now examine Eq. (10.147) closely. As expected from the characteristics in Fig. 10.27(c), the right-hand side is an odd (symmetric) function of $V_{in1} - V_{in2}$, dropping to zero for a zero input difference. But, can the difference under the square root vanish, too? That would suggest that $I_{D1} - I_{D2}$ falls to zero as $(V_{in1} - V_{in2})^2$ reaches $4I_{SS}/(\mu_n C_{ox} W/L)$, an effect not predicted by our qualitative sketches in Fig. 10.27(c). Furthermore, it appears that the argument of the square root becomes *negative* as $(V_{in1} - V_{in2})^2$ exceeds this value! How should these results be interpreted?

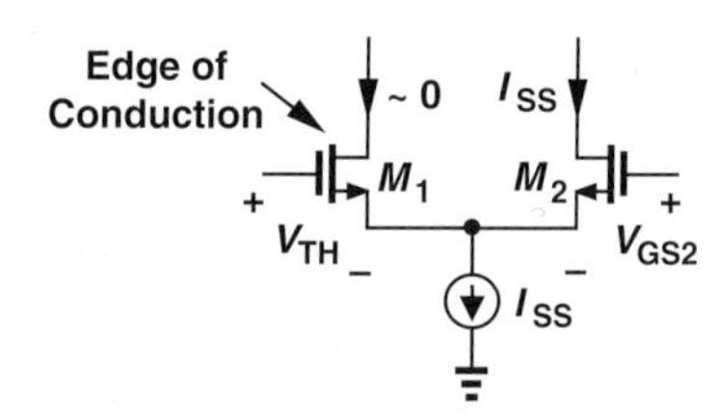

Figure 10.32 MOS differential pair with one device off.

Implicit in our foregoing derivations is the assumption both transistors are *on*. However, as $|V_{in1} - V_{in2}|$ rises, at some point M_1 or M_2 turns off, violating the above equations. We must therefore determine the input difference that places one of the transistors at the edge of conduction. This can be accomplished by equating Eqs. (10.145), (10.146), or (10.147) to I_{SS}, but this leads to lengthy algebra. Instead, we recognize from Fig. 10.32 that if, for example, M_1 approaches the edge of conduction, then its gate-source voltage falls to a value equal to V_{TH}. Also, the gate-source voltage of M_2 must be sufficiently large to accommodate a drain current of I_{SS}:

$$V_{GS1} = V_{TH} \tag{10.148}$$

$$V_{GS2} = V_{TH} + \sqrt{\frac{2I_{SS}}{\mu_n C_{ox}\frac{W}{L}}}. \tag{10.149}$$

It follows from Eq. (10.133) that

$$|V_{in1} - V_{in2}|_{max} = \sqrt{\frac{2I_{SS}}{\mu_n C_{ox}\frac{W}{L}}}, \tag{10.150}$$

where $|V_{in1} - V_{in2}|_{max}$ denotes the input difference that places one transistor at the edge of conduction. Equation (10.150) is invalid for input differences greater than this value.

Indeed, substituting from Eq. (10.150) in (10.147) also yields $|I_{D1} - I_{D2}| = I_{SS}$. We also note that $|V_{in1} - V_{in2}|_{max}$ can be related to the equilibrium overdrive [Eq. (10.111)] as follows:

$$|V_{in1} - V_{in2}|_{max} = \sqrt{2}(V_{GS} - V_{TH})_{equil.}. \qquad (10.151)$$

The above findings are very important and stand in contrast to the behavior of the bipolar differential pair and Eq. (10.83): the MOS pair steers *all* of the tail current[3] for $|V_{in1} - V_{in2}|_{max}$ whereas the bipolar counterpart only *approaches* this condition for a finite input difference. Equation (10.151) provides a great deal of intuition into the operation of the MOS pair. Specifically, we plot I_{D1} and I_{D2} as in Fig. 10.33(a), where $\Delta V_{in} = V_{in1} - V_{in2}$, arriving at the differential characteristics in Figs. 10.33(b) and (c). The circuit thus behaves linearly for small values of ΔV_{in} and becomes completely nonlinear for $\Delta V_{in} > \Delta V_{in,max}$. In other words, $\Delta V_{in,max}$ serves as an absolute bound on the input signal levels that have any effect on the output.

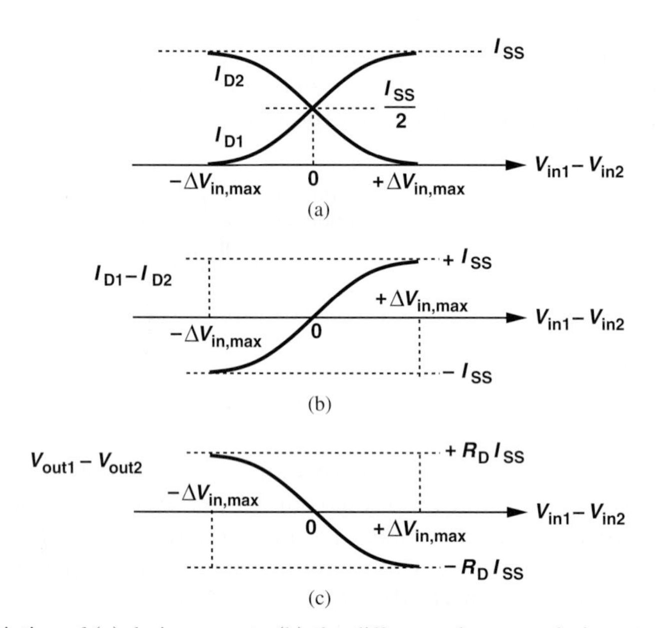

Figure 10.33 Variation of (a) drain currents, (b) the difference between drain currents, and (c) differential output voltage as a function of input.

Example 10-19 Examine the input/output characteristic of a MOS differential pair if (a) the tail current is doubled, or (b) the transistor aspect ratio is doubled.

Solution (a) Equation (10.150) suggests that doubling I_{SS} increases $\Delta V_{in,max}$ by a factor of $\sqrt{2}$. Thus, the characteristic of Fig. 10.33(c) *expands* horizontally. Furthermore, since $I_{SS}R_D$ doubles, the characteristic expands vertically as well. Figure 10.34(a) illustrates the result, displaying a greater slope.

[3]In reality, MOS devices carry a small current for $V_{GS} = V_{TH}$, making these observations only an approximate illustration.

(b) Doubling W/L lowers $\Delta V_{in,max}$ by a factor of $\sqrt{2}$ while maintaining $I_{SS}R_D$ constant. The characteristic therefore *contracts* horizontally [Fig. 10.34(b)], exhibiting a larger slope in the vicinity of $\Delta V_{in} = 0$.

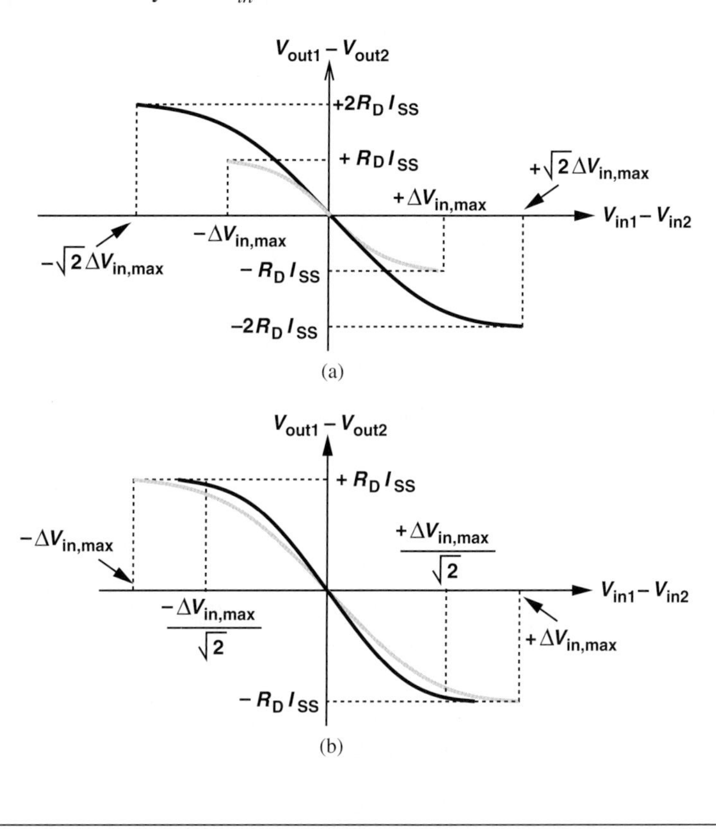

Figure 10.34

Exercise Repeat the above example if (a) the tail current is halved, or (b) the transistor aspect ratio is halved.

Example 10-20 Design an NMOS differential pair for a power budget of 3 mW and $\Delta V_{in,max} = 500$ mV. Assume $\mu_n C_{ox} = 100\ \mu\text{A/V}^2$ and $V_{DD} = 1.8$ V.

Solution The tail current must not exceed 3 mW/1.8 V = 1.67 mA. From Eq. (10.150), we write

$$\frac{W}{L} = \frac{2I_{SS}}{\mu_n C_{ox} \Delta V_{in,max}^2} \tag{10.152}$$

$$= 133.6. \tag{10.153}$$

The value of the load resistors is determined by the required voltage gain.

Exercise How does the above design change if the power budget is raised to 5 mW?

10.3.3 Small-Signal Analysis

The small-signal analysis of the MOS differential pair proceeds in a manner similar to that in Section 10.2.3 for the bipolar counterpart. The definition of "small" signals in this case can be seen from Eq. (10.147); if

$$|V_{in1} - V_{in2}|^2 \ll \frac{4I_{SS}}{\mu_n C_{ox}\frac{W}{L}}, \tag{10.154}$$

then

$$I_{D1} - I_{D2} \approx \frac{1}{2}\mu_n C_{ox}\frac{W}{L}(V_{in1} - V_{in2})\sqrt{\frac{4I_{SS}}{\mu_n C_{ox}\frac{W}{L}}} \tag{10.155}$$

$$= \sqrt{\mu_n C_{ox}\frac{W}{L}I_{SS}}(V_{in1} - V_{in2}). \tag{10.156}$$

Now, the differential inputs and outputs are *linearly* proportional, and the circuit operates linearly.

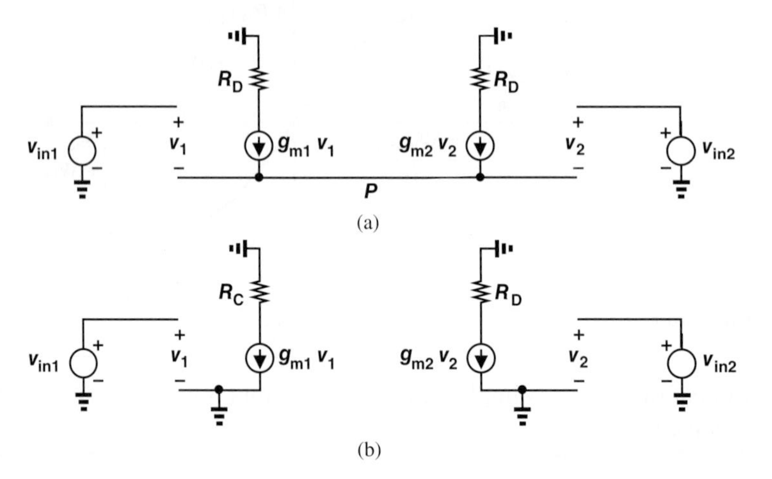

Figure 10.35 (a) Small-signal model of MOS differential pair, (b) simplified circuit.

We now use the small-signal model to prove that the tail node remains constant in the presence of small differential inputs. If $\lambda = 0$, the circuit reduces to that shown in Fig. 10.35(a), yielding

$$v_{in1} - v_1 = v_{in2} - v_2 \tag{10.157}$$

$$g_{m1}v_1 + g_{m2}v_2 = 0. \tag{10.158}$$

Assuming perfect symmetry, we have from Eq. (10.158)

$$v_1 = -v_2 \tag{10.159}$$

and for differential inputs, we require $v_{in1} = -v_{in2}$. Thus, Eq. (10.157) translates to

$$v_{in1} = v_1 \tag{10.160}$$

and hence

$$v_P = v_{in1} - v_1 \tag{10.161}$$

$$= 0. \tag{10.162}$$

Alternatively, we can simply utilize Eqs. (10.86)–(10.91) with the observation that $v_\pi/r_\pi = 0$ for a MOSFET, arriving at the same result.

With node P acting as a virtual ground, the concept of half circuit applies, leading to the simplified topology in Fig. 10.35(b). Here,

$$v_{out1} = -g_m R_D v_{in1} \tag{10.163}$$

$$v_{out2} = -g_m R_D v_{in2}, \tag{10.164}$$

and, therefore,

$$\frac{v_{out1} - v_{out2}}{v_{in1} - v_{in2}} = -g_m R_D. \tag{10.165}$$

Example 10-21 Prove that Eq. (10.156) can also yield the differential voltage gain.

Solution Since $V_{out1} - V_{out2} = -R_D(I_{D1} - I_{D2})$ and since $g_m = \sqrt{\mu_n C_{ox}(W/L)I_{SS}}$ (why?), we have from Eq. (10.156)

$$V_{out1} - V_{out2} = -R_D\sqrt{\mu_n C_{ox}\frac{W}{L}I_{SS}}(V_{in1} - V_{in2}) \tag{10.166}$$

$$= -g_m R_D(V_{in1} - V_{in2}). \tag{10.167}$$

This is, of course, to be expected. After all, small-signal operation simply means approximating the input/output characteristic [Eq. (10.147)] with a straight line [Eq. (10.156)] around an operating point (equilibrium).

Exercise Using the equation $g_m = 2I_D/(V_{GS} - V_{TH})$, express the above result in terms of the equilibrium overdrive voltage.

As with the bipolar circuits studied in Examples 10-10 and 10-14, the analysis of MOS differential topologies is greatly simplified if virtual grounds can be identified. The following examples reinforce this concept.

Example 10-22 Determine the voltage gain of the circuit shown in Fig. 10.36(a). Assume $\lambda \neq 0$.

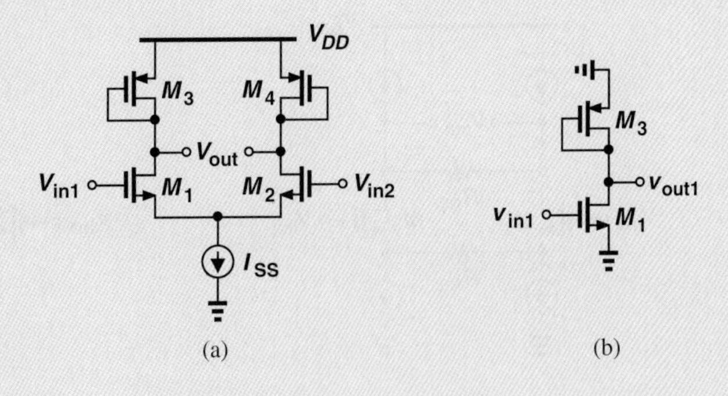

Figure 10.36

Solution Drawing the half circuit as in Fig. 10.36(b), we note that the total resistance seen at the drain of M_1 is equal to $(1/g_{m3})||r_{O3}||r_{O1}$. The voltage gain is therefore equal to

$$A_v = -g_{m1}\left(\frac{1}{g_{m3}}||r_{O3}||r_{O1}\right). \tag{10.168}$$

Exercise Repeat the above example if a resistance of value R_1 is inserted in series with the sources of M_3 and M_4.

Example 10-23 Assuming $\lambda = 0$, compute the voltage gain of the circuit illustrated in Fig. 10.37(a).

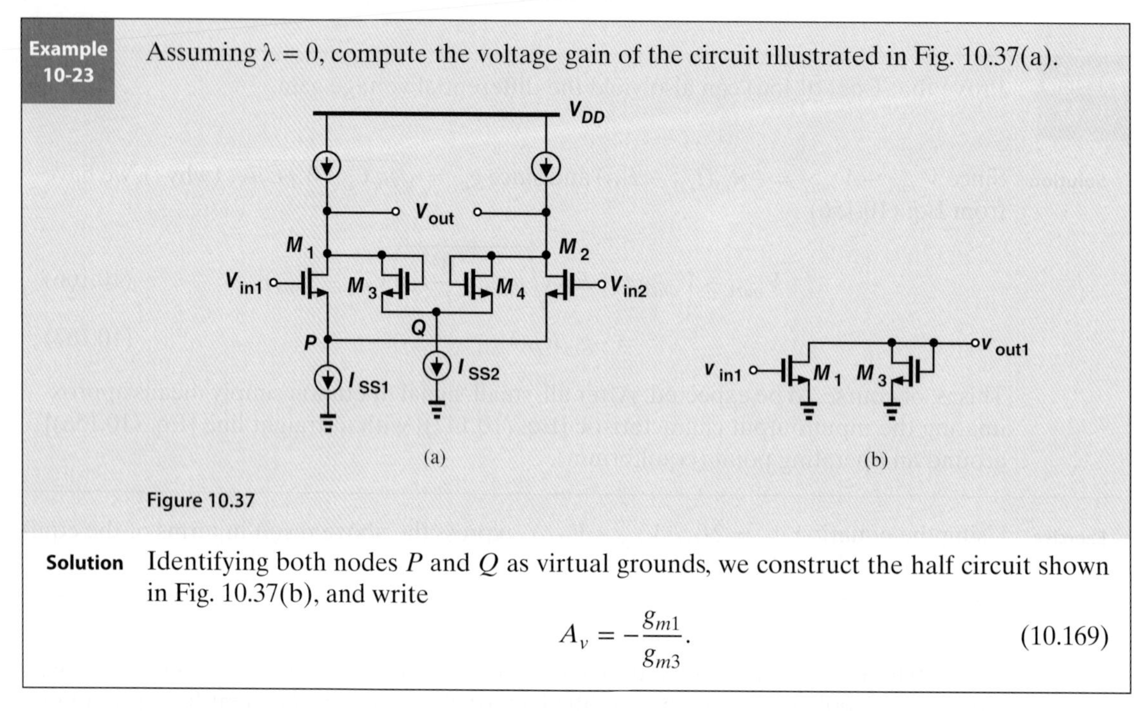

Figure 10.37

Solution Identifying both nodes P and Q as virtual grounds, we construct the half circuit shown in Fig. 10.37(b), and write

$$A_v = -\frac{g_{m1}}{g_{m3}}. \tag{10.169}$$

Exercise Repeat the above example if $\lambda \neq 0$.

Example 10-24 Assuming $\lambda = 0$, calculate the voltage gain of the topology shown in Fig. 10.38(a).

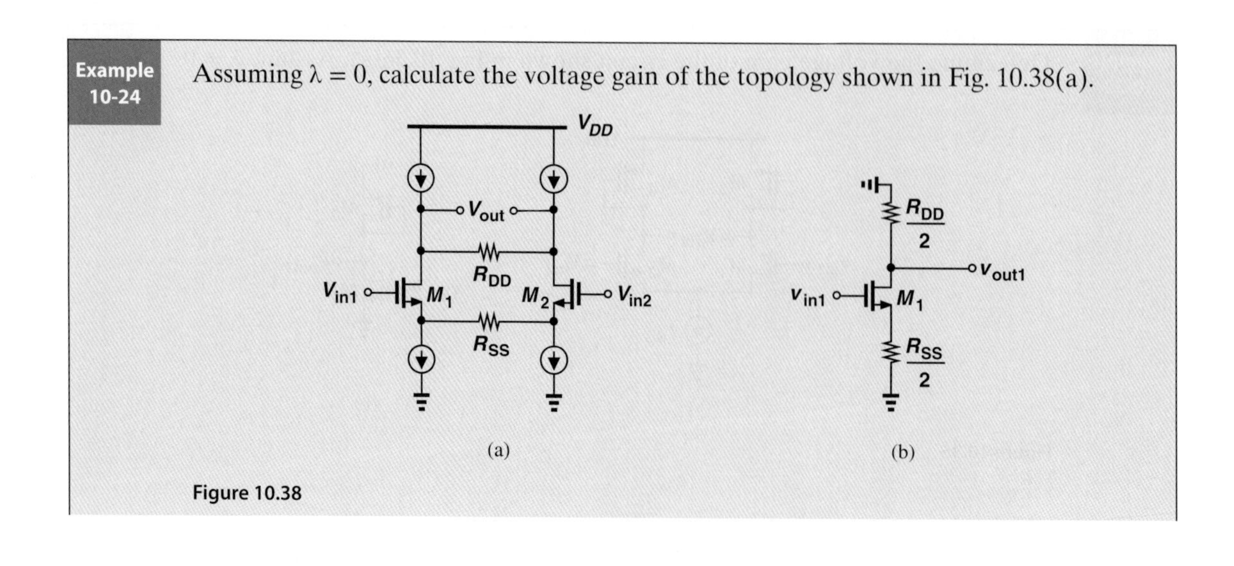

Figure 10.38

Solution Grounding the midpoint of R_{SS} and R_{DD}, we obtain the half circuit in Fig. 10.38(b), where

$$A_v = -\frac{\frac{R_{DD}}{2}}{\frac{R_{SS}}{2} + \frac{1}{g_m}}. \tag{10.170}$$

Exercise Repeat the above example if the load current sources are replaced with diode-connected PMOS devices.

10.4 CASCODE DIFFERENTIAL AMPLIFIERS

Recall from Chapter 9 that cascode stages provide a substantially higher voltage gain than simple CE and CS stages do. Noting that the differential gain of differential pairs is equal to the single-ended gain of their corresponding half circuits, we surmise that cascoding boosts the gain of differential pairs as well.

We begin our study with the structure depicted in Fig. 10.39(a), where Q_3 and Q_4 serve as cascode devices and I_1 and I_2 are ideal. Recognizing that the bases of Q_3 and Q_4 are at ac ground, we construct the half circuit shown in Fig. 10.39(b). Equation (9.51) readily gives the gain as

$$A_v = -g_{m1}[g_{m3}(r_{O1}||r_{\pi 3})r_{O3} + r_{O1}||r_{\pi 3}], \tag{10.171}$$

confirming that a differential cascode achieves a much higher gain.

The developments in Chapter 9 also suggest the use of *pnp* cascodes for current sources I_1 and I_2 in Fig. 10.39(a). As illustrated in Fig. 10.40(a), the resulting configuration can be analyzed with the aid of its half circuit, Fig. 10.40(b). Utilizing Eq. (9.61), we express the voltage gain as

$$A_v \approx -g_{m1}[g_{m3}r_{O3}(r_{O1}||r_{\pi 3})]||[g_{m5}r_{O5}(r_{O7}||r_{\pi 5})]. \tag{10.172}$$

Called a "telescopic cascode," the topology of Fig. 10.40(b) exemplifies the internal circuit of some operational amplifiers.

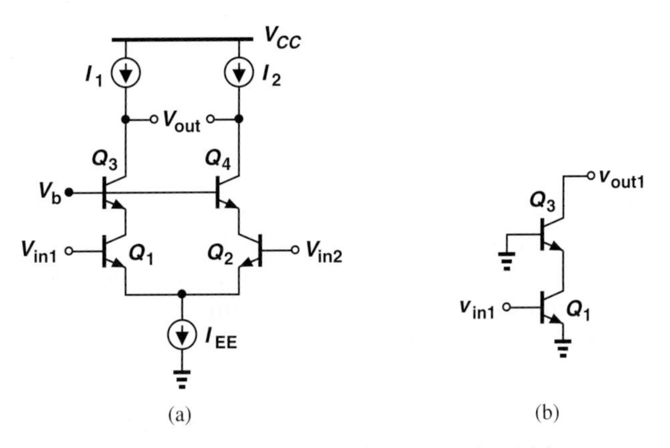

Figure 10.39 (a) Bipolar cascode differential pair, (b) half circuit of (a).

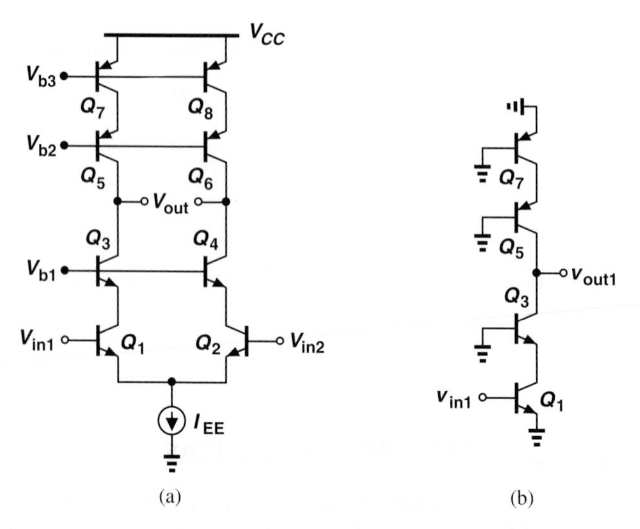

Figure 10.40 (a) Bipolar cascode differential pair with cascode loads, (b) half circuit of (a).

Example 10-25

Due to a manufacturing defect, a parasitic resistance has appeared between nodes A and B in the circuit of Fig. 10.41(a). Determine the voltage gain of the circuit.

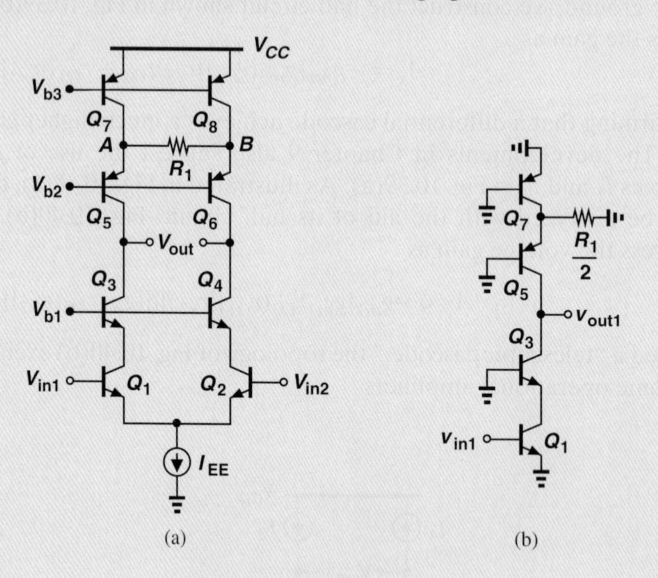

Figure 10.41

Solution The symmetry of the circuit implies that the midpoint of R_1 is a virtual ground, leading to the half circuit shown in Fig. 10.41(b). Thus, $R_1/2$ appears in parallel with r_{O7}, lowering the output impedance of the *pnp* cascode. Since the value of R_1 is not given, we cannot make approximations and must return to the original expression for the cascode output impedance, Eq. (9.1):

$$R_{op} = \left[1 + g_{m5}\left(r_{O7}||r_{\pi5}||\frac{R_1}{2}\right)\right] r_{O5} + r_{O7}||r_{\pi5}||\frac{R_1}{2}. \tag{10.173}$$

The resistance seen looking down into the *npn* cascode remains unchanged and approximately equal to $g_{m3}r_{O3}(r_{O1}||r_{\pi3})$. The voltage gain is therefore equal to

$$A_v = -g_{m1}[g_{m3}r_{O3}(r_{O1}||r_{\pi3})]||R_{op}. \tag{10.174}$$

Exercise If $\beta = 50$ and $V_A = 4$ V for all transistors and $I_{EE} = 1$ mA, what value of R_1 degrades the gain by a factor of two?

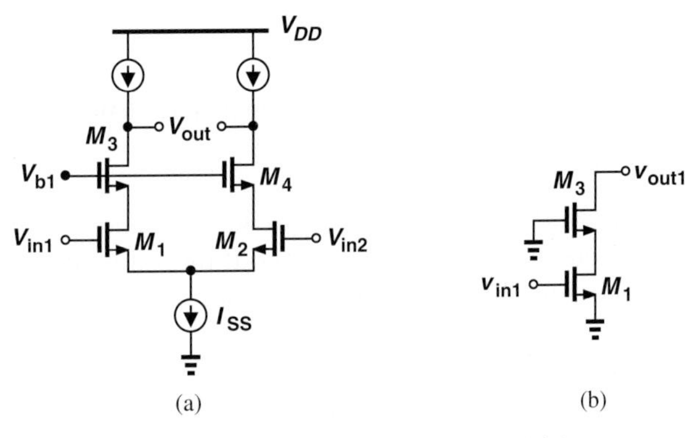

Figure 10.42 (a) MOS cascode differential pair, (b) half circuit of (a).

We now turn our attention to differential MOS cascodes. Following the above developments for bipolar counterparts, we consider the simplified topology of Fig. 10.42(a) and draw the half circuit as depicted in Fig. 10.42(b). From Eq. (9.69),

$$A_v \approx -g_{m3}r_{O3}g_{m1}r_{O1}. \tag{10.175}$$

As illustrated in Fig. 10.43(a), the complete CMOS telescopic cascode amplifier incorporates PMOS cascades as load current sources, yielding the half circuit shown in Fig. 10.43(b). It follows from Eq. (9.72) that the voltage gain is given by

$$A_v \approx -g_{m1}[(g_{m3}r_{O3}r_{O1})||(g_{m5}r_{O5}r_{O7})]. \tag{10.176}$$

Did you know?

Most stand-alone bipolar op amps do *not* use the telescopic cascode differential pair. One reason may be that the input common-mode range of this circuit is quite limited, a serious issue for general-purpose op amps. In analog integrated circuits, on the other hand, telescopic cascode op amps have been utilized extensively because they provide a high voltage gain and have a better high-frequency behavior than other op amp topologies. If you open up your WiFi or cell phone receiver, it is likely that you will see some analog filters and analog-to-digital converters using telescopic op amps.

Example 10-26 Due to a manufacturing defect, two equal parasitic resistances, R_1 and R_2, have appeared as shown in Fig. 10.44(a). Compute the voltage gain of the circuit.

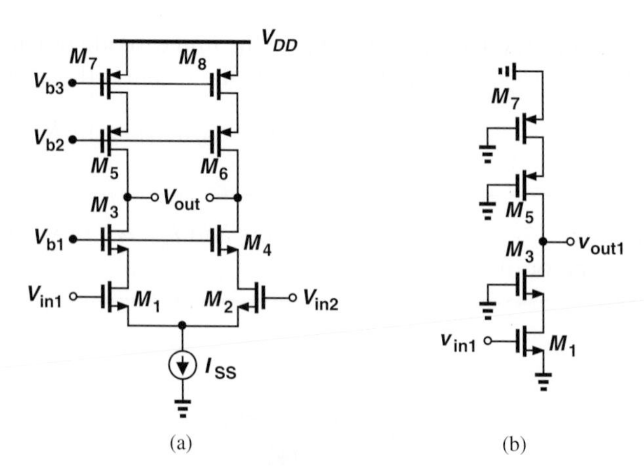

Figure 10.43 (a) MOS telescopic cascode amplifier, (b) half circuit of (a).

Solution Noting that R_1 and R_2 appear in parallel with r_{O5} and r_{O6}, respectively, we draw the half circuit as depicted in Fig. 10.44(b). Without the value of R_1 given, we must resort to the original expression for the output impedance, Eq. (9.3):

$$R_p = [1 + g_{m5}(r_{O5}||R_1)]r_{O7} + r_{O5}||R_1. \tag{10.177}$$

The resistance seen looking into the drain of the NMOS cascode can still be approximated as

$$R_n \approx g_{m3}r_{O3}r_{O1}. \tag{10.178}$$

The voltage gain is then simply equal to

$$A_v = -g_{m1}(R_p||R_n). \tag{10.179}$$

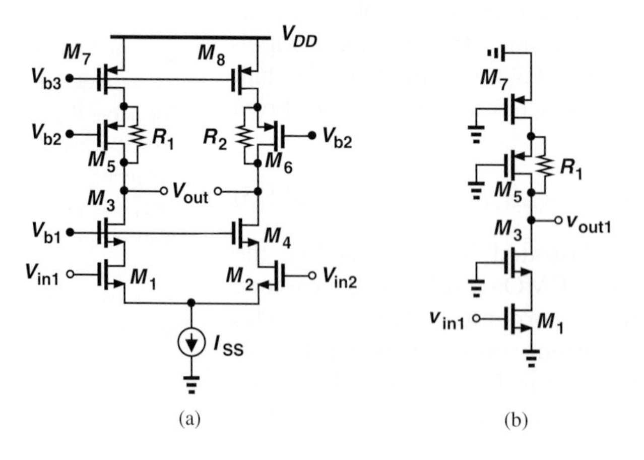

Figure 10.44

Exercise Repeat the above example if in addition to R_1 and R_2, a resistor of value R_3 appears between the sources of M_3 and M_4.

10.5 COMMON-MODE REJECTION

In our study of bipolar and MOS differential pairs, we have observed that these circuits produce *no* change in the output if the input CM level changes. The common-mode rejection property of differential circuits plays a critical role in today's electronic systems. As the reader may have guessed, in practice the CM rejection is not infinitely high. In this section, we examine the CM rejection in the presence of nonidealities.

The first nonideality relates to the output impedance of the tail current source. Consider the topology shown in Fig. 10.45(a), where R_{EE} denotes the output impedance of I_{EE}. What happens if the input CM level changes by a small amount? The symmetry requires that Q_1 and Q_2 still carry equal currents and $V_{out1} = V_{out2}$. But, since the base voltages of both Q_1 and Q_2 rise, so does V_P. In fact, noting that $V_{out1} = V_{out2}$, we can place a short circuit between the output nodes, reducing the topology to that shown in Fig. 10.45(b). That is, as far as node P is concerned, Q_1 and Q_2 operate as an emitter follower. As V_P increases, so does the current through R_{EE} and hence the collector currents of Q_1 and Q_2. Consequently, the output common-mode level falls. The change in the output CM level can be computed by noting that the stage in Fig. 10.45(b) resembles a degenerated CE stage. That is, from Chapter 5,

$$\frac{\Delta V_{out,CM}}{\Delta V_{in,CM}} = -\frac{\dfrac{R_C}{2}}{R_{EE} + \dfrac{1}{2g_m}} \tag{10.180}$$

$$= -\frac{R_C}{2R_{EE} + g_m^{-1}}, \tag{10.181}$$

where the term $2g_m$ represents the transconductance of the parallel combination of Q_1 and Q_2. This quantity is called the "common-mode gain." These observations apply to the MOS counterpart equally well. An alternative approach to arriving at Eq. (10.180) is outlined in Problem 10.65.

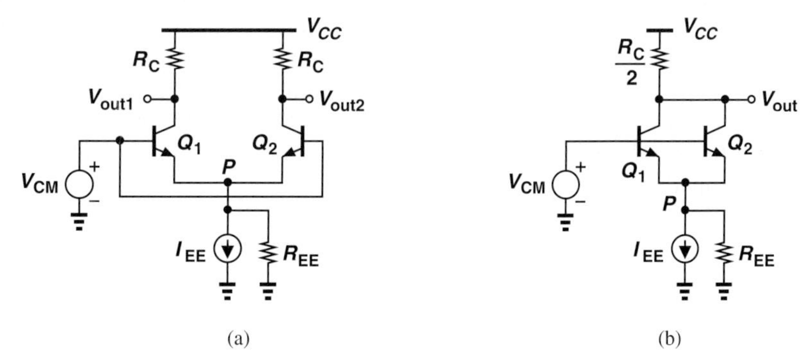

Figure 10.45 (a) CM response of differential pair in the presence of finite tail impedance, (b) simplified circuit of (a).

In summary, if the tail current exhibits a finite output impedance, the differential pair produces an output CM change in response to an input CM change. The reader may naturally wonder whether this is a serious issue. After all, so long as the quantity of interest is the *difference* between the outputs, a change in the output CM level introduces no corruption. Figure 10.46(a) illustrates such a situation. Here, two differential inputs, V_{in1} and V_{in2}, experience some common-mode noise, $V_{in,CM}$. As a result, the base voltages of Q_1 and Q_2 with respect to ground appear as shown in Fig. 10.46(b). With an ideal tail current source, the input CM variation would have no effect at the output, leading to the output waveforms shown in Fig. 10.46(b). On the other hand, with $R_{EE} < \infty$, the *single-ended* outputs are corrupted, but not the differential output [Fig. 10.46(c)].

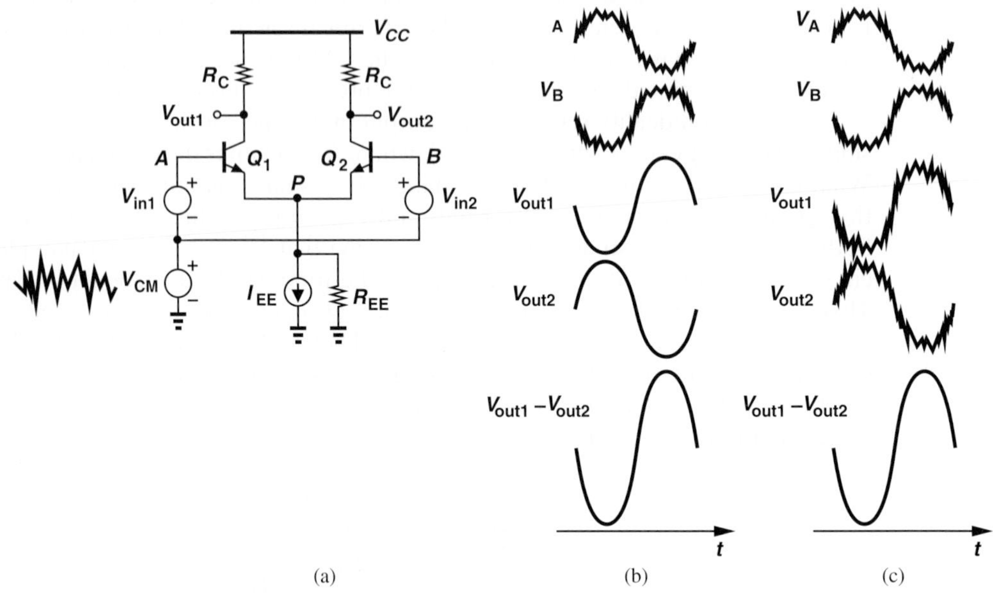

Figure 10.46 (a) Differential pair sensing input CM noise, (b) effect of CM noise at output with $R_{EE} = \infty$, (c) effect of CM noise at the output with $R_{EE} \neq \infty$.

In summary, the above study indicates that, in the presence of input CM noise, a finite CM gain does not corrupt the differential output and hence proves benign.[4] However, if the circuit suffers from *asymmetries* and a finite tail current source impedance, then the differential output is corrupted. During manufacturing, random "mismatches" appear between the two sides of the differential pair; for example, the transistors or the load resistors may display slightly different dimensions. Consequently, the change in the tail current due to an input CM variation may affect the *differential* output.

As an example of the effect of asymmetries, we consider the simple case of load resistor mismatch. Depicted in Fig. 10.47(a) for a MOS pair,[5] this imperfection leads to a difference between V_{out1} and V_{out2}. We must compute the change in I_{D1} and I_{D2} and multiply the result by R_D and $R_D + \Delta R_D$.

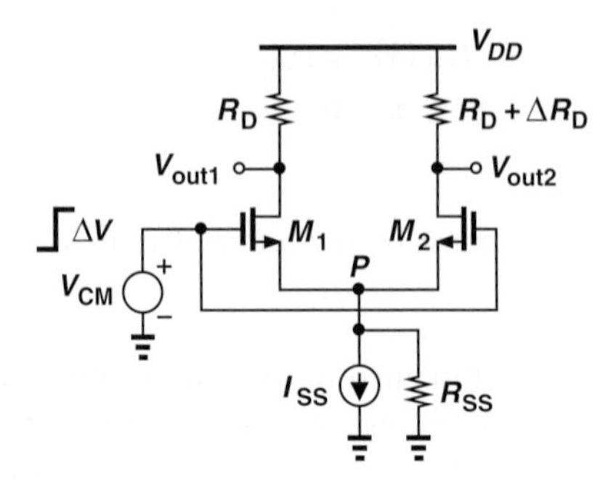

Figure 10.47 MOS pair with asymmetric loads.

[4]Interestingly, older literature has considered this effect troublesome.
[5]We have chosen a MOS pair here to show that the treatment is the same for both technologies.

How do we determine the change in I_{D1} and I_{D2}? Neglecting channel-length modulation, we first observe that

$$I_{D1} = \frac{1}{2}\mu_n C_{ox}\frac{W}{L}(V_{GS1} - V_{TH})^2 \tag{10.182}$$

$$I_{D2} = \frac{1}{2}\mu_n C_{ox}\frac{W}{L}(V_{GS2} - V_{TH})^2, \tag{10.183}$$

concluding that ΔI_{D1} must be equal to ΔI_{D2} because $V_{GS1} = V_{GS2}$ and hence $\Delta V_{GS1} = \Delta V_{GS2}$. In other words, the load resistor mismatch does not impact the symmetry of currents carried by M_1 and M_2.[6] Writing $\Delta I_{D1} = \Delta I_{D2} = \Delta I_D$ and $\Delta V_{GS1} = \Delta V_{GS2} = \Delta V_{GS}$, we recognize that both ΔI_{D1} and ΔI_{D2} flow through R_{SS}, creating a voltage change of $2\Delta I_D R_{SS}$ across it. Thus,

$$\Delta V_{CM} = \Delta V_{GS} + 2\Delta I_D R_{SS} \tag{10.184}$$

and, since $\Delta V_{GS} = \Delta I_D / g_m$,

$$\Delta V_{CM} = \Delta I_D \left(\frac{1}{g_m} + 2R_{SS}\right). \tag{10.185}$$

That is,

$$\Delta I_D = \frac{\Delta V_{CM}}{\dfrac{1}{g_m} + 2R_{SS}}. \tag{10.186}$$

Produced by each transistor, this current change flows through both R_D and $R_D + \Delta R_D$, thereby generating a differential output change of

$$\Delta V_{out} = \Delta V_{out1} - \Delta V_{out2} \tag{10.187}$$

$$= \Delta I_D R_D - \Delta I_D (R_D + \Delta R_D) \tag{10.188}$$

$$= -\Delta I_D \cdot \Delta R_D \tag{10.189}$$

$$= -\frac{\Delta V_{CM}}{\dfrac{1}{g_m} + 2R_{SS}}\Delta R_D. \tag{10.190}$$

It follows that

$$\left|\frac{\Delta V_{out}}{\Delta V_{CM}}\right| = \frac{\Delta R_D}{\dfrac{1}{g_m} + 2R_{SS}}. \tag{10.191}$$

(This result can also be obtained through small-signal analysis.) We say the circuit exhibits "common mode to differential mode (DM) conversion" and denote the above gain by A_{CM-DM}. In practice, we strive to minimize this corruption by maximizing the output impedance of the tail current source. For example, a bipolar current source may employ emitter degeneration and a MOS current source may incorporate a relatively long transistor. It is therefore reasonable to assume $R_{SS} \gg 1/g_m$ and

$$A_{CM-DM} \approx \frac{\Delta R_D}{2R_{SS}}. \tag{10.192}$$

[6]But with $\lambda \neq 0$, it would.

Example 10-27 Determine A_{CM-DM} for the circuit shown in Fig. 10.48. Assume $V_A = \infty$ for Q_1 and Q_2.

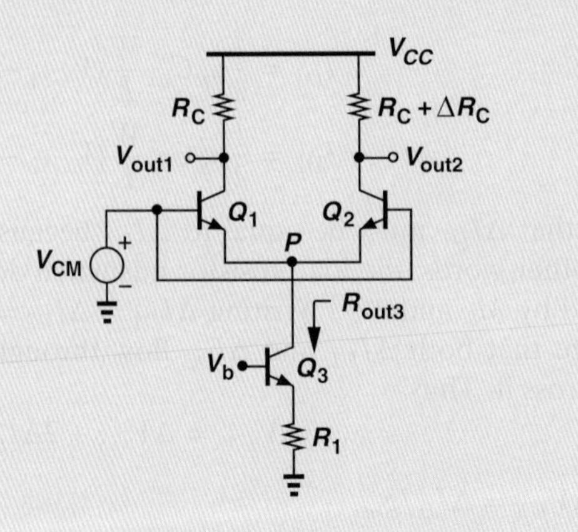

Figure 10.48

Solution Recall from Chapter 5 that emitter degeneration raises the output impedance to

$$R_{out3} = [1 + g_{m3}(R_1||r_{\pi 3})]r_{O3} + R_1||r_{\pi 3}. \tag{10.193}$$

Replacing this value for R_{SS} in Eq. (10.191) yields

$$A_{CM-DM} = \frac{\Delta R_C}{\dfrac{1}{g_{m1}} + 2\{[1 + g_{m3}(R_1||r_{\pi 3})]r_{O3} + R_1||r_{\pi 3}\}}. \tag{10.194}$$

Exercise Calculate the above result if $R_1 \to \infty$.

The mismatches between the transistors in a differential pair also lead to CM-DM conversion. This effect is beyond the scope of this book [1].

While undesirable, CM-DM conversion cannot be simply quantified by A_{CM-DM}. If the circuit provides a large *differential* gain, A_{DM}, then the *relative* corruption at the output is small. We therefore define the "common-mode rejection ratio" (CMRR) as

$$\text{CMRR} = \frac{A_{DM}}{A_{CM-DM}}. \tag{10.195}$$

Representing the ratio of "good" to "bad," CMRR serves as a measure of how much wanted signal and how much unwanted corruption appear at the output if the input consists of a differential component and common-mode noise.

Example 10-28 Calculate the CMRR of the circuit in Fig. 10.48.

Solution For small mismatches (e.g., 1%), $\Delta R_C \ll R_C$, and the differential gain is equal to $g_{m1}R_C$. Thus,

$$\text{CMRR} = \frac{g_{m1}R_C}{\Delta R_C}\left\{\frac{1}{g_{m1}} + 2[1 + g_{m3}(R_1||r_{\pi 3})]r_{O3} + 2(R_1||r_{\pi 3})\right\}. \tag{10.196}$$

Exercise Determine the CMRR if $R_1 \to \infty$.

10.6 DIFFERENTIAL PAIR WITH ACTIVE LOAD

In this section, we study an interesting combination of differential pairs and current mirrors that proves useful in many applications. To arrive at the circuit, let us first address a problem encountered in some cases.

Recall that the op amps used in Chapter 8 have a differential input but a *single-ended* output [Fig. 10.49(a)]. Thus, the internal circuits of such op amps must incorporate a stage that "converts" a differential input to a single-ended output. We may naturally consider the topology shown in Fig. 10.49(b) as a candidate for this task. Here, the output is sensed at node Y with respect to ground rather than with respect to node X.[7] Unfortunately, the voltage gain is now halved because the signal swing at node X is not used.

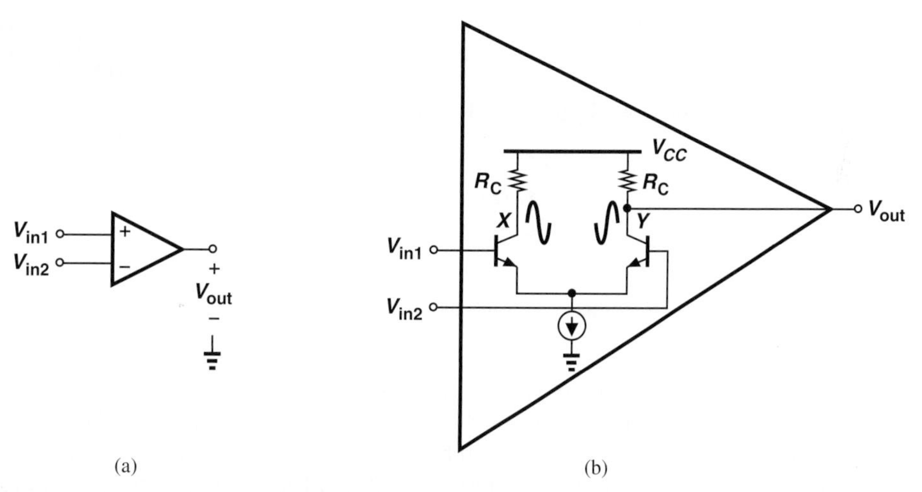

Figure 10.49 (a) Circuit with differential input and single-ended output, (b) possible implementation of (a).

We now introduce a topology that serves the task of "differential to single-ended" conversion while resolving the above issues. Shown in Fig. 10.50, the circuit employs a symmetric differential pair, Q_1-Q_2, along with a current-mirror load, Q_3-Q_4. (Transistors Q_3 and Q_4 are also identical.) The output is sensed with respect to ground.

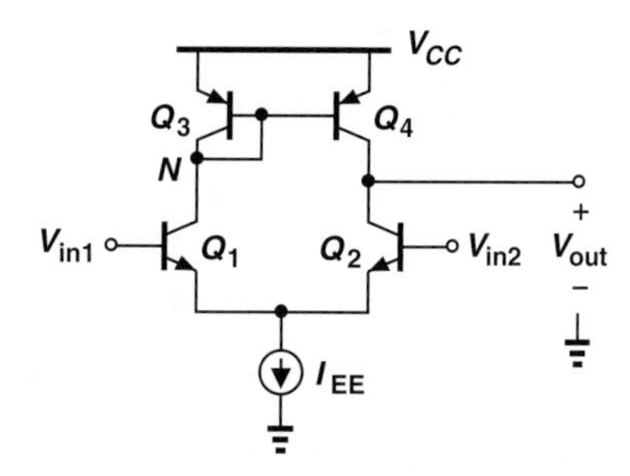

Figure 10.50 Differential pair with active load.

[7]In practice, additional stages precede this stage so as to provide a high gain.

10.6.1 Qualitative Analysis

It is instructive to first decompose the circuit of Fig. 10.50 into two sections: the input differential pair and the current-mirror load. As depicted in Fig. 10.51(a) (along with a fictitious load R_L), Q_1 and Q_2 produce equal and opposite changes in their collector currents in response to a differential change at the input, creating a voltage change of $\Delta I R_L$ across R_L. Now consider the circuit in Fig. 10.51(b) and suppose the current drawn from Q_3 increases from $I_{EE}/2$ to $I_{EE}/2 + \Delta I$. What happens? First, since the small-signal impedance seen at node N is approximately equal to $1/g_{m3}$, V_N changes by $\Delta I/g_{m3}$ (for small ΔI). Second, by virtue of current mirror action, the collector current of Q_4 also *increases* by ΔI. As a result, the voltage across R_L changes by $\Delta I R_L$.

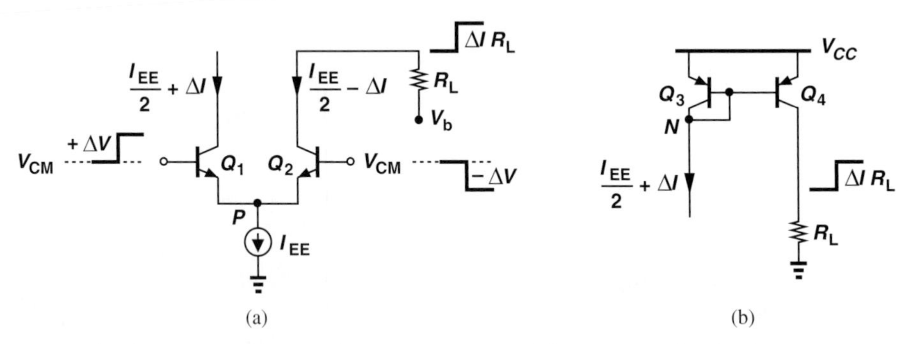

Figure 10.51 (a) Response of input pair to input change, (b) response of active load to current change.

In order to understand the detailed operation of the circuit, we apply small, differential changes at the input and follow the signals to the output (Fig. 10.52). The load resistor, R_L, is added to augment our intuition but it is not necessary for the actual operation. With the input voltage changes shown here, we note that I_{C1} increases by some amount ΔI and I_{C2} decreases by the same amount. Ignoring the role of Q_3 and Q_4 for the moment, we observe that the fall in I_{C2} translates to a *rise* in V_{out} because Q_2 draws less current from R_L. The output change can therefore be an amplified version of ΔV.

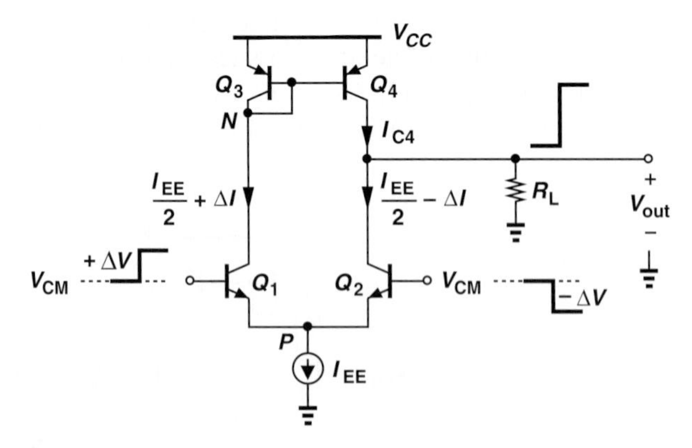

Figure 10.52 Detailed operation of pair with active load.

Let us now determine how the change in I_{C1} travels through Q_3 and Q_4. Neglecting the base currents of these two transistors, we recognize that the change in I_{C3} is also equal to ΔI. This change is copied into I_{C4} by virtue of the current mirror action. In other words, in response to the differential input shown in Fig. 10.52, I_{C1}, $|I_{C3}|$, and $|I_{C4}|$ *increase* by ΔI. Since Q_4 "injects" a greater current into the output node, V_{out} rises.

In summary, the circuit of Fig. 10.52 contains *two signal paths*, one through Q_1 and Q_2 and another through Q_1, Q_3, and Q_4 [Fig. 10.53(a)]. For a differential input change, each path experiences a current change, which translates to a voltage change at the output node. The key point here is that the two paths *enhance* each other at the output; in the above example, each path forces V_{out} to *increase*.

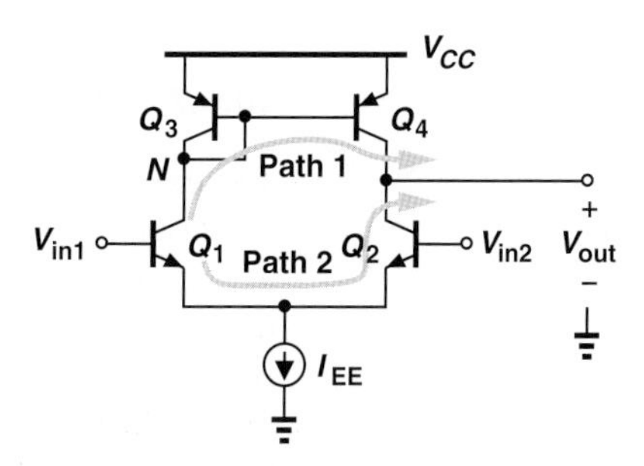

Figure 10.53 Signal paths in pair with active load.

Our initial examination of Q_3 and Q_4 in Fig. 10.52 indicates an interesting difference with respect to current mirrors studied in Chapter 9: here Q_3 and Q_4 carry *signals* in addition to bias currents. This also stands in contrast to the current-source loads in Fig. 10.54, where the base-emitter voltage of the load transistors remains *constant* and independent of signals. Called an "active load" to distinguish it from the load transistors in Fig. 10.54, the combination of Q_3 and Q_4 plays a critical role in the operation of the circuit.

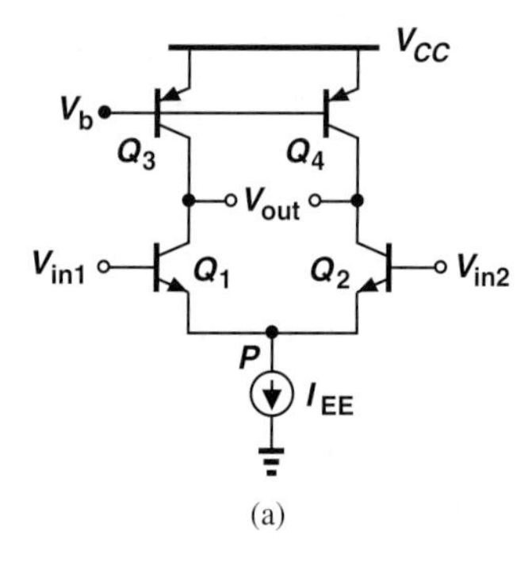

(a)

Figure 10.54 Differential pair with current-source loads.

The foregoing analysis directly applies to the CMOS counterpart, shown in Fig. 10.55. Specifically, in response to a small, differential input, I_{D1} rises to $I_{SS}/2 + \Delta I$ and I_{D2} falls to $I_{SS}/2 - \Delta I$. The change in I_{D2} tends to raise V_{out}. Also, the change in I_{D1} and I_{D3} is copied into I_{D4}, *increasing* $|I_{D4}|$ and raising V_{out}. (In this circuit, too, the current mirror transistors are identical.)

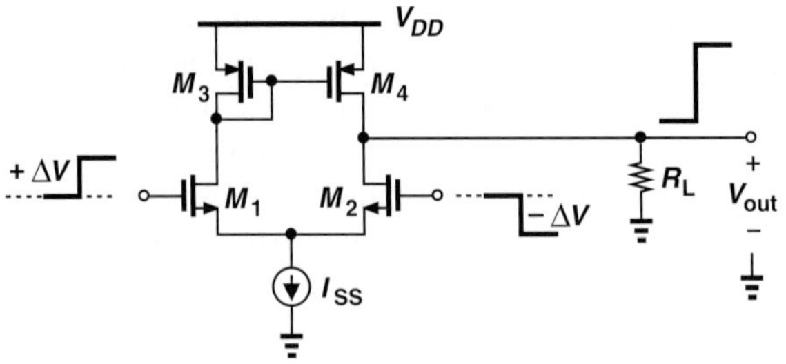

Figure 10.55 MOS differential pair with active load.

Did you know?
The active load was a significant breakthrough in op amp design. Earlier op amps employed a differential pair at the input but discarded the signal on one side at the final output. The first trace of active loads can be seen in the LM101A op amp designed by Robert Widlar at National Semiconductor in 1967. Widlar recognized that the active load avoids large resistor values (which would occupy a large chip area), saves voltage headroom, and provides a high gain. The differential pair with active load is also known as the "five-transistor OTA," where OTA stands for "operational transconductance amplifier" and was used to refer to CMOS op amps in olden days.

10.6.2 Quantitative Analysis

The existence of the signal paths in the differential to single-ended converter circuit suggests that the voltage gain of the circuit must be greater than that of a differential topology in which only *one* output node is sensed with respect to ground [e.g., Fig. 10.49(b)]. To confirm this conjecture, we wish to determine the small-signal single-ended output, v_{out}, divided by the small-signal differential input, $v_{in1} - v_{in2}$. We deal with a CMOS implementation here (Fig. 10.56) to demonstrate that both CMOS and bipolar versions are treated identically.

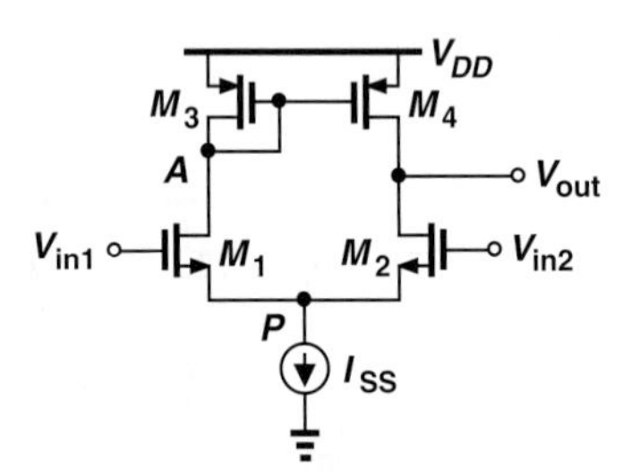

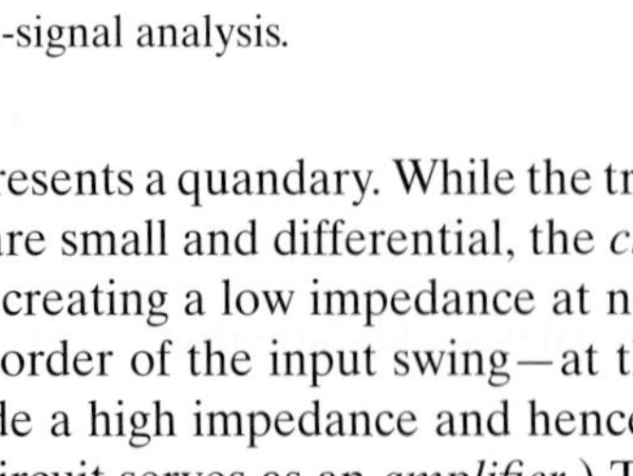

Figure 10.56 MOS pair for small-signal analysis.

The circuit of Fig. 10.56 presents a quandary. While the transistors themselves are symmetric and the input signals are small and differential, the *circuit* is asymmetric. With the diode-connected device, M_3, creating a low impedance at node A, we expect a relatively small voltage swing—on the order of the input swing—at this node. On the other hand, transistors M_2 and M_4 provide a high impedance and hence a large voltage swing at the output node. (After all, the circuit serves as an *amplifier*.) The asymmetry resulting from the very different voltage swings at the drains of M_1 and M_2 disallows grounding node P for small-signal analysis. We present two approaches to solving this circuit.

Approach I Without a half circuit available, the analysis can be performed through the use of a complete small-signal model of the amplifier. Referring to the equivalent circuit shown in Fig. 10.57, where the dashed boxes indicate each transistor, we perform the analysis in two steps. In the first step, we note that i_X and i_Y must add up to zero at node P and hence $i_X = -i_Y$. Also, $v_A = -i_X(g_{mP}^{-1}||r_{OP})$ and

$$-i_Y = \frac{v_{out}}{r_{OP}} + g_{mP}v_A \tag{10.197}$$

$$= \frac{v_{out}}{r_{OP}} - g_{mP}i_X\left(\frac{1}{g_{mP}}\middle\| r_{OP}\right) \tag{10.198}$$

$$= i_X. \tag{10.199}$$

Thus,

$$i_X = \frac{v_{out}}{r_{OP}\left[1 + g_{mP}\left(\frac{1}{g_{mP}}\middle\| r_{OP}\right)\right]}. \tag{10.200}$$

In the second step, we write a KVL around the loop consisting of all four transistors. The current through r_{ON} of M_1 is equal to $i_X - g_{mN}v_1$ and that through r_{ON} of M_2 equal to $i_Y - g_{mN}v_2$. It follows that

$$-v_A + (i_X - g_{mN}v_1)r_{ON} - (i_Y - g_{mN}v_2)r_{ON} + v_{out} = 0. \tag{10.201}$$

Since $v_1 - v_2 = v_{in1} - v_{in2}$ and $i_X = -i_Y$,

$$-v_A + 2i_Xr_{ON} - g_{mN}r_{ON}(v_{in1} - v_{in2}) + v_{out} = 0. \tag{10.202}$$

Substituting for v_A and i_X from above, we have

$$\frac{v_{out}}{r_{OP}\left[1 + g_{mP}\left(\frac{1}{g_{mP}}\middle\| r_{OP}\right)\right]}\left(\frac{1}{g_{mP}}\middle\| r_{OP}\right) + 2r_{ON}\frac{v_{out}}{r_{OP}\left[1 + g_{mP}\left(\frac{1}{g_{mP}}\middle\| r_{OP}\right)\right]} + v_{out} = g_{mN}r_{ON}(v_{in1} - v_{in2}). \tag{10.203}$$

Solving for v_{out} yields

$$\frac{v_{out}}{v_{in1} - v_{in2}} = g_{mN}r_{ON}\frac{r_{OP}\left[1 + g_{mP}\left(\frac{1}{g_{mP}}\middle\| r_{OP}\right)\right]}{2r_{ON} + 2r_{OP}}. \tag{10.204}$$

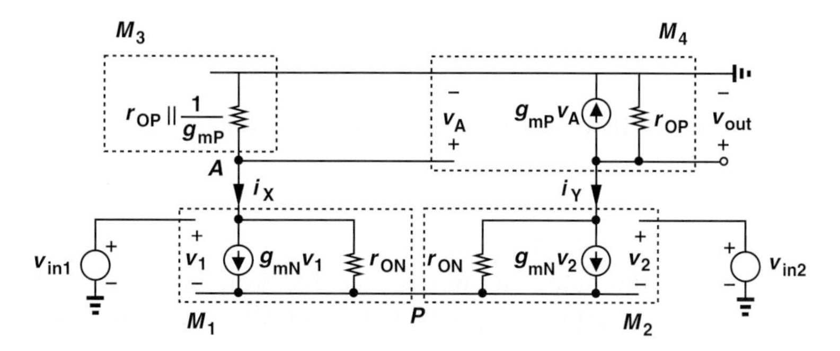

Figure 10.57 Small-signal equivalent circuit of differential pair with active load.

This is the exact expression for the gain. If $g_{mP}r_{OP} \gg 1$, then

$$\frac{v_{out}}{v_{in1} - v_{in2}} = g_{mN}(r_{ON}||r_{OP}). \quad (10.205)$$

The gain is indepedent of g_{mP} and equal to that of the fully-differential circuit. In other words, the use of the active load has restored the gain.

Approach II* In this approach, we decompose the circuit into sections that more easily lend themselves to analysis by inspection. As illustrated in Fig. 10.58(a), we first seek a Thevenin equivalent for the section consisting of v_{in1}, v_{in2}, M_1, and M_2, assuming v_{in1} and v_{in2} are differential. Recall that v_{Thev} is the voltage between A and B in the "open-circuit condition" [Fig. 10.58(b)]. Under this condition, the circuit *is* symmetric, resembling the topology of Fig. 10.18(a). Equation (10.97) thus yields

$$v_{Thev} = -g_{mN}r_{ON}(v_{in1} - v_{in2}), \quad (10.206)$$

where the subscript N refers to NMOS devices.

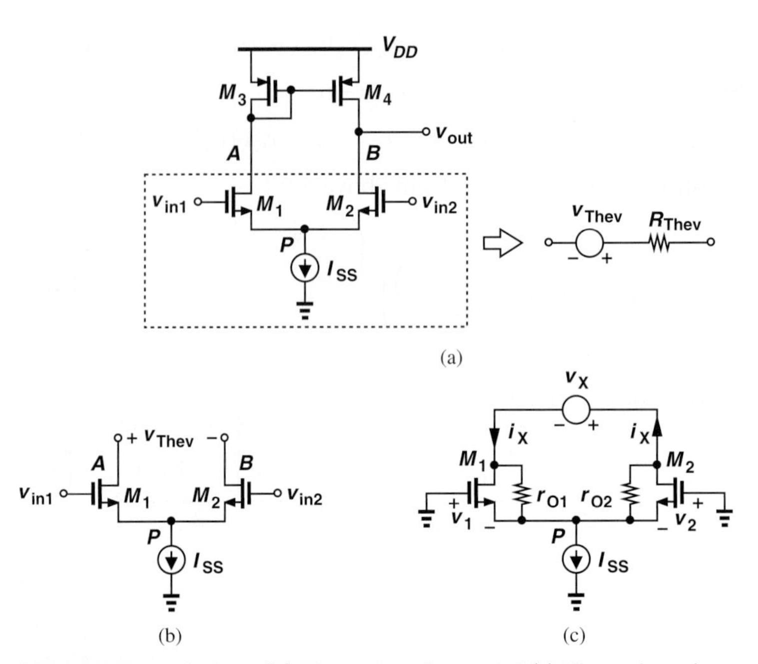

Figure 10.58 (a) Thevenin equivalent, (b) Thevenin voltage, and (c) Thevenin resistance of input pair.

To determine the Thevenin resistance, we set the inputs to zero and apply a voltage between the output terminals [Fig. 10.58(c)]. Noting that M_1 and M_2 have equal gate-source voltages ($v_1 = v_2$) and writing a KVL around the "output" loop, we have

$$(i_X - g_{m1}v_1)r_{O1} + (i_X + g_{m2}v_2)r_{O2} = v_X \quad (10.207)$$

*This section can be skipped in a first reading.

and hence

$$R_{Thev} = 2r_{ON}. \tag{10.208}$$

The reader is encouraged to obtain this result using half circuits as well.

Having reduced the input sources and transistors to a Thevenin equivalent, we now compute the gain of the overall amplifier. Figure 10.59 depicts the simplified circuit, where the diode-connected transistor M_3 is replaced with $(1/g_{m3})||r_{O3}$ and the output impedance of M_4 is drawn explicitly. The objective is to calculate v_{out} in terms of v_{Thev}. Since the voltage at node E with respect to ground is equal to $v_{out} + v_{Thev}$, we can view v_A as a divided version of v_E:

$$v_A = \frac{\frac{1}{g_{m3}} \Big\| r_{O3}}{\frac{1}{g_{m3}} \Big\| r_{O3} + R_{Thev}} (v_{out} + v_{Thev}). \tag{10.209}$$

Given by $g_{m4}v_A$, the small-signal drain current of M_4 must satisfy KCL at the output node:

$$g_{m4}v_A + \frac{v_{out}}{r_{O4}} + \frac{v_{out} + v_{Thev}}{\frac{1}{g_{m3}} \Big\| r_{O3} + R_{Thev}} = 0, \tag{10.210}$$

where the last term on the left-hand side represents the current flowing through R_{Thev}. It follows from Eqs. (10.209) and (10.210) that

$$\left(g_{m4} \frac{\frac{1}{g_{m3}} \Big\| r_{O3}}{\frac{1}{g_{m3}} \Big\| r_{O3} + R_{Thev}} + \frac{1}{\frac{1}{g_{m3}} \Big\| r_{O3} + R_{Thev}} \right) (v_{out} + v_{Thev}) + \frac{v_{out}}{r_{O4}} = 0. \tag{10.211}$$

Recognizing that $1/g_{m3} \ll r_{O3}$, and $1/g_{m3} \ll R_{Thev}$ and assuming $g_{m3} = g_{m4} = g_{mp}$ and $r_{O3} = r_{O4} = r_{OP}$, we reduce Eq. (10.211) to

$$\frac{2}{R_{Thev}}(v_{out} + v_{Thev}) + \frac{v_{out}}{r_{OP}} = 0. \tag{10.212}$$

Equations (10.206) and (10.212) therefore give

$$v_{out}\left(\frac{1}{r_{ON}} + \frac{1}{r_{OP}}\right) = \frac{g_{mN} r_{ON}(v_{in1} - v_{in2})}{r_{ON}} \tag{10.213}$$

and hence

$$\frac{v_{out}}{v_{in1} - v_{in2}} = g_{mN}(r_{ON}||r_{OP}). \tag{10.214}$$

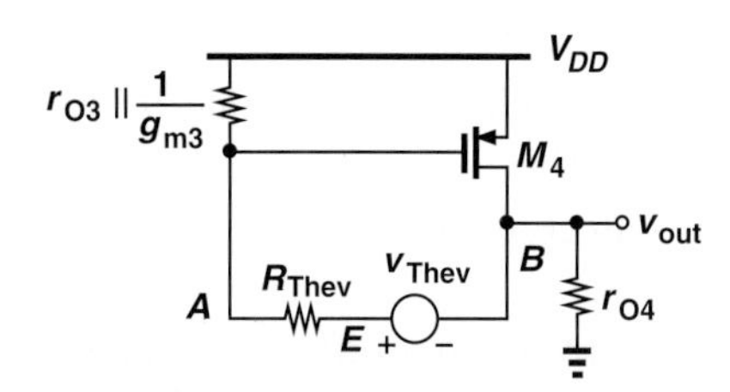

Figure 10.59 Simplified circuit for calculation of voltage gain.

The gain is independent of g_{mp}. Interestingly, the gain of this circuit is the same as the *differential* gain of the topology in Fig. 10.53(b). In other words, the path through the active load restores the gain even though the output is single-ended.

Example 10-29 In our earlier observations, we surmised that the voltage swing at node A in Fig. 10.58 is much less than that at the output. Prove this point.

Solution As depicted in Fig. 10.60, KCL at the output node indicates that the total current drawn by M_2 must be equal to $-v_{out}/r_{O4} - g_{m4}v_A$. This current flows through M_1 and hence through M_3, generating

$$v_A = -(v_{out}/r_{O4} + g_{m4}v_A)\left(\frac{1}{g_{m3}}\middle\|r_{O3}\right). \tag{10.215}$$

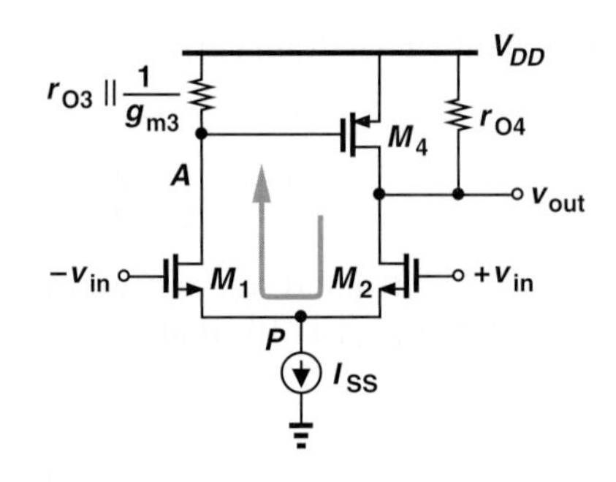

Figure 10.60

That is,

$$v_A \approx -\frac{v_{out}}{2g_{mP}r_{OP}}, \tag{10.216}$$

revealing that v_A is indeed much less than v_{out}.

Exercise Calculate the voltage gain from the differential input to node A.

10.7 CHAPTER SUMMARY

- Single-ended signals are voltages measured with respect to ground. A differential signal consists of two single-ended signals carried over two wires, with the two components beginning from the same dc (common-mode) level and changing by equal and opposite amounts.
- Compared with single-ended signals, differential signals are more immune to common-mode noise.
- A differential pair consists of two identical transistors, a tail current, and two identical loads.
- The transistor currents in a differential pair remain constant as the input CM level changes, i.e., the circuit "rejects" input CM changes.
- The transistor currents change in opposite directions if a differential input is applied, i.e., the circuit responds to differential inputs.

- For small, differential changes at the input, the tail node voltage of a differential pair remains constant and is thus considered a virtual ground node.
- Bipolar differential pairs exhibit a hyperbolic tangent input/output characteristic. The tail current can be mostly steered to one side with a differential input of about $4V_T$.
- For small-signal operation, the input differential swing of a bipolar differential pair must remain below roughly V_T. The pair can then be decomposed into two half circuits, each of which is simply a common-emitter stage.
- MOS differential pairs can steer the tail current with a differential input equal to $\sqrt{2I_{SS}/(\mu_n C_{ox} W/L)}$, which is $\sqrt{2}$ larger than the equilibrium overdrive of each transistor.
- Unlike their bipolar counterparts, MOS differential pairs can provide more or less linear characteristics depending on the choice of the device dimensions.
- The input transistors of a differential pair can be cascoded so as to achieve a higher voltage gain. Similarly, the loads can be cascoded to maximize the voltage gain.
- The differential output of a perfectly symmetric differential pair remains free from input CM changes. In the presence of asymmetries and a finite tail current source impedance, a fraction of the input CM change appears as a differential component at the output, corrupting the desired signal.
- The gain seen by the CM change normalized to the gain seen by the desired signal is called the common-mode rejection ration.
- It is possible to replace the loads of a differential pair with a current mirror so as to provide a single-ended output while maintaining the original gain. The circuit is called a differential pair with active load.

PROBLEMS

Sec. 10.1 General Considerations

SOL **10.1.** To calculate the effect of ripple at the output of the circuit in Fig. 10.1, we can assume V_{CC} is a small-signal "input" and determine the (small-signal) gain from V_{CC} to V_{out}. Compute this gain, assuming $V_A < \infty$.

10.2. Repeat Problem 10.1 for the circuit of Fig. 10.2(a), assuming $R_{C1} = R_{C2}$.

10.3. Repeat Problem 10.1 for the stages shown in Fig. 10.61. Assume $V_A < \infty$ and $\lambda > 0$.

10.4. In the circuit of Fig. 10.62, $I_1 = I_0 \cos \omega t + I_0$ and $I_2 = -I_0 \cos \omega t + I_0$. Plot the waveforms at X and Y and determine their peak-to-peak swings and common-mode level.

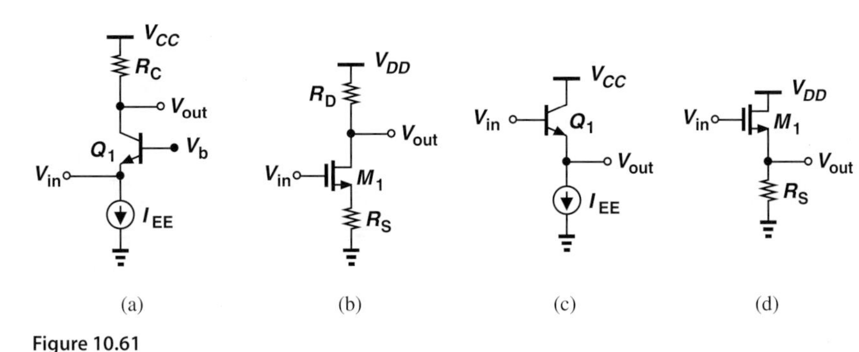

Figure 10.61

10.5. Repeat Problem 10.4 for the circuit depicted in Fig. 10.63. Also, plot the voltage at node P as a function of time.

SOL **10.6.** Repeat Problem 10.4 for the topology shown in Fig. 10.64.

10.7. Repeat Problem 10.4 for the topology shown in Fig. 10.65.

10.8. Repeat Problem 10.4, but assume $I_2 = -I_0 \cos \omega t + 0.8 I_0$. Can X and Y be considered true differential signals?

*__10.9.__ Assuming $I_1 = I_0 \cos \omega t + I_0$ and $I_2 = -I_0 \cos \omega t + I_0$, plot V_X and V_Y as a function of time for the circuits illustrated in Fig. 10.66. Assume I_0 is constant.

*__10.10.__ Assuming $V_1 = V_0 \cos \omega t + V_0$ and $V_2 = -V_0 \cos \omega t + V_0$, plot V_P as a function of time for the circuits shown in Fig. 10.67. Assume I_T is constant.

Sec. 10.2 Bipolar Differential Pair

10.11. Suppose in Fig. 10.9, V_{CC} rises by ΔV. Neglecting the Early effect, determine the change in V_X, V_Y, and $V_X - V_Y$. Explain why we say the circuit "rejects" supply noise.

10.12. In Fig. 10.9, I_{EE} experiences a change of ΔI. How do V_X, V_Y, and $V_X - V_Y$ change? SOL

10.13. Repeat Problem 10.12, but assuming that $R_{C1} = R_{C2} + \Delta R$. Neglect the Early effect.

10.14. Consider the circuit of Fig. 10.11(a) and assume $I_{EE} = 1$ mA. What is the maximum allowable value of R_C if Q_1 must remain in the active region?

10.15. In the circuit of Fig. 10.11(b), $R_C = 500\ \Omega$. What is the maximum allowable

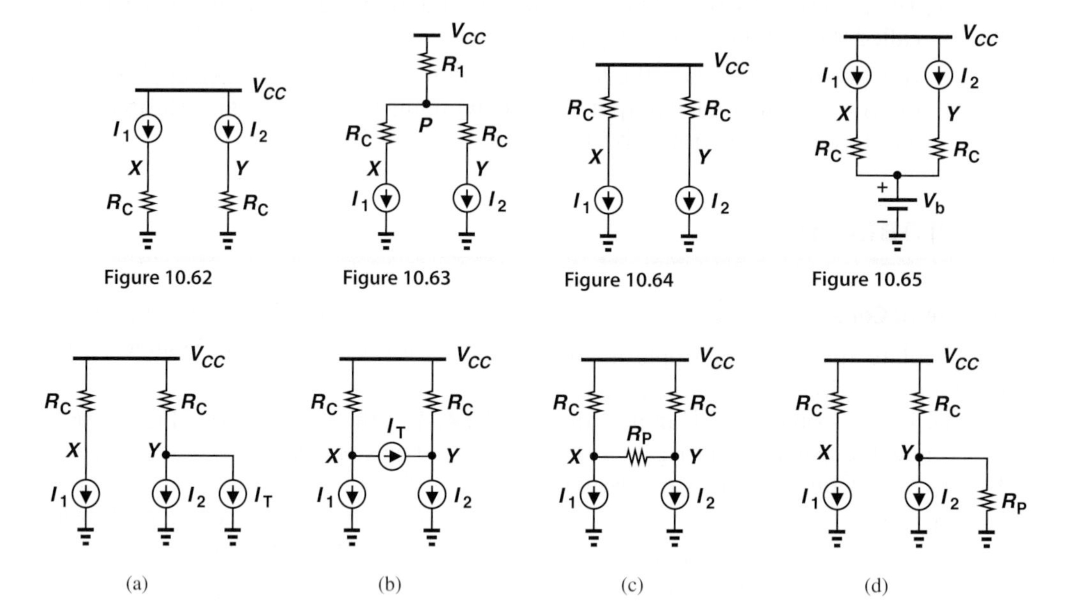

Figure 10.62 Figure 10.63 Figure 10.64 Figure 10.65

Figure 10.66

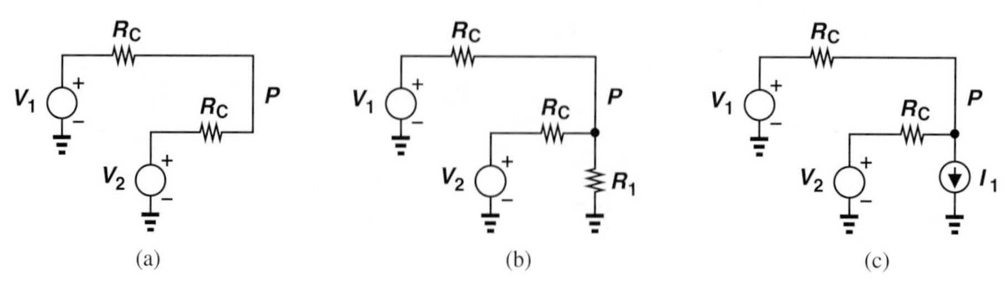

Figure 10.67

value of I_{EE} if Q_2 must remain in the active region?

10.16. Suppose $I_{EE} = 1$ mA and $R_C = 800\ \Omega$ in Fig. 10.11(a). Determine the region of operation of Q_1.

***10.17.** What happens to the characteristics depicted in Fig. 10.12 if (a) I_{EE} is halved, (b) V_{CC} rises by ΔV, or (c) R_C is halved?

Sec. 10.2.2 Large-Signal Analysis of Bipolar Pair

10.18. In the differential pair of Fig. 10.14, $I_{C1}/I_{C2} = 5$. What is the corresponding input differential voltage? With this voltage applied, how does I_{C1}/I_{C2} change if the temperature rises from 27° C to 100° C?

10.19. In the circuit of Fig. 10.14, the small-signal transconductance of Q_2 falls as $V_{in1} - V_{in2}$ rises because I_{C2} decreases. Using Eq. (10.63), determine the input difference at which the transconductance of Q_2 drops by a factor of 2.

10.20. Suppose the input differential signal applied to a bipolar differential pair must not change the transconductance (and hence the bias current) of each transistor by more than 10%. From Eq. (10.63), determine the maximum allowable input.

***10.21.** It is possible to define a differential transconductance for the bipolar differential pair of Fig. 10.14:

$$G_m = \frac{\partial(I_{C1} - I_{C2})}{\partial(V_{in1} - V_{in2})}. \qquad (10.217)$$

From Eqs. (10.63) and (10.65), compute G_m and plot the result as a function of $V_{in1} - V_{in2}$. What is the maximum value of G_m? At what value of $V_{in1} - V_{in2}$ does G_m drop by a factor of two with respect to its maximum value?

SOL ***10.22.** With the aid of Eq. (10.83), we can compute the small-signal voltage gain of the bipolar differential pair:

$$A_v = \frac{\partial(V_{out1} - V_{out2})}{\partial(V_{in1} - V_{in2})}. \qquad (10.218)$$

Determine the gain and compute its value if $V_{in1} - V_{in2}$ contains a dc component of 30 mV.

****10.23.** In Example 10-9, $R_C = 500\ \Omega$, $I_{EE} = 1$ mA, and $V_{CC} = 2.5$ V. Assume

$$V_{in1} = V_0 \sin \omega t + V_{CM} \qquad (10.219)$$

$$V_{in2} = -V_0 \sin \omega t + V_{CM}, \qquad (10.220)$$

where $V_{CM} = 1$ V denotes the input common-mode level.

(a) If $V_0 = 2$ mV, plot the output waveforms (as a function of time).

(b) If $V_0 = 50$ mV, determine the time t_1 at which one transistor carries 95% of the tail current. Plot the output waveforms.

10.24. Explain what happens to the characteristics shown in Fig. 10.15 if the ambient temperature goes from 27° C to 100° C.

****10.25.** The study in Example 10-9 suggests that a differential pair can convert a sinusoid to a square wave. Using the circuit parameters given in Problem 10.23, plot the output waveforms if $V_0 = 80$ mV or 160 mV. Explain why the output square wave becomes "sharper" as the input amplitude increases.

10.26. In Problem 10.25, estimate the slope of the output square waves for $V_0 = 80$ mV or 160 mV if $\omega = 2\pi \times (100$ MHz).

Sec. 10.2.3 Small-Signal Analysis of Bipolar Pair

10.27. Repeat the small-signal analysis of Fig. 10.17 for the circuit shown in Fig. 10.68. (First, prove that P is still a virtual ground.)

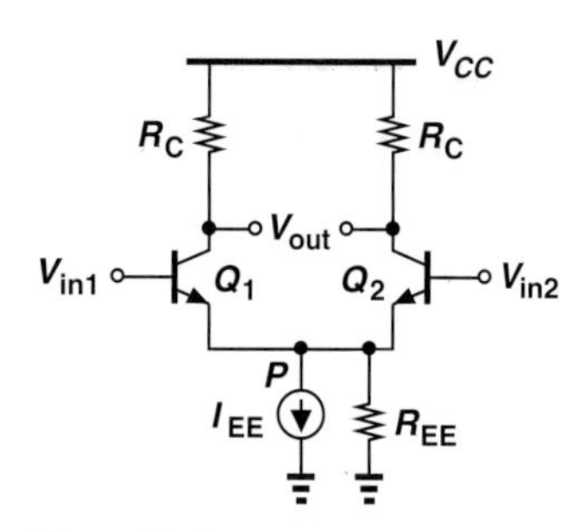

Figure 10.68

10.28. Using a small-signal model and including the output resistance of the transistors, prove that Eq. (10.91) holds in the presence of the Early effect.

***10.29.** In Fig. 10.69, $I_{EE} = 1$ mA and $V_A = 5$ V. Calculate the voltage gain of the circuit. Note that the gain is independent of the tail current.

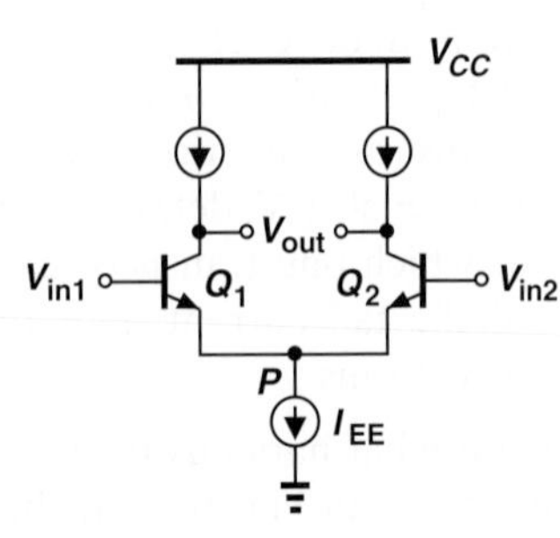

Figure 10.69

10.30. Consider the circuit shown in Fig. 10.70, where $I_{EE} = 2$ mA, $V_{A,n} = 5$ V $V_{A,p} = 4$ V. What value of $R_1 = R_2$ allows a voltage gain of 50?

10.31. The circuit of Fig. 10.70 must provide a gain of 50 with $R_1 = R_2 = 5$ kΩ. If $V_{A,n} = 5$ V and $V_{A,p} = 4$ V, calculate the required tail current.

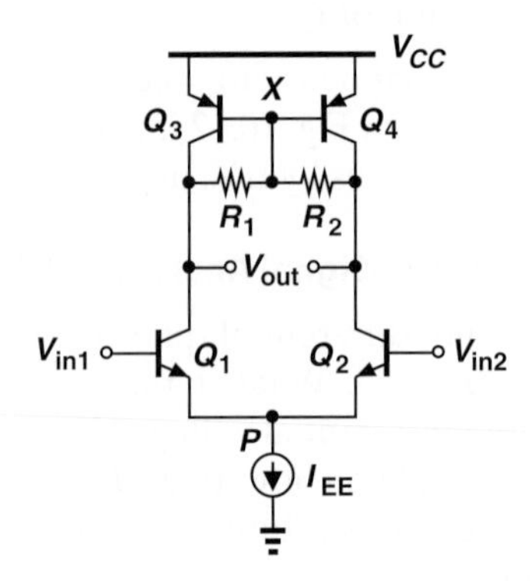

Figure 10.70

****10.32.** Assuming perfect symmetry and $V_A < \infty$, compute the differential voltage gain of each stage depicted in Fig. 10.71.

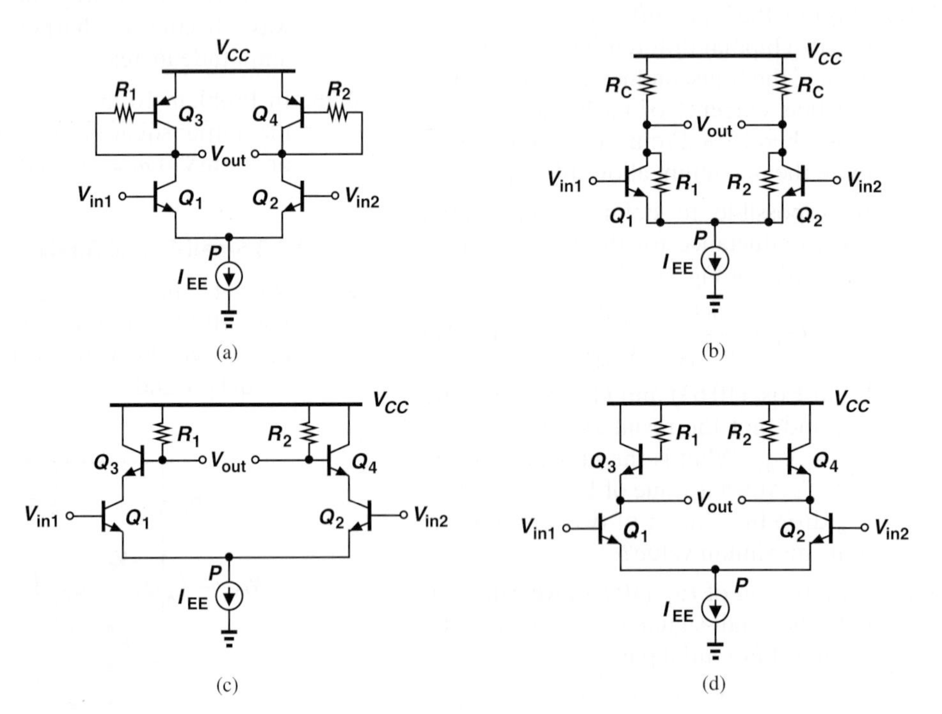

Figure 10.71

*10.33. Consider the differential pair illustrated in Fig. 10.72. Assuming perfect symmetry and $V_A = \infty$,
(a) Determine the voltage gain.
(b) Under what condition does the gain become *independent* of the tail currents? This is an example of a very linear circuit because the gain does not vary with the input or output signal levels.

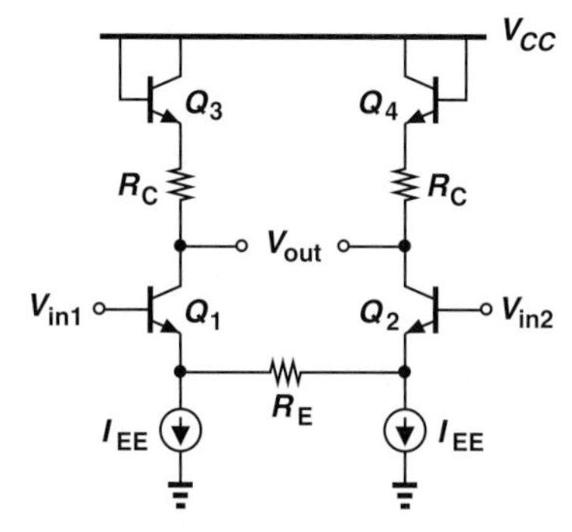

Figure 10.72

10.34. Assuming perfect symmetry and $V_A < \infty$, compute the differential voltage gain of each stage depicted in Fig. 10.73. You may need to compute the gain as $A_v = -G_m R_{out}$ in some cases.

Sec. 10.3 MOS Differential Pair

SOL 10.35. Consider the MOS differential pair of Fig. 10.26. What happens to the tail node voltage if (a) the width of M_1 and M_2 is doubled, (b) I_{SS} is doubled, (c) the gate oxide thickness is doubled?

10.36. In the MOS differential pair of Fig. 10.26, $V_{CM} = 1$ V, $I_{SS} = 1$ mA, and $R_D = 1$ kΩ. What is the minimum allowable supply voltage if the transistors must remain in saturation? Assume $V_{TH,n} = 0.5$ V.

10.37. The MOS differential pair of Fig. 10.26 must be designed for an equilibrium overdrive of 200 mV. If $\mu_n C_{ox} = 100\ \mu\text{A/V}^2$ and $W/L = 20/0.18$, what is the required value of I_{SS}?

10.38. For a MOSFET, the "current density" can be defined as the drain current divided by the device width for a given channel length. Explain how the equilibrium overdrive voltage of a MOS differential pair varies as a function of the current density.

*10.39. A MOS differential pair contains a parasitic resistance tied between its tail node and ground (Fig. 10.74). Without using the small-signal model, prove that P is still a virtual ground for small, differential inputs.

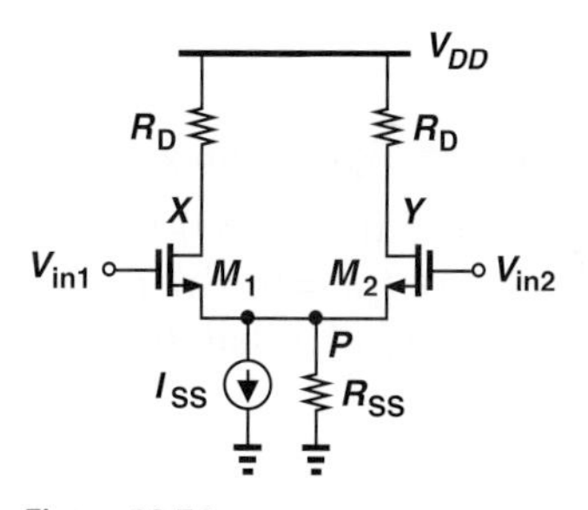

Figure 10.74

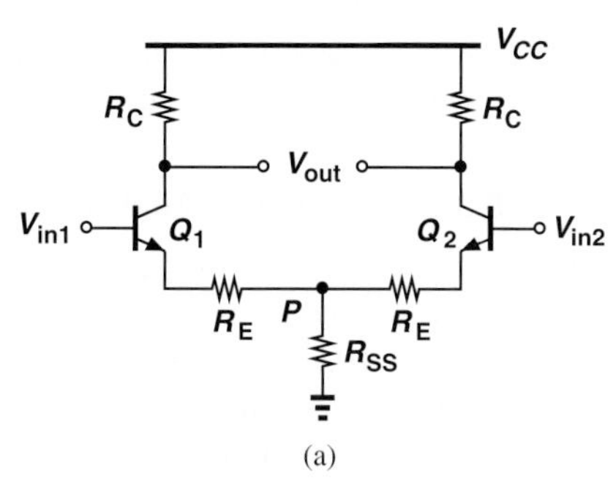

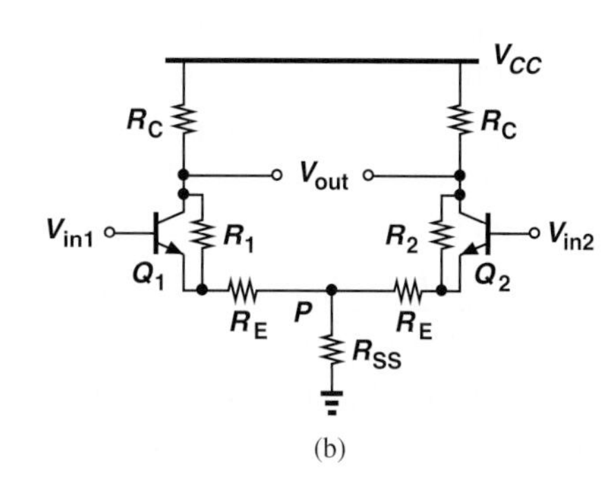

Figure 10.73

SOL **10.40.** In Fig. 10.27(a), $V_{in1} = 1.5$ V and $V_{in2} = 0.3$ V. Assuming M_2 is off, determine a condition among the circuit parameters that guarantees operation of M_1 in saturation.

10.41. An adventurous student constructs the circuit shown in Fig. 10.75 and calls it a "differential amplifier" because $I_D \propto (V_{in1} - V_{in2})$. Explain which aspects of our differential signals and amplifiers this circuit violates.

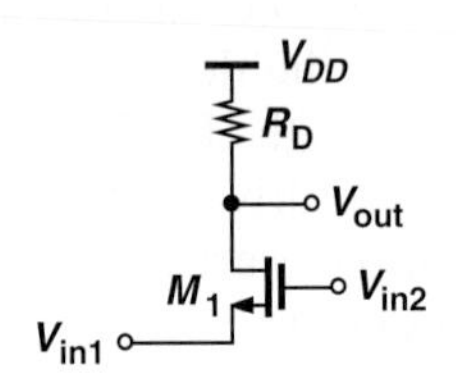

Figure 10.75

10.42. Repeat Example 10-16 for a supply voltage of 2 V. Formulate the trade-off between V_{DD} and W/L for a given output common-mode level.

Sec. 10.3.2 Large-Signal Analysis of MOS Pair

10.43. Examine Eq. (10.139) for the following cases: (a) $I_{D1} = 0$, (b) $I_{D1} = I_{SS}/2$, and (c) $I_{D1} = I_{SS}$. Explain the significance of these cases.

10.44. Prove that the right-hand side of Eq. (10.144) is always negative if the solution with the negative sign is considered.

10.45. From Eq. (10.147), determine the value of $V_{in1} - V_{in2}$ such that $I_{D1} - I_{D2} = I_{SS}$. Verify that is result is equal to $\sqrt{2}$ times the equilibrium overdrive voltage.

10.46. From Eq. (10.147), compute the small-signal transconductance of a MOS differential pair, defined as

$$G_m = \frac{\partial(I_{D1} - I_{D2})}{\partial(V_{in1} - V_{in2})}. \quad (10.221)$$

Plot the result as a function of $V_{in1} - V_{in2}$ and determine its maximum value.

****10.47.** Suppose a new type of MOS transistor has been invented that exhibits the following I-V characteristic:

$$I_D = \gamma(V_{GS} - V_{TH})^3, \quad (10.222)$$

where γ is a proportionality factor. Figure 10.76 shows a differential pair employing such transistors.

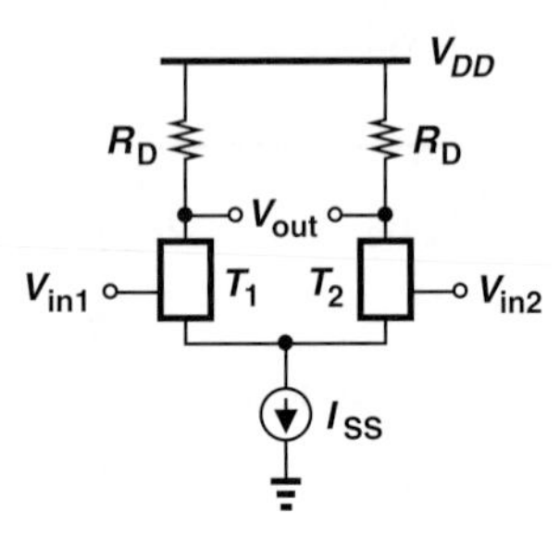

Figure 10.76

(a) What similarities exist between this circuit and the standard MOS differential pair?
(b) Calculate the equilibrium overdrive voltage of T_1 and T_2.
(c) At what value of $V_{in1} - V_{in2}$ does one transistor turn off?

***10.48.** Using the result obtained in Problem 10.46, calculate the value of $V_{in1} - V_{in2}$ at which the transconductance drops by a factor of 2.

***10.49.** Explain what happens to the characteristics shown in Fig. 10.33 if (a) the gate oxide thickness of the transistor is doubled, (b) the threshold voltage is halved, (c) I_{SS} and W/L are halved.

***10.50.** Assuming that the mobility of carriers falls at high temperatures, explain what happens to the characteristics of Fig. 10.33 as the temperature rises. SOL

Sec. 10.3.3 Small-Signal Analysis of MOS Pair

10.51. A student who has a single-ended voltage source constructs the circuit shown in Fig. 10.77, hoping to obtain differential outputs. Assume perfect symmetry but $\lambda = 0$ for simplicity.

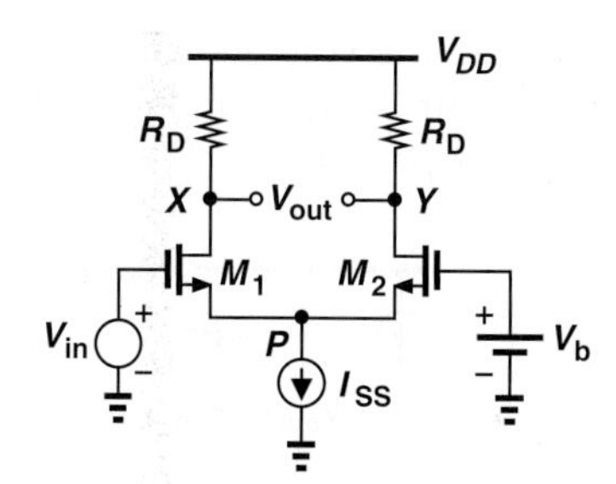

Figure 10.77

(a) Viewing M_1 as a common-source stage degenerated by the impedance seen at the source of M_2, calculate v_X in terms of v_{in}.
(b) Viewing M_1 as a source follower and M_2 as a common-gate stage, calculate v_Y in terms of v_{in}.
(c) Add the results obtained in (a) and (b) with proper polarities. If the voltage gain is defined as $(v_X - v_Y)/v_{in}$, how does it compare with the gain of differentially driven pairs?

10.52. Calculate the differential voltage gain of the circuits depicted in Fig. 10.78. Assume perfect symmetry and $\lambda > 0$.

*__10.53.__ Calculate the differential voltage gain of the circuits depicted in Fig. 10.79. Assume perfect symmetry and $\lambda > 0$. You may need to compute the gain as $A_v = -G_m R_{out}$ in some cases.

Sec. 10.4 Cascode Differential Pairs

10.54. The cascode differential pair of Fig. 10.39(a) must achieve a voltage gain of 4000. If Q_1–Q_4 are identical and $\beta = 100$, what is the minimum required Early voltage?

10.55. Due to a manufacturing error, a parasitic resistance, R_P, has appeared in the circuit of Fig. 10.80. Calculate the voltage gain.

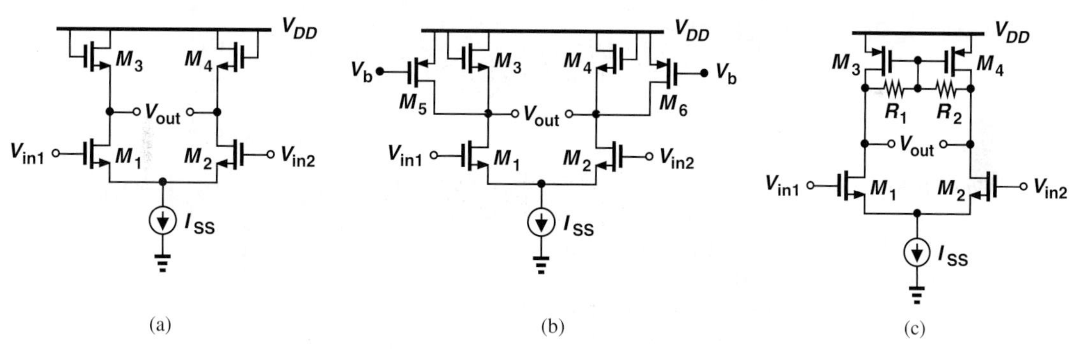

Figure 10.78

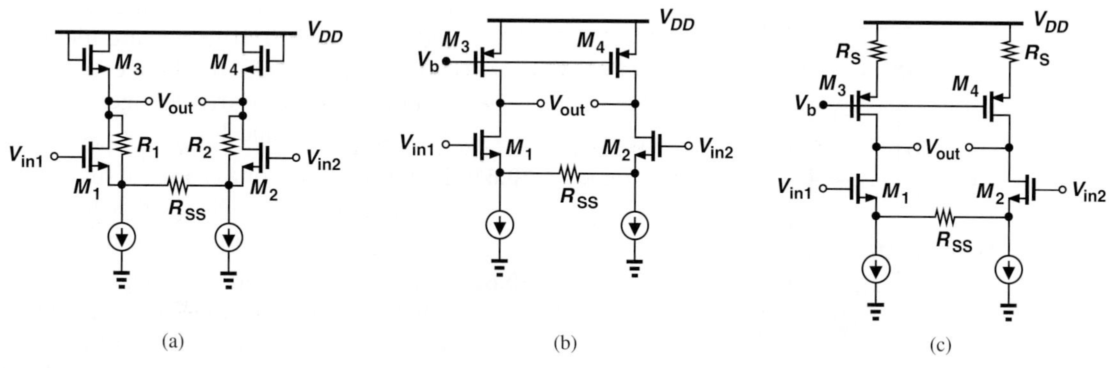

Figure 10.79

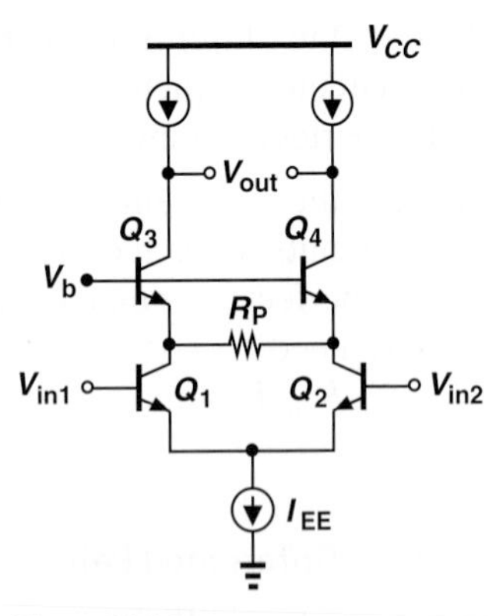

Figure 10.80

10.56. Repeat Problem 10.53 for the circuit shown in Fig. 10.81.

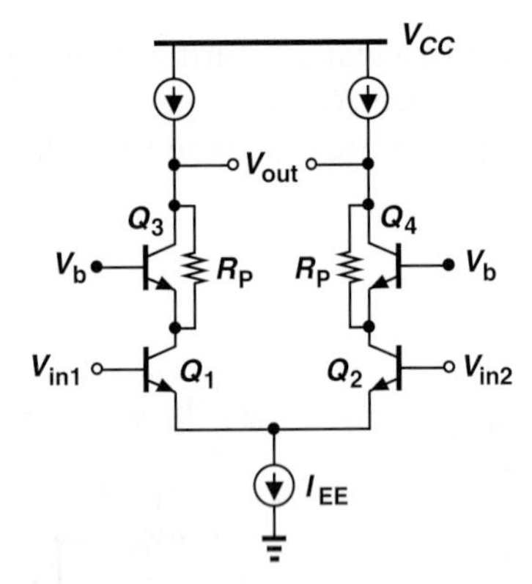

Figure 10.81

****10.57.** A student has mistakenly used *pnp* cascode transistors in a differential pair as illustrated in Fig. 10.82. Calculate the voltage gain of the circuit. (Hint: $A_v = -G_m R_{out}$.)

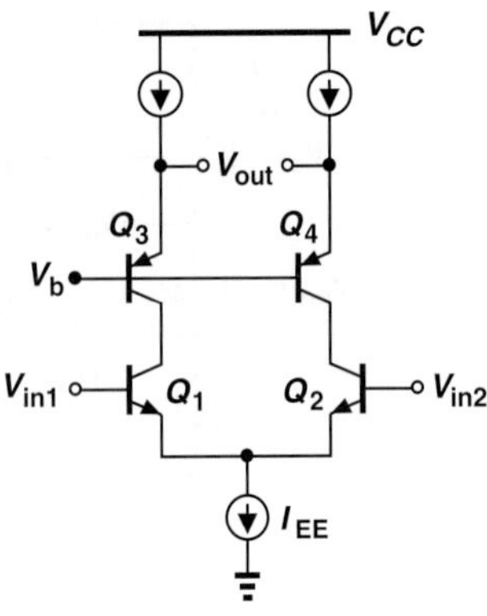

Figure 10.82

***10.58.** Calculate the voltage gain of the degenerated pair depicted in Fig. 10.83. (Hint: $A_v = -G_m R_{out}$.)

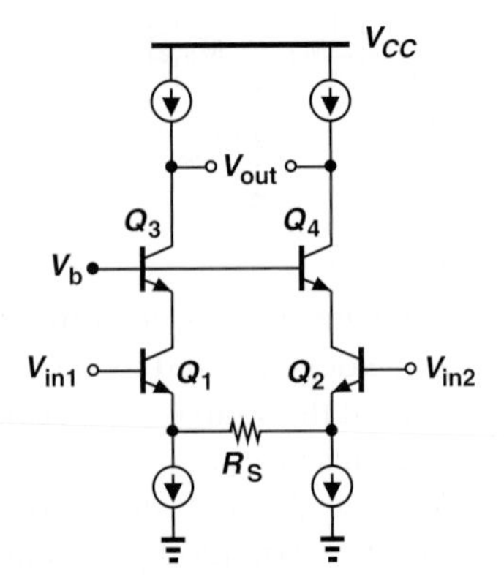

Figure 10.83

***10.59.** Realizing that the circuit of Fig. 10.82 suffers from a low gain, the student makes the modification shown in Fig. 10.84. Calculate the voltage gain of this topology. SOL

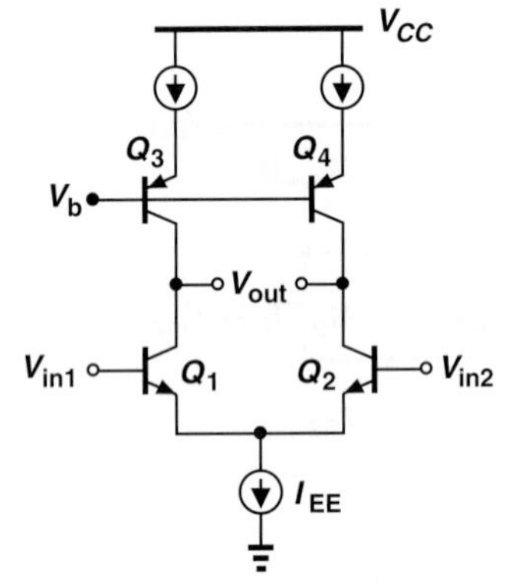

Figure 10.84

****10.60.** The telescopic cascode of Fig. 10.40 is to operate as an op amp having an open-loop gain of 800. If Q_1-Q_4 are identical and so are Q_5-Q_8, determine the minimum allowable Early voltage. Assume $\beta_n = 2\beta_p = 100$ and $V_{A,n} = 2V_{A,p}$.

10.61. Determine the voltage gain of the circuit depicted in Fig. 10.85. Is this topology considered a telescopic cascode?

10.62. The MOS cascode of Fig. 10.42(a) must provide a voltage gain of 300. If $W/L = 20/0.18$ for M_1-M_4 and $\mu_n C_{ox} = 100\ \mu A/V^2$, determine the required tail current. Assume $\lambda = 0.1\ V^{-1}$.

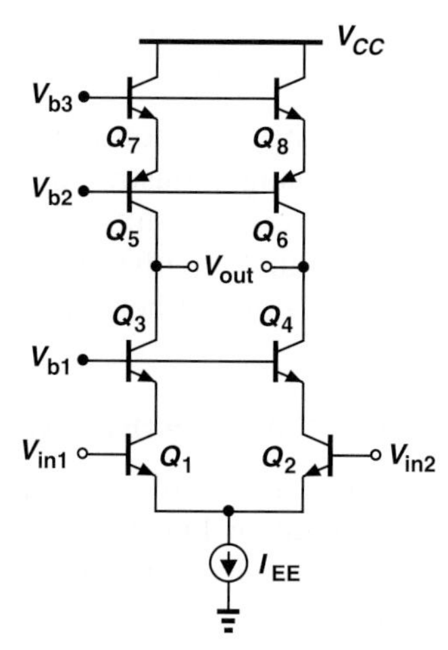

Figure 10.85

10.63. A student adventurously modifies a CMOS telescopic cascode as shown in Fig. 10.86, where the PMOS cascode transistors are replaced with NMOS devices. Assuming $\lambda > 0$, compute the voltage gain of the circuit. (Hint: The impedance seen looking into the source of M_5 or M_6 is *not* equal to $1/g_m || r_O$.)

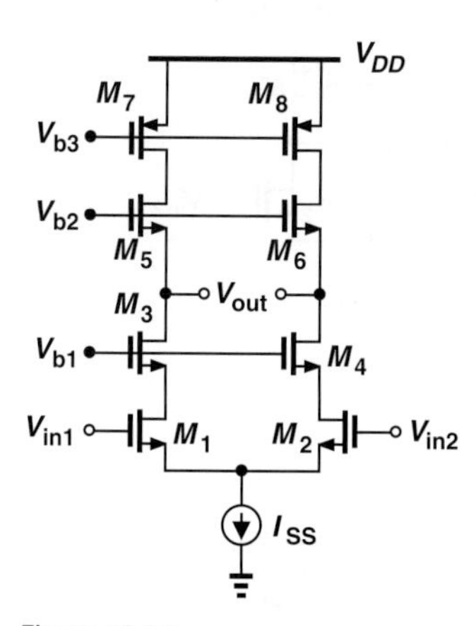

Figure 10.86

10.64. The MOS telescopic cascode of Fig. 10.43(a) is designed for a voltage gain of 200 with a tail current of 1 mA. If $\mu_n C_{ox} = 100\ \mu\text{A/V}^2$, $\mu_p C_{ox} = 50\ \mu\text{A/V}^2$, $\lambda_n = 0.1\ \text{V}^{-1}$, and $\lambda_p = 0.2\ \text{V}^{-1}$, determine $(W/L)_1 = \cdots = (W/L)_8$.

Sec. 10.5 Common-Mode Rejection

10.65. Consider the circuit of Fig. 10.45(a) and replace R_{EE} with two parallel resistors equal to $2R_{EE}$ places on the two sides of the current source. Now draw a vertical line of symmetry through the circuit and decompose it to two common-mode half circuits, each having a degeneration resistor equal to $2R_{EE}$. Prove that Eq. (10.180) still holds.

10.66. The bipolar differential pair depicted in Fig. 10.87 must exhibit a common-mode gain of less than 0.01. Assuming $V_A = \infty$ for Q_1 and Q_2 but $V_A < \infty$ for Q_3, prove that

$$R_C I_C < 0.02(V_A + V_T). \qquad (10.223)$$

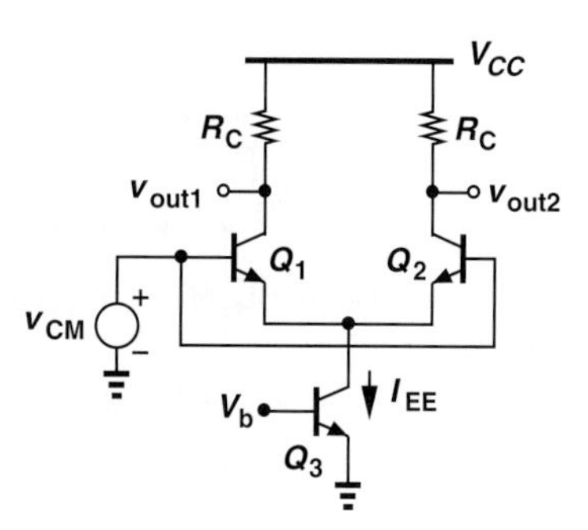

Figure 10.87

*10.67. Compute the common-mode gain of the MOS differential pair shown in Fig. 10.88. Assume $\lambda = 0$ for M_1 and M_2 but $\lambda \neq 0$ for M_3. Prove

$$A_{CM} = \frac{-R_D I_{SS}}{\dfrac{2}{\lambda} + (V_{GS} - V_{TH})_{eq.}}, \qquad (10.224)$$

where $(V_{GS} - V_{TH})_{eq.}$ denotes the equilibrium overdrive of M_1 and M_2.

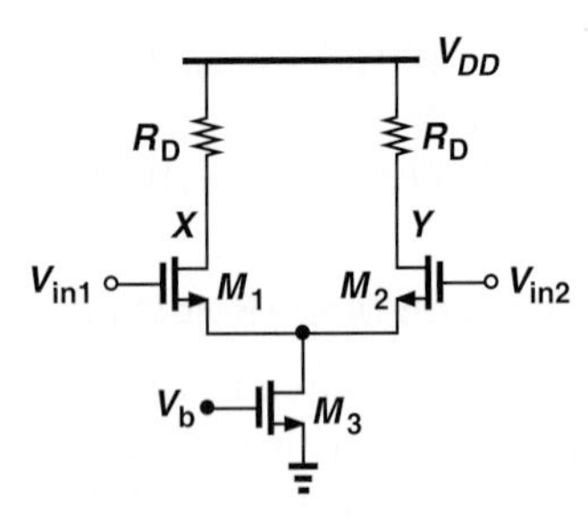

Figure 10.88

10.68. Calculate the common-mode gain of the circuit depicted in Fig. 10.89. Assume $\lambda > 0, g_m r_O \gg 1$, and use the relationship $A_v = -G_m R_{out}$.

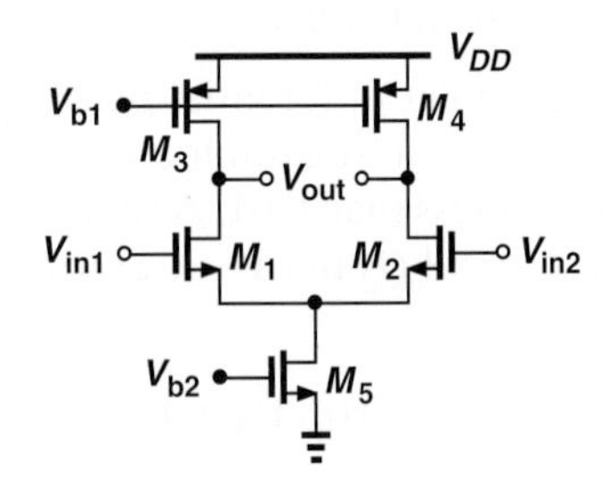

Figure 10.89

10.69. Repeat Problem 10.68 for the circuits shown in Fig. 10.90.

*__10.70.__ Compute the common-mode rejection ratio of the stages illustrated in Fig. 10.91 and compare the results. For simplicity, neglect channel-length modulation in M_1 and M_2 but not in other transistors.

Sec. 10.6 Differential Pair with Active Load

10.71. Determine the small-signal gain v_{out}/i_1 SOL in the circuit of Fig. 10.92 if $(W/L)_3 = N(W/L)_4$. Neglect channel-length modulation.

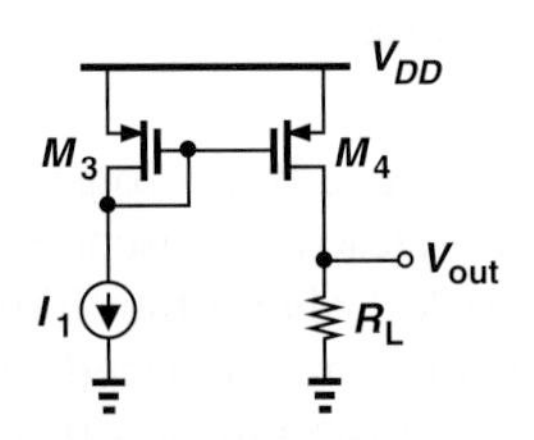

Figure 10.92

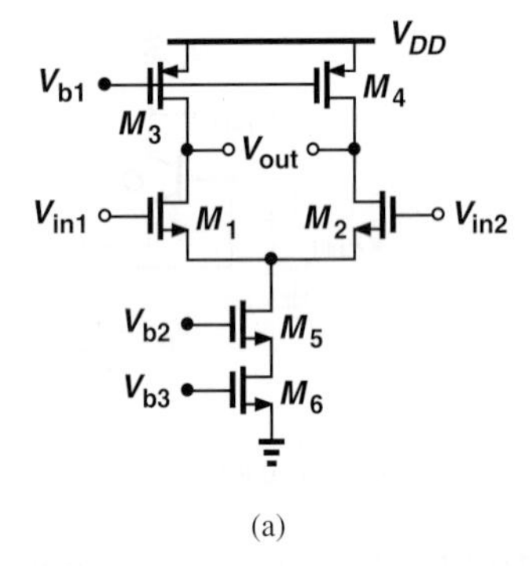

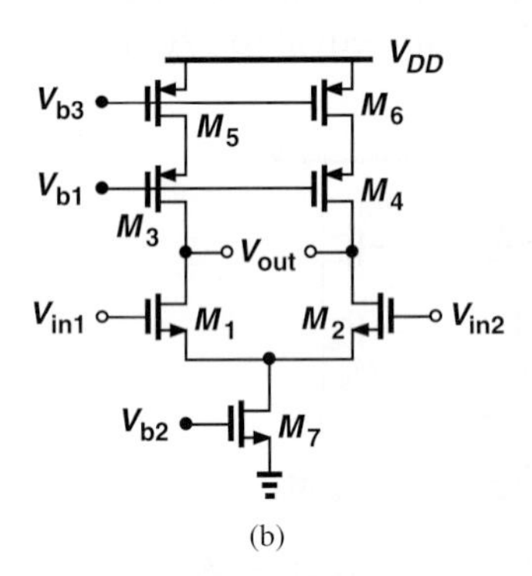

Figure 10.90

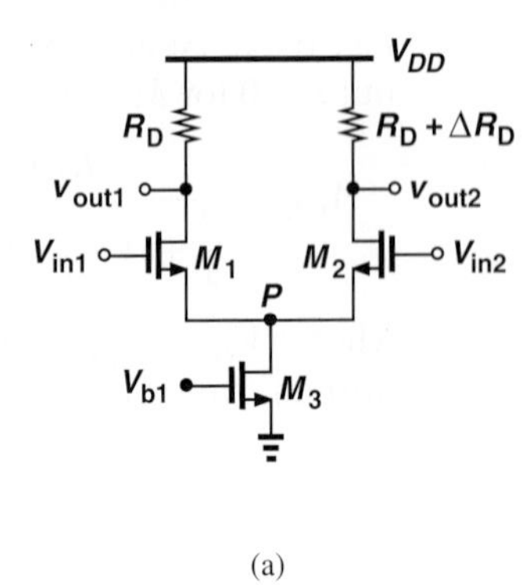

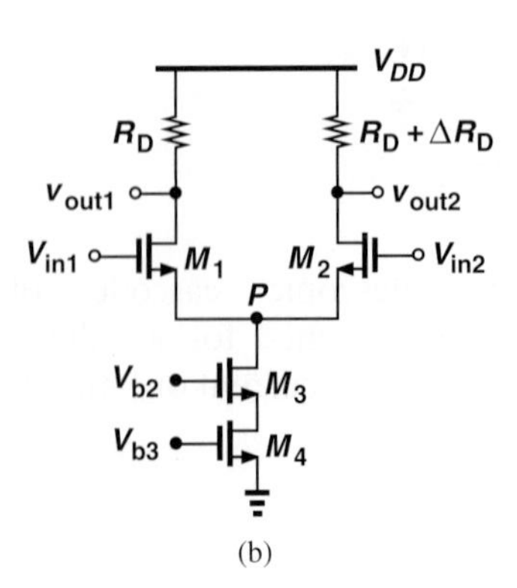

Figure 10.91

10.72. In the circuit shown in Fig. 10.93, I_1 changes from I_0 to $I_0 + \Delta I$ and I_2 from I_0 to $I_0 - \Delta I$. Neglecting channel-length modulation, calculate V_{out} before and after the change if
(a) $(W/L)_3 = (W/L)_4$.
(b) $(W/L)_3 = 2(W/L)_4$.

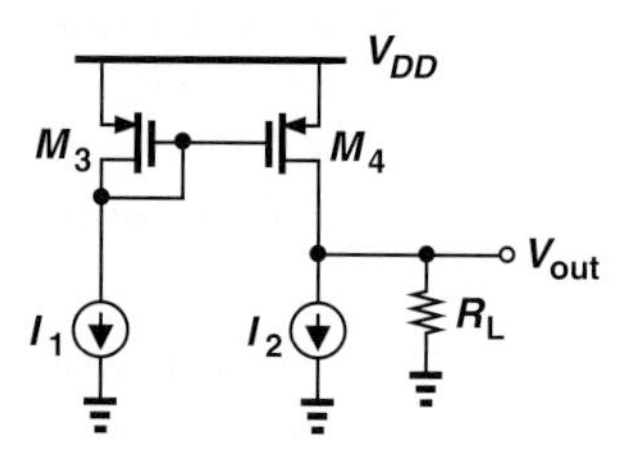

Figure 10.93

10.73. Neglecting channel-length modulation, compute the small-signal gains v_{out}/i_1 and v_{out}/i_2 in Fig. 10.94.

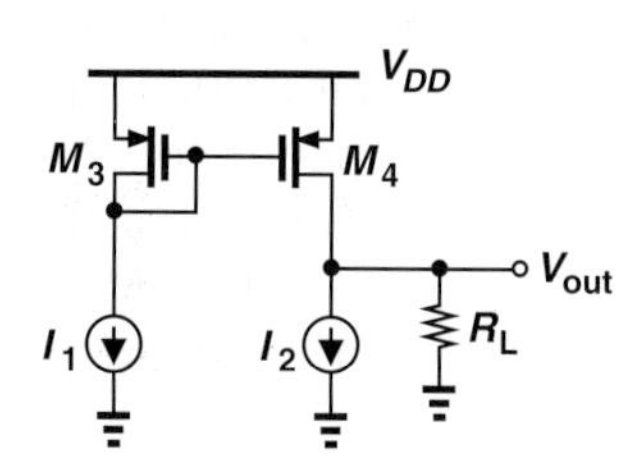

Figure 10.94

10.74. Consider the circuit of Fig. 10.95, where the inputs are tied to a common-mode level. Assume M_1 and M_2 are identical and so are M_3 and M_4.

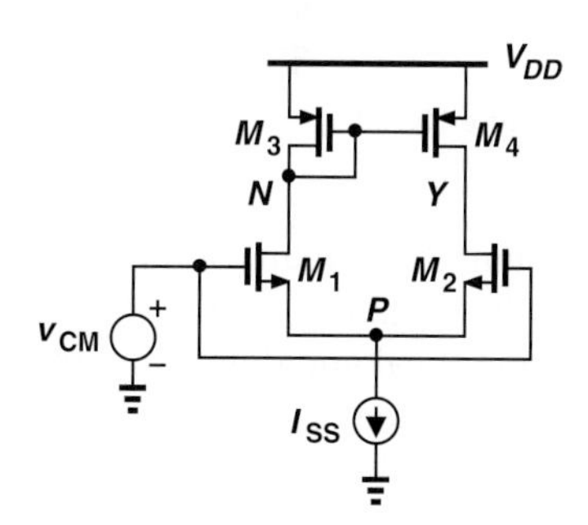

Figure 10.95

(a) Neglecting channel-length modulation, calculate the voltage at node N.
(b) Invoking symmetry, determine the voltage at node Y.
(c) What happens to the results obtained in (a) and (b) if V_{DD} changes by a small amount ΔV?

10.75. We wish to design the stage shown in Fig. 10.96 for a voltage gain of 100. If $V_{A,n} = 5$ V, what is the required Early voltage for the *pnp* transistors?

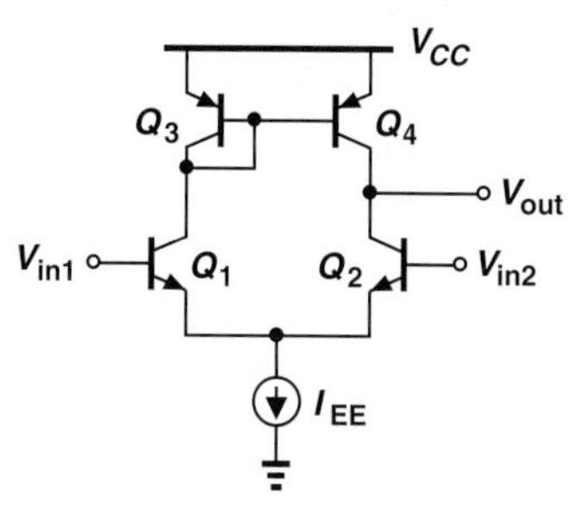

Figure 10.96

10.76. Repeat the analysis in Fig. 10.58 but by constructing a Norton equivalent for the input differential pair.

10.77. Determine the output impedance of the circuit shown in Fig. 10.56. Assume $g_m r_O \gg 1$.

10.78. Using the result obtained in Problem 10.77 and the relationship $A_v = -G_m R_{out}$, compute the voltage gain of the stage.

Design Problems

10.79. Design the bipolar differential pair of Fig. 10.7(a) for a voltage gain of 10 and a power budget of 2 mW. Assume $V_{CC} =$ 2.5 V and $V_A = \infty$.

10.80. The bipolar differential pair of Fig. 10.7(a) must operate with an input common-mode level of 1.2 V without driving the transistors into saturation. Design the circuit for maximum voltage gain and a power budget of 3 mW. Assume $V_{CC} = 2.5$ V.

10.81. The differential pair depicted in Fig. 10.97 must provide a gain of 5 and a power budget of 4 mW. Moreover, the gain of the circuit must change by less than 2% if the collector current of either transistor changes by 10%. Assuming $V_{CC} = 2.5$ V and $V_A = \infty$, design the circuit. (Hint: A 10% change in I_C leads to a 10% change in g_m.)

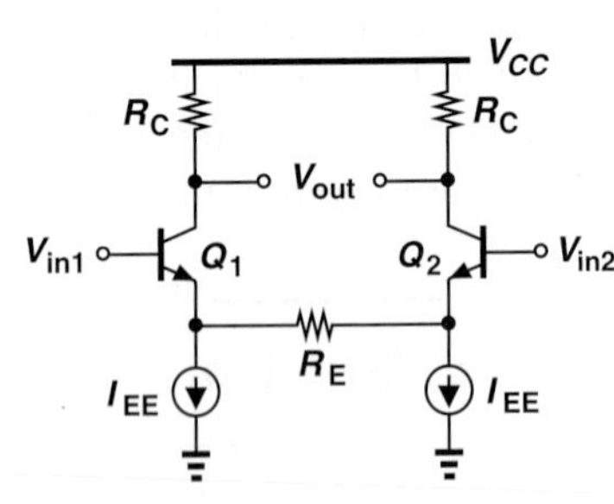

Figure 10.97

10.82. Design the circuit of Fig. 10.98 for a gain of 50 and a power budget of 1 mW. Assume $V_{A,n} = 6$ V and $V_{CC} = 2.5$ V.

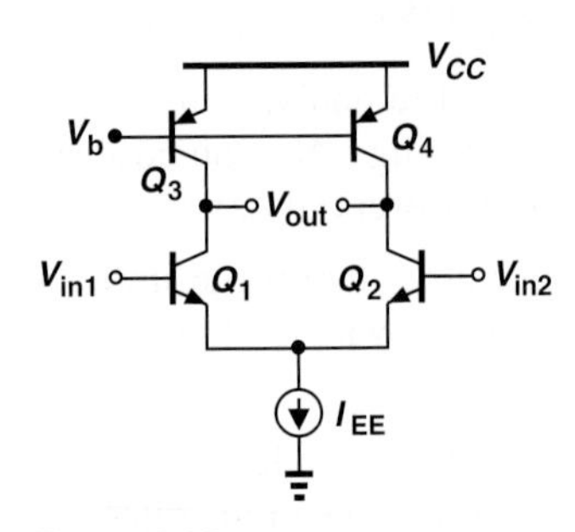

Figure 10.98

10.83. Design the circuit of Fig. 10.99 for a gain of 100 and a power budget of 1 mW. Assume $V_{A,n} = 10$ V, $V_{A,p} = 5$ V, and $V_{CC} = 2.5$ V. Also, $R_1 = R_2$.

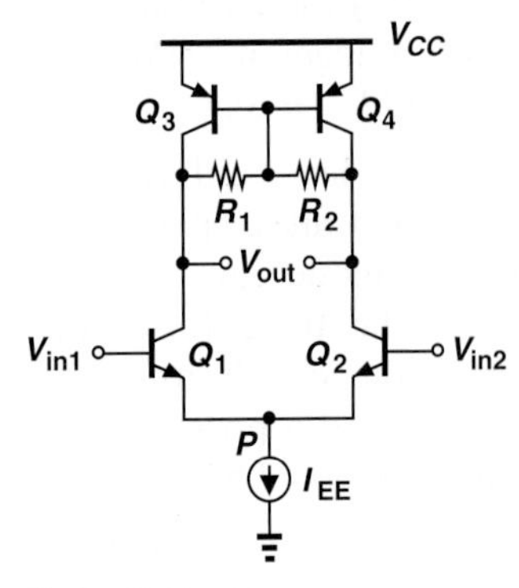

Figure 10.99

SOL **10.84.** Design the MOS differential pair of Fig. 10.31 for $\Delta V_{in,max} = 0.3$ V and a power budget of 3 mW. Assume $R_D = 500\ \Omega$, $\lambda = 0$, $\mu_n C_{ox} = 100\ \mu A/V^2$, and $V_{DD} = 1.8$ V.

10.85. Design the MOS differential pair of Fig. 10.31 for an equilibrium overdrive voltage of 100 mV and a power budget of 2 mW. Select the value of R_D to place the transistor at the edge of triode region for an input common-mode level of 1 V. Assume $\lambda = 0$, $\mu_n C_{ox} = 100\ \mu A/V^2$, $V_{TH,n} = 0.5$ V, and $V_{DD} = 1.8$ V. What is the voltage gain of the resulting design?

10.86. Design the MOS differential pair of Fig. 10.31 for a voltage gain of 5 and a power dissipation of 1 mW if the equilibrium overdrive must be at least 150 mV. Assume $\lambda = 0$, $\mu_n C_{ox} = 100\ \mu A/V^2$, and $V_{DD} = 1.8$ V.

10.87. The differential pair depicted in Fig. 10.100 must provide a gain of 40. Assuming the same (equilibrium) overdrive for all of the transistors and a power dissipation of 2 mW, design the circuit. Assume $\lambda_n = 0.1\ V^{-1}$, $\lambda_p = 0.2\ V^{-1}$, $\mu_n C_{ox} = 100\ \mu A/V^2$, $\mu_p C_{ox} = 100\ \mu A/V^2$, and $V_{DD} = 1.8$ V.

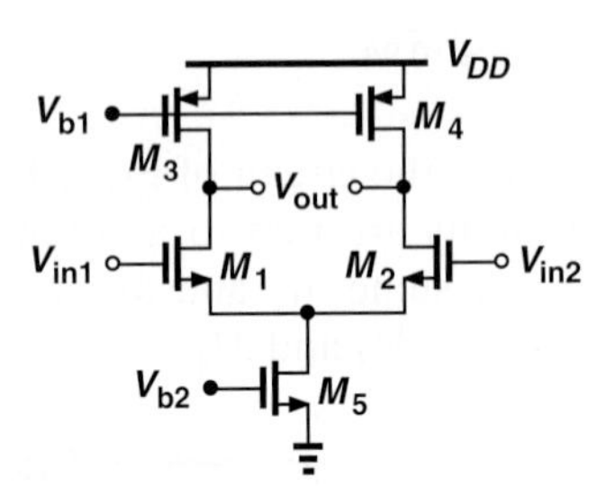

Figure 10.100

10.88. Design the circuit of Fig. 10.39(a) for a voltage gain of 4000. Assume Q_1-Q_4 are identical and determine the required Early voltage. Also, $\beta = 100$, $V_{CC} = 2.5$ V, and the power budget is 1 mW.

10.89. Design the telescopic cascode of Fig. 10.40(a) for a voltage gain of 2000. Assume Q_1-Q_4 are identical and so are Q_5-Q_8. Also, $\beta_n = 100$, $\beta_p = 50$, $V_{A,n} = 5$ V, $V_{CC} = 2.5$ V, and the power budget is 2 mW.

10.90. Design the telescopic cascode of Fig. 10.43(a) for a voltage gain of 600 and a power budget of 4 mW. Assume an (equilibrium) overdrive of 100 mV for the NMOS devices and 150 mV for the PMOS devices. If $V_{DD} = 1.8$ V, $\mu_n C_{ox} = 100\ \mu A/V^2$, $\mu_p C_{ox} = 50\ \mu A/V^2$, and $\lambda_n = 0.1\ V^{-1}$, determine the required value of λ_p. Assume M_1-M_4 are identical and so are M_5-M_8.

10.91. The differential pair of Fig. 10.101 must achieve a CMRR of 60 dB($= 1000$). Assume a power budget of 2 mW, a nominal differential voltage gain of 5, and neglecting channel-length modulation in M_1 and M_2, compute the minimum required λ for M_3. Assume $(W/L)_{1,2} = 10/0.18$, $\mu_n C_{ox} = 100\ \mu A/V^2$, $V_{DD} = 1.8$ V, and $\Delta R/R = 2\%$.

SOL **10.92.** Design the differential pair of Fig. 10.50 for a voltage gain of 200 and a power budget of 3 mW with a 2.5-V supply. Assume $V_{A,n} = 2V_{A,p}$.

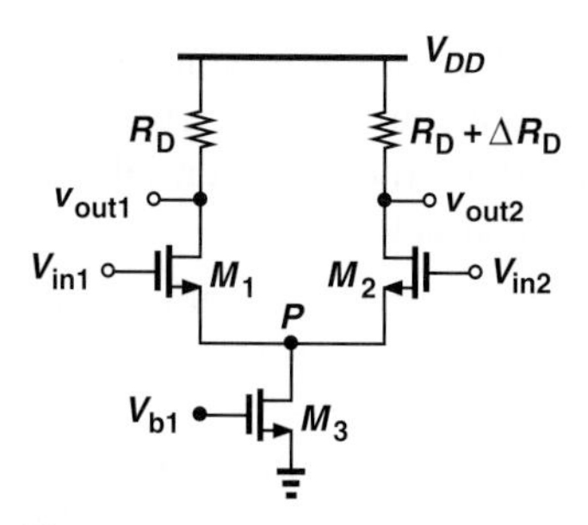

Figure 10.101

10.93. Design the circuit of Fig. 10.56 for a voltage gain of 20 and a power budget of 1 mW with $V_{DD} = 1.8$ V. Assume M_1 operates at the edge of saturation if the input common-mode level is 1 V. Also, $\mu_n C_{ox} = 2\mu_p C_{ox} = 100\ \mu A/V^2$, $V_{TH,n} = 0.5$ V, $V_{TH,p} = -0.4$ V, $\lambda_n = 0.5\lambda_p = 0.1\ V^{-1}$.

SPICE PROBLEMS

In the following problems, use the MOS device models given in Appendix A. For bipolar transistors, assume $I_{S,npn} = 5 \times 10^{-16}$ A, $\beta_{npn} = 100$, $V_{A,npn} = 5$ V, $I_{S,pnp} = 8 \times 10^{-16}$ A, $\beta_{pnp} = 50$, $V_{A,pnp} = 3.5$ V.

10.94. Consider the differential amplifier shown in Fig. 10.102, where the input CM level is equal to 1.2 V.

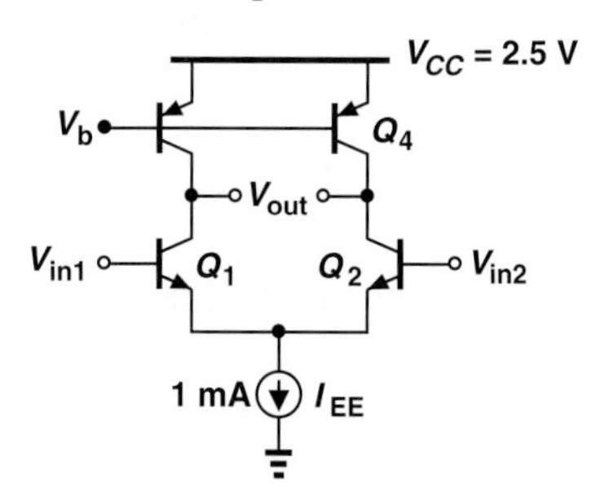

Figure 10.102

(a) Adjust the value of V_b so as to set the output CM level to 1.5 V.

(b) Determine the small-signal differential gain of the circuit. (Hint: To provide differential inputs, use an independent voltage source for one side and a voltage-dependent voltage source for the other.)

(c) What happens to the output CM level and the gain if V_b varies by ± 10 mV?

10.95. The differential amplifier depicted in Fig. 10.103 employs two current mirrors to establish the bias for the input and load devices. Assume $W/L = 10\ \mu m/0.18\ \mu m$ for M_1-M_6. The input CM level is equal to 1.2 V.

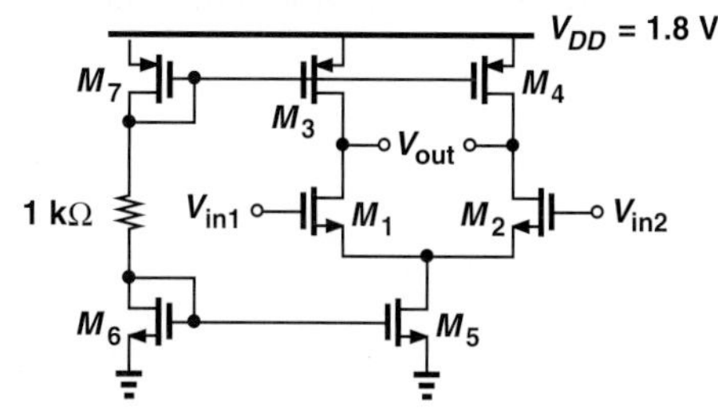

Figure 10.103

(a) Select $(W/L)_7$ so as to set the output CM level to 1.5 V. (Assume $L_7 = 0.18\ \mu m$.)
(b) Determine the small-signal differential gain of the circuit.
(c) Plot the differential input/output characteristic.

10.96. Consider the circuit illustrated in Fig. 10.104. Assume a small dc drop across R_1 and R_2.
(a) Select the input CM level to place Q_1 and Q_2 at the edge of saturation.
(b) Select the value of R_1 $(= R_2)$ such that these resistors reduce the differential gain by no more than 20%.

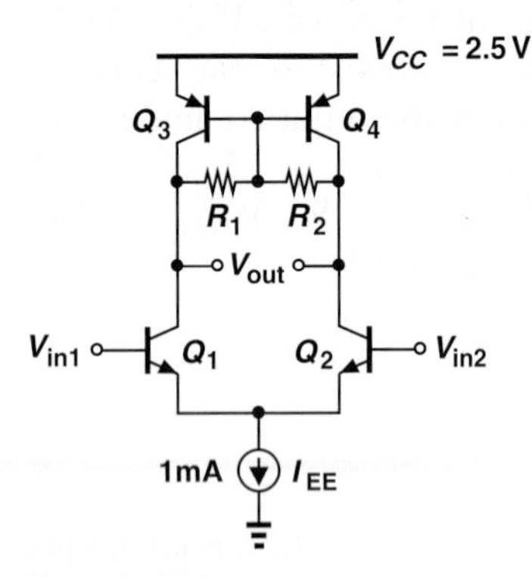

Figure 10.104

10.97. In the differential amplifier of Fig. 10.105, $W/L = 10\ \mu m/0.18\ \mu m$ for all of the transistors. Assume an input CM level of 1 V and $V_b = 1.5$ V.
(a) Select the value of I_1 so that the output CM level places M_3 and M_4 at the edge of saturation.
(b) Determine the small-signal differential gain.

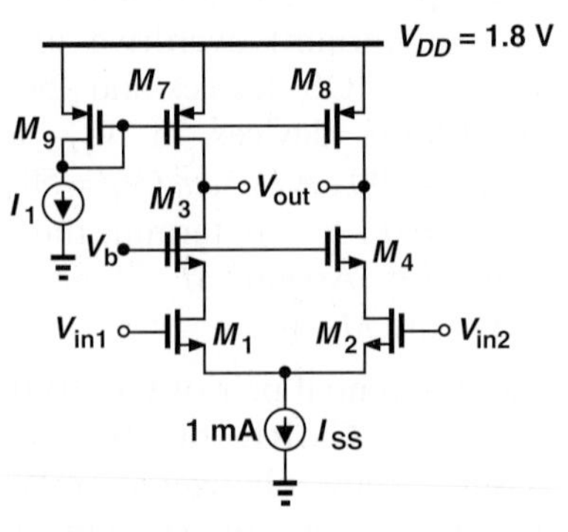

Figure 10.105

10.98. In the circuit of Fig. 10.106, $W/L = 10\ \mu m/0.18\ \mu m$ for M_1-M_4. Assume an input CM level of 1.2 V.

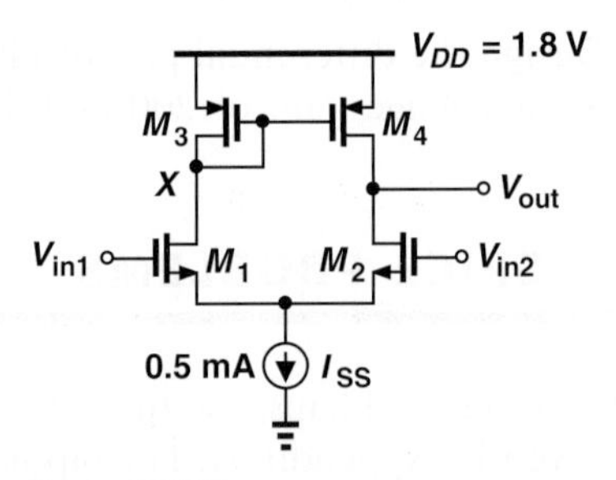

Figure 10.106

(a) Determine the output dc level and explain why it is equal to V_X?
(b) Determine the small-signal gains $v_{out}/(v_{in1} - v_{in2})$ and $v_X/(v_{in1} - v_{in2})$.
(c) Determine the change in the output dc level if W_4 changes by 5%.

REFERENCE

1. B. Razavi, *Design of Analog CMOS Integrated Circuits*, McGraw-Hill, 2001.

11

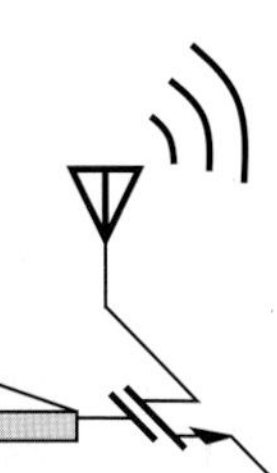

Frequency Response

Watch Companion YouTube Videos:

Razavi Electronics 2, Lec 17: Introduction to Frequency Response: Basic Concepts
https://www.youtube.com/watch?v=XbM1-WgGXQc

Razavi Electronics 2, Lec 18: Useful Frequency Response Concepts, Finding Poles by Inspection
https://www.youtube.com/watch?v=oC2otEX3i0I

Razavi Electronics 2, Lec 19: Miller Effect, High-Frequency Model of Bipolar Transistors
https://www.youtube.com/watch?v=sUObJPJQ8Os

Razavi Electronics 2, Lec 20: Examples of Capacitances in Bipolar Circuits, High-Freq. Model of MOSFETs
https://www.youtube.com/watch?v=MoORc8l0RtE

Razavi Electronics 2, Lec 21: Computation of Freq. Resp., Freq. Resp. of Common-Emitter/Source Stages
https://www.youtube.com/watch?v=0r3zsOlSsG8

Razavi Electronics 2, Lec 22: Dominant-Pole Approx., Resp. of CE/CS & Common-Base/Gate Stage
https://www.youtube.com/watch?v=anYWGLJIOB8

Razavi Electronics 2, Lec 23: Examples of High-Speed Circuits, Response of Common-Base/Gate Stage
https://www.youtube.com/watch?v=IntsUt7dUjQ

Razavi Electronics 2, Lec 24: Response of Emitter/Source Followers, Input & Output Impedances
https://www.youtube.com/watch?v=__kLV7tlPuE

Razavi Electronics 2, Lec 25: Output Imp. of Followers, Freq. Resp. of Cascodes and Diff. Pairs; ft
https://www.youtube.com/watch?v=Q8G_If19GrE

Razavi Electronics 2, Lec 26: Additional Examples of Frequency Response, Cascaded Stages
https://www.youtube.com/watch?v=z8aC8W8XflY

The need for operating circuits at increasingly higher speeds has always challenged designers. From radar and television systems in the 1940s to gigahertz microprocessors today, the demand to push circuits to higher frequencies has required a solid understanding of their speed limitations.

In this chapter, we study the effects that limit the speed of transistors and circuits, identifying topologies that better lend themselves to high-frequency operation. We also develop skills for deriving transfer functions of circuits, a critical task in the study of stability and frequency compensation (12). We assume bipolar transistors remain in the active mode and MOSFETs in the saturation region. The outline is shown below.

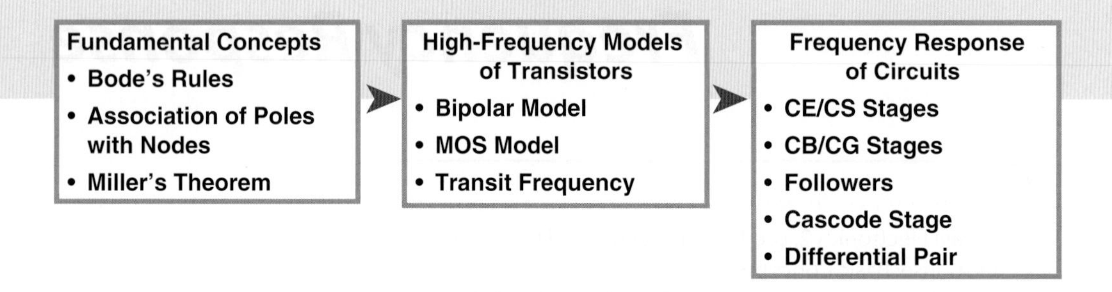

11.1 FUNDAMENTAL CONCEPTS

11.1.1 General Considerations

What do we mean by "frequency response?" As illustrated in Fig. 11.1(a), the idea is to apply a sinusoid at the input of the circuit and observe the output while the input frequency is varied. As exemplified by Fig. 11.1(a), the circuit may exhibit a high gain at low frequencies but a "roll-off" as the frequency increases. We plot the magnitude of the gain as in Fig. 11.1(b) to represent the circuit's behavior at all frequencies of interest. We may loosely call f_1 the useful bandwidth of the circuit. Before investigating the cause of this roll-off, we must ask: why is frequency response important? The following examples illustrate the issue.

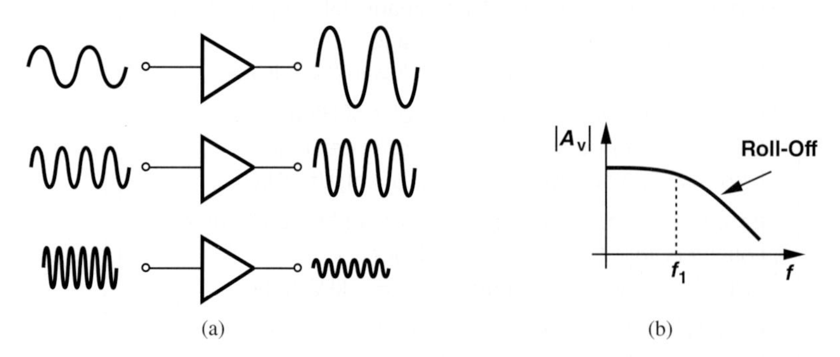

Figure 11.1 (a) Conceptual test of frequency response, (b) gain roll-off with frequency.

Example 11-1

Explain why people's voices over the phone sound different from their voices in face-to-face conversation.

Solution The human voice contains frequency components from 20 Hz to 20 kHz [Fig. 11.2(a)]. Thus, circuits processing the voice must accommodate this frequency range. Unfortunately, the phone system suffers from a limited bandwidth, exhibiting the frequency response shown in Fig. 11.2(b). Since the phone suppresses frequencies above 3.5 kHz, each person's voice is altered. In high-quality audio systems, on the other hand, the circuits are designed to cover the entire frequency range.

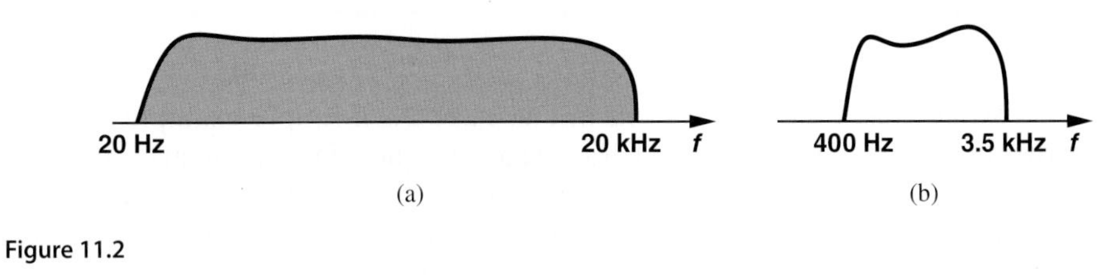

Figure 11.2

Exercise Whose voice does the phone system alter more, men's or women's?

Example 11-2

When you record your voice and listen to it, it sounds somewhat different from the way you hear it directly when you speak. Explain why?

Solution During recording, your voice propagates through the air and reaches the audio recorder. On the other hand, when you speak and listen to your own voice simultaneously, your voice propagates not only through the air but also from your mouth through your skull to your ear. Since the frequency response of the path through your skull is different from that through the air (i.e., your skull passes some frequencies more easily than others), the way you hear your own voice is different from the way other people hear your voice.

Exercise Explain what happens to your voice when you have a cold.

Example 11-3

Video signals typically occupy a bandwidth of about 5 MHz. For example, the graphics card delivering the video signal to the display of a computer must provide at least 5 MHz of bandwidth. Explain what happens if the bandwidth of a video system is insufficient.

Solution With insufficient bandwidth, the "sharp" edges on a display become "soft," yielding a fuzzy picture. This is because the circuit driving the display is not fast enough to abruptly change the contrast from, e.g., complete white to complete black from one pixel to the next. Figures 11.3(a) and (b) illustrate this effect for a high-bandwidth and low-bandwidth driver, respectively. (The display is scanned from left to right.)

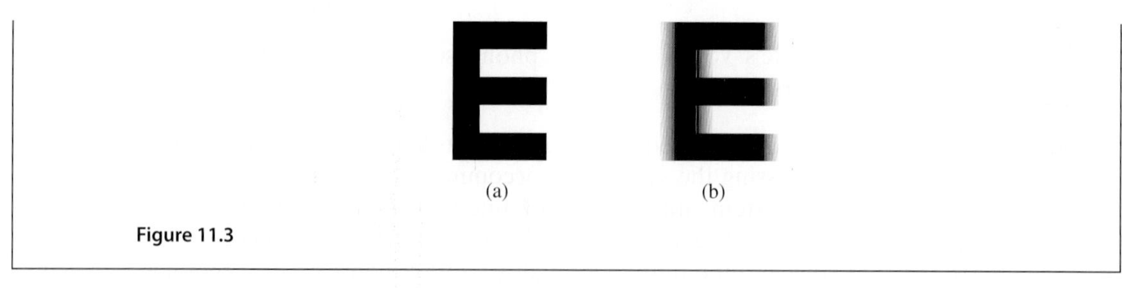

Figure 11.3

Exercise What happens if the display is scanned from top to bottom?

What causes the gain roll-off in Fig. 11.1? As a simple example, let us consider the low-pass filter depicted in Fig. 11.4(a). At low frequencies, C_1 is nearly open and the current through R_1 nearly zero; thus, $V_{out} = V_{in}$. As the frequency increases, the impedance of C_1 falls and the voltage divider consisting of R_1 and C_1 attenuates V_{in} to a greater extent. The circuit therefore exhibits the behavior shown in Fig. 11.4(b).

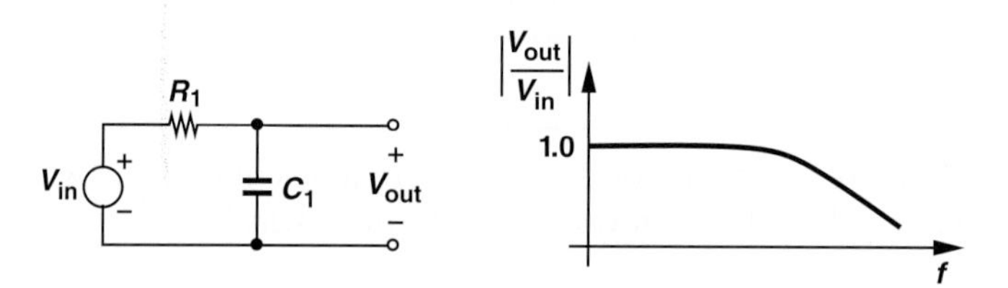

Figure 11.4 (a) Simple low-pass filter, and (b) its frequency response.

As a more interesting example, consider the common-source stage illustrated in Fig. 11.5(a), where a load capacitance, C_L, appears at the output. At low frequencies, the signal current produced by M_1 prefers to flow through R_D because the impedance of C_L, $1/(C_L s)$, remains high. At high frequencies, on the other hand, C_L "steals" some of the signal current and shunts it to ground, leading to a lower voltage swing at the output. In fact, from the small-signal equivalent circuit of Fig. 11.5(b),[1] we note that R_D and C_L are in parallel and hence:

$$V_{out} = -g_m V_{in} \left(R_D \parallel \frac{1}{C_L s} \right). \tag{11.1}$$

That is, as the frequency increases, the parallel impedance falls and so does the amplitude of V_{out}.[2] The voltage gain therefore drops at high frequencies.

The reader may wonder why we use *sinusoidal* inputs in our study of frequency response. After all, an amplifier may sense a voice or video signal that bears no resemblance to sinusoids. Fortunately, such signals can be viewed as a summation of many sinusoids with different frequencies (and phases). Thus, responses such as that in Fig. 11.5(b) prove useful so long as the circuit remains linear and hence superposition can be applied.

[1]Channel-length modulation is neglected here.
[2]We use upper-case letters for frequency-domain quantities (Laplace transforms) even though they denote small-signal values.

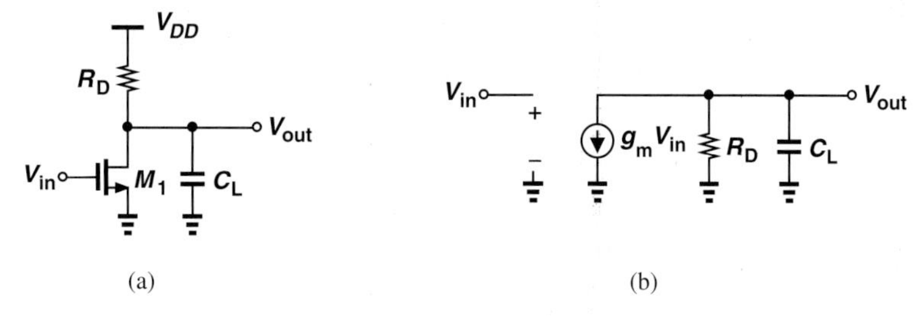

Figure 11.5 (a) CS stage with load capacitance, (b) small-signal model of the circuit.

11.1.2 Relationship Between Transfer Function and Frequency Response

We know from basic circuit theory that the transfer function of a circuit can be written as

$$H(s) = A_0 \frac{\left(1 - \frac{s}{\omega_{z1}}\right)\left(1 - \frac{s}{\omega_{z2}}\right)\cdots}{\left(1 - \frac{s}{\omega_{p1}}\right)\left(1 - \frac{s}{\omega_{p2}}\right)\cdots}, \tag{11.2}$$

where A_0 denotes the low-frequency gain because $H(s) \to A_0$ as $s \to 0$. The frequencies ω_{zj} and ω_{pj} represent the zeros and poles of the transfer function, respectively. If the input to the circuit is a sinusoid of the form $x(t) = A \cos(2\pi ft) = A \cos \omega t$, then the output can be expressed as

$$y(t) = A|H(j\omega)| \cos[\omega t + \angle H(j\omega)], \tag{11.3}$$

where $H(j\omega)$ is obtained by making the substitution $s = j\omega$. Called the "magnitude" and the "phase," $|H(j\omega)|$ and $\angle H(j\omega)$ respectively reveal the frequency response of the circuit. In this chapter, we are primarily concerned with the former. Note that f (in Hz) and ω (in radians per second) are related by a factor of 2π. For example, we may write $\omega = 5 \times 10^{10}$ rad/s $= 2\pi(7.96$ GHz$)$.

Example 11-4 Determine the transfer function and frequency response of the CS stage shown in Fig. 11.5(a).

Solution From Eq. (11.1), we have

$$H(s) = \frac{V_{out}}{V_{in}}(s) = -g_m \left(R_D \parallel \frac{1}{C_L s}\right) \tag{11.4}$$

$$= \frac{-g_m R_D}{R_D C_L s + 1}. \tag{11.5}$$

For a sinusoidal input, we replace $s = j\omega$ and compute the magnitude of the transfer function[3]:

$$\left|\frac{V_{out}}{V_{in}}\right| = \frac{g_m R_D}{\sqrt{R_D^2 C_L^2 \omega^2 + 1}}. \tag{11.6}$$

[3]The magnitude of the complex number $a + jb$ is equal to $\sqrt{a^2 + b^2}$.

As expected, the gain begins at $g_m R_D$ at low frequencies, rolling off as $R_D^2 C_L^2 \omega^2$ becomes comparable with unity. At $\omega = 1/(R_D C_L)$,

$$\left|\frac{V_{out}}{V_{in}}\right| = \frac{g_m R_D}{\sqrt{2}}. \tag{11.7}$$

Since $20 \log \sqrt{2} \approx 3$ dB, we say the −3 dB bandwidth of the circuit is equal to $1/(R_D C_L)$ (Fig. 11.6).

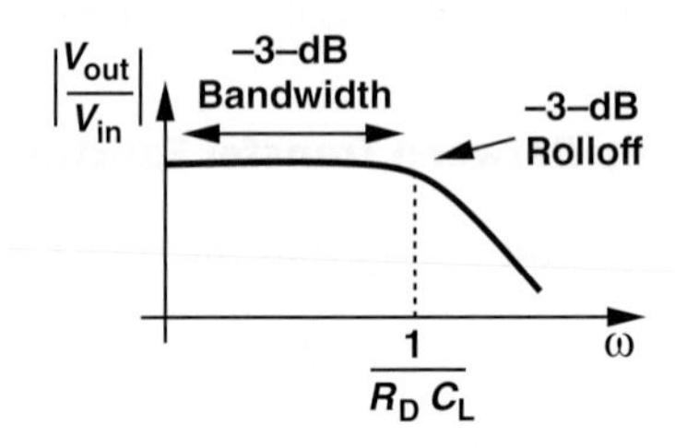

Figure 11.6

Exercise Derive the above results if $\lambda \neq 0$.

Example 11-5 Consider the common-emitter stage shown in Fig. 11.7. Derive a relationship among the gain, the −3 dB bandwidth, and the power consumption of the circuit. Assume $V_A = \infty$.

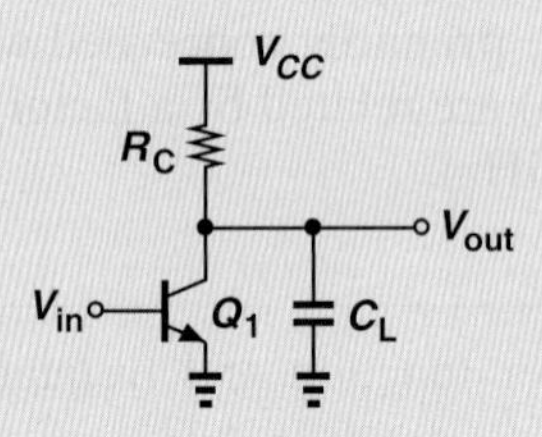

Figure 11.7

Solution In a manner similar to the CS topology of Fig. 11.5(a), the bandwidth is given by $1/(R_C C_L)$, the low-frequency gain by $g_m R_C = (I_C/V_T)R_C$, and the power consumption by $I_C \cdot V_{CC}$. For the highest performance, we wish to maximize both the gain and the bandwidth (and hence the product of the two) and minimize the power dissipation. We therefore define a "figure of merit" as

$$\frac{\text{Gain} \times \text{Bandwidth}}{\text{Power Consumption}} = \frac{\dfrac{I_C}{V_T} R_C \times \dfrac{1}{R_C C_L}}{I_C \cdot V_{CC}} \tag{11.8}$$

$$= \frac{1}{V_T \cdot V_{CC}} \frac{1}{C_L}. \tag{11.9}$$

Thus, the overall performance can be improved by lowering (a) the temperature[4]; (b) V_{CC} but at the cost of limiting the voltage swings; or (c) the load capacitance. In practice, the load capacitance receives the greatest attention. Equation (11.9) becomes more complex for CS stages (Problem 11.15).

Exercise Derive the above results if $V_A < \infty$.

Example 11-6 Explain the relationship between the frequency response and step response of the simple low-pass filter shown in Fig. 11.4(a).

Solution To obtain the transfer function, we view the circuit as a voltage divider and write

$$H(s) = \frac{V_{out}}{V_{in}}(s) = \frac{\dfrac{1}{C_1 s}}{\dfrac{1}{C_1 s} + R_1} \tag{11.10}$$

$$= \frac{1}{R_1 C_1 s + 1}. \tag{11.11}$$

The frequency response is determined by replacing s with $j\omega$ and computing the magnitude:

$$|H(s = j\omega)| = \frac{1}{\sqrt{R_1^2 C_1^2 \omega^2 + 1}}. \tag{11.12}$$

The -3 dB bandwidth is equal to $1/(R_1 C_1)$.

The circuit's response to a step of the form $V_0 u(t)$ is given by

$$V_{out}(t) = V_0 \left(1 - \exp\frac{-t}{R_1 C_1}\right) u(t). \tag{11.13}$$

The relationship between Eqs. (11.12) and (11.13) is that, as $R_1 C_1$ increases, the bandwidth *drops* and the step response becomes *slower*. Figure 11.8 plots this behavior, revealing that a narrow bandwidth results in a sluggish time response. This observation explains the effect seen in Fig. 11.3(b): since the signal cannot rapidly jump from low (white) to high (black), it spends some time at intermediate levels (shades of gray), creating "fuzzy" edges.

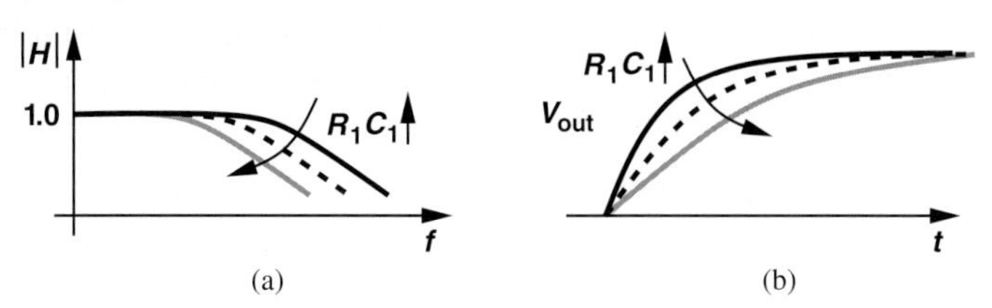

Figure 11.8

Exercise At what frequency does $|H|$ fall by a factor of two?

[4]For example, by immersing the circuit in liquid nitrogen ($T = 77$ K), but requiring that the user carry a tank around!

Bioengineering Application: Impedance Spectroscopy

Measuring the impedance of cells from a patient's body can provide information about his/her health. This type of characterization requires that the impedance be measured across a certain frequency range and is called "impedance spectroscopy" (IS). For example, neurodegenerative diseases can be studied with the aid of IS [1].

An implementation example of IS is described in [1]. As shown in Fig. 11.9(a), the system employs a waveform generator to apply a sinusoidal voltage to the sample, and a current sense circuit to measure the resulting current. The ratio of the voltage, V_1, and the current, I_1, represents the impedance, Z_{cell}. The frequency of the sinusoid can be varied from 1 Hz to 10 MHz.

The current sense circuit is depicted in Fig. 11.9(b), where I_1 flows through a virtual ground and R_F, generating a voltage, V_{out}. We have $Z_{cell} = V_1/I_1$ and $V_{out} = -I_1 R_F$. It follows that

$$Z_{cell} = -R_F \frac{V_1}{V_{out}}. \tag{11.14}$$

Since V_1 and V_{out} are known, Z_{cell} can be computed at each sinusoid frequency. The cell impedance typically behaves as a parallel RC circuit [1], causing $|Z_{cell}|$ to fall with frequency.

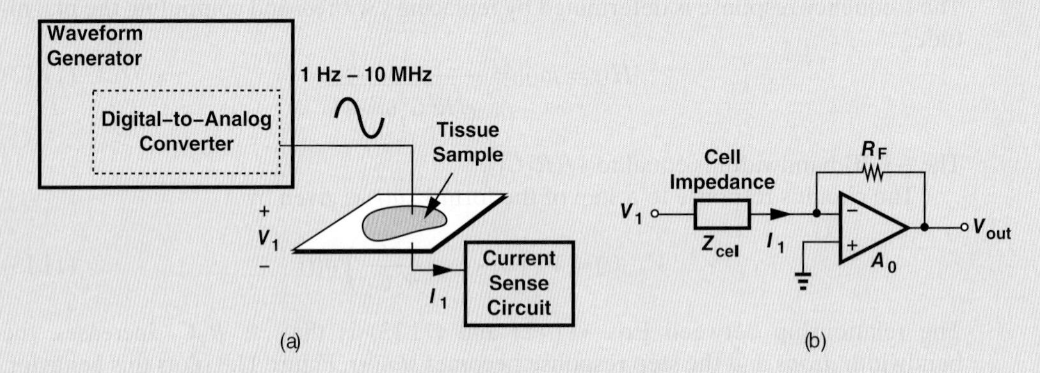

Figure 11.9 (a) System for performing impedance spectroscopy on tissue and (b) circuit implementation.

11.1.3 Bode's Rules

The task of obtaining $|H(j\omega)|$ from $H(s)$ and plotting the result is somewhat tedious. For this reason, we often utilize Bode's rules (approximations) to construct $|H(j\omega)|$ rapidly. Bode's rules for $|H(j\omega)|$ are as follows:

- As ω passes each pole frequency, the slope of $|H(j\omega)|$ *decreases* by 20 dB/dec. (A slope of 20 dB/dec simply means a tenfold change in H for a tenfold increase in frequency.)
- As ω passes each zero frequency, the slope of $|H(j\omega)|$ *increases* by 20 dB/dec.[5]

[5]Complex poles may result in sharp peaks in the frequency response, an effect neglected in Bode's approximation.

Example 11-7 Construct the Bode plot of $|H(j\omega)|$ for the CS stage depicted in Fig. 11.5(a).

Solution Equation (11.5) indicates a pole frequency of

$$|\omega_{p1}| = \frac{1}{R_D C_L}. \tag{11.15}$$

The magnitude thus begins at $g_m R_D$ at low frequencies and remains flat up to $\omega = |\omega_{p1}|$. At this point, the slope changes from zero to −20 dB/dec. Figure 11.10 illustrates the result. In contrast to Fig. 11.5(b), the Bode approximation ignores the 3 dB roll-off at the pole frequency—but it greatly simplifies the algebra. As evident from Eq. (11.6), for $R_D^2 C_L^2 \omega^2 \gg 1$, Bode's rule provides a good approximation.

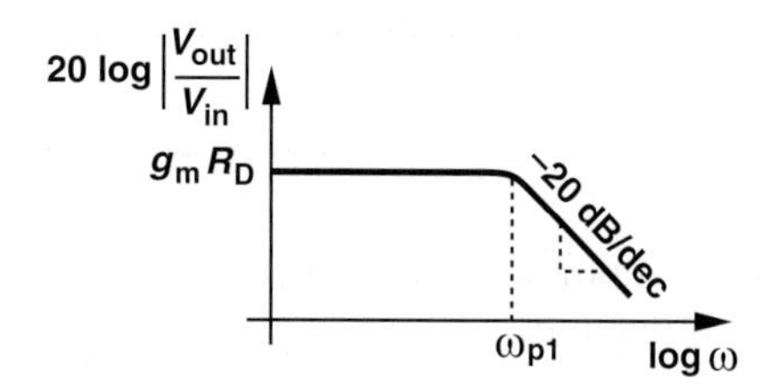

Figure 11.10

Exercise Construct the Bode plot for $g_m = (150\ \Omega)^{-1}$, $R_D = 2\ \text{k}\Omega$, and $C_L = 100$ fF.

11.1.4 Association of Poles with Nodes

The poles of a circuit's transfer function play a central role in the frequency response. The designer must therefore be able to identify the poles *intuitively* so as to determine which parts of the circuit appear as the "speed bottleneck."

The CS topology studied in Example 11-4 serves as a good example for identifying poles by inspection. Equation (11.5) reveals that the pole frequency is given by the inverse of the product of the total resistance seen between the output node and ground and the total capacitance seen between the output node and ground. Applicable to many circuits, this observation can be generalized as follows: if node j in the signal path exhibits a small-signal resistance of R_j to ground and a capacitance of C_j to ground, then it contributes a pole of magnitude $(R_j C_j)^{-1}$ to the transfer function.

Did you know?

Bode's rules are an example of "necessity is mother of invention." While working on electronic filters and equalizers at Bell Labs in the 1930s, Hendrik Bode, like other researchers in the field, faced the problem of determining the stability of complex circuits and systems. With no computers or even calculators available, the engineers of that era had to perform lengthy hand calculations for this purpose. Bode thus came up with his approximations of the frequency response and, as seen in Chapter 12, a simple method of studying the stability of feedback systems.

Example 11-8 Determine the poles of the circuit shown in Fig. 11.11. Assume $\lambda = 0$.

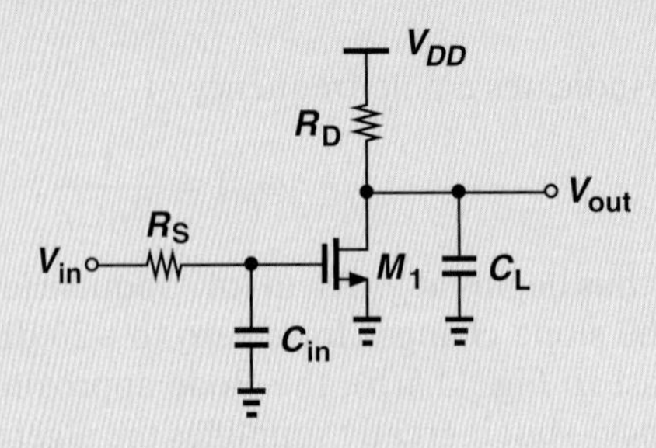

Figure 11.11

Solution Setting V_{in} to zero, we recognize that the gate of M_1 sees a resistance of R_S and a capacitance of C_{in} to ground. Thus,

$$|\omega_{p1}| = \frac{1}{R_S C_{in}}. \tag{11.16}$$

We may call ω_{p1} the "input pole" to indicate that it arises in the input network. Similarly, the "output pole" is given by

$$|\omega_{p2}| = \frac{1}{R_D C_L}. \tag{11.17}$$

Since the low-frequency gain of the circuit is equal to $-g_m R_D$, we can readily write the magnitude of the transfer function as:

$$\left|\frac{V_{out}}{V_{in}}\right| = \frac{g_m R_D}{\sqrt{\left(1+\omega^2/\omega_{p1}^2\right)\left(1+\omega^2/\omega_{p2}^2\right)}}. \tag{11.18}$$

Exercise If $\omega_{p1} = \omega_{p2}$, at what frequency does the gain drop by 3 dB?

Example 11-9 Compute the poles of the circuit shown in Fig. 11.12. Assume $\lambda = 0$.

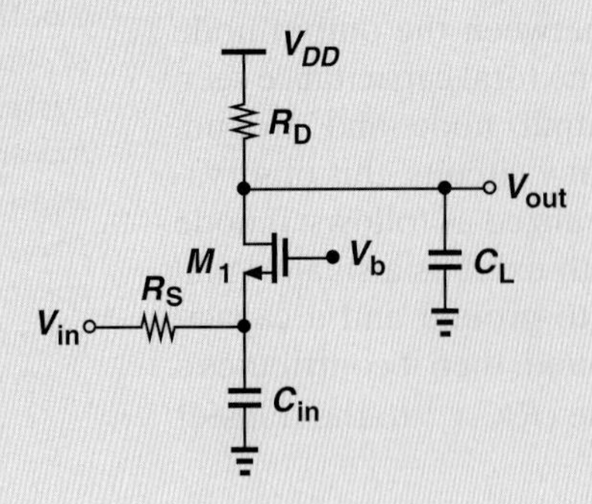

Figure 11.12

Solution With $V_{in} = 0$, the small-signal resistance seen at the source of M_1 is given by $R_S \parallel (1/g_m)$, yielding a pole at

$$\omega_{p1} = \frac{1}{\left(R_S \parallel \frac{1}{g_m}\right) C_{in}}. \tag{11.19}$$

The output pole is given by $\omega_{p2} = (R_D C_L)^{-1}$.

Exercise How do we choose the value of R_D such that the output pole frequency is ten times the input pole frequency?

The reader may wonder how the foregoing technique can be applied if a node is loaded with a "floating" capacitor, i.e., a capacitor whose other terminal is also connected to a node in the signal path (Fig. 11.13). In general, we cannot utilize this technique and must write the circuit's equations and obtain the transfer function. However, an approximation given by "Miller's theorem" can simplify the task in some cases.

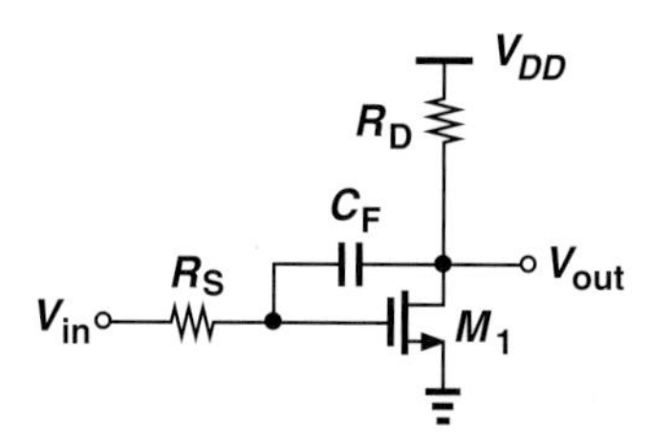

Figure 11.13 Circuit with floating capacitor.

11.1.5 Miller's Theorem

Our above study and the example in Fig. 11.13 make it desirable to obtain a method that "transforms" a floating capacitor to two *grounded* capacitors, thereby allowing association of one pole with each node. Miller's theorem is such a method. Miller's theorem, however, was originally conceived for another reason. In the late 1910s, John Miller had observed that parasitic capacitances appearing between the input and output of an amplifier may drastically lower the input impedance. He then proposed an analysis that led to the theorem.

Consider the general circuit shown in Fig. 11.14(a), where the floating impedance, Z_F, appears between nodes 1 and 2. We wish to transform Z_F to two grounded impedances as depicted in Fig. 11.14(b), while ensuring all of the currents and voltages in the circuit remain unchanged. To determine Z_1 and Z_2, we make two observations: (1) the current drawn by Z_F from node 1 in Fig. 11.14(a) must be equal to that drawn by Z_1 in Fig. 11.14(b); and (2) the current injected to node 2 in Fig. 11.14(a) must be equal to that injected by Z_2 in Fig. 11.14(b). (These requirements guarantee that the circuit does not "feel" the transformation.) Thus,

$$\frac{V_1 - V_2}{Z_F} = \frac{V_1}{Z_1} \tag{11.20}$$

$$\frac{V_1 - V_2}{Z_F} = -\frac{V_2}{Z_2}. \tag{11.21}$$

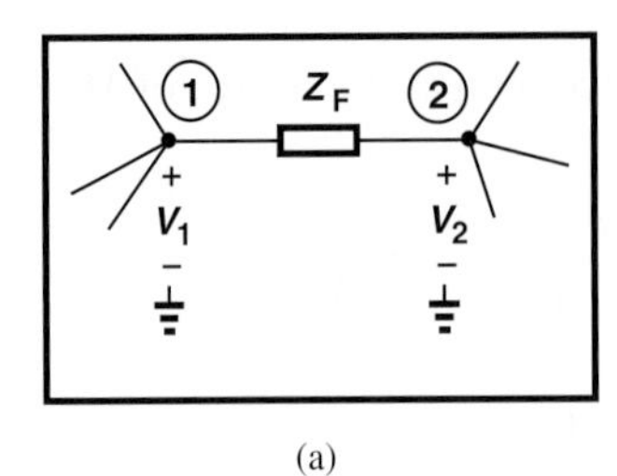

(a)

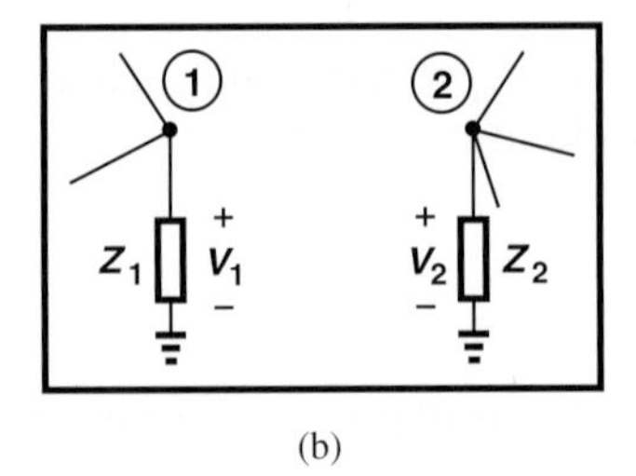

(b)

Figure 11.14 (a) General circuit including a floating impedance, (b) equivalent of (a) as obtained from Miller's theorem.

Denoting the voltage gain from node 1 to node 2 by A_v, we obtain

$$Z_1 = Z_F \frac{V_1}{V_1 - V_2} \tag{11.22}$$

$$= \frac{Z_F}{1 - A_v} \tag{11.23}$$

and

$$Z_2 = Z_F \frac{-V_2}{V_1 - V_2} \tag{11.24}$$

$$= \frac{Z_F}{1 - \dfrac{1}{A_v}}. \tag{11.25}$$

Called Miller's theorem, the results expressed by Eqs. (11.23) and (11.25) prove extremely useful in analysis and design. In particular, Eq. (11.23) suggests that the floating impedance is *reduced* by a factor of $1 - A_v$ when "seen" at node 1.

As an important example of Miller's theorem, let us assume Z_F is a single capacitor, C_F, tied between the input and output of an inverting amplifier [Fig. 11.15(a)]. Applying Eq. (11.23), we have

$$Z_1 = \frac{Z_F}{1 - A_v} \tag{11.26}$$

$$= \frac{1}{(1 + A_0)C_F s}, \tag{11.27}$$

where the substitution $A_v = -A_0$ is made. What type of impedance is Z_1? The $1/s$ dependence suggests a capacitor of value $(1 + A_0)C_F$, as if C_F is "amplified" by a factor of $1 + A_0$. In other words, a capacitor C_F tied between the input and output of an inverting amplifier with a gain of A_0 raises the *input capacitance* by an amount equal to $(1 + A_0)C_F$. We say such a circuit suffers from "Miller multiplication" of the capacitor.

The effect of C_F at the *output* can be obtained from Eq. (11.25):

$$Z_2 = \frac{Z_F}{1 - \dfrac{1}{A_v}} \tag{11.28}$$

$$= \frac{1}{\left(1 + \dfrac{1}{A_0}\right) C_F s}, \tag{11.29}$$

which is close to $(C_F s)^{-1}$ if $A_0 \gg 1$. Figure 11.15(b) summarizes these results.

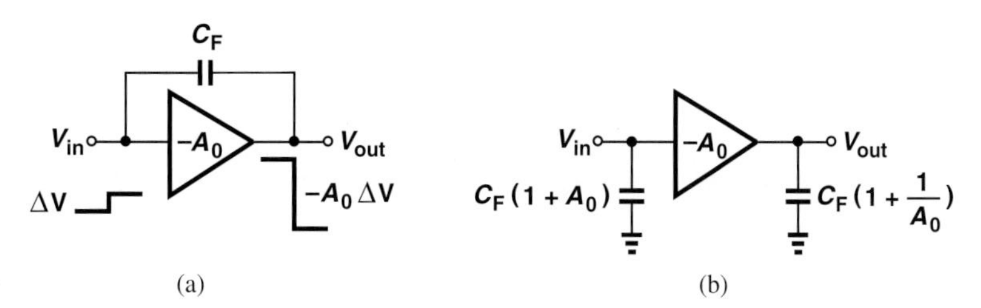

Figure 11.15 (a) Inverting circuit with floating capacitor, (b) equivalent circuit as obtained from Miller's theorem.

The Miller multiplication of capacitors can also be explained intuitively. Suppose the input voltage in Fig. 11.15(a) goes up by a small amount ΔV. The output then goes *down* by $A_0\Delta V$. That is, the voltage across C_F increases by $(1 + A_0)\Delta V$, requiring that the input provide a proportional charge. By contrast, if C_F were not a floating capacitor and its right plate voltage did not change, it would experience only a voltage change of ΔV and require less charge.

The above study points to the utility of Miller's theorem for conversion of floating capacitors to grounded capacitors. The following example demonstrates this principle.

Did you know?

The Miller effect was discovered by John Miller in 1919. In his original paper, Miller observes that "the apparent input capacity can become a number of times greater than the actual capacities between the tube electrodes." (Back then, capacitance and capacitor were called "capacity" and "condensor," respectively.) A curious effect with respect to Miller multiplication is that, if the amplifier gain is *positive* and greater than unity, then we obtain a *negative* input capacitance, $(1 - A_v)C_F$. Does this happen? Yes, indeed. Shown below is a circuit realizing a negative capacitance. Since the gain from V_{in1} to V_Y (and from V_{in2} to V_X) is positive, C_F can be multiplied by a negative number. This technique is used in many high-speed circuits to partially cancel the effect of undesired (positive) capacitances.

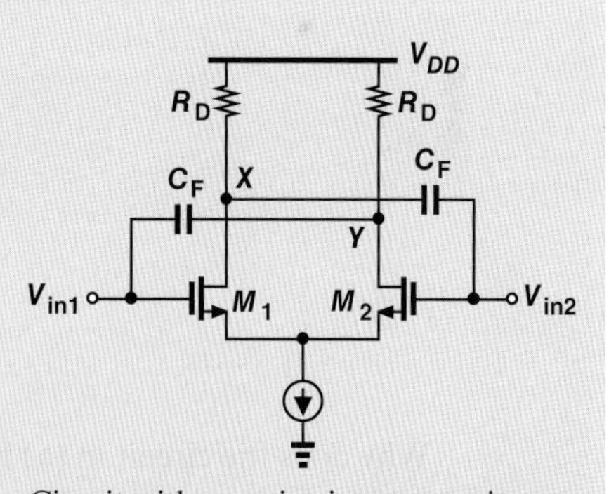

Circuit with negative input capacitance.

Example 11-10 Estimate the pole of the circuit shown in Fig. 11.16(a). Assume $\lambda = 0$.

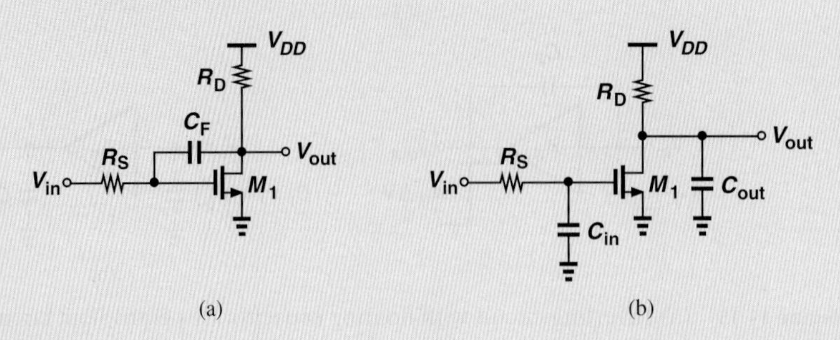

Figure 11.16

Solution Noting that M_1 and R_D constitute an inverting amplifier having a gain of $-g_mR_D$, we utilize the results in Fig. 11.15(b) to write:

$$C_{in} = (1 + A_0)C_F \tag{11.30}$$

$$= (1 + g_mR_D)C_F \tag{11.31}$$

and

$$C_{out} = \left(1 + \frac{1}{g_mR_D}\right) C_F, \tag{11.32}$$

thereby arriving at the topology depicted in Fig. 11.16(b). From our study in Example 11-8, we have:

$$\omega_{in} = \frac{1}{R_SC_{in}} \tag{11.33}$$

$$= \frac{1}{R_S(1 + g_mR_D)C_F} \tag{11.34}$$

and

$$\omega_{out} = \frac{1}{R_DC_{out}} \tag{11.35}$$

$$= \frac{1}{R_D\left(1 + \dfrac{1}{g_mR_D}\right) C_F}. \tag{11.36}$$

Why does the circuit in (a) have one pole but that in (b) two? This is explained below.

Exercise Calculate C_{in} if $g_m = (150\ \Omega)^{-1}$, $R_D = 2\ \text{k}\Omega$, and $C_F = 80$ fF.

The reader may find the above example somewhat inconsistent. Miller's theorem requires that the floating impedance and the voltage gain be computed *at the same frequency* whereas Example 11-10 uses the *low-frequency* gain, g_mR_D, even for the purpose of finding high-frequency poles. After all, we know that the existence of C_F lowers the voltage gain from the gate of M_1 to the output at high frequencies. Owing to this inconsistency, we call the procedure in Example 11-10 the "Miller approximation." Without this approximation, i.e., if A_0 is expressed in terms of circuit parameters at the frequency of interest, application of Miller's theorem would be no simpler than direct solution of the circuit's equations. Due to the approximation, the circuit in the above example exhibits two poles.

Another artifact of Miller's approximation is that it may eliminate a *zero* of the transfer function. We return to this issue in Section 11.4.3.

The general expression in Eq. (11.23) can be interpreted as follows: an impedance tied between the input and output of an inverting amplifier with a gain of A_v is lowered by a factor of $1 + A_v$ if seen at the input (with respect to ground). This reduction of impedance (hence increase in capacitance) is called "Miller effect." For example, we say Miller effect raises the input capacitance of the circuit in Fig. 11.16(a) to $(1 + g_m R_D)C_F$.

11.1.6 General Frequency Response

Our foregoing study indicates that capacitances in a circuit tend to lower the voltage gain at high frequencies. It is possible that capacitors reduce the gain at *low* frequencies as well. As a simple example, consider the high-pass filter shown in Fig. 11.17(a), where the voltage division between C_1 and R_1 yields

$$\frac{V_{out}}{V_{in}}(s) = \frac{R_1}{R_1 + \frac{1}{C_1 s}} \tag{11.37}$$

$$= \frac{R_1 C_1 s}{R_1 C_1 s + 1}, \tag{11.38}$$

and hence

$$\left|\frac{V_{out}}{V_{in}}\right| = \frac{R_1 C_1 \omega}{\sqrt{R_1^2 C_1^2 \omega_1^2 + 1}}. \tag{11.39}$$

Plotted in Fig. 11.17(b), the response exhibits a roll-off as the frequency of operation falls *below* $1/(R_1 C_1)$. As seen from Eq. (11.38), this roll-off arises because the zero of the transfer function occurs at the origin.

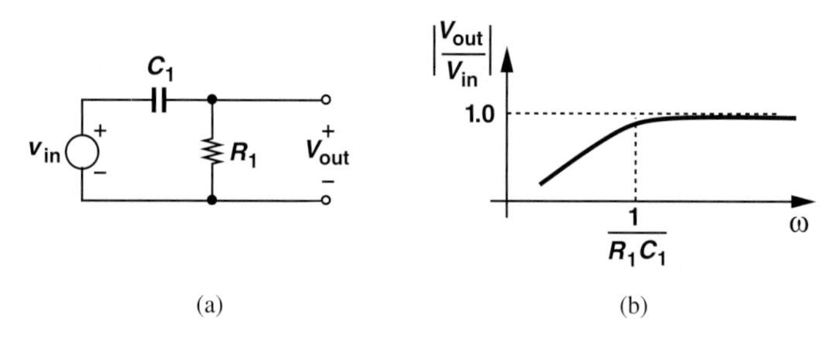

Figure 11.17 (a) Simple high-pass filter, and (b) its frequency response.

The low-frequency roll-off may prove undesirable. The following example illustrates this point.

Example 11-11 Figure 11.18 depicts a source follower used in a high-quality audio amplifier. Here, R_i establishes a gate bias voltage equal to V_{DD} for M_1, and I_1 defines the drain bias current. Assume $\lambda = 0$, $g_m = 1/(200\ \Omega)$, and $R_1 = 100\ \text{k}\Omega$. Determine the minimum required value of C_1 and the maximum tolerable value of C_L.

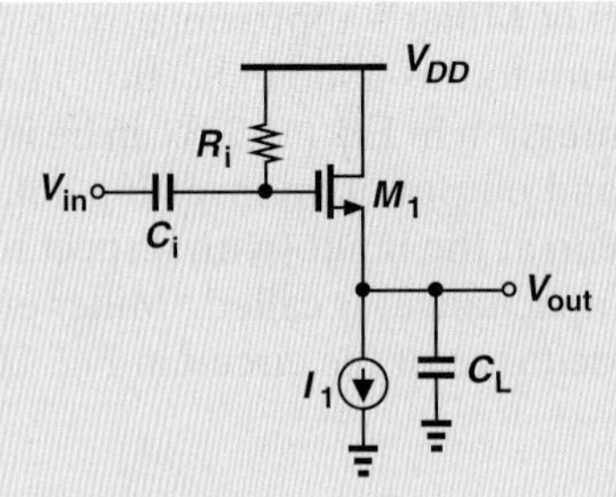

Figure 11.18

Solution Similar to the high-pass filter of Fig. 11.17, the input network consisting of R_i and C_i attenuates the signal at low frequencies. To ensure that audio components as low as 20 Hz experience a small attenuation, we set the corner frequency $1/(R_iC_i)$ to $2\pi \times$ (20 Hz), thus obtaining

$$C_i = 79.6\text{ nF}. \tag{11.40}$$

This value is, of course, much too large to be integrated on a chip. Since Eq. (11.39) reveals a 3-dB attenuation at $\omega = 1/(R_iC_i)$, in practice we must choose even a larger capacitor if a lower attenuation is desired.

The load capacitance creates a pole at the output node, lowering the gain at high frequencies. Setting the pole frequency to the upper end of the audio range, 20 kHz, and recognizing that the resistance seen from the output node to ground is equal to $1/g_m$, we have

$$\omega_{p,out} = \frac{g_m}{C_L} \tag{11.41}$$

$$= 2\pi \times (20\text{ kHz}), \tag{11.42}$$

and hence

$$C_L = 39.8\text{ nF}. \tag{11.43}$$

An efficient driver, the source follower can tolerate a very large load capacitance (for the audio band).

Exercise Repeat the above example if I_1 and the width of M_1 are halved.

Why did we use capacitor C_i in the above example? Without C_i, the circuit's gain would not fall at low frequencies, and we would not need perform the above calculations. Called a "coupling" capacitor, C_i allows the signal frequencies of interest to pass through the circuit while blocking the dc content of V_{in}. In other words, C_i isolates the bias conditions of the source follower from those of the *preceding* stage. Figure 11.19(a) illustrates an example in which a CS stage precedes the source follower. The coupling capacitor permits independent bias voltages at nodes X and Y. For example, V_Y can be chosen relatively low (placing M_2 near the triode region) to allow a large drop across R_D, thereby maximizing the voltage gain of the CS stage (why?).

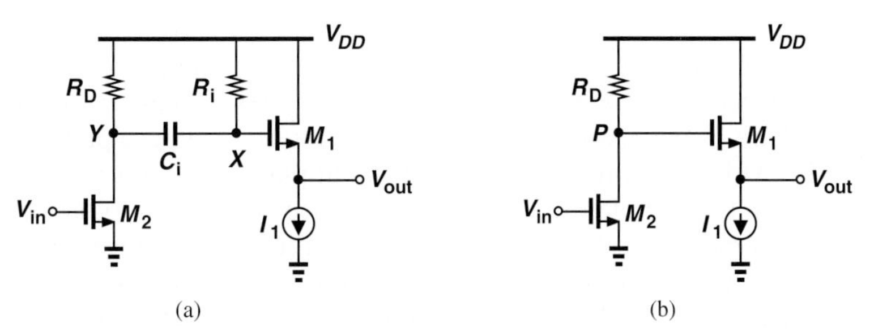

Figure 11.19 Cascade of CS stage and source follower with (a) capacitor coupling and (b) direct coupling.

To convince the reader that capacitive coupling proves essential in Fig. 11.19(a), we consider the case of "direct coupling" [Fig. 11.19(b)] as well. Here, to maximize the voltage gain, we wish to set V_P just above $V_{GS2} - V_{TH2}$, e.g., 200 mV. On the other hand, the gate of M_2 must reside at a voltage of at least $V_{GS1} + V_{I1}$, where V_{I1} denotes the minimum voltage required by I_1. Since $V_{GS1} + V_{I1}$ may reach 600-700 mV, the two stages are quite incompatible in terms of their bias points, necessitating capacitive coupling.

Capacitive coupling (also called "ac coupling") is more common in discrete circuit design due to the large capacitor values required in many applications (e.g., C_i in the above audio example). Nonetheless, many integrated circuits also employ capacitive coupling, especially at low supply voltages, if the necessary capacitor values are no more than a few picofarads.

Figure 11.20 shows a typical frequency response and the terminology used to refer to its various attributes. We call ω_L the lower corner or lower "cut-off" frequency and ω_H the upper corner or upper cut-off frequency. Chosen to accommodate the signal frequencies of interest, the band between ω_L and ω_H is called the "midband range" and the corresponding gain the "midband gain."

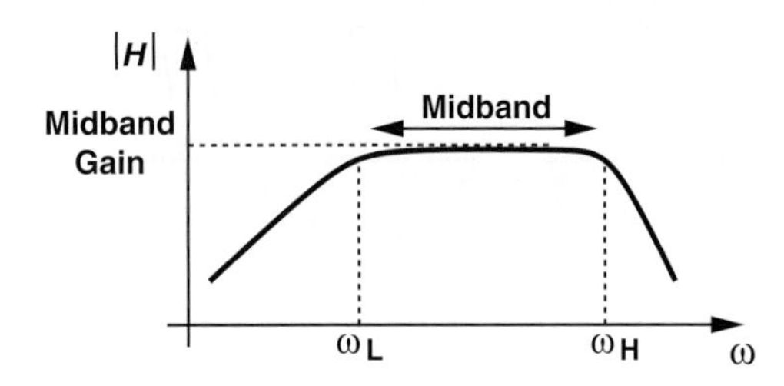

Figure 11.20 Typical frequency response.

Robotics Application: Accelerometers

Accelerometers measure the acceleration and are extensively used in robotics to determine the force of gravity, whether a robot is going uphill or downhill, or compute the robot's velocity and the distance that it travels.

Accelerometers operate by sensing changes in the capacitance of a device. Consider the conceptual structure depicted in Fig. 11.21(a), where conductive plates (pillars) 1 and 2 are installed on a nonconductive substrate. Suppose plate 2 is designed such that it tilts with acceleration [Fig. 11.21(b)], but plate 2 remains fixed. Thus, the capacitance between the two plates changes as the device accelerates to the left or to the right.

In practice, it is difficult to measure the small changes in C_{12}. We thus modify the structure to the "differential" capacitor" geometry shown in Fig. 11.21(c). Here, the first pate is flanked by plates 2 and 3, both of which can tilt. If these plates tilt to the right, then C_{12} decreases and C_{13} increases; the changes are approximately equal for a small tilt angle. We can now measure the *difference* between C_{12} and C_{13} as a quantity representing acceleration. As shown in Fig. 11.21(d), we drive plates 2 and 3 by differential sinusoids V_a and V_b, respectively, and examine the voltage on plate 1. Writing the impedance of each capacitor as $1/(Cs)$ and applying superposition to the capacitive voltage divider, the reader can show that

$$V_{out} = \frac{C_{12}}{C_{12} + C_{13}} V_a + \frac{C_{13}}{C_{12} + C_{13}} V_b. \tag{11.44}$$

With no acceleration, $C_{12} = C_{13}$ and $V_{out} = 0$ because $V_a = -V_b$. With acceleration, on the other hand, $C_{12} = C_0 + \Delta C$ and $C_{13} = C_0 - \Delta C$, yielding

$$V_{out} = \frac{\Delta C(V_a - V_b)}{C_{12} + C_{13}}. \tag{11.45}$$

Here, C_0 denotes the value when the two capacitors are equal. It follows that V_{out} assumes positive or negative values depending on the direction of the acceleration. Other types of accelerometers place the variable capacitance in an oscillator and measure the changes in the oscillation frequency.

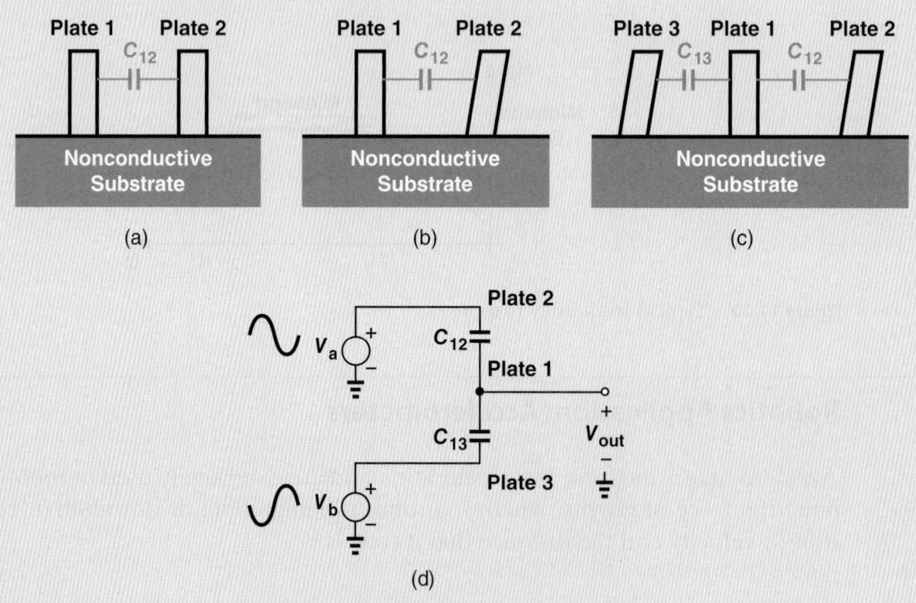

Figure 11.21 (a) Simple capacitor structure, (b) change in capacitance as a result of acceleration, (c) differential structure, and (d) measurement of capacitance change.

11.2 HIGH-FREQUENCY MODELS OF TRANSISTORS

The speed of many circuits is limited by the capacitances within each transistor. It is therefore necessary to study these capacitances carefully.

11.2.1 High-Frequency Model of Bipolar Transistor

Recall from Chapter 4 that the bipolar transistor consists of two *pn* junctions. The depletion region associated with the junctions[6] gives rise to a capacitance between base and emitter, denoted by C_{je}, and another between base and collector, denoted by C_μ [Fig. 11.22(a)]. We may then add these capacitances to the small-signal model to arrive at the representation shown in Fig. 11.22(b).

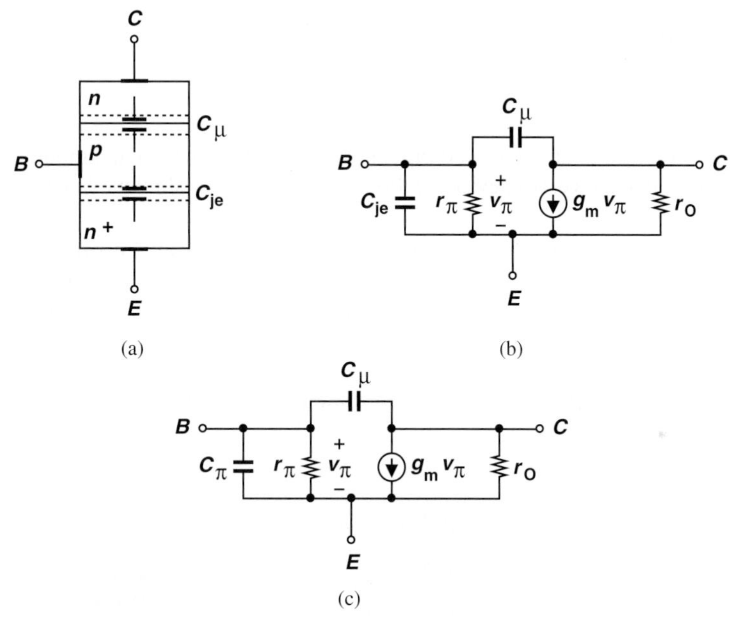

Figure 11.22 (a) Structure of bipolar transistor showing junction capacitances, (b) small-signal model with junction capacitances, (c) complete model accounting for base charge.

Unfortunately, this model is incomplete because the base-emitter junction exhibits another effect that must be taken into account. As explained in Chapter 4, the operation of the transistor requires a (nonuniform) charge profile in the base region to allow the diffusion of carriers toward the collector. In other words, if the transistor is suddenly turned on, proper operation does not begin until enough charge carriers enter the base region and *accumulate* so as to create the necessary profile. Similarly, if the transistor is suddenly turned off, the charge carriers stored in the base must be *removed* for the collector current to drop to zero.

The above phenomenon is quite similar to charging and discharging a capacitor: to change the collector current, we must change the base charge profile by injecting or removing some electrons or holes. Modeled by a second capacitor between the base and emitter,

[6]As mentioned in Chapter 4, both forward-biased and reversed-biased junctions contain a depletion region and hence a capacitance associated with it.

C_b, this effect is typically more significant than the depletion region capacitance. Since C_b and C_{je} appear in parallel, they are lumped into one and denoted by C_π [Fig. 11.22(c)].

In integrated circuits, the bipolar transistor is fabricated atop a grounded substrate [Fig. 11.23(a)]. The collector-substrate junction remains reverse-biased (why?), exhibiting a junction capacitance denoted by C_{CS}. The complete model is depicted in Fig. 11.23(b). We hereafter employ this model in our analysis. In modern integrated-circuit bipolar transistors, C_{je}, C_μ, and C_{CS} are on the order of a few femtofarads for the smallest allowable devices.

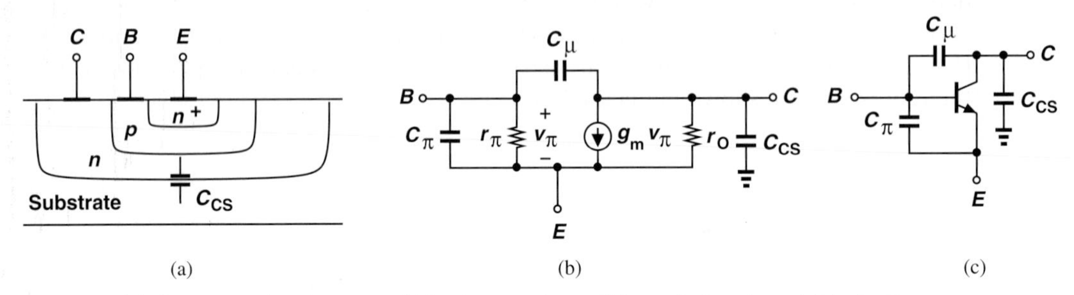

Figure 11.23 (a) Structure of an integrated bipolar transistor, (b) small-signal model including collector-substrate capacitance, (c) device symbol with capacitances shown explicitly.

In the analysis of frequency response, it is often helpful to first draw the transistor capacitances on the circuit diagram, simplify the result, and then construct the small-signal equivalent circuit. We may therefore represent the transistor as shown in Fig. 11.23(c).

Example 11-12 Identify all of the capacitances in the circuit shown in Fig. 11.24(a).

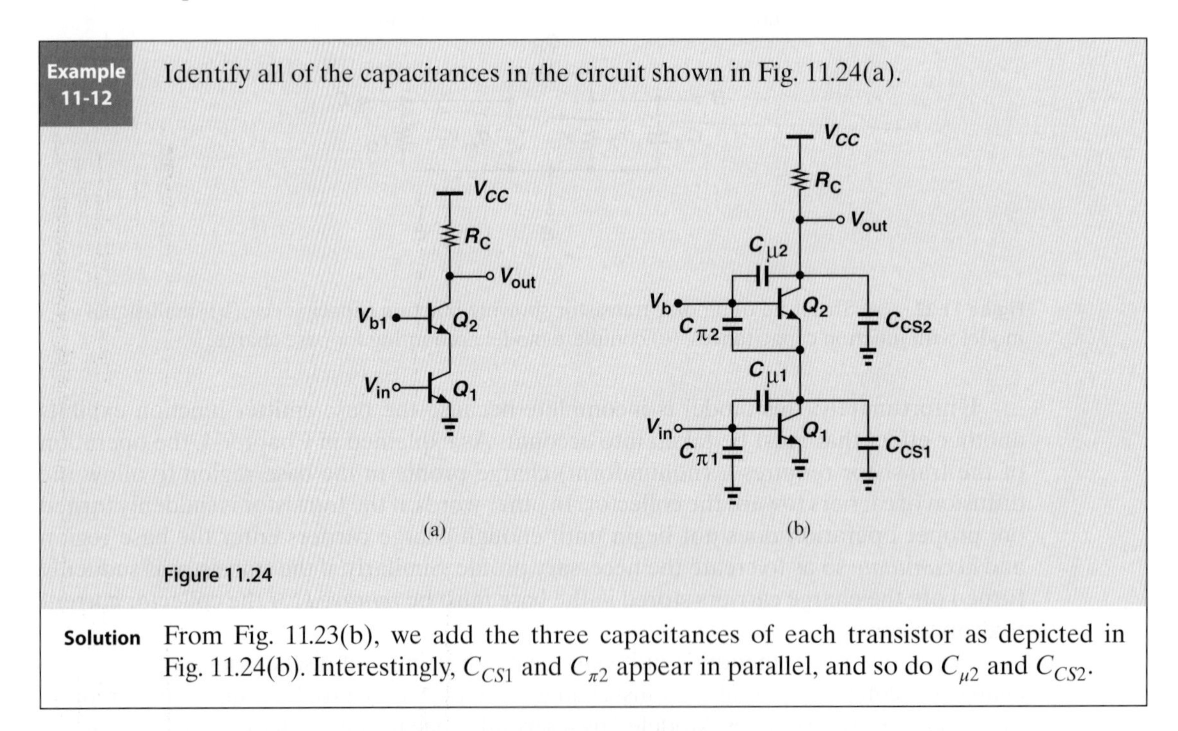

Figure 11.24

Solution From Fig. 11.23(b), we add the three capacitances of each transistor as depicted in Fig. 11.24(b). Interestingly, C_{CS1} and $C_{\pi 2}$ appear in parallel, and so do $C_{\mu 2}$ and C_{CS2}.

Exercise Construct the small-signal equivalent circuit of the above cascode.

11.2.2 High-Frequency Model of MOSFET

Our study of the MOSFET structure in Chapter 6 revealed several capacitive components. We now study these capacitances in the device in greater detail.

As illustrated in Fig. 11.25(a), the MOSFET displays three prominent capacitances: one between the gate and the channel (called the "gate oxide capacitance" and given by WLC_{ox}), and two associated with the reverse-biased source-bulk and drain-bulk junctions. The first component presents a modeling difficulty because the transistor model does not contain a "channel." We must therefore decompose this capacitance into one between the gate and the source and another between the gate and the drain [Fig. 11.25(b)]. The exact partitioning of this capacitance is beyond the scope of this book, but, in the saturation region, C_1 is about 2/3 of the gate-channel capacitance whereas $C_2 \approx 0$.

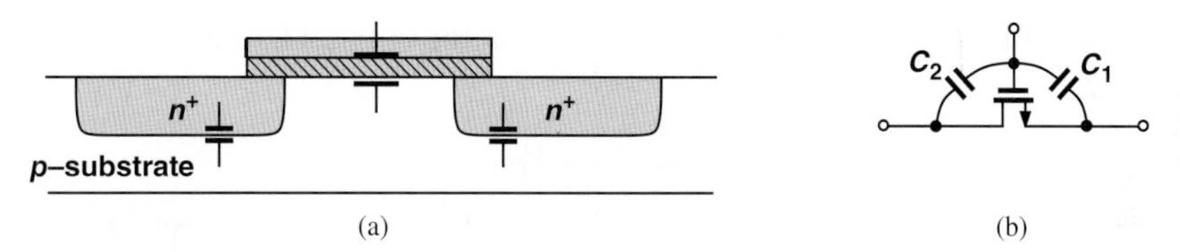

Figure 11.25 (a) Structure of MOS device showing various capacitances, (b) partitioning of gate-channel capacitance between source and drain.

Two other capacitances in the MOSFET become critical in some circuits. As shown in Fig. 11.26, these components arise from both the physical overlap of the gate with source/drain areas[7] and the fringe field lines between the edge of the gate and the top of the S/D regions. Called the gate-drain or gate-source "overlap" capacitance, this (symmetric) effect persists even if the MOSFET is off.

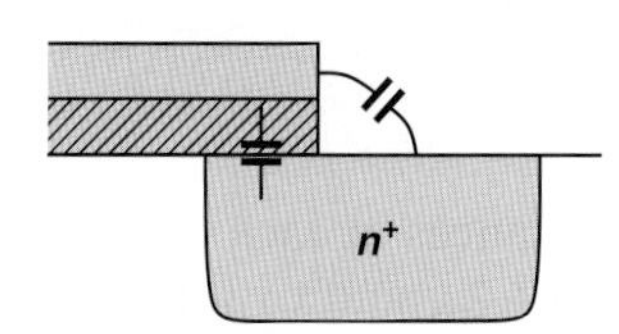

Figure 11.26 Overlap capacitance between gate and drain (or source).

We now construct the high-frequency model of the MOSFET. As depicted in Fig. 11.27(a), this representation consists of: (1) the capacitance between the gate and source, C_{GS} (including the overlap component); (2) the capacitance between the gate and drain (including the overlap component); and (3) the junction capacitances between the source and bulk and the drain and bulk, C_{SB} and C_{DB}, respectively. (We assume the bulk remains at ac ground.) As mentioned in Section 11.2.1, we often draw the capacitances on the transistor symbol [Fig. 11.27(b)] before constructing the small-signal model.

[7]As mentioned in Chapter 6, the S/D areas protrude under the gate during fabrication.

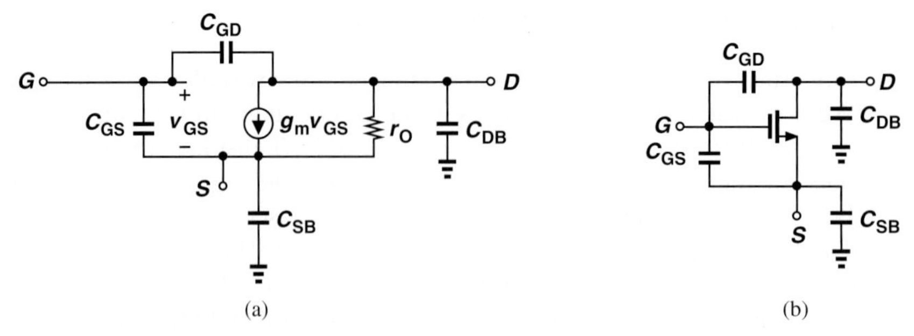

Figure 11.27 (a) High-frequency model of MOSFET, (b) device symbol with capacitances shown explicitly.

Example 11-13 Identify all of the capacitances in the circuit of Fig. 11.28(a).

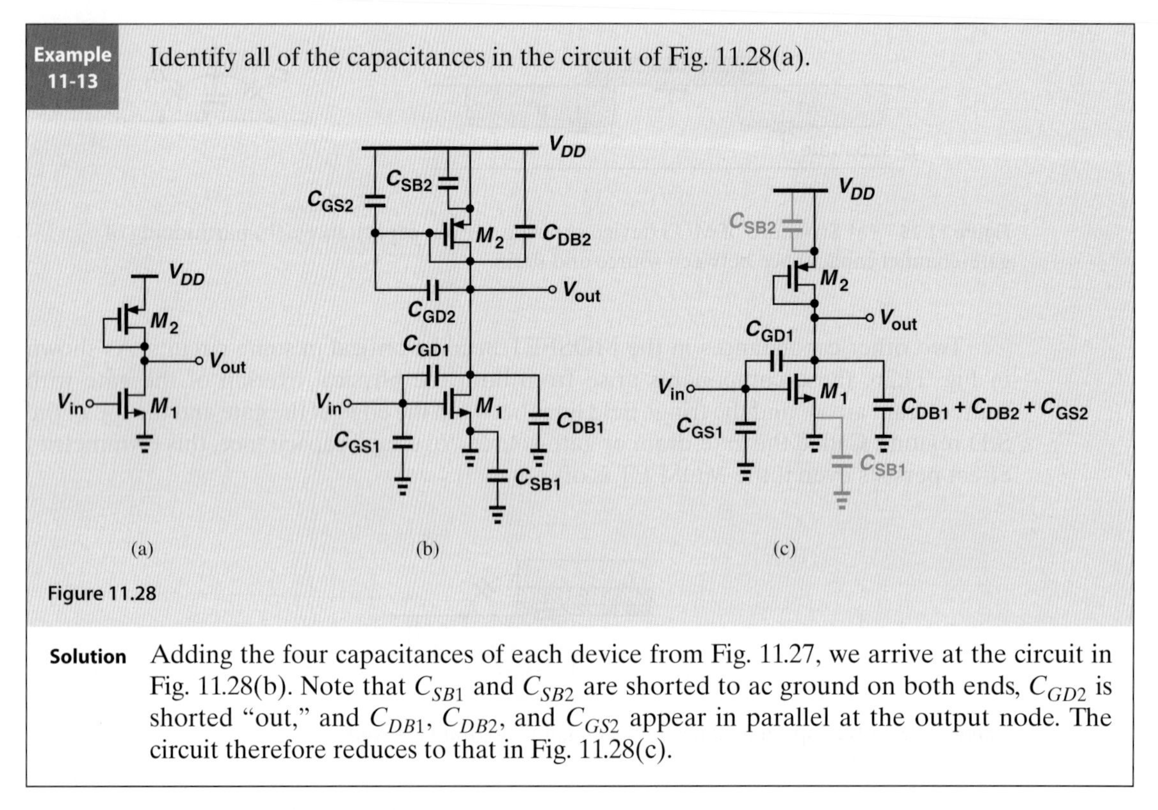

Figure 11.28

Solution Adding the four capacitances of each device from Fig. 11.27, we arrive at the circuit in Fig. 11.28(b). Note that C_{SB1} and C_{SB2} are shorted to ac ground on both ends, C_{GD2} is shorted "out," and C_{DB1}, C_{DB2}, and C_{GS2} appear in parallel at the output node. The circuit therefore reduces to that in Fig. 11.28(c).

Exercise Noting that M_2 is a diode-connected device, construct the small-signal equivalent circuit of the amplifier.

11.2.3 Transit Frequency

With various capacitances surrounding bipolar and MOS devices, is it possible to define a quantity that represents the ultimate speed of the transistor? Such a quantity would prove useful in comparing different types or generations of transistors as well as in predicting the performance of circuits incorporating the devices.

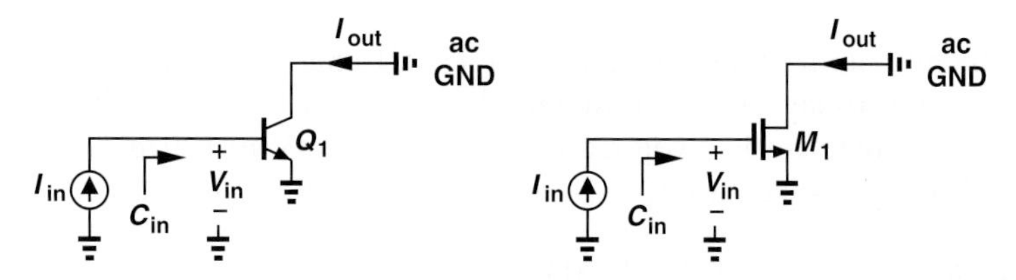

Figure 11.29 Conceptual setup for measurement of f_T of transistors.

A measure of the intrinsic speed of transistors[8] is the "transit" or "cut-off" frequency, f_T, defined as the frequency at which the small-signal *current gain* of the device falls to unity. As illustrated in Fig. 11.29 (without the biasing circuitry), the idea is to inject a sinusoidal current into the base or gate and measure the resulting collector or drain current while the input frequency, f_{in}, is increased. We note that, as f_{in} increases, the input capacitance of the device lowers the input impedance, Z_{in}, and hence the input voltage $V_{in} = I_{in}Z_{in}$ and the output current. We neglect C_μ and C_{GD} here (but take them into account in Problem 11.26). For the bipolar device in Fig. 11.29(a),

$$Z_{in} = \frac{1}{C_\pi s} \parallel r_\pi. \tag{11.46}$$

Since $I_{out} = g_m I_{in} Z_{in}$,

$$\frac{I_{out}}{I_{in}} = \frac{g_m r_\pi}{r_\pi C_\pi s + 1} \tag{11.47}$$

$$= \frac{\beta}{r_\pi C_\pi s + 1}. \tag{11.48}$$

At the transit frequency, $\omega_T (= 2\pi f_T)$, the magnitude of the current gain falls to unity:

$$r_\pi^2 C_\pi^2 \omega_T^2 = \beta^2 - 1 \tag{11.49}$$

$$\approx \beta^2. \tag{11.50}$$

That is,

$$\omega_T \approx \frac{g_m}{C_\pi}. \tag{11.51}$$

The transit frequency of MOSFETs is obtained in a similar fashion. We therefore write:

$$2\pi f_T \approx \frac{g_m}{C_\pi} \text{ or } \frac{g_m}{C_{GS}}. \tag{11.52}$$

Note that the collector-substrate or drain-bulk capacitance does not affect f_T owing to the ac ground established at the output.

Modern bipolar and MOS transistors boast f_T's above 100 GHz. Of course, the speed of complex circuits using such devices is quite lower.

[8]By "intrinsic" speed, we mean the performance of the device *by itself*, without any other limitations imposed or enhancements provided by the circuit.

Example 11-14 The minimum channel length of MOSFETs has been scaled from 1 μm in the late 1980s to 65 nm today. Also, the inevitable reduction of the supply voltage has reduced the gate-source overdive voltage from about 400 mV to 100 mV. By what factor has the f_T of MOSFETs increased?

Solution It can proved (Problem 11.28) that

$$2\pi f_T = \frac{3}{2}\frac{\mu_n}{L^2}(V_{GS} - V_{TH}). \tag{11.53}$$

Thus, the transit frequency has increased by approximately a factor of 59. For example, if $\mu_n = 400\ \text{cm}^2/(\text{V}\cdot\text{s})$, then 65 nm devices having an overdrive of 100 mV exhibit an f_T of 226 GHz.

Exercise Determine the f_T if the channel length is scaled down to 45 nm but the mobility degrades to $300\ \text{cm}^2/(\text{V}\cdot\text{s})$.

Did you know?

If the f_T of 65 nm MOSFETs is around 220 GHz, is it possible to operate such a device at a *higher* frequency? Yes, indeed. The key is to use inductors to cancel the effect of capacitors. Suppose as shown in Fig. (a), we place inductor L_1 in parallel with C_{GS}. (Realized as a metal spiral on the chip, the inductor has some resistance, R_S.) At the resonance frequency, $\omega_0 = 1/\sqrt{L_1 C_{GS}}$, the parallel combination reduces to a single resistor, almost as if M_1 had no gate-source capacitance! For this reason, the use of on-chip inductors has become common in high-frequency design. Figure (b) shows the chip photograph of a 300 GHz oscillator designed by the author in 65 nm technology. Such high frequencies find application in medical imaging.

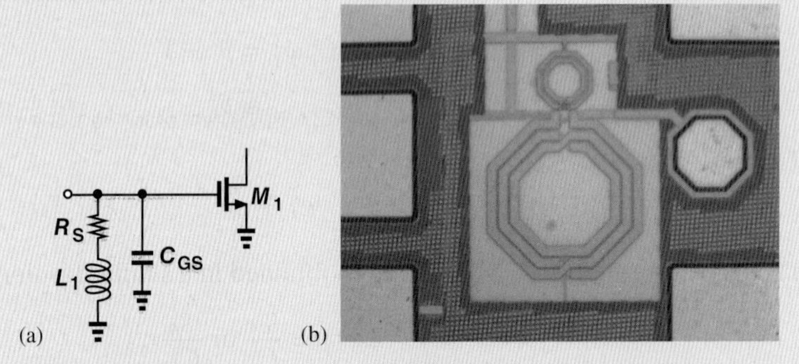

(a) Use of resonance to cancel transistor capacitance, (b) chip photograph of a 300 GHz CMOS oscillator.

11.3 ANALYSIS PROCEDURE

We have thus far seen a number of concepts and tools that help us study the frequency response of circuits. Specifically, we have observed that:

- The frequency response refers to the magnitude of the transfer function of a system.[9]

[9] In a more general case, the frequency response also includes the phase of the transfer function, as studied in Chapter 12.

- Bode's approximation simplifies the task of plotting the frequency response if the poles and zeros are known.
- In many cases, it is possible to associate a pole with each node in the signal path.
- Miller's theorem proves helpful in decomposing floating capacitors into grounded elements.
- Bipolar and MOS devices exhibit various capacitances that limit the speed of circuits.

In order to methodically analyze the frequency response of various circuits, we prescribe the following steps:

1. Determine which capacitors impact the low-frequency region of the response and compute the low-frequency cut-off. In this calculation, the transistor capacitances can be neglected as they typically impact only the high-frequency region.
2. Calculate the midband gain by replacing the above capacitors with short circuits while still neglecting the transistor capacitances.
3. Identify and add to the circuit the capacitances contributed by each transistor.
4. Noting ac grounds (e.g., the supply voltage or constant bias voltages), merge the capacitors that are in parallel and omit those that play no role in the circuit.
5. Determine the high-frequency poles and zeros by inspection or by computing the transfer function. Miller's theorem may prove useful here.
6. Plot the frequency response using Bode's rules or exact calculations.

We now apply this procedure to various amplifier topologies.

11.4 FREQUENCY RESPONSE OF CE AND CS STAGES

11.4.1 Low-Frequency Response

As mentioned in Section 11.1.6, the gain of amplifiers may fall at low frequencies due to certain capacitors in the signal path. Let us consider a general CS stage with its input bias network and an input coupling capacitor [Fig. 11.30(a)]. At low frequencies, the transistor capacitances negligibly affect the frequency response, leaving only C_i as the frequency-dependent component. We write $V_{out}/V_{in} = (V_{out}/V_X)(V_X/V_{in})$, neglect channel-length modulation, and note that both R_1 and R_2 are tied between X and ac ground. Thus, $V_{out}/V_X = -R_D/(R_S + 1/g_m)$ and

$$\frac{V_X}{V_{in}}(s) = \frac{R_1 \parallel R_2}{R_1 \parallel R_2 + \dfrac{1}{C_i s}} \tag{11.54}$$

$$= \frac{(R_1 \parallel R_2)\, C_i s}{(R_1 \parallel R_2)\, C_i s + 1}. \tag{11.55}$$

Similar to the high-pass filter of Fig. 11.17, this network attenuates the low frequencies, dictating that the lower cut-off be chosen below the lowest signal frequency, $f_{sig,\min}$ (e.g., 20 Hz in audio applications):

$$\frac{1}{2\pi[(R_1 \parallel R_2)\, C_i]} < f_{sig,\min}. \tag{11.56}$$

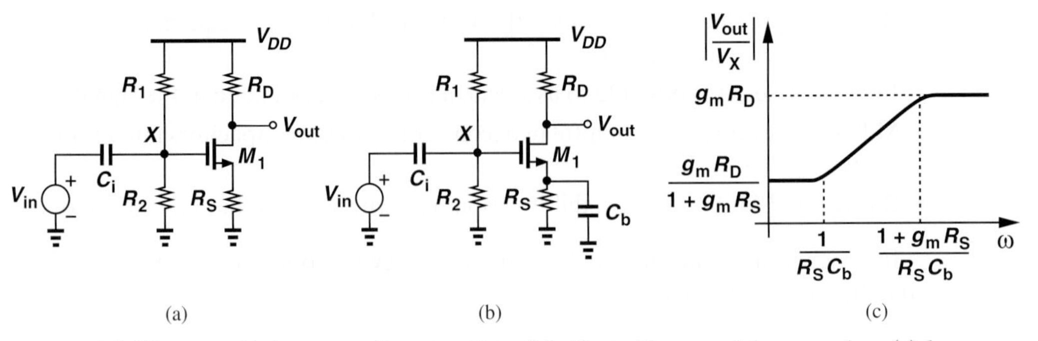

Figure 11.30 (a) CS stage with input coupling capacitor, (b) effect of bypassed degeneration, (c) frequency response with bypassed degeneration.

In applications demanding a greater midband gain, we place a "bypass" capacitor in parallel with R_S [Fig. 11.30(b)] so as to remove the effect of degeneration at midband frequencies. To quantify the role of C_b, we place its impedance, $1/(C_b s)$, in parallel with R_S in the midband gain expression:

$$\frac{V_{out}}{V_X}(s) = \frac{-R_D}{R_S \parallel \dfrac{1}{C_b s} + \dfrac{1}{g_m}} \tag{11.57}$$

$$= \frac{-g_m R_D (R_S C_b s + 1)}{R_S C_b s + g_m R_S + 1}. \tag{11.58}$$

Figure 11.30(c) shows the Bode plot of the frequency response in this case. At frequencies well below the zero, the stage operates as a degenerated CS amplifier, and at frequencies well above the pole, the circuit experiences no degeneration. Thus, the pole frequency must be chosen significantly smaller than the lowest signal frequency of interest.

The above analysis can also be applied to a CE stage. Both types exhibit low-frequency roll-off due to the input coupling capacitor and the degeneration bypass capacitor.

11.4.2 High-Frequency Response

Consider the CE and CS amplifiers shown in Fig. 11.31(a), where R_S may represent the output impedance of the preceding stage, i.e., it is not added deliberately. Identifying the capacitances of Q_1 and M_1, we arrive at the complete circuits depicted in Fig. 11.31(b), where the source-bulk capacitance of M_1 is grounded on both ends. The small-signal equivalents of these circuits differ by only r_π [Fig. 11.31(c)],[10] and can be reduced to one if V_{in}, R_S and r_π are replaced with their Thevenin equivalent [Fig. 11.31(d)]. In practice, $R_S \ll r_\pi$ and hence $R_{Thev} \approx R_S$. Note that the output resistance of each transistor would simply appear in parallel with R_L.

With this unified model, we now study the high-frequency response, first applying Miller's approximation to develop insight and then performing an accurate analysis to arrive at more general results.

[10]The Early effect and channel-length modulation are neglected here.

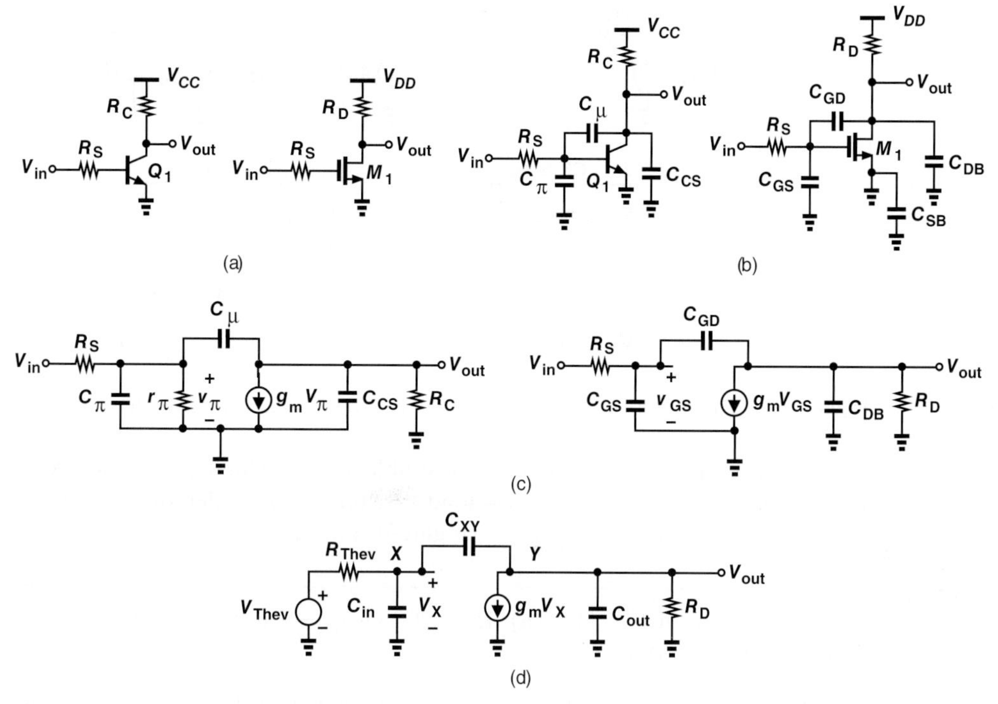

Figure 11.31 (a) CE and CS stages, (b) inclusion of transistor capacitances, (c) small-signal equivalents, (d) unified model of both circuits.

11.4.3 Use of Miller's Theorem

With C_{XY} tied between two floating nodes, we cannot simply associate one pole with each node. However, following Miller's approximation as in Example 11-10, we can decompose C_{XY} into two grounded components (Fig. 11.32):

$$C_X = (1 + g_m R_L)\, C_{XY} \tag{11.59}$$

$$C_Y = \left(1 + \frac{1}{g_m R_L}\right) C_{XY}. \tag{11.60}$$

Now, each node sees a resistance and capacitances only to ground. In accordance with our notations in Section 11.1, we write

$$|\omega_{p,in}| = \frac{1}{R_{Thev}[C_{in} + (1 + g_m R_L)C_{XY}]} \tag{11.61}$$

$$|\omega_{p,out}| = \frac{1}{R_L\left[C_{out} + \left(1 + \frac{1}{g_m R_L}\right) C_{XY}\right]}. \tag{11.62}$$

If $g_m R_L \gg 1$, the capacitance at the output node is simply equal to $C_{out} + C_{XY}$.

The intuition gained from the application of Miller's theorem proves invaluable. The input pole is approximately given by the source resistance, the base-emitter or gate-source capacitance, and the *Miller multiplication* of the base-collector or gate-drain capacitance.

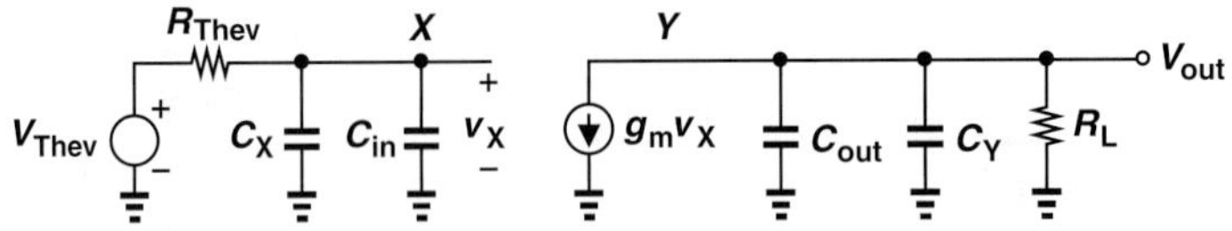

CE Stage	CS Stage
$V_{Thev} = V_{in}\dfrac{r_\pi}{r_\pi + R_S}$	$V_{Thev} = V_{in}$
$R_{Thev} = R_S \Vert r_\pi$	$R_{Thev} = R_S$
$C_X = C_\mu\,(1 + g_m R_L)$	$C_X = C_{GD}\,(1 + g_m R_L)$
$C_Y = C_\mu\,(1 + \dfrac{1}{g_m R_L})$	$C_Y = C_{GD}\,(1 + \dfrac{1}{g_m R_L})$

Figure 11.32 Parameters in unified model of CE and CS stages with Miller's approximation.

The Miller multiplication makes it undesirable to have a high gain in the circuit. The output pole is roughly determined by the load resistance, the collector-substrate or drain-bulk capacitance, and the base-collector or gate-drain capacitance.

Example 11-15 In the CE stage of Fig. 11.31(a), $R_S = 200\ \Omega$, $I_C = 1$ mA, $\beta = 100$, $C_\pi = 100$ fF, $C_\mu =$ 20 fF, and $C_{CS} = 30$ fF.

(a) Calculate the input and output poles if $R_L = 2\ \text{k}\Omega$. Which node appears as the speed bottleneck (limits the bandwidth)?

(b) Is it possible to choose R_L such that the output pole limits the bandwidth?

Solution (a) Since $r_\pi = 2.6\ \text{k}\Omega$, we have $R_{Thev} = 186\ \Omega$. Fig. 11.32 and Eqs. (11.61) and (11.62) thus give

$$|\omega_{p,in}| = 2\pi \times (516\ \text{MHz}) \tag{11.63}$$

$$|\omega_{p,out}| = 2\pi \times (1.59\ \text{GHz}). \tag{11.64}$$

We observe that the Miller effect multiplies C_μ by a factor of 78, making its contribution much greater than that of C_π. As a result, the input pole limits the bandwidth.

(b) We must seek such a value of R_L that yields $|\omega_{p,in}| > \omega_{p,out}|$:

$$\frac{1}{(R_S \parallel r_\pi)[C_\pi + (1 + g_m R_L)C_\mu]} > \frac{1}{R_L\left[C_{CS} + \left(1 + \dfrac{1}{g_m R_L}\right)C_\mu\right]}. \tag{11.65}$$

If $g_m R_L \gg 1$, then we have

$$[C_{CS} + C_\mu - g_m(R_S \parallel r_\pi)C_\mu]R_L > (R_S \parallel r_\pi)C_\pi. \tag{11.66}$$

With the values assumed in this example, the left-hand side is negative, implying that no solution exists. The reader can prove that this holds even if $g_m R_L$ is not much greater than unity. Thus, the input pole remains the speed bottleneck here.

Exercise Repeat the above example if $I_C = 2$ mA and $C_\pi = 180$ fF.

Example 11-16 An electrical engineering student designs the CS stage of Fig. 11.31(a) for a certain low-frequency gain and high-frequency response. Unfortunately, in the layout phase, the student uses a MOSFET half as wide as that in the original design. Assuming that the bias current is also halved, determine the gain and the poles of the circuit.

Solution Both the width and the bias current of the transistor are halved, and so is its transconductance (why?). The small-signal gain, g_mR_L, is therefore halved.

Reducing the transistor width by a factor of two also lowers all of the capacitances by the same factor. From Fig. 11.32 and Eqs. (11.61) and (11.62), we can express the poles as

$$|\omega_{p,in}| = \frac{1}{R_S\left[\frac{C_{in}}{2} + \left(1 + \frac{g_mR_L}{2}\right)\frac{C_{XY}}{2}\right]} \tag{11.67}$$

$$|\omega_{p,out}| = \frac{1}{R_L\left[\frac{C_{out}}{2} + \left(1 + \frac{2}{g_mR_L}\right)\frac{C_{XY}}{2}\right]}, \tag{11.68}$$

where C_{in}, g_m, C_{XY}, and C_{out} denote the parameters corresponding to the original device width. We observe that $\omega_{p,in}$ has risen in magnitude by more than a factor of two, and $\omega_{p,out}$ by approximately a factor of two (if $g_mR_L \gg 2$). In other words, the gain is halved and the bandwidth is roughly doubled, suggesting that the gain-bandwidth product is approximately constant.

Exercise What happens if both the width and the bias current are twice their nominal values?

11.4.4 Direct Analysis

The use of Miller's theorem in the previous section provides a quick and intuitive perspective on the performance. However, we must carry out a more accurate analysis so as to understand the limitations of Miller's approximation in this case.

The circuit of Fig. 11.31(d) contains two nodes and can therefore be solved by writing two KCLs. That is,[11]

$$\text{At Node } X: (V_{out} - V_X)C_{XY}s = V_XC_{in}s + \frac{V_X - V_{Thev}}{R_{Thev}} \tag{11.69}$$

$$\text{At Node } Y: (V_X - V_{out})C_{XY}s = g_mV_X + V_{out}\left(\frac{1}{R_L} + C_{out}s\right). \tag{11.70}$$

We compute V_X from Eq. (11.70):

$$V_X = V_{out}\frac{C_{XY}s + \frac{1}{R_L} + C_{out}s}{C_{XY}s - g_m} \tag{11.71}$$

[11] Recall that we denote frequency-domain quantities with upper-case letters.

and substitute the result in Eq. (11.69) to arrive at

$$V_{out}C_{XY}s - \left(C_{XY}s + C_{in}s + \frac{1}{R_{Thev}}\right)\frac{C_{XY}s + \frac{1}{R_L} + C_{out}s}{C_{XY}s - g_m}V_{out} = \frac{-V_{Thev}}{R_{Thev}}. \tag{11.72}$$

It follows that

$$\frac{V_{out}}{V_{Thev}}(s) = \frac{(C_{XY}s - g_m)R_L}{as^2 + bs + 1}, \tag{11.73}$$

where

$$a = R_{Thev}R_L(C_{in}C_{XY} + C_{out}C_{XY} + C_{in}C_{out}) \tag{11.74}$$

$$b = (1 + g_m R_L)C_{XY}R_{Thev} + R_{Thev}C_{in} + R_L(C_{XY} + C_{out}). \tag{11.75}$$

Note from Fig. 11.32 that for a CE stage, Eq. (11.73) must be multiplied by $r_\pi/(R_S + r_\pi)$ to obtain V_{out}/V_{in}—without affecting the location of the poles and the zero.

Let us examine the above results carefully. The transfer function exhibits a zero at

$$\omega_z = \frac{g_m}{C_{XY}}. \tag{11.76}$$

(The Miller approximation fails to predict this zero.) Since C_{XY} (i.e., the base-collector or the gate-drain overlap capacitance) is relatively small, the zero typically appears at very high frequencies and hence is unimportant.[12]

As expected, the system contains two poles given by the values of s that force the denominator to zero. We can solve the quadratic $as^2 + bs + 1 = 0$ to determine the poles but the results provide little insight. Instead, we first make an interesting observation in regard to the quadratic denominator: if the poles are given by ω_{p1} and ω_{p2}, we can write

$$as^2 + bs + 1 = \left(\frac{s}{\omega_{p1}} + 1\right)\left(\frac{s}{\omega_{p2}} + 1\right) \tag{11.77}$$

$$= \frac{s^2}{\omega_{p1}\omega_{p2}} + \left(\frac{1}{\omega_{p1}} + \frac{1}{\omega_{p2}}\right)s + 1. \tag{11.78}$$

Now suppose one pole is much farther from the origin than the other: $\omega_{p2} \gg \omega_{p1}$. (This is called the "dominant pole" approximation to emphasize that ω_{p1} dominates the frequency response.) Then, $\omega_{p1}^{-1} + \omega_{p2}^{-1} \approx \omega_{p1}^{-1}$, i.e.,

$$b = \frac{1}{\omega_{p1}}, \tag{11.79}$$

and from Eq. (11.75),

$$|\omega_{p1}| = \frac{1}{(1 + g_m R_L)C_{XY}R_{Thev} + R_{Thev}C_{in} + R_L(C_{XY} + C_{out})}. \tag{11.80}$$

How does this result compare with that obtained using the Miller approximation? Equation (11.80) does reveal the Miller effect of C_{XY} but it also contains the additional term $R_L(C_{XY} + C_{out})$ [which is close to the output time constant predicted by Eq. (11.62)].

[12] As explained in more advanced courses, this zero does become problematic in the internal circuitry of op amps.

To determine the "nondominant" pole, ω_{p2}, we recognize from Eqs. (11.78) and (11.79) that

$$|\omega_{p2}| = \frac{b}{a} \tag{11.81}$$

$$= \frac{(1 + g_m R_L)C_{XY}R_{Thev} + R_{Thev}C_{in} + R_L(C_{XY} + C_{out})}{R_{Thev}R_L(C_{in}C_{XY} + C_{out}C_{XY} + C_{in}C_{out})}. \tag{11.82}$$

Example 11-17 Using the dominant-pole approximation, compute the poles of the circuit shown in Fig. 11.33(a). Assume both transistors operate in saturation and $\lambda \neq 0$.

Solution Noting that C_{SB1}, C_{GS2}, and C_{SB2} do not affect the circuit (why?), we add the remaining capacitances as depicted in Fig. 11.33(b), simplifying the result as illustrated in Fig. 11.33(c), where

$$C_{in} = C_{GS1} \tag{11.83}$$

$$C_{XY} = C_{GD1} \tag{11.84}$$

$$C_{out} = C_{DB1} + C_{GD2} + C_{DB2}. \tag{11.85}$$

It follows from Eqs. (11.80) and (11.82) that

$$\omega_{p1} \approx \frac{1}{[1 + g_{m1}(r_{O1} \parallel r_{O2})]C_{XY}R_S + R_SC_{in} + (r_{O1} \parallel r_{O2})(C_{XY} + C_{out})} \tag{11.86}$$

$$\omega_{p2} \approx \frac{[1 + g_{m1}(r_{O1} \parallel r_{O2})]C_{XY}R_S + R_SC_{in} + (r_{O1} \parallel r_{O2})(C_{XY} + C_{out})}{R_S(r_{O1} \parallel r_{O2})(C_{in}C_{XY} + C_{out}C_{XY} + C_{in}C_{out})}. \tag{11.87}$$

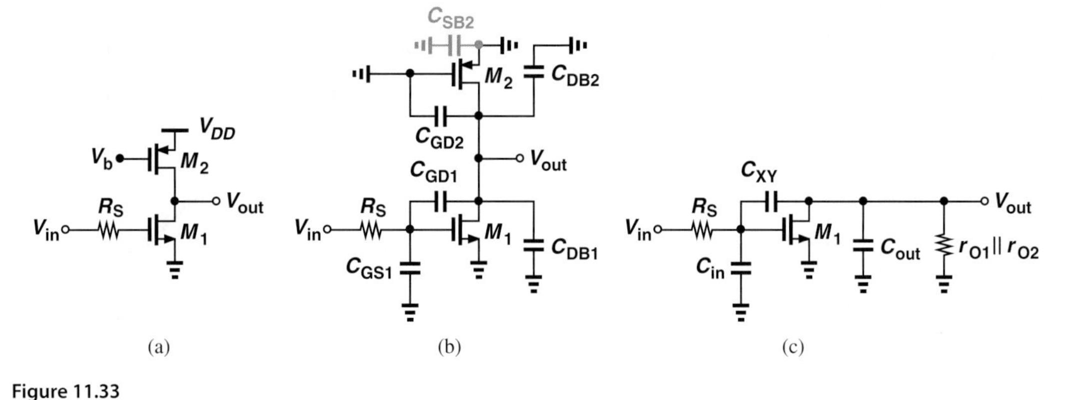

Figure 11.33

Exercise Repeat the above example if $\lambda \neq 0$.

Example 11-18 In the CS stage of Fig. 11.31(a), we have $R_S = 200\ \Omega$, $C_{GS} = 250$ fF, $C_{GD} = 80$ fF, $C_{DB} = 100$ fF, $g_m = (150\ \Omega)^{-1}$, $\lambda = 0$, and $R_L = 2\ \text{k}\Omega$. Plot the frequency response with the aid of (a) Miller's approximation, (b) the exact transfer function, (c) the dominant-pole approximation.

Solution (a) With $g_m R_L = 13.3$, Eqs. (11.61) and (11.62) yield

$$|\omega_{p,in}| = 2\pi \times (571 \text{ MHz}) \tag{11.88}$$

$$|\omega_{p,out}| = 2\pi \times (428 \text{ MHz}). \tag{11.89}$$

(b) The transfer function in Eq. (11.73) gives a zero at $g_m/C_{GD} = 2\pi \times (13.3 \text{ GHz})$. Also, $a = 2.12 \times 10^{-20}$ s^{-2} and $b = 6.39 \times 10^{-10}$ s. Thus,

$$|\omega_{p1}| = 2\pi \times (264 \text{ MHz}) \tag{11.90}$$

$$|\omega_{p2}| = 2\pi \times (4.53 \text{ GHz}). \tag{11.91}$$

Note the large error in the values predicted by Miller's approximation. This error arises because we have multiplied C_{GD} by the midband gain $(1 + g_m R_L)$ rather than the gain at high frequencies.[13]

(c) The results obtained in part (b) predict that the dominant-pole approximation produces relatively accurate results as the two poles are quite far apart. From Eqs. (11.80) and (11.82), we have

$$|\omega_{p1}| = 2\pi \times (249 \text{ MHz}) \tag{11.92}$$

$$|\omega_{p2}| = 2\pi \times (4.79 \text{ GHz}). \tag{11.93}$$

Figure 11.34 plots the results. The low-frequency gain is equal to 22 dB ≈ 13 and the −3 dB bandwidth predicted by the exact equation is around 250 MHz.

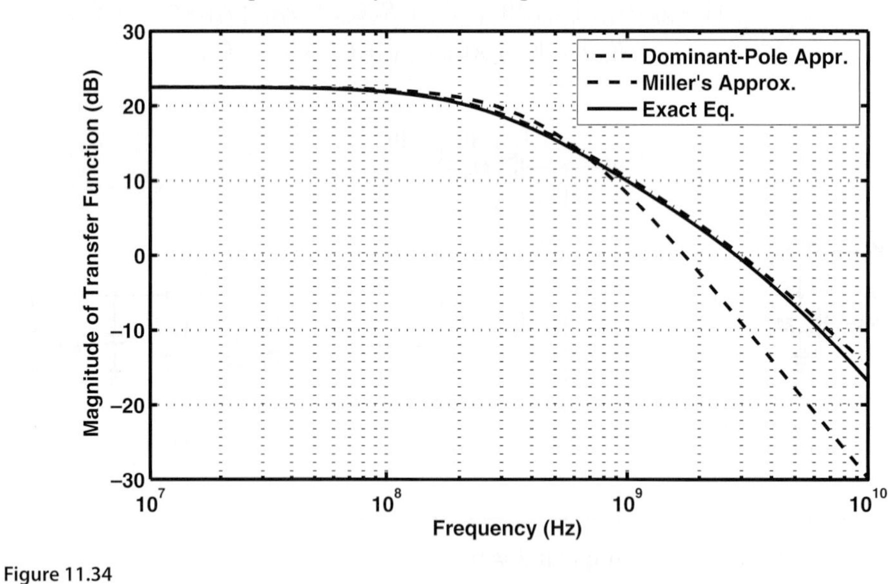

Figure 11.34

Exercise Repeat the above example if the device width (and hence its capacitances) and the bias current are halved.

[13]The large discrepancy between $|\omega_{p,out}|$ and $|\omega_{p2}|$ results from an effect called "pole splitting" and is studied in more advanced courses.

11.4.5 Input Impedance

The high-frequency input impedances of the CE and CS amplifiers determine the ease with which these circuits can be driven by other stages. Our foregoing analysis of the frequency response and particularly the Miller approximation readily yields this impedance.

As illustrated in Fig. 11.35(a), the input impedance of a CE stage consists of two parallel components: $C_\pi + (1 + g_m R_D)C_\mu$ and r_π.[14] That is,

$$Z_{in} \approx \frac{1}{[C_\pi + (1 + g_m R_D)C_\mu]s} \parallel r_\pi. \tag{11.94}$$

Similarly, the MOS counterpart exhibits an input impedance given by

$$Z_{in} \approx \frac{1}{[C_{GS} + (1 + g_m R_D)C_{GD}]s}. \tag{11.95}$$

With a high voltage gain, the Miller effect may substantially lower the input impedance at high frequencies.

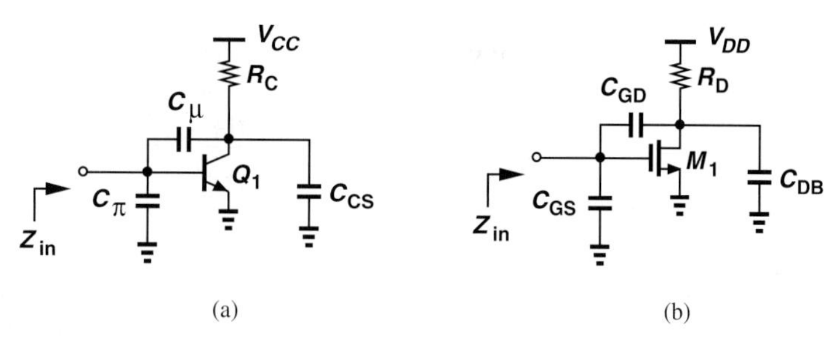

Figure 11.35 Input impedance of (a) CE and (b) CS stages.

Did you know?

Most RF receivers incorporate a common-source or common-emitter amplifier at their front end. This "low-noise" amplifier must present an input resistance of 50 Ω so as to "match" the impedance of the antenna [Fig. (a)]. But how could a CS stage have such a low input resistance? A clever technique is to add an inductor in series with the source of the transistor [Fig. (b)]. It can be shown that the input impedance is given by

$$Z_{in}(s) = \frac{1}{C_{GS}s} + L_1 s + \frac{L_1 g_m}{C_{GS}}.$$

Note that the last term is a real quantity, reprsenting a resistance. Proper choice of L_1, g_m, and C_{GS} provides a value of 50 Ω. Next time you turn on your cell phone or your GPS, you may be receiving an RF signal through an inductively-degenerated CS amplifier.

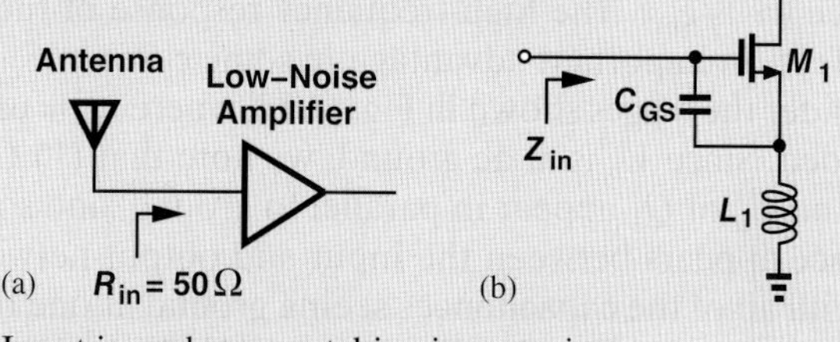

Input impedance matching in a receiver.

[14]In calculation of the input impedance, the output impedance of the preceding stage (denoted by R_S) is excluded.

11.5 FREQUENCY RESPONSE OF CB AND CG STAGES

11.5.1 Low-Frequency Response

As with CE and CS stages, the use of capacitive coupling leads to low-frequency roll-off in CB and CG amplifiers. Consider the CB circuit depicted in Fig. 11.36(a), where I_1 defines the bias current of Q_1 and V_b is chosen to ensure operation in the forward active region (V_b is less than the collector bias voltage). How large should C_i be? Since C_i appears in series with R_S, we replace R_S with $R_S + (C_i s)^{-1}$ in the midband gain expression, $R_C/(R_S + 1/g_m)$, and write the resulting transfer function as

$$\frac{V_{out}}{V_{in}}(s) = \frac{R_C}{R_S + (C_i s)^{-1} + 1/g_m} \tag{11.96}$$

$$= \frac{g_m R_C C_i s}{(1 + g_m R_S)C_i s + g_m}. \tag{11.97}$$

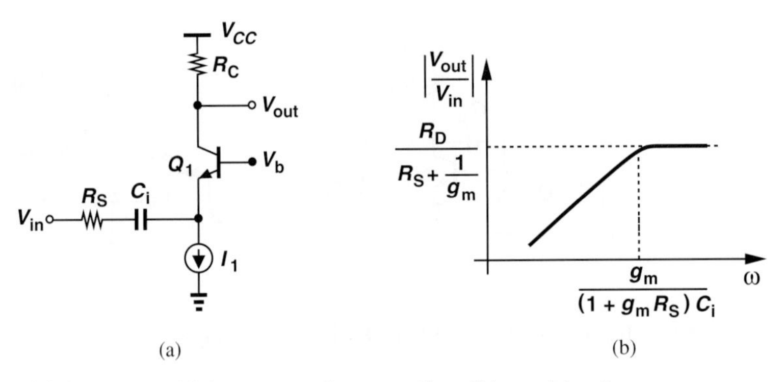

Figure 11.36 (a) CB stage with input capacitor coupling, (b) resulting frequency response.

Equation (11.96) implies that the signal does not "feel" the effect of C_i if $|(C_i s)^{-1}| \ll R_S + 1/g_m$. From another perspective, Eq. (11.97) yields the response shown in Fig. 11.36(b), revealing a pole at

$$|\omega_p| = \frac{g_m}{(1 + g_m R_S)C_i} \tag{11.98}$$

and suggesting that this pole must remain quite lower than the minimum signal frequency of interest. These two conditions are equivalent.

11.5.2 High-Frequency Response

We know from Chapters 5 and 7 that CB and CG stages exhibit a relatively low input impedance ($\approx 1/g_m$). The high-frequency response of these circuits does not suffer from Miller effect, an important advantage in some cases.

Consider the stages shown in Fig. 11.37, where $r_O = \infty$ and the transistor capacitances are included. Since V_b is at ac ground, we note that (1) C_π and $C_{GS} + C_{SB}$ go to ground; (2) C_{CS} and C_μ of Q_1 appear in parallel to ground, and so do C_{GD} and C_{DB} of M_1; (3) no capacitance appears between the input and output networks, avoiding the Miller effect. In fact, with all of the capacitances seeing ground at one of their terminals, we can readily associate one pole with each node. At node X, the total resistance seen to ground is given by $R_S \parallel (1/g_m)$, yielding

$$|\omega_{p,X}| = \frac{1}{\left(R_S \parallel \frac{1}{g_m}\right) C_X}, \tag{11.99}$$

where $C_X = C_\pi$ or $C_{GS} + C_{SB}$. Similarly, at Y,

$$|\omega_{p,Y}| = \frac{1}{R_L C_Y}, \tag{11.100}$$

where $C_Y = C_\mu + C_{CS}$ or $C_{GD} + C_{DB}$.

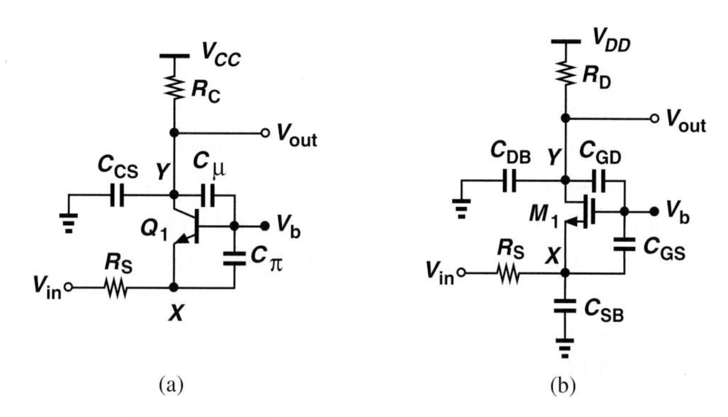

Figure 11.37 (a) CB and (b) CG stages including transistor capacitances.

It is interesting to note that the "input" pole magnitude is on the order of the f_T of the transistor: C_X is equal to C_π or roughly equal to C_{GS} while the resistance seen to ground is less than $1/g_m$. For this reason, the input pole of the CB/CG stage rarely creates a speed bottleneck.[15]

Example 11-19 Compute the poles of the circuit shown in Fig. 11.38(a). Assume $\lambda = 0$.

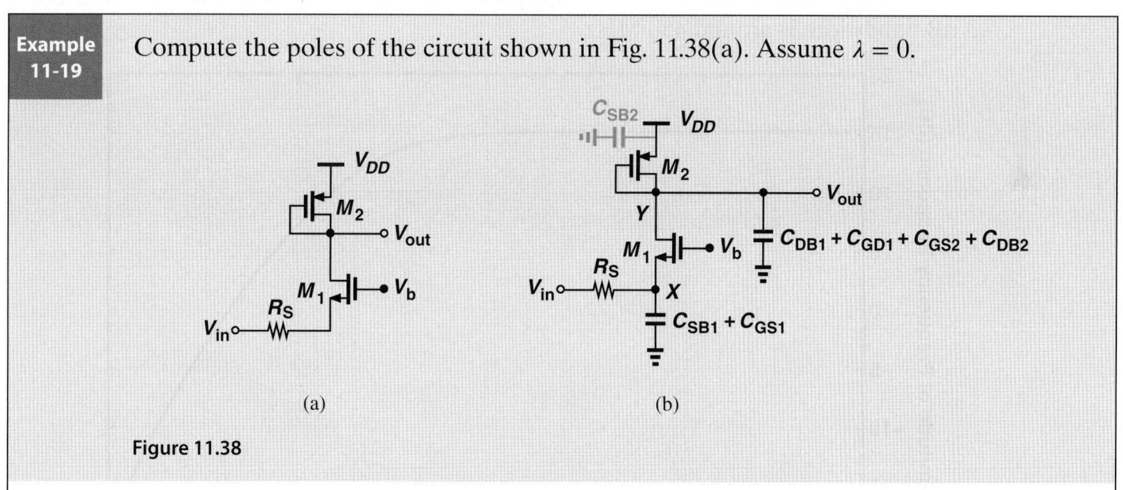

Figure 11.38

Solution Noting that C_{GD2} and C_{SB2} play no role in the circuit, we add the device capacitances as depicted in Fig. 11.38(b). The input pole is thus given by

$$|\omega_{p,X}| = \frac{1}{\left(R_S \left\| \frac{1}{g_{m1}} \right.\right)(C_{SB1} + C_{GD1})}. \tag{11.101}$$

[15]One exception is encountered in radio-frequency circuits (e.g., cellphones), where the input capacitance becomes undesirable.

Since the small-signal resistance at the output node is equal to $1/g_{m2}$, we have

$$|\omega_{p,Y}| = \frac{1}{\frac{1}{g_{m2}}(C_{DB1} + C_{GD1} + C_{GS2} + C_{DB2})}. \quad (11.102)$$

Exercise Repeat the above example if M_2 operates as a current source, i.e., its gate is connected to a constant voltage.

Example 11-20 The CS stage of Example 11-18 is reconfigured to a common-gate amplifier (with R_S tied to the source of the transistor). Plot the frequency response of the circuit.

Solution With the values given in Example 11-18 and noting that $C_{SB} = C_{DB}$,[16] we obtain from Eqs. (11.99) and (11.100),

$$|\omega_{p,in}| = 2\pi \times (5.31 \text{ GHz}) \quad (11.103)$$

$$|\omega_{p,out}| = 2\pi \times (442 \text{ MHz}). \quad (11.104)$$

With no Miller effect, the input pole has dramatically risen in magnitude. The output pole, however, limits the bandwidth. Also, the low-frequency gain is now equal to $R_D/(R_S + 1/g_m) = 5.7$, more than a factor of two lower than that of the CS stage. Figure 11.39 plots the result. The low-frequency gain is equal to 15 dB $\approx$ 5.7 and the −3 dB bandwidth is around 450 MHz.

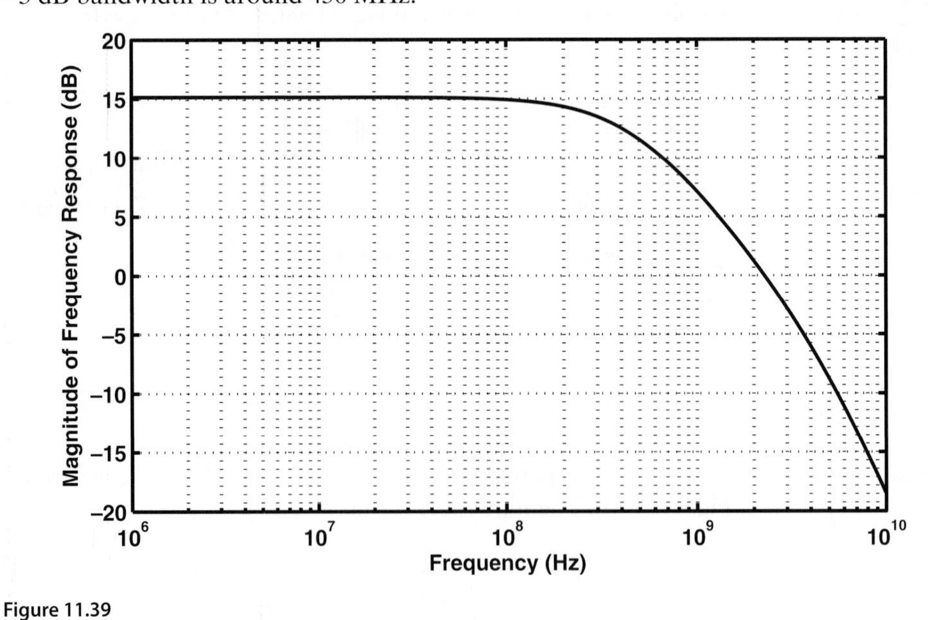

Figure 11.39

Exercise Repeat the above example if the CG amplifier drives a load capacitance of 150 fF.

[16] In reality, the junction capacitances C_{SB} and C_{DB} sustain different reverse bias voltages and are therefore not quite equal.

11.6 FREQUENCY RESPONSE OF FOLLOWERS

The low-frequency response of followers is similar to that studied in Example 11-11 and that of CE/CS stages. We thus study the high-frequency behavior here.

In Chapters 5 and 7, we noted that emitter and source followers provide a high input impedance and a relatively low output impedance while suffering from a sub-unity (positive) voltage gain. Emitter followers, and occasionally source followers, are utilized as buffers and their frequency characteristics are of interest.

Figure 11.40 illustrates the stages with relevant capacitances. The emitter follower is loaded with C_L to create both a more general case and greater similarity between the bipolar and MOS counterparts. We observe that each circuit contains two grounded capacitors and one floating capacitor. While the latter may be decomposed using Miller's approximation, the resulting analysis is beyond the scope of this book. We therefore perform a direct analysis by writing the circuit's equations. Since the bipolar and MOS versions in Fig. 11.40 differ by only r_π, we first analyze the emitter follower and subsequently let r_π (or β) approach infinity to obtain the transfer function of the source follower.

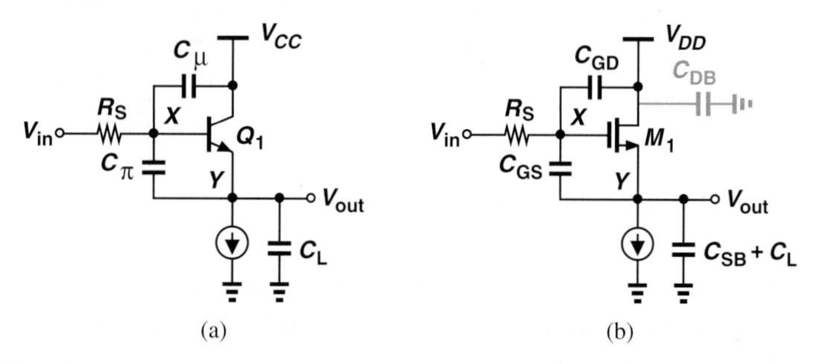

Figure 11.40 (a) Emitter follower and (b) source follower including transistor capacitances.

Consider the small-signal equivalent shown in Fig. 11.41. Recognizing that $V_X = V_{out} + V_\pi$ and the current through the parallel combination of r_π and C_π is given by $V_\pi/r_\pi + V_\pi C_\pi s$, we write a KCL at node X:

$$\frac{V_{out} + V_\pi - V_{in}}{R_S} + (V_{out} + V_\pi)C_\mu s + \frac{V_\pi}{r_\pi} + V_\pi C_\pi s = 0, \tag{11.105}$$

and another at the output node:

$$\frac{V_\pi}{r_\pi} + V_\pi C_\pi s + g_m V_\pi = V_{out} C_L s. \tag{11.106}$$

The latter gives

$$V_\pi = \frac{V_{out} C_L s}{\dfrac{1}{r_\pi} + g_m + C_\pi s}, \tag{11.107}$$

which, upon substitution in Eq. (11.105) and with the assumption $r_\pi \gg g_m^{-1}$, leads to

$$\frac{V_{out}}{V_{in}} = \frac{1 + \dfrac{C_\pi}{g_m} s}{as^2 + bs + 1}, \tag{11.108}$$

where

$$a = \frac{R_S}{g_m}(C_\mu C_\pi + C_\mu C_L + C_\pi C_L) \quad (11.109)$$

$$b = R_S C_\mu + \frac{C_\pi}{g_m} + \left(1 + \frac{R_S}{r_\pi}\right)\frac{C_L}{g_m}. \quad (11.110)$$

The circuit thus exhibits a zero at

$$|\omega_z| = \frac{g_m}{C_\pi}, \quad (11.111)$$

which, from Eq. (11.52), is near the f_T of the transistor. The poles of the circuit can be computed using the dominant-pole approximation described in Section 11.4.4. In practice, however, the two poles do not fall far from each other, necessitating direct solution of the quadratic denominator.

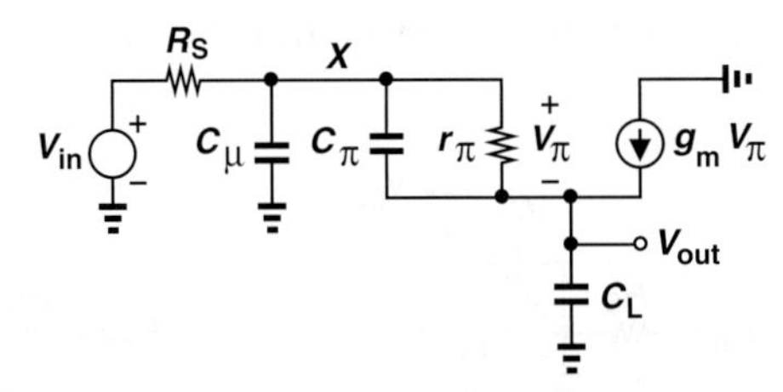

Figure 11.41 Small-signal equivalent of emitter follower.

The above results also apply to the source follower if $r_\pi \to \infty$ and corresponding capacitance substitutions are made (C_{SB} and C_L are in parallel):

$$\frac{V_{out}}{V_{in}} = \frac{1 + \frac{C_{GS}}{g_m}s}{as^2 + bs + 1}, \quad (11.112)$$

where

$$a = \frac{R_S}{g_m}[C_{GD}C_{GS} + C_{GD}(C_{SB} + C_L) + C_{GS}(C_{SB} + C_L)] \quad (11.113)$$

$$b = R_S C_{GD} + \frac{C_{GS} + C_{SB} + C_L}{g_m}. \quad (11.114)$$

Example 11-21 A source follower is driven by a resistance of 200 Ω and drives a load capacitance of 100 fF. Using the transistor parameters given in Example 11-18, plot the frequency response of the circuit.

Solution The zero occurs at $g_m/C_{GS} = 2\pi \times (4.24\ \text{GHz})$. To compute the poles, we obtain a and b from Eqs. (11.113) and (11.114), respectively:

$$a = 2.58 \times 10^{-21}\ \text{s}^{-2} \quad (11.115)$$

$$b = 5.8 \times 10^{-11}\ \text{s}. \quad (11.116)$$

The two poles are then equal to

$$\omega_{p1} = 2\pi[-1.79 \text{ GHz} + j(2.57 \text{ GHz})] \tag{11.117}$$

$$\omega_{p2} = 2\pi[-1.79 \text{ GHz} - j(2.57 \text{ GHz})]. \tag{11.118}$$

With the values chosen here, the poles are complex. Figure 11.42 plots the frequency response. The −3 dB bandwidth is approximately equal to 3.5 GHz.

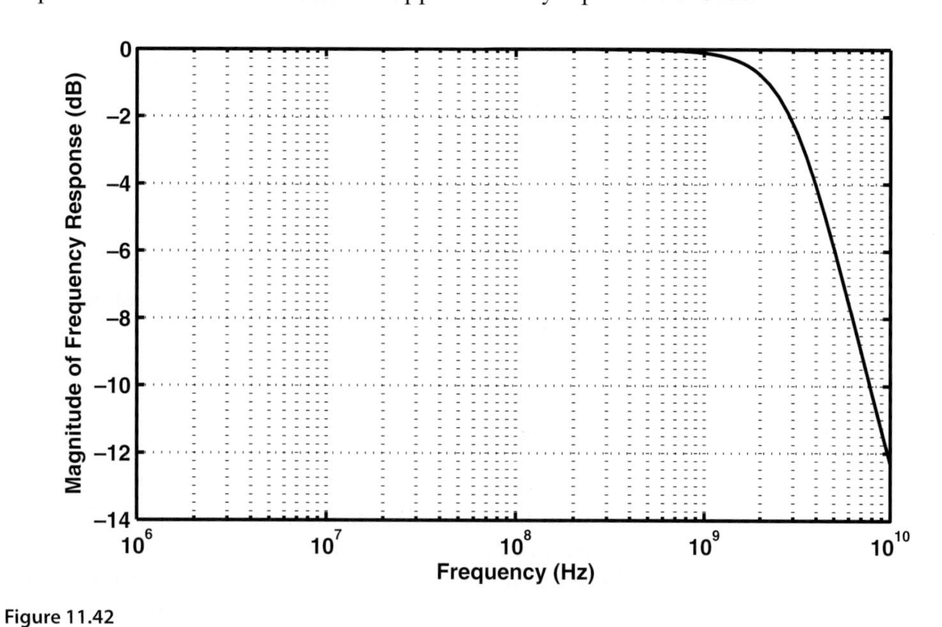

Figure 11.42

Exercise For what value of g_m do the two poles become real and equal?

Example 11-22 Determine the transfer function of the source follower shown in Fig. 11.43(a), where M_2 acts as a current source.

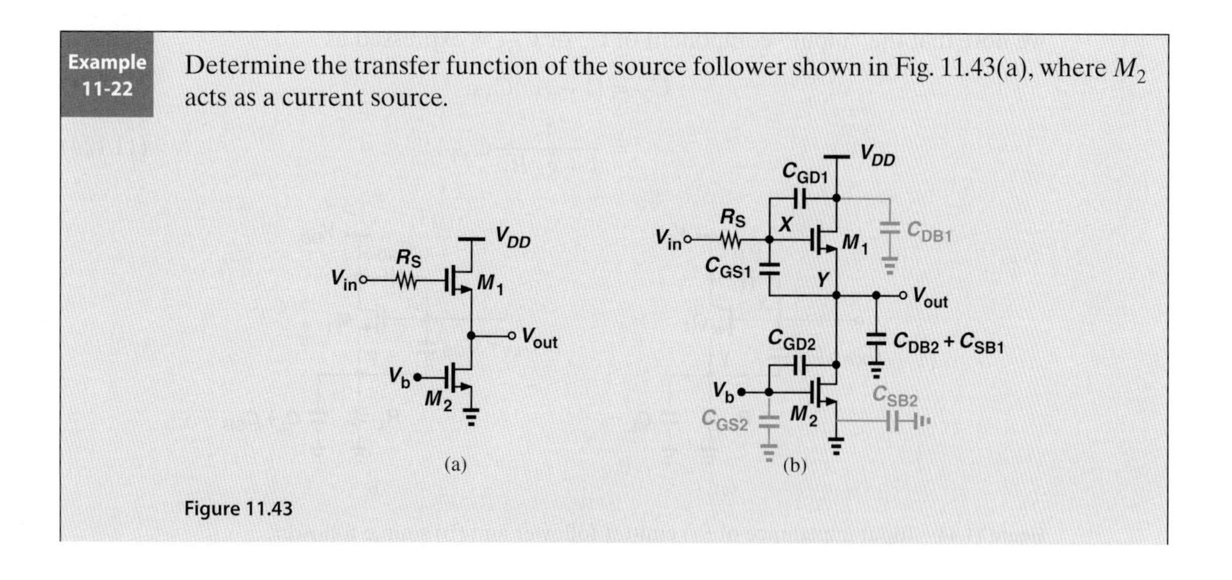

Figure 11.43

Solution Noting that C_{GS2} and C_{SB2} play no role in the circuit, we include the transistor capacitances as illustrated in Fig. 11.43(b). The result resembles that in Fig. 11.40, but with C_{GD2} and C_{DB2} appearing in parallel with C_{SB1}. Thus, Eq. (11.112) can be rewritten as

$$\frac{V_{out}}{V_{in}}(s) = \frac{1 + \frac{C_{GS1}}{g_{m1}}s}{as^2 + bs + 1}, \tag{11.119}$$

where

$$a = \frac{R_S}{g_{m1}}[C_{GD1}C_{GS1} + (C_{GD1} + C_{GS1})(C_{SB1} + C_{GD2} + C_{DB2})] \tag{11.120}$$

$$b = R_S C_{GD1} + \frac{C_{GD1} + C_{SB1} + C_{GD2} + C_{DB2}}{g_{m1}}. \tag{11.121}$$

Exercise Assuming M_1 and M_2 are identical and using the transistor parameters given in Example 11-18, calculate the pole frequencies.

11.6.1 Input and Output Impedances

In Chapter 5, we observed that the input resistance of the emitter follower is given by $r_\pi + (\beta + 1)R_L$, where R_L denotes the load resistance. Also, in Chapter 7, we noted that the source follower input resistance approaches infinity at low frequencies. We now employ an approximate but intuitive analysis to obtain the input capacitance of followers.

Consider the circuits shown in Fig. 11.44, where C_π and C_{GS} appear between the input and output and can therefore be decomposed using Miller's theorem. Since the low-frequency gain is equal to

$$A_v = \frac{R_L}{R_L + \frac{1}{g_m}}, \tag{11.122}$$

we note that the "input" component of C_π or C_{GS} is expressed as

$$C_X = (1 - A_v)C_{XY} \tag{11.123}$$

$$= \frac{1}{1 + g_m R_L}C_{XY}. \tag{11.124}$$

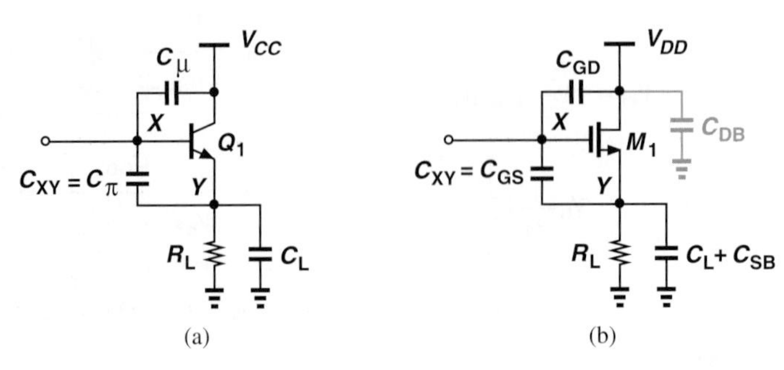

Figure 11.44 Input impedance of (a) emitter follower and (b) source follower.

Interestingly, the input capacitance of the follower contains only a *fraction* of C_π or C_{GS}, depending on how large $g_m R_L$ is. Of course, C_μ or C_{GD} directly adds to this value to yield the total input capacitance.

Example 11-23 Estimate the input capacitance of the follower shown in Fig. 11.45. Assume $\lambda \neq 0$.

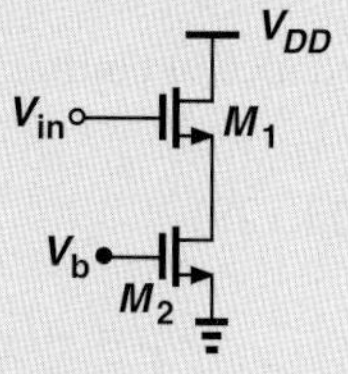

V_{DD}, V_{in}, M_1, V_b, M_2

Figure 11.45

Solution From Chapter 7, the low-frequency gain of the circuit can be written as

$$A_v = \frac{r_{O1} \parallel r_{O2}}{r_{O1} \parallel r_{O2} + \frac{1}{g_{m1}}}. \tag{11.125}$$

Also, from Fig. 11.44(b), the capacitance appearing between the input and output is equal to C_{GS1}, thereby providing

$$C_{in} = C_{GD1} + (1 - A_v)C_{GS1} \tag{11.126}$$

$$= C_{GD1} + \frac{1}{1 + g_{m1}(r_{O1} \parallel r_{O2})} C_{GS1}. \tag{11.127}$$

For example, if $g_{m1}(r_{O1} \parallel r_{O2}) \approx 10$, then only 9% of C_{GS1} appears at the input.

Exercise Repeat the above example if $\lambda = 0$.

Let us now turn our attention to the output impedance of followers. Our study of the emitter follower in Chapter 5 revealed that the output resistance is equal to $R_S/(\beta + 1) + 1/g_m$. Similarly, Chapter 7 indicated an output resistance of $1/g_m$ for the source follower. At high frequencies, these circuits display an interesting behavior.

Consider the followers depicted in Fig. 11.46(a), where other capacitances and resistances are neglected for the sake of simplicity. As usual, R_S represents the output resistance of a preceding stage or device. We first compute the output impedance of the emitter follower and subsequently let $r_\pi \to \infty$ to determine that of the source follower. From the equivalent circuit in Fig. 11.46(b), we have

$$(I_X + g_m V_\pi)\left(r_\pi \left\| \frac{1}{C_\pi s}\right.\right) = -V_\pi \tag{11.128}$$

and also

$$(I_X + g_m V_\pi)R_S - V_\pi = V_X. \tag{11.129}$$

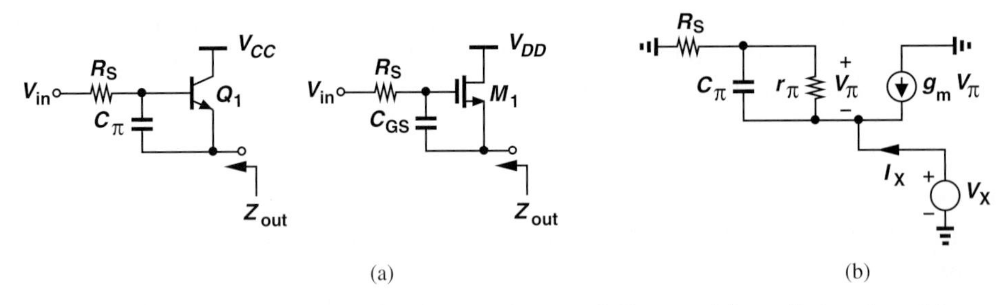

Figure 11.46 (a) Output impedance of emitter and source followers, (b) small-signal model.

Finding V_π from Eq. (11.128)

$$V_\pi = -I_X \frac{r_\pi}{r_\pi C_\pi s + \beta + 1} \tag{11.130}$$

and substituting in Eq. (11.129), we obtain

$$\frac{V_X}{I_X} = \frac{R_S r_\pi C_\pi s + r_\pi + R_S}{r_\pi C_\pi s + \beta + 1}. \tag{11.131}$$

As expected, at low frequencies $V_X/I_X = (r_\pi + R_S)/(\beta + 1) \approx 1/g_m + R_S/(\beta + 1)$. On the other hand, at very high frequencies, $V_X/I_X = R_S$, a meaningful result considering that C_π becomes a short circuit.

The two extreme values calculated above for the output impedance of the emitter follower can be used to develop greater insight. Plotted in Fig. 11.47, the magnitude of this impedance *falls* with ω if $R_S < 1/g_m + R_S/(\beta + 1)$ or *rises* with ω if $R_S > 1/g_m + R_S/(\beta + 1)$. In analogy with the impedance of capacitors and inductors, we say Z_{out} exhibits a capacitive behavior in the former case and an inductive behavior in the latter.

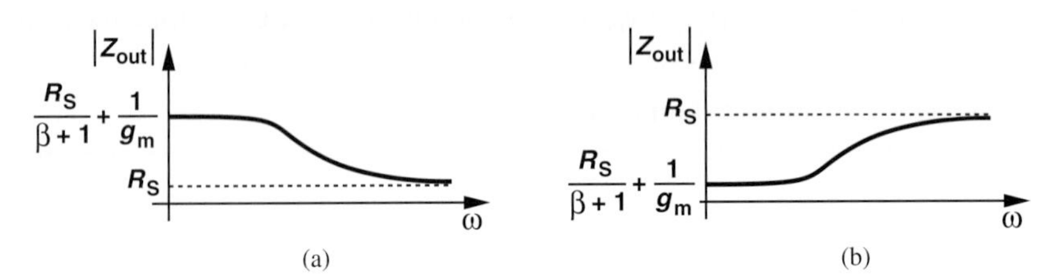

Figure 11.47 Output impedance of emitter follower as a function of frequency for (a) small R_S and (b) large R_S.

Which case is more likely to occur in practice? Since a follower serves to *reduce* the driving impedance, it is reasonable to assume that the follower low-frequency output impedance is *lower* than R_S.[17] Thus, the inductive behavior is more commonly encountered. (It is even possible that the inductive output impedance leads to oscillation if the follower sees a certain amount of load capacitance.)

The above development can be extended to source followers by factoring r_π from the numerator and denominator of Eq. (11.131) and letting r_π and β approach infinity:

$$\frac{V_X}{I_X} = \frac{R_S C_{GS} s + 1}{C_{GS} s + g_m}, \tag{11.132}$$

[17] If the follower output resistance is *greater* than R_S, then it is better to omit the follower!

where $(\beta + 1)/r_\pi$ is replaced with g_m, and C_π with C_{GD}. The plots of Fig. 11.47 are redrawn for the source follower in Fig. 11.48, displaying a similar behavior.

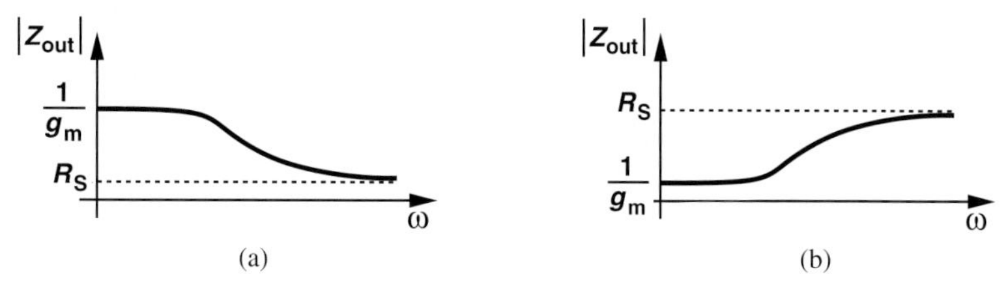

Figure 11.48 Output impedance of source follower as a function of frequency for (a) small R_S and (b) large R_S.

The inductive impedance seen at the output of followers proves useful in the realization of "active inductors."

Example 11-24 Figure 11.49 depicts a two-stage amplifier consisting of a CS circuit and a source follower. Assuming $\lambda \neq 0$ for M_1 and M_2 but $\lambda = 0$ for M_3, and neglecting all capacitances except C_{GS3}, compute the output impedance of the amplifier.

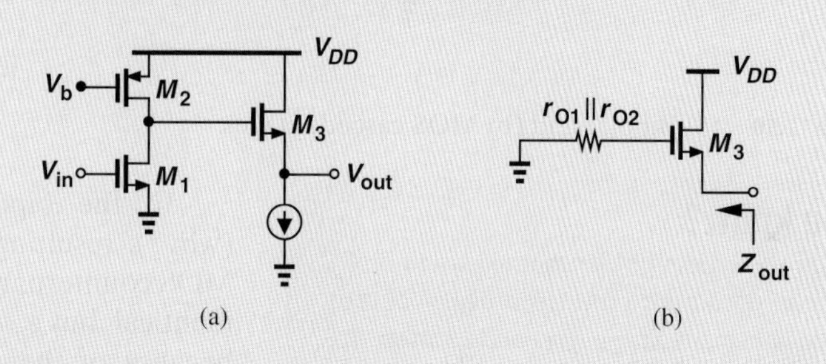

Figure 11.49

Solution The source impedance seen by the follower is equal to the output resistance of the CS stage, which is equal to $r_{O1} \parallel r_{O2}$. Assuming $R_S = r_{O1} \parallel r_{O2}$ in Eq. (11.132), we have

$$\frac{V_X}{I_X} = \frac{(r_{O1} \parallel r_{O2})C_{GS3}s + 1}{C_{GS3}s + g_{m3}}. \tag{11.133}$$

Exercise Determine Z_{out} in the above example if $\lambda \neq 0$ for M_1-M_3.

11.7 FREQUENCY RESPONSE OF CASCODE STAGE

Our analysis of the CE/CS stage in Section 11.4 and the CB/CG stage in Section 11.5 reveals that the former provides a relatively high input resistance but suffers from Miller effect whereas the latter exhibits a relatively low input resistance but is free from Miller effect. We wish to combine the desirable properties of the two topologies, obtaining a circuit with a relatively high input resistance and no or little Miller effect. Indeed, this thought process led to the invention of the cascode topology in the 1940s.

Consider the cascodes shown in Fig. 11.50. As mentioned in Chapter 9, this structure can be viewed as a CE/CS transistor, Q_1 or M_1, followed by a CB/CG device, Q_2 or M_2. As such, the circuit still exhibits a relatively high (for Q_1) or infinite (for M_1) input resistance while providing a voltage gain equal to $g_{m1}R_L$.[18] But, how about the Miller multiplication of $C_{\mu1}$ or C_{GD1}? We must first compute the voltage gain from node X to node Y. Assuming $r_O = \infty$ for all transistors, we recognize that the impedance seen at Y is equal to $1/g_{m2}$, yielding a small-signal gain of

$$A_{v,XY} = \frac{v_Y}{v_X} \tag{11.134}$$

$$= -\frac{g_{m1}}{g_{m2}}. \tag{11.135}$$

(a) (b)

Figure 11.50 (a) Bipolar and (b) MOS cascode stages.

In the bipolar cascode, $g_{m1} = g_{m2}$ (why?), resulting in a gain of -1. In the MOS counterpart, M_1 and M_2 need not be identical, but g_{m1} and g_{m2} are comparable because of their relatively weak dependence upon W/L. We therefore say the gain from X to Y remains near -1 in most practical cases, concluding that the Miller effect of $C_{XY} = C_{\mu1}$ or C_{GD1} is given by

$$C_X = (1 - A_{v,XY})C_{XY} \tag{11.136}$$

$$\approx 2C_{XY}. \tag{11.137}$$

This result stands in contrast to that expressed by Eq. (11.59), suggesting that the cascode transistor breaks the trade-off between the gain and the input capacitance due to Miller effect.

Let us continue our analysis and estimate the poles of the cascode topology with the aid of Miller's approximation. Illustrated in Fig. 11.51 is the bipolar

Did you know?

In addition to reducing Miller multiplication of C_μ, the cascode structure provides a higher output impedance, a higher voltage gain, and greater stability. By stability, we mean the tendency of the amplifier not to oscillate. Since C_μ or C_{GD} returns a fraction of the output signal to the input, it can cause instability in high-frequency amplifiers. For this reason, we commonly use the cascode topology at the front end of RF receivers, often with inductive degeneration as shown below.

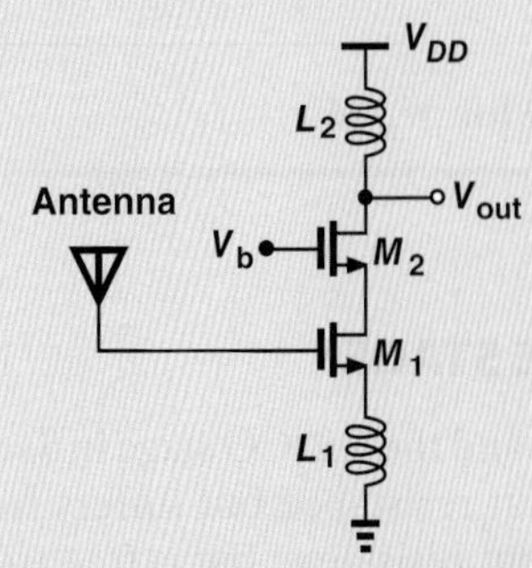

Use of cascode in a cellphone low-noise amplifier.

[18]The voltage division between R_S and $r_{\pi1}$ lowers the gain slightly in the bipolar circuit.

cascode along with the transistor capacitances. Note that the effect of $C_{\mu 1}$ at Y is also equal to $(1 - A_{v,XY}^{-1})C_{\mu 1} = 2C_{\mu 1}$. Associating one pole with each node gives

$$|\omega_{p,X}| = \frac{1}{(R_S \parallel r_{\pi 1})(C_{\pi 1} + 2C_{\mu 1})} \tag{11.138}$$

$$|\omega_{p,Y}| = \frac{1}{\frac{1}{g_{m2}}(C_{CS1} + C_{\pi 2} + 2C_{\mu 1})} \tag{11.139}$$

$$|\omega_{p,out}| = \frac{1}{R_L(C_{CS2} + C_{\mu 2})}. \tag{11.140}$$

It is interesting to note that the pole at node Y falls near the f_T of Q_2 if $C_{\pi 2} \gg C_{CS1} + 2C_{\mu 1}$. Even for comparable values of $C_{\pi 2}$ and $C_{CS1} + 2C_{\mu 1}$, we can say this pole is on the order of $f_T/2$, a frequency typically much higher than the signal bandwidth. For this reason, the pole at node Y often has negligible effect on the frequency response of the cascode stage.

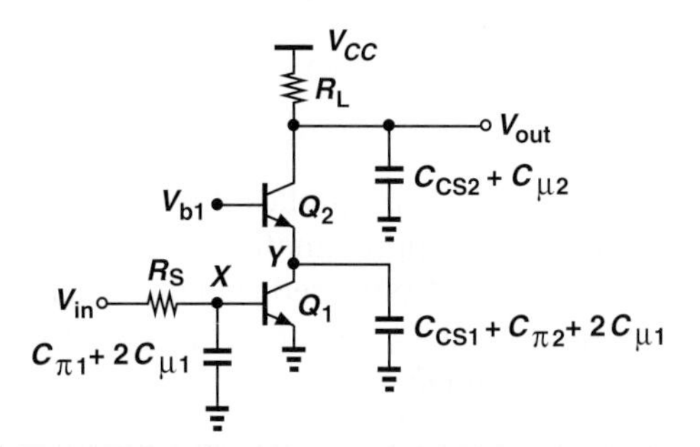

Figure 11.51 Bipolar cascode including transistor capacitances.

The MOS cascode is shown in Fig. 11.52 along with its capacitances after the use of Miller's approximation. Since the gain from X to Y in this case may not be equal to -1, we use the actual value, $-g_{m1}/g_{m2}$, to arrive at a more general solution. Associating one pole with each node, we have

$$|\omega_{p,X}| = \frac{1}{R_S\left[C_{GS1} + \left(1 + \frac{g_{m1}}{g_{m2}}\right)C_{GD1}\right]} \tag{11.141}$$

$$|\omega_{p,Y}| = \frac{1}{\frac{1}{g_{m2}}\left[C_{DB1} + C_{GS2} + \left(1 + \frac{g_{m2}}{g_{m1}}\right)C_{GD1} + C_{SB2}\right]} \tag{11.142}$$

$$|\omega_{p,out}| = \frac{1}{R_L(C_{DB2} + C_{GD2})}. \tag{11.143}$$

We note that $\omega_{p,Y}$ is still in the range of $f_T/2$ if C_{GS2} and $C_{DB1} + (1 + g_{m2}/g_{m1})C_{GD1}$ are comparable.

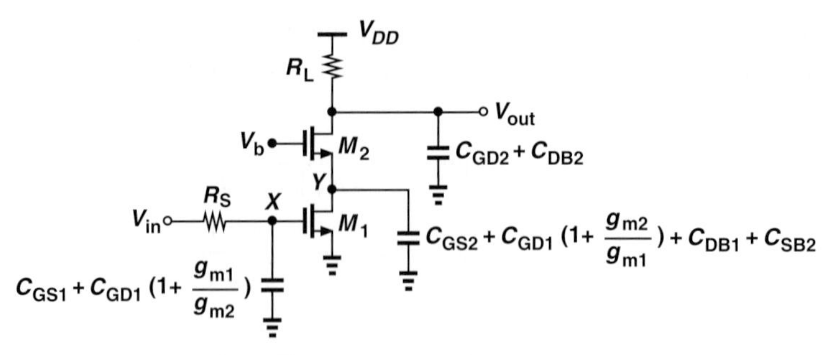

Figure 11.52 MOS cascode including transistor capacitances.

Example 11-25 The CS stage studied in Example 11-18 is converted to a cascode topology. Assuming the two transistors are identical, estimate the poles, plot the frequency response, and compare the results with those of Example 11-18. Assume $C_{DB} = C_{SB}$.

Solution Using the values given in Example 11-18, we write from Eqs. (11.141), (11.142), and (11.143):

$$|\omega_{p,X}| = 2\pi \times (1.95\ \text{GHz}) \tag{11.144}$$

$$|\omega_{p,Y}| = 2\pi \times (1.73\ \text{GHz}) \tag{11.145}$$

$$|\omega_{p,out}| = 2\pi \times (442\ \text{MHz}). \tag{11.146}$$

Note that the pole at node Y is significantly lower than $f_T/2$ in this particular example. Compared with the Miller approximation results obtained in Example 11-18, the input pole has risen considerably. Compared with the exact values derived in that example, the cascode bandwidth (442 MHz) is nearly twice as large. Figure 11.53 plots the frequency response of the cascode stage.

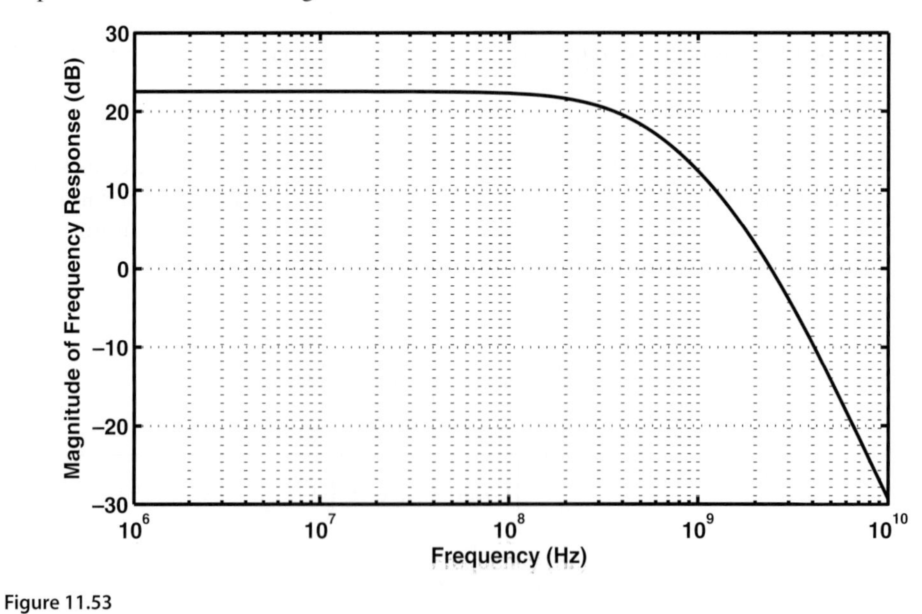

Figure 11.53

Exercise Repeat the above example if the width of M_2 and hence all of its capacitances are doubled. Assume $g_{m2} = (100\ \Omega)^{-1}$.

Example 11-26

In the cascode shown in Fig. 11.54, transistor M_3 serves as a constant current source, allowing M_1 to carry a larger current than M_2. Estimate the poles of the circuit, assuming $\lambda = 0$.

Solution Transistor M_3 contributes C_{GD3} and C_{DB3} to node Y, thus lowering the corresponding pole magnitude. The circuit contains the following poles:

$$|\omega_{p,X}| = \frac{1}{R_S\left[C_{GS1} + \left(1 + \frac{g_{m1}}{g_{m2}}\right)C_{GD1}\right]} \quad (11.147)$$

$$|\omega_{p,Y}| = \frac{1}{\frac{1}{g_{m2}}\left[C_{DB1} + C_{GS2} + \left(1 + \frac{g_{m2}}{g_{m1}}\right)C_{GD1} + C_{GD3} + C_{DB3} + C_{SB2}\right]} \quad (11.148)$$

$$|\omega_{p,out}| = \frac{1}{R_L(C_{DB2} + C_{GD2})}. \quad (11.149)$$

Note that $\omega_{p,X}$ also reduces in magnitude because the addition of M_3 lowers I_{D2} and hence g_{m2}.

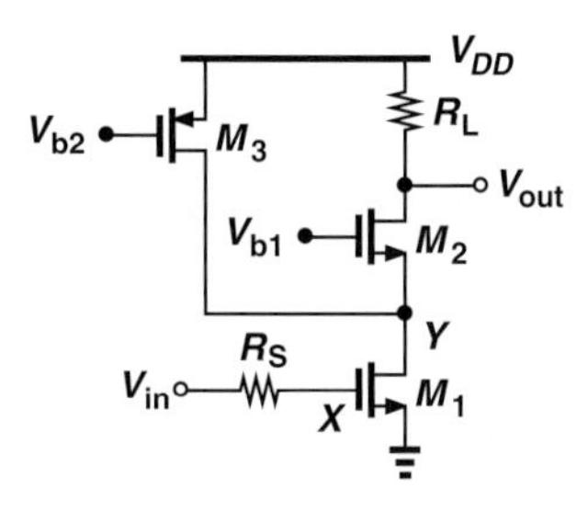

Figure 11.54

Exercise Calculate the pole frequencies in the above example using the transistor parameters given in Example 11-18 for M_1-M_3.

From our studies of the cascode topology in Chapter 9 and in this chapter, we identify two important, distinct attributes of this circuit: (1) the ability to provide a high output impedance and hence serve as a good current source and/or high-gain amplifier; (2) the reduction of the Miller effect and hence better high-frequency performance. Both of these properties are exploited extensively.

11.7.1 Input and Output Impedances

The foregoing analysis of the cascode stage readily provides estimates for the I/O impedances. From Fig. 11.51, the input impedance of the bipolar cascode is given by

$$Z_{in} = r_{\pi 1} \left\| \frac{1}{(C_{\pi 1} + 2C_{\mu 1})s} \right., \quad (11.150)$$

where Z_{in} does not include R_S. The output impedance is equal to

$$Z_{out} = R_L \left\| \frac{1}{(C_{\mu 2} + C_{CS2})s} \right., \quad (11.151)$$

where the Early effect is neglected. Similarly, for the MOS stage shown in Fig. 11.52, we have

$$Z_{in} = \frac{1}{\left[C_{GS1} + \left(1 + \frac{g_{m1}}{g_{m2}}\right) C_{GD1}\right] s} \tag{11.152}$$

$$Z_{out} = R_L \frac{1}{(C_{GD2} + C_{DB2})}, \tag{11.153}$$

where it is assumed $\lambda = 0$.

If R_L is large, the output resistance of the transistors must be taken into account. This calculation is beyond the scope of this book.

11.8 FREQUENCY RESPONSE OF DIFFERENTIAL PAIRS

The half-circuit concept introduced in Chapter 10 can also be applied to the high-frequency model of differential pairs, thus reducing the circuit to those studied above.

Figure 11.55(a) depicts two bipolar and MOS differential pairs along with their capacitances. For small differential inputs, the half circuits can be constructed as shown in Fig. 11.55(b). The transfer function is therefore given by Eq. (11.73):

$$\frac{V_{out}}{V_{Thev}}(s) = \frac{(C_{XY}s - g_m)R_L}{as^2 + bs + 1}, \tag{11.154}$$

where the same notation is used for various parameters. Similarly, the input and output impedances (from each node to ground) are equal to those in Eqs. (11.94) and (11.95), respectively.

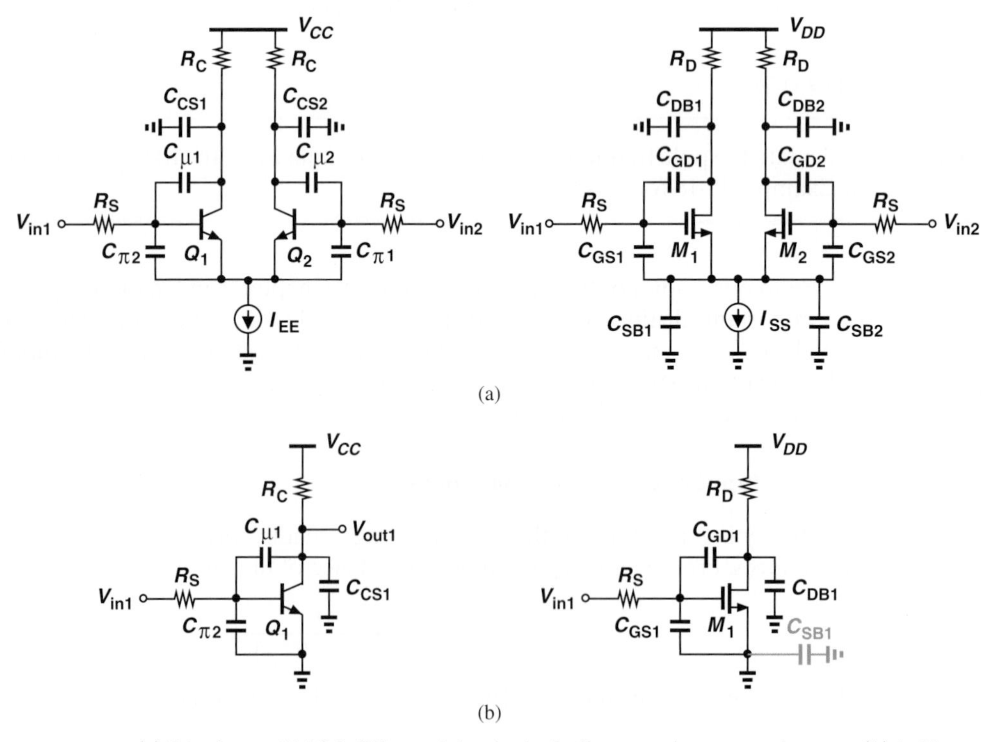

Figure 11.55 (a) Bipolar and MOS differential pairs including transistor capacitances, (b) half circuits.

Example 11-27 A differential pair employs cascode devices to lower the Miller effect [Fig. 11.56(a)]. Estimate the poles of the circuit.

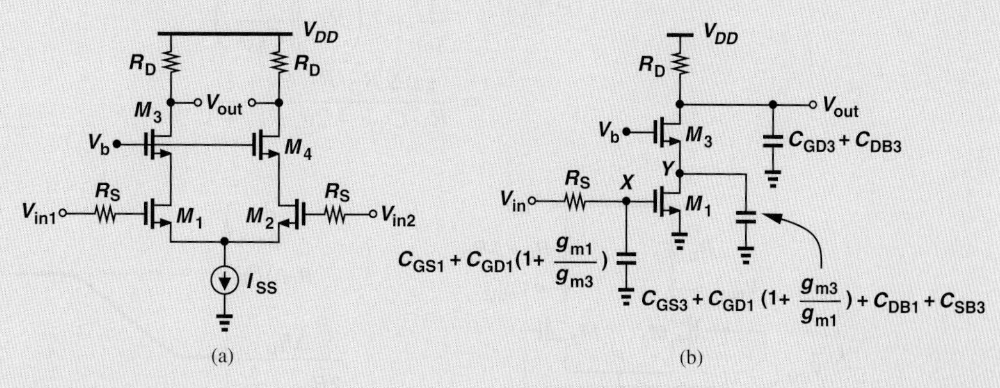

Figure 11.56

Solution Employing the half circuit shown in Fig. 11.56(b), we utilize the results obtained in Section 11.7:

$$|\omega_{p,X}| = \frac{1}{R_S\left[C_{GS1} + \left(1 + \frac{g_{m1}}{g_{m3}}\right)C_{GD1}\right]} \tag{11.155}$$

$$|\omega_{p,Y}| = \frac{1}{\frac{1}{g_{m3}}\left[C_{DB1} + C_{GS3} + \left(1 + \frac{g_{m3}}{g_{m1}}\right)C_{GD1}C_{SB3}\right]} \tag{11.156}$$

$$|\omega_{p,out}| = \frac{1}{R_L(C_{DB3} + C_{GD3})}. \tag{11.157}$$

Exercise Calculate the pole frequencies using the transistor parameters given in Example 11-18. Assume the width and hence the capacitances of M_3 are twice those of M_1. Also, $g_{m3} = \sqrt{2}g_{m1}$.

11.8.1 Common-Mode Frequency Response*

The CM response studied in Chapter 10 included no transistor capacitances. At high frequencies, capacitances may *raise* the CM gain (and lower the differential gain), thus degrading the common-mode rejection ratio.

Let us consider the MOS differential pair shown in Fig. 11.57(a), where a finite capacitance appears between node P and ground. Since C_{SS} shunts R_{SS}, we expect the total impedance between P and ground to fall at high frequencies, leading to a higher CM gain.

*This section can be skipped in a first reading.

In fact, we can simply replace R_{SS} with $R_{SS} \parallel [1/(C_{SS}s)]$ in Eq. (10.191):

$$\left|\frac{\Delta V_{out}}{\Delta V_{CM}}\right| = \frac{\Delta R_D}{\dfrac{1}{g_m} + 2\left(R_{SS} \left\| \dfrac{1}{C_{SS}s}\right.\right)} \tag{11.158}$$

$$= \frac{g_m \Delta R_D(R_{SS}C_S + 1)}{R_{SS}C_{SS}s + 2g_m R_{SS} + 1}. \tag{11.159}$$

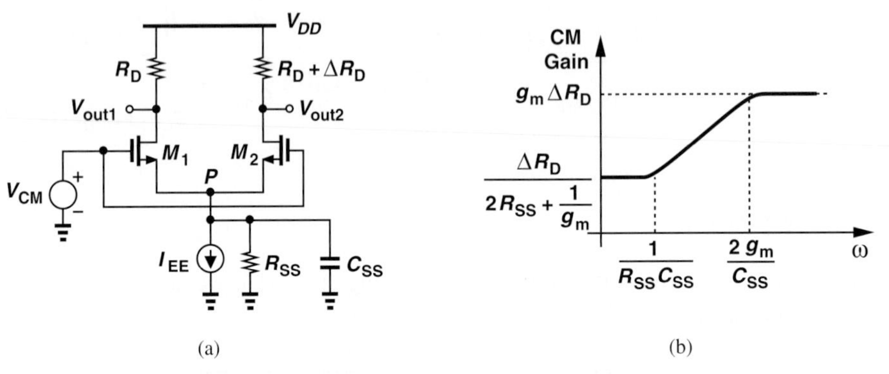

Figure 11.57 (a) Differential pair with parasitic capacitance at the tail node, (b) CM frequency response.

Since R_{SS} is typically quite large, $2g_m R_{SS} \gg 1$, yielding the following zero and pole frequencies:

$$|\omega_z| = \frac{1}{R_{SS}C_{SS}} \tag{11.160}$$

$$|\omega_p| = \frac{2g_m}{C_{SS}}, \tag{11.161}$$

and the Bode approximation plotted in Fig. 11.57(b). The CM gain indeed rises dramatically at high frequencies—by a factor of $2g_m R_{SS}$ (why?).

Figure 11.58 depicts the transistor capacitances that constitute C_{SS}. For example, M_3 is typically a wide device so that it can operate with a small V_{DS}, thereby adding large capacitances to node P.

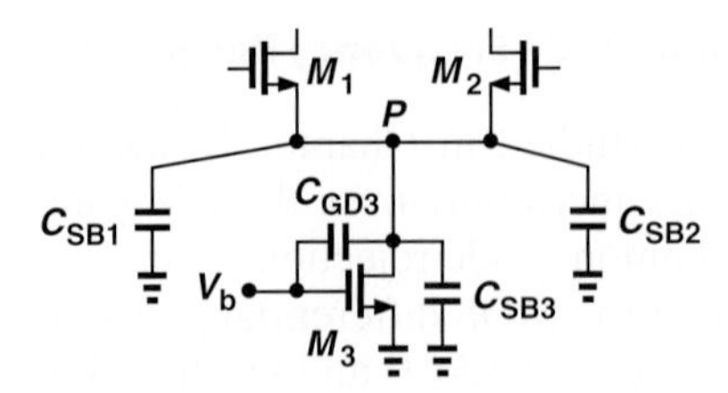

Figure 11.58 Transistor capacitance contributions to the tail node.

11.9 ADDITIONAL EXAMPLES

Example 11-28

The amplifier shown in Fig. 11.59(a) incorporates capacitive coupling both at the input and between the two stages. Determine the low-frequency cut-off of the circuit. Assume $I_S = 5 \times 10^{-16}$ A, $\beta = 100$, and $V_A = \infty$.

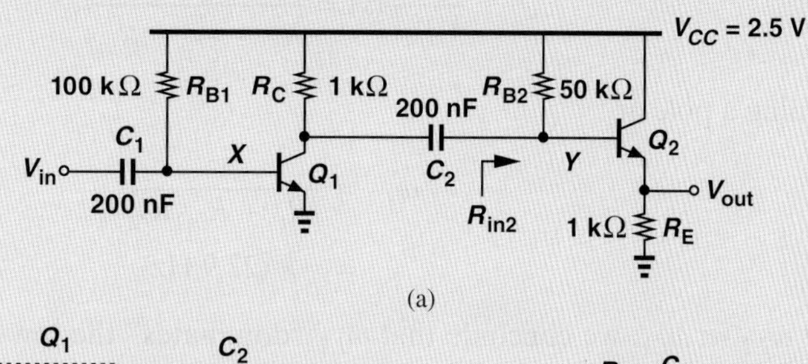

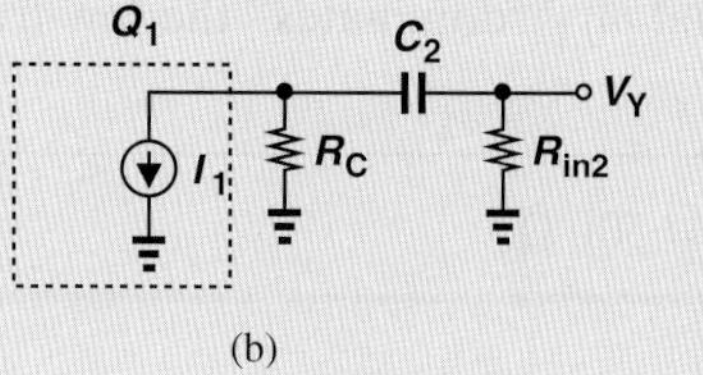

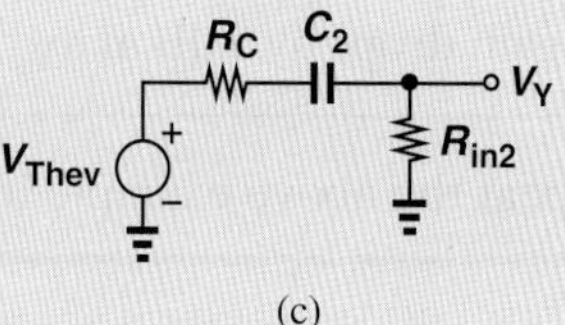

Figure 11.59

Solution We must first compute the operating point and small-signal parameters of the circuit. From Chapter 5, we begin with an estimate of for V_{BE1}, e.g., 800 mV, and express the base current of Q_1 as $(V_{CC} - V_{BE1})/R_{B1}$ and hence

$$I_{C1} = \beta \frac{V_{CC} - V_{BE1}}{R_{B1}} \tag{11.162}$$

$$= 1.7 \text{ mA}. \tag{11.163}$$

It follows that $V_{BE1} = V_T \ln(I_{C1}/I_{S1}) = 748$ mV and $I_{C1} = 1.75$ mA. Thus, $g_{m1} = (14.9\ \Omega)^{-1}$ and $r_{\pi 1} = 1.49$ kΩ. For Q_2, we have

$$V_{CC} = I_{B2}R_{B2} + V_{BE2} + R_E I_{C2}, \tag{11.164}$$

and therefore

$$I_{C2} = \frac{V_{CC} - V_{BE2}}{R_{B2}/\beta + R_E} \tag{11.165}$$

$$= 1.13 \text{ mA}, \tag{11.166}$$

where it is assumed $V_{BE2} \approx 800$ mV. Iteration yields $I_{C2} = 1.17$ mA. Thus, $g_{m2} = (22.2\ \Omega)^{-1}$ and $r_{\pi 2} = 2.22$ kΩ.

Let us now consider the first stage by itself. Capacitor C_1 forms a high-pass filter along with the input resistance of the circuit, R_{in1}, thus attenuating low frequencies. Since $R_{in1} = r_{\pi 2} \parallel R_{B1}$, the low-frequency cut-off of this stage is equal to

$$\omega_{L1} = \frac{1}{(r_{\pi 1} \parallel R_{B1})C_1} \tag{11.167}$$

$$= 2\pi \times (542 \text{ Hz}). \tag{11.168}$$

The second coupling capacitor also creates a high-pass response along with the input resistance of the second stage, $R_{in2} = R_{B2} \| [r_{\pi 2} + (\beta+1)R_E]$. To compute the cut-off frequency, we construct the simplified interface shown in Fig. 11.59(b) and determine V_Y/I_1. In this case, it is simpler to replace I_1 and R_C with a Thevenin equivalent, Fig. 11.59(c), where $V_{Thev} = -I_1 R_C$. We now have

$$\frac{V_Y}{V_{Thev}}(s) = \frac{R_{in2}}{R_C + \frac{1}{C_2 s} + R_{in2}}, \tag{11.169}$$

obtaining a pole at

$$\omega_{L2} = \frac{1}{(R_C + R_{in2})C_2} \tag{11.170}$$

$$= \pi \times (22.9 \text{ Hz}). \tag{11.171}$$

Since $\omega_{L2} \ll \omega_{L1}$, we conclude that ω_{L1} "dominates" the low-frequency response, i.e., the gain drops by 3 dB at ω_{L1}.

Exercise Repeat the above example if $R_E = 500\ \Omega$.

Example 11-29

The circuit of Fig. 11.60(a) is an example of amplifiers realized in integrated circuits. It consists of a degenerated stage and a self-biased stage, with moderate values for C_1 and C_2. Assuming M_1 and M_2 are identical and have the same parameters as those given in Example 11-18, plot the frequency response of the amplifier.

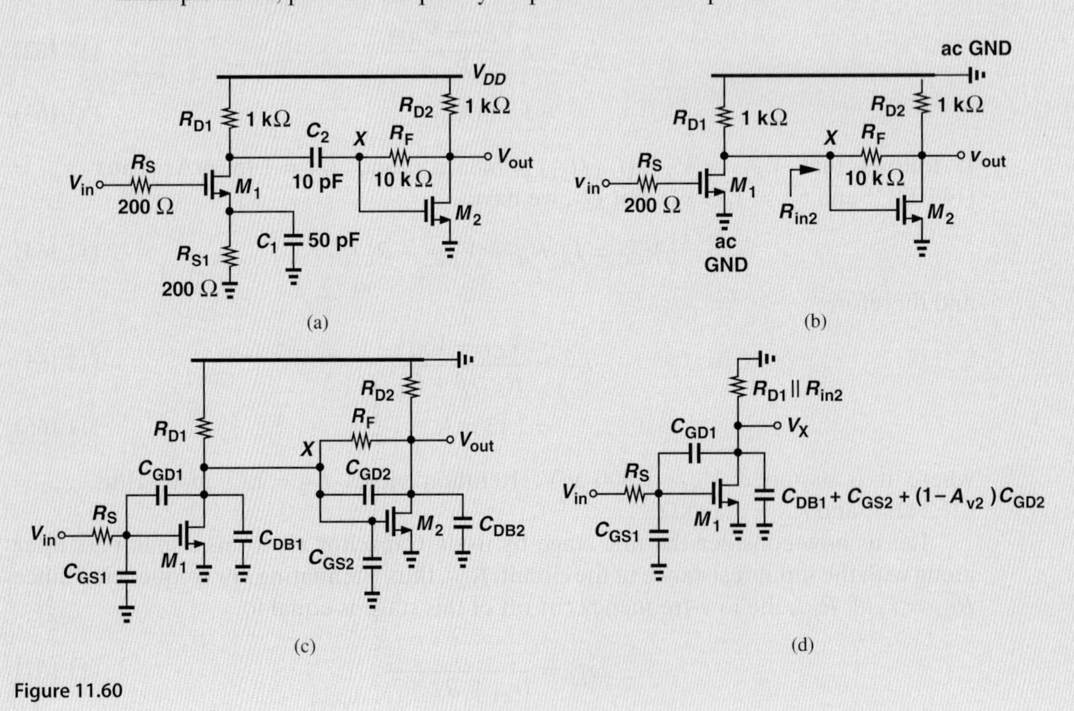

Figure 11.60

Solution **Low-Frequency Behavior** We begin with the low-frequency region and first consider the role of C_1. From Eq. (11.58) and Fig. 11.30(c), we note that C_1 contributes a low-frequency cut-off at

$$\omega_{L1} = \frac{g_{m1}R_{S1} + 1}{R_{S1}C_1} \tag{11.172}$$

$$= 2\pi \times (37.1 \text{ MHz}). \tag{11.173}$$

A second low-frequency cut-off is contributed by C_2 and the input resistance of the second stage, R_{in2}. This resistance can be calculated with the aid of Miller's theorem:

$$R_{in2} = \frac{R_F}{1 - A_{v2}}, \tag{11.174}$$

where A_{v2} denotes the voltage gain from X to the output. Since $R_F \gg R_{D2}$, we have $A_{v2} \approx -g_{m2}R_{D2} = -6.67$,[19] obtaining $R_{in2} = 1.30$ kΩ. Using an analysis similar to that in the previous example, the reader can show that

$$\omega_{L2} = \frac{1}{(R_{D1} + R_{in2})C_2} \tag{11.175}$$

$$= 2\pi \times (6.92 \text{ MHz}). \tag{11.176}$$

Since ω_{L1} remains well above ω_{L2}, the cut-off is dominated by the former.

Midband Behavior In the next step, we compute the midband gain. At midband frequencies, C_1 and C_2 act as a short circuit and the transistor capacitances play a negligible role, allowing the circuit to be reduced to that in Fig. 11.60(b). We note that $v_{out}/v_{in} = (v_X/v_{in})(v_{out}/v_X)$ and recognize that the drain of M_1 sees two resistances to ac ground: R_{D1} and R_{in2}. That is,

$$\frac{v_X}{v_{in}} = -g_{m1}(R_{D1} \parallel R_{in2}) \tag{11.177}$$

$$= -3.77. \tag{11.178}$$

The voltage gain from node X to the output is approximately equal to $-g_{m2}R_{D2}$ because $R_F \gg R_{D2}$.[20] The overall midband gain is therefore roughly equal to 25.1.

High-Frequency Behavior To study the response of the amplifier at high frequencies, we insert the transistor capacitances, noting that C_{SB1} and C_{SB2} play no role because the source terminals of M_1 and M_2 are at ac ground. We thus arrive at the simplified topology shown in Fig. 11.60(c), where the overall transfer function is given by $V_{out}/V_{in} = (V_X/V_{in})(V_{out}/V_X)$.

How do we compute V_X/V_{in} in the presence of the loading of the second stage? The two capacitances C_{DB1} and C_{GS2} are in parallel, but how about the effect of R_F and C_{GD2}? We apply Miller's approximation to both components so as to convert them to grounded elements. The Miller effect of R_F was calculated above to be equivalent to $R_{in2} = 1.3$ kΩ. The Miller multiplication of C_{GD2} is given by $(1 - A_{v2})C_{GD2} = 614$ fF.

[19] With this estimate of the gain, we can express the Miller effect of R_F at the output as $R_F/(1 - A_{v2}^{-1}) \approx 8.7$ kΩ, place this resistance in parallel with R_{L2}, and write $A_{v2} = -g_{m2}(R_{D2} \parallel 8.7 \text{ k}\Omega) = -5.98$. But we continue without this iteration for simplicity.

[20] If not, then the circuit must be solved using a complete small-signal equivalent.

The first stage can now be drawn as illustrated in Fig. 11.60(d), lending itself to the CS analysis performed in Section 11.4. The zero is given by $g_{m1}/C_{GD1} = 2\pi \times (13.3 \text{ GHz})$. The two poles can be calculated from Eqs. (11.73), (11.74), and (11.75):

$$|\omega_{p1}| = 2\pi \times (242 \text{ MHz}) \tag{11.179}$$

$$|\omega_{p2}| = 2\pi \times (2.74 \text{ GHz}). \tag{11.180}$$

The second stage contributes a pole at its output node. The Miller effect of C_{GD2} at the output is expressed as $(1 - A_{v2}^{-1})C_{GD2} \approx 1.15C_{GD2} = 92$ fF. Adding C_{DB2} to this value yields the output pole as

$$|\omega_{p3}| = \frac{1}{R_{L2}(1.15C_{GD2} + C_{DB2})} \tag{11.181}$$

$$= 2\pi \times (0.829 \text{ GHz}). \tag{11.182}$$

We observe that ω_{p1} dominates the high-frequency response. Figure 11.61 plots the overall response. The midband gain is about 26 dB $\approx$ 20, around 20% lower than the calculated result. This is primarily due to the use of Miller approximation for R_F. Also, the "useful" bandwidth can be defined from the lower −3 dB cut-off ($\approx$ 40 MHz) to the upper −3 dB cut-off ($\approx$ 300 MHz) and is almost one decade wide. The gain falls to unity at about 2.3 GHz.

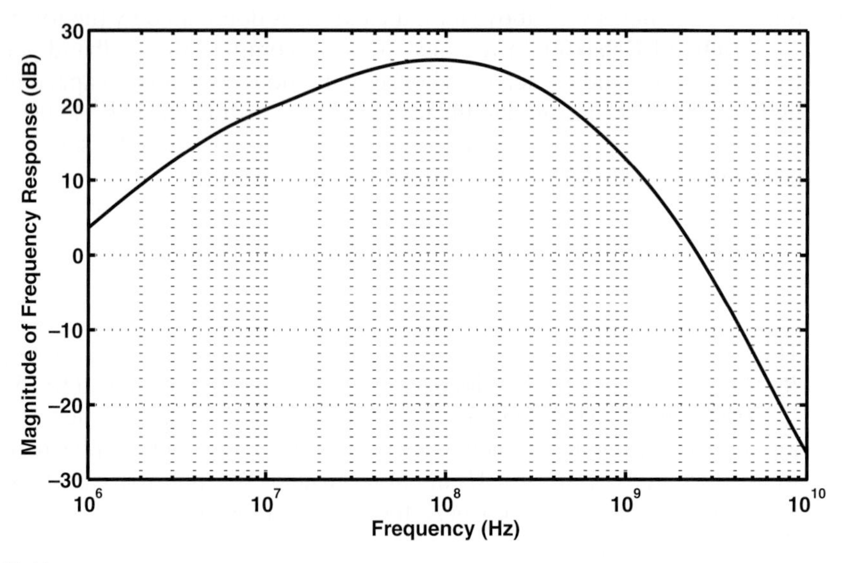

Figure 11.61

11.10 CHAPTER SUMMARY

- The speed of circuits is limited by various capacitances that the transistors and other components contribute to each node.
- The speed can be studied in the time domain (e.g., by applying a step) or in the frequency domain (e.g., by applying a sinusoid). The frequency response of a circuit corresponds to the latter test.

- As the frequency of operation increases, capacitances exhibit a lower impedance, reducing the gain. The gain thus rolls off at high signal frequencies.
- To obtain the frequency response, we must derive the transfer function of the circuit. The magnitude of the transfer function indicates how the gain varies with frequency.
- Bode's rules approximate the frequency response if the poles and zeros are known.
- A capacitance tied between the input and output of an inverting amplifier appears at the input with a factor equal to one minus the gain of the amplifier. This is called the Miller effect.
- In many circuits, it is possible to associate a pole with each node, i.e., calculate the pole frequency as the inverse of the product of the capacitance and resistance seen between the node and ac ground.
- Miller's theorem allows a floating impedance to be decomposed into to grounded impedances.
- Owing to coupling or degeneration capacitors, the frequency response may also exhibit roll-off as the frequency falls to very low values.
- Bipolar and MOS transistors contain capacitances between their terminals and from some terminals to ac ground. When solving a circuit, these capacitances must be identified and the resulting circuit simplified.
- The CE and CS stages exhibit a second-order transfer function and hence two poles. Miller's approximation indicates an input pole that embodies Miller multiplication of the base-collector or gate-drain capacitance.
- If the two poles of a circuit are far from each other, the "dominant-pole approximation" can be used to find a simple expression for each pole frequency.
- The CB and CG stages do not suffer from the Miller effect and achieve a higher speed than CE/CS stages, but their lower input impedance limits their applicability.
- Emitter and source followers provide a wide bandwidth. Their output impedance, however, can be inductive, causing instability in some cases.
- To benefit from the higher input impedance of CE/CS stages but reduce the Miller effect, a cascode stage can be used.
- The differential frequency response of differential pairs is similar to that of CE/CS stages.

PROBLEMS

Sec. 11.1.2 Transfer Function and Frequency Response

11.1. In the amplifier of Fig. 11.62, $R_D = 1\ \text{k}\Omega$ and $C_L = 1$ pF. Neglecting channel-length modulation and other capacitances, determine the frequency at which the gain falls by 10% (≈ 1 dB).

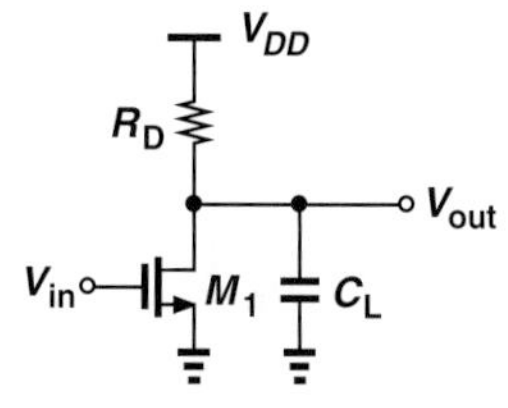

Figure 11.62

SOL **11.2.** In the circuit of Fig. 11.63, we wish to achieve a −3-dB bandwidth of 1 GHz with a load capacitance of 2 pF. What is the maximum (low-frequency) gain that can be achieved with a power dissipation of 2 mW? Assume $V_{CC} = 2.5$ V and neglect the Early effect and other capacitances.

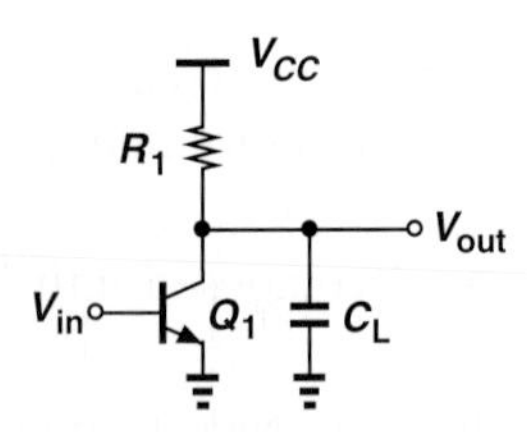

Figure 11.63

11.3. Determine the −3 dB bandwidth of the circuits shown in Fig. 11.64. Assume $V_A = \infty$ but $\lambda > 0$. Neglect other capacitances.

Sec. 11.1.3 Bode's Rules

11.4. Construct the Bode plot of $|V_{out}/V_{in}|$ for the stages depicted in Fig. 11.64.

11.5. A circuit contains two coincident (i.e., equal) poles at ω_{p1}. Construct the Bode plot of $|V_{out}/V_{in}|$.

11.6. An amplifier exhibits two poles at 100 MHz and 10 GHz and a zero at 1 GHz. Construct the Bode plot of $|V_{out}/V_{in}|$.

11.7. An ideal integrator contains a pole at the origin, i.e., $\omega_p = 0$. Construct the Bode plot of $|V_{out}/V_{in}|$. What is the gain of the circuit at arbitrarily *low* frequencies?

11.8. An ideal differentiator provides a zero at the origin, i.e., $\omega_z = 0$. Construct the Bode plot of $|V_{out}/V_{in}|$. What is the gain of the circuit at arbitrarily *high* frequencies?

11.9. Figure 11.65 illustrates a cascade of two identical CS stages. Neglecting channel-length modulation and other capacitances, construct the Bode plot of $|V_{out}/V_{in}|$. Note that $V_{out}/V_{in} = (V_X/V_{in})(V_{out}/V_X)$.

***11.10.** In Problem 11.9, derive the transfer function of the circuit, substitute $s = j\omega$, and obtain an expression for $|V_{out}/V_{in}|$. Determine the −3 dB bandwidth of the circuit.

***11.11.** Consider the circuit shown in Fig. 11.66. SOL Derive the transfer function assuming $\lambda > 0$ but neglecting other capacitances.

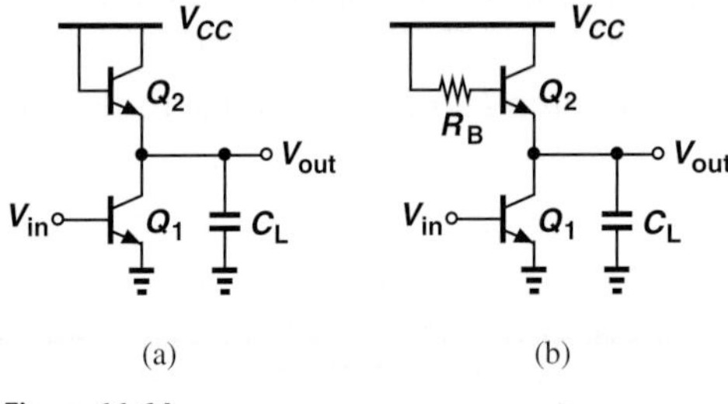

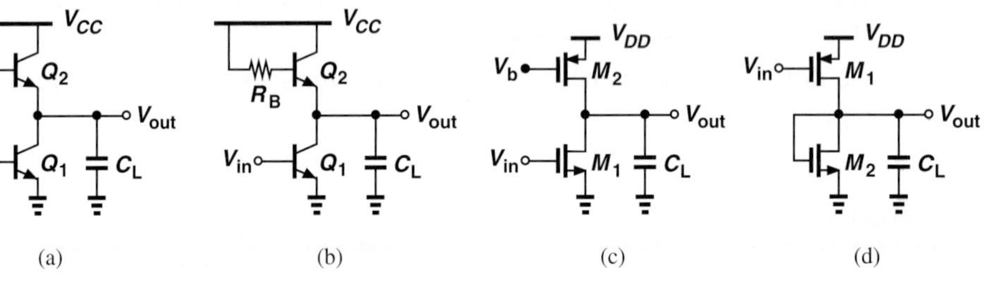

Figure 11.64

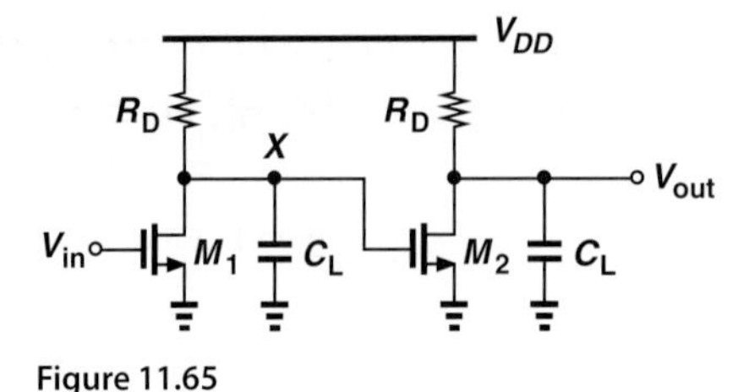

Figure 11.65

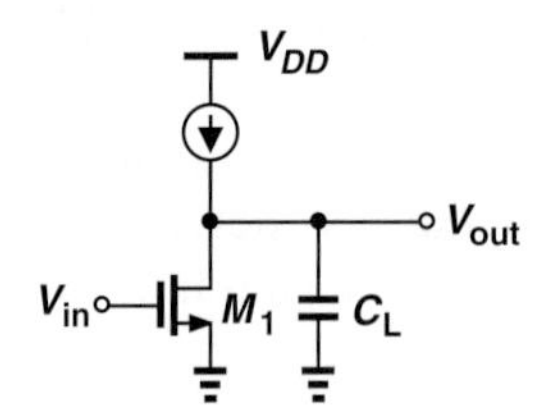

Figure 11.66

Explain why the circuit operates as an ideal integrator if $\lambda \to 0$.

*11.12. Due to a manufacturing error, a parasitic resistance R_p has appeared in series with the source of M_1 in Fig. 11.67. Assuming $\lambda = 0$ and neglecting other capacitances, determine the input and output poles of the circuit.

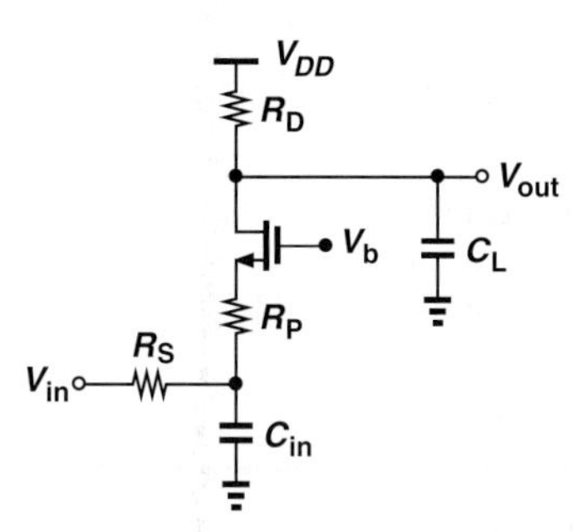

Figure 11.67

11.13. Repeat Problem 11.13 for the circuit shown in Fig. 11.68.

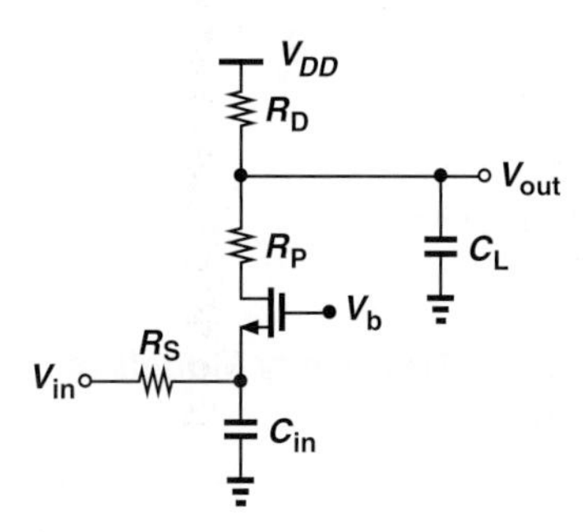

Figure 11.68

11.14. Repeat Problem 11.12 for the CS stage depicted in Fig. 11.69.

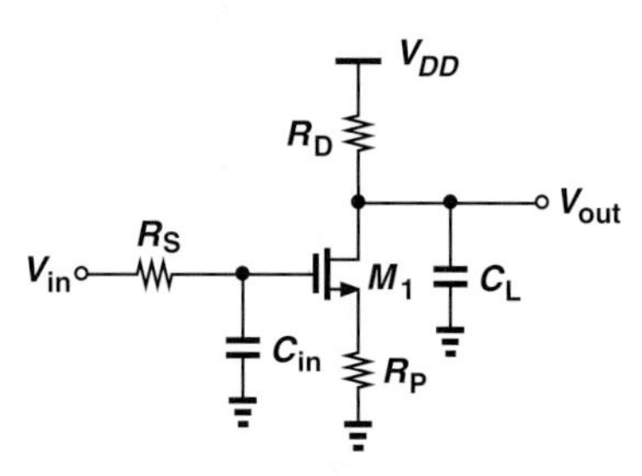

Figure 11.69

*11.15. Derive a relationship for the figure of merit defined by Eq. (11.8) for a CS stage. Consider only the load capacitance.

Sec. 11.1.5 Miller's Theorem

11.16. Apply Miller's theorem to resistor R_F in Fig. 11.70 and estimate the voltage gain of the circuit. Assume $V_A = \infty$ and R_F is large enough to allow the approximation $v_{out}/v_X = -g_m R_C$.

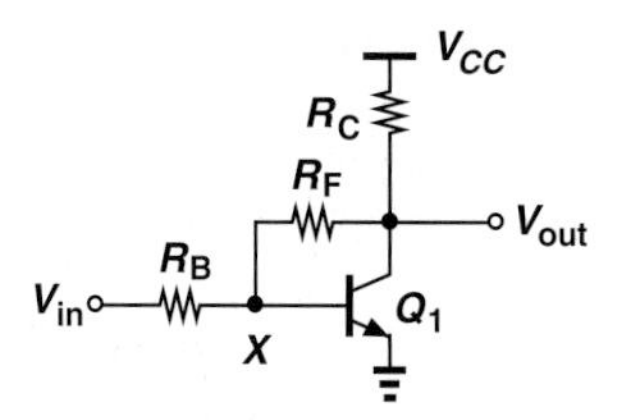

Figure 11.70

11.17. Repeat Problem 11.16 for the source follower in Fig. 11.71. Assume $\lambda = 0$ and R_F is large enough to allow the approximation $v_{out}/v_X = R_L/(R_L + g_m^{-1})$.

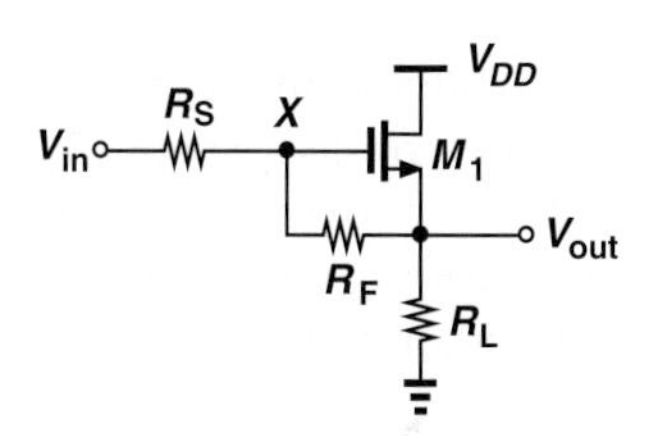

Figure 11.71

**11.18. Consider the common-base stage illustrated in Fig. 11.72, where the output resistance of Q_1 is drawn explicitly. Utilize Miller's theorem to estimate the gain. Assume r_O is large enough to allow the approximation $v_{out}/v_X = g_m R_C$.

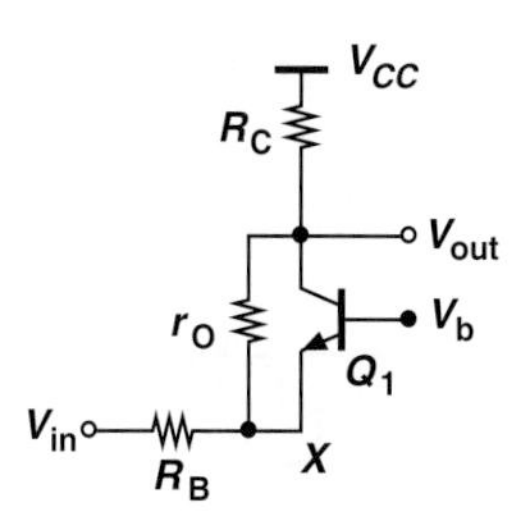

Figure 11.72

*11.19. Using Miller's theorem, estimate the input capacitance of the circuit depicted in Fig. 11.73. Assume $\lambda > 0$ but neglect other capacitances. What happens if $\lambda \to 0$?

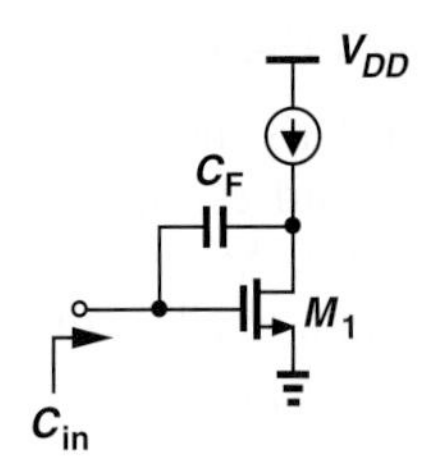

Figure 11.73

11.20. Repeat Problem 11.19 for the source follower shown in Fig. 11.74.

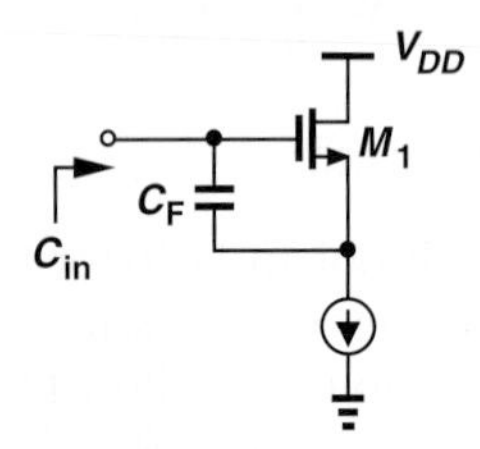

Figure 11.74

SOL *__11.21.__ Using Miller's theorem, explain how the common-base stage illustrated in Fig. 11.75 provides a *negative* input capacitance. Assume $V_A = \infty$ and neglect other capacitances.

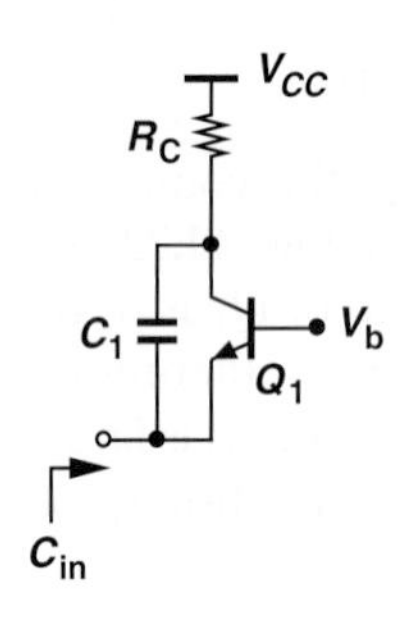

Figure 11.75

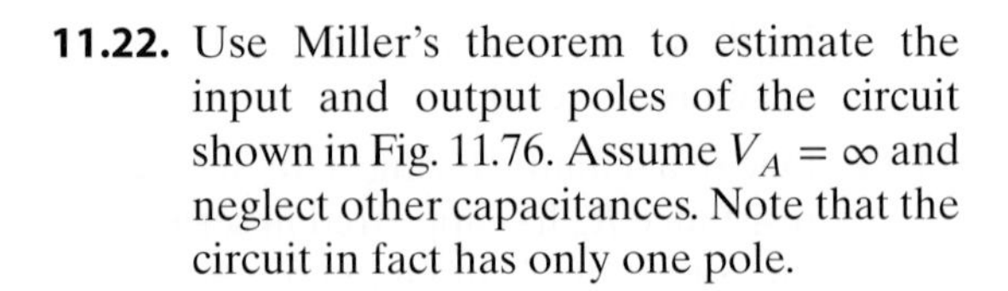

11.22. Use Miller's theorem to estimate the input and output poles of the circuit shown in Fig. 11.76. Assume $V_A = \infty$ and neglect other capacitances. Note that the circuit in fact has only one pole.

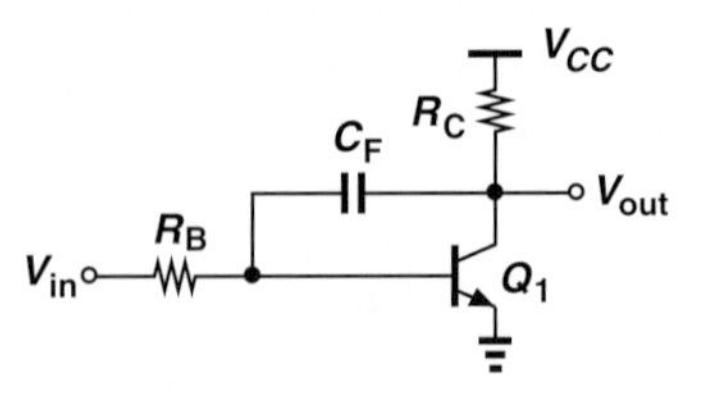

Figure 11.76

**__11.23.__ Repeat Problem 11.22 for the circuit in Fig. 11.77.

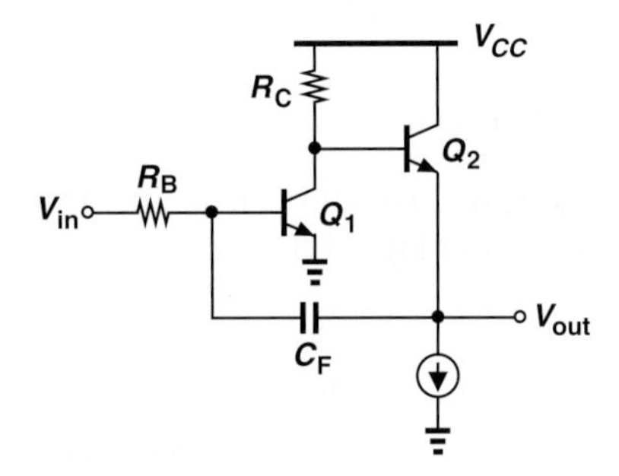

Figure 11.77

Sec. 11.2 High-Frequency Models of Transistors

11.24. For the bipolar circuits depicted in Fig. 11.78, identify all of the transistor capacitances and determine which ones are in parallel and which ones are grounded on both ends.

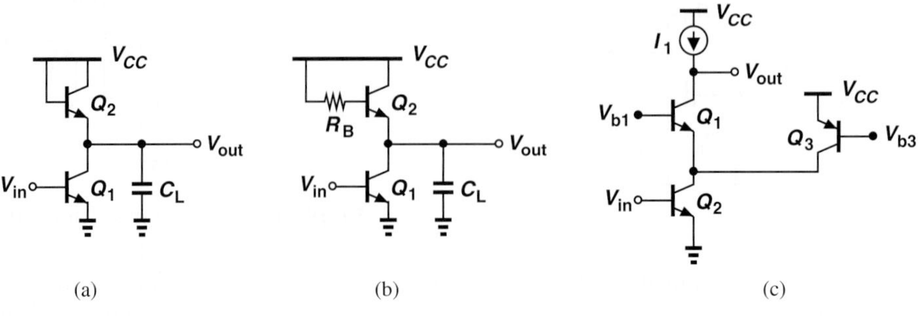

Figure 11.78

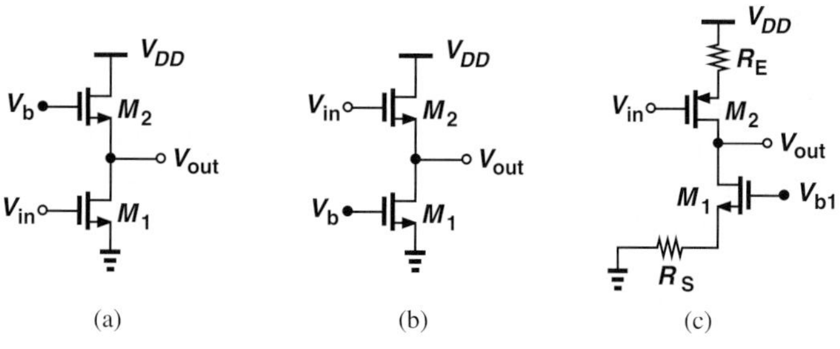

Figure 11.79

11.25. For the MOS circuits shown in Fig. 11.79, identify all of the transistor capacitances and determine which ones are in parallel and which ones are grounded on both ends.

11.26. In arriving at Eq. (11.52) for the f_T of transistors, we neglected C_μ and C_{GD}. Repeat the derivation without this approximation.

11.27. It can be shown that, if the minority carriers injected by the emitter into the base take τ_F seconds to cross the base region, then $C_b = g_m \tau_F$.

(a) Writing $C_\pi = C_b + C_{je}$, assuming that C_{je} is independent of the bias current, and using Eq. (11.52), derive an expression for the f_T of bipolar transistors in terms of the collector bias current.

(b) Sketch f_T as a function of I_C.

SOL *__11.28.__ It can be shown that $C_{GS} \approx (2/3)WLC_{ox}$ for a MOSFET operating in saturation. Using Eq. (11.52), prove that

$$2\pi f_T = \frac{3}{2}\frac{\mu_n}{L^2}(V_{GS} - V_{TH}). \quad (11.183)$$

Note that f_T increases with the overdrive voltage.

*__11.29.__ Having solved Problem 11.28 successfully, a student attempts a different substitution for g_m: $2I_D/(V_{GS} - V_{TH})$, arriving at

$$2\pi f_T = \frac{3}{2}\frac{2I_D}{WLC_{ox}}\frac{1}{V_{GS} - V_{TH}}. \quad (11.184)$$

This result suggests that f_T *decreases* as the overdrive voltage increases! Explain this apparent discrepancy between Eqs. (11.183) and (11.184).

*__11.30.__ Using Eq. (11.52) and the results of Problems 11.28 and 11.29, plot the f_T of a MOSFET (a) as a function of W for a constant I_D, (b) as a function of I_D for a constant W. Assume L remains constant in both cases.

*__11.31.__ Using Eq. (11.52) and the results of Problems 11.28 and 11.29, plot the f_T of a MOSFET (a) as a function of $V_{GS} - V_{TH}$ for a constant I_D, (b) as a function of I_D for a constant $V_{GS} - V_{TH}$. Assume L remains constant in both cases.

*__11.32.__ Using Eq. (11.52) and the results of Problems 11.28 and 11.29, plot the f_T of a MOSFET (a) as a function of W for a constant $V_{GS} - V_{TH}$, (b) as a function of $V_{GS} - V_{TH}$ for a constant W. Assume L remains constant in both cases.

*__11.33.__ In order to lower channel-length modulation in a MOSFET, we double the device length. (a) How should the device width be adjusted to maintain the same overdrive voltage and the same drain current? (b) How do these changes affect the f_T of the transistor?

*__11.34.__ We wish to halve the overdrive voltage of a transistor so as to provide a greater voltage headroom in a circuit. Determine the change in the f_T if (a) I_D is constant and W is increased, or (b) W is constant and I_D is decreased. Assume L is constant.

11.35. Using Miller's theorem, determine the input and output poles of the CE and CS stages depicted in Fig. 11.31(a) while including the output impedance of the transistors.

Sec. 11.4 Frequency Response of CE and CS Stage

11.36. The common-emitter stage of Fig. 11.80 employs a current-source load to achieve a high gain (at low frequencies). Assuming $V_A < \infty$ and using Miller's theorem, determine the input and output poles and hence the transfer function of the circuit.

Figure 11.80

11.37. Repeat Problem 11.36 for the stage shown in Fig. 11.81.

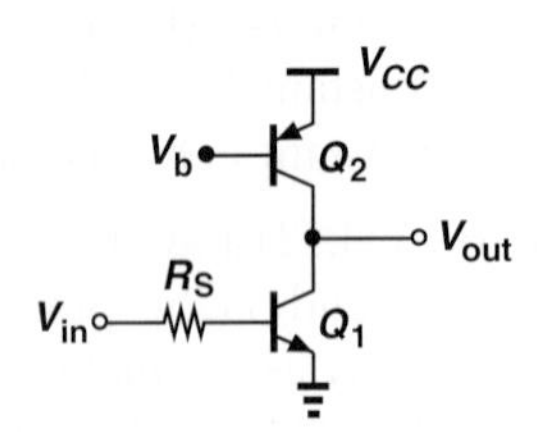

Figure 11.81

*__11.38.__ Assuming $\lambda > 0$ and using Miller's theorem, determine the input and output poles of the stages depicted in Fig. 11.82.

SOL **11.39.** In the CS stage of Fig. 11.31(a), $R_S = 200\ \Omega$, $R_D = 1\ \text{k}\Omega$, $I_{D1} = 1$ mA, $C_{GS} = 50$ fF, $C_{GD} = 10$ fF, $C_{DB} = 15$ fF, and $V_{GS} - V_{TH} = 200$ mV. Determine the poles of the circuit using (a) Miller's approximation, and (b) the transfer function given by Eq. (11.73). Compare the results.

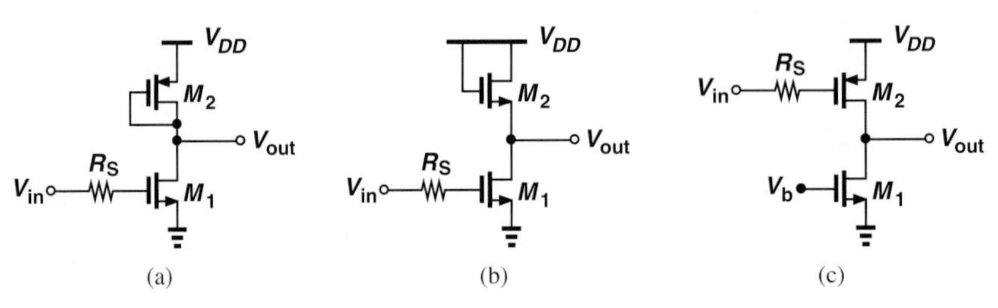

Figure 11.82

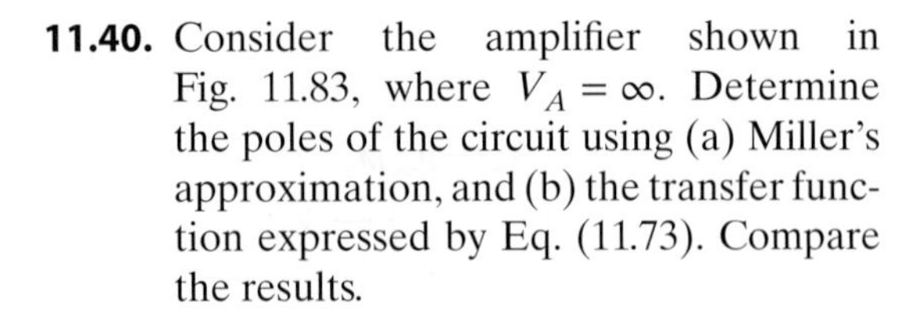

11.40. Consider the amplifier shown in Fig. 11.83, where $V_A = \infty$. Determine the poles of the circuit using (a) Miller's approximation, and (b) the transfer function expressed by Eq. (11.73). Compare the results.

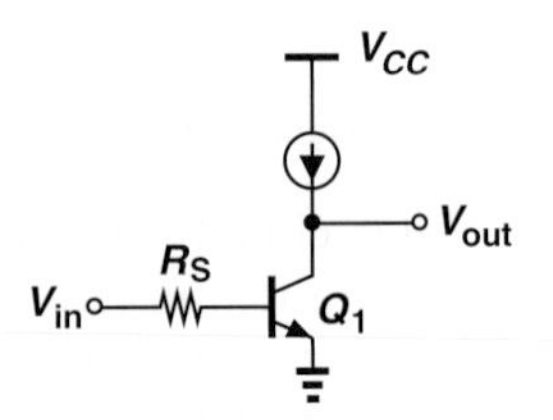

Figure 11.83

11.41. Repeat Problem 11.40 but use the dominant-pole approximation. How do the results compare?

*__11.42.__ The circuit depicted in Fig. 11.84 is called an "active inductor." Neglecting other capacitances and assuming $\lambda = 0$, compute Z_{in}. Use Bode's rule to plot $|Z_{in}|$ as a function of frequency and explain why it exhibits inductive behavior.

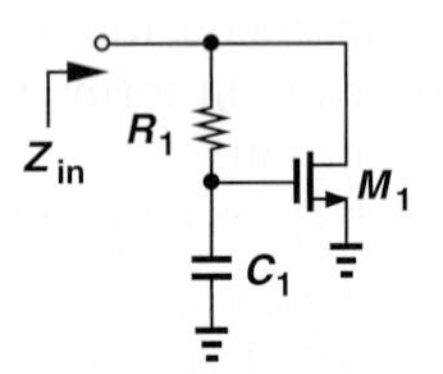

Figure 11.84

*__11.43.__ Determine the input and output impedances of the stage depicted in Fig. 11.85 without using Miller's theorem. Assume $V_A = \infty$.

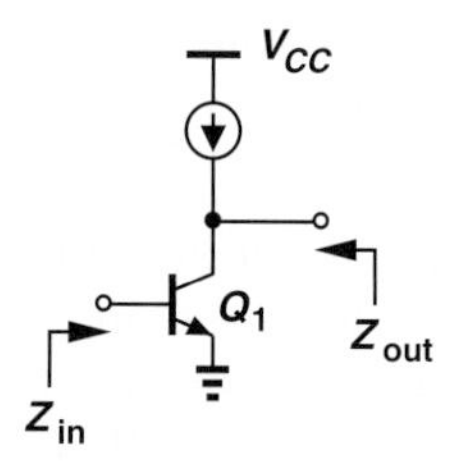

Figure 11.85

**11.44. Compute the transfer function of the circuit shown in Fig. 11.86 without using Miller's theorem. Assume $\lambda > 0$.

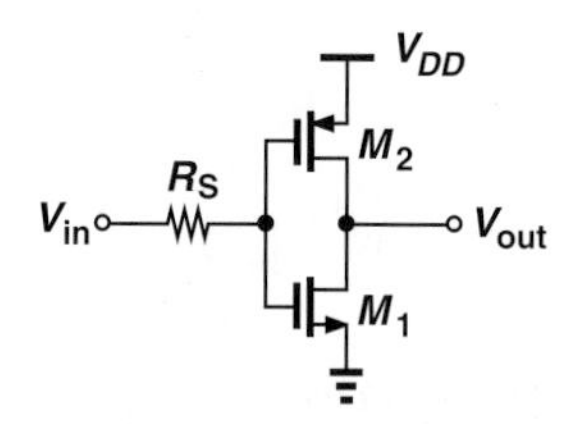

Figure 11.86

*11.45. Calculate the input impedance of the stage illustrated in Fig. 11.87 without using Miller's theorem. Assume $\lambda = 0$.

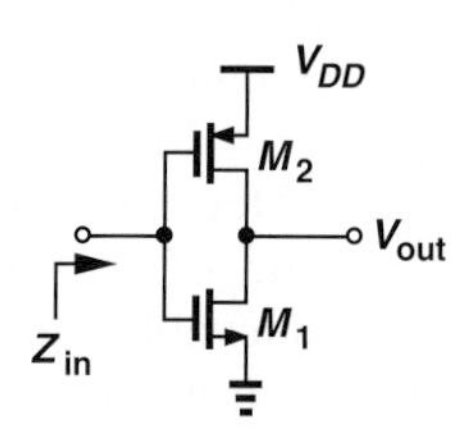

Figure 11.87

Sec. 11.5 Frequency Response of CB and CG Stages

11.46. Determine the transfer function of the circuits shown in Fig. 11.88. Assume $\lambda = 0$ for M_1.

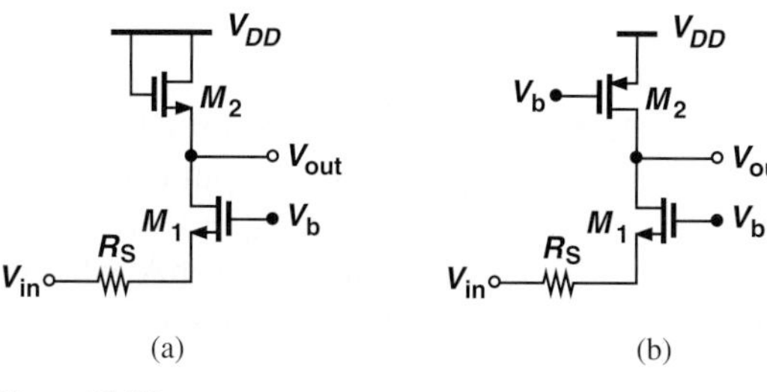

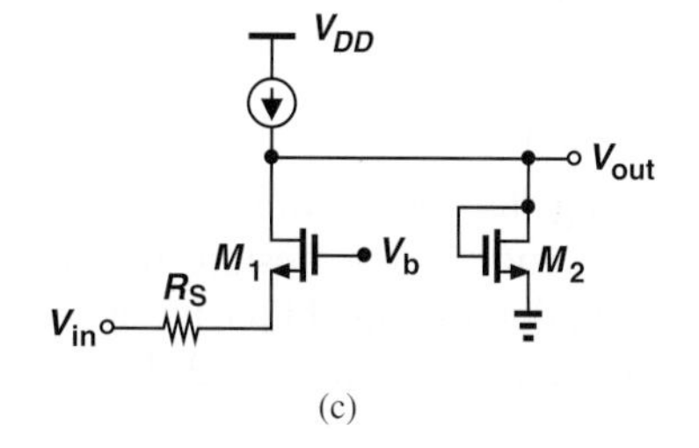

Figure 11.88

Sec. 11.6 Frequency Response of Followers

11.47. Consider the source follower shown in Fig. 11.89, where the current source is mistakenly replaced with a diode-connected device. Taking into account only C_{GS1}, compute the input capacitance of the circuit. Assume $\lambda \neq 0$. SOL

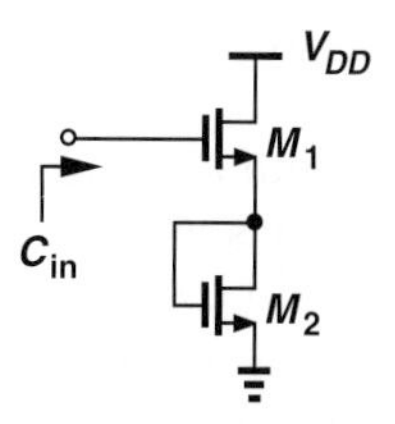

Figure 11.89

11.48. Determine the output impedance of the emitter follower depicted in Fig. 11.90, including C_μ. Sketch $|Z_{out}|$ as a function of frequency. Assume $V_A = \infty$.

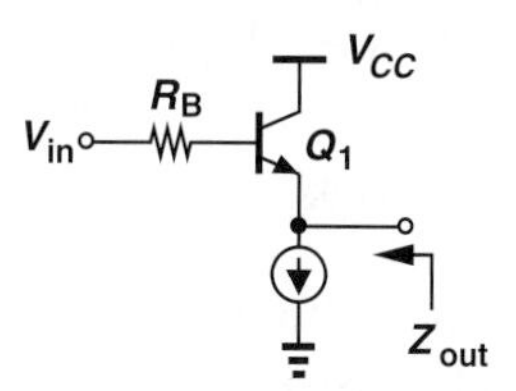

Figure 11.90

Sec. 11.7 Frequency Response of Cascode Stage

11.49. In the cascode of Fig. 11.91, Q_3 serves as a constant current source, providing 75% of the bias current of Q_1. Assuming $V_A = \infty$ and using Miller's theorem, determine the poles of the circuit. Is Miller's effect more or less significant here than in the standard cascode topology of Fig. 11.50(a)?

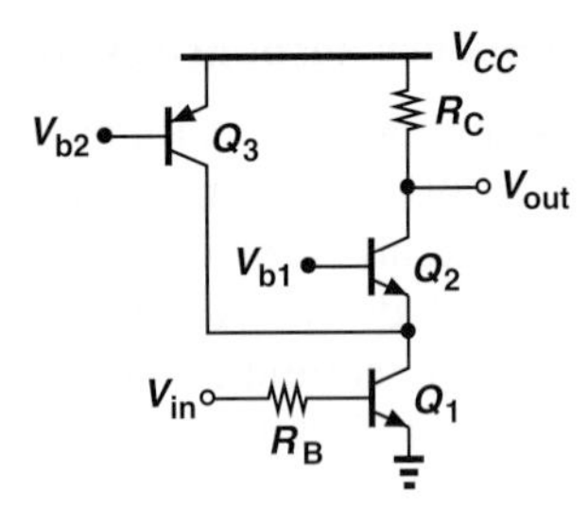

Figure 11.91

**11.50. Due to manufacturing error, a parasitic resistor R_p has appeared in the cascode stage of Fig. 11.92. Assuming $\lambda = 0$ and using Miller's theorem, determine the poles of the circuit.

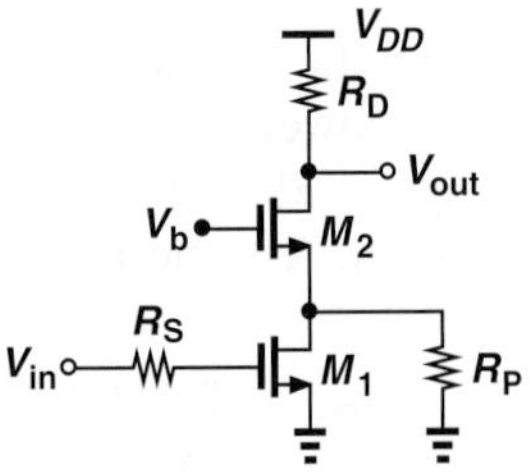

Figure 11.92

**11.51. In analogy with the circuit of Fig. 11.91, a student constructs the stage depicted in Fig. 11.93 but mistakenly uses an NMOS device for M_3. Assuming $\lambda = 0$ and using Miller's theorem, compute the poles of the circuit.

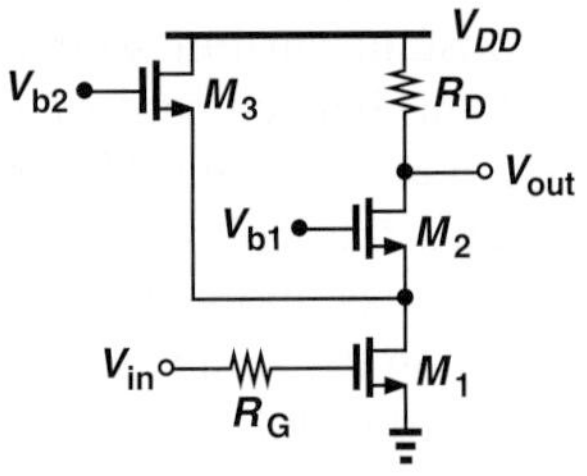

Figure 11.93

Design Problems

11.52. Using the results obtained in Problems 11.9 and 11.10, design the two-stage amplifier of Fig. 11.65 for a total voltage gain of 20 and a −3 dB bandwidth of 1 GHz. Assume each stage carries a bias current of 1 mA, $C_L = 50$ fF, and $\mu_n C_{ox} = 100 \mu A/V^2$.

11.53. We wish to design the CE stage of Fig. 11.94 for an input pole at 500 MHz and an output pole at 2 GHz. Assuming $I_C = 1$ mA, $C_\pi = 20$ fF, $C_\mu = 5$ fF, $C_{CS} = 10$ fF, and $V_A = \infty$, and using Miller's theorem, determine the values of R_B and R_C such that the (low-frequency) voltage gain is maximized. You may need to use iteration.

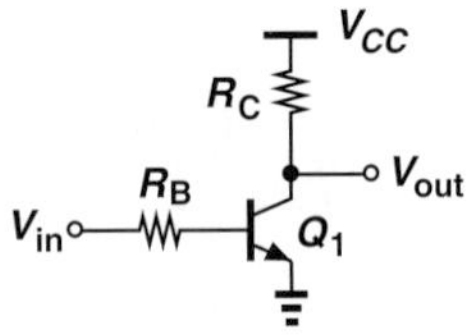

Figure 11.94

11.54. Repeat Problem 11.53 with the additional assumption that the circuit must drive a load capacitance of 20 fF.

11.55. We wish to design the common-base stage of Fig. 11.95 for a −3 dB bandwidth of 10 GHz. Assume $I_C = 1$ mA, $V_A = \infty$, $R_S = 50\ \Omega$, $C_\pi = 20$ fF, $C_\mu = 5$ fF, and $C_{CS} = 20$ fF. Determine the maximum allowable value of R_C and hence the maximum achievable gain. (Note that the input and output poles may affect the bandwidth.)

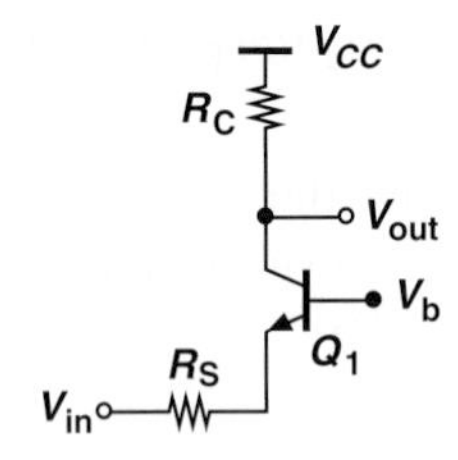

Figure 11.95

11.56. The emitter follower of Fig. 11.96 must be designed for an input capacitance of less than 50 fF. If $C_\mu = 10$ fF, $C_\pi = 100$ fF, $V_A = \infty$, and $I_C = 1$ mA, what is the minimum tolerable value of R_L?

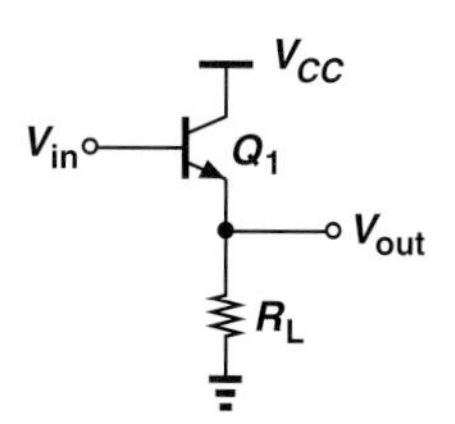

Figure 11.96

SOL **11.57.** An NMOS source follower must drive a load resistance of 100 Ω with a voltage gain of 0.8. If $I_D = 1$ mA, $\mu_n C_{ox} = 100\ \mu\text{A/V}^2$, $C_{ox} = 12\ \text{fF}/\mu\text{m}^2$, and $L = 0.18\ \mu\text{m}$, what is the minimum input capacitance that can be achieved? Assume $\lambda = 0$, $C_{GD} \approx 0$, $C_{SB} \approx 0$, and $C_{GS} = (2/3)WLC_{ox}$.

11.58. We wish to design the MOS cascode of Fig. 11.97 for an input pole of 5 GHz and an output pole of 10 GHz. Assume M_1 and M_2 are identical, $I_D = 0.5$ mA, $C_{GS} = (2/3)WLC_{ox}$, $C_{ox} = 12\ \text{fF}/\mu\text{m}^2$, $\mu_n C_{ox} = 100\ \mu\text{A/V}^2$, $\lambda = 0$, $L = 0.18\ \mu\text{m}$ and $C_{GD} = C_0 W$, where $C_0 = 0.2\ \text{fF}/\mu\text{m}$ denotes the gate-drain capacitance per unit width. Determine the maximum allowable values of R_G, R_D, and the voltage gain. Use Miller's approximation for C_{GD1}. Assume an overdrive voltage of 200 mV for each transistor.

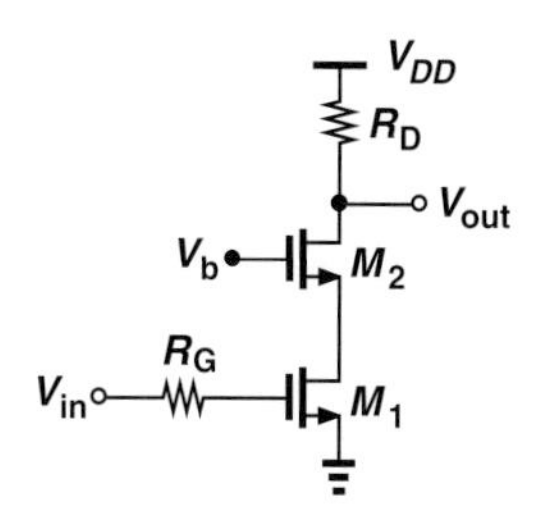

Figure 11.97

11.59. Repeat Problem 11.58 if $W_2 = 4W_1$ so as to reduce the Miller multiplication of C_{GD1}.

SPICE PROBLEMS

In the following problems, use the MOS device models given in Appendix A. For bipolar transistors, assume $I_{S,npn} = 5 \times 10^{-16}$ A, $\beta_{npn} = 100$, $V_{A,npn} = 5$ V, $I_{S,pnp} = 8 \times 10^{-16}$ A, $\beta_{pnp} = 50$, $V_{A,pnp} = 3.5$ V. Also, SPICE models the effect of charge storage in the base by a parameter called $\tau_F = C_b/g_m$. Assume $\tau_F(tf) = 20$ ps.

11.60. In the two-stage amplifier shown in Fig. 11.98, $W/L = 10\ \mu\text{m}/0.18\ \mu\text{m}$ for M_1-M_4.

(a) Select the input dc level to obtain an output dc level of 0.9 V.

(b) Plot the frequency response and compute the low-frequency gain and the −3 dB bandwidth.

(c) Repeat (a) and (b) for $W = 20\ \mu\text{m}$ and compare the results.

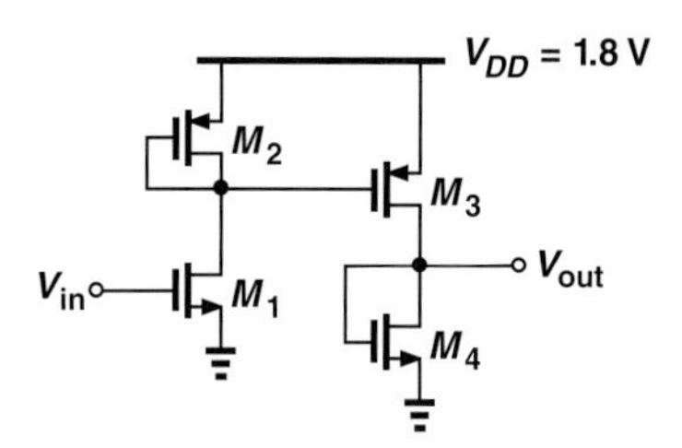

Figure 11.98

11.61. The circuit of Fig. 11.99 must drive a load capacitance of 100 fF.

(a) Select the input dc level to obtain an output dc level of 1.2 V.

(b) Plot the frequency response and compute the low-frequency gain and the −3 dB bandwidth.

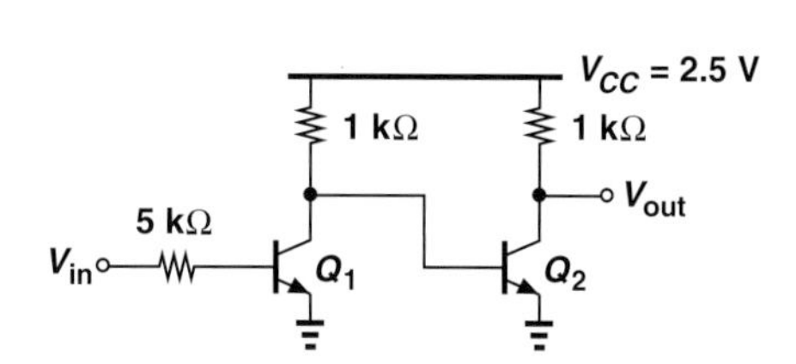

Figure 11.99

11.62. The self-biased stage depicted in Fig. 11.100 must drive a load capacitance of 50 fF with a maximum gain-bandwidth product (= midband gain × unity-gain bandwidth). Assuming $R_1 = 500\ \Omega$ and $L_1 = 0.18\,\mu\text{m}$, determine W_1, R_F, and R_D.

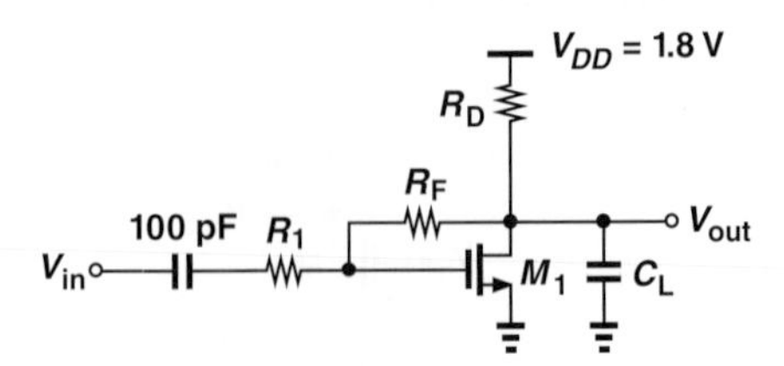

Figure 11.100

11.63. Repeat Problem 11.62 for the circuit shown in Fig. 11.101. (Determine R_F and R_C.)

11.64. The two-stage amplifier shown in Fig. 11.102 must achieve a maximum gain-bandwidth product while driving $C_L = 50$ fF. Assuming M_1-M_4 have a width of W And a length of 0.18 μm, determine R_F and W.

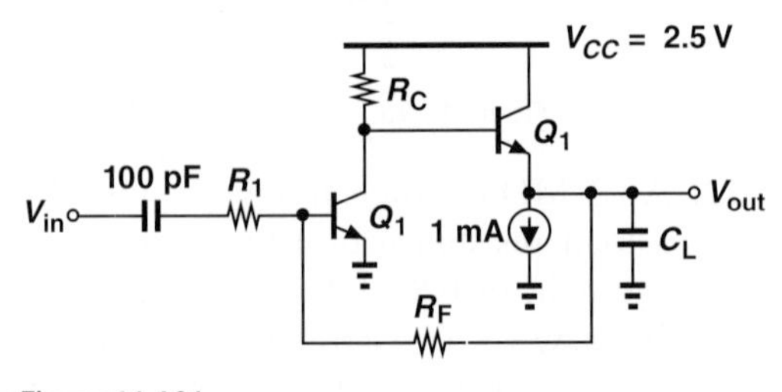

Figure 11.101

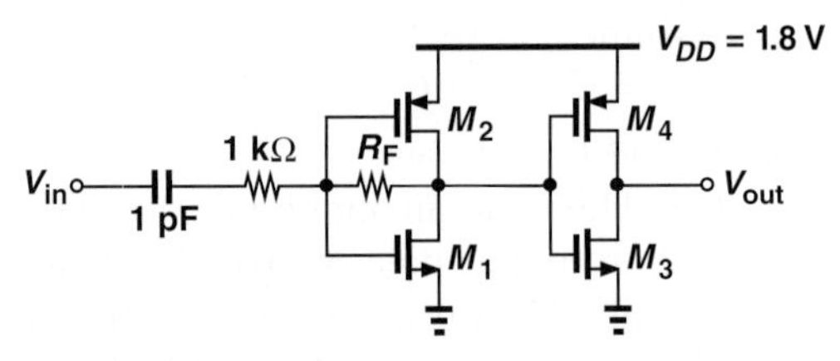

Figure 11.102

REFERENCE

1. V. Viswam, et al., "Impedance spectroscopy and electrophysiological imaging of cells with a high-density CMOS microelectrode array system," *IEEE Trans. Biomedical Circuits and Systems*, vol. 12, pp. 1356–1368, Dec. 2018.

12

Feedback

Watch Companion YouTube Videos:

Razavi Electronics 2, Lec 27: Introduction to Feedback, General Feedback System
https://www.youtube.com/watch?v=Ok-1kMqUnPk

Razavi Electronics 2, Lec 28: Feedback Examples, Concept of Loop Gain
https://www.youtube.com/watch?v=1TL0swYML8Q

Razavi Electronics 2, Lec 29: Application Examples of Feedback, Properties of Feedback
https://www.youtube.com/watch?v=98uAU2xG3EI

Razavi Electronics 2, Lec 30: A Closer Look at Properties of Feedback Systems
https://www.youtube.com/watch?v=zCyAsIiPRRs

Razavi Electronics 2, Lec 31: Foundations for Feedback Analysis: Types of Amplifiers
https://www.youtube.com/watch?v=lZtvEkFCtXQ

Razavi Electronics 2, Lec 32: Foundations for Feedback Analysis: Sense & Return Mechanisms
https://www.youtube.com/watch?v=udClJeiQKBI

Razavi Electronics 2, Lec 33: Feedback Circuit Examples, Sign of Feedback
https://www.youtube.com/watch?v=caoAI1EEXDk

Razavi Electronics 2, Lec 34: Four Feedback Topologies, Voltage-Voltage (Shunt-Series) Feedback
https://www.youtube.com/watch?v=GQ7T9MeH-60

Razavi Electronics 2, Lec 35: Examples of Voltage-Voltage Feedback
https://www.youtube.com/watch?v=jtjikoFRlK8

Razavi Electronics 2, Lec 36: Additional Examples of Voltage-Voltage Feedback
https://www.youtube.com/watch?v=S6Ks8btnmB8

Razavi Electronics 2, Lec 37: Voltage-Current (Shunt-Shunt) Feedback
https://www.youtube.com/watch?v=rKsORTxlxa8

Razavi Electronics 2, Lec 38: Examples of Voltage-Current Feedback, Current-Voltage-Feedback
https://www.youtube.com/watch?v=a2O59zdMYx4

Razavi Electronics 2, Lec 39: Application Examples of Feedback, More on Current-Voltage Feedback
https://www.youtube.com/watch?v=ZNWbV5G2wrs

Razavi Electronics 2, Lec 40: Current-Current Feedback, App. of Feedback in Power Management
https://www.youtube.com/watch?v=QzHxmqB8I1U

Razavi Electronics 2, Lec 41: Effect of Loading in Feedback, Opening the Loop Properly
https://www.youtube.com/watch?v=NRYMrheAAJI

Razavi Electronics 2, Lec 42: Accurate Analysis of Feedback Circuits
https://www.youtube.com/watch?v=8h0n65voDDU

Razavi Electronics 2, Lec 43: Introduction to Instability in Feedback Systems
https://www.youtube.com/watch?v=kC8FYL8gr3E

Razavi Electronics 2, Lec 44: Bode's Rules, Stability Condition, Circuit Examples
https://www.youtube.com/watch?v=UKf4tVoULIo

Razavi Electronics 2, Lec 45: Additional Stability Examples, Phase Margin, Freq. Compensation
https://www.youtube.com/watch?v=KQjOVSKydyM

Feedback is an integral part of our lives. Try touching your fingertips together with your eyes closed; you may not succeed the first time because you have broken a feedback loop that ordinarily "regulates" your motions. The regulatory role of feedback manifests itself in biological, mechanical, and electronic systems, allowing precise realization of "functions." For example, an amplifier targeting a precise gain of 2.00 is designed much more easily with feedback than without.

This chapter deals with the fundamentals of (negative) feedback and its application to electronic circuits. The outline is shown below.

General Considerations
- Elements of Feedback Systems
- Loop Gain
- Properties of Negative Feedback

Amplifiers and Sense/Return Methods
- Types of Amplifiers
- Amplifier Models
- Sense/Return Methods
- Polarity of Feedback

Analysis of Feedback Circuits
- Four Types of Feedback
- Effect of Finite I/O Impedances

Stability and Compensation
- Loop Instability
- Phase Margin
- Frequency Compensation

12.1 GENERAL CONSIDERATIONS

As soon as he reaches the age of 18, John eagerly obtains his driver's license, buys a used car, and begins to drive. Upon his parents' stern advice, John continues to observe the speed limit while noting that *every* other car on the highway drives faster. He then reasons that the speed limit is more of a "recommendation" and exceeding it by a small amount would not be harmful. Over the ensuing months, John gradually raises his speed so as to catch up with the rest of the drivers on the road, only to see flashing lights in his rear-view mirror one day. He pulls over to the shoulder of the road, listens to the sermon given by the police officer, receives a speeding ticket, and, dreading his parents' reaction, drives home—now strictly adhering to the speed limit.

John's story exemplifies the "regulatory" or "corrective" role of negative feedback. Without the police officer's involvement, John would probably continue to drive increasingly faster, eventually becoming a menace on the road.

Shown in Fig. 12.1, a negative feedback system consists of four essential components. (1) The "feedforward" system:[1] the main system, probably "wild" and poorly controlled. John, the gas pedal, and the car form the feedforward system, where the input is the amount of pressure that John applies to the gas pedal and the output is the speed of the car. (2) Output sense mechanism: a means of measuring the output. The police officer's radar serves this purpose here. (3) Feedback network: a network that generates a "feedback signal," X_F, from the sensed output. The police officer acts as the feedback network by reading the radar display, walking to John's car, and giving him a speeding ticket. The quantity $K = X_F/Y$ is called the "feedback factor." (4) Comparison or return mechanism: a means of subtracting the feedback signal from the input to obtain the "error," $E = X - X_F$. John makes this comparison himself, applying less pressure to the gas pedal—at least for a while.

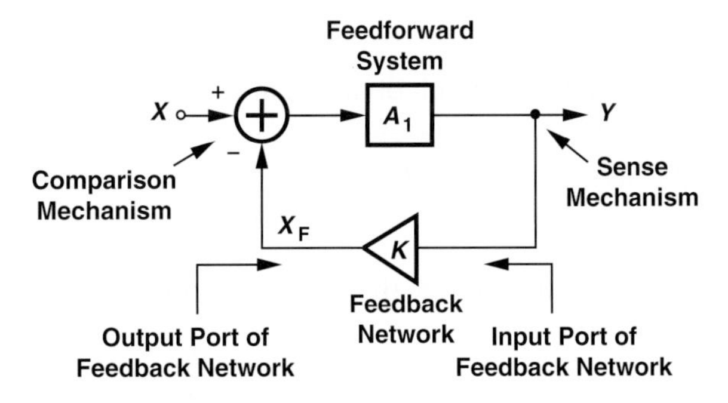

Figure 12.1 General feedback system.

The feedback in Fig. 12.1 is called "negative" because X_F is subtracted from X. Positive feedback, too, finds application in circuits such as oscillators and digital latches. If $K = 0$, i.e., no signal is fed back, then we obtain the "open-loop" system. If $K \neq 0$, we say the system operates in the "closed-loop" mode. As seen throughout this chapter, analysis of a feedback system requires expressing the closed-loop parameters in terms of the open-loop parameters. Note that the input port of the feedback network refers to that sensing the *output* of the forward system.

As our first step towards understanding the feedback system of Fig. 12.1, let us determine the closed-loop transfer function Y/X. Since $X_F = KY$, the error produced by the subtractor is equal to $X - KY$, which serves as the input of the forward system:

$$(X - KY)A_1 = Y. \tag{12.1}$$

[1]Also called the "forward" system.

That is,

$$\frac{Y}{X} = \frac{A_1}{1 + KA_1}. \tag{12.2}$$

This equation plays a central role in our treatment of feedback, revealing that negative feedback reduces the gain from A_1 (for the open-loop system) to $A_1/(1 + KA_1)$. The quantity $A_1/(1 + KA_1)$ is called the "closed-loop gain." Why do we deliberately lower the gain of the circuit? As explained in Section 12.2, the benefits accruing from negative feedback very well justify this reduction of the gain.

Example 12-1 Analyze the noninverting amplifier of Fig. 12.2 from a feedback point of view.

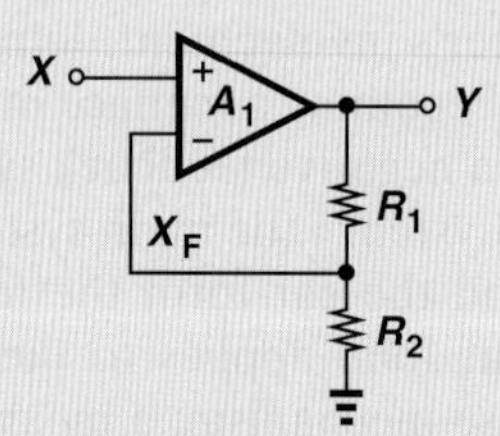

Figure 12.2

Solution The op amp A_1 performs two functions: subtraction of X and X_F and amplification. The network R_1 and R_2 also performs two functions: sensing the output voltage and providing a feedback factor of $K = R_2/(R_1 + R_2)$. Thus, Eq. (12.2) gives

$$\frac{Y}{X} = \frac{A_1}{1 + \frac{R_2}{R_1 + R_2} A_1}, \tag{12.3}$$

which is identical to the result obtained in Chapter 8.

Exercise Perform the above analysis if $R_2 = \infty$.

It is instructive to compute the error, E, produced by the subtractor. Since $E = X - X_F$ and $X_F = KA_1 E$,

$$E = \frac{X}{1 + KA_1}, \tag{12.4}$$

suggesting that the difference between the feedback signal and the input diminishes as KA_1 increases. In other words, the feedback signal becomes a close "replica" of the input (Fig. 12.3). This observation leads to a great deal of insight into the operation of feedback systems.

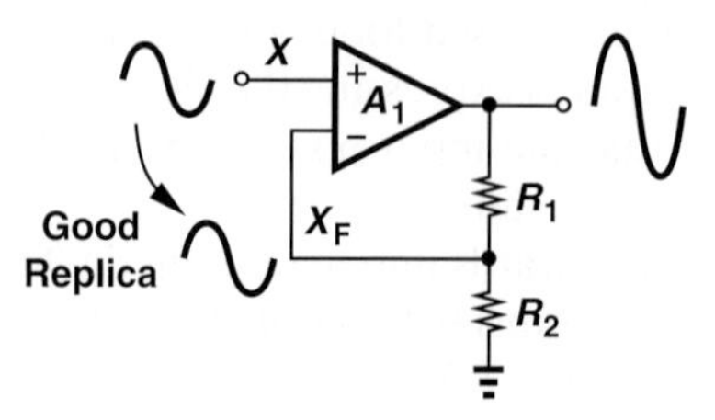

Figure 12.3 Feedback signal as a good replica of the input.

Example 12-2 Explain why in the circuit of Fig. 12.2, Y/X approaches $1 + R_1/R_2$ as $[R_2/(R_1 + R_2)]A_1$ becomes much greater than unity.

Solution If $KA_1 = [R_2/(R_1 + R_2)]A_1$ is large, X_F becomes almost identical to X, i.e., $X_F \approx X$. The voltage divider therefore requires that

$$Y\frac{R_2}{R_1 + R_2} \approx X \tag{12.5}$$

and hence

$$\frac{Y}{X} \approx 1 + \frac{R_1}{R_2}. \tag{12.6}$$

Of course, Eq. (12.3) yields the same result if $[R_2/(R_1 + R_2)]A_1 \gg 1$.

Exercise Repeat the above example if $R_2 = \infty$.

12.1.1 Loop Gain

In Fig. 12.1, the quantity KA_1, which is equal to product of the gain of the forward system and the feedback factor, determines many properties of the overall system. Called the "loop gain," KA_1 has an interesting interpretation. Let us set the input X to zero and "break" the loop at an arbitrary point, e.g., as depicted in Fig. 12.4(a). The resulting topology can be viewed as a system with an input M and an output N. Now, as shown in Fig. 12.4(b), let us apply a test signal at M and follow it through the feedback network, the subtractor, and the forward system to obtain the signal at N.[2] The input of A_1 is equal to $-KV_{test}$, yielding

$$V_N = -KV_{test}A_1 \tag{12.7}$$

and hence

$$KA_1 = -\frac{V_N}{V_{test}}. \tag{12.8}$$

In other words, if a signal "goes around the loop," it experiences a gain equal to $-KA_1$; hence the term "loop gain." It is important not to confuse the closed-loop gain, $A_1/(1 + KA_1)$, with the loop gain, KA_1.

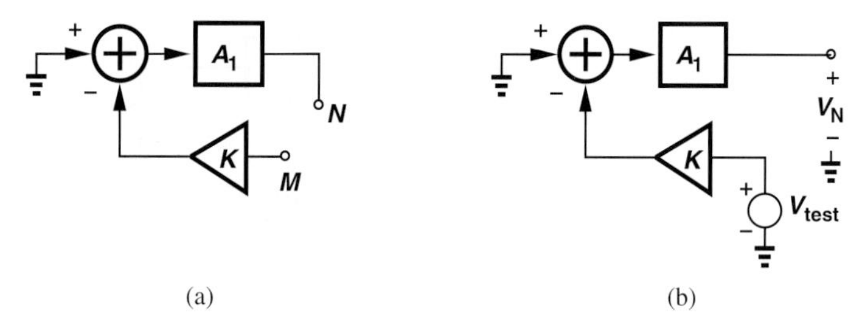

Figure 12.4 Computation of the loop gain by (a) breaking the loop and (b) applying a test signal.

[2]We use voltage quantities in this example, but other quantities work as well.

Example 12-3 Compute the loop gain of the feedback system of Fig. 12.1 by breaking the loop at the input of A_1.

Solution Illustrated in Fig. 12.5 is the system with the test signal applied to the input of A_1. The output of the feedback network is equal to KA_1V_{test}, yielding

$$V_N = -KA_1V_{test} \tag{12.9}$$

and hence the same result as in Eq. (12.8).

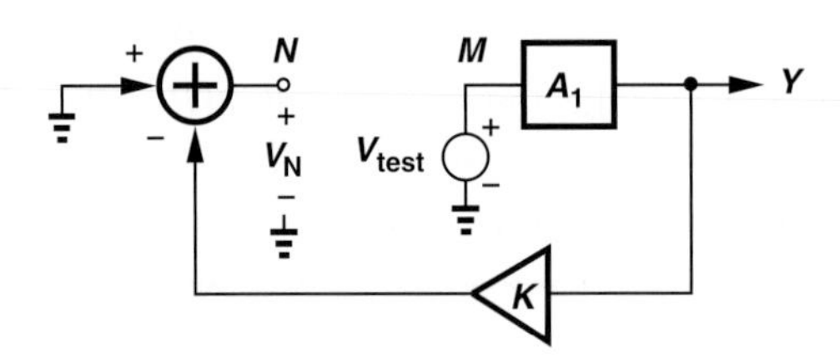

Figure 12.5

Exercise Compute the loop gain by breaking the loop at the input of the subtractor.

Did you know?

Negative feedback is a common phenomenon in nature. When you enter a bright room, your brain commands your pupils to become smaller. If you listen to very loud music for several hours, you feel hard of hearing afterwards because your ear has adjusted your hearing threshold in response to the volume. And if you try writing with your eyes closed, your handwriting will not be good because the feedback loop monitoring your pen strokes is broken.

The reader may wonder if an ambiguity exists with respect to the *direction* of the signal flow in the loop gain test. For example, can we modify the topology of Fig. 12.4(b) as shown in Fig. 12.6? This would mean applying V_{test} to the *output* of A_1 and expecting to observe a signal at its *input* and eventually at N. While possibly yielding a finite value, such a test does not represent the actual behavior of the circuit. In the feedback system, the signal flows from the input of A_1 to its output and from the input of the feedback network to its output.

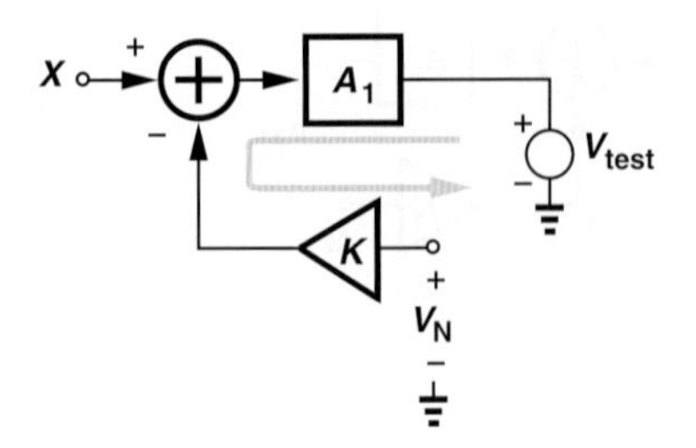

Figure 12.6 Incorrect method of applying test signal.

Bioengineering Application: Electrocardiography

Our studies in Chapters 8 and 10 suggest that the ECG signal can be sensed differentially by two noninverting amplifiers [Fig. 12.7(a)]. The 60-Hz common-mode noise is present in V_X, V_Y, V_A, and V_B, but not in $V_X - V_Y$ or in $V_A - V_B$.

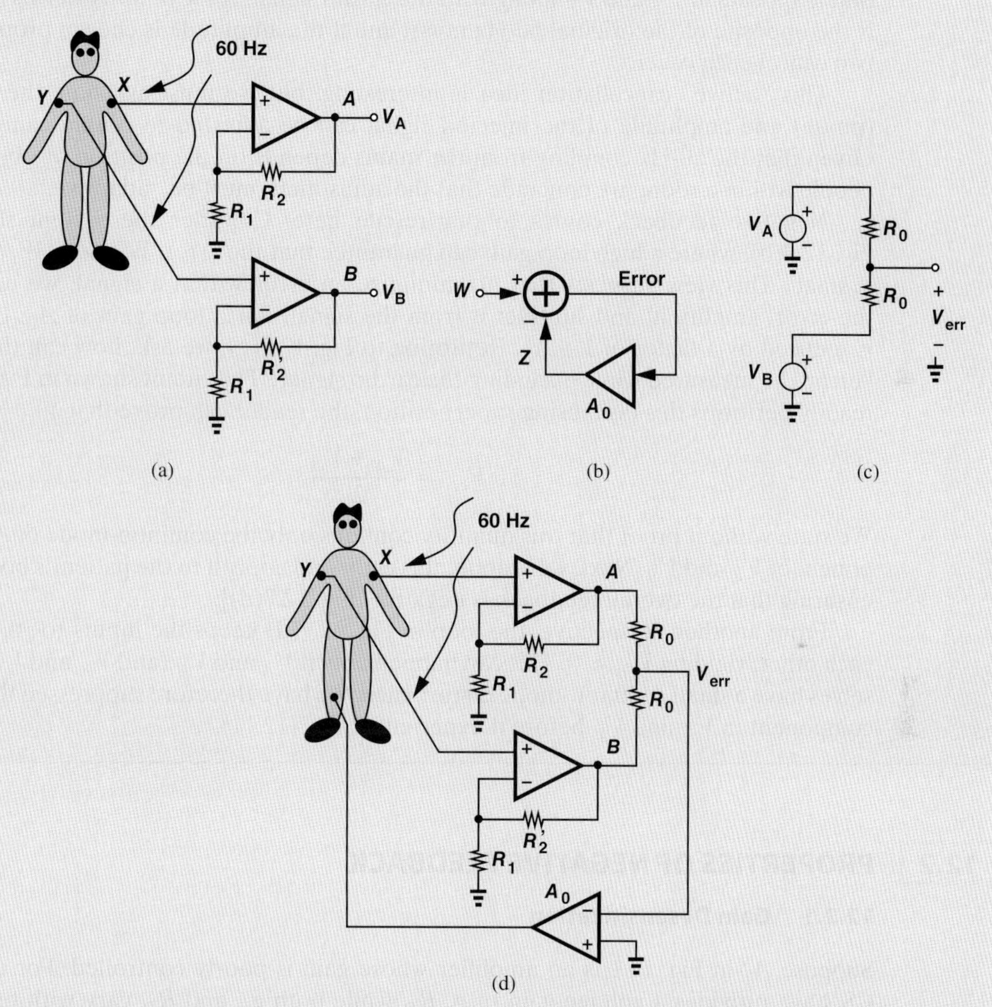

Figure 12.7 (a) Measuring the cardio signal by two noninverting amplifiers, (b) suppression of an error by negative feedback, (c) reconstruction of CM noise in V_A and V_B, and (d) negative-feedback loop suppressing CM noise.

In practice, the situation is more complex, and $V_A - V_B$ is not completely free from 60 Hz. Suppose, for example, that, due to manufacturing tolerances, R_2 and R_2' are not exactly equal. We then observe that

$$V_A - V_B = \left(1 + \frac{R_2}{R_1}\right) V_X - \left(1 + \frac{R_2'}{R_1}\right) V_Y \tag{12.10}$$

$$= \left(1 + \frac{R_2}{R_1}\right)(V_X - V_Y) + \frac{R_2 - R_2'}{R_1} V_Y. \tag{12.11}$$

Interestingly, and unfortunately, $V_A - V_B$ carries a faction of V_Y, which does contain the 60-Hz CM disturbance.

How do we suppress this residual noise in $V_A - V_B$? If we can magically reduce the CM noise in V_X and V_Y, we have less corruption in $V_A - V_B$ as well. This can be accomplished if we deliberately "inject" a 60-Hz waveform to the patient's body such that it appears in V_X and V_Y along with the cardio signal. If the *polarity* of this injection is the opposite of the original 60-Hz noise, and if the amplitude is chosen properly, the two cancel each other.

This "active" cancellation idea is interesting, but we must ask how the polarity (phase) and amplitude of the injected signal can be adjusted to obtain cancellation. Given that the 60-Hz coupling from the mains depends on the position of objects and people within a room, we conclude that the adjustment must be "adaptive."

Negative feedback comes to our rescue here. Consider the system shown in Fig. 12.7(b), where a high loop gain can guarantee that the error, $W - Z = W/(1 + A_0)$, is small. This view suggests that, to minimize an error within a signal, we can sense the error, amplify it, and subtract it from the signal. For a loop gain of A_0, the error is reduced by a factor of $1 + A_0$. Returning to Fig. 12.7(a), we ask, how can the 60-Hz "error" be measured while excluding the cardio signal? The circuit shown in Fig. 12.7(c) readily performs this task: using superposition, the reader can prove that

$$V_{err} = \frac{V_A + V_B}{2}. \tag{12.12}$$

We state without proof that this quantity contains only the common-mode 60-Hz component of V_A and V_B. Next, we amplify this error and apply it to the patient's body while ensuring that the overall feedback is negative [Fig. 12.7(d)].

From another perspective, the loop in Fig. 12.7(d) keeps the inputs to A_0 close to each other, yielding $V_{err} \approx 0$. This can happen only if V_A and V_B (and V_X and V_Y) themselves have a small common-mode disturbance. We have therefore suppressed the 60-Hz component in V_X and V_Y before it experiences Eq. (12.11).

12.2 PROPERTIES OF NEGATIVE FEEDBACK

12.2.1 Gain Desensitization

Suppose A_1 in Fig. 12.1 is an amplifier whose gain is poorly controlled. For example, a CS stage provides a voltage gain of $g_m R_D$ while both g_m and R_D vary with process and temperature; the gain thus may vary by as much as $\pm 20\%$. Also, suppose we require a voltage gain of 4.00.[3] How can we achieve such precision? Equation (12.2) points to a potential solution: if $KA_1 \gg 1$, we have

$$\frac{Y}{X} \approx \frac{1}{K}, \tag{12.13}$$

a quantity independent of A_1. From another perspective, Eq. (12.4) indicates that $KA_1 \gg 1$ leads to a small error, forcing X_F to be nearly equal to X and hence Y nearly equal to

[3]Some analog-to-digital converters (ADCs) require very precise voltage gains. For example, a 10-bit ADC may call for a gain of 2.000.

X/K. Thus, if K can be defined precisely, then A_1 impacts Y/X negligibly and a high precision in the gain is attained. The circuit of Fig. 12.2 exemplifies this concept very well. If $A_1R_2/(R_1+R_2) \gg 1$, then

$$\frac{Y}{X} \approx \frac{1}{K} \quad (12.14)$$

$$\approx 1 + \frac{R_1}{R_2}. \quad (12.15)$$

Why is R_1/R_2 more precisely defined than g_mR_D is? If R_1 and R_2 are made of the same material and constructed identically, then the variation of their value with process and temperature does not affect their ratio. As an example, for a closed-loop gain of 4.00, we choose $R_1 = 3R_2$ and implement R_1 as the series combination of three "unit" resistors equal to R_2. Illustrated in Fig. 12.8, the idea is to ensure that R_1 and R_2 "track" each other; if R_2 increases by 20%, so does each unit in R_1 and hence the total value of R_1, still yielding a gain of $1 + 1.2R_1/(1.2R_2) = 4$.

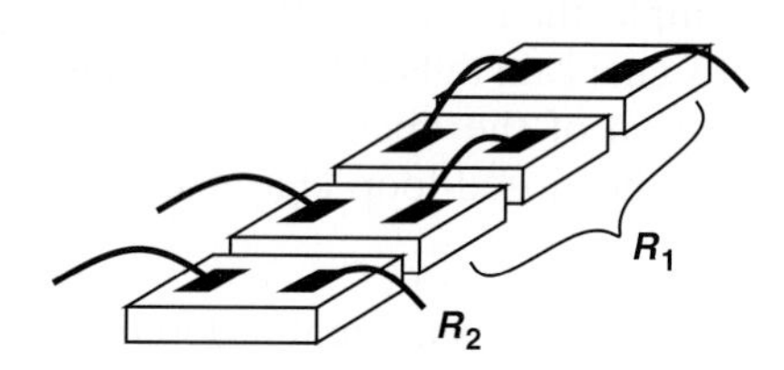

Figure 12.8 Construction of resistors for good matching.

Example 12-4 The circuit of Fig. 12.2 is designed for a nominal gain of 4. (a) Determine the actual gain if $A_1 = 1000$. (b) Determine the percentage change in the gain if A_1 drops to 500.

Solution For a nominal gain of 4, Eq. (12.15) implies that $R_1/R_2 = 3$. (a) The actual gain is given by

$$\frac{Y}{X} = \frac{A_1}{1 + KA_1} \quad (12.16)$$

$$= 3.984. \quad (12.17)$$

Note that the loop gain $KA_1 = 1000/4 = 250$. (b) If A_1 falls to 500, then

$$\frac{Y}{X} = 3.968. \quad (12.18)$$

Thus, the closed-loop gain changes by only $(3.984/3.968)/3.984 = 0.4\%$ if A_1 drops by factor of 2.

Exercise Determine the percentage change in the gain if A_1 falls to 200.

The above example reveals that the closed-loop gain of a feedback circuit becomes relatively independent of the open-loop gain so long as the loop gain, KA_1, remains sufficiently higher than unity. This property of negative feedback is called "gain desensitization."

We now see why we are willing to accept a reduction in the gain by a factor of $1 + KA_1$. We begin with an amplifier having a high, but poorly-controlled gain and apply negative feedback around it so as to obtain a better-defined, but inevitably lower gain. This concept was also extensively employed in the op amp circuits described in Chapter 8.

The gain desensitization property of negative feedback means that *any* factor that influences the open-loop gain has less effect on the closed-loop gain. Thus far, we have blamed only process and temperature variations, but many other phenomena change the gain as well.

- As the signal *frequency* rises, A_1 may fall, but $A_1/(1 + KA_1)$ remains relatively constant. We therefore expect that negative feedback *increases* the bandwidth (at the cost of gain).
- If the *load resistance* changes, A_1 may change; e.g., the gain of a CS stage depends on the load resistance. Negative feedback, on the other hand, makes the gain less sensitive to load variations.
- The signal *amplitude* affects A_1 because the forward amplifier suffers from nonlinearity. For example, the large-signal analysis of differential pairs in Chapter 10 reveals that the small-signal gain falls at large input amplitudes. With negative feedback, however, the variation of the open-loop gain due to nonlinearity manifests itself to a lesser extent in the closed-loop characteristics. That is, negative feedback improves the linearity.

We now study these properties in greater detail.

12.2.2 Bandwidth Extension

Let us consider a one-pole open-loop amplifier with a transfer function

$$A_1(s) = \frac{A_0}{1 + \dfrac{s}{\omega_0}}. \tag{12.19}$$

Here, A_0 denotes the low-frequency gain and ω_0 the -3 dB bandwidth. Noting from Eq. (12.2) that negative feedback lowers the low-frequency gain by a factor of $1 + KA_1$, we wish to determine the resulting bandwidth improvement. The closed-loop transfer function is obtained by substituting Eq. (12.19) for A_1 in Eq. (12.2):

$$\frac{Y}{X} = \frac{\dfrac{A_0}{1 + \dfrac{s}{\omega_0}}}{1 + K\dfrac{A_0}{1 + \dfrac{s}{\omega_0}}}. \tag{12.20}$$

Multiplying the numerator and the denominator by $1 + s/\omega_0$ gives

$$\frac{Y}{X}(s) = \frac{A_0}{1 + KA_0 + \dfrac{s}{\omega_0}} \tag{12.21}$$

$$= \frac{\dfrac{A_0}{1 + KA_0}}{1 + \dfrac{s}{(1 + KA_0)\omega_0}}. \tag{12.22}$$

In analogy with Eq. (12.19), we conclude that the closed-loop system now exhibits:

$$\text{Closed–Loop Gain} = \frac{A_0}{1 + KA_0} \tag{12.23}$$

$$\text{Closed–Loop Bandwidth} = (1 + KA_0)\omega_0. \tag{12.24}$$

In other words, the gain and bandwidth are scaled by the same factor but in opposite directions, displaying a *constant* product.

Example 12-5 Plot the closed-loop frequency response given by Eq. (12.22) for $K = 0$, 0.1, and 0.5. Assume $A_0 = 200$.

Solution For $K = 0$, the feedback vanishes and Y/X reduces to $A_1(s)$ as given by Eq. (12.19). For $K = 0.1$, we have $1 + KA_0 = 21$, noting that the gain decreases and the bandwidth increases by the same factor. Similarly, for $K = 0.5$, $1 + KA_0 = 101$, yielding a proportional reduction in gain and increase in bandwidth. The results are plotted in Fig. 12.9.

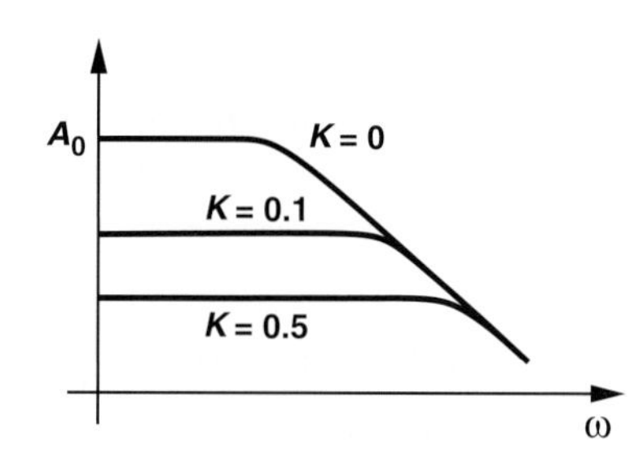

Figure 12.9

Exercise Repeat the above example for $K = 1$.

Example 12-6 Prove that the unity-gain bandwidth of the above system remains independent of K if $1 + KA_0 \gg 1$ and $K^2 \ll 1$.

Solution The magnitude of Eq. (12.22) is equal to

$$\left|\frac{Y}{X}(j\omega)\right| = \frac{\dfrac{A_0}{1 + KA_0}}{\sqrt{1 + \dfrac{\omega^2}{(1 + KA_0)^2\omega_0^2}}}. \tag{12.25}$$

Equating this result to unity and squaring both sides, we write

$$\left(\frac{A_0}{1 + KA_0}\right)^2 = 1 + \frac{\omega_u^2}{(1 + KA_0)^2\omega_0^2}, \tag{12.26}$$

where ω_u denotes the unity-gain bandwidth. It follows that

$$\omega_u = \omega_0\sqrt{A_0^2 - (1+KA_0)^2} \tag{12.27}$$

$$\approx \omega_0\sqrt{A_0^2 - K^2A_0^2} \tag{12.28}$$

$$\approx \omega_0 A_0, \tag{12.29}$$

which is equal to the gain-bandwidth product of the open-loop system. Figure 12.10 depicts the results.

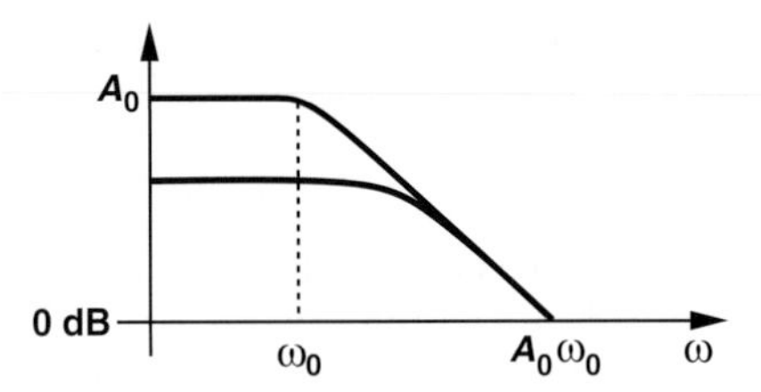

Figure 12.10

Exercise If $A_0 = 1000$, $\omega_0 = 2\pi \times (10\text{ MHz})$, and $K = 0.5$, calculate the unity-gain bandwidth from Eqs. (12.27) and (12.29) and compare the results.

12.2.3 Modification of I/O Impedances

As mentioned above, negative feedback makes the closed-loop gain less sensitive to the load resistance. This effect fundamentally arises from the modification of the *output impedance* as a result of feedback. Feedback modifies the *input* impedance as well. We will formulate these effects carefully in the following sections, but it is instructive to study an example at this point.

Example 12-7 Figure 12.11 depicts a transistor-level realization of the feedback circuit shown in Fig. 12.2. Assume $\lambda = 0$ and $R_1 + R_2 \gg R_D$ for simplicity. (a) Identify the four components of the feedback system. (b) Determine the open-loop and closed-loop voltage gain. (c) Determine the open-loop and closed-loop I/O impedances.

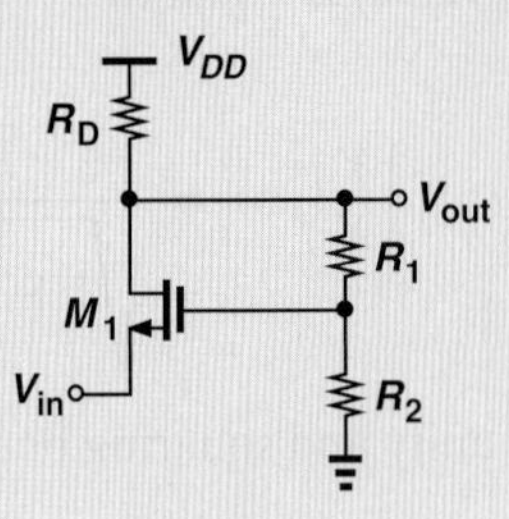

Figure 12.11

Solution (a) In analogy with Fig. 12.11, we surmise that the forward system (the main amplifier) consists of M_1 and R_D, i.e., a common-gate stage. Resistors R_1 and R_2 serve as both the sense mechanism and the feedback network, returning a signal equal to $V_{out}R_2/(R_1 + R_2)$ to the subtractor. Transistor M_1 itself operates as the subtractor because the small-signal drain current is proportional to the *difference* between the gate and source voltages:

$$i_D = g_m(v_G - v_S). \tag{12.30}$$

(b) The forward system provides a voltage gain equal to

$$A_0 \approx g_m R_D \tag{12.31}$$

because $R_1 + R_2$ is large enough that its loading on R_D can be neglected. The closed-loop voltage gain is thus given by

$$\frac{v_{out}}{v_{in}} = \frac{A_0}{1 + KA_0} \tag{12.32}$$

$$= \frac{g_m R_D}{1 + \frac{R_2}{R_1 + R_2} g_m R_D}. \tag{12.33}$$

We should note that the overall gain of this stage can also be obtained by simply solving the circuit's equations—as if we know nothing about feedback. However, the use of feedback concepts both provides a great deal of insight and simplifies the task as circuits become more complex.

(c) The open-loop I/O impedances are those of the CG stage:

$$R_{in,open} = \frac{1}{g_m} \tag{12.34}$$

$$R_{out,open} = R_D. \tag{12.35}$$

At this point, we do not know how to obtain the closed-loop I/O impedances in terms of the open-loop parameters. We therefore simply solve the circuit. From Fig. 12.12(a), we recognize that R_D carries a current approximately equal to i_X because $R_1 + R_2$ is assumed large. The drain voltage of M_1 is thus given by $i_X R_D$, leading to a gate voltage equal to $+i_X R_D R_2/(R_1 + R_2)$. Transistor M_1 generates a drain current proportional to v_{GS}:

$$i_D = g_m v_{GS} \tag{12.36}$$

$$= g_m \left(\frac{+i_X R_D R_2}{R_1 + R_2} - v_X \right). \tag{12.37}$$

Since $i_D = -i_X$, Eq. (12.37) yields

$$\frac{v_X}{i_X} = \frac{1}{g_m} \left(1 + \frac{R_2}{R_1 + R_2} g_m R_D \right). \tag{12.38}$$

That is, the input resistance *increases* from $1/g_m$ by a factor equal to $1 + g_m R_D R_2/(R_1 + R_2)$, the same factor by which the gain decreases.

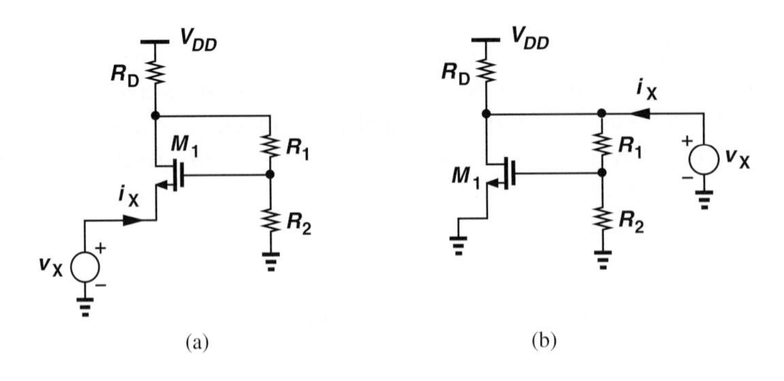

Figure 12.12

To determine the output resistance, we write from Fig. 12.12(b),

$$v_{GS} = \frac{R_2}{R_1 + R_2} v_X, \tag{12.39}$$

and hence

$$i_D = g_m v_{GS} \tag{12.40}$$

$$= g_m \frac{R_2}{R_1 + R_2} v_X. \tag{12.41}$$

Noting that, if $R_1 + R_2 \gg R_D$, then $i_X \approx i_D + v_X/R_D$, we obtain

$$i_X \approx g_m \frac{R_2}{R_1 + R_2} v_X + \frac{v_X}{R_D}. \tag{12.42}$$

It follows that

$$\frac{v_X}{i_X} = \frac{R_D}{1 + \dfrac{R_2}{R_1 + R_2} g_m R_D}. \tag{12.43}$$

The output resistance thus *decreases* by the "universal" factor $1 + g_m R_D R_2/(R_1 + R_2)$.

The above computation of I/O impedances can be greatly simplified if feedback concepts are employed. As exemplified by Eqs. (12.38) and (12.43), the factor $1 + KA_0 = 1 + g_m R_D R_2/(R_1 + R_2)$ plays a central role here. Our treatment of feedback circuits in this chapter will provide the foundation for this point.

Exercise In some applications, the input and output impedances of an amplifier must both be equal to 50 Ω. What relationship guarantees that the input and output impedances of the above circuit are equal?

The reader may raise several questions at this point. Do the input impedance and the output impedance always scale down and up, respectively? Is the modification of I/O impedances by feedback *desirable*? We consider one example here to illustrate a point and defer more rigorous answers to subsequent sections.

Example 12-8 The common-gate stage of Fig. 12.11 must drive a load resistance $R_L = R_D/2$. How much does the gain change (a) without feedback, (b) with feedback?

Solution (a) Without feedback [Fig. 12.13(a)], the CG gain is equal to $g_m(R_D||R_L) = g_m R_D/3$. That is, the gain drops by factor of three.

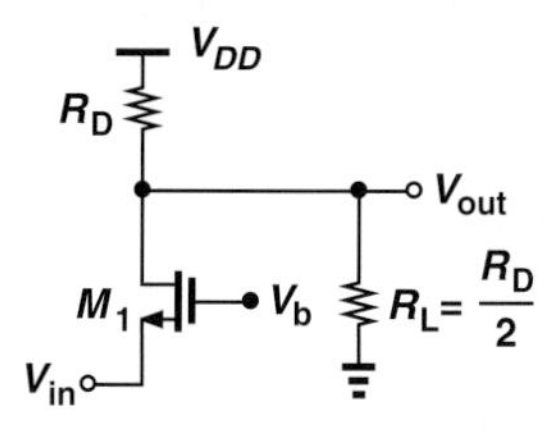

Figure 12.13

(b) With feedback, we use Eq. (12.33) but recognize that the open-loop gain has fallen to $g_m R_D/3$:

$$\frac{v_{out}}{v_{in}} = \frac{g_m R_D/3}{1 + \frac{R_2}{R_1 + R_2} g_m R_D/3} \tag{12.44}$$

$$= \frac{g_m R_D}{3 + \frac{R_2}{R_1 + R_2} g_m R_D}. \tag{12.45}$$

For example, if $g_m R_D R_2/(R_1 + R_2) = 10$, then this result differs from the "unloaded" gain expression in Eq. (12.33) by about 18%. Feedback therefore desensitizes the gain to load variations.

Exercise Repeat the above example for $R_L = R_D$.

12.2.4 Linearity Improvement

Consider a system having the input/output characteristic shown in Fig. 12.14(a). The nonlinearity observed here can also be viewed as the variation of the *slope* of the characteristic, i.e., the small-signal gain. For example, this system exhibits a gain of A_1 near $x = x_1$ and A_2 near $x = x_2$. If placed in a negative-feedback loop, the system provides a more uniform gain for different signal levels and, therefore, operates more linearly. In fact, as illustrated in Fig. 12.14(b) for the closed-loop system, we can write

$$\text{Gain at } x_1 = \frac{A_1}{1 + KA_1} \tag{12.46}$$

$$\approx \frac{1}{K}\left(1 - \frac{1}{KA_1}\right), \tag{12.47}$$

Did you know?

The large-signal behavior of circuits is generally nonlinear. For example, a low-quality audio amplifier driving a speaker suffers from nonlinearity as we increase the volume (i.e., the output signal swing); the result is the distinct distortion in the sound. A familiar type of nonlinearity is the type observed in differential pairs, e.g., the hyperbolic relationship obtained for bipolar implementations. In fact, the audio amplifier in your stereo system incorporates a differential pair at the input and other stages that can deliver a high output power [Fig. (a)]. To reduce distortion, the circuit is then placed in a negative feedback loop [Fig. (b)].

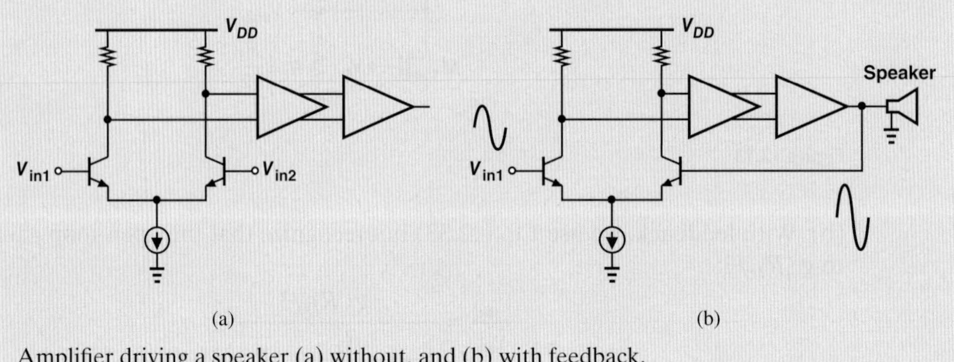

Amplifier driving a speaker (a) without, and (b) with feedback.

where it is assumed $KA_1 \gg 1$. Similarly,

$$\text{Gain at } x_2 = \frac{A_2}{1 + KA_2} \tag{12.48}$$

$$\approx \frac{1}{K}\left(1 - \frac{1}{KA_2}\right). \tag{12.49}$$

Thus, so long as KA_1 and KA_2 are large, the variation of the closed-loop gain with the signal level remains much less than that of the open-loop gain.

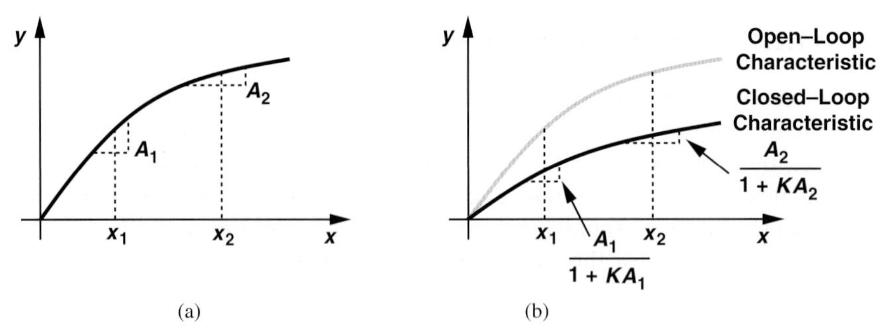

Figure 12.14 (a) Nonlinear open-loop characteristic of an amplifier, (b) improvement in linearity due to feedback.

All of the above attributes of negative feedback can also be considered a result of the minimal error property illustrated in Fig. 12.3. For example, if at different signal levels, the forward amplifier's gain varies, the feedback still ensures the feedback signal is a close replica of the input, and so is the output.

Robotics Application: Negative Feedback

Robots incorporate negative feedback for precise control of various motions. An interesting example arises in robotic hands, which are used to grasp objects and find a wide range of usage from "pick-and-place" functions in factories to surgery on humans.

When grasping an object, e.g., an egg, a robotic hand must apply enough pressure to keep the object from sliding out, but not so much as to crush it. This delicate operation is performed as illustrated in Fig. 12.15, where the robotic hand employs pressure sensors on its fingers. Each sensor behaves as a resistor, R_S, whose value depends on the pressure. The stage consisting of R_S, R_F, and the op amp generates

$$V_X = -\frac{R_F}{R_S}V_B, \tag{12.50}$$

i.e., a voltage that represents the pressure [1]. This result is digitized by an analog-to-digital converter (ADC) and subtracted from a reference value. The difference (the "error") controls the pressure applied by the finger to the object. This negative-feedback loop seeks to minimize the error between the ADC output and the reference, eventually settling with a certain pressure. Of course, the reference value must be chosen carefully.

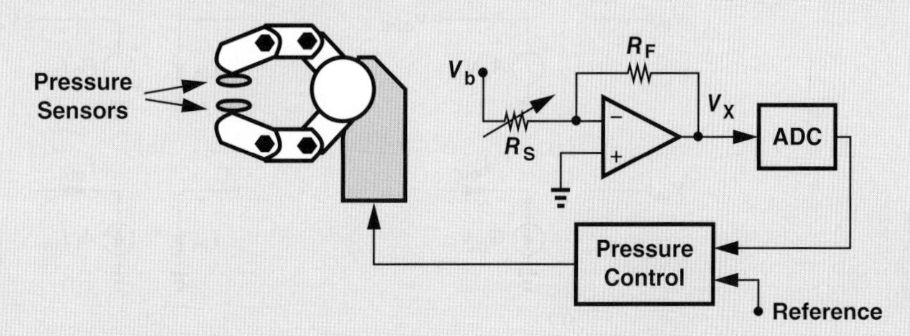

Figure 12.15 Negative-feedback loop for pressure control in a robotic arm.

12.3 TYPES OF AMPLIFIERS

The amplifiers studied thus far in this book sense and produce voltages. While less intuitive, other types of amplifiers also exist, i.e., those that sense and/or produce currents. Figure 12.16 depicts the four possible combinations along with their input and output impedances in the ideal case. For example, a circuit sensing a current must display a *low* input impedance to resemble a current meter. Similarly, a circuit generating an output current must achieve a *high* output impedance to approximate a current source. The reader is encouraged to confirm the other cases as well. The distinction among the four types of amplifiers becomes important in the analysis of feedback circuits. Note that the "current-voltage" and "voltage-current" amplifiers of Figs. 12.16(b) and (c) are commonly known as "transimpedance" and "transconductance" amplifiers, respectively.

12.3.1 Simple Amplifier Models

For our studies later in this chapter, it is beneficial to develop simple models for the four amplifier types. Depicted in Fig. 12.17 are the models for the ideal case. The voltage

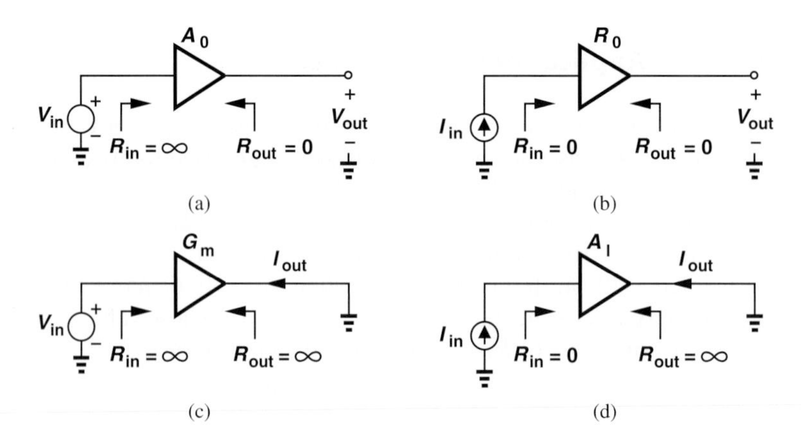

Figure 12.16 (a) Voltage, (b) transimpedance, (c) transconductance, and (d) current amplifiers.

amplifier in Fig. 12.17(a) provides an *infinite* input impedance so that it can sense voltages as an *ideal* voltmeter, i.e., without loading the preceding stage. Also, the circuit exhibits a *zero* output impedance so as to serve as an ideal voltage source, i.e., deliver $v_{out} = A_0 v_{in}$ regardless of the load impedance.

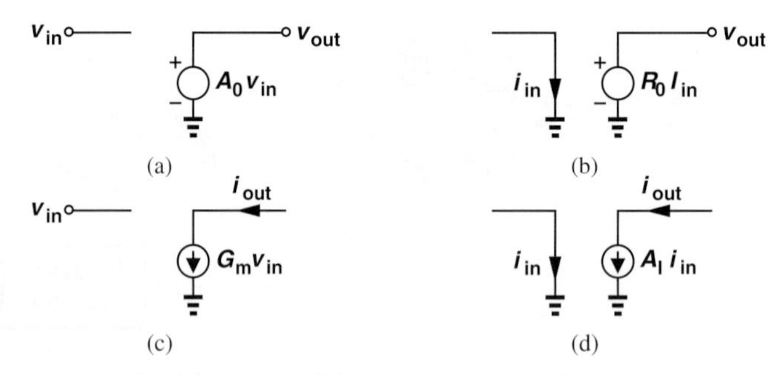

Figure 12.17 Ideal models for (a) voltage, (b) transimpedance, (c) transconductance, and (d) current amplifiers.

The transimpedance amplifier in Fig. 12.17(b) has a *zero* input impedance so that it can measure currents as an ideal current meter. Similar to the voltage amplifier, the output impedance is also zero if the circuit operates as an ideal voltage source. Note that the "transimpedance gain" of this amplifier, $R_0 = v_{out}/i_{in}$, has a dimension of resistance. For example, a transimpedance gain of 2 kΩ means a 1-mA change in the input current leads to a 2-V change at the output.

The I/O impedances of the topologies in Figs. 12.17(c) and (d) follow similar observations. It is worth noting that the amplifier of Fig. 12.17(c) has a "transconductance gain," $G_m = i_{out}/v_{in}$, with a dimension of transconductance.

In reality, the ideal models in Fig. 12.17 may not be accurate. In particular, the I/O impedances may not be negligibly large or small. Figure 12.18 shows more realistic models of the four amplifier types. Illustrated in Fig. 12.18(a), the voltage amplifier model contains an input resistance in *parallel* with the input port and an output resistance in *series* with the output port. These choices are unique and become clearer if we attempt other combinations. For example, if we envision the model as shown in Fig. 12.18(b), then the input and output impedances remain equal to infinity and zero, respectively, regardless of the values of R_{in} and R_{out}. (Why?) Thus, the topology of Fig. 12.18(a) serves as the only possible model representing finite I/O impedances.

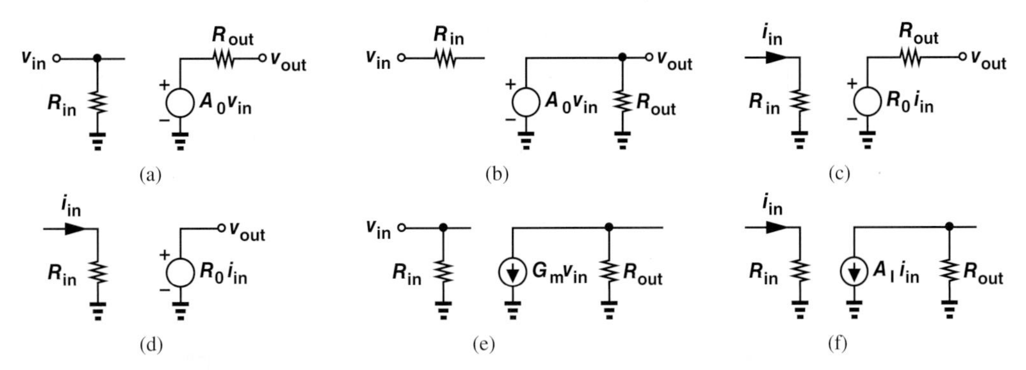

Figure 12.18 (a) Realistic model of voltage amplifier, (b) incorrect voltage amplifier model, (c) realistic model of transimpedance amplifier, (d) incorrect model of transimpedance amplifier, (e) realistic model of transconductance amplifier, (f) realistic model of current amplifier.

Figure 12.18(c) depicts a nonideal transimpedance amplifier. Here, the input resistance appears in *series* with the input. Again, if we attempt a model such as that in Fig. 12.18(d), the input resistance is zero. The other two amplifier models in Figs. 12.18(e) and (f) follow similar concepts.

12.3.2 Examples of Amplifier Types

It is instructive to study examples of the above four types. Figure 12.19(a) shows a cascade of a CS stage and a source follower as a "voltage amplifier." The circuit indeed provides a high input impedance (similar to a voltmeter) and a low output impedance (similar to a voltage source). Figure 12.19(b) depicts a cascade of a CG stage and a source follower as a transimpedance amplifier. Such a circuit displays low input and output impedances to serve as a "current sensor" and a "voltage generator." Figure 12.19(c) illustrates a single MOSFET as a transconductance amplifier. With high input and output impedances, the circuit efficiently senses voltages and generates currents. Finally, Fig. 12.19(d) shows a common-gate transistor as a current amplifier. Such a circuit must provide a low input impedance and a high output impedance.

Let us also determine the small-signal "gain" of each circuit in Fig. 12.19, assuming $\lambda = 0$ for simplicity. The voltage gain, A_0, of the cascade in Fig. 12.19(a) is equal to $-g_mR_D$ if $\lambda = 0$.[4] The gain of the circuit in Fig. 12.19(b) is defined as v_{out}/i_{in}, called the "transimpedance gain," and denoted by R_T. In this case, i_{in} flows through M_1 and R_D, generating a voltage equal to $i_{in}R_D$ at both the drain of M_1 and the source of M_2. That is, $v_{out} = i_{in}R_D$ and hence $R_T = R_D$.

For the circuit in Fig. 12.19(c), the gain is defined as i_{out}/v_{in}, called the "transconductance gain," and denoted by G_m. In this example, $G_m = g_m$. For the current amplifier in Fig. 12.19(d), the current gain, A_I, is equal to unity because the input current simply flows to the output.

[4]Recall from Chapter 7 that the gain of the source follower is equal to unity in this case.

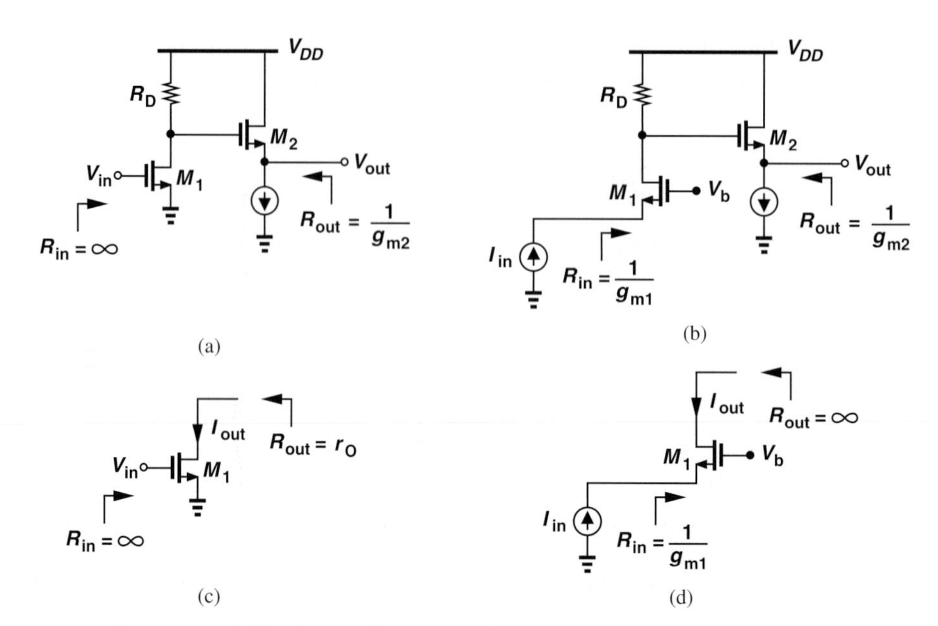

Figure 12.19 Examples of (a) voltage, (b) transimpedance, (c) transconductance, and (d) current amplifiers.

Example 12-9 With a current gain of unity, the topology of Fig. 12.19(d) appears hardly better than a piece of wire. What is the advantage of this circuit?

Solution The important property of this circuit lies in its input impedance. Suppose the current source serving as the input suffers from a large parasitic capacitance, C_p. If applied directly to a resistor R_D [Fig. 12.20(a)], the current would be wasted through C_p at high frequencies, exhibiting a −3 dB bandwidth of only $(R_D C_p)^{-1}$. On the other hand, the use of a CG stage [Fig. 12.20(b)] moves the input pole to g_m/C_p, a much higher frequency.

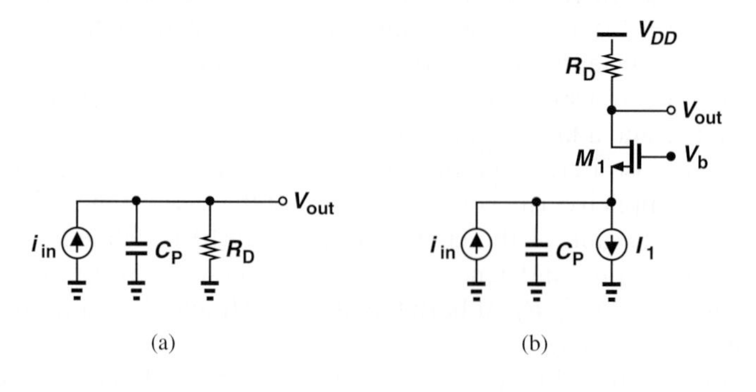

Figure 12.20

Exercise Determine the transfer function V_{out}/I_{in} for each of the above circuits.

12.4 SENSE AND RETURN TECHNIQUES

Recall from Section 12.1 that a feedback system includes means of sensing the output and "returning" the feedback signal to the input. In this section, we study such means so as to recognize them easily in a complex feedback circuit.

How do we measure the voltage across a port? We place a voltmeter in *parallel* with the port, and require that the voltmeter have a *high* input impedance so that it does not disturb the circuit [Fig. 12.21(a)]. By the same token, a feedback circuit sensing an output voltage must appear in parallel with the output and, ideally, exhibit an infinite impedance [Fig. 12.21(b)]. Shown in Fig. 12.21(c) is an example in which the resistive divider consisting of R_1 and R_2 senses the output voltage and generates the feedback signal, v_F. To approach the ideal case, $R_1 + R_2$ must be very large so that A_1 does not "feel" the effect of the resistive divider.

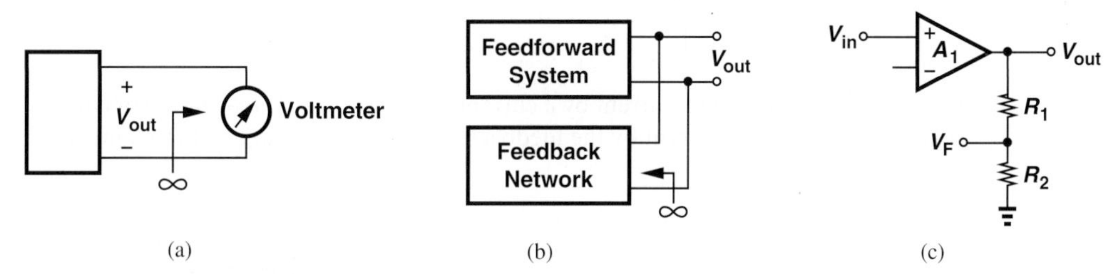

Figure 12.21 (a) Sensing a voltage by a voltmeter, (b) sensing the output voltage by the feedback network, (c) example of implementation.

How do we measure the current flowing through a wire? We *break* the wire and place a current meter in *series* with the wire [Fig. 12.22(a)]. The current meter in fact consists of a small resistor, so that it does not disturb the circuit, and a voltmeter that measures the voltage drop across the resistor [Fig. 12.22(b)]. Thus, a feedback circuit sensing an output current must appear in *series* with the output and, ideally, exhibit a zero impedance [Fig. 12.22(c)]. Depicted in Fig. 12.22(d) is an implementation of this concept. A resistor placed in series with the source of M_1 senses the output current, generating a proportional feedback voltage, V_F. Ideally, R_S is so small ($\ll 1/g_{m1}$) that the operation of M_1 remains unaffected.

To return a voltage or current to the input, we must employ a mechanism for adding or subtracting such quantities.[5] To add two voltage sources, we place them in *series* [Fig. 12.23(a)]. Thus, a feedback network returning a voltage must appear in series with the input signal [Fig. 12.23(b)], so that

$$v_e = v_{in} - v_F. \tag{12.51}$$

For example, as shown in Fig. 12.23(c), a differential pair can subtract the feedback voltage from the input. Alternatively, as mentioned in Example 12-7, a single transistor can operate as a voltage subtractor [Fig. 12.23(d)].

[5]Of course, only quantities having the same dimension can be added or subtracted. That is, a voltage cannot be added to a current.

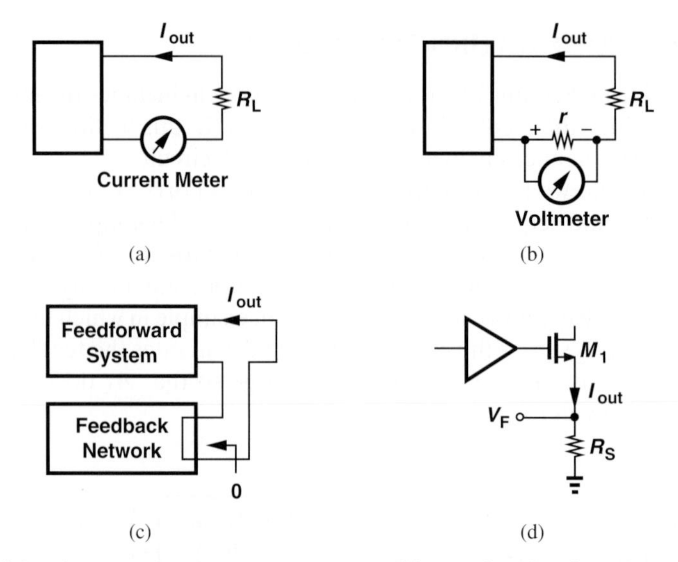

Figure 12.22 (a) Sensing a current by a current meter, (b) actual realization of current meter, (c) sensing the output current by the feedback network, (d) example of implementation.

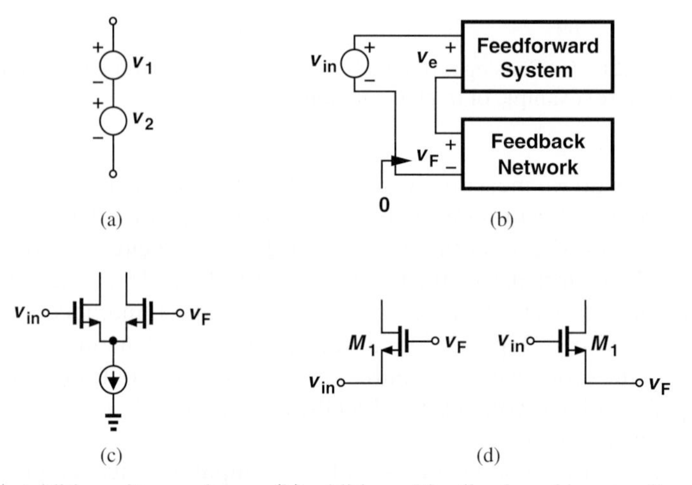

Figure 12.23 (a) Addition of two voltages, (b) addition of feedback and input voltages, (c) differential pair as a voltage subtractor, (d) single transistor as a voltage subtractor.

To add two current sources, we place them in *parallel* [Fig. 12.24(a)]. Thus, a feedback network returning a current must appear in parallel with the input signal, Fig. 12.24(b), so that

$$i_e = i_{in} - i_F. \tag{12.52}$$

For example, a transistor can return a current to the input [Fig. 12.24(c)]. So can a resistor if it is large enough to approximate a current source [Fig. 12.24(d)].

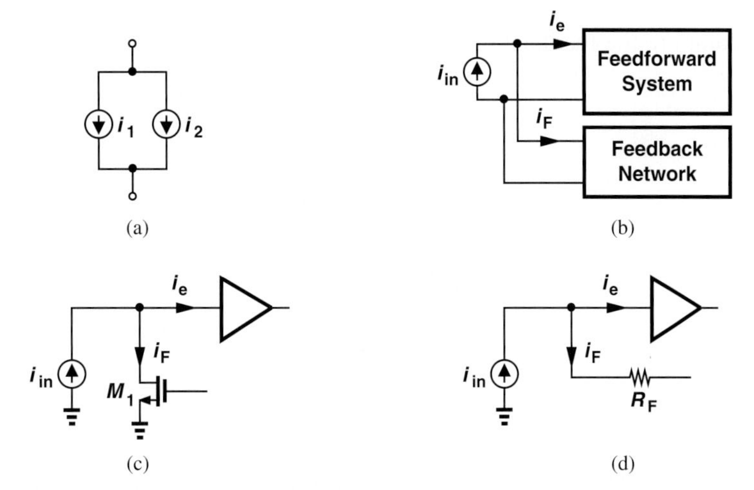

Figure 12.24 (a) Addition of two currents, (b) addition of feedback current and input current, (c) circuit realization, (d) another realization.

Example 12-10 Determine the types of sensed and returned signals in the circuit of Fig. 12.25.

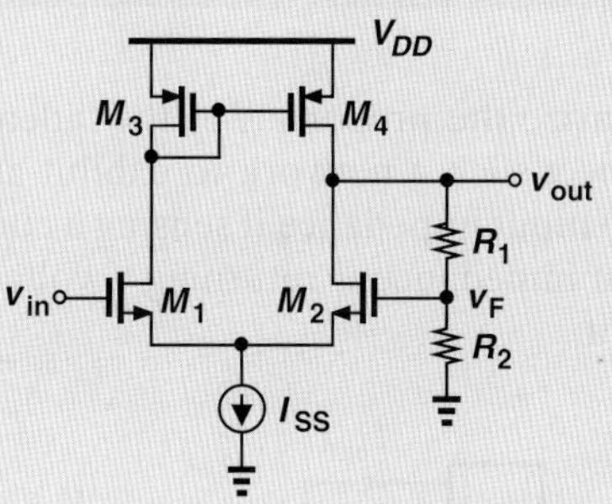

Figure 12.25

Solution This circuit is an implementation of the noninverting amplifier shown in Fig. 12.2. Here, the differential pair with the active load plays the role of an op amp. The resistive divider senses the output voltage and serves as the feedback network, producing $v_F = [R_2/(R_1 + R_2)]v_{out}$. Also, M_1 and M_2 operate as both part of the op amp (the forward system) and a voltage subtractor. The amplifier therefore combines the topologies in Figs. 12.21(c) and 12.23(c).

Exercise Repeat the above example if $R_2 = \infty$.

Example 12-11 Compute the feedback factor, K, for the circuit depicted in Fig. 12.26. Assume $\lambda = 0$.

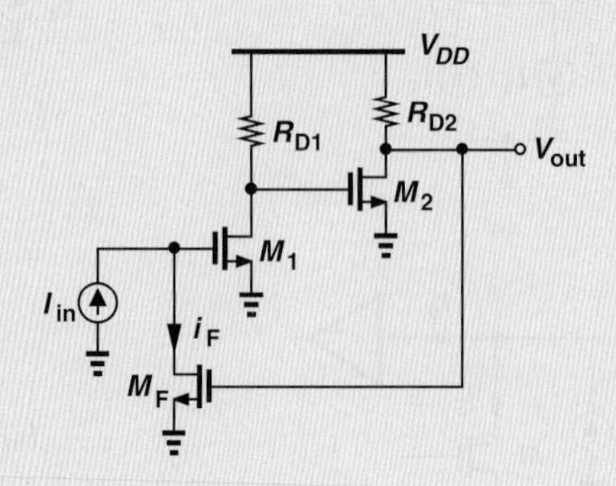

Figure 12.26

Solution Transistor M_F both senses the output voltage and returns a current to the input. The feedback factor is thus given by

$$K = \frac{i_F}{v_{out}} = g_{mF}, \tag{12.53}$$

where g_{mF} denotes the transconductance of M_F.

Exercise Calculate the feedback factor if M_F is degenerated by a resistor of value R_S.

Let us summarize the properties of the "ideal" feedback network. As illustrated in Fig. 12.27, we expect such a network to exhibit an infinite input impedance if sensing a voltage and a zero input impedance if sensing a current. Moreover, the network must provide a zero output impedance if returning a voltage and an infinite output impedance if returning a current.

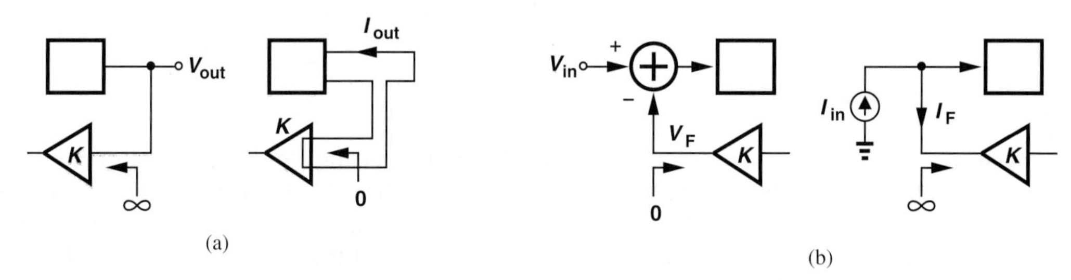

Figure 12.27 (a) Input impedance of ideal feedback networks for sensing voltage and current quantities, (b) output impedance of ideal feedback networks for producing voltage and current quantities.

12.5 POLARITY OF FEEDBACK

While the block diagram of a feedback system, e.g., Fig. 12.1, readily reveals the polarity of feedback, an actual circuit implementation may not. The procedure of determining this polarity involves three steps: (a) assume the input signal goes up (or down); (b) follow the change through the forward amplifier and the feedback network; (c) determine whether the returned quantity *opposes* or *enhances* the original "effect" produced by the input change. A simpler procedure is as follows: (a) set the input to zero; (b) break the loop; (c) apply a test signal, V_{test}, travel around the loop, examine the returned signal, V_{ret}, and determine the polarity of V_{ret}/V_{test}.

Example 12-12 Determine the polarity of feedback in the circuit of Fig. 12.28.

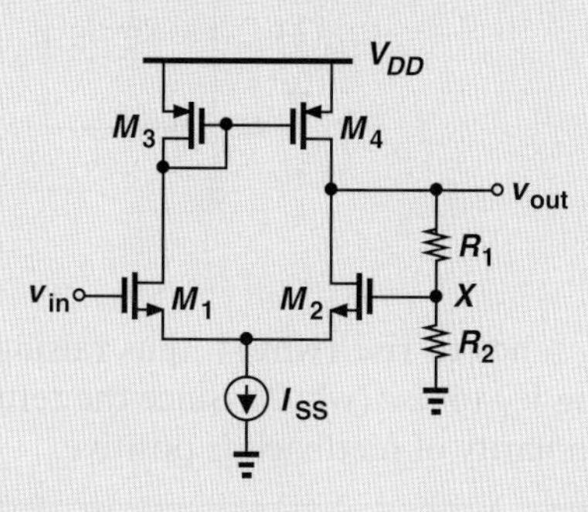

Figure 12.28

Solution If V_{in} goes up, I_{D1} tends to increase and I_{D2} tends to decrease. As a result, V_{out} and hence V_X tend to rise. The rise in V_X tends to increase I_{D2} and decrease I_{D1}, counteracting the effect of the change in V_{in}. The feedback is therefore negative. The reader is encouraged to apply the second procedure.

Exercise Suppose the top terminal of R_1 is tied to the drain of M_1 rather than the the drain of M_2. Determine the polarity of feedback.

Example 12-13 Determine the polarity of feedback in the circuit of Fig. 12.29.

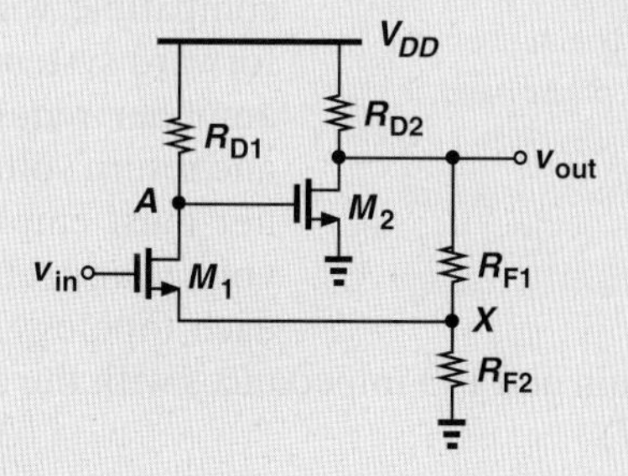

Figure 12.29

Solution If V_{in} goes up, I_{D1} tends to increase. Thus, V_A falls, V_{out} rises, and so does V_X. The rise in V_X tends to *reduce* I_{D1} (why?), thereby opposing the effect produced by V_{in}. The feedback is therefore negative.

Exercise Repeat the above example if M_2 is converted to a CG stage, i.e., its source is tied to node A and its gate to a bias voltage.

Example 12-14 Determine the polarity of feedback in the circuit of Fig. 12.30.

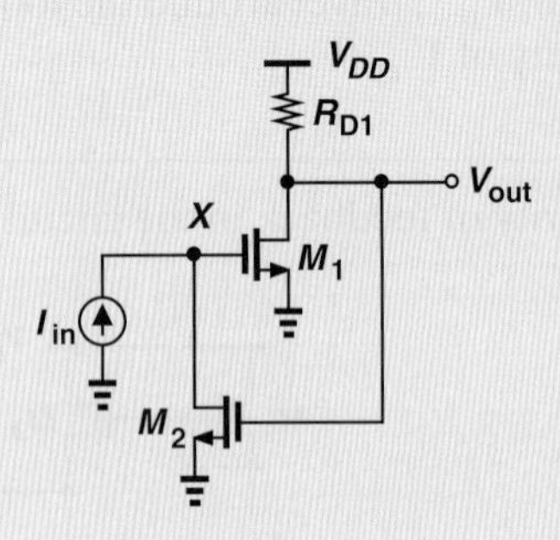

Figure 12.30

Solution If I_{in} goes up, V_X tends to rise (why?), thus raising I_{D1}. As a result, V_{out} falls and I_{D2} decreases, allowing V_X to *rise* (why?). Since the returned signal enhances the effect produced by I_{in}, the polarity of feedback is positive.

Exercise Repeat the above example if M_2 is a PMOS device (still operating as a CS stage). What happens if $R_D \to \infty$? Is this result expected?

12.6 FEEDBACK TOPOLOGIES

Our study of different types of amplifiers in Section 12.3 and sense and return mechanisms in Section 12.4 suggests that four feedback topologies can be constructed. Each topology includes one of four types of amplifiers as its forward system. The feedback network must, of course, sense and return quantities compatible with those produced and sensed by the forward system, respectively. For example, a voltage amplifier requires that the feedback network sense and return voltages, whereas a transimpedance amplifier must employ a feedback network that senses a voltage and returns a current. In this section, we study each topology and compute the closed-loop characteristics such as gain and I/O impedances with the assumption that the feedback network is ideal (Fig. 12.27).

Did you know?

While we focus on negative feedback in this book, positive feedback also finds its own applications. Suppose, for example, that a new ice cream store adopts a unique recipe and begins to attract many customers. If the store now adds one more scoop to each ice cream *for free*, then it will draw even more customers. That is, the store owner is providing a feedback signal that *enhances* the customer response (positive feedback). On the other hand, if the owner is greedy and *reduces* the size of the ice cream, then the store loses customers (negative feedback).

12.6.1 Voltage-Voltage Feedback

Illustrated in Fig. 12.31(a), this topology incorporates a voltage amplifier, requiring that the feedback network sense the output voltage and return a voltage to the subtractor. Recall from Section 12.4 that such a feedback network appears in *parallel* with the output and in *series* with the input,[6] ideally exhibiting an infinite input impedance and a zero output impedance.

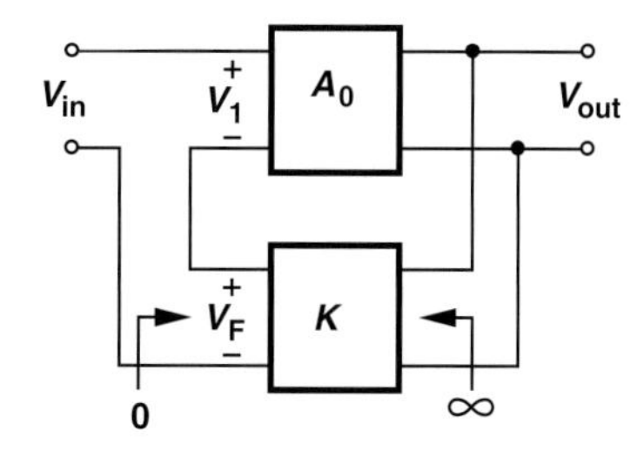

Figure 12.31 Voltage-voltage feedback.

We first calculate the closed-loop gain. Since

$$V_1 = V_{in} - V_F \tag{12.54}$$

$$V_{out} = A_0 V_1 \tag{12.55}$$

$$V_F = KV_{out}, \tag{12.56}$$

we have

$$V_{out} = A_0(V_{in} - KV_{out}), \tag{12.57}$$

and hence

$$\frac{V_{out}}{V_{in}} = \frac{A_0}{1 + KA_0}, \tag{12.58}$$

an expected result.

Example 12-15 Determine the closed-loop gain of the circuit shown in Fig. 12.32, assuming $R_1 + R_2$ is very large.

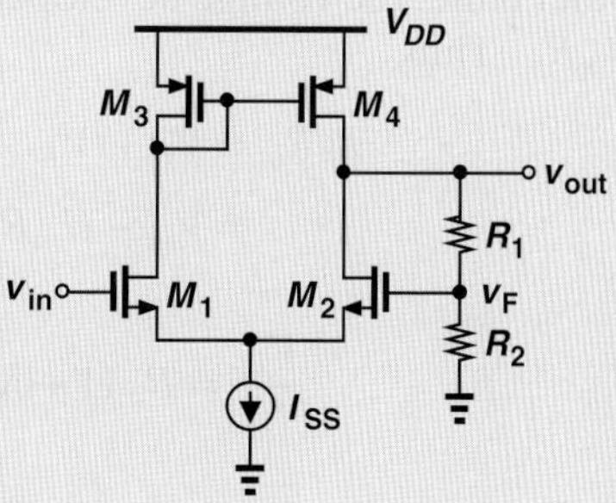

Figure 12.32

[6]For this reason, this type of feedback is also called the "series-shunt" topology, where the first term refers to the return mechanism at the input and the second term to the sense mechanism at the output.

Solution As evident from Examples 12-10 and 12-12, this topology indeed employs negative voltage-voltage feedback: the resistive network senses V_{out} with a high impedance (because $R_1 + R_2$ is very large), returning a voltage to the gate of M_2. As mentioned in Example 12-10, M_1 and M_2 serve as the input stage of the forward system and as a subtractor.

Noting that A_0 is the gain of the circuit consisting of M_1-M_4, we write from Chapter 10

$$A_0 = g_{mN}(r_{ON}||r_{OP}), \tag{12.59}$$

where the subscripts N and P refer to NMOS and PMOS devices, respectively.[7] With $K = R_2/(R_1 + R_2)$, we obtain

$$\frac{V_{out}}{V_{in}} = \frac{g_{mN}(r_{ON}||r_{OP})}{1 + \dfrac{R_2}{R_1 + R_2} g_{mN}(r_{ON}||r_{OP})}. \tag{12.60}$$

As expected, if the loop gain remains much greater than unity, then the closed-loop gain is approximately equal to $1/K = 1 + R_1/R_2$.

Exercise If $g_{mN} = 1/(100\ \Omega)$, $r_{ON} = 5\ \text{k}\Omega$, and $r_{OP} = 2\ \text{k}\Omega$, determine the required value of $R_2/(R_1 + R_2)$ for a closed loop gain of 4. Compare the result with the nominal value of $(R_2 + R_1)/R_2 = 4$.

In order to analyze the effect of feedback on the I/O impedances, we assume the forward system is a nonideal voltage amplifier (i.e., it exhibits finite I/O impedances) while the feedback network remains ideal. Depicted in Fig. 12.33 is the overall topology including a finite input resistance for the forward amplifier. Without feedback, of course, the entire input signal would appear across R_{in}, producing an input current of V_{in}/R_{in}.[8] With feedback, on the other hand, the voltage developed at the input of A_0 is equal to $V_{in} - V_F$ and also equal to $I_{in}R_{in}$. Thus,

$$I_{in}R_{in} = V_{in} - V_F \tag{12.61}$$

$$= V_{in} - (I_{in}R_{in})A_0K. \tag{12.62}$$

It follows that

$$\frac{V_{in}}{I_{in}} = R_{in}(1 + KA_0). \tag{12.63}$$

Interestingly, negative feedback around a voltage amplifier *raises* the input impedance by the universal factor of one plus the loop gain. This impedance modification brings the circuit closer to an ideal voltage amplifier.

[7]We observe that $R_1 + R_2$ must be much greater than $r_{ON}||r_{OP}$ for this to hold. This serves as the definition of $R_1 + R_2$ being "very large."

[8]Note that V_{in} and R_{in} carry equal currents because the feedback network must appear in *series* with the input [Fig. 12.23(a)].

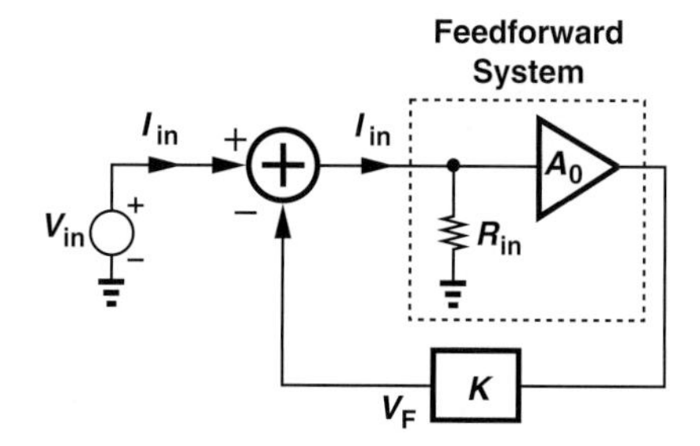

Figure 12.33 Calculation of input impedance.

Example 12-16 Determine the input impedance of the stage shown in Fig. 12.34(a) if $R_1 + R_2$ is very large.

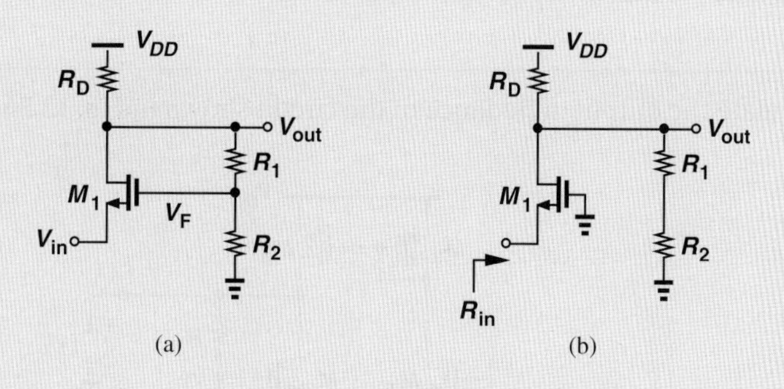

Figure 12.34

Solution We first open the loop to calculate R_{in} in Eq. (12.63). To open the loop, we break the gate of M_1 from the feedback signal and tie it to ground [Fig. 12.34(b)]:

$$R_{in} = \frac{1}{g_m}. \tag{12.64}$$

The closed-loop input impedance is therefore given by

$$\frac{V_{in}}{I_{in}} = \frac{1}{g_m}\left(1 + \frac{R_2}{R_1 + R_2} g_m R_D\right). \tag{12.65}$$

Exercise What happens if $R_2 \to \infty$? Is this result expected?

The effect of feedback on the output impedance can be studied with the aid of the diagram shown in Fig. 12.35, where the forward amplifier exhibits an output impedance of R_{out}. Expressing the error signal at the input of A_0 as $-V_F = -KV_X$, we write the output voltage of A_0 as $-KA_0V_X$ and hence

$$I_X = \frac{V_X - (-KA_0V_X)}{R_{out}}, \tag{12.66}$$

where the current drawn by the feedback network is neglected. Thus,

$$\frac{V_X}{I_X} = \frac{R_{out}}{1 + KA_0}, \tag{12.67}$$

revealing that negative feedback *lowers* the output impedance if the topology senses the output voltage. The circuit is now a better voltage amplifier—as predicted by our gain desensitization analysis in Section 12.2.

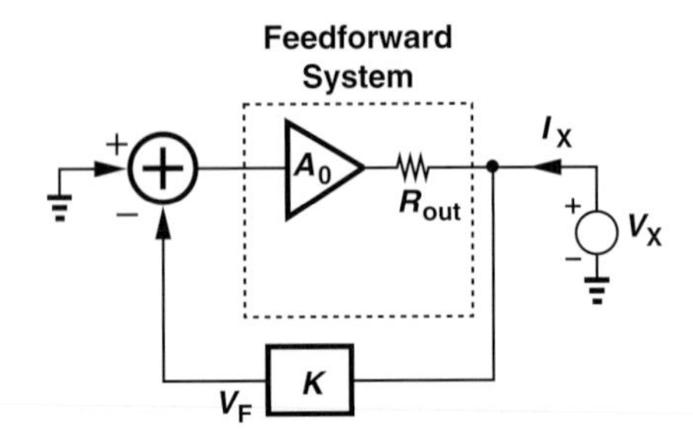

Figure 12.35 Calculation of output impedance.

Example 12-17 Calculate the output impedance of the circuit shown in Fig. 12.36 if $R_1 + R_2$ is very large.

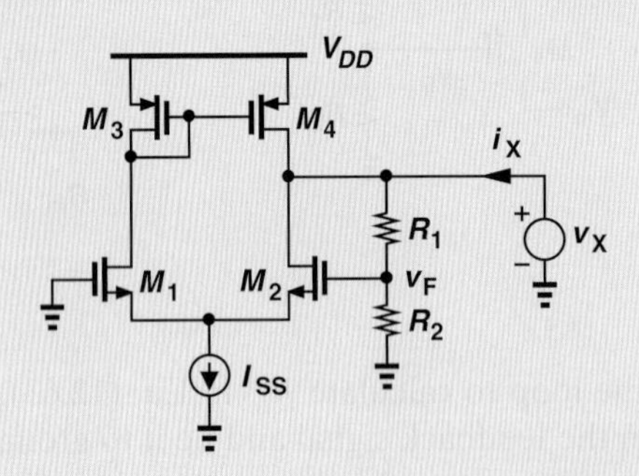

Figure 12.36

Solution Recall from Example 12-15 that the open-loop output impedance is equal to $r_{ON}||r_{OP}$ and $KA_0 = [R_2/(R_1 + R_2)]g_{mN}(r_{ON}||r_{OP})$. Thus, the closed-loop output impedance, $R_{out,closed}$, is given by

$$R_{out,closed} = \frac{r_{ON}||r_{OP}}{1 + \frac{R_2}{R_1 + R_2} g_{mN}(r_{ON}||r_{OP})}. \tag{12.68}$$

If the loop gain is much greater than unity,

$$R_{out,closed} \approx \left(1 + \frac{R_1}{R_2}\right)\frac{1}{g_{mN}}, \tag{12.69}$$

a value independent of r_{ON} and r_{OP}. In other words, while the open-loop amplifier suffers from a *high* output impedance, the application of negative feedback lowers R_{out} to a multiple of $1/g_{mN}$.

Exercise What happens if $R_2 \to \infty$? Can you prove this result by direct analysis of the circuit?

In summary, voltage-voltage feedback lowers the gain and the output impedance by $1 + KA_0$ and raises the input impedance by the same factor.

12.6.2 Voltage-Current Feedback

Depicted in Fig. 12.37, this topology employs a transimpedance amplifier as the forward system, requiring that the feedback network sense the output voltage and return a current to the subtractor. In our terminology, the first term in "voltage-current feedback" refers to the quantity *sensed* at the output, and the second, to the quantity returned to the input. (This terminology is not standard.) Also, recall from Section 12.4 that such a feedback network must appear in parallel with the output and with the input,[9] ideally providing both an infinite input impedance and an infinite output impedance (why?). Note that the feedback factor in this case has a dimension of *conductance* because $K = I_F/V_{out}$.

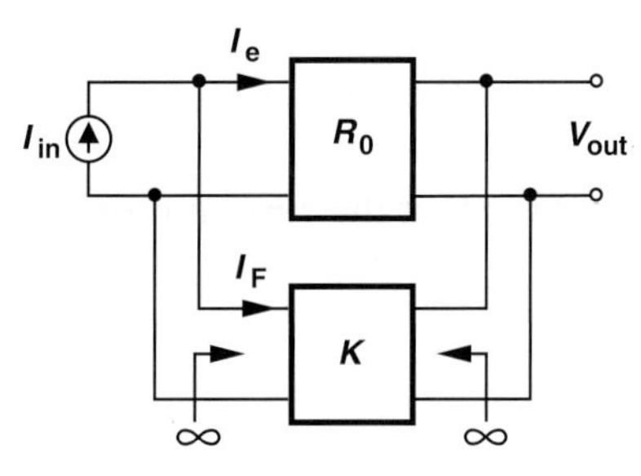

Figure 12.37 Voltage-current feedback.

Did you know?

Most electronic devices incorporate negative feedback in their "voltage regulators." Since the integrated circuits in a device such as a cell phone must operate with a tight supply voltage range, a regulator follows the battery so as to deliver a relatively constant voltage as the battery discharges and its voltage falls. The figure below shows an example, where Q_1, the resistor divider, and the op amp form a negative-feedback loop. To regulate the output voltage, V_{reg}, the loop divides it and compares the result with a reference voltage, V_{REF}. With a high loop gain, we must have $V_{reg}R_1/(R_1 + R_2) \approx V_{REF}$ and hence $V_{reg} \approx (1 + R_2/R_1)V_{REF}$. That is, the output is independent of V_{bat} and A_1. Of course, V_{REF} itself must remain constant. This voltage is provided by a Zener diode in discrete implementations or a "bandgap circuit" in integrated designs.

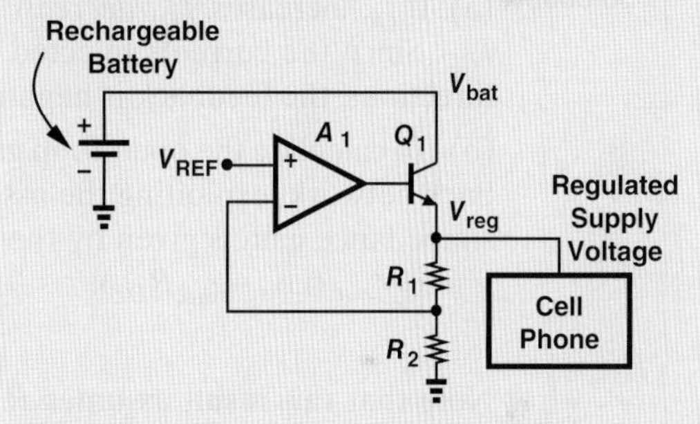

Voltage regulator using negative feedback.

We first compute the closed-loop gain, expecting to obtain a familiar result. Since $I_e = I_{in} - I_F$ and $V_{out} = I_eR_0$, we have

$$V_{out} = (I_{in} - I_F)R_0 \tag{12.70}$$

$$= (I_{in} - KV_{out})R_0, \tag{12.71}$$

and hence

$$\frac{V_{out}}{I_{in}} = \frac{R_0}{1 + KR_0}. \tag{12.72}$$

[9]For this reason, this type is also called "shunt-shunt" feedback.

Example 12-18

For the circuit shown in Fig. 12.38(a), assume $\lambda = 0$ and R_F is very large and (a) prove that the feedback is negative; (b) calculate the open-loop gain; (c) calculate the closed-loop gain.

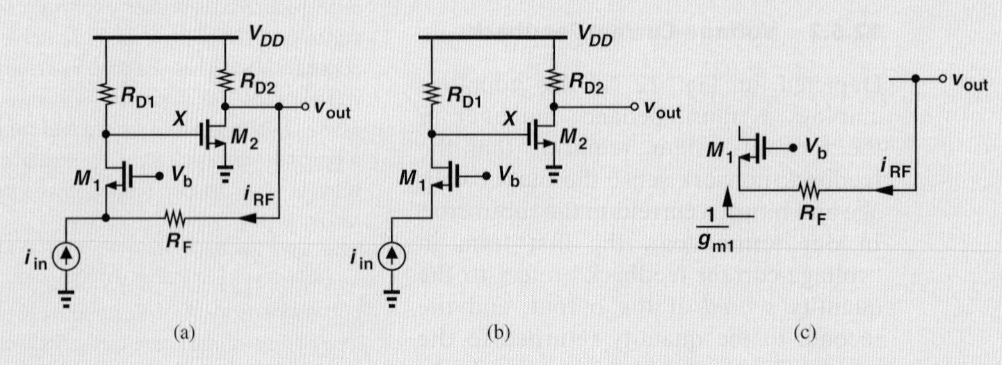

Figure 12.38

Solution

(a) If I_{in} increases, I_{D1} *decreases* and V_X rises. As a result, V_{out} falls, thereby reducing I_{RF}. Since the currents injected by I_{in} and R_F into the input node change in opposite directions, the feedback is negative.

(b) To calculate the open-loop gain, we consider the forward amplifier without the feedback network, exploiting the assumption that R_F is very large [Fig. 12.38(b)]. The transimpedance gain is given by the gain from I_{in} to V_X (i.e., R_{D1}) multiplied by that from V_X to V_{out} (i.e., $-g_{m2}R_{D2}$):

$$R_0 = R_{D1}(-g_{m2}R_{D2}). \tag{12.73}$$

Note that this result assumes $R_F \gg R_{D2}$ so that the gain of the second stage remains equal to $-g_{m2}R_{D2}$.

(c) To obtain the closed-loop gain, we first note that the current returned by R_F to the input is approximately equal to V_{out}/R_F if R_F is very large. To prove this, we consider a section of the circuit as in Fig. 12.38(c) and write

$$I_{RF} = \frac{V_{out}}{R_F + \dfrac{1}{g_{m1}}}. \tag{12.74}$$

Thus, if $R_F \gg 1/g_{m1}$, the returned current is approximately equal to V_{out}/R_F. (We say "R_F operates as a current source.") That is, $K = -1/R_F$, where the negative sign arises from the direction of the current *drawn* by R_F from the input node with respect to that in Fig. 12.37. Forming $1 + KR_0$, we express the closed-loop gain as

$$\left.\frac{V_{out}}{I_{in}}\right|_{closed} = \frac{-g_{m2}R_{D1}R_{D2}}{1 + \dfrac{g_{m2}R_{D1}R_{D2}}{R_F}}, \tag{12.75}$$

which reduces to $-R_F$ if $g_{m2}R_{D1}R_{D2} \gg R_F$.

It is interesting to note that the assumption that R_F is very large translates to two conditions in this example: $R_F \gg R_{D2}$ and $R_F \gg 1/g_{m1}$. The former arises from the output network calculations and the latter from the input network calculations. What happens if one or both of these assumptions are not valid? We deal with this (relatively common) situation in Section 12.7.

Exercise What is the closed-loop gain if $R_{D1} \to \infty$? How can this result be interpreted? (Hint: the infinite open-loop gain creates a virtual ground node at the source of M_1.)

We now proceed to determine the closed-loop I/O impedances. Modeling the forward system as an ideal transimpedance amplifier but with a finite input impedance R_{in} (Section 12.3), we construct the test circuit shown in Fig. 12.39. Since the current flowing through R_{in} is equal to V_X/R_{in} (why?), the forward amplifier produces an output voltage equal to $(V_X/R_{in})R_0$ and hence

$$I_F = K\frac{V_X}{R_{in}}R_0. \tag{12.76}$$

Writing a KCL at the input node thus yields

$$I_X - K\frac{V_X}{R_{in}}R_0 = \frac{V_X}{R_{in}} \tag{12.77}$$

and hence

$$\frac{V_X}{I_X} = \frac{R_{in}}{1 + KR_0}. \tag{12.78}$$

That is, a feedback loop returning current to the input *lowers* the input impedance by a factor of one plus the loop gain, bringing the circuit closer to an ideal "current sensor."

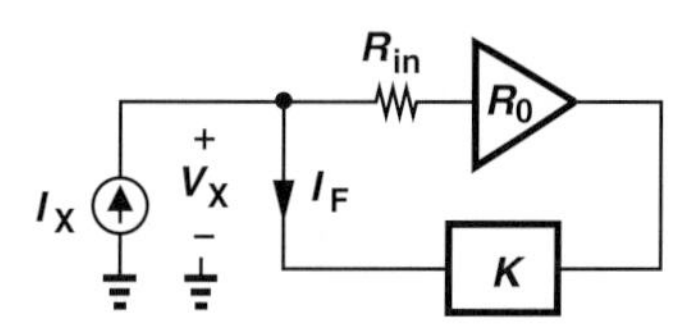

Figure 12.39 Calculation of input impedance.

Example 12-19 Determine the closed-loop input impedance of the circuit studied in Example 12-18.

Solution The open-loop amplifier shown in Fig. 12.38(b) exhibits an input impedance $R_{in} = 1/g_{m1}$ because R_F is assumed to be very large. With $1 + KR_0$ from the denominator of Eq. (12.75), we obtain

$$R_{in,closed} = \frac{1}{g_{m1}} \cdot \frac{1}{1 + \dfrac{g_{m2}R_{D1}R_{D2}}{R_F}}. \tag{12.79}$$

Exercise Explain what happens if $R_{D1} \to \infty$ and why.

From our study of voltage-voltage feedback in Section 12.6.1, we postulate that voltage-current feedback too lowers the output impedance because a feedback loop "regulating" the output voltage tends to stabilize it despite load impedance variations.

Drawing the circuit as shown in Fig. 12.40, where the input current source is set to zero and R_{out} models the open-loop output resistance, we observe that the feedback network produces a current of $I_F = KV_X$. Upon flowing through the forward amplifier, this current translates to $V_A = -KV_X R_0$ and hence

$$I_X = \frac{V_X - V_A}{R_{out}} \tag{12.80}$$

$$= \frac{V_X + KV_X R_0}{R_{out}}, \tag{12.81}$$

where the current drawn by the feedback network is neglected. Thus,

$$\frac{V_X}{I_X} = \frac{R_{out}}{1 + KR_0}, \tag{12.82}$$

an expected result.

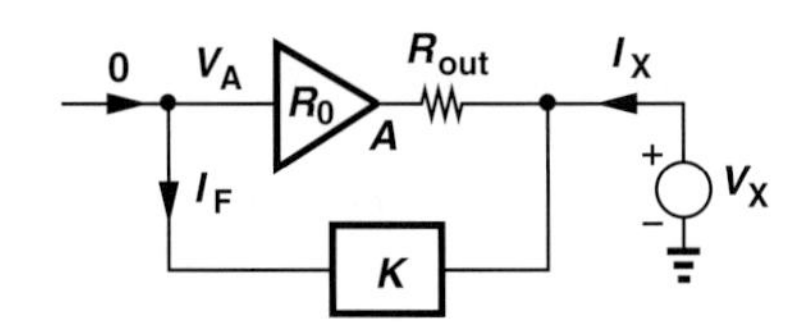

Figure 12.40 Calculation of output impedance.

Example 12-20 Calculate the closed-loop output impedance of the circuit studied in Example 12-18.

Solution From the open-loop circuit in Fig. 12.38(b), we have $R_{out} \approx R_{D2}$ because R_F is assumed very large. Writing $1 + KR_0$ from the denominator of Eq. (12.75) gives

$$R_{out,closed} = \frac{R_{D2}}{1 + \dfrac{g_{m2}R_{D1}R_{D2}}{R_F}}. \tag{12.83}$$

Exercise Explain what happens if $R_{D1} \to \infty$ and why.

12.6.3 Current-Voltage Feedback

Shown in Fig. 12.41(a), this topology incorporates a transconductance amplifier, requiring that the feedback network sense the output current and return a voltage to the subtractor.

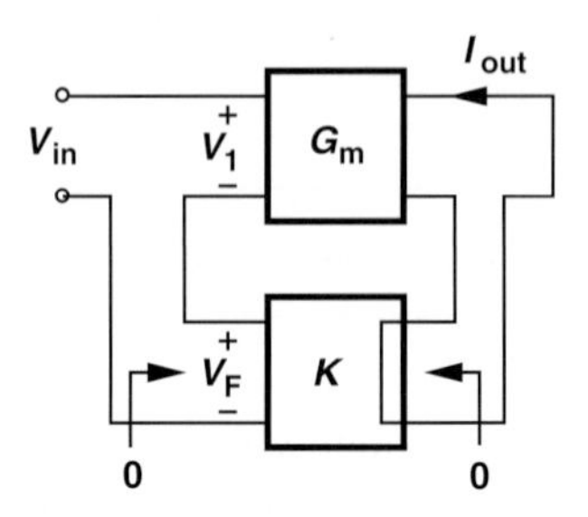

Figure 12.41 Current-voltage feedback.

Again, in our terminology, the first term in "current-voltage feedback" refers to the quantity sensed at the output, and the second, to the quantity returned to the input. Recall from Section 12.4 that such a feedback network must appear in series with the output and with the input,[10] ideally exhibiting zero input and output impedances. Note that the feedback factor in this case has a dimension of *resistance* because $K = V_F/I_{out}$.

Let us first confirm that the closed-loop gain is equal to the open-loop gain divided by one plus the loop gain. Since the forward system produces a current equal to $I_{out} = G_m(V_{in} - V_F)$ and since $V_F = KI_{out}$, we have

$$I_{out} = G_m(V_{in} - KI_{out}) \tag{12.84}$$

and hence

$$\frac{I_{out}}{V_{in}} = \frac{G_m}{1 + KG_m}. \tag{12.85}$$

Example 12-21

We wish to deliver a well-defined current to a laser diode as shown in Fig. 12.42(a),[11] but the transconductance of M_1 is poorly controlled. For this reason, we "monitor" the current by inserting a small resistor R_M in series, sensing the voltage across R_M, and returning the result to the input of an op amp [Fig. 12.42(b)]. Estimate I_{out} if the op amp provides a very high gain. Calculate the closed-loop gain for the implementation shown in Fig. 12.42(c).

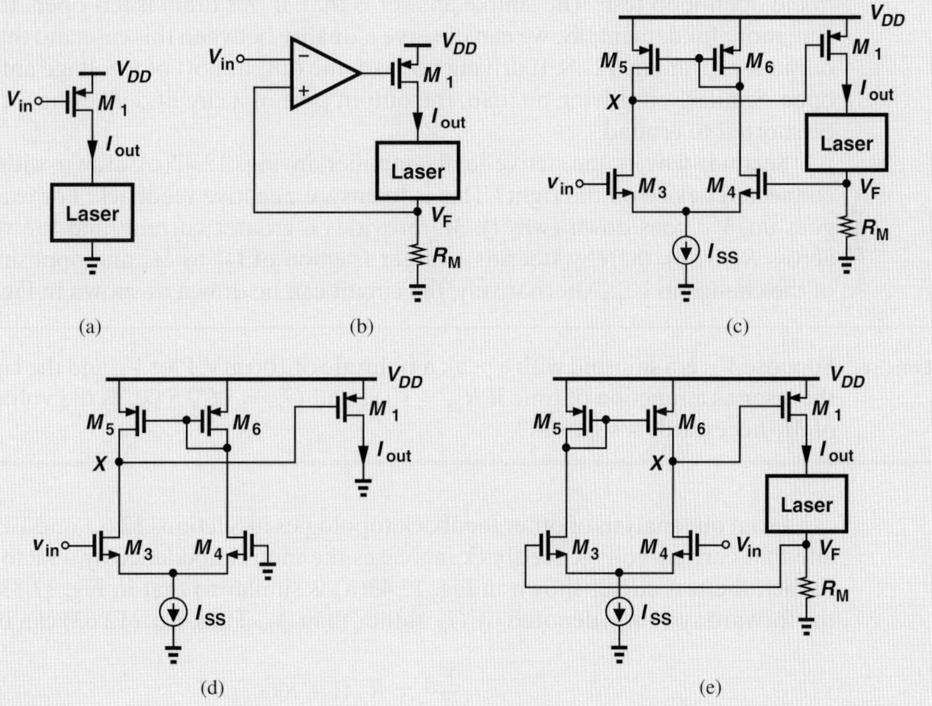

Figure 12.42

[10]For this reason, this type is also called "series-series" feedback.

[11]Laser diodes convert electrical signals to optical signals and are widely used in DVD players, long-distance communications, etc.

Solution If the gain of the op amp is very high, the difference between V_{in} and V_F is very small. Thus, R_M sustains a voltage equal to V_{in} and hence

$$I_{out} \approx \frac{V_{in}}{R_M}. \tag{12.86}$$

We now determine the open-loop gain of the transistor-level implementation in Fig. 12.42(c). The forward amplifier can be identified as shown in Fig. 12.42(d), where the gate of M_4 is grounded because the feedback signal (voltage) is set to zero. Since $I_{out} = -g_{m1}V_X$ (why?) and $V_X = -g_{m3}(r_{O3}||r_{O5})V_{in}$, we have

$$G_m = g_{m1}g_{m3}(r_{O3}||r_{O5}). \tag{12.87}$$

The feedback factor $K = V_F/I_{out} = R_M$. Thus,

$$\left.\frac{I_{out}}{V_{in}}\right|_{closed} = \frac{g_{m1}g_{m3}(r_{O3}||r_{O5})}{1 + g_{m1}g_{m3}(r_{O3}||r_{O5})R_M}. \tag{12.88}$$

Note that if the loop gain is much greater than unity, then

$$\left.\frac{I_{out}}{V_{in}}\right|_{closed} \approx \frac{1}{R_M}. \tag{12.89}$$

We must now answer two questions. First, why is the drain of M_1 *shorted* to ground in the open-loop test? The simple answer is that, if this drain is left open, then $I_{out} = 0$! But, more fundamentally, we can observe a duality between this case and that of voltage outputs, e.g., in Fig. 12.38. If driving no load, the output port of a voltage amplifier is left open. Similarly, if driving no load, the output port of a circuit delivering a current must be shorted to ground.

Second, why is the active-load amplifier in Fig. 12.42(c) drawn with the diode-connected device on the right? This is to ensure negative feedback. For example, if V_{in} goes up, V_X goes down (why?), M_1 provides a greater current, and the voltage drop across R_M rises, thereby steering a larger fraction of I_{SS} to M_4 and opposing the effect of the change in V_{in}. Alternatively, the circuit can be drawn as shown in Fig. 12.42(e).

Exercise Suppose V_{in} is a sinusoid with a peak amplitude of 100 mV. Plot V_F and the current through the laser as a function of time if $R_M = 10\ \Omega$ and $G_m = 1/(0.5\ \Omega)$. Is the voltage at the gate of M_1 necessarily a sinusoid?

From our analysis of other feedback topologies in Sections 12.6.1 and 12.6.2, we postulate that current-voltage feedback increases the input impedance by a factor of $1 + KG_m$. In fact, the test circuit shown in Fig. 12.43(a) is similar to that in Fig. 12.33—except that the forward system is denoted by G_m rather than A_0. Thus, Eq. (12.63) can be rewritten as

$$\frac{V_{in}}{I_{in}} = R_{in}(1 + KG_m). \tag{12.90}$$

The output impedance is calculated using the test circuit of Fig. 12.43(b). Note that, in contrast to the cases in Figs. 12.35 and 12.40, the test voltage source is inserted in *series* with the output port of the forward amplifier and the input port of the feedback network. The voltage developed at port A is equal to $-KI_X$ and the current drawn by the G_m stage

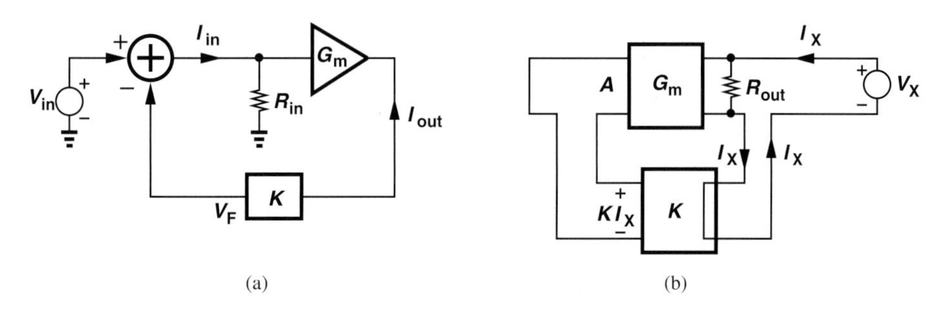

Figure 12.43 Calculation of (a) input and (b) output impedances.

equal to $-KG_mI_X$. Since the current flowing through R_{out} is given by V_X/R_{out}, a KCL at the output node yields

$$I_X = \frac{V_X}{R_{out}} - KG_mI_X \tag{12.91}$$

and hence

$$\frac{V_X}{I_X} = R_{out}(1 + KG_m). \tag{12.92}$$

Interestingly, a negative feedback loop sensing the output current *raises* the output impedance, bringing the circuit closer to an ideal current generator. As in other cases studied thus far, this occurs because negative feedback tends to regulate the output quantity that it senses.

Example 12-22 An alternative approach to regulating the current delivered to a laser diode is shown in Fig. 12.44(a). As in the circuit of Fig. 12.42(b), the very small resistor R_M monitors the current, generating a proportional voltage and feeding it back to the subtracting device, M_1. Determine the closed-loop gain and I/O impedances of the circuit.

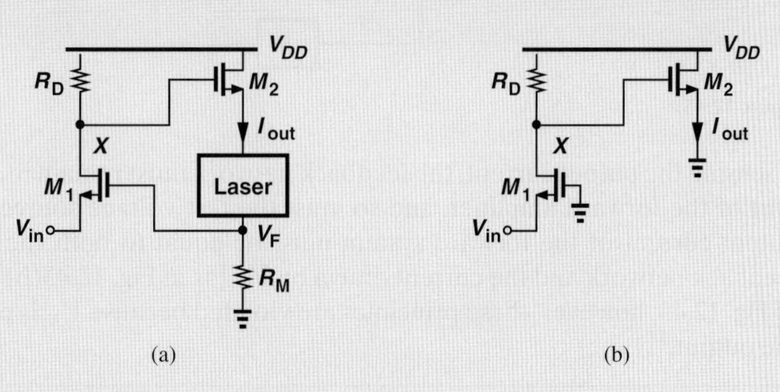

Figure 12.44

Solution Since R_M is very small, the open-loop circuit reduces to that shown in Fig. 12.44(b), where the gain can be expressed as

$$G_m = \frac{V_X}{V_{in}} \cdot \frac{I_{out}}{V_X} \tag{12.93}$$

$$= g_{m1}R_D \cdot g_{m2}. \tag{12.94}$$

The input impedance is equal to $1/g_{m1}$ and the output impedance equal to $1/g_{m2}$.[12] The feedback factor is equal to R_M, yielding

$$\left.\frac{I_{out}}{V_{in}}\right|_{closed} = \frac{g_{m1}g_{m2}R_D}{1+g_{m1}g_{m2}R_DR_M}, \tag{12.95}$$

which reduces to $1/R_M$ if the loop gain is much greater than unity. The input impedance rises by a factor of $1 + G_mR_M$:

$$R_{in,closed} = \frac{1}{g_{m1}}(1+g_{m1}g_{m2}R_DR_M), \tag{12.96}$$

and so does the output impedance (i.e., that seen by the laser):

$$R_{out,closed} = \frac{1}{g_{m2}}(1+g_{m1}g_{m2}R_DR_M). \tag{12.97}$$

Exercise If an input impedance of 500 Ω and an output impedance of 5 kΩ are desired, determine the required values of g_{m1} and g_{m2}. Assume $R_D = 1$ kΩ and $R_M = 100$ Ω.

Example 12-23 A student attempts to calculate the output impedance of the current-voltage feedback topology with the aid of circuit depicted in Fig. 12.45. Explain why this topology is an incorrect representation of the actual circuit.

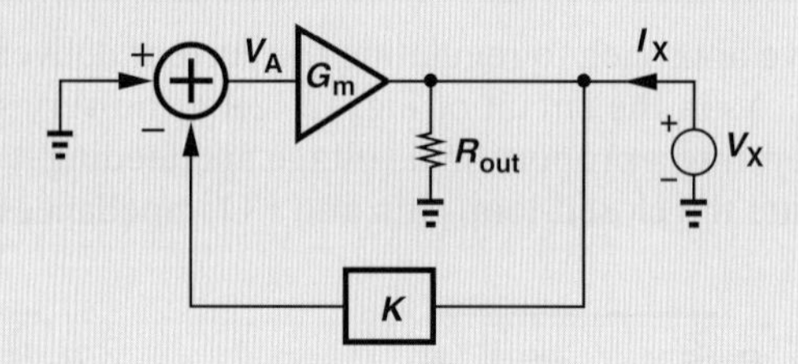

Figure 12.45

Solution If sensing the output current, the feedback network must remain in *series* with the output port of the forward amplifier, and so must the test voltage source. In other words, the output current of the forward system must be equal to both the input current of the feedback network and the current drawn by V_X [as in Fig. 12.43(b)]. In the arrangement of Fig. 12.45, however, these principles are violated because V_X is placed in parallel with the output.[13]

Exercise Apply the above (incorrect) test to the circuit of Fig. 12.44 and examine the results.

[12]To measure the output impedance, the test voltage source must be placed in series with the output wire.

[13]If the feedback network is ideal and hence has a zero input impedance, then V_X must supply an infinite current.

12.6.4 Current-Current Feedback

From the analysis of the first three feedback topologies, we predict that this type lowers the gain, raises the output impedance, and lowers the input impedance, all by a factor of one plus the loop gain.

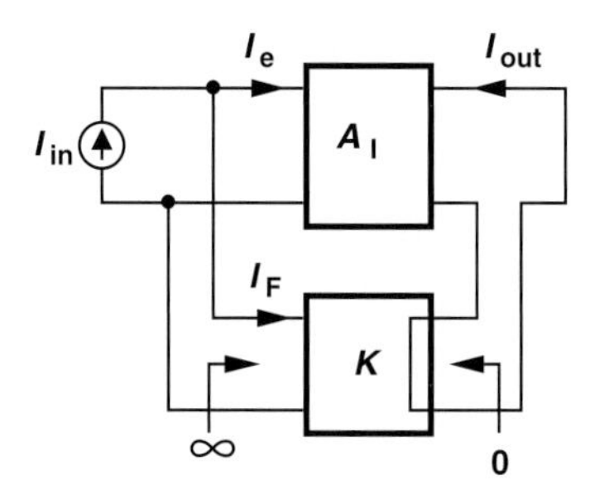

Figure 12.46 Current-current feedback.

As shown in Fig. 12.46, current-current feedback senses the output in series and returns the signal in parallel with the input. The forward system has a current gain of A_I and the feedback network a dimensionless gain of $K = I_F/I_{out}$. Given by $I_{in} - I_F$, the current entering the forward amplifier yields

$$I_{out} = A_I(I_{in} - I_F) \tag{12.98}$$

$$= A_I(I_{in} - KI_{out}) \tag{12.99}$$

and hence

$$\frac{I_{out}}{I_{in}} = \frac{A_I}{1 + KA_I}. \tag{12.100}$$

The input impedance of the circuit is calculated with the aid of the arrangement depicted in Fig. 12.47. As in the case of voltage-current feedback (Fig. 12.39), the input impedance of the forward amplifier is modeled by a series resistor, R_{in}. Since the current flowing through R_{in} is equal to V_X/R_{in}, we have $I_{out} = A_I V_X/R_{in}$ and hence $I_F = KA_I V_X/R_{in}$. A KCL at the input node therefore gives

$$I_X = \frac{V_X}{R_{in}} + I_F \tag{12.101}$$

$$= \frac{V_X}{R_{in}} + KA_I\frac{V_X}{R_{in}}. \tag{12.102}$$

That is,

$$\frac{V_X}{I_X} = \frac{R_{in}}{1 + KA_I}. \tag{12.103}$$

Did you know?

While less intuitive than voltage regulation, current regulation is in fact quite common, namely, in battery chargers. A battery charged with a constant *voltage* may have a short lifetime. (Imagine what happens if two ideal voltage sources of unequal values are placed in parallel!) For this reason, we prefer to charge batteries with a constant current. The figure below shows an example where op amp A_1 forces V_X to be close to the reference voltage, V_{REF}. We say R_S and A_1 monitor and stabilize the source current of M_1. The drain current of M_1 drives the battery. Of course, as the battery charges, its voltage eventually exceeds the permitted value, causing damage. The charger must therefore incorporate additional circuitry for overcharge protection.

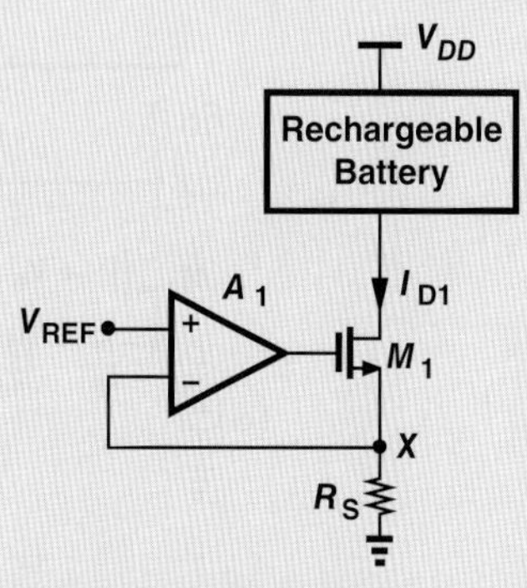

Current regulation in a battery charger.

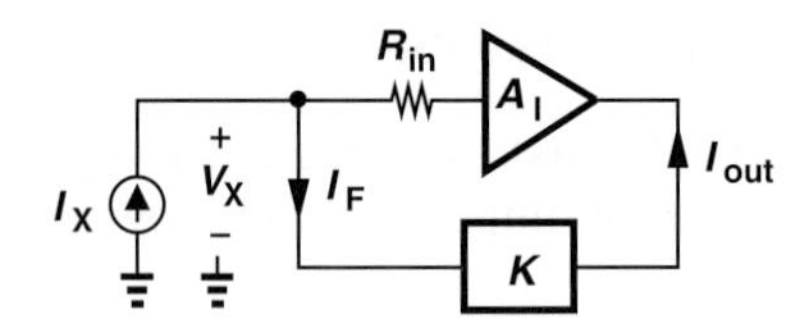

Figure 12.47 Calculation of input impedance.

For the output impedance, we utilize the test circuit shown in Fig. 12.48, where the input is left open and V_X is inserted in series with the output port. Since $I_F = KI_X$, the forward amplifier produces an output current equal to $-KA_I I_X$. Noting that R_{out} carries a current of V_X/R_{out} and writing a KCL at the output node, we have

$$I_X = \frac{V_X}{R_{out}} - KA_I I_X. \tag{12.104}$$

It follows that

$$\frac{V_X}{I_X} = R_{out}(1 + KA_I). \tag{12.105}$$

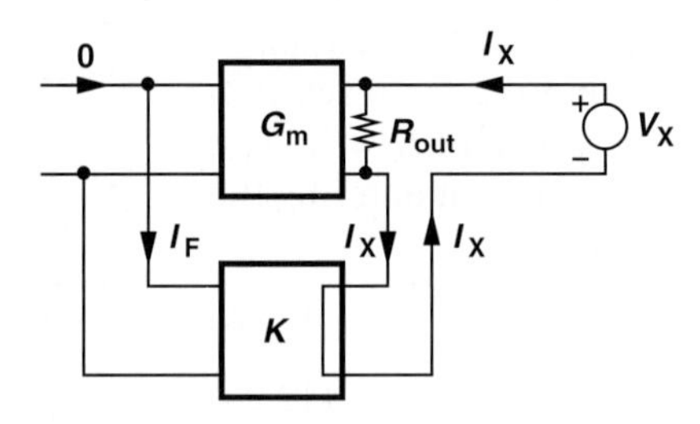

Figure 12.48 Calculation of output impedance.

Example 12-24

Consider the circuit shown in Fig. 12.49(a), where the output current delivered to a laser diode is regulated by negative feedback. Prove that the feedback is negative and compute the closed-loop gain and I/O impedances if R_M is very small and R_F very large.

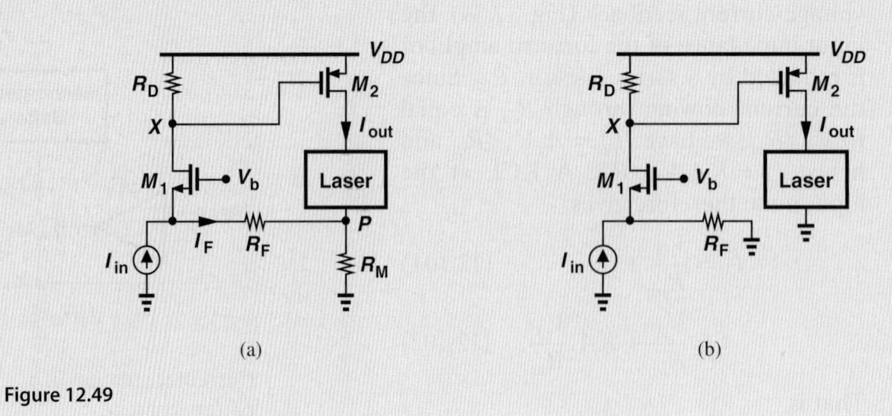

Figure 12.49

Solution Suppose I_{in} increases. Then, the source voltage of M_1 tends to rise, and so does its drain voltage (why?). As a result, the overdrive of M_2 decreases, I_{out} and hence V_P fall, and I_F *increases*, thereby lowering the source voltage of M_1. Since the feedback signal, I_F, opposes the effect produced by I_{in}, the feedback is negative.

We must now analyze the open-loop system. Since R_M is very small, we assume V_P remains near zero, arriving at the open-loop circuit depicted in Fig. 12.49(b). The assumption that R_F is very large ($\gg 1/g_{m1}$) indicates that almost all of I_{in} flows through M_1 and R_D, thus generating $V_X = I_{in}R_D$ and hence

$$I_{out} = -g_{m2}V_X \tag{12.106}$$

$$= -g_{m2}R_D I_{in}. \tag{12.107}$$

That is,

$$A_I = -g_{m2}R_D. \tag{12.108}$$

The input impedance is approximately equal to $1/g_{m1}$ and the output impedance is equal to r_{O2}.

To obtain the closed-loop parameters, we must compute the feedback factor, I_F/I_{out}. Recall from Example 12-18 that the current returned by R_F can be approximated as $-V_P/R_F$ if $R_F \gg 1/g_{m1}$. We also note that $V_P = I_{out}R_M$, concluding that

$$K = \frac{I_F}{I_{out}} \tag{12.109}$$

$$= \frac{-V_P}{R_F} \cdot \frac{1}{I_{out}} \tag{12.110}$$

$$= -\frac{R_M}{R_F}. \tag{12.111}$$

The closed-loop parameters are therefore given by:

$$A_{I,closed} = \frac{-g_{m2}R_D}{1 + g_{m2}R_D\dfrac{R_M}{R_F}} \tag{12.112}$$

$$R_{in,closed} = \frac{1}{g_{m1}} \cdot \frac{1}{1 + g_{m2}R_D\dfrac{R_M}{R_F}} \tag{12.113}$$

$$R_{out,closed} = r_{O2}\left(1 + g_{m2}R_D\frac{R_M}{R_F}\right). \tag{12.114}$$

Note that if $g_{m2}R_DR_M/R_F \gg 1$, then the closed-loop gain is simply given by $-R_F/R_M$.

Exercise Noting that $R_{out}|_{closed}$ is the impedance seen by the laser in the closed-loop circuit, construct a Norton equivalent for the entire circuit that drives the laser.

The effect of feedback on the input and output impedances of the forward amplifier is summarized in Fig. 12.50.

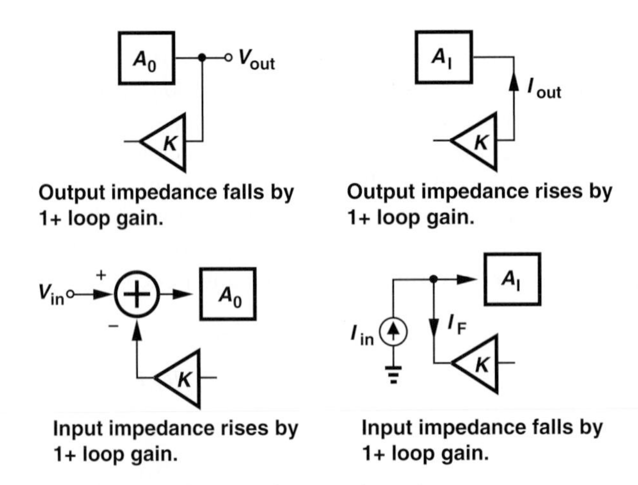

Figure 12.50 Effect of feedback on input and output impedances.

12.7 EFFECT OF NONIDEAL I/O IMPEDANCES

Our study of feedback topologies in Section 12.6 has been based on idealized models for the feedback network, always assuming that the I/O impedances of this network are very large or very small depending on the type of feedback. In practice, however, the finite I/O impedances of the feedback network may considerably alter the performance of the circuit, thereby necessitating analysis techniques to account for these effects. In such cases, we say the feedback network "loads" the forward amplifier and the "loading effects" must be determined.

Before delving into the analysis, it is instructive to understand the difficulty in the context of an example.

Example 12-25 Suppose in the circuit of Example 12-7, $R_1 + R_2$ is *not* much greater than R_D. How should we analyze the circuit?

Solution In Example 12-7, we constructed the open-loop circuit by simply neglecting the effect of $R_1 + R_2$. Here, on the other hand, $R_1 + R_2$ tends to reduce the open-loop gain because it appears in parallel with R_D. We therefore surmise that the open-loop circuit must be configured as shown in Fig. 12.51, with the open-loop gain given by

$$A_O = g_{m1}[R_D||(R_1 + R_2)], \tag{12.115}$$

and the output impedance

$$R_{out,open} = R_D||(R_1 + R_2). \tag{12.116}$$

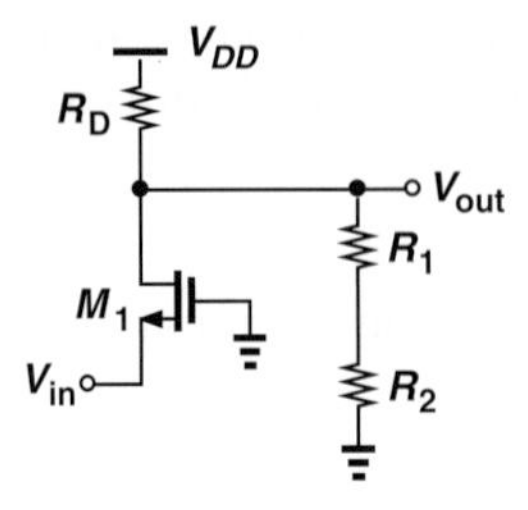

Figure 12.51

Other forward and feedback parameters are identical to those calculated in Example 12-7. Thus,

$$A_{v,closed} = \frac{g_{m1}[R_D||(R_1 + R_2)]}{1 + \frac{R_2}{R_1 + R_2} g_{m1}[R_D||(R_1 + R_2)]} \tag{12.117}$$

$$R_{in,closed} = \frac{1}{g_{m1}} \left\{ 1 + \frac{R_2}{R_1 + R_2} g_{m1}[R_D||(R_1 + R_2)] \right\} \tag{12.118}$$

$$R_{out,closed} = \frac{R_D||(R_1 + R_2)}{1 + \frac{R_2}{R_1 + R_2} g_{m1}[R_D||(R_1 + R_2)]}. \tag{12.119}$$

Exercise Repeat the above example if R_D is replaced with an ideal current source.

The above example easily lends itself to intuitive inspection. But many other circuits do not. To gain more confidence in our analysis and deal with more complex circuits, we must develop a systematic approach.

12.7.1 Inclusion of I/O Effects

We present a methodology here that allows the analysis of the four feedback topologies even if the I/O impedances of the forward amplifier or the feedback network depart from their ideal values. The methodology is based on a formal proof that is somewhat beyond the scope of this book and can be found in [2].

Our methodology proceeds in six steps:

1. Identify the forward amplifier.
2. Identify the feedback network.
3. Break the feedback network according to the rules described below.
4. Calculate the open-loop parameters.
5. Determine the feedback factor according to the rules described below.
6. Calculate the closed-loop parameters.

Rules for Breaking the Feedback Network The third step is carried out by "duplicating" the feedback network at both the input and the output of the overall system. Illustrated in Fig. 12.52, the idea is to "load" both the input and the output of the forward amplifier by proper copies of the feedback network. The copy tied to the output is called the "sense duplicate" and that connected to the input, the "return duplicate." We must also decide what to do with the output port of the former and the input port of the latter, i.e., whether to short or open these ports. This is accomplished through the use of the "termination" rules depicted in Fig. 12.53. For example, for voltage-voltage feedback [Fig. 12.53(a)], the output port of the sense replica is left *open* while the input of the return duplicate is shorted. Similarly, for voltage-current feedback [Fig. 12.53(b)], both the output port of the sense duplicate and the input port of the return duplicate are shorted.

The formal proof of these concepts is given in [2] but it is helpful to remember these rules based on the following intuitive (but not quite rigorous) observations. In an ideal situation, a feedback network sensing an output *voltage* is driven by a zero impedance,

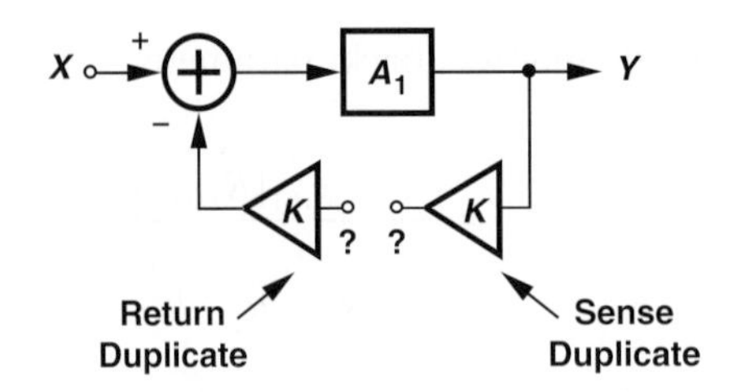

Figure 12.52 Method of breaking the feedback loop.

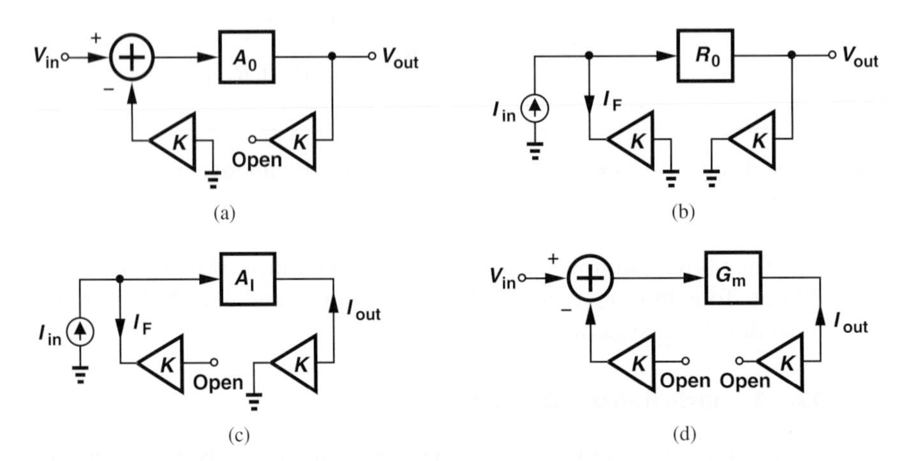

Figure 12.53 Proper termination of duplicates in (a) voltage-voltage, (b) voltage-current, (c) current-current, and (d) current-voltage feedback.

namely, the output impedance of the forward amplifier. Thus, the input port of the *return* duplicate is *shorted*. Moreover, a feedback network returning a voltage to the input ideally sees an infinite impedance, namely, the input impedance of the forward amplifier. Thus, the output port of the *sense* duplicate is left open. Similar observations apply to the other three cases.

Calculation of Feedback Factor The fifth step entails the calculation of the feedback factor, a task requiring the rules illustrated in Fig. 12.54. Depending on the type of feedback, the output port of the feedback network is shorted or opened, and the ratio of the output current or voltage to the input is defined as the feedback factor. For example, in a voltage-voltage feedback topology, the output port of the feedback network is open [Fig. 12.54(a)] and $K = V_2/V_1$.

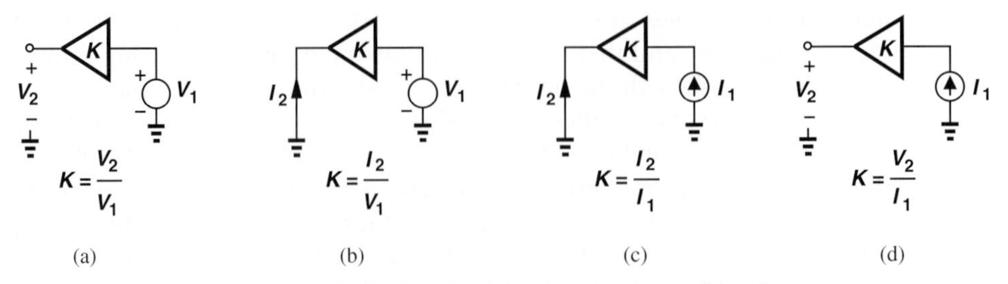

Figure 12.54 Calculation of feedback factor for (a) voltage-voltage, (b) voltage-current, (c) current-current, and (d) current-voltage feedback.

The proof of these rules is provided in [2], but an intuitive view can also be developed. First, the stimulus (voltage or current) applied to the *input* of the feedback network is of the same type as the quantity sensed at the *output* of the forward amplifier. Second, the output port of the feedback network is opened (shorted) if the returned quantity is a voltage (current)—just as in the case of the sense duplicates in Fig. 12.53. Of course, if the output port of the feedback network is left open, the quantity of interest is a voltage, V_2. Similarly, if the port is shorted, the quantity of interest is a current, I_2.

In order to reinforce the above principles, we reconsider the examples studied thus far in this chapter and determine the closed-loop parameters if I/O impedance effects are not negligible.

Example 12-26 Analyze the amplifier depicted in Fig. 12.55(a) if $R_1 + R_2$ is not much less than R_D.

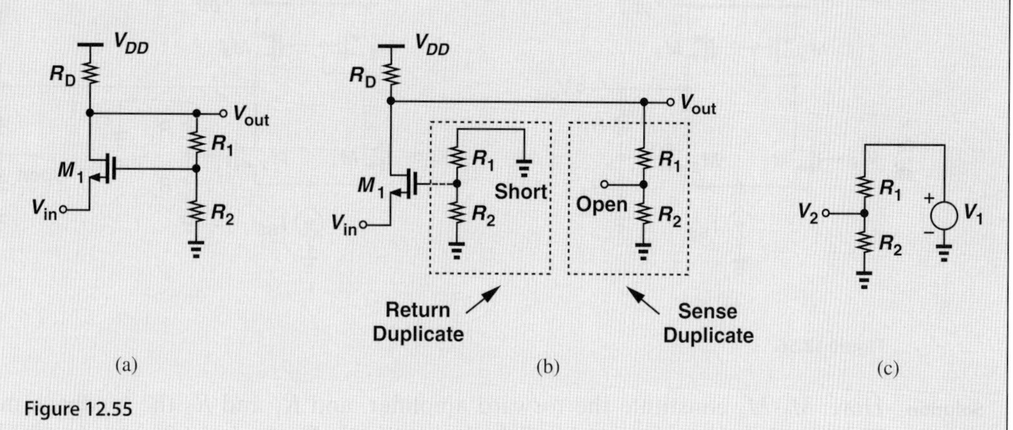

Figure 12.55

Solution We identify the forward system as M_1 and R_D, and the feedback network as R_1 and R_2. We construct the open-loop circuit according to Fig. 12.53(a), as shown in Fig. 12.55(b). Note that the feedback network appears twice. The sense duplicate output port is left open and the input port of the return duplicate is shorted. The open-loop parameters of this topology were computed in Example 12-25.

To determine the feedback factor, we follow the rule in Fig. 12.54(a) to form the circuit shown in Fig. 12.55(c), arriving at

$$K = \frac{V_2}{V_1} \tag{12.120}$$

$$= \frac{R_2}{R_1 + R_2}. \tag{12.121}$$

It follows that

$$KA_0 = \frac{R_2}{R_1 + R_2} g_{m1}[R_D || (R_1 + R_2)], \tag{12.122}$$

and hence

$$A_{v,closed} = \frac{g_{m1}[R_D || (R_1 + R_2)]}{1 + \frac{R_2}{R_1 + R_2} g_{m1}[R_D || (R_1 + R_2)]} \tag{12.123}$$

$$R_{in,closed} = \frac{1}{g_{m1}} \left\{ 1 + \frac{R_2}{R_1 + R_2} g_{m1}[R_D || (R_1 + R_2)] \right\} \tag{12.124}$$

$$R_{out,closed} = \frac{R_D||(R_1+R_2)}{1+\frac{R_2}{R_1+R_2}g_{m1}[R_D||(R_1+R_2)]}. \tag{12.125}$$

Obtained through our general methodology, these results agree with those found by inspection in Example 12-25.

Exercise Repeat the above analysis if R_D is replaced with an ideal current source.

Example 12-27 Analyze the circuit of Fig. 12.56(a) if $R_1 + R_2$ is not much greater than $r_{OP}||r_{ON}$.

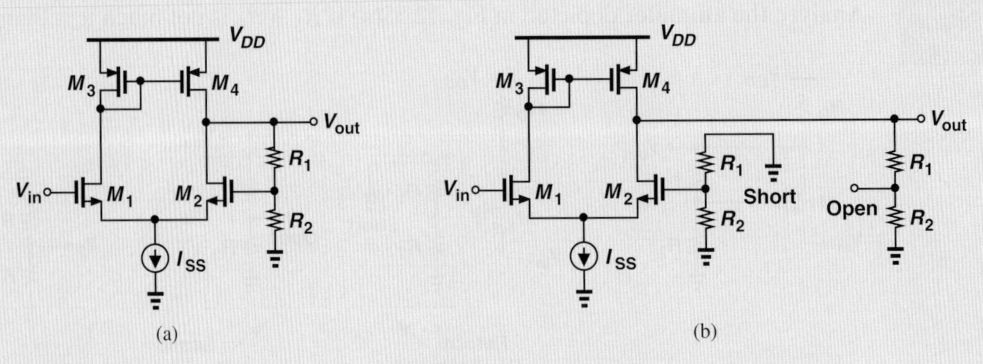

Figure 12.56

Solution Here, M_1-M_4 constitute the forward amplifier, and R_1 and R_2 the feedback network. The loop is broken in a manner similar to that in Example 12-26 because the type of feedback is the same [Fig. 12.56(b)]. The open-loop parameters are therefore given by

$$A_0 = g_{mN}[r_{ON}||r_{OP}||(R_1+R_2)] \tag{12.126}$$

$$R_{in,open} = \infty \tag{12.127}$$

$$R_{out,open} = r_{ON}||r_{OP}||(R_1+R_2). \tag{12.128}$$

The test circuit for calculation of the feedback factor is identical to that in Fig. 12.55(c), yielding

$$K = \frac{R_2}{R_1+R_2}. \tag{12.129}$$

It follows that

$$\left.\frac{V_{out}}{V_{in}}\right|_{closed} = \frac{g_{mN}[r_{ON}||r_{OP}||(R_1+R_2)]}{1+\frac{R_2}{R_1+R_2}g_{mN}[r_{ON}||r_{OP}||(R_1+R_2)]} \tag{12.130}$$

$$R_{in,closed} = \infty \tag{12.131}$$

$$R_{out,closed} = \frac{r_{ON}||r_{OP}||(R_1+R_2)}{1+\frac{R_2}{R_1+R_2}g_{mN}[r_{ON}||r_{OP}||(R_1+R_2)]}. \tag{12.132}$$

Exercise Repeat the above example if a load resistor of R_L is tied between the output of the circuit and ground.

Example 12-28 Analyze the circuit of Fig. 12.57(a).

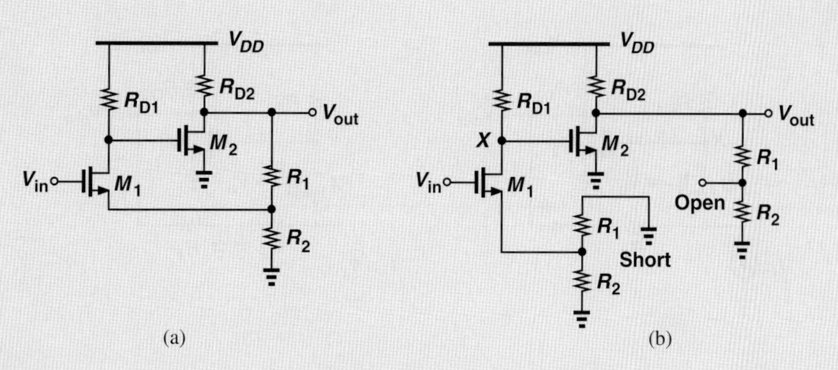

Figure 12.57

Solution We identify the forward system as M_1, R_{D1}, M_2, and R_{D2}. The feedback network consists of R_1 and R_2 and returns a voltage to the source of the subtracting transistor, M_1. In a manner similar to the above two examples, the open-loop circuit is constructed as shown in Fig. 12.57(b). Note that M_1 is now degenerated by $R_1||R_2$. Writing $A_0 = (V_X/V_{in})(V_{out}/V_X)$, we have

$$A_0 = \frac{-R_{D1}}{\dfrac{1}{g_m} + R_1||R_2} \cdot \{-g_{m2}[R_{D2}||(R_1 + R_2)]\} \tag{12.133}$$

$$R_{in,open} = \infty \tag{12.134}$$

$$R_{out,open} = R_{D2}||(R_1 + R_2). \tag{12.135}$$

As in the above example, the feedback factor is equal to $R_2/(R_1 + R_2)$, yielding

$$\left.\frac{V_{out}}{V_{in}}\right|_{closed} = \frac{A_0}{1 + \dfrac{R_2}{R_1 + R_2}A_0} \tag{12.136}$$

$$R_{in,closed} = \infty \tag{12.137}$$

$$R_{out,closed} = \frac{R_{D2}||(R_1 + R_2)}{1 + \dfrac{R_2}{R_1 + R_2}A_0}, \tag{12.138}$$

where A_0 is given by (12.133).

Exercise Repeat the above example if M_2 is degenerated by a resistor of value R_S.

Example 12-29 Analyze the circuit of Fig. 12.58(a), assuming that R_F is not very large.

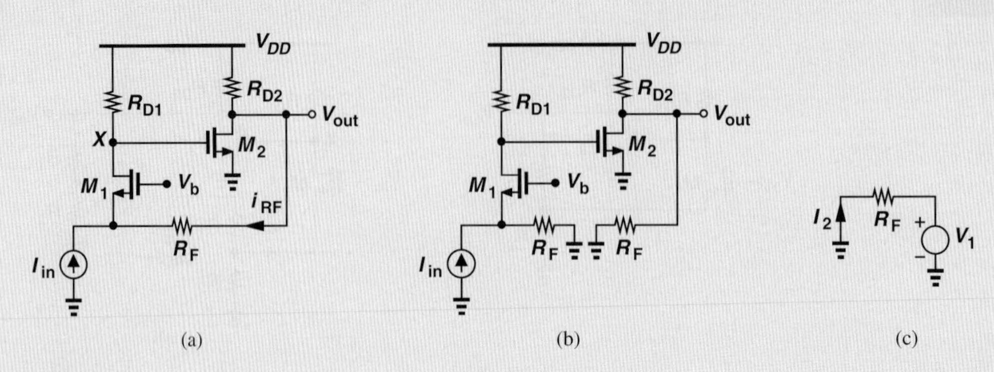

Figure 12.58

Solution As a voltage-current feedback topology, this circuit must be handled according to the rules in Figs. 12.53(b) and 12.54(b). The forward amplifier is formed by M_1, R_{D1}, M_2, and R_{D2}. The feedback network simply consists of R_F. The loop is opened as shown in Fig. 12.58(b), where, from Fig. 12.53(b), the output port of the sense duplicate is *shorted*. Since I_{in} splits between R_F and M_1, we have

$$V_X = I_{in}\frac{R_F R_{D1}}{R_F + \dfrac{1}{g_{m1}}}. \tag{12.139}$$

Noting that $R_0 = V_{out}/I_{in} = (V_X/I_{in})(V_{out}/V_X)$, we write

$$R_0 = \frac{R_F R_{D1}}{R_F + \dfrac{1}{g_{m1}}} \cdot [-g_{m2}(R_{D2}||R_F)]. \tag{12.140}$$

The open-loop input and output impedances are respectively given by

$$R_{in,open} = \frac{1}{g_{m1}}||R_F \tag{12.141}$$

$$R_{out,open} = R_{D2}||R_F. \tag{12.142}$$

To obtain the feedback factor, we follow the rule in Fig. 12.54(b) and construct the test circuit shown in Fig. 12.58(c), obtaining

$$K = \frac{I_2}{V_1} \tag{12.143}$$

$$= -\frac{1}{R_F}. \tag{12.144}$$

Note that both R_0 and K are negative here, yielding a positive loop gain and hence confirming that the feedback is negative. The closed-loop parameters are thus expressed as

$$\left.\frac{V_{out}}{I_{in}}\right|_{closed} = \frac{R_0}{1 - \dfrac{R_0}{R_F}} \tag{12.145}$$

$$R_{in,closed} = \frac{\dfrac{1}{g_{m1}}||R_F}{1 - \dfrac{R_0}{R_F}} \tag{12.146}$$

$$R_{out,closed} = \frac{R_{D2}||R_F}{1 - \dfrac{R_0}{R_F}}, \tag{12.147}$$

where R_0 is given by Eq. (12.140).

Exercise Repeat the above example if R_{D2} is replaced with an ideal current source.

Example 12-30 Analyze the circuit of Fig. 12.59(a), assuming R_M is not small, $r_{O1} < \infty$, and the laser diode has an impedance of R_L.

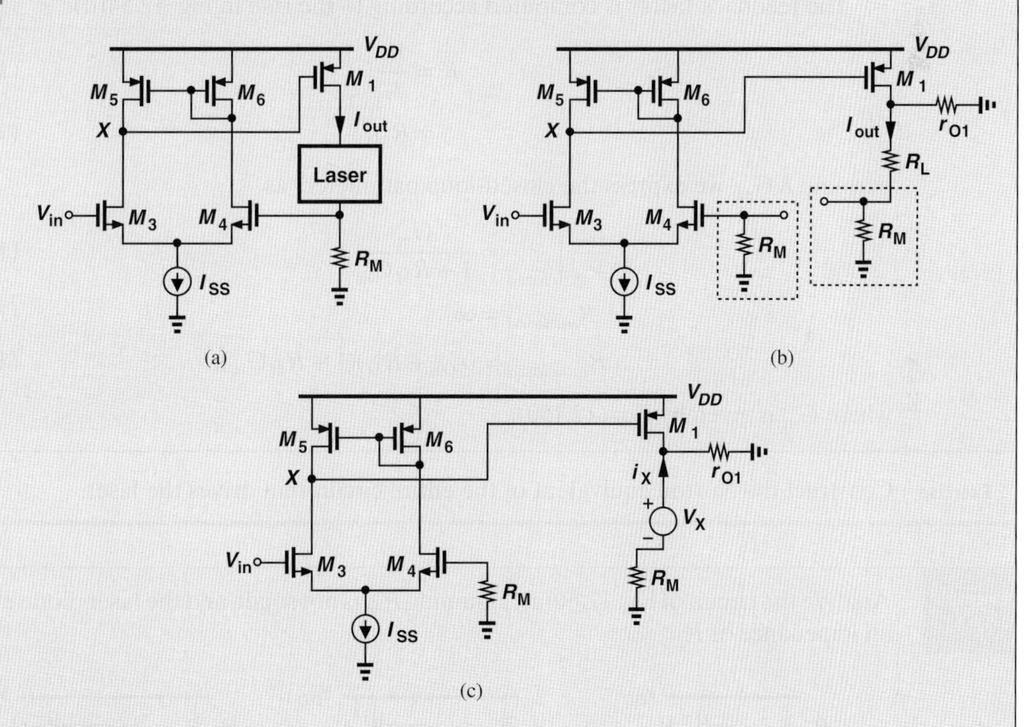

Figure 12.59

Solution This circuit employs current-voltage feedback and must be opened according to the rules shown in Figs. 12.53(d) and 12.54(d). The forward amplifier is formed by M_1 and M_3-M_6, and the feedback network consists of R_M. Depicted in Fig. 12.54(d), the open-loop circuit contains two instances of the feedback network, with the output port of the sense duplicate and the input port of the return duplicate left open. The open-loop gain $G_m = I_{out}/V_{in} = (V_X/V_{in})(I_{out}/V_X)$, and

$$\frac{V_X}{V_{in}} = -g_{m3}(r_{O3}||r_{O5}). \tag{12.148}$$

To calculate I_{out}/V_X, we note that the current produced by M_1 is divided between r_{O1} and $R_L + R_M$:

$$I_{out} = -\frac{r_{O1}}{r_{O1} + R_L + R_M} g_{m1} V_X, \tag{12.149}$$

where the negative sign arises because I_{out} flows *out* of the transistor. The open-loop gain is therefore equal to

$$G_m = \frac{g_{m3}(r_{O3}||r_{O5})g_{m1}r_{O1}}{r_{O1} + R_L + R_M}. \tag{12.150}$$

The output impedance is measured by replacing R_L with a test voltage source and measuring the small-signal current [Fig. 12.59(c)]. The top and bottom terminals of V_X respectively see an impedance of r_{O1} and R_M to ac ground; thus,

$$R_{in,open} = \frac{V_X}{I_X} \tag{12.151}$$

$$= r_{O1} + R_M. \tag{12.152}$$

The feedback factor is computed according to the rule in Fig. 12.54(d):

$$K = \frac{V_2}{I_1} \tag{12.153}$$

$$= R_M. \tag{12.154}$$

Forming KG_m, we express the closed-loop parameters as

$$\left.\frac{I_{out}}{V_{in}}\right|_{closed} = \frac{G_m}{1 + R_M G_m}, \tag{12.155}$$

$$R_{in,closed} = \infty \tag{12.156}$$

$$R_{out,closed} = (r_{O1} + R_M)(1 + R_M G_m), \tag{12.157}$$

where G_m is given by Eq. (12.150).

Exercise Construct the Norton equivalent of the entire circuit that drives the laser.

Example 12-31 Analyze the circuit of Fig. 12.60(a), assuming R_M is not small, and the laser diode exhibits an impedance of R_L.

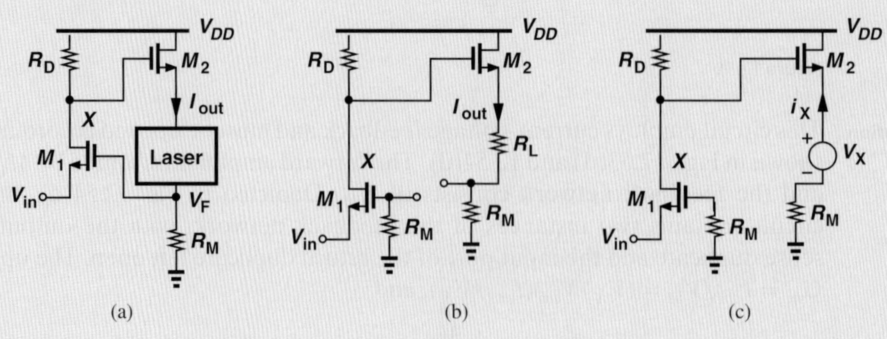

Figure 12.60

Solution The forward amplifier consisting of M_1, R_D, and M_2 senses a voltage and delivers a current to the load, and resistor R_M plays the role of the feedback network. In a manner similar to Example 12-30, we open the loop as shown in Fig. 12.60(b), where $G_m = I_{out}/V_{in} = (V_X/V_{in})(I_{out}/V_X)$. As a common-gate stage, M_1 and R_D yield $V_X/V_{in} = g_{m1}R_D$. To determine I_{out}, we first view M_2 as a source follower and calculate the voltage gain V_A/V_X from Chapter 7:

$$\frac{V_A}{V_X} = \frac{R_L + R_M}{R_L + R_M + \dfrac{1}{g_{m2}}}. \tag{12.158}$$

Thus,

$$I_{out} = \frac{V_A}{R_L + R_M} \tag{12.159}$$

$$= \frac{V_X}{R_L + R_M + \dfrac{1}{g_{m2}}}, \tag{12.160}$$

yielding the open-loop gain as

$$G_m = \frac{g_{m1}R_D}{R_L + R_M + \dfrac{1}{g_{m2}}}. \tag{12.161}$$

The open-loop input impedance is equal to $1/g_{m1}$. For the open-loop output impedance, we replace R_L with a test voltage source [Fig. 12.60(c)], obtaining

$$\frac{V_X}{I_X} = \frac{1}{g_{m2}} + R_M. \tag{12.162}$$

The feedback factor remains identical to that in Example 12-30, leading to the following expressions for the closed-loop parameters:

$$\left.\frac{I_{out}}{V_{in}}\right|_{closed} = \frac{G_m}{1 + R_M G_m} \tag{12.163}$$

$$R_{in,closed} = \frac{1}{g_{m1}}(1 + R_M G_m) \tag{12.164}$$

$$R_{out,closed} = \left(\frac{1}{g_{m2}} + R_M\right)(1 + R_M G_m), \tag{12.165}$$

where G_m is given by Eq. (12.161).

Exercise Repeat the above example if a resistor of value of R_1 is tied between the source of M_2 and ground.

Example 12-32 Analyze the circuit of Fig. 12.61(a), assuming R_F is not large, R_M is not small, and the laser diode is modeled by a resistance R_L. Also, assume $r_{O2} < \infty$.

Solution As a current-current feedback topology, the amplifier must be analyzed according to the rules illustrated in Figs. 12.53(c) and 12.54(c). The forward system consists of M_1, R_D, and M_2, and the feedback network includes R_M and R_F. The loop is opened as depicted in Fig. 12.61(b), where the output port of the sense duplicate is shorted because

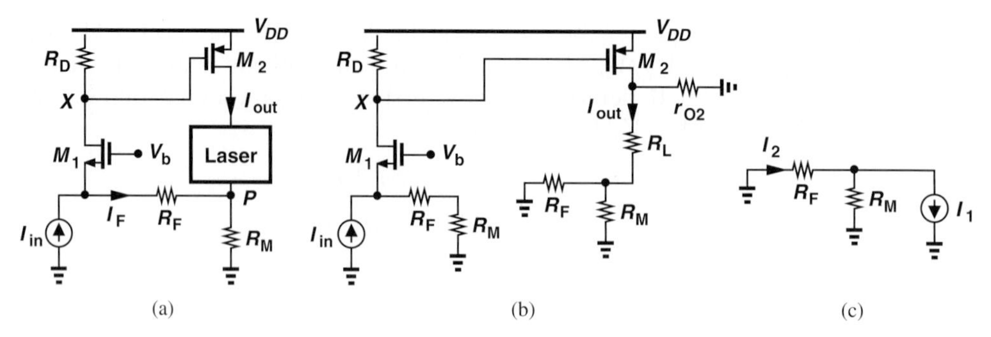

Figure 12.61

the feedback network returns a current to the input. Given by $(V_X/I_{in})(I_{out}/V_X)$, the open-loop gain is computed as

$$A_{I,open} = \frac{(R_F + R_M)R_D}{R_F + R_M + \dfrac{1}{g_{m1}}} \cdot \frac{-g_{m2}r_{O2}}{r_{O2} + R_L + R_M||R_F}, \tag{12.166}$$

where the two fractions account for the division of I_{in} between $R_F + R_M$ and M_1, and the division of I_{D2} between r_{O2} and $R_L + R_M||R_F$.

The open-loop I/O impedances are expressed as

$$R_{in,open} = \frac{1}{g_{m1}}||(R_F + R_M) \tag{12.167}$$

$$R_{out,open} = r_{O2} + R_F||R_M, \tag{12.168}$$

with the latter obtained in a manner similar to that depicted in Fig. 12.59(c).

To determine the feedback factor, we apply the rule of Fig. 12.54(c) as shown in Fig. 12.61(c), thereby obtaining

$$K = \frac{I_2}{I_1} \tag{12.169}$$

$$= -\frac{R_M}{R_M + R_F}. \tag{12.170}$$

The closed-loop parameters are thus given by

$$A_{I,closed} = \frac{A_{I,open}}{1 - \dfrac{R_M}{R_M + R_F}A_{I,open}} \tag{12.171}$$

$$R_{in,closed} = \frac{\dfrac{1}{g_{m1}}||(R_F + R_M)}{1 - \dfrac{R_M}{R_M + R_F}A_{I,open}} \tag{12.172}$$

$$R_{out,closed} = (r_{O2} + R_F||R_M)\left(1 - \frac{R_M}{R_M + R_F}A_{I,open}\right), \tag{12.173}$$

where $A_{I,open}$ is expressed by Eq. (12.166).

Exercise Construct the Norton equivalent of the entire circuit that drives the laser.

Example 12-33 Figure 12.62(a) depicts a circuit similar to that in Fig. 12.61(a), but the output of interest here is V_{out}. Analyze the amplifier and study the differences between the two.

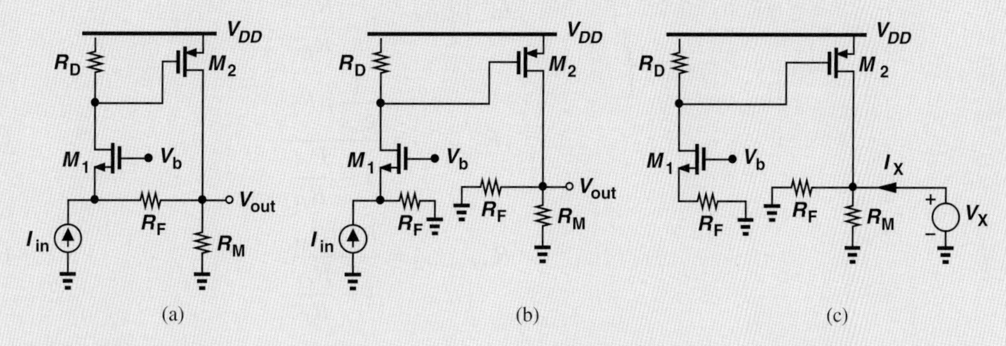

Figure 12.62

Solution The circuit incorporates voltage-current feedback. In contrast to the previous case, R_M now belongs to the forward amplifier rather than the feedback network. After all, M_2 would not be able to generate an output *voltage* without R_M. In fact, this circuit resembles the configuration of Fig. 12.58(a), except that the common-source stage employs a PMOS device.

Opening the loop with the aid of the rules in Fig. 12.53(b), we arrive at the topology in Fig. 12.62(b). Note that the return duplicate in this case (R_F) differs from that in Fig. 12.61(b) ($R_F + R_M$). The open-loop gain is then equal to

$$R_O = \frac{V_{out}}{I_{in}} \tag{12.174}$$

$$= \frac{R_F R_D}{R_F + \dfrac{1}{g_{m1}}}[-g_{m2}(R_F||R_M)], \tag{12.175}$$

and the open-loop input impedance is given by

$$R_{in,open} = \frac{1}{g_{m1}}||R_F. \tag{12.176}$$

The output impedance is computed as illustrated in Fig. 12.62(c):

$$R_{out,open} = \frac{V_X}{I_X} \tag{12.177}$$

$$= R_F||R_M. \tag{12.178}$$

If $r_{O2} < \infty$, then it simply appears in parallel with R_F and R_M in both (12.175) and (12.178).

The feedback factor also differs from that in Example 12-32 and is determined with the aid of Fig. 12.54(b):

$$K = \frac{I_2}{V_1} \tag{12.179}$$

$$= -\frac{1}{R_F}, \tag{12.180}$$

yielding the following expressions for the closed-loop parameters:

$$\left.\frac{V_{out}}{I_{in}}\right|_{closed} = \frac{R_O}{1 - \dfrac{R_O}{R_F}} \tag{12.181}$$

$$R_{in,closed} = \frac{\dfrac{1}{g_{m1}}||R_F}{1 - \dfrac{R_O}{R_F}} \tag{12.182}$$

$$R_{out,closed} = \frac{R_F||R_M}{1 - \dfrac{R_O}{R_F}}. \tag{12.183}$$

In contrast to Example 12-32, the output impedance in this case decreases by $1 - R_O/R_F$ even though the circuit topology remains unchanged. This is because the output impedance is measured very differently in the two cases.

Exercise Repeat the above example if M_2 is degenerated by a resistor of value R_S.

12.8 STABILITY IN FEEDBACK SYSTEMS

Our studies in this chapter have thus far revealed many important benefits of negative feedback. Unfortunately, if designed poorly, negative-feedback amplifiers may behave "badly" or even oscillate. We say the system is marginally stable or simply unstable. In this section, we reexamine our understanding of feedback so as to define the meaning of "behaving badly" and determine the sources of instability.

Did you know?

A simple feedback circuit that, in fact, creates quite a bit of confusion is the op-amp-based noninverting amplifier [Fig. (a)] or a "poor man's" realization thereof [Fig. (b)]. What type of feedback do we have here? The quantity sensed at the output is a voltage, but what is returned to the input? Since the input signal, V_{in}, is a voltage quantity, we may suspect that the feedback network (R_2) returns a voltage. However, for two voltages to add or subtract, they must be placed *in series*. The proper view here is to assume R_1 and R_2, respectively, convert V_{in} and V_{out} to current, and the resulting currents are summed at the virtual ground node, X. We study these commonly-used circuits in more advanced courses.

(a) (b)

Inverting amplifier with feedback.

12.8.1 Review of Bode's Rules

In our review of Bode's rules in Chapter 11, we noted that the slope of the magnitude of a transfer function decreases (increases) by 20 dB/dec as the frequency passes a pole (zero). We now review Bode's rule for plotting the phase of the transfer function:

The phase of a transfer function begins to decrease (increase) at one-tenth of the pole (zero) frequency, incurs a change of −45° (+45°) at the pole (zero) frequency, and experiences a total change of nearly −90° (+90°) at ten times the pole (zero) frequency.[14]

Since the phase begins to change at one-tenth of a pole or zero frequency, even poles or zeros that seem far may affect it significantly—a point of contrast to the behavior of the magnitude.

Example 12-34 Figure 12.63(a) depicts the magnitude response of an amplifier. Using Bode's rule, plot the phase response.[15]

Solution Plotted in Fig. 12.63(b), the phase begins to fall at $0.1\omega_{p1}$, reaches −45° at ω_{p1} and −90° at $10\omega_{p1}$, begins to rise at $0.1\omega_z$, reaches −45° at ω_z and approximately zero at $10\omega_z$, and finally begins to fall at $0.1\omega_{p2}$, reaching −45° at ω_{p2} and −90° at $10\omega_{p2}$. In this example, we have assumed that the pole and zero frequencies are so wide apart that $10\omega_{p1} < 0.1\omega_z$ and $10\omega_z < 0.1\omega_{p2}$. In practice, this may not hold, requiring more detailed calculations.

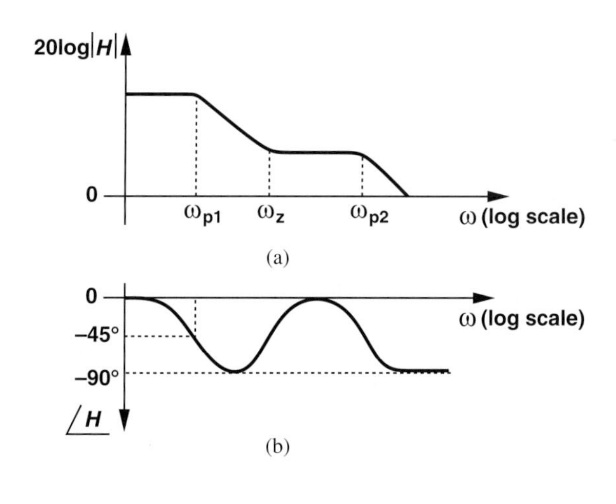

Figure 12.63

Exercise Repeat the above example if ω_{p1} falls between ω_z and ω_{p2}.

Amplifiers having multiple poles may become unstable if placed in a negative-feedback loop. The following example serves as our first step toward understanding this phenomenon.

[14]It is assumed that the poles and the zeros are located in the left half plane.

[15]In general, it may not be possible to construct the phase profile from the magnitude plot.

Example 12-35 Construct the magnitude and phase response of an amplifier having (a) one pole, (b) two poles, or (c) three poles.

Solution Figures 12.64(a)–(c) show the response for the three cases. The phase shift between the input and the output asymptotically approaches −90°, −180°, and −270° in one-pole, two-pole, and three-pole systems, respectively. An important observation here is that the three-pole system introduces a phase shift of −180° at a finite frequency ω_1, reversing the sign of an input sinusoid at this frequency [Fig. 12.64(d)]. For example, a 1-GHz sinusoid is shifted (delayed) by 0.5 ns.

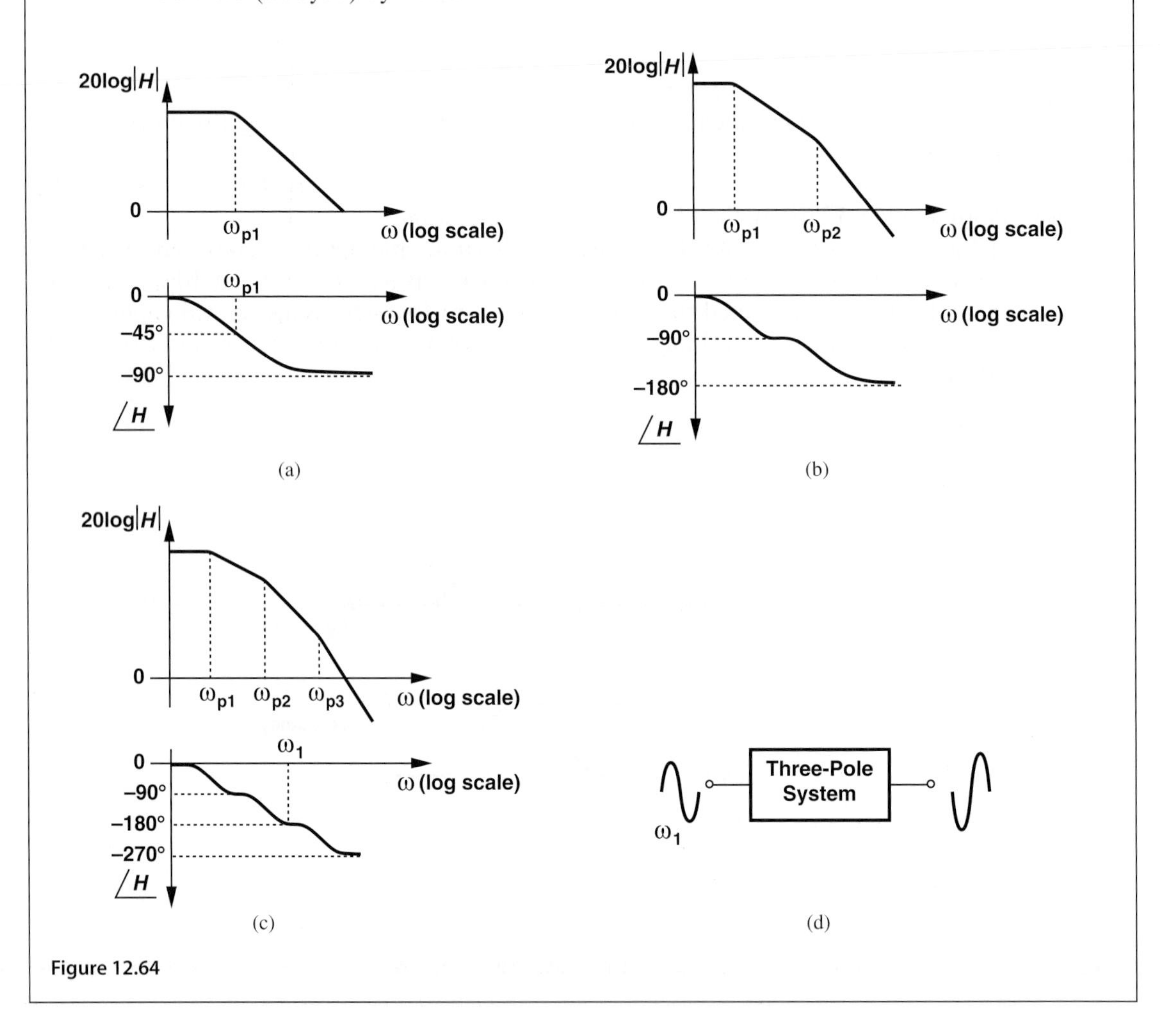

Figure 12.64

Exercise Repeat the above analysis for a three-pole system if $\omega_{p1} = \omega_{p2}$.

12.8.2 Problem of Instability

Suppose an amplifier having a transfer function $H(s)$ is placed in a negative feedback loop (Fig. 12.65). As with the cases studied in Section 12.1, we write the closed-loop transfer function as

$$\frac{Y}{X}(s) = \frac{H(s)}{1 + KH(s)}, \tag{12.184}$$

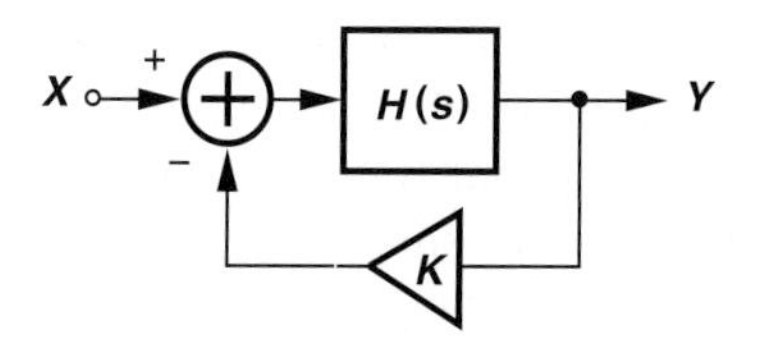

Figure 12.65 Negative feedback system.

where $KH(s)$ is sometimes called the "loop transmission" rather than the loop gain to emphasize its frequency dependence. Recall from Chapter 11 that for a sinusoidal input, $x(t) = A\cos\omega t$, we simply make the substitution $s = j\omega$ in the above transfer function. We assume the feedback factor, K, exhibits no frequency dependence. (For example, it is equal to the ratio of two resistors.)

An interesting question that arises here is, what happens if at a certain input frequency ω_1, the loop transmission, $KH(j\omega_1)$, becomes equal to -1? Then, the closed-loop system provides an infinite gain (even though the open-loop amplifier does not). To understand the consequences of infinite gain, we recognize that even an infinitesimally small input at ω_1 leads to a finite output component at this frequency. For example, the devices comprising the subtractor generate electronic "noise" containing all frequencies. A small noise component at ω_1 therefore experiences a very high gain and emerges as a large sinusoid at the output. We say the system oscillates at ω_1.[16]

It is also possible to understand the above oscillation phenomenon intuitively. Recall from Example 12-35 that a three-pole system introducing a phase shift of $-180°$ reverses the sign of the input signal. Now, if $H(s)$ in Fig. 12.65 contains three poles such that $\angle H = -180°$ at ω_1, then the feedback becomes *positive* at this frequency, thereby producing a feedback signal that *enhances* the input. Circulating around the loop, the signal may thus continue to grow in amplitude. In practice, the final amplitude remains bounded by the supply voltage or other "saturation" mechanisms in the circuit.

For analysis purposes, we express the condition $KH(j\omega_1) = -1$ in a different form. Viewing KH as a complex quantity, we recognize that this condition is equivalent to

$$|KH(j\omega_1| = 1 \tag{12.185}$$

$$\angle KH(j\omega_1) = -180°, \tag{12.186}$$

the latter confirming our above intuitive perspective. In fact, Eq. (12.186) guarantees positive feedback (sufficient delay) and Eq. (12.185) ensures sufficient loop gain for the circulating signal to grow. Called "Barkhausen's criteria" for oscillation, Eqs. (12.185) and (12.186) prove extremely useful in the study of stability.

Example 12-36 Explain why a two-pole system cannot oscillate.[17]

Solution As evident from the Bode plots in Fig. 12.64(b), the phase shift produced by such a system reaches $-180°$ only at $\omega = \infty$, where $|H| \to 0$. In other words, at no frequency are both Eqs. (12.185) and (12.186) satisfied.

Exercise What happens if one of the poles is at the origin?

[16] It can be proved that the circuit continues to produce a sinusoid at ω_1 even if the electronic noise of the devices ceases to exist. The term "oscillation" is thus justified.

[17] It is assumed that at least one of the poles is not at the origin.

In summary, a negative feedback system may become unstable if the forward amplifier introduces a phase shift of $-180°$ at a finite frequency, ω_1, *and* the loop transmission $|KH|$ is equal to unity at that frequency. These conditions become intuitive in the time domain as well. Suppose, as shown in Fig. 12.66(a), we apply a small sinusoid at ω_1 to the system and follow it around the loop as time progresses. The sinusoid incurs a sign reversal as it emerges at the output of the forward amplifier [Fig. 12.66(b)]. Assumed frequency-independent, the feedback factor simply scales the result by a factor of K, producing an inverted replica of the input at node A if $|KH(j\omega_1)| = 1$. This signal is now *subtracted* from the input, generating a sinusoid at node B with *twice* the amplitude [Fig. 12.66(c)]. The signal level thus continues to grow after each trip around the loop. On the other hand, if $|KH(j\omega_1)| < 1$, then the output cannot grow indefinitely.

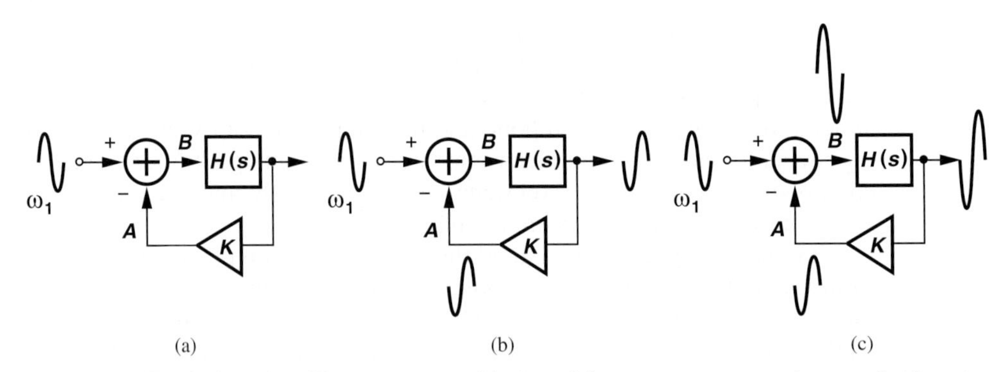

Figure 12.66 Evolution of oscillatory system with time: (a) a component at ω_1 is sensed at input, (b) the component returns to subtractor with a 180° phase shift, and (c) the subtractor enhances the signal at node B.

Example 12-37

A three-pole feedback system exhibits the frequency response depicted in Fig. 12.67. Does this system oscillate?

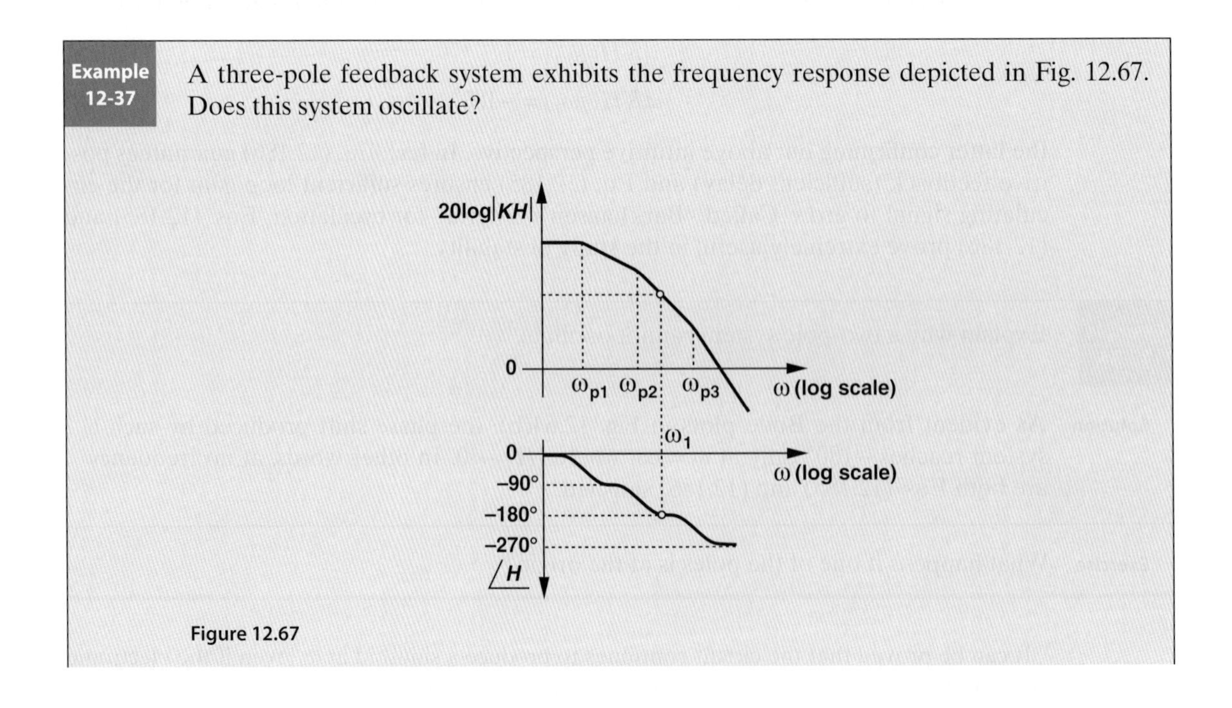

Figure 12.67

Solution Yes, it does. The loop gain at ω_1 is *greater* than unity, but we note from the analysis in Fig. 12.66 that indefinite signal growth still occurs, in fact more rapidly. After each trip around the loop, a sinusoid at ω_1 experiences a gain of $|KH| > 1$ and returns with opposite phase to the subtractor.

Exercise Suppose ω_{p1} is halved in value. Does the system still oscillate?

12.8.3 Stability Condition

Our foregoing investigation indicates that if $|KH(j\omega_1)| \geq 1$ and $\angle H(j\omega_1) = -180°$, then the negative feedback system oscillates. Thus, to avoid instability, we must ensure that these two conditions do not occur at the same frequency.

Figure 12.68 depicts two scenarios wherein the two conditions do not coincide. Are both of these systems stable? In Fig. 12.68(a), the loop gain at ω_1 *exceeds* unity (0 dB), still leading to oscillation. In Fig. 12.68(b), on the other hand, the system cannot oscillate at ω_1 (due to insufficient phase shift) or ω_2 (because of inadequate loop gain).

The frequencies at which the loop gain falls to unity or the phase shift reaches −180° play such a critical role as to deserve specific names. The former is called the "gain crossover frequency" (ω_{GX}) and the latter, the "phase crossover frequency" (ω_{PX}). In Fig. 12.68(b), for example, $\omega_{GX} = \omega_1$ and $\omega_{PX} = \omega_2$. The key point emerging from the two above scenarios is that stability requires that

$$\omega_{GX} < \omega_{PX}. \quad (12.187)$$

Did you know?

We have seen that a negative-feedback system can become unstable if the loop experiences a large phase shift. Since phase shift and *delay* are roughly equivalent, we can say excess delay causes instability. Two people walking past each other in a narrow hallway sometimes end up dancing to the left and to the right because of the slight delay they incur in correcting themselves. A microphone connected to a speaker system (as shown in the figure below) may whistle because the sound coming back from the speaker to the microphone through the air experiences delay. In this case, the feedback system consisting of the microphone, the amplifier, the speaker, and the return path through the air oscillates. You can change the frequency of oscillation (the pitch of the whistle) by adjusting the delay.

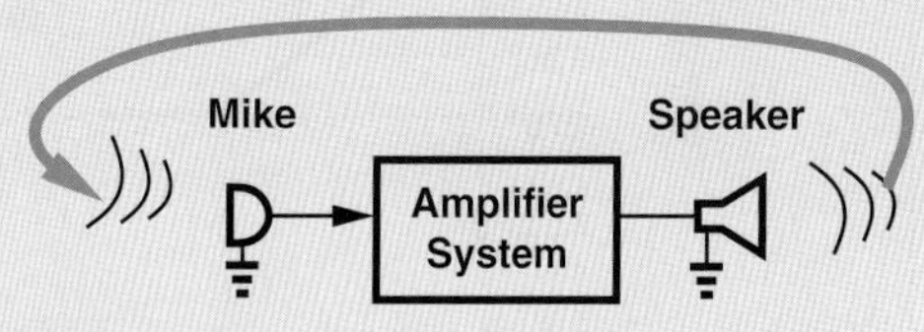

Unwanted feedback from speaker to microphone.

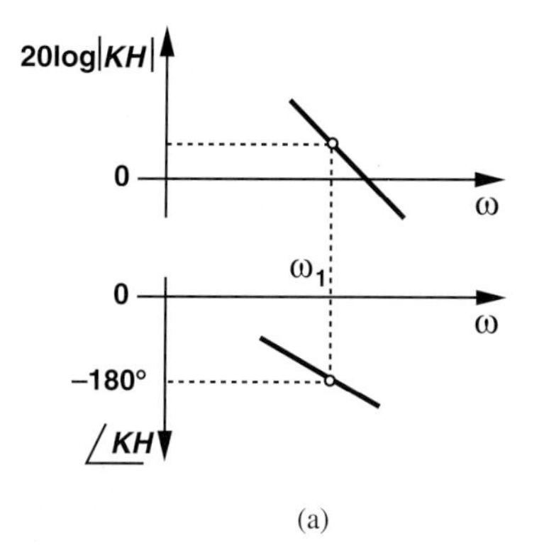

Figure 12.68 Systems with phase reaching −180° (a) before and (b) after the loop gain reaches unity.

In summary, to guarantee stability in negative-feedback systems, we must ensure that the loop gain falls to unity *before* the phase shift reaches −180° so that Barkhausen's criteria do not hold at the same frequency.

Example 12-38

We wish to apply negative feedback with $K = 1$ around the three-stage amplifier shown in Fig. 12.69(a). Neglecting other capacitances and assuming identical stages, plot the frequency response of the circuit and determine the condition for stability. Assume $\lambda = 0$.

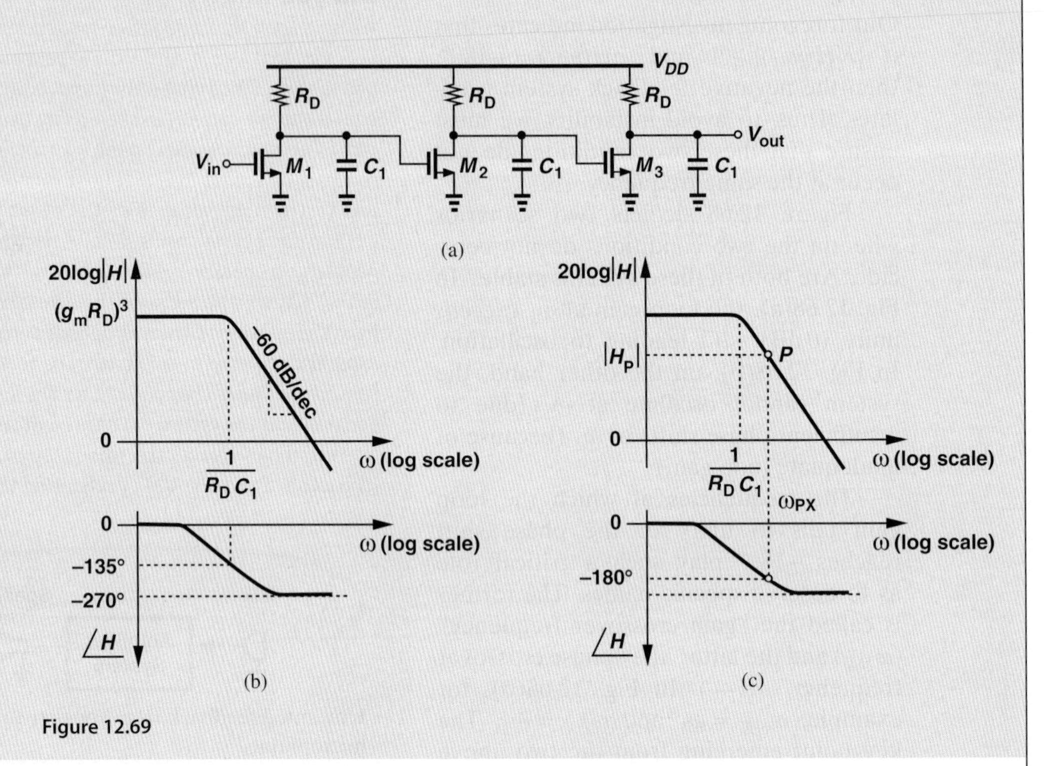

Figure 12.69

Solution The circuit exhibits a low-frequency gain of $(g_m R_D)^3$ and three *coincident* poles given by $(R_D C_1)^{-1}$. Thus, as depicted in Fig. 12.69(b), $|H|$ begins to fall at a rate of 60 dB/dec at $\omega_p = (R_D C_1)^{-1}$. The phase begins to change at one-tenth of this frequency,[18] reaches −135° at ω_p, and approaches −270° at $10\omega_p$.

To guarantee that a unity-feedback system incorporating this amplifier remains stable, we must ensure that $|KH|(= |H|)$ falls below unity at the phase crossover frequency. Illustrated in Fig. 12.69(c), the procedure entails identifying ω_{PX} on the phase response, finding the corresponding point, P, on the gain response, and requiring that $|H_P| < 1$.

[18]Strictly speaking, we note that the three coincident poles affect the phase at frequencies even below $0.1\omega_p$.

We now repeat the procedure analytically. The amplifier transfer function is given by the product of those of the three stages:[19]

$$H(s) = \frac{(g_m R_D)^3}{\left(1 + \frac{s}{\omega_p}\right)^3}, \tag{12.188}$$

where $\omega_p = (R_D C_1)^{-1}$. The phase of the transfer function can be expressed as[20]

$$\angle H(j\omega) = -3 \tan^{-1} \frac{\omega}{\omega_p}, \tag{12.189}$$

where the factor 3 accounts for the three coincident poles. The phase crossover occurs if $\tan^{-1}(\omega/\omega_p) = 60°$ and hence

$$\omega_{PX} = \sqrt{3}\omega_p. \tag{12.190}$$

The magnitude must remain less than unity at this frequency:

$$\frac{(g_m R_D)^3}{\left[\sqrt{1 + \left(\frac{\omega_{PX}}{\omega_p}\right)^2}\right]^3} < 1. \tag{12.191}$$

It follows that

$$g_m R_D < 2. \tag{12.192}$$

If the low-frequency gain of each stage exceeds 2, then a feedback loop around this amplifier with $K = 1$ becomes unstable.

Exercise Repeat the above example if the last stage incorporates a load resistance equal to $2R_D$.

Example 12-39 A common-source stage is placed in a unity-gain feedback loop as shown in Fig. 12.70. Explain why this circuit does not oscillate.

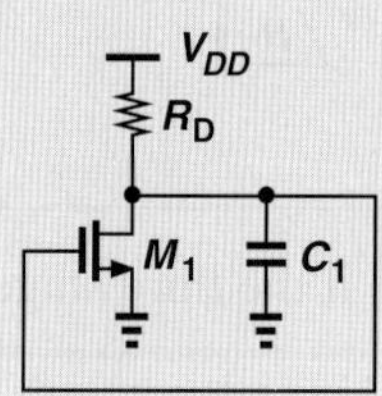

Figure 12.70

[19]For simplicity, we drop the negative sign in the gain of each stage here. The final result is still valid.

[20]Recall that the phase of a complex number $a + jb$ is given by $\tan^{-1}(b/a)$. Also, the phase of a product is equal to the sum of phases.

Solution Since the circuit contains only one pole, the phase shift cannot reach 180° at any frequency. The circuit is thus stable.

Exercise What happens if $R_D \to \infty$ and $\lambda \to 0$?

Example 12-40 Repeat Example 12-38 if the target value of K is $1/2$, i.e., the feedback is weaker.

Solution We plot $|KH| = 0.5|H|$ and $\angle KH = \angle H$ as shown in Fig. 12.71(a). Note the $|KH|$ plot is simply shifted down by 6 dB on a logarithmic scale. Starting from the phase crossover frequency, we determine the corresponding point, P, on $|KH|$ and require that $0.5|H_P| < 1$. Recognizing that Eqs.(12.189) and (12.190) still hold, we write

$$\frac{0.5(g_m R_D)^3}{\left[\sqrt{1+\left(\frac{\omega_{PX}}{\omega_p}\right)^2}\right]^3} < 1. \tag{12.193}$$

That is,

$$(g_m R_D)^3 < \frac{2^3}{0.5}. \tag{12.194}$$

Thus, the weaker feedback permits a greater open-loop gain.

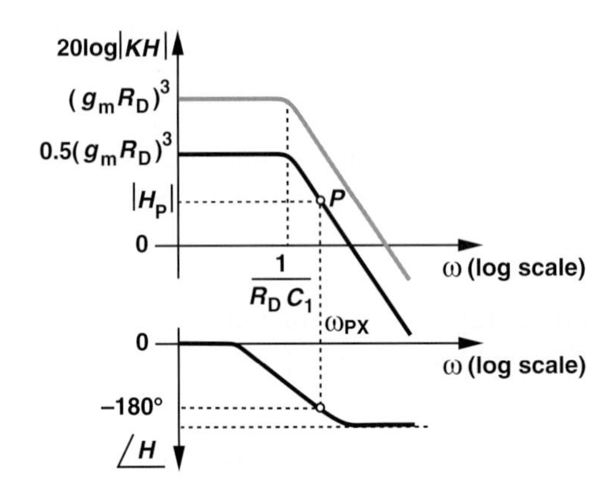

Figure 12.71

Exercise Repeat the above example if the third stage incorporates a load resistor of value $2R_D$.

12.8.4 Phase Margin

Our study of instability in negative-feedback systems reveals that ω_{GX} must remain below ω_{PX} to avoid oscillation. But by how much? We surmise that if $\omega_{GX} < \omega_{PX}$ but the difference between the two is small, then the feedback system displays an almost-oscillatory response. Shown in Fig. 12.72 are three cases illustrating this point. In Fig. 12.72(a),

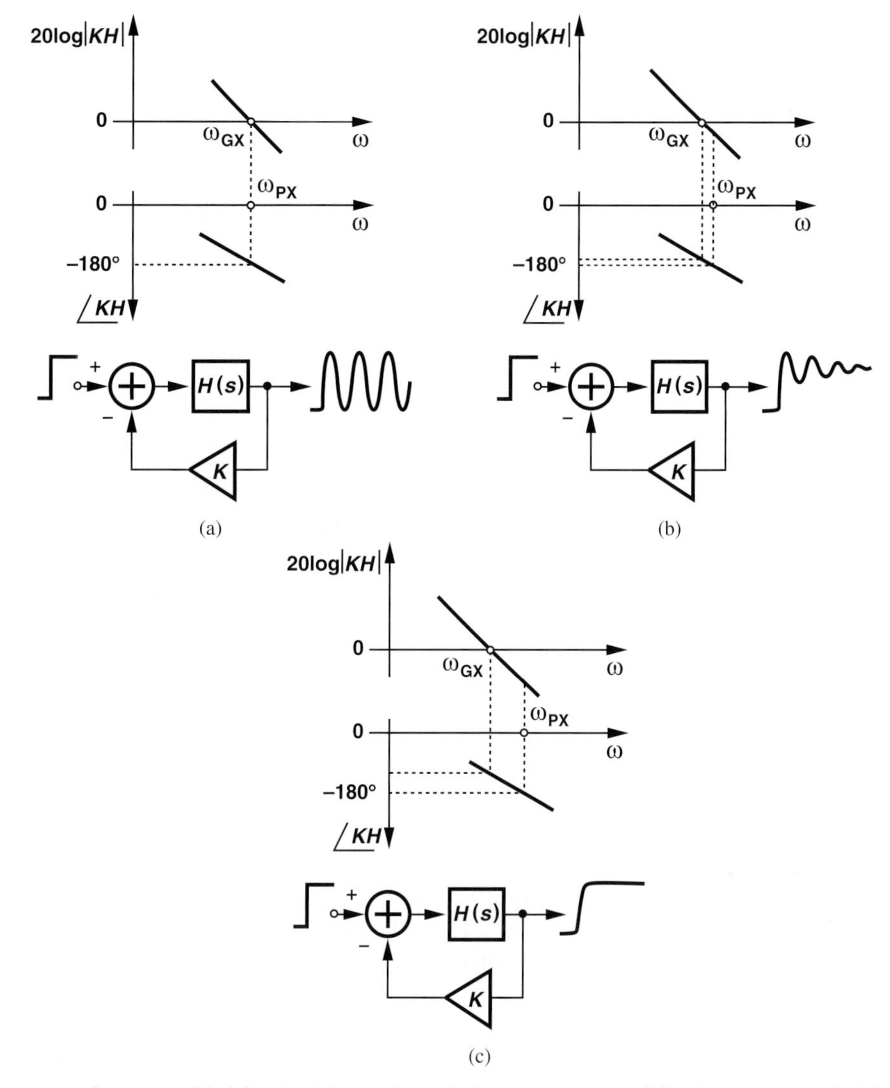

Figure 12.72 Systems with (a) coincident gain and phase crossovers, (b) gain crossover slightly below phase crossover, and (c) gain crossover well below phase crossover.

Barkhausen's criteria are met and the system produces oscillation in response to an input step. In Fig. 12.72(b), $\omega_{GX} < \omega_{PX}$, but the step response "rings" for a long time because the system is "marginally" stable and behaves "badly." We therefore postulate that a well-behaved system is obtained only if a sufficient "margin" is allowed between ω_{GX} and ω_{PX} [Fig. 12.72(c)]. Note that in this case, $\angle KH$ at ω_{GX} remains significantly more positive than $-180°$. We assume K has no phase shift and hence $\angle KH = \angle H$.

A measure commonly used to quantify the stability of feedback systems is the "phase margin" (PM). As exemplified by the cases in Fig. 12.72, the more stable a system is, the greater is the *difference* between $\angle H(\omega_{GX})$ and $-180°$. Indeed, this difference is called the phase margin:

$$\text{Phase Margin} = \angle H(\omega_{GX}) + 180°. \tag{12.195}$$

Example 12-41 Figure 12.73 plots the frequency response of a multipole amplifier. The magnitude drops to unity at the second pole frequency. Determine the phase margin of a feedback system employing this amplifier with $K = 1$.

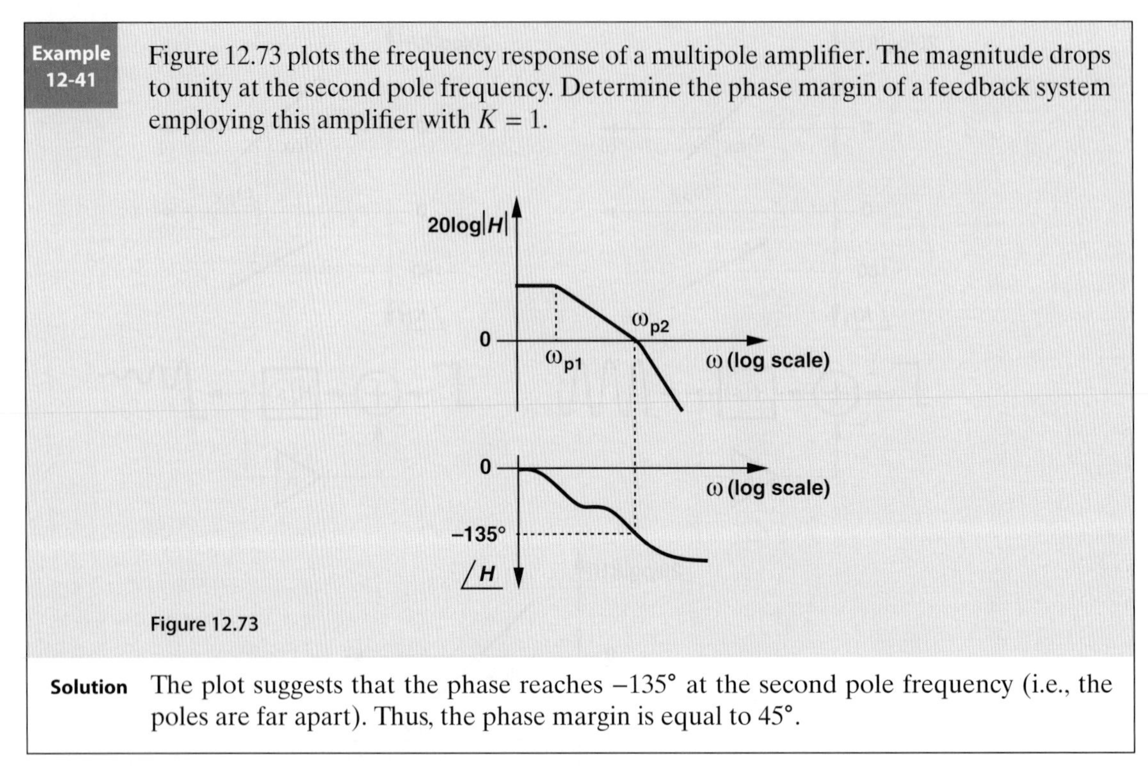

Figure 12.73

Solution The plot suggests that the phase reaches $-135°$ at the second pole frequency (i.e., the poles are far apart). Thus, the phase margin is equal to 45°.

Exercise Is the phase margin greater or less than 45° if $K = 0.5$?

How much phase margin is necessary? For a well-behaved response, we typically require a phase margin of 60°. Thus, the above example is not considered an acceptable design. In other words, the gain crossover must fall below the *second* pole.

12.8.5 Frequency Compensation

It is possible that after the design of an amplifier is completed, the phase margin proves inadequate. How is the circuit modified to improve the stability? For example, how do we make the three-stage amplifier of Example 12-38 stable if $K = 1$ and $g_m R_D > 2$? The solution is to make two of the poles unequal in magnitude. Called "frequency compensation," this task can be accomplished by shifting ω_{GX} toward the origin (without changing ω_{PX}). In other words, if $|KH|$ is forced to drop to unity at a lower frequency, then the phase margin increases [Fig. 12.74(a)].

How can ω_{GX} be shifted toward the origin? We recognize that if the *dominant* pole is translated to lower frequencies, so is ω_{GX}. Figure 12.74(b) illustrates an example where the first pole is shifted from ω_{p1} to ω'_{p1}, but other poles are constant. As a result, ω_{GX} decreases in magnitude.

What happens to the phase after compensation? As shown in Fig. 12.74(b), the low-frequency section of $\angle KH$ changes because ω_{p1} is moved to ω'_{P1}, but the critical section near ω_{GX} does not. Consequently, the phase margin increases.

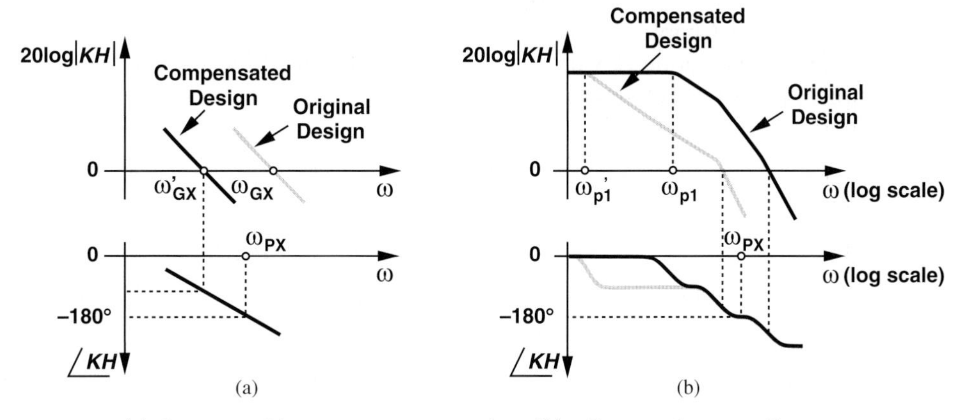

Figure 12.74 (a) Concept of frequency compensation, (b) effect on phase profile.

Example 12-42

The amplifier shown in Fig. 12.75(a) employs a cascode stage and a CS stage. Assuming that the pole at node B is dominant, sketch the frequency response and explain how the circuit can be "compensated."

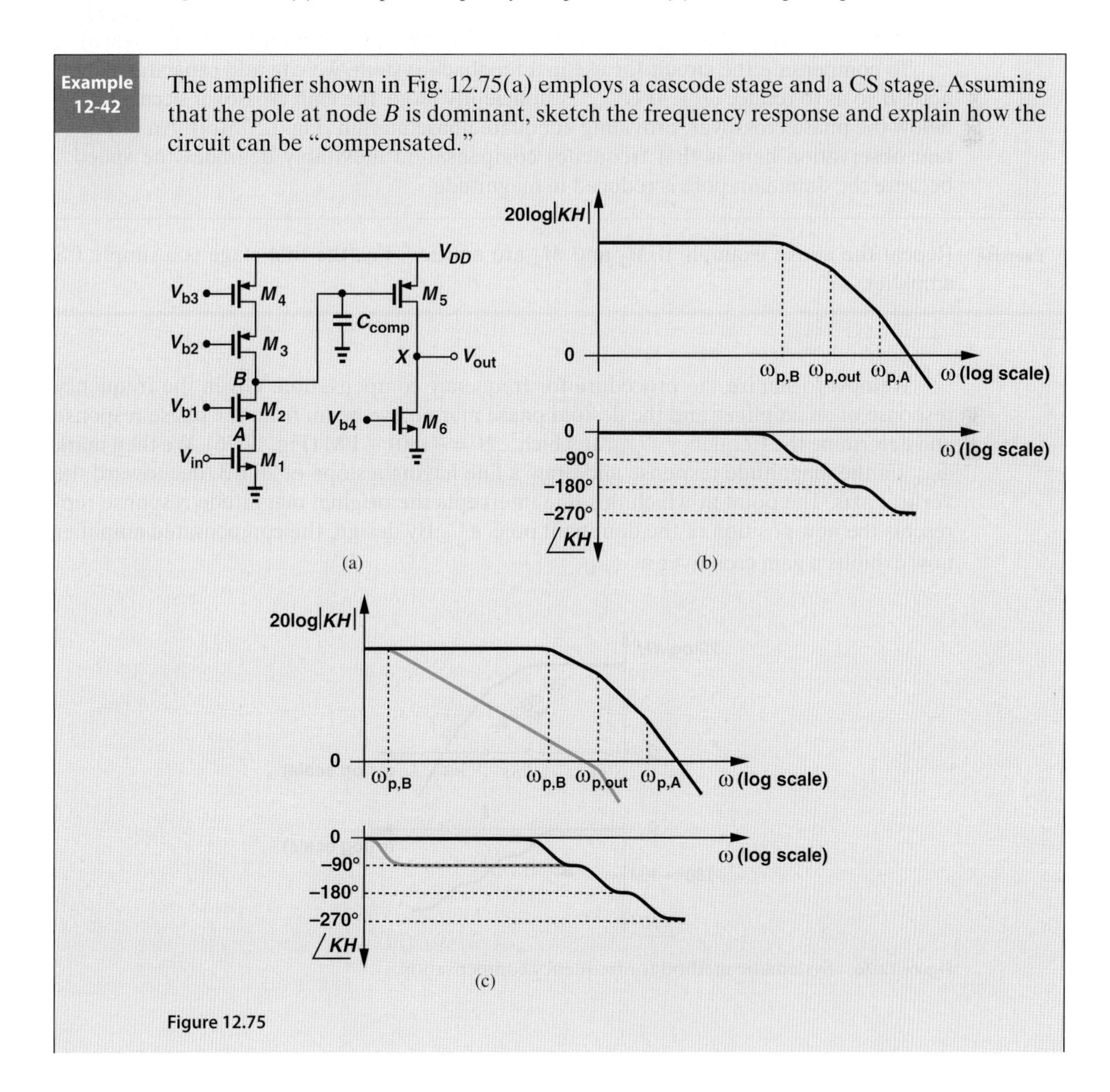

Figure 12.75

Solution Recall from Chapter 11 that the cascode stage exhibits one pole arising from node A and another from node B, with the latter falling much closer to the origin. We can express these poles respectively as

$$\omega_{p,A} \approx \frac{g_{m2}}{C_A} \tag{12.196}$$

$$\omega_{p,B} \approx \frac{1}{[(g_{m2}r_{O2}r_{O1})||(g_{m3}r_{O3}r_{O4})]C_B}, \tag{12.197}$$

where C_A and C_B denote the total capacitance seen at each node to ground. The third pole is associated with the output node:

$$\omega_{p,out} = \frac{1}{(r_{O5}||r_{O6})C_{out}}, \tag{12.198}$$

where C_{out} represents the total capacitance at the output node. We assume $\omega_{P,B} < \omega_{P,out} < \omega_{P,A}$. The frequency response of the amplifier is plotted in Fig. 12.75(b).

To compensate the circuit for use in a feedback system, we can add capacitance to node B so as to reduce $\omega_{p,B}$. If C_{comp} is sufficiently large, the gain crossover occurs well below the phase crossover, providing adequate phase margin [Fig. 12.75(c)]. An important observation here is that frequency compensation inevitably degrades the speed because the dominant pole is reduced in magnitude.

Exercise Repeat the above example if M_2 and M_3 are omitted, i.e., the first stage is a simple CS amplifier.

We now formalize the procedure for frequency compensation. Given the frequency response of an amplifier and the desired phase margin, we begin from the phase response and determine the frequency, ω_{PM}, at which $\angle H = -180° + \text{PM}$ (Fig. 12.76). We then mark ω_{PM} on the magnitude response and draw a line having a slope of 20 dB/dec toward the vertical axis. The point at which this line intercepts the original magnitude response represents the new position of the dominant pole, ω'_{p1}. By design, the compensated amplifier now exhibits a gain crossover at ω_{PM}.

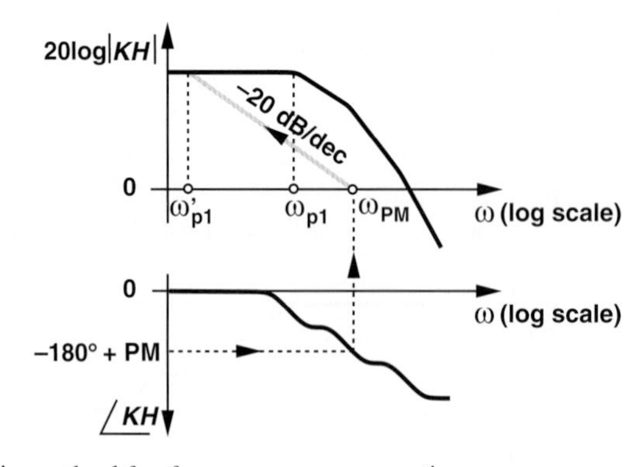

Figure 12.76 Systematic method for frequency compensation.

Example 12-43

A multipole amplifier exhibits the frequency response plotted in Fig. 12.77(a). Assuming the poles are far apart, compensate the amplifier for a phase margin of 45° with $K = 1$.

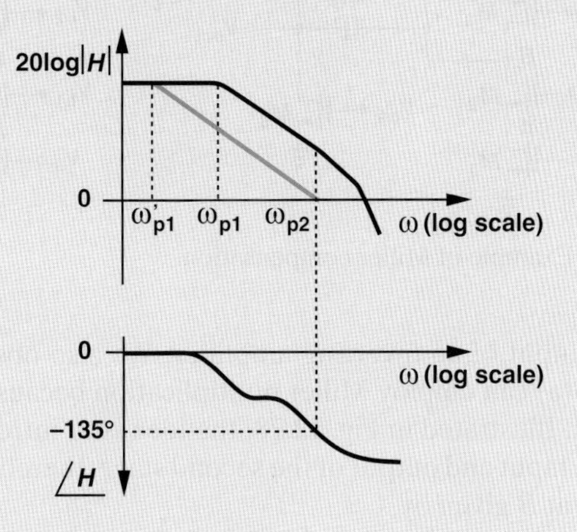

Figure 12.77

Solution Since the phase reaches $-135°$ at $\omega = \omega_{p2}$, in this example $\omega_{PM} = \omega_{p2}$. We thus draw a line with a slope of -20 dB/dec from ω_{p2} toward the vertical axis. The dominant pole must therefore be translated to ω'_{p1}. Since this phase margin is generally inadequate, in practice, $\omega_{PM} < \omega_{p2}$.

Exercise Repeat the above example for $K = 0.5$ and compare the results.

The reader may wonder why the line originating at ω_{PM} must rise at a slope of 20 dB/dec (rather than 40 or 60 dB/dec) toward the vertical axis. Recall from Examples 12-41 and 12-43 that, for adequate phase margin, the gain crossover must occur *below* the second pole. Thus, the magnitude response of the *compensated* amplifier does not "see" the second pole as it approaches ω_{GX}; i.e., the magnitude response has a slope of only -20 dB/dec.

Did you know?

It is interesting to know how one capacitor changed the fate of a product forever. A general-purpose, innovative op amp (LM101) introduced by Robert Widlar at National Semiconductor in 1967 employed Miller compensation but with an *external* 30 pF capacitor. Evidently, the compensation capacitor was left off-chip to provide greater flexibility for the user. In 1968, David Fullagar at Fairchild Semiconductor introduced another op amp very similar to the LM101, except that the compensation capacitor was integrated on the chip. This was the 741 op amp, which quickly became popular and has served the electronics industry for nearly half a century.

12.8.6 Miller Compensation

In Example 12-42, we noted that a capacitor can be tied from node B to ground to compensate the amplifier. The required value of this capacitor may be quite large, necessitating a large area on an integrated circuit. But recall from Miller's theorem in Chapter 11 that the apparent value of a capacitor increases if the device is connected between the

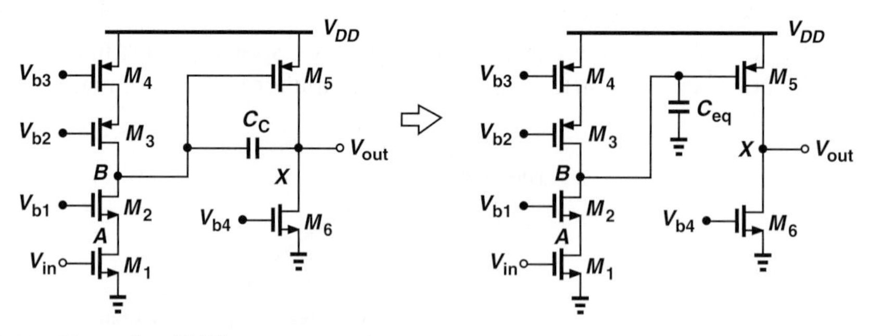

Figure 12.78 Example of Miller compensation.

input and output of an inverting amplifier. We also observe that the two-stage amplifier of Fig. 12.75(a) can employ Miller multiplication because the second stage provides some voltage gain. Illustrated in Fig. 12.78, the idea is to introduce the compensation capacitor between the input and output of the second stage, thereby creating an equivalent *grounded* capacitance at B given by

$$C_{eq} = (1 - A_v)C_c \tag{12.199}$$

$$= [1 + g_{m5}(r_{O5}||r_{O6})]C_c. \tag{12.200}$$

Called "Miller compensation," this technique reduces the required value of C_c by a factor of $1 + g_{m5}(r_{O5}||r_{O6})$.

Miller compensation entails a number of interesting side effects. For example, it shifts not only the dominant pole but also the output pole. Such phenomena are studied in more advanced texts, e.g., [2].

12.9 CHAPTER SUMMARY

- Negative feedback can be used to regulate the behavior of systems that are otherwise "untamed" and poorly controlled.
- A negative feedback system consists of four components: forward system, output sense mechanism, feedback network, and input comparison mechanism.
- The loop gain of a feedback system can, in principle, be obtained by breaking the loop, injecting a test signal, and calculating the gain as the signal goes around the loop. The loop gain determines many properties of feedback systems, e.g., gain, frequency response, and I/O impedances.
- The loop gain and closed-loop gain should not be confused with each other. The latter refers to the overall gain from the main input to the main output while the feedback loop is closed.
- If the loop gain is much greater than unity, the closed-loop gain becomes approximately equal to the inverse of the feedback factor.
- Making the closed-loop gain relatively independent of the open-loop gain, negative feedback provides many useful properties: it reduces the sensitivity of the gain to component variations, load variations, frequency variations, and signal level variations.
- Amplifiers can generally be viewed as one of four types with voltage or current inputs and outputs. In the ideal case, the input impedance of a circuit is infinite if it senses a

voltage or zero if it senses a current. Also, the output impedance of an ideal circuit is zero if it generates a voltage or infinite if it generates a current.

- Voltage quantities are sensed in parallel and current quantities in series. Voltage quantities are summed in series and current quantities in parallel.
- Depending on the type of the forward amplifier, four feedback topologies can be constructed. The closed-loop gain of each is equal to the open-loop gain divided by one plus the loop gain.
- A negative-feedback loop sensing and regulating the output voltage lowers the output impedance by a factor of one plus the loop gain, making the circuit a better voltage source.
- A negative-feedback loop sensing and regulating the output current raises the output impedance by a factor of one plus the loop gain, making the circuit a better current source.
- A negative-feedback loop returning a voltage to the input raises the input impedance by one plus the loop gain, making the circuit a better voltage sensor.
- A negative-feedback loop returning a current to the input lowers the input impedance by one plus the loop gain, making the circuit a better current sensor.
- If the feedback network departs from its ideal model, then it "loads" the forward amplifier characteristics. In this case, a methodical method must be followed that included the effect of finite I/O impedances.
- A high-frequency signal traveling through a forward amplifier experiences significant phase shift. With several poles, it is possible that the phase shift reaches 180°.
- A negative-feedback loop that introduces a large phase shift may become a positive-feedback loop at some frequency and begin to oscillate if the loop gain at that frequency is unity or higher.
- To avoid oscillation, the gain crossover frequency must fall below the phase crossover frequency.
- Phase margin is defined as 180° minus the phase of the loop transmission at the gain crossover frequency.
- To ensure a well-behaved time and frequeny response, a negative-feedback system must realize sufficient phase margin, e.g., 60°.
- If a feedback circuit suffers from insufficient phase margin, then it must be "frequency-compensated." The most common method is to lower the dominant pole so as to reduce the gain crossover frequency (without changing the phase profile). This typically requires adding a large capacitor from the dominant pole node to ground.
- To lower the dominant pole, one can exploit Miller multiplication of capacitors.

PROBLEMS

Sec. 12.1 General Considerations

12.1. Determine the transfer function, Y/X, for the systems shown in Fig. 12.79.

12.2. For the systems depicted in Fig. 12.79, compute the transfer function W/X.

12.3. For the systems depicted in Fig. 12.79, compute the transfer function E/X. SOL

12.4. Calculate the loop gain of the circuits illustrated in Fig. 12.80. Assume the op amp exhibits an open-loop gain of A_1, but is otherwise ideal. Also, $\lambda = 0$.

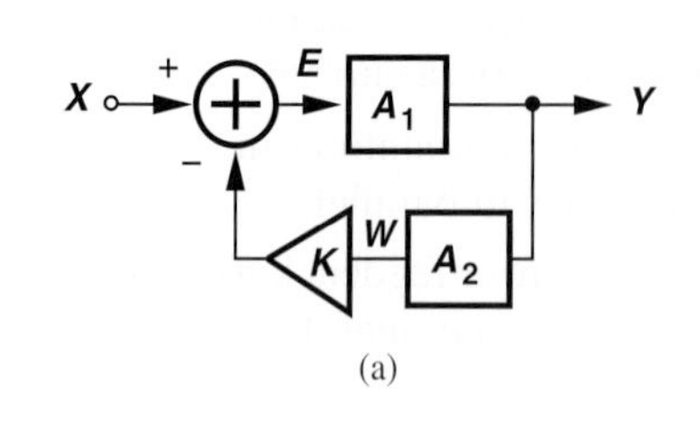

(a)

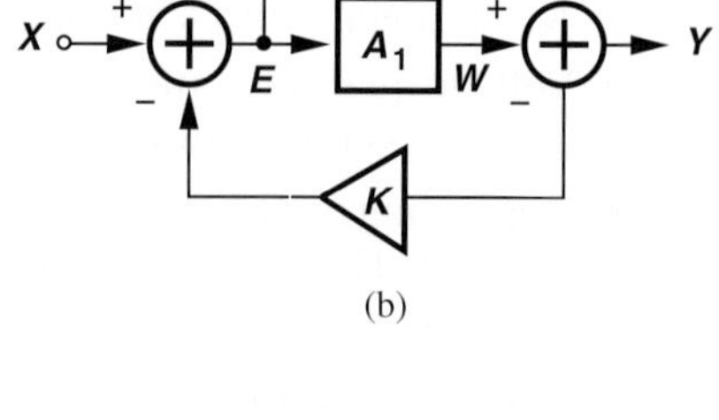

(b)

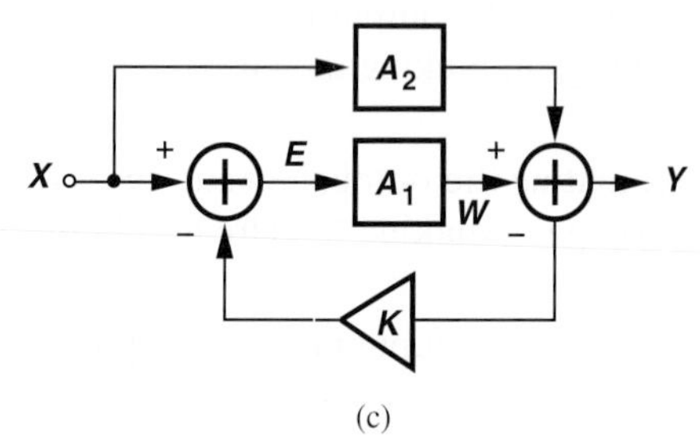

(c)

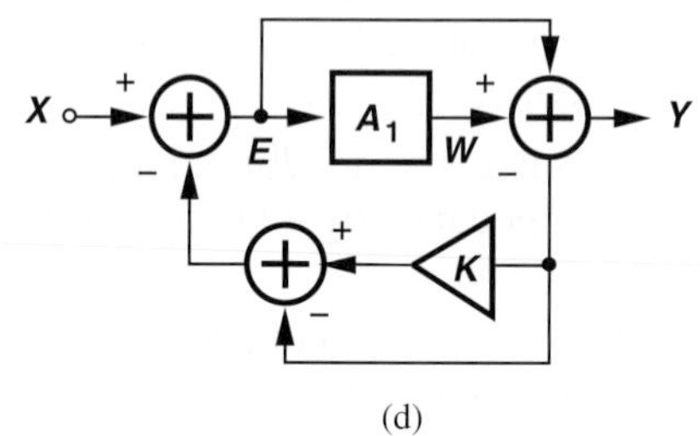

(d)

Figure 12.79

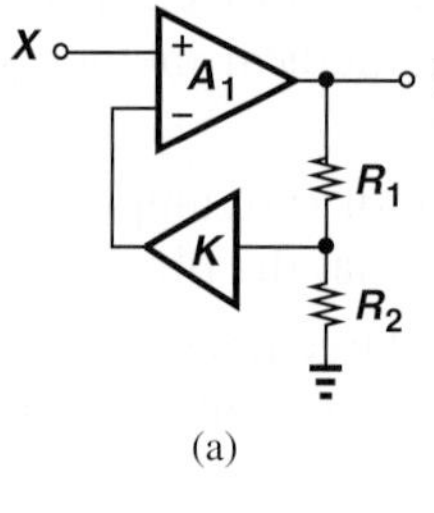

(a)

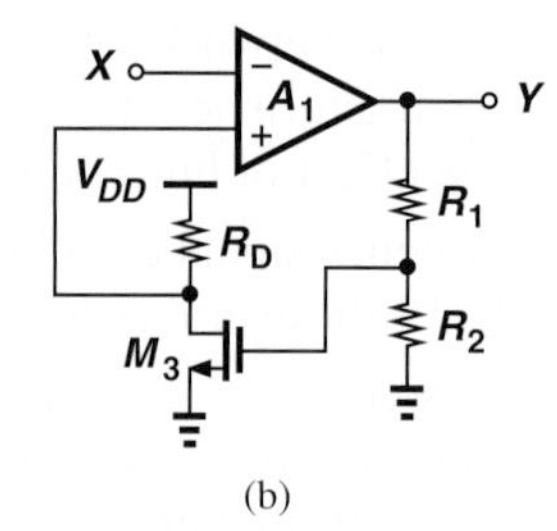

(b)

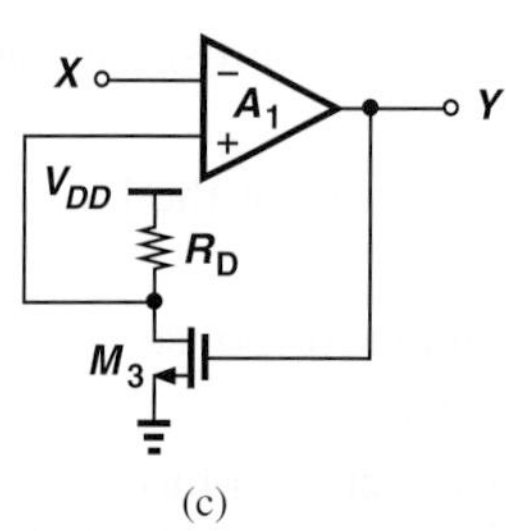

(c)

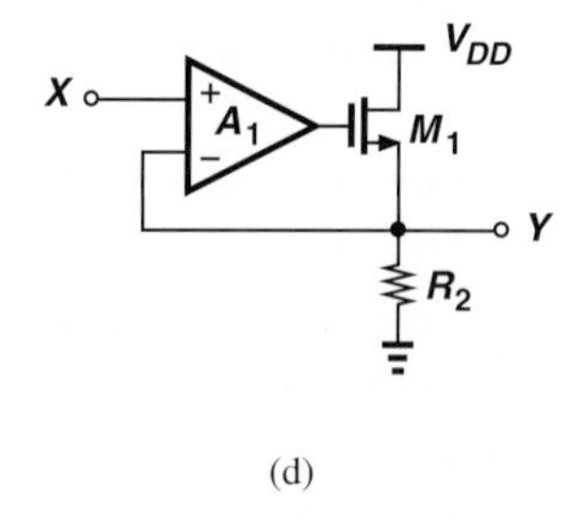

(d)

Figure 12.80

12.5. Using the results obtained in Problem 12.4, compute the closed-loop gain of the circuits shown in Fig. 12.80.

12.6. In the circuit of Fig. 12.3, the input is a sinusoid with a peak amplitude of 2 mV. If $A_1 = 500$ and $R_1/R_2 = 7$, determine the amplitude of the output waveform and the feedback waveform.

Sec. 12.2 Properties of Negative Feedback

12.7. Suppose the open-loop gain A_1 in Fig. 12.1 changes by 20%. Determine the minimum loop gain necessary to ensure the closed-loop gain changes by less than 1%.

12.8. Consider the feedback system shown in Fig. 12.81, where the common-source stage serves as the feedforward network. Assume $\mu_n C_{ox}$ may vary by $\pm 10\%$ and λ by $\pm 20\%$. What is the minimum loop gain necessary to guarantee that the closed-loop gain varies by less than $\pm 5\%$?

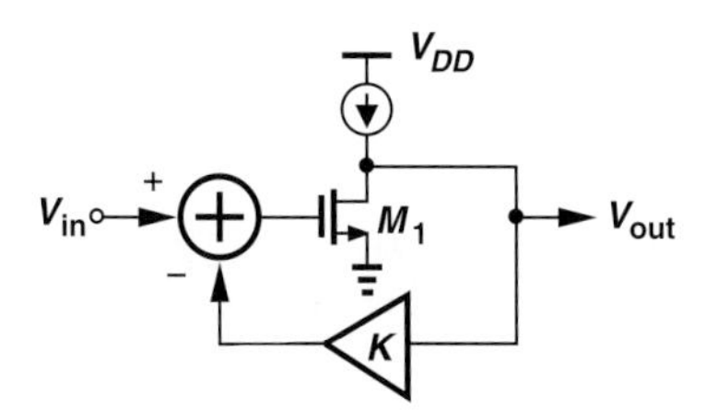

Figure 12.81

12.9. In some applications, we may define a "−1 dB bandwidth" as the frequency at which the gain falls by 10%. Determine the −1 dB bandwidth of the open-loop and closed-loop first-order systems described by Eqs. (12.19) and (12.22). Can we say the −1 dB bandwidth increases by $1 + KA_0$ as a result of feedback?

12.10. The circuit of Fig. 12.82 must achieve a closed-loop −3 dB bandwidth of B. Determine the required value of K. Neglect other capacitances and assume $\lambda > 0$.

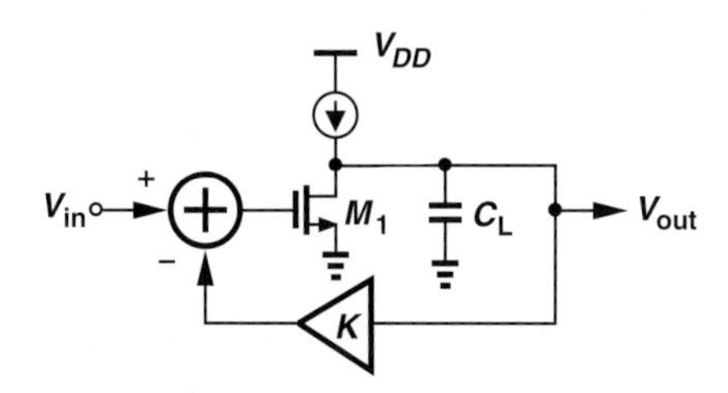

Figure 12.82

12.11. Repeat Example 12-7 for the circuit depicted in Fig. 12.83. Assume the impedance of C_1 and C_2 at the frequency of interest is much higher than R_D.

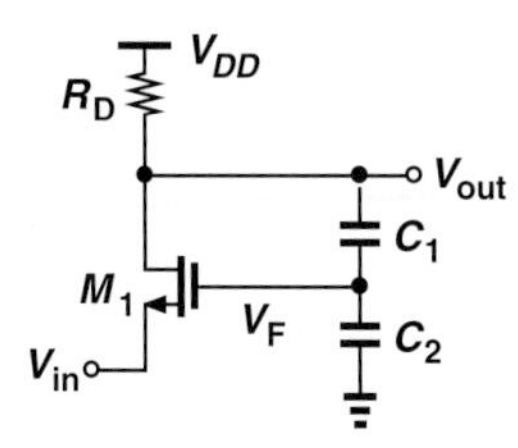

Figure 12.83

12.12. In Example 12-8, the closed-loop gain of the circuit must fall below its "unloaded" value by less than 10%. What is the lowest tolerable value of R_L?

12.13. In Fig. 12.14, $A_1 = 500$ and $A_2 = 420$. SOL What value of K guarantees that the closed-loop gains at x_1 and x_2 differ by less than 5%? What closed-loop gain is achieved under this condition?

***12.14.** The characteristic in Fig. 12.14(a) is sometimes approximated as

$$y = \alpha_1 x - \alpha_3 x^3, \qquad (12.201)$$

where α_1 and α_3 are constant.

(a) Determine the small-signal gain $\partial y/\partial x$ at $x = 0$ and $x = \Delta x$.

(b) Determine the closed-loop gain at $x = 0$ and $x = \Delta x$ for a feedback factor of K.

12.15. Using the developments in Fig. 12.18, draw the amplifier model for each stage in Fig. 12.19.

12.16. Determine the amplifier model for the circuit depicted in Fig. 12.84. Assume $\lambda > 0$.

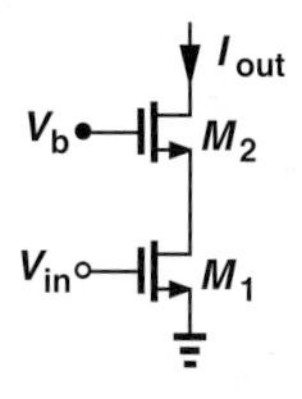

Figure 12.84

12.17. Repeat Problem 12.16 for the circuit in Example 12-7.

Sec. 12.4 Sense and Return Mechanisms

12.18. Identify the sense and return mechanisms in each amplifier depicted in Fig. 12.85.

12.19. Identify the sense and return mechanisms in each amplifier depicted in Fig. 12.86.

12.20. Identify the sense and return mechanisms in each amplifier depicted in Fig. 12.87.

***12.21.** Identify the sense and return mechanisms in each amplifier depicted in Fig. 12.88.

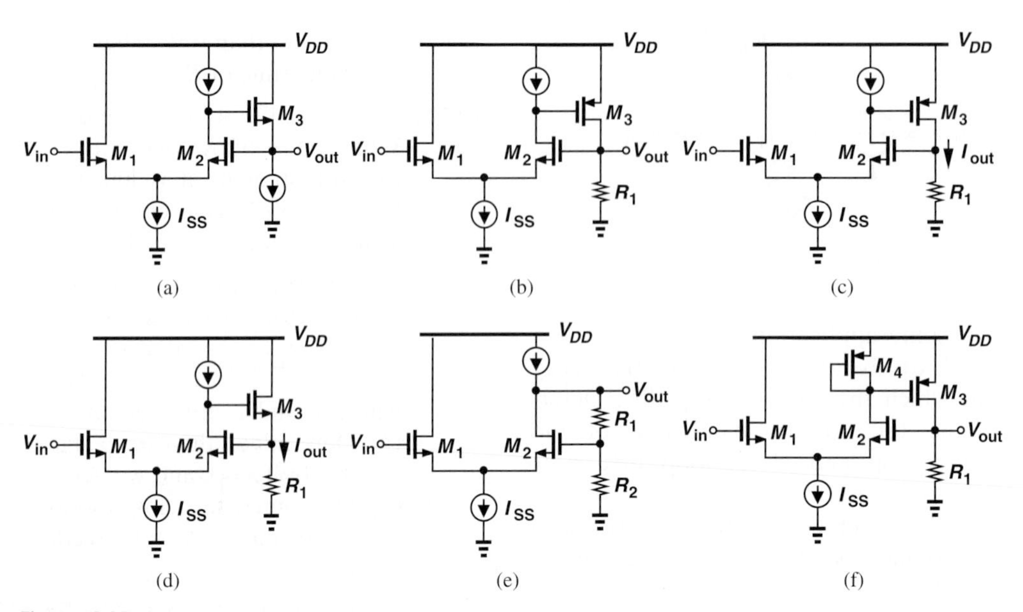

Figure 12.85

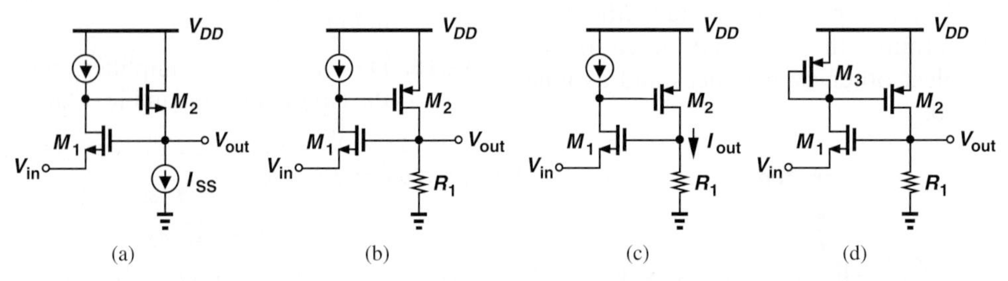

Figure 12.86

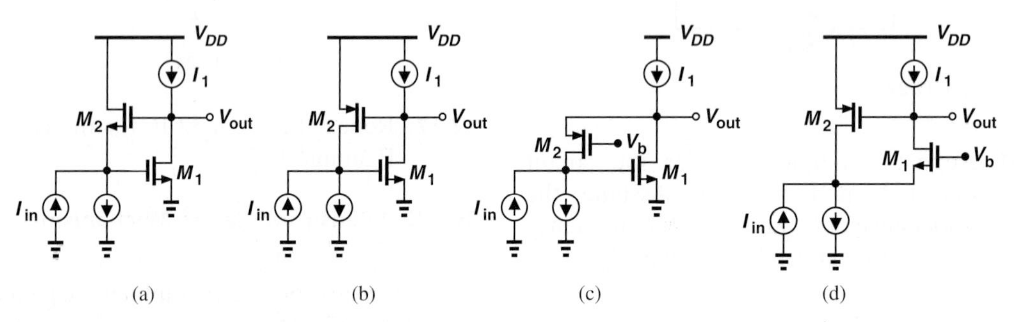

Figure 12.87

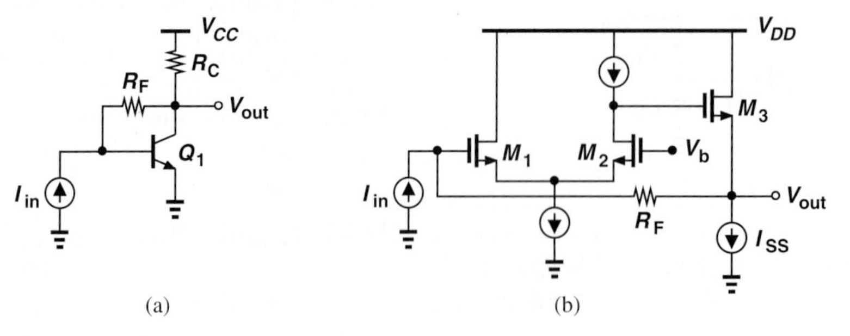

Figure 12.88

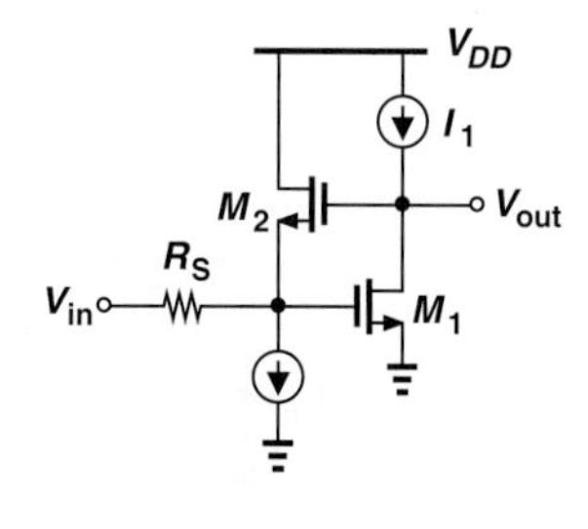

(a)

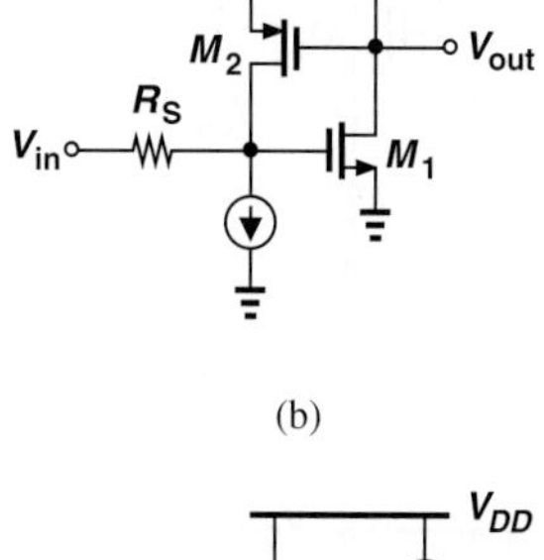

(b)

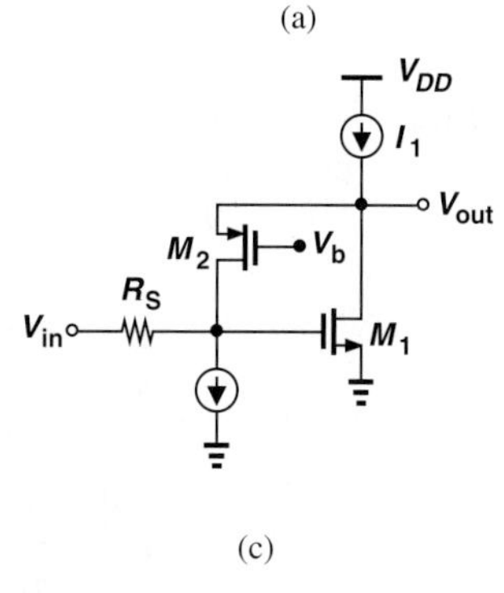

(c)

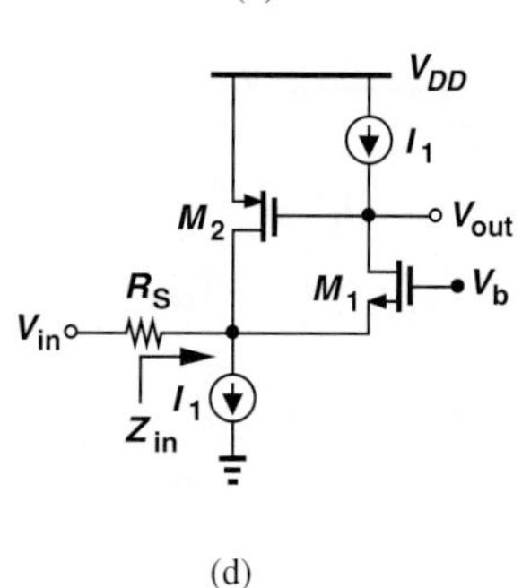

(d)

Figure 12.89

Sec. 12.5 Polarity of Feedback

*12.22. Determine the polarity of feedback in each of the stages illustrated in Fig. 12.89.

12.23. Determine the polarity of feedback in the circuit of Example 12-11.

SOL 12.24. Determine the polarity of feedback in the circuits depicted in Figs. 12.85–12.88.

Sec. 12.6.1 Voltage-Voltage Feedback

12.25. Consider the feedback circuit shown in Fig. 12.90, where $R_1 + R_2 \gg R_D$. Compute the closed-loop gain and I/O impedances of the circuit. Assume $\lambda \neq 0$.

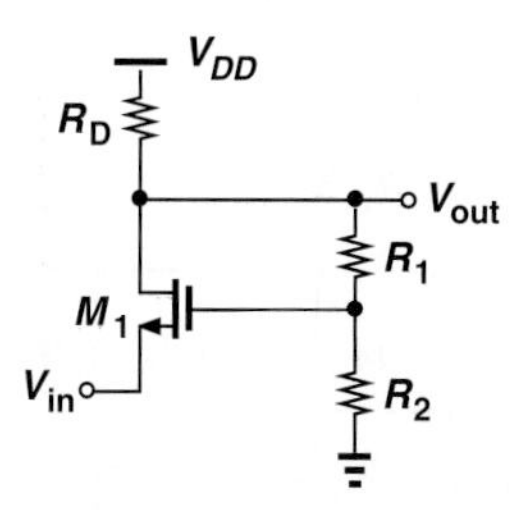

Figure 12.90

12.26. Repeat Problem 12.25 for the topology of Fig. 12.91. Assume C_1 and C_2 are very small and neglect other capacitances.

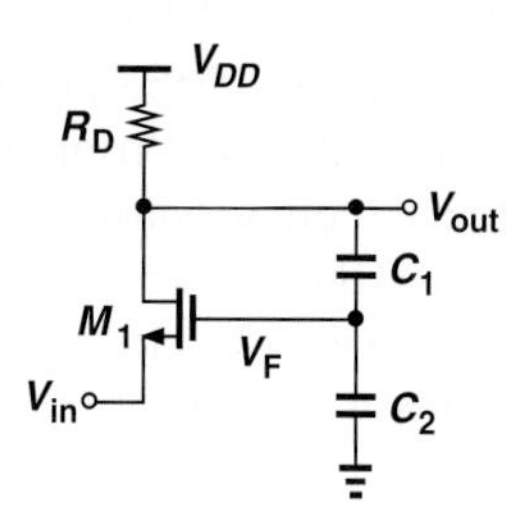

Figure 12.91

12.27. The amplifier shown in Fig. 12.92 provides a closed-loop gain close to unity but a very low output impedance. Assuming $\lambda > 0$, determine the closed-loop gain and output impedance and compare the results with those of a simple source follower.

*12.28. An adventurous student replaces the NMOS source follower in Fig. 12.92 with a PMOS common-source stage

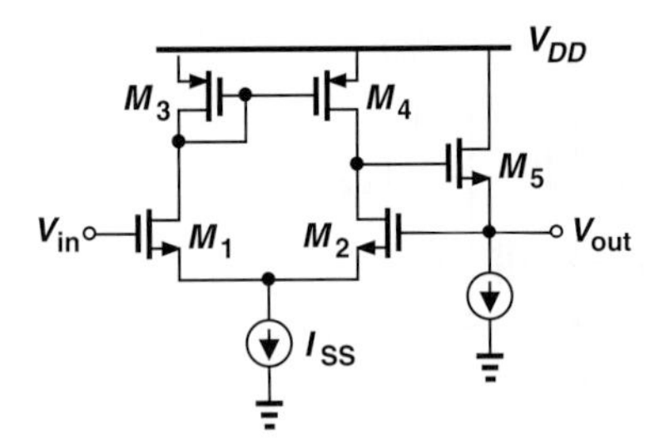

Figure 12.92

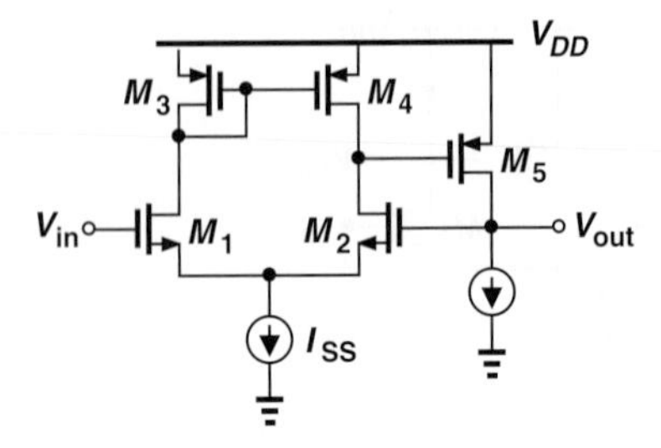

Figure 12.93

(Fig. 12.93). Unfortunately, the amplifier does not operate well.

(a) Prove by inspection that the feedback is positive.

(b) Breaking the loop at the gate of M_2, determine the loop gain and prove that the feedback is positive.

*__12.29.__ Having discovered the polarity of feedback, the student in Problem 12.28 modifies the circuit as shown in Fig. 12.94. Determine the closed-loop gain and I/O impedances of the circuit and compare the results with those obtained in Problem 12.27.

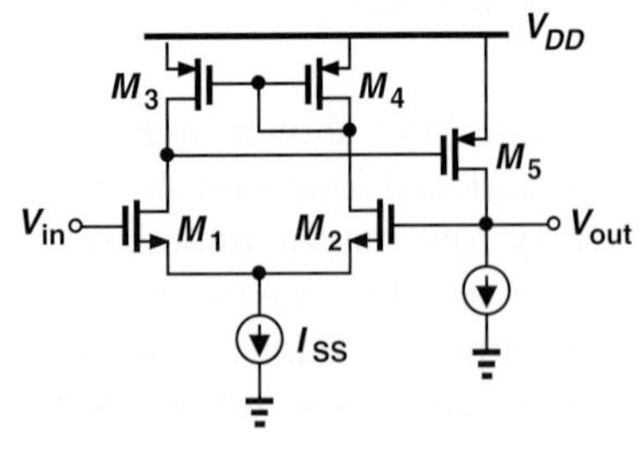

Figure 12.94

Sec. 12.6.2 Voltage-Current Feedback

12.30. Repeat Example 12-18 for the circuit illustrated in Fig. 12.95.

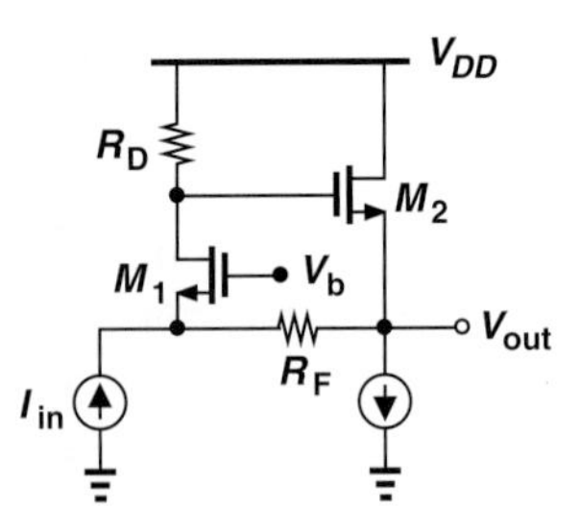

Figure 12.95

12.31. A student adventurously modifies the circuit of Example 12-18 to that shown in Fig. 12.96. Assume $\lambda = 0$.

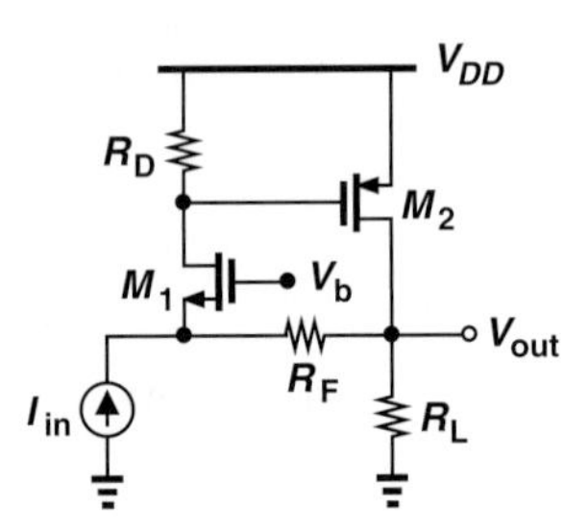

Figure 12.96

(a) Prove by inspection that the feedback is positive.

(b) Assuming R_F is very large and breaking the loop at the gate of M_2, calculate the loop gain and prove that the feedback is positive.

12.32. Determine the closed-loop I/O impedances of the circuit shown in Fig. 12.96.

**__12.33.__ The amplifier depicted in Fig. 12.97 consists of a common-gate stage (M_1 and R_D) and a feedback network (R_1, R_2, and M_2). Assuming $R_1 + R_2$ is very large and $\lambda = 0$, compute the closed-loop gain and I/O impedances. SOL

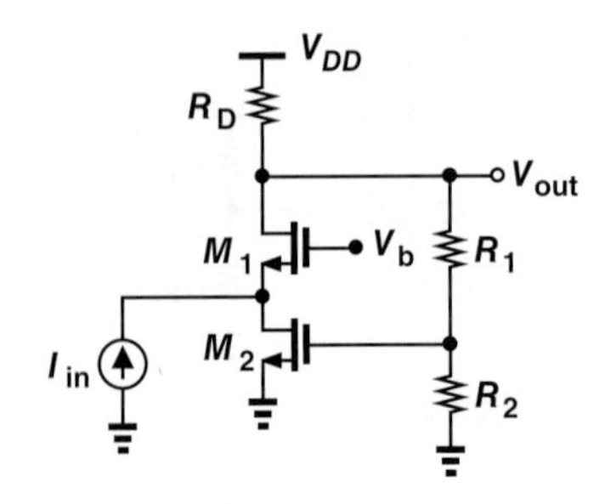

Figure 12.97

**12.34. Repeat Problem 12.33 for the circuit illustrated in Fig. 12.98. Assume C_1 and C_2 are very small and neglect other capacitances.

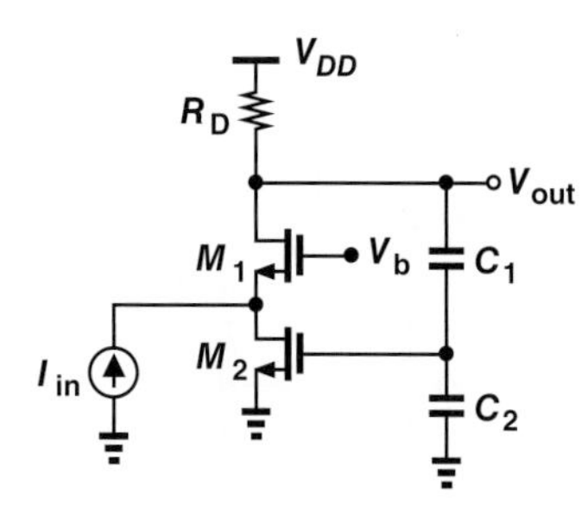

Figure 12.98

Sec. 12.6.3 Current-Voltage Feedback

12.35. A "laser diode" converts current to light (as in laser pointers). We wish to design a circuit that delivers a well-defined current to a laser diode. Shown in Fig. 12.99 is an example in which resistor R_M measures the current flowing through D_1 and amplifier A_1 subtracts the resulting voltage drop from V_{in}. Assume R_M is very small and $V_A = \infty$.

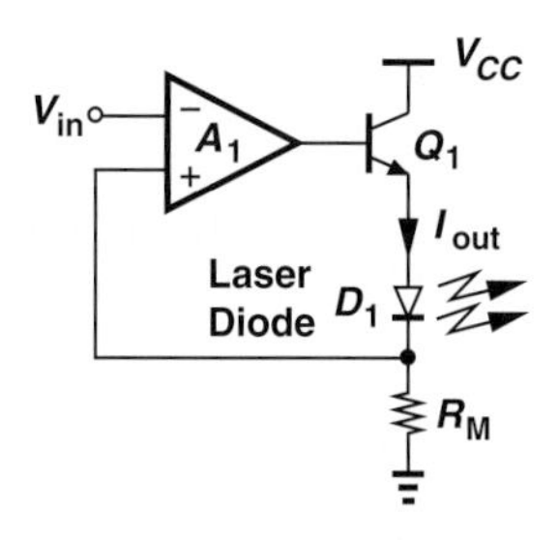

Figure 12.99

(a) Following the procedure used in Example 12-21, determine the open-loop gain.
(b) Calculate the loop gain and the closed-loop gain.

12.36. Following the procedure used in Example 12-22, compute the open-loop and closed-loop output impedances of the circuit depicted in Fig. 12.99.

*12.37. A student mistakenly replaces the common-emitter *pnp* device in Fig. 12.99 with an *npn* emitter follower (Fig. 12.100). Repeat Problems 12.35 and 12.36 for this circuit and compare the results.

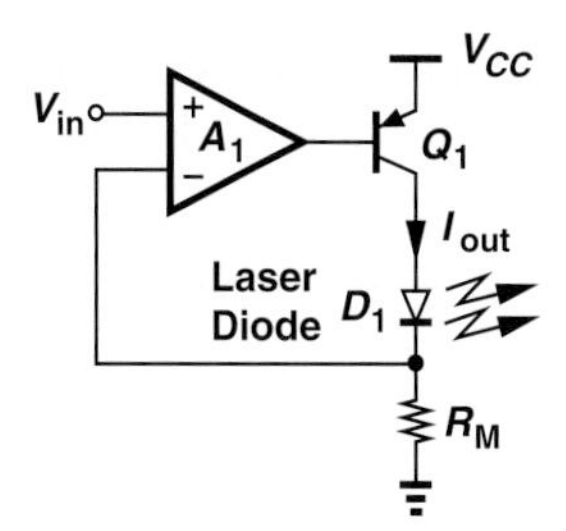

Figure 12.100

12.38. The amplifier A_1 in Fig. 12.100 can be realized as a common-base stage (Fig. 12.101). Repeat Problem 12.37 for this circuit. For simplicity, assume $\beta \to \infty$.

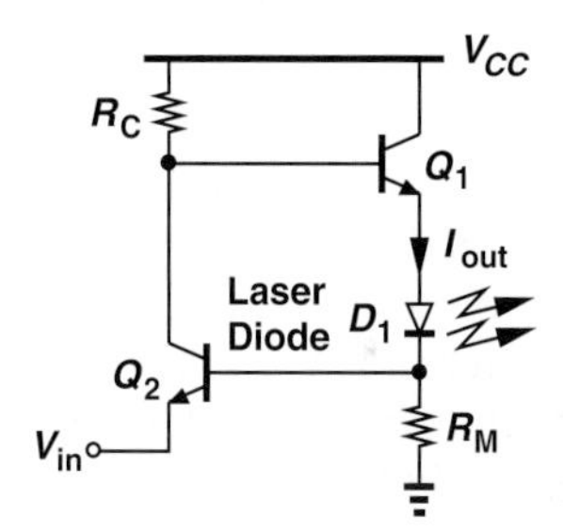

Figure 12.101

Sec. 12.6.4 Current-Current Feedback

12.39. A student has adventurously replaced the PMOS common-source stage in Fig. 12.49(a) with an NMOS source follower (Fig. 12.102).

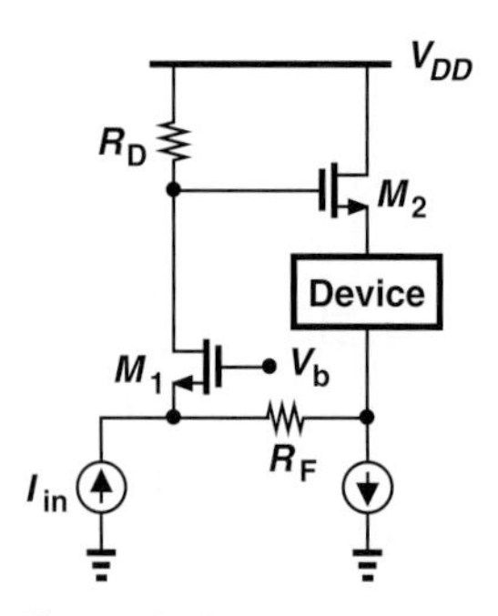

Figure 12.102

(a) Prove by inspection that the feedback is positive.
(b) Break the loop at the gate of M_2, determine the loop gain, and prove that the feedback is positive.

**12.40. Consider the feedback circuit depicted in Fig. 12.103. Assume $V_A = \infty$.

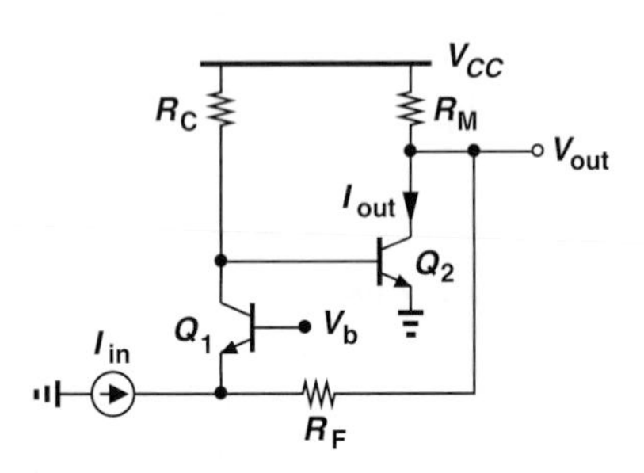

Figure 12.103

(a) Suppose the output quantity of interest is the collector current of Q_2, I_{out}. Assuming R_M is very small and R_F is very large, determine the closed-loop gain and I/O impedances of the circuit.
(b) Now, suppose the output quantity of interest is V_{out}. Assuming R_F is very large, compute the closed-loop gain and I/O impedances of the circuit.

Sec. 12.7 Effect of Finite I/O Impedances

12.41. The common-gate stage shown in Fig. 12.104 employs an ideal current source as its load, requiring that the loading introduced by R_1 and R_2 be taken into account. Repeat Example 12-26 for this circuit.

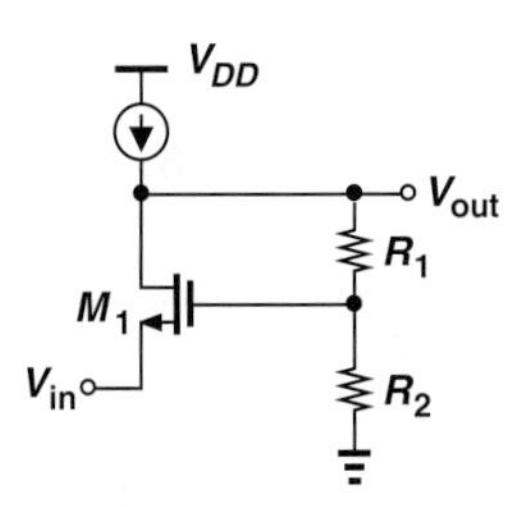

Figure 12.104

12.42. Figure 12.105 depicts the bipolar counterpart of the circuit studied in Example 12-26. Assuming $R_1 + R_2$ is not very large, $1 \ll \beta < \infty$ and $V_A = \infty$, determine the closed-loop gain and I/O impedances. SOL

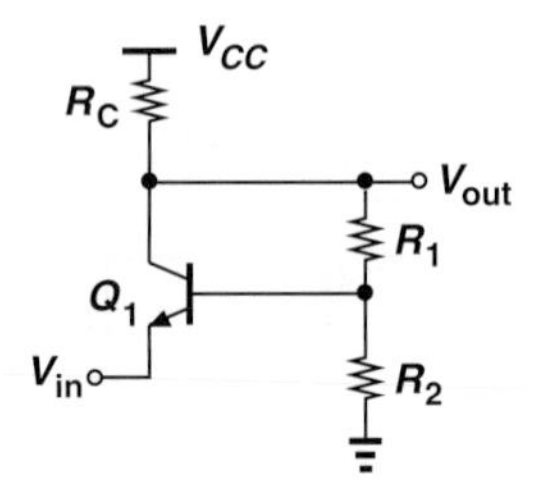

Figure 12.105

12.43. Repeat Problem 12.42 for the amplifier illustrated in Fig. 12.106.

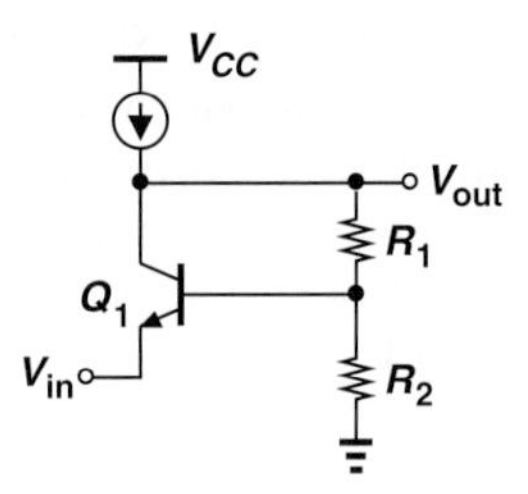

Figure 12.106

*12.44. Repeat Example 12-28 for the circuit shown in Fig. 12.107. Assume $\lambda = 0$.

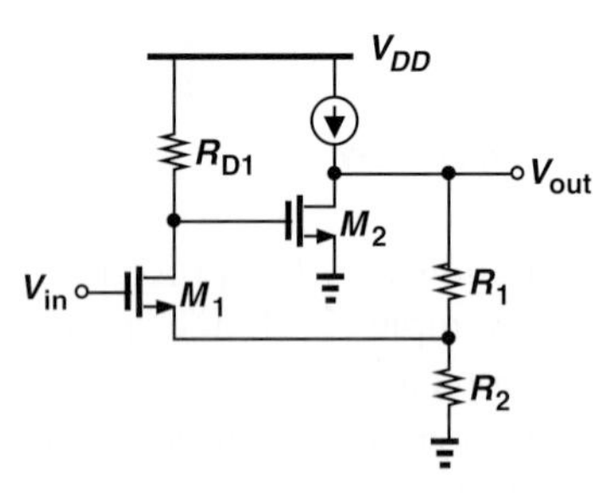

Figure 12.107

*12.45. Repeat Example 12-28 for the circuit shown in Fig. 12.108. Assume $V_A = \infty$.

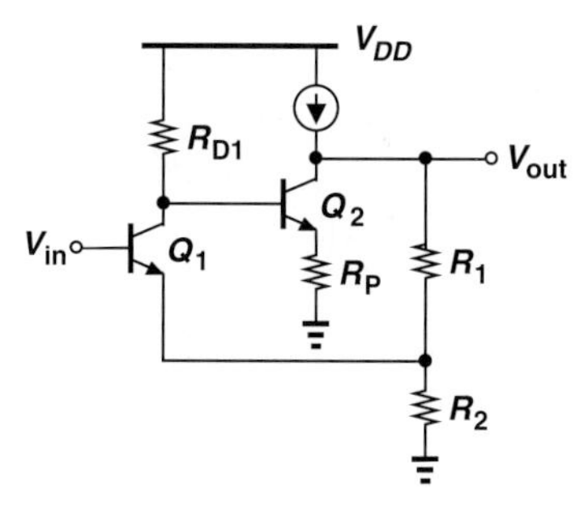

Figure 12.108

12.46. Assuming $V_A = \infty$, determine the closed-loop gain and I/O impedances of the amplifier depicted in Fig. 12.109. (For open-loop calculations, it is helpful to view Q_1 and Q_2 as a follower and a common-base stage, respectively.)

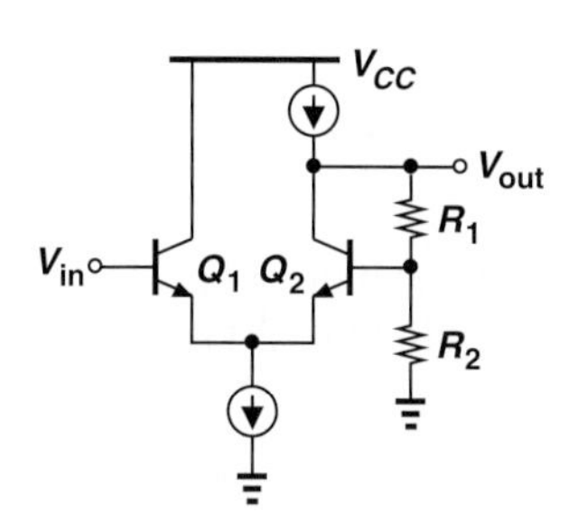

Figure 12.109

**12.47. Repeat Example 12-29 for the circuit illustrated in Fig. 12.110. Assume $\lambda > 0$.

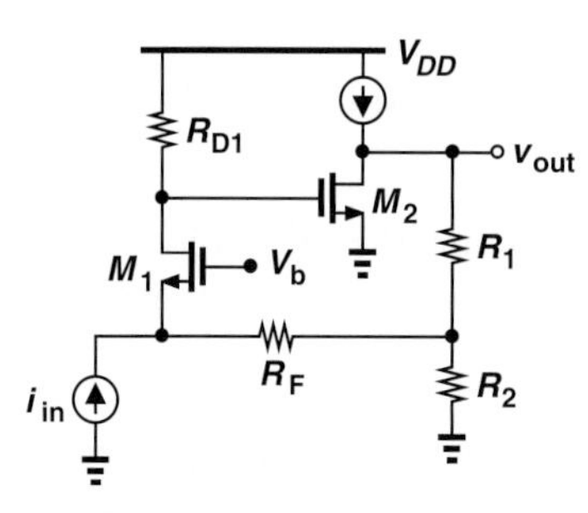

Figure 12.110

12.48. Repeat Example 12-29 for the bipolar transimpedance amplifier shown in Fig. 12.111. Assume $V_A = \infty$.

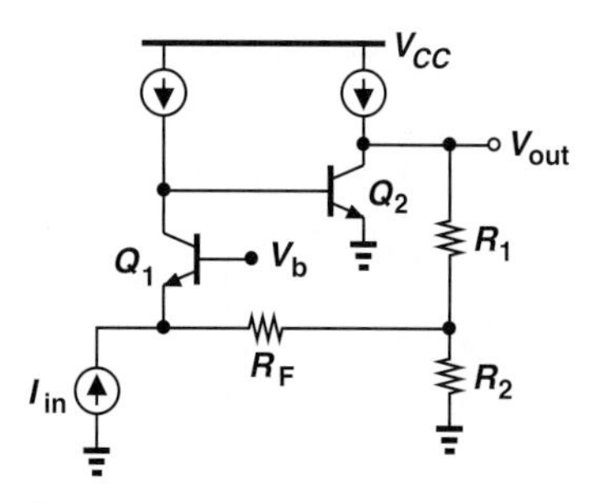

Figure 12.111

12.49. Figure 12.112 depicts a popular transimpedance amplifier topology. Repeat the analysis of Example 12-29 for this circuit. Assume $V_A < \infty$.

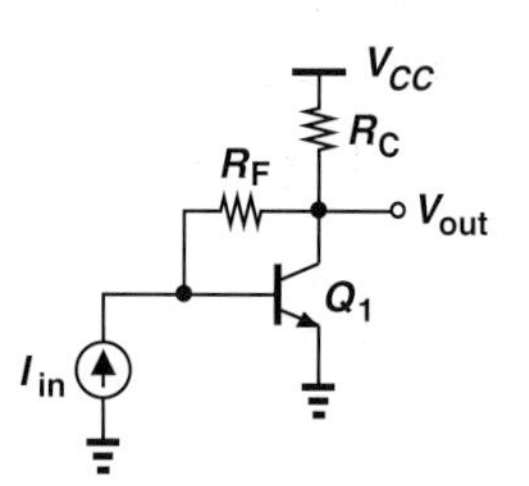

Figure 12.112

*12.50. The circuit of Fig. 12.112 can be improved by inserting an emitter follower at the output (Fig. 12.113). Assuming $V_A < \infty$, repeat Example 12-29 for this topology.

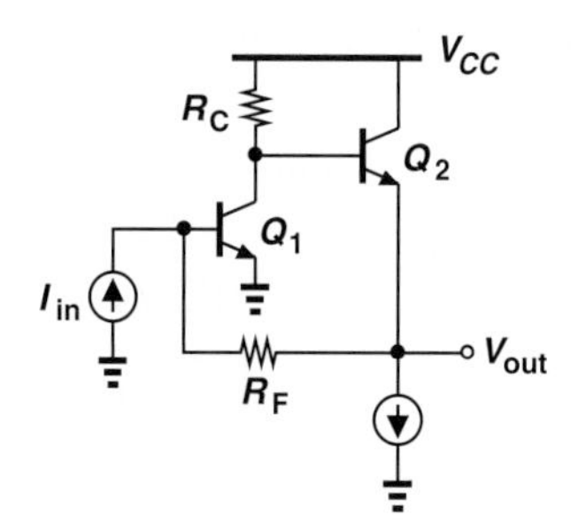

Figure 12.113

**12.51. Determine the closed-loop gain and I/O impedances of the circuits shown in Fig. 12.114, including the loading effects of each feedback network. Assume $\lambda = 0$.

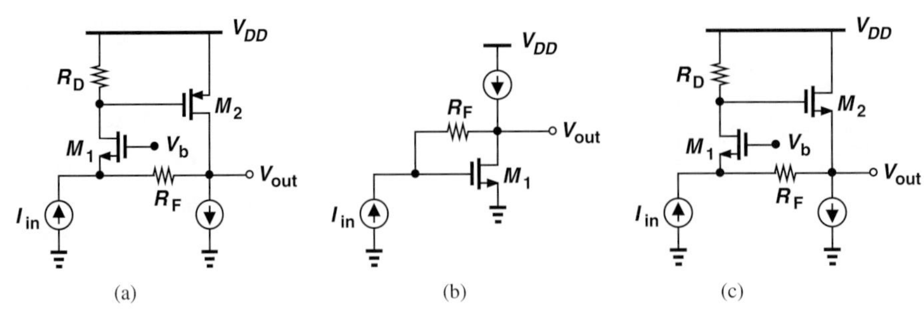

Figure 12.114

12.52. The circuit of Fig. 12.99, repeated in Fig. 12.115, employs a value for R_M that is not very small. Assuming $V_A < \infty$ and the diode exhibits an impedance of R_L, repeat the analysis of Example 12-30 for this circuit.

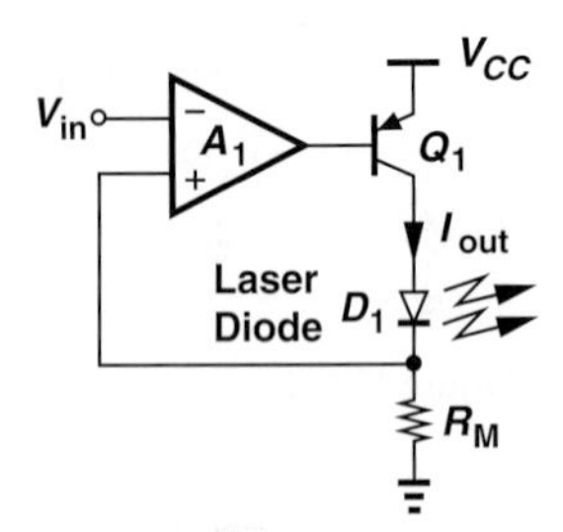

Figure 12.115

12.53. Repeat Example 12-32 for the circuit shown in Fig. 12.116. Note that R_M is replaced with a current source but the analysis proceeds in a similar manner.

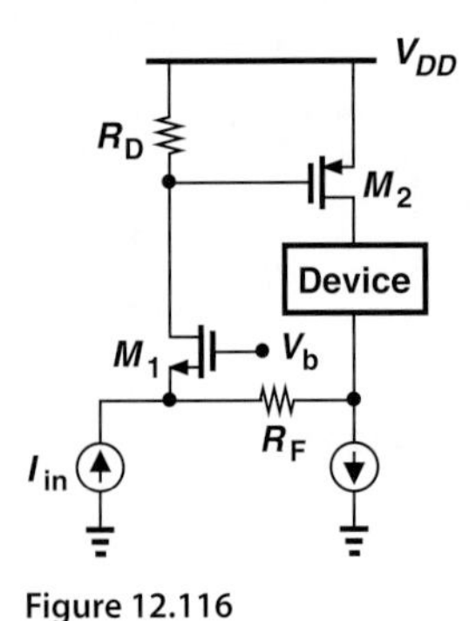

Figure 12.116

12.54. Repeat Problem 12.53 for the circuit illustrated in Fig. 12.117. Assume $V_A = \infty$.

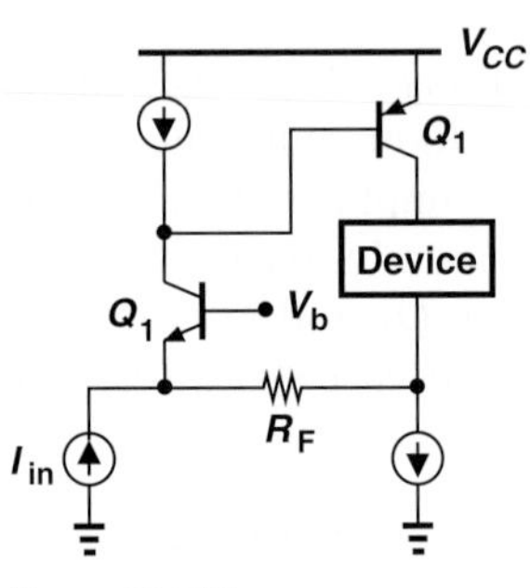

Figure 12.117

12.55. Repeat Problem 12.53 for the topology depicted in Fig. 12.118. Assume $V_A = \infty$.

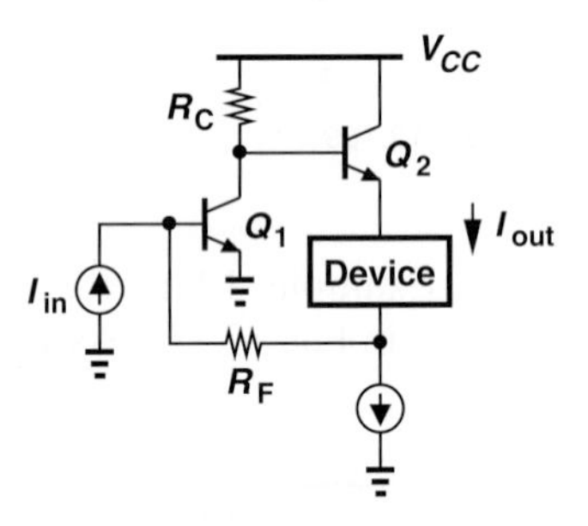

Figure 12.118

****12.56.** Compute the closed-loop gain and I/O impedances of the stages illustrated in Fig. 12.119.

Sec. 12.8.1 Review of Bode's Rules

12.57. Construct the Bode plots for the magnitude and phase of the following systems:

(a) $\omega_{p1} = 2\pi \times (10\ \text{MHz})$, $\omega_{p2} = 2\pi \times (120\ \text{MHz})$, $\omega_z = 2\pi \times (1\ \text{GHz})$.

(b) $\omega_z = 2\pi \times (10\ \text{MHz})$, $\omega_{p1} = 2\pi \times (120\ \text{MHz})$, $\omega_{p2} = 2\pi \times (1\ \text{GHz})$.

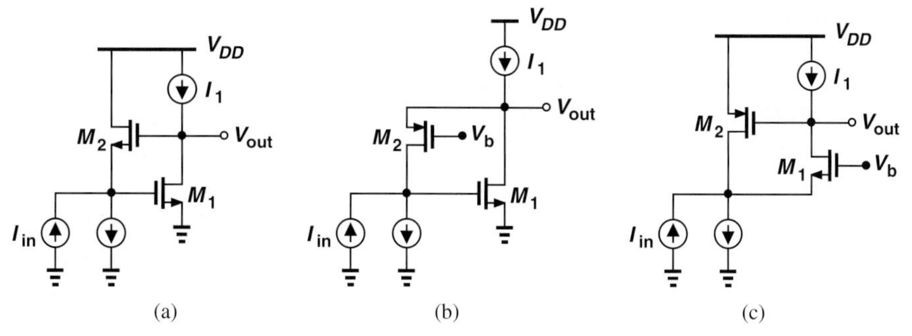

Figure 12.119

(c) $\omega_z = 0$, $\omega_{p1} = 2\pi \times (10 \text{ MHz})$, $\omega_{p2} = 2\pi \times (120 \text{ MHz})$.

(d) $\omega_{p1} = 0$, $\omega_z = 2\pi \times (10 \text{ MHz})$, $\omega_{p2} = 2\pi \times (120 \text{ MHz})$.

12.58. In the Bode plots of Fig. 12.63, explain qualitatively what happens as ω_z comes closer to ω_{p1} or ω_{p2}.

****12.59.** Assuming $\lambda = 0$ and without using Miller's theorem, determine the transfer function of the circuit depicted in Fig. 12.120 and construct its Bode plots.

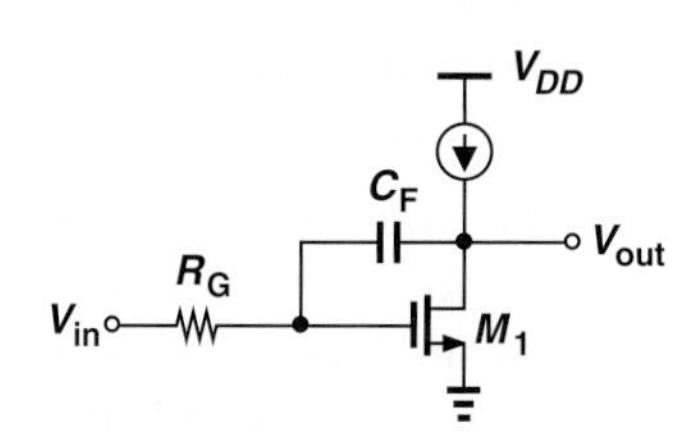

Figure 12.120

Sec. 12.8.2 Problem of Instability

12.60. In the system of Example 12-37, we gradually decrease the value of K without changing the position of the poles. Explain why decreasing K can make the system stable?

SOL ***12.61.** The three coincident poles in Example 12-38 do affect the phase even at $0.1\omega_p$. Calculate the phase of the transfer function at $\omega = 0.1\omega_p$.

***12.62.** Unlike a one-pole system, the magnitude response of the circuit in Example 12-38 falls by more than 3 dB at the pole frequency. Determine $|H|$ at ω_p. Can we say $|H|$ falls 9 dB due to three coincident poles?

12.63. Repeat Example 12-38 for $K = 0.1$.

12.64. Repeat Example 12-38 for four identical stages and compare the results.

12.65. Consider a one-pole circuit whose open-loop transfer function is given by

$$H(s) = \frac{A_0}{1 + \frac{s}{\omega_0}}. \tag{12.202}$$

Determine the phase margin of a feedback network using this circuit with $K = 1$.

12.66. Repeat Problem 12.65 for $K = 0.5$.

12.67. In each case illustrated in Fig. 12.72, what happens if K is reduced by a factor of 2?

12.68. Suppose the amplifier in Example 12-41 is described by

$$H(s) = \frac{A_0}{\left(1 + \frac{s}{\omega_{p1}}\right)\left(1 + \frac{s}{\omega_{p2}}\right)}, \tag{12.203}$$

where $\omega_{p2} \gg \omega_{p1}$. Compute the phase margin if the circuit is employed in a feedback system with $K = 0.5$.

12.69. Explain what happens to the characteristics illustrated in Fig. 12.74 if K drops by a factor of two. Assume ω_{p1} and ω'_{p1} remain constant.

****12.70.** Figure 12.121 depicts the amplifier of Example 12-38 with a compensation capacitor added to node X. Explain how the circuit can be compensated for a phase margin of 45°.

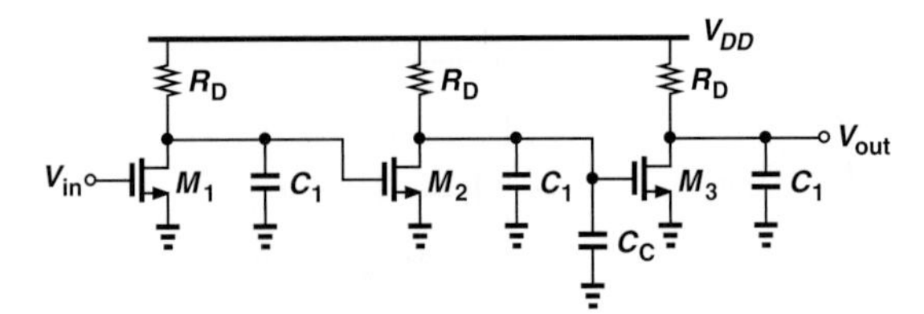

Figure 12.121

Design Problems

In the following problems, unless otherwise stated, assume $\mu_n C_{ox} = 2\mu_p C_{ox} = 100\ \mu A/V^2$ and $\lambda_n = 0.5\lambda_p = 0.1 V^{-1}$.

12.71. Design the circuit of Example 12-15 for an open-loop gain of 50 and a nominal closed-loop gain of 4. Assume $I_{SS} =$ 0.5 mA. Choose $R_1 + R_2 \approx 10(r_{O2}||r_{O4})$.

12.72. Design the circuit of Example 12-16 for an open-loop gain of 10, a closed-loop input impedance of 50 Ω, and a nominal closed-loop gain of 2. Calculate the closed-loop I/O impedances. Assume $R_1 + R_2 \approx 10R_D$.

SOL **12.73.** Design the transimpedance amplifier of Example 12-18 for an open-loop gain of 10 kΩ, a closed-loop gain of 1 kΩ, a closed-loop input impedance of 50 Ω, and a closed-loop output impedance of 200 Ω. Assume $R_{D1} = 1$ kΩ and R_F is very large.

12.74. Repeat Problem 12.73 for the circuit shown in Fig. 12.96.

12.75. We wish to design the transimpedance amplifier depicted in Fig. 12.103 for a closed-loop gain of 1 kΩ. Assume each transistor carries a collector bias current of 1 mA, $\beta = 100$, $V_A = \infty$, and R_F is very large.
(a) Determine the values of R_C and R_M for an open-loop gain of 20 kΩ and an open-loop output impedance of 500 Ω.
(b) Compute the required value of R_F.
(c) Calculate the closed-loop I/O impedances.

12.76. Design the circuit illustrated in Fig. 12.108 for an open-loop voltage gain of 20, an open-loop output impedance of 2 kΩ, and a closed-loop voltage gain of 4. Assume $\lambda = 0$. Is the solution unique? If not, how should the circuit parameters be chosen to minimize the power dissipation?

12.77. Design the circuit of Fig. 12.109 for a closed-loop gain of 2, a tail current of 1 mA, and minimum output impedance. Assume $\beta = 100$ and $V_A = \infty$. SOL

12.78. Design the transimpedance amplifier of Fig. 12.113 for a closed-loop gain of 1 kΩ an output impedance of 50 Ω. Assume each transistor is biased at a collector current of 1 mA and $V_A = \infty$.

SPICE PROBLEMS

In the following problems, use the MOS device models given in Appendix A. For bipolar transistors, assume $I_{S,npn} = 5 \times 10^{-16}$ A, $\beta_{npn} = 100$, $V_{A,npn} = 5$ V, $I_{S,pnp} = 8 \times 10^{-16}$ A, $\beta_{pnp} = 50$, $V_{A,pnp} = 3.5$ V. Also, SPICE models the effect of charge storage in the base by a parameter called $\tau_F = C_b/g_m$. Assume $\tau_F(tf) = 20$ ps.

12.79. Figure 12.122 shows a transimpedance amplifier often used in optical communications. Assume $R_F = 2$ kΩ.
(a) Select the value of R_C so that Q_1 carries a bias current of 1 mA.
(b) Estimate the loop gain.
(c) Determine the closed-loop gain and I/O impedances.
(d) Determine the change in the closed-loop gain if V_{CC} varies by ±10%.

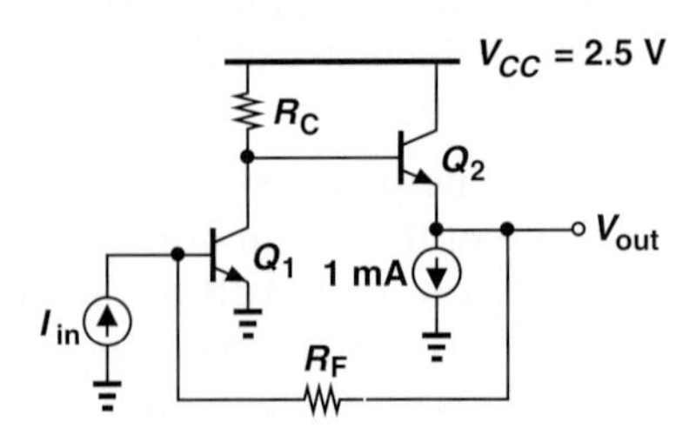

Figure 12.122

12.80. Figure 12.123 depicts another transimpedance amplifier, where the bias current of M_1 is defined by the mirror arrangement (M_2 and M_3). Assume $W/L = 20\ \mu m/0.18\ \mu m$ for M_1–M_3.

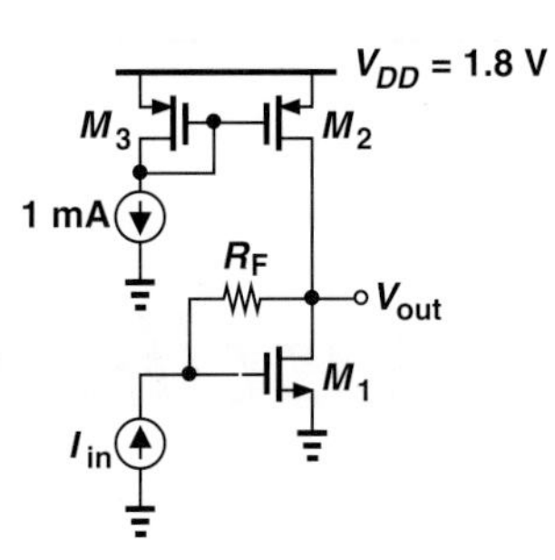

Figure 12.123

(a) What value of R_F yields a closed-loop gain of 1 kΩ?
(b) Determine the change in the closed-loop gain if V_{DD} varies by ±10%.
(c) Suppose the circuit drives a load capacitance of 100 fF. Verify that the input impedance exhibits *inductive* behavior and explain why.

12.81. In the circuit shown in Fig. 12.124, $W/L = 20\ \mu m/0.18\ \mu m$ for M_1 and M_2.

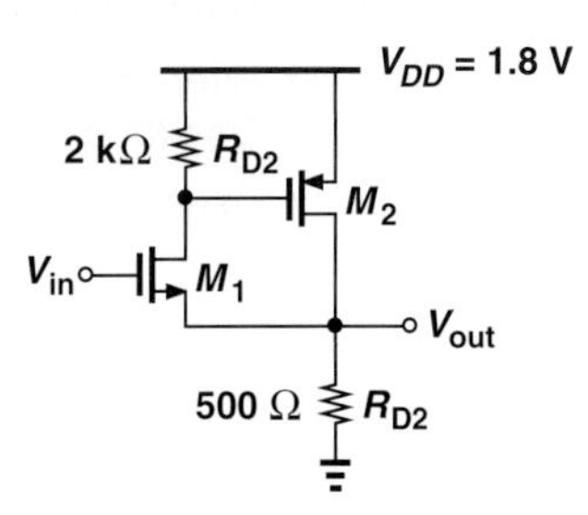

Figure 12.124

(a) Determine the circuit's operating points for an input dc level of 0.9 V.
(b) Determine the closed-loop gain and I/O impedances.

12.82. In the circuit of Fig. 12.125, the three stages provide a high gain, approximating an op amp. Assume $(W/L)_{1-6} = 10\ \mu m/0.18\ \mu m$.

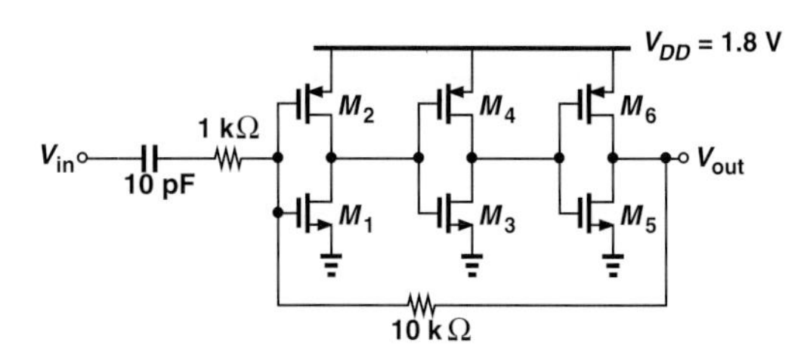

Figure 12.125

(a) Explain why the circuit is potentially unstable.
(b) Determine the step response and explain the circuit's behavior.
(c) Place a capacitor between nodes X and Y and adjust its value to obtain a well-behaved step response.
(d) Determine the gain error of the circuit with respect to the nominal value of 10.

12.83. In the three-stage amplifier of Fig. 12.126, $(W/L)_{1-7} = 20\ \mu m/0.18\ \mu m$.

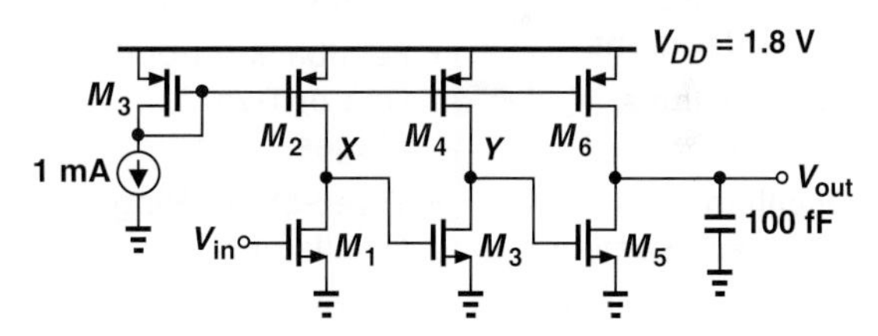

Figure 12.126

(a) Determine the phase margin.
(b) Place a capacitor between nodes X and Y so as to obtain a phase margin of 60°. What is the unity-gain bandwidth under this condition?
(c) Repeat (b) if the compensation capacitor is tied between X and ground and compare the results.

REFERENCES

1. A. Almassri, et al., "Pressure sensor: state of the art, design, and application for robotic hand," *J. Sensors*, vol. 2015, July 2015.
2. B. Razavi, *Design of Analog CMOS Integrated Circuits*, McGraw-Hill, 2001.

13

Oscillators

Most of our studies in previous chapters have focused on the analysis and design of amplifiers. In this chapter, we turn our attention to another important class of analog circuits, namely, oscillators. From your laptop computer to your cell phone, today's electronic devices use oscillators for numerous purposes, and pose interesting challenges. For example, the clock driving a 3 GHz microprocessor is generated by an on-chip oscillator running at 3 GHz. Also, a WiFi transceiver employs a 2.4 GHz or 5 GHz on-chip oscillator to generate a "carrier." Shown below is the outline of the chapter. The reader is encouraged to review Chapter 12 before delving into oscillators.

General Considerations	Ring Oscillators	LC Oscillators	Other Oscillators
• Barkhausen's Criteria • Oscillatory Feedback System • Startup Condition • Oscillator Topologies	• Feedback around One or Two Stages • Three-Stage Ring Oscillator • Internal Waveforms	• Parallel LC Tank • Cross-Coupled Oscillator • Colpitts Oscillator	• Phase Shift Oscillator • Wien-Bridge Oscillator • Crystal Oscillator

13.1 GENERAL CONSIDERATIONS

We know from previous chapters that an amplifier *senses* a signal and reproduces it at the output, perhaps with some gain. An oscillator, on the other hand, *generates* a signal, typically a periodic one. For example, the clock in a microprocessor resembles a square wave (Fig. 13.1).

How can a circuit generate a periodic output without an input? Let us return to our study of amplifier stability in Chapter 12 and recall that a negative-feedback circuit can oscillate if Barkhausen's criteria are met. That is, as shown in Fig. 13.2, we have

$$\frac{Y}{X}(s) = \frac{H(s)}{1 + H(s)}, \tag{13.1}$$

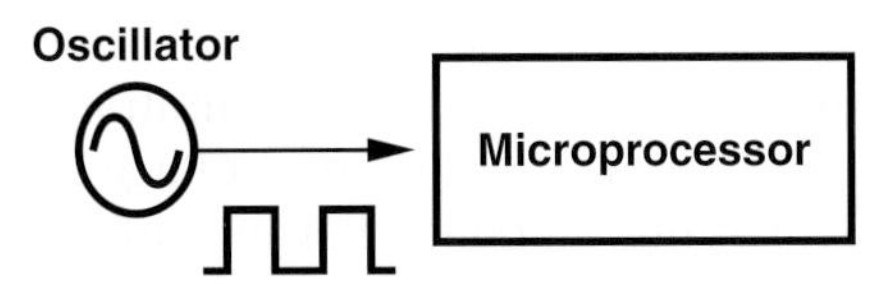

Figure 13.1 High-speed oscillator driving a microprocessor.

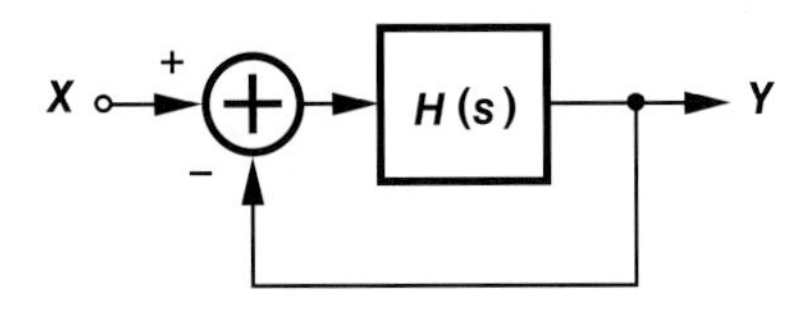

Figure 13.2 Feedback system for oscillation study.

which goes to infinity at a frequency of ω_1 if $H(s = j\omega_1) = -1$, or, equivalently, $|H(j\omega_1)| = 1$ and $\angle H(j\omega_1) = 180°$. We may therefore view an oscillator as a badly-designed feedback amplifier! The key point here is that the signal traveling around the loop experiences so much phase shift (i.e., delay) that, upon reaching the subtractor, it actually *enhances X*. With enough loop gain, the circuit continues to amplify X indefinitely, generating an infinitely large output waveform from a finite swing at X.

Did you know?

The most commonplace use of oscillators is in (electronic) watches. A crystal oscillator (studied later in this chapter) runs at a precise frequency of $32{,}768(= 2^{15})$ Hz. This frequency is then divided down by means of a 15-bit counter to generate a 1-Hz square waveform, providing the "time base." This waveform shows the seconds on the watch. It is also divided by 60 and another 60 to count the minutes and the hours, respectively. A great challenge in early electronic watches was to design these counters for a very low power consumption so that the watch battery would last a few months.

It is important not to confuse the *frequency-dependent* 180° phase shift stipulated by Barkhausen with the 180° phase shift necessary for negative feedback. As depicted in Fig. 13.3(a), the loop contains one net signal inversion (the negative sign at the input of the adder) so as to ensure negative feedback *and* another 180° of phase shift at ω_1. In other words, the *total* phase shift around the loop reaches 360° at ω_1 [Fig. 13.3(b)].

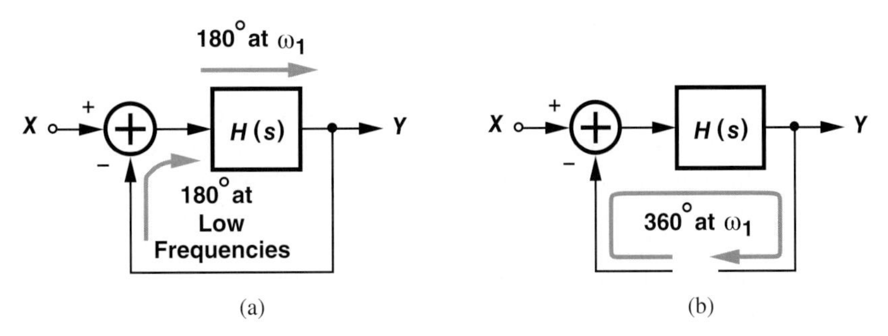

Figure 13.3 (a) Phase shift around an oscillator loop, (b) alternative view.

We must now answer two urgent questions. First, where does X come from? (We just stated that oscillators do not have an input.) In practice, X comes from the noise of the devices within the loop. Transistors and resistors in the oscillator produce noise at all frequencies, providing the "seed" for oscillation at ω_1. Second, does the output amplitude really go to infinity? No, in reality, saturation or nonlinear effects in the circuit limit the output swing. After all, if the supply voltage is 1.5 V, it would be difficult to produce a swing greater than this amount.[1] For example, consider the conceptual arrangement shown in Fig. 13.4, where a common-source stage provides amplification within $H(s)$. As the output swing grows, at some point M_1 enters the triode region and its transconductance falls. Consequently, the loop gain decreases, eventually approaching the barely acceptable value, unity.

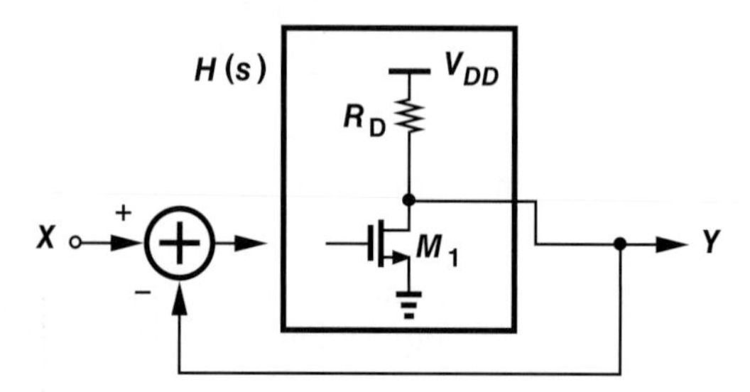

Figure 13.4 Feedback loop containing a common-source stage.

Example 13-1 An oscillator employs a differential pair [Fig. 13.5(a)]. Explain what limits the output amplitude.

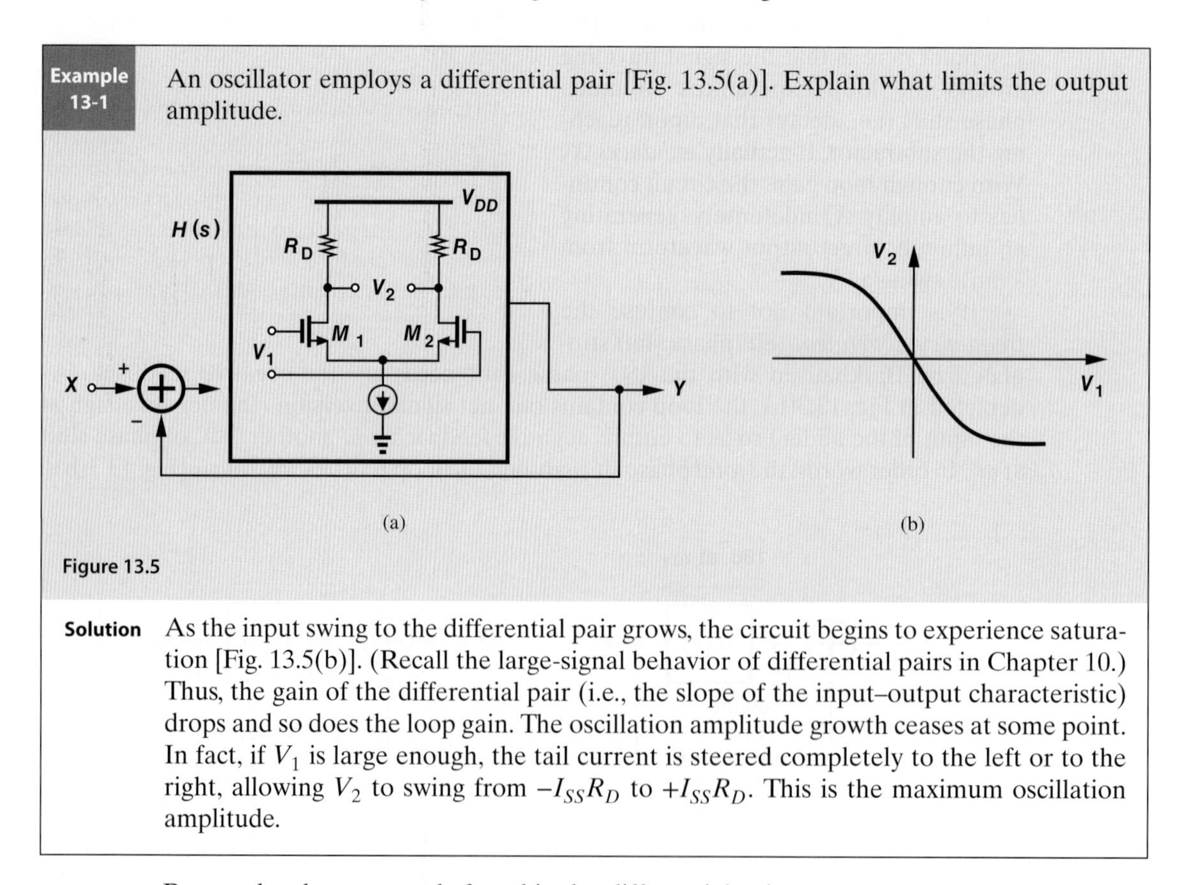

Figure 13.5

Solution As the input swing to the differential pair grows, the circuit begins to experience saturation [Fig. 13.5(b)]. (Recall the large-signal behavior of differential pairs in Chapter 10.) Thus, the gain of the differential pair (i.e., the slope of the input–output characteristic) drops and so does the loop gain. The oscillation amplitude growth ceases at some point. In fact, if V_1 is large enough, the tail current is steered completely to the left or to the right, allowing V_2 to swing from $-I_{SS}R_D$ to $+I_{SS}R_D$. This is the maximum oscillation amplitude.

Exercise Repeat the above example for a bipolar differential pair.

[1]But some oscillators can generate an output swing about twice the supply voltage.

TABLE 13.1 Summary of various oscillator topologies and their applications.

Oscillator Topology	Ring Oscillator	LC Oscillators: Cross-Coupled Oscillator	LC Oscillators: Colpitts Oscillator	Phase Shift Oscillator	Wien-Bridge Oscillator	Crystal Oscillator
Implementation	Integrated	Integrated	Discrete or Integrated	Discrete	Discrete	Discrete or Integrated
Typical Frequency Range	Up to Several Gigahertz	Up to Hundreds of Gigahertz	Up to Tens of Gigahertz	Up to a Few Megahertz	Up to a Few Megahertz	Up to About 100 MHz
Application	Microprocessors and Memories	Wireless Transceivers	Stand-Alone Oscillators	Prototype Design	Prototype Design	Precise Reference

Startup Condition From the first Barkhausen criterion, we may design the circuit for a unity loop gain at the desired oscillation frequency, ω_1. This is called the oscillation "startup condition." However, this choice places the circuit at the edge of failure: a slight change in the temperature, process, or supply voltage may drop the loop gain below 1. For this and other reasons, the loop gain is usually quite larger than unity. (In fact, the design typically begins with the required output voltage swing rather than the loop gain.)

What aspects of an oscillator design are important? Depending on the application, the specifications include the frequency of oscillation, output amplitude, power consumption, and complexity. In some cases, the "noise" in the output waveform is also critical.

Oscillators can be realized as either integrated or discrete circuits. The topologies are quite different in the two cases but still rely on Barkhausen's criteria. We study both types here. It is helpful to first take a glance at the various types of oscillators studied in this chapter. Table 13.1 summarizes the topologies and some of their attributes.

13.2 RING OSCILLATORS

Most microprocessors and memories incorporate CMOS "ring oscillators." As the name implies, the circuit consists of a number of stages in a ring, but to understand the underlying principles, we must take a few steps back.

Let us place a common-source amplifier in a negative-feedback loop and see if it oscillates. As shown in Fig. 13.6(a), we tie the output to the input. The feedback (at low frequencies) is negative because the stage has a voltage gain of $-g_m R_D$ (if $\lambda = 0$). Now, consider the small-signal model [Fig. 13.6(b)]. Does this circuit oscillate? Of the two Barkhausen's criteria, the loop gain requirement, $|H(j\omega_1)| = 1$, appears possible. But how about the phase requirement? Neglecting C_{GD1}, we observe that the circuit's capacitances merge into one at node X, forming a single (open-loop) pole with R_D: $\omega_{p,X} = -(R_D C_L)^{-1}$. Unfortunately, a single pole can provide a maximum phase shift of $-90°$ (at $\omega = \infty$). That is, the frequency-dependent phase shift of the open-loop transfer function, $H(s)$, does not exceed $-90°$, prohibiting oscillation. From the perspective of Fig. 13.3(b), the *total* phase shift around the loop cannot reach 360°.

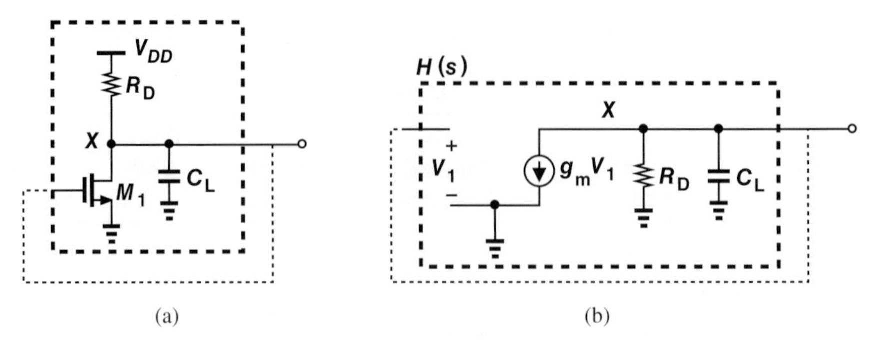

Figure 13.6 (a) Hypothetical oscillator using a single CS stage, (b) equivalent circuit of (a).

This brief analysis suggests that we should increase the delay or phase shift around the loop. For example, let us cascade *two* CS stages (Fig. 13.7). Now, the open-loop circuit contains two poles, exhibiting a maximum phase shift of 180°. Do we have an oscillator? No, not yet; each CS stage provides a phase shift of −90° only at $\omega = \infty$, but no *gain* at this frequency. That is, we still cannot meet both of Barkhausen's criteria at the same frequency.

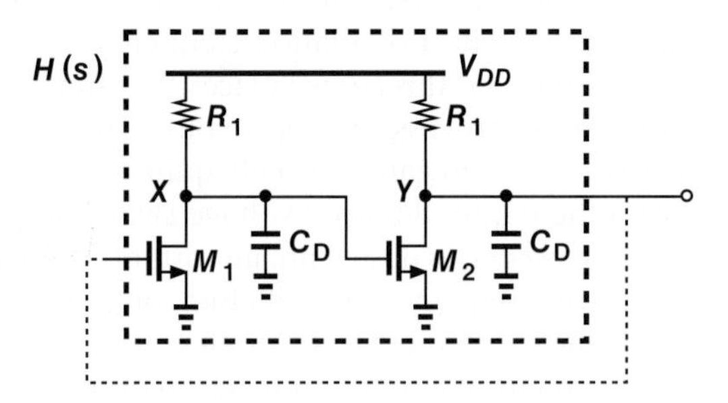

Figure 13.7 Feedback loop using two CS stages.

By now, we see the trend and postulate that we must insert one more CS stage in the loop, as shown in the "ring oscillator" of Fig. 13.8.[2] Each pole must provide a phase shift of only 60° (or −60°). Since the phase shift for a pole at $R_D C_D$ is equal to $-\tan^{-1}(R_D C_D \omega)$, we have

$$-\tan^{-1}(R_D C_D \omega_1) = -60°, \tag{13.2}$$

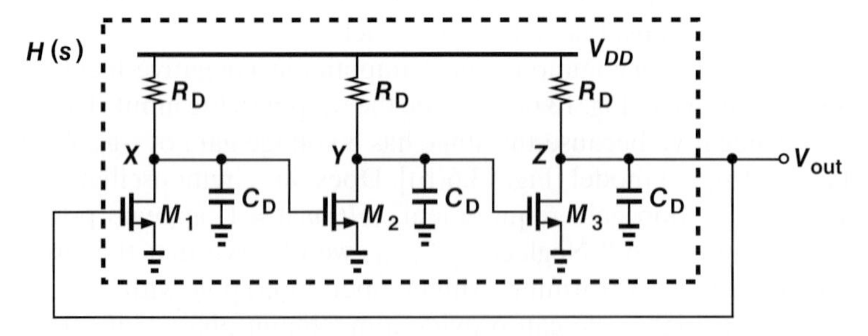

Figure 13.8 Simple three-stage ring oscillator.

[2]We neglect C_{GD} here.

obtaining an oscillation frequency of

$$\omega_1 = \frac{\sqrt{3}}{R_D C_D}. \tag{13.3}$$

The reader is encouraged to apply the perspective of Fig. 13.3(b) and prove that the total phase shift around the loop is 360° at ω_1. The startup condition is calculated by setting $H(s)$ at $s = j\omega_1$ equal to unity. If $\lambda = 0$, the transfer function of each stage is given by $-g_m R_D/(1 + R_D C_D s)$. We replace s with $j\omega_1$, find the magnitude of the transfer function, raise it to the third power (for three identical stages), and equate the result to 1:

$$\left(\frac{g_m R_D}{\sqrt{1 + R_D^2 C_D^2 \omega_1^2}}\right)^3 = 1. \tag{13.4}$$

It follows that

$$g_m R_D = \sqrt{1 + R_D^2 C_D^2 \omega_1^2} \tag{13.5}$$

$$= 2, \tag{13.6}$$

indicating that the *low-frequency* gain of each stage must exceed 2 to ensure oscillation startup.

Example 13-2 A student runs a transient SPICE simulation on the ring oscillator of Fig. 13.8 but observes that all three drain voltages are equal and the circuit does not oscillate. Explain why. Assume that the stages are identical.

Solution At the beginning of a transient simulation, SPICE computes the dc operating points for all of the devices. With identical stages, SPICE finds equal drain voltages as one solution of the network and retains it. Thus, the three drain voltages remain at the same value indefinitely. In the actual circuit, on the other hand, the electronic noise of the devices perturbs these voltages, initiating oscillation. (The transient simulation in SPICE does not include device noise.) In order to "kick" the circuit in SPICE, we can apply an initial condition of, say, 0 to node X. As a result, $V_Y = V_{DD}$, and $V_Z \approx 0$, forcing V_X to rise toward V_{DD}. Thus, SPICE cannot find an equilibrium point and is forced to allow oscillation.

Exercise What happens if the ring contains four identical stages and we apply an initial condition of zero to one of the nodes?

Another type of ring oscillator can be conceived as follows. Suppose we replace the load resistors in Fig. 13.8 with PMOS current sources as shown in Fig. 13.9(a). The circuit still satisfies our foregoing derivations if we substitute $r_{Op}||r_{On}$ for R_D.[3] But let us change each PMOS current source to an *amplifying* device by connecting its gate to the input

[3]We assume all of the transistors are in saturation, which is not quite correct when the drain voltage comes close to ground or V_{DD}.

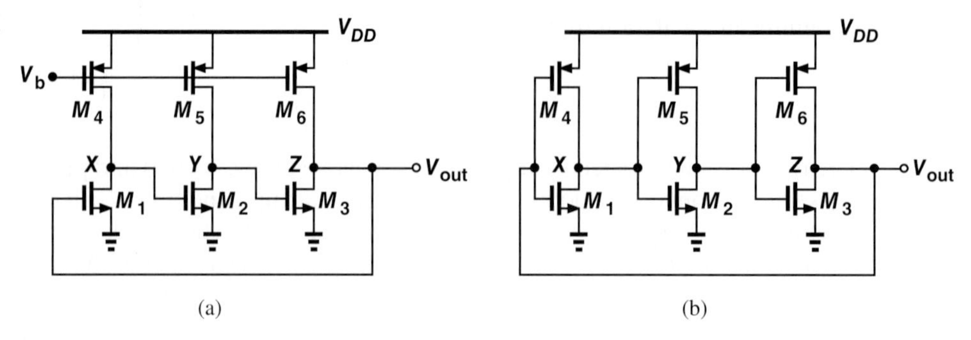

Figure 13.9 Ring oscillators using (a) CS stages with PMOS loads, (b) CMOS inverters.

of the corresponding stage [Fig. 13.9(b)]. Each stage is now a CMOS inverter (a basic building block in logic design), providing a voltage gain of $-(g_{mp} + g_{mn})(r_{Op}||r_{On})$ if both transistors are in saturation. This type of ring oscillator finds use in many applications. Note the transistors themselves contribute capacitance to each node, limiting the speed.

The operation of the inverter-based ring oscillator can also be studied from a different perspective. If V_X starts at zero, we have $V_Y = V_{DD}$ and $V_Z = 0$. Thus, the first stage wants to raise V_X to V_{DD}. Since each stage has some phase shift (delay), the circuit oscillates such that the three voltages toggle between 0 and V_{DD} consecutively (Fig. 13.10). First, V_X rises; after some delay, V_Y falls; after another delay, V_Z rises; finally, with some delay, V_X falls. If each inverter has a delay of T_D seconds, the overall oscillation period is equal to $6T_D$ and hence the output frequency is given by $1/(6T_D)$.

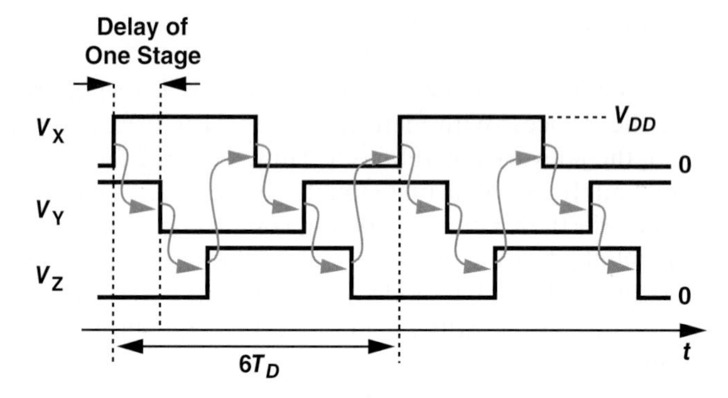

Figure 13.10 Ring oscillator waveforms.

Example 13-3 Can we cascade four inverters to implement a four-stage ring oscillator?

Solution No, we cannot. Consider the ring in Fig. 13.11 and suppose the circuit begins with $V_X = 0$. Thus, $V_Y = V_{DD}$, $V_Z = 0$, and $V_W = V_{DD}$. Since the first stage senses a *high input*, it happily retains its low output indefinitely. Note that all of the transistors are either off or in deep triode region (with zero drain current), yielding a zero loop gain and violating Barkhausen's first criterion. We say the circuit is "latched up." In general, a single-ended ring having an even number of inverters experiences latch-up.

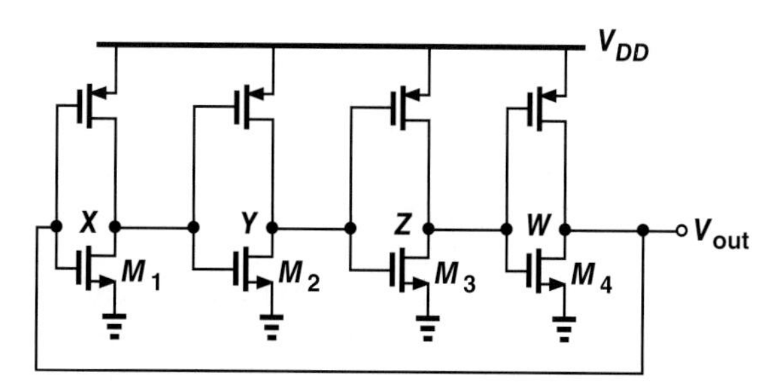

Figure 13.11

How fast can ring oscillators run? The gate delay in 40-nm CMOS technology is about 8 ps. Thus, a three-stage ring can oscillate at frequencies as high as 20 GHz. Their simplicity makes ring oscillators a popular choice in many integrated circuits. For example, memories, microprocessors, and some communication systems employ ring oscillators for on-chip clock generation.

Bioengineering Application: Proximity Vital Sign Detector

In some applications, it is desirable to detect people's respiration or heart beat without a physical contact to their body. For example, detecting humans behind walls in a fire or monitoring the breathing of infants without attaching devices to them prove extremely helpful. Similarly, such a "contactless" device can be placed inside a mobile phone so as to measure the user's vital signs.

A method of sensing chest movements for detection of respiration employs an oscillator [1]. Recall that the oscillation frequency of the cross-coupled or Colpitts topologies depends on the tank's capacitance. Shown in Fig. 13.12, the system in [1] incorporates the capacitance between the subject's chest and an antenna as part of the tank. As a result, chest movements translate to a change in the capacitance and hence in the oscillation frequency. The principal challenge here is that the capacitance change is extremely small, requiring that the frequency be measured with great accuracy.

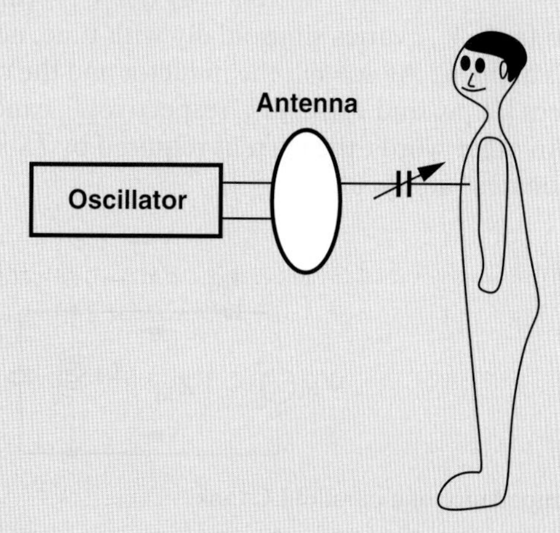

Figure 13.12 Detection of chest movement by measuring capacitance change.

13.3 LC OSCILLATORS

Another class of oscillators employs inductors and capacitors to define the oscillation frequency. Called "LC oscillators," these circuits can be realized in both integrated and discrete forms but with different topologies and design constraints. We begin with integrated LC oscillators.

Did you know?

A tough challenge in the design of integrated LC oscillators has been the realization of on-chip inductors. While discrete inductors can readily extend in three dimensions [Fig. (a)], integrated devices must be preferably based on a "planar" (two-dimensional) structure. Figure (b) shows a "spiral" inductor commonly employed in integrated circuits. The spiral is made of the metal layer (copper or aluminum) that is used for on-chip wiring.

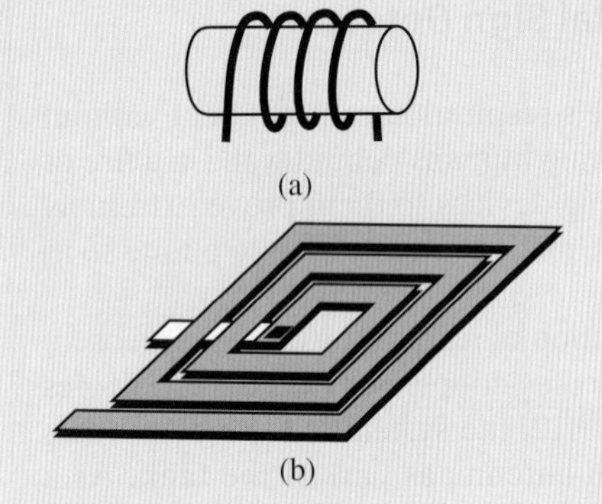

(a) Simple three-dimensional inductor, (b) integrated spiral inductor.

Why LC oscillators? Why not just ring oscillators? LC oscillators offer two advantages that have made them popular, especially in radio-frequency and wireless transceivers: they can operate faster than ring oscillators (the author has developed one that reaches 300 GHz in 65 nm CMOS technology), and they exhibit less noise (although we have not studied noise in this book). Unfortunately, LC oscillators are more difficult to design and occupy a larger chip area than ring oscillators. As our first step, let us return to some concepts from basic circuit theory.

13.3.1 Parallel LC Tanks

Shown in Fig. 13.13, an ideal parallel LC tank provides an impedance given by

$$Z_1(s) = (L_1 s)||\frac{1}{C_1 s} \tag{13.7}$$

$$= \frac{L_1 s}{L_1 C_1 s^2 + 1}. \tag{13.8}$$

For a sinusoidal input current or voltage, we have $s = j\omega$ and

$$Z_1(j\omega) = \frac{jL_1\omega}{1 - L_1 C_1 \omega^2}, \tag{13.9}$$

observing that the impedance goes to infinity at $\omega_1 = 1/\sqrt{L_1 C_1}$. That is, even though the voltage applied to the tank, V_{in}, varies sinusoidally with time, no net current flows into the tank. How does this happen? At $\omega = \omega_1$, the inductor and the capacitor exhibit equal and opposite impedances [$jL_1\omega_1$ and $1/(jC_1\omega_1)$, respectively], canceling each other and yielding an open circuit. In other words, the current required by L_1 is exactly provided by C_1. We say the tank "resonates" at $\omega = \omega_1$.

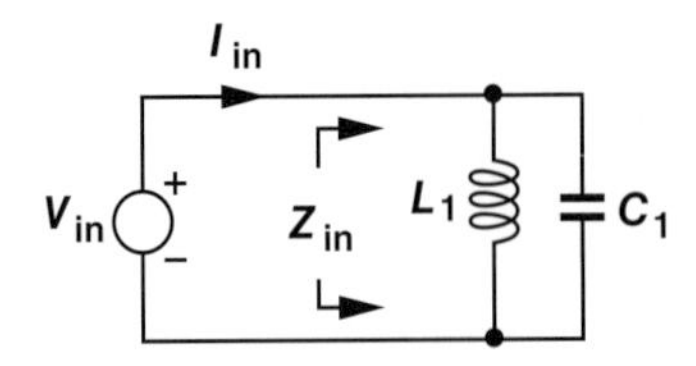

Figure 13.13 Impedance of a parallel LC tank.

Example 13-4 Sketch the magnitude and phase of $Z_1(j\omega)$ as a function of frequency.

Solution We have

$$|Z_1(j\omega)| = \frac{L_1\omega}{|1 - L_1C_1\omega^2|}, \tag{13.10}$$

obtaining the plot in Fig. 13.14(a). As for the phase, we note from Eq. (13.9) that if $\omega < \omega_1$, then $1 - L_1C_1\omega^2 > 0$ and $\angle Z_1(j\omega) = \angle(jL_1\omega) = +90°$ [Fig. 13.14(b)]. On the other hand, if $\omega > \omega_1$, then $1 - L_1C_1\omega^2 < 0$ and hence $\angle Z_1(j\omega) = \angle(jL_1\omega_1) - 180° = -90°$. We roughly say the tank has an inductive behavior for $\omega < \omega_1$ and a capacitive behavior for $\omega > \omega_1$.

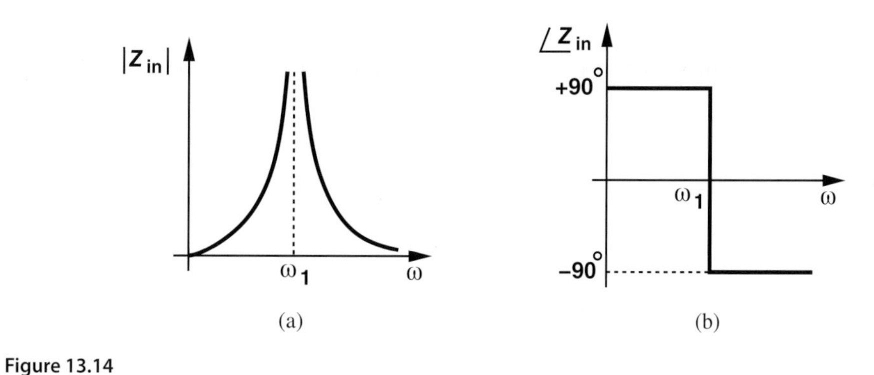

Figure 13.14

Exercise Determine $|Z_1|$ at $\omega = \omega_1/2$ and $\omega = 2\omega_1/2$.

In practice, the impedance of a parallel LC tank does not go to infinity at the resonance frequency. To understand this point, we recognize that the wire forming the inductor has a finite *resistance*. As illustrated in Fig. 13.15(a), when L_1 carries current, its wire resistance, R_1, heats up, dissipating energy. Thus, V_{in} must replenish this energy in every cycle, and $Z_2 < \infty$ even at resonance. The circuit is now called a "lossy tank" to emphasize the loss of energy within the inductor's resistance.

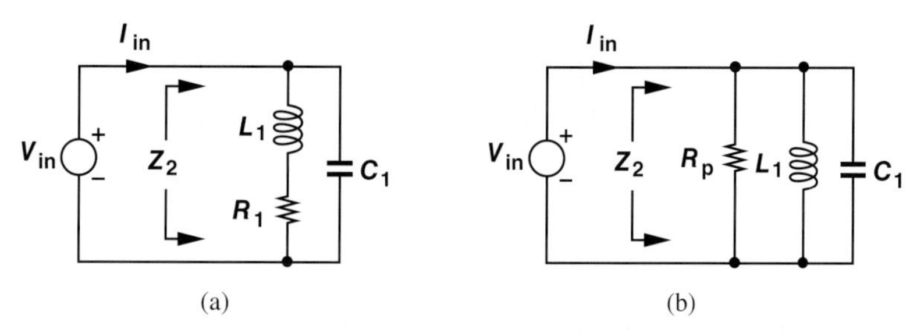

Figure 13.15 (a) Impedance of lossy tank, (b) equivalent circuit of (a).

In the analysis of LC oscillators, we prefer to model the loss of the tank by a parallel resistance, R_p [Fig. 13.15(b)]. Are the two circuits in Fig. 13.15 equivalent? They cannot be equivalent at all frequencies: at $\omega \approx 0$, L_1 is a short circuit and C_1 is open, yielding $Z_2 = R_1$ in Fig. 13.15(a) but $Z_2 = 0$ in Fig. 13.15(b). But for a narrow range around the resonance frequency, the two models can be equivalent. The proof and derivations are outlined in Problem 13.23 for the interested reader, but we present the final result here: for the two tanks to be approximately equivalent, we must have

$$R_p = \frac{L_1^2\omega^2}{R_1}. \quad (13.11)$$

Note that an ideal inductor exhibits $R_1 = 0$ and hence $R_p = \infty$. The following example illustrates how the parallel model simplifies the analysis.

Example 13-5 Plot the magnitude and phase of $Z_2(s)$ in Fig. 13.15(b) as a function of frequency.

Solution We have

$$Z_2(s) = R_p||(L_1s)||\frac{1}{C_1s} \quad (13.12)$$

$$= \frac{R_pL_1s}{R_pL_1C_1s^2 + L_1s + R_p}. \quad (13.13)$$

At $s = j\omega$,

$$Z_2(j\omega) = \frac{jR_pL_1\omega}{R_p(1 - L_1C_1\omega^2) + jL_1\omega}. \quad (13.14)$$

At $\omega_1 = 1/\sqrt{L_1C_1}$, we have $Z_2(j\omega_1) = R_p$, an expected result because the inductor and capacitor impedances still cancel each other. Since Z_2 reduces to a single resistance at ω_1, $\angle Z_2(j\omega_1) = 0$. We also note that (1) at very low frequencies, $jL_1\omega$ is very small, dominating the parallel combination, i.e., $Z_2 \approx jL_1\omega$, and (2) at very high frequencies, $1/(jC_1\omega)$ is very small and hence $Z_2 \approx 1/(jC_1\omega)$. Thus, $|Z_2|$ and $\angle Z_2$ follow the general behaviors shown in Fig. 13.16.

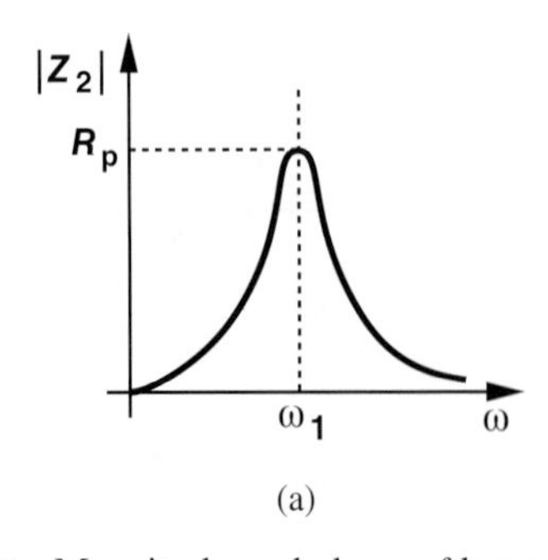

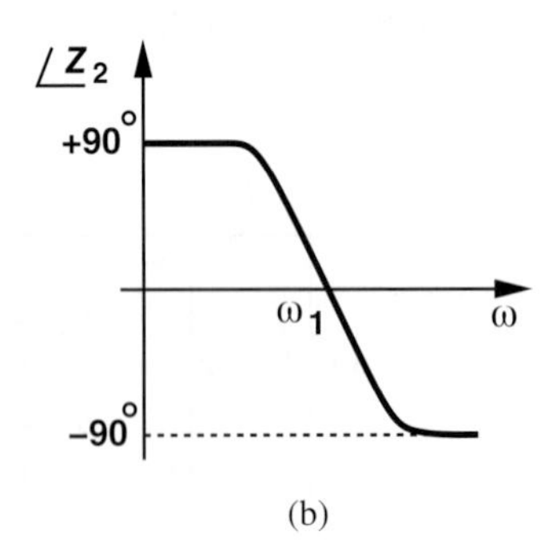

Figure 13.16 Magnitude and phase of lossy tank.

Exercise Determine $|Z_2|$ at $\omega = \omega_1/2$ and $\omega = 2\omega_1/2$.

Example 13-6 Suppose we apply an initial voltage of V_0 across the capacitor in an isolated parallel tank. Study the behavior of the circuit in the time domain if the tank is ideal or lossy.

Solution As illustrated in Fig. 13.17(a) for the ideal tank, the capacitor begins to discharge through the inductor, i.e., the electric energy is transformed to magnetic energy. When $V_{out} = 0$, only L_1 carries energy in the form of a current. This current now continues to charge C_1 toward $-V_0$. This transfer of energy between C_1 and L_1 repeats and the tank oscillates indefinitely.

With a lossy tank [Fig. 13.17(b)], on the other hand, a nonzero output voltage causes current flow through R_p and hence dissipation of energy. Thus, the tank loses some energy in every cycle, producing a decaying oscillatory output. To construct an oscillator, we must somehow cancel this decay.

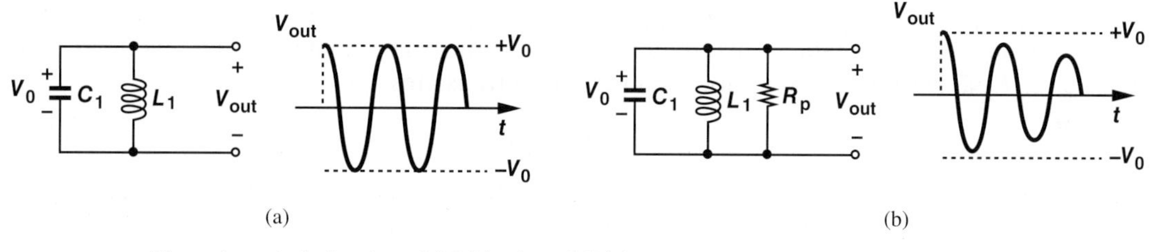

Figure 13.17 Time-domain behavior of (a) ideal, and (b) lossy tanks.

Exercise Calculate the maximum energy stored in L_1 in Fig. 13.17(a).

With our basic understanding of the parallel LC tank, we can now incorporate it in amplifying stages and oscillators.

13.3.2 Cross-Coupled Oscillator

In our study of CMOS amplifiers, we have considered common-source stages with resistor or current-source loads. Now, let us construct a common-source stage using a parallel LC tank as its load [Fig. 13.18(a)]. We wish to analyze the frequency response of this "tuned" amplifier. Denoting the tank impedance by Z_2 and neglecting channel-length modulation,[4] we have

$$\frac{V_{out}}{V_{in}} = -g_m Z_2(s), \tag{13.15}$$

where $Z_2(s)$ is given by Eq. (13.13). Using the plots of $|Z_2|$ and $\angle Z_2$ in Fig. 13.16, we can sketch $|V_{out}/V_{in}|$ and $\angle(V_{out}/V_{in})$ as shown in Fig. 13.18(b). Note $\angle(V_{out}/V_{in})$ is obtained by shifting $\angle Z_2$ by 180° (up or down) to account for the negative sign in $-g_m Z_2(s)$. The CS stage thus exhibits a gain that reaches a maximum of $g_m R_p$ at resonance and approaches zero at very low or very high frequencies. The phase shift at ω_1 is equal to 180° because the load reduces to a resistor at resonance.

Does the CS stage of Fig. 13.18(a) oscillate if we tie its output to its input? As illustrated in Fig. 13.3(b), the total phase shift around the loop must reach 360° at a finite frequency, but Fig. 13.18(b) reveals that this is not possible. We therefore insert one more

[4]The output resistance of M_1 can be simply absorbed in R_p.

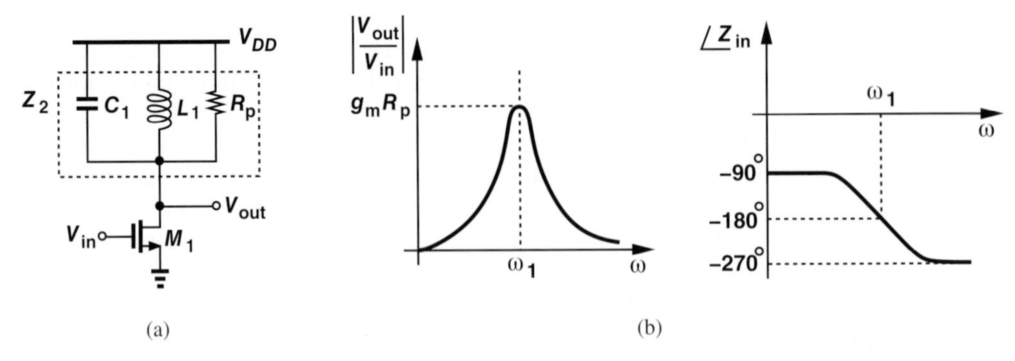

Figure 13.18 (a) CS stage with a tank load, (b) magnitude and phase plots of the stage.

CS stage in the loop and try again [Fig. 13.19(a)]. For a total phase shift of 360°, each stage must provide 180°, which is possible at $\omega = \omega_1$ in Fig. 13.18(b). Thus, the circuit oscillates at ω_1 if the loop gain at this frequency is sufficient. Since each stage has a voltage gain of g_mR_p at ω_1, Barkhausen's loop gain criterion translates to

$$(g_mR_p)^2 \geq 1. \tag{13.16}$$

Stated more accurately, the startup condition emerges as $g_m(R_p||r_O) \geq 1$. With identical stages, the oscillator of Fig. 13.19 generates differential signals at nodes X and Y [Fig. 13.19(b)] (why?), a useful property for integrated-circuit applications.

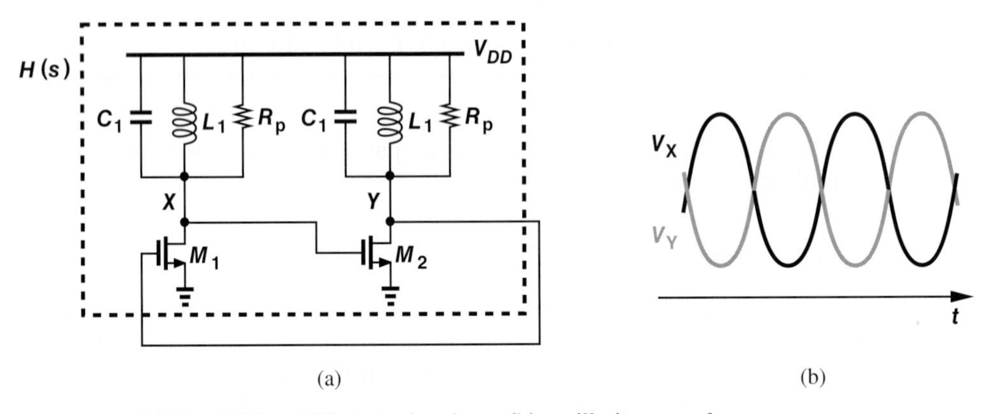

Figure 13.19 (a) Two LC-load CS stages in a loop, (b) oscillation waveforms.

A critical issue in the above topology is that the bias current of the transistors is poorly defined. Since no current mirror or other means of proper biasing are used, the drain currents of M_1 and M_2 vary with process, supply voltage, and temperature. For example, if the transistors' threshold voltage is lower than the nominal value, then the peak value of V_X yields a greater overdrive voltage for M_2 and hence a larger drain current. To resolve this issue, we first note that the gate of each device is tied to the drain of the other and redraw the circuit as shown in Fig. 13.20(a). Now, M_1 and M_2 almost resemble a differential pair whose output is fed back to its input. Let us then add a tail current source as illustrated in Fig. 13.20(b), ensuring that the total bias current of M_1 and M_2 is equal to I_{SS}. We usually redraw this circuit as shown in Fig. 13.20(c).

The "cross-coupled" oscillator of Fig. 13.20(c) is the most popular and robust LC oscillator used in integrated circuits. The carrier frequency in your cell phone and in its GPS receiver is very likely generated by such a topology.

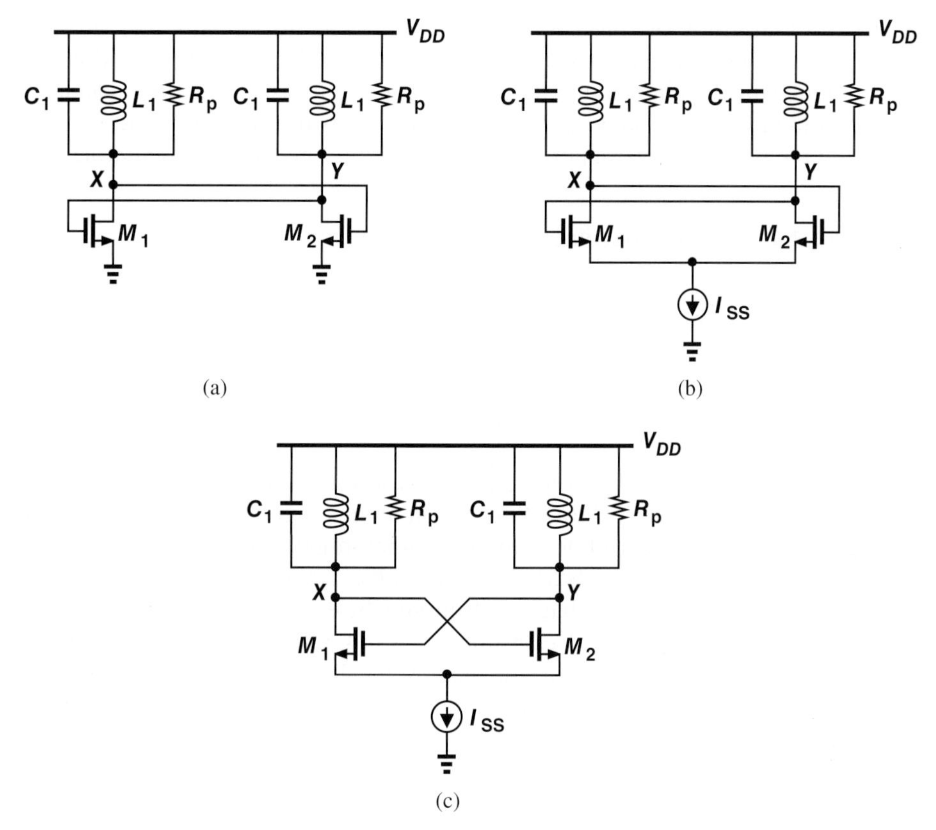

Figure 13.20 Different drawings of the cross-coupled oscillator.

Example 13-7 Plot the drain currents of M_1 and M_2 in Fig. 13.20(c) if the voltage swings at X and Y are large.

Solution Let us first consider the differential pair in Fig. 13.21(a). With large input voltage swings, the entire current is steered to the left or to the right [Fig. 13.21(b)]. The circuit of Fig. 13.20(b), too, exhibits the same behavior, producing drain currents that swing between zero and I_{SS} [Fig. 13.21(c)].

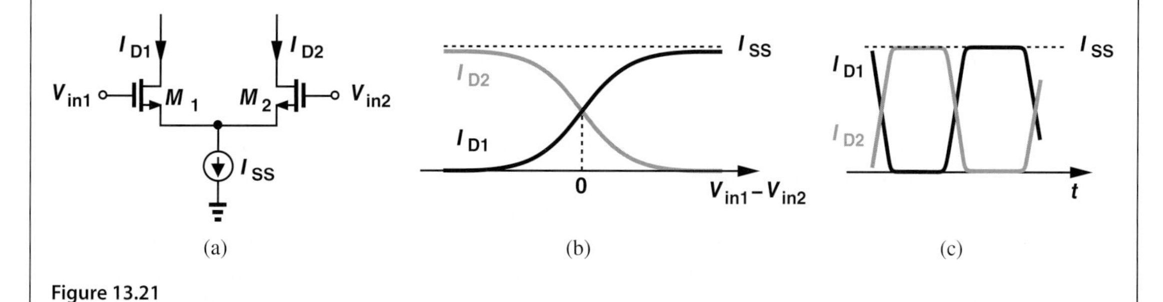

Figure 13.21

Exercise Redraw the cross-coupled oscillator with PMOS transistors.

Our study of ring and cross-coupled oscillators points to a general procedure for the analysis of oscillators: open the feedback loop (while including the effect of I/O impedances as in Chapter 12), determine the transfer function around the loop (similar to the loop gain), and equate the phase of this result to 360° and its magnitude to unity. In the next section, we apply this procedure to the Colpitts oscillator.

13.3.3 Colpitts Oscillator

The Colpitts topology employs only one transistor and finds wide application in discrete design. This is because (high-frequency) discrete transistors are more expensive than passive discrete devices. (In integrated circuits, on the other hand, transistors are the least expensive because they occupy the smallest area.) Since bipolar transistors are much more common in discrete design than are MOSFETs, we analyze a bipolar Colpitts oscillator here.

How can we construct an oscillator using only one transistor? We observed in Figs. 13.6(a) and 13.18(a) that a common-source (or common-emitter) stage cannot serve this purpose. But how about a common-gate (or common-base) stage? Depicted in Fig. 13.22(a), the Colpitts oscillator resembles a common-base topology whose output (the collector voltage) is fed back to its input (the emitter node). Current source I_1 defines the bias current of Q_1, and V_b ensures Q_1 is in the forward active region. As with the cross-coupled oscillator, resistor R_p models the loss of the inductor. This resistor can also model the input resistance of the subsequent stage, e.g., r_π if the oscillator drives a simple common-emitter stage.

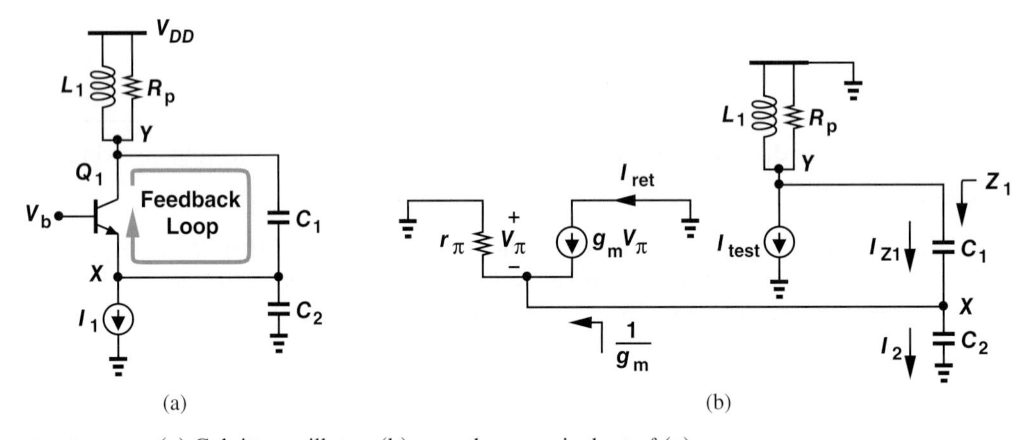

Figure 13.22 (a) Colpitts oscillator, (b) open-loop equivalent of (a).

In order to analyze the Colpitts oscillator, we wish to break the feedback loop. Neglecting the Early effect, we note that Q_1 in Fig. 13.22(a) operates as an ideal voltage-dependent current source, injecting its small-signal current into node Y. We therefore break the loop at the collector as shown in Fig. 13.22(b), where an independent current source I_{test} is drawn from Y, and the current returned by the transistor, I_{ret}, is measured as the quantity of interest. The transfer function I_{ret}/I_{test} must exhibit a phase of 360° and a magnitude of at least unity at the frequency of oscillation.

We observe that I_{test} is divided between $(L_1s)||R_p = L_1sR_p/(L_1s + R_p)$ and Z_1, which is given by

$$Z_1 = \frac{1}{C_1s} + \frac{1}{g_m} \bigg\| \frac{1}{C_2s} \tag{13.17}$$

$$= \frac{1}{C_1s} + \frac{1}{C_2s + g_m}. \tag{13.18}$$

That is, the current flowing through C_1 is equal to

$$I_{Z1} = -I_{test}\frac{\dfrac{L_1sR_p}{L_1s + R_p}}{\dfrac{L_1sR_p}{L_1s + R_p} + \dfrac{1}{C_1s} + \dfrac{1}{C_2s + g_m}}. \tag{13.19}$$

This current is now multiplied by the parallel combination of $1/(C_2s)$ and $1/g_m$ to yield V_X. Since $I_{ret} = g_mV_\pi = -g_mV_X$, we have

$$\frac{I_{ret}}{I_{test}}(s) = \frac{g_mR_pL_1C_1s^2}{L_1C_1C_2R_ps^3 + [g_mR_pL_1C_1 + L_1(C_1 + C_2)]s^2 + [g_mL_1 + R_p(C_1 + C_2)]s + g_mR_p}. \tag{13.20}$$

We now equate this transfer function to unity (which is equivalent to setting its phase to 360° and its magnitude to 1) and cross-multiply, obtaining:

$$L_1C_1C_2R_ps^3 + L_1(C_1 + C_2)s^2 + [g_mL_1 + R_p(C_1 + C_2)]s + g_mR_p = 0. \tag{13.21}$$

At the oscillation frequency, $s = j\omega_1$, both the real and imaginary parts of the right-hand side must drop to zero:

$$-L_1(C_1 + C_2)\omega_1^2 + g_mR_p = 0 \tag{13.22}$$

$$-L_1C_1C_2R_p\omega_1^3 + [g_mL_1 + R_p(C_1 + C_2)]\omega_1 = 0. \tag{13.23}$$

From the second equation, we obtain the oscillation frequency:

$$\omega_1^2 = \frac{(C_1 + C_2)}{L_1C_1C_2} + \frac{g_m}{R_pC_1C_2}. \tag{13.24}$$

The second term on the right is typically negligible, yielding

$$\omega_1^2 \approx \frac{1}{L_1\dfrac{C_1C_2}{C_1 + C_2}}. \tag{13.25}$$

That is, the oscillation occurs at the resonance of L_1 and the series combination of C_1 and C_2. Using this result in Eq. (13.22) gives the startup condition:

$$g_mR_p = \frac{(C_1 + C_2)^2}{C_1C_2}. \tag{13.26}$$

The transistor must thus provide sufficient transconductance to satisfy or exceed this requirement. Since the right-hand side is minimum if $C_1 = C_2$, we conclude that g_mR_p must be at least equal to 4.

Example 13-8 Compare the startup conditions of cross-coupled and Colpitts oscillators.

Solution We note from Eq. (13.16) that the cross-coupled topology requires a minimum $g_m R_p$ of 1, i.e., it can tolerate a lossier inductor than the Colpitts oscillator can. (Also, note that the Colpitts topology provides only a single-ended output.)

Exercise How much is the dc voltage at node Y in Fig. 13.22(a)? Can you sketch the oscillation waveform at this node?

Where is the output node in the oscillator of Fig. 13.22(a)? The output can be sensed at node Y, in which case the input resistance of the next stage (e.g., r_π) shunts R_p, requiring a greater g_m to satisfy the startup condition. Alternatively, the output can be sensed at the emitter (Fig. 13.23). This is usually preferable in discrete design because (1) discrete inductors have a low loss (a high equivalent R_p) and are therefore sensitive to resistive loading, and (2) with only R_{in} loading the emitter (and $R_p \to \infty$), the startup condition is modified to:

$$g_m R_{in} = 1 \tag{13.27}$$

Derived in Problem 13.37, this more relaxed condition simplifies the design of the oscillator. For example, the oscillator can drive a lower load resistance in this case than when the load is tied to the collector. It is important to note that most textbooks derive Eq. (13.27) as the startup condition, which holds only if R_p is very large (the inductor has a low loss).

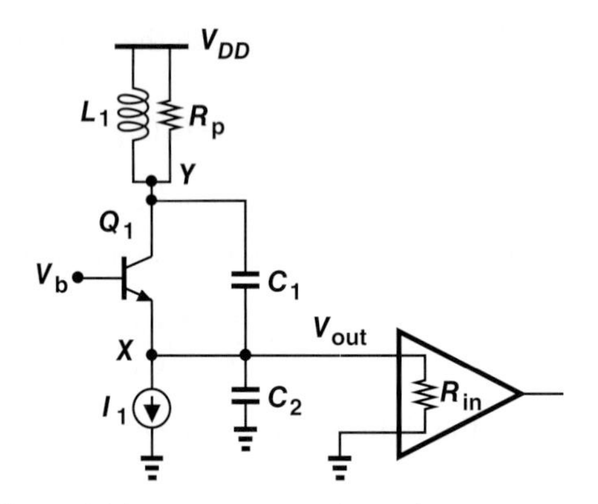

Figure 13.23 Colpitts oscillator driving next stage at its emitter.

13.4 PHASE SHIFT OSCILLATOR

In our development of ring oscillators in Section 13.2, we created sufficient phase shift by cascading three *active* stages. Alternatively, we can cascade passive sections along with a single amplifier to achieve the same goal. Shown in Fig. 13.24(a) is a "phase shift oscillator" based on this principle. We expect that the three RC sections can provide a phase shift of 180° at the frequency of interest even if the amplifier itself contributes negligible phase. Nonetheless, the signal attenuation introduced by the passive stages must be compensated by the amplifier to fulfill the startup condition.[5]

[5]Note that the bottom terminal of R_3 must in fact be tied to a bias voltage that is proper for the amplifier.

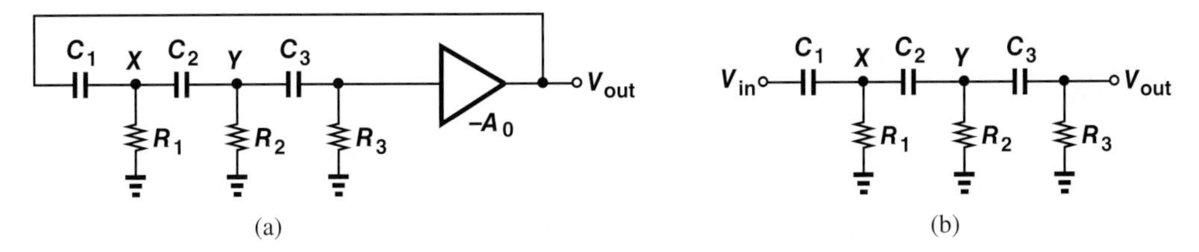

Figure 13.24 (a) Phase shift oscillator, (b) phase shift network.

Let us first compute the transfer function of the passive network shown in Fig. 13.24(b), assuming that $C_1 = C_2 = C_3 = C$ and $R_1 = R_2 = R_3 = R$. Beginning from the output, we write the current through R_3 as V_{out}/R and hence

$$V_Y = \frac{V_{out}}{R}\frac{1}{Cs} + V_{out}. \tag{13.28}$$

Dividing V_Y by R_2 and multiplying it by $1/(C_2 s)$, we have the voltage drop across C_2 and thus

$$V_X = \left(\frac{V_Y}{R_2} + \frac{V_{out}}{R_3}\right)\frac{1}{Cs} + V_Y \tag{13.29}$$

$$= \left(\frac{1}{RCs} + 1\right)^2 V_{out} + \frac{V_{out}}{RC_3}. \tag{13.30}$$

Finally,

$$V_{in} = \frac{V_{out}}{R^3C^3s^3} + \frac{5V_{out}}{R^2C^2s^2} + \frac{6V_{out}}{RCs} + V_{out} \tag{13.31}$$

and hence

$$\frac{V_{out}}{V_{in}} = \frac{(RCs)^3}{(RCs)^3 + 6(RCs)^2 + 5RCs + 1}. \tag{13.32}$$

At $s = j\omega_1$,

$$\angle\frac{V_{out}}{V_{in}} = 3 \times 90° - \tan^{-1}\frac{5RC\omega - (RC\omega)^3}{1 - 6R^2C^2\omega^2}. \tag{13.33}$$

For oscillation to occur at ω_1, this phase must reach 180°:

$$1 - 6R^2C^2\omega_1^2 = 0. \tag{13.34}$$

It follows that

$$\omega_1 = \frac{1}{\sqrt{6}RC}. \tag{13.35}$$

For the startup condition to hold, we multiply the magnitude of Eq. (13.32) by the gain of the amplifier and equate the result to unity:

$$\frac{(\sqrt{6})^{-3}A}{5/\sqrt{6} - (\sqrt{6})^{-3}} = 1. \tag{13.36}$$

That is, the gain of the amplifier must be at least:

$$A = 29. \tag{13.37}$$

The phase shift oscillator is occasionally used in discrete design as it requires only one amplifying stage. This topology does not find wide application in integrated circuits because its output noise is quite high.

Example 13-9 Design the phase shift oscillator using an op amp.

Solution We must configure the op amp as an *inverting* amplifier. Figure 13.25(a) shows an example. Here, however, resistor R_4 appears between node Z and a virtual ground, equivalently shunting resistor R_3. Thus, for our foregoing derivations to apply, we must choose $R_3||R_4 = R_2 = R_1 = R$. In fact, we may simply allow R_3 to be infinity and R_4 to be equal to R, arriving at the topology depicted in Fig. 13.25(b).

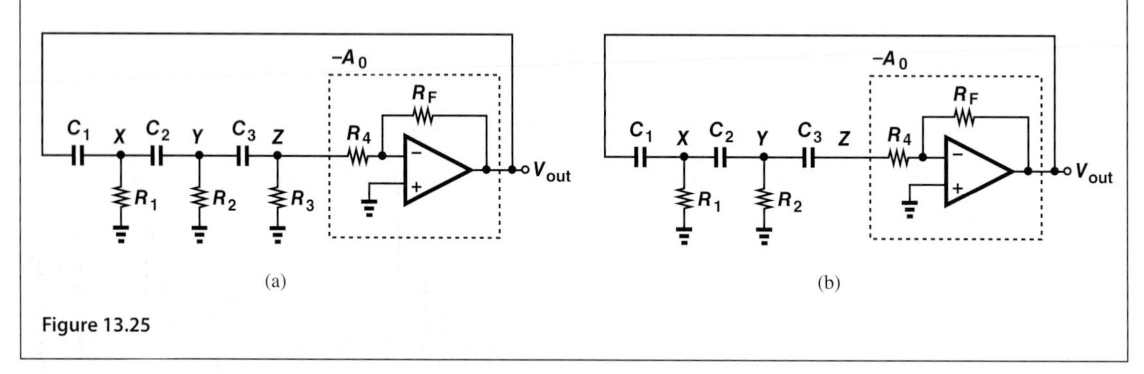

Figure 13.25

Exercise How should R_F/R_4 be chosen to obtain a loop gain of 1 at the frequency of oscillation?

What determines the oscillation amplitude in the circuit of Fig. 13.25(b)? If the loop gain at ω_1 is greater than unity, the amplitude grows until the op amp output swings from one supply rail to the other. Owing to the saturation of the op amp, the output waveform resembles a square wave rather than a sinusoid, an undesirable effect in some applications. Moreover, the saturation tends to slow down the op amp response, limiting the maximum oscillation frequency. For these reasons, one may opt to define ("stabilize") the oscillation amplitude by additional means. For example, as illustrated in Fig. 13.26(a), we can replace the feedback resistor with two "anti-parallel" diodes. The output now swings by one diode drop ($V_{D,on} = 700$ to $800\,\text{mV}$) below and above its average value.

The oscillation amplitude obtained above may prove inadequate in many applications. We must therefore modify the feedback network such that the diodes turn on only when V_{out} reaches a larger, predetermined value. To this end, we divide V_{out} down and feed the result to the diodes [Fig. 13.26(b)]. Assuming a constant-voltage model for D_1 and D_2, we observe that one diode turns on when

$$V_{out}\frac{R_{D2}}{R_{D2}+R_{D1}} = V_{D,on}, \tag{13.38}$$

and hence

$$V_{out} = \left(1+\frac{R_{D1}}{R_{D2}}\right)V_{D,on}. \tag{13.39}$$

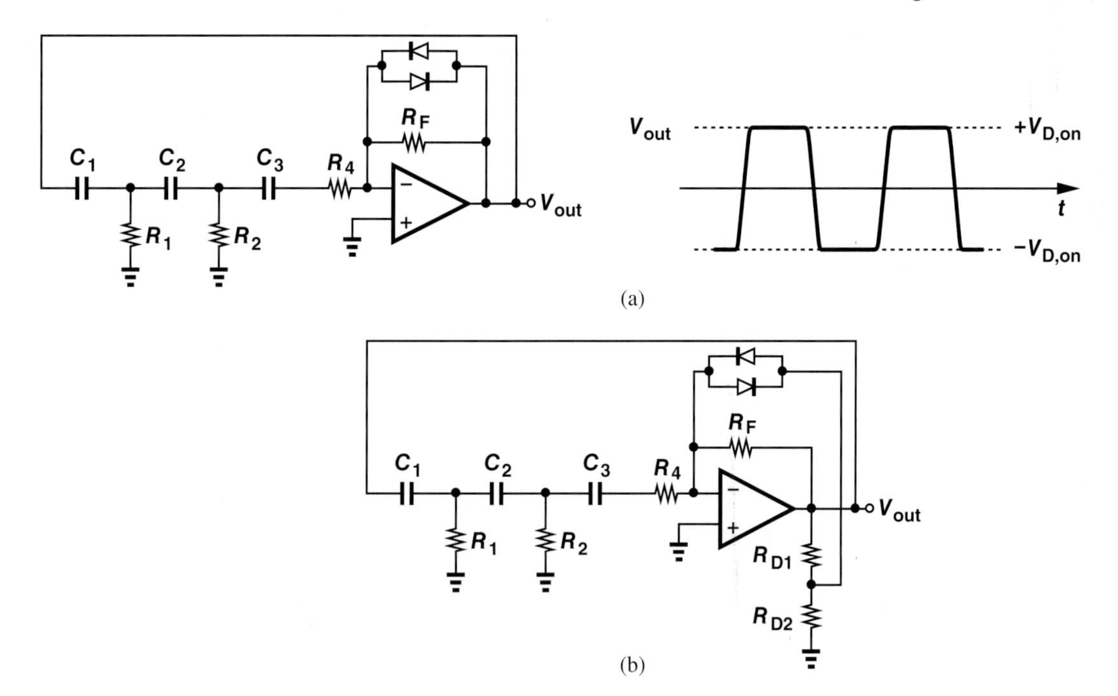

Figure 13.26 (a) Use of diodes to limit the output swing, (b) alternative topology providing larger output swing.

13.5 WIEN-BRIDGE OSCILLATOR

The Wien-bridge oscillator is another topology sometimes used in discrete design as it requires only one amplifying stage. Unlike the phase shift oscillator, however, the Wien-bridge configuration employs a passive feedback network with *zero* phase shift rather than 180° phase shift. The amplifier must therefore provide a *positive* gain so that the total phase shift at the frequency of oscillation is equal to zero (or 360°).

Let us first construct a simple passive network with zero phase shift at a *single* frequency. Shown in Fig. 13.27(a) is an example. If $R_1 = R_2 = R$ and $C_1 = C_2 = C$, we have

$$\frac{V_{out}}{V_{in}}(s) = \frac{\dfrac{R}{RCs+1}}{\dfrac{R}{RCs+1} + \dfrac{1}{Cs} + R} \tag{13.40}$$

$$= \frac{RCs}{R^2C^2s^2 + 3RCs + 1}. \tag{13.41}$$

The phase thus emerges as

$$\angle \frac{V_{out}}{V_{in}}(s = j\omega) = \frac{\pi}{2} - \tan^{-1}\frac{3RC\omega}{1 - R^2C^2\omega^2}, \tag{13.42}$$

falling to zero at

$$\omega_1 = \frac{1}{RC}. \tag{13.43}$$

> **Did you know?**
>
> In 1939, two young Stanford graduates named William Hewlett and David Packard used the Wien-bridge oscillator to design a sound generator for the soundtrack of the Disney movie Fantasia. History has it that Hewlett and Packard borrowed $500 from their advisor, Fredrick Terman, to construct and sell eight of these generators to Disney. Thus began the company known today as HP. For the first several decades, HP designed and manufactured only test equipment, e.g., oscilloscopes, signal generators, power supplies, etc.

We now place this network around an op amp as illustrated in Fig. 13.27(b). Denoting the gain of the non-inverting amplifier by A, we multiply the magnitude of Eq. (13.41) by A and equate the result to unity:

$$\left|\frac{ARCj\omega}{1 - R^2C^2\omega^2 + 3jRC\omega}\right| = 1. \quad (13.44)$$

At ω_1, this equation yields

$$A = 3. \quad (13.45)$$

That is, we choose $R_{F1} \geq 2R_{F2}$.

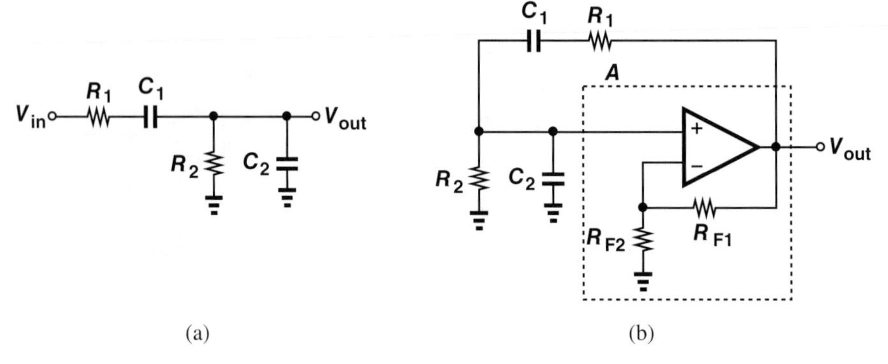

Figure 13.27 (a) Phase shift network, (b) Wien-bridge oscillator.

To avoid uncontrolled amplitude growth, the Wien-bridge oscillator can incorporate diodes in the gain definition network, R_{F1} and R_{F2}. As depicted in Fig. 13.28, two anti-parallel diodes can be inserted in series with R_{F1} so as to create strong feedback as $|V_{out}|$ exceeds $V_{D,on}$. If larger amplitudes are desired, resistor R_{F3} can be added to divide V_{out} and apply the result to the diodes.

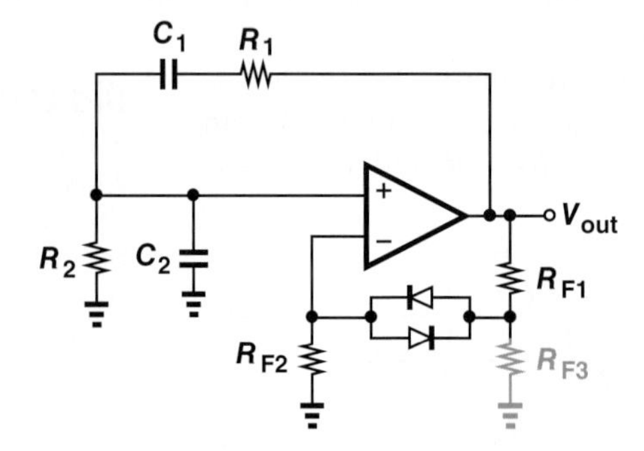

Figure 13.28 Addition of diodes to limit output swing of Wien-bridge oscillator.

Robotics Application: Capacitive Sensors

Capacitive sensors find application in human-robot interactions—e.g., in robot-assisted dressing of the elderly—and for enhancing safety in a robotic environment. The ability to perform "pretouch" sensing helps these sensors avoid collisions.

Capacitive sensors can detect a nearby object by measuring how much that object changes the capacitance of a given structure. Consider first the capacitor shown in

Fig. 13.29(a) and note the electric field lines going from plate A to plate B. If an object approaches the top side of the structure, it can change some of the field lines arriving at plate B [Fig. 13.29(b)], thereby changing the capacitance between A and B. We must measure this change.

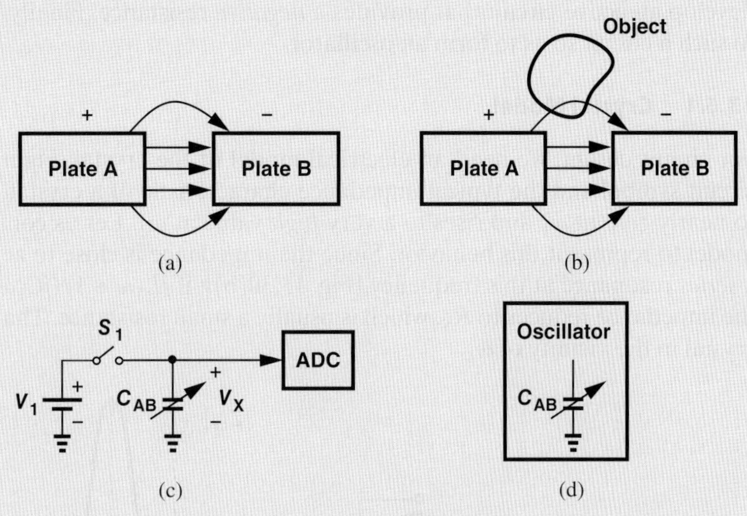

Figure 13.29 (a) A capacitor with its field lines, (b) change in field lines in proximity of an object, (c) use of an ADC to measure capacitance change, and (d) use of an oscillator to measure capacitance change.

One method of measuring the capacitance change is illustrated in Fig. 13.29(c). We first turn on S_1 and charge C_{AB} to V_1. Next, we turn S_1 off and monitor V_X by an analog-to-digital converter (ADC). If C_{AB} remains constant, then $V_X = V_1$. On the other hand, if C_{AB} changes, so does V_X because the charge on the capacitor is constant and $Q = CV$. Thus, by comparing V_X with V_1, the ADC decides how V_X and C_{AB} have changed. This operation is repeated hundreds of times per second to capture changes in C_{AB}.

Another method places C_{AB} within a oscillator [Fig. 13.29(d)] such that the oscillation frequency is a function of this capacitance. We note that ring, LC, phase-shift, and Wien-bridge oscillators satisfy this condition.

13.6 CRYSTAL OSCILLATORS

The oscillators studied thus far do not offer a precise output frequency. For example, as the temperature varies, so does the value of the capacitances in each circuit, creating a drift in the oscillation frequency. Many applications, on the other hand, demand a precise clock frequency. If the oscillator frequency in your watch departs from 2^{15} Hz by 0.1%, the time reading will be 10 minutes off after one week.

For high-precision applications, we employ "crystal oscillators." A crystal is made of a piezoelectric material such as quartz and it mechanically vibrates at a certain frequency if subjected to a voltage difference. Crystals are attractive as a frequency "reference" for three reasons. (1) Given by the physical dimensions of the crystal, the vibration frequency is extremely stable with temperature, varying by only a few parts per million (ppm) for a 1° change, (2) the crystal can be cut with relative ease in the factory so as to produce a precise vibration frequency, e.g., with an error of 10–20 ppm.[6] (3) Crystals exhibit a very low loss, behaving almost like an ideal LC tank. That is, an electric impulse applied to the crystal makes it vibrate for thousands of cycles before the oscillation decays.

Our treatment of crystal oscillators in this section proceeds as follows. First, we derive a circuit model for the crystal, concluding that it behaves as a lossy LC tank. Next, we develop an active circuit that provides a *negative* resistance. Finally, we attach the crystal to such a circuit so as to form an oscillator.

13.6.1 Crystal Model

For circuit design, we need an electrical model of the crystal. Figure 13.30(a) shows the circuit symbol and the typical impedance characteristic of a crystal. The impedance falls to nearly *zero* at ω_1 and rises to a very *high* value at ω_2. Let us construct an RLC circuit model to represent this behavior. Since the impedance is close to zero at ω_1, we envision a *series* resonance at this frequency [Fig. 13.30(b)]: if $jL_1\omega + 1/(jC_1\omega) = 0$ at $\omega = \omega_1$, then the impedance reduces to R_S, which is usually a small resistance. That is, Z_S can model the crystal in the vicinity of ω_1.

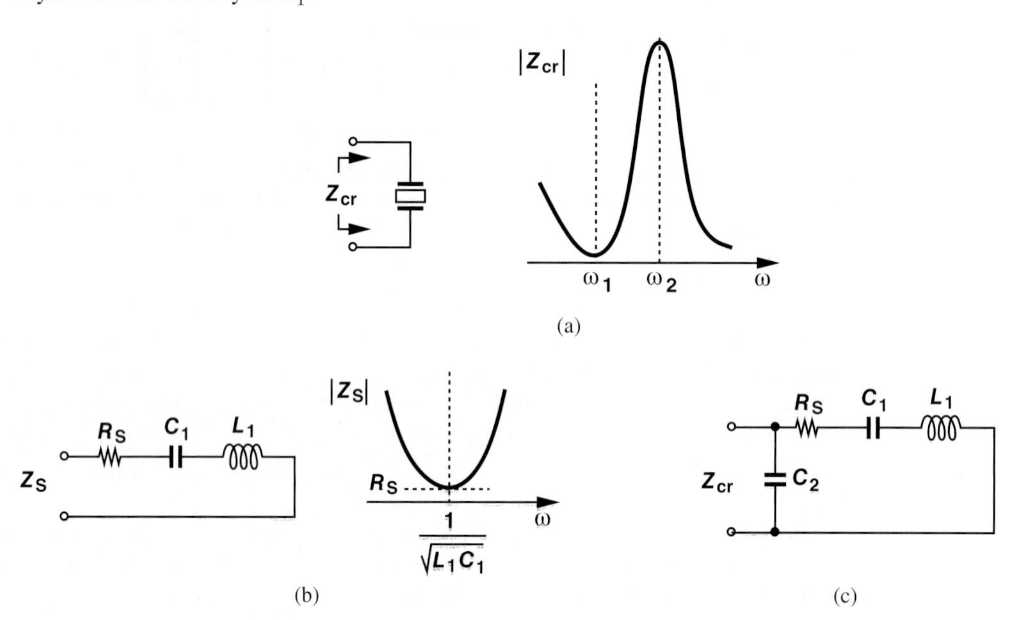

Figure 13.30 (a) Symbol and impedance of a crystal, (b) circuit model for series resonance, (c) complete model.

Around ω_2, the device experiences *parallel* resonance—as seen earlier in LC oscillators. We can therefore place a capacitance in parallel with Z_S as shown in Fig. 13.30(c). To determine ω_2 in terms of the circuit parameters, we neglect R_S and write

$$Z_{cr}(j\omega) = Z_S(j\omega)||\frac{1}{jC_2\omega} \tag{13.46}$$

$$\approx \frac{1 - L_1C_1\omega^2}{j\omega(C_1 + C_2 - L_1C_1C_2\omega^2)}. \tag{13.47}$$

[6]Crystals with an error of a few ppm are also available but at a higher cost.

We note that Z_{cr} goes to infinity at

$$\omega_2 = \frac{1}{\sqrt{L_1 \dfrac{C_1 C_2}{C_1 + C_2}}}. \tag{13.48}$$

In practice, ω_1 and ω_2 are very close, i.e.,

$$\frac{1}{\sqrt{L_1 C_1}} \approx \frac{1}{\sqrt{L_1 \dfrac{C_1 C_2}{C_1 + C_2}}} \tag{13.49}$$

and hence

$$C_1 \approx \frac{C_1 C_2}{C_1 + C_2}. \tag{13.50}$$

It follows that

$$C_2 \gg C_1. \tag{13.51}$$

Example 13-10 If $C_2 \gg C_1$, find a relation between the series and parallel resonance frequencies.

Solution We have

$$\frac{\omega_2}{\omega_1} = \sqrt{\frac{C_1 + C_2}{C_2}} \tag{13.52}$$

$$\approx 1 + \frac{C_1}{2C_2}. \tag{13.53}$$

Exercise Derive an expression for ω_2/ω_1 if R_S is not neglected.

13.6.2 Negative-Resistance Circuit

In order to arrive at a popular crystal oscillator topology, we must first devise a circuit that provides a *negative* (small-signal) input resistance. Consider the topology shown in Fig. 13.31(a), where the bias network of M_1 is omitted for simplicity. Let us obtain Z_{in} with the aid of the arrangement in Fig. 13.31(b), neglecting channel-length modulation and other capacitances. Upon flowing through C_A, I_X generates a gate-source voltage for M_1. Thus, the drain current is given by

$$I_1 = -\frac{I_X}{C_A s} g_m. \tag{13.54}$$

Since C_B carries a current equal to $I_X - I_1$, it sustains a voltage equal to $(I_X - I_1)/C_B s = [I_X + g_m I_X/(C_A s)]/(C_B s)$. Writing a KVL around C_A, V_X and C_B, we eventually obtain

$$V_X = \frac{I_X}{C_A s} + \frac{I_X}{C_B s} + \frac{g_m I_X}{C_A C_B s^2}. \tag{13.55}$$

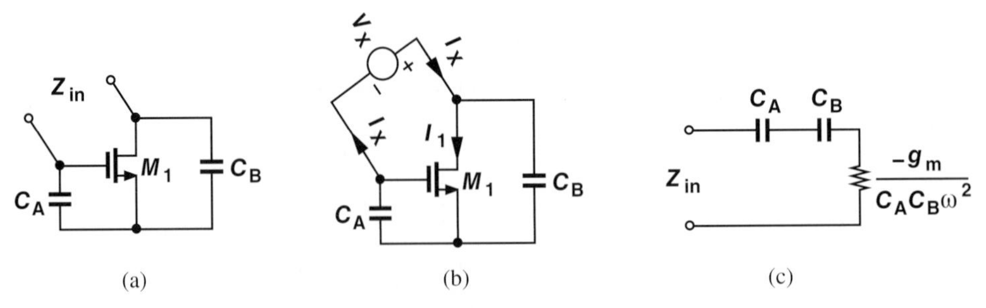

Figure 13.31 (a) Circuit providing negative resistance, (b) setup for impedance measurement, (c) equivalent impedance.

That is,

$$Z_{in}(s) = \frac{1}{C_A s} + \frac{1}{C_B s} + \frac{g_m}{C_A C_B s^2}. \tag{13.56}$$

For a sinusoidal input, $s = j\omega$, and

$$Z_{in}(j\omega) = \frac{1}{jC_A\omega} + \frac{1}{jC_B\omega} - \frac{g_m}{C_A C_B \omega^2}. \tag{13.57}$$

What do the three terms in this equation signify? The first two represent two capacitors in series. The third, on the other hand, is *real*, i.e., a resistance, and *negative* [Fig. 13.31(c)]. A small-signal negative resistance simply means that if the voltage across the device increases, the current through it *decreases*.

A negative resistance can help sustain oscillation. To understand this point, consider a lossy parallel LC tank [Fig. 13.32(a)]. As explained previously, an initial condition on the capacitor leads to a decaying oscillation because R_p dissipates energy in every cycle. Let us now place a negative resistance in parallel with R_p [Fig. 13.32(b)]. We choose $|-R_1| = R_p$, obtaining $(-R_1)||R_p = \infty$. Since R_1 and R_p cancel, the tank consisting of L_1 and C_1 sees no net loss, as if the tank were *lossless*. In other words, since the energy lost by R_p in every cycle is replenished by the active circuit, the oscillation continues indefinitely.

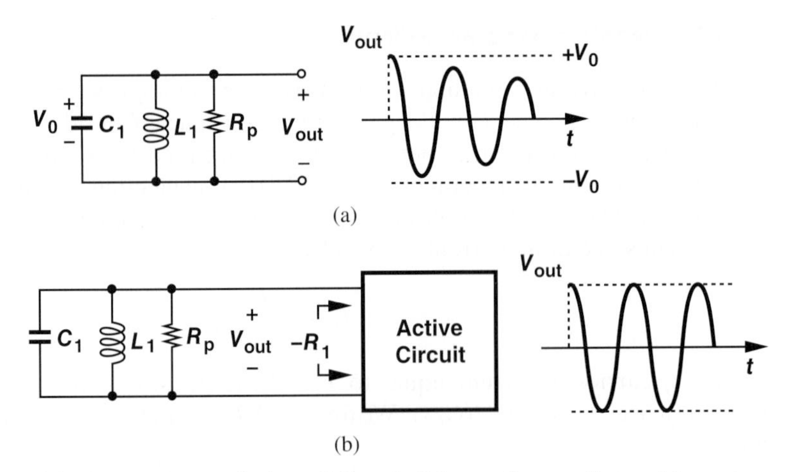

Figure 13.32 (a) Time response of a lossy LC tank, (b) use of a negative resistance to cancel the loss of the tank.

What happens if $|-R_1| < R_p$? Then, $(-R_1)||R_p$ is still negative, allowing the oscillation amplitude to grow until nonlinear mechanisms in the active circuit limit the amplitude (Section 13.1).

Did you know?

Crystal oscillators provide a very precise and stable output frequency. But what if we need a variable frequency? For example, the microprocessor in your laptop runs at a high clock frequency if heavy computation is necessary but switches to a low clock frequency to save power if there is little computation demand. How can such a clock be generated? This task is accomplished using a "phase-locked loop," a circuit that multiplies the crystal frequency by a programmable factor. For example, with a crystal frequency of 10 MHz and a multiplication factor ranging from 10 to 300, the microprocessor can run at a clock frequency of 100 MHz to 3 GHz.

13.6.3 Crystal Oscillator Implementation

We now attach a crystal to a negative resistance to form an oscillator [Fig. 13.33(a)]. Replacing the crystal with its electrical model and the negative-resistance circuit with its equivalent network, we arrive at Fig. 13.33(b). Of course, to benefit from the precise resonance frequency of the crystal, we must choose C_A and C_B so as to *minimize* their effect on the oscillation frequency. As evident from Fig. 13.33(b), this occurs if $C_AC_B/(C_A + C_B)$ is much *smaller* than the crystal impedance, Z_{cr}. However, if C_A and C_B are excessively large, then the negative resistance, $-g_m/(C_AC_B\omega^2)$, is not "strong" enough to cancel the crystal loss. In typical designs, C_A and C_B are chosen 10 to 20 times smaller than C_2.

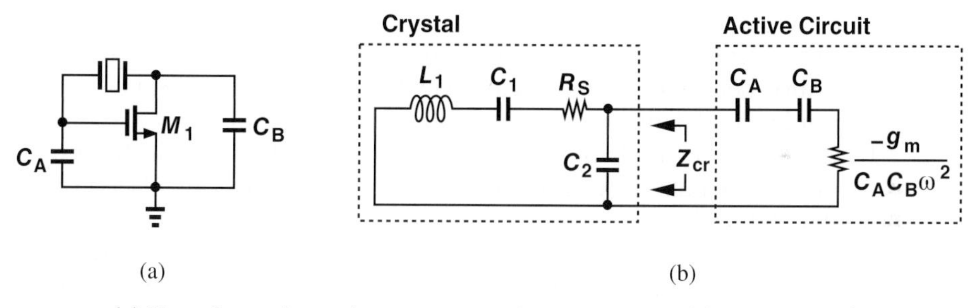

Figure 13.33 (a) Use of negative resistance to cancel loss of crystal, (b) equivalent circuit.

The analysis of the basic crystal oscillator in Fig. 13.33 is somewhat beyond the scope of this book and is outlined in Problem 13.51 for the interested reader. It can be shown that the circuit oscillates at the crystal's parallel resonance frequency if

$$L_1C_1\omega^2 - 1 \leq g_mR_S\frac{C_1C_2}{C_AC_B}. \tag{13.58}$$

The crystal data sheet specifies L_1, C_1, C_2, and R_S. The designer must choose C_A, C_B, and g_m properly.

We must now add bias elements to the circuit. Unlike parallel LC tanks, a crystal does not provide a path for the bias current or voltage. (Recall the series capacitance, C_1, in the crystal model.) For example, the stages in Fig. 13.34(a) do not operate properly because the drain bias current of M_A is zero and the gate bias voltage of M_B is not defined. We can add a feedback resistor as shown in Fig. 13.34(b) to realize a self-biased stage. Note that R_F must be very large (tens of kiloohms) to contribute negligible loss. We can replace R_D with a current source [Fig. 13.34(c)]. Now the current source can be transformed to an *amplifying* device if its gate is tied to node X [Fig. 13.34(d)].

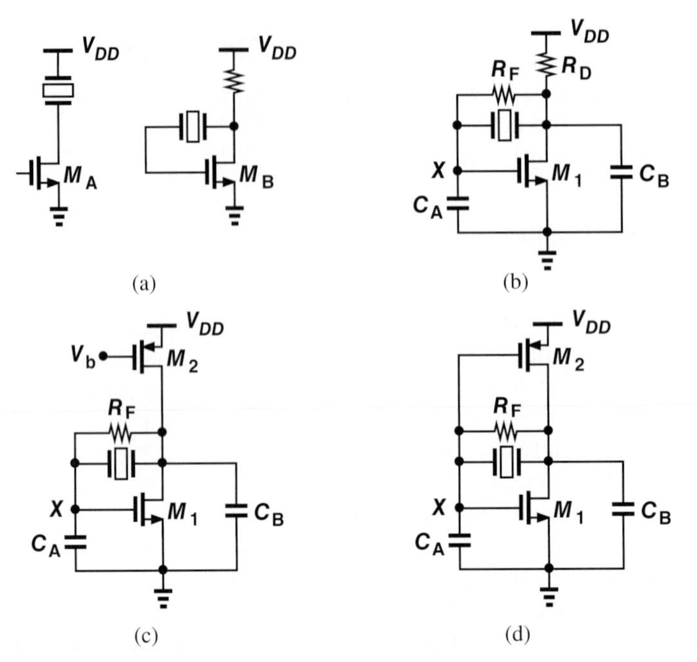

Figure 13.34 (a) Stages with no dc bias path, (b) simple biasing of a crystal oscillator, (c) biasing using a PMOS current source, (d) inverter-based crystal oscillator.

The circuit of Fig. 13.34(d) merits several remarks. First, both transistors are biased in saturation before oscillation begins (why?). Second, for small-signal operation, M_1 and M_2 appear in parallel, providing a total transconductance of $g_{m1} + g_{m2}$. Third, M_1 and M_2 can be viewed as a CMOS inverter (Chapter 16) that is biased at its trip point. This oscillator topology is popular in integrated circuits, with the inverter placed on the chip and the crystal off the chip.

The circuit of Fig. 13.34(d) may exhibit a tendency to oscillate at higher harmonics of the crystal's parallel resonance frequency. For example, if this resonance frequency is 20 MHz, the circuit may oscillate at 40 MHz. To avoid this issue, a low-pass filter must be inserted in the feedback loop so as to suppress the gain at higher frequencies. As illustrated in Fig. 13.35, we place resistor R_1 in series with the feedback network. The pole frequency, $1/(2\pi R_1 C_B)$, is typically chosen slightly above the oscillation frequency.

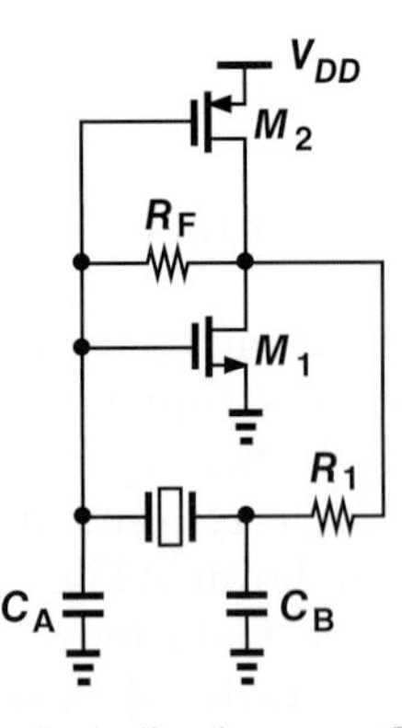

Figure 13.35 Complete crystal oscillator including low-pass filter to avoid higher modes.

In discrete circuit design, a high-speed CMOS inverter may not be available. An alternative topology using a single bipolar transistor can be derived from the circuit of Fig. 13.33(a) as shown in Fig. 13.36(a). To bias the transistor, we add a large resistor from the collector to the base and an inductor from the collector to V_{CC} [Fig. 13.36(b)]. We wish L_1 to provide the bias current of Q_1 but not affect the oscillation frequency. Thus, we choose L_1 large enough that $L_1\omega$ is a high impedance (approximately an open circuit). An inductor playing such a role is called a "radio-frequency choke" (RFC). Note that this circuit reduces to that in Fig. 13.36(a) if R_F and L_1 are large.

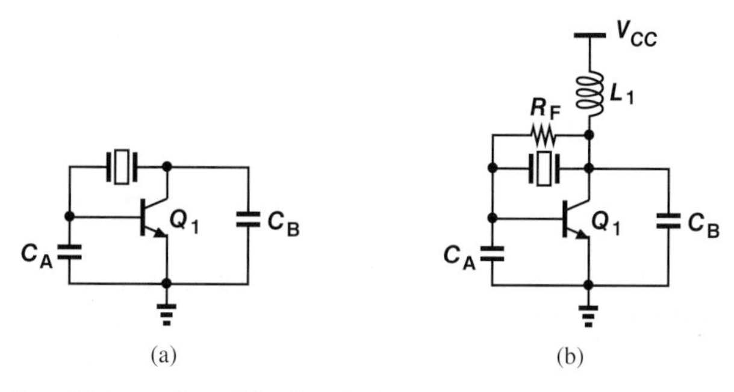

Figure 13.36 Crystal oscillator using a bipolar device.

13.7 CHAPTER SUMMARY

- An oscillator can be viewed as a negative-feedback system with so much phase shift (delay) in the loop that the feedback becomes positive at the oscillation frequency.
- The magnitude of the loop gain must exceed unity at the oscillation frequency. This is called the "startup condition."
- The voltage swing in an oscillator is determined by saturation or nonlinear behavior of the devices.
- Ring oscillators consist of multiple identical gain stages in a loop and find wide application in integrated circuits, e.g., microprocessors and memories.
- The impedance of a parallel LC tank exhibits a zero phase at resonance. A lossy tank reduces to a single resistor at this frequency.
- If two common-source stages having resonant loads are placed in a feedback loop, an oscillator is formed.
- The cross-coupled LC oscillator is extensively used in high-frequency integrated circuits, e.g., WiFi transceivers. This topology provides a differential output.
- The Colpitts LC oscillator employs a single transistor and finds application in high-frequency discrete design. This topology provides a single-ended output.
- For low to moderate frequencies, the phase shift and Wien-bridge oscillators are used in discrete design. They can be readily implemented by means of op amps.
- For precise and stable frequencies, crystal oscillators can be used. Such circuits serve as the "reference frequency" in many applications, e.g., microprocessors, memories, wireless transceivers, etc.

PROBLEMS

Sec. 13.1 Oscillation Conditions

13.1. A negative-feedback system is shown in Fig. 13.37. Under what conditions does the system oscillate?

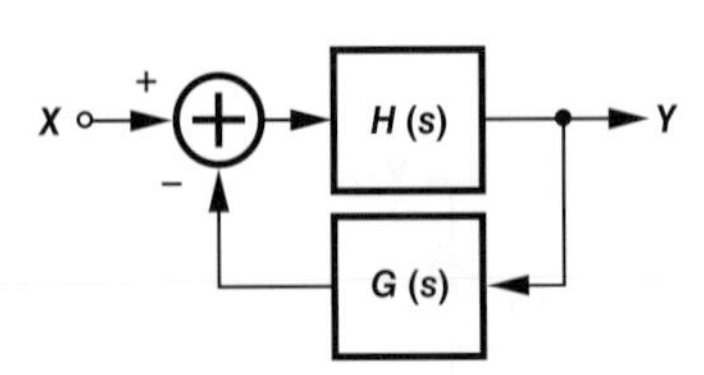

Figure 13.37

13.2. A negative-feedback system is shown in Fig. 13.38. Under what conditions does the system oscillate?

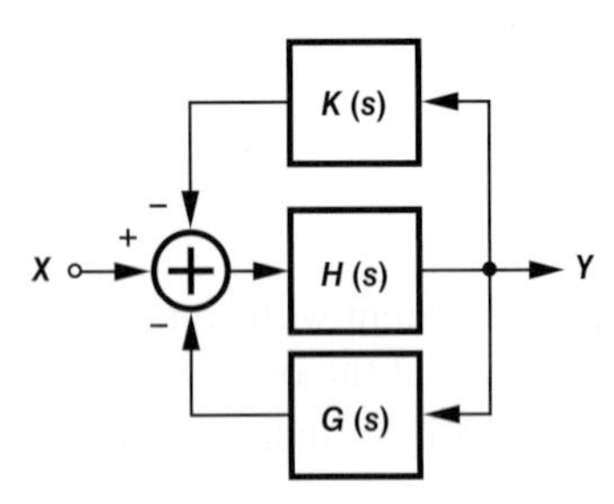

Figure 13.38

13.3. Consider the simple common-emitter stage shown in Fig. 13.39. Explain why this circuit does not oscillate.

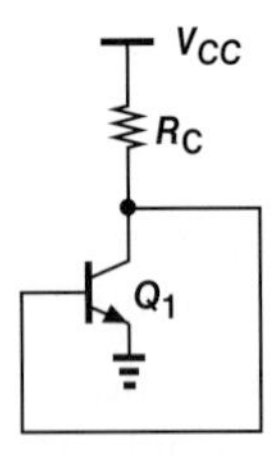

Figure 13.39

13.4. A differential pair is placed in a negative-feedback loop as shown in Fig. 13.40. Can this circuit oscillate? Explain.

***13.5.** Explain what happens if the polarity of feedback is changed in the circuit of Fig. 13.40, i.e., the gate of M_1 is tied to the drain of M_2 and vice versa.

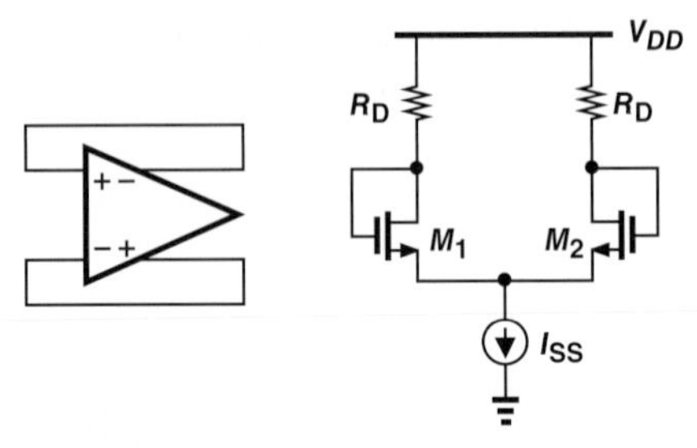

Figure 13.40

***13.6.** A differential pair followed by source followers is placed in a negative-feedback loop as illustrated in Fig. 13.41. Consider only the capacitances shown in the circuit. Can this circuit oscillate? Explain.

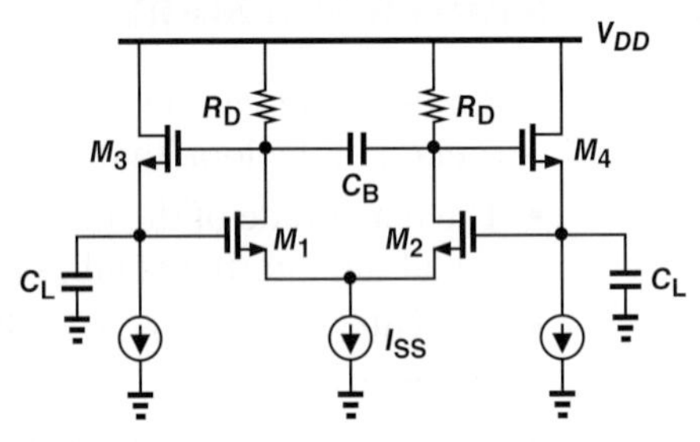

Figure 13.41

****13.7.** We insert two resistors in series with the gates of M_1 and M_2 in Fig. 13.41. Taking into account C_{GS1} and C_{GS2} in addition to the other four capacitors, explain whether the circuit can oscillate.

Sec. 13.2 Ring Oscillators

13.8. Suppose in the ring oscillator of Fig. 13.8, the value of R_D is doubled. How do the oscillation frequency and startup condition change?

***13.9.** Suppose in the ring oscillator of Fig. 13.8, the value of C_D is doubled. How do the oscillation frequency and startup condition change?

*13.10. In the ring oscillator of Fig. 13.8, we assume C_D arises from C_{GS} and neglect other capacitances. If the width and bias current of each transistor are doubled and R_D is halved, what happens to the oscillation frequency?

13.11. The supply voltage of the ring oscillator of Fig. 13.8 is gradually reduced. Explain why the oscillation eventually ceases.

**13.12. Derive the oscillation frequency and startup condition for the ring oscillator of Fig. 13.8 if the number of stages is increased to five.

*13.13. Derive the oscillation frequency and startup condition for the ring oscillator of Fig. 13.9(a). Consider only the C_{GS} of the NMOS transistors and assume all transistors are in saturation.

13.14. Suppose the bias voltage, V_b, in Fig. 13.9(a) is gradually raised to reduce the bias current of each stage. Does the circuit have a lower or higher tendency to oscillate? (Hint: as the bias current decreases, r_O rises more rapidly than g_m falls.)

**13.15. Derive the oscillation frequency and startup condition for the ring oscillator of Fig. 13.9(b). Consider the C_{GS} of both NMOS and PMOS transistors and assume all transistors are in saturation.

13.16. Draw the large-signal waveforms of Fig. 13.10 for a five-stage ring oscillator similar to the circuit of Fig. 13.9(b).

13.17. A ring oscillator is sometimes used to provide *multiple* outputs with different phases. What is the phase difference between consecutive nodes in the circuit of Fig. 13.9(b)? (Hint: consider the waveforms in Fig. 13.10.)

13.18. A ring oscillator employs N stages. What is the phase difference between the consecutive outputs of the circuit?

Sec. 13.3 LC Oscillators

13.19. Repeat the plots in Fig. 13.14 if both L_1 and C_1 in Fig. 13.13 are doubled.

13.20. In the circuit of Fig. 13.13, assume $V_{in}(t) = V_0 \cos \omega_{in} t$. Plot $I_{in}(t)$ if ω_{in} is slightly below ω_1 or above ω_1. (Hint: consider the magnitude and phase response in Fig. 13.14.)

13.21. Compute $Z_2(s)$ in the tank of Fig. 13.15(a). Compute the pole frequencies.

13.22. Determine the pole frequencies of $Z_2(s)$ in Eq. (13.13) and sketch their locations on the complex plane as R_p goes from infinity to a small value.

*13.23. In this problem, we wish to determine how the tank in Fig. 13.15(a) can be transformed to that in Fig. 13.15(b). Compute the impedance of each tank at a frequency $s = j\omega$ and equate the two impedances. Now, equate their real parts and do the same with their imaginary parts. Also, assume $L_1\omega/R_1 \gg 1$. (We say the inductor has a high quality factor, Q.) Determine the value of R_p.

13.24. Explain qualitatively what happens to the plots in Fig. 13.16 if R_p is doubled.

13.25. Sketch the instantaneous power dissipated by R_p in Fig. 13.17(b) as a function of time. Can you predict what we will obtain if we integrate the area under this plot?

13.26. In the CS stage of Fig. 13.18(a), $V_{in} = V_0 \cos \omega_1 t + V_1$, where $\omega_1 = 1/\sqrt{L_1 C_1}$ and V_1 is a bias value. Plot V_{out} as a function of time. (Hint: what is the dc value of the output when the input is just equal to V_1?)

13.27. Explain qualitatively what happens to the plots in Fig. 13.18(b) if R_p is doubled.

**13.28. In the circuit of Fig. 13.19(a), we break the loop at the gate of M_1 and apply an input as shown in Fig. 13.42. Suppose $V_{in} = V_0 \cos \omega_1 t + V_1$, where ω_1 is the resonance frequency of each tank and V_1 is a bias voltage. Plot the waveforms at nodes X and Y.

13.29. Suppose the LC oscillator of Fig. 13.19(a) is realized with *ideal* tanks, i.e., $R_p = \infty$. Taking channel-length modulation into account, determine the startup condition.

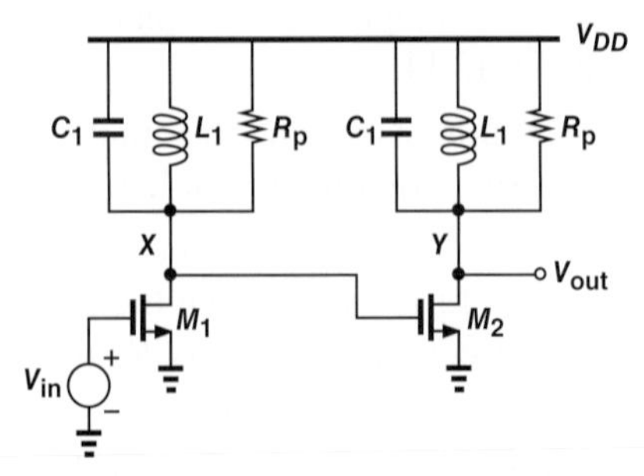

Figure 13.42

*13.30. Suppose the two tanks in the oscillator of Fig. 13.19(a) have slightly different resonance frequencies. Can you roughly predict the oscillation frequency? (Hint: consider the open-loop frequency response.)

13.31. Explain why the circuit of Fig. 13.20(c) does not oscillate if the tanks are replaced with resistors.

13.32. If we increase C_1 in the Colpitts oscillator of Fig. 13.22(a), do we relax or tighten the startup condition?

13.33. What happens if $C_2 = 0$ in the Colpitts oscillator of Fig. 13.22(a)?

**13.34. Repeat the analysis of the Colpitts oscillator but by breaking the loop at the emitter of Q_1. The equivalent circuit is shown in Fig. 13.43. Note that the loading seen at the emitter, $1/g_m$, is included in parallel with C_2.

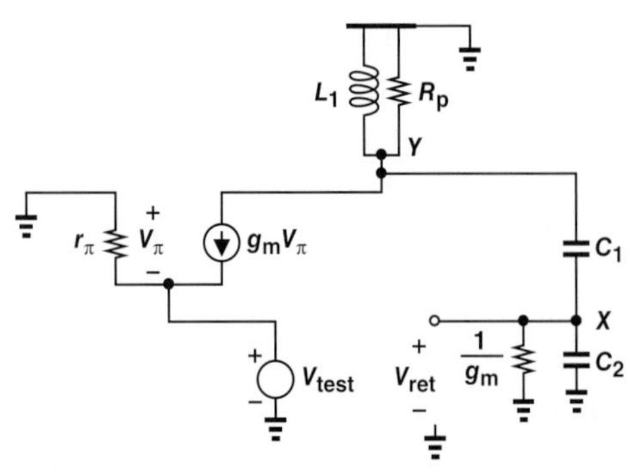

Figure 13.43

13.35. Repeat the analysis of the Colpitts oscillator while including r_O of Q_1 but assuming $R_p = \infty$.

13.36. Repeat the analysis of the Colpitts oscillator while assuming an output resistance, R_1, for I_1 but neglecting R_p.

13.37. Derive Eq. (13.27) if $R_p = \infty$. You can use the equivalent circuit of Fig. 13.22(b) and tie R_{in} from X to ground.

Sec. 13.4 and 13.5 Phase Shift and Wien Bridge Oscillators

13.38. In the oscillator of Fig. 13.24(a), we have $R_1 = R_2 = R_3 = R$, $C_1 = C_2 = C_3 = C$, and $V_{out} = V_0 \cos \omega_0 t$, where $\omega_0 = 1/(RC)$. Plot the waveforms at X and Y. Assume $A_0 = 2$.

13.39. A student decides to employ three *low-pass* sections to create the phase shift necessary in a phase shift oscillator (Fig. 13.44). If $R_1 = R_2 = R_3 = R$ and $C_1 = C_2 = C_3 = C$, repeat the analysis for this circuit.

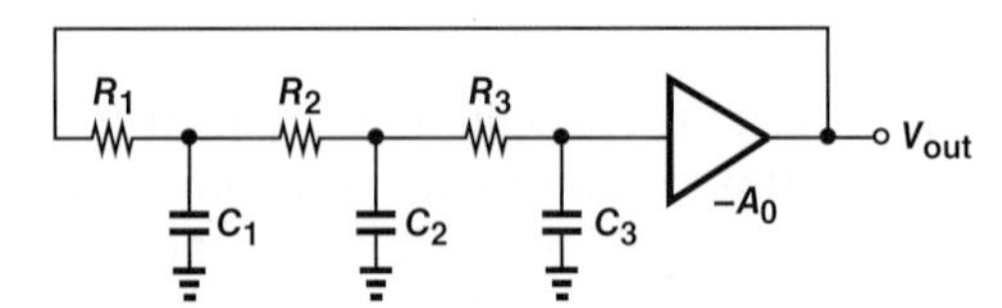

Figure 13.44

13.40. Compare the oscillation frequency and startup condition of the phase shift oscillator with those of a three-stage ring oscillator.

13.41. Suppose a phase shift oscillator incorporates *four* high-pass sections with equal resistors and capacitors. Derive the oscillation frequency and the startup condition for such a circuit.

*13.42. Can the circuit shown in Fig. 13.45 oscillate? Explain. Assume $R_1 = R_2 = R_3 = R$ and $C_1 = C_2 = C_3 = C$.

13.43. Consider the oscillator of Fig. 13.24(a) and assume the amplifier contains a pole, i.e., $A(s) = A_0/(1 + s/\omega_0)$. Also, assume the phase shift network contains only two high-pass sections with $R_1 = R_2 = R$ and $C_1 = C_2 = C$. Can ω_0 be chosen such that this circuit oscillates?

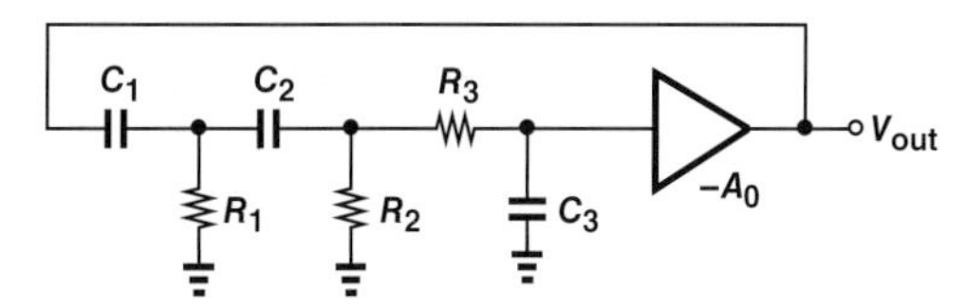

Figure 13.45

13.44. In the circuit of Fig. 13.27(a), $R_1 = R_2 = R$ and $C_1 = C_2 = C$. If $V_{in} = V_0 \cos \omega_0 t$, where $\omega_0 = 1/(RC)$, plot V_{out} as a function of time.

*__13.45.__ A student decides to modify the Wien oscillator of Fig. 13.27(b) as shown in Fig. 13.46. Can this circuit oscillate? Explain.

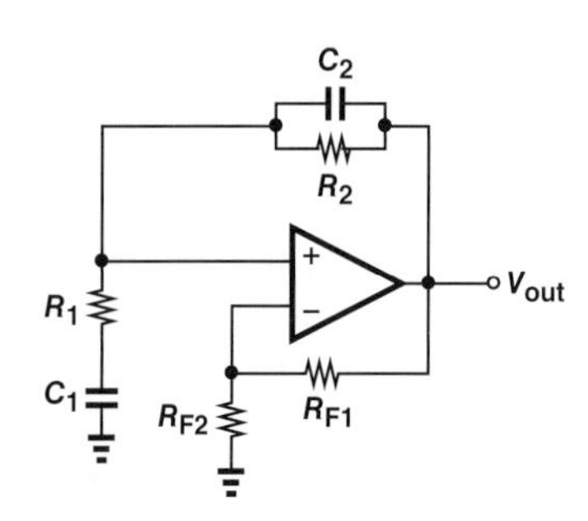

Figure 13.46

Sec. 13.6 Crystal Oscillators

13.46. Derive an expression for Z_S in Fig. 13.30(b) if $s = j\omega$.

13.47. Sketch the real and imaginary parts of Z_{in} in Fig. 13.31(c) as a function of frequency.

13.48. Suppose the negative-resistance circuit of Fig. 13.31(a) employs a bipolar transistor rather than a MOSFET. Determine Z_{in} and the equivalent circuit.

13.49. Determine Z_{in} in Fig. 13.31(a) if channel-length modulation is not neglected. Can you construct a simple equivalent circuit for Z_{in} such as that in Fig. 13.31(c)?

13.50. Suppose capacitor C_1 in Fig. 13.32(a) begins with an initial condition of V_0. Derive an equation for V_{out} assuming that R_p is large. (A large R_p means the tank has a high quality factor, Q.)

**__13.51.__ We wish to determine the startup condition for the crystal oscillator of Fig. 13.33(b).

(a) Prove that Z_{cr} is given by the following equation at the parallel resonance frequency:

$$Z_{cr}(j\omega_2) = \frac{L_1 C_1 \omega_2^2 - 1}{R_S C_1 C_2 \omega_2^2} + \frac{1}{jC_2\omega_2}. \tag{13.59}$$

(b) Now, cancel the real part of $Z_{cr}(j\omega_2)$ by the negative resistance and prove Eq. (13.58).

**__13.52.__ Repeat Problem 13.51 if channel-length modulation is not neglected.

Design Problems

In the following problems, unless otherwise stated, assume $\mu_n C_{ox} = 2\mu_p C_{ox} = 100\ \mu\text{A/V}^2$, $\lambda_n = 0.5$ $\lambda_p = 0.1\ \text{V}^{-1}$, $V_{THN} = 0.3$ V, and $V_{THP} = -0.35$ V.

13.53. In Fig. 13.8, $R_D = 1\ \text{k}\Omega$. Design the circuit for a power budget of 3 mW and a frequency of 1 GHz. Assume $V_{DD} = 1.5$ V and $\lambda = 0$.

13.54. In the circuit of Fig. 13.9(a), $V_{DD} - V_b = 0.6$ V.

(a) Choose W/L of the PMOS devices for a bias current of 1 mA.

(b) Choose W/L of the NMOS devices to meet the startup condition, $g_{mN}(r_{ON}||r_{OP}) = 2$.

13.55. (a) A 10 nH inductor has a series resistance of 10 Ω. Determine the equivalent parallel resistance, R_p, at 1 GHz.

(b) Design a 1 GHz cross-coupled oscillator using this inductor with a power budget of 2 mW and $V_{DD} = 1.5$ V. Choose W/L of the two transistors such that 95% of the tail current is steered to the left or to the right for $|V_X - V_Y| = 500$ mV. For simplicity, assume each transistor contributes a capacitance of $1.5 \times W$ fF, where W is in microns.

13.56. (a) A 20 nH inductor has a series resistance of 15 Ω. Compute the equivalent parallel resistance, R_p, at 2 GHz.

(b) Design a 2 GHz Colpitts oscillator using this inductor with a power budget of 2 mW and $V_{DD} = 1.5$ V. For simplicity, neglect the capacitances of the bipolar transistor and assume $C_1 = C_2$. Verify that the startup condition is met.

13.57. Design the phase shift oscillator of Fig. 13.25 for a frequency of 10 MHz, assuming an ideal op amp and $C_1 = C_2 = C_3 = 1$ nF.

13.58. Design the Wien bridge oscillator of Fig. 13.28(b) for a frequency of 10 MHz, assume an ideal op amp and $C_1 = C_2 = 1$ nF.

13.59. (a) A crystal with a parallel resonance frequency at 10 MHz has $C_2 = 100$ pF, $C_1 = 10$ pF [Fig. 13.30(c)]. Determine the value of L_1.

(b) Suppose the crystal series resistance is equal to 5 Ω. Design the oscillator of Fig. 13.34(d) for a frequency of 10 MHz. Neglect the transistor capacitances and assume $C_A = C_B = 20$ pF, $(W/L)_2 = 2(W/L)_1$, and $V_{DD} = 1.2$ V.

SPICE PROBLEMS

In the following problems, use the MOS device models given in Appendix A. For bipolar transistors, assume $I_S = 5 \times 10^{-16}$ A, $\beta = 100$, and $V_A = 5$ V. Also, assume a supply voltage of 1.8 V. (In SPICE, one node of the oscillators must be initialized near zero or V_{DD} to ensure startup.)

13.60. Simulate the oscillator of Fig. 13.8 with $W/L = 10/0.18$ and $C_D = 20$ fF. Choose the value of R_D so that the circuit barely oscillates. Compare the value of $g_m R_D$ with the theoretical minimum of 2. Plot the voltage swings at X, Y, and Z and measure their phase difference.

13.61. Repeat Problem 13.60 but choose R_D equal to four times the minimum acceptable value. How much is the voltage swing in this case? Does the frequency decrease by a factor of 4?

13.62. Simulate the oscillator of Fig. 13.9(a) with $(W/L)_N = 10/0.18$ and $(W/L)_P = 15/0.18$. Choose V_b to obtain a bias current of 0.5 mA in each branch.

(a) Measure the oscillation frequency.

(b) Now, change V_b by ± 100 mV and measure the oscillation frequency. Such a circuit is called a voltage-controlled oscillator (VCO).

13.63. Repeat Problem 13.62 with five stages in the ring and compare the oscillation frequencies with the previous case. Do they decrease by a ratio of 5 to 3?

13.64. Simulate the ring oscillator of Fig. 13.9(b) in two cases.

(a) Choose $(W/L)_P = 2(W/L)_N = 20/0.18$.

(b) Choose $(W/L)_P = 2(W/L)_N = 10/0.18$. Which case yields a higher oscillation frequency?

13.65. We wish to design the circuit of Fig. 13.9(b) for the highest oscillation frequency. Begin with $(W/L)_N = (W/L)_P = 5/0.18$ and decrease the width of the transistors in 0.5 μm steps. Plot the oscillation frequency as a function of W.

13.66. We can construct a *four*-stage ring oscillator if we employ differential pairs rather than inverters. Simulate a ring comprising four identical differential pairs with $W/L = 10/0.18$, a tail current of 0.5 mA, and a load resistor of 1 kΩ. Choose the feedback to be *negative* at low frequencies. Plot the waveforms provided by the four stages and measure their phase difference.

13.67. Simulate the cross-coupled oscillator of Fig. 13.20(c) with $W/L = 10/0.18$, $I_{SS} =$ 1 mA, and $L_1 = 10$ nH. Place a resistance of 10 Ω in series with each inductor (and exclude R_p) and add enough capacitance from X and Y to ground so as to obtain an oscillation frequency of 1 GHz. Plot the output voltages and the drain currents of M_1 and M_2 as a function of time. What is the minimum value of I_{SS} to sustain oscillation?

13.68. Design and simulate the Colpitts oscillator of Fig. 13.22(a) with $L_1 = 10$ nH, $V_b = 1.2$ V, and $I_1 = 1$ mA. Choose R_p such that $Q = R_p/(L_1\omega) = 10$ at 2 GHz. Also, select the value of $C_1 = C_2$ for an oscillation frequency of 2 GHz. Plot the waveforms at the collector and emitter of Q_1.

13.69. In Problem 13.68, reduce I_1 in steps of 0.1 mA until the oscillation ceases. Compare the minimum required I_1 with the theoretical value predicted by Eq. (13.26).

REFERENCE

1. Y. Hong, et al., "Noncontact proximity vital sign sensor based on PLL for sensitivity enhancement," *IEEE Trans. Biomedical Circuits and Systems*, vol. 8, pp. 584–593, Aug. 2014.

14

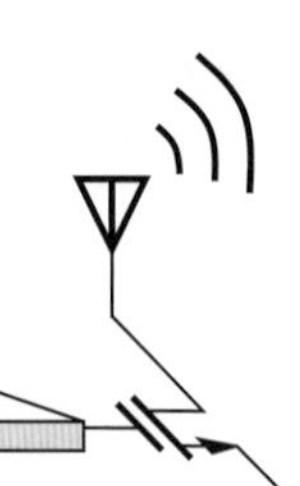

Output Stages and Power Amplifiers

The amplifier circuits studied in previous chapters aim to achieve a high gain with desirable input and output impedance levels. However, many applications require circuits that can deliver a high power to the load. For example, the cellphone described in Chapter 1 must drive the antenna with 1 W of power. As another example, typical stereo systems deliver tens or hundreds of watts of audio power to speakers. Such circuits are called "power amplifiers" (PAs).

This chapter deals with circuits that can provide a high output power. We first reexamine circuits studied in previous chapters to understand their shortcomings for this task. Next, we introduce the "push-pull" stage and various modifications to improve its performance. The chapter outline is shown below.

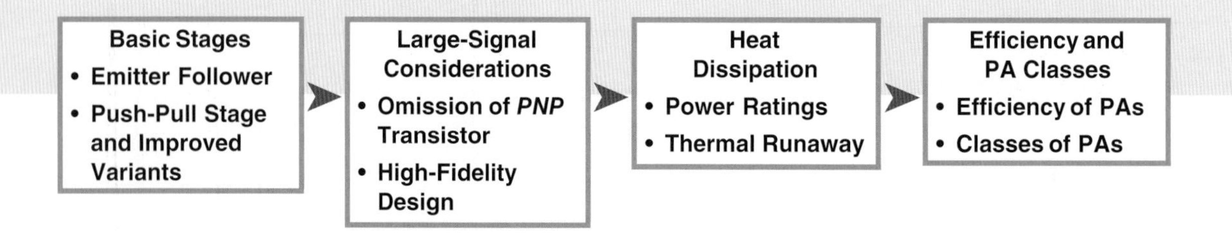

14.1 GENERAL CONSIDERATIONS

The reader may wonder why the amplifier stages studied in previous chapters are not suited to high-power applications. Suppose we wish to deliver 1 W to an 8-Ω speaker. Approximating the signal with a sinusoid of peak amplitude V_P, we express the power absorbed by the speaker as

$$P_{out} = \left(\frac{V_P}{\sqrt{2}}\right)^2 \cdot \frac{1}{R_L}, \tag{14.1}$$

where $V_P/\sqrt{2}$ denotes the root mean square (rms) value of the sinusoid and R_L represents the speaker impedance. For $R_L = 8\ \Omega$ and $P_{out} = 1$ W,

$$V_P = 4\ \text{V}. \tag{14.2}$$

Also, the peak current flowing through the speaker is given by $I_P = V_P/R_L = 0.5$ A.

We can make a number of important observations here. (1) The resistance that must be driven by the amplifier is much lower than the typical values (hundreds to thousands of ohms) seen in previous chapters. (2) The current levels involved in this example are much greater than the typical currents (milliamperes) encountered in previous circuits. (3) The voltage swings delivered by the amplifier can hardly be viewed as "small" signals, requiring a good understanding of the large-signal behavior of the circuit. (4) The power drawn from the supply voltage, at least 1 W, is much higher than our typical values. (5) A transistor carrying such high currents and sustaining several volts (e.g., between collector and emitter) dissipates a high power and, as a result, *heats up*. High-power transistors must therefore handle high currents and high temperature.[1]

Based on the above observations, we can predict the parameters of interest in the design of power stages:

1. "Distortion," i.e., the nonlinearity resulting from large-signal operation. A high-quality audio amplifier must achieve a very low distortion so as to reproduce music with high fidelity. In previous chapters, we rarely dealt with distortion.
2. "Power efficiency" or simply "efficiency," denoted by η and defined as

$$\eta = \frac{\text{Power Delivered to Load}}{\text{Power Drawn from Supply}}. \tag{14.3}$$

   For example, a cellphone power amplifier that consumes 3 W from the battery to deliver 1 W to the antenna provides $\eta \approx 33.3\%$. In previous chapters, the efficiency of circuits was of little concern because the absolute value of the power consumption was quite small (a few milliwatts).
3. "Voltage rating." As suggested by Eq. (14.1), higher power levels or load resistance values translate to large voltage swings and (possibly) high supply voltages. Also, the transistors in the output stage must exhibit breakdown voltages well above the output voltage swings.

14.2 EMITTER FOLLOWER AS POWER AMPLIFIER

With its relatively low output impedance, the emitter follower may be considered a good candidate for driving "heavy" loads, i.e., low impedances. As shown in Chapter 5, the small-signal gain of the follower is given by

$$A_v = \frac{R_L}{R_L + \dfrac{1}{g_m}}. \tag{14.4}$$

We may therefore surmise that for, say, $R_L = 8\ \Omega$, a gain near unity can be obtained if $1/g_m \ll R_L$, e.g., $1/g_m = 0.8\ \Omega$, requiring a collector bias current of 32.5 mA. We assume $\beta \gg 1$.

[1] And, in some applications, high voltages.

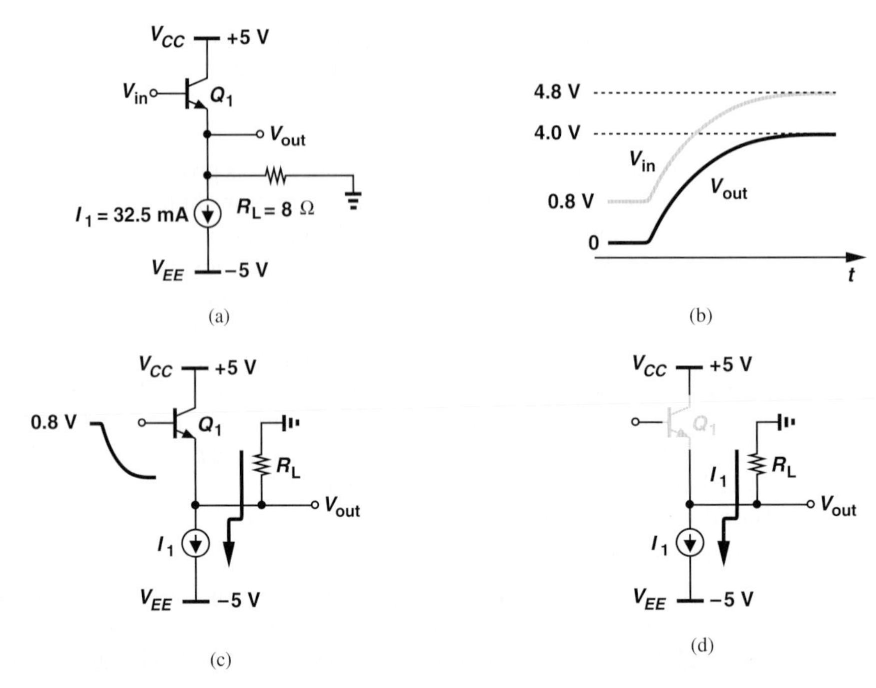

Figure 14.1 (a) Follower driving a heavy load, (b) input and output waveforms, (c) current path as input becomes more negative, (d) current path as input becomes more positive.

But, let us analyze the circuit's behavior in delivering *large* voltage swings (e.g. 4 V_P) to heavy loads. To this end, consider the follower shown in Fig. 14.1(a), where I_1 serves as the bias current source. To simplify the analysis, we assume the circuit operates from negative and positive power supplies, allowing V_{out} to be centered around zero. For $V_{in} \approx 0.8$ V, we have $V_{out} \approx 0$ and $I_C \approx 32.5$ mA. If V_{in} rises from 0.8 V to 4.8 V, the emitter voltage follows the base voltage with a relatively constant difference of 0.8 V, producing a 4-V swing at the output [Fig. 14.1(b)].

Now suppose V_{in} begins from +0.8 V and gradually goes down [Fig. 14.1(c)]. We expect V_{out} to go below zero and hence part of I_1 to flow from R_L. For example, if $V_{in} \approx 0.7$ V, then $V_{out} \approx -0.1$ V, and R_L carries a current of 12.5 mA. That is, $I_{C1} \approx I_{E1} = 20$ mA. Similarly, if $V_{in} \approx 0.6$ V, then $V_{out} \approx -0.2$ V, $I_{RL} \approx 25$ mA, and hence $I_{C1} \approx 7.5$ mA. In other words, the collector current of Q_1 continues to fall.

What happens as V_{in} becomes more negative? Does V_{out} still track V_{in}? We observe that for a sufficiently low V_{in}, the collector current of Q_1 drops to *zero* and R_L carries the *entire* I_1 [Fig. 14.1(d)]. For lower values of V_{in}, Q_1 remains off and $V_{out} = -I_1 R_L = -260$ mV.

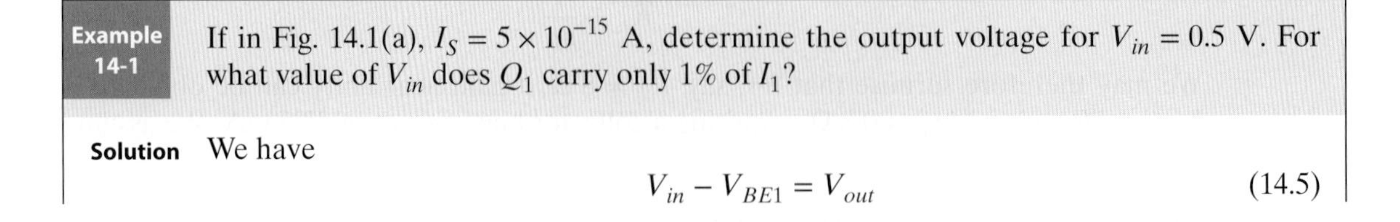

Example 14-1 If in Fig. 14.1(a), $I_S = 5 \times 10^{-15}$ A, determine the output voltage for $V_{in} = 0.5$ V. For what value of V_{in} does Q_1 carry only 1% of I_1?

Solution We have

$$V_{in} - V_{BE1} = V_{out} \tag{14.5}$$

and

$$\frac{V_{out}}{R_L} + I_1 = I_{C1}. \quad (14.6)$$

Since $V_{BE1} = V_T \ln(I_{C1}/I_S)$, Eqs. (14.5) and (14.6) can be combined to yield

$$V_{in} - V_T \ln\left[\left(\frac{V_{out}}{R_L} + I_1\right)\frac{1}{I_S}\right] = V_{out}. \quad (14.7)$$

Beginning with a guess $V_{out} = -0.2$ V and after a few iterations, we obtain

$$V_{out} \approx -211 \text{ mV}. \quad (14.8)$$

Note from Eq. (14.6) that $I_{C1} \approx 6.13$ mA.

To determine the value of V_{in} that yields $I_{C1} \approx 0.01 I_1 = 0.325$ mA, we eliminate V_{out} from Eqs. (14.5) and (14.6):

$$V_{in} = V_T \ln\frac{I_{C1}}{I_S} + (I_{C1} - I_1)R_L. \quad (14.9)$$

Setting $I_{C1} = 0.325$ mA, we obtain

$$V_{in} \approx 390 \text{ mV}. \quad (14.10)$$

Note from Eq. (14.5) that $V_{out} \approx -257$ mV under this condition.

Exercise Repeat the above example if $R_L = 16\ \Omega$ and $I_1 = 16$ mA.

Let us summarize our thoughts thus far. In the arrangement of Fig. 14.1(a), the output tracks the input[2] as V_{in} rises because Q_1 can carry both I_1 and the current drawn by R_L. On the other hand, as V_{in} falls, so does I_{C1}, eventually turning Q_1 off and leading to a *constant* output voltage even though the input changes. As illustrated in the waveforms of Fig. 14.2(a), the output is severely distorted. From another perspective, the input/output characteristic of the circuit, depicted in Fig. 14.2(b), begins to depart substantially from a straight line as V_{in} falls below approximately 0.4 V (from Example 14-2).

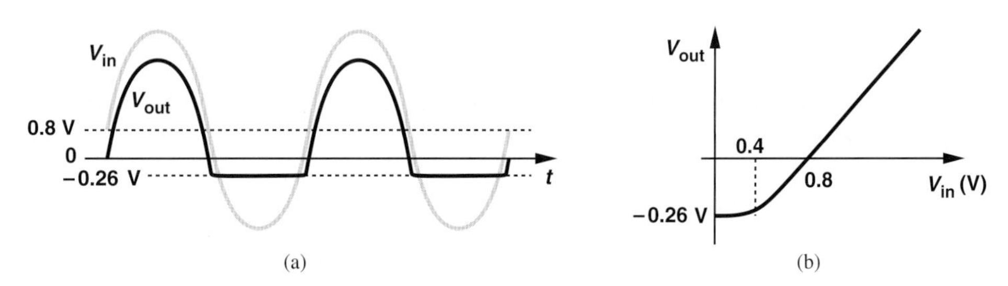

Figure 14.2 (a) Distortion in a follower, (b) input/output characteristic.

[2]The tracking may not be quite faithful because V_{BE} experiences some change, but we ignore this effect for now.

Our foregoing study reveals that the follower of Fig. 14.1(a) cannot deliver voltage swings as large as ±4 V to an 8-Ω speaker. How can we remedy the situation? Noting that $V_{out,min} = -I_1 R_L$, we can increase I_1 to greater than 50 mA so that for $V_{out} = -4$ V, Q_1 still remains on. This solution, however, yields a higher power dissipation and a lower efficiency.

14.3 PUSH-PULL STAGE

Considering the operation of the emitter follower in the previous section, we postulate that the performance can be improved if I_1 increases *only when needed*. In other words, we envision an arrangement wherein I_1 increases as V_{in} becomes *more negative* and vice versa. Shown in Fig. 14.3(a) is a possible realization of this idea. Here, the constant current source is replaced with a *pnp* emitter follower so that, as Q_1 begins to turn off, Q_2 "kicks in" and allows V_{out} to track V_{in}.

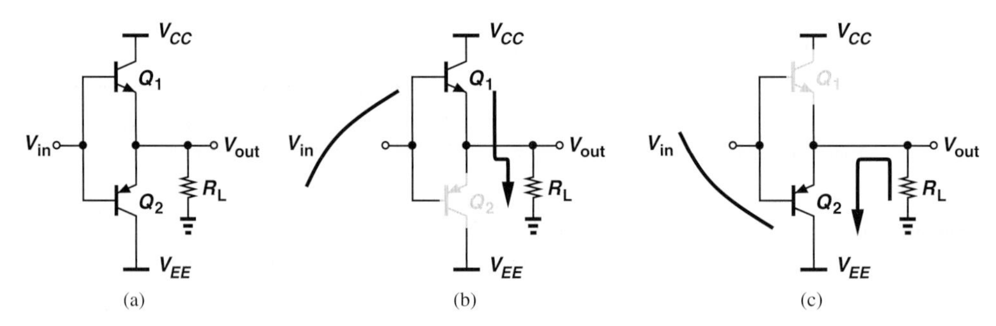

Figure 14.3 (a) Basic push-pull stage, (b) current path for sufficiently positive inputs, (c) current path for sufficiently negative inputs.

Called the "push-pull" stage, this circuit merits a detailed study. We note that if V_{in} is sufficiently positive, Q_1 operates as an emitter follower, $V_{out} = V_{in} - V_{BE1}$, and Q_2 remains *off* [Fig. 14.3(b)] because its base-emitter junction is reverse-biased. By symmetry, if V_{in} is sufficiently negative, the reverse occurs [Fig. 14.3(c)] and $V_{out} = V_{in} + |V_{BE2}|$. We say Q_1 "pushes" current into R_L in the former case and Q_2 "pulls" current from R_L in the latter.

Example 14-2 Sketch the input/output characteristic of the push-pull stage for very positive or very negative inputs.

Solution As noted above,

$$V_{out} = V_{in} + |V_{BE2}| \quad \text{for very negative inputs} \tag{14.11}$$

$$V_{out} = V_{in} - V_{BE1} \quad \text{for very positive inputs.} \tag{14.12}$$

That is, for negative inputs, Q_2 shifts the signal *up*, and for positive inputs, Q_1 shifts the signal *down*. Figure 14.4 plots the resulting characteristic.

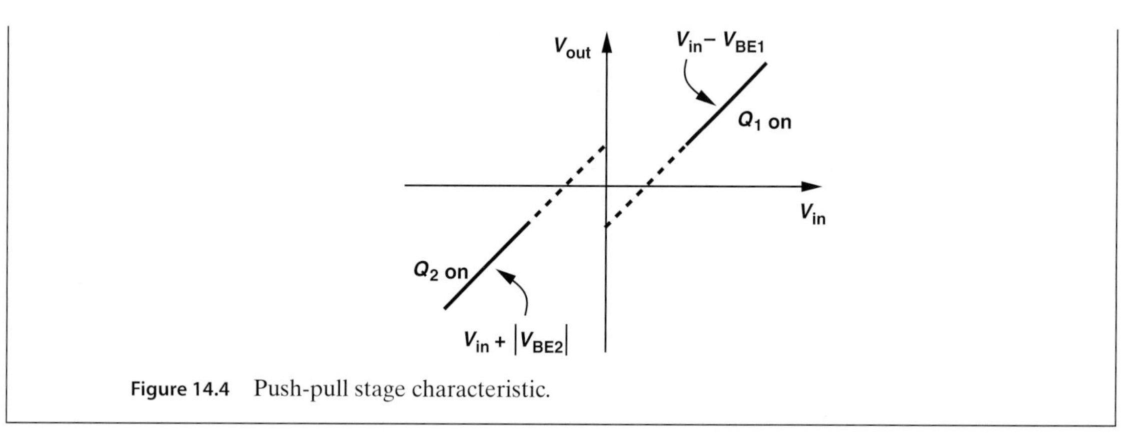

Figure 14.4 Push-pull stage characteristic.

Exercise Repeat the above example for a CMOS output stage.

What happens as V_{in} approaches zero? The rough characteristic in Fig. 14.4 suggests that the two segments cannot meet if they must remain linear. In other words, the overall characteristic inevitably incurs nonlinearity and resembles that shown in Fig. 14.5, exhibiting a "dead zone" around $V_{in} = 0$.

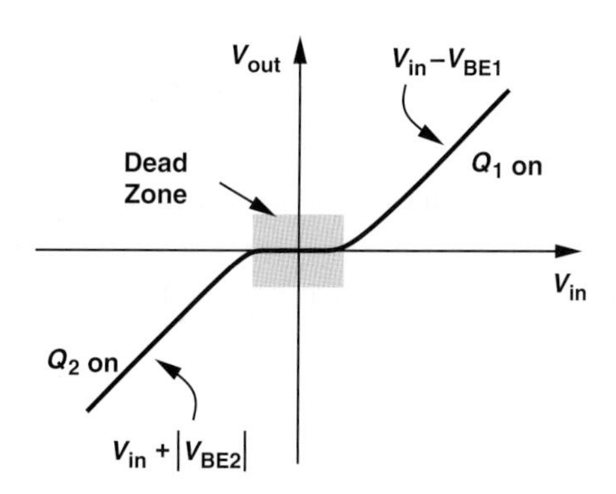

Figure 14.5 Push-pull stage characteristic with dead zone.

Why does the circuit suffer from a dead zone? We make two observations. First, Q_1 and Q_2 cannot be on simultaneously: for Q_1 to be on, $V_{in} > V_{out}$, but for Q_2, $V_{in} < V_{out}$. Second, if $V_{in} = 0$, V_{out} must also be zero. This can be proved by contradiction. For example, if $V_{out} > 0$ (Fig. 14.6), then the current V_{out}/R_L must be provided by Q_1 (from V_{CC}), requiring $V_{BE1} > 0$ and hence $V_{out} = V_{in} - V_{BE1} < 0$. That is, for $V_{in} = 0$, both transistors are off.

Now suppose V_{in} begins to increase from zero. Since V_{out} is initially at zero, V_{in} must reach at least $V_{BE} \approx 600 - 700$ mV before Q_1 turns on. The output therefore remains at zero for $V_{in} < 600$ mV, exhibiting the dead zone depicted in Fig. 14.5. Similar observations apply to the dead zone for $V_{in} < 0$.

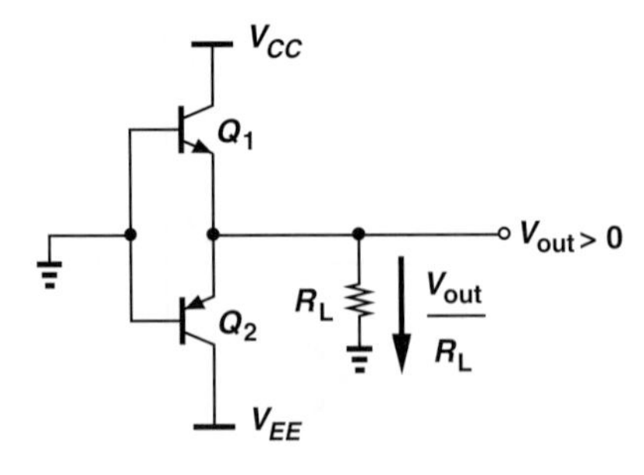

Figure 14.6 Push-pull stage with zero input voltage.

Example 14-3 Sketch the small-signal gain for the characteristic of Fig. 14.5 as a function of V_{in}.

Solution The gain (slope) is near unity for very negative or positive inputs, falling to zero in the dead zone. Figure 14.7 plots the result.

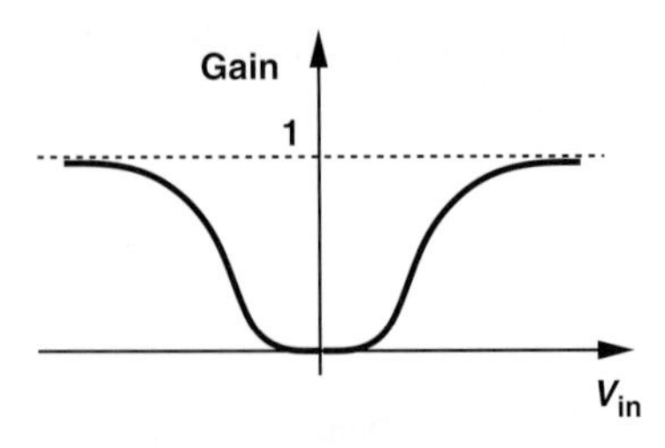

Figure 14.7 Gain of push-pull stage as a function of input.

Exercise Repeat the above example if R_L is replaced with an ideal current source.

In summary, the simple push-pull stage of Fig. 14.3(a) operates as a *pnp* or *npn* emitter follower for sufficiently negative or positive inputs, respectively, but turns off for -600 mV $< V_{in} < +600$ mV. The resulting dead zone substantially distorts the input signal.

Example 14-4 Suppose we apply a sinusoid with a peak amplitude of 4 V to the push-pull stage of Fig. 14.3(a). Sketch the output waveform.

Solution For V_{in} well above 600 mV, either Q_1 or Q_2 serves as an emitter follower, thus producing a reasonable sinusoid at the output. Under this condition, the plot in Fig. 14.5 indicates that $V_{out} = V_{in} + |V_{BE2}|$ or $V_{in} - V_{BE1}$. Within the dead zone, however, $V_{out} \approx 0$. Illustrated in Fig. 14.8, V_{out} exhibits distorted "zero crossings." We also say the circuit suffers from "crossover distortion."

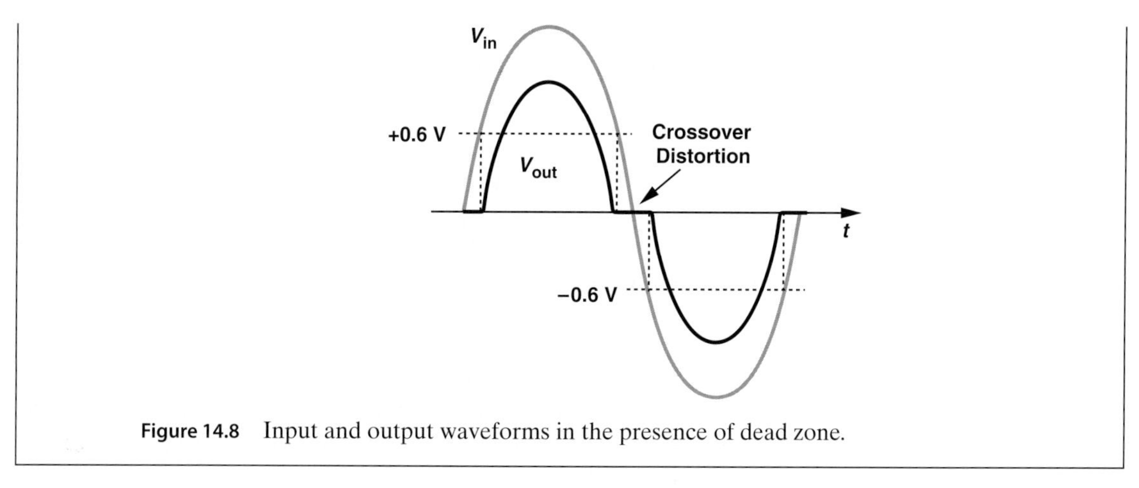

Figure 14.8 Input and output waveforms in the presence of dead zone.

Exercise Sketch the input and output waveforms if the push-pull stage incorporates NMOS and PMOS transistors with zero threshold voltage.

14.4 IMPROVED PUSH-PULL STAGE

14.4.1 Reduction of Crossover Distortion

In most applications, the distortion introduced by the simple push-pull stage of Fig. 14.3(a) proves unacceptable. We must therefore devise methods of reducing or eliminating the dead zone.

The distortion in the push-pull stage fundamentally arises from the input connections: since the bases of Q_1 and Q_2 in Fig. 14.3(a) are shorted together, the two

Did you know?

The ear is quite sensitive to the clipping or crossover distortion produced by followers or push-pull stages. With a sinusoidal input at frequency f_{in}, the output is no longer a pure sinusoid, exhibiting harmonics at 2 f_{in}, 3 f_{in}, etc. One may then say that the ear is sensitive to harmonics. In fact, many audio enthusiasts, especially guitarists, prefer power amplifiers that use vacuum tubes (also called "valves") because these devices have softer clipping characteristics than do transistors.

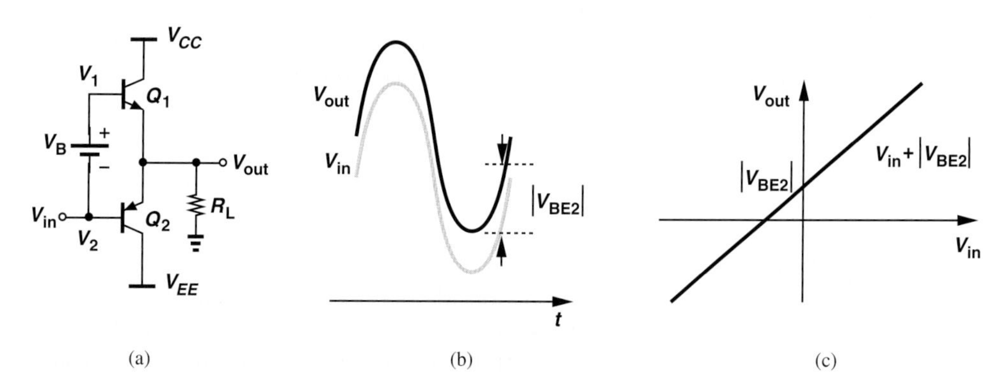

Figure 14.9 (a) Addition of voltage source to remove the dead zone, (b) input and output waveforms, (c) input/output characteristic.

transistors cannot remain on simultaneously around $V_{in} = 0$. We surmise that the circuit can be modified as shown in Fig. 14.9(a), where a battery of voltage V_B is inserted between the two bases. What is the required value of V_B? If Q_1 is to remain on, then $V_1 = V_{out} + V_{BE1}$. Similarly, if Q_2 is to remain on, then $V_2 = V_{out} - |V_{BE2}|$. Thus,

$$V_B = V_1 - V_2 \tag{14.13}$$

$$= V_{BE1} + |V_{BE2}|. \tag{14.14}$$

We say V_B must be approximately equal to $2V_{BE}$ (even though V_{BE1} and $|V_{BE2}|$ may not be equal).

With the connection of V_{in} to the base of Q_2, $V_{out} = V_{in} + |V_{BE2}|$; i.e., the output is a replica of the input but shifted up by $|V_{BE2}|$. If the base-emitter voltages of Q_1 and Q_2 are assumed constant, both transistors remain on for all input and output levels, yielding the waveforms depicted in Fig. 14.9(b). The dead zone is thus eliminated. The input/output characteristic is illustrated in Fig. 14.9(c).

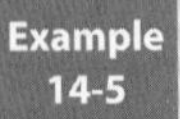

Study the behavior of the stage shown in Fig. 14.10(a). Assume $V_B \approx 2V_{BE}$.

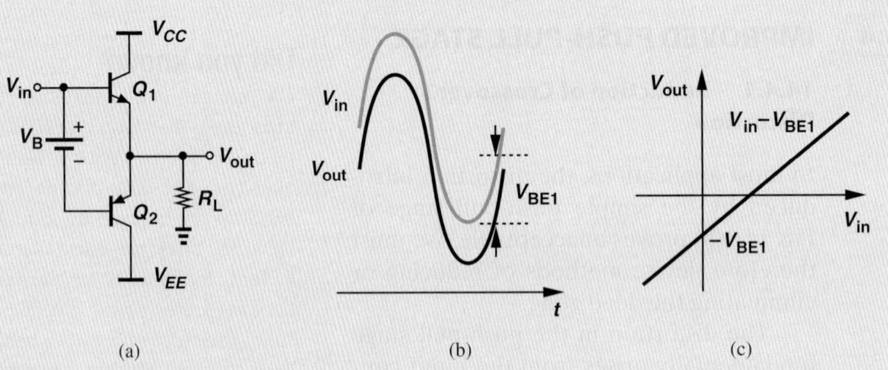

Figure 14.10 (a) Push-pull stage with input applied to base of Q_1, (b) input and output waveforms, (c) input/output characteristic.

Solution In this circuit, both transistors remain on simultaneously, and $V_{out} = V_{in} - V_{BE1}$. Thus, the output is a replica of the input but shifted down. Figures 14.10(b) and (c) plot the waveforms and the input/output characteristic, respectively.

Exercise What happens if $V_B \approx V_{BE}$?

We now determine how the battery V_B in Fig. 14.9(a) must be implemented. Since $V_B = V_{BE1} + |V_{BE2}|$, we naturally decide that two *diodes* placed in series can provide the required voltage drop, thereby arriving at the topology shown in Fig. 14.11(a). Unfortunately, the diodes carry no current here (why?), exhibiting a zero voltage drop. This difficulty is readily overcome by adding a current source on top [Fig. 14.11(b)]. Now, I_1 provides both the bias current of D_1 and D_2 and the base current of Q_1.

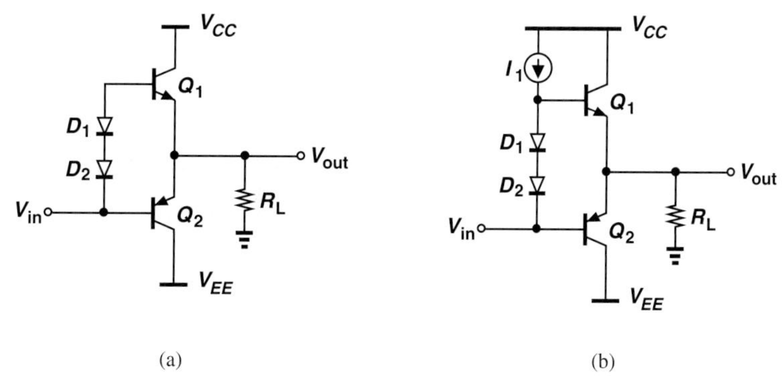

Figure 14.11 (a) Use of diodes as a voltage source, (b) addition of current source I_1 to bias the diodes.

Example 14-6 Determine the current flowing through the voltage source V_{in} in Fig. 14.11(b).

Solution The current flowing through D_1 and D_2 is equal to $I_1 - I_{B1}$ (Fig. 14.12). The voltage source must sink both this current and the base current of Q_2. Thus, the total current flowing through this source is equal to $I_1 - I_{B1} + |I_{B2}|$.

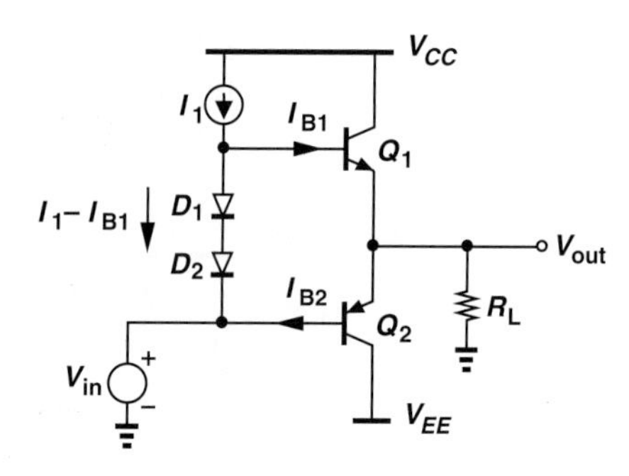

Figure 14.12 Circuit to examine base currents.

Exercise Sketch the current flowing through the voltage source as a function of V_{in} as V_{in} goes from −4 V to +4 V. Assume $\beta_1 = 25$, $\beta_2 = 15$, and $R_L = 8\ \Omega$.

Example 14-7 Under what condition are the base currents of Q_1 and Q_2 in Fig. 14.11(b) equal? Assume $\beta_1 = \beta_2 \gg 1$.

Solution We must seek the condition $I_{C1} = |I_{C2}|$. As depicted in Fig. 14.13, this means no current flows through R_L and $V_{out} = 0$. As V_{out} departs from zero, the current flowing through R_L is provided by either Q_1 or Q_2 and hence $I_{C1} \neq |I_{C2}|$ and $I_{B1} \neq |I_{B2}|$. Thus, the base currents are equal only at $V_{out} = 0$.

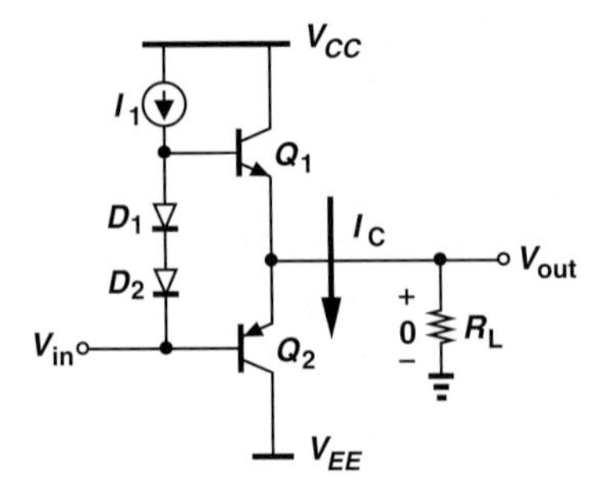

Figure 14.13 Stage with zero output voltage.

Exercise Repeat the above example if $\beta_1 = 2\beta_2$.

Did you know?

Can we use a push-pull stage at the output of op amps? Yes, many general-purpose op amps, such as the 741, include a push-pull stage in order to drive loads more efficiently than an emitter follower can. Integrated op amps, on the other hand, must operate with low supply voltages and avoid push-pull stages. After all, a 1-V power supply cannot accommodate two V_{BE}'s.

Example 14-8

Study the behavior of the circuit illustrated in Fig. 14.14, where I_2 absorbs both the bias current of D_2 and I_{B2}.

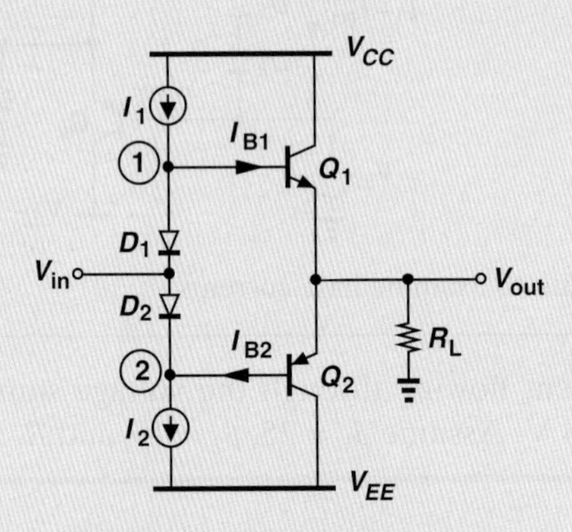

Figure 14.14 Stage with input applied to midpoint of diodes.

Solution Here, we have $V_1 = V_{in} + V_{D1}$ and $V_{out} = V_1 - V_{BE1}$. If $V_{D1} \approx V_{BE1}$, then $V_{out} \approx V_{in}$, exhibiting no level shift with respect to the input. Also, the current flowing through D_1 is equal to $I_1 - I_{B1}$ and that through D_2 equal to $I_2 - |I_{B2}|$. Thus, if $I_1 = I_2$ and $I_{B1} \approx I_{B2}$, the input voltage source need not sink or source a current for $V_{out} = 0$, a point of contrast with respect to the circuit of Fig. 14.12.

Exercise Sketch the current provided by the input source as a function of V_{in} as V_{in} goes from −4 V to +4 V. Assume $\beta_1 = 25$, $\beta_2 = 15$, and $R_L = 8\ \Omega$.

14.4.2 Addition of CE Stage

The two current sources in Fig. 14.14 can be realized with *pnp* and *npn* transistors as depicted in Fig. 14.15(a). We may therefore decide to apply the input signal to the base of one of the current sources so as to obtain a greater gain. Illustrated in Fig. 14.15(b), the idea is to employ Q_4 as a common-emitter stage, thus providing voltage gain from V_{in} to the base of Q_1 and Q_2.[3] The CE stage is called the "predriver."

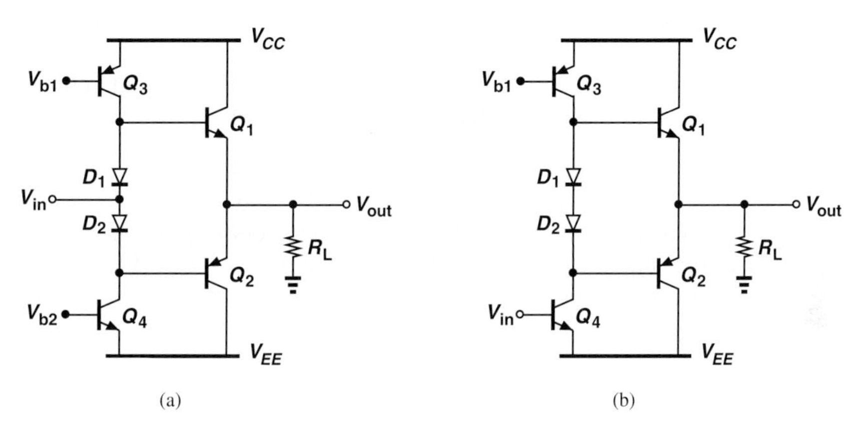

Figure 14.15 (a) Push-pull stage with realization of current sources, (b) stage with input applied to base of Q_4.

The push-pull circuit of Fig. 14.15(b) is used extensively in high-power output stages and merits a detailed analysis. We must first answer the following questions: (1) Given the bias currents of Q_3 and Q_4, how do we determine those of Q_1 and Q_2? (2) What is the overall voltage gain of the circuit in the presence of a load resistance R_L?

To answer the first question, we assume $V_{out} = 0$ for bias calculations and also $I_{C4} = I_{C3}$. If $V_{D1} = V_{BE1}$ and $V_{D2} = |V_{BE2}|$, then $V_A = 0$ (why?). With both V_{out} and V_A at zero, the circuit can be reduced to that shown in Fig. 14.16(a), revealing a striking resemblance to a current mirror. In fact, since

$$V_{D1} = V_T \ln \frac{|I_{C3}|}{I_{S,D1}}, \tag{14.15}$$

where the base current of Q_1 is neglected and $I_{S,D1}$ denotes the saturation current of D_1, and since $V_{BE1} = V_T \ln(I_{C1}/I_{S,Q1})$, we have

$$I_{C1} = \frac{I_{S,Q1}}{I_{S,D1}} |I_{C3}|. \tag{14.16}$$

To establish a well-defined value for $I_{S,Q1}/I_{S,D1}$, diode D_1 is typically realized as a diode-connected bipolar transistor [Fig. 14.16(b)] in integrated circuits. Note that a similar analysis can be applied to the bottom half of the circuit, namely, Q_4, D_2, and Q_2.

The second question can be answered with the aid of the simplified circuit shown in Fig. 14.17(a), where $V_A = \infty$ and $2r_D$ represents the total small-signal resistance of D_1

[3]If the dc level of V_{in} is close to V_{CC}, then V_{in} is applied to the base of Q_3 instead.

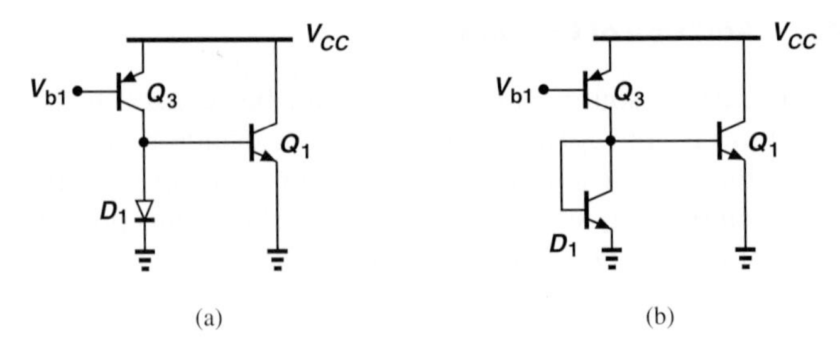

Figure 14.16 (a) Simplified diagram of a push-pull stage, (b) illustration of current mirror action.

and D_2. Let us assume for simplicity that $2r_D$ is relatively small and $v_1 \approx v_2$, further reducing the circuit to that illustrated in Fig. 14.17(b),[4] where

$$\frac{v_{out}}{v_{in}} = \frac{v_N}{v_{in}} \cdot \frac{v_{out}}{v_N}. \tag{14.17}$$

Now, Q_1 and Q_2 operate as two emitter followers *in parallel*, i.e., as a single transistor having an r_π equal to $r_{\pi 1}||r_{\pi 2}$ and a g_m equal to $g_{m1} + g_{m2}$ [Fig. 14.17(c)]. For this circuit, we have $v_{\pi 1} = v_{pi2} = v_N - v_{out}$ and

$$\frac{v_{out}}{R_L} = \frac{v_N - v_{out}}{r_{\pi 1}||r_{\pi 2}} + (g_{m1} + g_{m2})(v_N - v_{out}). \tag{14.18}$$

It follows that

$$\frac{v_{out}}{v_N} = \frac{1 + (g_{m1} + g_{m2})(r_{\pi 1}||r_{\pi 2})}{\dfrac{r_{\pi 1}||r_{\pi 2}}{R_L} + 1 + (g_{m1} + g_{m2})(r_{\pi 1}||r_{\pi 2})}. \tag{14.19}$$

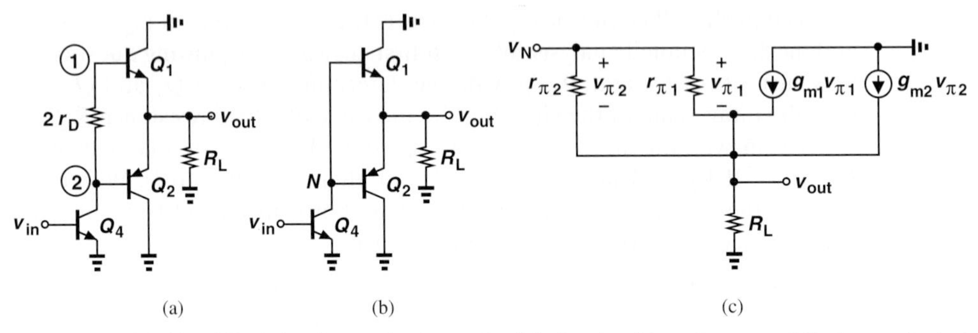

Figure 14.17 (a) Simplified circuit to calculate gain, (b) circuit with resistance of diodes neglected, (c) small-signal model.

Multiplying the numerator and denominator by R_L, dividing both by $1 + (g_{m1} + g_{m2}) \times (r_{\pi 1}||r_{\pi 2})$, and assuming $(g_{m1} + g_{m2})(r_{\pi 1}||r_{\pi 2}) \gg 1$, we obtain

$$\frac{v_{out}}{v_N} = \frac{R_L}{R_L + \dfrac{1}{g_{m1} + g_{m2}}}, \tag{14.20}$$

a result expected of a follower transistor having a transconductance of $g_{m1} + g_{m2}$.

[4]It is important to note that this representation is valid for signals but not for biasing.

To compute v_N/v_{in}, we must first derive the impedance seen at node N, R_N. From the circuit of Fig. 14.17(c), the reader can show that

$$R_N = (g_{m1} + g_{m2})(r_{\pi 1}||r_{\pi 2})R_L + r_{\pi 1}||r_{\pi 2}. \tag{14.21}$$

(Note that for $I_{C1} = I_{C2}$ and $\beta_1 = \beta_2$, this expression reduces to the input impedance of a simple emitter follower.) Consequently,

$$\frac{v_{out}}{v_{in}} = -g_{m4}[(g_{m1} + g_{m2})(r_{\pi 1}||r_{\pi 2})R_L + r_{\pi 1}||r_{\pi 2}]\frac{R_L}{R_L + \dfrac{1}{g_{m1} + g_{m2}}} \tag{14.22}$$

$$= -g_{m4}(r_{\pi 1}||r_{\pi 2})(g_{m1} + g_{m2})R_L. \tag{14.23}$$

Example 14-9 Calculate the output impedance of the circuit shown in Fig. 14.18(a). For simplicity, assume $2r_D$ is small.

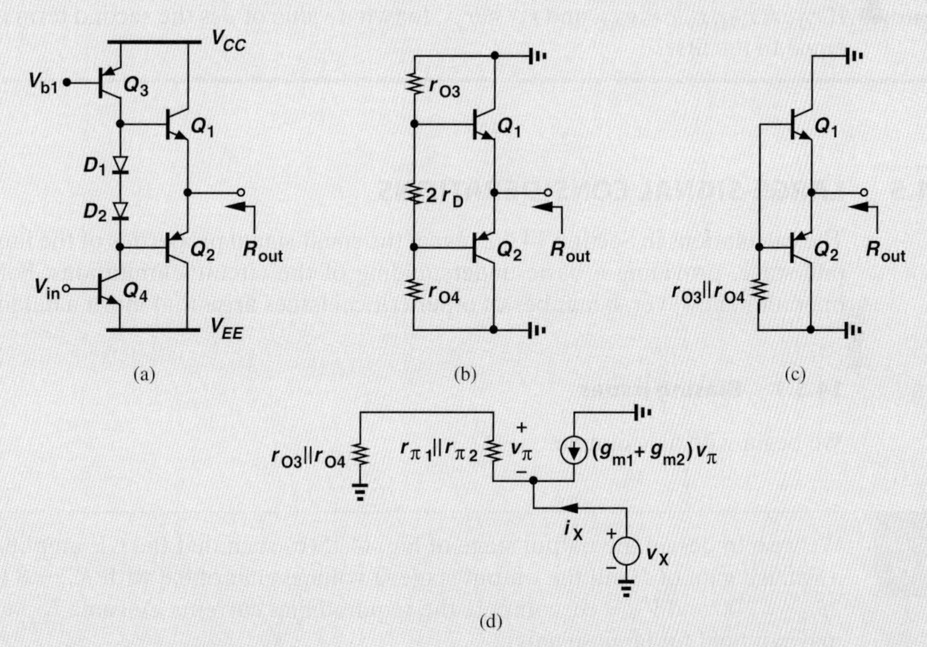

Figure 14.18 (a) Circuit for calculation of output impedance, (b) simplified diagram, (c) further simplification, (d) small-signal model.

Solution The circuit can be reduced to that in Fig. 14.18(b), and, with $2r_D$ negligible, to that in Fig. 14.18(c). Utilizing the composite model illustrated in Fig. 14.17(c), we obtain the small-signal equivalent circuit of Fig. 14.18(d), where $V_A = \infty$ for Q_1 and Q_2 but not for Q_3 and Q_4. Here, $r_{O3}||r_{O4}$ and $r_{pi1}||r_{pi2}$ act as a voltage divider:

$$v_\pi = -v_X\frac{r_{\pi 1}||r_{\pi 2}}{r_{\pi 1}||r_{\pi 2} + r_{O3}||r_{O4}}. \tag{14.24}$$

A KCL at the output node gives

$$i_X = \frac{v_X}{r_{\pi 1}||r_{\pi 2} + r_{O3}||r_{O4}} + (g_{m1} + g_{m2})v_X \frac{r_{\pi 1}||r_{\pi 2}}{r_{\pi 1}||r_{\pi 2} + r_{O3}||r_{O4}}. \quad (14.25)$$

It follows that

$$\frac{v_X}{i_X} = \frac{r_{\pi 1}||r_{\pi 2} + r_{O3}||r_{O4}}{1 + (g_{m1} + g_{m2})(r_{\pi 1}||r_{\pi 2})} \quad (14.26)$$

$$\approx \frac{1}{g_{m1} + g_{m2}} + \frac{r_{O3}||r_{O4}}{(g_{m1} + g_{m2})(r_{\pi 1}||r_{\pi 2})}, \quad (14.27)$$

if $(g_{m1} + g_{m2})(r_{\pi 1}||r_{\pi 2}) \gg 1$.

The key observation here is that the second term in Eq. (14.27) may raise the output impedance considerably. As a rough approximation, we assume $r_{O3} \approx r_{O4}$, $g_{m1} \approx g_{m2}$, and $r_{\pi 1} \approx r_{\pi 2}$, concluding that the second term is on the order of $(r_O/2)/\beta$. This effect becomes particularly problematic in discrete design because power transistors typically suffer from a low β.

Exercise If $r_{O3} \approx r_{O4}$, $g_{m1} \approx g_{m2}$, and $r_{\pi 1} \approx r_{\pi 2}$, for what value of β is the second term in Eq. (14.27) equal to the first?

14.5 LARGE-SIGNAL CONSIDERATIONS

The calculations in Section 14.4.2 reveal the small-signal properties of the improved push-pull stage, providing a basic understanding of the circuit's limitations. For large-signal operation, however, a number of other critical issues arise that merit a detailed study.

14.5.1 Biasing Issues

We begin with an example.

Example 14-10 We wish to design the output stage of Fig. 14.15(b) such that the CE amplifier provides a voltage gain of 5 and the output stage, a voltage gain of 0.8 with $R_L = 8\ \Omega$. If $\beta_{npn} = 2\beta_{pnp} = 100$ and $V_A = \infty$, compute the required bias currents. Assume $I_{C1} \approx I_{C2}$ (which may not hold for large signals).

Solution From Eq. (14.20) for $v_{out}/v_N = 0.8$, we have

$$g_{m1} + g_{m2} = \frac{1}{2\ \Omega}. \quad (14.28)$$

With $I_{C1} \approx I_{C2}$, $g_{m1} \approx g_{m2} \approx (4\ \Omega)^{-1}$ and hence $I_{C1} \approx I_{C2} \approx 6.5$ mA. Also, $r_{\pi 1}||r_{\pi 2} = 133\ \Omega$. Setting Eq. (14.20) equal to $-5 \times 0.8 = -4$, we have $I_{C4} \approx 195\ \mu$A. We thus bias Q_3 and Q_4 at 195 μA.

Exercise Repeat the above example if the second stage must provide a voltage gain of 2.

The above example entails moderate current levels in the milliampere range. But what happens if the stage must deliver a swing of, say, 4 V_P to the load? Each output transistor must now provide a peak current of 4 V/8 Ω = 500 mA. Does the design in Example 14-10 deliver such voltage and current swings without difficulty? Two issues must be considered here. First, a bipolar transistor carrying 500 mA requires a large emitter area, about 500 times the emitter area of a transistor capable of handling 1 mA.[5] Second, with a β of 100, the peak base current reaches as high as 5 mA! How is this base current provided? Transistor Q_1 receives maximum base current if Q_4 turns off so that the entire I_{C3} flows to the base of Q_1. Referring to the bias currents obtained in Example 14-10, we observe that the circuit can be simplified as shown in Fig. 14.19 for the peak of positive half cycles. With an I_{C3} of only 195 μA, the collector current of Q_1 cannot exceed roughly 100 × 195 μA = 19.5 mA, far below the desired value of 500 mA.

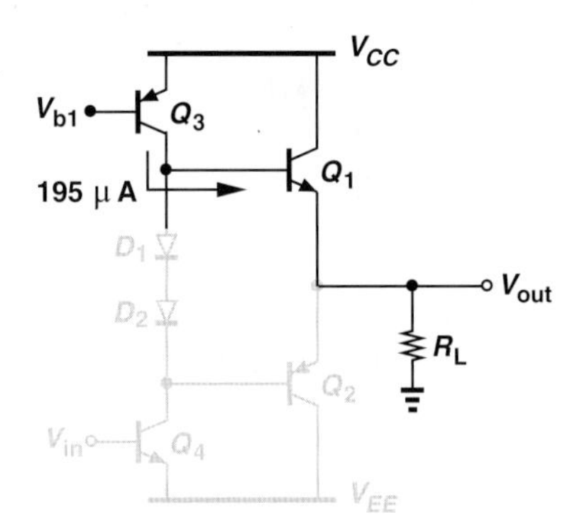

Figure 14.19 Calculation of maximum available current.

The key conclusion here is that, while achieving a small-signal gain of near unity with an 8-Ω load, the output stage can deliver an output swing of only 195 mA × 8 Ω = 156 mV_P. We must therefore provide a much higher base current, requiring proportionally higher bias currents in the predriver stage. In practice, power transistors suffer from a low β, e.g., 20, exacerbating this issue.

14.5.2 Omission of *PNP* Power Transistor

PNP power transistors typically suffer from both a low current gain and a low f_T, posing serious constraints on the design of output stages. Fortunately, it is possible to combine an *npn* device with a *pnp* transistor to improve the performance.

Consider the common-emitter *npn* transistor, Q_2, depicted in Fig. 14.20(a). We wish to modify the circuit so that Q_2 exhibits the characteristics of an *emitter follower*. To this end, we add the *pnp* device Q_3 as shown in Fig. 14.20(b) and prove that the Q_2-Q_3 combination operates as an emitter follower. With the aid of the small-signal equivalent circuit illustrated in Fig. 14.20(c) ($V_A = \infty$), and noting that the collector current of Q_3 serves as the base current of Q_2, and hence $g_{m2}v_{\pi 2} = -\beta_2 g_{m3}(v_{in} - v_{out})$, we write a KCL at the output node:

$$-g_{m3}(v_{in} - v_{out})\beta_2 + \frac{v_{out} - v_{in}}{r_{\pi 3}} - g_{m3}(v_{in} - v_{out}) = -\frac{v_{out}}{R_L}. \tag{14.29}$$

[5]For a given emitter area, if the collector current exceeds a certain level, "high-level injection" occurs, degrading the transistor performance, e.g., β.

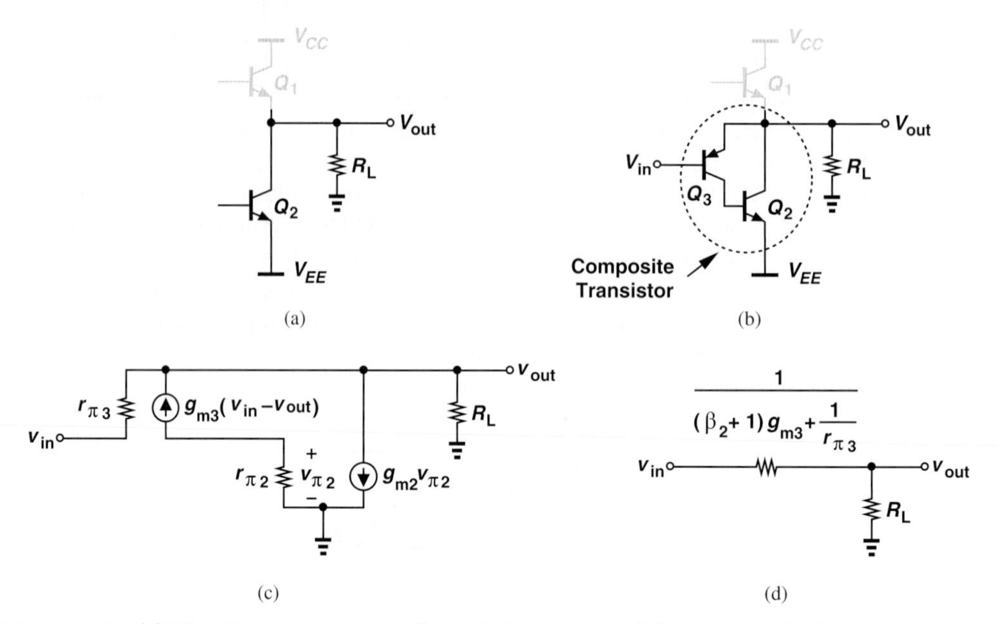

Figure 14.20 (a) Use of an *npn* transistor for pull-down action, (b) composite device, (c) small-signal model, (d) equivalent circuit.

Note that the first term on the left-hand side represents the collector current of Q_2. It follows that

$$\frac{v_{out}}{v_{in}} = \frac{R_L}{R_L + \frac{1}{(\beta_2 + 1)g_{m3} + \frac{1}{r_{\pi 3}}}}. \tag{14.30}$$

In analogy with the standard emitter follower (Chapter 5), we can view this result as voltage division between two resistances of values $[(\beta_2 + 1)g_{m3} + 1/r_{\pi 3}]^{-1}$ and R_L [Fig. 14.20(d)]. That is, the output resistance of the circuit (excluding R_L) is given by

$$R_{out} = \frac{1}{(\beta_2 + 1)g_{m3} + \frac{1}{r_{\pi 3}}} \tag{14.31}$$

$$\approx \frac{1}{(\beta_2 + 1)g_{m3}} \tag{14.32}$$

because $1/r_{\pi 3} = g_{m3}/\beta_3 \ll (\beta_2 + 1)g_{m3}$. If Q_3 alone operated as a follower, the output impedance would be quite higher ($1/g_{m3}$).

The results expressed by Eqs. (14.30) and (14.32) are quite interesting. The voltage gain of the circuit can approach unity if the output resistance of the Q_2-Q_3 combination, $[(\beta_2 + 1)g_{m3}]^{-1}$, is much less than R_L. In other words, the circuit acts as an emitter follower but with an output impedance that is lower by a factor of $\beta_2 + 1$.

Example 14-11 Compute the input impedance of the circuit shown in Fig. 14.20(c).

Solution Since the current drawn from the input is equal to $(v_{in} - v_{out})/r_{\pi 3}$, we have from Eq. (14.30)

$$i_{in} = \frac{1}{r_{\pi 3}}\left(v_{in} - v_{in}\frac{R_L}{R_L + \dfrac{1}{(\beta_2+1)g_{m3}}}\right), \tag{14.33}$$

where $1/r_{pi3}$ is neglected with respect to $(\beta_2 + 1)g_{m3}$. It follows that

$$\frac{v_{in}}{i_{in}} = \beta_3(\beta_2 + 1)R_L + r_{\pi 3}. \tag{14.34}$$

Interestingly, R_L is boosted by a factor of $\beta_3(\beta_2 + 1)$ as seen at the input—as if the Q_2-Q_3 combination provides a current gain of $\beta_3(\beta_2 + 1)$.

Exercise Calculate the output impedance if $r_{O3} < \infty$.

The circuit of Fig. 14.20(b) proves superior to a single *pnp* emitter follower. However, it also introduces an additional pole at the base of Q_2. Also, since Q_3 carries a small current, it may not be able to charge and discharge the large capacitance at this node. To alleviate these issues, a constant current source is typically added as shown in Fig. 14.21 so as to raise the bias current of Q_3.

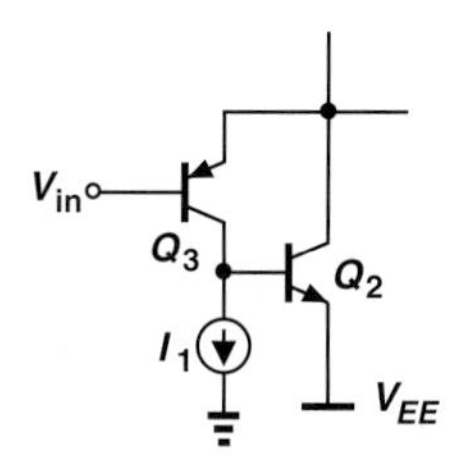

Figure 14.21 Addition of current source to improve speed of composite device.

Example 14-12 Compare the two circuits depicted in Fig. 14.22 in terms of the minimum allowable input voltage and the minimum achievable output voltage. (Bias components are not shown.) Assume the transistors do not enter saturation.

Solution In the emitter follower of Fig. 14.22(a), V_{in} can be as low as zero so that Q_2 operates at the edge of saturation. The minimum achievable output level is thus equal to $|V_{BE2}| \approx 0.8$ V.

In the topology of Fig. 14.22(b), V_{in} can be equal to the collector voltage of Q_3, which is equal to V_{BE2} with respect to ground. The output is then given by $V_{in} + |V_{BE3}| = V_{BE2} + |V_{BE3}| \approx 1.6$ V, a disadvantage of this topology. We say the circuit "wastes" one V_{BE} in voltage headroom.

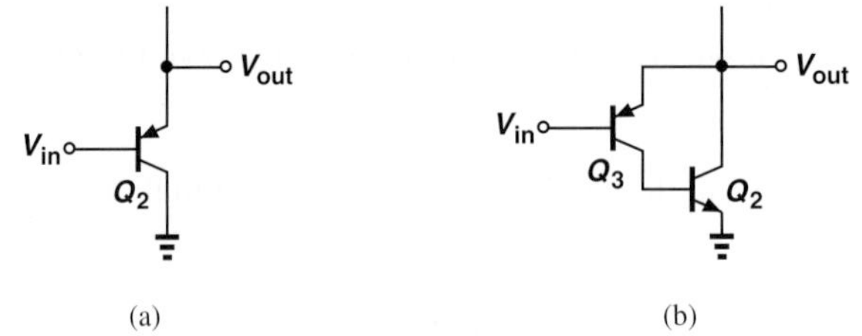

(a) (b)

Figure 14.22 Voltage headroom for (a) simple follower, (b) composite device.

Exercise Explain why Q_2 cannot enter saturation in this circuit.

14.5.3 High-Fidelity Design

Even with the diode branch present in Fig. 14.15(b), the output stage introduces some distortion in the signal. Specifically, since the collector currents of Q_1 and Q_2 vary considerably in each half cycle, so does their transconductance. As a result, the voltage division relationship governing the emitter follower, Eq. (14.20), exhibits an *input-dependent* behavior: as V_{out} becomes more positive, g_{m1} rises (why?) and A_v comes closer to unity. Thus, the circuit experiences nonlinearity.

In most applications, especially in audio systems, the distortion produced by the push-pull stage proves objectionable. For this reason, the circuit is typically embedded in a negative feedback loop to reduce the nonlinearity. Figure 14.23 illustrates a conceptual realization, where amplifier A_1, the output stage, and resistors R_1 and R_2 form a noninverting amplifier (Chapter 8), yielding $V_{out} \approx (1 + R_1/R_2)V_{in}$ and significantly lowering the distortion. However, owing to the multiple poles contributed by A_1 and the push-pull stage, this topology may become unstable, necessitating frequency compensation (Chapter 12).

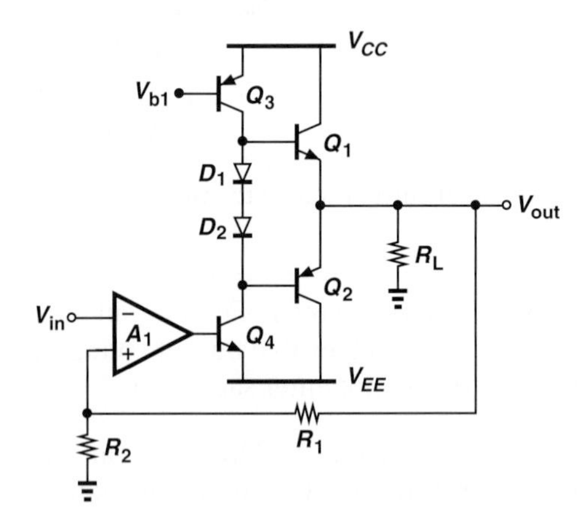

Figure 14.23 Reduction of distortion by feedback.

14.6 SHORT-CIRCUIT PROTECTION

Electronic devices and circuits may experience "hostile" conditions during handling, assembly, and usage. For example, a person attempting to connect a speaker to a stereo may accidentally short the amplifier output to ground while the stereo is on. The high currents flowing through the circuit under this condition may permanently damage the output transistors. Thus, a means of limiting the short-circuit current is necessary.

The principle behind short-circuit protection is to sense the output current (by a small series resistor) and reduce the base drive of the output transistors if this current exceeds a certain level. Shown in Fig. 14.24 is an example, where Q_S senses the voltage drop across r, "stealing" some of the base current of Q_1 as V_r approaches 0.7 V. For example, if $r = 0.25\ \Omega$, then the emitter current of Q_1 is limited to about 2.8 A.

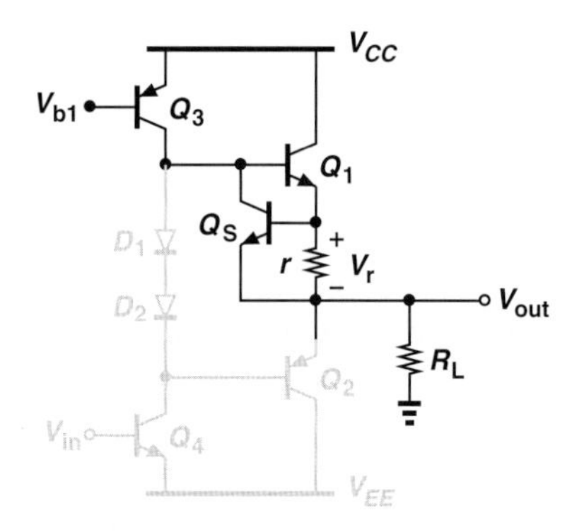

Figure 14.24 Short-circuit protection.

Did you know?

Many audio power amplifiers are implemented as "hybrids." A hybrid consists of a small board (called the "substrate") with the components mounted on it in chip form (rather than in packaged form). Shown below is the simplified circuit diagram of the STK4200, a 100-W audio PA manufactured by Sanyo. We recognize most of the circuit configurations here. Transistors Q_1-Q_4 form a differential pair with active load, serving as a simple op amp. Transistor Q_5 realizes the second gain stage, and C_c performs Miller compensation. Devices Q_6 and Q_7 along with R_1 and R_2 comprise a "V_{BE} multiplier," creating a dc voltage difference of about four V_{BE}'s between A and B. Finally, Q_8-Q_{11} and D_1 operate as a push-pull stage. The circuit is placed in a negative feedback loop in a manner similar to a noninverting amplifier.

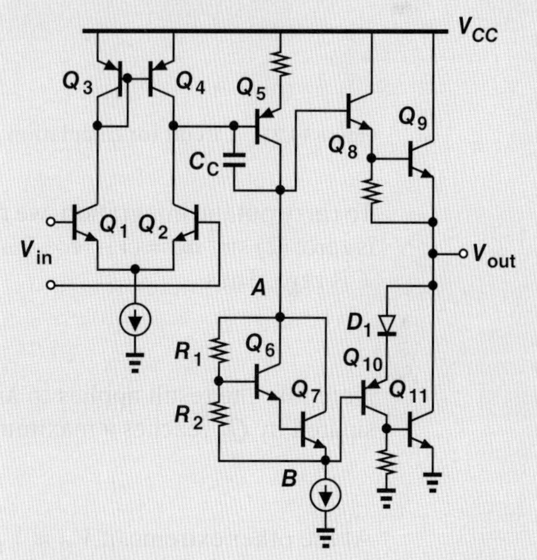

Simplified circuit of high-power audio amplifier STK4200.

The protection scheme of Fig. 14.24 suffers from several drawbacks. First, resistor r directly raises the output impedance of the circuit. Second, the voltage drop across r under normal operating condition, e.g., 0.5–0.6 V, reduces the maximum output voltage swing. For example, if the base voltage of Q_1 approaches V_{CC}, then $V_{out} = V_{CC} - V_{BE1} - V_r \approx V_{CC} - 1.4$ V.

14.7 HEAT DISSIPATION

Since the output transistors in a power amplifier carry a finite current and sustain a finite voltage for part of the period, they consume power and hence heat up. If the junction temperature rises excessively, the transistor may be irreversibly damaged. Thus, the "power rating" (the maximum allowable power dissipation) of each transistor must be chosen properly in the design process.

14.7.1 Emitter Follower Power Rating

Let us first compute the power dissipated by Q_1 in the simple emitter follower of Fig. 14.25, assuming that the circuit delivers a sinusoid of $V_P \sin \omega t$ to a load resistance R_L. Recall from Section 14.2 that $I_1 \geq V_P/R_L$ to ensure V_{out} can reach $-V_P$. The instantaneous power dissipated by Q_1 is given by $I_C \cdot V_{CE}$ and its average value (over one period) equals:

$$P_{av} = \frac{1}{T}\int_0^T I_C \cdot V_{CE}\, dt, \tag{14.35}$$

where $T = 2\pi/\omega$. Since $I_C \approx I_E = I_1 + V_{out}/R_L$ and $V_{CE} = V_{CC} - V_{out} = V_{CC} - V_P \sin \omega t$, we have

$$P_{av} = \frac{1}{T}\int_0^T \left(I_1 + \frac{V_P \sin \omega t}{R_L}\right)(V_{CC} - V_P \sin \omega t)\, dt. \tag{14.36}$$

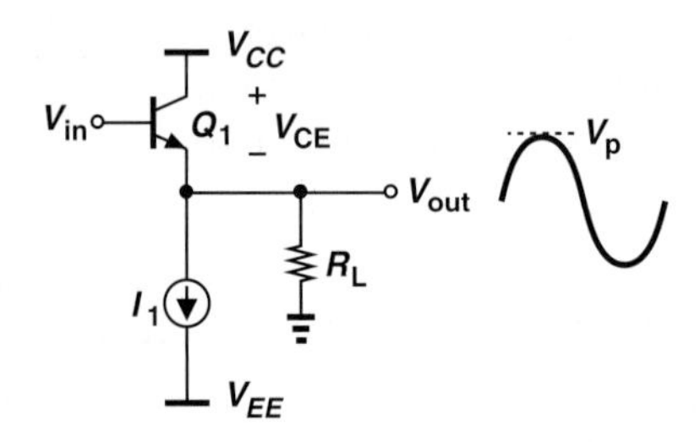

Figure 14.25 Circuit for calculation of follower power dissipation.

To carry out the integration, we note that (1) the average value of $\sin \omega t$ over one period T is zero; (2) $\sin^2 \omega t = (1 - \cos 2\omega t)/2$; and (3) the average value of $\cos 2\omega t$ over one period T is zero. Thus,

$$P_{av} = I_1\left(V_{CC} - \frac{V_P}{2}\right). \tag{14.37}$$

Note that the result applies to any type of transistor (why?). Interestingly, the power dissipated by Q_1 reaches a maximum in the *absence* of signals, i.e., with $V_P = 0$:

$$P_{av,\max} = I_1 V_{CC}. \tag{14.38}$$

At the other extreme, if $V_P \approx V_{CC}$,[6] then

$$P_{av} \approx I_1 \frac{V_{CC}}{2}. \tag{14.39}$$

Example 14-13 Calculate the power dissipated by the current source I_1 in Fig. 14.25.

Solution The current source sustains a voltage equal to $V_{out} - V_{EE} = V_P \sin \omega t - V_{EE}$. Thus,

$$P_{I1} = \frac{1}{T}\int_0^T I_1(V_P \sin \omega t - V_{EE})\, dt \tag{14.40}$$

$$= -I_1 V_{EE}. \tag{14.41}$$

[6]Here, V_{BE} is neglected with respect to V_P.

The value is, of course, positive because $V_{EE} < 0$ to accommodate negative swings at the output.

Exercise Explain why the power delivered by V_{EE} is equal to that dissipated by I_1.

14.7.2 Push-Pull Stage Power Rating

We now determine the power dissipated by the output transistors in the push-pull stage (Fig. 14.26). To simplify our calculations, we assume that each transistor carries a negligible current around $V_{out} = 0$ and turns off for half of the period. If $V_{out} = V_P \sin \omega t$, then $I_{RL} = (V_P/R_L) \sin \omega t$ but only for half of the cycle. Also, the collector-emitter voltage of Q_1 is given by $V_{CC} - V_{out} = V_{CC} - V_P \sin \omega t$. The average power dissipated in Q_1 is therefore equal to

$$P_{av} = \frac{1}{T} \int_0^{T/2} V_{CE} \cdot I_C dt \tag{14.42}$$

$$= \frac{1}{T} \int_0^{T/2} (V_{CC} - V_P \sin \omega t) \left(\frac{V_P}{R_L} \sin \omega t \right) dt, \tag{14.43}$$

where $T = 1/\omega$, and β is assumed large enough to allow the approximation $I_C \approx I_E$. Expanding the terms inside the integral and noting that

$$\int_0^{T/2} \cos 2\omega t \, dt = 0, \tag{14.44}$$

we have

$$P_{av} = \frac{1}{T} \int_0^{T/2} \frac{V_{CC} V_P}{R_L} \sin \omega t \, dt - \frac{1}{T} \int_0^{T/2} \frac{V_P^2}{2R_L} dt \tag{14.45}$$

$$= \frac{V_{CC} V_P}{\pi R_L} - \frac{V_P^2}{4R_L} \tag{14.46}$$

$$= \frac{V_P}{R_L} \left(\frac{V_{CC}}{\pi} - \frac{V_P}{4} \right). \tag{14.47}$$

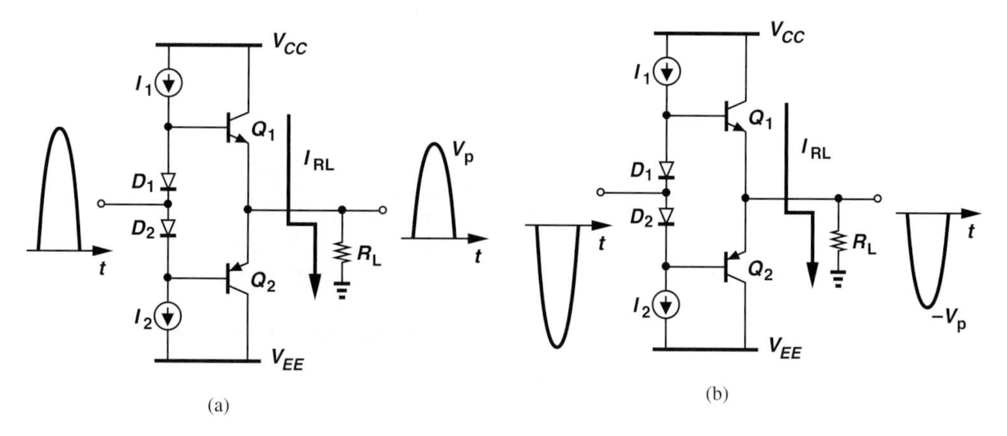

Figure 14.26 Push-pull stage during (a) positive half cycle and (b) negative half cycle.

For example, if $V_P = 4$ V, $R_L = 8\ \Omega$, and $V_{CC} = 6$ V, then Q_1 dissipates 455 mW. Transistor Q_2 also consumes this amount of power if $|V_{EE}| = V_{CC}$.

Equation (14.47) indicates that for $V_P \approx 0$ or $V_P \approx 4V_{CC}/\pi$, the power dissipated in Q_1 approaches zero, suggesting that P_{av} must reach a maximum between these extremes. Differentiating P_{av} with respect to V_P and equating the result to zero, we have $V_P = 2V_{CC}/\pi$ and hence

$$P_{av,\max} = \frac{V_{CC}^2}{\pi^2 R_L}. \tag{14.48}$$

Example 14-14 A student observes from Eq. (14.47) that $P_{av} = 0$ if $V_P = 4V_{CC}/\pi$, concluding that this choice of peak swing is the best because it minimizes the power "wasted" by the transistor. Explain the flaw in the student's reasoning.

Solution With a supply voltage of V_{CC}, the circuit cannot deliver a peak swing of $4V_{CC}/\pi (> V_{CC})$. It is thus impossible to approach $P_{av} = 0$.

Exercise Compare the power dissipated in Q_1 with that delivered to R_L for $V_P = 2V_{CC}/\pi$.

Did you know?

Heat sinks are not just necessary for power transistors. Any other semiconductor device that dissipates a high power may require a heat sink to avoid rising to high temperatures. For example, some microprocessors consume 100 W and must be attached to heat sinks. In fact, one or two decades ago, there was a doom's day prediction that the advance of microprocessors would eventually stop due to our limited ability to remove heat from them. Fortunately, sophisticated heat sinks were introduced that could solve this problem.

The problem of heat dissipation becomes critical for power levels greater than a few hundred milliwatts. The physical size of transistors is quite small, e.g., 1 mm × 1 mm × 0.5 mm, and so is the surface area through which the heat can exit. Of course, from the perspective of device capacitances and cost, the transistor(s) must not be enlarged just for the purpose of heat dissipation. It is therefore desirable to employ other means that increase the conduction of the heat. Called a "heat sink" and shown in Fig. 14.27, one such means is formed as a metal structure (typically aluminum) with a large surface area and attached to the transistor or chip package. The idea is to "sink" the heat from the package and subsequently dissipate it through a much larger surface area.

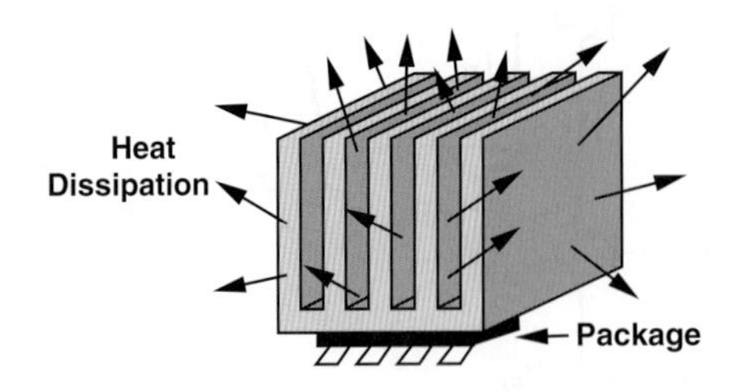

Figure 14.27 Example of heat sink.

14.7.3 Thermal Runaway

As described above, the output transistors in a power amplifier experience elevated temperatures. Even in the presence of a good heat sink, the push-pull stage is susceptible to a phenomenon called "thermal runaway" that can damage the devices.

To understand this effect, let us consider the conceptual stage depicted in Fig. 14.28(a), where the battery $V_B \approx 2V_{BE}$ eliminates the dead zone and $V_{out} = 0$. What happens as the junction temperature of Q_1 and Q_2 rises? It can be proved that, for a given base-emitter voltage, the collector current increases with temperature. Thus, with a constant V_B, Q_1 and Q_2 carry increasingly larger currents, dissipating greater power. The higher dissipation in turn further raises the junction temperature and hence the collector currents, etc. The resulting positive feedback continues until the transistors are damaged.

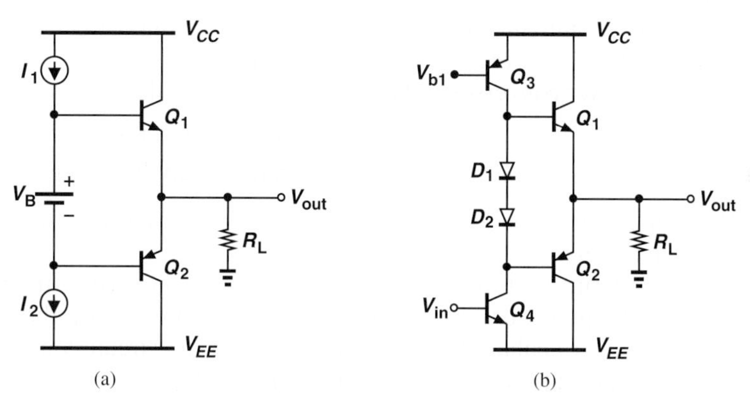

Figure 14.28 (a) Runaway in the presence of constant voltage shift, V_B, (b) use of diodes to avoid runaway.

Interestingly, the use of diode biasing [Fig. 14.28(b)] can prohibit thermal runaway. If the diodes experience the same temperature change as the output transistors, then $V_{D1} + V_{D2}$ *decreases* as the temperature rises (because their bias current is relatively constant), thereby stabilizing the collector currents of Q_1 and Q_2. From another perspective, since D_1 and Q_1 form a current mirror, I_{C1} is a constant multiple of I_1 if D_1 and Q_1 remain at the same temperature. More accurately, we have for D_1 and D_2:

$$V_{D1} + V_{D2} = V_T \ln \frac{I_{D1}}{I_{S,D1}} + V_T \ln \frac{I_{D2}}{I_{S,D2}} \tag{14.49}$$

$$= V_T \ln \frac{I_{D1} I_{D2}}{I_{S,D1} I_{S,D2}}. \tag{14.50}$$

Similarly, for Q_1 and Q_2:

$$V_{BE1} + V_{BE2} = V_T \ln \frac{I_{C1}}{I_{S,Q1}} + V_T \ln \frac{I_{C2}}{I_{S,Q2}} \tag{14.51}$$

$$= V_T \ln \frac{I_{C1} I_{C2}}{I_{S,Q1} I_{S,Q2}}. \tag{14.52}$$

Equating (14.50) and (14.52) and assuming the same value of V_T (i.e., the same temperature) for both expressions, we write

$$\frac{I_{D1}I_{D2}}{I_{S,D1}I_{S,D2}} = \frac{I_{C1}I_{C2}}{I_{S,Q1}I_{S,Q2}}. \tag{14.53}$$

Since $I_{D1} \approx I_{D2} \approx I_1$, $I_{C1} \approx I_{C2}$, we observe that I_{C1} and I_{C2} "track" I_1 so long as the I_S values (which are temperature-dependent) also track.

14.8 EFFICIENCY

Since power amplifiers draw large amounts of power from the supply voltage, their "efficiency" proves critical in most applications. In a cellphone, for example, a PA delivering 1 W to the antenna may pull several watts from the battery, a value comparable to the power dissipation of the rest of the circuits in the phone.

The "power conversion efficiency" of a PA, η, is defined as

$$\eta = \frac{\text{Power Delivered to Load}}{\text{Power Drawn from Supply Voltage}}. \tag{14.54}$$

Thus, an efficiency of 30% in the above cellphone translates to a power drain of 3.33 W from the battery.

It is instructive to compute the efficiency of the two output stages studied in this chapter. The procedure consists of three steps: (1) calculate the power delivered to the load, P_{out}; (2) calculate the power dissipated in the circuit components (e.g., the output transistors), P_{ckt}; (3) determine $\eta = P_{out}/(P_{out} + P_{ckt})$.

14.8.1 Efficiency of Emitter Follower

With the results obtained in Section 14.7.1 the efficiency of emitter followers can be readily calculated. Recall that the power dissipated by Q_1 is equal to

$$P_{av} = I_1\left(V_{CC} - \frac{V_P}{2}\right) \tag{14.55}$$

or that consumed by I_1 is

$$P_{I1} = -I_1 V_{EE}. \tag{14.56}$$

If $V_{EE} = -V_{CC}$, the total power "wasted" in the circuit is given by

$$P_{ckt} = I_1\left(2V_{CC} - \frac{V_P}{2}\right). \tag{14.57}$$

It follows that

$$\eta = \frac{P_{out}}{P_{out} + P_{ckt}} \tag{14.58}$$

$$= \frac{\dfrac{V_P^2}{2R_L}}{\dfrac{V_P^2}{2R_L} + I_1\left(2V_{CC} - \dfrac{V_P}{2}\right)}. \tag{14.59}$$

For proper operation, I_1 must be at least equal to V_P/R_L, yielding

$$\eta = \frac{V_P}{4V_{CC}}. \tag{14.60}$$

That is, the efficiency reaches a maximum of 25% as V_P approaches V_{CC}.[7] Note that this result holds only if $I_1 = V_P/R_L$.

Example 14-15 An emitter follower designed to deliver a peak swing of V_P operates with an output swing of $V_P/2$. Determine the efficiency of the circuit.

Solution Since the circuit is originally designed for an output swing of V_P, we have $V_{CC} = -V_{EE} \approx V_P$ and $I_1 = V_P/R_L$. Replacing V_P with $V_{CC}/2$ and I_1 with V_{CC}/R_L in (14.59), we have

$$\eta = \frac{1}{15}. \tag{14.61}$$

This low efficiency results because both the supply voltages and I_1 are "overdesigned."

Exercise At what peak swing does the efficiency reach 20%?

The maximum efficiency of 25% proves inadequate in many applications. For example, a stereo amplifier delivering 50 W to a speaker would consume 150 W in the output stage, necessitating very large (and expensive) heat sinks.

14.8.2 Efficiency of Push-Pull Stage

In Section 14.7.2, we determined that each of Q_1 and Q_2 in Fig. 14.26 consumes a power of

$$P_{av} = \frac{V_P}{R_L}\left(\frac{V_{CC}}{\pi} - \frac{V_P}{4}\right). \tag{14.62}$$

Thus,

$$\eta = \frac{\dfrac{V_P^2}{2R_L}}{\dfrac{V_P^2}{2R_L} + \dfrac{2V_P}{R_L}\left(\dfrac{V_{CC}}{\pi} - \dfrac{V_P}{4}\right)} \tag{14.63}$$

$$= \frac{\pi}{4}\frac{V_P}{V_{CC}}. \tag{14.64}$$

The efficiency thus reaches a maximum of $\pi/4 = 78.5\%$ for $V_P \approx V_{CC}$, a much more attractive result than that of the emitter follower. For this reason, push-pull stages are very common in many applications, e.g., audio amplifiers.

[7]This is only an approximation because V_{CE} or the voltage across I_1 cannot go to zero.

Example 14-16 Calculate the efficiency of the stage depicted in Fig. 14.26. Assume $I_1(=I_2)$ is chosen so as to allow a peak swing of V_P at the output. Also, $V_{CC}=-V_{EE}$.

Solution Recall from Section 14.3 that I_1 must be at least equal to $(V_P/R_L)/\beta$. Thus, the branch consisting of I_1, D_1, D_2, and I_2 consumes a power of $2V_{CC}(V_P/R_L)/\beta$, yielding an overall efficiency of:

$$\eta = \frac{\dfrac{V_P^2}{2R_L}}{\dfrac{2V_PV_{CC}}{\pi R_L}+\dfrac{2V_PV_{CC}}{\beta R_L}} \tag{14.65}$$

$$= \frac{1}{4}\frac{V_P}{\dfrac{V_{CC}}{\pi}+\dfrac{V_P}{\beta}}. \tag{14.66}$$

We should note the approximation made here: with the diode branch present, we can no longer assume that each output transistor is on for only half of the cycle. That is, Q_1 and Q_2 consume slightly greater power, leading to a lower η.

Exercise If $V_P \le V_{CC}$ and $\beta \gg \pi$, what is the maximum efficiency that can be achieved in this circuit?

14.9 POWER AMPLIFIER CLASSES

The emitter follower and push-pull stages studied in this chapter exhibit distinctly different properties: in the former, the transistor conducts current throughout the entire cycle, and the efficiency is low; in the latter, each transistor is on for about half of the cycle, and the efficiency is high. These observations lead to different "PA classes."

An amplifier in which each transistor is on for the entire cycle is called a "class A" stage [Fig. 14.29(a)]. Exemplified by the emitter follower studied in Section 14.2, class A circuits suffer from a low efficiency but provide a higher linearity than other classes.

A "class B stage" is one in which each transistor conducts for half of the cycle [Fig. 14.29(b)]. The simple push-pull circuit of Fig. 14.3(a) is an example of class B stages.[8] The efficiency in this case reaches $\pi/4 = 78.5\%$, but the distortion is rather high.

As a compromise between linearity and efficiency, PAs are often configured as "class AB" stages, wherein each transistor remains on for greater than half a cycle [Fig. 14.29(c)]. The modified push-pull stage of Fig. 14.11(a) serves as an example of class AB amplifiers.

Many other classes of PAs have been invented and used in various applications. Examples include classes C, D, E, and F. The reader is referred to more advanced texts [1].

[8]The dead zone in this stage in fact allows conduction for slightly *less* than half cycle for each transistor.

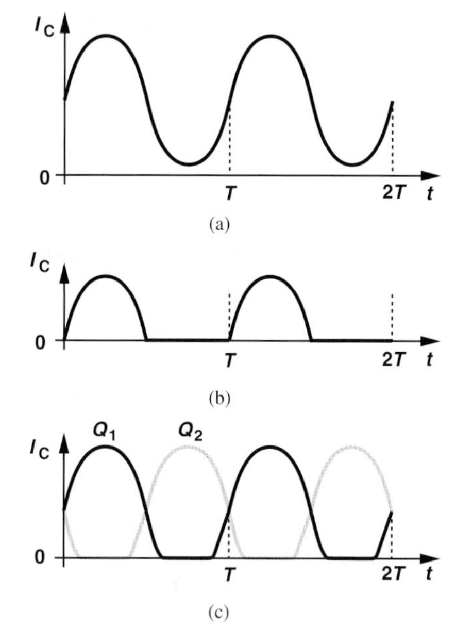

Figure 14.29 Collector waveforms for (a) class A, (b) class B, and (c) class C operation.

14.10 CHAPTER SUMMARY

- Power amplifiers deliver high power levels and large signal swings to relatively low load impedances.
- Both the distortion and efficiency of power amplifiers are critical parameters.
- While providing a low small-signal output impedance, emitter followers operate poorly under large-signal conditions.
- A push-pull stage consists of an *npn* follower and a *pnp* follower. Each device conducts for about half of the input cycle, improving the efficiency.
- A simple push-pull stage suffers from a dead zone, across which neither transistor conducts and the small-signal gain falls to zero.
- The crossover distortion resulting from the dead zone can be reduced by biasing the push-pull transistors for a small quiescent current.
- With two diodes placed between the bases of the push-pull transistors and a CE amplifier preceding this stage, the circuit can provide a high output power with moderate distortion.
- The output *pnp* transistors is sometimes replaced with a composite *pnp-npn* structure that provides a higher current gain.
- In low-distortion applications, the output stage may be embedded in a negative feedback loop to suppress the nonlinearity.
- A push-pull stage operating at high temperatures may suffer from thermal runaway, whereby the elevated temperatures allow the output transistors to draw higher currents, which in turn makes them dissipate even more.

- The power efficiency of emitter followers rarely reaches 25%, whereas it can approach 79% for push-pull stages.
- Power amplifiers can operate in different classes depending on across what fraction of the input cycle the transistor conducts. These classes include class A and class B.

PROBLEMS

Unless otherwise stated, assume $V_{CC} = +5$ V, $V_{EE} = -5$ V, $V_{BE,on} = 0.8$ V, $I_S = 6 \times 10^{-17}$, $V_A = \infty$, $R_L = 8\ \Omega$, and $\beta \gg 1$ in the following problems.

Sec. 14.2 Emitter Follower as Power Amplifier

14.1. Consider the emitter follower shown in Fig. 14.30. We wish to deliver a power of 0.5 W to $R_L = 8\ \Omega$.

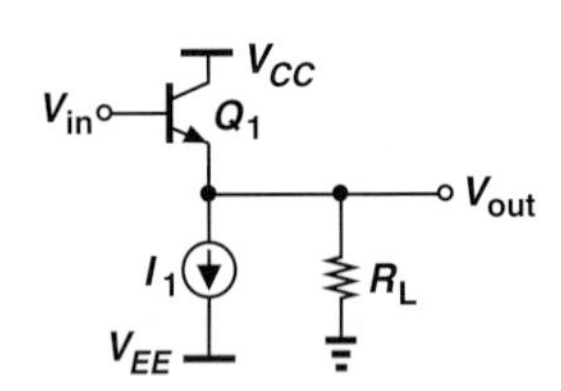

Figure 14.30

(a) Determine I_1 for a small-signal voltage gain of 0.8.

(b) Writing $g_m = I_C/V_T$, calculate the voltage gain as V_{in} reaches its positive peak value (corresponding to an output power of 0.5 W). The change in voltage gain represents nonlinearity.

14.2. For the emitter follower of Fig. 14.30, we can express the voltage gain as

$$A_v = \frac{I_C R_L}{I_C R_L + V_T}. \quad (14.67)$$

Recall from Section 14.2 that $I_1 \geq V_P/R_L$, where V_P denotes the peak voltage delivered to R_L

(a) Assuming $I_1 = V_P/R_L$ and $V_P \gg V_T$, determine an expression for A_V if the swings are small.

(b) Now assume the output reaches a peak of V_P. Calculate the small-signal voltage gain in this region and obtain the change with respect to the result in part (a).

14.3. A student designs the emitter follower of Fig. 14.30 for a small-signal voltage gain of 0.7 and a load resistance of 4 Ω. For a sinusoidal input, estimate the largest average power that can be delivered to the load without turning Q_1 off.

14.4. Suppose the follower of Fig. 14.30 is designed for a small-signal voltage gain of A_v. Determine the maximum power that can be delivered to the load without turning Q_1 off. SOL

14.5. Due to a manufacturing error, the load of an emitter follower is tied between the output and V_{CC} (Fig. 14.31). Assume $I_{S1} = 5 \times 10^{-17}$ A, $R_L = 8\ \Omega$ and $I_1 = 20$ mA.

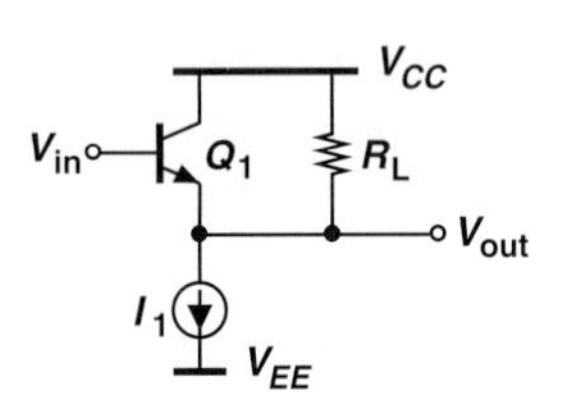

Figure 14.31

(a) Calculate the bias current of Q_1 for $V_{in} = 0$.

(b) For what value of V_{in} does Q_1 carry only 1% of I_1?

14.6. The emitter follower of Fig. 14.30 senses a sinusoidal input with a peak amplitude of 1 V. Assume $I_{S1} = 6 \times 10^{-17}$ A, $R_L = 8\ \Omega$ and $I_1 = 25$ mA.

(a) Calculate V_{BE} for $V_{in} = +1$ V and $V_{in} = -1$ V. (This change is a measure of the nonlinearity.)

(b) Noting that $V_{out} = V_{in} - V_{BE}$, sketch the output waveform.

14.7. In the circuit of Problem 14.6, determine the maximum input swing for which V_{BE} changes by less than 10 mV from the positive peak to the negative peak. Determine the ratio of ΔV_{BE} and peak-to-peak output swing as a measure of the nonlinearity.

Sec. 14.3 Push-Pull Stage

14.8. For the push-pull stage of Fig. 14.3(a), sketch the base current of Q_1 as a function of V_{in}.

14.9. Consider the push-pull stage depicted in Fig. 14.32, where a current source, I_1, is tied from the output node to ground.

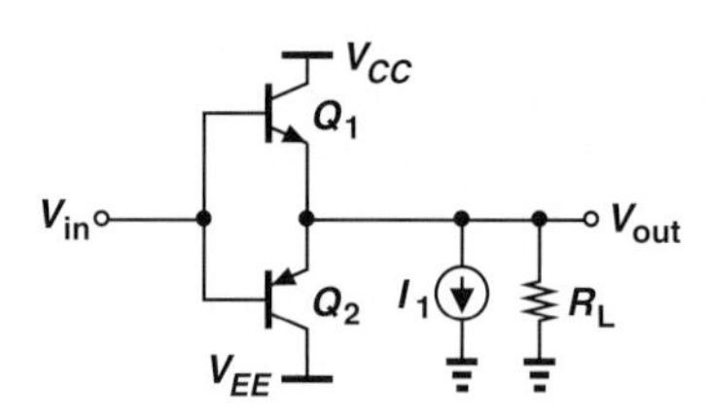

Figure 14.32

(a) Suppose $V_{in} = 0$. Determine a relationship between I_1 and R_L to guarantee that Q_1 is on, i.e., $V_{BE1} \approx$ 800 mV.

(b) With the condition obtained in (a), calculate the input voltage at which Q_2 turns on, i.e., $V_{BE2} \approx 800$ mV.

14.10. Explain how I_1 in Fig. 14.32 alters the input/output characteristic and the dead zone.

SOL *__14.11.__ The circuit shown in Fig. 14.33 precedes the output *npn* device with an emitter follower. Sketch the input/output characteristic and estimate the width of the dead zone.

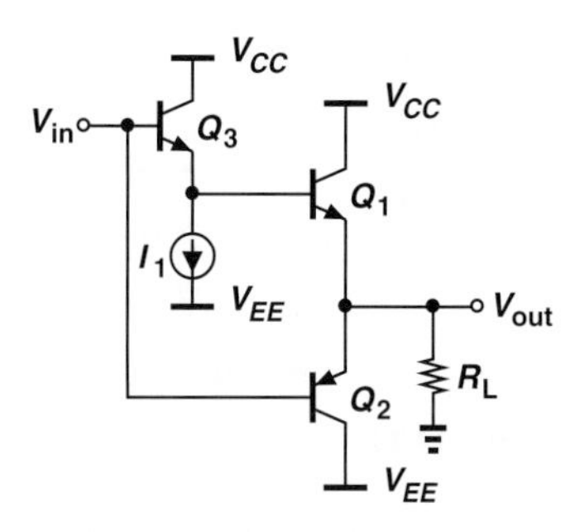

Figure 14.33

*__14.12.__ Repeat Problem 14.11 for the stage depicted in Fig. 14.34.

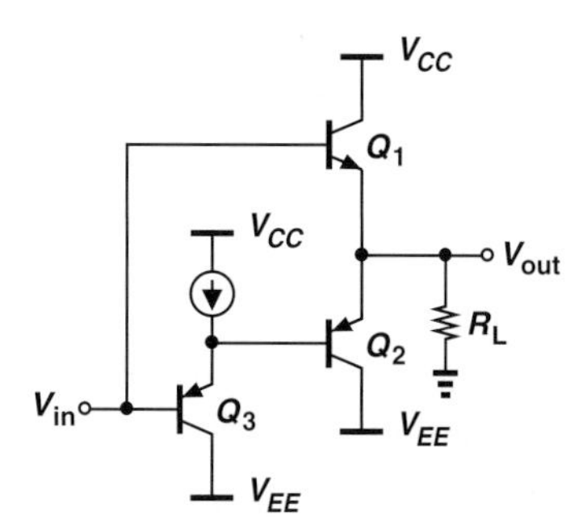

Figure 14.34

14.13. Figure 14.35 shows a CMOS realization of the push-pull stage.

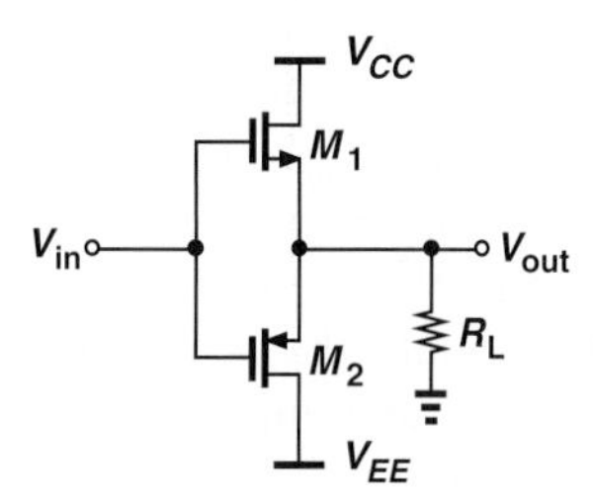

Figure 14.35

(a) Sketch the input/output characteristic of the circuit.

(b) Determine the small-signal voltage gain for the positive and negative inputs outside the dead zone.

14.14. A large sinusoidal input is applied to the circuit of Fig. 14.33. Sketch the output waveform.

14.15. Repeat Problem 14.14 for the circuit of Fig. 14.34.

Sec. 14.4 Improved Push-Pull Stage

14.16. Consider the push-pull stage illustrated in Fig. 14.36, where $V_B \approx V_{BE}$ (rather than $2V_{BE}$).

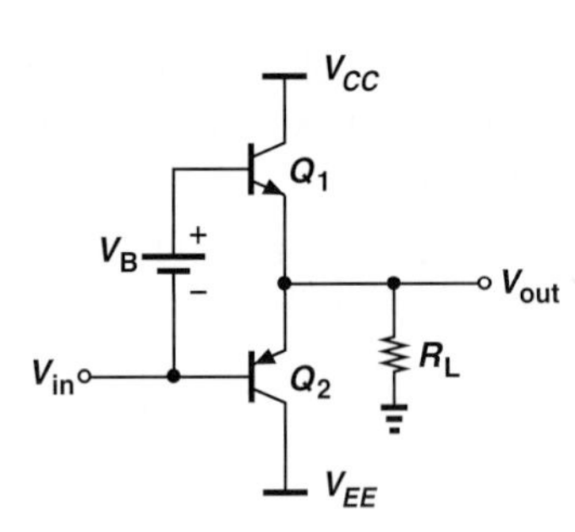

Figure 14.36

(a) Sketch the input/output characteristic.

(b) Sketch the output waveform for a sinusoidal input.

14.17. In the push-pull stage of Fig. 14.36, $I_{S1} = 5 \times 10^{-17}$ A and $I_{S2} = 8 \times 10^{-17}$ A. Calculate the value of V_B so as to establish a bias current of 5 mA in Q_1 and Q_2 (for $V_{out} = 0$).

14.18. Suppose the design in Problem 14.17 operates with a peak input swing of 2 V and $R_L = 8\ \Omega$.

(a) Calculate the small-signal voltage gain for $V_{out} \approx 0$.

(b) Use the gain obtained in (a) to estimate the output voltage swing.

(c) Estimate the peak collector current of Q_1 assuming that Q_2 still carries 5 mA.

SOL **14.19.** The stage of Fig. 14.36 is designed with $V_B \approx 2V_{BE}$ to suppress crossover distortion. Sketch the collector currents of Q_1 and Q_2 as a function of V_{in}.

14.20. Consider the circuit shown in Fig. 14.37, where V_B is placed in series with the emitter of Q_1. Sketch the input/output characteristic.

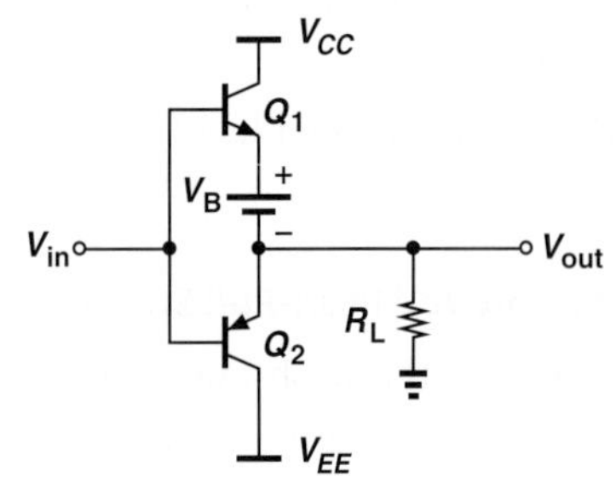

Figure 14.37

14.21. In the circuit of Fig. 14.11(b), we have $V_{BE1} + |V_{BE2}| = V_{D1} + V_{D2}$. Under what condition can we write $I_{C1}I_{C2} = I_{D1}I_{D2}$?

14.22. The circuit of Fig. 14.11(b) is designed with $I_1 = 1$ mA and $I_{S,Q1} = I_{S,Q2} = 16I_{S,D1} = 16I_{S,D2}$. Calculate the bias current of Q_1 and Q_2 (for $V_{out} = 0$). (Hint: $V_{BE1} + |V_{BE2}| = V_{D1} + V_{D2}$.)

14.23. The stage of Fig. 14.11(b) must be designed for a bias current of 5 mA in Q_1 and Q_2 (for $V_{out} = 0$). If $I_{S,Q1} = I_{S,Q2} = 8I_{S,D1} = 8I_{S,D2}$, determine the required value of I_1. (Hint: $V_{BE1} + |V_{BE2}| = V_{D1} + V_{D2}$.)

14.24. In the output stage of Fig. 14.11(b), $I_1 = 2$ mA, $I_{S,Q1} = 8I_{S,D1}$, and $I_{S,Q2} = 16I_{S,D2}$. Determine the bias current of Q_1 and Q_2 (for $V_{out} = 0$). (Hint: $V_{BE1} + |V_{BE2}| = V_{D1} + V_{D2}$.)

*__14.25.__ A critical problem in the design of the push-pull stage shown in Fig. 14.11(b) is the temperature difference between the diodes and the output transistors because the latter consume much greater power and tend to rise to a higher temperature. Noting that $V_{BE1} + |V_{BE2}| = V_{D1} + V_{D2}$, explain how a temperature difference introduces an error in the bias currents of Q_1 and Q_2.

*__14.26.__ Determine the small-signal voltage gain of the stage depicted in Fig. 14.38. Assume I_1 is an ideal current source. Neglect the incremental resistance of D_1 and D_2. SOL

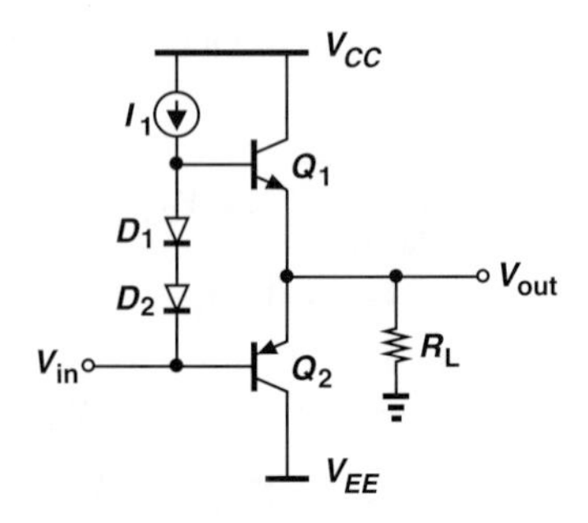

Figure 14.38

**__14.27.__ Repeat Problem 14.26 but do not neglect the incremental resistance of D_1 and D_2.

14.28. The output stage of Fig. 14.38 must achieve a small-signal voltage gain of 0.8. Determine the required bias current of Q_1 and Q_2. Neglect the incremental resistance of D_1 and D_2.

14.29. Compute the small-signal input impedance of the output stage depicted in Fig. 14.38 if the incremental resistance of D_1 and D_2 is not neglected. From the result, determine the condition under which this resistance is neglected.

14.30. The stage of Fig. 14.15(b) is designed with a bias current of 1 mA in Q_3 and Q_4 and 10 mA in Q_1 and Q_2. Assuming $\beta_1 = 40$, $\beta_2 = 20$, and $R_L = 8\ \Omega$, calculate the small-signal voltage gain if the incremental resistance of D_1 and D_2 is neglected.

*__14.31.__ Noting that $g_{m1} \approx g_{m2}$ in Fig. 14.15(b), prove that Eq. (14.23) reduces to

$$\frac{v_{out}}{v_{in}} = -\frac{2\beta_1\beta_2}{\beta_1 + \beta_2} g_{m4} R_L. \qquad (14.68)$$

Interestingly, the gain remains independent of the bias current of Q_1 and Q_2.

14.32. The output stage of Fig. 14.15(b) must provide a small-signal voltage gain of 4. Assuming $\beta_1 = 40$, $\beta_2 = 20$, and $R_L = 8\ \Omega$, determine the required bias current of Q_3 and Q_4. Neglect the incremental resistance of D_1 and D_2.

SOL **__14.33.__ Consider Eq. (14.27) and note that $g_{m1} \approx g_{m2} = g_m$. Prove that the output impedance can be expressed as

$$\frac{v_X}{i_X} \approx \frac{1}{2g_m} + \frac{r_{O3}||r_{O4}}{2\beta_1\beta_2}(\beta_1 + \beta_2). \quad (14.69)$$

14.34. The push-pull stage of Fig. 14.15(b) employs a bias current of 1 mA in Q_3 and Q_4 and 8 mA in Q_1 and Q_2. If Q_3 and Q_4 suffer from the Early effect and $V_{A3} = 10$ V and $V_{A4} = 15$ V,

(a) Calculate the small-signal output impedance of the circuit if $\beta_1 = 40$ and $\beta_2 = 20$.

(b) Using the result obtained in (a), determine the voltage gain if the stage drives a load resistance of 8 Ω.

14.35. The circuit of Fig. 14.15(b) employs a bias current of 1 mA in Q_3 and Q_4. If $\beta_1 = 40$ and $\beta_2 = 20$, calculate the maximum current that Q_1 and Q_2 can deliver to the load.

14.36. We wish to deliver a power of 0.5 W to an 8-Ω load. Determine the minimum required bias current of Q_3 and Q_4 in Fig. 14.15(b) if $\beta_1 = 40$ and $\beta_2 = 20$.

14.37. The push-pull stage of Fig. 14.15 delivers an average power of 0.5 W to $R_L = 8\ \Omega$ with $V_{CC} = 5$ V. Compute the average power dissipated in Q_1.

14.38. The push-pull stage of Fig. 14.15 incorporates transistors with a maximum power rating of 0.75 W. If $V_{CC} = 5$ V, what is the largest power that the circuit can deliver to an 8-Ω load?

14.39. Repeat Problem 14.38 but assume that V_{CC} can be chosen freely.

Sec. 14.5 Large-Signal Considerations

14.40. Consider the composite stage shown in Fig. 14.39. Assume $I_1 = 5$ mA, $\beta_1 = 40$, and $\beta_2 = 50$. Calculate the base current of Q_2.

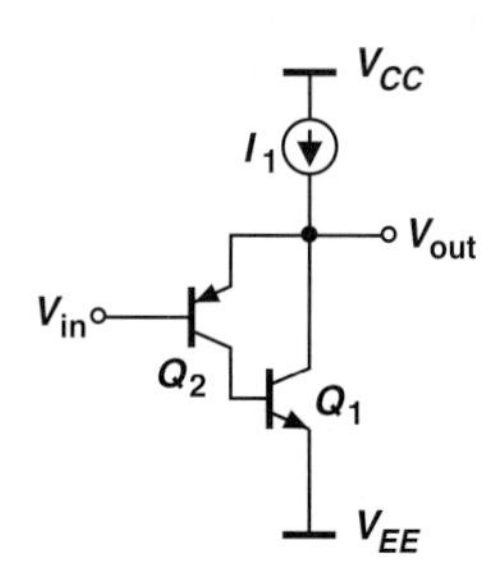

Figure 14.39

14.41. In Problem 14.40, $V_{in} = 0.5$ V. Determine the output voltage if $I_{S2} = 6 \times 10^{-17}$ A. SOL

14.42. In Problem 14.40, calculate the input and output impedances of the circuit.

14.43. In the circuit of Fig. 14.39, determine the value of I_1 so as to obtain an output impedance of 1 Ω. Assume $\beta_1 = 40$ and $\beta_2 = 50$.

14.44. An emitter follower delivers a peak swing of 0.5 V to an 8-Ω load with $V_{CC} =$ 2 V. If the bias current is 70 mA, calculate the power efficiency of the circuit.

14.45. In a realistic emitter follower design, the peak swing reaches only $V_{CC} - V_{BE}$. Determine the efficiency in this case.

14.46. Repeat Problem 14.45 for a push-pull stage.

*__14.47.__ A push-pull stage is designed to deliver a peak swing of V_P to a load resistance of R_L. What is the efficiency if the circuit delivers a swing of only $V_P/2$?

14.48. A push-pull stage operating from $V_{CC} =$ 3 V delivers a power of 0.2 W to an 8-Ω load. Determine the efficiency of the circuit.

Design Problems

14.49. We wish to design the emitter follower of Fig. 14.30 for a power of 1 W delivered to $R_L = 8\ \Omega$. Determine I_1 and the power rating of Q_1.

14.50. The emitter follower of Fig. 14.30 must be designed to drive $R_L = 4\ \Omega$ with a voltage gain of 0.8. Determine I_1, the maximum output voltage swing, and the power rating of Q_1.

SOL **14.51.** Consider the push-pull stage shown in Fig. 14.38. Determine the bias current of Q_1 and Q_2 so as to obtain a voltage gain of 0.6 with $R_L = 8\ \Omega$. Neglect the incremental resistance of D_1 and D_2.

14.52. The push-pull stage of Fig. 14.38 must deliver a power of 1 W to $R_L = 8\ \Omega$. Determine the minimum allowable supply voltage if $|V_{BE}| \approx 0.8$ V and the minimum value of I_1 if $\beta_1 = 40$.

14.53. Suppose the transistors in the stage of Fig. 14.38 exhibit a maximum power rating of 2 W. What is the largest power that the circuit can deliver to an 8-Ω load?

14.54. Repeat Problem 14.53 for a 4-Ω load.

14.55. We wish to design the push-pull amplifier for Fig. 14.15(b) for a small-signal voltage gain of 4 with $R_L = 8\ \Omega$. If $\beta_1 = 40$ and $\beta_2 = 20$, compute the bias current of Q_3 and Q_4. What is the maximum current that Q_1 can deliver to R_L?

14.56. Repeat Problem 14.55 with $R_L = 4\ \Omega$ and compare the results.

14.57. The push-pull stage of Fig. 14.15(b) must be designed for an output power of 2 W and $R_L = 8\ \Omega$. Assume $|V_{BE}| \approx 0.8$ V, $\beta_1 = 40$, and $\beta_2 = 20$.
(a) Determine the minimum required supply voltage if Q_3 and Q_4 must remain in the active region.
(b) Calculate the minimum required bias current of Q_3 and Q_4.
(c) Determine the average power dissipated in Q_1 while the circuit delivers 2 W to the load.
(d) Compute the overall efficiency of the circuit, taking into account the bias current of Q_3 and Q_4.

14.58. A stereo system requires a push-pull stage similar to that in Fig. 14.15(b) with a voltage gain of 5 and an output power of 5 W. Assume $R_L = 4\ \Omega$, $\beta_1 = 40$, and $\beta_2 = 20$.
(a) Calculate the bias current of Q_3 and Q_4 to achieve the required voltage gain. Does the result satisfy the required output power?
(b) Determine the bias current of Q_3 and Q_4 to achieve the required output power. What is the resulting voltage gain?

SPICE PROBLEMS

In the following problems, use the MOS device models given in Appendix A. For bipolar transistors, assume $I_{S,npn} = 5 \times 10^{-16}$ A, $\beta_{npn} = 100$, $V_{A,npn} = 5$ V, $I_{S,pnp} = 8 \times 10^{-16}$ A, $\beta_{pnp} = 50$, $V_{A,pnp} = 3.5$ V. Also, SPICE models the effect of charge storage in the base by a parameter called $\tau_F = C_b/g_m$. Assume $\tau_F(tf) = 20$ ps.

14.59. The emitter follower shown in Fig. 14.40 must deliver a power of 50 mW to an 8-Ω speaker at a frequency of 5 kHz.

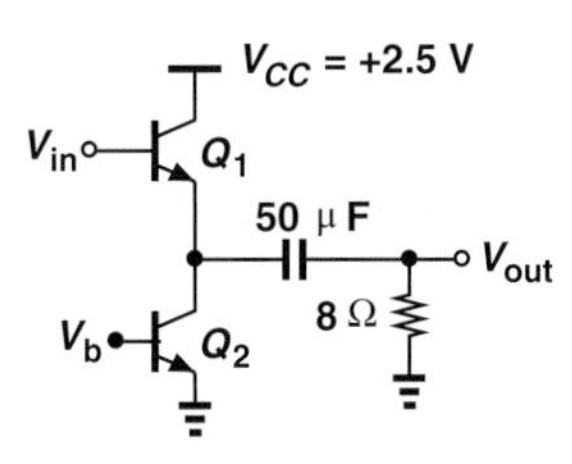

Figure 14.40

(a) Determine the minimum required supply voltage.
(b) Determine the minimum bias current of Q_2.
(c) Using the values obtained in (a) and (b) and a current mirror to bias Q_2, examine the output waveform. What supply voltage and bias current yield a relatively pure sinusoid?

14.60. Repeat Problem 14.59 for the source follower of Fig. 14.41, where $(W/L)_{1-2} = 300\ \mu m/0.18\ \mu m$ and compare the results.

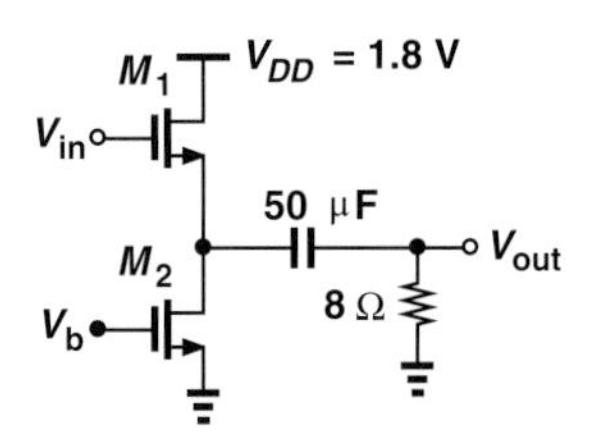

Figure 14.41

14.61. Plot the input/output characteristic of the circuit shown in Fig. 14.42 for $-2\text{ V} < V_{in} < +2\text{ V}$. Also, plot the output waveform for an input sinusoid having a peak amplitude of 2 V. How are these results changed if the load resistance is raised to 16 Ω.

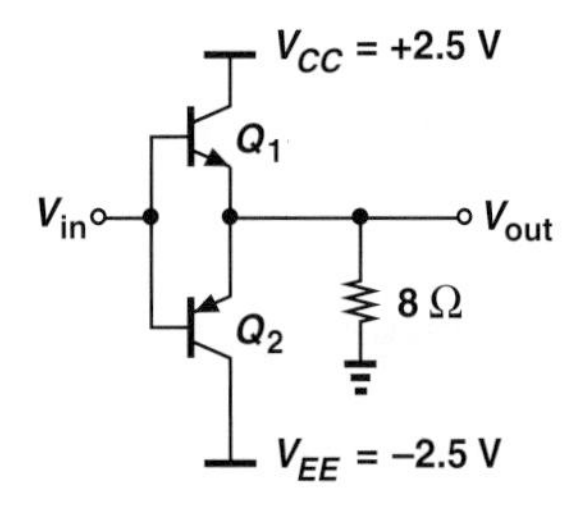

Figure 14.42

14.62. In the push-pull stage of Fig. 14.43, Q_3 and Q_4 operate as diodes.

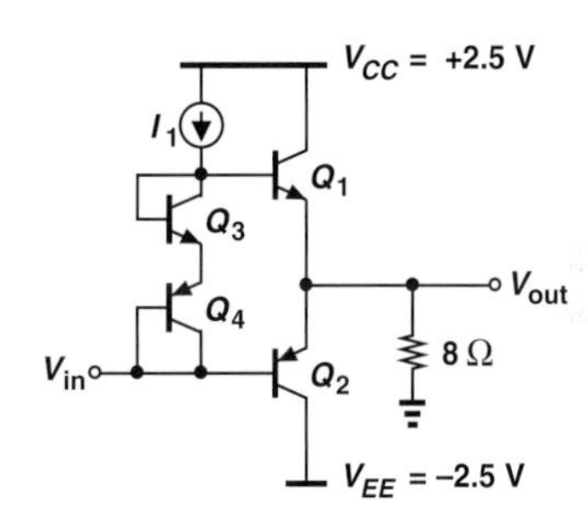

Figure 14.43

(a) Select I_1 so that the circuit can deliver a peak swing of 1.5 V to the load.
(b) Under these conditions, examine the output waveform and explain what happens.
(c) SPICE allows scaling of bipolar transistors as follows: `q1 col bas emi sub bimod m=16`, where $m = 16$ denotes a 16-fold increase

in the size of the transistor (as if 16 unit transistors were placed in parallel). Using $m = 16$ for Q_1 and Q_6, repeat part (b).

14.63. The feedback push-pull stage of Fig. 14.44 must deliver a peak swing of 1.5 V to an 8-Ω load.

(a) What is the minimum required value of I_1?

(b) Using an input dc level of −1.7 V and a scaling factor of 16 for Q_1 and Q_2 (as in Problem 14.62), examine the output waveform.

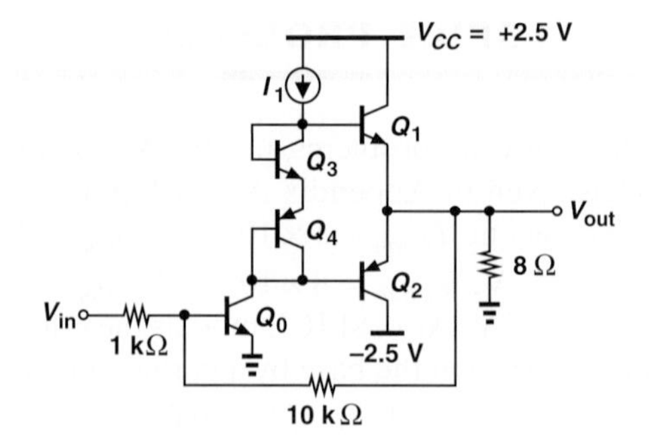

Figure 14.44

REFERENCE

1. B. Razavi, *RF Microelectronics*, Upper Saddle River, NJ: Prentice Hall, 1998.

15

Analog Filters

Our treatment of microelectronics thus far has mostly concentrated on the problem of amplification. Another important function vastly used in electronic systems is "filtering." For example, a cellphone incorporates filters to suppress "interferers" that are received in addition to the desired signal. Similarly, a high-fidelity audio system must employ filters to eliminate the 60 Hz (50 Hz) ac line interference. This chapter provides an introduction to analog filters. The outline is shown below.

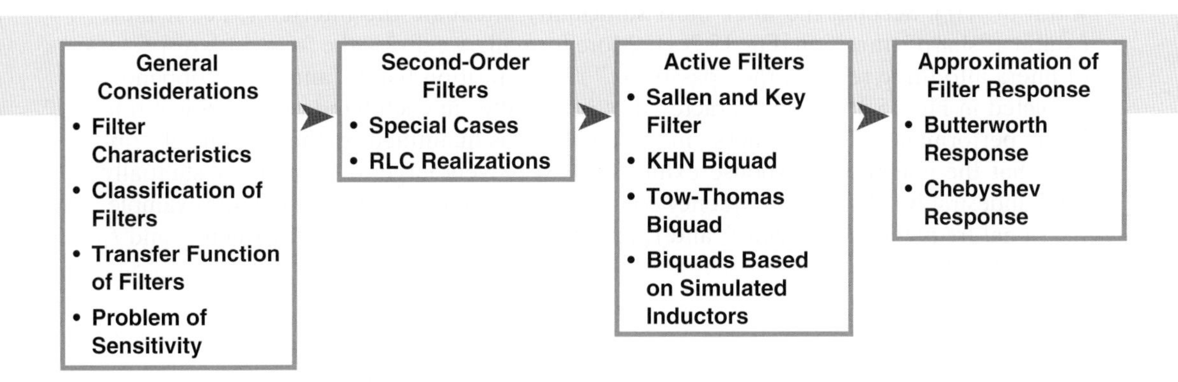

15.1 GENERAL CONSIDERATIONS

In order to define the performance parameters of filters, we first take a brief look at some applications. Suppose a cellphone receives a desired signal, $X(f)$, with a bandwidth of 200 kHz at a center frequency of 900 MHz [Fig. 15.1(a)]. As mentioned in Chapter 1, the receiver may translate this spectrum to zero frequency and subsequently "detect" the signal.

Now, let us assume that, in addition to $X(f)$, the cellphone receives a large interferer centered at 900 MHz + 200 kHz [Fig. 15.1(b)].[1] After translation to zero center frequency,

[1]This is called the "adjacent channel."

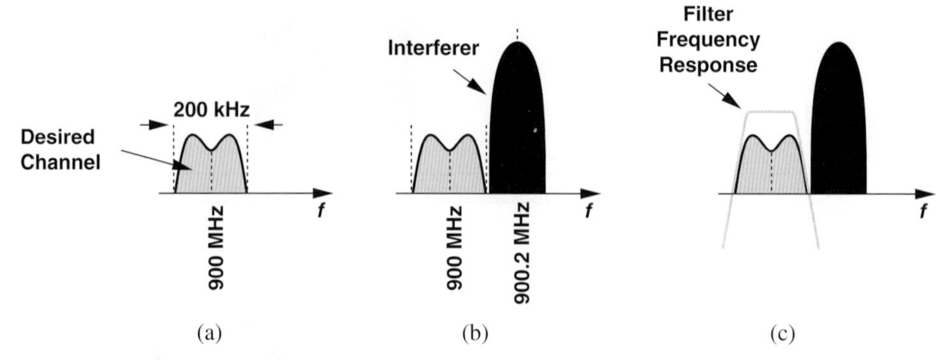

Figure 15.1 (a) Desired channel in a receiver, (b) large interferer, (c) use of filter to suppress the interferer.

the desired signal is still accompanied by the large interferer and cannot be detected properly. We must therefore "reject" the interferer by means of a filter [Fig. 15.1(c)].

15.1.1 Filter Characteristics

Which characteristics of the above filter are important here? First, the filter must not affect the desired signal; i.e., it must provide a "flat" frequency response across the bandwidth of $X(f)$. Second, the filter must sufficiently attenuate the interferer; i.e., it must exhibit a "sharp" transition [Fig. 15.2(a)]. More formally, we divide the frequency response of filters into three regions: the "passband," the "transition band," and the "stopband." Depicted in Fig. 15.2(b), the characteristics of the filter in each band play a critical role in the performance. The "flatness" in the passband is quantified by the amount of "ripple" that the magnitude response exhibits. If excessively large, the ripple substantially (and undesirably) alters the frequency contents of the signal. In Fig. 15.2(b), for example, the signal frequencies between f_2 and f_3 are attenuated whereas those between f_3 and f_4 are amplified.

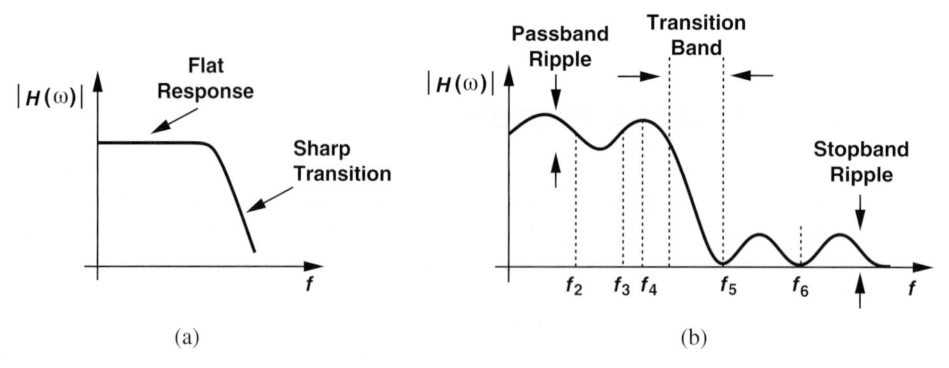

Figure 15.2 (a) Generic and (b) ideal filter characteristics.

The width of the transition band determines how much of the interferer remains alongside the signal, i.e., the inevitable corruption inflicted upon the signal by the interferer. For this reason, the transition band must be sufficiently narrow, i.e., the filter must provide sufficient "selectivity."

The stopband "attenuation" and ripple also impact the performance. The attenuation must be large enough to suppress the interferer to well below the signal level. The ripple in this case proves less critical than that in the passband, but it simply subtracts from the stopband attenuation. In Fig. 15.2(b), for example, the stopband attenuation is degraded between f_5 and f_6 as a result of the ripple.

Example 15-1 In a wireless application, the interferer in the adjacent channel may be 25 dB higher than the desired signal. Determine the required stopband attenuation of the filter in Fig. 15.2(b) if the signal power must exceed the interferer power by 15 dB for proper detection.

Solution As illustrated in Fig. 15.3, the filter must suppress the interferer by 40 dB, requiring the same amount of stopband attenuation.

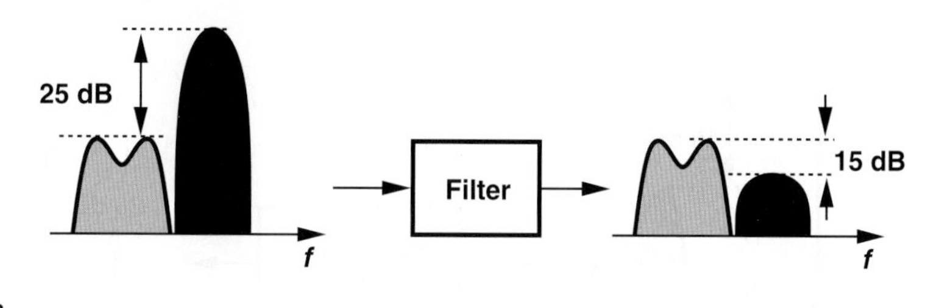

Figure 15.3

Exercise Suppose there are two interferers in two adjacent channels, each one 25 dB higher than the desired signal. Determine the stopband attenuation if the signal power must exceed each interferer by 18 dB.

In addition to the above characteristics, other parameters of analog filters such as linearity, noise, power dissipation, and complexity must also be taken into account. These issues are described in [1].

15.1.2 Classification of Filters

Filters can be categorized according to their various properties. We study a few classifications of filters in this section.

One classification of filters relates to the frequency band that they "pass" or "reject." The example illustrated in Fig. 15.2(b) is called a "low-pass" filter as it passes low-frequency signals and rejects high-frequency components. Conversely, one can envision a "high-pass" filter, wherein low-frequency signals are rejected (Fig. 15.4).

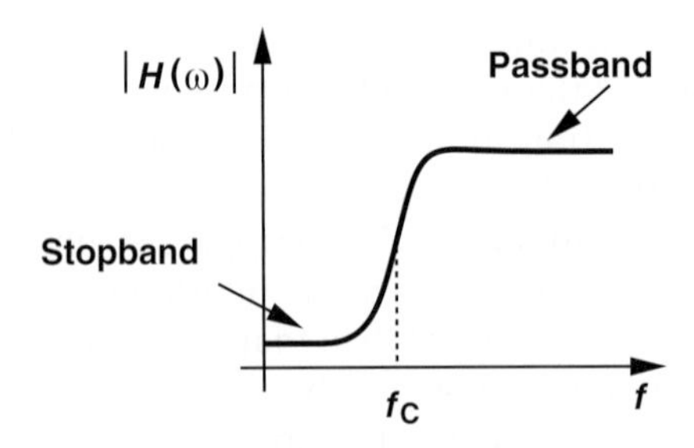

Figure 15.4 High-pass filter frequency response.

Example 15-2 We wish to amplify a signal in the vicinity of 1 kHz but the circuit board and wires pick up a strong 60 Hz component from the line electricity. If this component is 40 dB higher than the desired signal, what filter stopband attenuation is necessary to ensure the signal level remains 20 dB above the interferer level?

Solution As shown in Fig. 15.5, the high-pass filter must provide a stopband attenuation of 60 dB at 60 Hz.

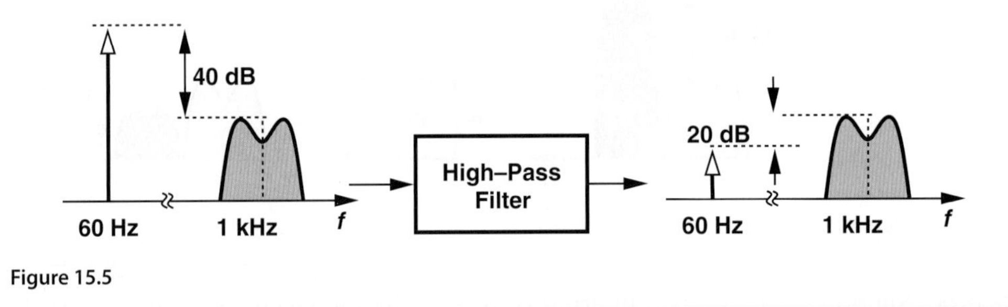

Figure 15.5

Exercise A signal in the audio frequency range is accompanied by an interferer at 100 kHz. If the interferer is 30 dB above the signal level, what stopband attenuation is necessary if the signal must be 20 dB above the interferer?

Some applications call for a "bandpass" filter, i.e., one that rejects both low- and high-frequency signals and passes a band in between (Fig. 15.6). The example below illustrates the need for such filters.

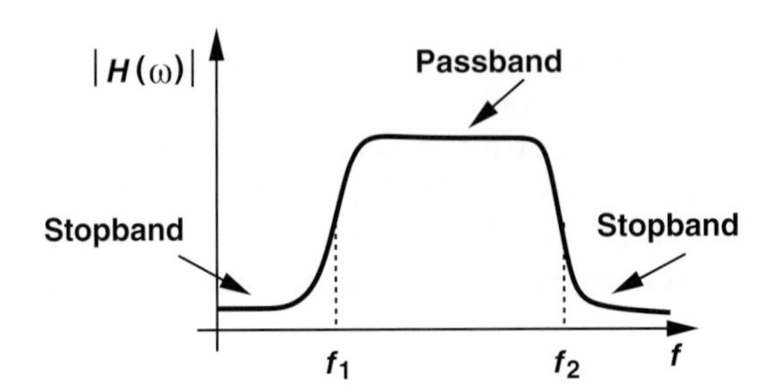

Figure 15.6

Example 15-3 Receivers designed for the Global Positioning System (GPS) operate at a frequency of approximately 1.5 GHz. Determine the interferers that may corrupt a GPS signal and the type of filters necessary to suppress them.

Solution The principal sources of interference in this case are cellphones operating in the 900-MHz and 1.9 GHz bands.[2] A bandpass filter is therefore required to reject these interferers (Fig. 15.7).

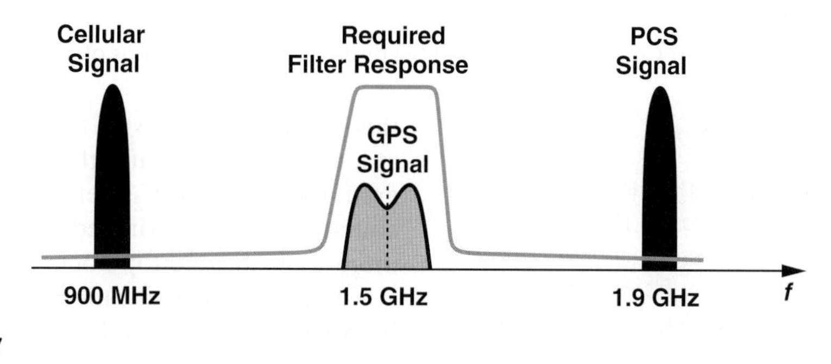

Figure 15.7

Exercise Bluetooth transceivers operate at 2.4 GHz. What type of filter is required to avoid corrupting Bluetooth signals by PCS signals?

Figure 15.8 summarizes four types of filters, including a "band-reject" response that suppresses components between f_1 and f_2.

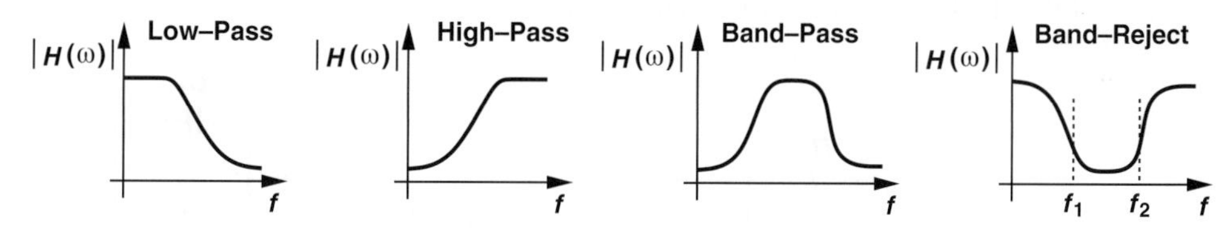

Figure 15.8 Summary of filter responses.

Another classification of analog filters concerns their circuit implementation and includes "continuous-time" and "discrete-time" realizations. The former type is exemplified by the familiar RC circuit depicted in Fig. 15.9(a), where C_1 exhibits a lower impedance as the frequency increases, thus attenuating high frequencies. The realization in Fig. 15.9(b) replaces R_1 with a "switched-capacitor" network. Here, C_2 is periodically switched between two nodes having voltages V_1 and V_2. We prove that this network acts as a resistor tied between the two nodes—an observation first made by James Maxwell in the 19th century.

In each cycle, C_2 stores a charge of $Q_1 = C_2V_1$ while connected to V_1 and $Q_2 = C_2V_2$ while tied to V_2. For example, if $V_1 > V_2$, C_2 absorbs charge from V_1 and delivers it to V_2, thus approximating a resistor. We also observe that the equivalent value of

[2]The former is called the "cellular band" and the latter, the "PCS band," where PCS stands for Personal Communication System.

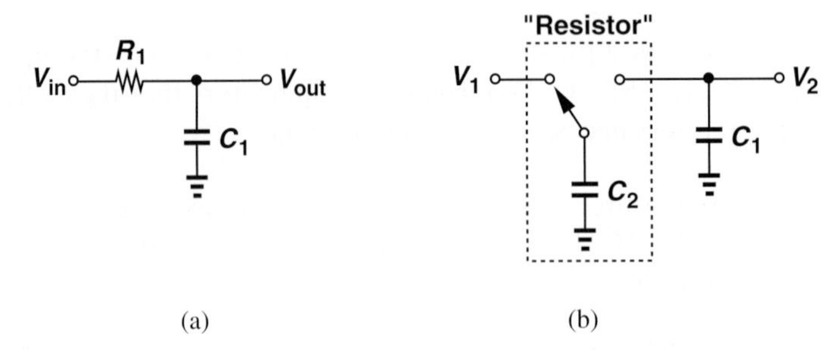

Figure 15.9 (a) Continuous-time and (b) discrete-time realizations of a low-pass filter.

Did you know?

The work on filters predates the vacuum tube and transistor developments. Even in the late 19th century, filters were considered for use in telegraph and telephone systems. Today, filters find application in almost all electronic devices. For example, our body picks up the 60 Hz (or 50 Hz) line voltage from the wirings around us, making it difficult for doctors to measure our brain waves. The equipment sensing these waves must therefore incorporate a band-reject filter to remove the undesired line voltage component.

this resistor *decreases* as the switching is performed at a higher rate because the amount of charge delivered from V_1 to V_2 per unit time increases. Of course, practical switched-capacitor filters employ more sophisticated topologies.

The third classification of filters distinguishes between "passive" and "active" implementations. The former incorporates only passive devices such as resistors, capacitors, and inductors, whereas the latter also employs amplifying components such as transistors and op amps.

The concepts studied in Chapter 8 readily provide examples of passive and active filters. A low-pass filter can be realized as the passive circuit in Fig. 15.10(a) or the active topology (integrator) in Fig. 15.10(b). Active filters provide much more flexibility in the design and find wide application in many electronic systems. Table 15.1 summarizes these classifications.

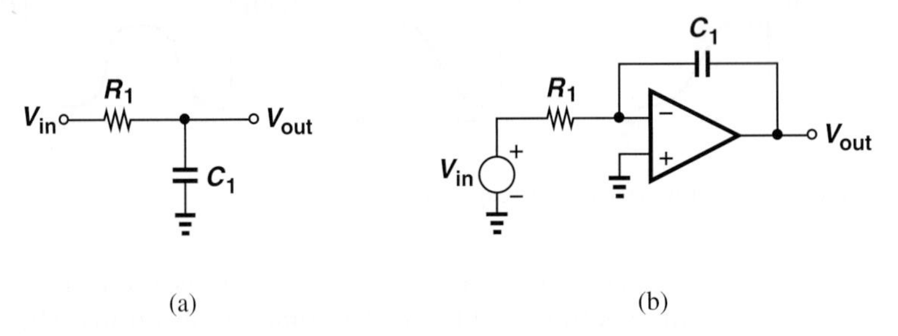

Figure 15.10 (a) Passive and (b) active realizations of a low-pass filter.

15.1.3 Filter Transfer Function

The foregoing examples of filter applications point to the need for a sharp transition (a high selectivity) in many cases. This is because (1) the interferer frequency is close to the desired signal band and/or (2) the interferer level is quite higher than the desired signal level.

How do we achieve a high selectivity? The simple low-pass filter of Fig. 15.11(a) exhibits a slope of only −20 dB/dec beyond the passband, thus providing only a tenfold suppression as the frequency increases by a factor of ten. We therefore postulate that *cascading* two such stages may sharpen the slope to −40 dB/dec, providing a suppression of 100 times for a tenfold increase in frequency [Fig. 15.11(b)]. In other words, increasing the "order" of the transfer function can improve the selectivity of the filter.

TABLE 15.1 Classifications of filters.

	Low-Pass	High-Pass	Band-Pass	Band-Reject
Frequency Response				
Continuous-Time and Discrete-Time				
Passive and Active				

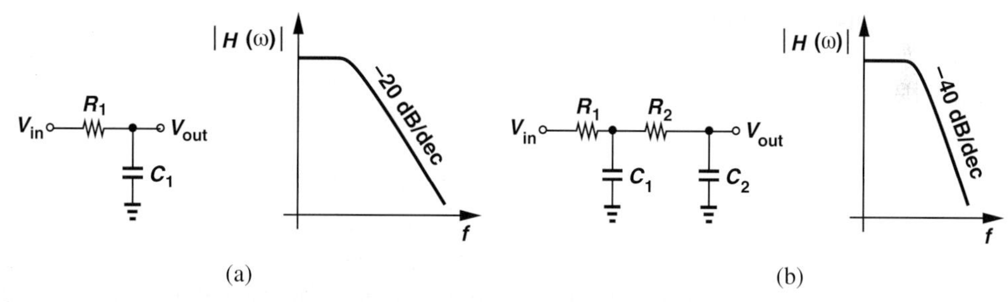

Figure 15.11 (a) First-order filter along with its frequency response, (b) addition of another RC section to sharpen the selectivity.

The selectivity, ripple, and other attributes of a filter are reflected in its transfer function, $H(s)$:

$$H(s) = \alpha \frac{(s - z_1)(s - z_2) \cdots (s - z_m)}{(s - p_1)(s - p_2) \cdots (s - p_n)}, \tag{15.1}$$

where z_k and p_k (real or complex) denote the zero and pole frequencies, respectively.

It is common to express z_k and p_k as $\sigma + j\omega$, where σ represents the real part and ω the imaginary part. One can then plot the poles and zeros on the complex plane.

Example 15-4

Construct the pole-zero diagram for the circuits shown in Fig. 15.12.

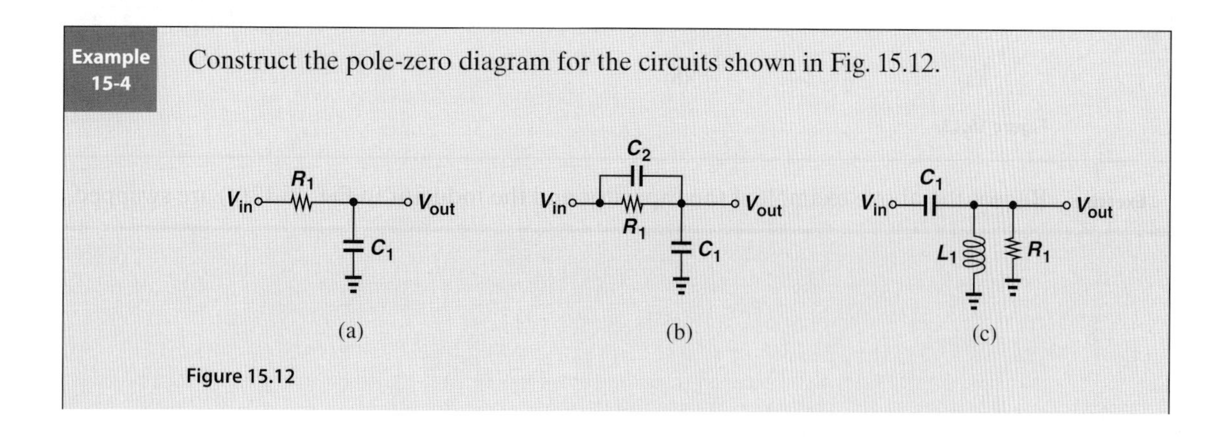

Figure 15.12

Solution For the circuit in Fig. 15.12(a), we have

$$H_a(s) = \frac{1}{R_1C_1s+1}, \tag{15.2}$$

obtaining a real pole at $-1/(R_1C_1)$. For the topology in Fig. 15.12(b),

$$H_b(s) = \frac{\frac{1}{C_1s}}{\frac{1}{C_1s}+R_1||\frac{1}{C_2s}} \tag{15.3}$$

$$= \frac{R_1C_2s+1}{R_1(C_1+C_2)s+1}. \tag{15.4}$$

The circuit therefore contains a zero at $-1/(R_1C_2)$ and a pole at $-1/[R_1(C_1+C_2)]$. Note that the zero arises from C_2. The arrangement in Fig. 15.12(c) provides the following transfer function:

$$H_c(s) = \frac{(L_1s)||R_1}{(L_1s)||R_1+\frac{1}{C_1s}} \tag{15.5}$$

$$= \frac{L_1C_1R_1s^2}{R_1L_1C_1s^2+L_1s+R_1}. \tag{15.6}$$

The circuit exhibits a zero at zero frequency and two poles that may be real or complex depending on whether $L^2 - 4R_1^2L_1C_1$ is positive or negative. Figure 15.13 summarizes our findings for the three circuits, where we have assumed $H_c(s)$ contains complex poles.

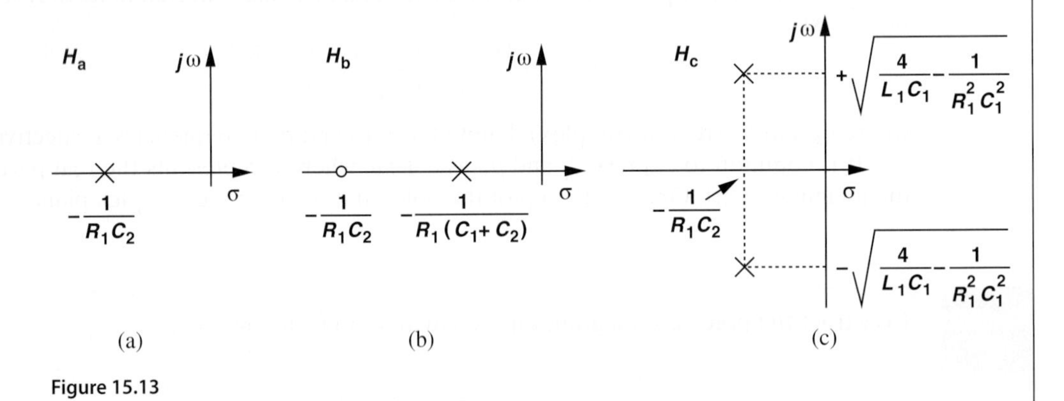

Figure 15.13

Exercise Repeat the above example if the capacitor and the inductor in Fig. 15.12(c) are swapped.

Example 15-5 Explain why the poles of the circuits in Fig. 15.12 must lie in the left half plane.

Solution Recall that the impulse response of a system contains terms such as $\exp(p_k t) = \exp(\sigma_k t)\ \exp(j\omega_k t)$. If $\sigma_k > 0$, these terms grow indefinitely with time while oscillating at a frequency of ω_k [Fig. 15.14(a)]. If $\sigma_k = 0$, such terms still introduce oscillation at ω_k [Fig. 15.14(b)]. Thus, we require $\sigma_k < 0$ for the system to remain stable [Fig. 15.14(c)].

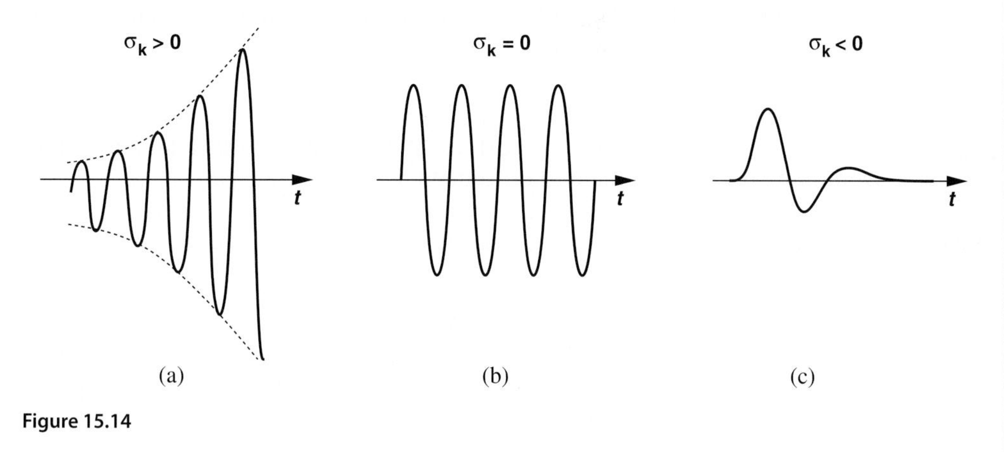

Figure 15.14

Exercise Redraw the above waveforms if ω_k is doubled.

It is instructive to make several observations in regard to Eq. (15.1). (1) The order of the numerator, m, cannot exceed that of the denominator; otherwise, $H(s) \to \infty$ as $s \to \infty$, an unrealistic situation. (2) For a physically-realizable transfer function, complex zeros or poles must occur in conjugate pairs, e.g., $z_1 = \sigma_1 + j\omega_1$ and $z_2 = \sigma_1 - j\omega_1$. (3) If a zero is located on the $j\omega$ axis, $z_{1,2} = \pm j\omega_1$, then $H(s)$ drops to zero at a sinusoidal input frequency of ω_1 (Fig. 15.15). This is because the numerator contains a product such as $(s - j\omega_1)(s + j\omega_1) = s^2 + \omega_1^2$, which vanishes at $s = j\omega_1$. In other words, imaginary zeros force $|H|$ to zero, thereby providing significant attenuation in their vicinity. For this reason, imaginary zeros are placed only in the *stopband*.

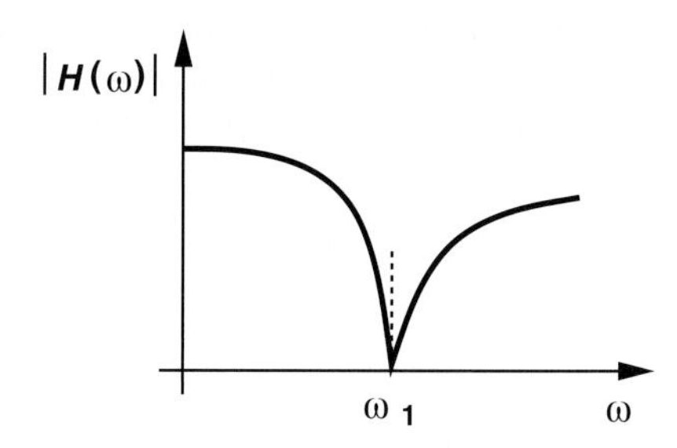

Figure 15.15 Effect of imaginary zero on the frequency response.

15.1.4 Problem of Sensitivity

The frequency response of analog filters naturally depends on the values of their constituent components. In the simple filter of Fig. 15.10(a), for example, the −3 dB corner frequency is given by $1/(R_1C_1)$. Such dependencies lead to errors in the cut-off frequency and other parameters in two situations: (a) the value of components varies with process and temperature (in integrated circuits), or (b) the *available* values of components deviate from those required by the design (in discrete implementations).[3]

We must therefore determine the change in each filter parameter in terms of a given change (tolerance) in each component value.

Example 15-6 In the low-pass filter of Fig. 15.10(a), resistor R_1 experiences a (small) change of ΔR_1. Determine the error in the corner frequency, $\omega_0 = 1/(R_1C_1)$.

Solution For small changes, we can utilize derivatives:

$$\frac{d\omega_0}{dR_1} = \frac{-1}{R_1^2C_1}. \tag{15.7}$$

Since we are usually interested in the *relative* (percentage) error in ω_0 in terms of the relative change in R_1, we write Eq. (15.7) as

$$\frac{d\omega_0}{\omega_0} = -\frac{dR_1}{\omega_0 \cdot R_1^2C_1} \tag{15.8}$$

$$= -\frac{dR_1}{R_1} \cdot \frac{1}{\omega_0 R_1 C_1} \tag{15.9}$$

$$= -\frac{dR_1}{R_1}. \tag{15.10}$$

For example, a +5% change in R_1 translates to a −5% error in ω_0.

Exercise Repeat the above example if C_1 experiences a small change of ΔC.

The above example leads to the concept of "sensitivity," i.e., how sensitive each filter parameter is with respect to the value of each component. Since in the first-order circuit, $|d\omega_0/\omega_0| = |dR_1/R_1|$, we say the sensitivity of ω_0 with respect to R_1 is unity in this example. More formally, the sensitivity of parameter P with respect to the component value C is defined as

$$S_C^P = \frac{\frac{dP}{P}}{\frac{dC}{C}}. \tag{15.11}$$

[3]For example, a particular design requires a 1.15-kΩ resistor but the closest available value is 1.2 kΩ.

Sensitivities substantially higher than unity are undesirable as they make it difficult to obtain a reasonable approximation of the required transfer function in the presence of component variations.

Example 15-7 Calculate the sensitivity of ω_0 with respect to C_1 for the low-pass filter of Fig. 15.10(a).

Solution Since

$$\frac{d\omega_0}{dC_1} = -\frac{1}{R_1 C_1^2}, \tag{15.12}$$

we have

$$\frac{d\omega_0}{\omega_0} = -\frac{dC_1}{C_1} \tag{15.13}$$

and hence

$$S_{C1}^{\omega 0} = -1. \tag{15.14}$$

Exercise Calculate the sensitivity of the pole frequency of the circuit in Fig. 15.12(b) with respect to R_1.

15.2 FIRST-ORDER FILTERS

As our first step in the analysis of filters, we consider first-order realizations, described by the transfer function

$$H(s) = \alpha \frac{s - z_1}{s - p_1}. \tag{15.15}$$

The circuit of Fig. 15.12(b) and its transfer function in Eq. (15.4) exemplify this type of filter. Depending on the relative values of z_1 and p_1, a low-pass or high-pass characteristic results, as illustrated in the plots of Fig. 15.16. Note that the stopband attenuation factor is given by z_1/p_1.

Let us consider the passive circuit of Fig. 15.12(b) as a candidate for realization of the above transfer function. We note that, since $z_1 = -1/(R_1 C_2)$ and $p_1 = -1/[R_1(C_1 + C_2)]$, the zero always falls *above* the pole, allowing only the response shown in Fig. 15.16(b).

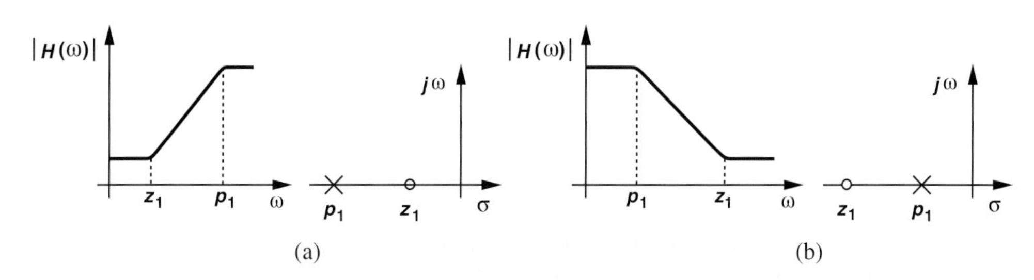

Figure 15.16 First-order (a) high-pass and (b) low-pass filters.

Example 15-8 Determine the response of the circuit depicted in Fig. 15.17(a).

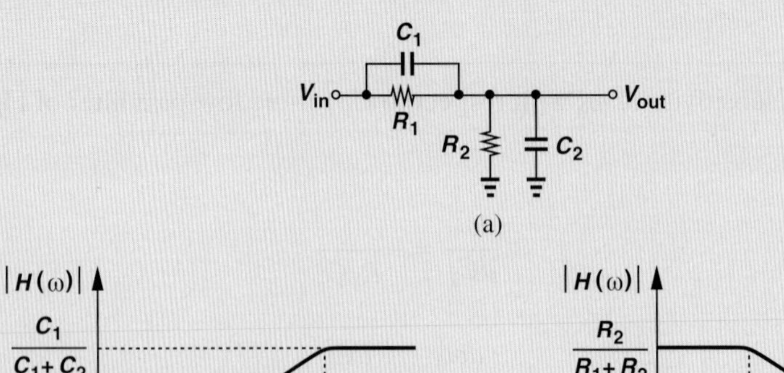

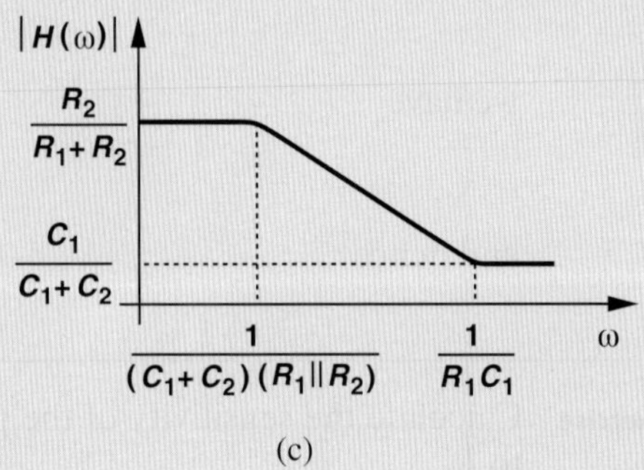

Figure 15.17

Solution We have

$$\frac{V_{out}}{V_{in}}(s) = \frac{R_2 || \frac{1}{C_2 s}}{R_2 || \frac{1}{C_2 s} + R_1 || \frac{1}{C_1 s}} \tag{15.16}$$

$$= \frac{R_2(R_1 C_1 s + 1)}{R_1 R_2 (C_1 + C_2)s + R_1 + R_2}. \tag{15.17}$$

The circuit contains a zero at $-1/(R_1C_1)$ and a pole at $-[(C_1 + C_2)R_1||R_2]^{-1}$. Depending on the component values, the zero may lie below or above the pole. Specifically, for the zero frequency to be lower:

$$\frac{1}{R_1 C_1} < \frac{R_1 + R_2}{(C_1 + C_2)R_1 R_2} \tag{15.18}$$

and hence

$$1 + \frac{C_2}{C_1} < 1 + \frac{R_1}{R_2}. \tag{15.19}$$

That is,

$$R_2 C_2 < R_1 C_1. \tag{15.20}$$

Figures 15.17(b) and (c) plot the response for the two cases $R_2C_2 < R_1C_1$ and $R_2C_2 > R_1C_1$, respectively. Note that $V_{out}/V_{in} = R_2/(R_1 + R_2)$ at $s = 0$ because the capacitors act as open circuit. Similarly, $V_{out}/V_{in} = C_1/(C_1 + C_2)$ at $s = \infty$ because the impedance of the capacitors becomes much smaller than R_1 and R_2 and hence the determining factor.

Exercise Design the circuit for a high-pass response with a zero frequency of 50 MHz and a pole frequency of 100 MHz. Use capacitors no larger than 10 pF.

Example 15-9 Figure 15.18(a) shows the active counterpart of the filter depicted in Fig. 15.17(a). Compute the response of the circuit. Assume the gain of the op amp is large.

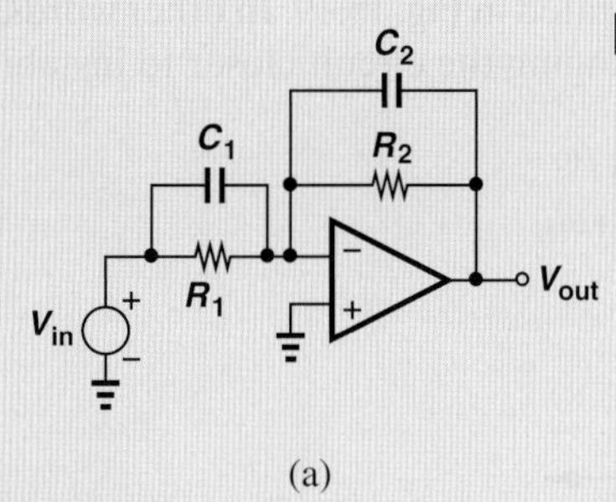

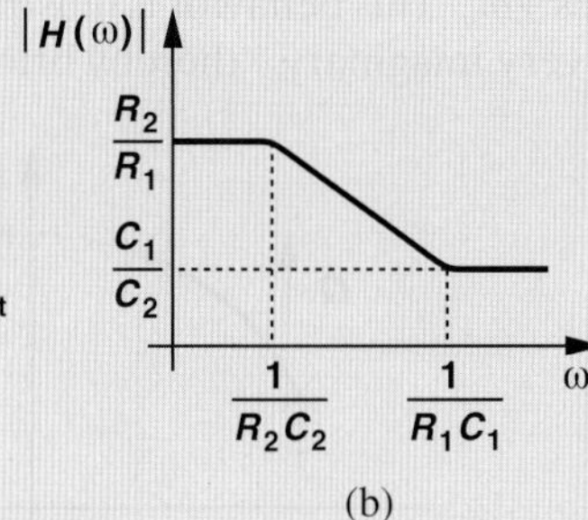

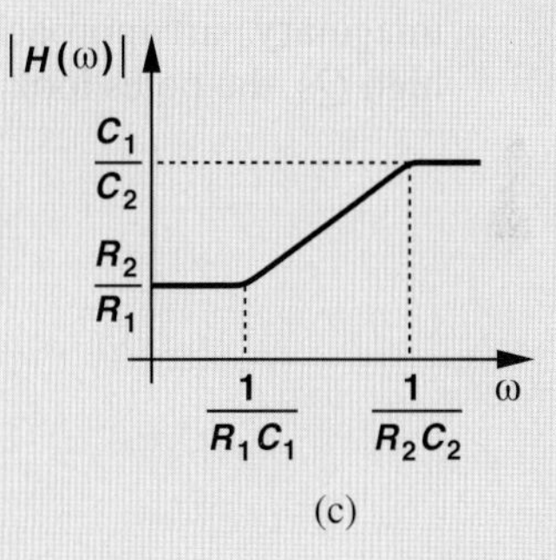

Figure 15.18

Solution We have from Chapter 8

$$\frac{V_{out}}{V_{in}}(s) = \frac{-\left(R_2 || \frac{1}{C_2 s}\right)}{R_1 || \frac{1}{C_1 s}} \tag{15.21}$$

$$= -\frac{R_2}{R_1} \cdot \frac{R_1 C_1 s + 1}{R_2 C_2 s + 1}. \tag{15.22}$$

As expected, at $s = 0$, $V_{out}/V_{in} = -R_2/R_1$ and at $s = \infty$, $V_{out}/V_{in} = -C_1/C_2$. Figures 15.18(b) and (c) plot the response for the two cases $R_1C_1 < R_2C_2$ and $R_1C_1 > R_2C_3$, respectively.

Exercise Is it possible for the pole frequency to be five times the zero frequency while the passband gain is ten times the stopband gain?

The first-order filters studied above provide only a slope of –20 dB/dec in the transition band. For a sharper attenuation, we must seek circuits of higher order.

15.3 SECOND-ORDER FILTERS

The general transfer function of second-order filters is given by the "biquadratic" equation:

$$H(s) = \frac{\alpha s^2 + \beta s + \gamma}{s^2 + \frac{\omega_n}{Q}s + \omega_n^2}. \tag{15.23}$$

Unlike the numerator, the denominator is expressed in terms of quantities ω_n and Q because they signify important aspects of the response. We begin our study by calculating the pole frequencies. Since most second-order filters incorporate complex poles, we assume $(\omega_n/Q)^2 - 4\omega_n^2 < 0$, obtaining

$$p_{1,2} = -\frac{\omega_n}{2Q} \pm j\omega_n\sqrt{1 - \frac{1}{4Q^2}}. \tag{15.24}$$

Note that as the "quality factor" of the poles, Q, increases, the real part decreases while the imaginary part approaches $\pm\omega_n$. This behavior is illustrated in Fig. 15.19. In other words, for high Qs, the poles look "very imaginary," thereby bringing the circuit closer to instability.

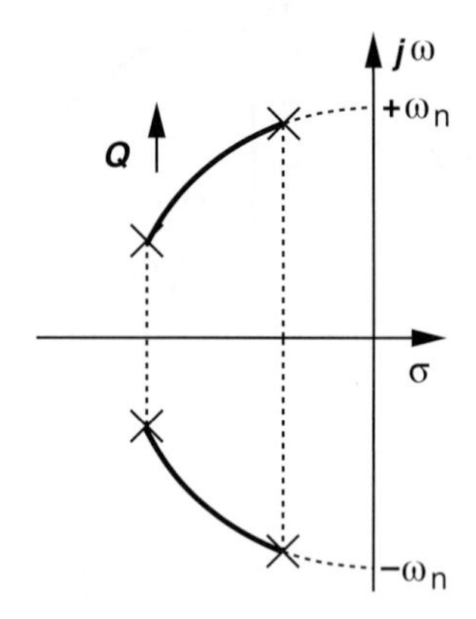

Figure 15.19 Variation of poles as a function of Q.

15.3.1 Special Cases

It is instructive to consider a few special cases of the biquadratic transfer function that prove useful in practice. First, suppose $\alpha = \beta = 0$ so that the circuit contains only poles[4] and operates as a low-pass filter (why?). The magnitude of the transfer function is then obtained by making the substitution $s = j\omega$ in Eq. (15.23) and expressed as

$$|H(j\omega)|^2 = \frac{\gamma^2}{(\omega_n^2 - \omega^2)^2 + \left(\frac{\omega_n}{Q}\omega\right)^2}. \tag{15.25}$$

Note that $|H(j\omega)|$ provides a slope of −40 dB/dec beyond the passband (i.e., if $\omega \gg \omega_n$). It can be shown (Problem 15.11) that the response is (a) free from

Did you know?

Is peaking in the frequency response of a filter desirable? Generally, it is not. Imagine an audio system using a 20 Hz to 20 kHz filter exhibits unwanted significant peaking at some intermediate frequency. Then, that frequency is amplified considerably more than others, bothering the ear. We should remark, however, that the desired frequency response in audio systems is usually *not* flat. In fact, the "equalizer" controls on stereo systems adjust the gain in different parts of the frequency response, allowing the user to tailor the sound for different types of music. The equalizer consists of a number of adjustable bandpass filters spanning the audio frequency range.

[4] Since $H(s) \to 0$ at $s = \infty$, we say the circuit exhibits two zeros at infinity.

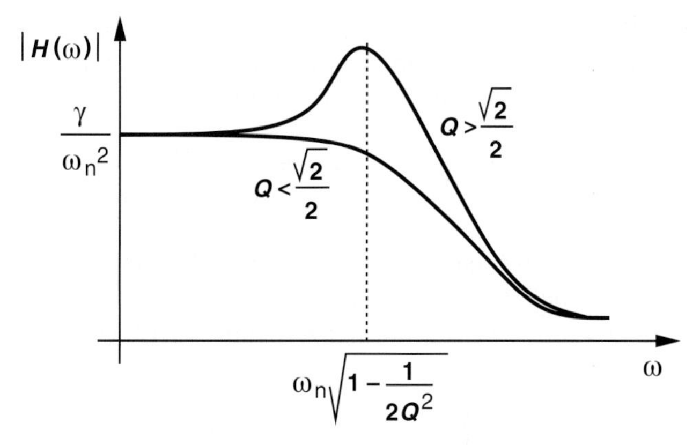

Figure 15.20 Frequency response of second-order system for different values of Q.

peaking if $Q \leq \sqrt{2}/2$; and (b) reaches a peak at $\omega_n\sqrt{1-1/(2Q^2)}$ if $Q > \sqrt{2}/2$ (Fig. 15.20). In the latter case, the peak magnitude normalized to the passband magnitude is equal to $Q/\sqrt{1-(4Q^2)^{-1}}$.

Example 15-10 Suppose a second-order LPF is designed with $Q = 3$. Estimate the magnitude and frequency of the peak in the frequency response.

Solution Since $2Q^2 = 18 \gg 1$, we observe that the normalized peak magnitude is $Q/\sqrt{1-1/(4Q^2)} \approx Q \approx 3$ and the corresponding frequency is $\omega_n\sqrt{1-1/(2Q^2)} \approx \omega_n$. The behavior is plotted in Fig. 15.21.

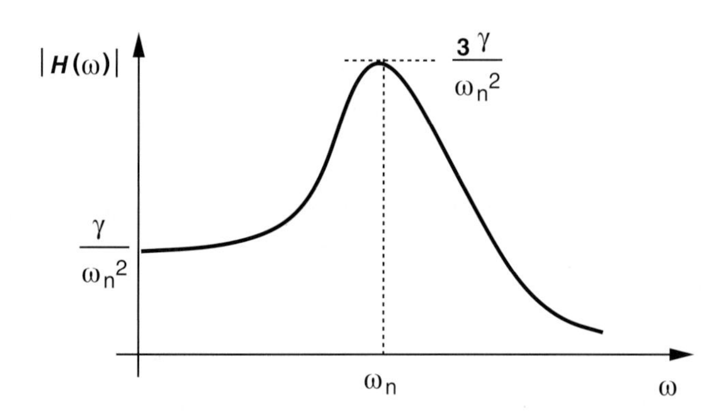

Figure 15.21

Exercise Repeat the above example for $Q = 1.5$.

How does the transfer function in Eq. (15.23) provide a *high-pass* response? In a manner similar to the first-order realization in Fig. 15.12(b), the zero(s) must fall *below* the poles. For example, with two zeros at the origin:

$$H(s) = \frac{\alpha s^2}{s^2 + \frac{\omega_n}{Q}s + \omega_n^2}, \tag{15.26}$$

we note that $H(s)$ approaches zero as $s \to 0$ and a constant value, α, as $s \to \infty$, thus providing a high-pass behavior (Fig. 15.22). As with the low-pass counterpart, the circuit exhibits a peak if $Q > \sqrt{2}/2$ with a normalized value of $Q/\sqrt{1-1/(4Q^2)}$ but at a frequency of $\omega_n/\sqrt{1-1/(2Q^2)}$.

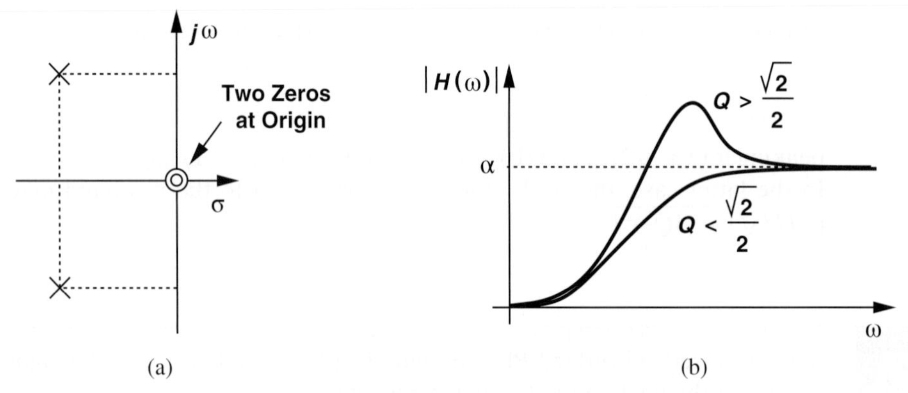

Figure 15.22 (a) Pole and zero locations and (b) frequency response of a second-order high pass filter.

Example 15-11 Explain why a high-pass response cannot be obtained if the biquadratic equation contains only one zero.

Solution Let us express such a case as

$$H(s) = \frac{\beta s + \gamma}{s^2 + \frac{\omega_n}{Q}s + \omega_n^2}. \tag{15.27}$$

Since $H(s) \to 0$ as $s \to \infty$, the system cannot operate as a high-pass filter.

Exercise Calculate the magnitude of $H(s)$.

A second-order system can also provide a band-pass response. Specifically, if

$$H(s) = \frac{\beta s}{s^2 + \frac{\omega_n}{Q}s + \omega_n^2}, \tag{15.28}$$

then, the magnitude approaches zero for both $s \to 0$ and $s \to \infty$, reaching a maximum in between (Fig. 15.23). It can be proved that the peak occurs at $\omega = \omega_n$ and has a value of $\beta Q/\omega_n$.

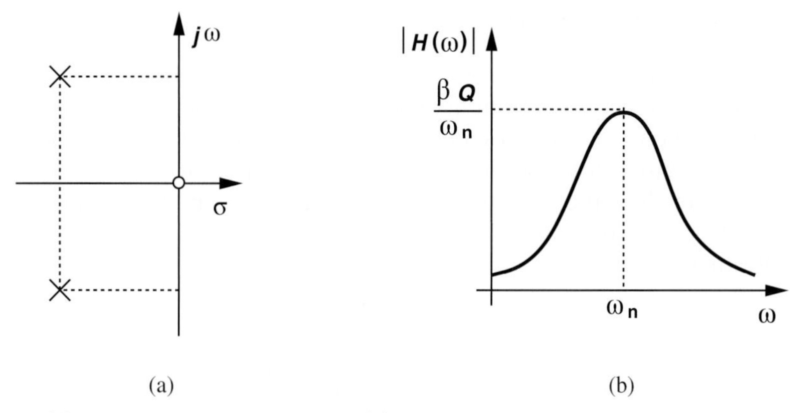

Figure 15.23 (a) Pole and zero locations and (b) frequency response of a second-order band-pass filter.

Example 15-12 Determine the −3 dB bandwidth of the response expressed by Eq. (15.28).

Solution As shown in Fig. 15.24, the response reaches $1/\sqrt{2}$ times its peak value at frequencies ω_1 and ω_2, exhibiting a bandwidth of $\omega_2 - \omega_1$. To calculate ω_1 and ω_2, we equate the squared magnitude to $(\beta Q/\omega_n)^2(1/\sqrt{2})^2$:

$$\frac{\beta^2\omega^2}{(\omega_n^2 - \omega^2)^2 + \left(\dfrac{\omega_n}{Q}\omega\right)^2} = \frac{\beta^2 Q^2}{2\omega_n^2}, \tag{15.29}$$

obtaining

$$\omega_{1,2} = \omega_0 \left[\sqrt{1 + \frac{1}{4Q^2}} \pm \frac{1}{2Q}\right]. \tag{15.30}$$

The total −3 dB bandwidth spans ω_1 to ω_2 and is equal to ω_o/Q. We say the "normalized" bandwidth is given by $1/Q$; i.e., the bandwidth trades with Q.

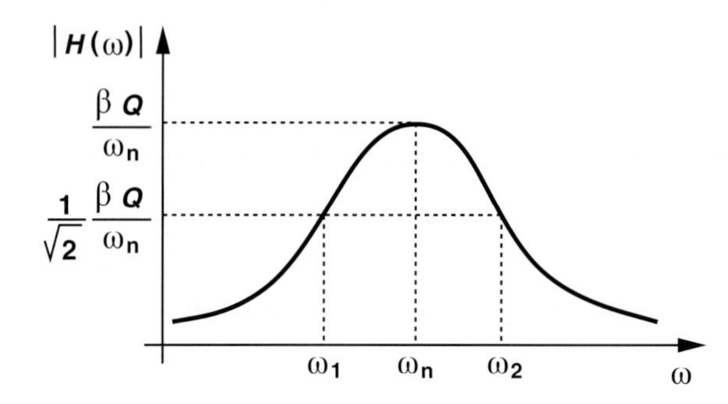

Figure 15.24

Exercise For what value of Q is ω_2 twice ω_1?

15.3.2 RLC Realizations

It is possible to implement the second-order transfer function in Eq. (15.23) by means of resistors, capacitors, and inductors. Such RLC realizations (a) find practical applications in low-frequency discrete circuits or high-frequency integrated circuits, and (b) prove useful as a *procedure* for designing *active* filters. We therefore study their properties here and determine how they can yield low-pass, high-pass, and band-pass responses.

Consider the parallel LC combination (called a "tank") depicted in Fig. 15.25(a). Writing

$$Z_1 = (L_1 s)||\frac{1}{C_1 s} \tag{15.31}$$

$$= \frac{L_1 s}{L_1 C_1 s^2 + 1}, \tag{15.32}$$

we note that the impedance contains a zero at the origin and two *imaginary* poles at $\pm j/\sqrt{L_1 C_1}$ [Fig. 15.25(b)]. We also examine the magnitude of the impedance by replacing s with $j\omega$:

$$|Z_1| = \frac{L_1 \omega}{1 - L_1 C_1 \omega^2}. \tag{15.33}$$

The magnitude thus begins from zero for $\omega = 0$, goes to *infinity* at $\omega_0 = 1/\sqrt{L_1 C_1}$, and returns to zero at $\omega = \infty$ [Fig. 15.25(c)]. The infinite impedance at ω_0 arises simply because the impedances of L_1 and C_1 cancel each other while operating in parallel.

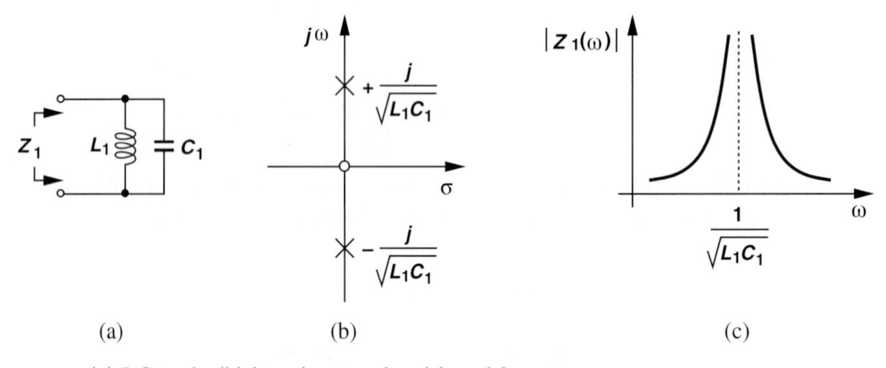

Figure 15.25 (a) LC tank, (b) imaginary poles, (c) and frequency response.

Example 15-13 Explain intuitively why the impedance of the tank goes to zero at $\omega = 0$ and $\omega = \infty$.

Solution At $\omega = 0$, L_1 operates as a short circuit. Similarly, at $\omega = \infty$, C_1 becomes a short.

Exercise Explain why the impedance has a zero at the origin.

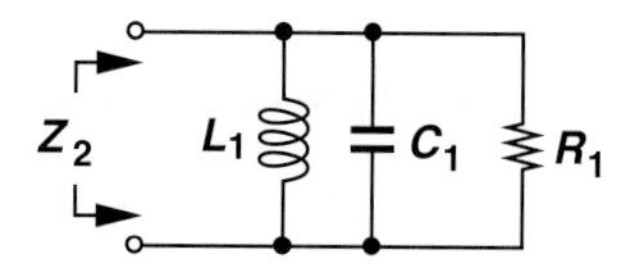

Figure 15.26 Lossy tank.

Now let us turn our attention to the parallel RLC tank depicted in Fig. 15.26(a). We can obtain Z_2 by replacing R_1 in parallel with Z_1 in Eq. (15.32):

$$Z_2 = R_1 || \frac{L_1 s}{L_1 C_1 s^2 + 1} \tag{15.34}$$

$$= \frac{R_1 L_1 s}{R_1 L_1 C_1 s^2 + L_1 s + R_1}. \tag{15.35}$$

The impedance still contains a zero at the origin due to the inductor. To compute the poles, we can factor $R_1 L_1 C_1$ from the denominator, thus obtaining a form similar to the denominator of Eq. (15.23):

$$R_1 L_1 C_1 s^2 + L_1 s + R_1 = R_1 L_1 C_1 \left(s^2 + \frac{1}{R_1 C_1} s + \frac{1}{L_1 C_1} \right) \tag{15.36}$$

$$= R_1 L_1 C_1 \left(s^2 + \frac{\omega_n}{Q} s + \omega_n^2 \right), \tag{15.37}$$

where $\omega_n = 1/\sqrt{L_1 C_1}$ and $Q = R_1 C_1 \omega_n = R_1 \sqrt{C_1/L_1}$. It follows from Eq. (15.24) that

$$p_{1,2} = -\frac{\omega_n}{2Q} \pm j\omega_n \sqrt{1 - \frac{1}{4Q^2}} \tag{15.38}$$

$$= -\frac{1}{2R_1 C_1} \pm j \frac{1}{\sqrt{L_1 C_1}} \sqrt{1 - \frac{L_1}{4R_1^2 C_1}}. \tag{15.39}$$

These results hold for complex poles, i.e., if

$$4R_1^2 > \frac{L_1}{C_1} \tag{15.40}$$

or

$$Q > \frac{1}{2}. \tag{15.41}$$

On the other hand, if R_1 decreases and $4R_1^2 < L_1/C_1$, we obtain real poles:

$$p_{1,2} = -\frac{\omega_n}{2Q} \pm \omega_n \sqrt{\frac{1}{4Q^2} - 1} \tag{15.42}$$

$$= -\frac{1}{2R_1 C_1} \pm \frac{1}{\sqrt{L_1 C_1}} \sqrt{\frac{L_1}{4R_1^2 C_1} - 1}. \tag{15.43}$$

So long as the excitation of the circuit does not alter its topology,[5] the poles are given by Eq. (15.39) or (15.43), a point that proves useful in the choice of filter structures.

[5]The "topology" of a circuit is obtained by setting all independent sources to zero.

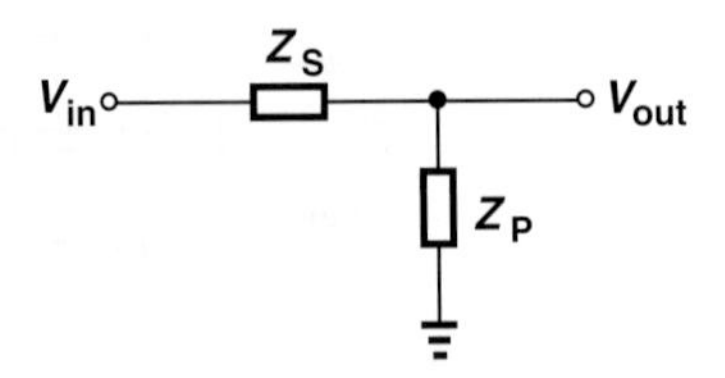

Figure 15.27 Voltage divider using general impedances.

Before studying different RLC filters, it is instructive to make several observations. Consider the voltage divider shown in Fig. 15.27, where a series impedance Z_S and a parallel impedance Z_P yield

$$\frac{V_{out}}{V_{in}}(s) = \frac{Z_P}{Z_S + Z_P}. \tag{15.44}$$

We note that (a) if, at high frequencies, Z_P goes to zero and/or Z_S goes to infinity,[6] then the circuit operates as a low-pass filter; (b) if, at low frequencies, Z_P goes to zero and/or Z_S goes to infinity, then the circuit serves as a high-pass filter; (c) if Z_S remains constant but Z_P falls to zero at both low and high frequencies, then the topology yields a band-pass response. These cases are conceptually illustrated in Fig. 15.28.

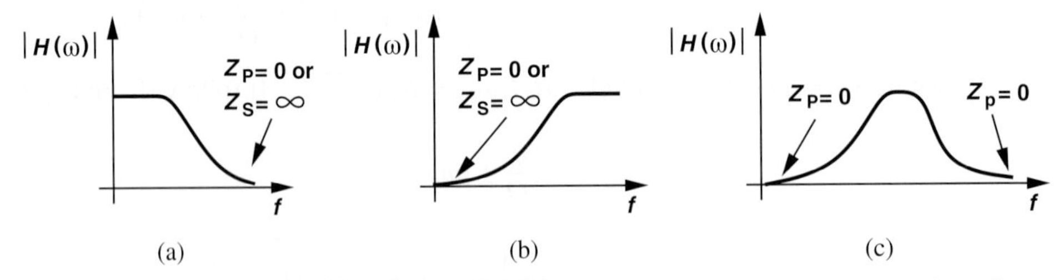

Figure 15.28 (a) Low-pass, (b) high-pass, and (c) bandpass responses obtained from the voltage divider of Fig. 15.27.

Low-Pass Filter Following the observation depicted in Fig. 15.28(a), we construct the circuit shown in Fig. 15.29, where

$$Z_S = L_1 s \rightarrow \infty \text{ as } s \rightarrow \infty \tag{15.45}$$

$$Z_P = \frac{1}{C_1 s} || R_1 \rightarrow 0 \text{ as } s \rightarrow \infty. \tag{15.46}$$

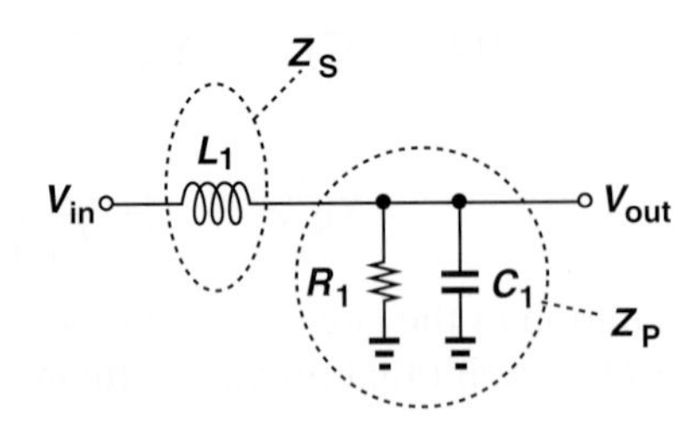

Figure 15.29 Low-pass filter obtained from Fig. 15.27.

[6]We assume Z_S and Z_P do not go to zero or infinity simultaneously.

This arrangement provides a low-pass response having the same poles as those given by Eq. (15.39) or (15.43) because for $V_{in} = 0$, it reduces to the topology of Fig. 15.26. Furthermore, the transition beyond the passband exhibits a *second-order* roll-off because both $Z_S \rightarrow \infty$ *and* $Z_P \rightarrow 0$. The reader can show that

$$\frac{V_{out}}{V_{in}}(s) = \frac{R_1}{R_1 C_1 L_1 s^2 + L_1 s + R_1}. \tag{15.47}$$

Example 15-14 Explain how the transfer function of Eq. (15.47) can provide a voltage gain greater than unity.

Solution If the Q of the network is sufficiently high, the frequency response exhibits "peaking," i.e., a gain of greater than unity in a certain frequency range. With a constant numerator, the transfer function provides this effect if the denominator falls to a local minimum. Writing the squared magnitude of the denominator as

$$|D|^2 = (R_1 - R_1 C_1 L_1 \omega^2)^2 + L_1^2 \omega^2 \tag{15.48}$$

and taking its derivative with respect to ω^2, we have

$$\frac{d|D|^2}{d(\omega^2)} = 2(-R_1 C_1 L_1)(R_1 - R_1 C_1 L_1 \omega^2) + L_1^2. \tag{15.49}$$

The derivative goes to zero at

$$\omega_a^2 = \frac{1}{L_1 C_1} - \frac{1}{2R_1^2 C_1^2}. \tag{15.50}$$

For a solution to exist, we require that

$$2R_1^2 \frac{C_1}{L_1} > 1 \tag{15.51}$$

or

$$Q > \frac{1}{\sqrt{2}}. \tag{15.52}$$

Comparison with Eq. (15.41) reveals the poles are complex here. The reader is encouraged to plot the resulting frequency response for different values of R_1 and prove that the peak value increases as R_1 decreases.

Exercise Compare the gain at ω_a with that at $1/\sqrt{L_1 C_1}$.

The peaking phenomenon studied in the above example proves undesirable in many applications as it disproportionately amplifies some frequency components of the signal. Viewed as ripple in the passband, peaking must remain below approximately 1 dB (10%) in such cases.

Example 15-15 Consider the low-pass circuit shown in Fig. 15.30 and explain why it is less useful than that of Fig. 15.29.

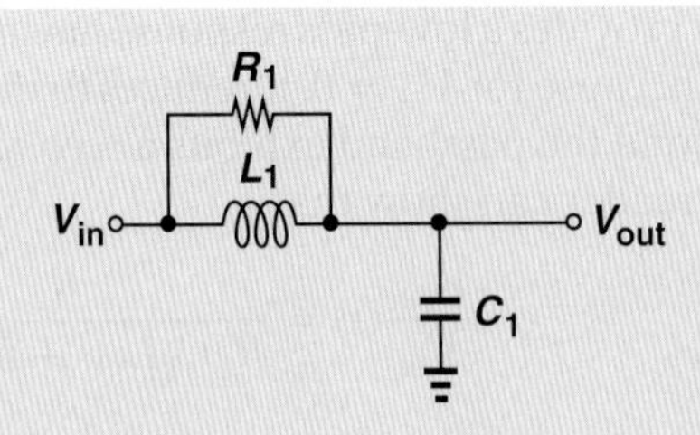

Figure 15.30

Solution This circuit satisfies the conceptual illustration in Fig. 15.28(a) and hence operates as a low-pass filter. However, at high frequencies, the parallel combination of L_1 and R_1 is dominated by R_1 because $L_1\omega \to \infty$, thereby reducing the circuit to R_1 and C_1. The filter thus exhibits a roll-off less sharp than the second-order response of the previous design.

Exercise What type of frequency response is obtained if L_1 and C_1 are swapped?

High-Pass Filter To obtain a high-pass response, we swap L_1 and C_1 in Fig. 15.29, arriving at the arrangement depicted in Fig. 15.31. Satisfying the principle illustrated in Fig. 15.28(b), the circuit acts as a second-order filter because as $s \to 0$, C_1 approaches an open circuit *and* L_1 a short circuit. The transfer function is given by

$$\frac{V_{out}}{V_{in}}(s) = \frac{(L_1 s)||R_1}{(L_1 s)||R_1 + \dfrac{1}{C_1 s}} \tag{15.53}$$

$$= \frac{L_1 C_1 R_1 s^2}{L_1 C_1 R_1 s^2 + L_1 s + R_1}. \tag{15.54}$$

The filter therefore contains two zeros at the origin. As with the low-pass counterpart, this circuit can exhibit peaking in its frequency response.

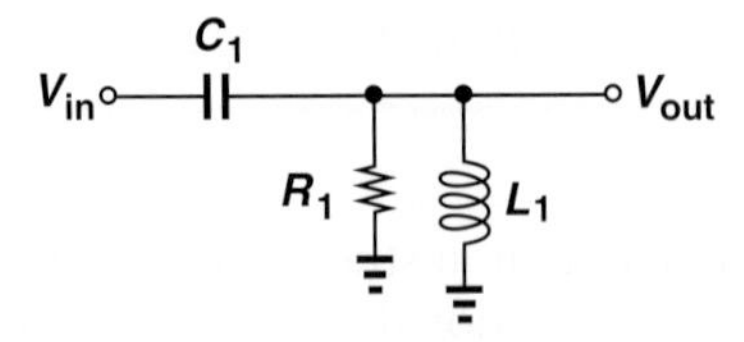

Figure 15.31 High-pass filter obtained from Fig. 15.27.

Band-Pass Filter From our observation in Fig. 15.28(c), we postulate that Z_P must contain both a capacitor and an inductor so that it approaches zero as $s \to 0$ or $s \to \infty$. Depicted in Fig. 15.32 is a candidate. Note that at $\omega = 1/\sqrt{L_1 C_1}$, the parallel combination

Figure 15.32 Band-pass filter obtained from Fig. 15.27.

of L_1 and C_1 acts as an open circuit, yielding $|V_{out}/V_{in}| = 1$. The transfer function is given by

$$\frac{V_{out}}{V_{in}}(s) = \frac{(L_1 s)||\frac{1}{C_1 s}}{(L_1 s)||\frac{1}{C_1 s} + R_1} \tag{15.55}$$

$$= \frac{L_1 s}{L_1 C_1 R_1 s^2 + L_1 s + R_1}. \tag{15.56}$$

15.4 ACTIVE FILTERS

Our study of second-order systems in the previous section has concentrated on passive RLC realizations. However, passive filters suffer from a number of drawbacks; e.g., they constrain the type of transfer function that can be implemented, and they may require bulky inductors. In this section, we introduce active implementations that provide second- or higher-order responses. Most active filters employ op amps to allow simplifying idealizations and hence a systematic procedure for the design of the circuit. For example, the op-amp-based integrator studied in Chapter 8 and repeated in Fig. 15.10(a) serves as an *ideal* integrator only when incorporating an ideal op amp, but it still provides a reasonable approximation with a practical op amp. (Thus, the term "integrator" is a simplifying idealization.)

An important concern in the design of active filters stems from the number of op amps required as it determines the power dissipation and even cost of the circuit. We therefore consider second-order realizations using one, two, or three op amps.

Did you know?

We have seen that inductors can form filters along with capacitors and resistors. Discrete inductors are realized as a spiral made of metal wire, just as the electrical symbol suggests. But how about integrated inductors? Shown in the figure below, a common structure is a "flat" spiral made of the metal wires that are used as interconnects on the chip. The inductance of this structure is given by the lateral dimension, *d* (and the number of turns), and can hardly exceed a few tens of nanohenries. For this reason, on-chip inductors are suited to only high-frequency operation, roughly above 500 MHz. RF transceivers, for example, use many on-chip inductors for resonance with parasitic capacitances.

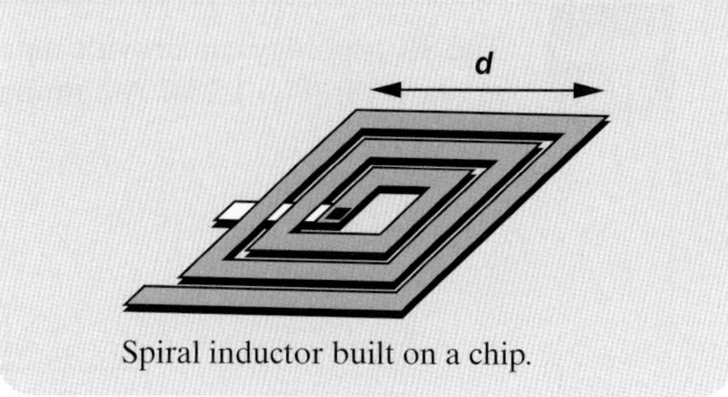

Spiral inductor built on a chip.

15.4.1 Sallen and Key Filter

The low-pass Sallen and Key (SK) filter employs one op amp to provide a second-order transfer function (Fig. 15.33). Note that the op amp simply serves as a unity-gain buffer,

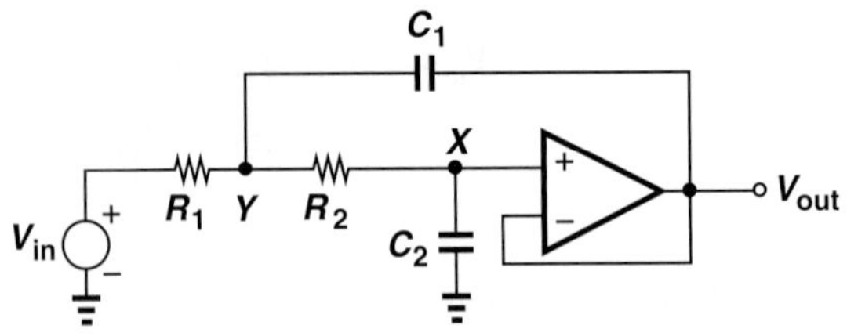

Figure 15.33 Basic Sallen and Key Filter.

thereby providing maximum bandwidth. Assuming an ideal op amp, we have $V_X = V_{out}$. Also, since the op amp draws no current, the current flowing through R_2 is equal to $V_X C_2 s = V_{out} C_2 s$, yielding

$$V_Y = R_2 C_2 s V_{out} + V_{out} \tag{15.57}$$

$$= V_{out}(1 + R_2 C_2 s). \tag{15.58}$$

Writing a KCL at node Y thus gives

$$\frac{V_{out}(1 + R_2 C_2 s) - V_{in}}{R_1} + V_{out} C_2 s + [V_{out}(1 + R_2 C_2 s) - V_{out}]C_1 s = 0 \tag{15.59}$$

and hence

$$\frac{V_{out}}{V_{in}}(s) = \frac{1}{R_1 R_2 C_1 C_2 s^2 + (R_1 + R_2)C_2 s + 1}. \tag{15.60}$$

To obtain a form similar to that in Eq. (15.23), we divide the numerator and the denominator by $R_1 R_2 C_1 C_2$ and define

$$Q = \frac{1}{R_1 + R_2}\sqrt{R_1 R_2 \frac{C_1}{C_2}} \tag{15.61}$$

$$\omega_n = \frac{1}{\sqrt{R_1 R_2 C_1 C_2}}. \tag{15.62}$$

Example 15-16 The SK topology can provide a passband voltage gain of greater than unity if configured as shown in Fig. 15.34. Assuming an ideal op amp, determine the transfer function of the circuit.

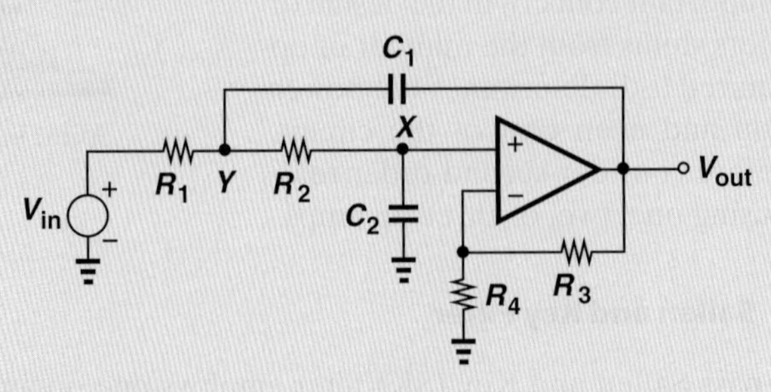

Figure 15.34 Sallen and Key filter with in-band gain.

Solution Returning to our derivations above, we note that now $(1 + R_3/R_4)V_X = V_{out}$, and the current flowing through R_2 is given by $V_X C_2 s = V_{out} C_2 s/(1 + R_3/R_4)$. It follows that

$$V_Y = \frac{V_{out}}{1 + \frac{R_3}{R_4}} + \frac{R_2 V_{out} C_2 s}{1 + \frac{R_3}{R_4}} \tag{15.63}$$

$$= V_{out} \frac{1 + R_2 C_2 s}{1 + \frac{R_3}{R_4}}. \tag{15.64}$$

A KCL at node Y thus yields

$$\frac{1}{R_1}\left(V_{out}\frac{1 + R_2C_2s}{1 + \frac{R_3}{R_4}} - V_{in}\right) + \left(V_{out}\frac{1 + R_2C_2s}{1 + \frac{R_3}{R_4}} - V_{out}\right)C_1 s + \frac{V_{out}C_2 s}{1 + \frac{R_3}{R_4}} = 0 \tag{15.65}$$

and hence

$$\frac{V_{out}}{V_{in}}(s) = \frac{1 + \frac{R_3}{R_4}}{R_1R_2C_1C_2s^2 + \left(R_1C_2 + R_2C_2 - R_1\frac{R_3}{R_4}C_1\right)s + 1}. \tag{15.66}$$

Interestingly, the value of ω_n remains unchanged.

Exercise Repeat the above analysis if a resistor of value R_0 is tied between node Y and ground.

Example 15-17 A common implementation of the SK filter assumes $R_1 = R_2$ and $C_1 = C_2$. Does such a filter contain complex poles? Consider the general case depicted in Fig. 15.34.

Solution From Eq. (15.66), we have

$$\frac{1}{Q} = \sqrt{\frac{R_1C_2}{R_2C_1}} + \sqrt{\frac{R_2C_2}{R_1C_1}} - \sqrt{\frac{R_1C_1}{R_2C_2}}\frac{R_3}{R_4}, \tag{15.67}$$

which, for $R_1 = R_2$ and $C_1 = C_2$, reduces to

$$\frac{1}{Q} = 2 - \frac{R_3}{R_4}. \tag{15.68}$$

That is,

$$Q = \frac{1}{2 - \frac{R_3}{R_4}}, \tag{15.69}$$

suggesting that Q begins from $1/2$ if $R_3/R_4 = 0$ (unity-gain feedback) and rises as R_3/R_4 approaches 2. The poles begin with real, equal values for $R_3/R_4 = 0$ and become complex for $R_3/R_4 > 0$ (Fig. 15.35).

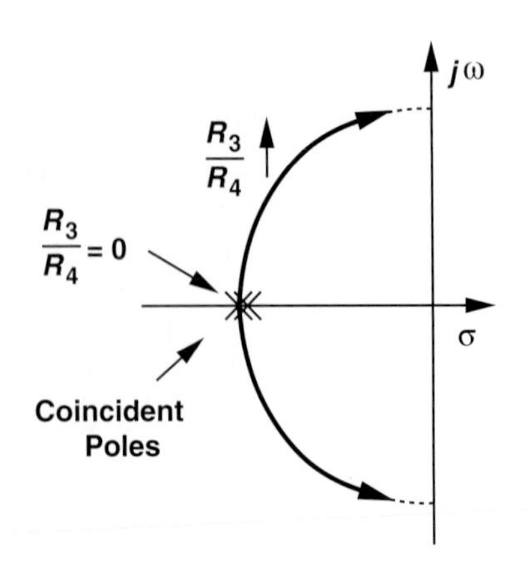

Figure 15.35

Exercise Calculate the pole frequencies if $R_3 = R_4$.

Sensitivity Analysis With so many components, how is the SK filter designed for a desired frequency response? An important objective in choosing the values is to minimize the sensitivities of the circuit. Considering the topology shown in Fig. 15.34 and defining $K = 1 + R_3/R_4$,[7] we compute the sensitivity of ω_n and Q with respect to the resistor and capacitor values.

From Eq. (15.66), we have $\omega_n = 1/\sqrt{R_1R_2C_1C_2}$ and hence:

$$\frac{d\omega_n}{dR_1} = -\frac{1}{2} \cdot \frac{1}{R_1\sqrt{R_1R_2C_1C_2}}. \tag{15.70}$$

That is,

$$\frac{d\omega_n}{\omega_n} = -\frac{1}{2}\frac{dR_1}{R_1} \tag{15.71}$$

and

$$S_{R1}^{\omega n} = -\frac{1}{2}. \tag{15.72}$$

This means a 1% error in R_1 translates to a 0.5% error in ω_n. Similarly,

$$S_{R2}^{\omega n} = S_{C1}^{\omega n} = S_{C2}^{\omega n} = -\frac{1}{2}. \tag{15.73}$$

For the Q sensitivities, we first rewrite Eq. (15.67) in terms of $K = 1 + R_3/R_4$:

$$\frac{1}{Q} = \sqrt{\frac{R_1C_2}{R_2C_1}} + \sqrt{\frac{R_2C_2}{R_1C_1}} - (K-1)\sqrt{\frac{R_1C_1}{R_2C_2}}. \tag{15.74}$$

Differentiating the right-hand side with respect to Q and the left-hand side with respect to R_1 yields:

$$\frac{-dQ}{Q^2} = \frac{dR_1}{2\sqrt{R_1}}\sqrt{\frac{C_2}{R_2C_1}} - \frac{dR_1}{2R_1\sqrt{R_1}}\sqrt{\frac{R_2C_2}{C_1}} - (K-1)\frac{dR_1}{2\sqrt{R_1}}\sqrt{\frac{C_1}{R_2C_2}} \tag{15.75}$$

$$= \frac{dR_1}{2R_1}\left[\sqrt{\frac{R_1C_2}{R_2C_1}} - \sqrt{\frac{R_2C_2}{R_1C_1}} - (K-1)\sqrt{\frac{R_1C_1}{R_2C_2}}\right]. \tag{15.76}$$

[7]In filter literature, the letter K denotes the gain of the filter and should not be confused with the feedback factor (Chapter 12).

It follows that

$$S_{R1}^{Q} = -\frac{1}{2}\left[\sqrt{\frac{R_1C_2}{R_2C_1}} - \sqrt{\frac{R_2C_2}{R_1C_1}} - (K-1)\sqrt{\frac{R_1C_1}{R_2C_2}}\right]Q. \tag{15.77}$$

The expression in the square brackets is similar to that in Eq. (15.74), except for a change in the sign of the second term. Adding and subtracting $2\sqrt{R_2C_2}/\sqrt{R_1C_1}$ to this expression and substituting for Q from Eq. (15.74), we arrive at

$$S_{R1}^{Q} = -\frac{1}{2} + Q\sqrt{\frac{R_2C_2}{R_1C_1}}. \tag{15.78}$$

Following the same procedure, the reader can show that:

$$S_{R2}^{Q} = -S_{R1}^{Q} \tag{15.79}$$

$$S_{C1}^{Q} = -S_{C2}^{Q} = -\frac{1}{2} + Q\left(\sqrt{\frac{R_1C_2}{R_2C_1}} + \sqrt{\frac{R_2C_2}{R_1C_1}}\right) \tag{15.80}$$

$$S_{K}^{Q} = QK\sqrt{\frac{R_1C_1}{R_2C_2}}. \tag{15.81}$$

Example 15-18 Determine the Q sensitivities of the SK filter for the common choice $R_1 = R_2 = R$ and $C_1 = C_2 = C$.

Solution From Eq. (15.74), we have

$$Q = \frac{1}{3-K} \tag{15.82}$$

and hence

$$S_{R1}^{Q} = -S_{R2}^{Q} = -\frac{1}{2} + \frac{1}{3-K} \tag{15.83}$$

$$S_{C1}^{Q} = -S_{C2}^{Q} = -\frac{1}{2} + \frac{2}{3-K} \tag{15.84}$$

$$S_{K}^{Q} = \frac{K}{3-K}. \tag{15.85}$$

Interestingly, for $K = 1$, the sensitivity to R_1 and R_2 vanishes and

$$\left|S_{C1}^{Q}\right| = \left|S_{C2}^{Q}\right| = \left|S_{K}^{Q}\right| = \frac{1}{2}. \tag{15.86}$$

The choice of equal component values *and* $K = 1$ thus leads to low sensitivities but also a limited Q and hence only a moderate transition slope. Moreover, the circuit provides no voltage gain in the passband.

Exercise Repeat the above example if $R_1 = 2R_2$.

In applications requiring a high Q and/or a high K, one can choose unequal resistors or capacitors so as to maintain reasonable sensitivities. The following example illustrates this point.

Example 15-19 An SK filter must be designed for $Q = 2$ and $K = 2$. Determine the choice of filter components for minimum sensitivities.

Solution For $S^Q_{k2} = 0$, we must have

$$Q\sqrt{\frac{R_2C_2}{R_1C_1}} = +\frac{1}{2} \tag{15.87}$$

and hence

$$\sqrt{\frac{R_2C_2}{R_1C_1}} = \frac{1}{4}. \tag{15.88}$$

For example, we can choose $R_1 = 4R_2$ and $C_1 = 4C_2$. But, how about the other sensitivities? For $S^Q_{C1} = -S^Q_{C2}$ to vanish,

$$\sqrt{\frac{R_1C_2}{R_2C_1}} + \sqrt{\frac{R_2C_2}{R_1C_1}} = \frac{1}{4}, \tag{15.89}$$

a condition in conflict with Eq. (15.88) because it translates to $\sqrt{R_1C_2}/\sqrt{R_2C_1} = 0$. In fact, we can combine Eqs. (15.78) and (15.80) to write

$$S^Q_{R1} + Q\sqrt{\frac{R_1C_2}{R_2C_1}} = S^Q_{C1}, \tag{15.90}$$

thereby observing that the two sensitivities cannot vanish simultaneously. Moreover, the term $\sqrt{R_2C_2}/\sqrt{R_1C_1}$ plays opposite roles in S^Q_{R1} and S^Q_K, leading to

$$S^Q_K = \frac{Q^2K}{S^Q_{R1} + \frac{1}{2}}. \tag{15.91}$$

That is, lowering S^Q_{R1} tends to *raise* S^Q_K.

The foregoing observations indicate that some compromise must be made to achieve reasonable (not necessarily minimum) sensitivities. For example, we choose

$$S^Q_{R1} = 1 \Rightarrow \sqrt{\frac{R_2C_2}{R_1C_1}} = \frac{3}{4} \tag{15.92}$$

$$S^Q_{C1} = \frac{5}{4} \Rightarrow \sqrt{\frac{R_1C_1}{R_2C_2}} = \frac{1}{8} \tag{15.93}$$

$$S^Q_K = \frac{8}{1.5} \tag{15.94}$$

The sensitivity to K is quite high and unacceptable in discrete design. In integrated circuits, on the other hand, K (typically the ratio of two resistors) can be controlled very accurately, thus allowing a large value of S^Q_K.

Exercise Can you choose sensitivities with respect to R_1 and C_1 such that S^Q_K remains below 2?

15.4.2 Integrator-Based Biquads

It is possible to realize the biquadratic transfer function of Eq. (15.23) by means of integrators. To this end, let us consider a special case where $\beta = \gamma = 0$:

$$\frac{V_{out}}{V_{in}}(s) = \frac{\alpha s^2}{s^2 + \dfrac{\omega_n}{Q}s + \omega_n^2}. \tag{15.95}$$

Cross-multiplying and rearranging the terms, we have

$$V_{out}(s) = \alpha V_{in}(s) - \frac{\omega_n}{Q} \cdot \frac{1}{s} V_{out}(s) - \frac{\omega_n^2}{s^2} V_{out}(s). \tag{15.96}$$

This expression suggests that V_{out} can be created as the sum of three terms: a scaled version of the input, an *integrated* version of the output, and a *doubly-integrated* version of the output. Figure 15.36(a) illustrates how V_{out} is generated by means of two integrators and a voltage adder. Utilizing the topologies introduced in Chapter 8, we readily arrive at the circuit realization depicted in Fig. 15.36(b). Note that the inherent signal inversion in each integrator necessitates returning V_X to the *noninverting* input of the adder and V_Y to the *inverting* input. Since

$$V_X = -\frac{1}{R_1 C_1 s} V_{out} \tag{15.97}$$

and

$$V_Y = -\frac{1}{R_2 C_2 s} V_X \tag{15.98}$$

$$= \frac{1}{R_1 R_2 C_1 C_2 s^2} V_{out}, \tag{15.99}$$

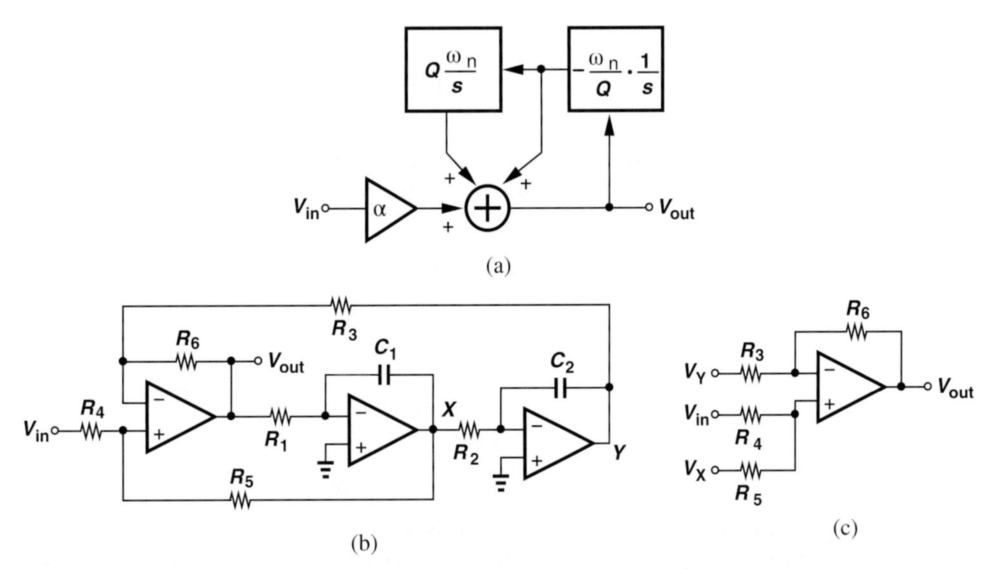

Figure 15.36 (a) Flow diagram showing the generation of V_{out} as a weighted sum of three terms, (b) realization of (a), (c) simplified diagram for calculating V_{out}.

we obtain from Fig. 15.36(c) the weighted sum of V_{in}, V_X, and V_Y as

$$V_{out} = \frac{V_{in}R_5 + V_X R_4}{R_4 + R_5}\left(1 + \frac{R_6}{R_3}\right) - V_Y \frac{R_6}{R_3} \tag{15.100}$$

$$= \frac{R_5}{R_4 + R_5}\left(1 + \frac{R_6}{R_3}\right)V_{in} - \frac{R_4}{R_4 + R_5}\frac{1}{R_1 C_1 s}V_{out} - \frac{R_6}{R_3}\frac{1}{R_1 R_2 C_1 C_2 s^2}V_{out}. \tag{15.101}$$

Equating similar terms in Eqs. (15.96) and (15.101) yields

$$\alpha = \frac{R_5}{R_4 + R_5}\left(1 + \frac{R_6}{R_3}\right) \tag{15.102}$$

$$\frac{\omega_n}{Q} = \frac{R_4}{R_4 + R_5} \cdot \frac{1}{R_1 C_1} \tag{15.103}$$

$$\omega_n^2 = \frac{R_6}{R_3} \cdot \frac{1}{R_1 R_2 C_1 C_2}. \tag{15.104}$$

It is thus possible to select the component values so as to obtain a desired transfer function.

Called the "KHN biquad" after its inventors, Kerwin, Huelsman, and Newcomb, the topology of Fig. 15.36(b) proves quite versatile. In addition to providing the high-pass transfer function of Eq. (15.95), the circuit can also serve as a low-pass and a band-pass filter. Specifically,

$$\frac{V_X}{V_{in}} = \frac{V_{out}}{V_{in}} \cdot \frac{V_X}{V_{out}} \tag{15.105}$$

$$= \frac{\alpha s^2}{s^2 + \frac{\omega_n}{Q}s + \omega_n^2} \cdot \frac{-1}{R_1 C_1 s}, \tag{15.106}$$

which is a band-pass function. Also,

$$\frac{V_Y}{V_{in}} = \frac{V_{out}}{V_{in}} \cdot \frac{V_Y}{V_{out}} \tag{15.107}$$

$$= \frac{\alpha s^2}{s^2 + \frac{\omega_n}{Q}s + \omega_n^2} \cdot \frac{1}{R_1 R_2 C_1 C_2 s^2}, \tag{15.108}$$

which provides a low-pass response.

Perhaps the most important attribute of the KHN biquad is its low sensitivities to component values. It can be shown that the sensitivity of ω_n with respect to all values is equal to 0.5 and

$$\left|S^Q_{R1,R2,C1,C2}\right| = 0.5 \tag{15.109}$$

$$\left|S^Q_{R4,R5}\right| = \frac{R_5}{R_4 + R_5} < 1 \tag{15.110}$$

$$\left|S^Q_{R3,R6}\right| = \frac{Q}{2}\frac{|R_3 - R_6|}{1 + \frac{R_5}{R_4}}\sqrt{\frac{R_2 C_2}{R_3 R_6 R_1 C_1}}. \tag{15.111}$$

Interesting, if $R_3 = R_6$, then $S^Q_{R3,R6}$ vanishes.

The use of three op amps in the feedback loop of Fig. 15.36(b) raises concern regarding the stability of the circuit because each op amp contributes several poles. Careful simulations are necessary to avoid oscillation.

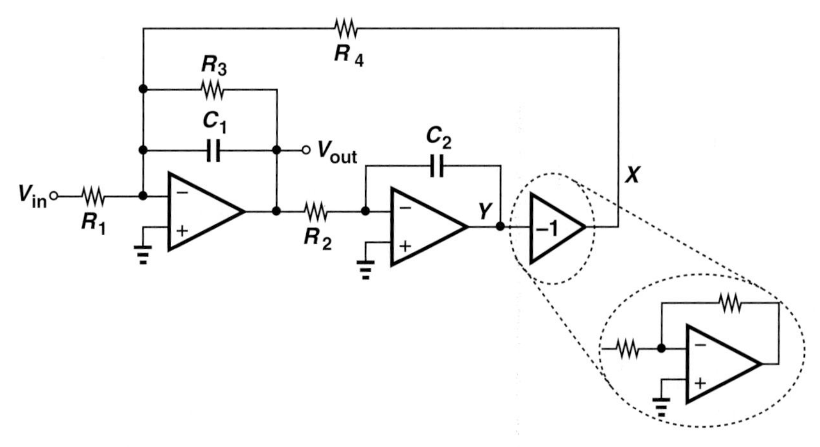

Figure 15.37 Tow-Thomas biquad.

Another type of biquad developed by Tow and Thomas is shown in Fig. 15.37. Here, the adder and the first integrator are *merged*, and resistor R_3 is introduced to create *lossy* integration. (Without R_3, the loop consisting of two ideal integrators oscillates.) Noting that $V_Y = -V_{out}/(R_2C_2s)$ and $V_X = -V_Y$, we sum the currents flowing through R_4 and R_1 and multiply the result by the parallel impedances of R_3 and C_1:

$$\left(\frac{V_{out}}{R_2C_2s}\cdot\frac{1}{R_4}+\frac{V_{in}}{R_1}\right)\frac{R_3}{R_3C_1s+1}=-V_{out}. \tag{15.112}$$

It follows that

$$\frac{V_{out}}{V_{in}}=-\frac{R_2R_3R_4}{R_1}\frac{C_2s}{R_2R_3R_4C_1C_2s^2+R_2R_4C_2s+R_3}, \tag{15.113}$$

which provides a band-pass response. The output at Y exhibits a low-pass behavior:

$$\frac{V_Y}{V_{in}}=\frac{R_3R_4}{R_1}\frac{1}{R_2R_3R_4C_1C_2s^2+R_2R_4C_2s+R_3}. \tag{15.114}$$

It can be shown that the sensitivities of the Tow-Thomas biquad with respect to the component values are equal to 0.5 or 1. An important advantage of this topology over the KHN biquad is accrued in integrated circuit design, where *differential* integrators obviate the need for the inverting stage in the loop, thus saving one op amp. Illustrated in Fig. 15.38, the idea is to swap the differential outputs of the second integrator to establish negative feedback.

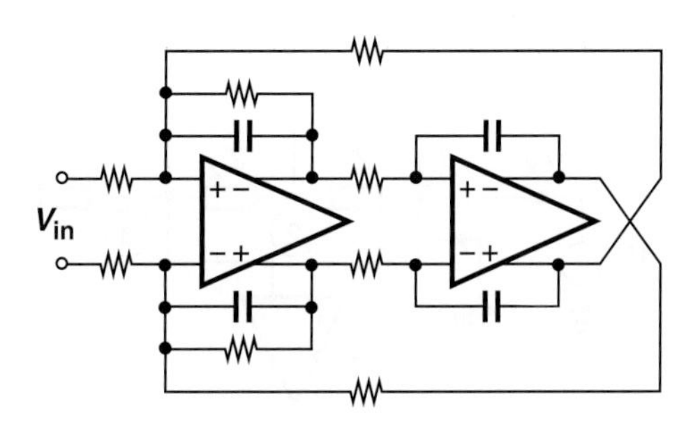

Figure 15.38 Differential Tow-Thomas filter.

Example 15-20 Prove that ω_n and Q of the Tow-Thomas filter can be adjusted (tuned) independently.

Solution From Eq. (15.114), we have

$$\omega_n = \frac{1}{\sqrt{R_2 R_4 C_1 C_2}} \tag{15.115}$$

and

$$Q^{-1} = \frac{1}{R_3}\sqrt{\frac{R_2 R_4 C_2}{C_1}}. \tag{15.116}$$

It is therefore possible to adjust ω_n by R_2 or R_4 and Q by R_3. As expected, if $R_3 = \infty$, then $Q = \infty$ and the circuit contains two purely imaginary poles.

Exercise A Tow-Thomas filter exhibits $\omega_n = 2\pi \times (10 \text{ MHz})$ and $Q = 5$. Is it possible to have $R_1 = R_2 = R_3$ and $C_1 = C_2$?

15.4.3 Biquads Using Simulated Inductors

Recall from Section 15.3.2 that second-order RLC circuits can provide low-pass, high-pass, or band-pass responses, but their usage in integrated circuits is limited because of the difficulty in building high-value, high-quality on-chip inductors. We may therefore ask: is it possible to emulate the behavior of an inductor by means of an active (inductorless) circuit?

Consider the circuit shown in Fig. 15.39, where general impedances $Z_1 - Z_5$ are placed in series and the feedback loops provided by the two (ideal) op amps force $V_1 - V_3$ and $V_3 - V_5$ to zero:

$$V_1 = V_3 = V_5 = V_X. \tag{15.117}$$

That is, the op amps establish a current of V_X/Z_5 through Z_5. This current flows through Z_4, yielding

$$V_4 = \frac{V_X}{Z_5} Z_4 + V_X. \tag{15.118}$$

Did you know?
In integrated circuits, active filters are much more common than passive filters. This is for two reasons: (1) passive filters tend to attenuate the signal, especially if they must provide high selectivity (i.e., a sharp transition in their frequency response); and (2) passive filters rely on high-Q inductors, which are difficult to realize on-chip (because the resistance of the integrated wires is relatively high). As a result, active filters using op amps, resistors, and capacitors are a more viable solution for integrated systems. Of course, the operation frequency of these filters is limited by the bandwidth of the op amps, posing research-worthy challenges.

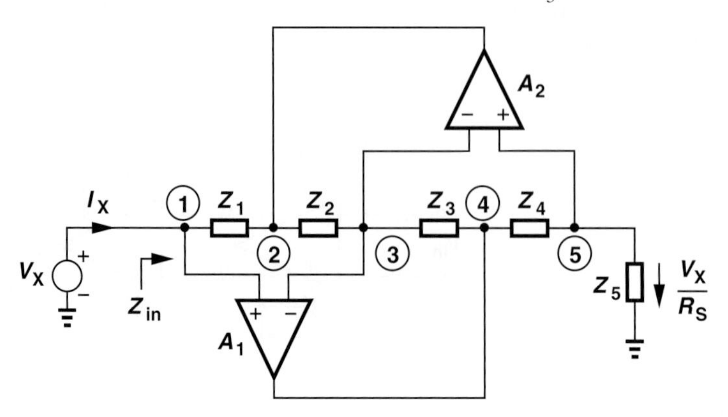

Figure 15.39 General impedance converter.

The current flowing through Z_3 (and hence through Z_2) is given by

$$I_{Z_3} = \frac{V_4 - V_3}{Z_3} \tag{15.119}$$

$$= \frac{V_X}{Z_5} \cdot \frac{Z_4}{Z_3}. \tag{15.120}$$

The voltage at node 2 is thus equal to

$$V_2 = V_3 - Z_2 I_{Z3} \tag{15.121}$$

$$= V_X - Z_2 \cdot \frac{V_X}{Z_5} \cdot \frac{Z_4}{Z_3}. \tag{15.122}$$

Finally,

$$I_X = \frac{V_X - V_2}{Z_1} \tag{15.123}$$

$$= V_X \frac{Z_2 Z_4}{Z_1 Z_3 Z_5} \tag{15.124}$$

and hence

$$Z_{in} = \frac{Z_1 Z_3}{Z_2 Z_4} Z_5. \tag{15.125}$$

The above result suggests that the circuit can "convert" Z_5 to a different type of impedance if $Z_1 - Z_4$ are chosen properly. For example, if $Z_5 = R_X$, $Z_2 = (Cs)^{-1}$, and $Z_1 = Z_3 = Z_4 = R_Y$ (Fig. 15.40), we have

$$Z_{in} = R_X R_Y Cs, \tag{15.126}$$

which is an inductor of value $R_X R_Y C$ (why?). We say the circuit converts a resistor to an inductor, i.e., it "simulates" an inductor.[8]

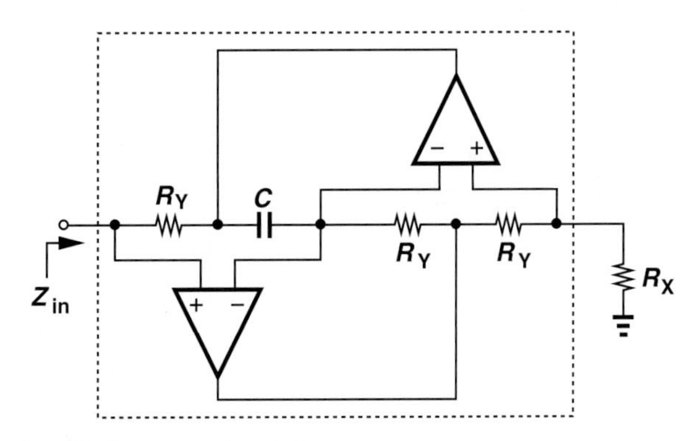

Figure 15.40 Example of inductance simulation.

[8]In today's terminology, we might call this an "emulated" inductor to avoid confusion with circuit simulation programs. But the term "simulated" has been used in this context since the 1960s.

Example 15-21 From Eq. (15.125), determine another possible combination of components that yields a simulated inductor.

Solution It is possible to choose $Z_4 = (Cs)^{-1}$ and the remaining passive elements to be resistors: $Z_1 = Z_2 = Z_3 = R_Y$. Thus,

$$Z_{in} = R_X R_Y Cs. \tag{15.127}$$

The resulting topology is depicted in Fig. 15.41.

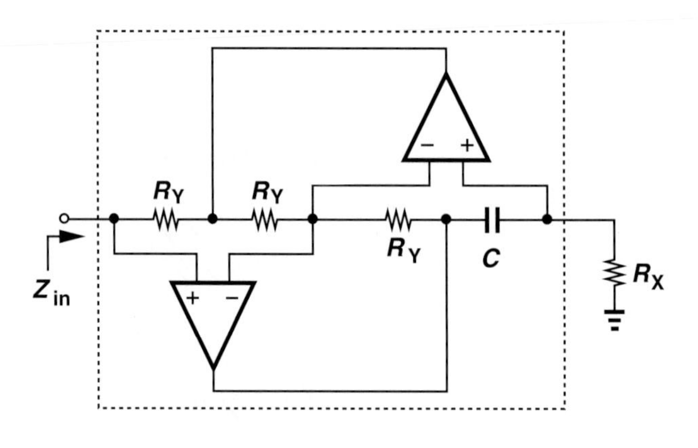

Figure 15.41

Exercise Is there yet another possible combination that yields a simulated inductor?

Introduced by Antoniou, the "general impedance converter" (GIC) in Fig. 15.39 and its descendants in Figs. 15.40 and 15.41 prove useful in transforming a passive RLC filter to an active counterpart. For example, as depicted in Fig. 15.42, a high-pass active section is obtained by replacing L_1 with a simulated inductor.

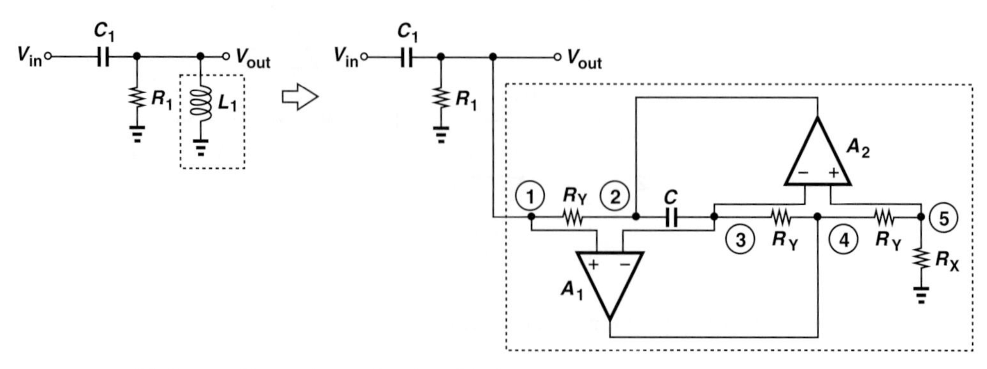

Figure 15.42 High-pass filter using a simulated inductor.

Example 15-22 Prove that node 4 in Fig. 15.42 can also serve as an output.

Solution Since $V_{out} = V_1 = V_3 = V_5$, the current flowing through R_X is equal to V_{out}/R_X, yielding

$$V_4 = \frac{V_{out}}{R_X} R_Y + V_{out} \tag{15.128}$$

$$= V_{out}\left(1 + \frac{R_Y}{R_X}\right). \tag{15.129}$$

Thus, V_4 is simply an amplified version of V_{out}. Driven by op amp A_1, this port exhibits a lower output impedance than does node 1, and is often utilized as the output of the circuit.

Exercise Determine the transfer function from V_{in} to V_3.

How is a *low-pass* filter derived? From the RLC network of Fig. 15.29, we note the need for a *floating* (rather than grounded) inductor and attempt to create such a device as shown in Fig. 15.43(a). Can this circuit serve as a floating inductor? A simple test is to tie a voltage source to node P and determine the Thevenin equivalent as seen from node Q [Fig. 15.43(b)]. To compute the open-circuit voltage, V_{Thev}, recall that the op amps force V_5 to be equal to V_P (= V_{in}). Since no current flows through R_X,

$$V_{Thev} = V_{in}. \tag{15.130}$$

To obtain Z_{Thev}, we set V_{in} to zero and apply a voltage to the left terminal of R_X [Fig. 15.43(c)]. Since $V_5 = V_1 = 0$, $I_X = V_X/R_X$ and hence

$$Z_{Thev} = R_X. \tag{15.131}$$

Unfortunately, the network operates as a simple resistor rather than a floating inductor! Fortunately, the impedance converter of Fig. 15.39 can overcome this difficulty. Consider the special case illustrated in Fig. 15.44(a), where $Z_1 = (Cs)^{-1}$, $Z_3 = (Cs)^{-1}||R_X$, and $Z_2 = Z_4 = Z_5 = R_X$. From Eq. (15.125), we have

$$Z_{in} = \frac{\frac{1}{Cs} \cdot \frac{R_X}{R_X Cs + 1} \cdot R_X}{R_X^2} \tag{15.132}$$

$$= \frac{1}{Cs(R_X Cs + 1)}. \tag{15.133}$$

This impedance may be viewed as a "super capacitor" because it is equal to the product of two capacitive components: $(Cs)^{-1}$ and $(R_X Cs + 1)^{-1}$.

Now, let us study the circuit depicted in Fig. 15.44(b):

$$\frac{V_{out}}{V_{in}} = \frac{Z_{in}}{Z_{in} + R_1} \tag{15.134}$$

$$= \frac{1}{R_1 R_X C^2 s^2 + R_1 Cs + 1}. \tag{15.135}$$

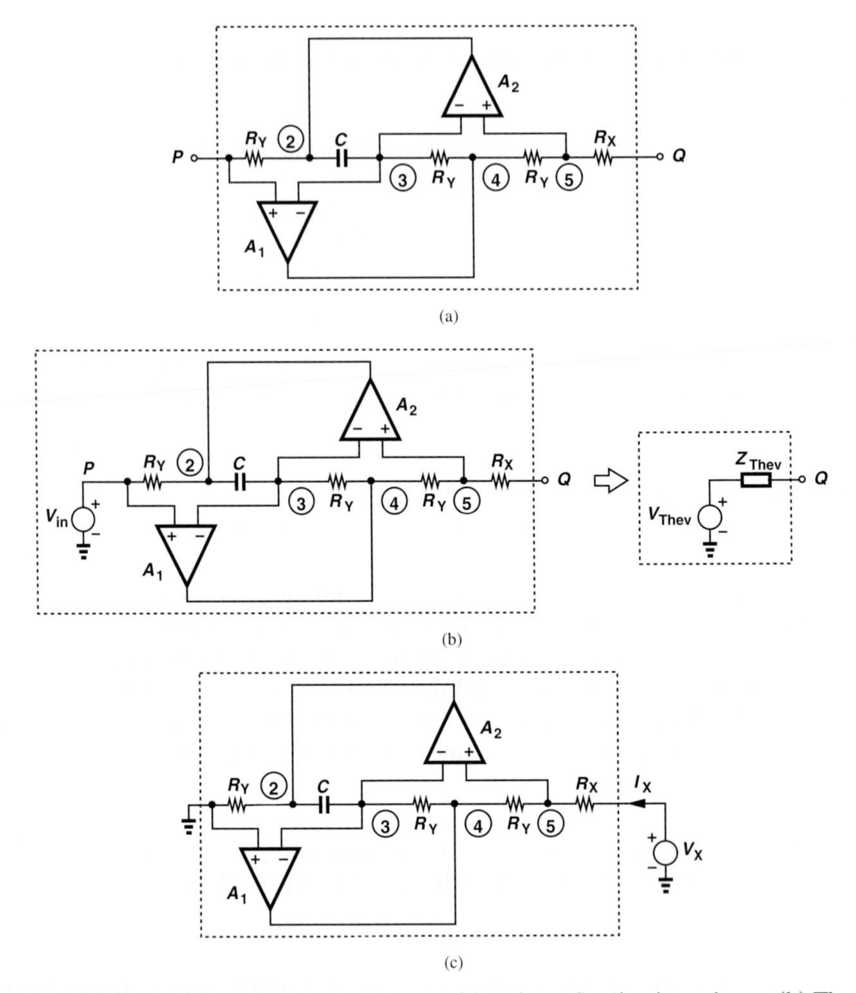

Figure 15.43 (a) General impedance converter considered as a floating impedance, (b) Thevenin equivalent, (c) test circuit for obtaining the output impedance.

This topology thus provides a second-order low-pass response. From Example 15-22, we note that node 4 serves as a better output port.

Example 15-23 Excited by the versatility of the general impedance converter, a student constructs the circuit shown in Fig. 15.45 as an alternative to that in Fig. 15.44. Explain why this topology is less useful.

Solution Employing $Z_3 = R_X$ and $Z_5 = (Cs)^{-1}||R_X$, this configuration provides the same transfer function as Eq. (15.135). However, V_4 is no longer a scaled version of V_{out}:

$$V_4 = \left[V_{out}\left(\frac{1}{R_X} + Cs\right)\right] R_X + V_{out} \tag{15.136}$$

$$= V_{out}(2 + R_X Cs). \tag{15.137}$$

Thus, the output can be sensed at only node 1, suffering from a relatively high impedance.

Exercise Determine the transfer function from V_{in} to V_5.

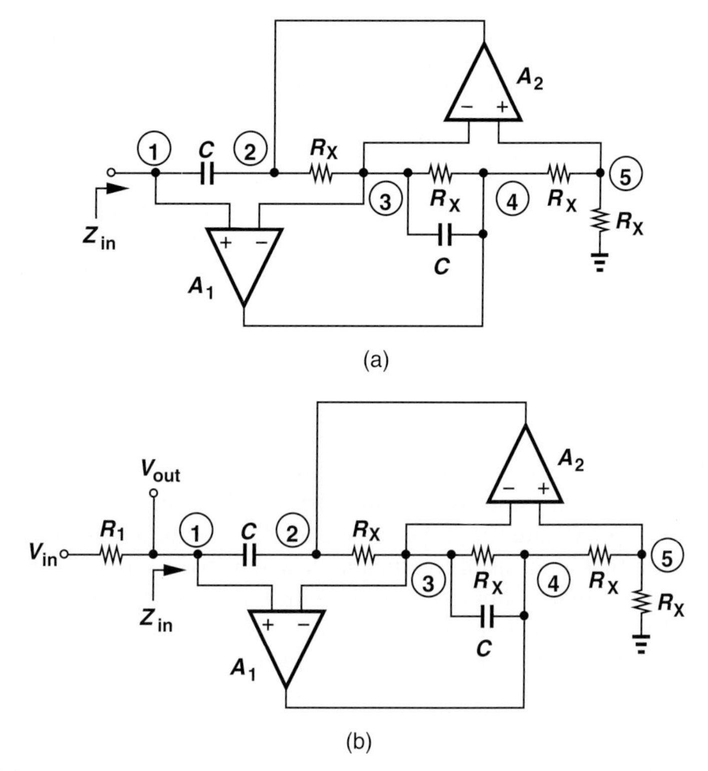

Figure 15.44 (a) General impedance converter producing a "super capacitor," (b) second-order low-pass filter obtained from (a).

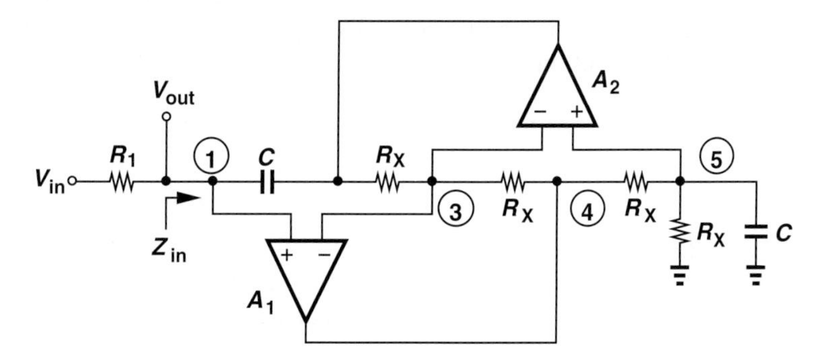

Figure 15.45

It can be proved that the sensitivities of the general impedance converter and the resulting filters with respect to component values are equal to 0.5 or 1. Such circuits therefore prove useful in both discrete and integrated design.

15.5 APPROXIMATION OF FILTER RESPONSE

How does the design of a filter begin? Based on the expected levels of the desired signal and the interferers, we decide on the required stopband attenuation. Next, depending on how close the interferer frequency is to the desired signal frequency, we choose a slope for the transition band. Finally, depending on the nature of the desired signal (audio, video, etc.), we select the tolerable passband ripple (e.g., 0.5 dB). We thus arrive at a "template" such as that shown in Fig. 15.46 for the frequency response of the filter.

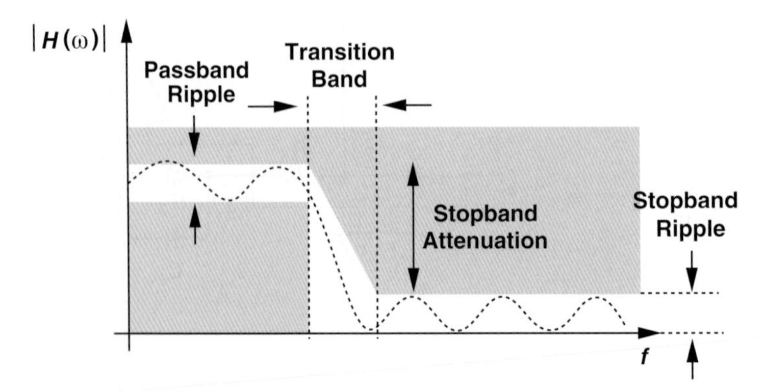

Figure 15.46 Frequency response template.

With the template in hand, how do we determine the required transfer function? This task is called the "approximation problem," by which we mean a transfer function is chosen to approximate the response dictated by the template. We prefer to select a transfer function that lends itself to efficient, low-sensitivity circuit realization.

A multitude of approximations with various trade-offs exist that prove useful in practice. Examples include "Butterworth," "Chebyshev," "elliptic," and "Bessel" responses. Most filters suffer from a trade-off between the passband ripple and the transition band slope. We study the first two types here and refer the reader to texts on filter design [1] for others.

15.5.1 Butterworth Response

The Butterworth response completely avoids ripple in the passband (and the stopband) at the cost of the transition band slope. This type of response only stipulates the *magnitude* of the transfer function as:

$$|H(j\omega)| = \frac{1}{\sqrt{1 + \left(\dfrac{\omega}{\omega_0}\right)^{2n}}}, \tag{15.138}$$

where n denotes the order of the filter.[9]

It is instructive to examine Eq. (15.138) carefully and understand its properties. First, we observe that the −3 dB bandwidth is calculated as:

$$\frac{1}{\sqrt{1 + \left(\dfrac{\omega_{-3\mathrm{dB}}}{\omega_0}\right)^{2n}}} = \frac{1}{\sqrt{2}} \tag{15.139}$$

and hence

$$\omega_{-3\mathrm{dB}} = \omega_0. \tag{15.140}$$

Interestingly, the −3 dB bandwidth remains independent of the order of the filter. Second, as n increases, the response assumes a sharper transition band and a greater passband flatness. Third, the response exhibits no ripple (local maxima or minima) because the first derivative of Eq. (15.138) with respect to ω vanishes only at $\omega = 0$. Figure 15.47 illustrates these points.

[9]This is called a "maximally-flat" response because the first $2n - 1$ derivatives of Eq. (15.138) vanish at $\omega = 0$.

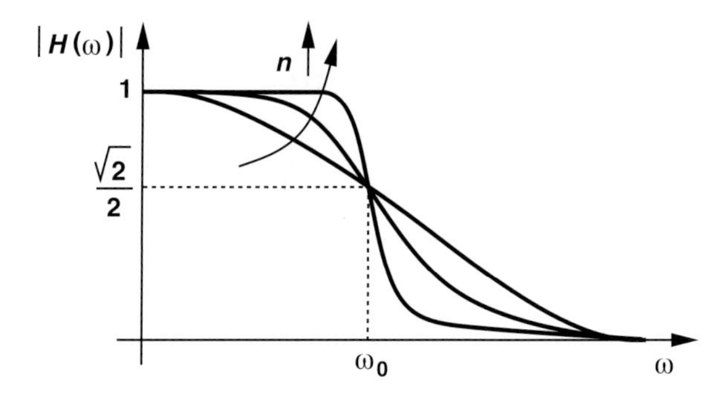

Figure 15.47 Butterworth response.

Example 15-24 A low-pass filter must provide a passband flatness of 0.45 dB for $f < f_1 = 1$ MHz and a stopband attenuation of 9 dB at $f_2 = 2$ MHz. Determine the order of a Butterworth filter satisfying these requirements.

Solution Figure 15.48 shows the template of the desired response. Noting that $|H(f_1 = 1\text{ MHz})| = 0.95$ (≈ -0.45 dB) and $|H(f_2 = 2\text{ MHz})| = 0.355$ (≈ -9 dB),

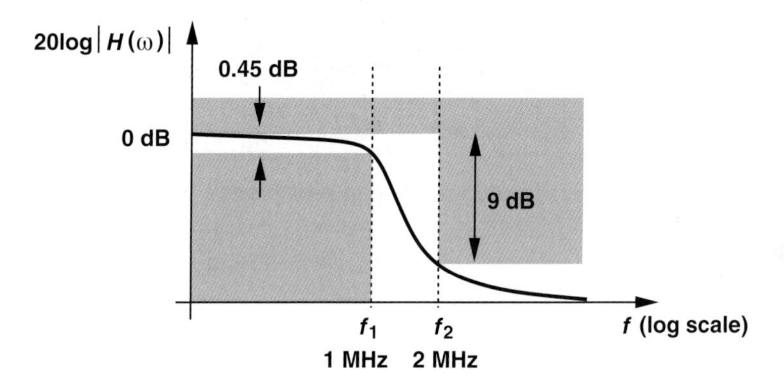

Figure 15.48

we construct two equations with two unknowns:

$$\frac{1}{1+\left(\dfrac{2\pi f_1}{\omega_0}\right)^{2n}} = 0.95^2 \tag{15.141}$$

$$\frac{1}{1+\left(\dfrac{2\pi f_2}{\omega_0}\right)^{2n}} = 0.355^2. \tag{15.142}$$

The former yields

$$\omega_0^{2n} = \frac{(2\pi f_1)^{2n}}{0.108}, \tag{15.143}$$

which, upon substitution in the latter, leads to

$$\left(\frac{f_2}{f_1}\right)^{2n} = 64.2. \tag{15.144}$$

Since $f_2 = 2f_1$, the smallest n that satisfies the requirement is 3. With $n = 3$, we obtain ω_0 from Eq. (15.141): $\omega_0 = 2\pi \times (1.45\ \text{MHz})$.

Exercise If the order of the filter must not exceed 2, how much attenuation can be obtained from f_1 to f_2?

Given filter specifications and hence a template, we can readily choose ω_0 and n in Eq. (15.138) to arrive at an acceptable Butterworth approximation. But how do we translate Eq. (15.138) to a *transfer function* and hence a circuit realization? Equation (15.138) suggests that the corresponding transfer function contains no zeros. To obtain the poles, we make a reverse substitution, $\omega = s/j$, and set the denominator to zero:

$$1 + \left(\frac{s}{j\omega_0}\right)^{2n} = 0. \tag{15.145}$$

That is,

$$s^{2n} + (-1)^n \omega_0^{2n} = 0. \tag{15.146}$$

This polynomial has $2n$ roots given by

$$p_k = \omega_0 \exp\frac{j\pi}{2} \exp\left(j\frac{2k-1}{2n}\pi\right),\quad k = 1, 2, \ldots, 2n, \tag{15.147}$$

but only the roots having a negative real part are acceptable (why?):

$$p_k = \omega_0 \exp\frac{j\pi}{2} \exp\left(j\frac{2k-1}{2n}\pi\right),\quad k = 1, 2, \ldots, n. \tag{15.148}$$

How are these poles located in the complex plane? As an example, suppose $n = 2$. Then,

$$p_1 = \omega_0 \exp\left(j\frac{3\pi}{4}\right) \tag{15.149}$$

$$p_2 = \omega_0 \exp\left(j\frac{5\pi}{4}\right). \tag{15.150}$$

As shown in Fig. 15.49(a), the poles are located at $\pm 135°$, i.e., their real and imaginary parts are equal in magnitude. For larger values of n, each pole falls on a circle of radius ω_0 and bears an angle of π/n with respect to the next pole [Fig. 15.49(b)].

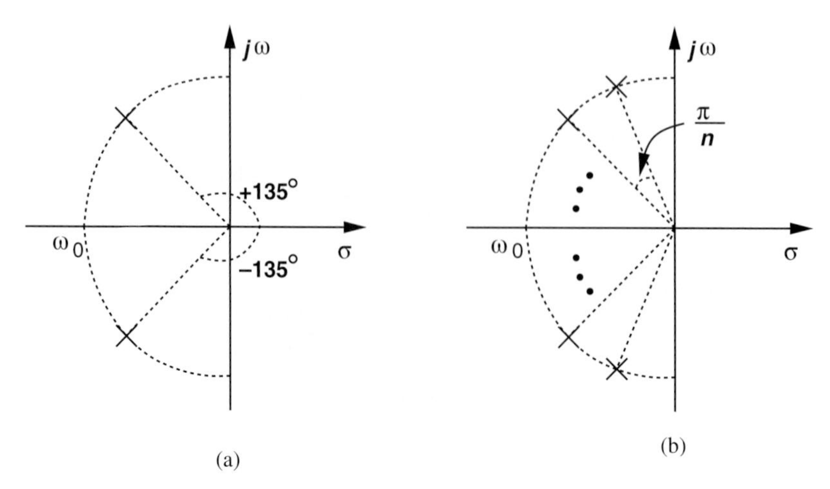

Figure 15.49 Locations of poles for (a) second-order, and (b) nth-order Butterworth filter.

Having obtained the poles, we now express the transfer function as

$$H(s) = \frac{(-p_1)(-p_2)\cdots(-p_n)}{(s-p_1)(s-p_2)\cdots(s-p_n)} \tag{15.151}$$

where the factor in the numerator is included to yield $H(s=0)=1$.

Example 15-25 Using a Sallen and Key topology as the core, design a Butterworth filter for the response derived in Example 15-24.

Solution With $n=3$ and $\omega_0 = 2\pi \times (1.45\text{ MHz})$, the poles appear as shown in Fig. 15.50(a). The complex conjugate poles p_1 and p_3 can be created by a second-order SK filter, and the real pole p_2 by a simple RC section. Since

$$p_1 = 2\pi \times (1.45\text{ MHz}) \times \left(\cos\frac{2\pi}{3} + j\sin\frac{2\pi}{3}\right) \tag{15.152}$$

$$p_3 = 2\pi \times (1.45\text{ MHz}) \times \left(\cos\frac{2\pi}{3} - j\sin\frac{2\pi}{3}\right), \tag{15.153}$$

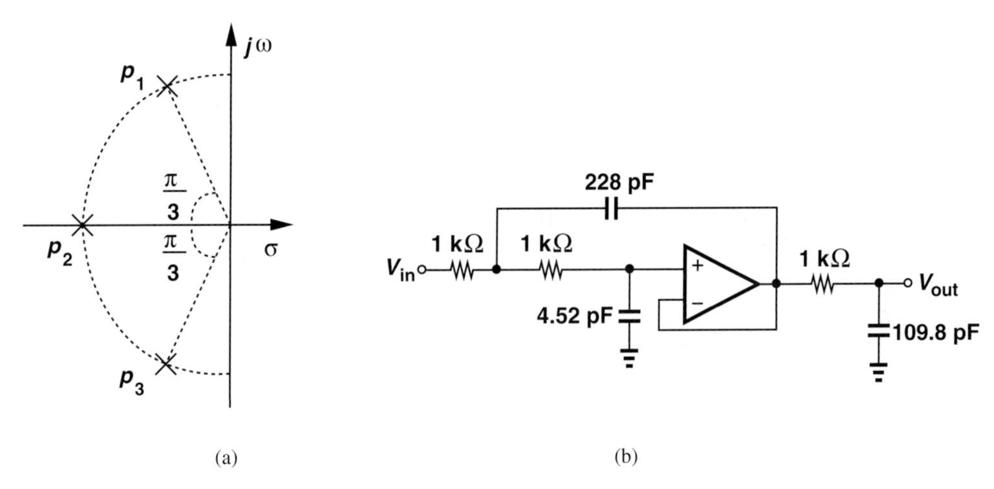

Figure 15.50

the SK transfer function can be written as

$$H_{SK}(s) = \frac{(-p_1)(-p_3)}{(s-p_1)(s-p_3)} \tag{15.154}$$

$$= \frac{[2\pi \times (1.45\text{ MHz})]^2}{s^2 - [4\pi \times (1.45\text{ MHz})\cos(2\pi/3)]s + [2\pi \times (1.45\text{ MHz})]^2}. \tag{15.155}$$

That is,

$$\omega_n = 2\pi \times (1.45\text{ MHz}) \tag{15.156}$$

$$Q = \frac{1}{2\cos\frac{2\pi}{3}} = 1. \tag{15.157}$$

In Eq. (15.61), we choose $R_1 = R_2$ and $C_1 = 4C_2$ to obtain $Q = 1$. From Eq. (15.62), to obtain $\omega_n = 2\pi \times (1.45\ \text{MHz}) = \left(\sqrt{4R_1^2C_2^2}\right)^{-1} = (2R_1C_2)^{-1}$, we have some freedom, e.g., $R_1 = 1\text{k}\Omega$ and $C_2 = 54.9$ pF. The reader is encouraged to verify that this design achieves low sensitivities.

The real pole, p_2, is readily created by an RC section:

$$\frac{1}{R_3C_3} = 2\pi \times (1.45\ \text{MHz}). \tag{15.158}$$

For example, $R_3 = 1\ \text{k}\Omega$ and $C_3 = 109.8$ pF. Figure 15.50(b) shows the resulting design.

Exercise If the 228 pF capacitor incurs an error of 10%, determine the error in the value of f_1.

The Butterworth response is employed only in rare cases where no ripple in the passband can be tolerated. We typically allow a small ripple (e.g., 0.5 dB) so as to exploit responses that provide a sharper transition slope and hence a greater stopband attenuation. The Chebyshev response is one such example.

15.5.2 Chebyshev Response

The Chebyshev response provides an "equiripple" passband behavior, i.e., with equal local maxima (and equal local minima). This type of response specifies the magnitude of the transfer function as:

$$|H(j\omega)| = \frac{1}{\sqrt{1 + \epsilon^2 C_n^2\left(\dfrac{\omega}{\omega_0}\right)}}, \tag{15.159}$$

where ϵ sets the amount of ripple and $C_n^2(\omega/\omega_0)$ denotes the "Chebyshev polynomial" of nth order. We consider ω_0 as the "bandwidth" of the filter. These polynomials are expressed recursively as

$$C_1\left(\frac{\omega}{\omega_0}\right) = \frac{\omega}{\omega_0} \tag{15.160}$$

$$C_2\left(\frac{\omega}{\omega_0}\right) = 2\left(\frac{\omega}{\omega_0}\right)^2 - 1 \tag{15.161}$$

$$C_3\left(\frac{\omega}{\omega_0}\right) = 4\left(\frac{\omega}{\omega_0}\right)^3 - 3\frac{\omega}{\omega_0} \tag{15.162}$$

$$C_{n+1}\left(\frac{\omega}{\omega_0}\right) = 2\frac{\omega}{\omega_0}C_n\left(\frac{\omega}{\omega_0}\right) - C_{n-1}\left(\frac{\omega}{\omega_0}\right) \tag{15.163}$$

or, alternatively, as

$$C_n\left(\frac{\omega}{\omega_0}\right) = \cos\left(n\cos^{-1}\frac{\omega}{\omega_0}\right) \quad \omega < \omega_0 \tag{15.164}$$

$$= \cosh\left(n\cosh^{-1}\frac{\omega}{\omega_0}\right) \quad \omega > \omega_0. \tag{15.165}$$

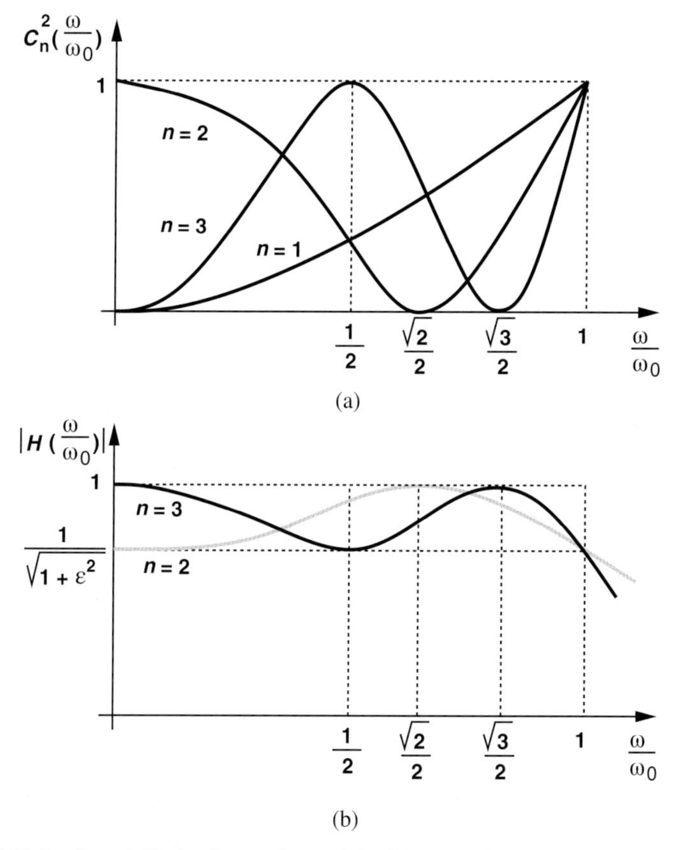

Figure 15.51 (a) Behavior of Chebyshev polynomials, (b) second- and third-order Chebyshev responses.

As illustrated in Fig. 15.51(a), higher-order polynomials experience a greater number of fluctuations between 0 and 1 in the range of $0 \leq \omega/\omega_0 \leq 1$, and monotonically rise thereafter. Scaled by ϵ^2, these fluctuations lead to n ripples in the passband of $|H|$ [Fig. 15.51(b)].

Example 15-26 Suppose the filter required in Example 15-24 is realized with a third-order Chebyshev response. Determine the attenuation at 2 MHz.

Solution For a flatness (ripple) of 0.45 dB in the passband:

$$\frac{1}{\sqrt{1+\epsilon^2}} = 0.95, \tag{15.166}$$

and hence

$$\epsilon = 0.329. \tag{15.167}$$

Also, $\omega_0 = 2\pi \times (1\ \text{MHz})$ because the response departs from unity by 0.45 dB at this frequency. It follows that

$$|H(j\omega)| = \frac{1}{\sqrt{1 + 0.329^2 \left[4\left(\frac{\omega}{\omega_0}\right)^3 - 3\frac{\omega}{\omega_0}\right]^2}}. \tag{15.168}$$

At $\omega_2 = 2\pi \times (2\text{ MHz})$,

$$|H(j\omega_2)| = 0.116 \tag{15.169}$$

$$= -18.7\text{ dB}. \tag{15.170}$$

Remarkably, the stopband attenuation improves by 9.7 dB if a Chebyshev response is employed.

Exercise How much attenuation can be obtained if the order if raised to four?

Let us summarize our understanding of the Chebyshev response. As depicted in Fig. 15.52, the magnitude of the transfer function in the passband is given by

$$|H_{PB}(j\omega)| = \frac{1}{\sqrt{1 + \epsilon^2 \cos^2\left(n \cos^{-1}\dfrac{\omega}{\omega_0}\right)}}, \tag{15.171}$$

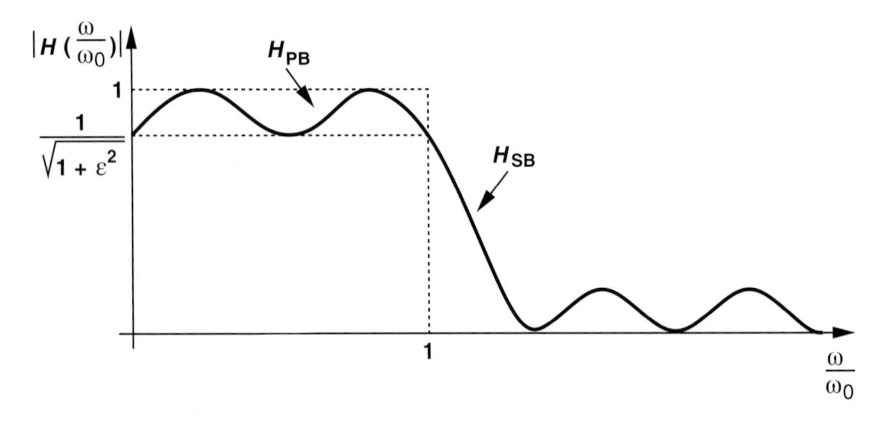

Figure 15.52 General Chebyshev response.

exhibiting a peak-to-peak ripple of

$$\text{Ripple}|_{\text{dB}} = 20 \log \sqrt{1 + \epsilon^2}. \tag{15.172}$$

In the stopband,

$$|H_{SB}(j\omega)| = \frac{1}{\sqrt{1 + \epsilon^2 \cosh^2\left(n \cosh^{-1}\dfrac{\omega}{\omega_0}\right)}}, \tag{15.173}$$

revealing the attenuation at frequencies greater than ω_0. In practice, we must determine n so as to obtain required values of ω_0, ripple, and stopband attenuation.

Example 15-27 A Chebyshev filter must provide a passband ripple of 1 dB across a bandwidth of 5 MHz and an attenuation of 30 dB at 10 MHz. Determine the order of the filter.

Solution We set ω_0 to $2\pi \times (5\text{ MHz})$ and write

$$1\text{ dB} = 20\log\sqrt{1+\epsilon^2}, \tag{15.174}$$

arriving at

$$\epsilon = 0.509. \tag{15.175}$$

Now, we equate Eq. (15.173) to 0.0316 (= −30 dB) at $\omega = 2\omega_0$:

$$\frac{1}{\sqrt{1+0.509^2\cosh^2(n\cosh^{-1}2)}} = 0.0316. \tag{15.176}$$

Since $\cosh^{-1} 2 \approx 1.317$, Eq. (15.176) yields

$$\cosh^2(1.317n) = 3862 \tag{15.177}$$

and hence

$$n > 3.66. \tag{15.178}$$

We must therefore select $n = 4$.

Exercise If the order is limited to three, how much attenuation can be obtained at 10 MHz?

With the order of the response determined, the next step in the design is to obtain the poles and hence the transfer function. It can be shown [1] that the poles are given by

$$p_k = -\omega_0 \sin\frac{(2k-1)\pi}{2n}\sinh\left(\frac{1}{n}\sinh^{-1}\frac{1}{\epsilon}\right) + j\omega_0\cos\frac{(2k-1)\pi}{2n}\cosh\left(\frac{1}{n}\sinh^{-1}\frac{1}{\epsilon}\right)$$

$$k = 1, 2, \ldots, n. \tag{15.179}$$

(The poles, in fact, reside on an ellipse.) The transfer function is then expressed as

$$H(s) = \frac{(-p_1)(-p_2)\cdots(-p_n)}{(s-p_1)(s-p_2)\cdots(s-p_n)}. \tag{15.180}$$

Example 15-28 Using two SK stages, design a filter that satisfies the requirements in Example 15-27.

Solution With $\epsilon = 0.509$ and $n = 4$, we have

$$p_k = -0.365\omega_0\sin\frac{(2k-1)\pi}{8} + 1.064j\omega_0\cos\frac{(2k-1)\pi}{8}, \quad k = 1, 2, 3, 4, \tag{15.181}$$

which can be grouped into two sets of conjugate poles

$$p_{1,4} = -0.365\omega_0\sin\frac{\pi}{8} \pm 1.064j\omega_0\cos\frac{\pi}{8} \tag{15.182}$$

$$= -0.140\omega_0 \pm 0.983j\omega_0 \tag{15.183}$$

$$p_{2,3} = -0.365\omega_0\sin\frac{3\pi}{8} \pm 1.064j\omega_0\cos\frac{3\pi}{8} \tag{15.184}$$

$$= -0.337\omega_0 \pm 0.407\,j\omega_0. \tag{15.185}$$

Figure 15.53(a) plots the pole locations. We note that p_1 and p_4 fall close to the imaginary axis and thus exhibit a relatively high Q. The SK stage for p_1 and p_4 is characterized by the following transfer function:

$$H_{SK1}(s) = \frac{(-p_1)(-p_4)}{(s-p_1)(s-p_4)} \tag{15.186}$$

$$= \frac{0.986\omega_0^2}{s^2 + 0.28\omega_0 s + 0.986\omega_0^2}, \tag{15.187}$$

indicating that

$$\omega_{n1} = 0.993\omega_0 = 2\pi \times (4.965 \text{ MHz}) \tag{15.188}$$

$$Q_1 = 3.55. \tag{15.189}$$

Equation (15.61) suggests that such a high Q can be obtained only if C_1/C_2 is large. For example, with $R_1 = R_2$, we require

$$C_1 = 50.4\, C_2 \tag{15.190}$$

and hence

$$\frac{1}{\sqrt{50.4 R_1 C_2}} = 2\pi \times (4.965 \text{ MHz}). \tag{15.191}$$

If $R_1 = 1\text{ k}\Omega$, then $C_2 = 4.52$ pF. (For discrete implementations, this value of C_2 is excessively small, necessitating that R_1 be scaled down by a factor of, say, 5.)

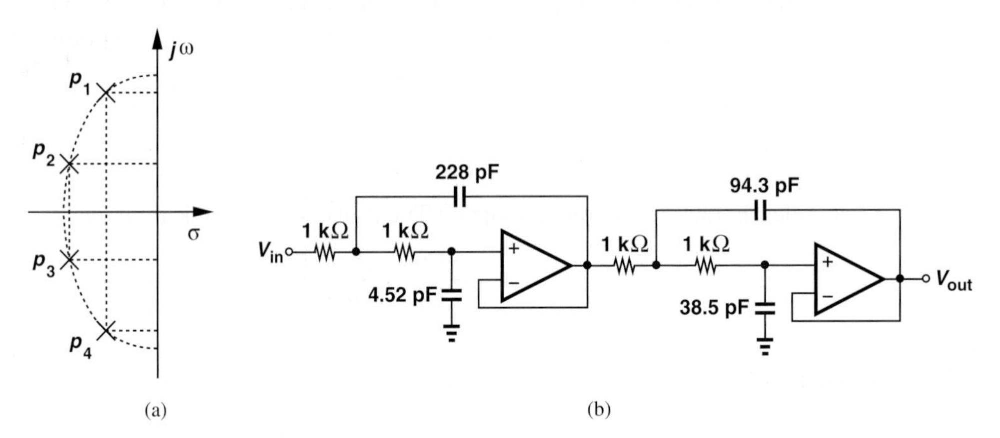

Figure 15.53

Similarly, the SK stage for p_2 and p_3 satisfies

$$H_{SK2}(s) = \frac{(-p_2)(-p_3)}{(s-p_2)(s-p_3)} \tag{15.192}$$

$$= \frac{0.279\omega_0^2}{s^2 + 0.674\omega_0 s + 0.279\omega_0^2}, \tag{15.193}$$

and hence

$$\omega_{n2} = 0.528\omega_0 = 2\pi \times (2.64 \text{ MHz}) \tag{15.194}$$

$$Q_2 = 0.783. \tag{15.195}$$

If $R_1 = R_2 = 1\ \text{k}\Omega$, then (15.61) and (15.62) translate to

$$C_1 = 2.45 C_2 \quad (15.196)$$

$$= 2.45 \times (38.5\ \text{pF}). \quad (15.197)$$

Figure 15.53(b) shows the overall design. The reader is encouraged to compute the sensitivities.

Exercise Repeat the above example if capacitor values must exceed 50 pF.

15.6 CHAPTER SUMMARY

- Analog filters prove essential in removing unwanted frequency components that may accompany a desired signal.
- The frequency response of a filter consists of a passband, a stopband, and a transition band between the two. The passband and stopband may exhibit some ripple.
- Filters can be classified as low-pass, high-pass, band-pass, or band-reject topologies. They can be realized as continuous-time or discrete-time configurations, and as passive or active circuits.
- The frequency response of filters has dependencies on various component values and, therefore, suffers from "sensitivity" to component variations. A well-designed filter ensures a small sensitivity with respect to each component.
- First-order passive or active filters can readily provide a low-pass or high-pass response, but their transition band is quite wide and stopband attenuation only moderate.
- Second-order filters have a greater stopband attenuation and are widely used. For a well-behaved frequency and time response, the Q of these filters is typically maintained below $\sqrt{2}2$.
- Continuous-time passive second-order filters employ RLC sections, but they become impractical at very low frequencies (because of large physical size of inductors and capacitors).
- Active filters employ op amps, resistors, and capacitors to create the desired frequency response. The Sallen and Key topology is an example.
- Second-order active (biquad) sections can be based on integrators. Examples include the KHN biquad and the Tow-Thomas biquad.
- Biquads can also incorporate simulated inductors, which are derived from the "general impedance converter" (GIC). The GIC can yield large inductor or capacitor values through the use of two op amps.
- The desired filter response must in practice be approximated by a realizable transfer function. Possible transfer functions include Butterworth and Chebyshev responses.
- The Butterworth response contains n complex poles on a circle and exhibits a maximally-flat behavior. It is suited to applications that are intolerant of any ripple in the passband.
- The Chebyshev response provides a sharper transition than Butterworth at the cost of some ripple in the passband and stopbands. It contains n complex poles on an ellipse.

PROBLEMS

Sec. 15.1 General Considerations

15.1. Determine the type of response (low-pass, high-pass, or band-pass) provided by each network depicted in Fig. 15.54.

15.2. Derive the transfer function of each network shown in Fig. 15.54 and determine the poles and zeros.

SOL **15.3.** We wish to realize a transfer function of the form

$$\frac{V_{out}}{V_{in}}(s) = \frac{1}{(s+a)(s+b)}, \qquad (15.198)$$

where a and b are real and positive. Which one of the networks illustrated in Fig. 15.54 can satisfy this transfer function?

***15.4.** In some applications, the input to a filter may be provided in the form of a current. Compute the transfer function, V_{out}/I_{in}, of each of the circuits depicted in Fig. 15.55 and determine the poles and zeros.

Sec. 15.2 First-Order Filters

15.5. For the high-pass filter depicted in Fig. 15.56, determine the sensitivity of the pole and zero frequencies with respect to R_1 and C_1.

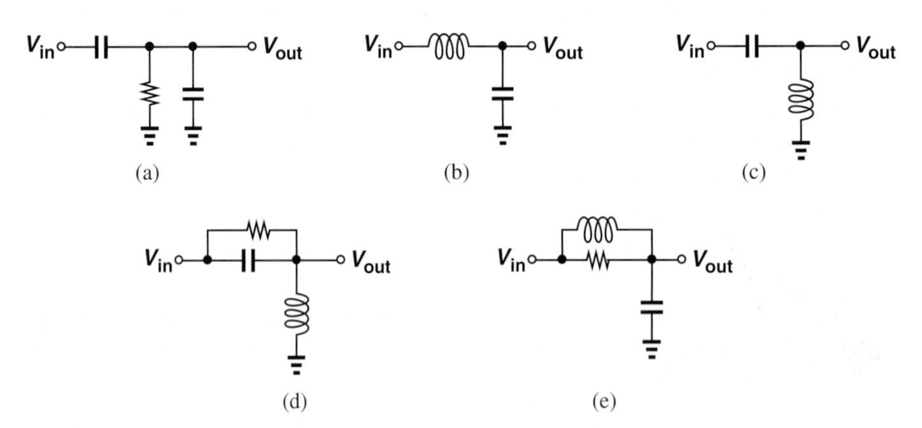

Figure 15.54

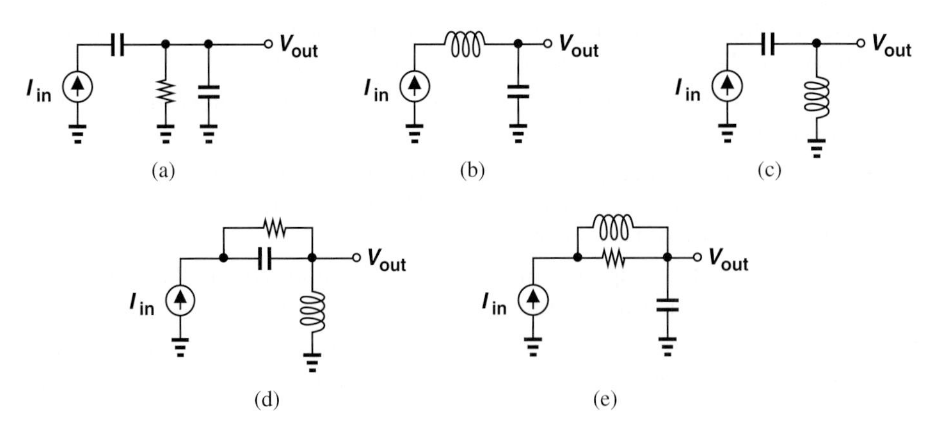

Figure 15.55

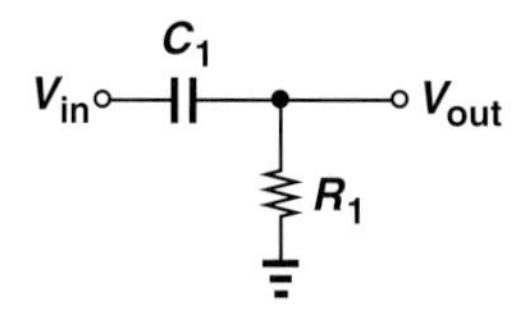

Figure 15.56

15.6. Consider the filter shown in Fig. 15.57. Compute the sensitivity of the pole and zero frequencies with respect to C_1, C_2, and R_1.

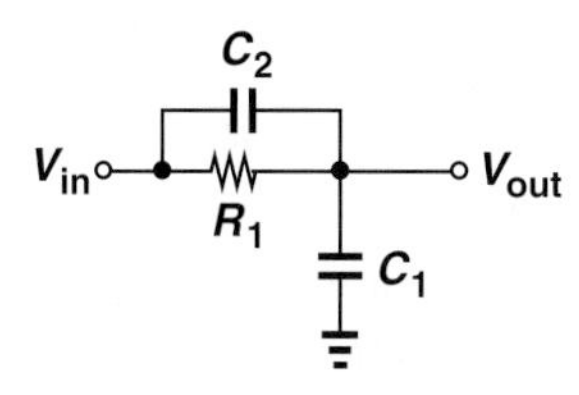

Figure 15.57

15.7. We wish to achieve a pole sensitivity of 5% in the circuit illustrated in Fig. 15.58. If R_1 exhibits a variability of 3%, what is the maximum tolerance of L_1?

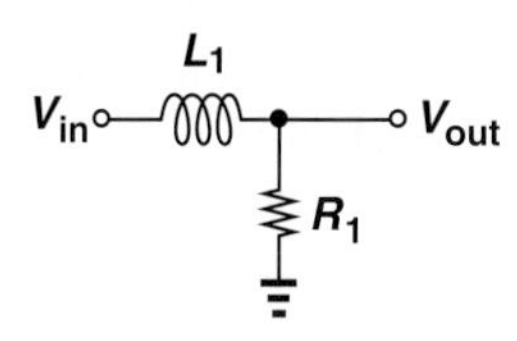

Figure 15.58

Sec. 15.3 Second-Order Filters

15.8. The low-pass filter of Fig. 15.59 is designed to contain two real poles.

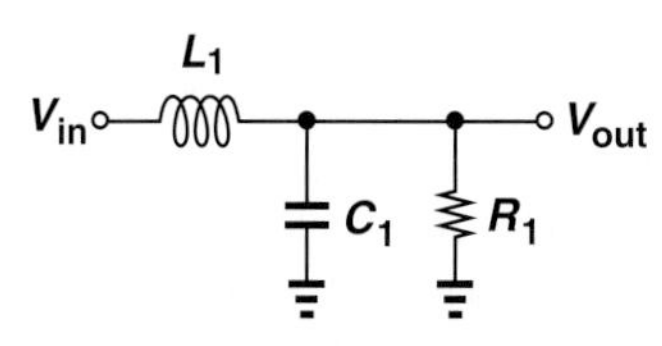

Figure 15.59

(a) Derive the transfer function.
(b) Compute the poles and the condition that guarantees they are real.
(c) Calculate the pole sensitivities to R_1, C_1, and L_1.

15.9. Explain what happens to the transfer functions of the circuits in Figs. 15.17(a) and 15.18(a) if the pole and zero coincide.

15.10. For what value of Q do the poles of the biquadratic transfer function, (15.23), coincide? What are the resulting pole frequencies? SOL

15.11. Prove that the response expressed by Eq. (15.25) exhibits no peaking (no local maximum) if $Q \leq \sqrt{2}/2$.

****15.12.** Prove that the response expressed by Eq. (15.25) reaches a normalized peak of $Q/\sqrt{1-(4Q^2)^{-1}}$ if $Q > \sqrt{2}/2$. Sketch the response for $Q = 2, 4$, and 8.

***15.13.** Prove that the response expressed by Eq. (15.28) reaches a normalized peak of Q/ω_n at $\omega = \omega_n$. Sketch the response for $Q = 2, 4$, and 8.

15.14. Consider the parallel RLC tank depicted in Fig. 15.26. Plot the location of the poles of the circuit in the complex plane as R_1 goes from very small values to very large values while L_1 and C_1 remain constant.

***15.15.** Repeat Problem 15.14 if R_1 and L_1 remain constant and C_1 varies from very small values to very large values.

15.16. With the aid of the observations made for Eq. (15.25), determine a condition for the low-pass filter of Fig. 15.29 to exhibit a peaking of 1 dB (10%).

Sec. 15.4.1 Sallen and Key Filter

15.17. Determine the poles of the Sallen and Key filter shown in Fig. 15.33 and plot their location in the complex plane as (a) R_1 varies from zero to ∞, (b) R_2 varies from zero to ∞, (c) C_1 varies from zero to ∞, or (d) C_2 varies from zero to ∞. In each case, assume other component values remain constant.

***15.18.** A student mistakenly configures a Sallen and Key filter as shown in Fig. 15.60. Determine the transfer function and explain why this is not a useful circuit. SOL

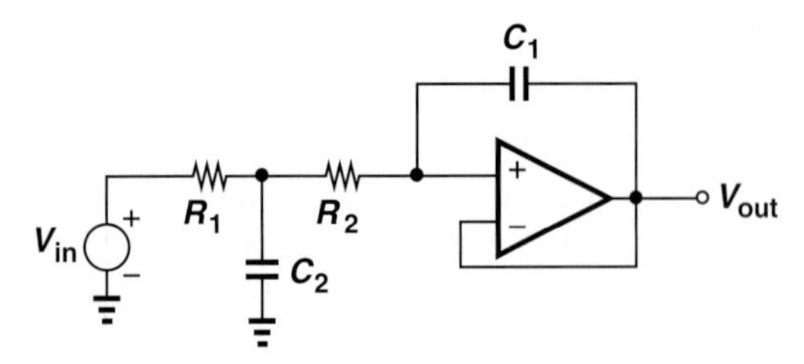

Figure 15.60

15.19. The Sallen and Key filter of Fig. 15.34 must be designed with $K = 4$ and $C_1 = C_2$. How should R_1/R_2 be chosen to yield $Q = 4$? What is the resulting Q sensitivity to R_1?

15.20. A Sallen and Key filter with $K = 1$ must exhibit a peaking of only 1 dB in its response. Determine the relationship required among the component values.

15.21. The Sallen and Key filter of Fig. 15.33 exhibits $S_{R1}^{Q} = 2$. If $C_2 = C_1$, plot Q as a function of $\sqrt{R_2/R_1}$ and determine the acceptable range of values of Q and $\sqrt{R_2/R_1}$.

***15.22.** Figure 15.61 shows a high-pass Sallen and Key filter. Derive the transfer function and determine Q and ω_n.

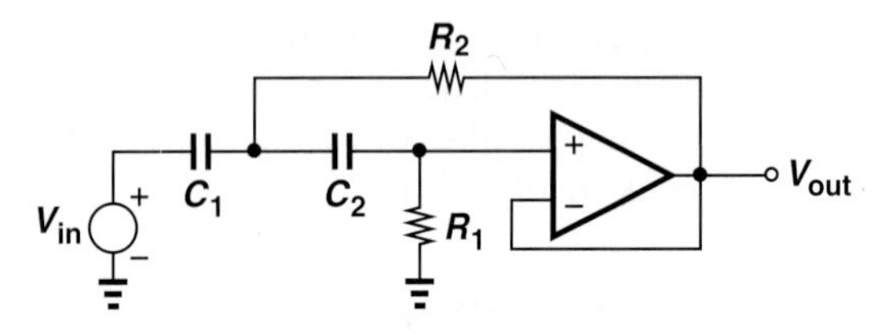

Figure 15.61

****15.23.** From the results obtained in Problem 15.22, compute the Q sensitivities of the circuit.

Sec. 15.4.2 Integrator-Based Biquads

15.24. It is possible to realize the transfer function of Eq. (15.95) by means of differentiators rather than integrators. Noting that the factor s in the frequency domain translates to d/dt in the time domain, construct a block diagram such as that shown in Fig. 15.36(a) but using only differentiators. (Due to amplification of noise at high frequencies, this implementation is less popular.)

15.25. The KHN biquad of Fig. 15.36(b) must provide a band-pass response with $Q = 2$ and $\omega_n = 2\pi \times (2\ \text{MHz})$. If $R_6 = R_3$, $R_1 = R_2$, and $C_1 = C_2$, determine the resistor and capacitor values subject to the restrictions $10\ \text{pF} < C < 1\ \text{nF}$ and $1\ \text{k}\Omega < R < 50\ \text{k}\Omega$.

15.26. From Eqs. (15.103) and (15.104), derive an expression for Q and explain why the sensitivity to R_3 and R_6 vanishes if $R_3 = R_6$. SOL

15.27. The KHN biquad of Fig. 15.36(b) must be designed for a low-pass response with a low-frequency gain $\alpha = 2$. Explain why this is impossible if $S_{R3,R6}^{Q}$ must be zero.

15.28. A KHN biquad must exhibit a peaking of only 1 dB in its low-pass response. Determine the relationship required among the component values. Assume $R_3 = R_6$ and the gain is unity.

***15.29.** A student mistakenly omits resistor R_5 from the KHN biquad of Fig. 15.36(b). Derive the resulting transfer function V_{out}/V_{in} and determine α, Q, and ω_n.

****15.30.** Determine the sensitivities of the Tow-Thomas filter shown in Fig. 15.37 with respect to resistor and capacitor values.

15.31. Equation (15.114) implies that the low-frequency gain of the Tow-Thomas filter is equal to R_4/R_1. Setting C_1 and C_2 to zero in Fig. 15.37, explain intuitively why this result makes sense.

***15.32.** The Tow-Thomas filter of Fig. 15.37 must be designed for a low-pass response having a peaking of 1 dB and a bandwidth $\omega_n = 2\pi \times (10\ \text{MHz})$. If $R_3 = 1\text{k}\Omega$, $R_2 = R_4$, and $C_1 = C_2$, determine the values of R_2 and C_1.

15.33. The transfer function in Eq. (15.114) reveals that resistor R_1 affects the low-frequency gain of the Tow-Thomas filter but not the frequency response. Replacing V_{in} and R_1 in Fig. 15.37 with a Norton SOL equivalent, explain intuitively why this result makes sense.

Sec. 15.4.3 Biquads Using Simulated Inductors

**15.34. For the general impedance converter of Fig. 15.39, determine all possible combinations of Z_1-Z_5 that yield an inductive behavior for Z_{in}. Assume each of Z_1-Z_5 consists of only one resistor or one capacitor. (Note that a solution is not acceptable if it does not provide a dc path to each input of the op amps.)

15.35. Repeat Problem 15.34 if a capacitive behavior for Z_{in} is required.

15.36. In Example 15-23, the parallel RC branch tied between node 5 and ground is replaced with a series branch $R_X + (Cs)^{-1}$. Determine the resulting transfer function V_{out}/V_{in}.

*15.37. Select the components in Fig. 15.39 such that the circuit provides a large capacitive impedance, i.e., it multiplies the value of a capacitor by a large number.

Sec. 15.5 Approximation of Filter Response

15.38. We wish to design a Butterworth filter with a roll-off of 1 dB at $\omega = 0.9\omega_0$. Determine the required order.

15.39. Using Eq. (15.138), plot the roll-off of a Butterworth response at $\omega = 0.9\omega_0$ as a function of n. Express the roll-off (on the vertical axis) in decibels.

15.40. Repeat Problem 15.39 for $\omega = 1.1\omega_0$. What order is required to obtain an attenuation of 20 dB at this frequency?

15.41. Suppose the filter of Example 15-24 receives an interferer at 5 MHz. How much attenuation does the filter provide?

15.42. A low-pass Butterworth filter must provide a passband flatness of 0.5 dB for $f < f_1 = 1$ MHz. If the order of the filter must not exceed 5, what is the greatest stopband attenuation at $f_2 = 2$ MHz?

*15.43. Explain why the poles expressed by Eq. (15.148) lie on a circle.

15.44. Repeat Example 15-25 but with an KHN biquad.

15.45. Repeat Example 15-25 but with a Tow-Thomas filter.

*15.46. Plot the Chebyshev response expressed by Eq. (15.159) for $n = 4$ and $\epsilon = 0.2$. Estimate the locations of the local maxima and minima in the passband.

15.47. A Chebyshev filter must provide an attenuation of 25 dB at 5 MHz. If the order of the filter must not exceed 5, what is the minimum ripple that can be achieved across a bandwidth of 2 MHz? SOL

15.48. Repeat Problem 15.47 for an order of 6 and compare the results.

15.49. Repeat Example 15-28 but with two KHN biquads.

15.50. Repeat Example 15-28 but with two Tow-Thomas biquads.

Design Problems

15.51. Design the first-order filter of Fig. 15.18(a) for a high-pass response so that the circuit attenuates an interferer at 1 MHz by 10 dB and passes frequencies above 5 MHz with a gain close to unity.

15.52. Design the passive filter of Fig. 15.29 for a −3 dB bandwidth of approximately 100 MHz, a peaking of 1 dB, and an inductance value less than 100 nH. SOL

15.53. Design the SK filter of Fig. 15.34 for $\omega_n = 2\pi \times (50\text{ MHz})$, $Q = 1.5$, and low-frequency gain of 2. Assume capacitor values must fall in the range of 10 pF to 100 pF.

15.54. Design a low-pass SK filter for a −3 dB bandwidth of 30 MHz with sensitivities no greater than unity. Assume a low-frequency gain of 2.

15.55. Design the KHN biquad of Fig. 15.36(b) for a bandpass response so that it provides a peak gain of unity at 10 MHz and an attenuation of 13 dB at 3 MHz and 33 MHz. Assume $R_3 = R_6$.

15.56. The design obtained in Problem 15.55 also provides low-pass and high-pass outputs. Determine the −3 dB corner frequencies for these two transfer functions.

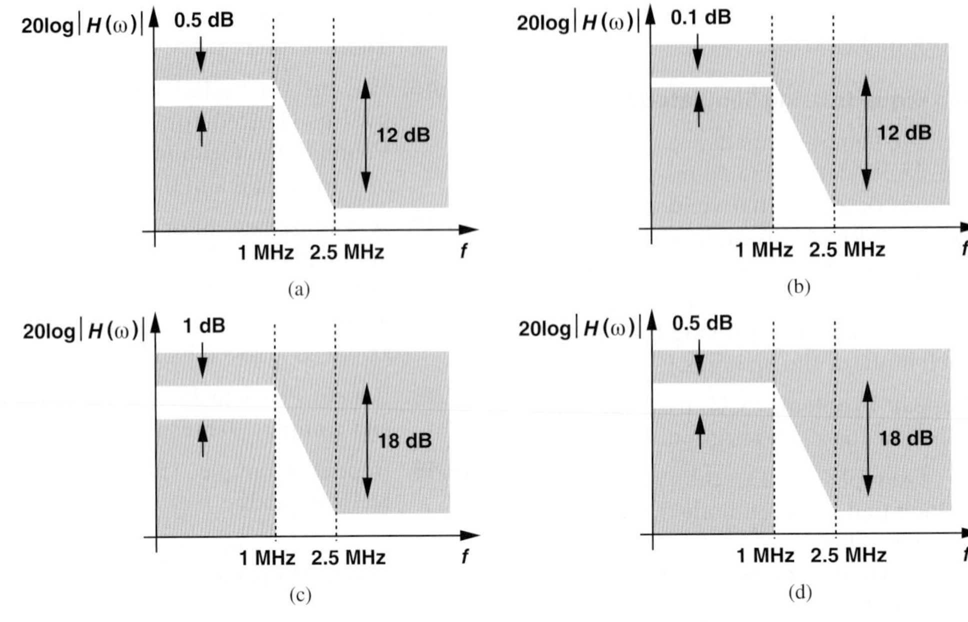

Figure 15.62

15.57. Repeat Problem 15.55 for the Tow-Thomas biquad shown in Fig. 15.37.

15.58. Design the active high-pass filter of Fig. 15.42 for a −3 dB corner frequency of 3.69 MHz and an attenuation of 13.6 dB at 2 MHz. Assume a peaking of 1 dB at 7 MHz.

15.59. Design the low-pass filter of Fig. 15.44(b) for a −3 dB bandwidth of 16.4 MHz and an attenuation of 6 dB at 20 MHz. Assume a peaking of 0.5 dB at 8 MHz.

15.60. For each frequency response template shown in Fig. 15.62, determine a Butterworth and a Chebyshev transfer function.

15.61. Following the methodology outlined in Examples 15-25 and 15-28, design filters for the Butterworth and Chebyshev transfer functions obtained in Problem 15.62.

15.62. Repeat Problem 15.61 but with Tow-Thomas biquads (and, if necessary, first-order RC sections).

SPICE PROBLEMS

15.63. Figure 15.63 shows the Butterworth filter designed in Example 15-25.

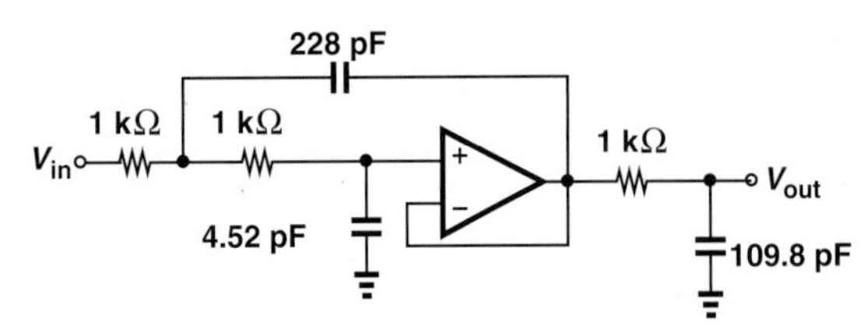

Figure 15.63

(a) Simulate the circuit with an op amp gain of 500 and determine if it meets the template specified in Example 15-24.

(b) Repeat (a) if the op amp exhibits an (open-loop) output resistance of 10 kΩ. (The output resistance can be modeled by inserting a 10 kΩ in series with the voltage-dependent source.)

(c) Repeat (b) if the op amp exhibits a single (open-loop) pole at 500 kHz. (The pole can be modeled by allowing a capacitor to form a low-pass filter with the 10 kΩ resistor.)

15.64. Repeat Problem 15.63 for the design obtained in Example 15-28.

15.65. (a) Repeat Example 15-28 with a cascade of two KHN biquads.

(b) Using SPICE, determine the minimum required op amp bandwidth if the overall response must exhibit a peaking no higher than 3 dB. Assume an op amp gain of 1000 and model the bandwidth as explained in Problem 15.63.

(c) Repeat (b) for the SK realization obtained in Example 15-28 and compare the results.

15.66. We must select an op amp for the SK design in Example 15-25. Suppose two types of op amps are available: one with an output resistance of 5 kΩ and a single pole at 200 MHz, and another with an output resistance of 10 kΩ and single pole at 100 MHz. Use SPICE to determine which op amp yields smaller peaking.

15.67. Consider the SK design in Example 15-25. Suppose the op amp provides an openloop gain of 1000 but is otherwise ideal.

(a) Does the response meet the template in Example 15-24 if all three resistors experience a change of +10%?

(b) Does the response meet the template in Example 15-24 if all three capacitors experience a change of +10%?

(c) What is the maximum tolerable error in the value of the resistors?

REFERENCE

1. R. Schaumann and M. E. van Valkenberg, *Design of Analog Filters*, Oxford University Press, 2001.

16

Digital CMOS Circuits

It is virtually impossible to find electronic devices in our daily lives that do not contain digital circuits. From watches and cameras to computers and cellphones, digital circuits account for more than 80% of the semiconductor market. Examples include microprocessors, memories, and digital signal processing ICs.

This chapter serves as an introduction to the analysis and design of digital CMOS (complementary metal oxide semiconductor) circuits. The objective is to provide a detailed transistor-level understanding of logical gates so as to prepare the reader for courses on digital circuit design. The outline is shown below.

General Considerations
- Static Characteristics
- Dynamic Characteristics

CMOS Inverter
- Voltage Transfer Characteristic
- Dynamic Behavior
- Power Dissipation

Other CMOS Gates
- NOR Gate
- NAND Gate

16.1 GENERAL CONSIDERATIONS

In the past five decades, digital circuits have evolved dramatically, going from a few gates per chip in the 1960s to hundreds of millions of transistors per chip today. Very early generations incorporated only resistors and diodes and were called "resistor-diode logic" (RDL). These were followed by bipolar realizations such as "transistor-transistor logic" (TTL) and "emitter-coupled logic" (ECL). But it was the advent of CMOS technology and the unique properties of digital CMOS circuits that led to the explosive growth of digital systems. We will study and appreciate these properties in this chapter.

Recall from basic logic design that digital systems employ building blocks such as gates, latches, and flipflops. For example, gates can form a "combinational" circuit that operates as a binary-Gray decoder. Similarly, gates and flipflops can comprise a "sequential" circuit that serves as a counter or a "finite-state machine." In this chapter, we delve into the internal design of some of these building blocks and analyze their limitations. In particular, we address three important questions:

1. What limits the speed of a digital gate?
2. How much power does a gate consume while running at a certain speed?
3. How much "noise" can a gate tolerate while producing a valid output?

These questions play a critical role in the design of digital systems. The first reveals how microprocessor speeds have risen from a few hundred megahertz to several gigahertz in past ten years. The second helps predict how much power a microprocessor drains from the battery of a laptop computer. The third illustrates how reliably a gate operates in the presence of nonidealities in the system.

16.1.1 Static Characterization of Gates

Unlike many of the amplifying stages studied in this book, logical gates always operate with *large* signals. In digital CMOS circuits, a logical ONE is represented by a voltage equal to the supply, V_{DD}, and a logical ZERO by zero volt. Thus, the inputs and outputs of gates swing between zero and V_{DD} as different states are processed.

How do we characterize the large-signal behavior of a circuit? Recall from Chapter 3 that we can construct the input/output characteristic by varying the input across the entire allowable range (e.g., 0 to V_{DD}) and computing the corresponding output. Also called the "voltage transfer characteristic" (VTC),[1] the result illustrates the operation of the gate in great detail, revealing departures from the ideal case.

As an example, consider a NOT gate whose logical operation is expressed as $X = \overline{A}$. Called an "inverter" and denoted by the symbol shown in Fig. 16.1(a), such a gate must ideally behave as depicted in Fig. 16.1(b). For $V_{in} = 0$, the output remains at a logical ONE, $V_{out} = V_{DD}$. For $V_{in} = V_{DD}$, the output provides a logical zero, $V_{out} = 0$. As V_{in} goes from 0 to V_{DD}, V_{out} abruptly changes its state at some value of the input, V_1.

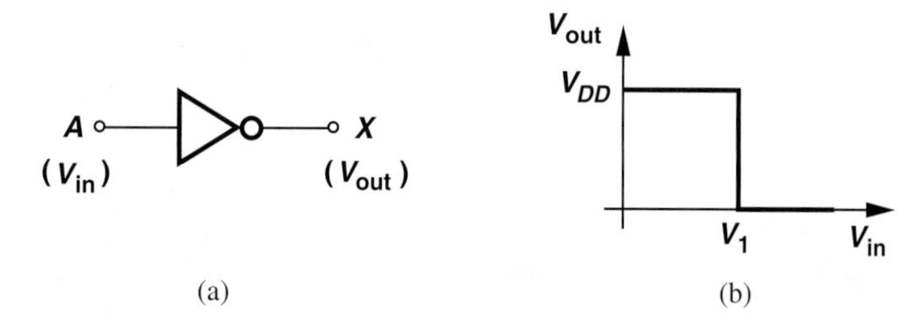

Figure 16.1 (a) Inverter, (b) ideal characteristic.

Example 16-1 Explain why a common-source stage can operate as an inverter.

Solution In the CS stage shown in Fig. 16.2(a), if $V_{in} = 0$, then M_1 is off, the voltage drop across R_D is zero, and hence $V_{out} = V_{DD}$. On the other hand, if $V_{in} = V_{DD}$, M_1 draws a relatively

[1]The term "transfer" should not create confusion with the *transfer function*.

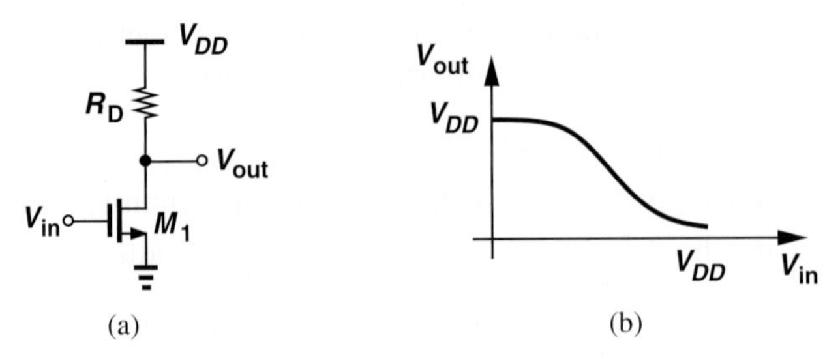

Figure 16.2 (a) CS stage, (b) input/output characteristic.

large current from R_D and $V_{out} = V_{DD} - I_D R_D$ can be near zero. Thus, as sketched in Fig. 16.2(b), the input/output characteristic resembles that of an inverter.

Exercise What happens if R_D is replaced by a PMOS current source?

Example 16-2 A common-source stage operates as a inverter. Determine the VTC for such a realization.

Solution Consider the CS stage shown in Fig. 16.3(a). We vary V_{in} from 0 to V_{DD} and plot the corresponding output. For $V_{in} \leq V_{TH}$, M_1 remains off and $V_{out} = V_{DD}$ (logical ONE). As V_{in} exceeds V_{TH}, M_1 turns on and V_{out} begins to fall:

$$V_{out} = V_{DD} - I_D R_D \tag{16.1}$$

$$= V_{DD} - \frac{1}{2}\mu_n C_{ox}\frac{W}{L}R_D(V_{in} - V_{TH})^2, \tag{16.2}$$

where channel-length modulation is neglected. As the input increases further, V_{out} drops, eventually driving M_1 into the triode region for $V_{out} \leq V_{in} - V_{TH}$ and hence:

$$V_{DD} - \frac{1}{2}\mu_n C_{ox}\frac{W}{L}R_D(V_{in1} - V_{TH})^2 \leq V_{in1} - V_{TH}. \tag{16.3}$$

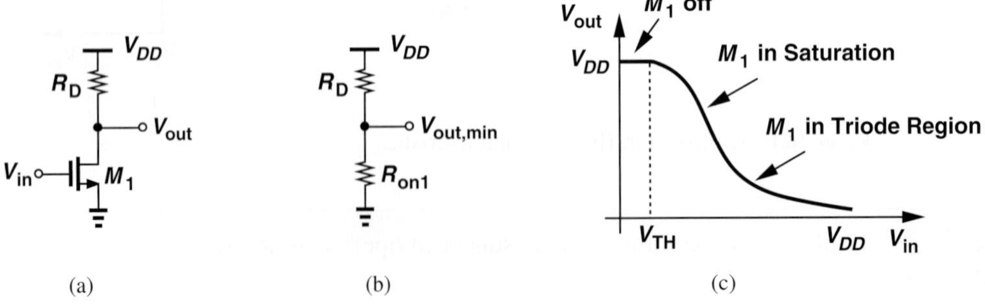

Figure 16.3 (a) CS stage, (b) equivalent circuit for M_1 in deep triode region, (c) input/output characteristic.

From this equation, the value of V_{in} that places M_1 at the edge of triode region can be calculated. As V_{in} exceeds this value, V_{out} continues to decrease, reaching its lowest level for $V_{in} = V_{DD}$:

$$V_{out,min} = V_{DD} - R_D I_{D,max} \tag{16.4}$$

$$= V_{DD} - \frac{1}{2}\mu_n C_{ox}\frac{W}{L}R_D[2(V_{DD} - V_{TH})V_{out,min} - V_{out,min}^2]. \tag{16.5}$$

Equation (16.5) can be solved to obtain $V_{out,min}$. If we neglect the second term in the square brackets, then

$$V_{out,min} \approx -\frac{V_{DD}}{1 + \mu_n C_{ox}\frac{W}{L}R_D(V_{DD} - V_{TH})}. \tag{16.6}$$

This is, of course, equivalent to viewing M_1 as a resistor of value $R_{on1} = [\mu_n C_{ox}(W/L)R_D(V_{DD} - V_{TH})]^{-1}$ and hence $V_{out,min}$ a result of voltage division between R_D and R_{on1} [Fig. 16.3(b)]. Figure 16.3(c) plots the VTC, illustrating the regions of operation. In this role, the CS stage is also called an "NMOS inverter."

Exercise Repeat the above example if R_D is replaced with a PMOS current source.

Can the characteristic of Fig. 16.1(b) be realized in practice? We recognize that V_{out} changes by an amount equal to V_{DD} for an infinitesimally small change in V_{in} around V_1, i.e., the voltage gain of the circuit is *infinite* at this point. In reality, as illustrated in Example 16-2, the gain remains finite, thereby producing a gradual transition from high to low (Fig. 16.4). We may call the range $V_0 < V_{in} < V_2$ the "transition region."

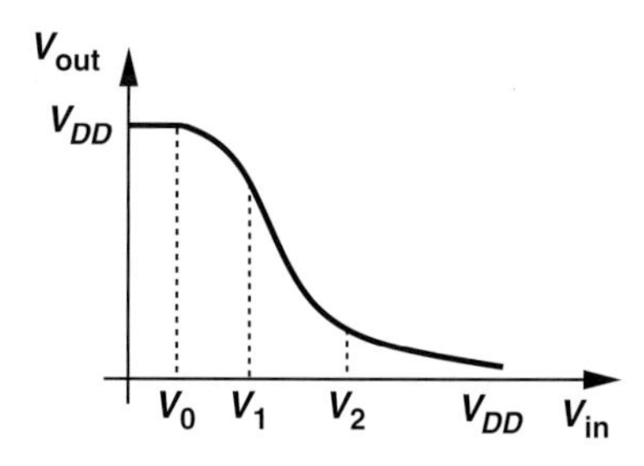

Figure 16.4 Characteristic with finite gain.

Example 16-3 An inverter must exhibit a transition region only 50 mV wide. If the supply voltage is 1.8 V, estimate the gain of the circuit in this region.

Solution Since a 50-mV change at the input results in a change of approximately 1.8 V at the output, the voltage gain is equal to $1.8/0.05 = 36$.

Exercise What happens to the transition region if the width of the NMOS transistor is increased?

The reader may wonder why the gradual transition in Fig. 16.4 may prove problematic. After all, if the input jumps between 0 and V_{DD}, the output still provides valid logical levels. In reality, however, the input may *not* reach exactly 0 or V_{DD}. For example, a logical zero may appear as +100 mV rather than 0 V. Such "degradation" of the logical levels arises from a multitude of phenomena in a large integrated circuit, but a simple example can illustrate this effect.

Example 16-4

The supply voltage, V_{DD}, is distributed on a microprocessor chip through a wide metal line 15 mm long [Fig. 16.5(a)]. Called the "power bus," this line carries a current of 5 A and suffers from a resistance of 25 mΩ. If inverter Inv_1 produces a logical ONE given by the local value of V_{DD}, determine the degradation in this level as sensed by inverter Inv_2.

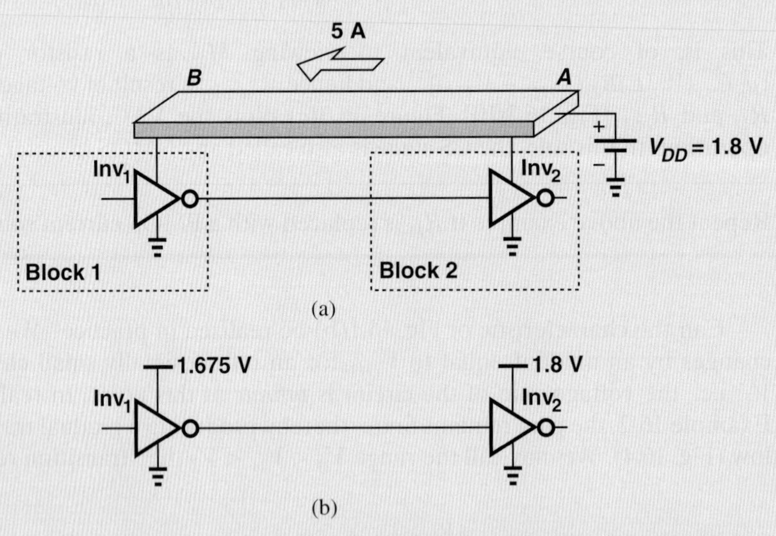

Figure 16.5 (a) Two inverters separated by a long distance on a chip, (b) equivalent supply voltages.

Solution The power bus experiences a voltage drop of 5 A × 25 mΩ = 125 mV from point *A* to point *B*, thereby allowing a logical ONE of only 1.8 V − 0.125 V = 1.675 V at the output of Inv_1 [Fig. 16.5(b)]. As a result, Inv_2 senses a high level that is degraded by 125 mV with respect to its own supply voltage, 1.8 V.

Exercise Repeat the above example if the width of the power bus is halved.

Did you know?

Early generations of MOS technology offered only NMOS or PMOS devices. The fabrication of both NMOS and PMOS transistors on the same chip faced technical and economical issues that would be resolved some years later. With only one device type available, the logic gates were in fact similar to the common-source stage studied here, except that the resistor was replaced with another transistor. The first Intel microprocessor, the 4004, was implemented in PMOS technology in 1971 and contained 2300 transistors. The problem with PMOS or NMOS gates was that they consumed power even in the idle state ("static power"). It was the invention of CMOS logic that revolutionized the semiconductor industry.

How much degradation can we tolerate in the input levels applied to a gate? Consider the situation depicted in Fig. 16.6, where both the low and high levels of the input, V_0 and V_2, respectively, depart considerably from their ideal values. Mapping these levels to the output, we observe that V_{out} also exhibits degraded logical levels. In a chain of gates, such successive degradations may make the system very "fragile" and even completely corrupt the states.

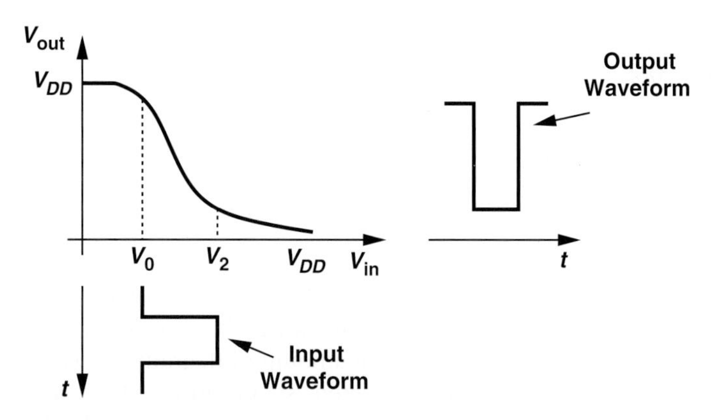

Figure 16.6 Degradation of output levels in an inverter.

Example 16-5 Sketch the small-signal voltage gain for the characteristic shown in Fig. 16.4 as a function of V_{in}.

Solution The slope of the VTC begins from zero, becomes more negative above V_0, and approaches zero again for $V_{in} > V_2$. Figure 16.7 plots the result.

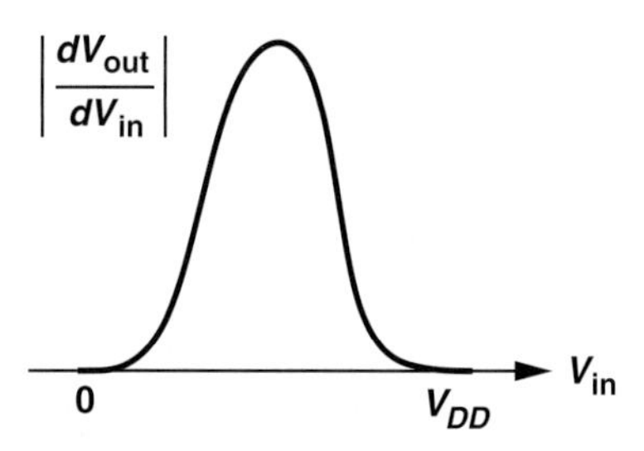

Figure 16.7

Exercise Is this plot necessarily symmetric? Use an CS stage as an example.

Example 16-6 Prove that the magnitude of the small-signal gain obtained in Example 16-5 must exceed unity at some point.

Solution Superimposing a line with a slope of –1 on the VTC as shown in Fig. 16.8, we note that the slope of the VTC is sharper than unity across part of the transition region. This is because the transition region spans a range narrower than 0 to V_{DD}.

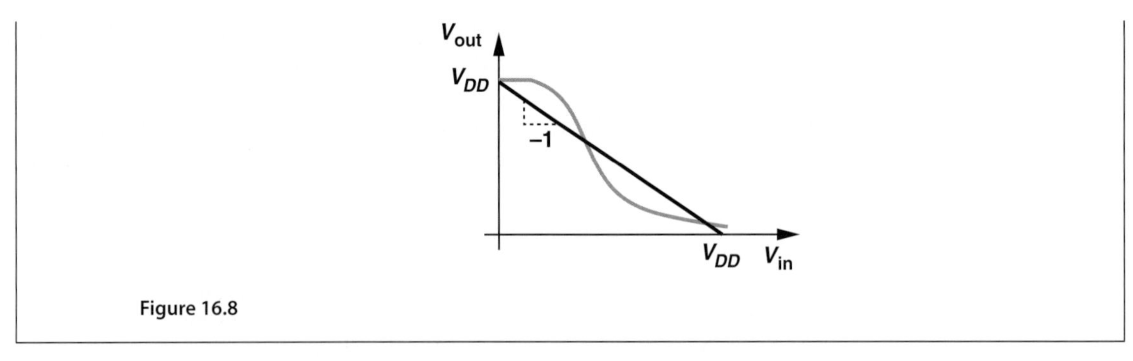

Figure 16.8

Exercise An inverter exhibits a gain of about 2 in its transition region. How wide is the transition region?

Noise Margin In order to quantify the robustness of a gate with respect to the degradation of the input logical levels, we introduce the concept of "noise margin" (NM). A rough definition is: NM is the maximum amount of degradation (noise) at the input that can be tolerated before the output is affected "significantly." What do we mean by "significantly?" We postulate that the output remains relatively unaffected if the *gain* of the circuit remains below unity, thus arriving at the following definition:

The noise margin is the maximum departure from the ideal logical level that places the gate at a small-signal voltage gain of unity.

The procedure for calculating NM is straightforward: we construct the VTC and determine the input level at which the small-signal gain reaches unity. The difference between this level and the ideal logical level yields the NM. Of course, we associate a noise margin with the input low level, NM_L, and another with the input high level, NM_H. Figure 16.9 summarizes these concepts. The two input voltages are denoted by V_{IL} and V_{IH}, respectively.

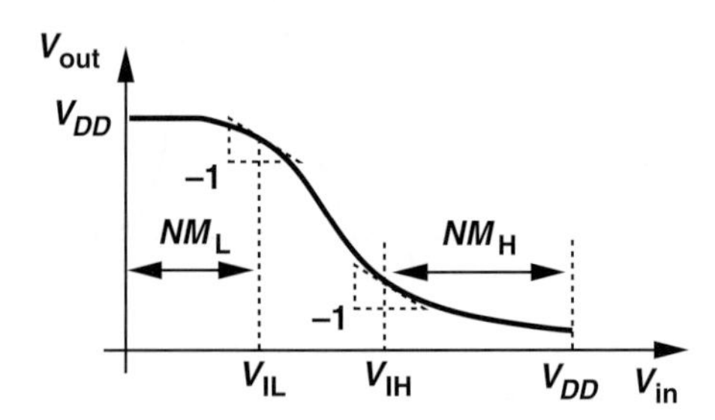

Figure 16.9 Illustration of noise margins.

Example 16-7 A common-source stage operates as an NMOS inverter. Compute the noise margins.

Solution We can adopt one of two approaches here. First, since the small-signal gain of the stage is equal to $-g_m R_D$ and since $g_m = \mu_n C_{ox}(W/L)(V_{GS} - V_{TH})$, we have

$$\mu_n C_{ox}\frac{W}{L}(V_{IL} - V_{TH})R_D = 1, \tag{16.7}$$

and hence

$$V_{IL} = \frac{1}{\mu_n C_{ox} \frac{W}{L} R_D} + V_{TH}. \tag{16.8}$$

In the second approach, we directly differentiate both sides of Eq. (16.2) with respect to V_{in}:

$$\frac{\partial V_{out}}{\partial V_{in}} = -\mu_n C_{ox} \frac{W}{L} R_D (V_{IL} - V_{TH}) \tag{16.9}$$

$$= -1 \tag{16.10}$$

and hence

$$NM_L = V_{IL} = \frac{1}{\mu_n C_{ox} \frac{W}{L} R_D} + V_{TH}. \tag{16.11}$$

That is, the input must exceed V_{TH} by $(\mu_n C_{ox} R_D W/L)^{-1}$ for the circuit to reach the unity-gain point.

As V_{in} drives M_1 into the triode region, the transconductance of M_1 and hence the voltage gain of the circuit begin to fall. Since in Chapter 6, we did not derive a small-signal model for MOSFETs operating in the triode region, we continue with the second approach:

$$V_{out} = V_{DD} - R_D I_D \tag{16.12}$$

$$= V_{DD} - \frac{1}{2} \mu_n C_{ox} \frac{W}{L} R_D [2(V_{in} - V_{TH}) V_{out} - V_{out}^2]. \tag{16.13}$$

We must equate the slope of this characteristic to −1 to determine NM_H:

$$\frac{\partial V_{out}}{\partial V_{in}} = -\frac{1}{2} \mu_n C_{ox} \frac{W}{L} R_D \left[2V_{out} + 2(V_{in} - V_{TH}) \frac{\partial V_{out}}{\partial V_{in}} - 2V_{out} \frac{\partial V_{out}}{\partial V_{in}} \right]. \tag{16.14}$$

With $\partial V_{out}/\partial V_{in} = -1$, Eq. (16.14) yields

$$V_{out} = \frac{1}{2\mu_n C_{ox} \frac{W}{L} R_D} + \frac{V_{in} - V_{TH}}{2}. \tag{16.15}$$

If this value of V_{out} is substituted in Eq. (16.13), the required value of V_{in} (V_{IH} in Fig. 16.9) can be obtained. Thus, $NM_H = V_{DD} - V_{IH}$.

Exercise If $R_D = 1\ \text{k}\Omega$, $\mu_n C_{ox} = 100\ \mu\text{A/V}^2$, $W/L = 10$, $V_{TH} = 0.5$ V, and $V_{DD} = 1.8$ V, calculate the high and low noise margins.

Example 16-8 As suggested by Eq. (16.6), the output low level of an NMOS inverter is always degraded. Derive a relationship to guarantee that this degradation remains below $0.05 V_{DD}$.

Solution Equating Eq. (16.6) to $0.05V_{DD}$, we have

$$\mu_n C_{ox}\frac{W}{L}R_D(V_{DD}-V_{TH})=19 \tag{16.16}$$

and hence

$$R_D=\frac{19}{\mu_n C_{ox}\frac{W}{L}(V_{DD}-V_{TH})}. \tag{16.17}$$

Note that the right-hand side is equal to 19 times the on-resistance of M_1. Thus, R_D must remain above $19R_{on1}$.

Exercise Repeat the above example if the degradation can be as high as $0.1V_{DD}$.

16.1.2 Dynamic Characterization of Gates

The input/output characteristic of a gate proves useful in determining the degradations that the circuit can tolerate in its input levels. Another important aspect of a gate's performance is its speed. How do we quantify the speed of a logical gate? Since the gate operates with large signals at the input and output and hence experiences heavy nonlinearity, the concepts of transfer function and bandwidth are not meaningful here. Instead, we must define the speed according to the role of gates in digital systems. An example serves us well at this point.

Example 16-9 The input to an NMOS inverter jumps from V_{DD} to 0 at $t=0$ [Fig. 16.10(a)]. If the circuit sees a load capacitance of C_L, how long does the output take to reach within 5% of the ideal high level? Assume V_{out} can be approximated by Eq. (16.6) when M_1 is on.

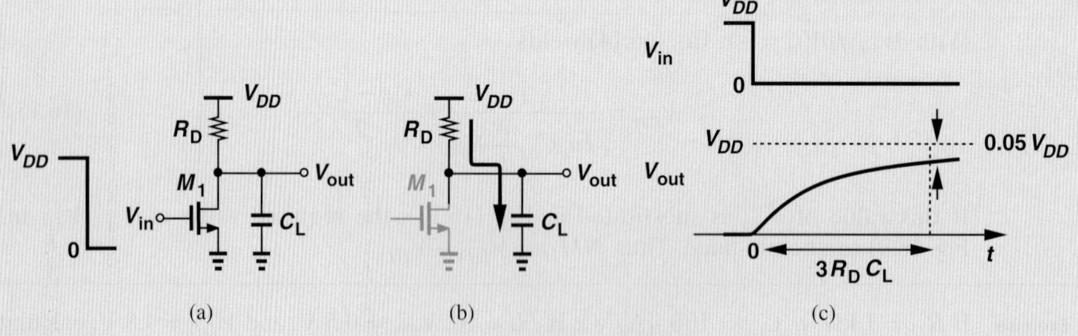

Figure 16.10 (a) NMOS inverter experiencing a step input, (b) charging path for C_L, (c) input and output waveforms.

Solution At $t=0^-$, M_1 is on, establishing an initial condition across C_L equal to

$$V_{out}(0^-)=\frac{V_{DD}}{1+\mu_n C_{ox}\frac{W}{L}R_D(V_{DD}-V_{TH})}. \tag{16.18}$$

At $t = 0^+$, the circuit reduces to that shown in Fig. 16.10(b), where C_L charges toward V_{DD} through R_D. We therefore have

$$V_{out}(t) = V_{out}(0^-) + [V_{DD} - V_{out}(0^-)]\left(1 - \exp\frac{-t}{R_D C_L}\right) \qquad t > 0 \tag{16.19}$$

(This equation is constructed so that the first term denotes the initial value if we choose $t = 0$, and the sum of the first and second terms yields the final value if we select $t = \infty$.) The time required for the output to reach within 95% of V_{DD}, $T_{95\%}$, is obtained from

$$0.95V_{DD} = V_{out}(0^-) + [V_{DD} - V_{out}(0^-)]\left(1 - \exp\frac{-T_{95\%}}{R_D C_L}\right). \tag{16.20}$$

It follows that

$$T_{95\%} = -R_D C_L \ln\frac{0.05V_{DD}}{V_{DD} - V_{out}(0^-)}. \tag{16.21}$$

If we can assume $V_{DD} - V_{out}(0^-) \approx V_{DD}$, then

$$T_{95\%} \approx 3R_D C_L. \tag{16.22}$$

In other words, the output takes about three time constants to reach a voltage close to the ideal high level [Fig. 16.10(c)]. Unlike ideal gates used in basic logic design, this inverter exhibits a finite transition time at the output.

Exercise How many time constants does the output take to reach within 90% of its ideal value?

The foregoing example reveals a fundamental limitation: in the presence of a load capacitance, a logical gate cannot respond *immediately* to an input. The circuit of Fig. 16.10(a) takes roughly three time constants to produce a reliable level at the output and, as such, suffers from a "delay." That is, the speed of gates is limited by the finite transition time at the output and the resulting delay.

Playing a critical role in high-speed digital design, the transition time and the delay must be defined carefully. As illustrated in Fig. 16.11(a), we define the output "risetime," T_R, as the time required for the output to go from 10% of V_{DD} to 90% of V_{DD}.[2] Similarly, the output "falltime," T_F, is defined as the time required for the output to go from 90% of V_{DD} to 10% of V_{DD}. In general, T_R and T_F may not be equal.

Since the *input* to a gate is produced by another gate and hence suffers from a finite transition time, the delay of the gate must be characterized with a realistic input waveform rather than the abrupt step in Fig. 16.11(a). We therefore apply a step with a typical risetime at the input and define the propagation delay as the difference between the time points at which the input and the output cross $V_{DD}/2$ [Fig. 16.11(b)]. Since the output rise and fall times may not be equal, a low-to-high delay, T_{PLH}, and a high-to-low delay, T_{PHL}, are necessary to characterize the speed. In today's CMOS technology, gate delays as little as 10 ps can be obtained.

[2]This definition applies only if the low and high levels are equal to 0 and V_{DD}, respectively.

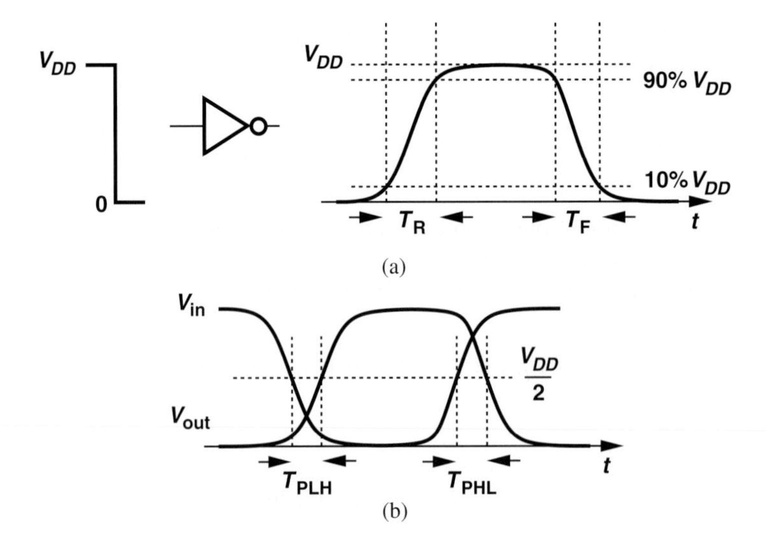

Figure 16.11 Definition of (a) rise and fall times, and (b) propagation delays.

The reader may wonder about the nature of the load capacitance in Example 16-9. If the gate drives only another stage on the chip, this capacitance arises from two sources: the input capacitance of the subsequent gate(s) and the capacitance associated with the "interconnect" (on-chip wire) that carries the signal from one circuit to another.

Example 16-10

An NMOS inverter drives an identical stage as depicted in Fig. 16.12. We say the first gate sees a "fanout" of unity. Assuming a 5% degradation in the output low level (Example 16-8), determine the time constant at node X when V_X goes from low to high. Assume $C_X \approx WLC_{ox}$.

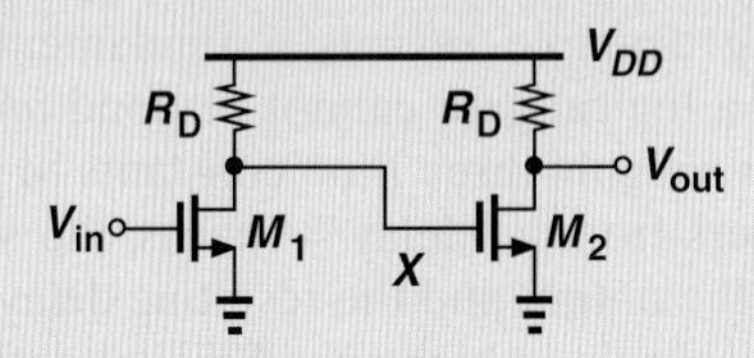

Figure 16.12 Cascade of inverters.

Solution Recall from Example 16-9 that this time constant is simply equal to $R_D C_X$. Assuming $R_D = 19R_{on1}$, we write

$$\tau = R_D C_X \tag{16.23}$$

$$= \frac{19}{\mu_n C_{ox}\frac{W}{L}(V_{DD} - V_{TH})} \cdot WLC_{ox} \tag{16.24}$$

$$= \frac{19L^2}{\mu_n(V_{DD} - V_{TH})}. \tag{16.25}$$

Exercise Suppose the width of M_2 is doubled while M_1 remains unchanged. Calculate the time constant.

Example 16-11 In Example 16-4, the wire connecting the output of Inv_1 to the input of Inv_2 exhibits a capacitance of 50×10^{-18} F (50 aF)[3] per micron of length. What is the interconnect capacitance driven by Inv_1?

Solution For 15,000 microns, we have

$$C_{int} = 15{,}000 \times 50 \times 10^{-18}\ \text{F} \tag{16.26}$$

$$= 750\ \text{fF}. \tag{16.27}$$

To appreciate the significance of this value, let us calculate the gate capacitance of a small MOSFET, e.g., with $W = 0.5\ \mu$m, $L = 0.18\ \mu$m, and $C_{ox} = 13.5\ \text{fF}/\mu\text{m}^2$:

$$C_{GS} \approx WLC_{ox} \tag{16.28}$$

$$\approx 1.215\ \text{fF}. \tag{16.29}$$

In other words, Inv_1 sees a load equivalent to a fanout of $750\ \text{fF}/1.215\ \text{fF} \approx 617$, as if it drives 640 gates.

Exercise What is the equivalent fanout if the width of the wire is halved?

16.1.3 Power-Speed Trade-Off

Integrated circuits containing millions of gates can consume a very high power (tens of watts). The power dissipation proves critical for several reasons. First, it determines the battery lifetime in portable applications such as laptop computers and cellphones. Second, it tends to raise the temperature of the chip, degrading the performance of the transistor.[4] Third, it requires special (expensive) packages that can conduct the heat away from the chip.

How does a gate consume power? Let us consider the NMOS inverter of Fig. 16.13 as an example. If $V_{in} = 0$, M_1 is off. On the other hand, if $V_{in} = V_{DD}$, M_1 draws a current equal to

$$I_D = \frac{V_{DD} - V_{out,min}}{R_D}, \tag{16.30}$$

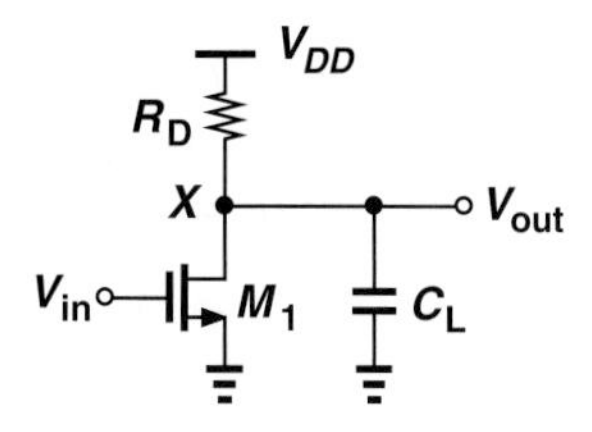

Figure 16.13 NMOS inverter driving a load capacitance.

[3]The abbreviation for 10^{-18} is called "ato."
[4]For example, the mobility of MOS devices falls as the temperature rises.

which, from Eq. (16.6), translates to

$$I_D = \frac{\mu_n C_{ox} \frac{W}{L}(V_{DD} - V_{TH})V_{DD}}{1 + \mu_n C_{ox} \frac{W}{L} R_D (V_{DD} - V_{TH})}. \tag{16.31}$$

Alternatively,

$$I_D = \frac{V_{DD}}{R_D + R_{on1}}. \tag{16.32}$$

The gate thus consumes a power of $I_D \cdot V_{DD}$ while the output is low. (If $R_D \gg R_{on1}$, then $I_D V_{DD} \approx V_{DD}^2 / R_D$.) Now, recall from Example 16-9 that the output risetime of the gate is determined by the time constant $R_D C_L$. We therefore observe a direct trade-off between the power dissipation and the speed: a high value of R_D reduces the power dissipation but yields a longer delay. In fact, we may define a figure of merit as the *product* of the power dissipation and the time constant:

$$(I_D V_{DD}) \cdot (R_D C_X) = \frac{V_{DD}^2}{R_D + R_{on1}} \cdot (R_D C_X). \tag{16.33}$$

As noted in Example 16-8, typically $R_D \gg R_{on1}$ and hence,

$$(I_D V_{DD}) \cdot (R_D C_X) \approx V_{DD}^2 C_X. \tag{16.34}$$

In digital design, the figure of merit is defined as the product of the power dissipation, P, and the *gate delay* rather than the output time constant. This is because the nonlinear operation of gates often prohibits the use of a single time constant to express the output transition behavior. As such, the figure of merit is called the "power-delay product" (PDP). Since T_{PHL} and T_{PLH} may not be equal, we define PDP with respect to the average of the two:

$$PDP = P \cdot \frac{T_{PHL} + T_{PLH}}{2}. \tag{16.35}$$

Note that PDP has dimension of energy, i.e., it indicates how much energy is consumed for a logical operation.

Example 16-12 Consider the cascade of identical NMOS inverters studied in Example 16-10. Assuming T_{PLH} is roughly equal to three time constants, determine the power-delay product for the low-to-high transitions at node X.

Solution Expressing the power dissipation as $I_D V_{DD} \approx V_{DD}^2 / R_D$, we have

$$PDP = (I_D V_{DD})(3 R_D C_X) \tag{16.36}$$

$$= 3 V_{DD}^2 C_X \tag{16.37}$$

$$= 3 V_{DD}^2 W L C_{ox}. \tag{16.38}$$

For example, if $V_{DD} = 1.8$ V, $W = 0.5$ μm, $L = 0.18$ μm, and $C_{ox} = 13$ fF/μm^2, then $PDP = 1.14 \times 10^{-14}$J $= 11.4$ fJ.

Exercise How much average power is consumed if the circuit runs at a frequency of 1 GHz?

16.2 CMOS INVERTER

Perhaps the most elegant and the most important circuit invention in CMOS technology, the CMOS inverter forms the foundation for modern digital VLSI systems. In this section, we study the static and dynamic properties of this circuit.

16.2.1 Initial Thoughts

We have seen in Section 16.1.1 that the NOT (inverter) function can be realized by a common-source stage, Fig. 16.3(a). As formulated in Examples 16-8 and 16-9, this circuit faces the following issues: (1) the load resistance, R_D, must be chosen much greater than the on-resistance of the transistor; (2) the value of R_D creates a trade-off between speed and power dissipation; (3) the inverter consumes a power of roughly V_{DD}^2/R_D so long as the output remains low. Of particular concern in large digital circuits is the last effect, called "static power dissipation" because the inverter consumes energy even though it is not switching. For example, in a VLSI chip containing one million gates, half of the outputs may be low at a given point in time, thereby demanding a power dissipation of $5 \times 10^5 \times V_{DD}^2/R_D$. If $V_{DD} = 1.8$ V and $R_D = 10$ kΩ, this amounts to 162 W of *static* power consumption!

The foregoing drawbacks of the NMOS inverter fundamentally arise from the "passive" nature of the load resistor, called the "pull-up" device here. Since R_D presents a *constant* resistance between V_{DD} and the output node, (1) M_1 must "fight" R_D while establishing a low level at the output and hence R_{on1} must remain much smaller than R_D [Fig. 16.14(a)]; (2) after M_1 turns off, only R_D can pull the output node up toward V_{DD} [Fig. 16.14(b)]; (3) the circuit draws a current of approximately V_{DD}/R_D from the supply when the output is low [Fig. 16.14(c)]. We therefore seek a more efficient realization that employs an "intelligent" pull-up device.

Let us ask, how should the ideal pull-up device behave in an inverter? When M_1 turns off, the pull-up device must connect the output node to V_{DD}, preferably with a low resistance [Fig. 16.15(a)]. On the other hand, when M_2 turns on, the pull-up device must turn

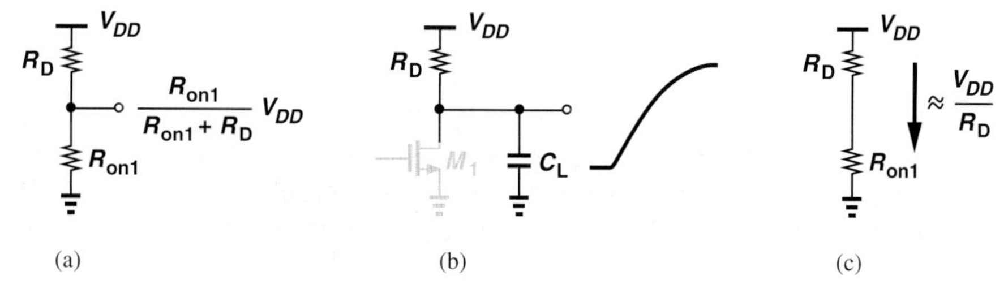

Figure 16.14 (a) Degradation of output level in an NMOS inverter, (b) risetime limitation due to R_D, (c) static power consumed during output low level.

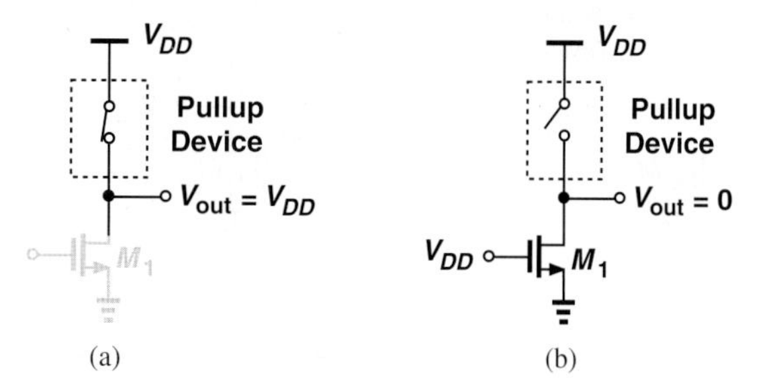

Figure 16.15 Use of active pullup device for (a) high output and (b) low output.

off [Fig. 16.15(b)] so that no current can flow from V_{DD} to ground (and V_{out} is exactly equal to zero). This latter property also reduces the *falltime* at the output, as illustrated in the following example.

Example 16-13

Consider the two inverter implementations depicted in Fig. 16.16. Suppose V_{in} jumps from 0 to V_{DD} at $t = 0$ and the pull-up device in Fig. 16.16(b) turns off at the same time. Compare the output falltimes of the two circuits if M_1 and C_L are identical in the two cases.

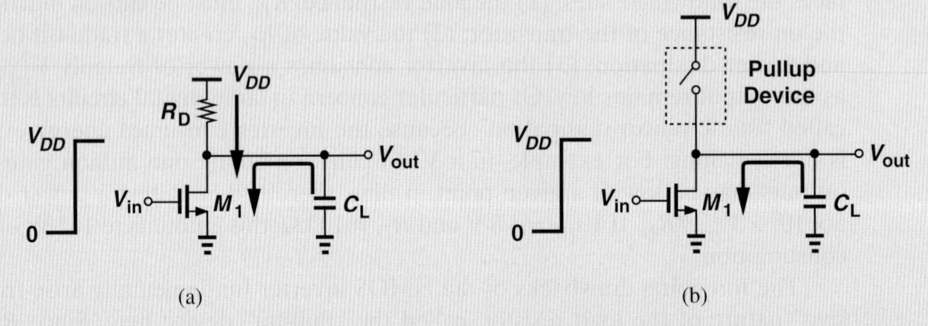

Figure 16.16 Comparison of (a) NMOS inverter and (b) inverter using an active pull-up device.

Solution In Fig. 16.16(a), M_1 must absorb *two* currents: one carried by R_D and another required to discharge C_L. In Fig. 16.16(b), on the other hand, I_{D1} simply discharges C_L because the pull-up device is turned off. As a consequence, V_{out} falls more rapidly in the topology of Fig. 16.16(b).

Exercise For each circuit, determine the energy consumed by M_1 as V_{out} falls from V_{DD} to zero.

In summary, we wish the pull-up device in Fig. 16.15 to turn on when M_1 turns off and vice versa. Is it possible to employ a transistor for this purpose and turn it on and off by the *input voltage* [Fig. 16.17(a)]? We recognized that the transistor must turn on when V_{in} is *low*, postulating that a PMOS device is necessary [Fig. 16.17(b)]. Called the "CMOS inverter," this topology benefits from "cooperation" between the NMOS device and the PMOS device: when M_1 wishes to pull V_{out} low, M_2 turns *off*, and vice versa.

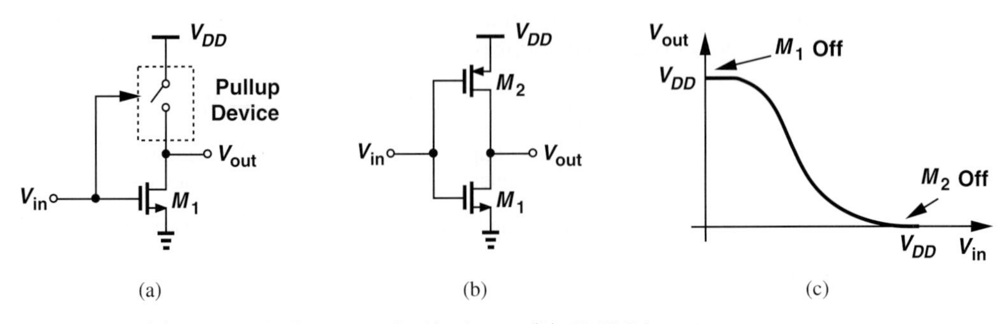

Figure 16.17 (a) Pull-up device controlled by input, (b) CMOS inverter.

It is important to note that, by virtue of the "active" pull-up device, the CMOS inverter indeed avoids the drawbacks of the NMOS implementation: (1) the output low level is *exactly* equal to zero because $V_{in} = V_{DD}$ ensures that M_2 remains off; (2) the circuit consumes *zero* static power for both high and low output levels. Figure 16.17(c) shows a rough sketch of the input/output characteristic, emphasizing that $V_{out} = 0$ for $V_{in} = V_{DD}$. Throughout this chapter, we denote the aspect ratios of the NMOS and PMOS transistors in an inverter by $(W/L)_1$ and $(W/L)_2$, respectively.

Did you know?

The CMOS inverter was proposed in 1963 by Sah and Wanlass of Fairchild. Unlike previous logic families, the CMOS inverter consumed nearly zero power in its idle state, opening the door for low-power digital integrated circuits. Interestingly, among the first applications of CMOS logic were digital watches, manufactured in the 1970s. In 1976, RCA introduced the 1802 microprocessor, evidently the first CMOS realization. Owing to its low power consumption (and its tolerance of cosmic radiation), this processor was used in the Galileo probe to Jupiter.

16.2.2 Voltage Transfer Characteristic

We begin our in-depth study of the CMOS inverter with its static characteristics. We must vary V_{in} from zero to V_{DD} and plot the corresponding output voltage. Note that the two transistors carry equal currents under all conditions (so long as the inverter is not loaded by any other circuit). Suppose $V_{in} = 0$ [Fig. 16.18(a)]. Then, M_1 is off and M_2 is on. How can M_2 remain on while $|I_{D2}| = I_{D1} = 0$? This is possible only if M_2 sustains a zero drain-source voltage. That is,

$$|I_{D2}| = \frac{1}{2}\mu_p C_{ox}\left(\frac{W}{L}\right)_2 [2(V_{DD} - |V_{TH2}|)|V_{DS2}| - V_{DS}^2] = 0 \tag{16.39}$$

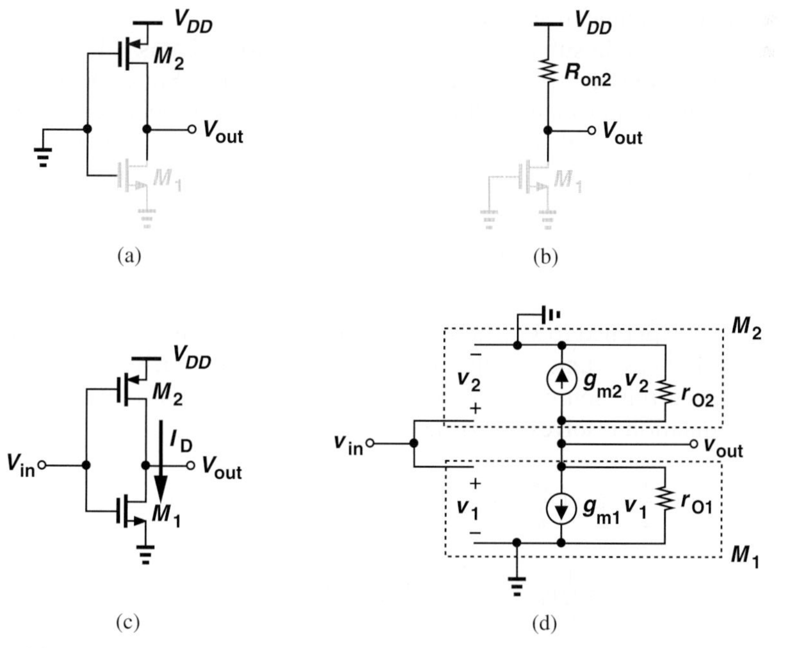

Figure 16.18 (a) CMOS inverter sensing a low input, (b) equivalent circuit, (c) supply current when both transistors are one, (d) small-signal model.

requires that

$$V_{DS2} = 0 \tag{16.40}$$

and hence

$$V_{out} = V_{DD}. \tag{16.41}$$

From another perspective, M_2 operates as a resistor of value

$$R_{on2} = \frac{1}{\mu_p C_{ox}\left(\frac{W}{L}\right)_2 (V_{DD} - |V_{TH2}|)}, \tag{16.42}$$

pulling the output node to V_{DD} [Fig. 16.18(b)].

As V_{in} rises, the gate-source overdrive of M_2 *decreases* and its on-resistance increases. But, for $V_{in} < V_{TH1}$, M_1 remains off and $V_{out} = V_{DD}$. As V_{in} exceeds V_{TH1} slightly, M_1 turns on, drawing a current from V_{DD} through the on-resistance of M_2 [Fig. 16.18(c)]. Since V_{out} is still close to V_{DD}, M_1 operates in saturation and M_2 still resides in the triode region. Equating the drain currents of the two, we have

$$\frac{1}{2}\mu_n C_{ox}\left(\frac{W}{L}\right)_1 (V_{in} - V_{TH1})^2$$
$$= \frac{1}{2}\mu_p C_{ox}\left(\frac{W}{L}\right)_2 [2(V_{DD} - V_{in} - |V_{TH2}|)(V_{DD} - V_{out}) - (V_{DD} - V_{out})^2], \tag{16.43}$$

where channel-length modulation is neglected. This quadratic equation can be solved in terms of $V_{DD} - V_{out}$ to express the behavior of V_{out} as a function of V_{in}. But from a qualitative point of view, we observe that V_{out} continues to fall as V_{in} rises because both I_{D1} and the channel resistance of M_2 increase.

If V_{out} falls sufficiently, M_2 enters saturation. That is, if $V_{out} = V_{in} + |V_{TH2}|$, then M_2 is about to exit the triode region. But how about M_1? Since the drain voltage of $M_1 (= V_{out})$ is *higher* than its gate voltage (V_{in}), this device still operates in saturation. To obtain the inverter VTC in this region, we equate the drain currents again and neglect channel-length modulation:

$$\frac{1}{2}\mu_n C_{ox}\left(\frac{W}{L}\right)_1 (V_{in} - V_{TH1})^2 = \frac{1}{2}\mu_p C_{ox}\left(\frac{W}{L}\right)_2 (V_{DD} - V_{in} - |V_{TH2}|)^2. \tag{16.44}$$

What happened to V_{out} here?! Equation (16.44) is meaningless as it does not contain V_{out} and implies a unique value for V_{in}. This quandary arises because we have allowed two *ideal* current sources to fight each other at the output node. Inclusion of channel-length modulation resolves this issue:

$$\frac{1}{2}\mu_n C_{ox}\left(\frac{W}{L}\right)_1 (V_{in} - V_{TH1})^2 (1 + \lambda_1 V_{out})$$
$$= \frac{1}{2}\mu_p C_{ox}\left(\frac{W}{L}\right)_2 (V_{DD} - V_{in} - |V_{TH2}|)^2 [1 + \lambda_2 (V_{DD} - V_{out})]. \tag{16.45}$$

It follows that

$$V_{out} = \frac{\mu_p\left(\frac{W}{L}\right)_2(V_{DD} - V_{in} - |V_{TH2}|)^2 - \mu_n\left(\frac{W}{L}\right)_1(V_{in} - V_{TH1})^2}{\lambda_2\mu_p\left(\frac{W}{L}\right)_2(V_{DD} - V_{in} - |V_{TH2}|)^2 + \lambda_1\mu_n\left(\frac{W}{L}\right)_1(V_{in} - V_{TH1})^2}. \tag{16.46}$$

To gain more insight and prove that V_{out} changes sharply here, let us compute the small-signal gain of the inverter in this region. Operating is saturation, each transistor can be modeled as a voltage-dependent current source with a finite output impedance [Fig. 16.18(d)]. Since $v_2 = v_1 = v_{in}$, a KCL at the output node yields

$$\frac{v_{out}}{v_{in}} = -(g_{m1} + g_{m2})(r_{O1}\|r_{O2}), \tag{16.47}$$

indicating that the voltage gain is on the order of the intrinsic gain of a MOSFET. Thus, for a small change in V_{in}, we expect a large change in V_{out}.

Figure 16.19(a) summarizes our findings this far. The output remains at V_{DD} for $V_{in} < V_{TH1}$, begins to fall as V_{in} exceeds V_{TH1}, and experiences a sharp drop when M_2 enters saturation. The input level at which $V_{out} = V_{in}$ is called the "trip point" (also called the "switching threshold") of the inverter [Fig. 16.19(b)]. Both transistors are in saturation at this point (why?). The trip point is denoted by V_M. Also, the maximum and minimum values of a gate output are denoted by V_{OH} and V_{OL}, respectively.

As the input goes beyond the trip point, $V_{in} - V_{out}$ eventually exceeds V_{TH1}, thereby driving M_1 into the triode region. The transconductance of M_1 therefore falls and so does

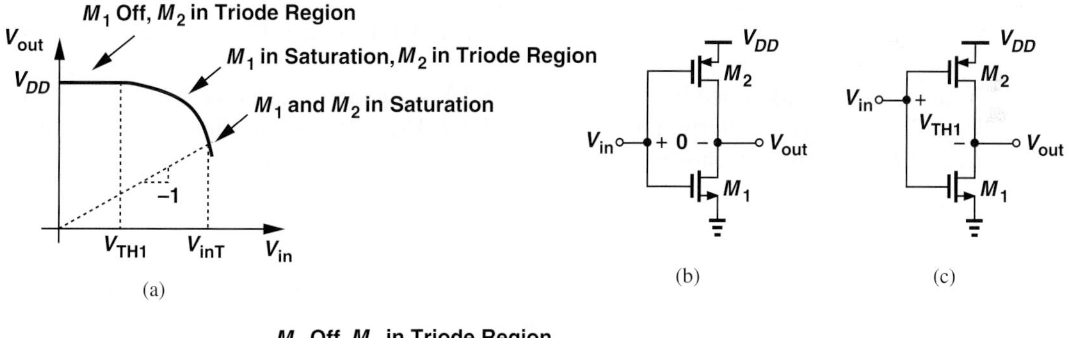

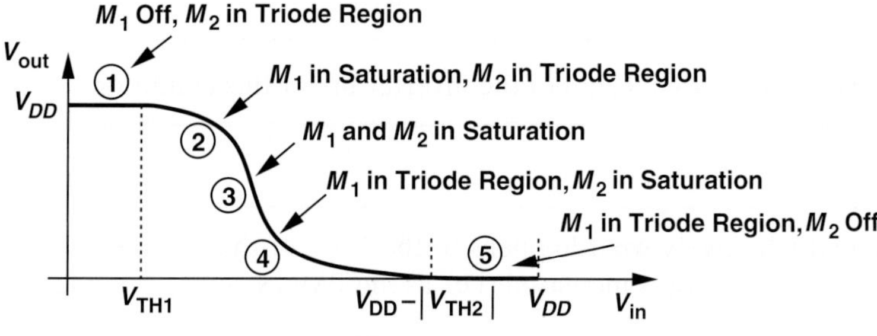

Figure 16.19 (a) Behavior of CMOS inverter for $V_{in} \leq V_{inT}$, (b) CMOS inverter at trip point, (c) M_1 at the edge of saturation, (d) overall characteristic.

the small-signal gain of the circuit [Fig. 16.19(c)]. We now have

$$\frac{1}{2}\mu_n C_{ox}\left(\frac{W}{L}\right)_1 [2(V_{in} - V_{TH1})V_{out} - V_{out}^2] = \frac{1}{2}\mu_p C_{ox}\left(\frac{W}{L}\right)_2 (V_{DD} - V_{in} - |V_{TH2}|)^2, \tag{16.48}$$

where channel-length modulation is neglected. From this quadratic equation, V_{out} can be expressed in terms of V_{in}, but we expect a more gradual slope due to the operation of M_1 in the triode region.

Finally, as V_{in} rises to $V_{DD} - |V_{TH2}|$, M_2 turns off, allowing $V_{out} = 0$. In this region, M_1 acts as a resistor carrying a zero current. Figure 16.19(d) plots the overall VTC, identifying different regions of operations by numbers.

Example 16-14 Determine a relationship between $(W/L)_1$ and $(W/L)_2$ that sets the trip point of the CMOS inverter to $V_{DD}/2$, thus providing a "symmetric" VTC.

Solution Replacing both V_{in} and V_{out} with $V_{DD}/2$ in Eq. (16.45), we have

$$\mu_n C_{ox}\left(\frac{W}{L}\right)_1\left(\frac{V_{DD}}{2} - V_{TH1}\right)^2\left(1 + \lambda_1\frac{V_{DD}}{2}\right)$$
$$= \mu_p C_{ox}\left(\frac{W}{L}\right)_2\left(\frac{V_{DD}}{2} - |V_{TH2}|\right)^2\left(1 + \lambda_2\frac{V_{DD}}{2}\right), \tag{16.49}$$

and hence

$$\frac{\left(\frac{W}{L}\right)_1}{\left(\frac{W}{L}\right)_2} = \frac{\mu_p\left(\frac{V_{DD}}{2} - |V_{TH2}|\right)^2\left(1 + \lambda_2\frac{V_{DD}}{2}\right)}{\mu_n\left(\frac{V_{DD}}{2} - V_{TH1}\right)^2\left(1 + \lambda_1\frac{V_{DD}}{2}\right)}. \tag{16.50}$$

In practice, the difference between $|V_{TH2}|$ and V_{TH1} can be neglected with respect to $V_{DD}/2 - |V_{TH1,2}|$. Similarly, $1 + \lambda_1 V_{DD}/2 \approx 1 + \lambda_2 V_{DD}/2$. Also, in digital design, both L_1 and L_2 are typically chosen equal to the minimum allowable value. Thus,

$$\frac{W_1}{W_2} \approx \frac{\mu_p}{\mu_n}. \tag{16.51}$$

Since the PMOS mobility is about one-third to one-half of the NMOS mobility, M_2 is typically twice to three times as wide as M_1.

Exercise What is the small-signal gain of the inverter under this condition?

Example 16-15 Explain qualitatively what happens to the VTC of the CMOS inverter as the width of the PMOS transistor is increased (i.e., as the PMOS device is made "stronger").

Solution Let us first consider the transition region around the trip point, where both M_1 and M_2 operate in saturation. As the PMOS device is made stronger, the circuit requires a *higher* input voltage to establish $I_{D1} = |I_{D2}|$. This is evident from Eq. (16.45): for $V_{out} = V_{DD}/2$, as $(W/L)_2$ increases, V_{in} must also increase so that $(V_{DD} - V_{in} - |V_{TH2}|)^2$ on the right-hand side decreases and $(V_{in} - V_{TH1})^2$ on the left-hand side increases. Consequently, the characteristic is shifted to the right (why?). (What happens to the small-signal gain near the trip point?)

Exercise What happens to the VTC of the CMOS inverter if the PMOS device experiences resistive degeneration?

Noise Margins Recall from Example 16-6 that a digital inverter always exhibits a small-signal voltage gain greater than unity in some region of the input/output characteristic. Since the gain of a CMOS inverter falls to zero near $V_{in} = 0$ and $V_{in} = V_{DD}$ (why?), we expect a gain of (negative) unity at two points between 0 and V_{DD}.

To determine the noise margin for logical low levels, we focus on region 2 in Fig. 16.19(d). With M_2 in the triode region, the voltage gain is relatively low and likely to assume a magnitude of unity somewhere. How do we express the gain of the circuit here? In a manner similar to Example 16-7, we directly differentiate both sides of Eq. (16.43) with respect to V_{in}:

$$2\mu_n\left(\frac{W}{L}\right)_1(V_{in} - V_{TH1}) = \mu_p\left(\frac{W}{L}\right)_2\left[-2(V_{DD} - V_{out}) - 2(V_{DD} - V_{in} - |V_{TH2}|)\frac{\partial V_{out}}{\partial V_{in}} + 2(V_{DD} - V_{out})\frac{\partial V_{out}}{\partial V_{in}}\right]. \quad (16.52)$$

The input level, V_{IL}, at which the gain reaches −1 can be solved by assuming $\partial V_{out}/\partial V_{in} = -1$:

$$\mu_n\left(\frac{W}{L}\right)_1(V_{IL} - V_{TH1}) = \mu_p\left(\frac{W}{L}\right)_2(2V_{OH} - V_{IL} - |V_{TH2}| - V_{DD}), \quad (16.53)$$

where V_{OH} denotes the corresponding output level. Obtaining V_{OH} from Eq. (16.53), substituting in Eq. (16.43), and carrying out some lengthy algebra, we arrive at

$$V_{IL} = \frac{2\sqrt{a}(V_{DD} - V_{TH1} - |V_{TH2}|)}{(a-1)\sqrt{a+3}} - \frac{V_{DD} - aV_{TH1} - |V_{TH2}|}{a-1}, \quad (16.54)$$

where

$$a = \frac{\mu_n\left(\frac{W}{L}\right)_1}{\mu_p\left(\frac{W}{L}\right)_2}. \quad (16.55)$$

Example 16-16 Recall from Example 16-14 that a symmetric VTC results if $a = 1$, $V_{TH1} = |V_{TH2}|$, and $\lambda_1 = \lambda_2$. Compute V_{IL} for this case.

Solution Choice of $a = 1$ in Eq. (16.54) yields $V_{IL} = \infty - \infty$. We can use L'Hopital's rule by first writing Eq. (16.54) as

$$V_{IL} = \frac{2\sqrt{a}(V_{DD} - 2V_{TH1}) - \sqrt{a+3}[V_{DD} - (a+1)V_{TH1}]}{(a-1)\sqrt{a+3}}, \tag{16.56}$$

where it is assumed $V_{TH1} = |V_{TH2}|$. Differentiating the numerator and the denominator with respect to a and substituting 1 for a, we have

$$V_{IL} = \frac{3}{8}V_{DD} + \frac{1}{4}V_{TH1}. \tag{16.57}$$

For example, if $V_{DD} = 1.8$ V and $V_{TH1} = 0.5$ V, then $V_{IL} = 0.8$ V.

Exercise Explain why V_{IL} must always exceed V_{TH1}.

We now turn our attention to NM_H and differentiate both sides of Eq. (16.48) with respect to V_{in}:

$$\mu_n\left(\frac{W}{L}\right)_1\left[2V_{out} + 2(V_{in} - V_{TH1})\frac{\partial V_{out}}{\partial V_{in}} - 2V_{out}\frac{\partial V_{out}}{\partial V_{in}}\right]$$

$$= 2\mu_p\left(\frac{W}{L}\right)_2(V_{in} - V_{DD} - |V_{TH2}|). \tag{16.58}$$

Again, we assume $\partial V_{out}/\partial V_{in} = -1$, $V_{in} = V_{IH}$, and $V_{out} = V_{OL}$, obtaining

$$V_{IH} = \frac{2a(V_{DD} - V_{TH1} - |V_{TH2}|)}{(a-1)\sqrt{1+3a}} - \frac{V_{DD} - aV_{TH1} - |V_{TH2}|}{a-1}. \tag{16.59}$$

The reader can prove that for $a = 1$, $V_{TH1} = |V_{TH2}|$, and $\lambda_1 = \lambda_2$,

$$NM_H = NM_L = \frac{3}{8}V_{DD} + \frac{1}{4}V_{TH1}. \tag{16.60}$$

Example 16-17 Compare the noise margins expressed by Eq. (16.60) with those of an *ideal* inverter.

Solution An ideal inverter is characterized by the behavior illustrated in Fig. 16.1(b), where the small-signal gain goes abruptly from zero to infinity at the trip point. With a symmetric VTC,

$$NM_{H,\text{ideal}} = NM_{L,\text{ideal}} = \frac{V_{DD}}{2}. \tag{16.61}$$

This value is greater than that in Eq. (16.60) because V_{TH1} and $|V_{TH2}|$ are typically less than $V_{DD}/2$ (and the gain in the transition region less than infinity).

Exercise Determine the reduction in the noise margins of an ideal inverter if its transition region gain is equal to 5. Assume a symmetric VTC.

Example 16-18 Explain what happens if V_{TH1} and $|V_{TH2}|$ in a CMOS inverter exceed $V_{DD}/2$.

Solution Consider the operation of the circuit for $V_{in} = V_{DD}/2$. In this case, *both* transistors are off, allowing the output node to "float." For this and speed reasons (explained in the next section), the threshold voltage is typically chosen to be less than $V_{DD}/4$.

Exercise What happens if $V_{TH1} = V_{DD}/4$ but $|V_{TH2}| = 3V_{DD}/4$?

16.2.3 Dynamic Characteristics

As explained in Section 16.1.2, the dynamic behavior of gates is related to the rate at which their output can change from one logical level to another. We now analyze the response of a CMOS inverter to a step input while the circuit drives a finite load capacitance. Our study of the NMOS inverter in Section 16.1.2 and the contrasts drawn in Section 16.2.1 prove useful here.

Qualitative Study Let us first understand qualitatively how a CMOS inverter charges and discharges a load capacitance. Suppose, as depicted in Fig. 16.20(a), V_{in} jumps from V_{DD} to 0 at $t = 0$ and V_{out} begins to from 0. Transistor M_1 turns off and transistor M_2 turns on in saturation, charging C_L toward V_{DD}. With the relatively constant current provided by M_2, V_{out} rises linearly until M_2 enters the triode region and hence supplies a smaller current. The output voltage continues to rise, almost as if M_2 acts as a resistor, eventually approaching V_{DD} and forcing the drain current of M_2 to zero. Figure 16.20(b) sketches the behavior of the output.

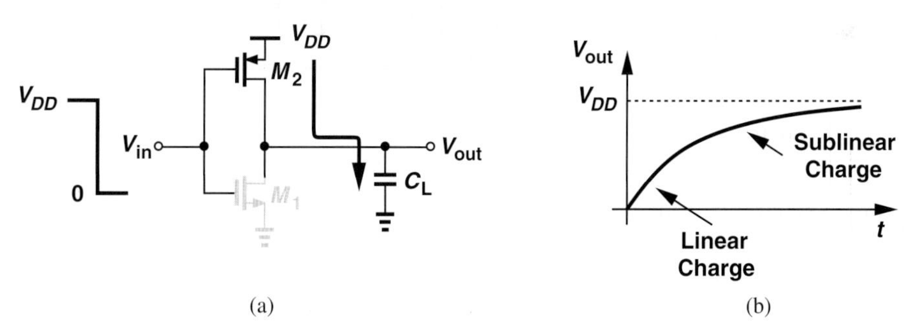

Figure 16.20 (a) CMOS inverter charging a load capacitance, (b) output waveform.

Example 16-19 Sketch the drain current of M_2 as a function of time.

Solution The current begins at a high (saturated) value and begins to fall as V_{out} exceeds $|V_{TH2}|$ (why?). Thereafter, the current continues to drop as V_{out} approaches V_{DD} and hence V_{DS2} falls to zero. Figure 16.21 plots the result.

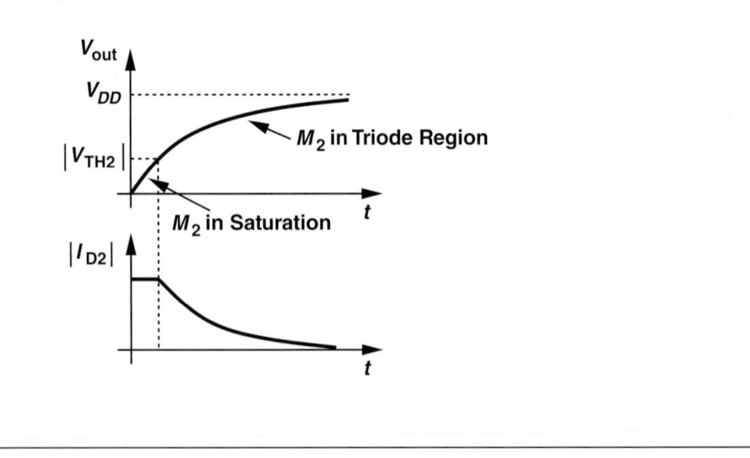

Figure 16.21

Exercise Sketch the supply current as a function of time.

Example 16-20 Sketch the output waveform of Fig. 16.20(b) for different values of $(W/L)_2$.

Solution As $(W/L)_2$ increases, so does the current drive of M_2 (in both saturation and triode regions). The circuit therefore exhibits a faster rising transition, as illustrated in Fig. 16.22. Of course, for very large values of W_2, the capacitance contributed by M_2 itself at the output node becomes comparable with C_L, and the speed improves to a lesser extent.

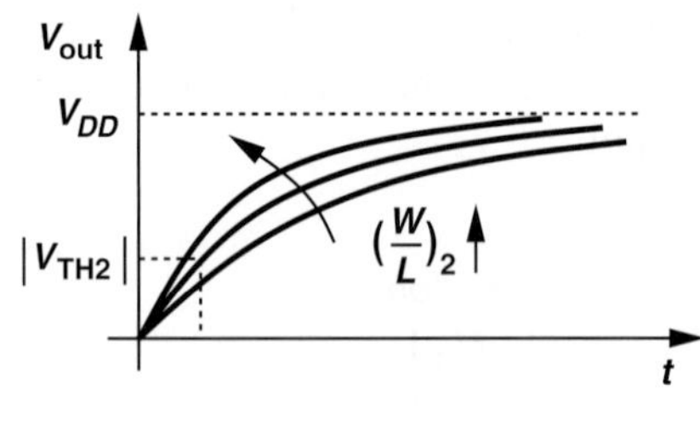

Figure 16.22

Exercise Sketch the drain current of M_2 for different values of $(W/L)^2$.

Did you know?

Intel's first microprocessor, the 4004, operated with a maximum clock frequency of 750 kHz, whereas today's processors reach nearly 4 GHz. This 5300x improvement in speed is primarily due to the scaling of device channel lengths from 10 μm to about 30 nm ($\approx$ 330x). Today's gate delays are around 8 ps. But doesn't 8 ps imply much higher speeds than 4 GHz? The key point here is that this value is observed in a chain of identical inverters. As we consider greater fanouts and, more importantly, the effect of long wires on the chip, the delays become substantially longer.

How about the output discharge behavior? As shown in Fig. 16.23(a), if the input steps from 0 to V_{DD} at $t = 0$, M_2 turns off, M_1 turns on, beginning to discharge C_L from V_{DD} toward 0. Transistor M_1 operates in saturation until V_{out} falls by V_{TH1} below the gate voltage ($= V_{DD}$), upon which I_{D1} begins to decrease, slowing down the discharge. Plotted in Fig. 16.23(b), V_{out} then gradually approaches zero.

Quantitative Analysis With the insights developed above, we can now quantify the rising and falling transitions at the output of the CMOS inverter, thereby arriving at the propagation delays. We neglect channel-length modulation here.

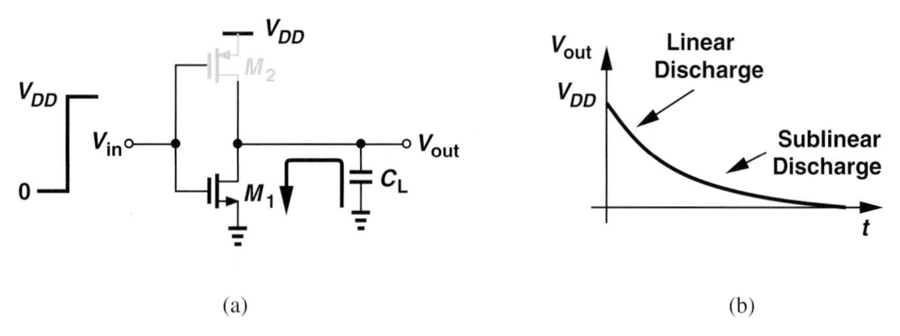

Figure 16.23 (a) CMOS inverter discharging a load capacitance, (b) output waveform.

Recall from Fig. 16.20 that, after the input falls to zero, M_2 begins to charge C_L with a constant current, given by

$$|I_{D2}| = \frac{1}{2}\mu_p C_{ox}\left(\frac{W}{L}\right)_2 (V_{DD} - |V_{TH2}|)^2, \tag{16.62}$$

producing

$$V_{out}(t) = \frac{|I_{D2}|}{C_L}t \tag{16.63}$$

$$= \frac{1}{2}\mu_p \frac{C_{ox}}{C_L}\left(\frac{W}{L}\right)_2 (V_{DD} - |V_{TH2}|)^2 t. \tag{16.64}$$

Transistor M_2 enters the triode region for $V_{out} = |V_{TH2}|$ at a time given by

$$T_{PLH1} = \frac{2|V_{TH2}|C_L}{\mu_p C_{ox}\left(\frac{W}{L}\right)_2 (V_{DD} - |V_{TH2}|)^2}. \tag{16.65}$$

Thereafter, M_2 operates in the triode region, yielding

$$|I_{D2}| = C_L \frac{dV_{out}}{dt}, \tag{16.66}$$

and hence

$$\frac{1}{2}\mu_p C_{ox}\left(\frac{W}{L}\right)_2 \left[2(V_{DD} - |V_{TH2}|)(V_{DD} - V_{out} - (V_{DD} - V_{out})^2\right] = C_L \frac{dV_{out}}{dt}. \tag{16.67}$$

Rearranging the terms gives

$$\frac{dV_{out}}{2(V_{DD} - |V_{TH2}|)(V_{DD} - V_{out}) - (V_{DD} - V_{out})^2} = \frac{1}{2}\mu_p \frac{C_{ox}}{C_L}\left(\frac{W}{L}\right)_2 dt. \tag{16.68}$$

Defining $V_{DD} - V_{out} = u$ and noting that

$$\int \frac{du}{au - u^2} = \frac{1}{a} \ln \frac{u}{a - u}, \tag{16.69}$$

we have

$$\frac{-1}{2(V_{DD} - |V_{TH2}|)} \ln \frac{V_{DD} - V_{out}}{V_{DD} - 2|V_{TH2}| + V_{out}}\Bigg|_{Vout=|VTH2|}^{Vout=VDD/2} = \frac{1}{2}\mu_p \frac{C_{ox}}{C_L}\left(\frac{W}{L}\right)_2 T_{PLH2}, \tag{16.70}$$

where the origin of time is chosen to coincide with $t = T_{PLH1}$ for simplicity, and T_{PLH2} denotes the time required for V_{out} to go from $|V_{TH2}|$ to $V_{DD}/2$. It follows that

$$T_{PLH2} = \frac{C_L}{\mu_p C_{ox}\left(\frac{W}{L}\right)_2 [V_{DD} - |V_{TH2}|)} \ln\left(3 - 4\frac{|V_{TH2}|}{V_{DD}}\right). \tag{16.71}$$

Interestingly, the denominator of Eq. (16.71) represents the inverse of the on-resistance of M_2 when it operates in the deep triode region. Thus,

$$T_{PLH2} = R_{on2} C_L \ln\left(3 - 4\frac{|V_{TH2}|}{V_{DD}}\right). \tag{16.72}$$

If $4|V_{TH2}| \approx V_{DD}$, this result reduce s to $T_{PLH2} = R_{on2} C_L \ln 2$—as if C_L charges up through a *constant* resistance equal to R_{on2}. The overall propagation delay is therefore given by

$$T_{PLH} = T_{PLH1} + T_{PLH2} \tag{16.73}$$

$$= R_{on2} C_L \left[\frac{2|V_{TH2}|}{V_{DD} - |V_{TH2}|} + \ln\left(3 - 4\frac{|V_{TH2}|}{V_{DD}}\right)\right]. \tag{16.74}$$

An important observation here is that T_{PLH} decreases as V_{DD} increases (why?). Also, for $|V_{TH2}| \approx V_{DD}/4$, the two terms inside the square brackets are nearly equal.

Example 16-21 A student decides to avoid the foregoing derivation of T_{PLH2} through the use of an average current for M_2. That is, I_{D2} is approximated as a constant value equal to the average between its initial value, $(1/2)\mu_p C_{ox}(W/L)_2(V_{DD} - |V_{TH2}|)^2$, and its final value, 0. Determine the resulting T_{PLH2} and compare with that expressed by Eq. (16.72).

Solution The average current is equal to $(1/4)\mu_p C_{ox}(W/L)_2(V_{DD} - |V_{TH2}|)^2$, yielding:

$$T_{PLH2} = \frac{C_L}{\mu_p C_{ox}\left(\frac{W}{L}\right)_2 (V_{DD} - |V_{TH2}|)} \cdot \frac{V_{DD}/2 - (V_{DD} - |V_{TH2}|)}{V_{DD} - |V_{TH2}|}. \tag{16.75}$$

Assuming $|V_{TH2}|$ is roughly equal to $V_{DD}/4$ and hence $V_{DD}/(V_{DD} - |V_{TH2}|) \approx 4/3$, we have

$$T_{PLH2} \approx \frac{4}{3} R_{on2} C_L, \tag{16.76}$$

about 50% greater than that obtained above.

Exercise What happens if $|V_{TH2}| \approx V_{DD}/3$?

The calculation of T_{PHL} follows the same procedure as above. Specifically, after the input jumps from 0 to V_{DD} [Fig. 16.23(a)], M_2 turns off and M_1 draws a current of $(1/2)\mu_n C_{ox}(W/L)_1(V_{DD} - V_{TH1}|)^2$. The time required for M_1 to enter the triode region is thus given by

$$T_{\mathrm{PHL1}} = \frac{2V_{TH1} C_L}{\left(\mu_n C_{ox}\frac{W}{L}\right)_1 (V_{DD} - V_{TH1})^2}. \tag{16.77}$$

After this point in time,

$$\frac{1}{2}\mu_n C_{ox}\left(\frac{W}{L}\right)_1 [2(V_{DD} - V_{TH1})V_{out} - V_{out}^2] = -C_L \frac{dV_{out}}{dt}, \tag{16.78}$$

where the negative sign on the right accounts for the flow of the current *out* of the capacitor. Using Eq. (16.69) to solve this differential equation and bearing in mind that $V_{out}(t = 0) = V_{DD} - V_{TH1}$, we obtain

$$\frac{-1}{2(V_{DD} - |V_{TH1}|)} \ln \frac{V_{out}}{2(V_{DD} - V_{TH2}) - V_{out}}\Bigg|_{V_{out}=VDD-VTH1}^{V_{out}=VDD/2} = \frac{1}{2}\mu_n \frac{C_{ox}}{C_L}\left(\frac{W}{L}\right)_1 T_{PLH2}. \tag{16.79}$$

It follows that

$$T_{\mathrm{PHL2}} = R_{on1} C_L \ln\left(3 - 4\frac{V_{TH1}}{V_{DD}}\right), \tag{16.80}$$

which, of course, has the same form as Eq. (16.72). Also, the total delay is given by

$$T_{PHL} = T_{\mathrm{PHL1}} + T_{\mathrm{PHL2}} \tag{16.81}$$

$$= R_{on1} C_L \left[\frac{2V_{TH1}}{V_{DD} - V_{TH1}} + \ln\left(3 - 4\frac{V_{TH1}}{V_{DD}}\right)\right]. \tag{16.82}$$

Example 16-22 Compare the two terms inside the square brackets in Eq. (16.82) as V_{TH1} varies from zero to $V_{DD}/2$.

Solution For $V_{TH1} = 0$, the first term is equal to 0 and and the second equal to $\ln 3 \approx 1.1$. As V_{TH1} increases, the two terms converge, both reaching 0.684 for $V_{TH1} = 0.255V_{DD}$. Finally, for $V_{TH1} = V_{DD}/2$, the first term rises to 2 and the second falls to 0. Figure 16.24 plots each term and the sum of the two, suggesting that low thresholds improve the speed.

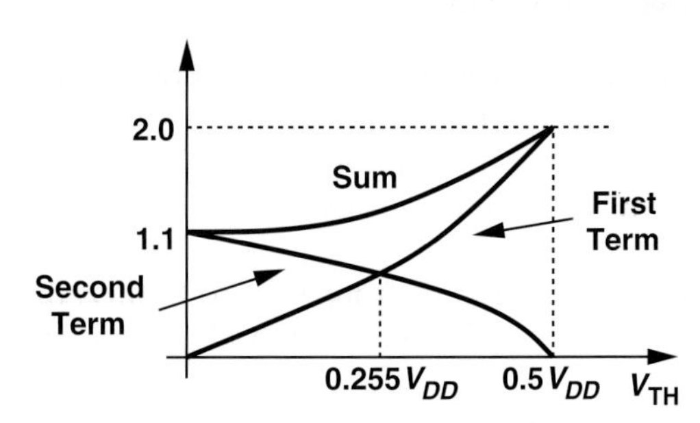

Figure 16.24

Exercise Repeat the above example if V_{TH1} varies from 0 to $3V_{DD}/4$.

Example 16-23 Due to a manufacturing error, an inverter is constructed as shown in Fig. 16.25, where M_1' appears in series with M_1 and is identical to M_1. Explain what happens to the output falltime. For simplicity, view M_1 and M_1' as resistors when they are on.

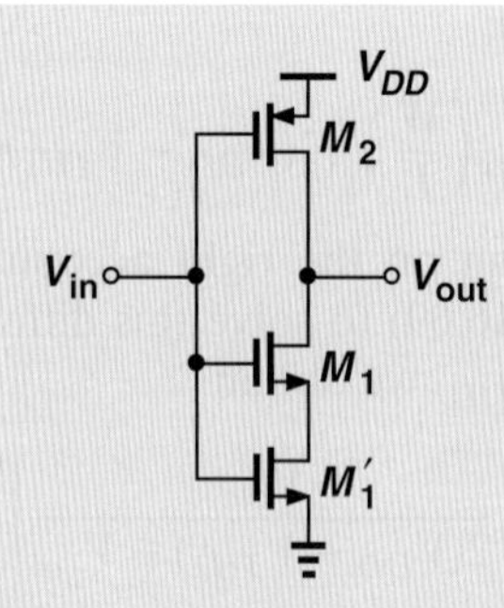

Figure 16.25

Solution Placing the two on-resistances in series, we have

$$R_{on1}||R'_{on1} = \frac{1}{\mu_n C_{ox}\left(\frac{W}{L}\right)_1 (V_{DD} - V_{TH1})} + \frac{1}{\mu_n C_{ox}\left(\frac{W}{L}\right)'_1 (V_{DD} - V'_{TH1})} \quad (16.83)$$

$$= \frac{2}{\mu_n C_{ox}\left(\frac{W}{L}\right)_1 (V_{DD} - V_{TH1})} \quad (16.84)$$

$$= 2R_{on1}. \quad (16.85)$$

Thus, the falltime is doubled.

Exercise What happens if M_1' is twice as wide as M_1?

16.2.4 Power Dissipation

Having determined the propagation delays of the CMOS inverter, we now turn our attention to the power dissipation of the circuit. Unlike the NMOS inverter, this type of logic consumes no static power. We therefore need only study the behavior of the circuit during transitions and determine the "dynamic" power dissipation. Let us first assume abrupt transitions at the input.

If the input voltage jumps from V_{DD} to 0, then the PMOS device charges the load capacitance toward V_{DD} [Fig. 16.26(a)]. As V_{out} approaches V_{DD}, the energy stored in C_L is equal to

$$E_1 = \frac{1}{2} C_L V_{DD}^2. \quad (16.86)$$

This energy is supplied by M_2 from V_{DD}. On the other hand, if V_{in} steps from 0 to V_{DD}, then the NMOS transistor discharges C_L toward zero [Fig. 16.26(b)]. That is, the energy E_1 is removed from C_L and dissipated by M_1 in the discharge process. This cycle repeats for every pair of falling and rising transitions at the input.

In summary, for every pair of falling and rising transitions at the input of the inverter, C_L acquires and loses an energy of $(1/2)C_L V_{DD}^2$. For a periodic input, we may then surmise that the circuit consumes an average power of $(1/2)C_L V_{DD}^2/T_{in}$, where T_{in} denotes the input period. Unfortunately, this result is incorrect. In addition to delivering energy to C_L, the PMOS transistor in Fig. 16.26(a) also consumes power because it carries a finite

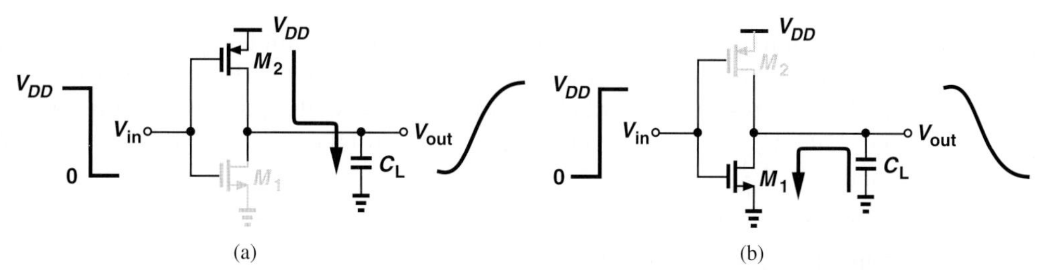

Figure 16.26 Power consumed in transistors during (a) charge and (b) discharge of load capacitance.

current while sustaining a finite voltage. In other words, the total energy drawn from V_{DD} in Fig. 16.26(a) consists of that stored on C_L plus that dissipated in M_2.

How do we compute the energy consumed by M_2? We first observe that (a) the instantaneous power dissipated in M_2 is given by $|V_{DS2} \parallel I_{D2}| = (V_{DD} - V_{out})|I_{D2}|$, and (b) this transistor charges the load capacitor and hence $|I_{D2}| = C_L dV_{out}/dt$. To calculate the energy lost in M_2, we must integrate the instantaneous power dissipation with respect to time:

$$E_2 = \int_{t=0}^{\infty} (V_{DD} - V_{out}) \left(C_L \frac{dV_{out}}{dt} \right) dt, \tag{16.87}$$

which reduces to

$$E_2 = C_L \int_{Vout=0}^{VDD} (V_{DD} - V_{out}) dV_{out} \tag{16.88}$$

$$= \frac{1}{2} C_L V_{DD}^2. \tag{16.89}$$

Interestingly, the energy consumed by M_2 is equal to that stored on C_L. Thus, the total energy drawn from V_{DD} is

$$E_{tot} = E_1 + E_2 \tag{16.90}$$

$$= C_L V_{DD}^2. \tag{16.91}$$

It follows that, for a periodic input with frequency f_{in}, the average power drawn from V_{DD} is equal to

$$P_{av} = f_{in} C_L V_{DD}^2. \tag{16.92}$$

Example 16-24

In the circuit of Fig. 16.27, $V_{out} = 0$ at $t = 0$. Compute the energy drawn from the supply as V_{out} reaches V_{DD}.

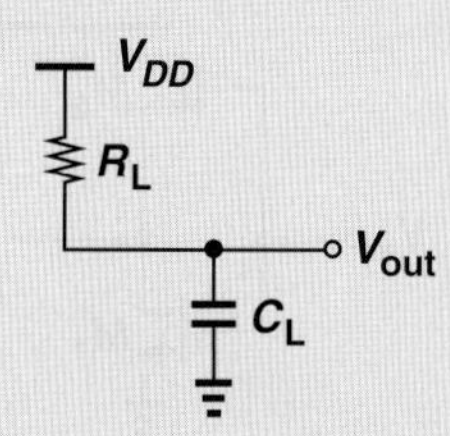

Figure 16.27

Solution We note that the derivation leading to Eq. (16.91) is completely general and independent of the I/V characteristics of the device that charges C_L. In other words, the circuit of Fig. 16.27 stores an energy of $(1/2)C_L V_{DD}^2$ on the load capacitor and consumes an energy of $(1/2)C_L V_{DD}^2$ in R_1 while charging C_L. The total energy supplied by V_{DD} is therefore equal to $C_L V_{DD}^2$.

Exercise Compute the energy consumed by R_L.

Did you know?

The CMOS inverter and the logic family derived from it became popular due to their nearly zero static power consumption. But the word "nearly" has become problematic in the past decade: each gate draws a finite leakage current in its idle state because a MOS transistor with zero V_{GS} is not quite off. With tens or hundreds of millions of gates, the static current has reached an appreciable portion of the overall current drawn by microprocessors from their supply voltage. In fact, in the early 2000s it was predicted that the leakage current would *exceed* the dynamic current by mid- to late 2000s. Fortunately, innovations in MOS technology and circuit design have avoided that.

Equation (16.92) plays a central role in CMOS logic design, elegantly expressing the dependence of P_{av} on the data rate, the load capacitance, and the supply voltage. The square dependence on V_{DD} calls for the *reduction* of the supply voltage, whereas Eqs. (16.74) and (16.82) for the propagation delays favor *raising* V_{DD}.

Power-Delay Product As mentioned in Section 16.1.3, the power-delay product represents the trade-off between the power dissipation and the speed. With the aid of Eqs. (16.35), (16.74), and (16.82) and assuming that T_{PHL} and T_{PLH} are roughly equal, we write

$$PDP = R_{on1} C_L^2 V_{DD}^2 \left[\frac{2V_{TH}}{V_{DD} - V_{TH}} + \ln\left(3 - 4\frac{V_{TH}}{V_{DD}}\right) \right]. \tag{16.93}$$

Interestingly, the PDP is proportional to C_L^2, underlining the importance of minimizing capacitances in the circuit.

Example 16-25 In the absence of long interconnects, C_L in Fig. 16.26 arises only from transistor capacitances. Consider a cascade of two identical inverters, Fig. 16.28, where the PMOS device is three times as wide as the NMOS transistor to provide a symmetric VTC. For simplicity, assume the capacitance at node X is equal to $4WLC_{ox}$. Also, $V_{THN} = |V_{THP}| \approx V_{DD}/4$. Compute the PDP.

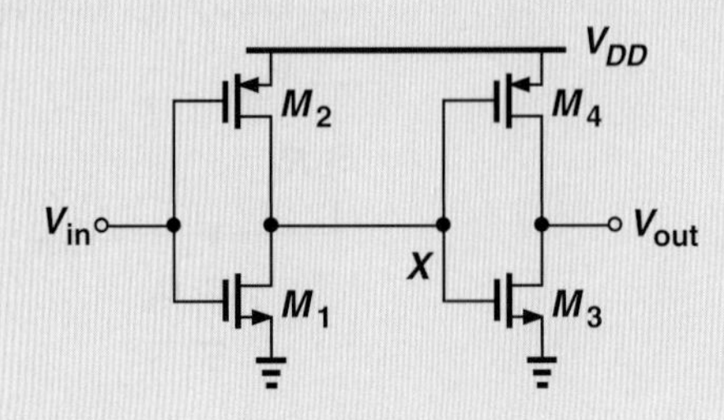

Figure 16.28

Solution We have

$$R_{on} = \frac{1}{\mu_n C_{ox}\left(\frac{W}{L}\right)(V_{DD} - V_{TH})} \tag{16.94}$$

$$\approx \frac{4}{3}\frac{1}{\mu_n C_{ox}\left(\frac{W}{L}\right)V_{DD}}. \tag{16.95}$$

Also, from Example 16-22, the two terms in the square brackets in Eqs. (16.74) and (16.82) add up to 1.36. Thus, Eq. (16.93) reduces to

$$PDP = \frac{7.25WL^2 C_{ox} f_{in} V_{DD}^2}{\mu_n}. \tag{16.96}$$

Exercise Suppose the withs of all four transistors are doubled. Does the delay of the first inverter change? How about the power dissipated per transition? From these observations, explain why PDP is linearly proportional to W.

Crowbar Current In our study of the dynamic power consumption, we have assumed abrupt transitions at the input. In practice, however, the input suffers from a finite transition time, thereby leading to another dissipation component.

Recall from the VTC of Fig. 16.19(d) that *both* transistors in an inverter are on in regions 2, 3, and 4. That is, if the input lies in the range $[V_{TH1}\ V_{DD} - |V_{TH2}|]$, then M_2 draws a current from V_{DD} and M_1 passes this current to ground—as if a direct path conducts current from V_{DD} to ground [Fig. 16.29(a)]. Called the "crowbar current," this component arises each time the input swings from one rail to the other with a finite transition time. As illustrated in Fig. 16.29(b), the circuit draws a crowbar current from t_1 to t_2.

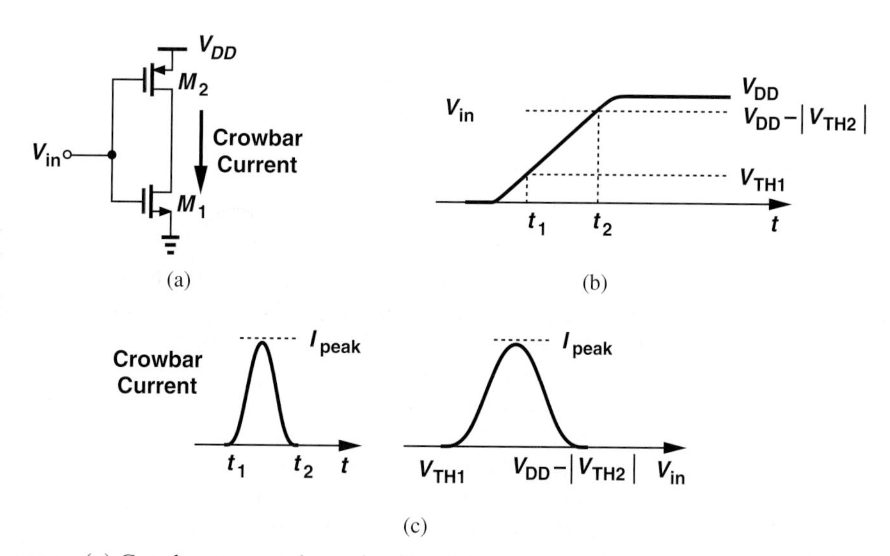

Figure 16.29 (a) Crowbar current drawn by CMOS inverter, (b) time period during which crowbar current is drawn, (c) crowbar current as a function of time and V_{in}.

How does the crowbar current very between t_1 and t_2 in Fig. 16.29(b)? For V_{in} slightly above V_{TH1}, M_1 is barely on, drawing only a small current. As V_{in} approaches the trip point of the inverter, both transistors enter saturation and the crowbar current reaches a maximum. Finally, as V_{in} reaches $V_{DD} - |V_{TH2}|$, the crowbar current returns to zero. Figure 16.29(c) plots the behavior of this current as a function of t and V_{in}. The peak value is obtained by assuming $V_{in} = V_{out} = V_{DD}/2$ in either side of Eq. (16.45):

$$I_{peak} = \frac{1}{2}\mu_n C_{ox}\left(\frac{W}{L}\right)_1\left(\frac{V_{DD}}{2} - V_{TH1}\right)^2\left(1 + \lambda_1\frac{V_{DD}}{2}\right). \tag{16.97}$$

16.3 CMOS NOR AND NAND GATES

The CMOS inverter serves as the foundation for realizing other logical gates. In this section, we study NOR and NAND gates, both of which find wide application.

16.3.1 NOR Gate

Recall from basic logic design that the OR operation, $A + B$, produces a high output if at least one input is high. The NOR gate, $\overline{A + B}$, thus generates a *low* output if at least one input is high.

How should a CMOS inverter be modified to serve as a NOR gate? First, we need two sets of NMOS and PMOS devices that are controlled by the two inputs. Second, considering the NMOS section first, we note that if one of the NMOS gates is *high*, the output (the drain voltage) must remain low. We then surmise that the NMOS section can be realized as shown in Fig. 16.30, recognizing that, if A or B is high, the corresponding transistor is on, pulling V_{out} to zero. This, of course, occurs only if the remainder of the circuit (the PMOS section) "cooperates," as observed for the inverter in Section 16.2.1.

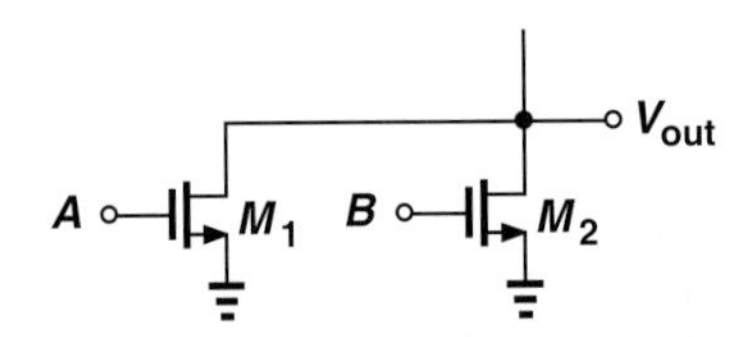

Figure 16.30 NMOS section of a NOR gate.

Example 16-26 Excited by the simple realization in Fig. 16.30, a student decides that the PMOS section should incorporate a similar topology, thus arriving at the circuit depicted in Fig. 16.31(a). Explain why this configuration does *not* operate as a NOR gate.

Solution Recall from Section 16.2.1 that cooperation between the NMOS and PMOS sections means that when one is on, the other must remain off. Unfortunately, the circuit of Fig. 16.31(a) fails to satisfy this principle. Specifically, if A is high and B is low, then *both* M_1 and M_4 are on [Fig. 16.31(b)], "fighting" each other and producing an ill-defined logical output. (Also, the circuit draws significant static power from V_{DD}.)

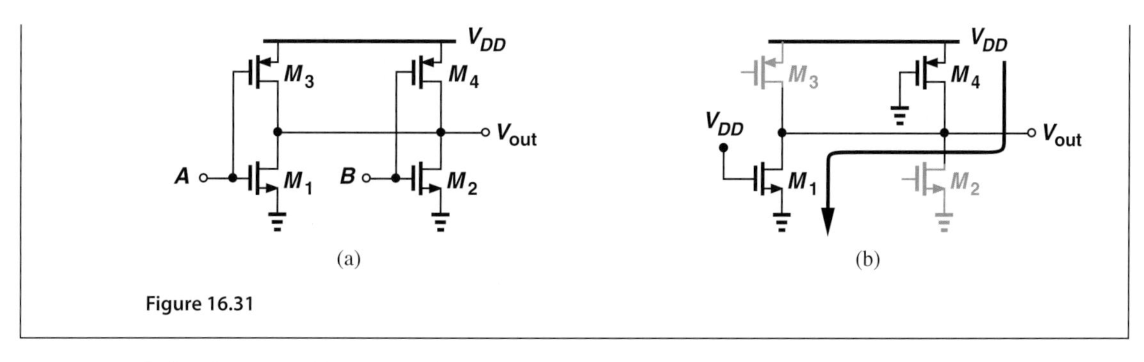

Figure 16.31

Exercise What happens if M_4 is omitted?

The above example reveals that the PMOS section must remain off if A or B (or both) are high. Moreover, if both inputs are low, the PMOS section must be *on* so as to ensure V_{out} is pulled up to V_{DD}. Shown in Fig. 16.32(a) is such a circuit, blocking the path from V_{DD} to V_{out} if one of the inputs is high (why?), but raising V_{out} to V_{DD} if *both* inputs are low. The operation, of course, remains unchanged if A and B are swapped.

Figure 16.32(b) depicts the overall CMOS NOR implementation. The reader is encouraged to verify the operation for all four input logical combinations and prove that the circuit consumes no static power.

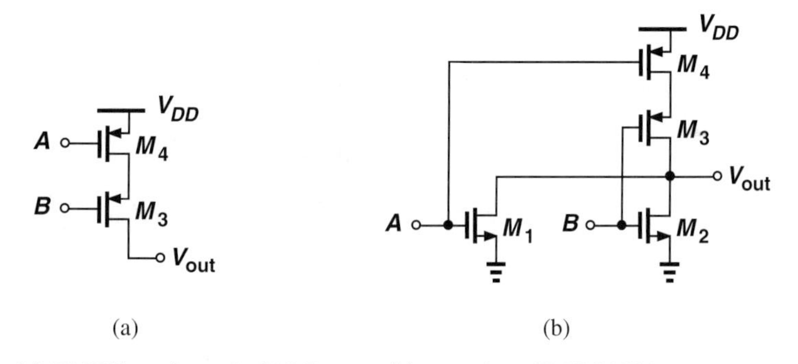

Figure 16.32 (a) PMOS section of a NOR gate, (b) complete CMOS NOR gate.

The reader may wonder why we did not attempt to implement an OR gate. As evident from the foregoing development, the evolution of the circuit from a CMOS inverter inherently contains an inversion. If an OR gate is necessary, the topology of Fig. 16.32(b) can be followed by an inverter.

Example 16-27 Construct a three-input NOR gate.

Solution We expand the NMOS section of Fig. 16.30 and the PMOS section of Fig. 16.32(a) so as to accommodate three inputs. The result is depicted in Fig. 16.33.

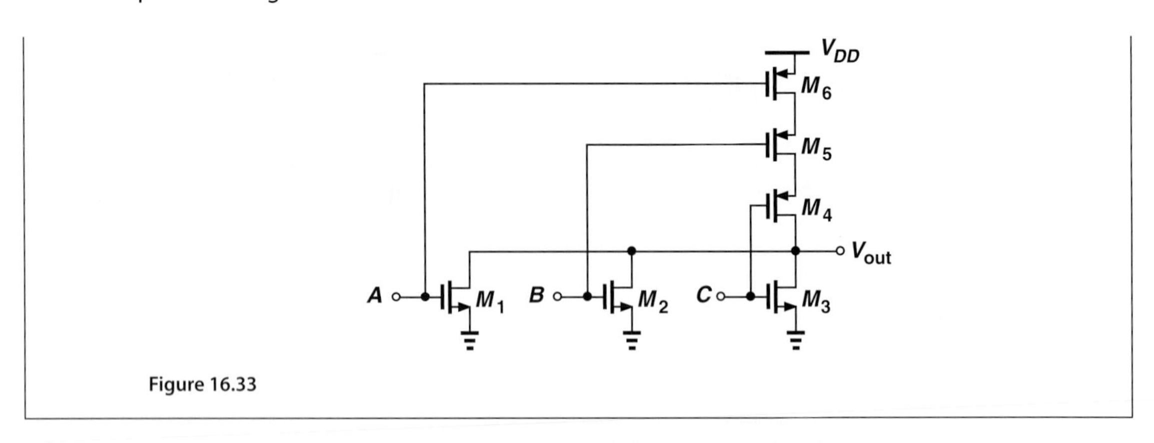

Figure 16.33

Exercise Study the behavior of the circuit if M_3 is accidentally omitted.

The principal drawback of the CMOS NOR gate stems from the use of PMOS devices *in series*. Recall that the low mobility of holes requires a proportionally wider PMOS transistor to obtain a symmetric VTC and, more importantly, equal rise and fall times. Viewing the transistors in a two-input NOR gate as resistors for simplicity, we observe that the PMOS section suffers from *twice* the resistance of each PMOS device (Example 16-23), creating a slow rising transition at the output (Fig. 16.34). If wider PMOS transistors are employed to reduce R_{on}, then their gate capacitance ($\approx WLC_{ox}$) increases, thereby loading the *preceding* stage. The situation worsens as the number of inputs to the gate increases.

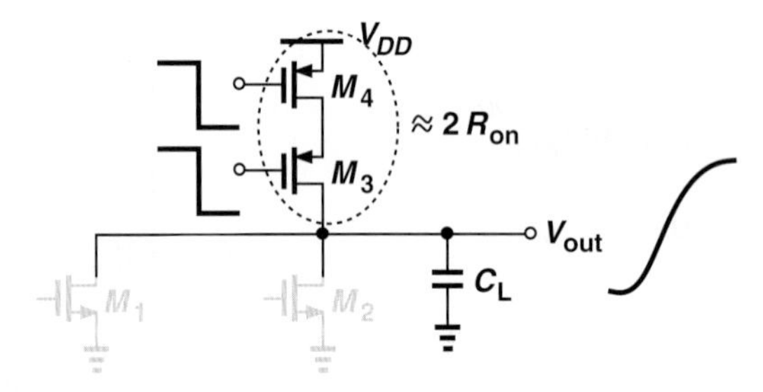

Figure 16.34 PMOS devices in series charging a load capacitance.

Example 16-28 Select the relative widths of the transistors in the three-input NOR gate of Fig. 16.33 for equal rise and fall times. Assume $\mu_n \approx 2\mu_p$ and equal channel lengths.

Solution The series combination of the three PMOS devices must present a resistance equal to that of an NMOS transistor. If $W_1 = W_2 = W_3 = W$, then we must choose

$$W_4 = W_5 = W_6 = 6W, \tag{16.98}$$

so as to ensure that each PMOS device exhibits an on-resistance equal to one-third of that of each NMOS transistor. Note that the gate presents a capacitance of about $7WLC_{ox}$ at each input, quite larger than that of an inverter ($\approx 3WLC_{ox}$).

Exercise Repeat the above example if $\mu_n \approx 3\mu_p$.

16.3.2 NAND Gate

The developments in Section 16.3.1 for the NOR gate can readily be extended to create a NAND gate. Since an NAND operation, $\overline{A \cdot B}$, produces a zero output if *both* inputs are high, we construct the NMOS section as shown in Fig. 16.35(a), where M_1 or M_2 *blocks* the path from V_{out} to ground unless both A and B remain high. The PMOS section, on other hand, must pull V_{out} to V_{DD} if at least one of the inputs is low, and is thus realized as shown in Fig. 16.35(b). Figure 16.35(c) depicts the overall NAND gate. This circuit, too, consumes zero static power.

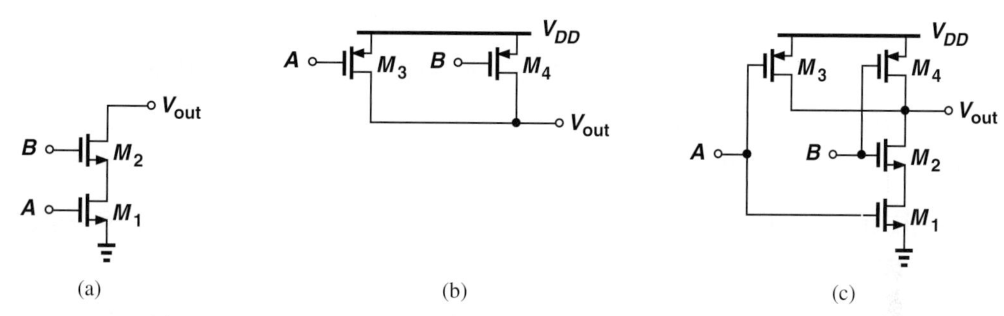

Figure 16.35 (a) NMOS section of a NAND gate, (b) PMOS section of a NAND gate, (c) complete CMOS NAND gate.

In contrast to the NOR gate, the NAND gate places NMOS devices in series, thus suffering less severely from speed limitation of PMOS transistors. The following example illustrates this point.

Example 16-29 Design a three-input NAND gate and determine the relative widths of the transistors for equal rise and fall times. Assume $\mu_n \approx 2\mu_p$ and equal channel lengths.

Solution Figure 16.36 shows the realization of the gate. With three NMOS transistors in series, we select a width of $3W$ for M_1-M_3 so that the total series resistance is equivalent to one device having a width of W. Each PMOS device must therefore have a width of $2W$. Consequently, the capacitance seen at each input is roughly equal to $5WLC_{ox}$, about 30% less than that of the NOR gate in Example 16-28.

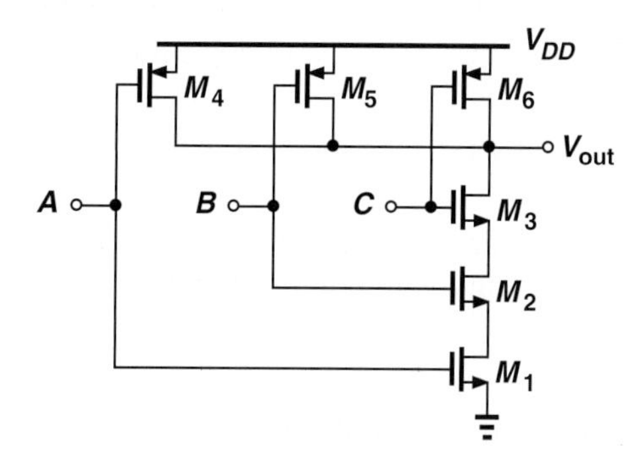

Figure 16.36

Exercise Repeat the above example if $\mu_n \approx 3\mu_p$.

In CMOS logic, the PMOS and NMOS sections are called "dual" of each other. In fact, given one section, we can construct the other according to the following rule: convert each series branch to parallel branches and vice versa.

Example 16-30 Determine the PMOS dual of the circuit shown in Fig. 16.37(a) and determine the logical function performed by the overall CMOS realization.

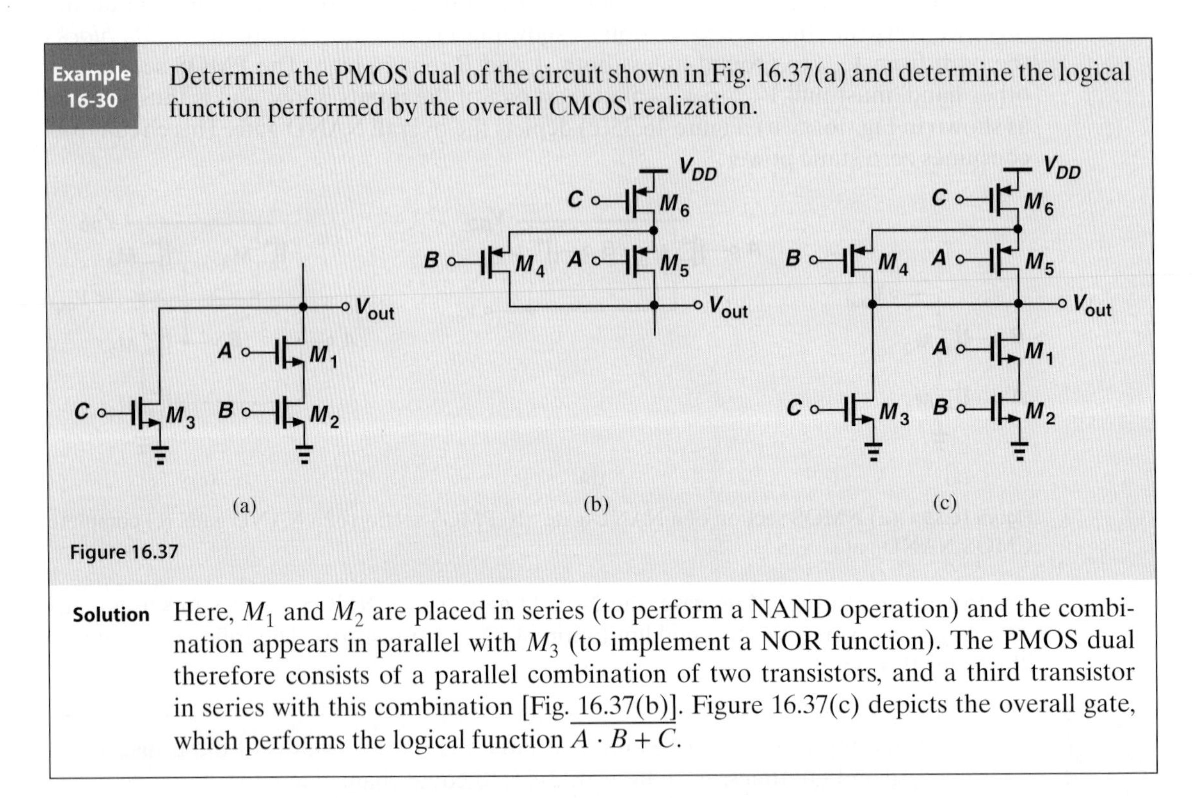

Figure 16.37

Solution Here, M_1 and M_2 are placed in series (to perform a NAND operation) and the combination appears in parallel with M_3 (to implement a NOR function). The PMOS dual therefore consists of a parallel combination of two transistors, and a third transistor in series with this combination [Fig. 16.37(b)]. Figure 16.37(c) depicts the overall gate, which performs the logical function $\overline{A \cdot B + C}$.

Exercise Suppose M_3 is accidentally omitted. Study the behavior of the gate.

16.4 CHAPTER SUMMARY

- Digital CMOS circuits account for more than 80% of the semiconductor market.
- The speed, power dissipation, and noise immunity of digital gates are critical parameters.
- The input/output characteristic of a gate reveals its immunity to noise or degraded logical levels.
- Noise margin is defined as the voltage degradation on the high or low levels that places the signal at the unity-gain point of the input/output characteristic.
- The speed of gates is given by the drive capability of the transistors and the capacitances contributed by transistors and interconnecting wires.
- The power-speed trade-off of gates is quantified by the power-delay product.
- The CMOS inverter is an essential building block in digital design. It consumes no power in the absence of signal transitions.
- The NMOS and PMOS devices in an inverter provide "active" pull-down and pull-up currents and hence enhance each other's operation.

- The average power dissipated by a CMOS inverter is equal to $f_{in} C_L V_{DD}^2$.
- Based on the CMOS inverter, other gates such as NOR and NAND gates can be derived. These gates also have zero static power.

PROBLEMS

Unless otherwise stated, in the following problems assume $V_{DD} = 1.8$ V, $\mu_n C_{ox} = 100\ \mu\text{A/V}^2$, $\mu_p C_{ox} = 50\ \mu\text{A/V}^2$, $V_{TH,N} = 0.4$ V, $V_{TH,P} = -0.5$ V, $\lambda_N = 0$, and $\lambda_P = 0$.

Sec. 16.1.1 Static Characterization of Codes

16.1. In the CS stage of Example 16-2, we have $R_D = 10$ kΩ and $(W/L)_1 = 3/0.18$. Calculate the output low level when $V_{in} = V_{DD}$.

16.2. The CS stage of Example 16-2 must achieve an output low level no higher than 100 mV. If $R_D = 5$ kΩ, determine the minimum required value of $(W/L)_1$.

16.3. Consider the PMOS common-source stage shown in Fig. 16.38. We wish to utilize this circuit as a logical inverter. Compute the low and high output levels if $(W/L)_1 = 20/0.18$ and $R_D = 5$ kΩ. Assume the input swings from zero to V_{DD}.

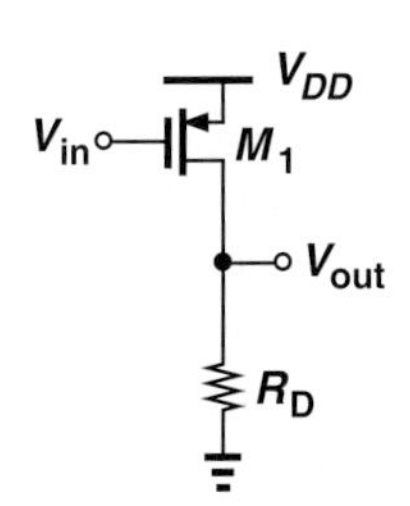

Figure 16.38

*16.4.** Some IC technologies provide no high-quality resistors. We may thus replace the resistor in a CS stage with a MOS realization as shown in Fig. 16.39. Here, M_2 approximates a pull-up resistor. Assume $(W/L)_1 = 3/0.18$ and $(W/L)_2 = 2/0.18$.

(a) Suppose $V_{in} = V_{DD}$. Assuming M_2 is in saturation, calculate the output low level. Is this assumption valid?

(b) Determine the trip point of this inverter, i.e., the input level at which $V_{out} = V_{in}$.

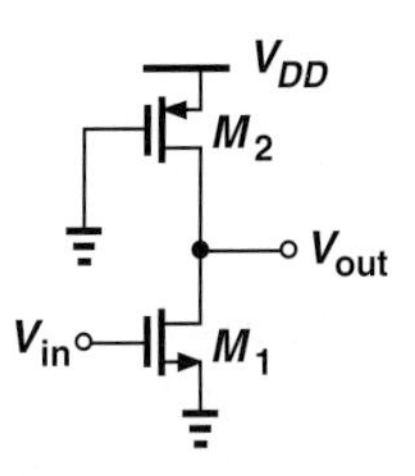

Figure 16.39

16.5. In the inverter of Fig. 16.39, the output low level must remain below 100 mV. If $(W/L)_2 = 3/0.18$, determine the minimum required value of $(W/L)_1$. SOL

16.6. The inverter of Fig. 16.39 must provide an output low level no higher than 80 mV. If $(W/L)_1 = 2/0.18$, what is the maximum allowable value of $(W/L)_2$?

16.7. In an NMOS inverter, $(W/L)_1 = 5/0.18$ and $R_D = 2$ kΩ. Calculate the noise margins.

*16.8.** Due to a manufacturing error, an NMOS inverter has been reconfigured as shown in Fig. 16.40.

(a) Determine the output for $V_{in} = 0$ and $V_{in} = V_{DD}$. Does the circuit invert?

(b) Can a trip point be obtained for this circuit?

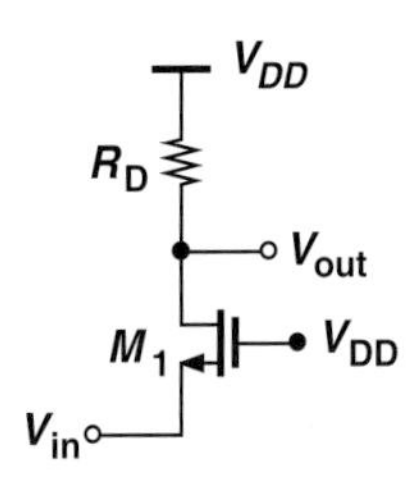

Figure 16.40

16.9. In Problem 16.7, we double the value of W/L or R_D. Determine what happens to the noise margins in each case.

16.10. A more conservative definition of noise margins would use the input level at which the small-signal gain reaches −0.5 (rather than −1). For an NMOS inverter with $(W/L)_1 = 5/0.18$ and $R_D = 2\ \text{k}\Omega$, compute such noise margins and compare the results with those obtained in Problem 16.7.

16.11. Consider the inverter shown in Fig. 16.39, assuming $(W/L)_1 = 4/0.18$ and $(W/L)_2 = 9/0.18$. Calculate the noise margins.

****16.12.** Consider the cascade of identical NMOS inverters depicted in Fig. 16.41. If $R_D = 5\ \text{k}\Omega$, determine $(W/L)_{1,2}$ such that the output low level of M_1 (for $V_{in} = V_{DD}$) is equal to NM_L of the second inverter. (In this situation, the output of the first inverter is degraded so much as to place the second stage at the point of unity gain.)

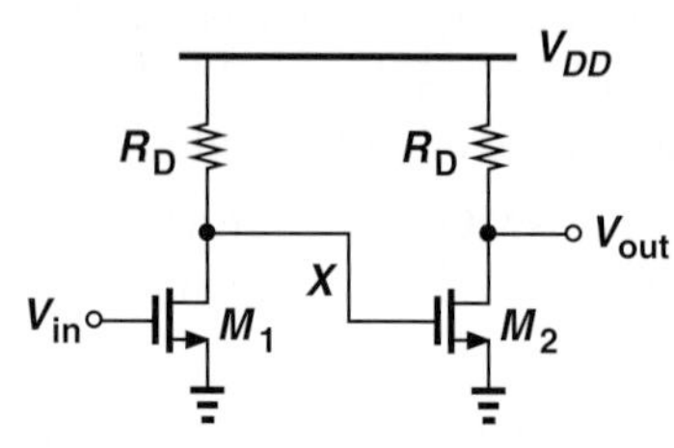

Figure 16.41

SOL ****16.13.** Two inverters having the characteristics shown in Fig. 16.42 are placed in a cascade. Sketch the overall VTC of the cascade if (a) inverter A precedes inverter B, or (b) inverter B precedes inverter A.

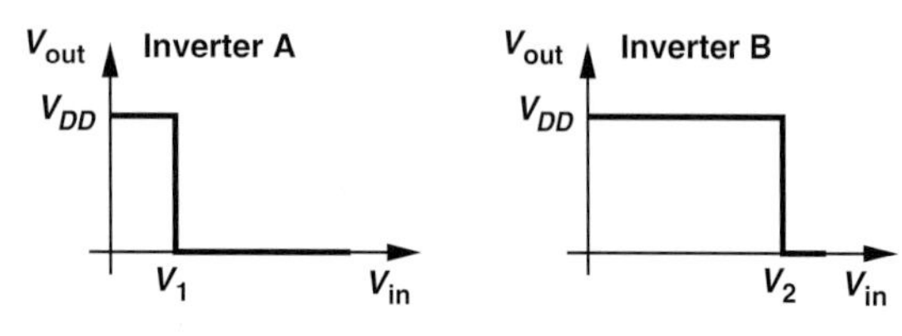

Figure 16.42

Sec. 16.1.2 Dynamic Characterization of Gates

16.14. An inverter is constructed as illustrated in Fig. 16.43, where R_{on} denotes the on-resistance of the switch. Assume $R_{on} \ll R_2$ so that the output low level is degraded negligibly.

(a) Compute the time required for the output to reach 95% of V_{DD} if S_1 turns off at $t = 0$.

(b) Compute the time required for the output to reach 5% of V_{DD} if S_1 turns on at $t = 0$. How does the result compare with that in (a)?

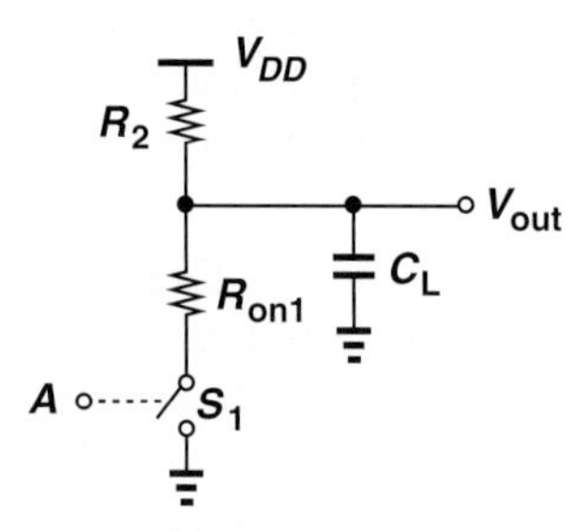

Figure 16.43

16.15. An NMOS inverter with a load capacitance of 100 fF exhibits an output low level of 50 mV and an output risetime of 200 ps. Compute the load resistor and $(W/L)_1$ if the risetime is given by three output time constants.

16.16. An NMOS inverter must drive a load capacitance of 50 fF with an output risetime of 100 ps. Assuming the risetime is given by three output time constants, determine the maximum load resistor value.

16.17. An NMOS inverter must drive a load capacitance of 100 fF while drawing a supply current of less than 1 mA when the output is low. What is the fastest risetime that the circuit can achieve? Assume the output low level is nearly zero.

Sec. 16.2 CMOS Inverter

16.18. In a CMOS inverter, $(W/L)_1 = 2/0.18$ and $(W/L)_2 = 3/0.18$. Determine the trip point of the circuit and the supply current drawn at this point.

16.19. For the inverter of Problem 16.18, calculate the small-signal voltage gain at the trip point if $\lambda_N = 0.1\text{V}^{-1}$ and $\lambda_P = 0.2\text{V}^{-1}$.

16.20. Explain qualitatively what happens to the VTC of a CMOS inverter as the *length* of M_1 or M_2 is increased.

16.21. A CMOS inverter employs $(W/L)_1 = 3/0.18$ and $(W/L)_2 = 7/0.18$. Derive expressions for the VTC in each region of Fig. 16.19(d) and plot the result.

SOL **16.22.** A CMOS inverter must provide a trip point equal to 0.5 V. Determine the required value of $(W/L)_1/(W/L)_2$. Note that a *low* trip point necessitates a strong NMOS device.

16.23. We often approximate the trip point of a CMOS inverter with the input voltage that places both transistors in saturation.
(a) Explain why this is a reasonable approximation if the inverter exhibits a high voltage gain around the trip point.
(b) Assuming $(W/L)_1 = 3/0.18$ and $(W/L)_2 = 7/0.18$, determine the minimum and maximum input voltages at which both transistors operate in saturation and calculate the difference between each and the trip point. Is the difference small?

***16.24.** Explain why a CMOS inverter with the device parameters given at the beginning of this problem set cannot achieve a trip point of 0.3 V.

16.25. Figure 16.44 shows three circuits along with three VTCs. Match the VTC with its corresponding circuit.

16.26. Due to a manufacturing error, a parasitic resistor $R_P = 2\ \text{K}\Omega$ has appeared in the inverter of Fig. 16.45. If $(W/L)_1 = 3/0.18$ and $(W/L)_2 = 5/0.18$, calculate the low and high output levels and the trip point.

16.27. Calculate the small-signal voltage gain of the circuit in Problem 16.26 at the trip point with and without R_P.

16.28. Calculate the noise margins for a CMOS inverter if $(W/L)_1 = 5/0.18$ and $(W/L)_2 = 11/0.18$.

16.29. Determine $(W/L)_1/(W/L)_2$ for a CMOS inverter if $NM_L = 0.6$ V. (Hint: solve the resulting equation by iteration.)

***16.30.** Consider Eq. (16.54) for V_{IL} $(= NM_L)$. Sketch the noise margin as a varies from 0 to infinity. Explain the results intuitively for very small and very large values of a.

16.31. Repeat Problem 16.30 for NM_H.

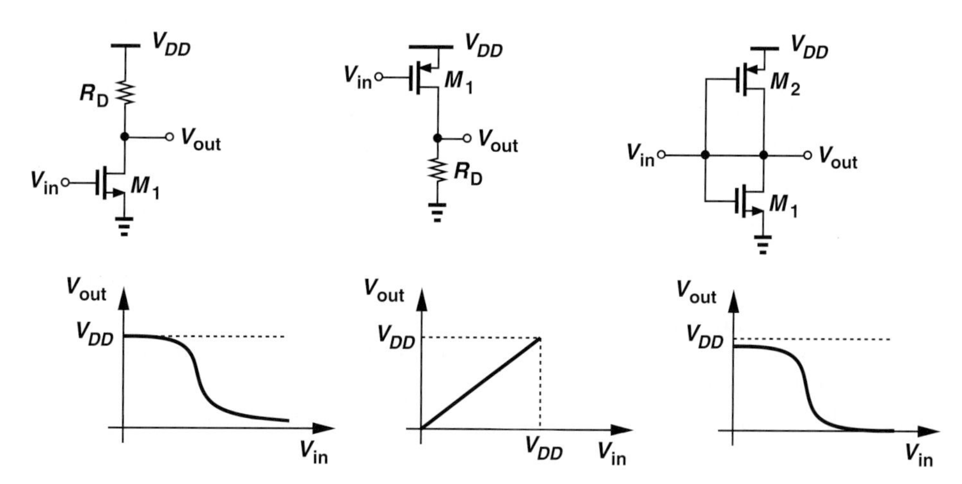

Figure 16.44

16.32. Calculate the noise margins for the circuit in Problem 16.26.

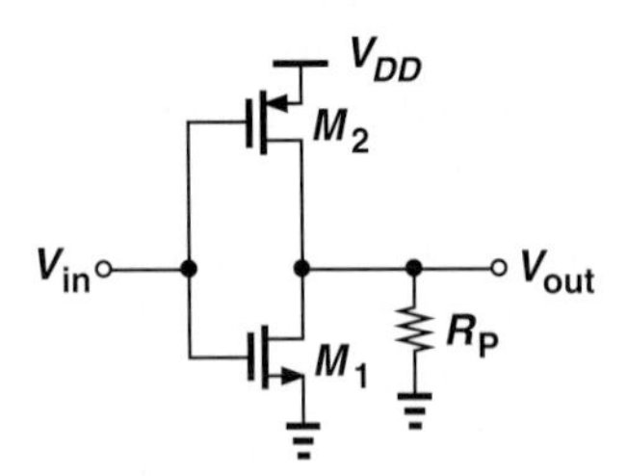

Figure 16.45

16.33. Consider the circuit shown in Fig. 16.20(a), where $V_{out}(t=0)=0$. If $(W/L)_2 = 6/0.18$ and $C_L = 50$ fF, determine the time it takes for the output to reach $V_{DD}/2$.

SOL **16.34.** Repeat Problem 16.33 for the time it takes the output to reach $0.95V_{DD}$ and compare the results.

16.35. In the circuit depicted in Fig. 16.23(a), the input jumps from 0 to V_1 at $t = 0$. Assuming $V_{out}(t=0) = V_{DD}$, $(W/L)_1 = 1/0.18$, and $C_L = 30$ fF, determine the time it takes the output to fall to $V_{DD}/2$ if (a) $V_1 = V_{DD}$ and (b) $V_1 = V_{DD}/2$.

16.36. Repeat Problem 16.35 for the time it takes the output to fall to $0.05V_{DD}$ and compare the results.

16.37. A CMOS inverter with $(W/L)_1 = 1/0.18$ and $(W/L)_2 = 3/0.18$ drives a load capacitance of 80 fF. Calculate T_{PHL} and T_{PLH}.

16.38. Suppose the supply voltage in Problem 16.37 is raised by 10%. By how much do T_{PHL} and T_{PLH} decrease?

16.39. Repeat Problem 16.38 with $V_{DD} = 0.9$ V and compare the results. Note the significant increase in T_{PHL} and T_{PLH}.

16.40. In Eq. (16.82), suppose $V_{TH1} = 0.4$ V. For what supply voltage do the two terms in the square brackets become equal? How should the supply voltage be chosen to make the first term 10% of the second?

16.41. A CMOS inverter must achieve symmetric propagation delays equal to 80 ps while driving a load capacitance of 50 fF. Determine $(W/L)_1$ and $(W/L)_2$. SOL

16.42. A CMOS inverter with $(W/L)_1 = 1/0.18$ exhibits a T_{PHL} of 100 ps with $C_L = 80$ fF. Determine the supply voltage.

** **16.43.** We have received a CMOS inverter with unknown device dimensions and thresholds. Tests indicate $T_{PHL} = 120$ ps with $C_L = 90$ fF and $V_{DD} = 1.8$ V, and $T_{PHL} = 160$ ps with $C_L = 90$ fF and V, $V_{DD} = 1.5$ V. Determine $(W/L)_1$ and V_{TH1}.

16.44. In Eq. (16.82), the argument of the logarithm becomes negative if $V_{DD} < 4V_{TH1}/3$. Explain intuitively why this happens.

16.45. A 1-kΩ resistor charges a capacitance of 100 fF from 0 V to V_{DD}. Determine the energy dissipated in the resistor.

16.46. A digital circuit contains one million gates, and runs with a clock frequency of 2 GHz. Assuming that, on the average, 20% of the gates switch in every clock cycle, and the average load capacitance seen by each gate is 20 fF, determine the average power dissipation. Neglect the crowbar current. (Note that the result is unrealistically low because the crowbar current is neglected.)

16.47. An inverter using very wide transistors is used as a "clock buffer" in a microprocessor to deliver a 2-GHz clock to various flipflops. Suppose the buffer drives five million transistors with an average width of 1 μm. If the gate length is 0.18 μm, $C_{ox} = 10$ fF/μm^2 and the gate capacitance is approximated by WLC_{ox}, determine the power dissipated by the clock buffer. Neglect the crowbar current (even though it is not negligible).

16.48. The supply voltage of an inverter increases by 10%. If $(W/L)_1 = 2/0.18$ and $(W/L)_2 = 4/0.18$, determine the change in the peak crowbar current.

**16.49. Approximating the crowbar current waveform in Fig. 16.29(c) with an isosceles triangle, calculate the average power dissipation resulting from this mechanism. Assume an operation frequency of f_1.

Sec. 16.3 CMOS NOR and NAND Gates

SOL **16.50.** A CMOS NOR gate drives a load capacitance of 20 fF. Suppose the input waveforms are as shown in Fig. 16.46, each having a frequency of $f_1 = 500$ MHz. Calculate the power dissipated by the gate. Neglect the crowbar current.

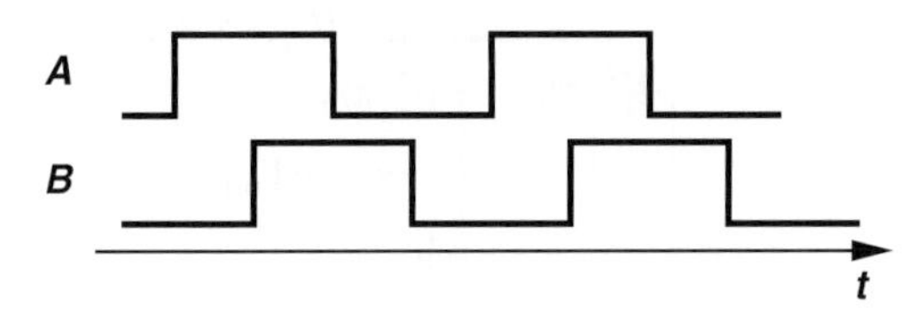

Figure 16.46

16.51. Repeat Problem 16.50 for a NAND gate.

16.52. For each NMOS section shown in Fig. 16.47, draw the dual PMOS section, construct the overall CMOS gate, and determine the logical function performed by the gate.

Design Problems

16.53. Design an NMOS inverter (i.e., determine R_D and W/L) for a static power budget of 0.5 mW and an output low level of 100 mV.

16.54. Design an NMOS inverter (i.e., determine R_D and W/L) for a static power budget of 0.25 mW and $NM_L =$ 600 mV.

16.55. Design an NMOS inverter (i.e., determine R_D and W/L) for an output low level of 100 mV and a power budget of 0.25 mW.

16.56. Determine $(W/L)_{1,2}$ for a CMOS inverter such that the trip point is equal to 0.8 V and the current drawn from V_{DD} at this point is equal to 0.5 mA. Assume $\lambda_n = 0.1\text{V}^{-1}$ and $\lambda_p = 0.2\text{V}^{-1}$. SOL

16.57. Is it possible to design a CMOS inverter such that $NM_L = NM_H = 0.7$ V if $V_{TH1} \neq |V_{TH2}|$? Explain why?

16.58. Determine $(W/L)_{1,2}$ for a CMOS inverter such that $T_{PLH} = T_{PHL} = 100$ ps while the circuit drives a load capacitance of 50 fF.

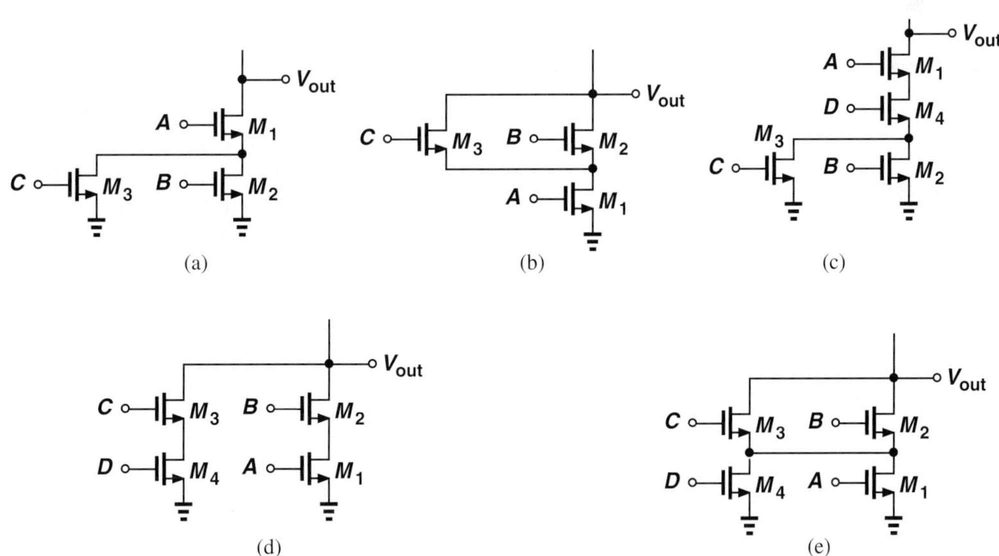

Figure 16.47

SPICE PROBLEMS

In the following problems, use the MOS device models given in Appendix A.

16.59. The inverter of Fig. 16.48 must provide a trip point at 0.8 V. If $(W/L)_1 = 0.6\ \mu\text{m}/0.18\ \mu\text{m}$, determine $(W/L)_2$. Also, plot the supply current as a function of V_{in} for $0 < V_{in} < 1.8$ V.

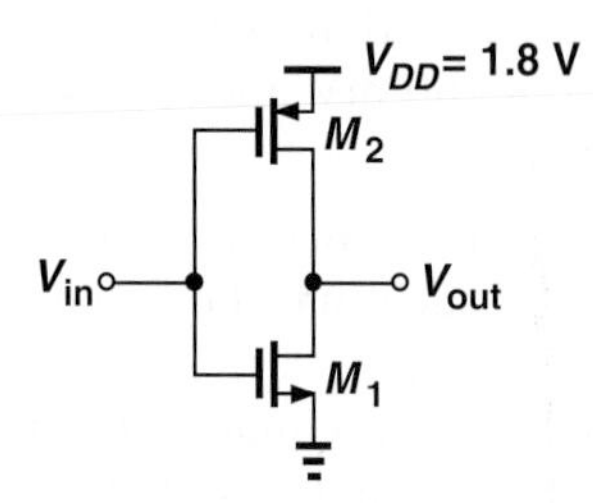

Figure 16.48

16.60. The inverter cascade shown in Fig. 16.49 drives a load capacitance of 100 fF. Assume $W_1 = 0.5W_2 = 0.6\ \mu\text{m}$, $W_3 = 0.5W_4$, and $L = 0.18\ \mu\text{m}$ for all four devices.

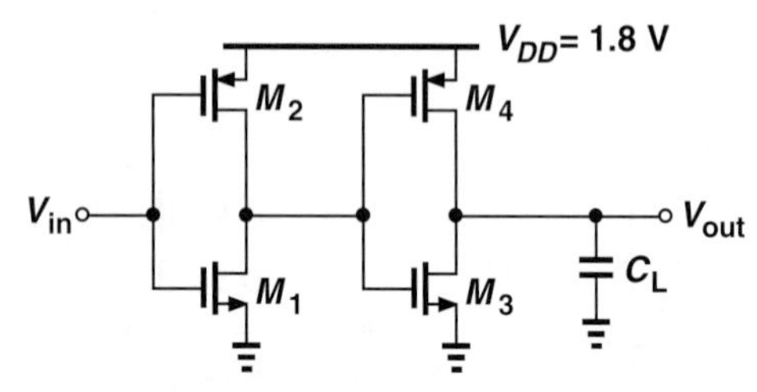

Figure 16.49

(a) Determine the optimum choice of W_3 (and W_4) if the total delay from V_{in} to V_{out} must be minimized. What is the delay contribution of each stage?

(b) Determine the average power dissipation of the circuit at a frequency of 500 MHz.

16.61. Consider a CMOS NAND gate with its inputs shorted together so as to form an inverter (Fig. 16.50). We wish to determine the delay of this circuit with a fanout of four; i.e., if it is loaded by a similar stage that incorporates devices whose width is scaled up by a factor of four. Use SPICE to compute this delay.

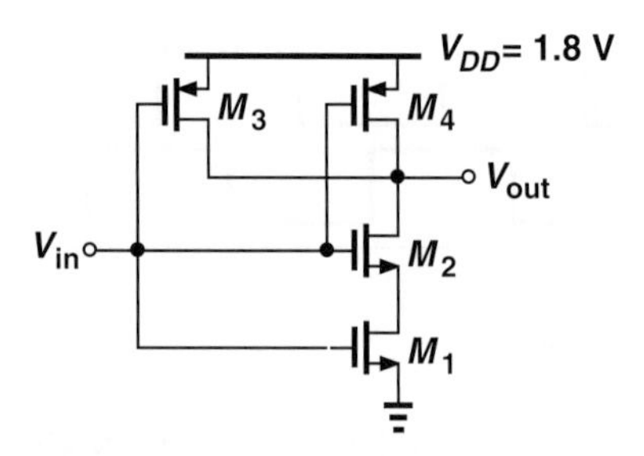

Figure 16.50

16.62. Repeat Problem 16.61 for a NOR gate and compare the results.

16.63. The circuit depicted in Fig. 16.51 is called a "ring oscillator." Assuming $V_{DD} = 1.8$ V and $W/L = 2\ \mu\text{m}/0.18\ \mu\text{m}$ for the NMOS devices, select W/L for the PMOS transistors such that the frequency of oscillation is maximized. (To start the oscillation in SPICE, you must apply an initial condition to one of the nodes, e.g., ic v(x) = 0.)

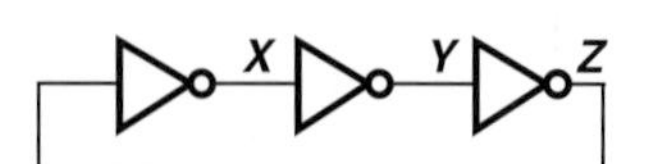

Figure 16.51

17

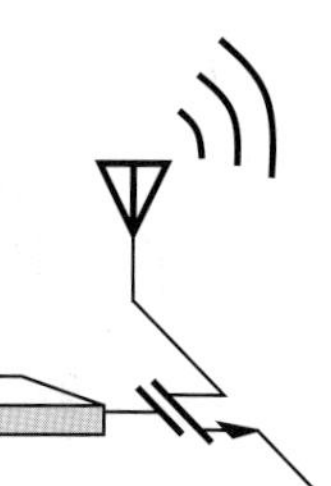

CMOS Amplifiers*

With the physics and operation of MOS transistors described in Chapter 6, we now deal with amplifier circuits employing such devices. While the field of microelectronics involves much more than amplifiers, our study of cellphones and digital cameras in Chapter 1 indicates the extremely wide usage of amplification, motivating us to master the analysis and design of such building blocks. This chapter proceeds as follows.

General Concepts
- Input and Output Impedances
- Biasing
- DC and Small-Signal Analysis

Operating Point Analysis
- Simple Biasing
- Source Degeneration
- Self-Biasing
- Biasing of PMOS Devices

Amplifier Topologies
- Common-Source Stage
- Common-Gate Stage
- Source Follower

Building the foundation for the remainder of this book, this chapter is quite long. Most of the concepts introduced here are invoked again in Chapter 5 (bipolar amplifiers). The reader is therefore encouraged to take frequent breaks and absorb the material in small doses.

17.1 GENERAL CONSIDERATIONS

Recall from Chapter 6 that a voltage-controlled current source along with a load resistor can form an amplifier. In general, an amplifier produces an output (voltage or current) that is a magnified version of the input (voltage or current). Since most electronic circuits both sense and produce voltage quantities,[1] our discussion primarily centers around "voltage amplifiers" and the concept of "voltage gain," v_{out}/v_{in}.

*This chapter is written for courses that deal with CMOS circuits *before* bipolar circuits.
[1]Exceptions are described in Chapter 12.

What other aspects of an amplifier's performance are important? Three parameters that readily come to mind are (1) power dissipation (e.g., because it determines the battery lifetime in a cellphone or a digital camera); (2) speed (e.g., some amplifiers in a cellphone or analog-to-digital converters in a digital camera must operate at high frequencies); (3) noise (e.g., the front-end amplifier in a cellphone or a digital camera processes small signals and must introduce negligible noise of its own).

17.1.1 Input and Output Impedances

In addition to the above parameters, the input and output (I/O) impedances of an amplifier play a critical role in its capability to interface with preceding and following stages. To understand this concept, let us first determine the I/O impedances of an *ideal* voltage amplifier. At the input, the circuit must operate as a voltmeter, i.e., sense a voltage without disturbing (loading) the preceding stage. The ideal input impedance is therefore infinite. At the output, the circuit must behave as a voltage source, i.e., deliver a constant signal level to any load impedance. Thus, the ideal output impedance is equal to zero.

In reality, the I/O impedances of a voltage amplifier may considerably depart from the ideal values, requiring attention to the interface with other stages. The following example illustrates the issue.

Example 17-1

An amplifier with a voltage gain of 10 senses a signal generated by a microphone and applies the amplified output to a speaker [Fig. 17.1(a)]. Assume the microphone can be modeled with a voltage source having a 10 mV peak-to-peak signal and a series resistance of 200 Ω. Also assume the speaker can be represented by an 8 Ω resistor.

(a) Determine the signal level sensed by the amplifier if the circuit has an input impedance of 2 kΩ or 500 Ω.

(b) Determine the signal level delivered to the speaker if the circuit has an output impedance of 10 Ω or 2 Ω.

Solution (a) Figure 17.1(b) shows the interface between the microphone and the amplifier. The voltage sensed by the amplifier is therefore given by

$$v_1 = \frac{R_{in}}{R_{in} + R_m} v_m. \quad (17.1)$$

For $R_{in} = 2\text{ k}\Omega$,

$$v_1 = 0.91 v_m, \quad (17.2)$$

only 9% less than the microphone signal level. On the other hand, for $R_{in} = 500\ \Omega$,

$$v_1 = 0.71 v_m, \quad (17.3)$$

i.e., nearly 30% loss. It is therefore desirable to maximize the input impedance in this case.

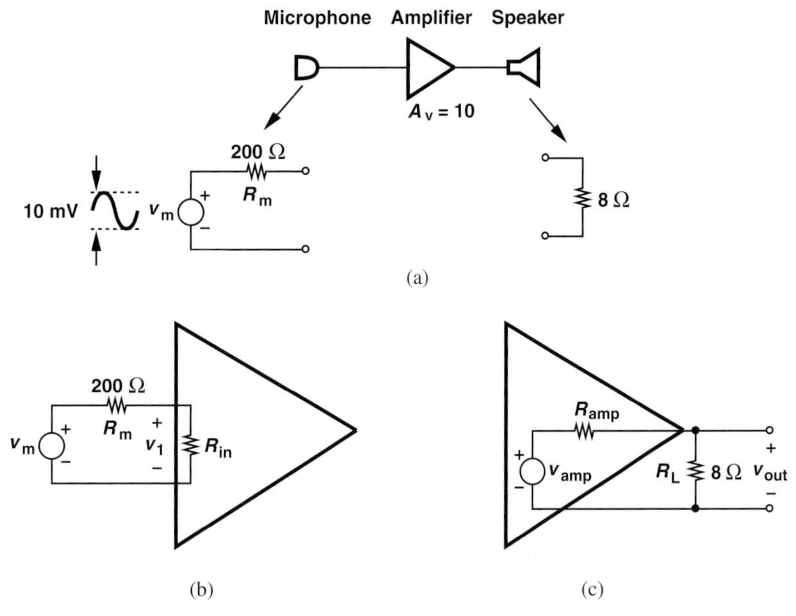

Figure 17.1 (a) Simple audio system, (b) signal loss due to amplifier input impedance, (c) signal loss due to amplifier output impedance.

(b) Drawing the interface between the amplifier and the speaker as in Fig. 17.1(c), we have

$$v_{out} = \frac{R_L}{R_L + R_{amp}} v_{amp}. \tag{17.4}$$

For $R_{amp} = 10\ \Omega$,

$$v_{out} = 0.44 v_{amp}, \tag{17.5}$$

a substantial attenuation. For $R_{amp} = 2\ \Omega$,

$$v_{out} = 0.8 v_{amp}. \tag{17.6}$$

Thus, the output impedance of the amplifier must be minimized.

Exercise Prove that the power delivered to R_L is maximized if $R_{amp} = R_L$ for a given value of R_L.

The importance of I/O impedances encourages us to carefully prescribe the method of measuring them. As with the impedance of two-terminal devices such as resistors and capacitors, the input (output) impedance is measured between the input (output) nodes of the circuit while all independent sources in the circuit are set to zero.[2] Illustrated in Fig. 17.2, the method involves applying a voltage source to the two nodes (also called "port") of interest, measuring the resulting current, and defining v_X/i_X as the impedance.

[2]Recall that a zero voltage source is replaced by a short and a zero current source by an open.

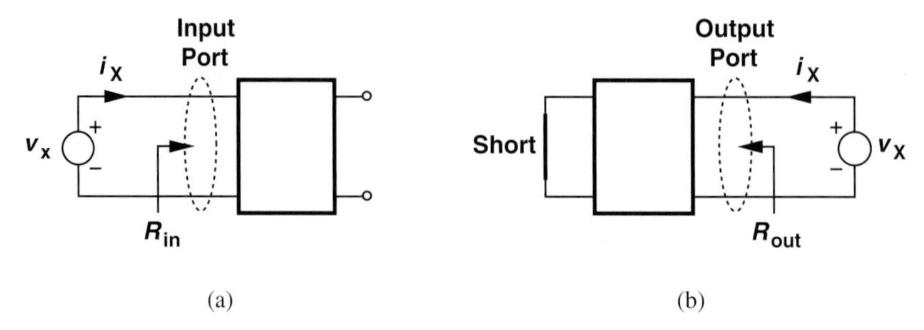

Figure 17.2 Measurement of (a) input and (b) output impedances.

Also shown are arrows to denote "looking into" the input or output port and the corresponding impedance.

The reader may wonder why the output port in Fig. 17.2(a) is left open whereas the input port in Fig. 17.2(b) is shorted. Since a voltage amplifier is driven by a voltage source during normal operation, and since all independent sources must be set to zero, the input port in Fig. 17.2(b) must be shorted to represent a zero voltage source. That is, the procedure for calculating the output impedance is identical to that used for obtaining the Thevenin impedance of a circuit (Chapter 1). In Fig. 17.2(a), on the other hand, the output remains open because it is not connected to any external sources.

Affecting the transfer of signals from one stage to the next, the I/O impedances are usually regarded as small-signal quantities—with the tacit assumption that the signal levels are indeed small. For example, the input impedance is obtained by applying a small change in the input voltage and measuring the resulting change in the input current. The small-signal models of semiconductor devices therefore prove crucial here.

Example 17-2 Assuming that the transistor operates in the saturation region, determine the input impedance of the circuit shown in Fig. 17.3(a).

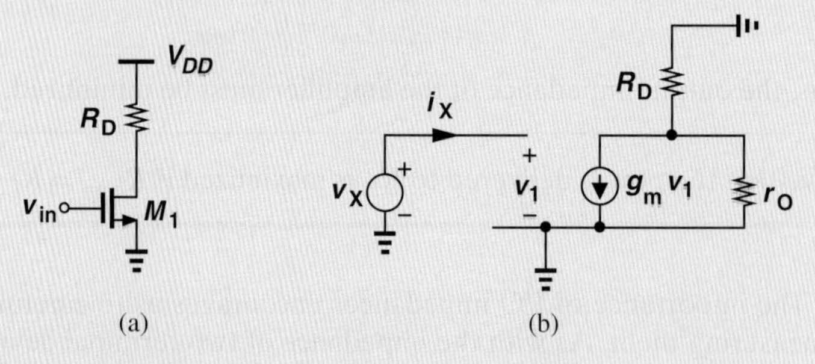

Figure 17.3 (a) Simple amplifier stage, (b) small-signal model.

Solution Constructing the small-signal equivalent circuit depicted in Fig. 17.3(b), we note that the gate draws no current (at low frequencies) and the input impedance is simply given by

$$\frac{v_x}{i_x} = \infty. \tag{17.7}$$

Exercise Compute the input impedance if $R_D = 0$ or $\lambda = 0$.

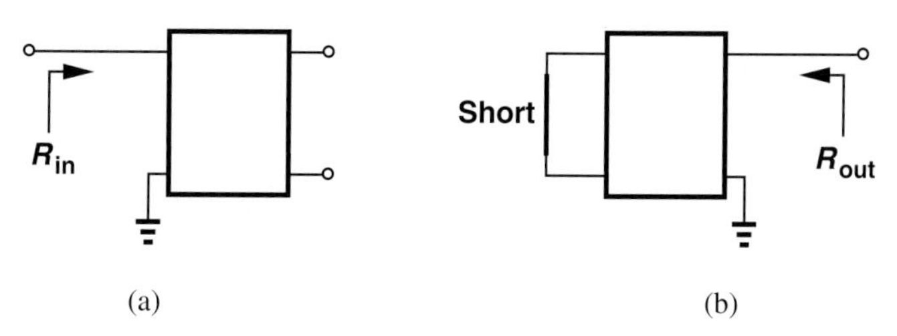

Figure 17.4 Concept of impedance seen at a node.

To simplify the notations and diagrams, we often refer to the impedance seen at a *node* rather than between two nodes (i.e., at a port). As illustrated in Fig. 17.4, such a convention simply assumes that the other node is the ground, i.e., the test voltage source is applied between the node of interest and ground.

Example 17-3 Calculate the impedance seen looking into the drain of M_1 in Fig. 17.5(a).

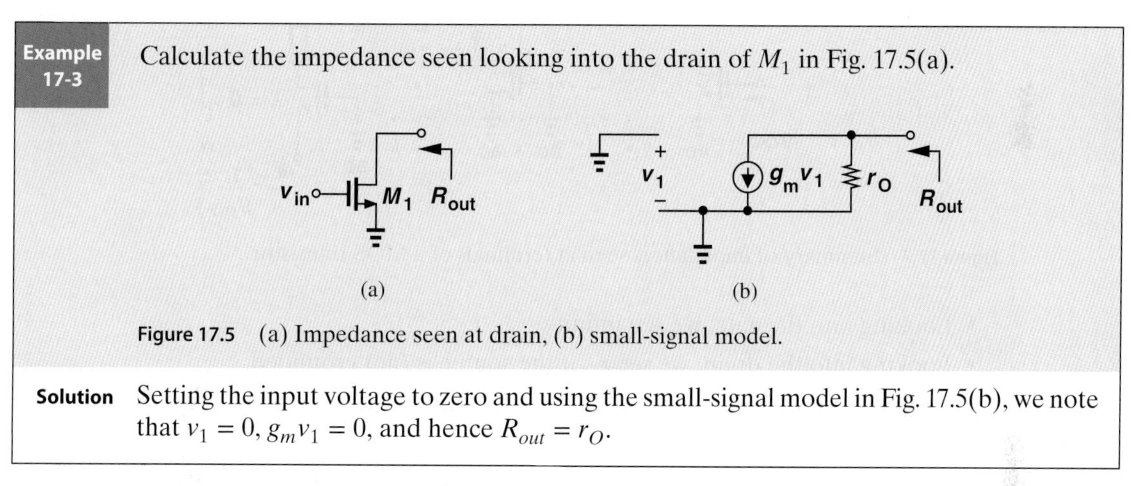

Figure 17.5 (a) Impedance seen at drain, (b) small-signal model.

Solution Setting the input voltage to zero and using the small-signal model in Fig. 17.5(b), we note that $v_1 = 0$, $g_m v_1 = 0$, and hence $R_{out} = r_O$.

Exercise What is the output impedance if a resistor is placed in series with the gate of M_1?

Example 17-4 Calculate the impedance seen at the source of M_1 in Fig. 17.6(a). Neglect channel-length modulation for simplicity.

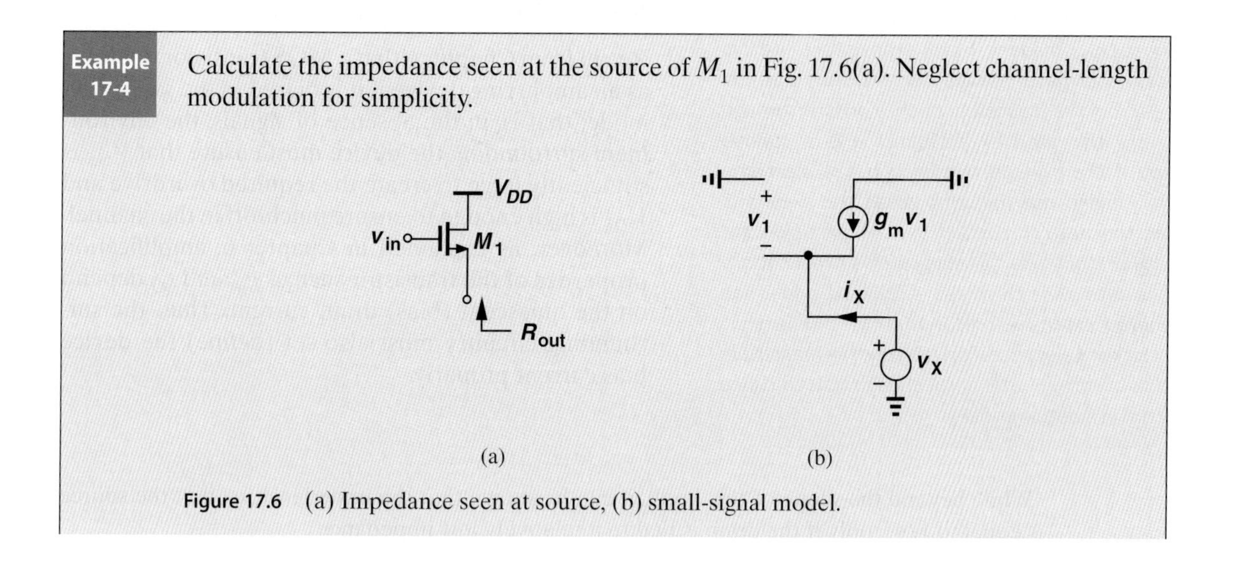

Figure 17.6 (a) Impedance seen at source, (b) small-signal model.

Solution Setting the input voltage to zero and replacing V_{DD} with ac ground, we arrive at the small-signal circuit shown in Fig. 17.6(b). Interestingly, $v_1 = -v_X$ and

$$g_m v_1 = -i_X. \tag{17.8}$$

That is,

$$\frac{v_X}{i_X} = \frac{1}{g_m}. \tag{17.9}$$

Exercise Does the above result change if a resistor is placed in series with the drain terminal of M_1?

The preceding three examples provide three important rules that will be used throughout this book (Fig. 17.7):

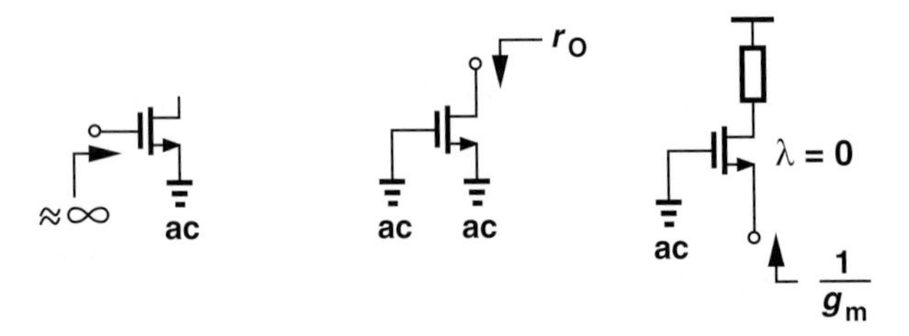

Figure 17.7 Summary of impedances seen at terminals of a MOS transistor.

- Looking into the gate, we see infinity.
- Looking into the drain, we see r_O if the source is (ac) grounded.
- Looking into the source, we see $1/g_m$ if the gate is (ac) grounded and channel-length modulation is neglected.

It is imperative that the reader master these rules and be able to apply them in more complex circuits.[3]

Did you know?

The observation that the (low-frequency) impedance seen at the gate is infinite comes from the assumption that the gate current is zero. Interestingly, modern MOS transistors exhibit a noticeable gate leakage current. This current arises from the "tunneling" of the carriers through the very thin gate oxide. Tunneling is a curious effect: if the oxide is thin enough, the electrons in the channel may *randomly* cross the oxide and enter the gate. Predicted by the particle-wave duality of electrons, tunneling grows exponentially as the gate oxide thickness is reduced, and it has prompted extensive research on MOSFETs that use thicker dielectrics. Fortunately, the magnitude of this current is still small enough not to cause concern in most amplifiers.

17.1.2 Biasing

Recall from Chapter 6 that a MOS transistor operates as an amplifying device if it is biased in the saturation mode; that is, in the absence of signals, the environment surrounding the device must ensure that V_{GS} is sufficiently large to create the required overdrive and V_{DS} is high enough to ensure pinch-off in the channel. Moreover, as explained in Chapter 6, amplification properties of the transistor such as g_m and r_O depend on the quiescent (bias) drain current. Thus, the surrounding circuitry must also set (define) the device bias current properly.

[3] While beyond the scope of this book, it can be shown that the impedance seen at the source is equal to $1/g_m$ only if the drain is tied to a relatively low impedance.

17.1.3 DC and Small-Signal Analysis

The foregoing observations lead to a procedure for the analysis of amplifiers (and many other circuits). First, we compute the operating (quiescent) conditions (terminal voltages and currents) of each transistor in the absence of signals. Called the "dc analysis" or "bias analysis," this step determines both the region of operation (saturation or triode) and the small-signal parameters of each device. Second, we perform "small-signal analysis," i.e., study the response of the circuit to small signals (superimposed on bias levels) and compute quantities such as the voltage gain and I/O impedances. As an example, Fig. 17.8 illustrates the bias and signal components of a voltage and a current.

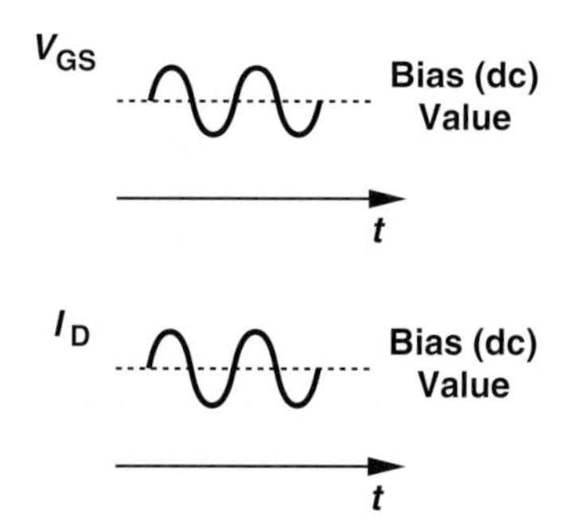

Figure 17.8 Bias and signal levels for a MOS transistor.

It is important to bear in mind that small-signal analysis deals with only (small) *changes* in voltages and currents in a circuit around their quiescent values. Thus, as mentioned in Chapter 6, all *constant* sources, i.e., voltage and current sources that do not vary with time, must be set to zero for small-signal analysis. For example, the supply voltage is constant and, while establishing proper bias points, plays no role in the response to small signals. We therefore ground all constant voltage sources[4] and open all constant current sources while constructing the small-signal equivalent circuit. From another point of view, the two steps described above follow the superposition principle: first, we determine the effect of constant voltages and currents while signal sources are set to zero, and second, we analyze the response to signal sources while constant sources are set to zero. Figure 17.9 summarizes these concepts.

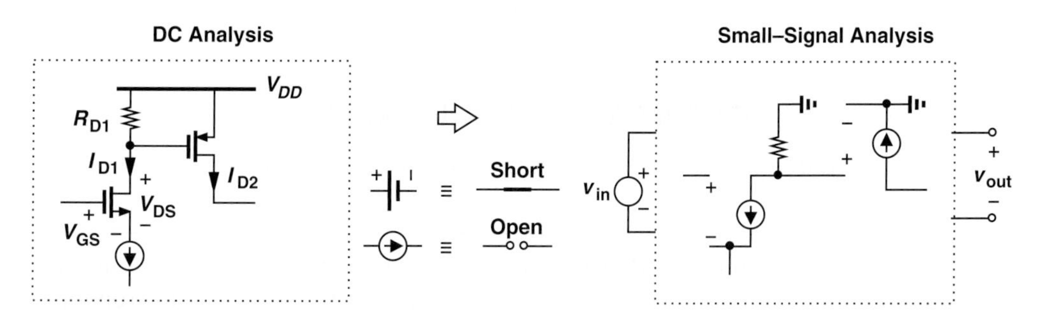

Figure 17.9 Steps in a general circuit analysis.

[4]We say all constant voltage sources that are tied to ground are replaced by an "ac ground."

We should remark that the *design* of amplifiers follows a similar procedure. First, the circuitry around the transistor is designed to establish proper bias conditions and hence the necessary small-signal parameters. Second, the small-signal behavior of the circuit is studied to verify the required performance. Some iteration between the two steps may often be necessary so as to converge toward the desired behavior.

How do we differentiate between small-signal and large-signal operations? In other words, under what conditions can we represent the devices with their small-signal models? If the signal perturbs the bias point of the device only negligibly, we say the circuit operates in the small-signal regime. In Fig. 17.8, for example, the change in I_D due to the signal must remain small. This criterion is justified because the amplifying properties of the transistor such as g_m and r_O are considered *constant* in small-signal analysis even though they in fact vary as the signal perturbs I_D. That is, a *linear* representation of the transistor holds only if the small-signal parameters themselves vary negligibly. The definition of "negligibly" depends somewhat on the circuit and the application, but as a rule of thumb, we consider 10% variation in the drain current as the upper bound for small-signal operation.

In drawing circuit diagrams hereafter, we will employ some simplified notations and symbols. Illustrated in Fig. 17.10 is an example where the battery serving as the supply voltage is replaced with a horizontal bar labeled V_{DD}.[5] Also, the input voltage source is simplified to one node called v_{in}, with the understanding that the other node is ground.

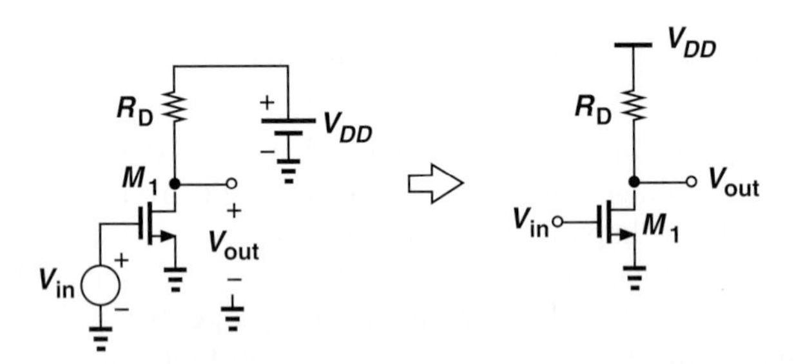

Figure 17.10 Notation for supply voltage.

In this chapter, we begin with the DC analysis and design of MOS stages, developing skills to determine or create bias conditions. This phase of our study requires no knowledge of signals and hence the input and output ports of the circuit. Next, we introduce various amplifier topologies and examine their small-signal behavior.

17.2 OPERATING POINT ANALYSIS AND DESIGN

It is instructive to begin our treatment of operating points with an example.

Example 17-5 A student familiar with MOS devices constructs the circuit shown in Fig. 17.11 and attempts to amplify the signal produced by a microphone. The microphone generates

[5]The subscript *DD* indicates supply voltage feeding the drain.

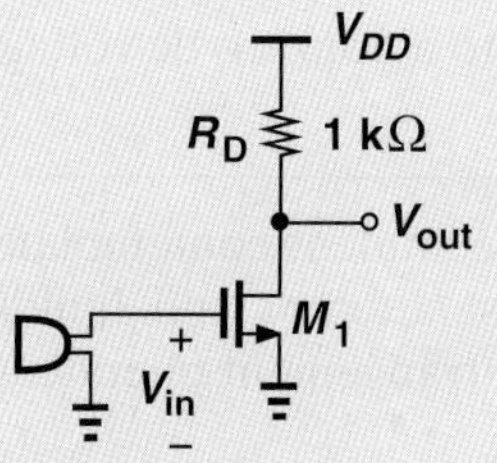

Figure 17.11 Amplifier driven directly by a microphone.

an output signal having a peak value of 20 mV with a zero dc (average) level. Explain what has happened.

Solution Unfortunately, the student has forgotten to bias the transistor. Since the microphone does not produce a dc output, a peak input of 20 mV fails to turn the transistor on. Consequently, the transistor carries no drain current and hence its transconductance is zero. The circuit thus generates no output signal.

Exercise Do we expect reasonable amplification if the threshold voltage of the device is zero?

As mentioned in Section 17.1.2, biasing seeks to fulfill two objectives: ensure operation in the saturation region, and set the drain current to the value required in the application. Let us return to the above example for a moment.

Example 17-6 Having realized the bias problem, the student in Example 17-5 modifies the circuit as shown in Fig. 17.12, connecting the gate to V_{DD} to allow dc biasing for the gate. Explain why the student needs to learn more about biasing.

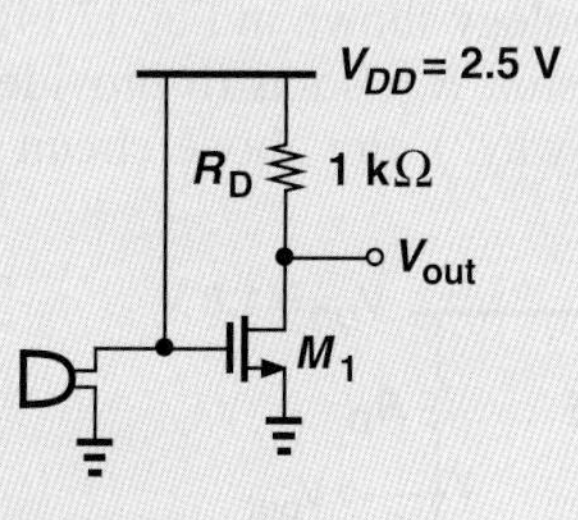

Figure 17.12 Amplifier with gate tied to V_{DD}.

Solution The fundamental issue here is that the signal generated by the microphone is *shorted* to V_{DD}. Acting as an ideal voltage source, V_{DD} maintains the gate voltage at a *constant* value, prohibiting any change introduced by the microphone. Since V_{GS} remains constant, so does V_{out}, leading to no amplification.

Exercise What happens if a resistor is placed in series with the source of M_1?

The preceding examples suggest that the gate can be neither tied to a zero dc voltage nor directly connected to a nonzero dc voltage. So what can we do?!

Example 17-7 Quite disappointed, the student contemplates transferring to the biology department. But he makes one last attempt and builds the circuit shown in Fig. 17.13, where $V_B = 0.75$ V. Can this arrangement operate as an amplifier?

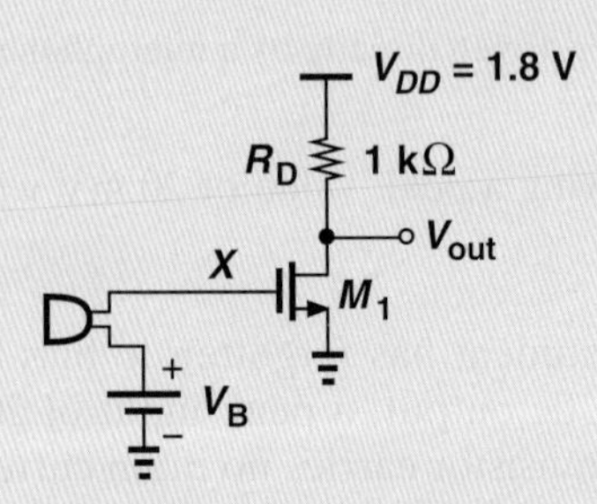

Figure 17.13 Amplifier with gate bias.

Solution Yes, it can. In the absence of voice, the microphone produces a zero output and hence $V_{GS} = V_B$. Proper choice of V_B can ensure operation in the saturation region with the required drain current (and the required transconductance).

Unfortunately, this arrangement necessitates one battery for each amplifier stage. (Different stages may operate with different gate-source voltages.) We must therefore seek a simple replacement for the battery. Nonetheless, the student has come close enough and need not quit electrical engineering.

Exercise Does the gate voltage depend on the value of R_D?

17.2.1 Simple Biasing

Now consider the topology shown in Fig. 17.14(a), where the gate is tied to V_{DD} through a relatively large resistor, R_G, so as to provide the gate bias voltage. With zero current flowing through R_G (why?), the above circuit yields $V_{GS} = V_{DD}$, a relatively large and

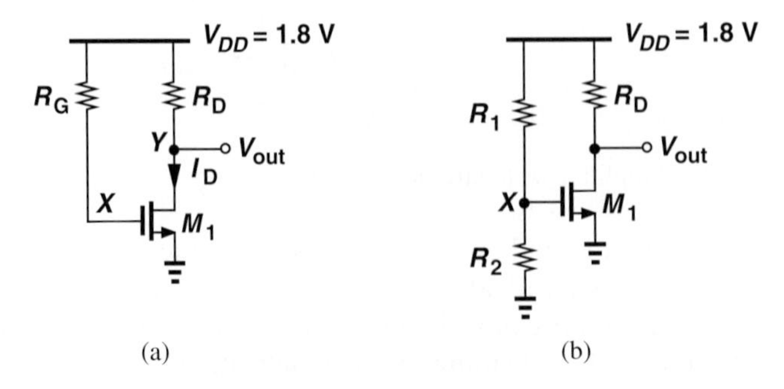

Figure 17.14 Use of (a) R_G, or (b) resistor divider to bias the gate.

fixed value. Most amplifier designs, on the other hand, require flexibility in the choice of V_{GS}. This issue can readily be remedied as depicted in Fig. 17.14(b), where

$$V_{GS} = \frac{R_2}{R_1 + R_2} V_{DD}. \tag{17.10}$$

Our objective is to analyze this circuit and determine its bias current and voltages.

We begin by assuming M_1 operates in the saturation region (and check this assumption at the end). We also neglect channel-length modulation in bias calculations. From the square-law characteristic,

$$I_D = \frac{1}{2}\mu_n C_{ox} \frac{W}{L}(V_{GS} - V_{TH})^2 \tag{17.11}$$

$$= \frac{1}{2}\mu_n C_{ox} \frac{W}{L}\left(\frac{R_2}{R_1 + R_2} V_{DD} - V_{TH}\right)^2. \tag{17.12}$$

Thus, proper choice of the resistor divider ratio and W/L can establish the required bias current.

We must also compute the drain-source voltage and determine whether the device is indeed in saturation. Noting that R_D carries a current equal to I_D and hence sustains a voltage of $R_D I_D$, we write a KVL around the supply voltage and the output branch:

$$V_{DS} = V_{DD} - R_D I_D. \tag{17.13}$$

For operation in saturation, the drain voltage must be no more than one threshold below the gate voltage, $V_{DS} \geq V_{GS} - V_{TH}$:

$$V_{DD} - R_D I_D \geq V_{GS} - V_{TH} \tag{17.14}$$

$$\geq \frac{R_2}{R_1 + R_2} V_{DD} - V_{TH}. \tag{17.15}$$

Example 17-8

Determine the bias current of M_1 in Fig. 17.14(b) assuming $V_{TH} = 0.5$ V, $\mu_n C_{ox} = 100\ \mu\text{A/V}^2$, $W/L = 5/0.18$, $\lambda = 0$, $R_1 = 20$ kΩ, and $R_2 = 15$ kΩ. What is the maximum allowable value of R_D for M_1 to remain in saturation?

Solution Since $V_{GS} = 0.771$ V, we have

$$I_D = 102\ \mu\text{A}. \tag{17.16}$$

The drain voltage can fall to as low as $V_{GS} - V_{TH} = 0.271$ V. Thus, the voltage drop across R_D can reach 1.529 V, allowing a maximum R_D of 15 kΩ.

Exercise If W is doubled, what is the maximum allowable value of R_D for M_1 to remain in saturation?

The reader may wonder why we concerned ourselves with the maximum value of R_D in the above example. As seen later in this chapter, the voltage gain of such a stage is proportional to R_D, demanding that it be maximized.

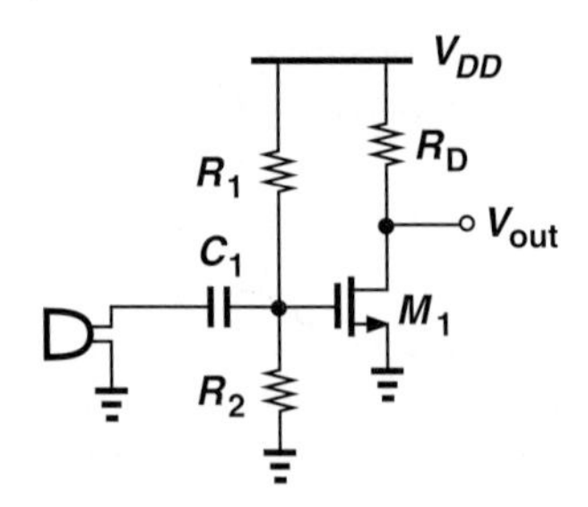

Figure 17.15 Use of capacitive coupling to isolate bias from microphone.

Did you know?

A serious issue in integrated circuits has been the implementation of large capacitors. A typical structure consists of two parallel conductors (e.g., metal layers used as interconnects on the chip). The difficulty is that integrated capacitors of greater than a few picofarads consume a large area, making the chip expensive. Imagine what happens if an audio application requires a 100 nF capacitor: the capacitor area would reach 6 to 10 square millimeters, much greater than the rest of the circuit. For this reason, many circuit techniques have been invented to avoid the use of large on-chip capacitors.

The reader may also wonder how exactly the microphone would be connected to the above amplifier. To ensure that only the *signal* is applied to the circuit and the zero dc level of the microphone does not disrupt the biasing, we must insert a device in series with the microphone that passes signals but blocks dc levels, i.e., a capacitor (Fig. 17.15). This technique is a direct result of the superposition perspective that we invoked earlier: for bias calculations we assume no signals are present and hence the capacitor is an open circuit, whereas for small-signal calculations we ignore the bias values and consider the capacitor a short circuit. Since the impedance of the capacitor, $1/(C_1 s)$, is inversely proportional to its value and the frequency of operation, we must choose C_1 large enough that it behaves nearly as a short circuit at the lowest frequency of interest. We return to this concept later in this chapter.

17.2.2 Biasing with Source Degeneration

In some applications, a resistor may be placed in series with the source of the transistor, thereby providing "source degeneration." Illustrated in Fig. 17.16 is such a topology, where the gate voltage is defined by R_1 and R_2. We assume M_1 operates in saturation and neglect channel-length modulation. Noting that the gate current is zero, we have

$$V_X = \frac{R_2}{R_1 + R_2} V_{DD}. \tag{17.17}$$

Since $V_X = V_{GS} + I_D R_S$,

$$\frac{R_2}{R_1 + R_2} V_{DD} = V_{GS} + I_D R_S. \tag{17.18}$$

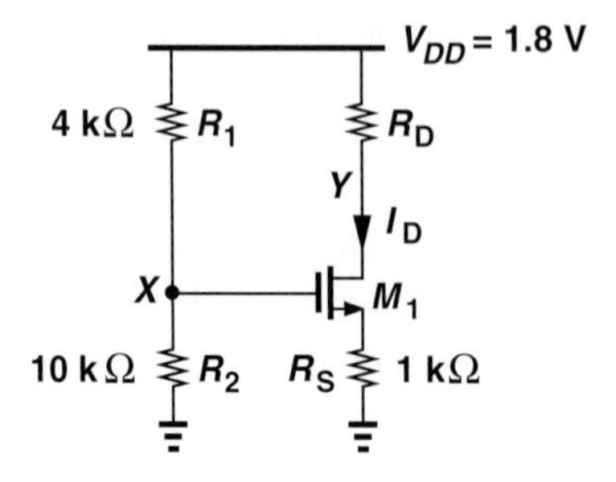

Figure 17.16 MOS stage with biasing.

Also,

$$I_D = \frac{1}{2}\mu_n C_{ox}\frac{W}{L}(V_{GS} - V_{TH})^2. \tag{17.19}$$

Equations (17.18) and (17.19) can be solved to obtain I_D and V_{GS}, either by iteration or by finding I_D from Eq. (17.18) and replacing for it in Eq. (17.19):

$$\left(\frac{R_2}{R_1 + R_2}V_{DD} - V_{GS}\right)\frac{1}{R_S} = \frac{1}{2}\mu_n C_{ox}\frac{W}{L}(V_{GS} - V_{TH})^2. \tag{17.20}$$

That is

$$V_{GS} = -(V_1 - V_{TH}) + \sqrt{(V_1 - V_{TH})^2 - V_{TH}^2 + \frac{2R_2}{R_1 + R_2}V_1 V_{DD}}, \tag{17.21}$$

$$= -(V_1 - V_{TH}) + \sqrt{V_1^2 + 2V_1\left(\frac{R_2 V_{DD}}{R_1 + R_2} - V_{TH}\right)}, \tag{17.22}$$

where

$$V_1 = \frac{1}{\mu_n C_{ox}\frac{W}{L}R_S}. \tag{17.23}$$

This value of V_{GS} can then be substituted in Eq. (17.18) to obtain I_D. Of course, V_Y must exceed $V_X - V_{TH}$ to ensure operation in the saturation region.

Example 17-9

Determine the bias current of M_1 in Fig. 17.16 assuming $V_{TH} = 0.5$ V, $\mu_n C_{ox} = 100\ \mu A/V^2$, $W/L = 5/0.18$, and $\lambda = 0$. What is the maximum allowable value of R_D for M_1 to remain in saturation?

Solution We have

$$V_X = \frac{R_2}{R_1 + R_2}V_{DD} \tag{17.24}$$

$$= 1.286\ \text{V}. \tag{17.25}$$

With an initial guess $V_{GS} = 1$ V, the voltage drop across R_S can be expressed as $V_X - V_{GS} = 286$ mV, yielding a drain current of 286 μA. From Eq. (17.19), we have

$$V_{GS} = V_{TH} + \sqrt{\frac{2I_D}{\mu_n C_{ox}\frac{W}{L}}} \tag{17.26}$$

$$= 0.954\ \text{V}. \tag{17.27}$$

Consequently,

$$I_D = \frac{V_X - V_{GS}}{R_S} \tag{17.28}$$

$$= 332\ \mu\text{A}, \tag{17.29}$$

and hence

$$V_{GS} = 0.989\ \text{V}. \tag{17.30}$$

This gives $I_D = 297\ \mu$A.

As seen from the iterations, the solutions converge slowly. We may therefore utilize the exact result in Eq. (17.22) to avoid lengthy calculations. Since $V_1 = 0.36$ V,

$$V_{GS} = 0.974 \text{ V} \tag{17.31}$$

and

$$I_D = \frac{V_X - V_{GS}}{R_S} \tag{17.32}$$

$$= 312\ \mu\text{A}. \tag{17.33}$$

The maximum allowable value of R_D is obtained if $V_Y = V_X - V_{TH} = 0.786$ V. That is,

$$R_D = \frac{V_{DD} - V_Y}{I_D} \tag{17.34}$$

$$= 3.25\ \text{k}\Omega. \tag{17.35}$$

Exercise Calculate the bias current if R_S is doubled.

Example 17-10 In the circuit of Example 17-9, assume M_1 is in saturation and $R_D = 2.5$ kΩ and compute (a) the maximum allowable value of W/L and (b) the minimum allowable value of R_S (with $W/L = 5/0.18$). Assume $\lambda = 0$.

Solution (a) As W/L becomes larger, M_1 can carry a larger current for a given V_{GS}. With $R_D = 2.5$ kΩ and $V_X = 1.286$ V, the maximum allowable value of I_D is given by

$$I_D = \frac{V_{DD} - V_Y}{R_D} \tag{17.36}$$

$$= 406\ \mu\text{A}. \tag{17.37}$$

The voltage drop across R_S is then equal to 406 mV, yielding $V_{GS} = 1.286$ V − 0.406 V = 0.88 V. In other words, M_1 must carry a current of 406 μA with $V_{GS} = 0.88$ V:

$$I_D = \frac{1}{2}\mu_n C_{ox}\frac{W}{L}(V_{GS} - V_{TH})^2 \tag{17.38}$$

$$406\ \mu\text{A} = (50\ \mu\text{A/V}^2)\frac{W}{L}(0.38\ \text{V})^2; \tag{17.39}$$

thus,

$$\frac{W}{L} = 56.2. \tag{17.40}$$

(b) With $W/L = 5/0.18$, the minimum allowable value of R_S gives a drain current of 406 μA. Since

$$V_{GS} = V_{TH} + \sqrt{\frac{2I_D}{\mu_n C_{ox}\frac{W}{L}}} \tag{17.41}$$

$$= 1.041\ \text{V}, \tag{17.42}$$

the voltage drop across R_S is equal to $V_X - V_{GS} = 245$ mV. It follows that

$$R_S = \frac{V_X - V_{GS}}{I_D} \quad (17.43)$$

$$= 604\ \Omega. \quad (17.44)$$

Exercise Repeat the above example if $V_{DD} = 1.5$ V.

17.2.3 Self-Biased Stage

Another biasing scheme commonly used in discrete and integrated circuits is shown in Fig. 17.17. Called "self-biased" because the gate voltage is provided from the drain, this stage exhibits many interesting and useful attributes. The circuit can be analyzed by noting that the voltage drop across R_G is zero and M_1 is always in saturation (why?). Thus,

$$I_D R_D + V_{GS} + R_S I_D = V_{DD}. \quad (17.45)$$

Finding V_{GS} from this equation and substituting it in Eq. (17.19), we have

$$I_D = \frac{1}{2}\mu_n C_{ox}\frac{W}{L}[V_{DD} - (R_S + R_D)I_D - V_{TH}]^2, \quad (17.46)$$

where channel-length modulation is neglected. It follows that

$$(R_S + R_D)^2 I_D^2 - 2\left[(V_{DD} - V_{TH})(R_S + R_D) + \frac{1}{\mu_n C_{ox}\frac{W}{L}}\right] I_D + (V_{DD} - V_{TH})^2 = 0. \quad (17.47)$$

The value of I_D can be obtained by solving the quadratic equation. With I_D known, V_{GS} can also be computed.

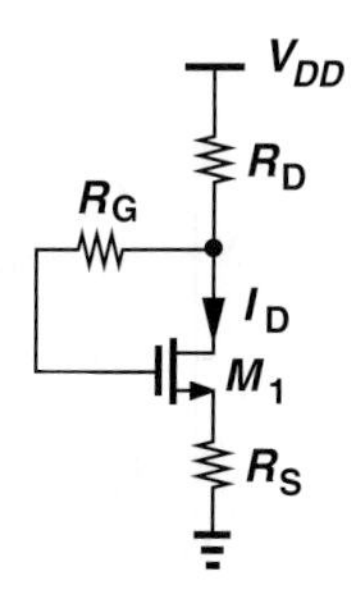

Figure 17.17 Self-biased MOS stage.

Example 17-11 Calculate the drain current of M_1 in Fig. 17.18 if $\mu_n C_{ox} = 100\ \mu\text{A/V}^2$, $V_{TH} = 0.5$ V, and $\lambda = 0$. What value of R_D is necessary to reduce I_D by a factor of two?

Solution Equation (17.47) gives

$$I_D = 556\ \mu\text{A}. \quad (17.48)$$

To reduce I_D to 278 μA, we solve Eq. (17.47) for R_D:

$$R_D = 2.867\ \text{k}\Omega. \tag{17.49}$$

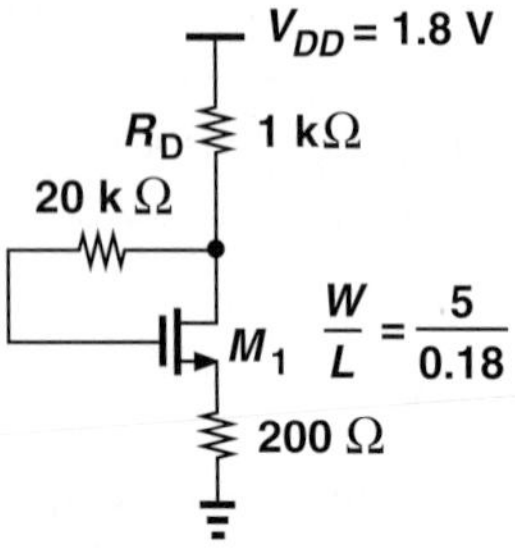

Figure 17.18 Example of self-biased MOS stage.

Exercise If W is quadrupled, does the bias current increase by a factor of four?

17.2.4 Biasing of PMOS Transistors

The dc biasing schemes considered thus far employ only NMOS devices. Similar ideas apply to PMOS circuits as well but attention must be paid to polarities and voltage relationships that ensure the device operates in saturation. The following examples reinforce these points.

Example 17-12 Determine the bias current of M_1 in Fig. 17.19 assuming $V_{TH} = -0.5$ V, $\mu_p C_{ox} = 50\ \mu\text{A/V}^2$, $W/L = 5/0.18$, $\lambda = 0$, $R_1 = 20$ kΩ, and $R_2 = 15$ kΩ. What is the maximum allowable value of R_D for M_1 to remain in saturation?

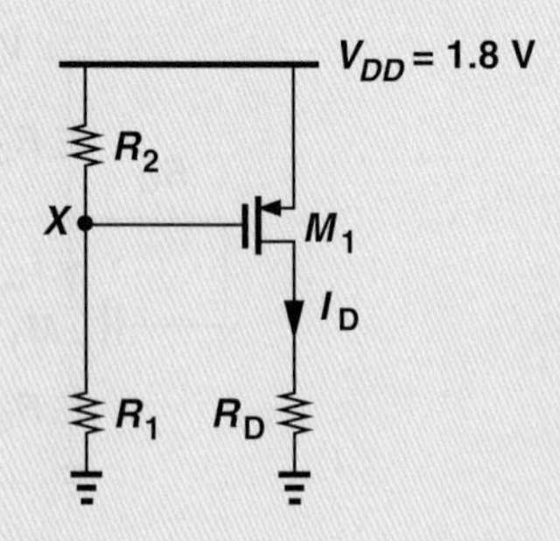

Figure 17.19 PMOS stage with biasing.

Solution The gate-source voltage of the transistor is given by the voltage drop across R_2:

$$V_{GS} = -\frac{R_2}{R_1 + R_2} V_{DD} \tag{17.50}$$

$$= -0.771\ \text{V} \tag{17.51}$$

The drain current is equal to

$$I_D = \frac{1}{2}\mu_p C_{ox}\frac{W}{L}(V_{GS} - V_{TH})^2 \tag{17.52}$$

$$= 56\ \mu\text{A}. \tag{17.53}$$

For M_1 to remain in saturation, its drain voltage can reach at most one V_{TH} *above* its gate voltage. That is, R_D can sustain a maximum voltage of $V_X + |V_{TH}| = 1.529$ V. The maximum allowable value of R_D is thus equal to 27.3 kΩ.

Exercise What is the bias current if $W/L = 10/0.18$?

Example 17-13 Determine the bias current of the self-biased stage shown in Fig. 17.20. Assume $V_{TH} = -0.5$ V, $\mu_p C_{ox} = 50\ \mu\text{A/V}^2$, $W/L = 5/0.18$, and $\lambda = 0$.

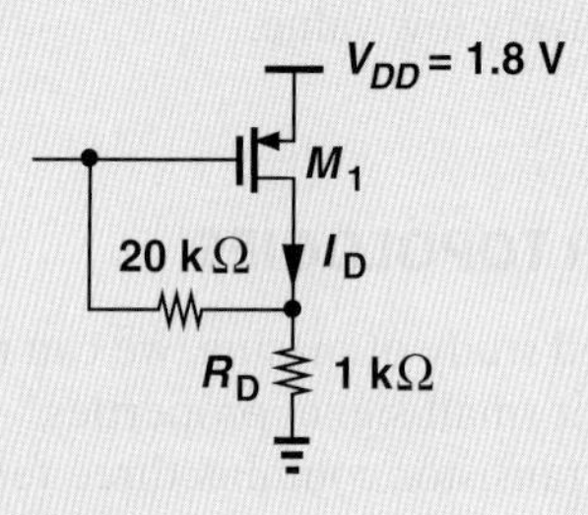

Figure 17.20 PMOS stage with self-biasing.

Solution Since $I_D R_D + |V_{GS}| = V_{DD}$, we have

$$I_D = \frac{1}{2}\mu_p C_{ox}\frac{W}{L}(V_{DD} - I_D R_D - |V_{TH}|)^2. \tag{17.54}$$

Solution of this quadratic equation for I_D gives

$$I_D = 418\ \mu\text{A}. \tag{17.55}$$

Note that the device is always in saturation.

Exercise What supply voltage gives a bias current of 200 μA?

17.2.5 Realization of Current Sources

MOS transistors operating in saturation can act as current sources. As illustrated in Fig. 17.21(a), an NMOS device serves as a current source with one terminal tied to ground, i.e., it draws current from node X to ground. On the other hand, a PMOS transistor [Fig. 17.21(b)] draws current from V_{DD} to node Y. If $\lambda = 0$, these currents remain independent of V_X or V_Y (so long as the transistors are in saturation).

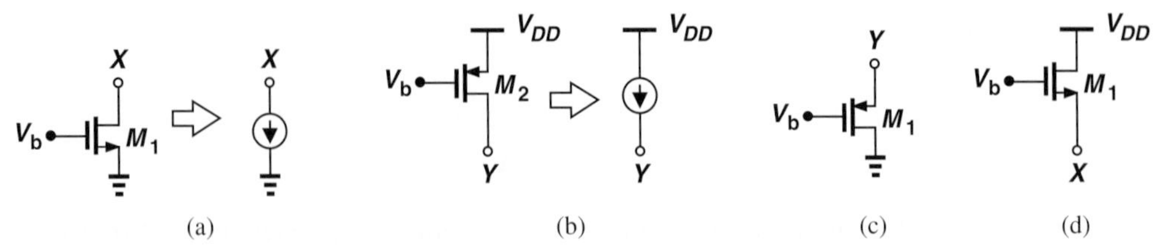

Figure 17.21 (a) NMOS device operating as a current source, (b) PMOS device operating as a current source, (c) PMOS topology not operating as a current source, (d) NMOS topology not operating as a current source.

It is important to understand that only the *drain* terminal of a MOSFET can draw a dc current and still present a high impedance. Specifically, NMOS or PMOS devices configured as shown in Figs. 17.21(c) and (d) do *not* operate as current sources because variation of V_X or V_Y directly changes the gate-source voltage of each transistor, thus changing the drain current considerably. To gain another perspective, the reader is encouraged to prove that the small-signal resistance seen at node Y or X (to ac ground) in these two circuits is equal to $1/g_m$, a relatively low value.

17.3 CMOS AMPLIFIER TOPOLOGIES

Following our detailed study of biasing, we can now delve into different amplifier topologies and examine their small-signal properties.[6]

Since the MOS transistor contains three terminals,[7] we may surmise that three possibilities exist for applying the input signal to the device, as conceptually illustrated in Figs. 17.22(a)–(c). Similarly, the output signal can be sensed from any of the terminals (with respect to ground) [Figs. 17.22(d)–(f)], leading to nine possible combinations of input and output networks and hence nine amplifier topologies.

However, as seen in Chapter 6, MOSFETs operating in the saturation region respond to gate-source voltage variations by varying their drain current. This property rules out the input connection shown in Fig. 17.22(c) because here V_{in} does not affect the gate or source voltages. Also, the topology in Fig. 17.22(f) proves of no value as V_{out} cannot be varied by varying the drain current. The number of possibilities therefore falls to four. But we note that the input and output connections in Figs. 17.22(b) and (e) remain incompatible because V_{out} would be sensed at the *input* node (the source) and the circuit would provide no function.

The above observations reveal three possible amplifier topologies. We study each carefully, seeking to compute its gain and input and output impedances. In all cases, the MOSFETs operate in saturation. The reader is encouraged to review Examples (17-2)–(17-4) and the three resulting rules illustrated in Fig. 17.7 before proceeding further.

[6]While beyond the scope of this book, the large-signal behavior of amplifiers also becomes important in many applications.
[7]The substrate also acts as a terminal through body effect, but we neglect this phenomenon in this book.

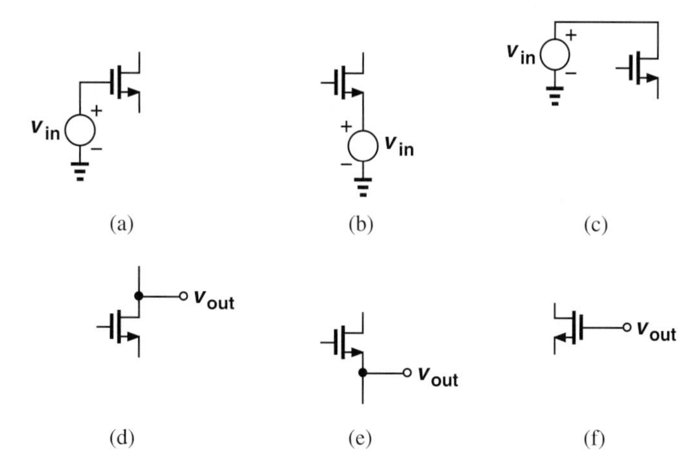

Figure 17.22 Possible input and output connections to a MOS transistor.

17.4 COMMON-SOURCE TOPOLOGY

If the input signal is applied to the gate [Fig. 17.22(a)] and the output signal is sensed at the drain [Fig. 17.22(d)], the circuit is called a "common-source" (CS) stage (Fig. 17.23). We have encountered and analyzed this circuit in different contexts without giving it a name. The term "common-source" is used because the source terminal is grounded and hence appears *in common* to the input and output ports. Nevertheless, we identify the stage based on the input and output connections (to the gate and from the drain, respectively) so as to avoid confusion in more complex topologies.

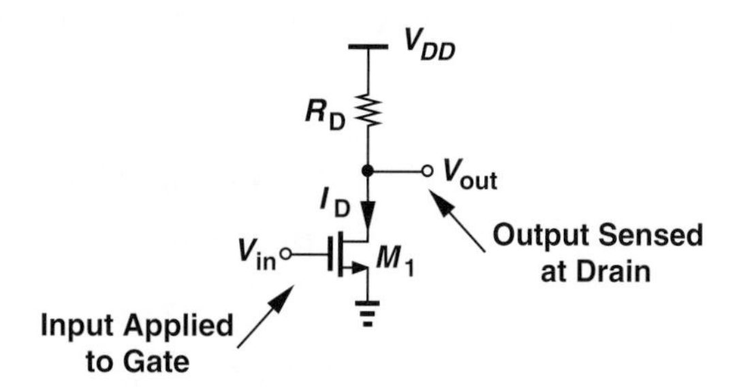

Figure 17.23 Common-source stage.

We deal with the CS amplifier in two phases: (a) analysis of the CS core to understand its fundamental properties, and (b) analysis of the CS stage including the bias circuitry as a more realistic case.

Analysis of CS Core Recall from the definition of transconductance that a small increment of ΔV applied to the base of M_1 in Fig. 17.23 increases the drain current by $g_m\Delta V$ and hence the voltage drop across R_D by $g_m\Delta VR_D$. In order to examine the amplifying properties of the CS stage, we construct the small-signal equivalent of the circuit, shown in Fig. 17.24. As explained in Chapter 6, the supply voltage node, V_{DD}, acts as an ac ground because its value remains constant with time. We neglect channel-length modulation for now.

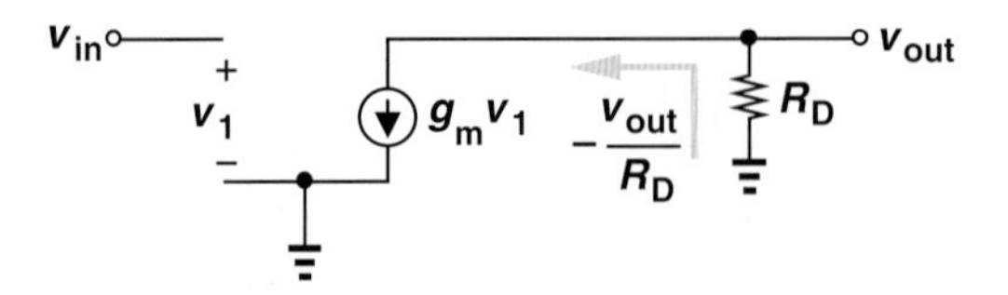

Figure 17.24 Small-signal model of CS stage.

Let us first compute the small-signal voltage gain $A_v = v_{out}/v_{in}$. Beginning from the output port and writing a KCL at the drain node, we have

$$-\frac{v_{out}}{R_D} = g_m v_1, \tag{17.56}$$

and $v_1 = v_{in}$. It follows that

$$A_v = -g_m R_D. \tag{17.57}$$

Equation (17.57) embodies two interesting and important properties of the CS stage. First, the small-signal gain is *negative* because raising the gate voltage and hence the drain current in Fig. 17.23 *lowers* V_{out}. Second, A_v is proportional to $g_m = \sqrt{2\mu_n C_{ox}(W/L)I_D}$ (i.e., the device aspect ratio and the drain bias current) and the drain resistor, R_D.

It is worth noting that the voltage gain of the stage is limited by the supply voltage. A higher drain bias current or a larger R_D dictates a greater voltage drop across R_D, but this drop cannot exceed V_{DD}.

Example 17-14 Calculate the small-signal voltage gain of the CS stage shown in Fig. 17.25 if $I_D = 1$ mA, $\mu_n C_{ox} = 100\ \mu\text{A/V}^2$, $V_{TH} = 0.5$ V, and $\lambda = 0$. Verify that M_1 operates in saturation.

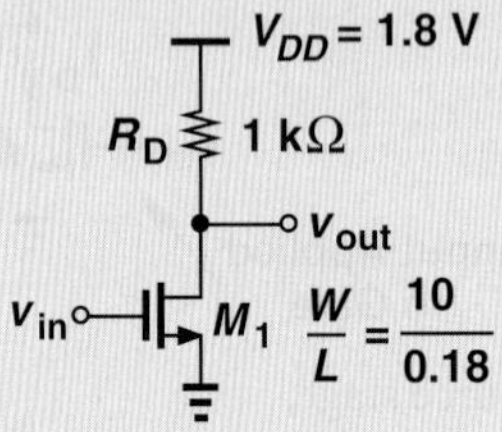

Figure 17.25 Example of CS stage.

Solution We have

$$g_m = \sqrt{2\mu_n C_{ox}\frac{W}{L}I_D} \tag{17.58}$$

$$= \frac{1}{300\ \Omega}. \tag{17.59}$$

Thus,

$$A_v = -g_m R_D \tag{17.60}$$

$$= 3.33. \tag{17.61}$$

To check the operation region, we first determine the gate-source voltage:

$$V_{GS} = V_{TH} + \sqrt{\frac{2I_D}{\mu_n C_{ox}\frac{W}{L}}} \quad (17.62)$$

$$= 1.1 \text{ V}. \quad (17.63)$$

The drain voltage is equal to $V_{DD} - R_D I_D = 0.8$ V. Since $V_{GS} - V_{TH} = 0.6$ V, the device indeed operates in saturation and has a margin of 0.2 V with respect to the triode region. For example, if R_D is doubled with the intention of doubling A_v, then M_1 enters the triode region and its transconductance drops.

Exercise Suppose we double W and choose R_D to place M_1 at the edge of saturation. Is the resulting gain higher or lower than the above value?

Example 17-15 Design a CS core with $V_{DD} = 1.8$ V and a power budget, P, of 1 mW while achieving a voltage gain of 5. Assume $V_{TH} = 0.5$ V, $\mu_n C_{ox} = 100\ \mu\text{A/V}^2$, $W/L = 5/0.18$, and $\lambda = 0$.

Solution Since $P = I_D V_{DD}$, the core can draw a maximum bias current of 556 μA. With such a current, the transistor transconductance reaches $g_m = \sqrt{2\mu_n C_{ox}(W/L)I_D} = 1/(569\ \Omega)$, requiring $R_D = 2845\ \Omega$ for a gain of 5.

Does such choice of bias current and load resistor conform to the supply voltage? That is, does M_1 operate in saturation? We must first calculate $V_{GS} = V_{TH} + \sqrt{2I_D/(\mu_n C_{ox} W/L)}$. With $I_D = 556\ \mu$A, $V_{GS} = 1.133$ V. On the other hand, the drain bias voltage is equal to $V_{DD} - R_D I_D = 0.218$ V. Unfortunately, the drain voltage is lower than $V_{GS} - V_{TH} = 0.633$ V, prohibiting M_1 from operation in saturation! That is, no solution exists.

Exercise Is there a solution if the power budget is reduced to 1 mW?

Writing the voltage gain as

$$A_v = -\sqrt{2\mu_n C_{ox}\frac{W}{L} I_D} R_D, \quad (17.64)$$

we may surmise that the gain can become arbitrarily large if W/L is increased indefinitely. In reality, however, as the device width increases (while the drain current remains constant), an effect called "subthreshold conduction" arises, which limits the transconductance. This effect is beyond the scope of this book, but the reader should bear in mind that the transconductance of a MOSFET cannot be increased arbitrarily by increasing only W/L.

Let us now calculate the I/O impedances of the CS stage. Using the equivalent circuit depicted in Fig. 17.26(a), we write

$$R_{in} = \frac{v_X}{i_X} \quad (17.65)$$

$$= \infty. \quad (17.66)$$

The very high input impedance proves essential in many applications.

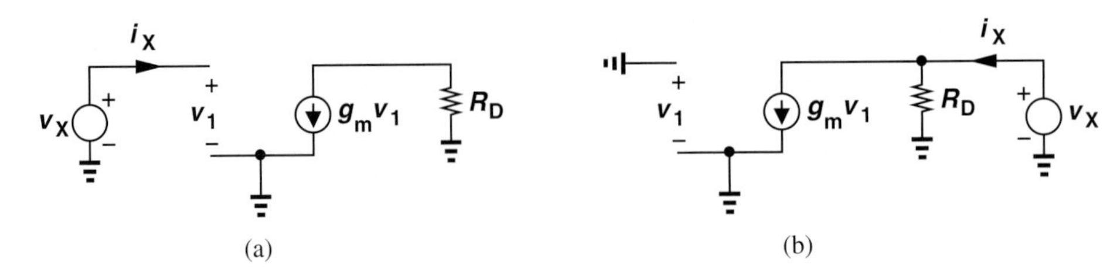

Figure 17.26 (a) Input and (b) output impedance calculation of CS stage.

The output impedance is obtained from Fig. 17.26(b), where the input voltage source is set to zero (replaced with a short). Since $v_1 = 0$, the dependent current source also vanishes, leaving R_D as the only component seen by v_X. In other words,

$$R_{out} = \frac{v_X}{i_X} \tag{17.67}$$

$$= R_D. \tag{17.68}$$

The output impedance therefore trades with the voltage gain, $-g_m R_D$.

Inclusion of Channel-Length Modulation Equation (17.57) suggests that the voltage gain of the CS stage can be increased indefinitely if $R_D \to \infty$ while g_m remains constant. From an intuitive point of view, a given change in the input voltage and hence the drain current gives rise to an increasingly larger output swing as R_D increases.

In reality, however, channel-length modulation limits the voltage gain even if R_D approaches infinity. Since achieving a high gain proves critical in circuits such as operational amplifiers, we must reexamine the above derivations in the presence of channel-length modulation.

Figure 17.27 depicts the small-signal equivalent circuit of the CS stage including the transistor output resistance. Note that r_O appears in parallel with R_D, allowing us to rewrite (17.57) as

$$A_v = -g_m (R_D || r_O). \tag{17.69}$$

We also recognize that the input impedance remains equal to infinity whereas the output impedance falls to

$$R_{out} = R_D || r_O. \tag{17.70}$$

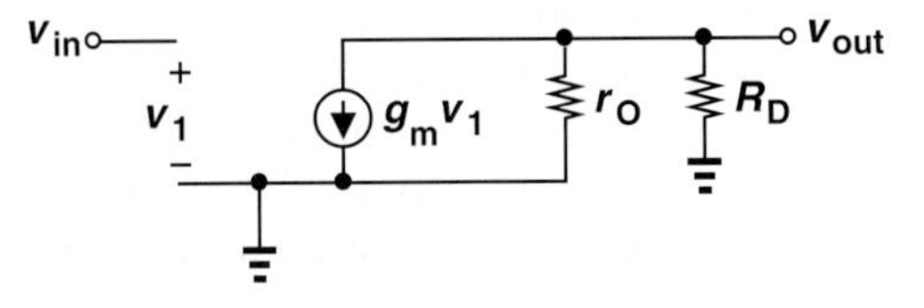

Figure 17.27 CS stage including channel-length modulation.

Example 17-16 The voltage gain of a CS stage drops by a factor of two when channel-length modulation is included. What is the λ of the transistor?

Solution The twofold reduction indicates that $r_O = R_D$. From Chapter 6, we know that $r_O = 1/(\lambda I_D)$, where I_D is the bias current. It follows that

$$\lambda = \frac{1}{I_D R_D}. \tag{17.71}$$

Exercise What value of λ drops the gain by 10% when channel-length modulation is included?

What happens as $R_D \to \infty$? The following example illustrates this case.

Example 17-17 Assuming M_1 operates in saturation, determine the voltage gain of the circuit depicted in Fig. 17.28(a) and plot the result as a function of the transistor channel length while other parameters remain constant.

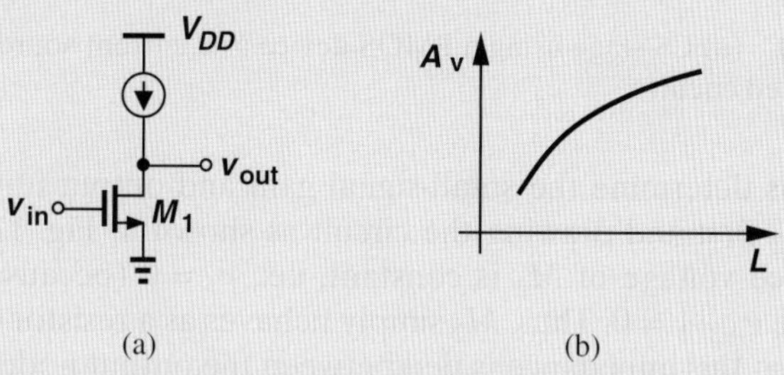

Figure 17.28 (a) CS stage with ideal current source as a load, (b) gain as a function of device channel length.

Solution The ideal current source presents an infinite small-signal resistance, allowing the use of Eq. (17.69) with $R_D = \infty$:

$$A_v = -g_m r_O. \tag{17.72}$$

Called the "intrinsic gain," this is the highest voltage gain that a single transistor can provide and its value falls in the range of 10–30 for MOS devices. Writing $g_m = \sqrt{2\mu_n C_{ox}(W/L)I_D}$ and $r_O = (\lambda I_D)^{-1}$, we have

$$|A_v| = \frac{\sqrt{2\mu_n C_{ox}\frac{W}{L}}}{\lambda\sqrt{I_D}}. \tag{17.73}$$

This result may imply that $|A_v|$ falls as L increases, but recall from Chapter 6 that $\lambda \propto L^{-1}$:

$$|A_v| \propto \sqrt{\frac{2\mu_n C_{ox} WL}{I_D}}. \tag{17.74}$$

Consequently, $|A_v|$ increases with L [Fig. 17.28(b)] and decreases with I_D.

Exercise What happens to the gain if both W and I_D are doubled? This is equivalent to placing two transistors in parallel.

17.4.1 CS Stage with Current-Source Load

As seen in the above example, the gain of a CS stage can be maximized by replacing the load resistor with a current source. But, how is the current source implemented? The observations made in relation to Fig. 17.21(b) suggest the use of a PMOS device as the load for an NMOS CS amplifier [Fig. 17.29(a)].

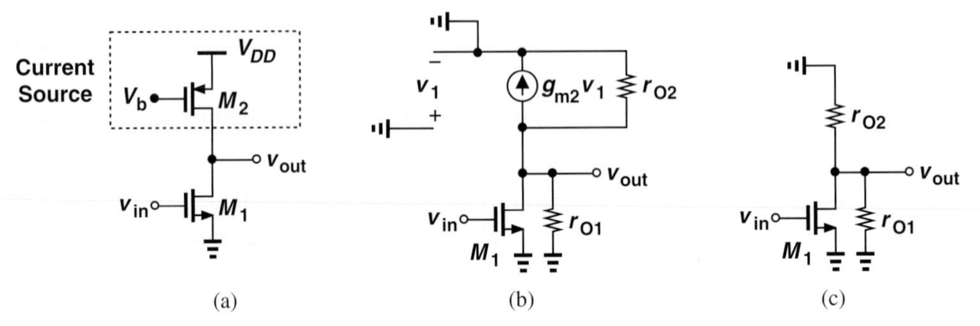

Figure 17.29 (a) CS stage using a PMOS device as a current source, (b) small-signal model of M_2, (c) simplified circuit.

Let us determine the small-signal gain and output impedance of the circuit. Focusing on M_2 first and drawing the circuit as shown in Fig. 17.29(b), we recognize that the gate-source voltage of M_2 is constant, i.e., $v_1 = 0$ (because v_1 denotes changes in V_{GS}) and hence $g_{m2}v_1 = 0$. Thus, M_2 simply behaves as a resistor equal to its output impedance. Noting that this impedance is tied between the output node and ac ground, we redraw the circuit as depicted in Fig. 17.29(c), where the output resistance of M_1 is shown explicitly. This stage is indeed similar to a standard CS amplifier except that it sees two parallel resistors from the output node to ac ground. Equations (17.69) and (17.70) respectively give

$$A_v = -g_{m1}(r_{O1}||r_{O2}) \tag{17.75}$$

$$R_{out} = r_{O1}||r_{O2}. \tag{17.76}$$

We conclude that inclusion of a realistic current source drops the gain from the intrinsic value ($g_m r_O$) to the value given above—by roughly a factor of two.

Example 17-18

Figure 17.30(a) shows a PMOS CS stage using an NMOS current source load. Compute the voltage gain of the circuit.

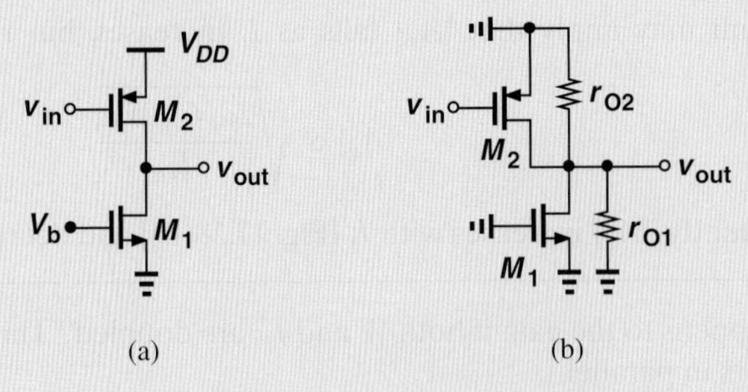

Figure 17.30 (a) CS stage using an NMOS device as current source, (b) simplified circuit.

Solution Here, M_2 acts as the "input" device and M_1 as the current source. Thus, M_2 generates a small-signal current equal to $g_{m2}v_{in}$, which, as shown in Fig. 17.30(b), flows through $r_{O1}||r_{O2}$, producing $v_{out} = -g_{m2}v_{in}(r_{O1}||r_{O2})$. Thus,

$$A_v = -g_{m2}(r_{O1}||r_{O2}). \tag{17.77}$$

Exercise For a given bias current, does the gain increase if L_1 increases?

The reader may have noticed that we use small-signal equivalent circuits only sparingly. Our objective is to learn "analysis by inspection," i.e., we wish to determine the properties of a circuit by just looking at it! This skill becomes increasingly more important as we deal with more complex circuits.

17.4.2 CS Stage with Diode-Connected Load

As explained in Chapter 6, a MOSFET whose gate and drain are shorted acts as a two-terminal device with an impedance of $(1/g_m)||r_O \approx 1/g_m$. In some applications, we may use a diode-connected MOSFET as the drain load. Illustrated in Fig. 17.31(a), such a topology exhibits only a moderate gain due to the relatively low impedance of the diode-connected device. With $\lambda = 0$, M_2 exhibits a small-signal resistance equal to $1/g_{m2}$, and $A_v = -g_m R_D$ yields

$$A_v = -g_{m1} \cdot \frac{1}{g_{m2}} \tag{17.78}$$

$$= -\frac{\sqrt{2\mu_n C_{ox}(W/L)_1 I_D}}{\sqrt{2\mu_n C_{ox}(W/L)_2 I_D}} \tag{17.79}$$

$$= -\sqrt{\frac{(W/L)_1}{(W/L)_2}}. \tag{17.80}$$

Interestingly, the gain is given by the dimensions of M_1 and M_2 and remains independent of process parameters μ_n and C_{ox} and the drain current, I_D.

A more accurate expression for the gain of the stage in Fig. 17.31(a) must take channel-length modulation into account. As depicted in Fig. 17.31(b), the resistance seen at the drain is now equal to $(1/g_{m2})||r_{O2}||r_{O1}$, and hence

$$A_v = -g_{m1}\left(\frac{1}{g_{m2}}||r_{O2}||r_{O1}\right). \tag{17.81}$$

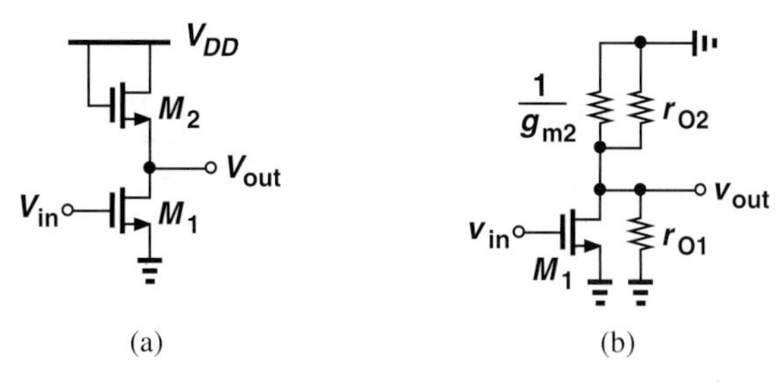

Figure 17.31 (a) MOS stage using a diode-connected load, (b) simplified circuit of (a).

Similarly, the output resistance of the stage is given by

$$R_{out} = \frac{1}{g_{m2}}||r_{O2}||r_{O1}. \tag{17.82}$$

Example 17-19

Determine the voltage gain of the circuit shown in Fig. 17.32 if $\lambda \neq 0$.

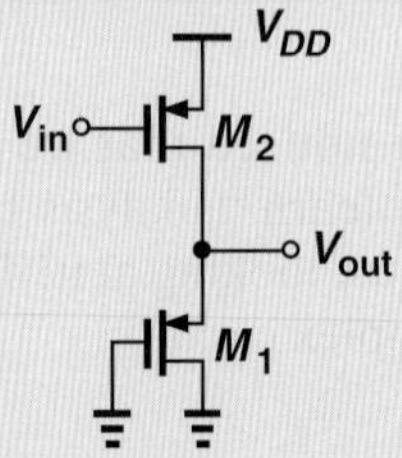

Figure 17.32 CS stage with diode-connected PMOS device.

Solution This stage is similar to that in Fig. 17.31(a), but with NMOS devices changed to PMOS transistors: M_1 serves as a common-source device and M_2 as a diode-connected load. Thus,

$$A_v = -g_{m2}\left(\frac{1}{g_{m1}}||r_{O1}||r_{O2}\right). \tag{17.83}$$

Exercise Calculate the gain if the gate of M_1 is tied to a constant voltage equal to 0.5 V.

17.4.3 CS Stage with Source Degeneration

In many applications, the CS core is modified as shown in Fig. 17.33(a), where a resistor R_S appears in series with the source. Called "source degeneration," this technique improves the "linearity" of the circuit and provides many other interesting properties that are studied in more advanced courses.

As with the CS core, we intend to determine the voltage gain and I/O impedances of the circuit, assuming M_1 is biased properly. Before delving into a detailed analysis, it is instructive to make some qualitative observations. Suppose the input signal raises the gate voltage by ΔV [Fig. 17.33(b)]. If R_S were zero, then the gate-source voltage would also

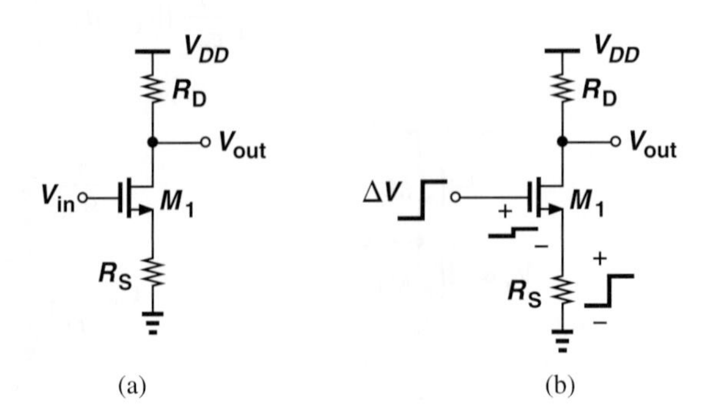

Figure 17.33 (a) CS stage with degeneration, (b) effect of input voltage change.

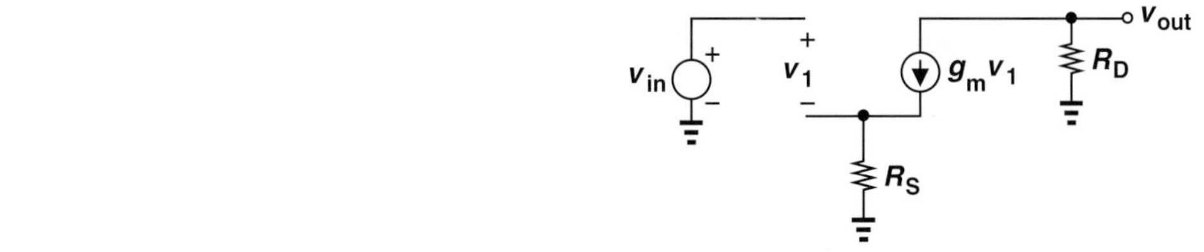

Figure 17.34 Small-signal model of CS stage with source degeneration.

increase by ΔV, producing a drain current change of $g_m\Delta V$. But with $R_S \neq 0$, some fraction of ΔV appears across R_S, thus leaving a change in V_{GS} that is *less* than ΔV. Consequently, the drain current change is also less than $g_m\Delta V$. We therefore expect that the voltage gain of the degenerated stage is *lower* than that of the CS core with no degeneration. While undesirable, the reduction in gain is incurred to improve other aspects of the performance.

We now quantify the foregoing observations by analyzing the small-signal behavior of the circuit. Depicted in Fig. 17.34 is the small-signal equivalent circuit, where V_{DD} is replaced with an ac ground and channel-length modulation is neglected. To determine v_{out}/v_{in}, we first write a KCL at the output node,

$$g_m v_1 = -\frac{v_{out}}{R_D}, \tag{17.84}$$

obtaining

$$v_1 = -\frac{v_{out}}{g_m R_D}. \tag{17.85}$$

We also recognize that the current through R_S is equal to $g_m v_1$. Thus, the voltage drop across R_S is given by $g_m v_1 R_S$. Since the voltage drop across R_S and v_1 must add up to v_{in}, we have

$$v_{in} = v_1 + g_m v_1 R_S \tag{17.86}$$

$$= \frac{-v_{out}}{g_m R_D}(1 + g_m R_S). \tag{17.87}$$

It follows that

$$\frac{v_{out}}{v_{in}} = -\frac{g_m R_D}{1 + g_m R_S}. \tag{17.88}$$

As predicted earlier, the magnitude of the voltage gain is lower than $g_m R_D$ for $R_S \neq 0$.

To arrive at an interesting interpretation of Eq. (17.88), we divide the numerator and denominator by g_m,

$$A_v = -\frac{R_D}{\frac{1}{g_m} + R_S}. \tag{17.89}$$

It is helpful to memorize this result as "the gain of the degenerated CS stage is equal to the total load resistance seen at the drain (to ground) divided by $1/g_m$ plus the total resistance tied from the source to ground." (In verbal descriptions, we often ignore the negative sign in the gain, with the understanding that it must be included.) This and similar interpretations throughout this book serve as powerful tools for analysis by inspection—often obviating the need for drawing small-signal circuits.

Example 17-20 Compute the voltage gain of the circuit shown in Fig. 17.35(a) if $\lambda = 0$.

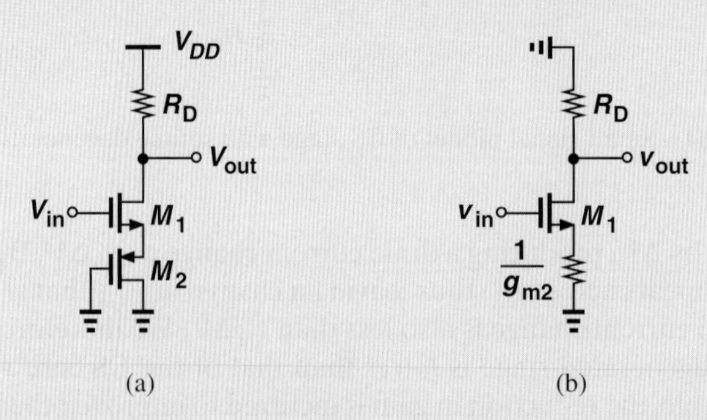

Figure 17.35 (a) Example of CS stage with degeneration, (b) simplified circuit.

Solution Transistor M_2 serves as a diode-connected device, presenting an impedance of $1/g_{m2}$ [Fig. 17.35(b)]. The gain is therefore given by Eq. (17.89) if R_S is replaced with $1/g_{m2}$:

$$A_v = -\frac{R_D}{\frac{1}{g_{m1}} + \frac{1}{g_{m2}}}. \tag{17.90}$$

Exercise Repeat the above example assuming $\lambda \neq 0$ for M_2.

The input impedance of the degenerated CS stage is infinite. But, how about the output impedance? Considering the equivalent circuit shown in Fig. 17.36, we note that the current flowing through R_S is equal to $g_m v_1$, and v_1 and the voltage across R_S must add up to zero (why?):

$$g_m v_1 R_S + v_1 = 0. \tag{17.91}$$

That is, $v_1 = 0$, indicating that all of i_x flows through R_D. It follows that

$$\frac{v_X}{i_X} = R_D, \tag{17.92}$$

the same as that of an undegenerated stage.

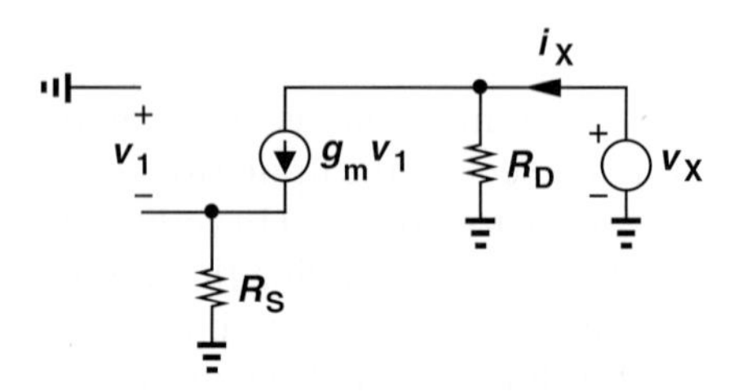

Figure 17.36 Equivalent circuit for calculation of output impedance.

Example 17-21 A CS stage incorporates a MOS device having a transconductance of $1/(200\ \Omega)$. If the circuit provides a voltage gain of 8 without degeneration and 4 with degeneration, calculate R_S and the output impedance. Assume $\lambda = 0$.

Solution With no degeneration, $g_m R_D = 8$ and hence $R_D = 1.6\ \text{k}\Omega$. With degeneration,

$$\frac{g_m R_D}{1 + g_m R_S} = 4, \tag{17.93}$$

yielding $R_D = 1/g_m = 200\ \Omega$.

Exercise What is the voltage gain if R_S reduced to $50\ \Omega$?

Did you know?

Source degeneration finds use in many circuits, e.g., high-speed "equalizers." Suppose we wish to transmit a data rate of 1 gigabit per second through an ethernet (CAT-5) cable. Since the cable attenuates high frequencies, [Fig. (a)], we follow it with a circuit that *amplifies* these frequencies by a larger amount than low frequencies. Such a circuit is called an equalizer and can be implemented as a degenerated CS stage [Fig. (b)]. Here, a capacitor is placed in parallel with R_S; as the frequency increases, the capacitor's impedance falls, lowering the amount of degeneration and raising the voltage gain. We face this issue in communication between a computer and a server or between two chips on a computer board.

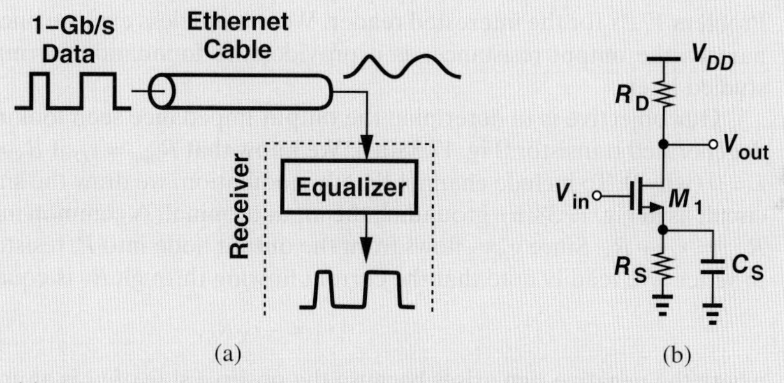

(a) Lossy cable carrying high-speed data, (b) equalizer circuit using capacitive degeneration.

The degenerated CS stage can be analyzed from a different perspective to provide more insight. Let us place the transistor and the source resistor in a black box still having three terminals [Fig. 17.37(a)]. For small-signal operation, we can view the box as a new transistor (or "active" device) and model its behavior by new values of transconductance and impedances. Denoted by G_m to avoid confusion with g_m of M_1, the equivalent transconductance is obtained from Fig. 17.37(b). Since the current flowing through R_S is equal to i_{out} and since the voltage across this resistor and v_1 must add up to v_{in}, we have $v_1 + g_m v_1 R_S = v_{in}$ and hence $v_1 = v_{in}/(1 + g_m R_S)$. Thus,

$$i_{out} = g_m v_1 \tag{17.94}$$

$$= g_m \frac{v_{in}}{1 + g_m R_S}, \tag{17.95}$$

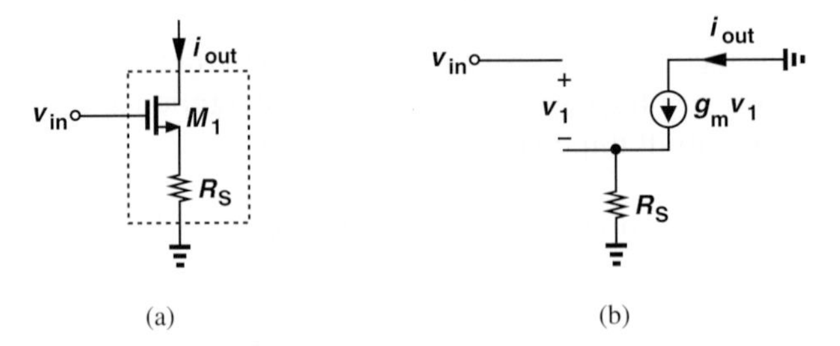

Figure 17.37 (a) Degenerated MOSFET viewed as a black box, (b) small-signal equivalent.

and hence

$$G_m = \frac{i_{out}}{v_{in}} \tag{17.96}$$

$$= \frac{g_m}{1 + g_m R_S}. \tag{17.97}$$

For example, the voltage gain of the stage with a load resistance of R_D is given by $-G_m R_D$.

An interesting property of the degenerated CS stage is that its voltage gain becomes relatively independent of the transistor transconductance and hence bias current if $g_m R_S \gg 1$. From Eq. (17.89), we note that $A_v \rightarrow -R_D/R_S$ under this condition.

Effect of Transistor Output Resistance The analysis of the degenerated CS stage has thus far neglected channel-length modulation effect. Somewhat beyond the scope of this book, the derivation of the circuit properties in the presence of this effect is outlined in Problem 17.31 for the interested reader. We nonetheless explore one aspect of the circuit, namely, the output resistance, as it provides the foundation for many other topologies studied later.

Our objective is to determine the output impedance seen looking into the drain of a degenerated transistor [Fig. 17.38(a)]. We know that $R_{out} = r_O$ if $R_S = 0$. Also, $R_{out} = \infty$ if $\lambda = 0$ (why?). To include channel-length modulation, we draw the small-signal equivalent circuit as in Fig. 17.38(b), grounding the input terminal. A common mistake here is to write $R_{out} = r_O + R_S$. Since $g_m v_1$ flows from the output node into P, resistors r_O and R_S are not in series. We readily note that the current flowing through R_S is equal to i_X. Thus,

$$v_1 = -i_X R_S, \tag{17.98}$$

where the negative sign arises because the positive side of v_1 is at ground. We also recognize that r_O carries a current of $i_X - g_m v_1$ and hence sustains a voltage of $(i_X - g_m v_1) r_O$. Adding this voltage to that across R_S $(= -v_1)$ and equating the result to v_X, we obtain

$$v_X = (i_X - g_m v_1) r_O - v_1 \tag{17.99}$$

$$= (i_X + g_m i_X R_S) r_O + i_X R_S. \tag{17.100}$$

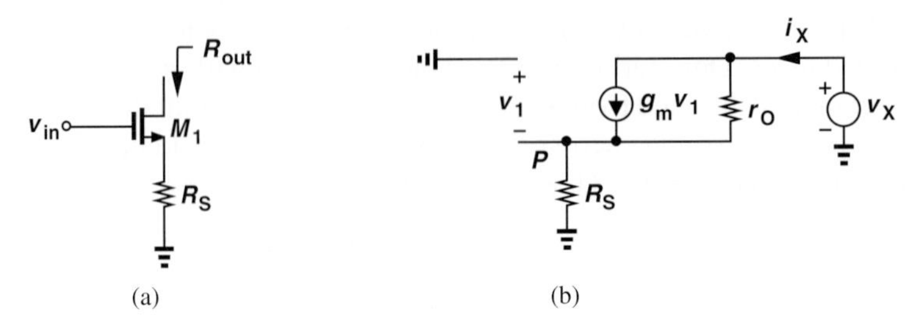

Figure 17.38 (a) Output impedance of degenerated stage, (b) equivalent circuit.

It follows that

$$R_{out} = (1 + g_m R_S) r_O + R_S \tag{17.101}$$

$$= r_O + (g_m r_O + 1) R_S. \tag{17.102}$$

Recall from (17.72) that the intrinsic gain of the transistor, $g_m r_O \gg 1$, and hence

$$R_{out} \approx r_O + g_m r_O R_S \tag{17.103}$$

$$\approx r_O (1 + g_m R_S). \tag{17.104}$$

Interestingly, source degeneration *raises* the output impedance from r_O to the above value, i.e., by a factor of $1 + g_m R_S$.

The reader may wonder if the increase in the output resistance is desirable or undesirable. The "boosting" of output resistance as a result of degeneration proves extremely useful in circuit design, leading to amplifiers with a higher gain as well as creating more ideal current sources. These concepts are studied in Chapter 9.

In the analysis of circuits, we sometimes draw the transistor output resistance explicitly to emphasize its significance (Fig. 17.39). This representation, of course, assumes M_1 itself does not contain another r_O.

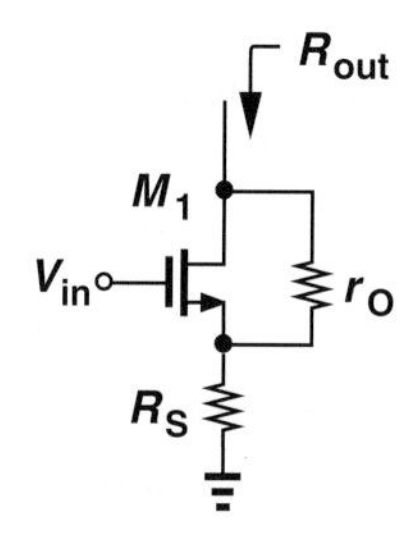

Figure 17.39 Stage with explicit depiction of r_O.

Example 17-22

We wish to design an NMOS current source having a value of 1 mA and an output resistance of 20 kΩ. Assume $\mu_n C_{ox} = 100\ \mu A/V^2$ and $\lambda = 0.25\ V^{-1}$. Compute the device aspect ratio and the degeneration resistance if the minimum allowable V_{DS} is 0.3 V.

Solution Setting the transistor overdrive voltage to 0.3 V (so that it can tolerate a minimum V_{DS} of the same value), we write:

$$g_m = \frac{2I_D}{V_{GS} - V_{TH}} \tag{17.105}$$

$$= \frac{1}{150\ \Omega}. \tag{17.106}$$

Also, since $g_m = \mu_n C_{ox}(W/L)(V_{GS} - V_{TH})$, we have $W/L = 222$. Moreover, $r_O = 1/(\lambda I_D) = 4$ kΩ. For the output resistance to reach 20 kΩ,

$$(1 + g_m R_S) r_O + R_S = 20\ \text{k}\Omega, \tag{17.107}$$

and hence

$$R_S = 578\ \Omega. \tag{17.108}$$

Exercise What is the output resistance if $I_D = 1$ mA, $V_{DS} = 0.15$ V and $R_S = 200\ \Omega$?

Example 17-23

Compute the output resistance of the circuit in Fig. 17.40(a) if M_1 and M_2 are identical.

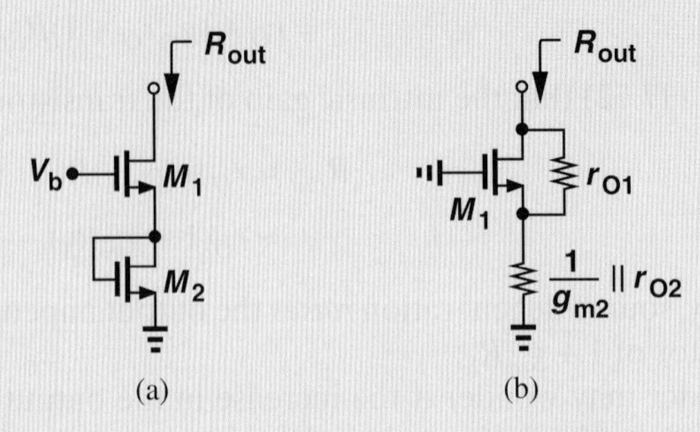

Figure 17.40 (a) Example of CS stage with degeneration, (b) simplified circuit.

Solution The diode-connected device M_2 can be represented by a small-signal resistance of $(1/g_{m2})||r_{O2} \approx 1/g_{m2}$. Transistor M_1 is degenerated by this resistance, and from Eq. (17.101):

$$R_{out} = r_{O1}\left(1 + g_{m1}\frac{1}{g_{m2}}\right) + \frac{1}{g_{m2}} \tag{17.109}$$

which, since $g_{m1} = g_{m2} = g_m$, reduces to

$$R_{out} = 2r_{O1} + \frac{1}{g_m} \tag{17.110}$$

$$\approx 2r_{O1}. \tag{17.111}$$

Exercise Suppose $W_2 = 2W_1$. Calculate the output impedance.

The procedure of progressively simplifying a circuit until it resembles a known topology proves extremely critical in our work. Called "analysis by inspection," this method obviates the need for complex small-signal models and lengthy calculations. The reader is encouraged to attempt the above example using the small-signal model of the overall circuit to appreciate the efficiency and insight provided by our intuitive approach.

Example 17-24

Determine the output resistance of the circuit in Fig. 17.41(a) and compare the result with that in the above example. Assume M_1 and M_2 are in saturation.

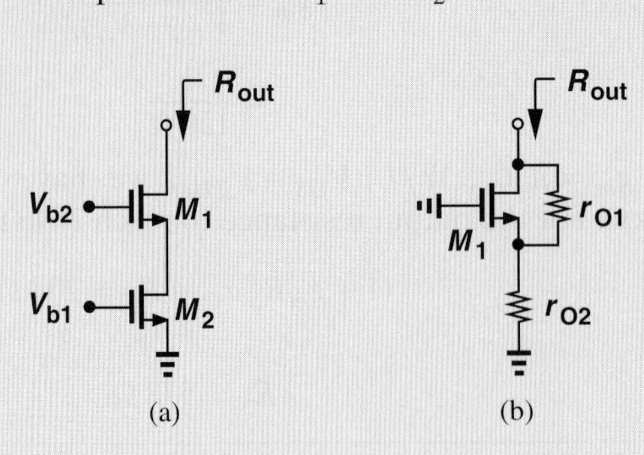

Figure 17.41 (a) Example of CS stage with degeneration, (b) simplified circuit.

Solution With its gate-source voltage fixed, transistor M_2 operates as a current source, introducing a resistance of r_{O2} from the source of M_1 to ground [Fig. 17.41(b)].

Equation (17.101) can therefore be written as

$$R_{out} = (1 + g_{m1}r_{O1})r_{O2} + r_{O1} \quad (17.112)$$

$$\approx g_{m1}r_{O1}r_{O2} + r_{O1}. \quad (17.113)$$

Assuming $g_{m1}r_{O2} \gg 1$ (which is valid in practice), we have

$$R_{out} \approx g_{m1}r_{O1}r_{O2}. \quad (17.114)$$

We observe that this value is significantly higher than that in (17.111).

Exercise Determine the output impedance if a resistor of value R_1 is inserted in series with the drain of M_2.

CS Stage with Biasing Having learned the small-signal properties of the common-source amplifier and its variants, we now study a more general case wherein the circuit contains a bias network as well. We begin with simple biasing schemes described earlier and progressively add complexity (and more robust performance) to the circuit. Let us begin with an example.

Example 17-25 A student familiar with the CS stage and basic biasing constructs the circuit shown in Fig. 17.42 to amplify the signal produced by a microphone. Unfortunately, M_1 carries no current, failing to amplify. Explain the cause of this problem.

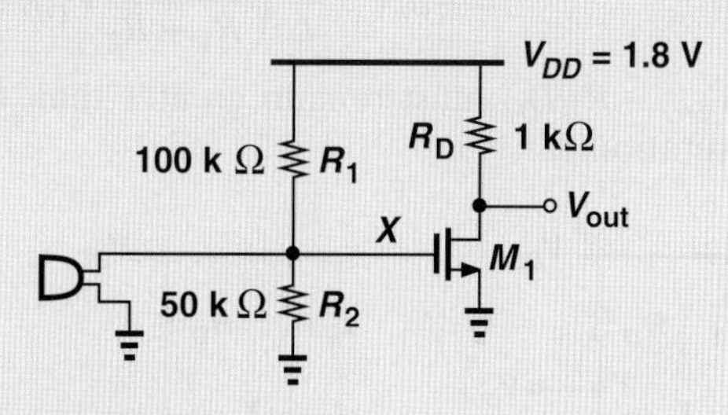

Figure 17.42 Microphone amplifier.

Solution Many microphones exhibit a small low-frequency resistance (e.g., < 100 Ω). If used in this circuit, such a microphone creates a low resistance from the gate of M_1 to ground, providing a very low gate voltage. For example, a microphone resistance of 100 Ω yields

$$V_X = \frac{100\ \Omega||50\ \text{k}\Omega}{100\ \text{k}\Omega + 100\ \Omega||50\ \text{k}\Omega} \times 2.5\ \text{V} \quad (17.115)$$

$$\approx 2.5\ \text{mV}. \quad (17.116)$$

Thus, the microphone low-frequency resistance disrupts the bias of the amplifier.

Exercise Can we resolve this issue by tying the source of M_1 to a negative voltage?

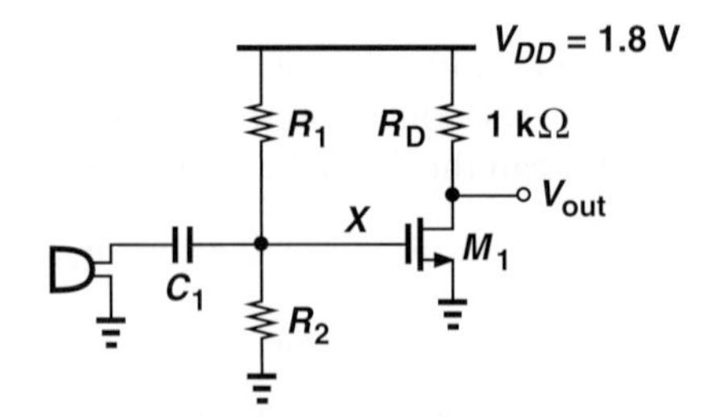

Figure 17.43 Capacitive coupling at the input of microphone amplifier.

How should the circuit of Fig. 17.42 be fixed? Since only the *signal* generated by the microphone is of interest, a series capacitor can be inserted as depicted in Fig. 17.43 so as to isolate the dc biasing of the amplifier from the microphone. That is, the bias point of M_1 remains independent of the resistance of the microphone because C_1 carries no bias current. The value of C_1 is chosen so that it provides a relatively low impedance (almost a short circuit) for the frequencies of interest. We say C_1 is a "coupling" capacitor and the input of this stage is "ac-coupled" or "capacitively coupled." Many circuits employ capacitors to isolate the bias conditions from "undesirable" effects. More examples clarify this point later.

The foregoing observation suggests that the methodology illustrated in Fig. 17.9 must include an additional rule: replace all capacitors with an open circuit for dc analysis and a short circuit for small-signal analysis.

Let us begin with the CS stage depicted in Fig. 17.44(a). For bias calculations, the signal source is set to zero and C_1 is opened, leading to Fig. 17.44(b). As shown earlier, we have

$$I_D = \frac{1}{2}\mu_n C_{ox}\frac{W}{L}(V_{GS} - V_{TH})^2 \tag{17.117}$$

$$= \frac{1}{2}\mu_n C_{ox}\frac{W}{L}\left(\frac{R_2}{R_1 + R_2}V_{DD} - V_{TH}\right)^2. \tag{17.118}$$

Also, $V_{DS}(= V_{DD} - R_D I_D)$ must remain greater than the overdrive voltage so that M_1 operates in saturation.

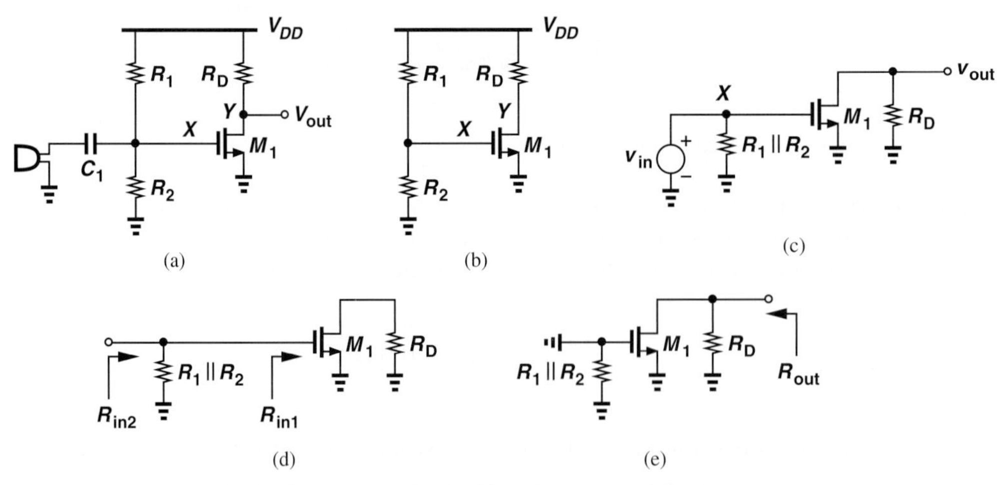

Figure 17.44 (a) Capacitive coupling at the input of a CS stage, (b) simplified stage for bias calculation, (c) simplified stage for small-signal calculation, (d) simplified circuit for input impedance calculation, (e) simplified circuit for output impedance calculation.

With the bias current known, the small-signal parameters g_m and r_O can be calculated. We now turn our attention to small-signal analysis, considering the simplified circuit of Fig. 17.44(c). Here, C_1 is replaced with a short and V_{DD} with ac ground, but M_1 is maintained as a symbol. Note that R_1 and R_2 now appear in parallel. We attempt to solve the circuit by inspection: if unsuccessful, we will resort to using a small-signal model for M_1 and writing KVLs and KCLs.

The circuit of Fig. 17.44(c) resembles the CS core illustrated in Fig. 17.23, except for R_1 and R_2. Interestingly, these resistors have no effect on the voltage at node X so long as v_{in} remains an ideal voltage source; i.e., $v_X = v_{in}$ regardless of the values of R_1 and R_2. Since the voltage gain from the gate to the drain is given by $v_{out}/v_X = -g_m R_D$, we have

$$\frac{v_{out}}{v_{in}} = -g_m R_D. \tag{17.119}$$

If $\lambda \neq 0$, then

$$\frac{v_{out}}{v_{in}} = -g_m(R_D||r_O). \tag{17.120}$$

However, the input impedance is affected by R_1 and R_2 [Fig. 17.44(d)]. Recall from Fig. 17.7 that the impedance seen looking into the gate, R_{in1}, is infinity. Here, R_1 and R_2 simply appear in parallel with R_{in1}, yielding

$$R_{in2} = R_1||R_2. \tag{17.121}$$

Thus, the bias resistors lower the input impedance.

To determine the output impedance, we set the input source to zero [Fig. 17.44(e)]. Comparing this circuit with that in Fig. 17.27, we recognize that R_{out} remains unchanged:

$$R_{out} = R_D||r_O. \tag{17.122}$$

Example 17-26

Having learned about ac coupling, the student in Example 17-25 modifies the design to that shown in Fig. 17.45 and attempts to drive a speaker. Unfortunately, the circuit still fails. Explain why.

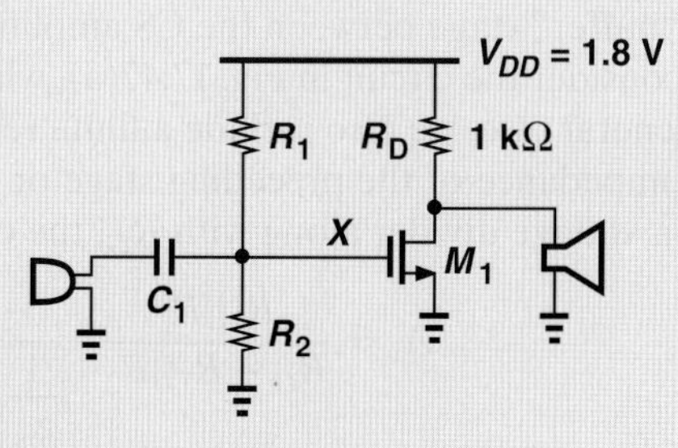

Figure 17.45 Amplifier with direct connection of speaker.

Solution Typical speakers incorporate a solenoid (inductor) to actuate a membrane. The solenoid exhibits a very low dc resistance, e.g., less than 1 Ω. Thus, the speaker in Fig. 17.45 shorts the drain to ground, driving M_1 into deep triode region.

Exercise Does the circuit behave better if the bottom terminal of the speaker is attached to V_{DD} rather than to ground?

Example 17-27 The student applies ac coupling to the output as well [Fig. 17.46(a)] and measures the quiescent points to ensure proper biasing. The drain bias voltage is 1.0 V and M_1 operates in the saturation region. However, the student still observes no gain in the circuit. If $g_m = (100\ \Omega)^{-1}$ and the speaker exhibits an impedance of $R_{sp} = 8\ \Omega$ in the audio range, explain why the circuit provides no gain.

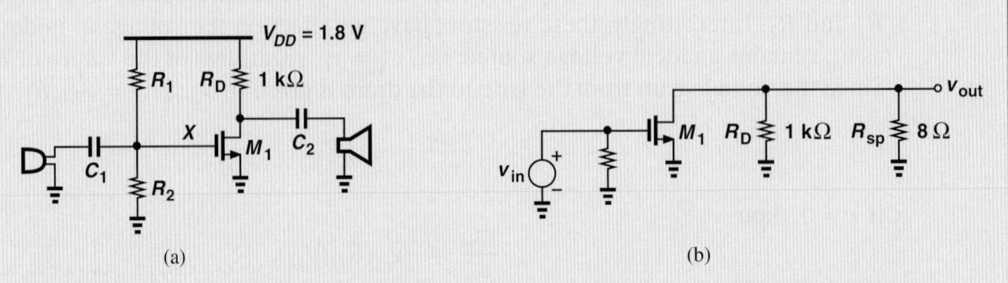

Figure 17.46 (a) Amplifier with capacitive coupling at the input and output, (b) simplified small-signal model.

Solution From the small-signal circuit of Fig. 17.46(b), we recognize that the equivalent load resistance is equal to $R_D||R_{sp} \approx 8\ \Omega$. Thus,

$$|A_v| = g_m(R_D||R_{sp}) = 0.08 \tag{17.123}$$

Exercise What speaker impedance drops the gain of the CS stage by only a factor of two?

The design in Fig. 17.46(a) exemplifies an improper interface between an amplifier and a load: the output impedance is so much higher than the load impedance that the connection of the load to the amplifier drops the gain drastically.

How can we remedy the problem of loading here? Since the voltage gain is proportional to g_m, we can bias M_1 at a much higher current to raise the gain. Alternatively, we can interpose a "buffer" stage between the CS amplifier and the speaker (Section 17.4.5).

Let us now consider the circuit in Fig. 17.47(a), where the transistor is degenerated by R_S. As a more general case, we also include a finite resistance, R_G in series with the input (i.e., the output impedance of the preceding stage or signal source) [Fig. 17.47(b)]. Since R_1 and R_2 form a voltage divider along with R_G, the overall voltage gain falls to

$$A_v = \frac{R_1||R_2}{R_G + R_1||R_2} \cdot \frac{-R_D}{\dfrac{1}{g_m} + R_S}, \tag{17.124}$$

where λ is assumed to be zero.

It is possible to utilize degeneration for biasing but eliminate its effect on the small-signal performance by means of a bypass capacitor [Fig. 17.47(c)]. If the capacitor operates as a short circuit at the frequency of interest, then the source of M_1 remains at ac ground and the amplification is free from degeneration:

$$A_v = -\frac{R_1||R_2}{R_G + R_1||R_2} g_m R_D. \tag{17.125}$$

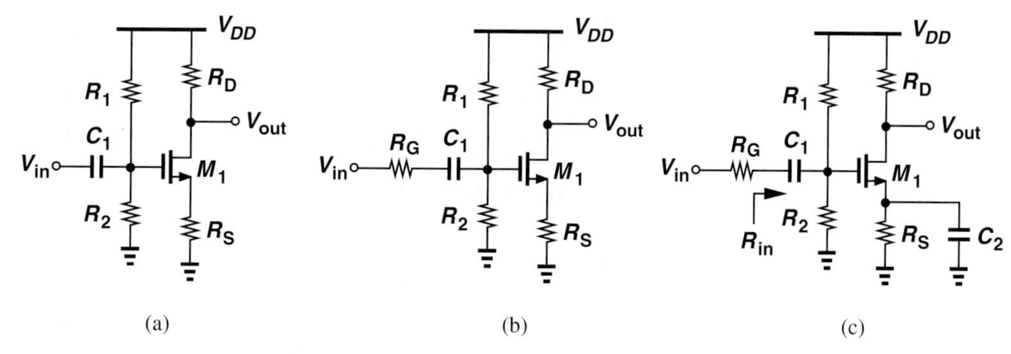

Figure 17.47 (a) CS stage with input coupling capacitor, (b) inclusion of R_G, (c) use of bypass capacitor.

Example 17-28 Design the CS stage of Fig. 17.47(c) for a voltage gain of 5, an input impedance of 50 kΩ, and a power budget of 5 mW. Assume $\mu_n C_{ox} = 100\ \mu\text{A/V}^2$, $V_{TH} = 0.5$ V, $\lambda = 0$, and $V_{DD} = 1.8$ V. Also, assume a voltage drop of 400 mV across R_S.

Solution The power budget along with $V_{DD} = 1.8$ V implies a maximum supply current of 2.78 mA. As an initial guess, we allocate 2.7 mA to M_1 and the remaining 80 μA to R_1 and R_2. It follows that

$$R_S = 148\ \Omega. \tag{17.126}$$

As with typical design problems, the choice of g_m and R_D is somewhat flexible so long as $g_m R_D = 5$. However, with I_D known, we must ensure a reasonable value for V_{GS}, e.g., $V_{GS} = 1$ V. This choice yields

$$g_m = \frac{2I_D}{V_{GS} - V_{TH}} \tag{17.127}$$

$$= \frac{1}{92.6\ \Omega}, \tag{17.128}$$

and hence

$$R_D = 463\ \Omega. \tag{17.129}$$

Writing

$$I_D = \frac{1}{2}\mu_n C_{ox}\frac{W}{L}(V_{GS} - V_{TH})^2 \tag{17.130}$$

gives

$$\frac{W}{L} = 216. \tag{17.131}$$

With $V_{GS} = 1$ V and a 400 mV drop across R_S, the gate voltage reaches 1.4 V, requiring that

$$\frac{R_2}{R_1 + R_2}V_{DD} = 1.4\ \text{V}, \tag{17.132}$$

which, along with $R_{in} = R_1||R_2 = 50$ kΩ, yields

$$R_1 = 64.3\ \text{k}\Omega \tag{17.133}$$

$$R_2 = 225\ \text{k}\Omega. \tag{17.134}$$

We must now check to verify that M_1 indeed operates in saturation. The drain voltage is given by $V_{DD} - I_D R_D = 1.8\text{ V} - 1.25\text{ V} = 0.55\text{ V}$. Since the gate voltage is equal to 1.4 V, the gate-drain voltage difference exceeds V_{TH}, driving M_1 into the triode region!

How did our design procedure lead to this result? For the given I_D, we have chosen an excessively large R_D, i.e., an excessively small g_m (because $g_m R_D = 5$), even though V_{GS} is reasonable. We must therefore increase g_m so as to allow a lower value for R_D. For example, suppose we halve R_D and double g_m by increasing W/L by a factor of four:

$$\frac{W}{L} = 864 \tag{17.135}$$

$$g_m = \frac{1}{46.3\ \Omega}. \tag{17.136}$$

The corresponding gate-source voltage is obtained from Eq. (17.127):

$$V_{GS} = 250\text{ mV}, \tag{17.137}$$

yielding a gate voltage of 650 mV.

Is M_1 in saturation? The drain voltage is equal to $V_{DD} - R_D I_D = 1.17$ V, a value higher than the gate voltage minus V_{TH}. Thus, M_1 operates in saturation.

Exercise Repeat the above example for a power budget of 3 mW and $V_{DD} = 1.2$ V.

Figure 17.48 summarizes the concepts studied in this section.

17.4.4 Common-Gate Topology

Following our extensive study of the CS stage, we now turn our attention to the "common-gate" (CG) topology. Nearly all of the concepts described for the CS configuration apply here as well. We therefore follow the same train of thought, but at a slightly faster pace.

Given the amplification capabilities of the CS stage, the reader may wonder why we study other amplifier topologies. As we will see, other configurations provide different circuit properties that are preferable to those of the CS stage in some applications. The reader is encouraged to review Examples 17-2–17-4, their resulting rules illustrated in Fig. 17.7, and the possible topologies in Fig. 17.22 before proceeding further.

Figure 17.49 shows the CG stage. The input is applied to the source and the output is sensed at the drain. Biased at a proper voltage, the gate acts as ac ground and hence as a node "common" to the input and output ports. As with the CS stage, we first study the core and subsequently add the biasing elements.

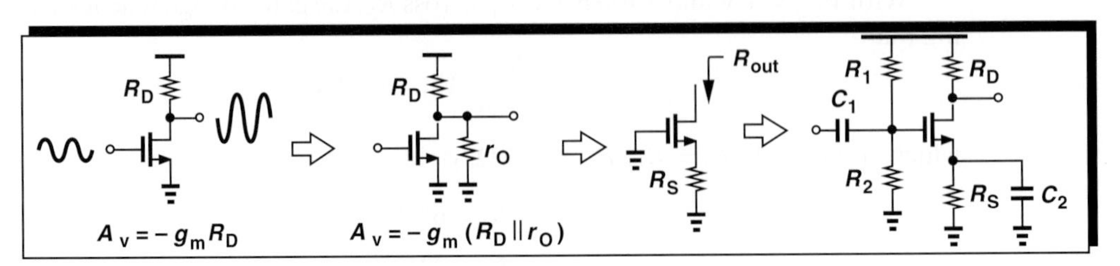

Figure 17.48 Summary of concepts studied thus far.

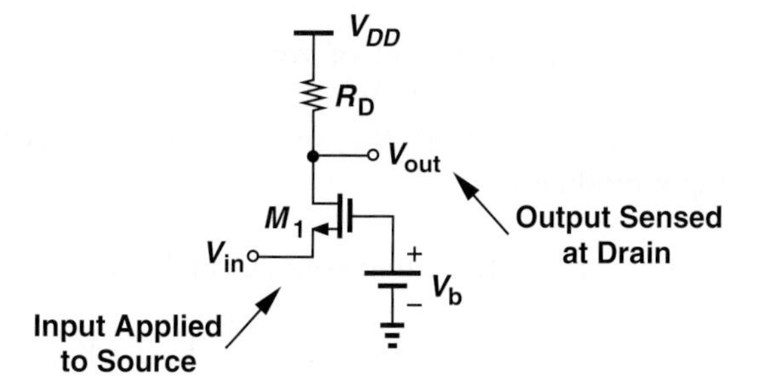

Figure 17.49 Common-gate stage.

Analysis of CG Core How does the CG stage of Fig. 17.50(a) respond to an input signal?[8] If V_{in} goes up by a small amount ΔV, the gate-source voltage of M_1 *decreases* by the same amount because the gate voltage is fixed. Consequently, the drain current falls by $g_m \Delta V$, allowing V_{out} to *rise* by $g_m \Delta V R_D$. We therefore surmise that the small-signal voltage gain is equal to

$$A_v = g_m R_D. \tag{17.138}$$

Interestingly, this expression is identical to the gain of the CS topology. Unlike the CS stage, however, this circuit exhibits a *positive* gain because an increase in V_{in} leads to an increase in V_{out}.

Let us confirm the above results with the aid of the small-signal equivalent depicted in Fig. 17.50(b), where channel-length modulation is neglected. Beginning with the output node, we equate the current flowing through R_D to $g_m v_1$:

$$-\frac{v_{out}}{R_D} = g_m v_1, \tag{17.139}$$

obtaining $v_1 = -v_{out}/(g_m R_D)$. Considering the input node next, we recognize that $v_1 = -v_{in}$. It follows that

$$\frac{v_{out}}{v_{in}} = g_m R_D. \tag{17.140}$$

As with the CS stage, the CG topology suffers from trade-offs among the gain, the voltage headroom, and the I/O impedances. The following example illustrates these trade-offs.

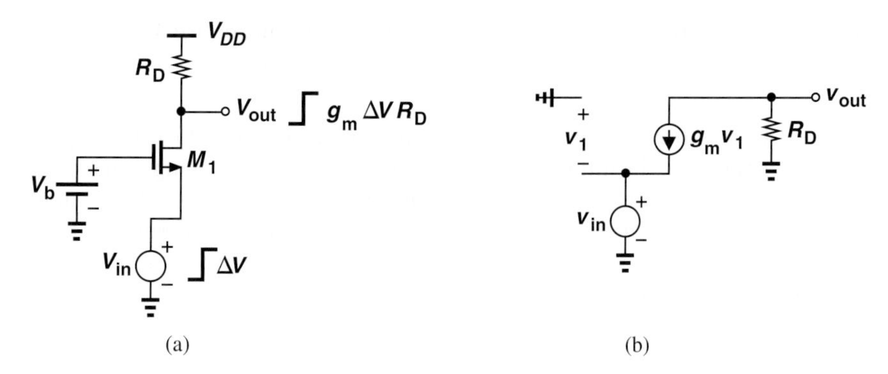

Figure 17.50 (a) Response of CG stage to small input change, (b) small-signal model.

[8]Note that the topologies of Figs. 17.49 and 17.50(a) are identical even though M_1 is drawn differently.

Example 17-29 A microphone having a dc level of zero drives a CG stage biased at $I_D = 0.5$ mA. If $W/L = 50$, $\mu_n C_{ox} = 100\ \mu\text{A/V}^2$, $V_{TH} = 0.5$ V, and $V_{DD} = 1.8$ V, determine the maximum allowable value of R_D and hence the maximum voltage gain. Neglect channel-length modulation.

Solution With W/L known, the gate-source voltage can be determined from

$$I_D = \frac{1}{2}\mu_n C_{ox}\frac{W}{L}(V_{GS} - V_{TH})^2 \tag{17.141}$$

as

$$V_{GS} = 0.947 \text{ V}. \tag{17.142}$$

For M_1 in Fig. 17.50(a) to remain in saturation,

$$V_{DD} - I_D R_D > V_b - V_{TH} \tag{17.143}$$

and hence

$$R_D < 2.71 \text{ k}\Omega. \tag{17.144}$$

Also, the above values of W/L and I_D yield $g_m = (447\ \Omega)^{-1}$ and

$$A_v \leq 6.06. \tag{17.145}$$

Figure 17.51 summarizes the allowable signal levels in this design. The gate voltage can be generated using a resistive divider similar to that in Fig. 17.47(a).

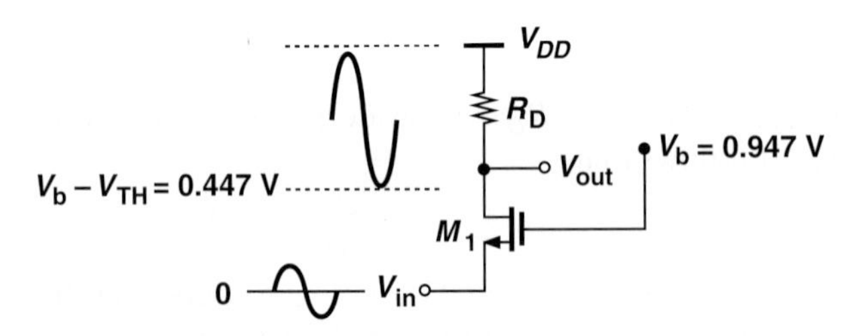

Figure 17.51 Signal levels in CG stage.

Exercise If a gain of 10 is required, what value should be chosen for W/L?

Let us now compute the I/O impedances of the CG topology so as to understand its capabilities in interfacing with preceding and following stages. The rules illustrated in Fig. 17.7 prove extremely useful here, obviating the need for small-signal equivalent circuits. Shown in Fig. 17.52(a), the simplified ac circuit reveals that R_{in} is simply the impedance seen looking into the source with the gate at ac ground. From the rules in Fig. 17.7, we have

$$R_{in} = \frac{1}{g_m} \tag{17.146}$$

if $\lambda = 0$. The input impedance of the CG stage is therefore relatively *low*, e.g., no more than a few hundred ohms.

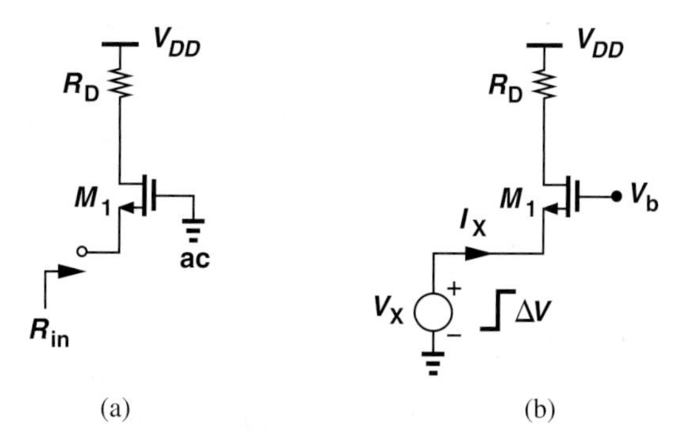

Figure 17.52 (a) Input impedance of CG stage, (b) response to a small change in input.

The input impedance of the CG stage can also be determined intuitively [Fig. 17.52(b)]. Suppose a voltage source V_X tied to the source of M_1 changes by a small amount ΔV. The gate-source voltage therefore changes by the same amount, leading to a change in the drain current equal to $g_m \Delta V$. Since the drain current flows through the input source, the current supplied by V_X also changes by $g_m \Delta V$. Consequently, $R_{in} = \Delta V_X / \Delta I_X = 1/g_m$.

Does an amplifier with a low input impedance find any practical use? Yes, indeed. For example, many stand-alone high-frequency amplifiers are designed with an input resistance of 50 Ω to provide "impedance matching" between modules in a cascade and the transmission lines (traces on a printed-circuit board) connecting the modules (Fig. 17.53).[9]

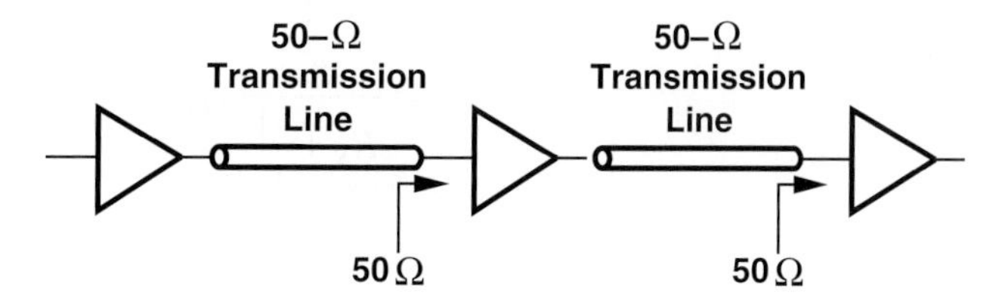

Figure 17.53 System using transmission lines.

The output impedance of the CG stage is computed with the aid of Fig. 17.54, where the input voltage source is set to zero. We note that $R_{out} = R_{out1} || R_D$, where R_{out1} is the impedance seen at the drain with the source grounded. From the rules of Fig. 17.7, we have $R_{out1} = r_O$ and hence

$$R_{out} = r_O || R_D \tag{17.147}$$

or

$$R_{out} = R_D \text{ if } \lambda = 0. \tag{17.148}$$

[9]If the input impedance of each stage is not matched to the characteristic impedance of the preceding transmission line, then "reflections" occur, corrupting the signal or at least creating dependence on the length of the lines.

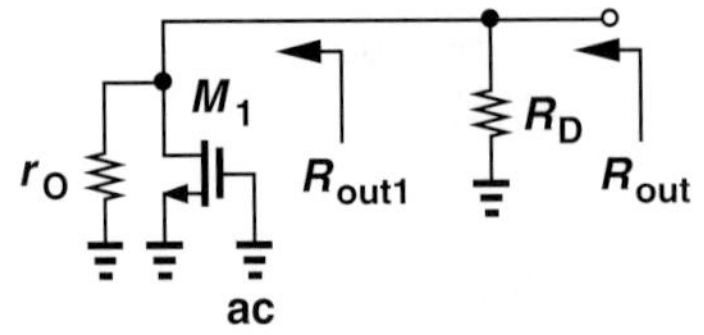

Figure 17.54 Output impedance of CG stage.

Example 17-30 A common-gate amplifier is designed for an input impedance of R_{in} and an output impedance of R_{out}. Neglecting channel-length modulation, determine the voltage gain of the circuit.

Solution Since $R_{in} = 1/g_m$ and $R_{out} = R_D$, we have

$$A_v = \frac{R_{out}}{R_{in}}. \tag{17.149}$$

Exercise Explain why we sometimes say the "current gain" of a CG stage is equal to unity.

It is instructive to study the behavior of the CG topology in the presence of a finite source resistance. Shown in Fig. 17.55, such a circuit suffers from signal attenuation from the input to node X, thereby providing a smaller voltage gain. More specifically, since the impedance seen looking into the source of M_1 (with the gate grounded) is equal to $1/g_m$ (for $\lambda = 0$), we have

$$v_X = \frac{\frac{1}{g_m}}{R_S + \frac{1}{g_m}} v_{in} \tag{17.150}$$

$$= \frac{1}{1 + g_m R_S} v_{in}. \tag{17.151}$$

We also recall from Eq. (17.140) that the gain from the source to the output is given by

$$\frac{v_{out}}{v_X} = g_m R_D. \tag{17.152}$$

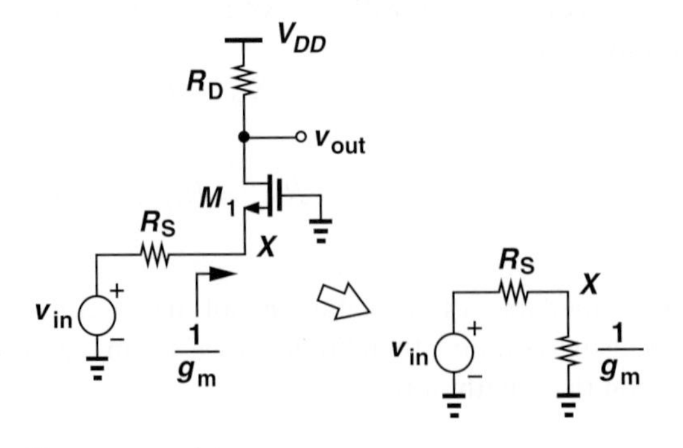

Figure 17.55 CG stage with source resistance.

It follows that

$$\frac{v_{out}}{v_{in}} = \frac{g_m R_D}{1 + g_m R_S} \tag{17.153}$$

$$= \frac{R_D}{\frac{1}{g_m} + R_S}, \tag{17.154}$$

a result identical to that of the CS stage (except for a negative sign) if R_S is viewed as a source degeneration resistor.

As with the CS stage, we may desire to analyze the CS topology in the general case: with source degeneration, $\lambda \neq 0$, and a resistance in series with the gate [Fig. 17.56(a)]. This analysis is somewhat beyond the scope of this book. Nevertheless, it is instructive to compute the output impedance. As illustrated in Fig. 17.56(b), R_{out} is equal to R_D in parallel with the impedance seen looking into the drain, R_{out1}. But R_{out1} is identical to the output resistance of a source-degenerated *common-source* stage:

$$R_{out1} = (1 + g_m R_S) r_O + R_S. \tag{17.155}$$

It follows that

$$R_{out} = R_D || [(1 + g_m R_S) r_O + R_S]. \tag{17.156}$$

The reader may have recognized that the output impedance of the CG stage is equal to that of the CS stage. Is this true in general? Recall that the output impedance is determined by setting the input source to zero. In other words, when calculating R_{out}, we have no knowledge of the input terminal of the circuit, as illustrated in Fig. 17.57 for CS and CG stages. It is therefore no coincidence that the output impedances are identical *if* the same assumptions are made for both circuits (e.g., identical values of λ and source degeneration).

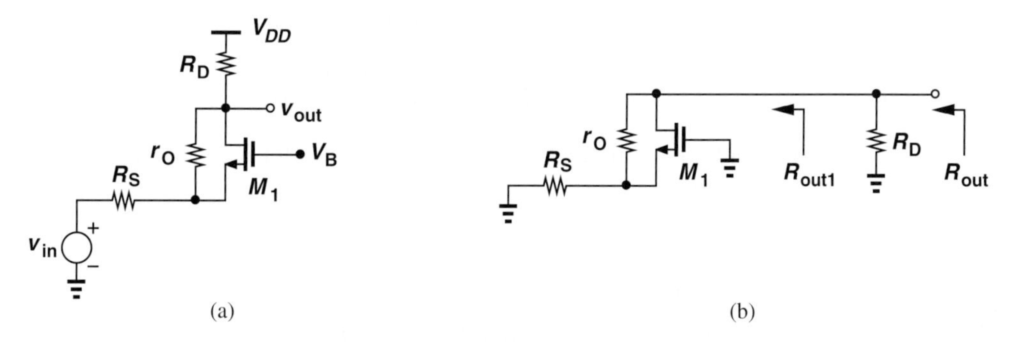

Figure 17.56 (a) General CG stage, (b) output impedance seen at different nodes.

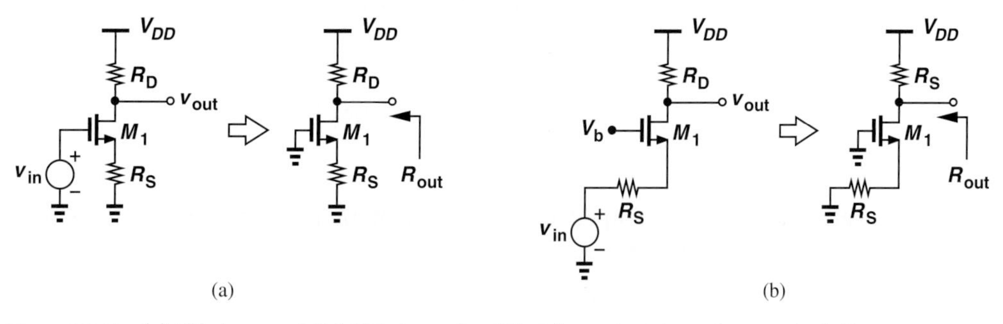

Figure 17.57 (a) CS stage and (b) CG stage simplified for output impedance calculation.

Example 17-31 For the circuit shown in Fig. 17.58(a), calculate the voltage gain if $\lambda = 0$ and the output impedance if $\lambda > 0$.

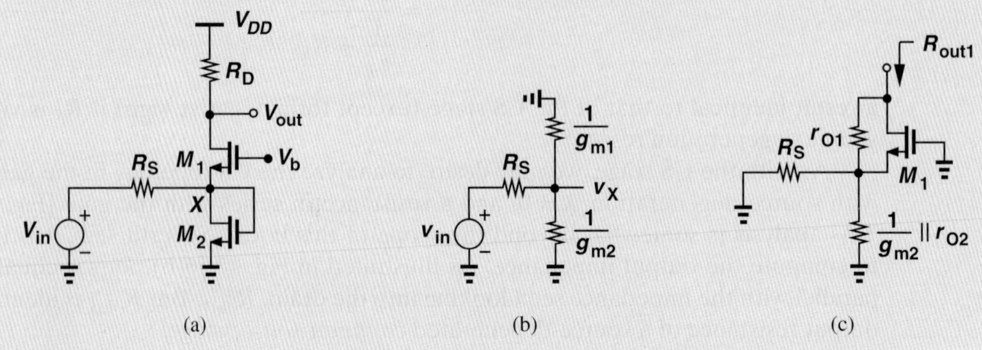

Figure 17.58 (a) Example of CG stage, (b) equivalent input network, (c) calculation of output resistance.

Solution We first compute v_X/v_{in} with the aid of the equivalent circuit depicted in Fig. 17.58(b):

$$\frac{v_X}{v_{in}} = \frac{\frac{1}{g_{m2}}||\frac{1}{g_{m1}}}{\frac{1}{g_{m2}}||\frac{1}{g_{m1}} + R_S} \tag{17.157}$$

$$= \frac{1}{1 + (g_{m1} + g_{m2})R_S}. \tag{17.158}$$

Noting that $v_{out}/v_X = g_{m1}R_D$ (why?), we have

$$\frac{v_{out}}{v_{in}} = \frac{g_{m1}R_D}{1 + (g_{m1} + g_{m2})R_S}. \tag{17.159}$$

To compute the output impedance, we first consider R_{out1}, as shown in Fig. 17.58(c), which is equal to

$$R_{out1} = (1 + g_{m1}r_{O1})\left(\frac{1}{g_{m2}}||r_{O2}||R_S\right) + r_{O1} \tag{17.160}$$

$$\approx g_{m1}r_{O1}\left(\frac{1}{g_{m2}}||R_S\right) + r_{O1}. \tag{17.161}$$

The overall output impedance is then given by

$$R_{out} = R_{out1}||R_D \tag{17.162}$$

$$\approx \left[g_{m1}r_{O1}\left(\frac{1}{g_{m2}}||R_S\right) + r_{O1}\right]||R_D. \tag{17.163}$$

Exercise Calculate the output impedance if the gate of M_2 is tied to a constant voltage.

CG Stage with Biasing Having learned the small-signal properties of the CG core, we now extend our analysis to the circuit including biasing. An example proves instructive at this point.

Example 17-32

A student decides to incorporate ac coupling at the input of a CG stage to ensure the bias is not affected by the signal source, drawing the design as shown in Fig. 17.59. Explain why this circuit does not work.

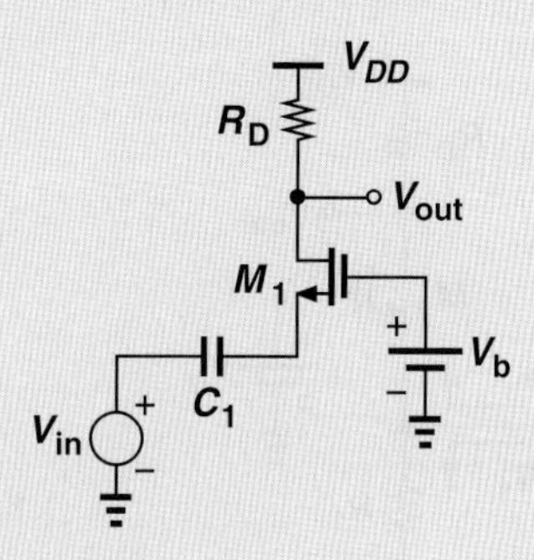

Figure 17.59 Cg stage lacking bias current.

Solution Unfortunately, the design provides no dc path for the source current of M_1, forcing a zero bias current and hence a zero transconductance.

Exercise Is the issue resolved if a capacitor is tied from the source of M_1 to ground?

Example 17-33

Somewhat embarrassed, the student quickly connects the source to ground so that $V_{GS} = V_b$ and a reasonable bias current can be established (Fig. 17.60). Explain why "haste makes waste."

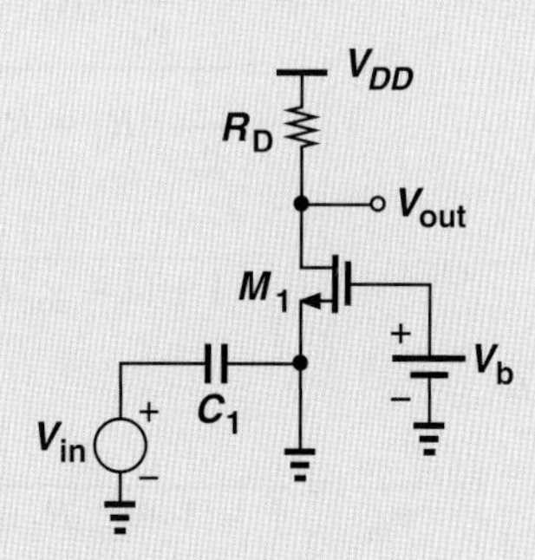

Figure 17.60 CG stage with source shorted to ground.

Solution The student has shorted the *signal* to ac ground. That is, the source voltage is equal to zero regardless of the value of v_{in}, yielding $v_{out} = 0$.

Exercise Does a signal current flow through C_1?

The foregoing examples imply that the source can remain neither open nor shorted to ground, thereby requiring some bias element. Shown in Fig. 17.61(a) is an example where R_1 provides a path for the bias current at the cost of lowering the input impedance. We recognize that R_{in} now consists of two *parallel* components: (1) $1/g_m$, seen looking "up" into the source and (2) R_1, seen looking "down." Thus,

$$R_{in} = \frac{1}{g_m}||R_1. \tag{17.164}$$

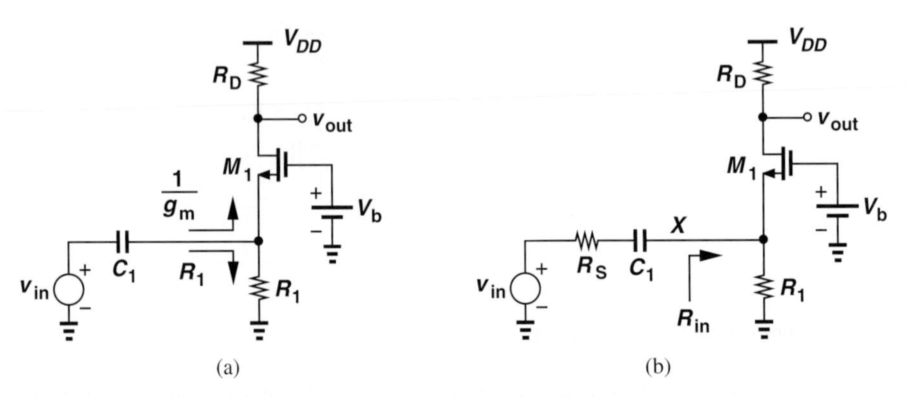

Figure 17.61 (a) CG stage with biasing, (b) inclusion of signal generator resistance.

The reduction in R_{in} manifests itself if the signal generator exhibits a finite output resistance. Depicted in Fig. 17.61(b), such a circuit attenuates the signal, lowering the overall voltage gain. Following the analysis illustrated in Fig. 17.55, we can write

$$\frac{v_X}{v_{in}} = \frac{R_{in}}{R_{in} + R_S} \tag{17.165}$$

$$= \frac{\frac{1}{g_m}||R_1}{\frac{1}{g_m}||R_1 + R_S} \tag{17.166}$$

$$= \frac{1}{1 + (1 + g_m R_1)R_S/R_1}. \tag{17.167}$$

Since $v_{out}/v_X = g_m R_D$,

$$\frac{v_{out}}{v_{in}} = \frac{1}{1 + (1 + g_m R_1)R_S/R_1} \cdot g_m R_D. \tag{17.168}$$

As usual, we have preferred solution by inspection over drawing the small-signal equivalent.

The reader may see a contradiction in our thoughts: on the one hand, we view the low input impedance of the CG stage a *useful* property; on the other hand, we consider the reduction of the input impedance due to R_1 *undesirable*. To resolve this apparent contradiction, we must distinguish between the two components $1/g_m$ and R_1, noting that the latter shunts the input source current to ground, thus "wasting" the signal. As shown in Fig. 17.62, i_{in} splits two ways, with only i_2 reaching R_D and contributing to the output signal.

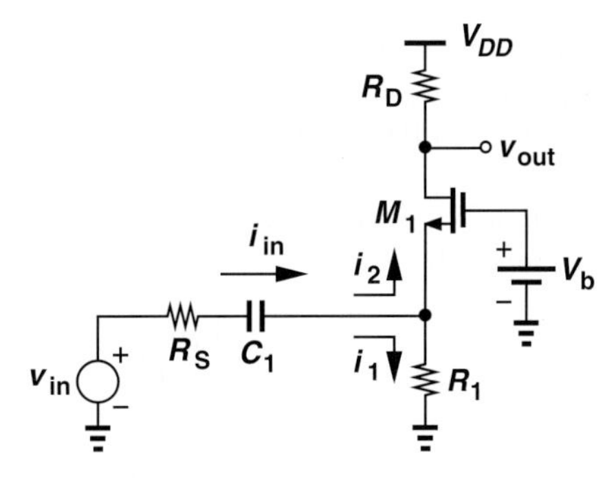

Figure 17.62 Small-signal input current components in a CG stage.

If R_1 decreases while $1/g_m$ remains constant, then i_2 also falls.[10] Thus, reduction of R_{in} due to R_1 is undesirable. By contrast, if $1/g_m$ decreases while R_1 remains constant, then i_2 rises. For R_1 to affect the input impedance negligibly, we must have

$$R_1 \gg \frac{1}{g_m}. \tag{17.169}$$

How is the gate voltage, V_b, generated? We can employ a resistive divider similar to that used in the CS stage. Shown in Fig. 17.63 is such a topology.

Example 17-34

Design the common-gate stage of Fig. 17.63 for the following parameters: $v_{out}/v_{in} = 5$, $R_S = 0$, $R_1 = 500\,\Omega$, $1/g_m = 50\,\Omega$, power budget = 2 mW, $V_{DD} = 1.8$ V. Assume $\mu_n C_{ox} = 100\,\mu\text{A/V}^2$, $V_{TH} = 0.5$ V, and $\lambda = 0$.

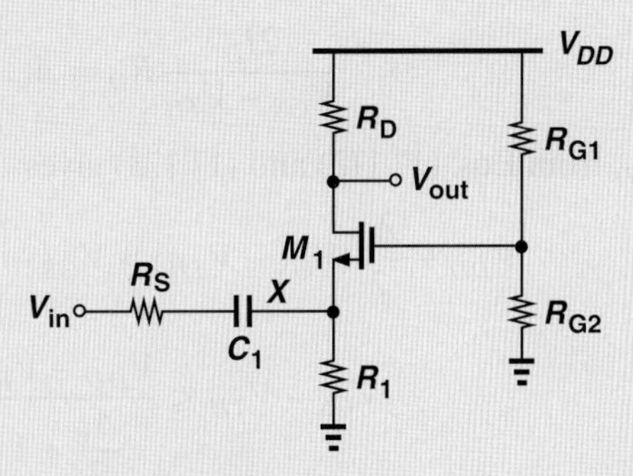

Figure 17.63 CG stage with gate bias network.

Solution From the power budget, we obtain a total supply current of 1.11 mA. Allocating 10 μA to the voltage divider, R_{G1} and R_{G2}, we leave 1.1 mA for the drain current of M_1. Thus, the voltage drop across R_1 is equal to 550 mV.

We must now compute two interrelated parameters: W/L and R_D. A larger value of W/L yields a greater g_m, allowing a lower value of R_D. As in Example 17-28, we choose an initial value for V_{GS} to arrive at a reasonable guess for W/L. For example, if $V_{GS} = 0.8$ V, then $W/L = 244$, and $g_m = 2I_D/(V_{GS} - V_{TH}) = (136.4\,\Omega)^{-1}$, dictating $R_D = 682\,\Omega$ for $v_{out}/v_{in} = 5$.

[10] In the extreme case, $R_1 = 0$ (Example 17-33) and $i_2 = 0$.

Let us determine whether M_1 operates in saturation. The gate voltage is equal to V_{GS} plus the drop across R_1, amounting to 1.35 V. On the other hand, the drain voltage is given by $V_{DD} - I_D R_D = 1.05$ V. Since the drain voltage exceeds $V_G - V_{TH}$, M_1 is indeed in saturation.

The resistive divider consisting of R_{G1} and R_{G2} must establish a gate voltage equal to 1.35 V while drawing 10 μA:

$$\frac{V_{DD}}{R_{G1} + R_{G2}} = 10\ \mu\text{A} \tag{17.170}$$

$$\frac{R_{G2}}{R_{G1} + R_{G2}} V_{DD} = 1.35\ \text{V}. \tag{17.171}$$

It follows that $R_{G1} = 45$ kΩ and $R_{G2} = 135$ kΩ.

Exercise If W/L cannot exceed 100, what voltage gain can be achieved?

Example 17-35 Suppose in Example 17-34, we wish to minimize W/L (and hence transistor capacitances). What is the minimum acceptable value of W/L?

Solution For a given I_D, as W/L decreases, $V_{GS} - V_{TH}$ increases. Thus, we must first compute the maximum allowable V_{GS}. We impose the condition for saturation as

$$V_{DD} - I_D R_D > V_{GS} + V_{R1} - V_{TH}, \tag{17.172}$$

where V_{R1} denotes the voltage drop across R_1, and set $g_m R_D$ to the required gain:

$$\frac{2I_D}{V_{GS} - V_{TH}} R_D = A_v. \tag{17.173}$$

Eliminating R_D from Eqs. (17.172) and (17.173) gives:

$$V_{DD} - \frac{A_v}{2}(V_{GS} - V_{TH}) > V_{GS} - V_{TH} + V_{R1} \tag{17.174}$$

and hence

$$V_{GS} - V_{TH} < \frac{V_{DD} - V_{R1}}{\dfrac{A_v}{2} + 1}. \tag{17.175}$$

In other words,

$$W/L > \frac{2I_D}{\mu_n C_{ox}\left(2\dfrac{V_{DD} - V_{R1}}{A_v + 2}\right)^2}. \tag{17.176}$$

It follows that

$$W/L > 172.5. \tag{17.177}$$

Exercise Repeat the above example for $A_v = 10$.

17.4.5 Source Follower

Another important circuit topology is the source follower (also called the "common-drain" stage). The reader is encouraged to review Examples 17-2–17-3, the rules illustrated in Fig. 17.7, and the possible topologies in Fig. 17.22 before proceeding further. For the sake of brevity, we may also use the term "follower" to refer to source followers in this chapter.

Shown in Fig. 17.64, the source follower senses the input at the gate of the transistor and produces the output at the source. The drain is tied to V_{DD} and hence ac ground. We first study the core and subsequently add the biasing elements.

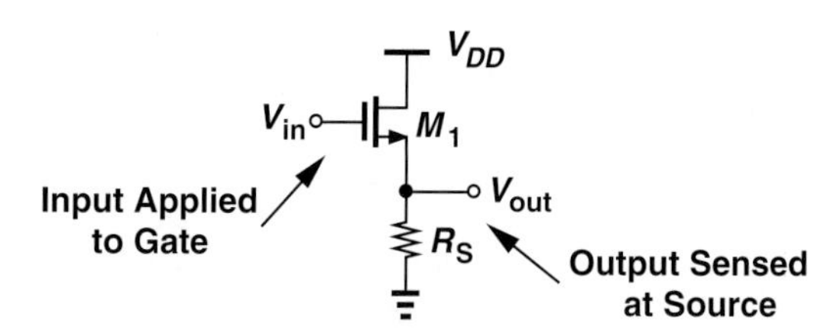

Figure 17.64 Source follower.

Source Follower Core How does the follower in Fig. 17.65(a) respond to a change in V_{in}? If V_{in} rises by a small amount ΔV_{in}, the gate-source voltage of M_1 tends to increase, raising the drain current. The higher drain current translates to a greater drop across R_S and hence a *higher* V_{out}. (If we mistakenly assume that V_{out} falls, then V_{GS} must rise and so must I_D, requiring that V_{out} go up.) Since V_{out} changes in the same direction as V_{in}, we expect the voltage gain to be positive. Note that V_{out} is always lower than V_{in} by an amount equal to V_{GS}, and the circuit is said to provide "level shift."

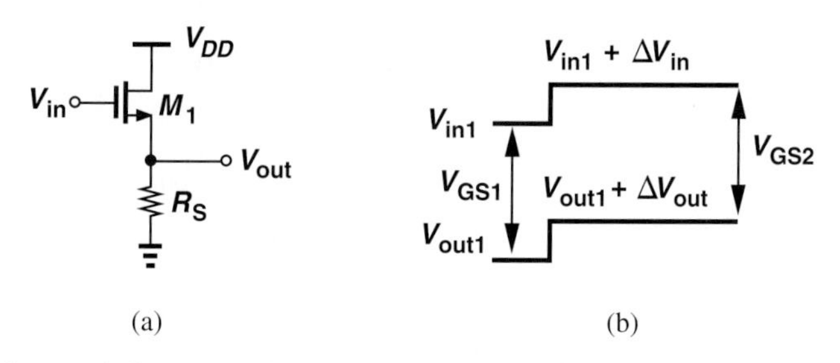

Figure 17.65 (a) Source follower sensing an input change, (b) response of the circuit.

Another interesting and important observation here is that the change in V_{out} cannot be larger than the change in V_{in}. Suppose V_{in} increases from V_{in1} to $V_{in1} + \Delta V_{in}$ and V_{out} from V_{out1} to $V_{out1} + \Delta V_{out}$ [Fig. 17.65(b)]. If the output changes by a *greater* amount than the input, $\Delta V_{out} > \Delta V_{in}$, then V_{GS2} must be *less* than V_{GS1}. But this means the drain current also decreases and so does $I_D R_S = V_{out}$, contradicting the assumption that V_{out} has increased. Thus, $\Delta V_{out} < \Delta V_{in}$, implying that the follower exhibits a voltage gain less than unity.[11]

The reader may wonder if an amplifier with a sub-unity gain has any practical value. As explained later, the input and output impedances of the source follower make it a particularly useful circuit for some applications.

[11] In an extreme case described later, the gain becomes equal to unity.

Let us now derive the small-signal properties of the follower, first assuming $\lambda = 0$. Shown in Fig. 17.66, the equivalent circuit yields

$$g_m v_1 = \frac{v_{out}}{R_S}. \tag{17.178}$$

We also have $v_{in} = v_1 + v_{out}$. Thus,

$$\frac{v_{out}}{v_{in}} = \frac{R_S}{R_S + \frac{1}{g_m}}. \tag{17.179}$$

The voltage gain is therefore positive and less than unity.

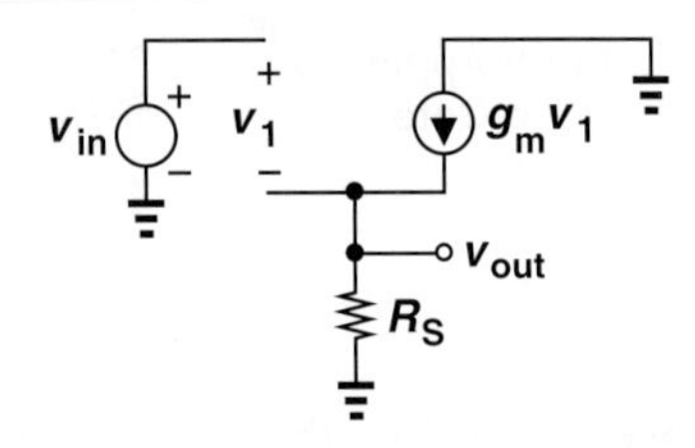

Figure 17.66 Small-signal model of source follower.

Example 17-36 In integrated circuits, the follower is typically realized as shown in Fig. 17.67. Determine the voltage gain if the current source is ideal and $V_A = \infty$.

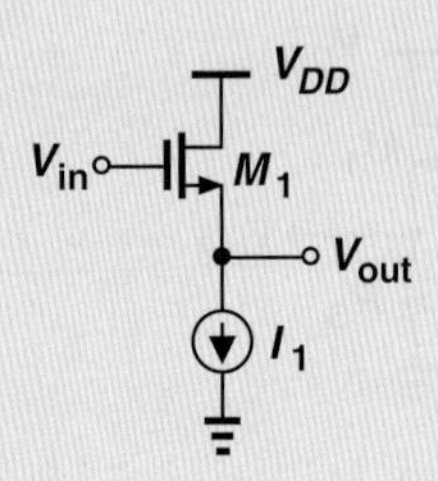

Figure 17.67 Follower with current source.

Solution Since the source resistor is replaced with an ideal current source, the value of R_S in Eq. (17.179) must tend to infinity, yielding

$$A_v = 1. \tag{17.180}$$

This result can also be derived intuitively. A constant current source flowing through M_1 requires that V_{GS} remain constant. Writing $V_{out} = V_{in} - V_{GS}$, we recognize that V_{out} exactly *follows* V_{in} if V_{GS} is constant.

Exercise Calculate the gain if $\lambda \neq 0$.

Equation (17.179) suggests that the source follower acts as a voltage divider, a perspective that can be reinforced by an alternative analysis. Suppose, as shown in Fig. 17.68(a), we wish to model v_{in} and M_1 by a Thevenin equivalent. The Thevenin voltage is given by the open-circuit output voltage produced by M_1 [Fig. 17.68(b)], as if M_1 operates with $R_S = \infty$ (Example 17-36). Thus, $v_{Thev} = v_{in}$. The Thevenin resistance is obtained by setting the input to zero [Fig. 17.68(c)] and is equal to $1/g_m$. The circuit of Fig. 17.68(a) therefore reduces to that shown in Fig. 17.68(d), confirming operation as a voltage divider.

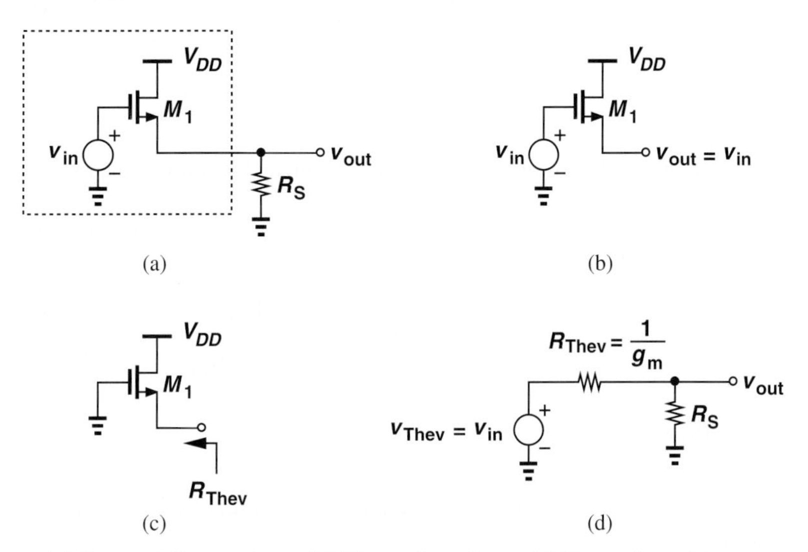

Figure 17.68 (a) Source follower stage, (b) Thevenin voltage, (c) Thevenin resistance, (d) simplified circuit.

Figure 17.69(a) depicts the small-signal equivalent of the source follower, including channel-length modulation. Recognizing that r_O appears in parallel with R_S and viewing the circuit as a voltage divider [Fig. 17.69(b)], we have

$$\frac{v_{out}}{v_{in}} = \frac{r_O||R_S}{\frac{1}{g_m} + r_O||R_S}. \tag{17.181}$$

It is desirable to maximize R_S and r_O.

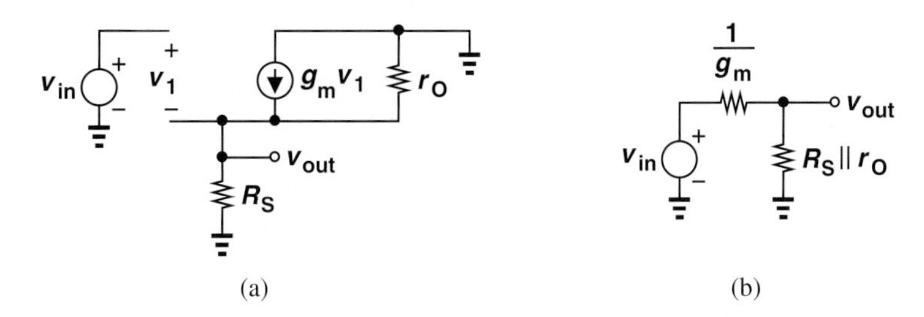

Figure 17.69 (a) Small-signal equivalent of source follower, (b) simplified circuit.

Example 17-37 A source follower is realized as shown in Fig. 17.70(a), where M_2 serves as a current source. Calculate the voltage gain of the circuit.

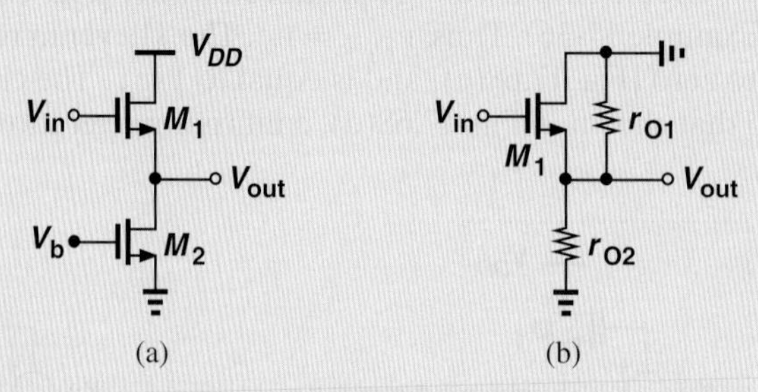

Figure 17.70 (a) Follower with ideal current source, (b) simplified circuit.

Solution Since M_2 simply presents an impedance of r_{O2} from the output node to ac ground [Fig. 17.70(b)], we substitute $R_S = r_{O2}$ in Eq. (17.181):

$$A_v = \frac{r_{O1}||r_{O2}}{\frac{1}{g_{m1}} + r_{O1}||r_{O2}}. \tag{17.182}$$

If $r_{O1}||r_{O2} \gg 1/g_{m1}$, then $A_v \approx 1$.

Exercise Repeat the above example if a resistance of value R_S is placed in series with the source of M_2.

Example 17-38 Design a source follower to drive a 50-Ω load with a voltage gain of 0.5 and a power budget of 10 mW. Assume $\mu_n C_{ox} = 100\ \mu A/V^2$, $V_{TH} = 0.5$ V, $\lambda = 0$, and $V_{DD} = 1.8$ V.

Solution With $R_S = 50\ \Omega$ and $r_O = \infty$ in Fig. 17.64, we have

$$A_v = \frac{R_S}{\frac{1}{g_m} + R_S} = 0.5 \tag{17.183}$$

and hence

$$g_m = \frac{1}{50\ \Omega}. \tag{17.184}$$

The power budget and supply voltage yield a maximum supply current of 5.56 mA. Using this value for I_D in $g_m = \sqrt{2\mu_n C_{ox}(W/L)I_D}$ gives

$$W/L = 360. \tag{17.185}$$

Exercise What voltage gain can be achieved if the power budget is raised to 15 mW?

In order to appreciate the usefulness of source followers, let us compute their input and output impedances. The input impedance is very high at low frequencies, making the circuit a good "voltmeter;" i.e., the follower can sense a voltage without disturbing it. As illustrated in Fig. 17.71, the output impedance consists of the resistance seen looking up into the source in parallel with that seen looking down into R_S. With $\lambda \neq 0$, the former is equal to $(1/g_m)||r_O$, yielding

$$R_{out} = \frac{1}{g_m}||r_O||R_S. \tag{17.186}$$

The circuit thus presents a relatively low output impedance. Followers can serve as good "buffers," e.g., between a CS stage and a low-impedance load (as in Example 17-27).

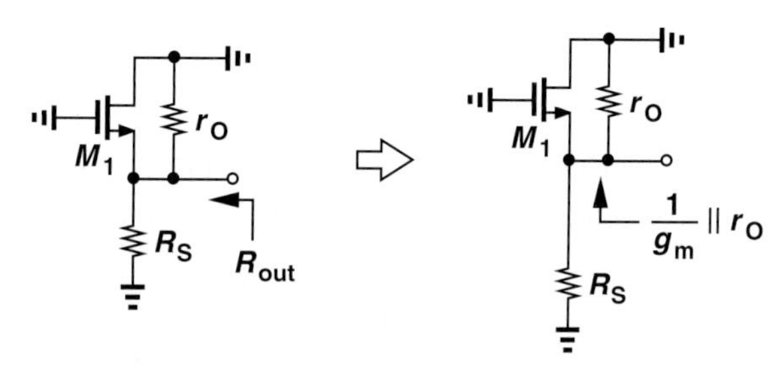

Figure 17.71 Output resistance of source follower.

Source Follower with Biasing The biasing of source followers entails defining the terminal voltages and the drain current. Figure 17.72 depicts an example where R_G establishes a dc voltage equal to V_{DD} at the gate of M_1 (why?) and R_S sets the drain bias current. Note that M_1 operates in saturation because the gate and drain voltages are equal. Also, the input impedance of the circuit has dropped from infinity to R_G.

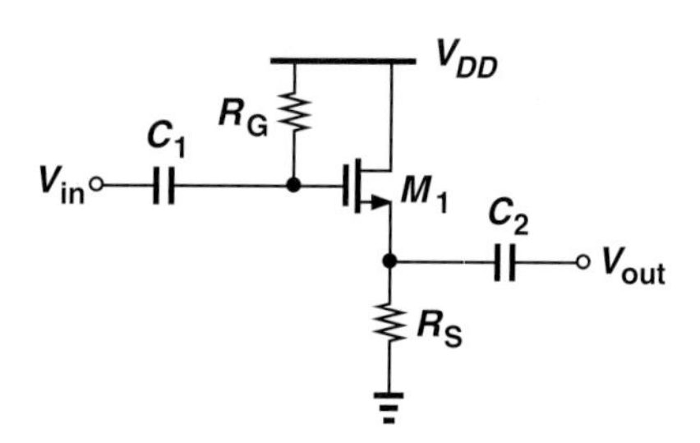

Figure 17.72 Source follower with biasing and coupling capacitors.

Let us compute the bias current of the circuit. With a zero voltage drop across R_G, we have

$$V_{GS} + I_D R_S = V_{DD}. \tag{17.187}$$

Neglecting channel-length modulation, we write

$$I_D = \frac{1}{2}\mu_n C_{ox}\frac{W}{L}(V_{GS} - V_{TH})^2 \tag{17.188}$$

$$= \frac{1}{2}\mu_n C_{ox}\frac{W}{L}(V_{DD} - I_D R_S - V_{TH})^2. \tag{17.189}$$

The resulting quadratic equation can be solved to obtain I_D.

Example 17-39 Design the source follower of Fig. 17.72 for a drain current of 1 mA and a voltage gain of 0.8. Assume $\mu_n C_{ox} = 100\ \mu\text{A/V}^2$, $V_{TH} = 0.5$ V, $\lambda = 0$, $V_{DD} = 1.8$ V, and $R_G = 50$ kΩ.

Solution The unknowns in this problem are V_{GS}, W/L, and R_S. The following three equations can be formed:

$$I_D = \frac{1}{2}\mu_n C_{ox}\frac{W}{L}(V_{GS} - V_{TH})^2 \tag{17.190}$$

$$I_D R_S + V_{GS} = V_{DD} \tag{17.191}$$

$$A_v = \frac{R_S}{\frac{1}{g_m} + R_S}. \tag{17.192}$$

If g_m is written as $2I_D/(V_{GS} - V_{TH})$, then Eqs. (17.191) and (17.192) do not contain W/L and can be solved to determine V_{GS} and R_S. We write Eq. (17.192) as

$$A_v = \frac{R_S}{\frac{V_{GS} - V_{TH}}{2I_D} + R_S} \tag{17.193}$$

$$= \frac{2I_D R_S}{V_{GS} - V_{TH} + 2I_D R_S} \tag{17.194}$$

$$= \frac{2I_D R_S}{V_{DD} - V_{TH} + I_D R_S}. \tag{17.195}$$

Thus,

$$R_S = \frac{V_{DD} - V_{TH}}{I_D}\frac{A_v}{2 - A_v} \tag{17.196}$$

$$= 867\ \Omega. \tag{17.197}$$

and

$$V_{GS} = V_{DD} - I_D R_S \tag{17.198}$$

$$= V_{DD} - (V_{DD} - V_{TH})\frac{A_v}{2 - A_v} \tag{17.199}$$

$$= 0.933\ \text{V}. \tag{17.200}$$

It follows from Eq. (17.190) that

$$\frac{W}{L} = 107. \tag{17.201}$$

Exercise What voltage gain can be achieved if W/L cannot exceed 50?

Equation (17.189) reveals that the bias current of the source follower varies with the supply voltage. To avoid this effect, integrated circuits bias the follower by means of a current source (Fig. 17.73).

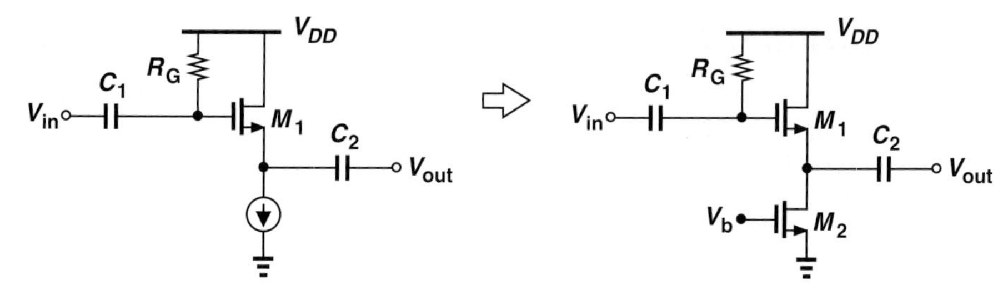

Figure 17.73 Source follower with biasing.

Robotics Application: CMOS Cameras

Vision-based robots rely on cameras to "see" the environment and act accordingly. Examples include robots that must perform inspection of parts in a factory or navigate a terrain while avoiding collision with other objects.

As explained in Chapter 1, a digital camera consists of a matrix of pixels, each of which receives light and, by means of a reverse-biased photodiode, converts it to current. How do we process this result? CMOS devices prove essential here. As shown in Fig. 17.74, we first set the voltage across the photodiode, D_1, to V_{DD} when S_0 is on. Next, we turn off S_0 and allow the carriers generated within D_1 by light to discharge C_1. This voltage change is sensed by a source follower and made available to a column that connects to other pixels. When S_j is turned on, the voltage provided by pixel j is measured, digitized, and stored by the column processing circuit. This switch then turns off and the next pixel is sensed.

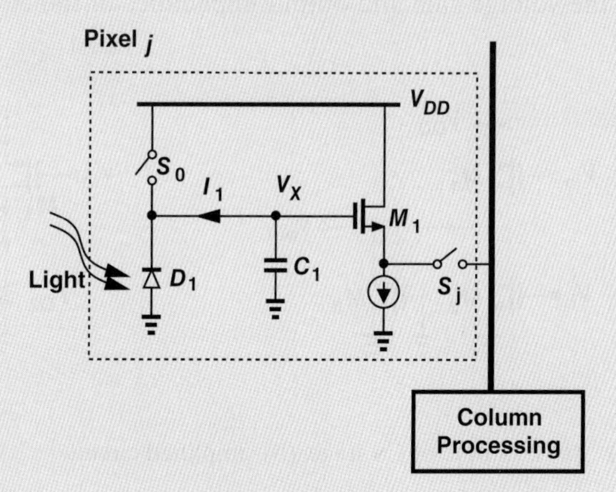

Figure 17.74 (a) Pixel electronics in a CMOS camera.

The switches and the follower greatly benefit from CMOS technology. Can we use an emitter follower instead of a source follower here? A bipolar transistor would draw its base current from C_1, discharging the capacitor and introducing an error in V_X.

17.5 ADDITIONAL EXAMPLES

This chapter has created a foundation for amplifier design, emphasizing that a proper bias point must be established to define the small-signal properties of each circuit. Depicted in Fig. 17.75, the three amplifier topologies studied here exhibit different gains and I/O impedances, each serving a specific application.

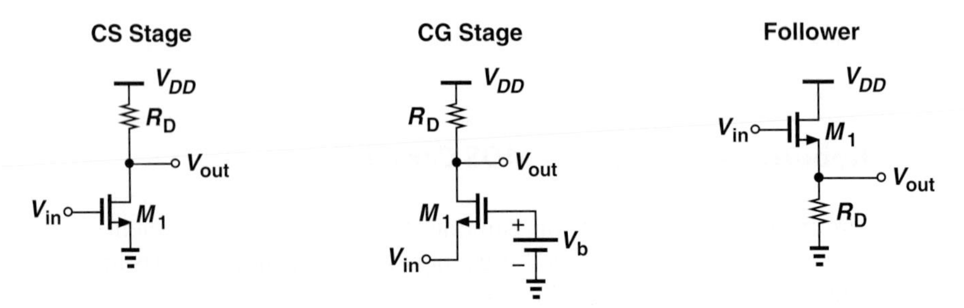

Figure 17.75 Summary of MOS amplifier topologies.

In this section, we consider a number of challenging examples, seeking to improve our circuit analysis techniques. As usual, our emphasis is on solution by inspection and hence intuitive understanding of the circuits. We assume various capacitors used in each circuit have a negligible impedance at the signal frequencies of interest.

Example 17-40 Calculate the voltage gain and output impedance of the circuit shown in Fig. 17.76(a).

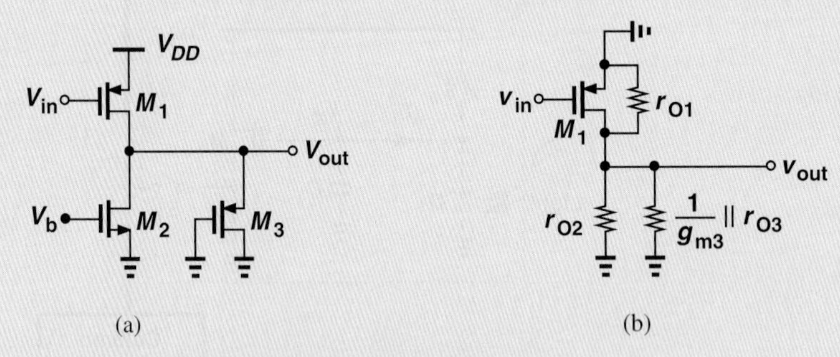

Figure 17.76 (a) Example of CS stage, (b) simplified circuit.

Solution We identify M_1 as a common-source device because it senses the input at its gate and generates the output at its drain. Transistors M_2 and M_3 therefore act as the load, with the former serving as a current source and the latter as a diode-connected device. Thus, M_2 can be replaced with a small-signal resistance equal to r_{O2}, and M_3 with another equal to $(1/g_{m3})||r_{O3}$. The circuit now reduces to that depicted in Fig. 17.76(b), yielding

$$A_v = -g_{m1}\left(\frac{1}{g_{m3}}||r_{O1}||r_{O2}||r_{O3}\right) \tag{17.202}$$

and

$$R_{out} = \frac{1}{g_{m3}}||r_{O1}||r_{O2}||r_{O3}. \qquad (17.203)$$

Note that $1/g_{m3}$ is dominant in both expressions.

Exercise Repeat the above example if M_2 is converted to a diode-connected device.

Example 17-41 Compute the voltage gain of the circuit shown in Fig. 17.77(a). Neglect channel-length modulation in M_1.

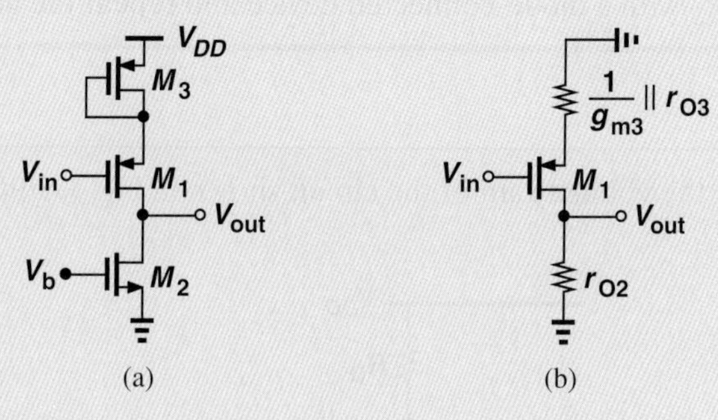

Figure 17.77 (a) Example of CS stage, (b) simplified circuit.

Solution Operating as a CS stage and degenerated by the diode-connected device M_3, transistor M_1 drives the current-source load, M_2. Simplifying the amplifier to that in Fig. 17.77(b), we have

$$A_v = -\frac{r_{O2}}{\frac{1}{g_{m1}} + \frac{1}{g_{m3}}||r_{O3}}. \qquad (17.204)$$

Exercise Repeat the above example if the gate of M_3 is tied to a constant voltage.

Example 17-42 Determine the voltage gain of the amplifiers illustrated in Fig. 17.78. For simplicity, assume $r_{O1} = \infty$ in Fig. 17.78(b).

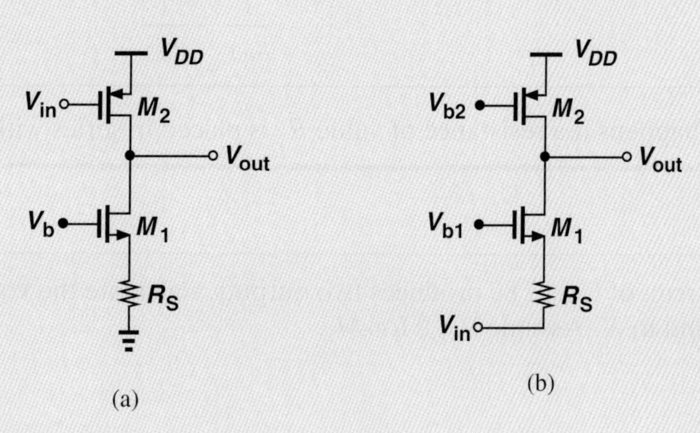

Figure 17.78 Examples of (a) CS and (b) CG stages.

Solution Degenerated by R_S, transistor M_1 in Fig. 17.78(a) presents an impedance of $(1+g_{m1}r_{O1})R_S+r_{O1}$ to the drain of M_2. Thus, the total impedance seen at the drain is equal to $[(1+g_{m1}r_{O1})R_S+r_{O1}]||r_{O2}$, giving a voltage gain of

$$A_{v1} = -g_{m2}\{[(1+g_{m1}r_{O1})R_S+r_{O1}]||r_{O1}\}. \tag{17.205}$$

In Fig. 17.78(b), M_1 operates as a common-gate stage and M_2 as the load, yielding:

$$A_{v2} = \frac{r_{O2}}{\frac{1}{g_{m1}}+R_S}. \tag{17.206}$$

Exercise Replace R_S with a diode-connected device and repeat the analysis.

Example 17-43 Calculate the voltage gain of the circuit shown in Fig. 17.79(a) if $\lambda = 0$.

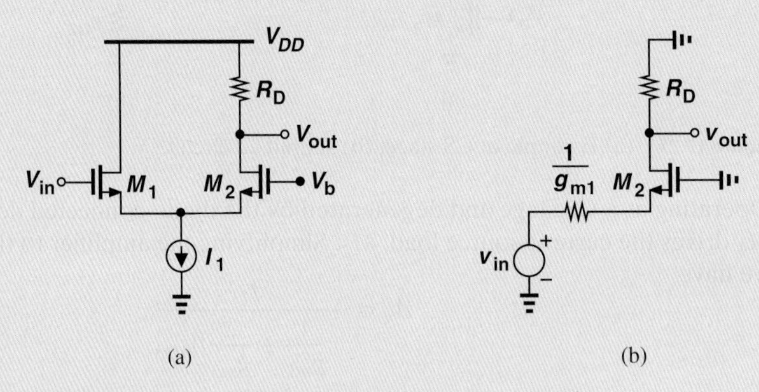

Figure 17.79 (a) Example of a composite stage, (b) simplified circuit.

Solution In this circuit, M_1 operates as a source follower and M_2 as a CG stage (why?). A simple method of analyzing the circuit is to replace v_{in} and M_1 with a Thevenin equivalent. From Fig. 17.68, we derive the model depicted in Fig. 17.79(b). Thus,

$$A_v = \frac{R_D}{\frac{1}{g_{m1}}+\frac{1}{g_{m2}}}. \tag{17.207}$$

Exercise What happens if a resistance of value R_1 is placed in series with the drain of M_1?

Example 17-44 The circuit of Fig. 17.80 produces two outputs. Calculate the voltage gain from the input to Y and to X. Assume $\lambda = 0$ for M_1.

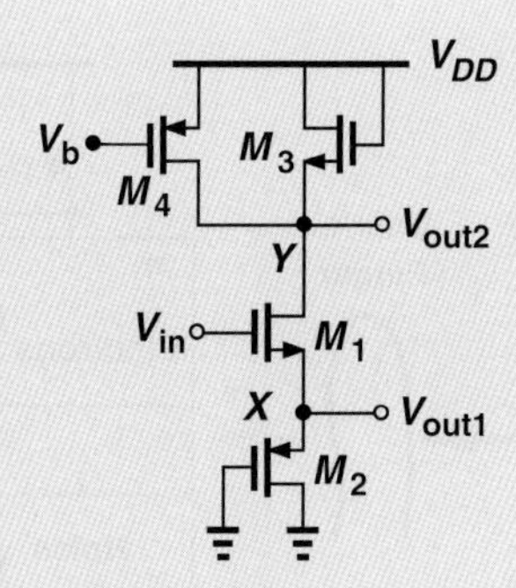

Figure 17.80 Example of composite stage.

Solution For V_{out1}, the circuit serves as a source follower. The reader can show that if $r_{O1} = \infty$, then M_3 and M_4 do not affect the source follower operation. Exhibiting a small-signal impedance of $(1/g_{m2})||r_{O2}$, transistor M_2 acts as a load for the follower, yielding from Eq. (17.181)

$$\frac{v_{out1}}{v_{in}} = \frac{\frac{1}{g_{m2}}||r_{O2}}{\frac{1}{g_{m2}}||r_{O2} + \frac{1}{g_{m1}}}. \tag{17.208}$$

For V_{out2}, M_1 operates as a degenerated CS stage with a drain load consisting of the diode-connected device M_3 and the current source M_4. This load impedance is equal to $(1/g_{m3})||r_{O3}||r_{O4}$, resulting in

$$\frac{v_{out2}}{v_{in}} = -\frac{\frac{1}{g_{m3}}||r_{O3}||r_{O4}}{\frac{1}{g_{m1}} + \frac{1}{g_{m2}}||r_{O2}}. \tag{17.209}$$

Exercise Which one of the two gains is higher? Explain intuitively why.

Bioengineering Application: Health Monitoring

Various devices have been introduced that can continuously monitor and record a user's vital signs. Such gathering of data proves useful for any individual: a doctor looking over several *years* of a patient's health history can more easily make a diagnosis than by simply asking the patient some questions.

It is possible to monitor a user's heart rate, blood pressure, and oxygen saturation by examining the blood flow in his/her finger. As illustrated in Fig. 17.81, an LED transmits light through the finger, and an array of photodiodes collects the light on the other side [1]. The intensity of this light has a large, relatively constant component (which we can view as a "bias" value) and a small, time-varying one that results from blood flow in the vessels.

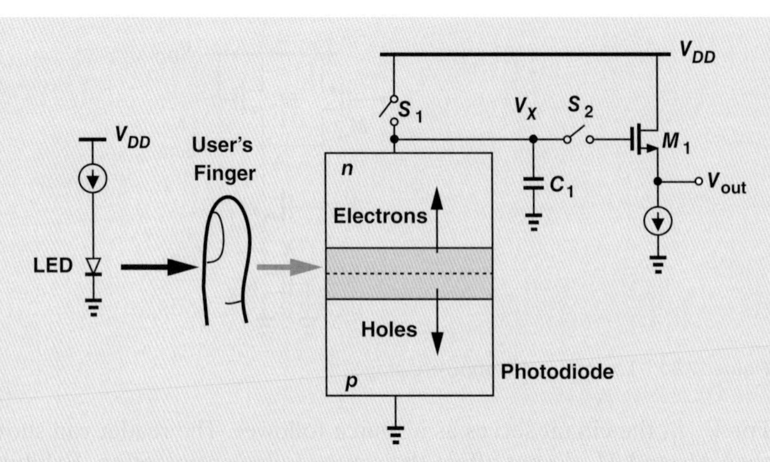

Figure 17.81 Monitoring vital signs by examining blood flow in the user's finger.

The circuit operates as follows [1]. First, S_1 is on, precharging the photodiode capacitance to V_{DD}. Next, S_1 turns off, S_2 turns on, and the photodiode current discharges the capacitance. The change in V_X is proportional to the light intensity and is sensed by the source follower. Tens of these photodiodes and their associated circuits are necessary to capture a sufficiently large area of the finger.

17.6 CHAPTER SUMMARY

- The impedances seen looking into the gate, drain, and source of a MOSFET are equal to infinity, r_O (with source grounded), and $1/g_m$ (with gate grounded), respectively.
- In order to obtain the required small-signal MOS parameters such as g_m and r_O, the transistor must be "biased," i.e., carry a certain drain current and sustain certain gate-source and drain-source voltages. Signals simply perturb these conditions.
- Biasing techniques establish the required gate voltage by means of a resistive path to the supply rails or the output node (self-biasing).
- With a single transistor, only three amplifier topologies are possible: common-source and common-gate stages and source followers.
- The CS stage provides a moderate voltage gain, a high input impedance, and a moderate output impedance.
- Source degeneration improves the linearity but lowers the voltage gain.
- Source degeneration raises the output impedance of CS stages considerably.
- The CG stage provides a moderate voltage gain, a low input impedance, and a moderate output impedance.
- The voltage gain expressions for CS and CG stages are similar but for a sign.
- The source follower provides a voltage gain less than unity, a high input impedance, and a low output impedance, serving as a good voltage buffer.

PROBLEMS

In the following problems, unless otherwise stated, assume $\mu_n C_{ox} = 200\ \mu\text{A/V}^2$, $\mu_p C_{ox} = 100\ \mu\text{A/V}^2$, $\lambda = 0$, and $V_{TH} = 0.4$ V for NMOS devices and -0.4 V for PMOS devices.

Sec. 17.2.1 Simple Biasing

17.1. In the circuit of Fig. 17.82, determine the maximum allowable value of W/L if M_1 must remain in saturation. Assume $\lambda = 0$.

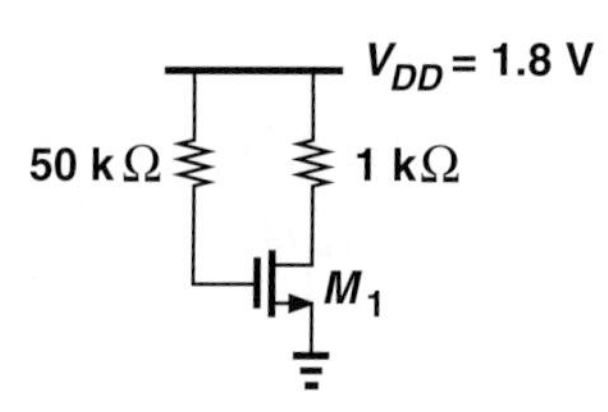

Figure 17.82

17.2. We wish to design the circuit of Fig. 17.83 for a drain current of 1 mA. If $W/L = 20/0.18$, compute R_1 and R_2 such that the input impedance is at least 20 kΩ.

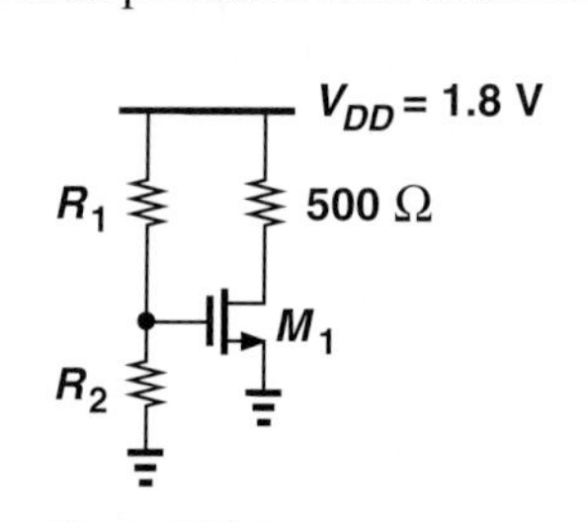

Figure 17.83

Sec. 17.2.2 Biasing with Source Degeneration

17.3. Consider the circuit shown in Fig. 17.84. Calculate the maximum transconductance that M_1 can provide (without going into the triode region).

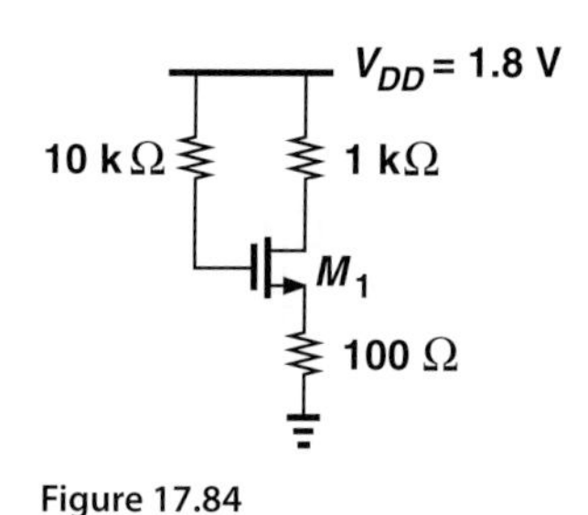

Figure 17.84

17.4. The circuit of Fig. 17.85 must be designed for a voltage drop of 200 mV across R_S. SOL

(a) Calculate the minimum allowable value of W/L if M_1 must remain in saturation.

(b) What are the required values of R_1 and R_2 if the input impedance must be at least 30 kΩ?

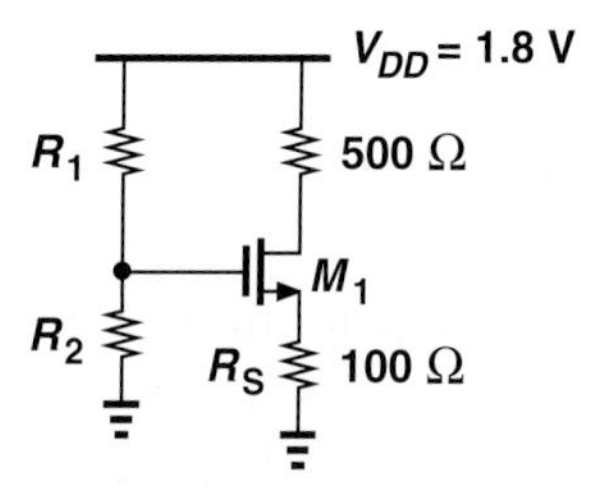

Figure 17.85

17.5. Consider the circuit depicted in Fig. 17.86, where $W/L = 20/0.18$. Assuming the current flowing through R_2 is one-tenth of I_{D1}, calculate the values of R_1 and R_2 so that $I_{D1} = 0.5$ mA.

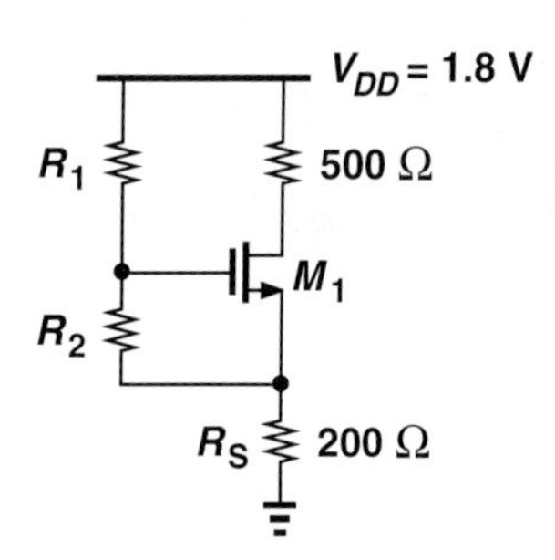

Figure 17.86

Sec. 17.2.3 Self-Biased Stage

17.6. The self-biased stage of Fig. 17.87 must be designed for a drain current of 1 mA. If M_1 is to provide a transconductance of $1/(100\ \Omega)$, calculate the required value of R_D.

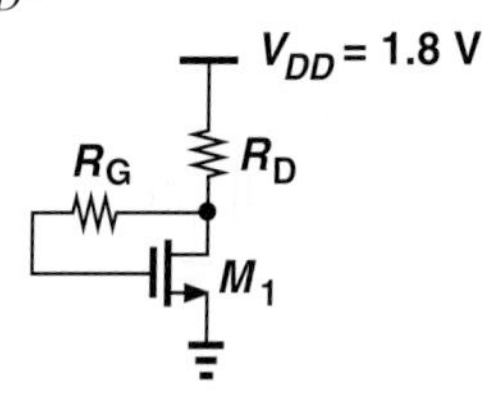

Figure 17.87

17.7. We wish to design the stage in Fig. 17.88 for a drain current of 0.5 mA. If $W/L = 50/0.18$, calculate the values of R_1 and R_2 such that these resistors carry a current equal to one-tenth of I_{D1}.

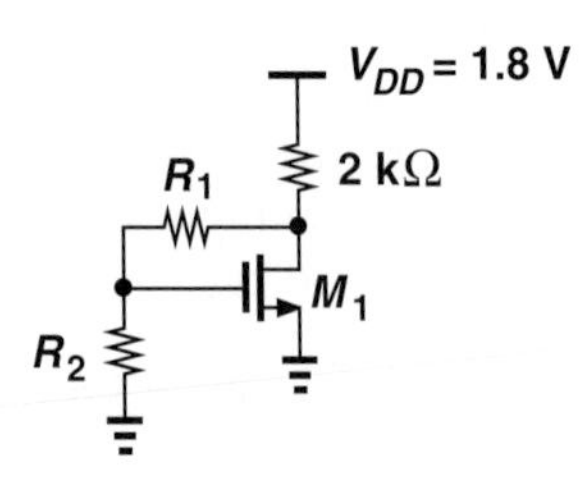

Figure 17.88

17.8. Due to a manufacturing error, a parasitic resistor, R_P, has appeared in the circuit of Fig. 17.89. We know that circuit samples free from this error exhibit $V_{GS} = V_{DS}$ whereas defective samples exhibit $V_{GS} = V_{DS} + V_{TH}$. Determine the values of W/L and R_P.

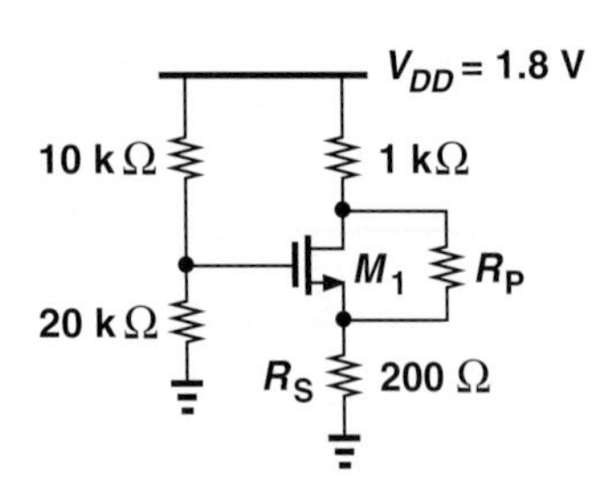

Figure 17.89

***17.9.** Due to a manufacturing error, a parasitic resistor, R_P, has appeared in the circuit of Fig. 17.90. We know that circuit samples free from this error exhibit $V_{GS} = V_{DS} + 100$ mV whereas defective samples exhibit $V_{GS} = V_{DS} + 50$ mV. Determine the values of W/L and R_P.

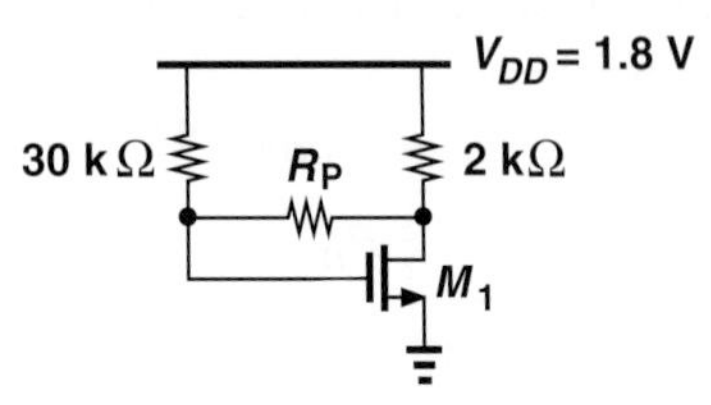

Figure 17.90

Sec. 17.2.5 Realization of Current Sources

17.10. An NMOS current source must be designed for an output resistance of 20 kΩ and an output current of 0.5 mA. What is the maximum tolerable value of λ? SOL

17.11. In the circuit of Fig. 17.91, M_1 and M_2 have lengths equal to 0.25 μm and $\lambda = 0.1\ \text{V}^{-1}$. Determine W_1 and W_2 such that $I_X = 2I_Y = 1$ mA. Assume $V_{DS1} = V_{DS2} = V_B = 0.8$ V. What is the output resistance of each current source?

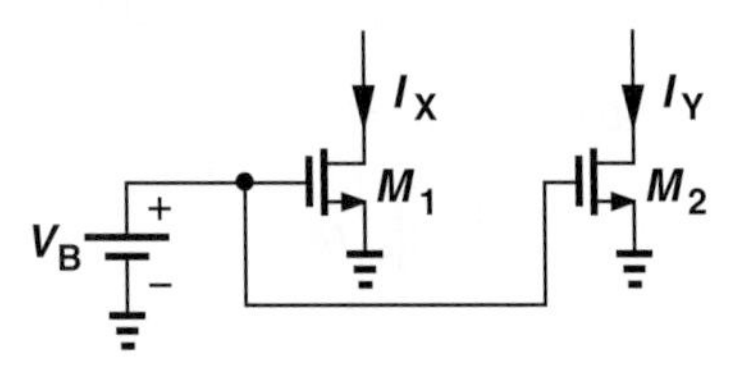

Figure 17.91

17.12. The two current sources in Fig. 17.92 must be designed for $I_X = I_Y = 0.5$ mA. If $V_{B1} = 1$ V, $V_{B2} = 1.2$ V, $\lambda = 0.1\ \text{V}^{-1}$, and $L_1 = L_2 = 0.25$ μm, calculate W_1 and W_2. Compare the output resistances of the two current sources.

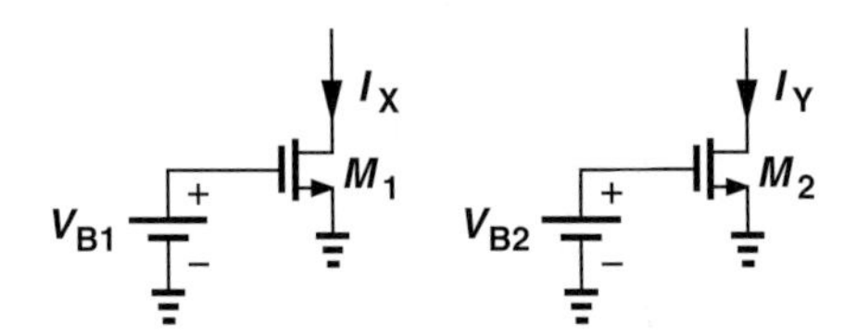

Figure 17.92

17.13. A student mistakenly uses the circuit of Fig. 17.93 as a current source. If $W/L = 10/0.25$, $\lambda = 0.1\ \text{V}^{-1}$, $V_{B1} = 0.2$ V, and V_X has a dc level of 1.2 V, calculate the impedance seen at the source of M_1.

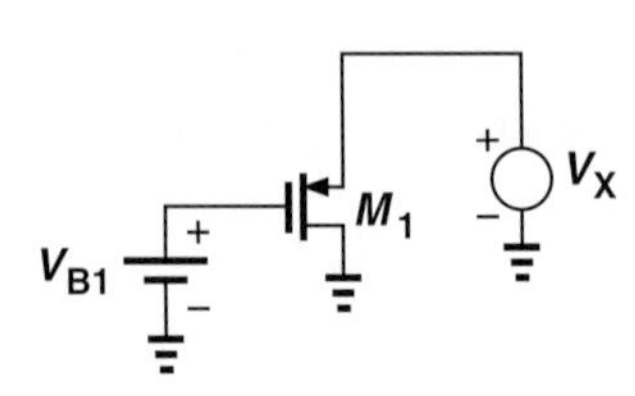

Figure 17.93

17.14. Consider the circuit shown in Fig. 17.94, where $(W/L)_1 = 10/0.18$ and $(W/L)_2 = 30/0.18$. if $\lambda = 0.1\ V^{-1}$, calculate V_B such that $V_X = 0.9$ V.

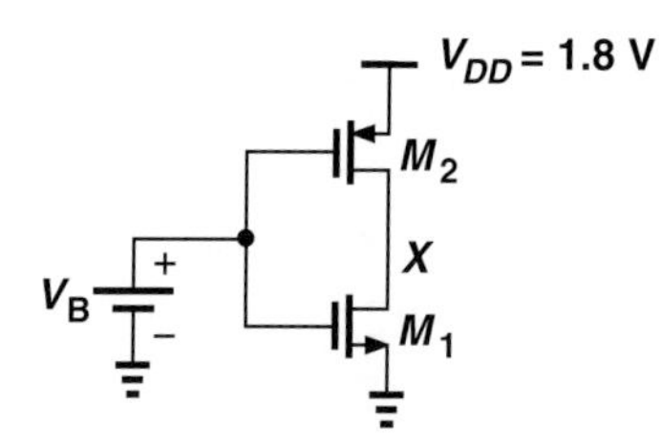

Figure 17.94

17.15. In the circuit of Fig. 17.95, M_1 and M_2 serve as current sources. Calculate I_X and I_Y if $V_B = 1$ V and $W/L = 20/0.25$. How are the output resistances of M_1 and M_2 related?

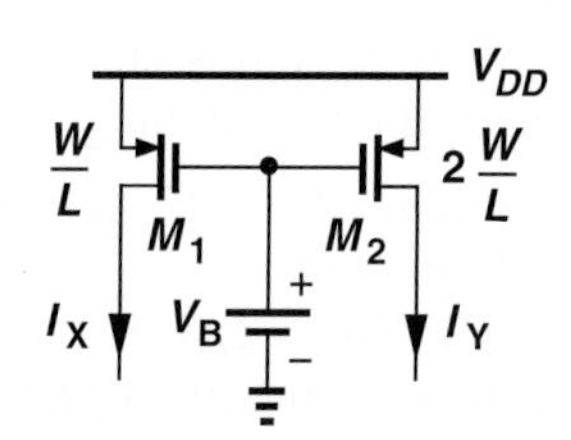

Figure 17.95

*__17.16.__ In the circuit of Fig. 17.96, $(W/L)_1 = 5/0.18$, $(W/L)_2 = 10/0.18$, $\lambda_1 = 0.1\ V^{-1}$, and $\lambda_2 = 0.15\ V^{-1}$.

(a) Determine V_B such that $I_{D1} = |I_{D2}| = 0.5$ mA for $V_X = 0.9$ V.

(b) Now sketch I_X as a function of V_X as V_X goes from 0 to V_{DD}.

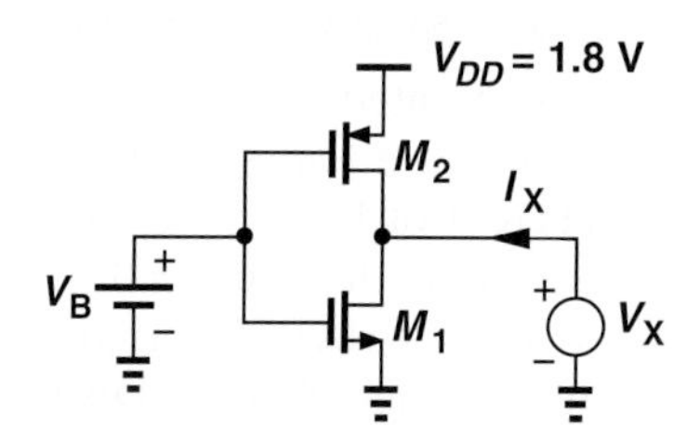

Figure 17.96

Sec. 17.4 Common-Source Stage

SOL **17.17.** In the common-source stage of Fig. 17.97, $W/L = 30/0.18$ and $\lambda = 0$.

(a) What gate voltage yields a drain current of 0.5 mA? (Verify that M_1 operates in saturation.)

(b) With such a drain bias current, calculate the voltage gain of the stage.

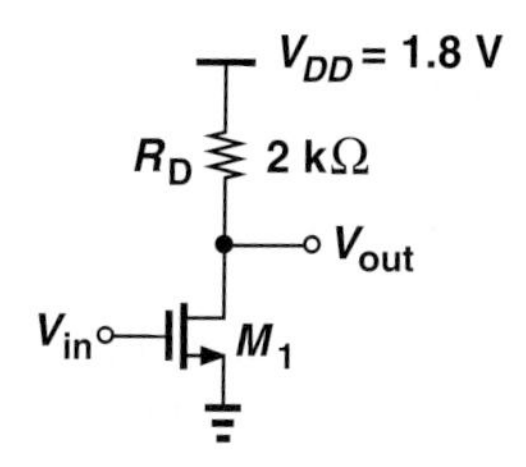

Figure 17.97

17.18. The circuit of Fig. 17.97 is designed with $W/L = 20/0.18$, $\lambda = 0$, and $I_D = 0.25$ mA.

(a) Compute the required gate bias voltage.

(b) With such a gate voltage, how much can W/L be increased while M_1 remains in saturation? What is the maximum voltage gain that can be achieved as W/L increases?

17.19. We wish to design the stage of Fig. 17.98 for a voltage gain of 5 with $W/L \leq 20/0.18$. Determine the required value of R_D if the power dissipation must not exceed 1 mW.

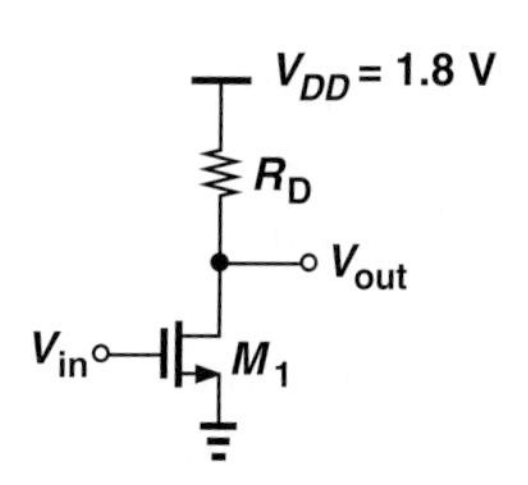

Figure 17.98

17.20. The CS stage of Fig. 17.99 must provide a voltage gain of 10 with a bias current of 0.5 mA. Assume $\lambda_1 = 0.1\ V^{-1}$, and $\lambda_2 = 0.15\ V^{-1}$.

(a) Compute the required value of $(W/L)_1$.

(b) If $(W/L)_2 = 20/0.18$, calculate the required value of V_B.

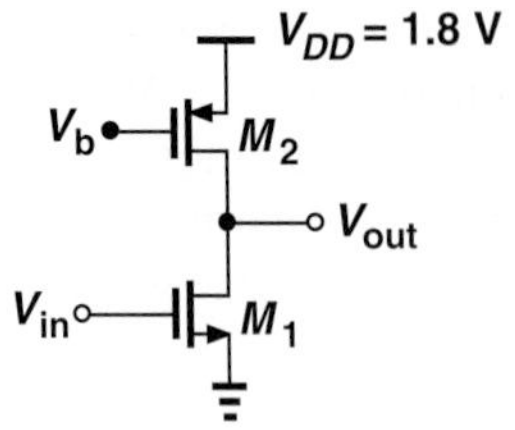

Figure 17.99

17.21. In the stage of Fig. 17.99, M_2 has a long length so that $\lambda_2 \ll \lambda_1$. Calculate the voltage gain if $\lambda_1 = 0.1\ \mathrm{V}^{-1}$, $(W/L)_1 = 20/0.18$, and $I_D = 1$ mA.

****17.22.** The circuit of Fig. 17.99 is designed for a bias current of I_1 with certain dimensions for M_1 and M_2. If the width and the length of both transistors are doubled, how does the voltage gain change? Consider two cases: (a) the bias current remains constant, or (b) the bias current is doubled.

17.23. The CS stage depicted in Fig. 17.100 must achieve a voltage gain of 15 at a bias current of 0.5 mA. If $\lambda_1 = 0.15\ \mathrm{V}^{-1}$ and $\lambda_2 = 0.05\ \mathrm{V}^{-1}$, determine the required value of $(W/L)_2$.

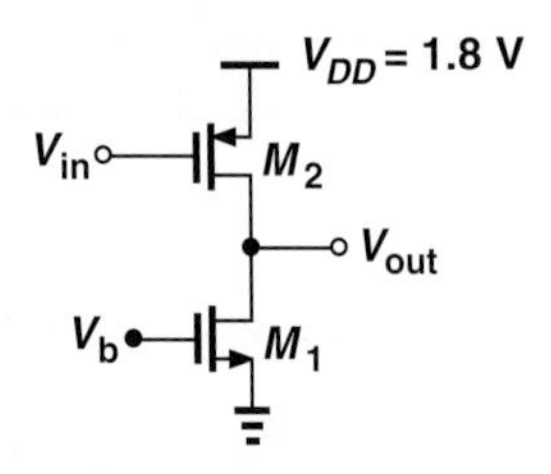

Figure 17.100

17.24. Explain which one of the topologies shown in Fig. 17.101 is preferred.

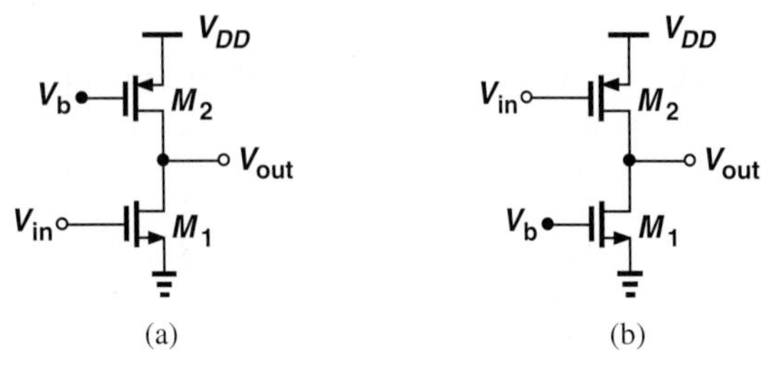

Figure 17.101

17.25. We wish to design the circuit shown in Fig. 17.102 for a voltage gain of 3. If $(W/L)_1 = 20/0.18$, determine $(W/L)_2$. Assume $\lambda = 0$.

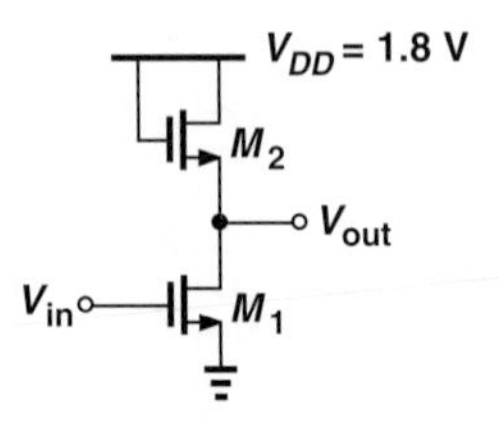

Figure 17.102

17.26. In the circuit of Fig. 17.102, $(W/L)_1 = 10/0.18$ and $I_{D1} = 0.5$ mA.

(a) If $\lambda = 0$, determine $(W/L)_2$ such that M_1 operates at the edge of saturation.

(b) Now calculate the voltage gain.

(c) Explain why this choice of $(W/L)_2$ yields the maximum gain.

17.27. The CS stage of Fig. 17.102 must achieve a voltage gain of 5. SOL

(a) If $(W/L)_2 = 2/0.18$, compute the required value of $(W/L)_1$.

(b) What is the maximum allowable bias current if M_1 must operate in saturation?

***17.28.** If $\lambda \neq 0$, determine the voltage gain of the stages shown in Fig. 17.103.

Sec. 17.4.3 CS Stage with Source Degeneration

17.29. In the circuit of Fig. 17.104, determine the gate voltage such that M_1 operates at the edge of saturation. Assume $\lambda = 0$.

17.30. The degenerated CS stage of Fig. 17.104 must provide a voltage gain of 4 with a bias current of 1 mA. Assume a drop of 200 mV across R_S and $\lambda = 0$.

(a) If $R_D = 1\ \mathrm{k\Omega}$, determine the required value of W/L. Does the transistor operate in saturation for this choice of W/L?

(b) If $W/L = 50/0.18$, determine the required value of R_D. Does the transistor operate in saturation for this choice of R_D?

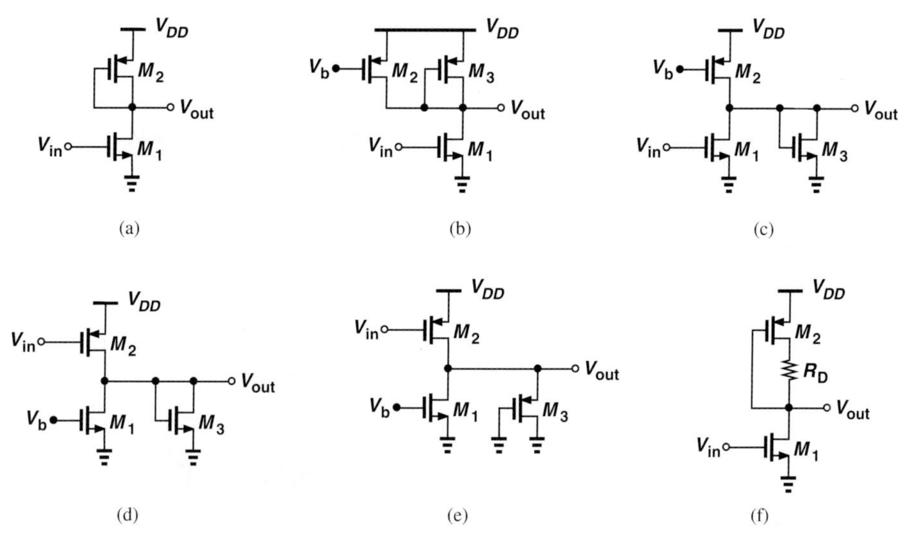

Figure 17.103

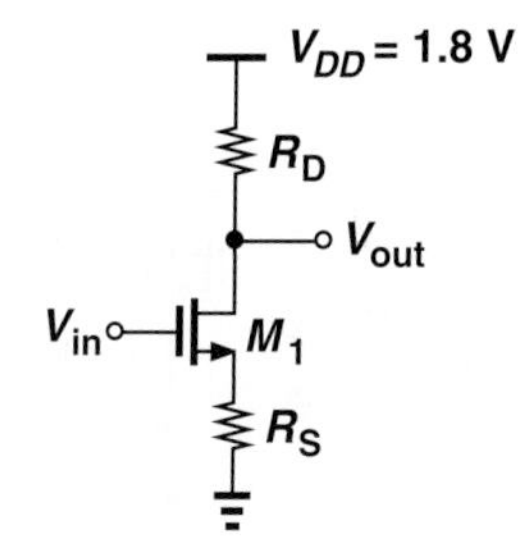

Figure 17.104

17.31. Consider a degenerated CS stage with $\lambda > 0$. Assuming $g_m r_O \gg 1$, calculate the voltage gain of the circuit.

17.32. Calculate the voltage gain of the circuits depicted in Fig. 17.105. Assume $\lambda = 0$.

17.33. Determine the output impedance of each circuit shown in Fig. 17.106. Assume $\lambda \neq 0$.

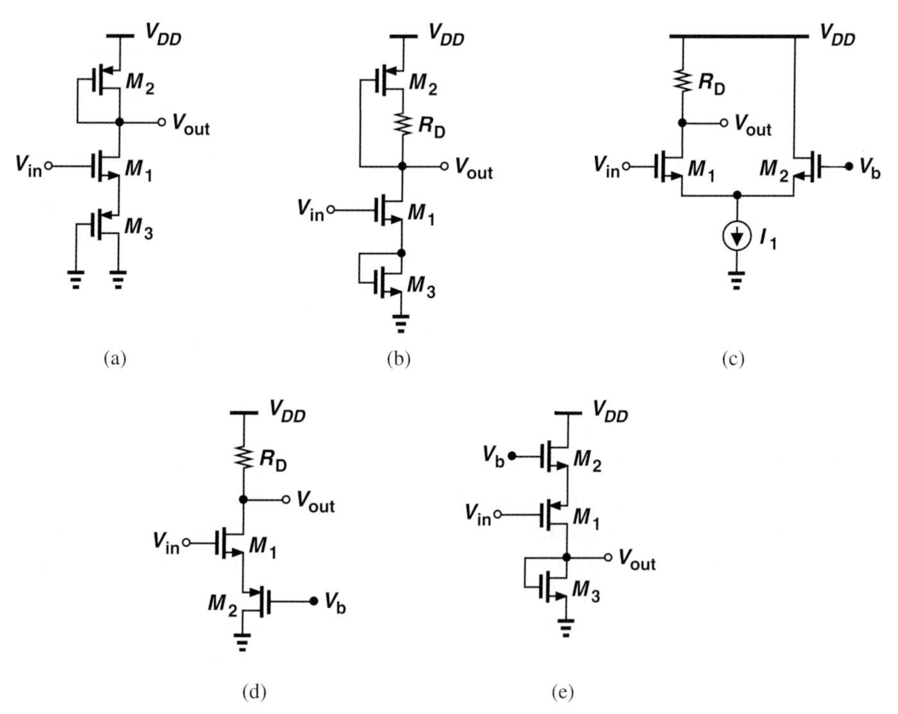

Figure 17.105

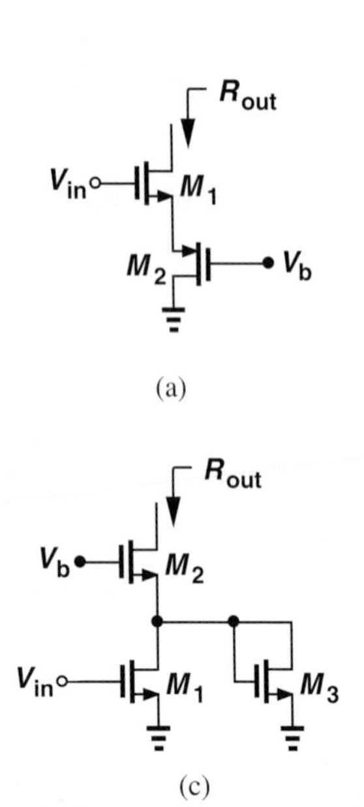

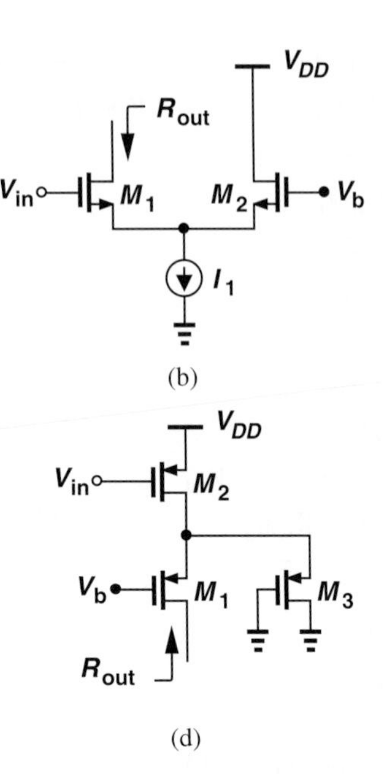

Figure 17.106

SOL **17.34.** The CS stage of Fig. 17.107 carries a bias current of 1 mA. If $R_D = 1\ \text{k}\Omega$ and $\lambda = 0.1\ \text{V}^{-1}$, compute the required value of W/L for a gate voltage of 1 V. What is the voltage gain of the circuit?

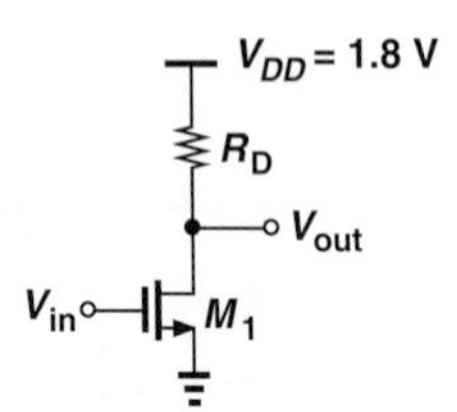

Figure 17.107

17.35. Repeat Problem 17.34 with $\lambda = 0$ and compare the results.

***17.36.** An adventurous student decides to try a new circuit topology wherein the input is applied to the drain and the output is sensed at the source (Fig. 17.108). Assume $\lambda \neq 0$, determine the voltage gain of the circuit and discuss the result.

17.37. In the common-source stage depicted in Fig. 17.109, the drain current of M_1 is defined by the ideal current source I_1 and remains independent of R_1 and R_2 (why?). Suppose $I_1 = 1\ \text{mA}$, $R_D = 500\ \Omega$, $\lambda = 0$, and C_1 is very large.

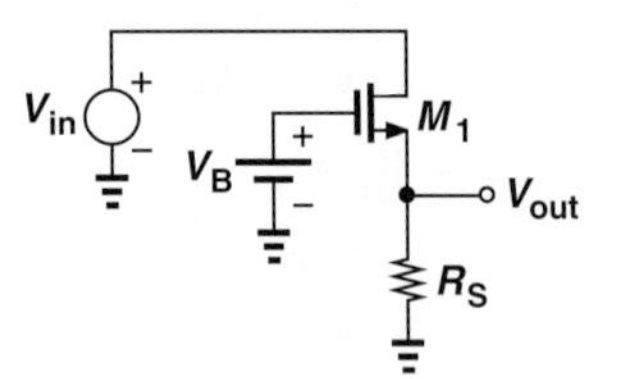

Figure 17.108

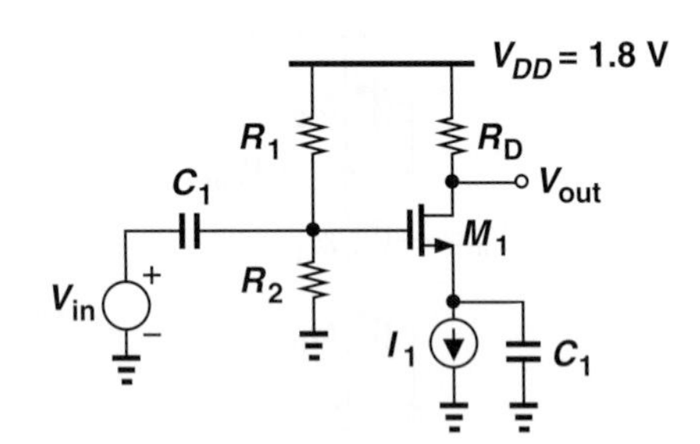

Figure 17.109

(a) Compute the value of W/L to obtain a voltage gain of 5.

(b) Choose the values of R_1 and R_2 to place the transistor 200 mV away from the triode region while $R_1 + R_2$ draws no more than 0.1 mA from the supply.

(c) With the values found in (b), what happens if W/L is twice that found in (a)? Consider both the bias conditions (e.g., whether M_1 comes closer to the triode region) and the voltage gain.

17.38. Consider the CS stage shown in Fig. 17.110, where I_1 defines the bias current of M_1 and C_1 is very large.
(a) If $\lambda = 0$ and $I_1 = 1$ mA, what is the maximum allowable value of R_D?
(b) With the value found in (a), determine W/L to obtain a voltage gain of 5.

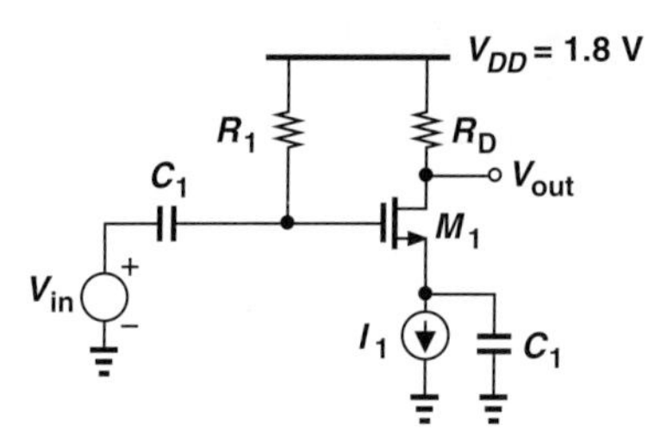

Figure 17.110

Sec. 17.4.4 Common-Gate Topology

17.39. The common-gate stage shown in Fig. 17.111 must provide a voltage gain of 4 and an input impedance of 50 Ω. If $I_D = 0.5$ mA, and $\lambda = 0$, determine the values of R_D and W/L.

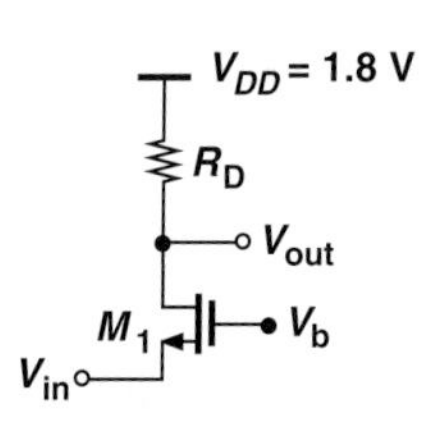

Figure 17.111

17.40. Suppose in Fig. 17.111, $I_D = 0.5$ mA, $\lambda = 0$, and $V_b = 1$ V. Determine the values of W/L and R_D for an input impedance of 50 Ω and maximum voltage gain (while M_1 remains in saturation).

17.41. A CG stage with a source resistance of R_S employs a MOSFET with $\lambda > 0$. Assuming $g_m r_{O \gg 1}$, calculate the voltage gain of the circuit.

17.42. The CG stage depicted in Fig. 17.112 must provide an input impedance of 50 Ω and an output impedance of 500 Ω. Assume $\lambda = 0$. SOL
(a) What is the maximum allowable value of I_D?
(b) With the value obtained in (a), calculate the required value of W/L.
(c) Compute the voltage gain.

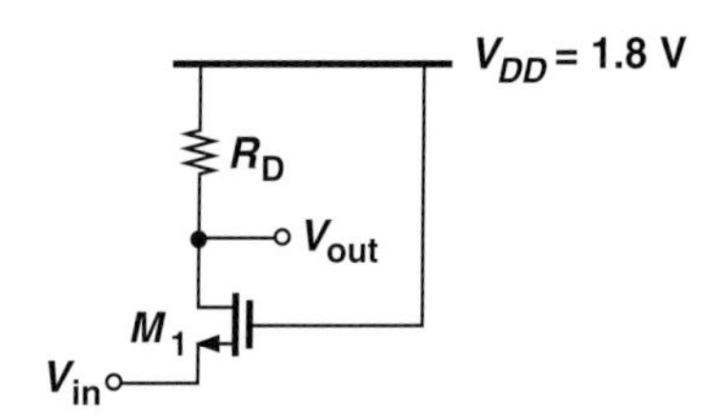

Figure 17.112

17.43. The CG amplifier shown in Fig. 17.113 is biased by means of $I_1 = 1$ mA. Assume $\lambda = 0$ and C_1 is very large.
(a) What value of R_D places the transistor M_1 100 mV away from the triode region?
(b) What is the required W/L if the circuit must provide a voltage gain of 5 with the value of R_D obtained in (a)?

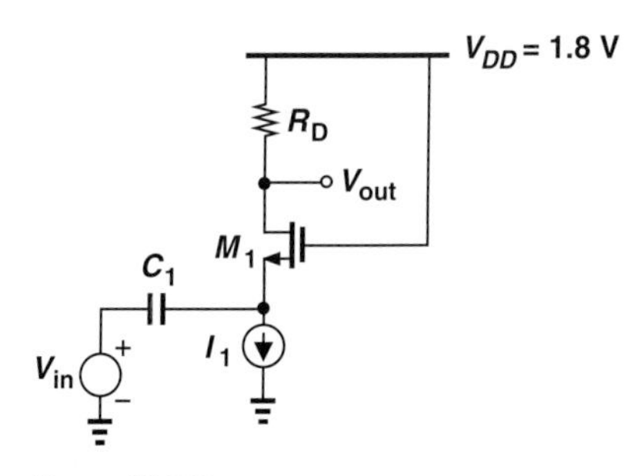

Figure 17.113

17.44. ** Determine the voltage gain of each stage depicted in Fig. 17.114. Assume $\lambda = 0$.

17.45. * Consider the circuit of Fig. 17.115, where a common-source stage (M_1 and R_{D1}) is followed by a common-gate stage (M_2 and R_{D2}).
(a) Writing $v_{out}/v_{in} = (v_X/v_{in})(v_{out}/v_X)$ and assuming $\lambda = 0$, compute the overall voltage gain.

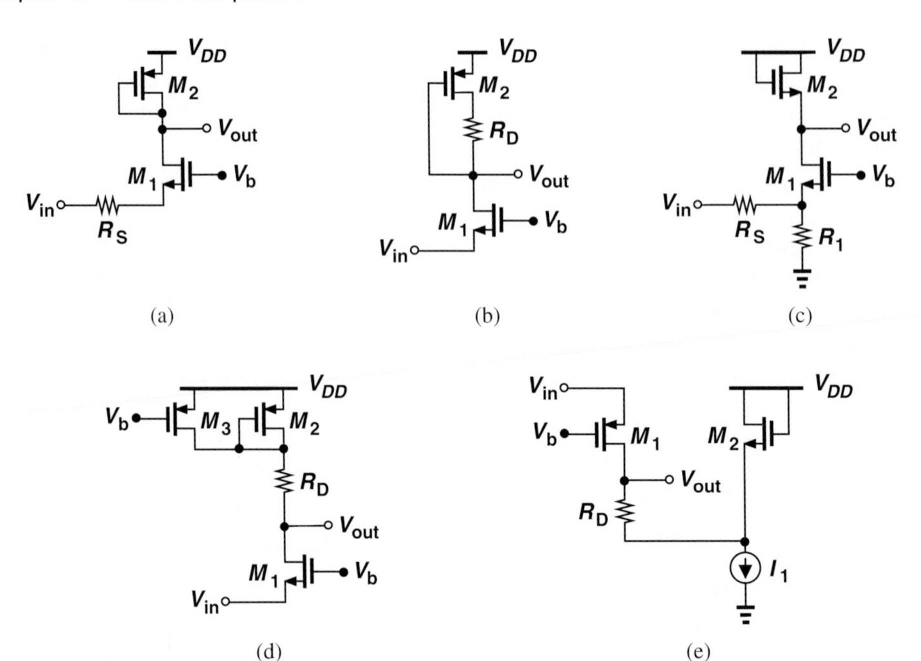

Figure 17.114

(b) Simplify the result obtained in (a) if $R_{D1} \to \infty$. Explain why this result is to be expected.

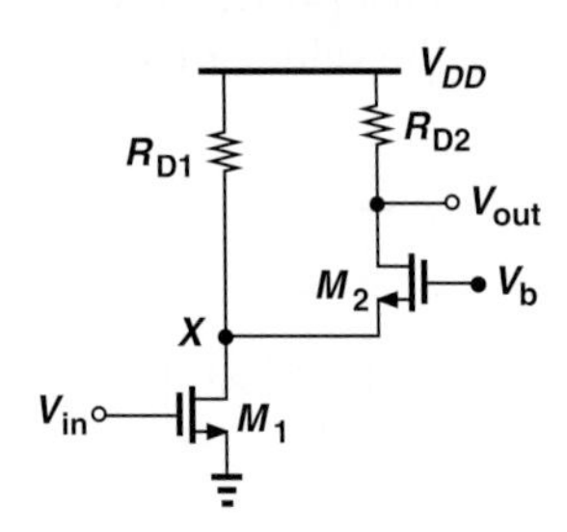

Figure 17.115

17.46. Repeat Problem 17.45 for the circuit shown in Fig. 17.116.

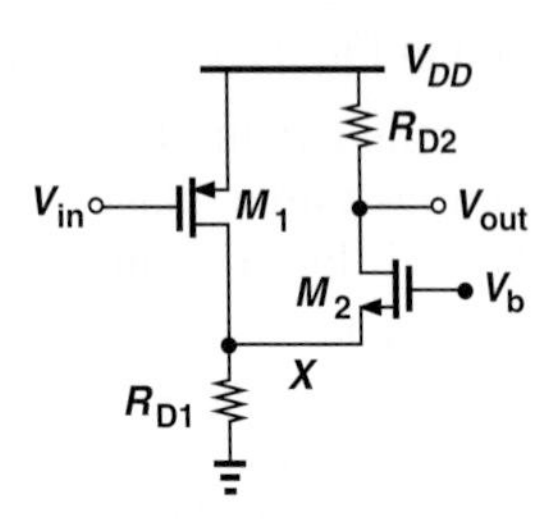

Figure 17.116

17.47. Calculate the voltage gain of the stage depicted in Fig. 17.117. Assume $\lambda = 0$ and the capacitors are very large.

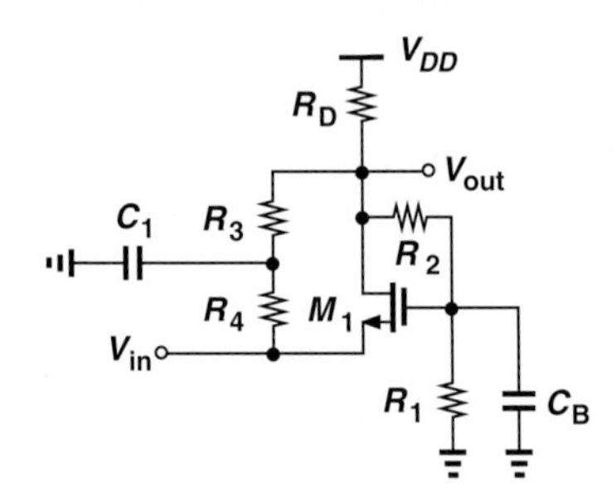

Figure 17.117

*__17.48.__ Assuming $\lambda = 0$, calculate the voltage gain of the circuit shown in Fig. 17.118. Explain why this stage is *not* a common-gate amplifier.

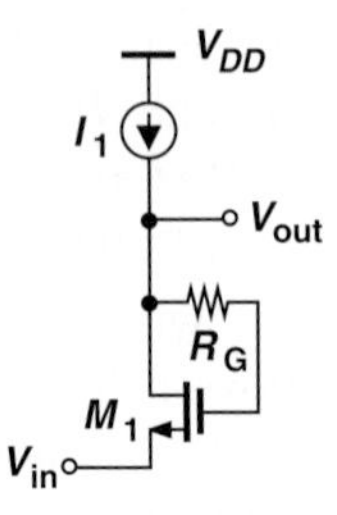

Figure 17.118

Sec. 17.4.5 Source Follower

17.49. The source follower shown in Fig. 17.119 is biased through R_G. Calculate the voltage gain if $W/L = 20/0.18$ and $\lambda = 0.1\ \text{V}^{-1}$.

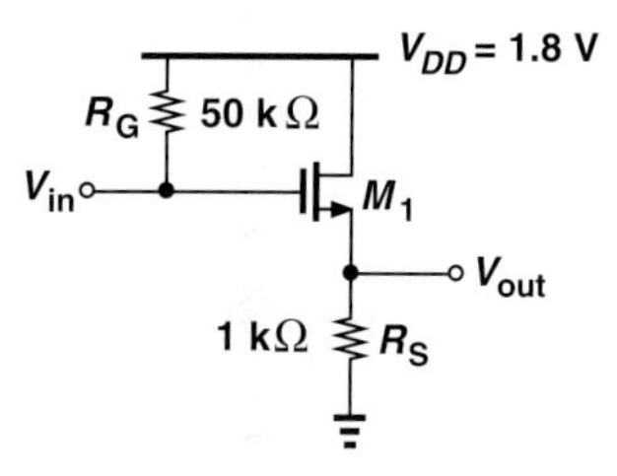

Figure 17.119

SOL **17.50.** We wish to design the source follower shown in Fig. 17.120 for a voltage gain of 0.8. If $W/L = 30/0.18$ and $\lambda = 0$, determine the required gate bias voltage.

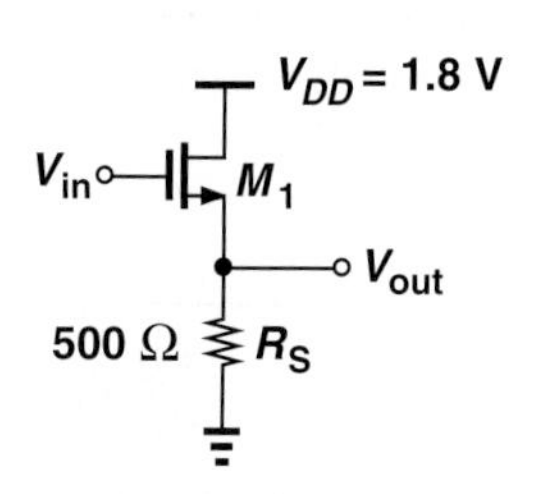

Figure 17.120

17.51. The source follower of Fig. 17.120 is to be designed with a maximum bias gate voltage of 1.8 V. Compute the required value of W/L for a voltage gain of 0.8 if $\lambda = 0$.

17.52. The source follower depicted in Fig. 17.121 employs a current source. Determine the values of I_1 and W/L if the circuit must provide an output impedance less than 100 Ω with $V_{GS} = 0.9$ V. Assume $\lambda = 0$.

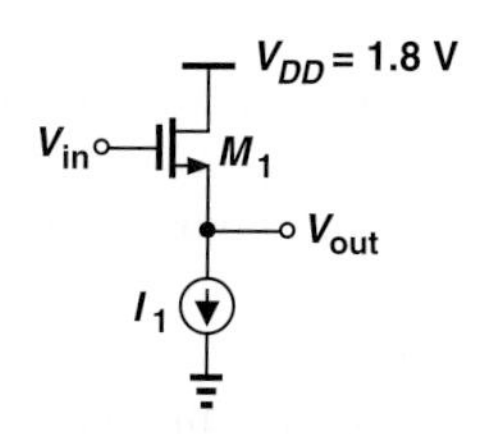

Figure 17.121

17.53. The circuit of Fig. 17.121 must exhibit an output impedance of less than 50 Ω with a power budget of 2 mW. Determine the required value of W/L. Assume $\lambda = 0$.

17.54. We wish to design the source follower of Fig. 17.122 for a voltage gain of 0.8 with a power budget of 3 mW. Compute the required value of W/L. Assume C_1 is very large and $\lambda = 0$.

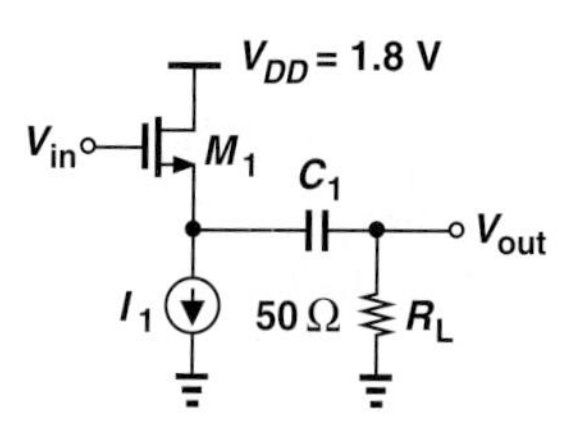

Figure 17.122

**17.55. Determine the voltage gain of the stages shown in Fig. 17.123. Assume $\lambda \neq 0$.

*17.56. Consider the circuit shown in Fig. 17.124, where a source follower (M_1 and I_1) precedes a common-gate stage (M_2 and R_D).

(a) Writing $v_{out}/v_{in} = (v_X/v_{in})(v_{out}/v_X)$, compute the overall voltage gain.

(b) Simplify the result obtained in (a) if $g_{m1} = g_{m2}$.

Design Problems

In the following problems, unless otherwise stated, assume $\lambda = 0$.

17.57. Design the CS stage shown in Fig. 17.125 for a voltage gain of 5 and an output impedance of 1 kΩ. Bias the transistor so that it operates 100 mV away from the triode region. Assume the capacitors are very large and $R_D = 10$ kΩ. SOL

17.58. The CS amplifier of Fig. 17.125 must be designed for a voltage gain of 5 with a power budget of 2 mW. If $R_D I_D = 1$ V, determine the required value of W/L. Make the same assumptions as those in Problem 17.57.

17.59. We wish to design the CS stage of Fig. 17.125 for maximum voltage gain but with $W/L \le 50/0.18$ and a maximum

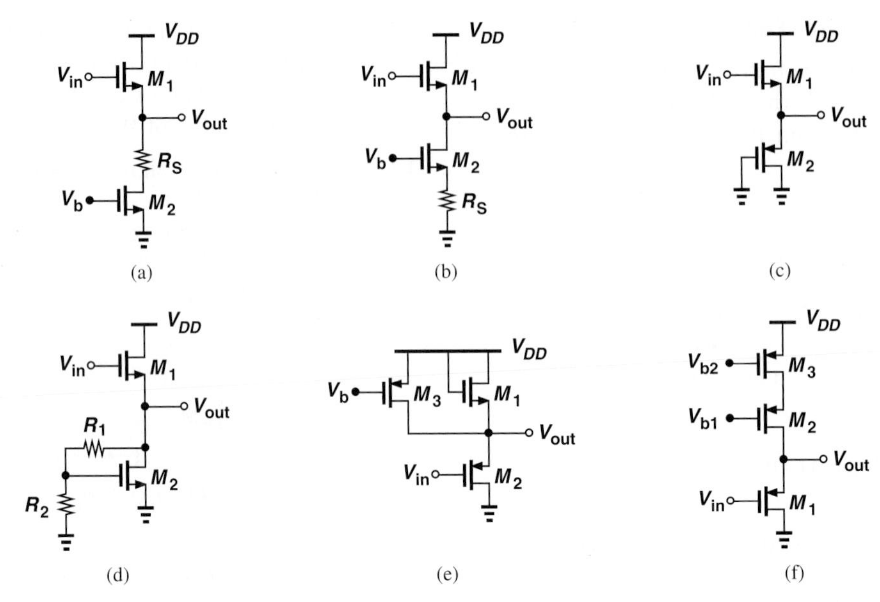

Figure 17.123

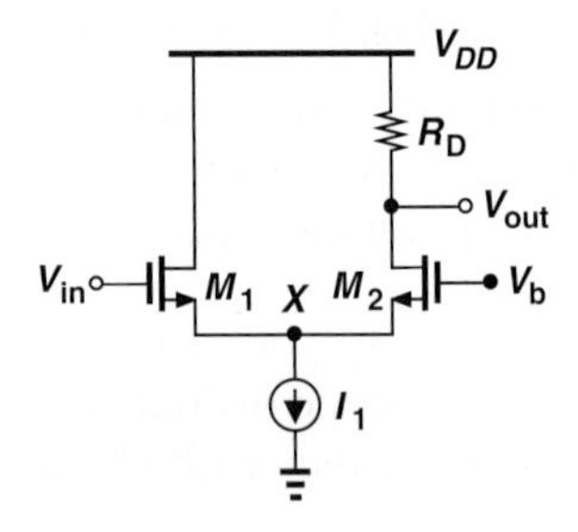

Figure 17.124

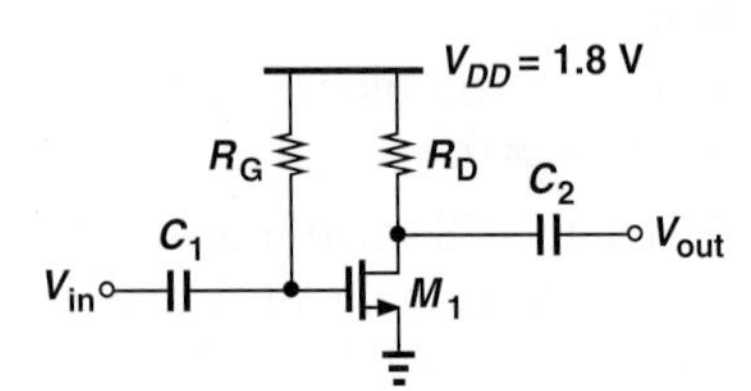

Figure 17.125

output impedance of 500 Ω. Determine the required current. Make the same assumptions as those in Problem 17.57.

17.60. The degenerated stage depicted in Fig. 17.126 must provide a voltage gain of 4 with a power budget of 2 mW while the voltage drop across R_S is equal to 200 mV. If the overdrive voltage of the transistor must not exceed 300 mV and $R_1 + R_2$ must consume less than than 5% of the allocated power, design the circuit. Make the same assumptions as those in Problem 17.57.

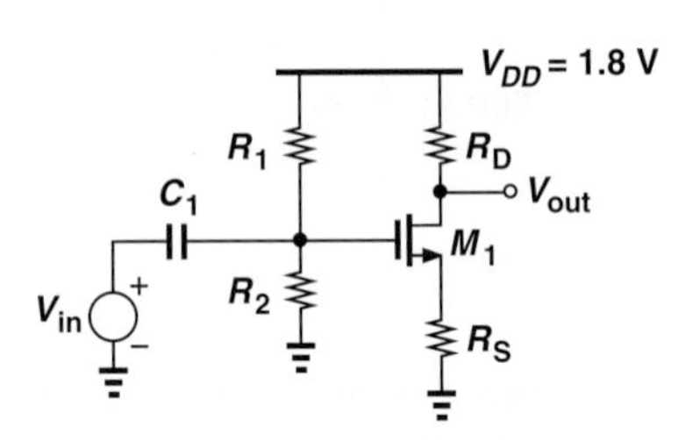

Figure 17.126

17.61. Design the circuit of Fig. 17.126 for a voltage gain of 5 and a power budget of 6 mW. Assume the voltage drop across R_S is equal to the overdrive voltage of the transistor and $R_D = 200\ \Omega$.

17.62. The circuit shown in Fig. 17.127 must provide a voltage gain of 6, with C_S serving as a low impedance at the frequencies of interest. Assuming a power budget of 2 mW and an input impedance of 20 kΩ, design the circuit such that M_1 operates 200 mV away from the triode region. Select the values of C_1 and C_S so that their impedance is negligible at 1 MHz.

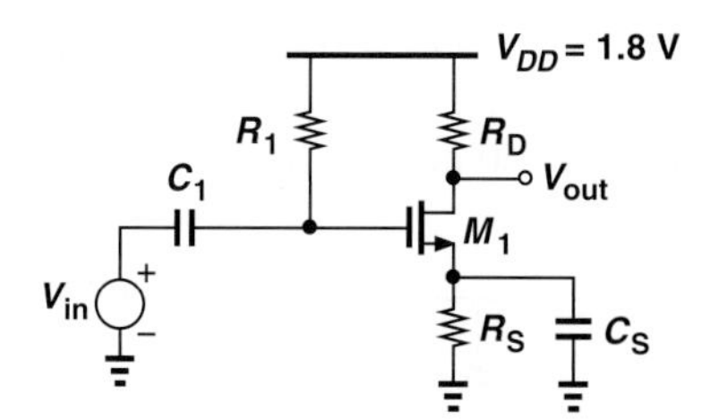

Figure 17.127

17.63. In the circuit of Fig. 17.128, M_2 serves as a current source. Design the stage for a voltage gain of 20 and a power budget of 2 mW. Assume $\lambda = 0.1\ \mathrm{V}^{-1}$ for both transistors and the maximum allowable level at the output is 1.5 V (i.e., M_2 must remain in saturation if $V_{out} \le 1.5$ V).

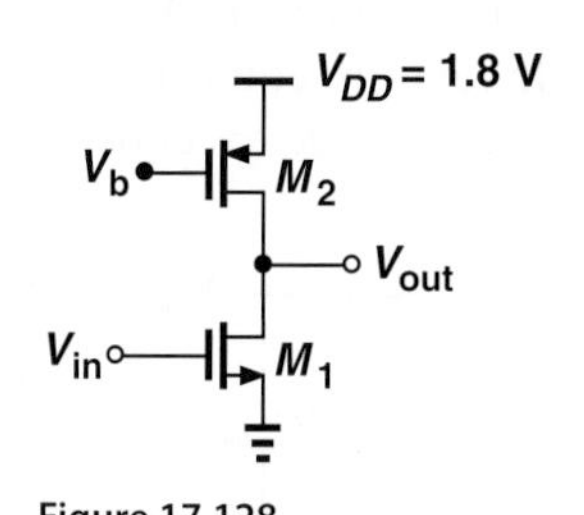

Figure 17.128

17.64. Consider the circuit shown in Fig. 17.129, where C_B is very large and $\lambda_n = 0.5\lambda_p = 0.1\ \mathrm{V}^{-1}$.

(a) Calculate the voltage gain.

(b) Design the circuit for a voltage gain of 15 and a power budget of 3 mW. Assume $R_G \approx 10(r_{O1}||r_{O2})$ and the dc level of the output must be equal to $V_{DD}/2$.

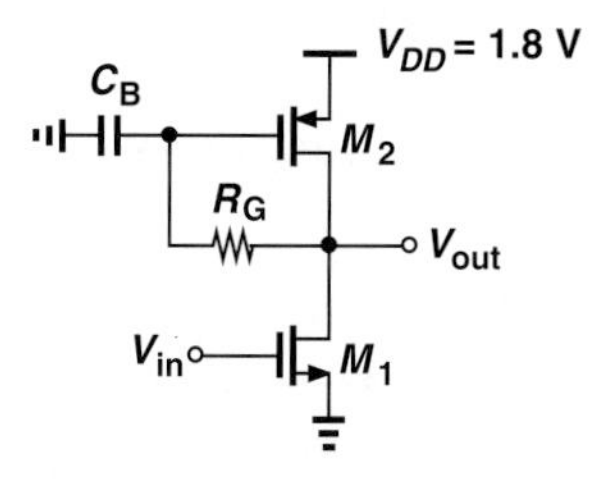

Figure 17.129

17.65. The CS stage of Fig. 17.130 incorporates a degenerated PMOS current source. The degeneration must raise the output impedance of the current source to about $10r_{O1}$ such that the voltage gain remains nearly equal to the intrinsic gain of M_1. Assume $\lambda = 0.1\ \mathrm{V}^{-1}$ for both transistors and a power budget of 2 mW.

(a) If $V_B = 1$ V, determine the values of $(W/L)_2$ and R_S so that the impedance seen looking into the drain of M_2 is equal to $10r_{O1}$.

(b) Determine $(W/L)_1$ to achieve a voltage gain of 30.

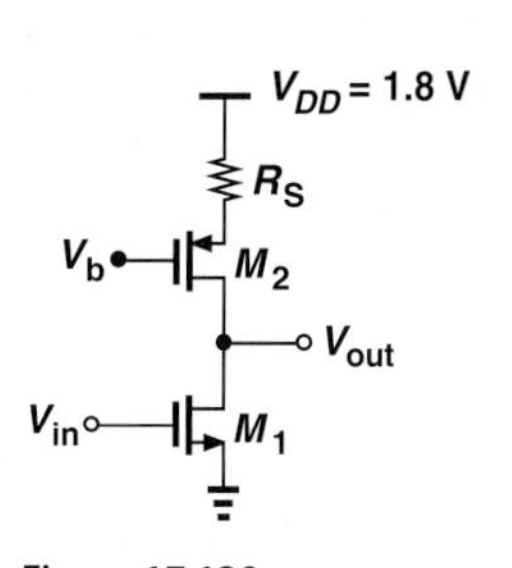

Figure 17.130

17.66. Assuming a power budget of 1 mW and an overdrive of 200 mV for M_1, design the circuit shown in Fig. 17.131 for a voltage gain of 4. SOL

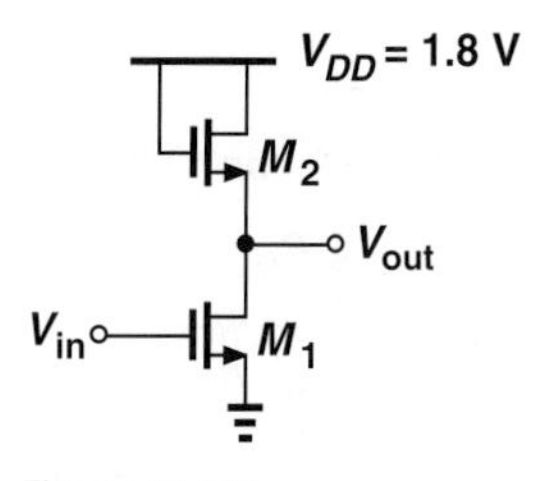

Figure 17.131

17.67. Design the common-gate stage depicted in Fig. 17.132 for an input impedance of

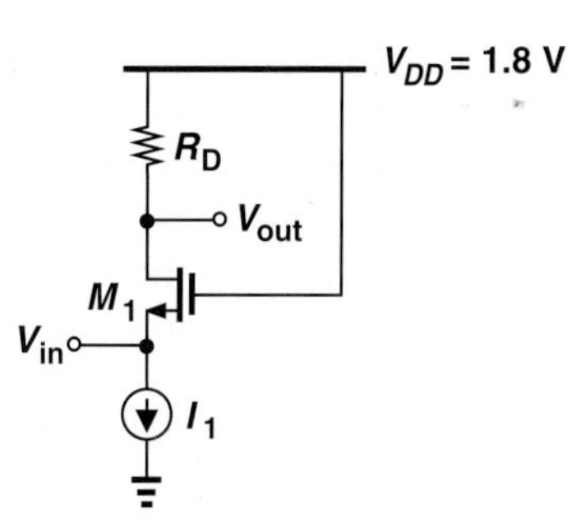

Figure 17.132

50 Ω and a voltage gain of 5. Assume a power budget of 3 mW.

17.68. Design the circuit of Fig. 17.133 such that M_1 operates 100 mV away from the triode region while providing a voltage gain of 4. Assume a power budget of 2 mW.

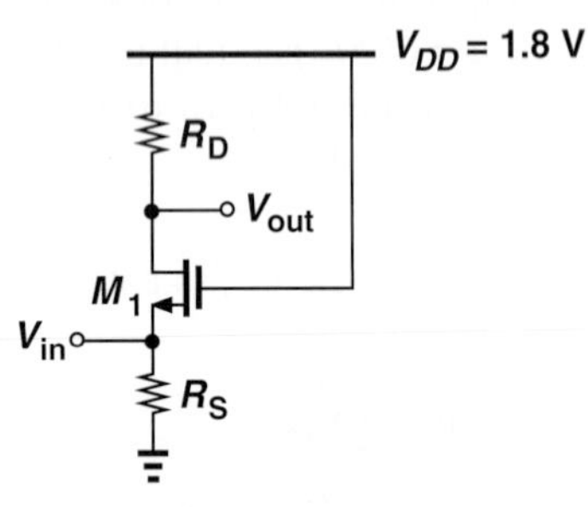

Figure 17.133

17.69. Figure 17.134 shows a self-biased common-gate stage, where $R_G \approx 10R_D$ and C_G serves as a low impedance so that the voltage gain is still given by $g_m R_D$. Design the circuit for a power budget of 5 mW and a voltage gain of 5. Assume $R_S \approx 10/g_m$ so that the input impedance remains approximately equal to $1/g_m$.

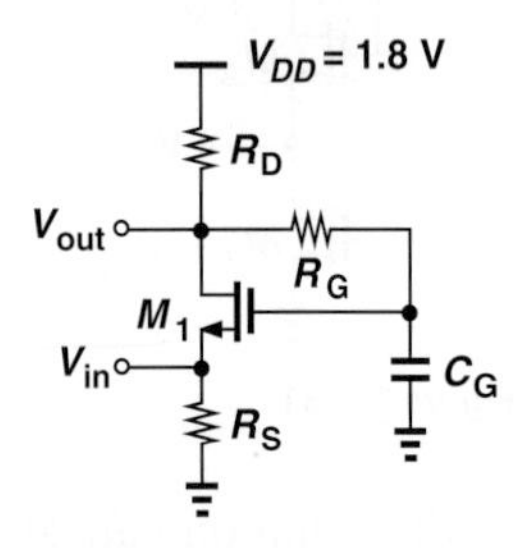

Figure 17.134

17.70. Design the CG stage shown in Fig. 17.135 such that it can accommodate an output swing of 500 mV_{pp}, i.e., V_{out} can fall below its bias value by 250 mV without driving M_1 into the triode region. Assume a voltage gain of 4 and an input impedance of 50 Ω. Select $R_S \approx 10/g_m$ and $R_1 + R_2 = 20$ kΩ. (Hint: since M_1 is biased 250 mV away from the triode region, we have $R_S I_D + V_{GS} - V_{TH} + 250\text{ mV} = V_{DD} - I_D R_D$.)

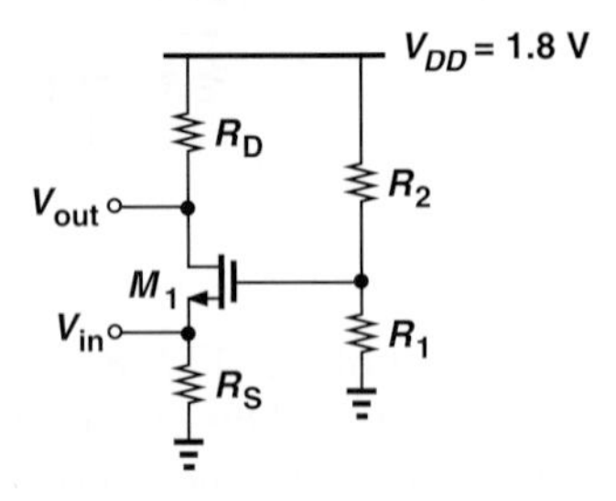

Figure 17.135

17.71. Design the source follower depicted in Fig. 17.136 for a voltage gain of 0.8 and a power budget of 2 mW. Assume the output dc level is equal to $V_{DD}/2$ and the input impedance exceeds 10 kΩ.

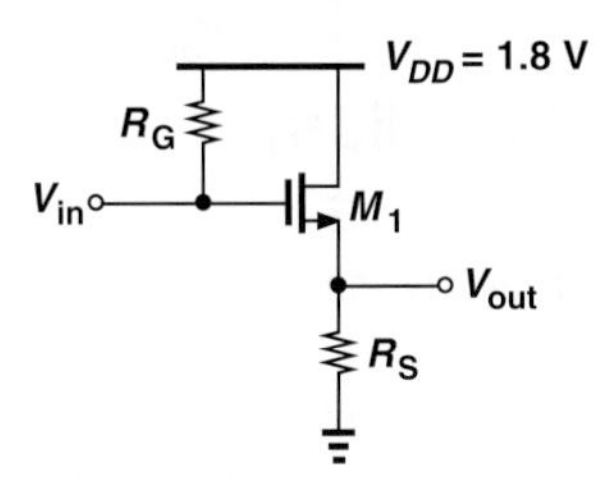

Figure 17.136

17.72. Consider the source follower shown in Fig. 17.137. The circuit must provide a voltage gain of 0.8 at 100 MHz while consuming 3 mW. Design the circuit such that the dc voltage at node X is equal to $V_{DD}/2$. Assume the input impedance exceeds 20 kΩ.

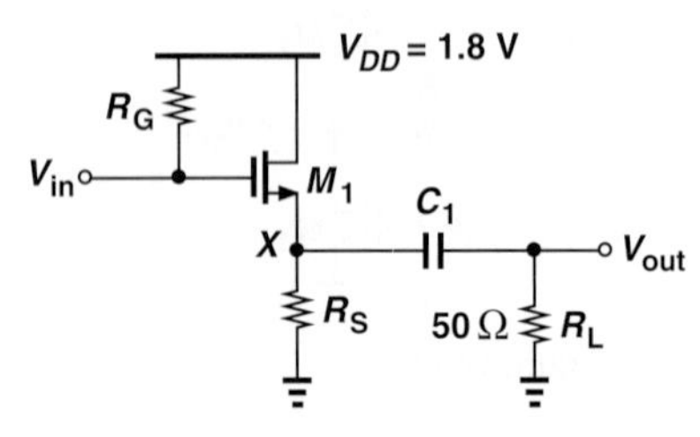

Figure 17.137

17.73. In the source follower of Fig. 17.138, M_2 serves as a current source. The circuit must operate with a power budget of 3 mW, a voltage gain of 0.9, and a minimum allowable output of 0.3 V (i.e., M_2 must remain in saturation if $V_{DS2} \geq 0.3$ V). Assuming $\lambda = 0.1\ \text{V}^{-1}$ for both transistors, design the circuit.

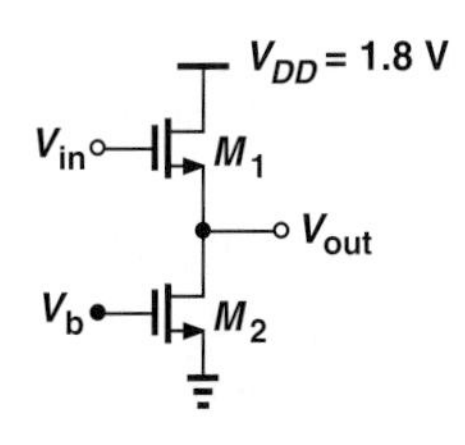

Figure 17.138

SPICE PROBLEMS

In the following problems, use the MOS models and source/drain dimensions given in Appendix A. Assume the substrates of NMOS and PMOS devices are tied to ground and V_{DD}, respectively.

17.74. In the circuit of Fig. 17.139, I_1 is an ideal current source equal to 1 mA.
(a) Using hand calculations, determine $(W/L)_1$ such that $g_{m1} = (100\ \Omega)^{-1}$.
(b) Select C_1 for an impedance of $\approx 100\ \Omega$ ($\ll 1$ kΩ) at 50 MHz.
(c) Simulate the circuit and obtain the voltage gain and output impedance at 50 MHz.
(d) What is the change in the gain if I_1 varies by $\pm 20\%$?

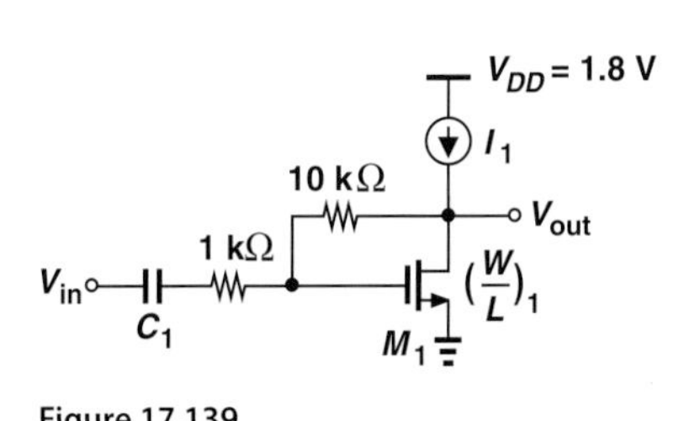

Figure 17.139

17.75. The source follower of Fig. 17.140 employs a bias current source, M_2.
(a) What value of V_{in} places M_2 at the edge of saturation?
(b) What value of V_{in} places M_1 at the edge of saturation?
(c) Determine the voltage gain if V_{in} has a dc value of 1.5 V.
(d) What is the change in the gain if V_b changes by ± 50 mV?

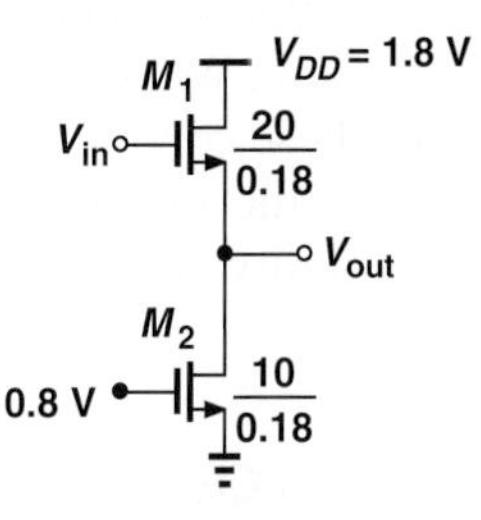

Figure 17.140

17.76. Figure 17.141 depicts a cascade of a source follower and a common-gate stage. Assume $V_b = 1.2$ V and $(W/L)_1 = (W/L)_2 = 10\ \mu\text{m}/0.18\ \mu\text{m}$.
(a) Determine the voltage gain if V_{in} has a dc value of 1.2 V.
(b) Verify that the gain drops if the dc value of V_{in} is higher or lower than 1.2 V.
(c) What dc value at the input reduces the gain by 10% with respect to that obtained in (a)?

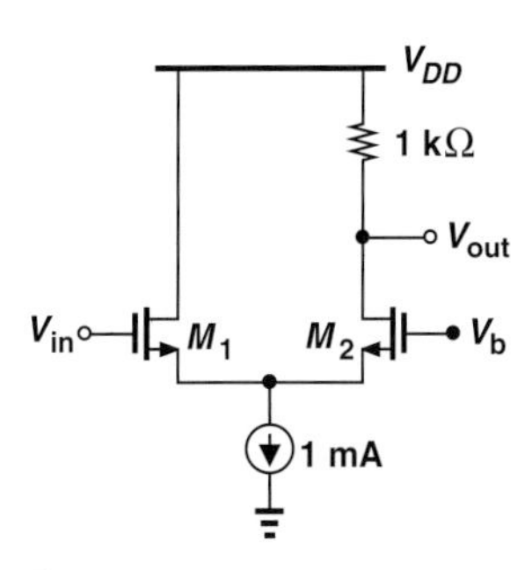

Figure 17.141

17.77. Consider the CS stage shown in Fig. 17.142, where M_2 operates as a resistor.

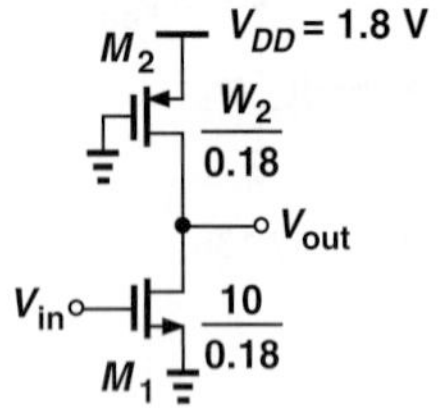

Figure 17.142

(a) Determine W_2 such that an input dc level of 0.8 V yields an output dc level of 1 V. What is the voltage gain under these conditions?

(b) What is the change in the gain if the mobility of the NMOS device varies by ±10%? Can you explain this result using the expressions derived in Chapter 6 for the transconductance?

17.78. Repeat Problem 17.77 for the circuit illustrated in Fig. 17.143 and compare the sensitivities to the mobility.

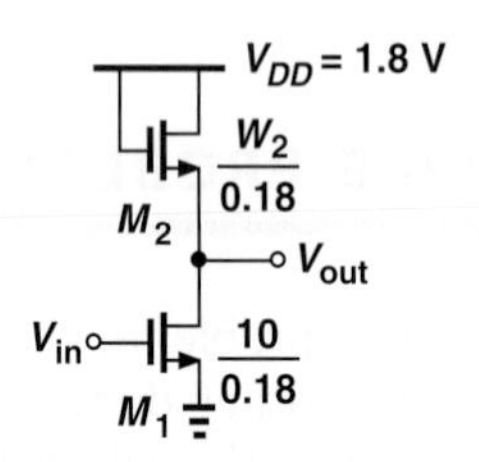

Figure 17.143

REFERENCE

1. A. Caizzone, et al., "A 2.6-μW monolithic CMOS photoplethysmographic (PPG) sensor operating with 2-μW LED power for continuous health monitoring," *IEEE Trans. Biomedical Circuits and Systems*, vol. 13, pp. 1243–1253, Dec. 2019.

A

Introduction to SPICE

The circuits encountered in microelectronics may contain a few devices or a few million devices.[1] How do we analyze and design these circuits? As the number of devices in a circuit increases, hand analysis becomes more difficult, eventually reaching a point where other methods are required. For example, one can *build* a prototype using discrete components and observe its behavior. However, discrete devices provide a poor approximation of modern integrated circuits. Furthermore, even for a few hundred devices, discrete prototypes become prohibitively complex.

Today's microelectronics employs simulation programs extensively. A versatile tool used to predict the behavior of circuits is Simulation Program with Integrated Circuit Emphasis (SPICE). While originally developed as a public-domain tool (at University of California, Berkeley), SPICE has evolved into commercial tools such as PSPICE, HSPICE, etc., most of which retain the same format. This appendix provides a tutorial overview of SPICE, enabling the reader to perform basic simulations. More details can be found in [1].

A.1 SIMULATION PROCEDURE

Suppose we have the circuit shown in Fig. A.1(a) and wish to use SPICE to study its frequency response. That is, we wish to verify that the response is relatively flat for $f < 1/(2\pi R_1 C_1) \approx 15.9$ MHz and begins to roll off thereafter [Fig. A.1(b)]. To this end, we apply a sinusoidal voltage to the input and vary its frequency from, say, 1 MHz to 50 MHz.

The procedure consists of two steps: (1) define the circuit in a language (format) that SPICE understands, and (2) use an appropriate command to tell SPICE to determine the frequency response. Let us begin with the first step. This step itself consists of three tasks.

1. Label each node in the circuit. Figure A.1(c) depicts an example, where the labels "in" and "out" refer to the input and output nodes, respectively. The common (ground) node *must* be called "0" in SPICE. While arbitrary, the labels chosen for other nodes

[1] Recent microprocessors contain one billion MOS transistors.

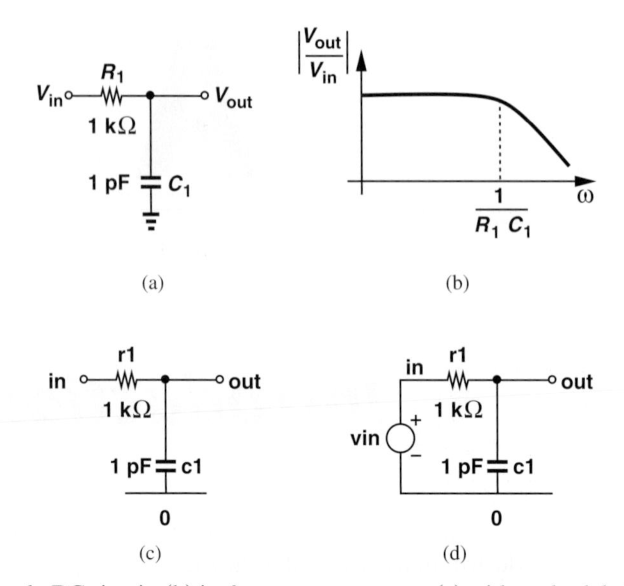

Figure A.1 (a) Simple RC circuit, (b) its frequency response, (c) with nodes labeled, (d) with elements labeled.

should carry some information about their respective nodes so as to facilitate reading the SPICE description of the circuit.

2. Label each element in the circuit. Defining the type of the element (resistors, capacitors, etc.), each of these labels must begin with a *specific letter* so that SPICE recognizes the element. For example, resistor labels must begin with r, capacitor labels with c, inductor labels with l, diode labels with d, and voltage sources with v.[2] Our simple circuit now appears as shown in Fig. A.1(d).
3. Construct the "netlist," i.e., a precise description of each element along with the nodes to which it is tied. The netlist consists of text lines, each describing one element, with the following format for two-terminal devices:

```
elementlabel node1 node2 value
```

From the example in Fig. A.1(d), we begin the netlist with:

```
r1 in out 1k
c1 out 0 1p
```

Note that the units are specified as a single letter (k for 10^3, p for 10^{-12}, etc.). For the input voltage source, we write

```
vin in 0 ac 1
```

where ac denotes our desire to determine the frequency (ac) response and hence designates V_{in} as a sinusoidal voltage source whose frequency will be varied. The value 1

[2]SPICE does not distinguish between lower-case and upper-case letters.

at the end represents the peak amplitude of the sinusoid. Also note that the first node, "in," is assumed to be the positive terminal of the voltage source.

The netlist must also include the "type of analysis" that we wish SPICE to perform. In our example, SPICE must vary the frequency from one value to another, e.g., 1 MHz to 50 MHz. The corresponding command appears as

```
.ac dec 200 1meg 50meg
```

Note that each "command" line begins with a period. The first entry, "ac," requests SPICE to perform an "ac analysis," i.e., determine the frequency response. The second and third entries, "dec 200," tell SPICE to simulate the circuit at 200 frequency values in every decade of frequency (e.g., from 1 MHz to 10 MHz). The last two entries, "1meg 50meg," set the lower and upper values of the frequency range, respectively. Note that "meg" denotes 10^6 and should not be confused with "m," which stands for 10^3.

We need two more lines to complete our netlist. The first line of the file is called the "title" and carries no information for SPICE. For example, the title line may read "My Amplifier." Note that SPICE always ignores the first line of the file, encountering errors if you forget to include the title. The last line of the file must be a ".end" command. Our netlist now appears as:

```
Test Circuit for Frequency Response
r1 in out 1k
c1 out 0 10p
vin in 0 ac 1
.ac dec 200 1meg 50 meg
.end
```

Note that, except for the first and last lines, the order of other lines in the netlist is unimportant.

What do we *do* with the above netlist? We must "run" SPICE on this file, which we call, for example, test.sp. Depending on the operating system, running SPICE may entail clicking on an icon in a graphics interface or simply typing:

```
spice test.sp
```

After the simulation is successfully run, various node voltages can be plotted using the graphics interface that accompanies SPICE.

Figure A.2 summarizes the SPICE simulation procedure. The definition of (voltage or current) sources in the netlist must be consistent with the type of analysis. In the above example, the input voltage source definition contains the entry "ac" so that SPICE applies the frequency sweep to V_{in} rather than to other sources.

At this point, the reader may raise many questions: How are other elements defined in the netlist? How are the units specified? Is the order of the node labels in the netlist important? How are other types of analysis specified? We answer these questions in the following sections.

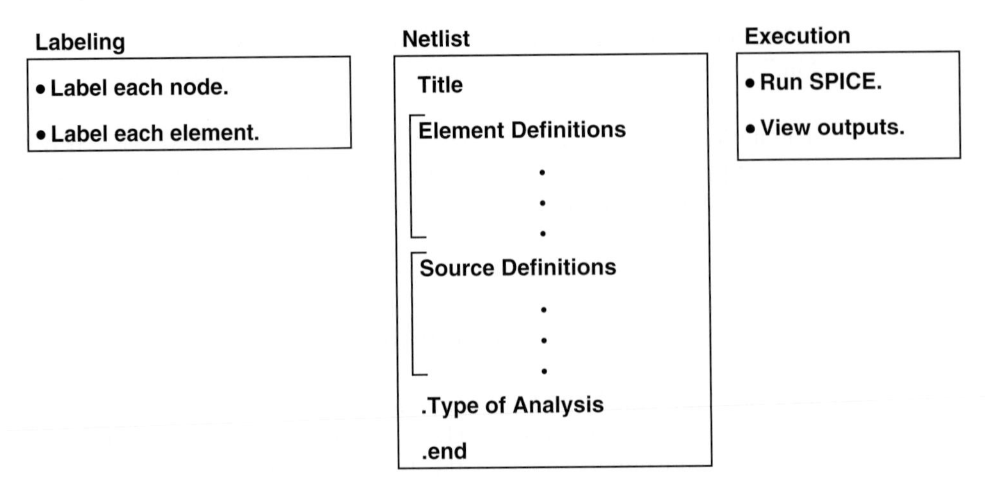

Figure A.2 Simulation procedure.

A.2 TYPES OF ANALYSIS

In addition to frequency response, other aspects of circuits may also be of interest. This section provides the (voltage or current) source descriptions and commands necessary to perform other types of analysis.

A.2.1 Operating Point Analysis

In many electronic circuits, we must first determine the bias conditions of the devices. SPICE performs such an analysis with the .op command. The following example illustrates the procedure.

Example A-1

Determine the currents flowing through R_3 and R_4 in Fig. A.3(a).

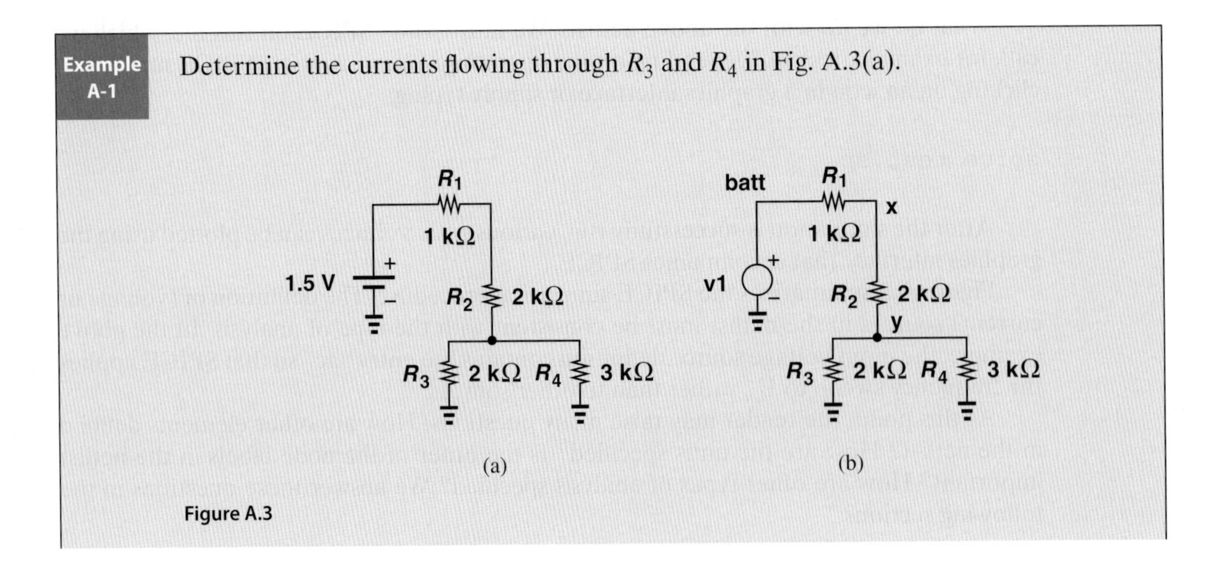

Figure A.3

Solution We label the nodes as shown in Fig. A.3(b) and construct the netlist as follows:

```
Simple Resistive Network
v1 batt 0 1.5
r1 batt x 1k
r2 x y 2k
r3 y 0 2k
r4 y 0 3k
.op
.end
```

SPICE predicts a current of 0.214 mA through R_3 and 0.143 mA through R_4.

A.2.2 Transient Analysis

Suppose we wish to study the pulse response of the RC section shown in Fig. A.1(d). Called "transient analysis," this type of simulation requires changing the vin and .ac lines while maintaining the same netlist descriptions for R_1 and C_1. The voltage source must now be specified as

```
                   V1  V2  Tdel  Tr  Tf  Tw
vin in 0 pulse(0   1   0     1n  2n  5n)
```

where $V_1, \ldots, T_w$ are defined as depicted in Fig. A.4(a).[3] We say V_{in} is a pulse that goes from 0 V to 1 V with zero delay (T_{del}), a rising transition of 1 ns (T_R), a falling transition of 2 ns (T_F), and a width of 5 ns (T_w). Note that the first node, "in," is assumed to be the positive terminal of the voltage source.

How do we tell SPICE to perform a transient analysis? The command is as follows:

```
.tran 0.2n 10n
```

where 0.2n indicates the increments ("time steps") that SPICE must use in calculating the response, and 10n the total time of interest [Fig. A.4(b)].

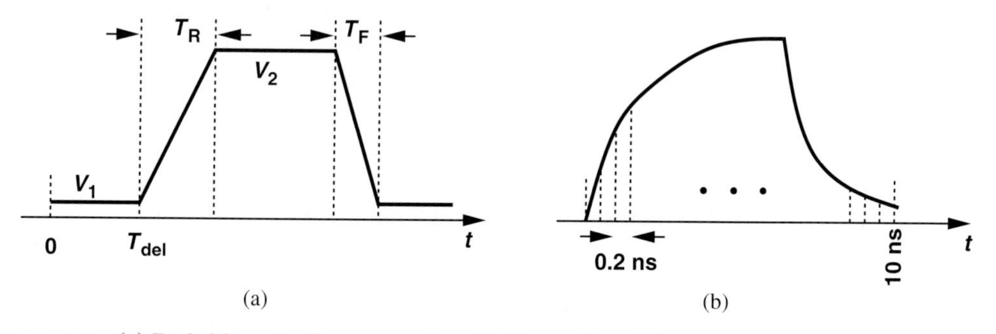

Figure A.4 (a) Definition of pulse parameters, (b) illustration of time step.

[3]The parentheses following the pulse description are for clarity and not essential.

The overall netlist now appears as:

```
Pulse Response Example
r1 in out 1k
c1 out 0 10p
vin in 0 pulse(0 1 0 1n 2n 5n)
.tran 0.2n 10n
.end
```

Example A-2 Construct a SPICE netlist for the pulse response of the circuit shown in Fig. A.5(a).

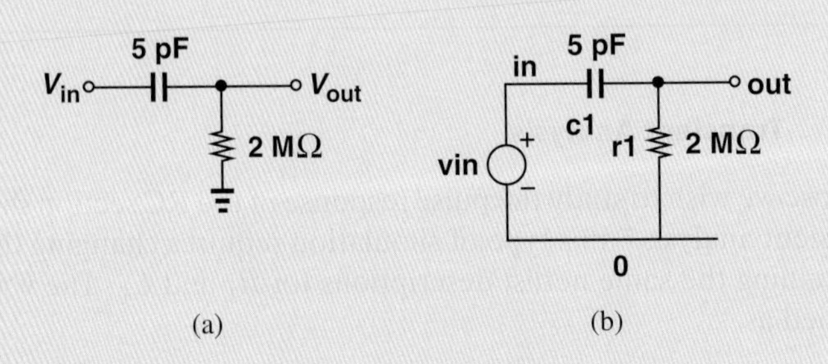

Figure A.5

Solution We begin with labeling the nodes and the elements [Fig. A.5(b)]. Given the time constant $R_1C_1 = 10\,\mu s$, we postulate that the rising and falling transitions of the input pulse can be as long as approximately 1 μs and still appear "abrupt" to the circuit. For the pulsewidth, we choose 30 μs to allow the output to "settle." We therefore have

```
High-Pass Filter Pulse Response
c1 in out 5p
r1 out 0 2meg
vin in 0 pulse(0 1 1u 1u 30u)
.tran 0.2u 60u
.end
```

(The letter u in the pulse description denotes 10^{-6}.) Note that the timestep is chosen sufficiently smaller than the pulse transition times, and the overall transient time long enough to reveal the response after the input falls to zero.

Example A-3 Revise the SPICE netlist constructed in Example A-2 so as to observe the *step* response of the circuit.

Solution We wish V_{in} to jump to 1 V and remain at this level. The pulse description, however, requires a pulsewidth value. Thus, we choose the pulsewidth sufficiently larger than our "observation window."

```
vin in 0 pulse(0 1 1u 1u 1)
.tran 0.2u 30u
```

A pulsewidth of 1 s proves quite versatile for step response analyses because most of our circuits exhibit a much faster response. Note that the overall transient time is now 30 μs, just long enough to show the response to the input rising edge.

Example A-4

Construct a SPICE netlist for the step response of the circuit depicted in Fig. A.6(a).

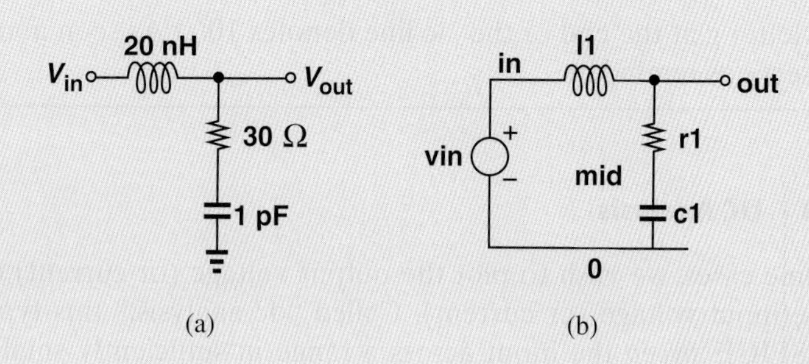

Figure A.6

Solution We begin with labeling the nodes and the elements [Fig. A.6(b)]. How do we choose the transition time of the step? Ignoring the damping behavior of the circuit for now, we may consider $R_1/(2L_1) = 1.5$ ns as the time constant of the response and hence choose the transition time to be about 150 ps. The netlist is as follows:

```
My RLC Circuit
l1 in out 20n
r1 out mid 30
c1 mid 0 1p
vin in 0 pulse(0 1 0 150p 150p 1)
.tran 25p 500p
.end
```

Note that the falling transition time is unimportant here.

Example A-5

Suppose we wish to determine the frequency response of the RLC circuit illustrated in Fig. A.6(a). Revise the netlist accordingly.

Solution We must often study both the transient and the ac response of circuits. For convenience, only *one* file should serve both purposes. Fortunately, SPICE allows us to "comment out" lines of the file by inserting a * at the beginning of each line. We therefore repeat the netlist from the above example, comment out the lines related to transient analysis, and add the lines necessary for ac analysis:

```
My RLC Circuit
l1 in out 20n
r1 out mid 30
c1 mid 0 1p
*vin in 0 pulse(0 1 0 150p 150p 1)
*.tran 25p 500p
*Added next two lines for ac analysis.
vin in 0 ac 1
.ac dec 100 1meg 1g
.end
```

(The letter g at the end of the .ac line denotes 10^9.) As seen above, comment lines can also serve as reminders.

A.2.3 DC Analysis

In some cases, we wish to plot the output voltage (or current) of a circuit as a function of the input voltage (or current). Called "dc analysis," this type of simulation requires that SPICE *sweep* the input across a range in sufficiently small steps. For example, we may write

```
vin   in    0   dc 1
                Lower     Upper     Step
                End       End       Size
.dc    vin      0.5       2         1m
```

The vin description specifies the type as dc with a nominal value of 1 V.[4] The dc sweep command begins with .dc and specifies vin as the source that must be swept. The following two entries denote the lower and upper ends of the range, respectively, and the last entry indicates the step size.

Example A-6

Construct a netlist to plot V_{out} as a function of V_{in} for the circuit shown in Fig. A.7(a). Assume an input range of −1 V to +1 V with 2-mV steps.

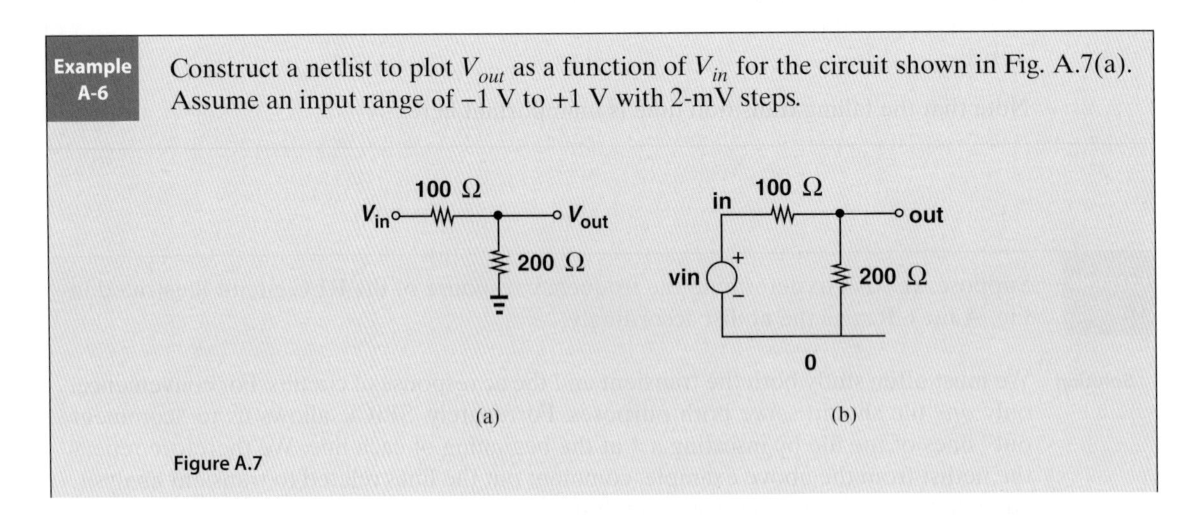

Figure A.7

[4]This nominal value is arbitrary and unimportant in dc analysis.

Solution We label the nodes and the elements as illustrated in Fig. A.7(b). The netlist can be written as:

```
Voltage Divider
r1 in out 100
r2 out 0 200
vin in 0 dc 1
.dc vin -1 +1 2m
.end
```

Note that the values of r1 and r2 are not followed by a unit so that SPICE assumes they are expressed in ohms.

A.3 ELEMENT DESCRIPTIONS

In our study of SPICE netlists thus far, we have seen descriptions of resistors, capacitors, inductors, and voltage sources. In this section, we consider the descriptions of elements such as current sources, diodes, bipolar transistors, and MOSFETs.

A.3.1 Current Sources

The definition of current sources for various types of analysis follows those of voltage sources, with the understanding that the current flows *out of* the first node and *into* the second node specified in the description. For example, the current source in Fig. A.8(a) is expressed as

```
iin in 0 ac 1
```

for ac analysis.

If the circuit is configured as shown in Fig. A.8(b), then we must write

```
iin 0 in ac 1
```

Similarly, for pulse response, the current source in Fig. A.8(a) can be expressed as

```
iin in 0 pulse(0 1m 0 0.1n 0.1n 5n)
```

where the current jumps from 0 to 1 mA with zero delay and a rising transition time of 0.1 ns.

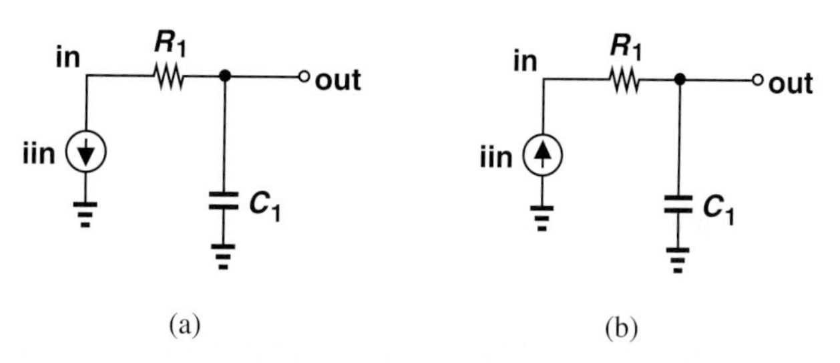

Figure A.8 Circuits for illustrating the polarity of current sources in SPICE.

Example A-7 Study the response of the circuit depicted in Fig. A.9(a) to an input current step.

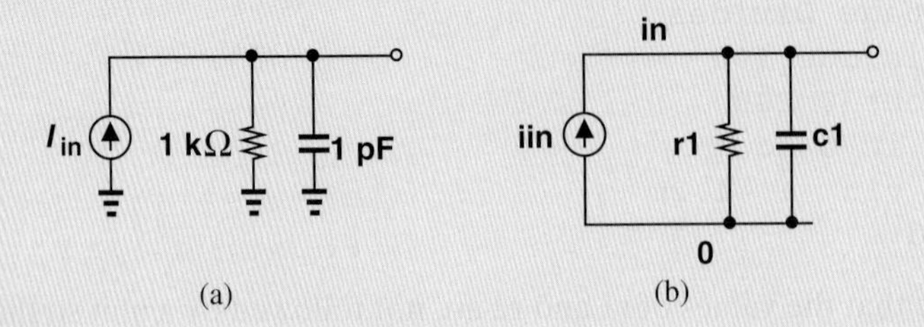

Figure A.9

Solution Labeling the circuit as shown in Fig. A.9(b) and noting a time constant of 1 ns, we write

```
Step Response Example
r1 in 0 1k
c1 in 0 1p
iin 0 in pulse(0 1m 0 0.1n 0.1n 1)
.tran 20p 3n
.end
```

A.3.2 Diodes

Unlike passive elements studied thus far, diodes cannot be specified by a "value." Rather, the equation $I_D = I_S[\exp(V_D/V_T) - 1]$ suggests that the value of I_S must be provided. Thus, in the example illustrated in Fig. A.10, we have

```
        Anode   Cathode    Is
d1       in       out     is=1f
```

where the element name begins with d to denote a diode, the first node indicates the anode, and the second represents the cathode. The last entry specifies the value of I_S as 1×10^{-15} A.[5]

In some cases, a reverse-biased diode may serve as a voltage-dependent capacitor, requiring that the value of the junction capacitance be specified. Recall that

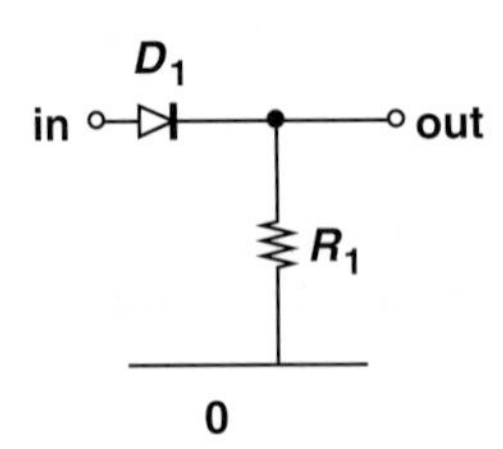

Figure A.10 Simple diode circuit.

[5]Note that f stands for femto and *not* for farad. That is, a capacitor expressed as 1f in SPICE description assumes a value of 1 fF.

$C_j = C_{j0}/\sqrt{1+|V_R|/V_0}$, where $V_R < 0$ is the reverse-bias voltage. We must therefore provide the values of C_{j0} and V_0 to SPICE.

The above diode line may then evolve to

```
d1 in out is=1f, cjo=1p, vj=0.7
```

(Note that the third letter in cjo is an o rather than a zero.) SPICE recognizes vj as V_0 for diodes.

Example A-8

Determine the step response of the circuit shown in Fig. A.11 if V_{in} jumps from 0 to 1 V and D_1 satisfies the parameters given above.

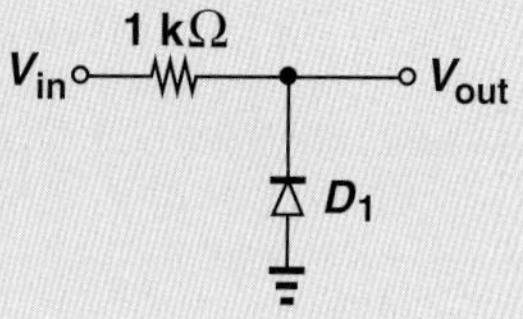

Figure A.11

Solution The voltage dependence of the junction capacitance of D_1 makes the analysis of this circuit difficult. For V_{out} near zero, D_1 experiences a small reverse bias, exhibiting a capacitance close to C_{j0}. As V_{out} rises, however, the capacitance falls, and so does the *time constant* of the circuit. Thus, SPICE proves quite useful here.

Labeling the circuit in our mind, we write the netlist as:

```
Step Response Example
r1 in out 1k
d1 out 0 is=1f, cjo=1p, vj=0.7
vin in 0 pulse(0 1 0 0.1n 0.1n 1)
.tran 25p 3n
.end
```

As we encounter more sophisticated devices, the number of parameters that must be specified for their SPICE description increases, thereby making the task of netlist construction cumbersome and error-prone. For example, today's MOSFETs require *hundreds* of parameters in their SPICE descriptions. To avoid repeating the parameters for each element, SPICE allows the definition of "models." For example, the above diode line can be written as

```
d1 in out mymodel
.model mymodel d (is=1f, cjo=1p, vj=0.7)
```

Upon reaching the fourth entry in the diode line, SPICE recognizes that this is not a *value*, but a model name and hence seeks a .model command that defines the details of "mymodel." The letter "d" in the .model line specifies a diode model. As seen below, this letter is replaced with "npn" for an *npn* bipolar transistor and "nmos" for an NMOS device.

Example A-9 Plot the input/output characteristic of the circuit shown in Fig. A.12. Assume D_1 and D_2 follow the above diode model.

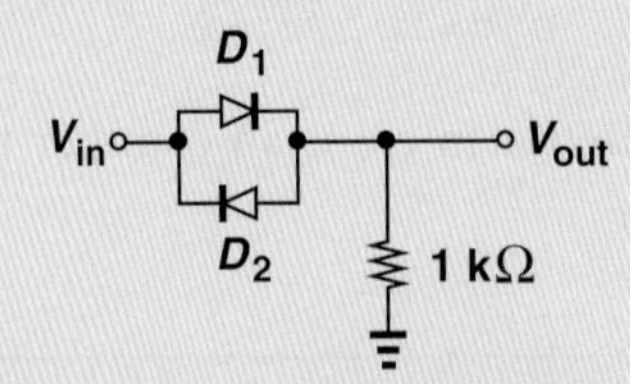

Figure A.12

Solution Labeling the circuit in our mind, we write the netlist as:

```
Diode Circuit
d1 in out mymodel
d2 out in mymodel
r1 out 0 1k
vin in 0 dc 1
.dc vin -3 +3 2m
.end
```

A.3.3 Bipolar Transistors

The definition of bipolar transistors requires special attention to the order of the terminals. Consider the example shown in Fig. A.13, where Q_1 is expressed as:

```
      Collector  Base   Emitter  Substrate  Model
q1      out       in      emi        0      bimod
```

where the device name begins with the letter q to indicate a bipolar transistor, and the first four nodes represent the collector, base, emitter, and substrate terminals, respectively. (In most cases, the substrate of *npn* transistors is tied to ground.) As with diodes, the parameters of the transistor are expressed in a model called, for example, bimod:

```
.model bimod npn (beta=100, is=10f)
```

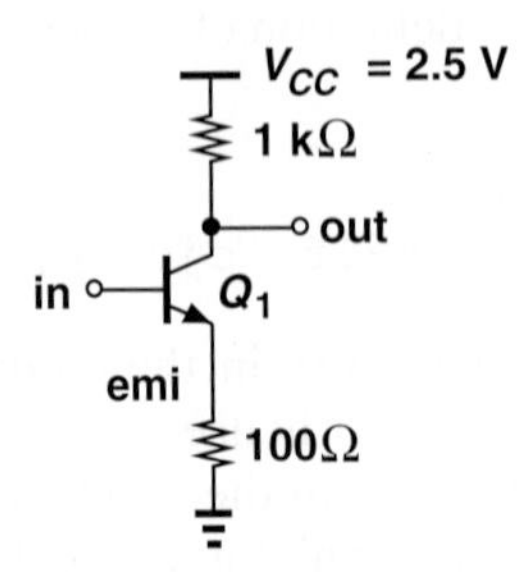

Figure A.13 Common-emitter stage.

Example A-10 Construct the SPICE netlist for the circuit of Fig. A.13. Assume the input must be swept from 0.8 V to 0.9 V.

Solution The netlist is as follows:

```
Simple CE Stage
q1 out in emi 0 bimod
remi emi 0 100
rout out vcc 1k
vcc vcc 0 2.5
vin in 0 dc 1
.dc 0.8 0.9 1m
.model bimod npn (beta=100, is=10f)
.end
```

Two observations prove useful here. (1) The two resistors are labeled according to the nodes to which they are attached. This approach allows us to find each resistor more readily than if it is simply labeled by a number, e.g., r1. (2) In the above netlist, the term "vcc" refers to two *distinct* entities: a voltage source (the first entry on the vcc line), and a node (the second entry on the vcc line).

The model of a bipolar transistor can contain high-frequency effects. For example, the base-emitter and base-collector junction capacitances are denoted by cje and cjc, respectively. The effect of charge storage in the base region is represented by a transit time, tf (equivalent to τ_F). Also, for integrated bipolar transistors, the collector-substrate junction capacitance, cjs, must be specified. Thus, a more complete model may read:

```
.model newmod npn (beta=100, is=10f, cje=5f, cjc=6f,
cjs=10f, tf=5p)
```

Modern bipolar transistor models contain hundreds of parameters.

Example A-11 Construct the netlist for the circuit shown in Fig. A.14(a), and obtain the frequency response from 100 MHz to 10 GHz. Use the above transistor model.

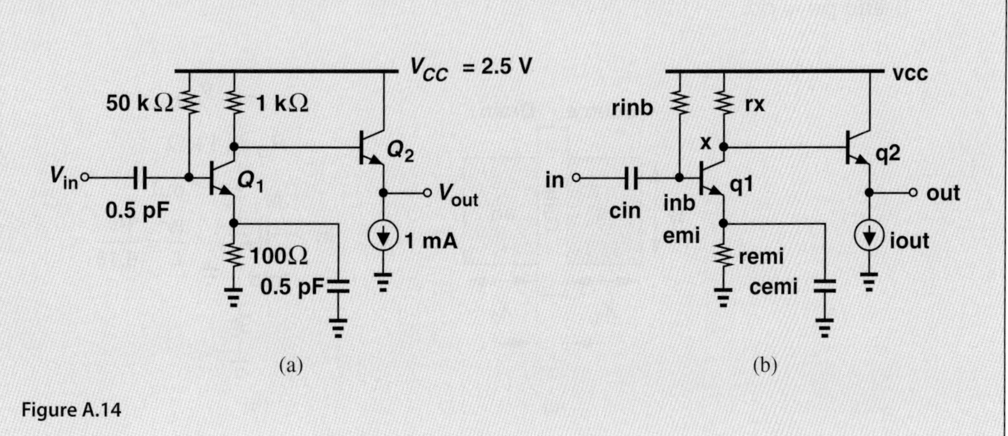

Figure A.14

Solution Labeling the circuit as depicted in Fig. A.14(b), we write

```
Two-Stage Amp
cin in inb 0.5p
rinb inb vcc 50k
q1 x inb emi 0 newmod
rx x vcc 1k
remi emi 0 2k
cemi emi 0 0.5p
q2 vcc x out 0 newmod
iout out 0 1m
vcc vcc 0 2.5
vin in 0 ac 1
.ac dec 100 100meg 10g
.model newmod npn (beta=100, is=10f, cje=5f, cjc=6f,
cjs=10f, tf=5p)
.end
```

A.3.4 MOSFETs

The definition of MOSFETs is somewhat similar to that of bipolar transistors but contains more details regarding the *dimensions* of the device. Unlike bipolar transistors, MOSFETs are both biased and "sized" so as to achieve certain small-signal properties. For example, both the transconductance and the output resistance of MOSFETs depend on the channel length.

In order to understand how the device dimensions are specified, we first consider the top view illustrated in Fig. A.15(a). In addition to the channel width and length, we must also provide the source/drain dimensions so that SPICE can calculate the associated capacitances. To this end, we specify the "area" and "perimeter" of the source and drain junctions. Denoted by "as" and "ps" for the source, respectively (and "ad" and "pd" for the drain), the area and perimeter are computed as follows: $\texttt{as} = X_1 \cdot W$, $\texttt{ps} = 2X_1 + 2W$, $\texttt{ad} = X_2 \cdot W$, $\texttt{pd} = 2X_2 + 2W$. In most cases, $X_1 = X_2$ and hence $\texttt{as} = \texttt{ad}$ and $\texttt{ps} = \texttt{pd}$.

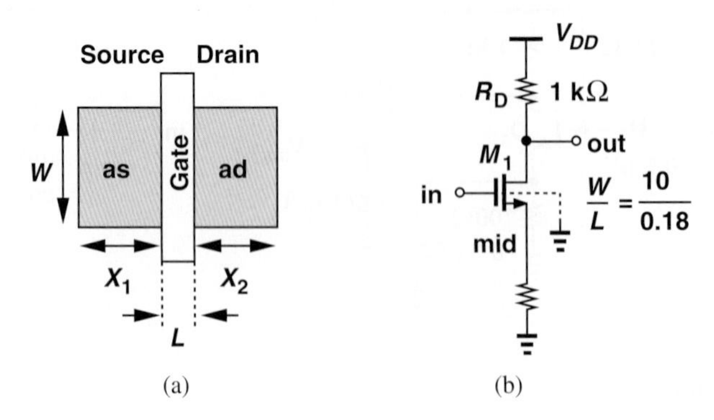

Figure A.15 (a) Top view of a MOSFET, (b) a common-source stage.

The value of $X_{1,2}$ is determined by "design rules" for each specific technology. As a rule of thumb, we assume $X_{1,2} \approx 3L_{min}$, where L_{min} denotes the minimum allowable channel length (e.g., 0.18 μm). In this section, we assume $X_{1,2} = 0.6$ μm.

Now consider the example shown in Fig. A.15(b), where the dashed line attached to M_1 indicates its substrate. Before considering the dimensions, we have:

```
        Drain   Gate   Source   Substrate   Model
m1       out    in      mid         0        nmos
```

As with the bipolar transistor, the terminal names appear in a certain order: drain, gate, source, and substrate. Now we add the dimensions:

```
    Drain Gate Source Substrate  Model
m1   out   in    mid        0     nmos   w=10u l=0.18u as=6p
+ps=21.2u ad=6p pd=21.2u
```

(The + sign allows continuing a line on the next.) The order of the dimensions is unimportant, but it is helpful to maintain a consistent pattern throughout the netlist so as to make it more readable. Note that

```
as=6p
```

denotes an area of 6×10^{-12} m^2.

The model of the MOSFET must provide various parameters of the transistor, e.g., mobility (uo), gate oxide thickness (tox), threshold voltage (vth), channel-length modulation coefficient (lambda), etc. For example,

```
.model mymod nmos (uo=360, tox=0.4n, vth=0.5, lambda=0.4)
```

Note that the default unit of mobility is cm^2/s, whereas the units of other parameters are based on the metric system. For example,

```
tox=0.4n
```

translates to 0.4×10^{-9} m = 40 Å.

Example A-12

Figure A.16(a) shows a two-stage amplifier. Construct a SPICE netlist to plot the input/output characteristic of the circuit. The substrate connections are not shown with the understanding that the default is ground for NMOS devices and V_{DD} for PMOS transistors.

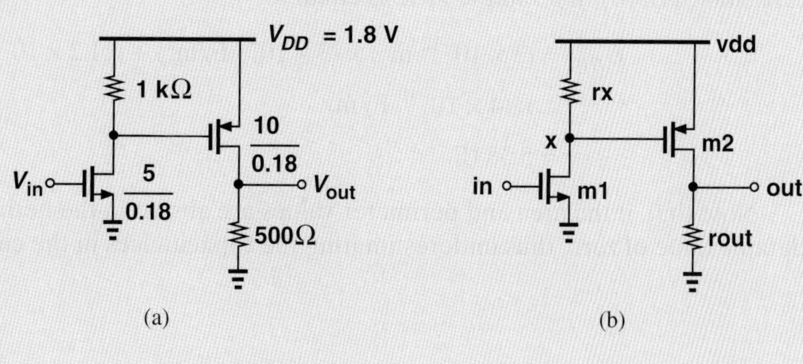

Figure A.16

Solution Labeling the circuit as depicted in Fig. A.16(b), we write

```
MOS Amplifier
m1 x in 0 0 nmos w=5u l=0.18u as=3p ps=11.2u ad=3p
pd=11.2u
rx x vdd 1k
m2 out x vdd vdd pmos w=10u l=0.8u as=6p ps=21.2u ad=6p
pd=21.2u
rout out 0 500
vdd vdd 0 1.8
vin in 0 dc 1
.dc vin 0 1 1m
.model mymod nmos (uo=360, tox=0.4n, vth=0.5, lambda=0.4)
.end
```

For high-frequency analysis, we must specify the junction capacitance of the source and drain areas. As illustrated in Fig. A.17, this capacitance is partitioned into two components: the "area" capacitance, C_j, and the "sidewall" capacitance, C_{jsw}. This separation is necessary because the values of C_j and C_{jsw} (e.g., per unit area) are typically unequal.

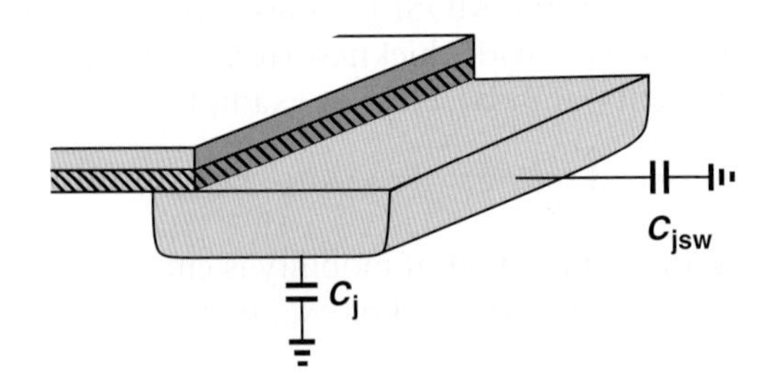

Figure A.17 Area and sidewall capacitances.

In SPICE, the above capacitance components are defined differently. The area capacitance is specified per unit area, e.g., $C_j = 3 \times 10^{-4}$ F/m^2 (= 0.3 fF/μm^2), whereas the sidewall capacitance is defined per *unit width*, e.g., $C_{jsw} = 4 \times 10^{-10}$ F/m (= 0.4 fF/μm). With these specifications, SPICE simply calculates the overall junction capacitance as $C_j \cdot ad + C_{jsw} \cdot pd$. For example, with the above values of C_j and C_{jsw}, the drain junction capacitance of M_1 in Example A-12 is equal to:

$$C_{DB1} = (3 \times 10^{-12}\ \text{m}^2) \times (3 \times 10^{-4}\ \text{F/m}^2) + (11.2 \times 10^{-6}\ \text{m}) \times (4 \times 10^{-10}\ \text{F/m}) \tag{A.1}$$

$$= 5.38\ \text{fF}. \tag{A.2}$$

Note that, if the area and perimeter values are absent in the netlist, SPICE may use a default value of zero, thus underestimating the capacitances in the circuit.

The source/drain junction capacitances exhibit a voltage dependence that may not follow the square-root equation associated with "abrupt" *pn* junctions. SPICE allows an equation of the form

$$C = \frac{C_0}{\left(1 + \frac{V_R}{\phi_B}\right)^m}, \tag{A.3}$$

where C_0 denotes the value for zero voltage across the junction, and m typically falls in the range of 0.3 to 0.4. Thus for C_j and C_{jsw}, we specify

```
(cjo, mj)
```

and

```
(cjswo, mjsw)
```

A more complete MOS model may therefore appear as:

```
.model mymod nmos (level=1, uo=360, tox=0.4n, vth=0.5,
lambda=0.4, +cjo=3e-4, mj=0.35, cjswo=40n, mjswo=0.3)
```

where the "level" denotes a certain complexity for the model. In practice, higher levels with many more parameters are used. Similarly, a PMOS model may be constructed as follows:

```
.model mymod2 pmos (level=1, uo=150, tox=0.4n, vth=-0.55,
+lambda=0.5, cjo=3.5e-4, mj=0.35, cjswo=35n, mjswo=0.3)
```

A.4 OTHER ELEMENTS AND COMMANDS

A.4.1 Dependent Sources

In addition to the independent voltage and current sources studied above, we may need to incorporate dependent sources in simulations. For example, as mentioned in Chapter 8, op amps can be viewed as voltage-dependent voltage sources. Similarly, a MOSFET acts as a voltage-dependent current source.

Consider the arrangement shown in Fig. A.18, where the voltage source tied between nodes C and D is equal to three times the voltage difference between nodes A and B. For simplicity, we call (A, B) the "input nodes," (C, D) the "output nodes," and the factor of 3, the "gain." Such a voltage-dependent voltage source is expressed as

```
      Output            Input     DC       Gain
      Nodes             Nodes    Value
e1    c d    poly(1)     a b       0        3
```


Figure A.18 Voltage-dependent voltage source.

Note the element name begins with the letter "e" to signify a voltage-dependent voltage source. The next two entries are the output nodes, with the first representing the positive terminal. The entry poly(1) indicates a first-order polynomial relationship between V_{CD} and V_{AB}. Next, the controlling (input) nodes are specified, and the zero is entered to denote a zero additional dc voltage. Finally, the gain is specified. In a more general case, this expression can realize $V_{CD} = \alpha + \beta V_{AB}$, where α is the dc value (zero in the above example) and β is the gain (3 in the previous example).

Example A-13 The circuit of Fig. A.19(a) employs an op amp with a gain of 500. Construct a SPICE netlist for the circuit.

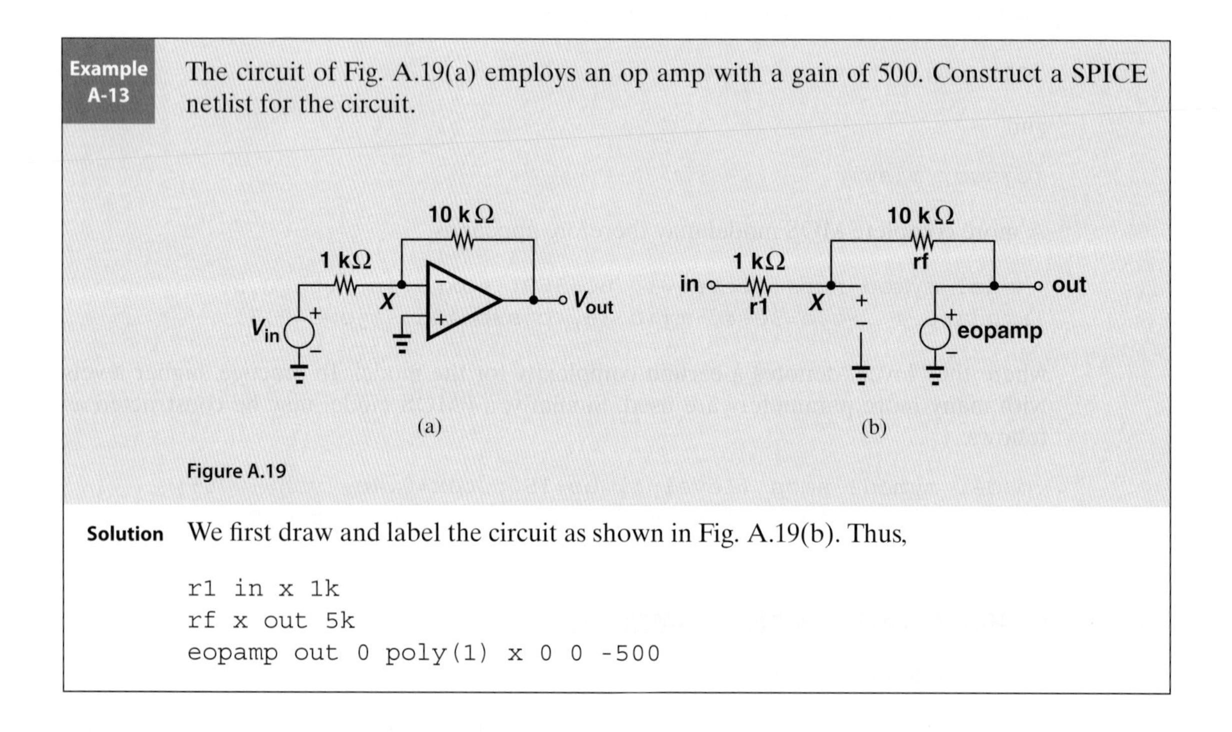

Figure A.19

Solution We first draw and label the circuit as shown in Fig. A.19(b). Thus,

```
r1 in x 1k
rf x out 5k
eopamp out 0 poly(1) x 0 0 -500
```

For the voltage-dependent current source depicted in Fig. A.20, the description is as follows:

```
g1 c d poly(1) a b 0 0.05
```

where the letter g denotes a voltage-dependent current source and the gain is specified as $1/(20\ \Omega) = 0.05\ \Omega^{-1}$.

Current-controlled voltage and current sources are also described in a similar manner, but they are rarely used.

A
B
C
D
$G_m V_{AB}$
$G_m = (20\ \Omega)^{-1}$

Figure A.20 Voltage-dependent current source.

A.4.2 Initial Conditions

In the transient analysis of circuits, we may wish to specify an initial voltage at a node with respect to ground. This is accomplished using the .ic command:

```
.ic v(x)=0.5
```

This example sets the initial voltage at node X to 0.5 V.

Reference

1. G. Roberts and A. S. Sedra, *SPICE*, Oxford University Press, 1997.

Input and Output Impedances

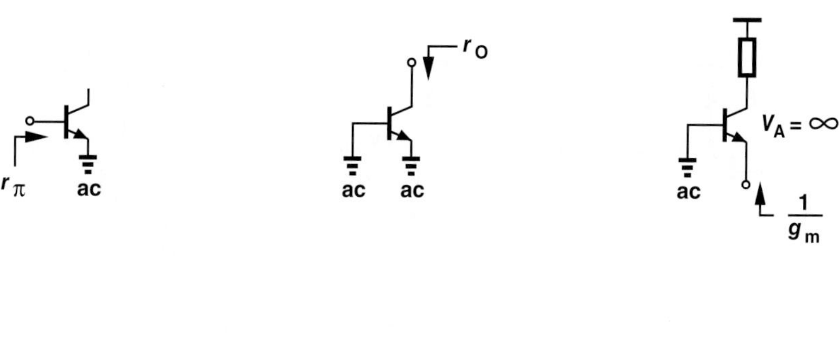

$r_\pi + (\beta + 1)R_E$

$V_A = \infty$

R_E

ac

$(1 + g_m r_O)(R_E \| r_\pi) + r_O$

ac

R_E

ac

R_B

$V_A = \infty$

ac

$\frac{1}{g_m} + \frac{R_B}{\beta + 1}$

$\approx \infty$

ac

r_O

ac ac

$\lambda = 0$

ac

$\frac{1}{g_m}$

$(1 + g_m r_O)R_S + r_O$

ac

R_S

ac

Voltage Gain Equations

V_{CC}, R_C, v_{out}, v_{in}

$$A_v = -g_m(R_C \| r_O)$$

V_{CC}, R_C, v_{out}, v_{in}, $V_A = \infty$, R_E

$$A_v = -\frac{R_C}{\frac{1}{g_m} + R_E}$$

V_{CC}, R_C, v_{out}, $V_A = \infty$, V_b, v_{in}, R_E

$$A_v = \frac{R_C}{\frac{1}{g_m} + R_E}$$

V_{CC}, v_{in}, v_{out}, R_E

$$A_v = \frac{R_E \| r_O}{\frac{1}{g_m} + R_E \| r_O}$$

V_{DD}, R_D, v_{out}, v_{in}

$$A_v = -g_m(R_D \| r_O)$$

V_{DD}, R_D, v_{out}, v_{in}, $\lambda = 0$, R_S

$$A_v = -\frac{R_D}{\frac{1}{g_m} + R_S}$$

V_{DD}, R_D, v_{out}, $\lambda = 0$, V_b, v_{in}, R_S

$$A_v = \frac{R_D}{\frac{1}{g_m} + R_S}$$

V_{DD}, v_{in}, v_{out}, R_S

$$A_v = \frac{R_S \| r_O}{\frac{1}{g_m} + R_S \| r_O}$$

Index

D

K

L

M

S

W

Z